DERIVATIVES AND INTEGRALS

Basic Differentiation Rules

1. $\dfrac{d}{dx}[cu] = cu'$

2. $\dfrac{d}{dx}[u \pm v] = u' \pm v'$

3. $\dfrac{d}{dx}[uv] = uv' + vu'$

4. $\dfrac{d}{dx}\left[\dfrac{u}{v}\right] = \dfrac{vu' - uv'}{v^2}$

5. $\dfrac{d}{dx}[c] = 0$

6. $\dfrac{d}{dx}[u^n] = nu^{n-1}u'$

7. $\dfrac{d}{dx}[x] = 1$

8. $\dfrac{d}{dx}[|u|] = \dfrac{u}{|u|}(u'), \quad u \neq 0$

9. $\dfrac{d}{dx}[\ln u] = \dfrac{u'}{u}$

10. $\dfrac{d}{dx}[e^u] = e^u u'$

11. $\dfrac{d}{dx}[\log_a u] = \dfrac{u'}{(\ln a)u}$

12. $\dfrac{d}{dx}[a^u] = (\ln a)a^u u'$

13. $\dfrac{d}{dx}[\sin u] = (\cos u)u'$

14. $\dfrac{d}{dx}[\cos u] = -(\sin u)u'$

15. $\dfrac{d}{dx}[\tan u] = (\sec^2 u)u'$

16. $\dfrac{d}{dx}[\cot u] = -(\csc^2 u)u'$

17. $\dfrac{d}{dx}[\sec u] = (\sec u \tan u)u'$

18. $\dfrac{d}{dx}[\csc u] = -(\csc u \cot u)u'$

19. $\dfrac{d}{dx}[\arcsin u] = \dfrac{u'}{\sqrt{1 - u^2}}$

20. $\dfrac{d}{dx}[\arccos u] = \dfrac{-u'}{\sqrt{1 - u^2}}$

21. $\dfrac{d}{dx}[\arctan u] = \dfrac{u'}{1 + u^2}$

22. $\dfrac{d}{dx}[\operatorname{arccot} u] = \dfrac{-u'}{1 + u^2}$

23. $\dfrac{d}{dx}[\operatorname{arcsec} u] = \dfrac{u'}{|u|\sqrt{u^2 - 1}}$

24. $\dfrac{d}{dx}[\operatorname{arccsc} u] = \dfrac{-u'}{|u|\sqrt{u^2 - 1}}$

25. $\dfrac{d}{dx}[\sinh u] = (\cosh u)u'$

26. $\dfrac{d}{dx}[\cosh u] = (\sinh u)u'$

27. $\dfrac{d}{dx}[\tanh u] = (\operatorname{sech}^2 u)u'$

28. $\dfrac{d}{dx}[\coth u] = -(\operatorname{csch}^2 u)u'$

29. $\dfrac{d}{dx}[\operatorname{sech} u] = -(\operatorname{sech} u \tanh u)u'$

30. $\dfrac{d}{dx}[\operatorname{csch} u] = -(\operatorname{csch} u \coth u)u'$

31. $\dfrac{d}{dx}[\sinh^{-1} u] = \dfrac{u'}{\sqrt{u^2 + 1}}$

32. $\dfrac{d}{dx}[\cosh^{-1} u] = \dfrac{u'}{\sqrt{u^2 - 1}}$

33. $\dfrac{d}{dx}[\tanh^{-1} u] = \dfrac{u'}{1 - u^2}$

34. $\dfrac{d}{dx}[\coth^{-1} u] = \dfrac{u'}{1 - u^2}$

35. $\dfrac{d}{dx}[\operatorname{sech}^{-1} u] = \dfrac{-u'}{u\sqrt{1 - u^2}}$

36. $\dfrac{d}{dx}[\operatorname{csch}^{-1} u] = \dfrac{-u'}{|u|\sqrt{1 + u^2}}$

Basic Integration Formulas

1. $\displaystyle\int kf(u)\,du = k\int f(u)\,du$

2. $\displaystyle\int [f(u) \pm g(u)]\,du = \int f(u)\,du \pm \int g(u)\,du$

3. $\displaystyle\int du = u + C$

4. $\displaystyle\int a^u\,du = \left(\dfrac{1}{\ln a}\right)a^u + C$

5. $\displaystyle\int e^u\,du = e^u + C$

6. $\displaystyle\int \sin u\,du = -\cos u + C$

7. $\displaystyle\int \cos u\,du = \sin u + C$

8. $\displaystyle\int \tan u\,du = -\ln|\cos u| + C$

9. $\displaystyle\int \cot u\,du = \ln|\sin u| + C$

10. $\displaystyle\int \sec u\,du = \ln|\sec u + \tan u| + C$

11. $\displaystyle\int \csc u\,du = -\ln|\csc u + \cot u| + C$

12. $\displaystyle\int \sec^2 u\,du = \tan u + C$

13. $\displaystyle\int \csc^2 u\,du = -\cot u + C$

14. $\displaystyle\int \sec u \tan u\,du = \sec u + C$

15. $\displaystyle\int \csc u \cot u\,du = -\csc u + C$

16. $\displaystyle\int \dfrac{du}{\sqrt{a^2 - u^2}} = \arcsin \dfrac{u}{a} + C$

17. $\displaystyle\int \dfrac{du}{a^2 + u^2} = \dfrac{1}{a}\arctan \dfrac{u}{a} + C$

18. $\displaystyle\int \dfrac{du}{u\sqrt{u^2 - a^2}} = \dfrac{1}{a}\operatorname{arcsec} \dfrac{|u|}{a} + C$

TRIGONOMETRY

Definition of the Six Trigonometric Functions

Right triangle definitions, where $0 < \theta < \pi/2$.

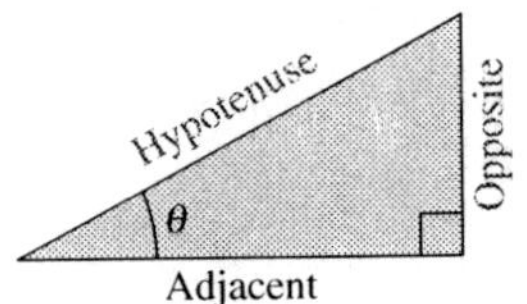

$$\sin \theta = \frac{\text{opp}}{\text{hyp}} \qquad \csc \theta = \frac{\text{hyp}}{\text{opp}}$$

$$\cos \theta = \frac{\text{adj}}{\text{hyp}} \qquad \sec \theta = \frac{\text{hyp}}{\text{adj}}$$

$$\tan \theta = \frac{\text{opp}}{\text{adj}} \qquad \cot \theta = \frac{\text{adj}}{\text{opp}}$$

Circular function definitions, where θ is any angle.

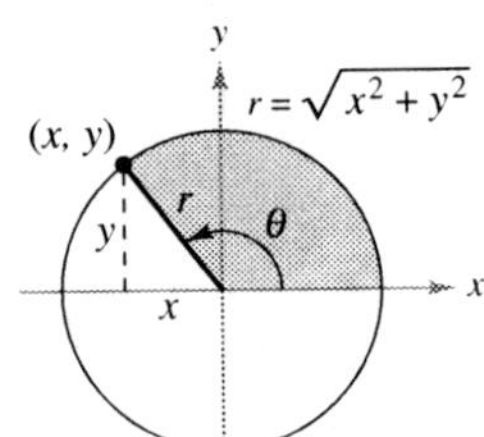

$$\sin \theta = \frac{y}{r} \qquad \csc \theta = \frac{r}{y}$$

$$\cos \theta = \frac{x}{r} \qquad \sec \theta = \frac{r}{x}$$

$$\tan \theta = \frac{y}{x} \qquad \cot \theta = \frac{x}{y}$$

Reciprocal Identities

$$\sin x = \frac{1}{\csc x} \qquad \sec x = \frac{1}{\cos x} \qquad \tan x = \frac{1}{\cot x}$$

$$\csc x = \frac{1}{\sin x} \qquad \cos x = \frac{1}{\sec x} \qquad \cot x = \frac{1}{\tan x}$$

Tangent and Cotangent Identities

$$\tan x = \frac{\sin x}{\cos x} \qquad \cot x = \frac{\cos x}{\sin x}$$

Pythagorean Identities

$$\sin^2 x + \cos^2 x = 1$$

$$1 + \tan^2 x = \sec^2 x \qquad 1 + \cot^2 x = \csc^2 x$$

Cofunction Identities

$$\sin\left(\frac{\pi}{2} - x\right) = \cos x \qquad \cos\left(\frac{\pi}{2} - x\right) = \sin x$$

$$\csc\left(\frac{\pi}{2} - x\right) = \sec x \qquad \tan\left(\frac{\pi}{2} - x\right) = \cot x$$

$$\sec\left(\frac{\pi}{2} - x\right) = \csc x \qquad \cot\left(\frac{\pi}{2} - x\right) = \tan x$$

Reduction Formulas

$$\sin(-x) = -\sin x \qquad \cos(-x) = \cos x$$

$$\csc(-x) = -\csc x \qquad \tan(-x) = -\tan x$$

$$\sec(-x) = \sec x \qquad \cot(-x) = -\cot x$$

Sum and Difference Formulas

$$\sin(u \pm v) = \sin u \cos v \pm \cos u \sin v$$

$$\cos(u \pm v) = \cos u \cos v \mp \sin u \sin v$$

$$\tan(u \pm v) = \frac{\tan u \pm \tan v}{1 \mp \tan u \tan v}$$

Double-Angle Formulas

$$\sin 2u = 2 \sin u \cos u$$

$$\cos 2u = \cos^2 u - \sin^2 u = 2 \cos^2 u - 1 = 1 - 2 \sin^2 u$$

$$\tan 2u = \frac{2 \tan u}{1 - \tan^2 u}$$

Power-Reducing Formulas

$$\sin^2 u = \frac{1 - \cos 2u}{2}$$

$$\cos^2 u = \frac{1 + \cos 2u}{2}$$

$$\tan^2 u = \frac{1 - \cos 2u}{1 + \cos 2u}$$

Sum-to-Product Formulas

$$\sin u + \sin v = 2 \sin\left(\frac{u + v}{2}\right) \cos\left(\frac{u - v}{2}\right)$$

$$\sin u - \sin v = 2 \cos\left(\frac{u + v}{2}\right) \sin\left(\frac{u - v}{2}\right)$$

$$\cos u + \cos v = 2 \cos\left(\frac{u + v}{2}\right) \cos\left(\frac{u - v}{2}\right)$$

$$\cos u - \cos v = -2 \sin\left(\frac{u + v}{2}\right) \sin\left(\frac{u - v}{2}\right)$$

Product-to-Sum Formulas

$$\sin u \sin v = \frac{1}{2}[\cos(u - v) - \cos(u + v)]$$

$$\cos u \cos v = \frac{1}{2}[\cos(u - v) + \cos(u + v)]$$

$$\sin u \cos v = \frac{1}{2}[\sin(u + v) + \sin(u - v)]$$

$$\cos u \sin v = \frac{1}{2}[\sin(u + v) - \sin(u - v)]$$

Math 2200
Calculus with Analytics

University of Georgia edition

Ron Larson | Bruce H. Edwards

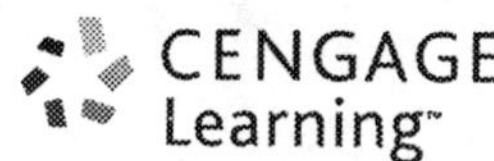

Australia • Brazil • Japan • Korea • Mexico • Singapore • Spain • United Kingdom • United States

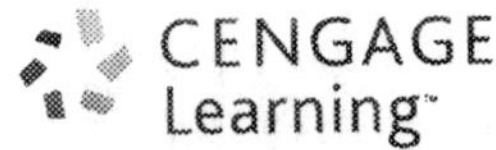

Math 2200
Calculus with Analytics: University of Georgia edition

Executive Editors:
 Maureen Staudt
 Michael Stranz

Senior Project Development Manager:
 Linda deStefano

Marketing Specialist:
 Courtney Sheldon

Senior Production/Manufacturing Manager:
 Donna M. Brown

PreMedia Manager:
 Joel Brennecke

Sr. Rights Acquisition Account Manager:
 Todd Osborne

Cover Image:
 Getty Images*

Calculus of a Single Variable Early Transcendental Functions, Fifth Edition
Ron Larson | Bruce H. Edwards

This book contains select works from existing Cengage Learning resources and was produced by Cengage Learning Custom Solutions for collegiate use. As such, those adopting and/or contributing to this work are responsible for editorial content accuracy, continuity and completeness.

Compilation © 2011 Cengage Learning

ISBN-13: 978-1-133-22873-8

ISBN-10: 1-133-22873-9

Cengage Learning
5191 Natorp Boulevard
Mason, Ohio 45040
USA
Cengage Learning is a leading provider of customized learning solutions with office locations around the globe, including Singapore, the United Kingdom, Australia, Mexico, Brazil, and Japan. Locate your local office at:
international.cengage.com/region.

Cengage Learning products are represented in Canada by Nelson Education, Ltd.
For your lifelong learning solutions, visit **www.cengage.com/custom.**
Visit our corporate website at **www.cengage.com.**

Printed in the United States of America

Contents

ADDITIONAL APPENDICES

Appendix F   Business and Economic Applications (Online)

1 Preparation for Calculus

This chapter reviews several concepts that will help you prepare for your study of calculus. These concepts include sketching the graphs of equations and functions, and fitting mathematical models to data. It is important to review these concepts before moving on to calculus.

In this chapter, you should learn the following.

- How to identify the characteristics of equations and sketch their graphs. (1.1)
- How to find and graph equations of lines, including parallel and perpendicular lines, using the concept of slope. (1.2)
- How to evaluate and graph functions and their transformations. (1.3)
- How to fit mathematical models to real-life data sets. (1.4)
- How to determine whether a function has an inverse function. The properties of inverse trigonometric functions. (1.5)
- The properties of the natural exponential and natural logarithmic functions. (1.6)

Jeremy Walker/Getty Images

In 2006, China surpassed the United States as the world's biggest emitter of carbon dioxide, the main greenhouse gas. Given the carbon dioxide concentrations in the atmosphere for several years, can older mathematical models still accurately predict future atmospheric concentrations compared with more recent models? (See Section 1.1, Example 6.)

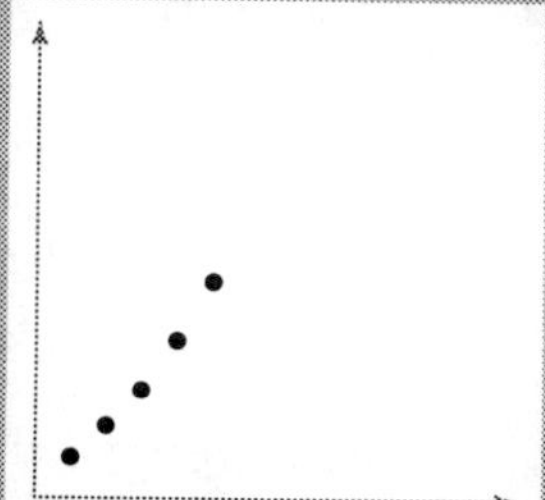 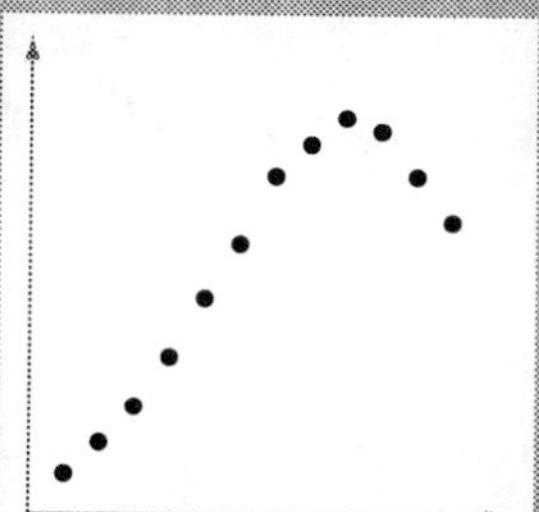 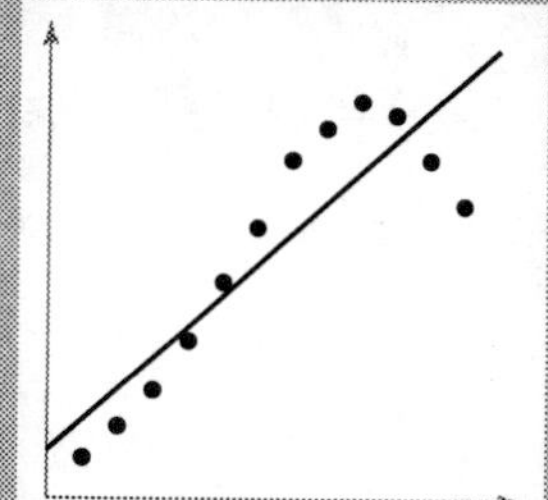 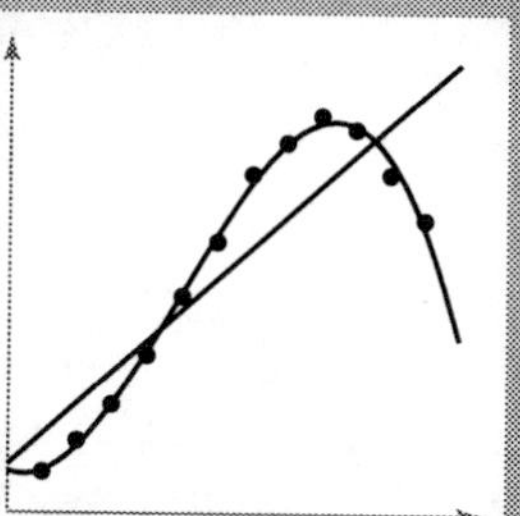

Mathematical models are commonly used to describe data sets. These models can be represented by many different types of functions, such as linear, quadratic, cubic, rational, and trigonometric functions. (See Section 1.4.)

1.1 Graphs and Models

- Sketch the graph of an equation.
- Find the intercepts of a graph.
- Test a graph for symmetry with respect to an axis and the origin.
- Find the points of intersection of two graphs.
- Interpret mathematical models for real-life data.

The Graph of an Equation

In 1637, the French mathematician René Descartes revolutionized the study of mathematics by joining its two major fields—algebra and geometry. With Descartes's coordinate plane, geometric concepts could be formulated analytically and algebraic concepts could be viewed graphically. The power of this approach is such that within a century, much of calculus had been developed.

The same approach can be followed in your study of calculus. That is, by viewing calculus from multiple perspectives—*graphically*, *analytically*, and *numerically*—you will increase your understanding of core concepts.

Consider the equation $3x + y = 7$. The point $(2, 1)$ is a **solution point** of the equation because the equation is satisfied (is true) when 2 is substituted for x and 1 is substituted for y. This equation has many other solutions, such as $(1, 4)$ and $(0, 7)$. To find other solutions systematically, solve the original equation for y.

$$y = 7 - 3x \qquad \text{Analytic approach}$$

Then construct a table of values by substituting several values for x.

x	0	1	2	3	4
y	7	4	1	-2	-5

Numerical approach

From the table, you can see that $(0, 7)$, $(1, 4)$, $(2, 1)$, $(3, -2)$, and $(4, -5)$ are solutions of the original equation $3x + y = 7$. Like many equations, this equation has an infinite number of solutions. The set of all solution points is the **graph** of the equation, as shown in Figure 1.1.

NOTE Even though we refer to the sketch shown in Figure 1.1 as the graph of $3x + y = 7$, it really represents only a *portion* of the graph. The entire graph would extend beyond the page. ∎

In this course, you will study many sketching techniques. The simplest is point plotting—that is, you plot points until the basic shape of the graph seems apparent.

EXAMPLE 1 Sketching a Graph by Point Plotting

Sketch the graph of $y = x^2 - 2$.

Solution First construct a table of values. Then plot the points shown in the table.

x	-2	-1	0	1	2	3
y	2	-1	-2	-1	2	7

Finally, connect the points with a *smooth curve*, as shown in Figure 1.2. This graph is a **parabola**. It is one of the conics you will study in Chapter 10. ∎

RENÉ DESCARTES (1596–1650)

Descartes made many contributions to philosophy, science, and mathematics. The idea of representing points in the plane by pairs of real numbers and representing curves in the plane by equations was described by Descartes in his book *La Géométrie*, published in 1637.

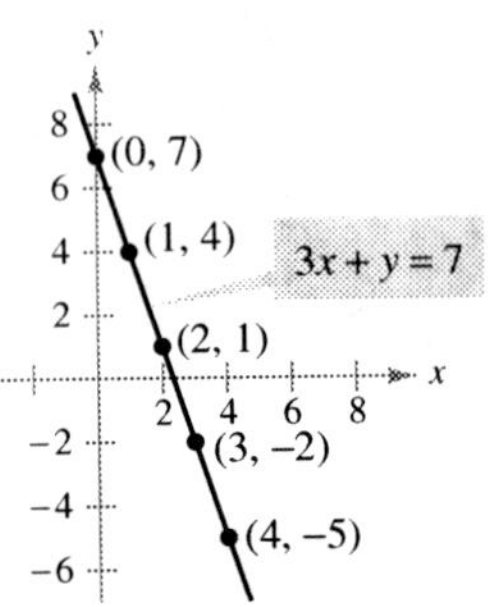

Graphical approach: $3x + y = 7$
Figure 1.1

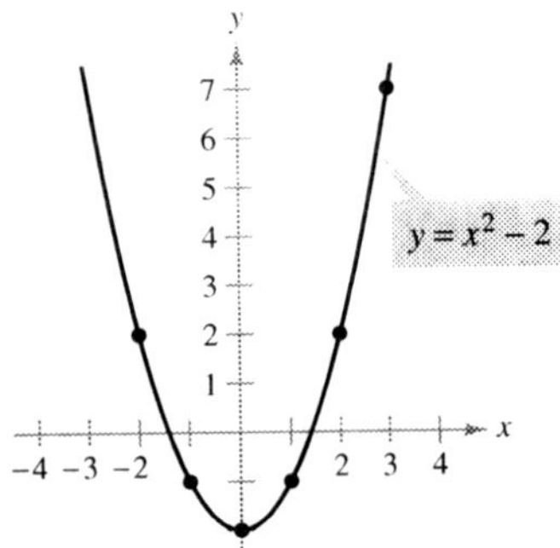

The parabola $y = x^2 - 2$
Figure 1.2

One disadvantage of point plotting is that to get a good idea about the shape of a graph, you may need to plot many points. With only a few points, you could misrepresent the graph. For instance, suppose that to sketch the graph of

$$y = \tfrac{1}{30}x(39 - 10x^2 + x^4)$$

you plotted only five points:

$$(-3, -3), (-1, -1), (0, 0), (1, 1), \text{ and } (3, 3)$$

as shown in Figure 1.3(a). From these five points, you might conclude that the graph is a line. This, however, is not correct. By plotting several more points, you can see that the graph is more complicated, as shown in Figure 1.3(b).

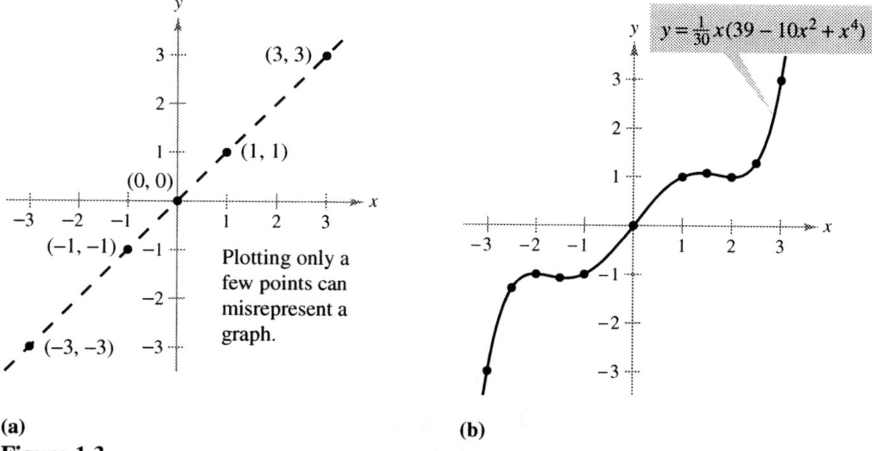

(a)   (b)

Figure 1.3

> **EXPLORATION**
>
> ***Comparing Graphical and Analytic Approaches***   Use a graphing utility to graph each equation. In each case, find a viewing window that shows the important characteristics of the graph.
>
> **a.** $y = x^3 - 3x^2 + 2x + 5$
>
> **b.** $y = x^3 - 3x^2 + 2x + 25$
>
> **c.** $y = -x^3 - 3x^2 + 20x + 5$
>
> **d.** $y = 3x^3 - 40x^2 + 50x - 45$
>
> **e.** $y = -(x + 12)^3$
>
> **f.** $y = (x - 2)(x - 4)(x - 6)$
>
> A purely graphical approach to this problem would involve a simple "guess, check, and revise" strategy. What types of things do you think an analytic approach might involve? For instance, does the graph have symmetry? Does the graph have turns? If so, where are they?
>
> As you proceed through Chapters 2, 3, and 4 of this text, you will study many new analytic tools that will help you analyze graphs of equations such as these.

TECHNOLOGY Technology has made sketching of graphs easier. Even with technology, however, it is possible to misrepresent a graph badly. For instance, each of the graphing utility screens in Figure 1.4 shows a portion of the graph of

$$y = x^3 - x^2 - 25.$$

From the screen on the left, you might assume that the graph is a line. From the screen on the right, however, you can see that the graph is not a line. So, whether you are sketching a graph by hand or using a graphing utility, you must realize that different "viewing windows" can produce very different views of a graph. In choosing a viewing window, your goal is to show a view of the graph that fits well in the context of the problem.

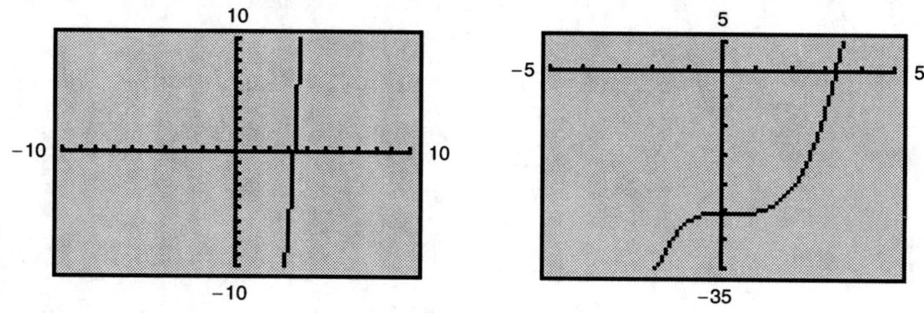

Graphing utility screens of $y = x^3 - x^2 - 25$

Figure 1.4

NOTE In this text, the term *graphing utility* means either a graphing calculator or computer graphing software such as *Maple*, *Mathematica*, or the *TI-89*.

Intercepts of a Graph

Two types of solution points that are especially useful in graphing an equation are those having zero as their *x*- or *y*-coordinate. Such points are called **intercepts** because they are the points at which the graph intersects the *x*- or *y*-axis. The point $(a, 0)$ is an **x-intercept** of the graph of an equation if it is a solution point of the equation. To find the *x*-intercepts of a graph, let *y* be zero and solve the equation for *x*. The point $(0, b)$ is a **y-intercept** of the graph of an equation if it is a solution point of the equation. To find the *y*-intercepts of a graph, let *x* be zero and solve the equation for *y*.

NOTE Some texts denote the *x*-intercept as the *x*-coordinate of the point $(a, 0)$ rather than the point itself. Unless it is necessary to make a distinction, we will use the term *intercept* to mean either the point or the coordinate. ■

It is possible for a graph to have no intercepts, or it might have several. For instance, consider the four graphs shown in Figure 1.5.

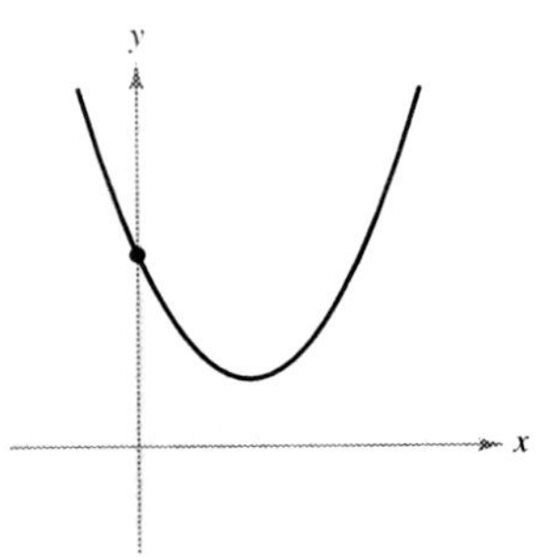

No *x*-intercepts
One *y*-intercept
Figure 1.5

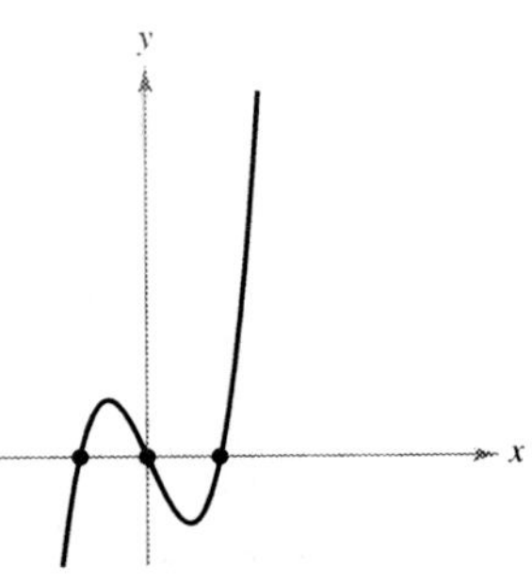

Three *x*-intercepts
One *y*-intercept

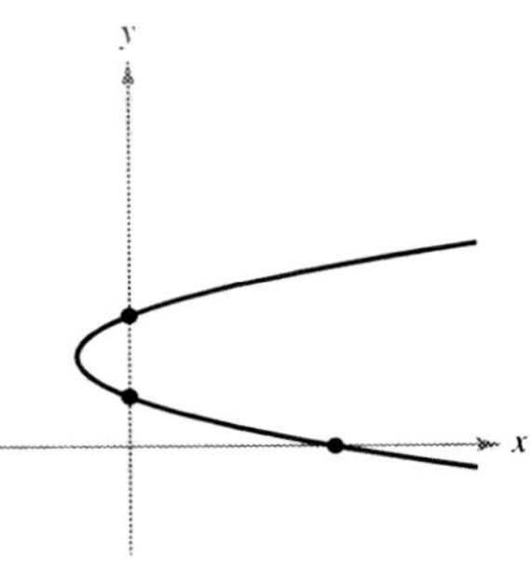

One *x*-intercept
Two *y*-intercepts

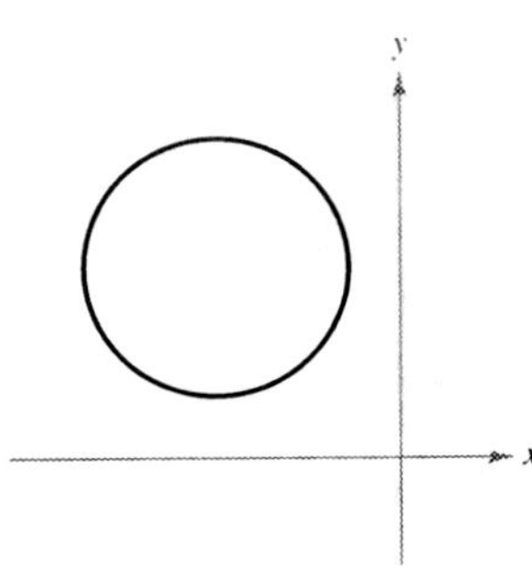

No intercepts

EXAMPLE 2   Finding *x*- and *y*-intercepts

Find the *x*- and *y*-intercepts of the graph of $y = x^3 - 4x$.

Solution To find the *x*-intercepts, let *y* be zero and solve for *x*.

$$x^3 - 4x = 0 \qquad \text{Let } y \text{ be zero.}$$
$$x(x - 2)(x + 2) = 0 \qquad \text{Factor.}$$
$$x = 0, 2, \text{ or } -2 \qquad \text{Solve for } x.$$

Because this equation has three solutions, you can conclude that the graph has three *x*-intercepts:

$$(0, 0), (2, 0), \text{ and } (-2, 0). \qquad \text{x-intercepts}$$

To find the *y*-intercepts, let *x* be zero. Doing so produces $y = 0$. So, the *y*-intercept is

$$(0, 0). \qquad \text{y-intercept}$$

(See Figure 1.6.) ■

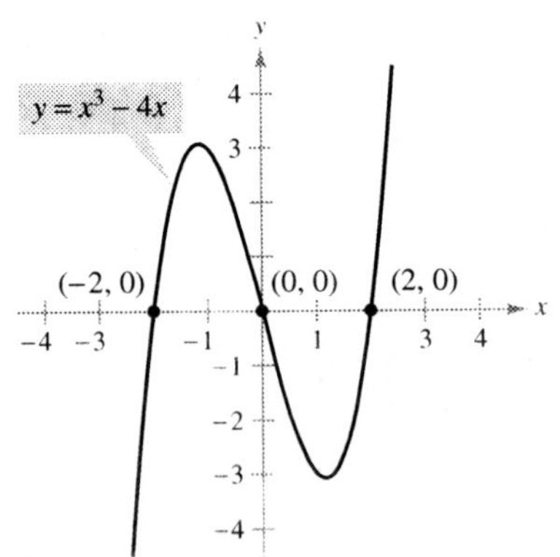

Intercepts of a graph
Figure 1.6

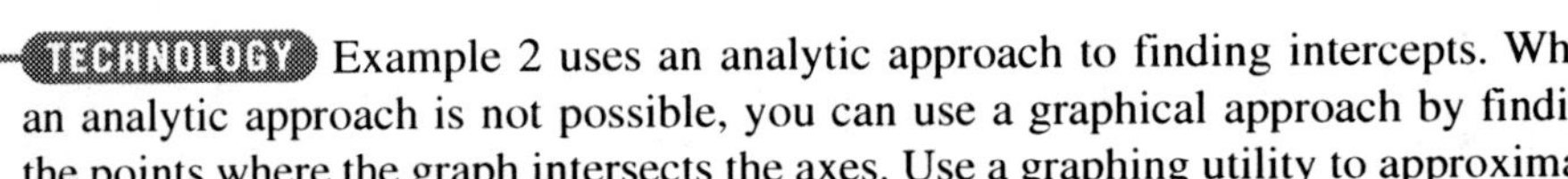

TECHNOLOGY Example 2 uses an analytic approach to finding intercepts. When an analytic approach is not possible, you can use a graphical approach by finding the points where the graph intersects the axes. Use a graphing utility to approximate the intercepts.

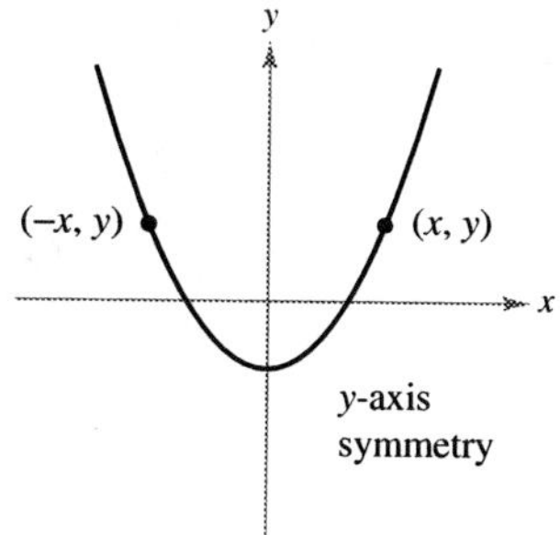

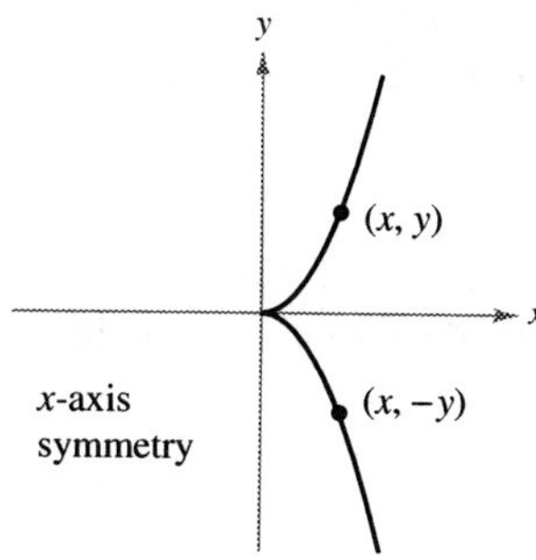

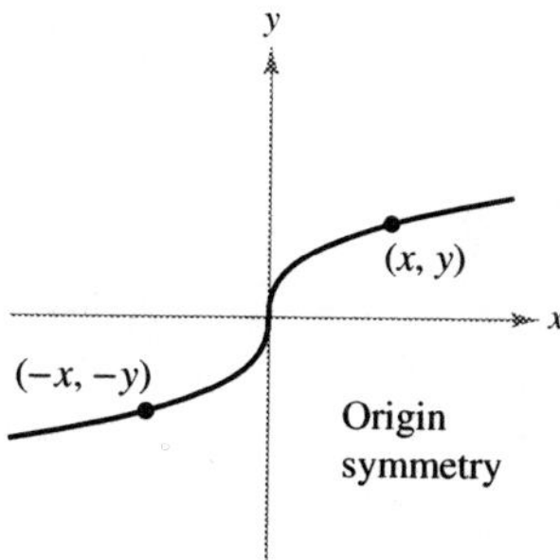

Figure 1.7

Symmetry of a Graph

Knowing the symmetry of a graph *before* attempting to sketch it is useful because you need only half as many points to sketch the graph. The following three types of symmetry can be used to help sketch the graphs of equations (see Figure 1.7).

1. A graph is **symmetric with respect to the y-axis** if, whenever (x, y) is a point on the graph, $(-x, y)$ is also a point on the graph. This means that the portion of the graph to the left of the y-axis is a mirror image of the portion to the right of the y-axis.

2. A graph is **symmetric with respect to the x-axis** if, whenever (x, y) is a point on the graph, $(x, -y)$ is also a point on the graph. This means that the portion of the graph above the x-axis is a mirror image of the portion below the x-axis.

3. A graph is **symmetric with respect to the origin** if, whenever (x, y) is a point on the graph, $(-x, -y)$ is also a point on the graph. This means that the graph is unchanged by a rotation of $180°$ about the origin.

TESTS FOR SYMMETRY

1. The graph of an equation in x and y is symmetric with respect to the y-axis if replacing x by $-x$ yields an equivalent equation.

2. The graph of an equation in x and y is symmetric with respect to the x-axis if replacing y by $-y$ yields an equivalent equation.

3. The graph of an equation in x and y is symmetric with respect to the origin if replacing x by $-x$ and y by $-y$ yields an equivalent equation.

The graph of a polynomial has symmetry with respect to the y-axis if each term has an even exponent (or is a constant). For instance, the graph of $y = 2x^4 - x^2 + 2$ has symmetry with respect to the y-axis. Similarly, the graph of a polynomial has symmetry with respect to the origin if each term has an odd exponent, as illustrated in Example 3.

EXAMPLE 3　Testing for Symmetry

Test the graph of $y = 2x^3 - x$ for symmetry with respect to the y-axis and to the origin.

Solution

y-axis Symmetry:

$$y = 2x^3 - x \qquad \text{Write original equation.}$$
$$y = 2(-x)^3 - (-x) \qquad \text{Replace } x \text{ by } -x.$$
$$y = -2x^3 + x \qquad \text{Simplify. It is not an equivalent equation.}$$

Origin Symmetry:

$$y = 2x^3 - x \qquad \text{Write original equation.}$$
$$-y = 2(-x)^3 - (-x) \qquad \text{Replace } x \text{ by } -x \text{ and } y \text{ by } -y.$$
$$-y = -2x^3 + x \qquad \text{Simplify.}$$
$$y = 2x^3 - x \qquad \text{Equivalent equation}$$

Because replacing both x by $-x$ and y by $-y$ yields an equivalent equation, you can conclude that the graph of $y = 2x^3 - x$ is symmetric with respect to the origin, as shown in Figure 1.8.

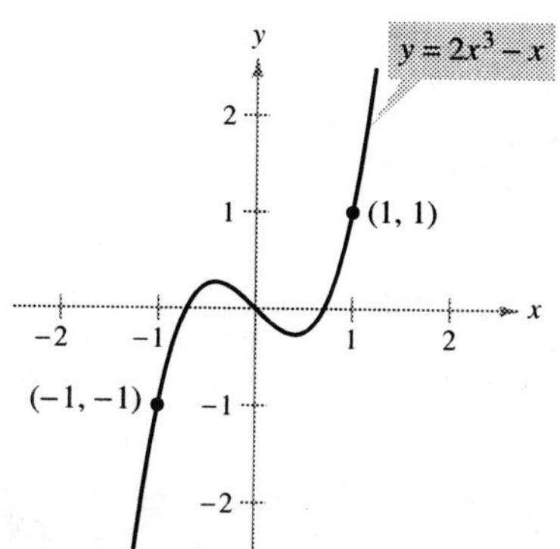

Origin symmetry
Figure 1.8

EXAMPLE 4   Using Intercepts and Symmetry to Sketch a Graph

Sketch the graph of $x - y^2 = 1$.

Solution   The graph is symmetric with respect to the x-axis because replacing y by $-y$ yields an equivalent equation.

$$x - y^2 = 1 \qquad \text{Write original equation.}$$
$$x - (-y)^2 = 1 \qquad \text{Replace } y \text{ by } -y.$$
$$x - y^2 = 1 \qquad \text{Equivalent equation}$$

This means that the portion of the graph below the x-axis is a mirror image of the portion above the x-axis. To sketch the graph, first sketch the portion above the x-axis. Then reflect in the x-axis to obtain the entire graph, as shown in Figure 1.9.   ■

TECHNOLOGY   Graphing utilities are designed so that they most easily graph equations in which y is a function of x (see Section 1.3 for a definition of **function**). To graph other types of equations, you need to split the graph into two or more parts *or* you need to use a different graphing mode. For instance, to graph the equation in Example 4, you can split it into two parts.

$$y_1 = \sqrt{x - 1} \qquad \text{Top portion of graph}$$
$$y_2 = -\sqrt{x - 1} \qquad \text{Bottom portion of graph}$$

Points of Intersection

A **point of intersection** of the graphs of two equations is a point that satisfies both equations. You can find the point(s) of intersection of two graphs by solving their equations simultaneously.

EXAMPLE 5   Finding Points of Intersection

Find all points of intersection of the graphs of $x^2 - y = 3$ and $x - y = 1$.

Solution   Begin by sketching the graphs of both equations on the *same* rectangular coordinate system, as shown in Figure 1.10. Having done this, it appears that the graphs have two points of intersection. You can find these two points, as follows.

$$y = x^2 - 3 \qquad \text{Solve first equation for } y.$$
$$y = x - 1 \qquad \text{Solve second equation for } y.$$
$$x^2 - 3 = x - 1 \qquad \text{Equate } y\text{-values.}$$
$$x^2 - x - 2 = 0 \qquad \text{Write in general form.}$$
$$(x - 2)(x + 1) = 0 \qquad \text{Factor.}$$
$$x = 2 \text{ or } -1 \qquad \text{Solve for } x.$$

The corresponding values of y are obtained by substituting $x = 2$ and $x = -1$ into either of the original equations. Doing this produces two points of intersection:

$$(2, 1) \quad \text{and} \quad (-1, -2). \qquad \text{Points of intersection} \qquad ■$$

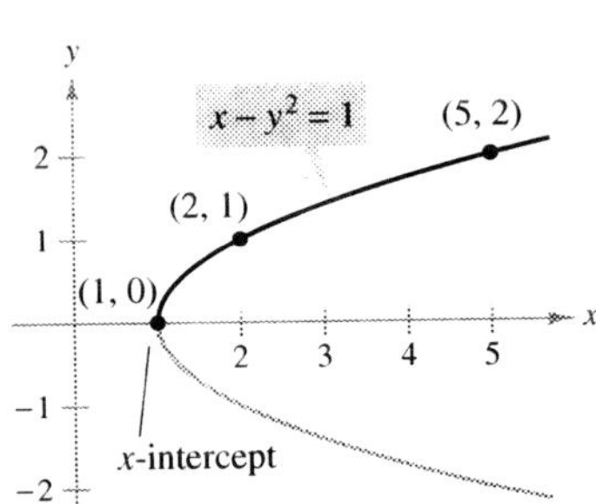

Figure 1.9

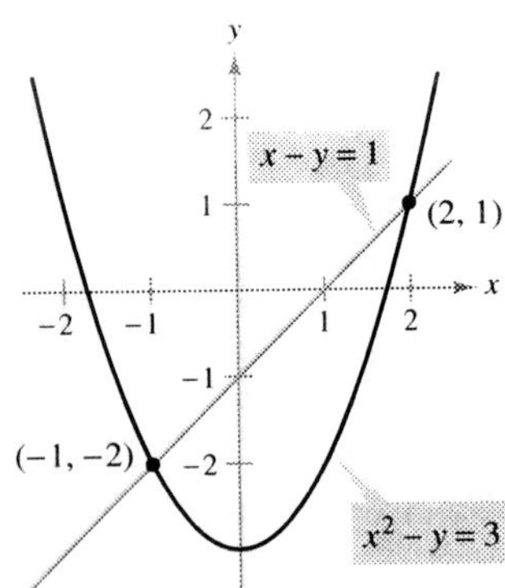

Two points of intersection
Figure 1.10

STUDY TIP   You can check the points of intersection in Example 5 by substituting into *both* of the original equations or by using the *intersect* feature of a graphing utility.

The icon ⟳ indicates that you will find a CAS Investigation on the book's website. The CAS Investigation is a collaborative exploration of this example using the computer algebra systems Maple *and* Mathematica.

Mathematical Models

Real-life applications of mathematics often use equations as **mathematical models.** In developing a mathematical model to represent actual data, you should strive for two (often conflicting) goals—accuracy and simplicity. That is, you want the model to be simple enough to be workable, yet accurate enough to produce meaningful results. Section 1.4 explores these goals more completely.

EXAMPLE 6 Comparing Two Mathematical Models

The Mauna Loa Observatory in Hawaii records the carbon dioxide concentration y (in parts per million) in Earth's atmosphere. The January readings for various years are shown in Figure 1.11. In the July 1990 issue of *Scientific American*, these data were used to predict the carbon dioxide level in Earth's atmosphere in the year 2035. The article used the quadratic model

$$y = 316.2 + 0.70t + 0.018t^2 \qquad \text{Quadratic model for 1960–1990 data}$$

where $t = 0$ represents 1960, as shown in Figure 1.11(a).

The data shown in Figure 1.11(b) represent the years 1980 through 2007 and can be modeled by

$$y = 304.1 + 1.64t \qquad \text{Linear model for 1980–2007 data}$$

where $t = 0$ represents 1960. What was the prediction given in the *Scientific American* article in 1990? Given the new data for 1990 through 2007, does this prediction for the year 2035 seem accurate?

The Mauna Loa Observatory in Hawaii has been measuring the increasing concentration of carbon dioxide in Earth's atmosphere since 1958. Carbon dioxide is the main greenhouse gas responsible for global climate warming.

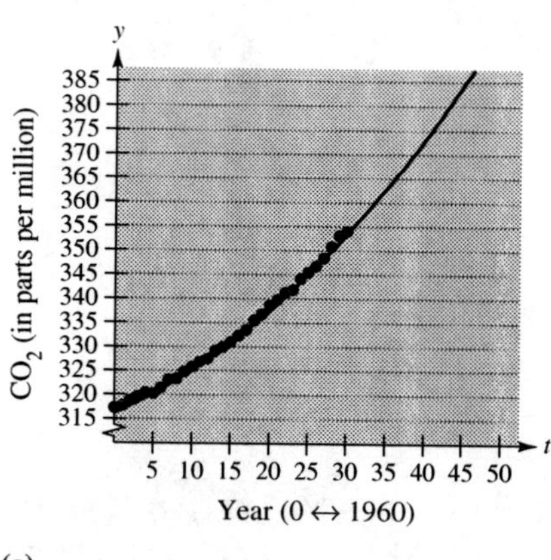

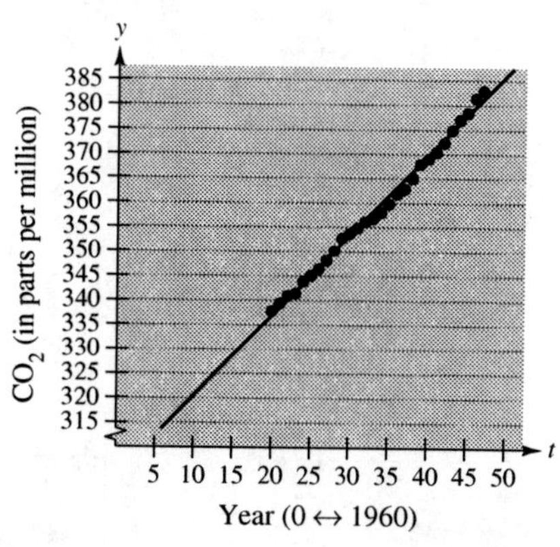

Figure 1.11

Solution   To answer the first question, substitute $t = 75$ (for 2035) into the quadratic model.

$$y = 316.2 + 0.70(75) + 0.018(75)^2 = 469.95 \qquad \text{Quadratic model}$$

So, the prediction in the *Scientific American* article was that the carbon dioxide concentration in Earth's atmosphere would reach about 470 parts per million in the year 2035. Using the linear model for the 1980–2007 data, the prediction for the year 2035 is

$$y = 304.1 + 1.64(75) = 427.1. \qquad \text{Linear model}$$

So, based on the linear model for 1980–2007, it appears that the 1990 prediction was too high.

NOTE   The models in Example 6 were developed using a procedure called *least squares regression* (see Section 13.9). The quadratic and linear models have correlations given by $r^2 = 0.997$ and $r^2 = 0.994$, respectively. The closer r^2 is to 1, the "better" the model.

1.1 Exercises

See www.CalcChat.com for worked-out solutions to odd-numbered exercises.

In Exercises 1–4, match the equation with its graph. [The graphs are labeled (a), (b), (c), and (d).]

(a) 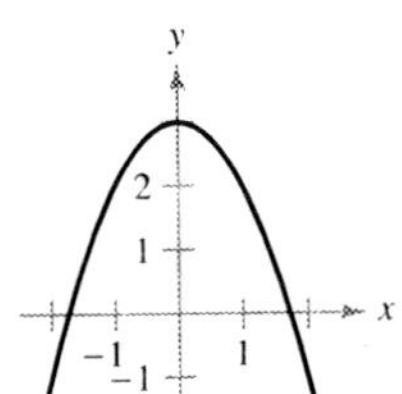

(b) 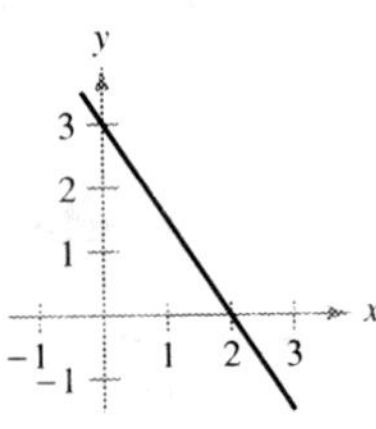

(c) 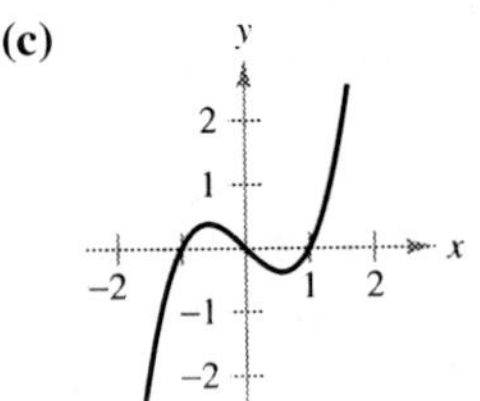

(d) 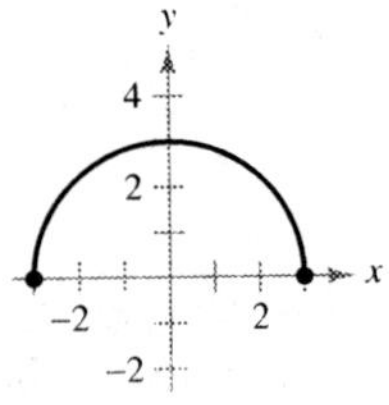

1. $y = -\frac{3}{2}x + 3$

2. $y = \sqrt{9 - x^2}$

3. $y = 3 - x^2$

4. $y = x^3 - x$

In Exercises 5–14, sketch the graph of the equation by point plotting.

5. $y = \frac{1}{2}x + 2$

6. $y = 5 - 2x$

7. $y = 4 - x^2$

8. $y = (x - 3)^2$

9. $y = |x + 2|$

10. $y = |x| - 1$

11. $y = \sqrt{x} - 6$

12. $y = \sqrt{x + 2}$

13. $y = \dfrac{3}{x}$

14. $y = \dfrac{1}{x + 2}$

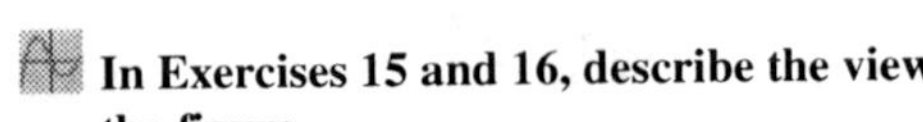 **In Exercises 15 and 16, describe the viewing window that yields the figure.**

15. $y = x^3 + 4x^2 - 3$

16. $y = |x| + |x - 16|$

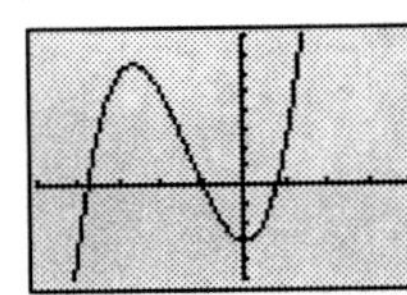

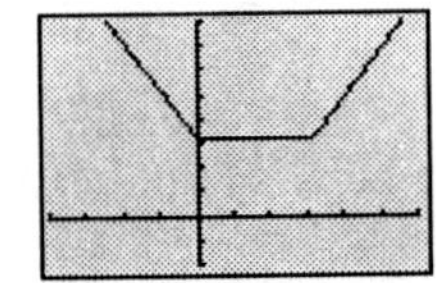

In Exercises 17 and 18, use a graphing utility to graph the equation. Move the cursor along the curve to approximate the unknown coordinate of each solution point accurate to two decimal places.

17. $y = \sqrt{5 - x}$ (a) $(2, y)$ (b) $(x, 3)$

18. $y = x^5 - 5x$ (a) $(-0.5, y)$ (b) $(x, -4)$

In Exercises 19–28, find any intercepts.

19. $y = 2x - 5$

20. $y = 4x^2 + 3$

21. $y = x^2 + x - 2$

22. $y^2 = x^3 - 4x$

23. $y = x\sqrt{16 - x^2}$

24. $y = (x - 1)\sqrt{x^2 + 1}$

25. $y = \dfrac{2 - \sqrt{x}}{5x}$

26. $y = \dfrac{x^2 + 3x}{(3x + 1)^2}$

27. $x^2y - x^2 + 4y = 0$

28. $y = 2x - \sqrt{x^2 + 1}$

In Exercises 29–40, test for symmetry with respect to each axis and to the origin.

29. $y = x^2 - 6$

30. $y = x^2 - x$

31. $y^2 = x^3 - 8x$

32. $y = x^3 + x$

33. $xy = 4$

34. $xy^2 = -10$

35. $y = 4 - \sqrt{x + 3}$

36. $xy - \sqrt{4 - x^2} = 0$

37. $y = \dfrac{x}{x^2 + 1}$

38. $y = \dfrac{x^2}{x^2 + 1}$

39. $y = |x^3 + x|$

40. $|y| - x = 3$

In Exercises 41–58, sketch the graph of the equation. Identify any intercepts and test for symmetry.

41. $y = 2 - 3x$

42. $y = -\frac{3}{2}x + 6$

43. $y = \frac{1}{2}x - 4$

44. $y = \frac{2}{3}x + 1$

45. $y = 9 - x^2$

46. $y = x^2 + 3$

47. $y = (x + 3)^2$

48. $y = 2x^2 + x$

49. $y = x^3 + 2$

50. $y = x^3 - 4x$

51. $y = x\sqrt{x + 5}$

52. $y = \sqrt{25 - x^2}$

53. $x = y^3$

54. $x = y^2 - 4$

55. $y = \dfrac{8}{x}$

56. $y = \dfrac{10}{x^2 + 1}$

57. $y = 6 - |x|$

58. $y = |6 - x|$

In Exercises 59–62, use a graphing utility to graph the equation. Identify any intercepts and test for symmetry.

59. $y^2 - x = 9$

60. $x^2 + 4y^2 = 4$

61. $x + 3y^2 = 6$

62. $3x - 4y^2 = 8$

In Exercises 63–70, find the points of intersection of the graphs of the equations.

63. $x + y = 8$
    $4x - y = 7$

64. $3x - 2y = -4$
    $4x + 2y = -10$

65. $x^2 + y = 6$
    $x + y = 4$

66. $x = 3 - y^2$
    $y = x - 1$

The symbol *indicates an exercise in which you are instructed to use graphing technology or a symbolic computer algebra system. The solutions of other exercises may also be facilitated by use of appropriate technology.*

67. $x^2 + y^2 = 5$
$x - y = 1$

68. $x^2 + y^2 = 25$
$-3x + y = 15$

69. $y = x^3$
$y = x$

70. $y = x^3 - 4x$
$y = -(x + 2)$

In Exercises 71–74, use a graphing utility to find the points of intersection of the graphs. Check your results analytically.

71. $y = x^3 - 2x^2 + x - 1$
$y = -x^2 + 3x - 1$

72. $y = x^4 - 2x^2 + 1$
$y = 1 - x^2$

73. $y = \sqrt{x + 6}$
$y = \sqrt{-x^2 - 4x}$

74. $y = -|2x - 3| + 6$
$y = 6 - x$

75. Modeling Data The table shows the Consumer Price Index (CPI) for selected years. *(Source: Bureau of Labor Statistics)*

Year	1975	1980	1985	1990	1995	2000	2005
CPI	53.8	82.4	107.6	130.7	152.4	172.2	195.3

(a) Use the regression capabilities of a graphing utility to find a mathematical model of the form $y = at^2 + bt + c$ for the data. In the model, y represents the CPI and t represents the year, with $t = 5$ corresponding to 1975.

(b) Use a graphing utility to plot the data and graph the model. Compare the data with the model.

(c) Use the model to predict the CPI for the year 2010.

76. Modeling Data The table shows the numbers of cellular phone subscribers (in millions) in the United States for selected years. *(Source: Cellular Telecommunications and Internet Association)*

Year	1990	1993	1996	1999	2002	2005
Number	5	16	44	86	141	208

(a) Use the regression capabilities of a graphing utility to find a mathematical model of the form $y = at^2 + bt + c$ for the data. In the model, y represents the number of subscribers and t represents the year, with $t = 0$ corresponding to 1990.

(b) Use a graphing utility to plot the data and graph the model. Compare the data with the model.

(c) Use the model to predict the number of cellular phone subscribers in the United States in the year 2015.

77. Break-Even Point Find the sales necessary to break even ($R = C$) if the cost C of producing x units is

$$C = 5.5\sqrt{x} + 10{,}000 \qquad \text{Cost equation}$$

and the revenue R from selling x units is

$$R = 3.29x. \qquad \text{Revenue equation}$$

78. Copper Wire The resistance y in ohms of 1000 feet of solid copper wire at 77°F can be approximated by the model

$$y = \frac{10{,}770}{x^2} - 0.37, \quad 5 \le x \le 100$$

where x is the diameter of the wire in mils (0.001 in.). Use a graphing utility to graph the model. If the diameter of the wire is doubled, the resistance is changed by about what factor?

WRITING ABOUT CONCEPTS

In Exercises 79 and 80, write an equation whose graph has the given property. (There may be more than one correct answer.)

79. The graph has intercepts at $x = -4$, $x = 3$, and $x = 8$.

80. The graph has intercepts at $x = -\frac{3}{2}$, $x = 4$, and $x = \frac{5}{2}$.

81. (a) Prove that if a graph is symmetric with respect to the x-axis and to the y-axis, then it is symmetric with respect to the origin. Give an example to show that the converse is not true.

(b) Prove that if a graph is symmetric with respect to one axis and to the origin, then it is symmetric with respect to the other axis.

CAPSTONE

82. Match the equation or equations with the given characteristic.

(i) $y = 3x^3 - 3x$ (ii) $y = (x + 3)^2$ (iii) $y = 3x - 3$
(iv) $y = \sqrt[3]{x}$ (v) $y = 3x^2 + 3$ (vi) $y = \sqrt{x + 3}$

(a) Symmetric with respect to the y-axis

(b) Three x-intercepts

(c) Symmetric with respect to the x-axis

(d) $(-2, 1)$ is a point on the graph

(e) Symmetric with respect to the origin

(f) Graph passes through the origin

True or False? **In Exercises 83–86, determine whether the statement is true or false. If it is false, explain why or give an example that shows it is false.**

83. If $(-4, -5)$ is a point on a graph that is symmetric with respect to the x-axis, then $(4, -5)$ is also a point on the graph.

84. If $(-4, -5)$ is a point on a graph that is symmetric with respect to the y-axis, then $(4, -5)$ is also a point on the graph.

85. If $b^2 - 4ac > 0$ and $a \ne 0$, then the graph of $y = ax^2 + bx + c$ has two x-intercepts.

86. If $b^2 - 4ac = 0$ and $a \ne 0$, then the graph of $y = ax^2 + bx + c$ has only one x-intercept.

In Exercises 87 and 88, find an equation of the graph that consists of all points (x, y) having the given distance from the origin. (For a review of the Distance Formula, see Appendix C.)

87. The distance from the origin is twice the distance from $(0, 3)$.

88. The distance from the origin is $K (K \ne 1)$ times the distance from $(2, 0)$.

1.2 Linear Models and Rates of Change

- Find the slope of a line passing through two points.
- Write the equation of a line given a point and the slope.
- Interpret slope as a ratio or as a rate in a real-life application.
- Sketch the graph of a linear equation in slope-intercept form.
- Write equations of lines that are parallel or perpendicular to a given line.

The Slope of a Line

The **slope** of a nonvertical line is a measure of the number of units the line rises (or falls) vertically for each unit of horizontal change from left to right. Consider the two points (x_1, y_1) and (x_2, y_2) on the line in Figure 1.12. As you move from left to right along this line, a vertical change of

$$\Delta y = y_2 - y_1 \qquad \text{Change in } y$$

units corresponds to a horizontal change of

$$\Delta x = x_2 - x_1 \qquad \text{Change in } x$$

units. (Δ is the Greek uppercase letter *delta*, and the symbols Δy and Δx are read "delta y" and "delta x.")

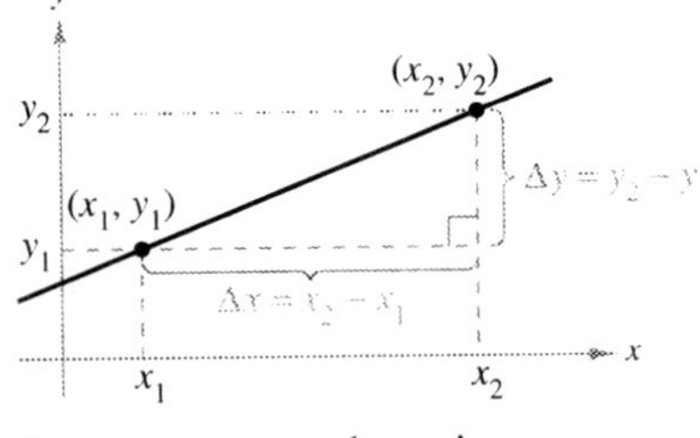

$\Delta y = y_2 - y_1 = \text{change in } y$
$\Delta x = x_2 - x_1 = \text{change in } x$
Figure 1.12

DEFINITION OF THE SLOPE OF A LINE

The **slope** m of the nonvertical line passing through (x_1, y_1) and (x_2, y_2) is

$$m = \frac{\Delta y}{\Delta x} = \frac{y_2 - y_1}{x_2 - x_1}, \quad x_1 \neq x_2.$$

Slope is not defined for vertical lines.

NOTE When using the formula for slope, note that

$$\frac{y_2 - y_1}{x_2 - x_1} = \frac{-(y_1 - y_2)}{-(x_1 - x_2)} = \frac{y_1 - y_2}{x_1 - x_2}.$$

So, it does not matter in which order you subtract *as long as* you are consistent and both "subtracted coordinates" come from the same point.

Figure 1.13 shows four lines: one has a positive slope, one has a slope of zero, one has a negative slope, and one has an "undefined" slope. In general, the greater the absolute value of the slope of a line, the steeper the line is. For instance, in Figure 1.13, the line with a slope of -5 is steeper than the line with a slope of $\frac{1}{5}$.

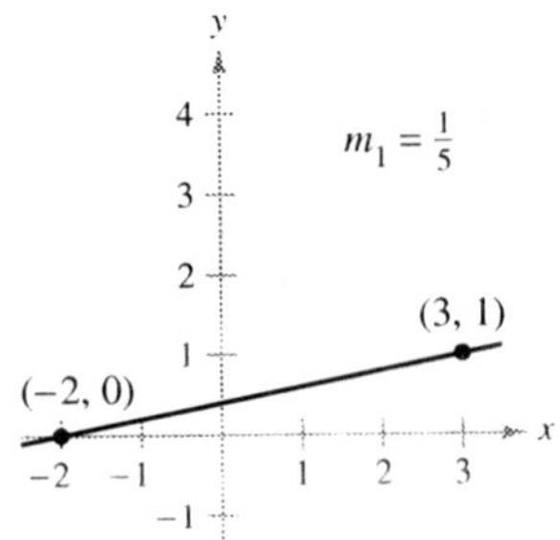

If m is positive, then the line rises from left to right.

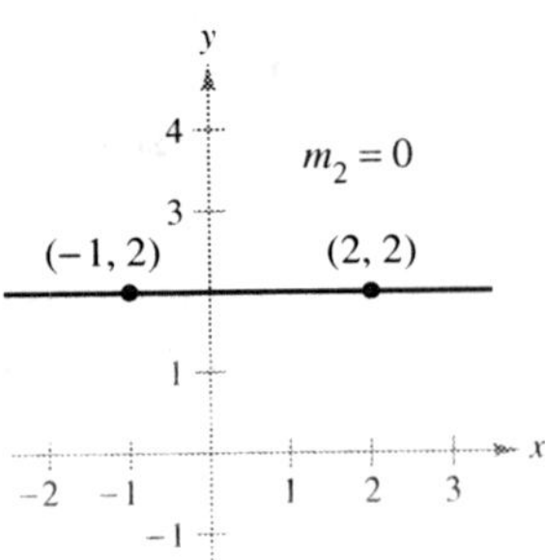

If m is zero, then the line is horizontal.

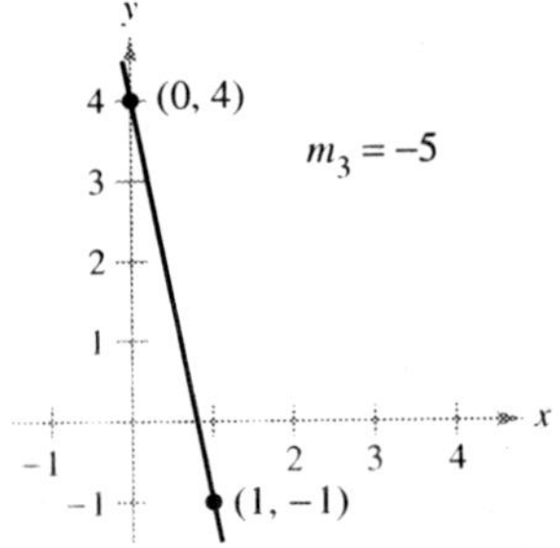

If m is negative, then the line falls from left to right.

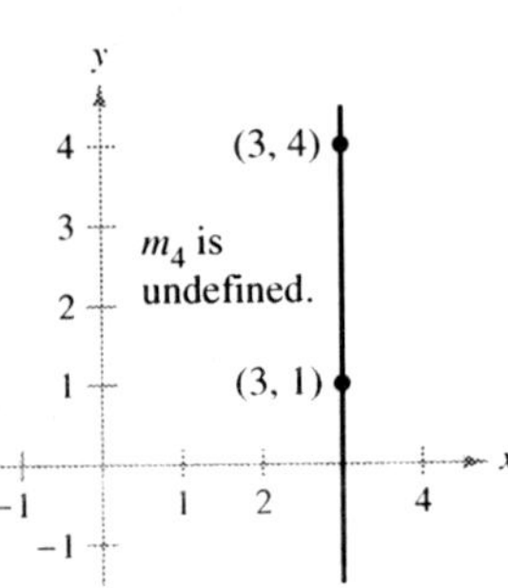

If m is undefined, then the line is vertical.

Figure 1.13

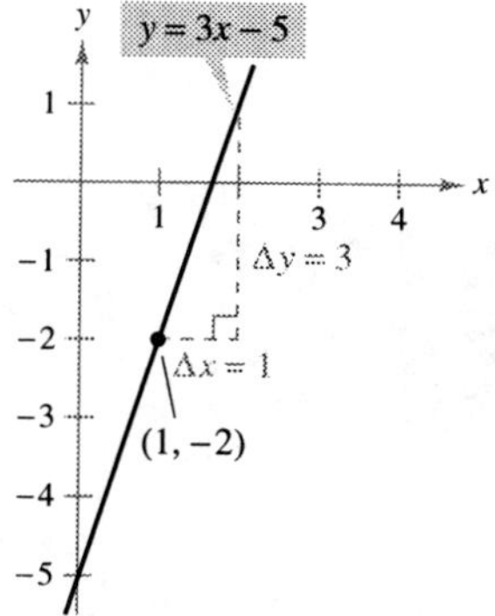
The line with a slope of 3 passing through
the point $(1, -2)$
Figure 1.15

EXPLORATION

Investigating Equations of Lines
Use a graphing utility to graph
each of the linear equations.
Which point is common to all
seven lines? Which value in the
equation determines the slope of
each line?

a. $y - 4 = -2(x + 1)$

b. $y - 4 = -1(x + 1)$

c. $y - 4 = -\frac{1}{2}(x + 1)$

d. $y - 4 = 0(x + 1)$

e. $y - 4 = \frac{1}{2}(x + 1)$

f. $y - 4 = 1(x + 1)$

g. $y - 4 = 2(x + 1)$

Use your results to write an
equation of a line passing through
$(-1, 4)$ with a slope of m.

Equations of Lines

Any two points on a nonvertical line can be used to calculate its slope. This can be
verified from the similar triangles shown in Figure 1.14. (Recall that the ratios of
corresponding sides of similar triangles are equal.)

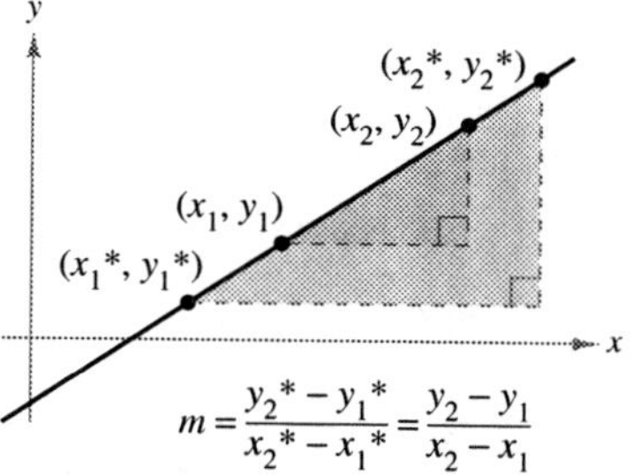

$$m = \frac{y_2{}^* - y_1{}^*}{x_2{}^* - x_1{}^*} = \frac{y_2 - y_1}{x_2 - x_1}$$

Any two points on a nonvertical line can be
used to determine its slope.
Figure 1.14

You can write an equation of a nonvertical line if you know the slope of the line
and the coordinates of one point on the line. Suppose the slope is m and the point is
(x_1, y_1). If (x, y) is any other point on the line, then

$$\frac{y - y_1}{x - x_1} = m.$$

This equation, involving the two variables x and y, can be rewritten in the form
$y - y_1 = m(x - x_1)$, which is called the **point-slope form of the equation of a line.**

POINT-SLOPE FORM OF THE EQUATION OF A LINE

An equation of the line with slope m passing through the point (x_1, y_1) is
given by

$$y - y_1 = m(x - x_1).$$

EXAMPLE 1   Finding an Equation of a Line

Find an equation of the line that has a slope of 3 and passes through the point $(1, -2)$.

Solution

$$\begin{aligned}
y - y_1 &= m(x - x_1) &&\text{Point-slope form} \\
y - (-2) &= 3(x - 1) &&\text{Substitute } -2 \text{ for } y_1, 1 \text{ for } x_1, \text{ and } 3 \text{ for } m. \\
y + 2 &= 3x - 3 &&\text{Simplify.} \\
y &= 3x - 5 &&\text{Solve for } y.
\end{aligned}$$

(See Figure 1.15.)

◼

NOTE Remember that only nonvertical lines have a slope. Consequently, vertical lines
cannot be written in point-slope form. For instance, the equation of the vertical line passing
through the point $(1, -2)$ is $x = 1$.

◼

Ratios and Rates of Change

The slope of a line can be interpreted as either a *ratio* or a *rate*. If the x- and y-axes have the same unit of measure, the slope has no units and is a **ratio.** If the x- and y-axes have different units of measure, the slope is a rate or **rate of change.** In your study of calculus, you will encounter applications involving both interpretations of slope.

EXAMPLE 2   Population Growth and Engineering Design

a. The population of Colorado was 3,827,000 in 1995 and 4,665,000 in 2005. Over this 10-year period, the average rate of change of the population was

$$\text{Rate of change} = \frac{\text{change in population}}{\text{change in years}}$$

$$= \frac{4,665,000 - 3,827,000}{2005 - 1995}$$

$$= 83,800 \text{ people per year.}$$

If Colorado's population continues to increase at this same rate for the next 10 years, it will have a 2015 population of 5,503,000 (see Figure 1.16).   (*Source: U.S. Census Bureau*)

b. In tournament water-ski jumping, the ramp rises to a height of 6 feet on a raft that is 21 feet long, as shown in Figure 1.17. The slope of the ski ramp is the ratio of its height (the rise) to the length of its base (the run).

$$\text{Slope of ramp} = \frac{\text{rise}}{\text{run}} \qquad \text{\small Rise is vertical change, run is horizontal change.}$$

$$= \frac{6 \text{ feet}}{21 \text{ feet}}$$

$$= \frac{2}{7}$$

In this case, note that the slope is a ratio and has no units.

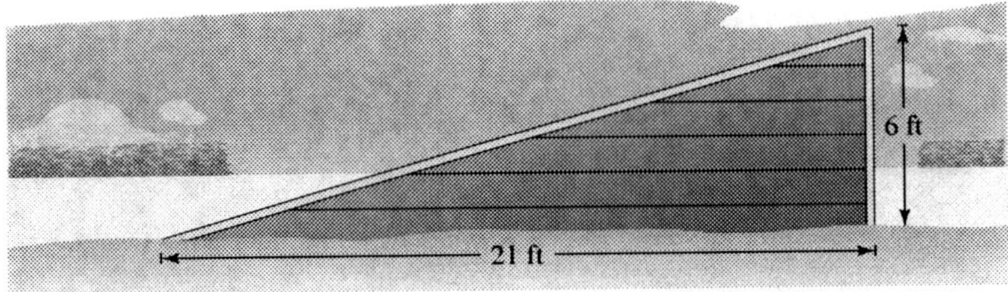

Dimensions of a water-ski ramp
Figure 1.17

The rate of change found in Example 2(a) is an **average rate of change.** An average rate of change is always calculated over an interval. In this case, the interval is $[1995, 2005]$. In Chapter 3 you will study another type of rate of change called an *instantaneous rate of change.*

Population of Colorado
Figure 1.16

Graphing Linear Models

Many problems in analytic geometry can be classified in two basic categories: (1) Given a graph, what is its equation? and (2) Given an equation, what is its graph? The point-slope equation of a line can be used to solve problems in the first category. However, this form is not especially useful for solving problems in the second category. The form that is better suited to sketching the graph of a line is the **slope-intercept form of the equation of a line.**

SLOPE-INTERCEPT FORM OF THE EQUATION OF A LINE

The graph of the linear equation

$$y = mx + b$$

is a line having a *slope* of m and a *y-intercept* at $(0, b)$.

EXAMPLE 3 Sketching Lines in the Plane

Sketch the graph of each equation.

a. $y = 2x + 1$ **b.** $y = 2$ **c.** $3y + x - 6 = 0$

Solution

a. Because $b = 1$, the y-intercept is $(0, 1)$. Because the slope is $m = 2$, you know that the line rises two units for each unit it moves to the right, as shown in Figure 1.18(a).

b. Because $b = 2$, the y-intercept is $(0, 2)$. Because the slope is $m = 0$, you know that the line is horizontal, as shown in Figure 1.18(b).

c. Begin by writing the equation in slope-intercept form.

$$3y + x - 6 = 0 \qquad \text{Write original equation.}$$
$$3y = -x + 6 \qquad \text{Isolate } y\text{-term on the left.}$$
$$y = -\frac{1}{3}x + 2 \qquad \text{Slope-intercept form}$$

In this form, you can see that the y-intercept is $(0, 2)$ and the slope is $m = -\frac{1}{3}$. This means that the line falls one unit for every three units it moves to the right, as shown in Figure 1.18(c).

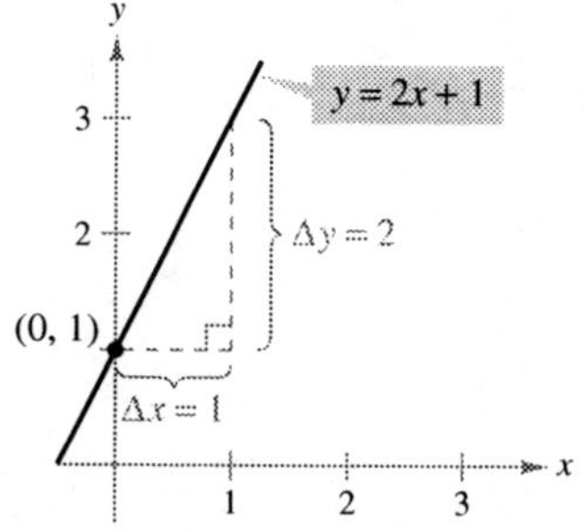

(a) $m = 2$; line rises

Figure 1.18

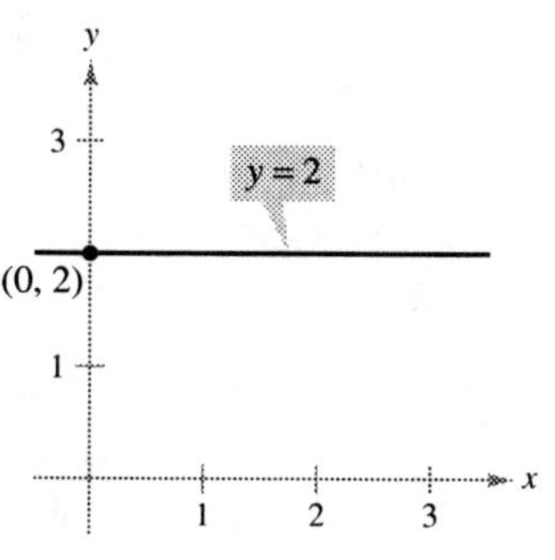

(b) $m = 0$; line is horizontal

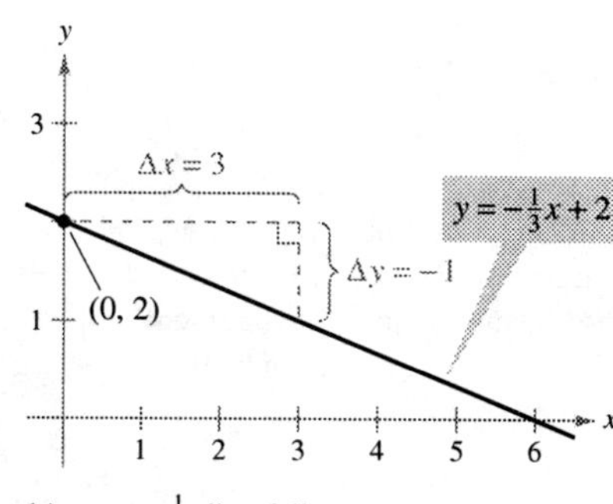

(c) $m = -\frac{1}{3}$; line falls

Because the slope of a vertical line is not defined, its equation cannot be written in the slope-intercept form. However, the equation of *any* line can be written in the **general form**

$$Ax + By + C = 0 \qquad \text{General form of the equation of a line}$$

where A and B are not *both* zero. For instance, the vertical line given by $x = a$ can be represented by the general form $x - a = 0$.

SUMMARY OF EQUATIONS OF LINES

1. General form: $Ax + By + C = 0$
2. Vertical line: $x = a$
3. Horizontal line: $y = b$
4. Point-slope form: $y - y_1 = m(x - x_1)$
5. Slope-intercept form: $y = mx + b$

Parallel and Perpendicular Lines

The slope of a line is a convenient tool for determining whether two lines are parallel or perpendicular, as shown in Figure 1.19. Specifically, nonvertical lines with the same slope are parallel and nonvertical lines whose slopes are negative reciprocals are perpendicular.

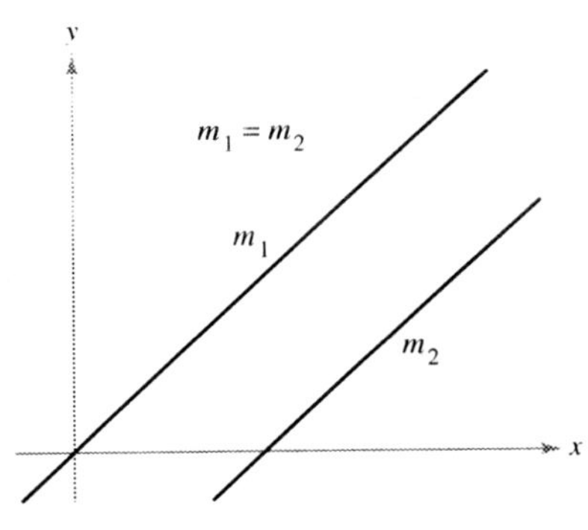

Parallel lines

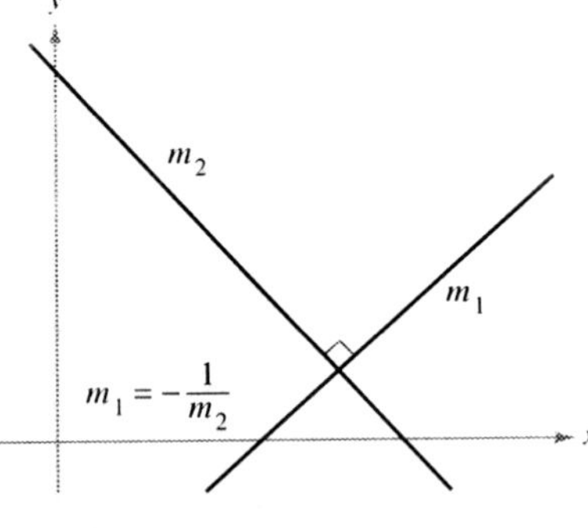

Perpendicular lines

Figure 1.19

STUDY TIP In mathematics, the phrase "if and only if" is a way of stating two implications in one statement. For instance, the first statement at the right could be rewritten as the following two implications.

a. If two distinct nonvertical lines are parallel, then their slopes are equal.

b. If two distinct nonvertical lines have equal slopes, then they are parallel.

PARALLEL AND PERPENDICULAR LINES

1. Two distinct nonvertical lines are **parallel** if and only if their slopes are equal—that is, if and only if

$$m_1 = m_2.$$

2. Two nonvertical lines are **perpendicular** if and only if their slopes are negative reciprocals of each other—that is, if and only if

$$m_1 = -\frac{1}{m_2}.$$

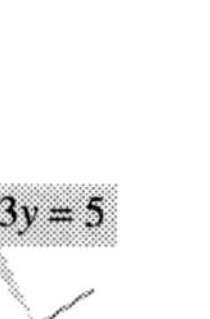

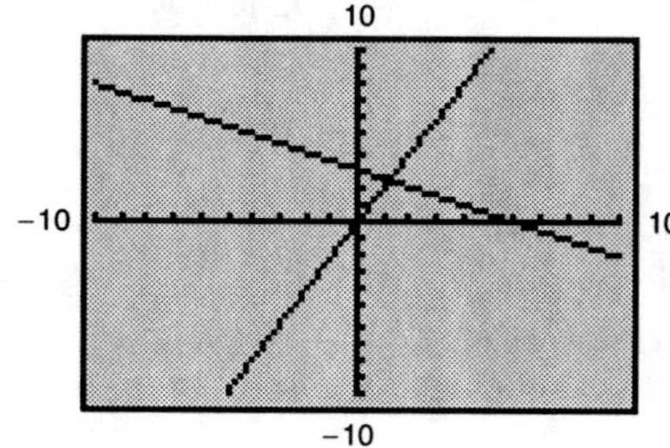

Lines parallel and perpendicular to
$2x - 3y = 5$
Figure 1.20

EXAMPLE 4 Finding Parallel and Perpendicular Lines

Find the general forms of the equations of the lines that pass through the point $(2, -1)$ and are

a. parallel to the line $2x - 3y = 5$.   **b.** perpendicular to the line $2x - 3y = 5$.

(See Figure 1.20.)

Solution   By writing the linear equation $2x - 3y = 5$ in slope-intercept form, $y = \frac{2}{3}x - \frac{5}{3}$, you can see that the given line has a slope of $m = \frac{2}{3}$.

a. The line through $(2, -1)$ that is parallel to the given line also has a slope of $\frac{2}{3}$.

$$y - y_1 = m(x - x_1) \qquad \text{Point-slope form}$$
$$y - (-1) = \tfrac{2}{3}(x - 2) \qquad \text{Substitute.}$$
$$3(y + 1) = 2(x - 2) \qquad \text{Simplify.}$$
$$2x - 3y - 7 = 0 \qquad \text{General form}$$

Note the similarity to the original equation.

b. Using the negative reciprocal of the slope of the given line, you can determine that the slope of a line perpendicular to the given line is $-\frac{3}{2}$. So, the line through the point $(2, -1)$ that is perpendicular to the given line has the following equation.

$$y - y_1 = m(x - x_1) \qquad \text{Point-slope form}$$
$$y - (-1) = -\tfrac{3}{2}(x - 2) \qquad \text{Substitute.}$$
$$2(y + 1) = -3(x - 2) \qquad \text{Simplify.}$$
$$3x + 2y - 4 = 0 \qquad \text{General form}$$

TECHNOLOGY PITFALL The slope of a line will appear distorted if you use different tick-mark spacing on the x- and y-axes. For instance, the graphing calculator screens in Figures 1.21(a) and 1.21(b) both show the lines given by

$$y = 2x \quad \text{and} \quad y = -\tfrac{1}{2}x + 3.$$

Because these lines have slopes that are negative reciprocals, they must be perpendicular. In Figure 1.21(a), however, the lines don't appear to be perpendicular because the tick-mark spacing on the x-axis is not the same as that on the y-axis. In Figure 1.21(b), the lines appear perpendicular because the tick-mark spacing on the x-axis is the same as on the y-axis. This type of viewing window is said to have a *square setting*.

(a) Tick-mark spacing on the x-axis is not the same as tick-mark spacing on the y-axis.

(b) Tick-mark spacing on the x-axis is the same as tick-mark spacing on the y-axis.

Figure 1.21

1.2　Exercises

See www.CalcChat.com for worked-out solutions to odd-numbered exercises.

In Exercises 1–6, estimate the slope of the line from its graph. To print an enlarged copy of the graph, go to the website *www.mathgraphs.com*.

1. 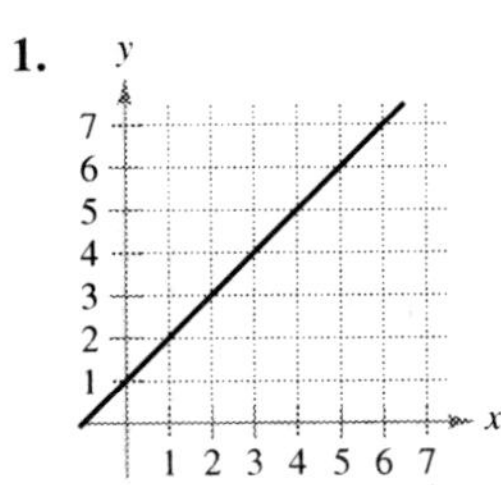

2. 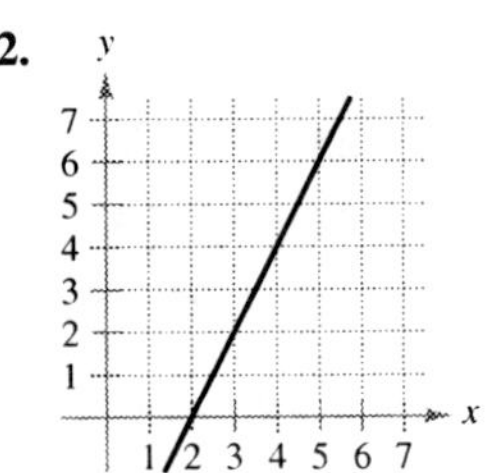

3. 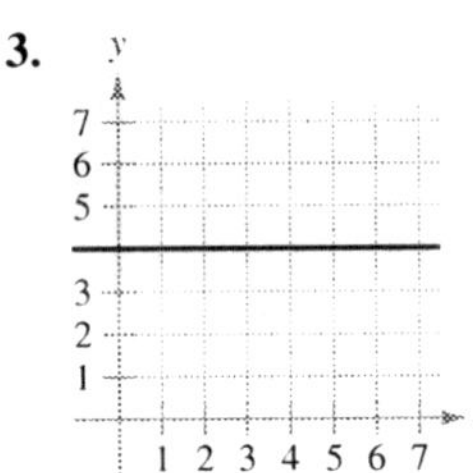

4. 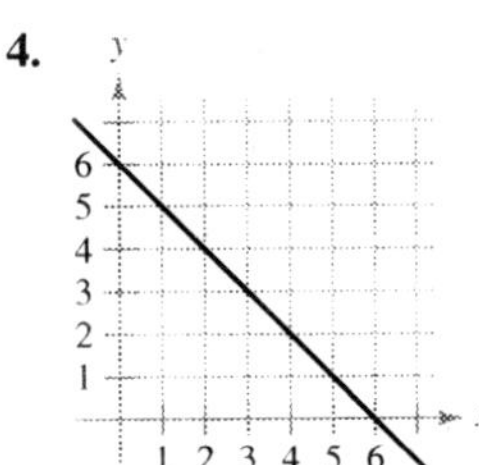

5. 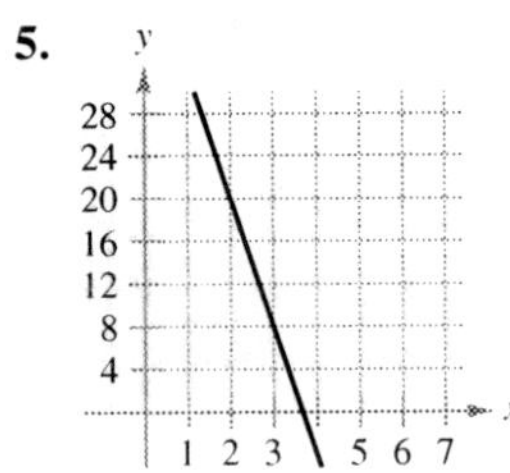

6. 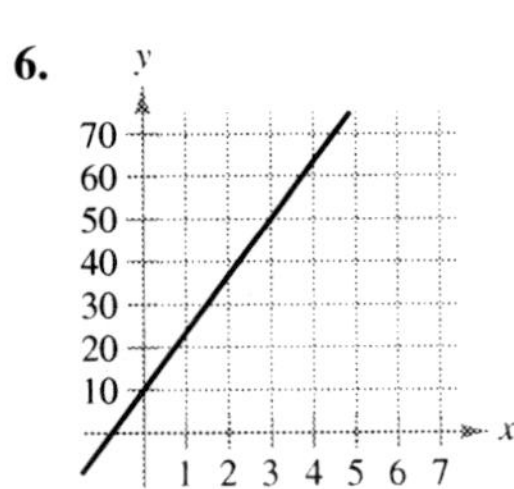

In Exercises 7 and 8, sketch the lines through the point with the given slopes. Make the sketches on the same set of coordinate axes.

	Point	Slopes			
7.	$(3, 4)$	(a) 1	(b) -2	(c) $-\frac{3}{2}$	(d) Undefined
8.	$(-2, 5)$	(a) 3	(b) -3	(c) $\frac{1}{3}$	(d) 0

In Exercises 9–14, plot the pair of points and find the slope of the line passing through them.

9. $(3, -4), (5, 2)$

10. $(1, 1), (-2, 7)$

11. $(4, 6), (4, 1)$

12. $(3, -5), (5, -5)$

13. $\left(-\frac{1}{2}, \frac{2}{3}\right), \left(-\frac{3}{4}, \frac{1}{6}\right)$

14. $\left(\frac{7}{8}, \frac{3}{4}\right), \left(\frac{5}{4}, -\frac{1}{4}\right)$

In Exercises 15–18, use the point on the line and the slope of the line to find three additional points that the line passes through. (There is more than one correct answer.)

	Point	Slope		Point	Slope
15.	$(6, 2)$	$m = 0$	**16.**	$(-4, 3)$	m is undefined.
17.	$(1, 7)$	$m = -3$	**18.**	$(-2, -2)$	$m = 2$

19. *Conveyor Design*　A moving conveyor is built to rise 1 meter for each 3 meters of horizontal change.

(a) Find the slope of the conveyor.

(b) Suppose the conveyor runs between two floors in a factory. Find the length of the conveyor if the vertical distance between floors is 10 feet.

20. *Rate of Change*　Each of the following is the slope of a line representing daily revenue y in terms of time x in days. Use the slope to interpret any change in daily revenue for a one-day increase in time.

(a) $m = 800$　　(b) $m = 250$　　(c) $m = 0$

21. *Modeling Data*　The table shows the populations y (in millions) of the United States for 2000 through 2005. The variable t represents the time in years, with $t = 0$ corresponding to 2000. (*Source: U.S. Bureau of the Census*)

t	0	1	2	3	4	5
y	282.4	285.3	288.2	291.1	293.9	296.6

(a) Plot the data by hand and connect adjacent points with a line segment.

(b) Use the slope of each line segment to determine the year when the population increased least rapidly.

22. *Modeling Data*　The table shows the rate r (in miles per hour) that a vehicle is traveling after t seconds.

t	5	10	15	20	25	30
r	57	74	85	84	61	43

(a) Plot the data by hand and connect adjacent points with a line segment.

(b) Use the slope of each line segment to determine the interval when the vehicle's rate changed most rapidly. How did the rate change?

In Exercises 23–28, find the slope and the y-intercept (if possible) of the line.

23. $y = 4x - 3$

24. $-x + y = 1$

25. $x + 5y = 20$

26. $6x - 5y = 15$

27. $x = 4$

28. $y = -1$

In Exercises 29–34, find an equation of the line that passes through the point and has the given slope. Sketch the line.

	Point	Slope		Point	Slope
29.	$(0, 3)$	$m = \frac{3}{4}$	**30.**	$(-5, -2)$	m is undefined.
31.	$(0, 0)$	$m = \frac{2}{3}$	**32.**	$(0, 4)$	$m = 0$
33.	$(3, -2)$	$m = 3$	**34.**	$(-2, 4)$	$m = -\frac{3}{5}$

In Exercises 35–44, find an equation of the line that passes through the points, and sketch the line.

35. $(0, 0)$, $(4, 8)$

36. $(0, 0)$, $(-1, 5)$

37. $(2, 1)$, $(0, -3)$

38. $(-2, -2)$, $(1, 7)$

39. $(2, 8)$, $(5, 0)$

40. $(-3, 6)$, $(1, 2)$

41. $(6, 3)$, $(6, 8)$

42. $(1, -2)$, $(3, -2)$

43. $\left(\frac{1}{2}, \frac{7}{2}\right)$, $\left(0, \frac{3}{4}\right)$

44. $\left(\frac{7}{8}, \frac{3}{4}\right)$, $\left(\frac{5}{4}, -\frac{1}{4}\right)$

45. Find an equation of the vertical line with x-intercept at 3.

46. Show that the line with intercepts $(a, 0)$ and $(0, b)$ has the following equation.

$$\frac{x}{a} + \frac{y}{b} = 1, \quad a \neq 0, b \neq 0$$

In Exercises 47–50, use the result of Exercise 46 to write an equation of the line in general form.

47. x-intercept: $(2, 0)$

   y-intercept: $(0, 3)$

48. x-intercept: $\left(-\frac{2}{3}, 0\right)$

   y-intercept: $(0, -2)$

49. Point on line: $(1, 2)$

   x-intercept: $(a, 0)$

   y-intercept: $(0, a)$

   $(a \neq 0)$

50. Point on line: $(-3, 4)$

   x-intercept: $(a, 0)$

   y-intercept: $(0, a)$

   $(a \neq 0)$

In Exercises 51–58, sketch a graph of the equation.

51. $y = -3$

52. $x = 4$

53. $y = -2x + 1$

54. $y = \frac{1}{3}x - 1$

55. $y - 2 = \frac{3}{2}(x - 1)$

56. $y - 1 = 3(x + 4)$

57. $2x - y - 3 = 0$

58. $x + 2y + 6 = 0$

 **59.** *Square Setting* Use a graphing utility to graph the lines $y = 2x - 3$ and $y = -\frac{1}{2}x + 1$ in each viewing window. Compare the graphs. Do the lines appear perpendicular? Are the lines perpendicular? Explain.

(a)
```
Xmin = -5
Xmax = 5
Xscl = 1
Ymin = -5
Ymax = 5
Yscl = 1
```

(b)
```
Xmin = -6
Xmax = 6
Xscl = 1
Ymin = -4
Ymax = 4
Yscl = 1
```


60. A line is represented by the equation $ax + by = 4$.

   (a) When is the line parallel to the x-axis?

   (b) When is the line parallel to the y-axis?

   (c) Give values for a and b such that the line has a slope of $\frac{5}{8}$.

   (d) Give values for a and b such that the line is perpendicular to $y = \frac{2}{5}x + 3$.

   (e) Give values for a and b such that the line coincides with the graph of $5x + 6y = 8$.

In Exercises 61–66, write the general forms of the equations of the lines through the point (a) parallel to the given line and (b) perpendicular to the given line.

	Point	*Line*		*Point*	*Line*
61.	$(-7, -2)$	$x = 1$	**62.**	$(-1, 0)$	$y = -3$
63.	$(2, 1)$	$4x - 2y = 3$	**64.**	$(-3, 2)$	$x + y = 7$
65.	$\left(\frac{3}{4}, \frac{7}{8}\right)$	$5x - 3y = 0$	**66.**	$(4, -5)$	$3x + 4y = 7$

Rate of Change **In Exercises 67–70, you are given the dollar value of a product in 2008 *and* the rate at which the value of the product is expected to change during the next 5 years. Write a linear equation that gives the dollar value V of the product in terms of the year t. (Let $t = 0$ represent 2000.)**

	2008 Value	*Rate*
67.	$1850	$250 increase per year
68.	$156	$4.50 increase per year
69.	$17,200	$1600 decrease per year
70.	$245,000	$5600 decrease per year

 **In Exercises 71 and 72, use a graphing utility to graph the parabolas and find their points of intersection. Find an equation of the line through the points of intersection and graph the line in the same viewing window.**

71. $y = x^2$

   $y = 4x - x^2$

72. $y = x^2 - 4x + 3$

   $y = -x^2 + 2x + 3$

In Exercises 73 and 74, determine whether the points are collinear. (Three points are *collinear* if they lie on the same line.)

73. $(-2, 1)$, $(-1, 0)$, $(2, -2)$

74. $(0, 4)$, $(7, -6)$, $(-5, 11)$

In Exercises 75–77, find the coordinates of the point of intersection of the given segments. Explain your reasoning.

75.

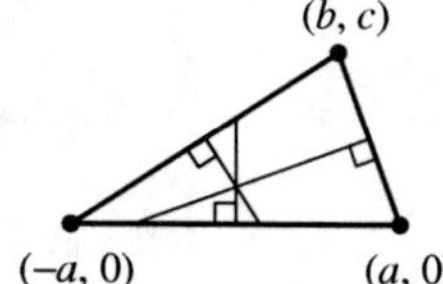

Perpendicular bisectors

76.

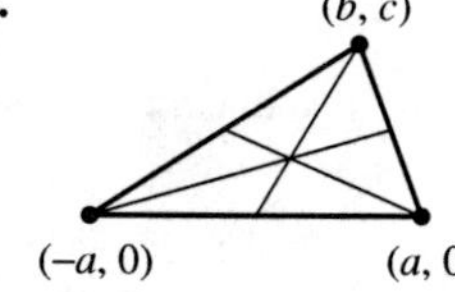

Medians

77.

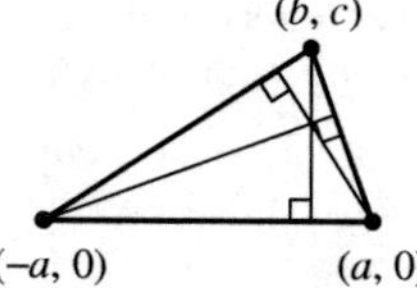

Altitudes

78. Show that the points of intersection in Exercises 75, 76, and 77 are collinear.

79. *Temperature Conversion*   Find a linear equation that expresses the relationship between the temperature in degrees Celsius C and degrees Fahrenheit F. Use the fact that water freezes at $0°C$ ($32°F$) and boils at $100°C$ ($212°F$). Use the equation to convert $72°F$ to degrees Celsius.

80. *Reimbursed Expenses*   A company reimburses its sales representatives \$175 per day for lodging and meals plus 48¢ per mile driven. Write a linear equation giving the daily cost C to the company in terms of x, the number of miles driven. How much does it cost the company if a sales representative drives 137 miles on a given day?

81. *Career Choice*   An employee has two options for positions in a large corporation. One position pays \$14.50 per hour *plus* an additional unit rate of \$0.75 per unit produced. The other pays \$11.20 per hour *plus* a unit rate of \$1.30.

 (a) Find linear equations for the hourly wages W in terms of x, the number of units produced per hour, for each option.

 (b) Use a graphing utility to graph the linear equations and find the point of intersection.

 (c) Interpret the meaning of the point of intersection of the graphs in part (b). How would you use this information to select the correct option if the goal were to obtain the highest hourly wage?

82. *Straight-Line Depreciation*   A small business purchases a piece of equipment for \$875. After 5 years the equipment will be outdated, having no value.

 (a) Write a linear equation giving the value y of the equipment in terms of the time x, $0 \le x \le 5$.

 (b) Find the value of the equipment when $x = 2$.

 (c) Estimate (to two-decimal-place accuracy) the time when the value of the equipment is \$200.

83. *Apartment Rental*   A real estate office manages an apartment complex with 50 units. When the rent is \$780 per month, all 50 units are occupied. However, when the rent is \$825, the average number of occupied units drops to 47. Assume that the relationship between the monthly rent p and the demand x is linear. (*Note:* The term *demand* refers to the number of occupied units.)

 (a) Write a linear equation giving the demand x in terms of the rent p.

 (b) *Linear extrapolation* Use a graphing utility to graph the demand equation and use the *trace* feature to predict the number of units occupied if the rent is raised to \$855.

 (c) *Linear interpolation* Predict the number of units occupied if the rent is lowered to \$795. Verify graphically.

84. *Modeling Data*   An instructor gives regular 20-point quizzes and 100-point exams in a mathematics course. Average scores for six students, given as ordered pairs (x, y), where x is the average quiz score and y is the average test score, are $(18, 87)$, $(10, 55)$, $(19, 96)$, $(16, 79)$, $(13, 76)$, and $(15, 82)$.

 (a) Use the regression capabilities of a graphing utility to find the least squares regression line for the data.

 (b) Use a graphing utility to plot the points and graph the regression line in the same viewing window.

 (c) Use the regression line to predict the average exam score for a student with an average quiz score of 17.

 (d) Interpret the meaning of the slope of the regression line.

 (e) The instructor adds 4 points to the average test score of everyone in the class. Describe the changes in the positions of the plotted points and the change in the equation of the line.

85. *Tangent Line*   Find an equation of the line tangent to the circle $x^2 + y^2 = 169$ at the point $(5, 12)$.

86. *Tangent Line*   Find an equation of the line tangent to the circle $(x - 1)^2 + (y - 1)^2 = 25$ at the point $(4, -3)$.

Distance   **In Exercises 87–92, find the distance between the point and line, or between the lines, using the formula for the distance between the point (x_1, y_1) and the line $Ax + By + C = 0$.**

$$\text{Distance} = \frac{|Ax_1 + By_1 + C|}{\sqrt{A^2 + B^2}}$$

87. Point: $(0, 0)$

 Line: $4x + 3y = 10$

88. Point: $(2, 3)$

 Line: $4x + 3y = 10$

89. Point: $(-2, 1)$

 Line: $x - y - 2 = 0$

90. Point: $(6, 2)$

 Line: $x = -1$

91. Line: $x + y = 1$

 Line: $x + y = 5$

92. Line: $3x - 4y = 1$

 Line: $3x - 4y = 10$

93. Show that the distance between the point (x_1, y_1) and the line $Ax + By + C = 0$ is

$$\text{Distance} = \frac{|Ax_1 + By_1 + C|}{\sqrt{A^2 + B^2}}.$$

94. Write the distance d between the point $(3, 1)$ and the line $y = mx + 4$ in terms of m. Use a graphing utility to graph the equation. When is the distance 0? Explain the result geometrically.

95. Prove that the diagonals of a rhombus intersect at right angles. (A rhombus is a quadrilateral with sides of equal lengths.)

96. Prove that the figure formed by connecting consecutive midpoints of the sides of any quadrilateral is a parallelogram.

97. Prove that if the points (x_1, y_1) and (x_2, y_2) lie on the same line as (x_1^*, y_1^*) and (x_2^*, y_2^*), then

$$\frac{y_2^* - y_1^*}{x_2^* - x_1^*} = \frac{y_2 - y_1}{x_2 - x_1}.$$

 Assume $x_1 \ne x_2$ and $x_1^* \ne x_2^*$.

98. Prove that if the slopes of two nonvertical lines are negative reciprocals of each other, then the lines are perpendicular.

True or False?   **In Exercises 99 and 100, determine whether the statement is true or false. If it is false, explain why or give an example that shows it is false.**

99. The lines represented by $ax + by = c_1$ and $bx - ay = c_2$ are perpendicular. Assume $a \ne 0$ and $b \ne 0$.

100. It is possible for two lines with positive slopes to be perpendicular to each other.

1.3　Functions and Their Graphs

- **Use function notation to represent and evaluate a function.**
- **Find the domain and range of a function.**
- **Sketch the graph of a function.**
- **Identify different types of transformations of functions.**
- **Classify functions and recognize combinations of functions.**

Functions and Function Notation

A **relation** between two sets X and Y is a set of ordered pairs, each of the form (x, y), where x is a member of X and y is a member of Y. A **function** from X to Y is a relation between X and Y having the property that any two ordered pairs with the same x-value also have the same y-value. The variable x is the **independent variable,** and the variable y is the **dependent variable.**

Many real-life situations can be modeled by functions. For instance, the area A of a circle is a function of the circle's radius r.

$$A = \pi r^2 \qquad \text{\textit{A is a function of r.}}$$

In this case r is the independent variable and A is the dependent variable.

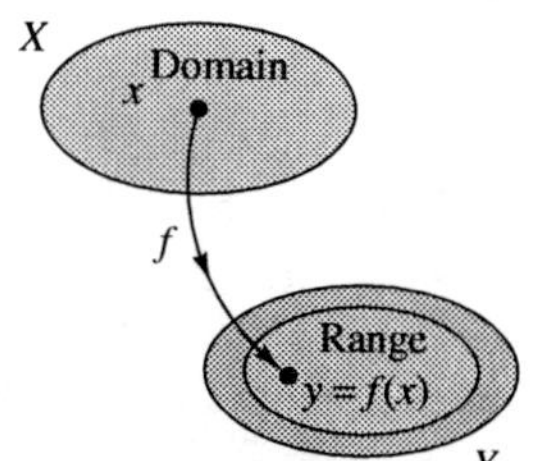

A real-valued function f of a real variable x
Figure 1.22

DEFINITION OF A REAL-VALUED FUNCTION OF A REAL VARIABLE

Let X and Y be sets of real numbers. A **real-valued function f of a real variable x** from X to Y is a correspondence that assigns to each number x in X exactly one number y in Y.

The **domain** of f is the set X. The number y is the **image** of x under f and is denoted by $f(x)$, which is called the **value of f at x.** The **range** of f is a subset of Y and consists of all images of numbers in X (see Figure 1.22).

Functions can be specified in a variety of ways. In this text, however, we will concentrate primarily on functions that are given by equations involving the dependent and independent variables. For instance, the equation

$$x^2 + 2y = 1 \qquad \text{\textit{Equation in implicit form}}$$

defines y, the dependent variable, as a function of x, the independent variable. To **evaluate** this function (that is, to find the y-value that corresponds to a given x-value), it is convenient to isolate y on the left side of the equation.

$$y = \frac{1}{2}(1 - x^2) \qquad \text{\textit{Equation in explicit form}}$$

Using f as the name of the function, you can write this equation as

$$f(x) = \frac{1}{2}(1 - x^2). \qquad \text{\textit{Function notation}}$$

FUNCTION NOTATION

The word *function* was first used by Gottfried Wilhelm Leibniz in 1694 as a term to denote any quantity connected with a curve, such as the coordinates of a point on a curve or the slope of a curve. Forty years later, Leonhard Euler used the word *function* to describe any expression made up of a variable and some constants. He introduced the notation $y = f(x)$.

The original equation, $x^2 + 2y = 1$, **implicitly** defines y as a function of x. When you solve the equation for y, you are writing the equation in **explicit** form.

Function notation has the advantage of clearly identifying the dependent variable as $f(x)$ while at the same time telling you that x is the independent variable and that the function itself is "f." The symbol $f(x)$ is read "f of x." Function notation allows you to be less wordy. Instead of asking "What is the value of y that corresponds to $x = 3$?" you can ask, "What is $f(3)$?"

In an equation that defines a function, the role of the variable x is simply that of a placeholder. For instance, the function given by

$$f(x) = 2x^2 - 4x + 1$$

can be described by the form

$$f(\blacksquare) = 2(\blacksquare)^2 - 4(\blacksquare) + 1$$

where parentheses are used instead of x. To evaluate $f(-2)$, simply place -2 in each set of parentheses.

$$
\begin{aligned}
f(-2) &= 2(-2)^2 - 4(-2) + 1 && \text{Substitute } -2 \text{ for } x. \\
&= 2(4) + 8 + 1 && \text{Simplify.} \\
&= 17 && \text{Simplify.}
\end{aligned}
$$

NOTE Although f is often used as a convenient function name and x as the independent variable, you can use other symbols. For instance, the following equations all define the same function.

$$
\begin{aligned}
f(x) &= x^2 - 4x + 7 && \text{Function name is } f, \text{ independent variable is } x. \\
f(t) &= t^2 - 4t + 7 && \text{Function name is } f, \text{ independent variable is } t. \\
g(s) &= s^2 - 4s + 7 && \text{Function name is } g, \text{ independent variable is } s.
\end{aligned}
$$

EXAMPLE 1 Evaluating a Function

For the function f defined by $f(x) = x^2 + 7$, evaluate each of the following.

a. $f(3a)$ **b.** $f(b - 1)$ **c.** $\dfrac{f(x + \Delta x) - f(x)}{\Delta x}, \quad \Delta x \neq 0$

Solution

$$
\begin{aligned}
\textbf{a.} \quad f(3a) &= (3a)^2 + 7 && \text{Substitute } 3a \text{ for } x. \\
&= 9a^2 + 7 && \text{Simplify.}
\end{aligned}
$$

$$
\begin{aligned}
\textbf{b.} \quad f(b - 1) &= (b - 1)^2 + 7 && \text{Substitute } b - 1 \text{ for } x. \\
&= b^2 - 2b + 1 + 7 && \text{Expand binomial.} \\
&= b^2 - 2b + 8 && \text{Simplify.}
\end{aligned}
$$

$$
\begin{aligned}
\textbf{c.} \quad \frac{f(x + \Delta x) - f(x)}{\Delta x} &= \frac{[(x + \Delta x)^2 + 7] - (x^2 + 7)}{\Delta x} \\
&= \frac{x^2 + 2x\Delta x + (\Delta x)^2 + 7 - x^2 - 7}{\Delta x} \\
&= \frac{2x\Delta x + (\Delta x)^2}{\Delta x} \\
&= \frac{\Delta x(2x + \Delta x)}{\Delta x} \\
&= 2x + \Delta x, \quad \Delta x \neq 0
\end{aligned}
$$

STUDY TIP In calculus, it is important to specify clearly the domain of a function or expression. For instance, in Example 1(c), the two expressions

$$\frac{f(x + \Delta x) - f(x)}{\Delta x} \quad \text{and} \quad 2x + \Delta x, \quad \Delta x \neq 0$$

are equivalent because $\Delta x = 0$ is excluded from the domain of each expression. Without a stated domain restriction, the two expressions would not be equivalent.

NOTE The expression in Example 1(c) is called a *difference quotient* and has a special significance in calculus. You will learn more about this in Chapter 3.

The Domain and Range of a Function

The domain of a function may be described explicitly, or it may be described *implicitly* by an equation used to define the function. The implied domain is the set of all real numbers for which the equation is defined, whereas an explicitly defined domain is one that is given along with the function. For example, the function given by

$$f(x) = \frac{1}{x^2 - 4}, \quad 4 \le x \le 5$$

has an explicitly defined domain given by $\{x \colon 4 \le x \le 5\}$. On the other hand, the function given by

$$g(x) = \frac{1}{x^2 - 4}$$

has an implied domain that is the set $\{x \colon x \ne \pm 2\}$.

EXAMPLE 2 Finding the Domain and Range of a Function

a. The domain of the function

$$f(x) = \sqrt{x - 1}$$

is the set of all x-values for which $x - 1 \ge 0$, which is the interval $[1, \infty)$. To find the range, observe that $f(x) = \sqrt{x - 1}$ is never negative. So, the range is the interval $[0, \infty)$, as indicated in Figure 1.23(a).

b. The domain of the tangent function, shown in Figure 1.23(b),

$$f(x) = \tan x$$

is the set of all x-values such that

$$x \ne \frac{\pi}{2} + n\pi, \quad n \text{ is an integer.} \qquad \text{Domain of tangent function}$$

The range of this function is the set of all real numbers. For a review of the characteristics of this and other trigonometric functions, see Appendix C.

EXAMPLE 3 A Function Defined by More than One Equation

Determine the domain and range of the function.

$$f(x) = \begin{cases} 1 - x, & \text{if } x < 1 \\ \sqrt{x - 1}, & \text{if } x \ge 1 \end{cases}$$

Solution Because f is defined for $x < 1$ and $x \ge 1$, the domain is the entire set of real numbers. On the portion of the domain for which $x \ge 1$, the function behaves as in Example 2(a). For $x < 1$, the values of $1 - x$ are positive. So, the range of the function is the interval $[0, \infty)$. (See Figure 1.24.) ∎

A function from X to Y is **one-to-one** if to each y-value in the range there corresponds exactly one x-value in the domain. For instance, the function given in Example 2(a) is one-to-one, whereas the functions given in Examples 2(b) and 3 are not one-to-one. A function from X to Y is **onto** if its range consists of all of Y.

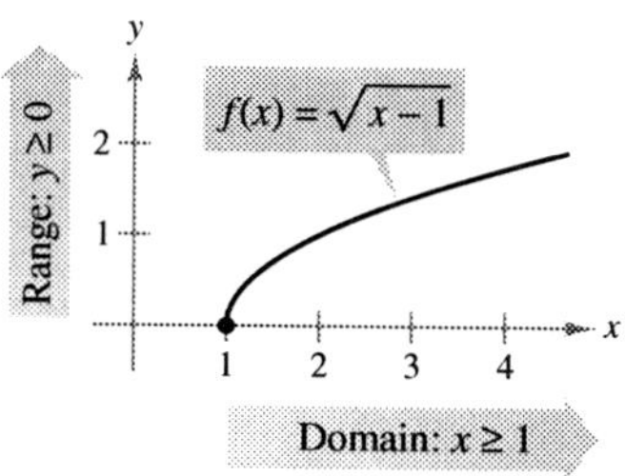

(a) The domain of f is $[1, \infty)$ and the range is $[0, \infty)$.

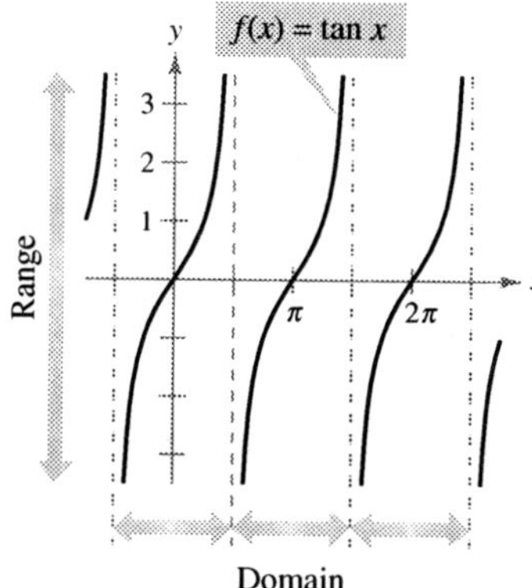

(b) The domain of f is all x-values such that $x \ne \frac{\pi}{2} + n\pi$ and the range is $(-\infty, \infty)$.

Figure 1.23

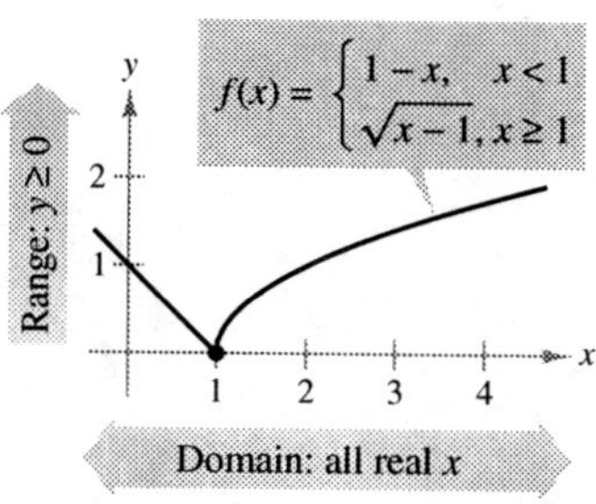

The domain of f is $(-\infty, \infty)$ and the range is $[0, \infty)$.

Figure 1.24

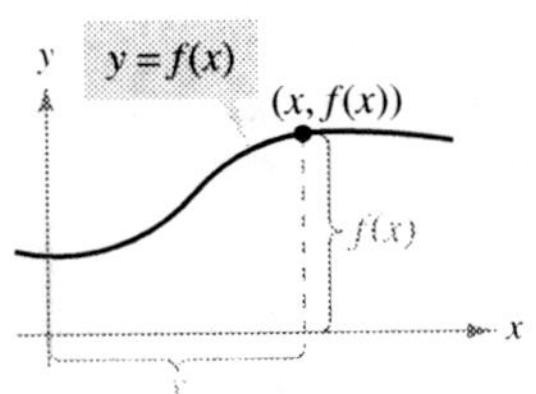

The graph of a function
Figure 1.25

The Graph of a Function

The graph of the function $y = f(x)$ consists of all points $(x, f(x))$, where x is in the domain of f. In Figure 1.25, note that

$$x = \text{the directed distance from the } y\text{-axis}$$

$$f(x) = \text{the directed distance from the } x\text{-axis}.$$

A vertical line can intersect the graph of a function of x at most *once*. This observation provides a convenient visual test, called the **Vertical Line Test,** for functions of x. That is, a graph in the coordinate plane is the graph of a function of x if and only if no vertical line intersects the graph at more than one point. For example, in Figure 1.26(a), you can see that the graph does not define y as a function of x because a vertical line intersects the graph twice. In Figures 1.26(b) and (c), the graphs do define y as a function of x.

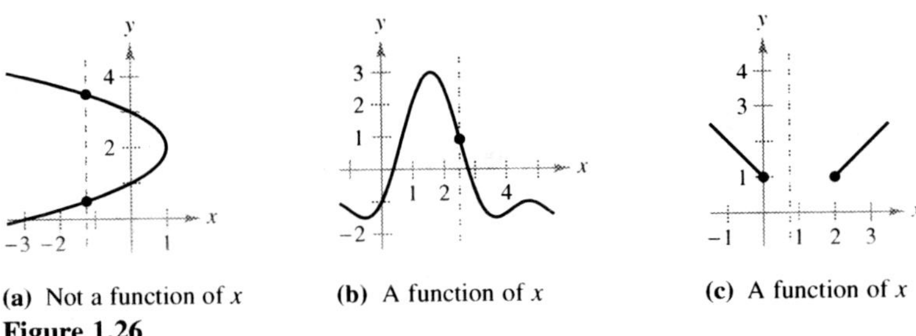

(a) Not a function of x **(b)** A function of x **(c)** A function of x

Figure 1.26

Figure 1.27 shows the graphs of eight basic functions. You should be able to recognize these graphs. (Graphs of the other four basic trigonometric functions are shown in Appendix C.)

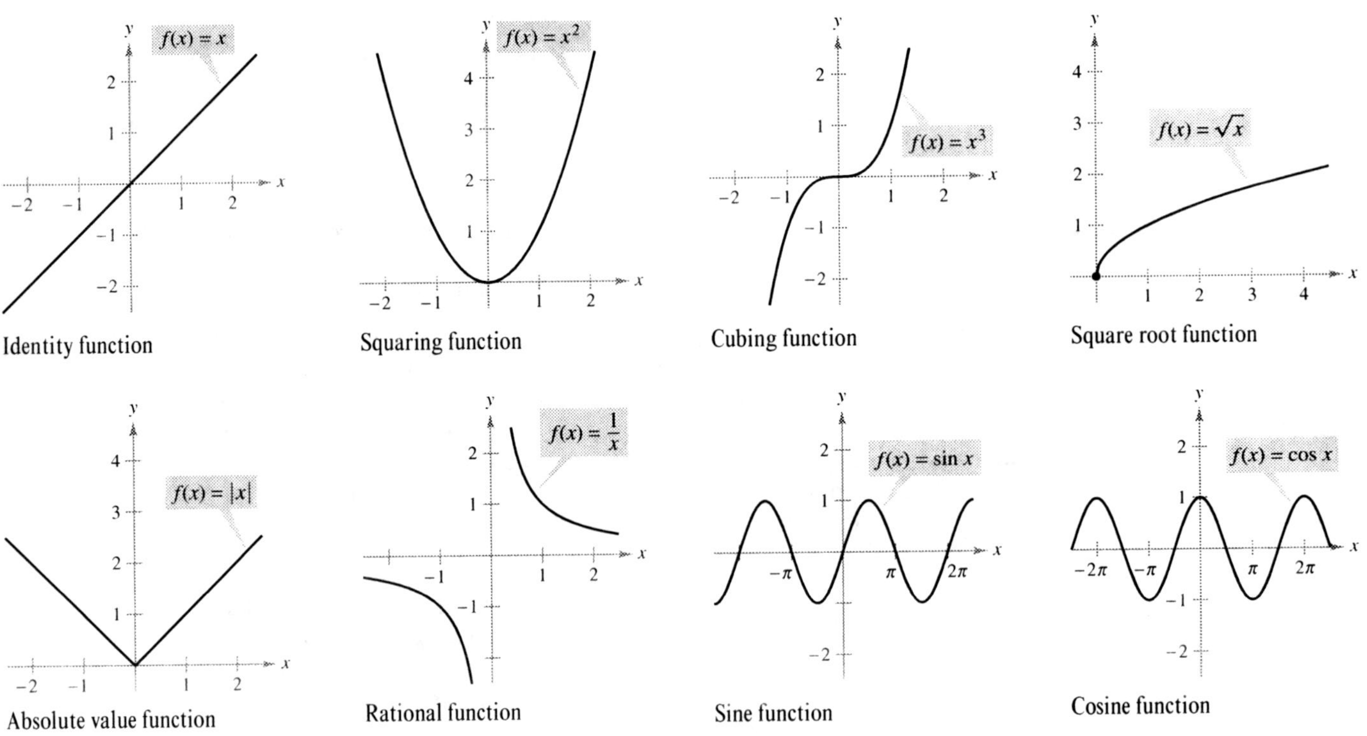

The graphs of eight basic functions
Figure 1.27

Transformations of Functions

Some families of graphs have the same basic shape. For example, compare the graph of $y = x^2$ with the graphs of the four other quadratic functions shown in Figure 1.28.

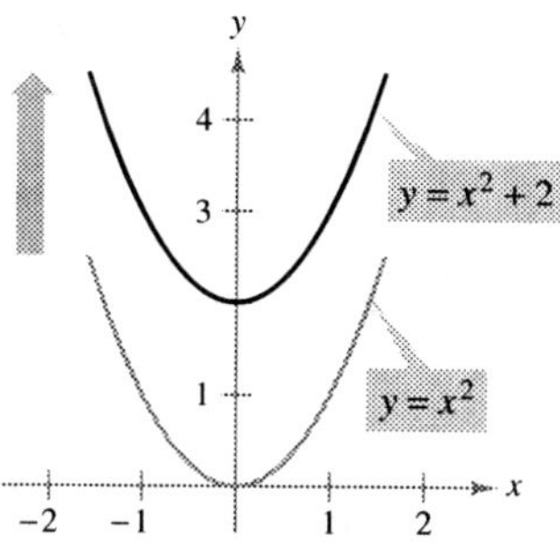

(a) Vertical shift upward

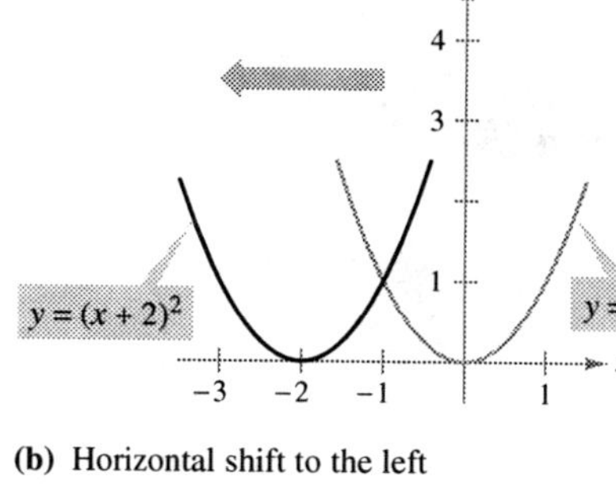

(b) Horizontal shift to the left

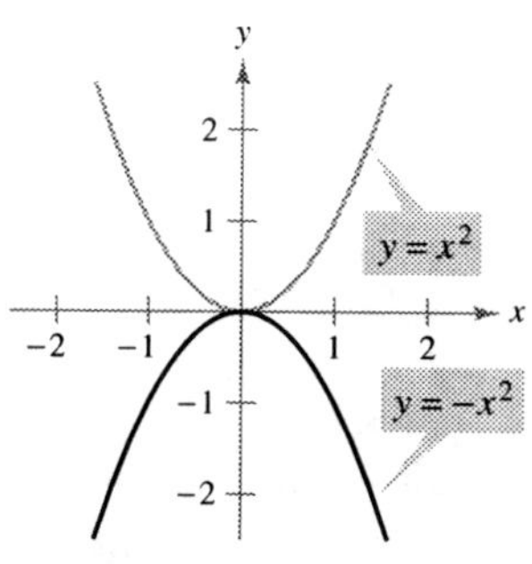

(c) Reflection

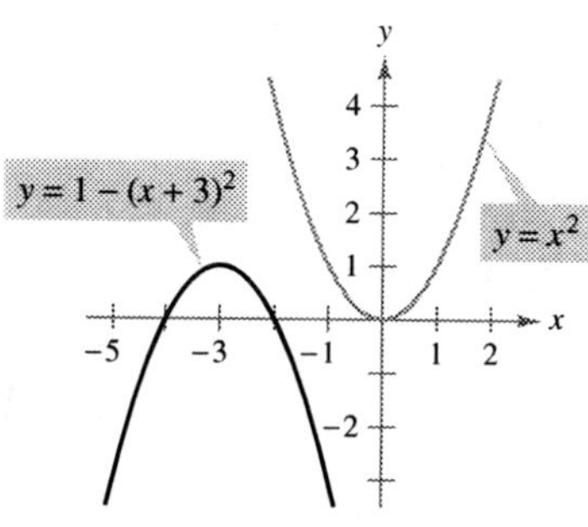

(d) Shift left, reflect, and shift upward

Figure 1.28

Each of the graphs in Figure 1.28 is a **transformation** of the graph of $y = x^2$. The three basic types of transformations illustrated by these graphs are vertical shifts, horizontal shifts, and reflections. Function notation lends itself well to describing transformations of graphs in the plane. For instance, if $f(x) = x^2$ is considered to be the original function in Figure 1.28, the transformations shown can be represented by the following equations.

$$y = f(x) + 2 \qquad \text{Vertical shift up two units}$$
$$y = f(x + 2) \qquad \text{Horizontal shift to the left two units}$$
$$y = -f(x) \qquad \text{Reflection about the } x\text{-axis}$$
$$y = -f(x + 3) + 1 \qquad \text{Shift left three units, reflect about } x\text{-axis, and shift up one unit}$$

BASIC TYPES OF TRANSFORMATIONS $(c > 0)$

Original graph:	$y = f(x)$
Horizontal shift c units to the **right:**	$y = f(x - c)$
Horizontal shift c units to the **left:**	$y = f(x + c)$
Vertical shift c units **downward:**	$y = f(x) - c$
Vertical shift c units **upward:**	$y = f(x) + c$
Reflection (about the x-axis):	$y = -f(x)$
Reflection (about the y-axis):	$y = f(-x)$
Reflection (about the origin):	$y = -f(-x)$

LEONHARD EULER (1707–1783)

In addition to making major contributions to almost every branch of mathematics, Euler was one of the first to apply calculus to real-life problems in physics. His extensive published writings include such topics as shipbuilding, acoustics, optics, astronomy, mechanics, and magnetism.

■ **FOR FURTHER INFORMATION** For more on the history of the concept of a function, see the article "Evolution of the Function Concept: A Brief Survey" by Israel Kleiner in *The College Mathematics Journal*. To view this article, go to the website *www.matharticles.com*.

Classifications and Combinations of Functions

The modern notion of a function is derived from the efforts of many seventeenth- and eighteenth-century mathematicians. Of particular note was Leonhard Euler, to whom we are indebted for the function notation $y = f(x)$. By the end of the eighteenth century, mathematicians and scientists had concluded that many real-world phenomena could be represented by mathematical models taken from a collection of functions called **elementary functions.** Elementary functions fall into three categories.

1. Algebraic functions (polynomial, radical, rational)
2. Trigonometric functions (sine, cosine, tangent, and so on)
3. Exponential and logarithmic functions

You can review the trigonometric functions in Appendix C. The other nonalgebraic functions, such as the inverse trigonometric functions and the exponential and logarithmic functions, are introduced in Sections 1.5 and 1.6.

The most common type of algebraic function is a **polynomial function**

$$f(x) = a_n x^n + a_{n-1} x^{n-1} + \cdots + a_2 x^2 + a_1 x + a_0$$

where n is a nonnegative integer. The numbers a_i are **coefficients,** with a_n the **leading coefficient** and a_0 the **constant term** of the polynomial function. If $a_n \neq 0$, then n is the degree of the polynomial function. The zero polynomial $f(x) = 0$ is not assigned a degree. It is common practice to use subscript notation for coefficients of general polynomial functions, but for polynomial functions of low degree, the following simpler forms are often used. (Note that $a \neq 0$.)

Zeroth degree:	$f(x) = a$	Constant function
First degree:	$f(x) = ax + b$	Linear function
Second degree:	$f(x) = ax^2 + bx + c$	Quadratic function
Third degree:	$f(x) = ax^3 + bx^2 + cx + d$	Cubic function

Although the graph of a polynomial function can have several turns, eventually the graph will rise or fall without bound as x moves to the right or left. Whether the graph of

$$f(x) = a_n x^n + a_{n-1} x^{n-1} + \cdots + a_2 x^2 + a_1 x + a_0$$

eventually rises or falls can be determined by the function's degree (odd or even) and by the leading coefficient a_n, as indicated in Figure 1.29. Note that the dashed portions of the graphs indicate that the **Leading Coefficient Test** determines *only* the right and left behavior of the graph.

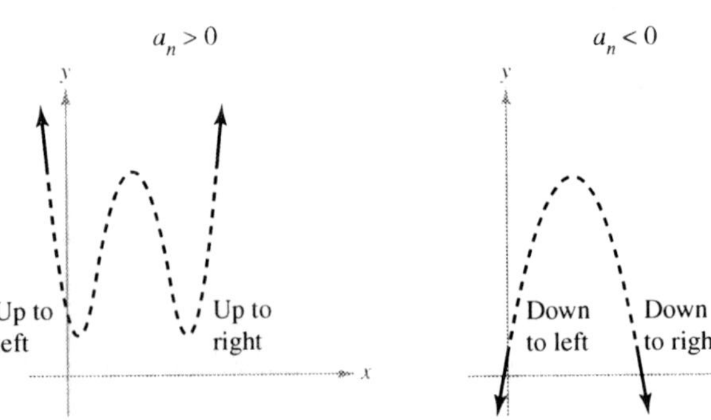

Graphs of polynomial functions of even degree ($n \geq 2$)

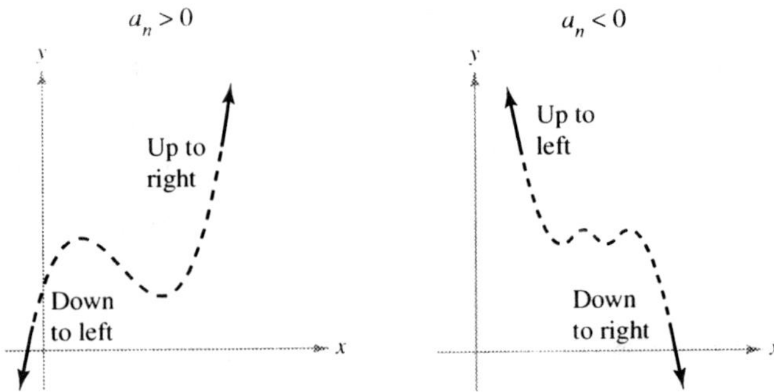

Graphs of polynomial functions of odd degree

The Leading Coefficient Test for polynomial functions
Figure 1.29

Just as a rational number can be written as the quotient of two integers, a **rational function** can be written as the quotient of two polynomials. Specifically, a function f is rational if it has the form

$$f(x) = \frac{p(x)}{q(x)}, \quad q(x) \neq 0$$

where $p(x)$ and $q(x)$ are polynomials.

Polynomial functions and rational functions are examples of **algebraic functions.** An algebraic function of x is one that can be expressed as a finite number of sums, differences, multiples, quotients, and radicals involving x^n. For example, $f(x) = \sqrt{x + 1}$ is algebraic. Functions that are not algebraic are **transcendental.** For instance, the trigonometric functions are transcendental.

Two functions can be combined in various ways to create new functions. For example, given $f(x) = 2x - 3$ and $g(x) = x^2 + 1$, you can form the functions shown.

$$(f + g)(x) = f(x) + g(x) = (2x - 3) + (x^2 + 1) \qquad \text{Sum}$$
$$(f - g)(x) = f(x) - g(x) = (2x - 3) - (x^2 + 1) \qquad \text{Difference}$$
$$(fg)(x) = f(x)g(x) = (2x - 3)(x^2 + 1) \qquad \text{Product}$$
$$(f/g)(x) = \frac{f(x)}{g(x)} = \frac{2x - 3}{x^2 + 1} \qquad \text{Quotient}$$

You can combine two functions in yet another way, called **composition.** The resulting function is called a **composite function.**

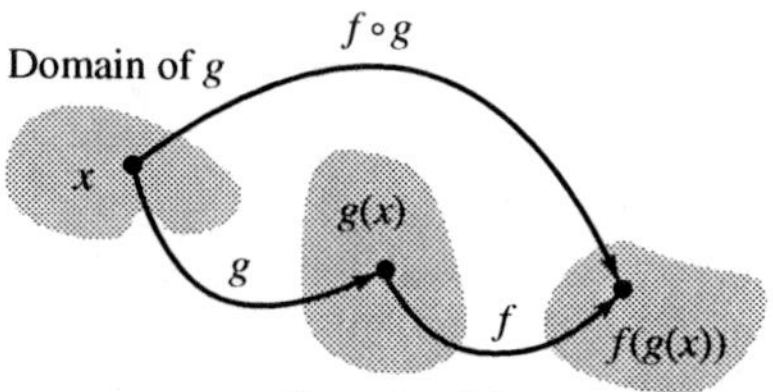

The domain of the composite function $f \circ g$
Figure 1.30

DEFINITION OF COMPOSITE FUNCTION

Let f and g be functions. The function given by $(f \circ g)(x) = f(g(x))$ is called the **composite** of f with g. The domain of $f \circ g$ is the set of all x in the domain of g such that $g(x)$ is in the domain of f (see Figure 1.30).

The composite of f with g may not be equal to the composite of g with f.

EXAMPLE 4   Finding Composites of Functions

Given $f(x) = 2x - 3$ and $g(x) = \cos x$, find the following.

a. $f \circ g$ **b.** $g \circ f$

Solution

a. $(f \circ g)(x) = f(g(x))$ Definition of $f \circ g$

          $= f(\cos x)$ Substitute $\cos x$ for $g(x)$.

          $= 2(\cos x) - 3$ Definition of $f(x)$

          $= 2 \cos x - 3$ Simplify.

b. $(g \circ f)(x) = g(f(x))$ Definition of $g \circ f$

          $= g(2x - 3)$ Substitute $2x - 3$ for $f(x)$.

          $= \cos(2x - 3)$ Definition of $g(x)$

Note that $(f \circ g)(x) \neq (g \circ f)(x)$.

In Section 1.1, an x-intercept of a graph was defined to be a point $(a, 0)$ at which the graph crosses the x-axis. If the graph represents a function f, the number a is a **zero** of f. In other words, *the zeros of a function f are the solutions of the equation $f(x) = 0$.* For example, the function $f(x) = x - 4$ has a zero at $x = 4$ because $f(4) = 0$.

In Section 1.1 you also studied different types of symmetry. In the terminology of functions, a function is **even** if its graph is symmetric with respect to the y-axis, and is **odd** if its graph is symmetric with respect to the origin. The symmetry tests in Section 1.1 yield the following test for even and odd functions.

TEST FOR EVEN AND ODD FUNCTIONS

The function $y = f(x)$ is **even** if $f(-x) = f(x)$.

The function $y = f(x)$ is **odd** if $f(-x) = -f(x)$.

NOTE Except for the constant function $f(x) = 0$, the graph of a function of x cannot have symmetry with respect to the x-axis because it then would fail the Vertical Line Test for the graph of the function. ∎

EXAMPLE 5 Even and Odd Functions and Zeros of Functions

Determine whether each function is even, odd, or neither. Then find the zeros of the function.

a. $f(x) = x^3 - x$ **b.** $g(x) = 1 + \cos x$

Solution

a. This function is odd because

$$f(-x) = (-x)^3 - (-x) = -x^3 + x = -(x^3 - x) = -f(x).$$

The zeros of f are found as shown.

$$x^3 - x = 0 \qquad \text{Let } f(x) = 0.$$
$$x(x^2 - 1) = 0 \qquad \text{Factor.}$$
$$x(x - 1)(x + 1) = 0 \qquad \text{Factor.}$$
$$x = 0, 1, -1 \qquad \text{Zeros of } f$$

See Figure 1.31(a).

b. This function is even because

$$g(-x) = 1 + \cos(-x) = 1 + \cos x = g(x). \qquad \cos(-x) = \cos(x)$$

The zeros of g are found as shown.

$$1 + \cos x = 0 \qquad \text{Let } g(x) = 0.$$
$$\cos x = -1 \qquad \text{Subtract 1 from each side.}$$
$$x = (2n + 1)\pi, \ n \text{ is an integer.} \qquad \text{Zeros of } g$$

See Figure 1.31(b). ∎

NOTE Each of the functions in Example 5 is either even or odd. However, some functions, such as $f(x) = x^2 + x + 1$, are neither even nor odd. ∎

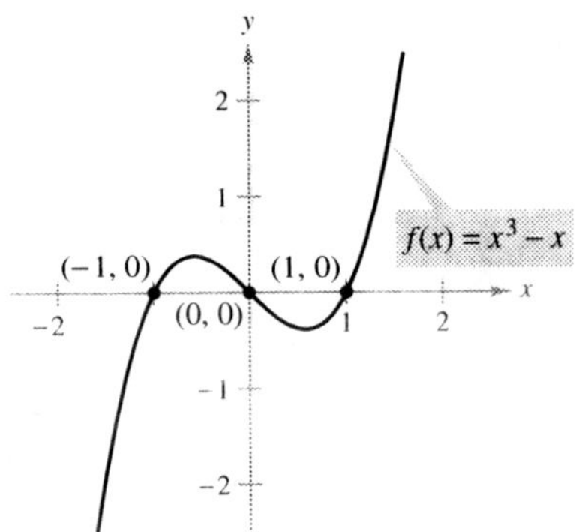

(a) Odd function

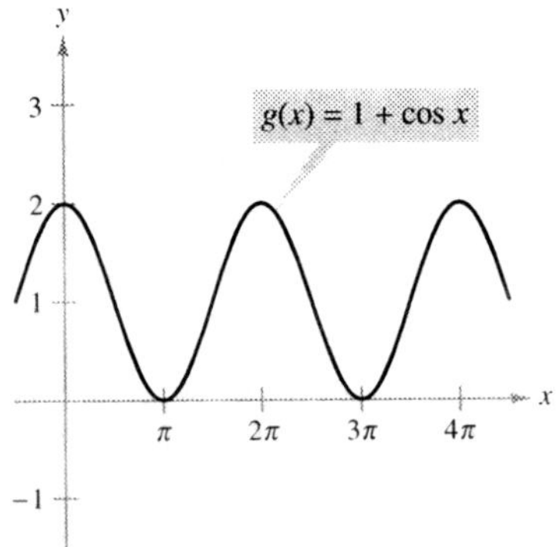

(b) Even function

Figure 1.31

1.3 Exercises

See www.CalcChat.com for worked-out solutions to odd-numbered exercises.

In Exercises 1 and 2, use the graphs of f and g to answer the following.

(a) Identify the domains and ranges of f and g.

(b) Identify $f(-2)$ and $g(3)$.

(c) For what value(s) of x is $f(x) = g(x)$?

(d) Estimate the solution(s) of $f(x) = 2$.

(e) Estimate the solutions of $g(x) = 0$.

1. 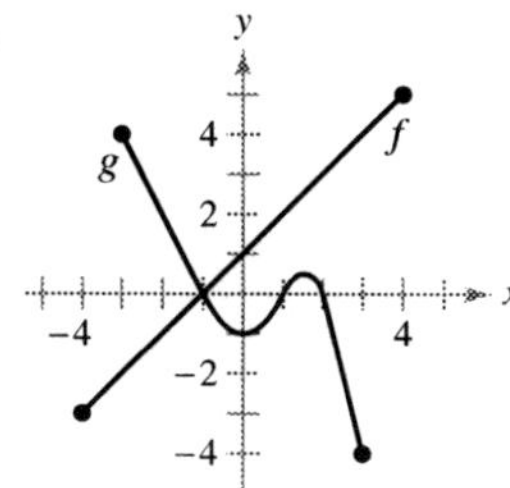   **2.** 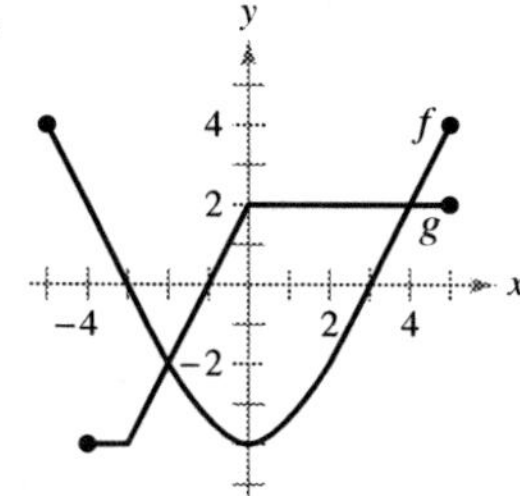

In Exercises 3–12, evaluate (if possible) the function at the given value(s) of the independent variable. Simplify the results.

3. $f(x) = 7x - 4$

  (a) $f(0)$

  (b) $f(-3)$

  (c) $f(b)$

  (d) $f(x - 1)$

4. $f(x) = \sqrt{x + 5}$

  (a) $f(-4)$

  (b) $f(11)$

  (c) $f(-8)$

  (d) $f(x + \Delta x)$

5. $g(x) = 5 - x^2$

  (a) $g(0)$

  (b) $g(\sqrt{5})$

  (c) $g(-2)$

  (d) $g(t - 1)$

6. $g(x) = x^2(x - 4)$

  (a) $g(4)$

  (b) $g(\frac{3}{2})$

  (c) $g(c)$

  (d) $g(t + 4)$

7. $f(x) = \cos 2x$

  (a) $f(0)$

  (b) $f(-\pi/4)$

  (c) $f(\pi/3)$

8. $f(x) = \sin x$

  (a) $f(\pi)$

  (b) $f(5\pi/4)$

  (c) $f(2\pi/3)$

9. $f(x) = x^3$

$$\frac{f(x + \Delta x) - f(x)}{\Delta x}$$

10. $f(x) = 3x - 1$

$$\frac{f(x) - f(1)}{x - 1}$$

11. $f(x) = \dfrac{1}{\sqrt{x - 1}}$

$$\frac{f(x) - f(2)}{x - 2}$$

12. $f(x) = x^3 - x$

$$\frac{f(x) - f(1)}{x - 1}$$

In Exercises 13–20, find the domain and range of the function.

13. $f(x) = 4x^2$

14. $g(x) = x^2 - 5$

15. $g(x) = \sqrt{6x}$

16. $h(x) = -\sqrt{x + 3}$

17. $f(t) = \sec \dfrac{\pi t}{4}$

18. $h(t) = \cot t$

19. $f(x) = \dfrac{3}{x}$

20. $g(x) = \dfrac{2}{x - 1}$

In Exercises 21–26, find the domain of the function.

21. $f(x) = \sqrt{x} + \sqrt{1 - x}$

22. $f(x) = \sqrt{x^2 - 3x + 2}$

23. $g(x) = \dfrac{2}{1 - \cos x}$

24. $h(x) = \dfrac{1}{\sin x - \dfrac{1}{2}}$

25. $f(x) = \dfrac{1}{|x + 3|}$

26. $g(x) = \dfrac{1}{|x^2 - 4|}$

In Exercises 27–30, evaluate the function as indicated. Determine its domain and range.

27. $f(x) = \begin{cases} 2x + 1, & x < 0 \\ 2x + 2, & x \geq 0 \end{cases}$

  (a) $f(-1)$ (b) $f(0)$ (c) $f(2)$ (d) $f(t^2 + 1)$

28. $f(x) = \begin{cases} x^2 + 2, & x \leq 1 \\ 2x^2 + 2, & x > 1 \end{cases}$

  (a) $f(-2)$ (b) $f(0)$ (c) $f(1)$ (d) $f(s^2 + 2)$

29. $f(x) = \begin{cases} |x| + 1, & x < 1 \\ -x + 1, & x \geq 1 \end{cases}$

  (a) $f(-3)$ (b) $f(1)$ (c) $f(3)$ (d) $f(b^2 + 1)$

30. $f(x) = \begin{cases} \sqrt{x + 4}, & x \leq 5 \\ (x - 5)^2, & x > 5 \end{cases}$

  (a) $f(-3)$ (b) $f(0)$ (c) $f(5)$ (d) $f(10)$

In Exercises 31–38, sketch a graph of the function and find its domain and range. Use a graphing utility to verify your graph.

31. $f(x) = 4 - x$

32. $g(x) = \dfrac{4}{x}$

33. $h(x) = \sqrt{x - 6}$

34. $f(x) = \frac{1}{4}x^3 + 3$

35. $f(x) = \sqrt{9 - x^2}$

36. $f(x) = x + \sqrt{4 - x^2}$

37. $g(t) = 3 \sin \pi t$

38. $h(\theta) = -5 \cos \dfrac{\theta}{2}$

WRITING ABOUT CONCEPTS

39. The graph of the distance that a student drives in a 10-minute trip to school is shown in the figure. Give a verbal description of characteristics of the student's drive to school.

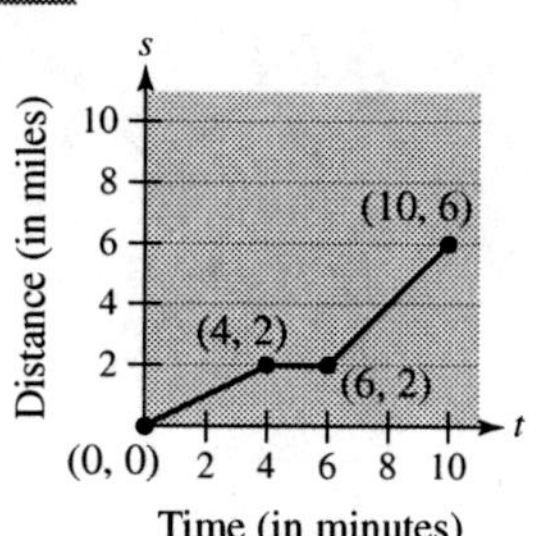

WRITING ABOUT CONCEPTS (continued)

40. A student who commutes 27 miles to attend college remembers, after driving a few minutes, that a term paper that is due has been forgotten. Driving faster than usual, the student returns home, picks up the paper, and once again starts toward school. Sketch a possible graph of the student's distance from home as a function of time.

In Exercises 41–44, use the Vertical Line Test to determine whether y is a function of x. To print an enlarged copy of the graph, go to the website *www.mathgraphs.com*.

41. $x - y^2 = 0$

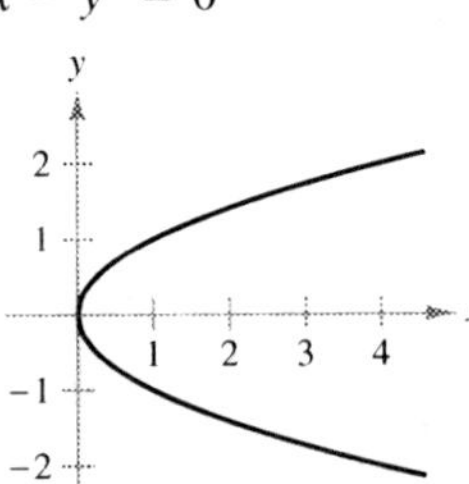

42. $\sqrt{x^2 - 4} - y = 0$

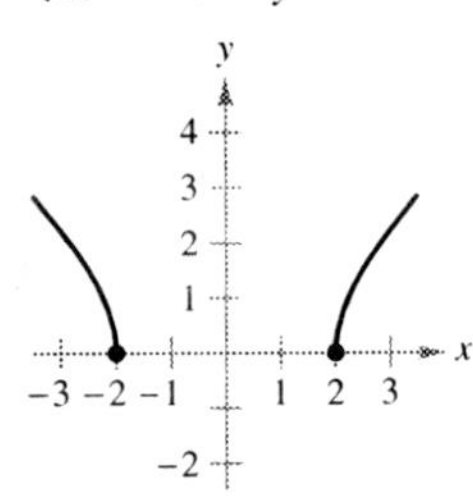

43. $y = \begin{cases} x + 1, & x \le 0 \\ -x + 2, & x > 0 \end{cases}$

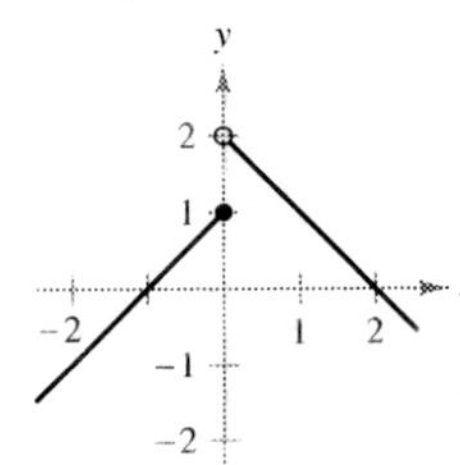

44. $x^2 + y^2 = 4$

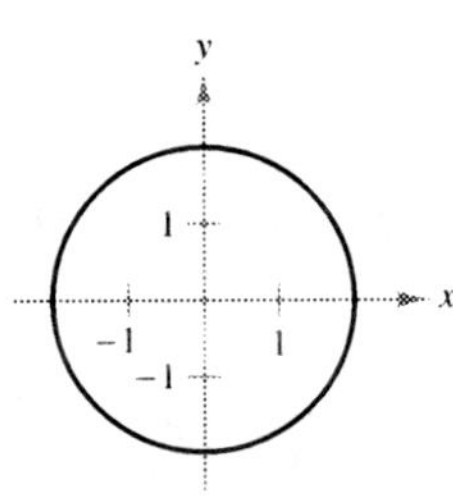

In Exercises 45–48, determine whether y is a function of x.

45. $x^2 + y^2 = 16$

46. $x^2 + y = 16$

47. $y^2 = x^2 - 1$

48. $x^2 y - x^2 + 4y = 0$

In Exercises 49–54, use the graph of $y = f(x)$ to match the function with its graph.

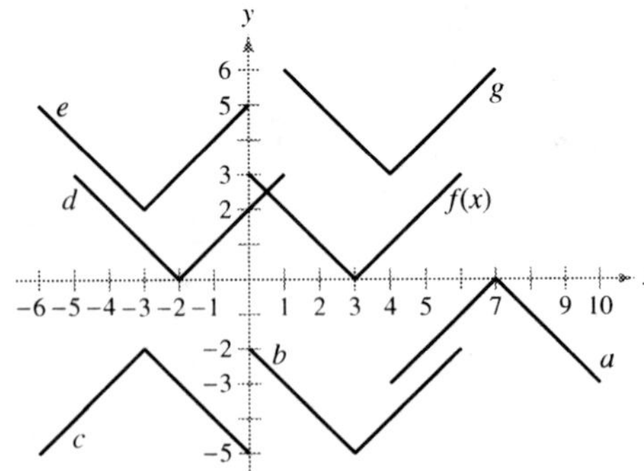

49. $y = f(x + 5)$

50. $y = f(x) - 5$

51. $y = -f(-x) - 2$

52. $y = -f(x - 4)$

53. $y = f(x + 6) + 2$

54. $y = f(x - 1) + 3$

55. Use the graph of f shown in the figure to sketch the graph of each function. To print an enlarged copy of the graph, go to the website *www.mathgraphs.com*.

(a) $f(x + 3)$ (b) $f(x - 1)$

(c) $f(x) + 2$ (d) $f(x) - 4$

(e) $3f(x)$ (f) $\frac{1}{4}f(x)$

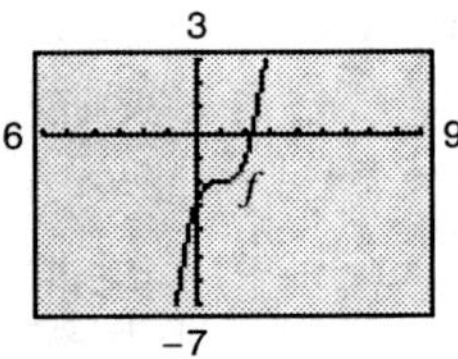

56. Use the graph of f shown in the figure to sketch the graph of each function. To print an enlarged copy of the graph, go to the website *www.mathgraphs.com*.

(a) $f(x - 4)$ (b) $f(x + 2)$

(c) $f(x) + 4$ (d) $f(x) - 1$

(e) $2f(x)$ (f) $\frac{1}{2}f(x)$

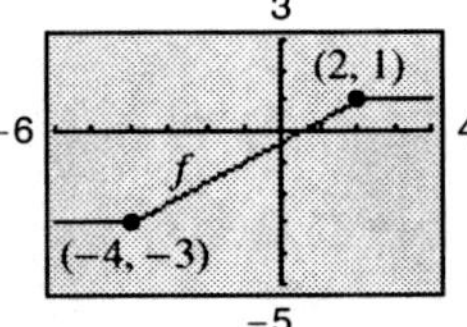

57. Use the graph of $f(x) = \sqrt{x}$ to sketch the graph of each function. In each case, describe the transformation.

(a) $y = \sqrt{x} + 2$ (b) $y = -\sqrt{x}$ (c) $y = \sqrt{x - 2}$

58. Specify a sequence of transformations that will yield each graph of h from the graph of the function $f(x) = \sin x$.

(a) $h(x) = \sin\left(x + \dfrac{\pi}{2}\right) + 1$ (b) $h(x) = -\sin(x - 1)$

59. Given $f(x) = \sqrt{x}$ and $g(x) = x^2 - 1$, evaluate each expression.

(a) $f(g(1))$ (b) $g(f(1))$ (c) $g(f(0))$

(d) $f(g(-4))$ (e) $f(g(x))$ (f) $g(f(x))$

60. Given $f(x) = \sin x$ and $g(x) = \pi x$, evaluate each expression.

(a) $f(g(2))$ (b) $f\left(g\left(\dfrac{1}{2}\right)\right)$ (c) $g(f(0))$

(d) $g\left(f\left(\dfrac{\pi}{4}\right)\right)$ (e) $f(g(x))$ (f) $g(f(x))$

In Exercises 61–64, find the composite functions $(f \circ g)$ and $(g \circ f)$. What is the domain of each composite function? Are the two composite functions equal?

61. $f(x) = x^2$
    $g(x) = \sqrt{x}$

62. $f(x) = x^2 - 1$
    $g(x) = \cos x$

63. $f(x) = \dfrac{3}{x}$
    $g(x) = x^2 - 1$

64. $f(x) = \dfrac{1}{x}$
    $g(x) = \sqrt{x + 2}$

65. Use the graphs of f and g to evaluate each expression. If the result is undefined, explain why.

(a) $(f \circ g)(3)$ (b) $g(f(2))$

(c) $g(f(5))$ (d) $(f \circ g)(-3)$

(e) $(g \circ f)(-1)$ (f) $f(g(-1))$

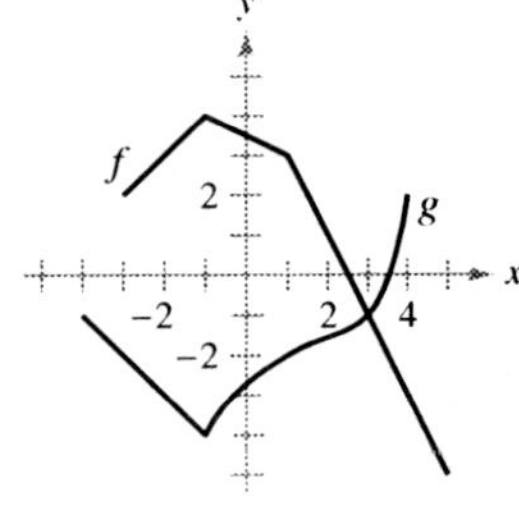

66. *Ripples*   A pebble is dropped into a calm pond, causing ripples in the form of concentric circles. The radius (in feet) of the outer ripple is given by $r(t) = 0.6t$, where t is the time in seconds after the pebble strikes the water. The area of the circle is given by the function $A(r) = \pi r^2$. Find and interpret $(A \circ r)(t)$.

Think About It   **In Exercises 67 and 68, $F(x) = f \circ g \circ h$. Identify functions for f, g, and h. (There are many correct answers.)**

67. $F(x) = \sqrt{2x - 2}$ **68.** $F(x) = -4 \sin(1 - x)$

In Exercises 69–72, determine whether the function is even, odd, or neither. Use a graphing utility to verify your result.

69. $f(x) = x^2(4 - x^2)$ **70.** $f(x) = \sqrt[3]{x}$

71. $f(x) = x \cos x$ **72.** $f(x) = \sin^2 x$

Think About It   **In Exercises 73 and 74, find the coordinates of a second point on the graph of a function f if the given point is on the graph and the function is (a) even and (b) odd.**

73. $\left(-\frac{3}{2}, 4\right)$ **74.** $(4, 9)$

75. The graphs of f, g, and h are shown in the figure. Decide whether each function is even, odd, or neither.

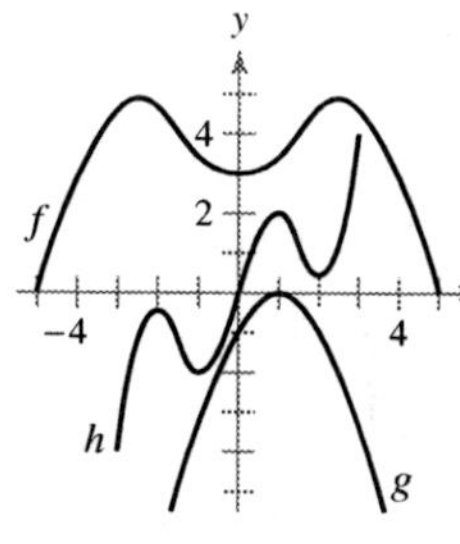

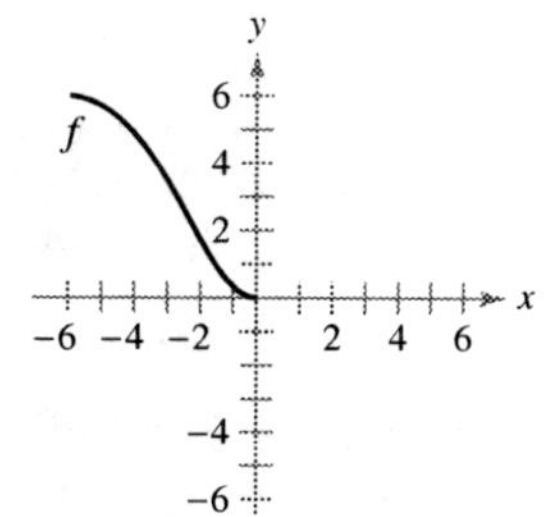

Figure for 75   **Figure for 76**

76. The domain of the function f shown in the figure is $-6 \le x \le 6$.

(a) Complete the graph of f given that f is even.

(b) Complete the graph of f given that f is odd.

Writing Functions   **In Exercises 77–80, write an equation for a function that has the given graph.**

77. Line segment connecting $(-2, 4)$ and $(0, -6)$

78. Line segment connecting $(3, 1)$ and $(5, 8)$

79. The bottom half of the parabola $x + y^2 = 0$

80. The bottom half of the circle $x^2 + y^2 = 36$

In Exercises 81–84, sketch a possible graph of the situation.

81. The speed of an airplane as a function of time during a 5-hour flight

82. The height of a baseball as a function of horizontal distance during a home run

83. The amount of a certain brand of sneaker sold by a sporting goods store as a function of the price of the sneaker

84. The value of a new car as a function of time over a period of 8 years

85. Find the value of c such that the domain of
$$f(x) = \sqrt{c - x^2}$$
is $[-5, 5]$.

86. Find all values of c such that the domain of
$$f(x) = \frac{x + 3}{x^2 + 3cx + 6}$$
is the set of all real numbers.

87. *Graphical Reasoning*   An electronically controlled thermostat is programmed to lower the temperature during the night automatically (see figure). The temperature T in degrees Celsius is given in terms of t, the time in hours on a 24-hour clock.

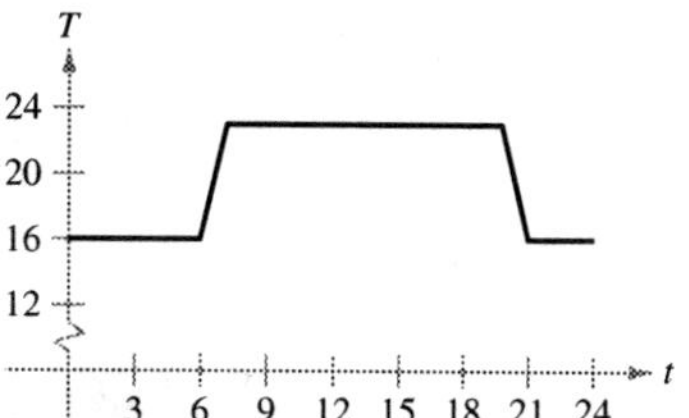

(a) Approximate $T(4)$ and $T(15)$.

(b) The thermostat is reprogrammed to produce a temperature $H(t) = T(t - 1)$. How does this change the temperature? Explain.

(c) The thermostat is reprogrammed to produce a temperature $H(t) = T(t) - 1$. How does this change the temperature? Explain.

CAPSTONE

88. Water runs into a vase of height 30 centimeters at a constant rate. The vase is full after 5 seconds. Use this information and the shape of the vase shown to answer the questions if d is the depth of the water in centimeters and t is the time in seconds (see figure).

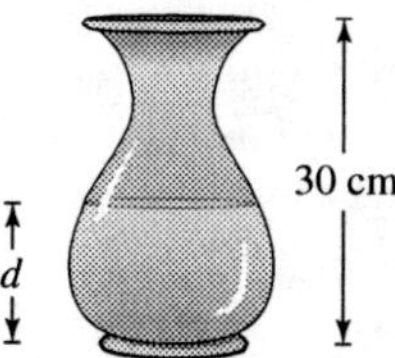

(a) Explain why d is a function of t.

(b) Determine the domain and range of the function.

(c) Sketch a possible graph of the function.

(d) Use the graph in part (c) to approximate $d(4)$. What does this represent?

89. *Modeling Data* The table shows the average numbers of acres per farm in the United States for selected years. (*Source: U.S. Department of Agriculture*)

Year	1955	1965	1975	1985	1995	2005
Acreage	258	340	420	441	438	444

(a) Plot the data, where A is the acreage and t is the time in years, with $t = 5$ corresponding to 1955. Sketch a freehand curve that approximates the data.

(b) Use the curve in part (a) to approximate $A(20)$.

90. *Automobile Aerodynamics* The horsepower H required to overcome wind drag on a certain automobile is approximated by

$$H(x) = 0.002x^2 + 0.005x - 0.029, \quad 10 \le x \le 100$$

where x is the speed of the car in miles per hour.

(a) Use a graphing utility to graph H.

(b) Rewrite the power function so that x represents the speed in kilometers per hour. [Find $H(x/1.6)$.]

91. *Think About It* Write the function

$$f(x) = |x| + |x - 2|$$

without using absolute value signs. (For a review of absolute value, see Appendix C.)

92. *Writing* Use a graphing utility to graph the polynomial functions $p_1(x) = x^3 - x + 1$ and $p_2(x) = x^3 - x$. How many zeros does each function have? Is there a cubic polynomial that has no zeros? Explain.

93. Prove that the function is odd.

$$f(x) = a_{2n+1}x^{2n+1} + \cdots + a_3x^3 + a_1x$$

94. Prove that the function is even.

$$f(x) = a_{2n}x^{2n} + a_{2n-2}x^{2n-2} + \cdots + a_2x^2 + a_0$$

95. Prove that the product of two even (or two odd) functions is even.

96. Prove that the product of an odd function and an even function is odd.

97. *Volume* An open box of maximum volume is to be made from a square piece of material 24 centimeters on a side by cutting equal squares from the corners and turning up the sides (see figure).

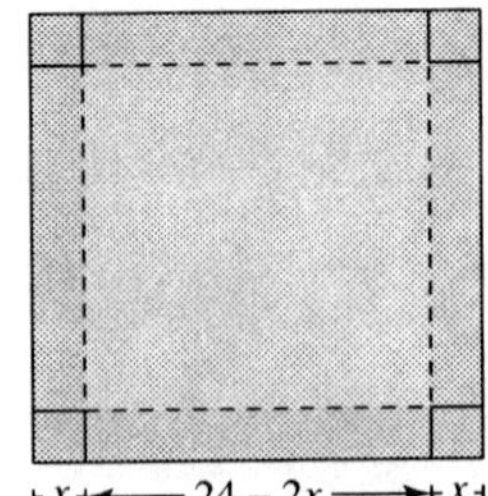

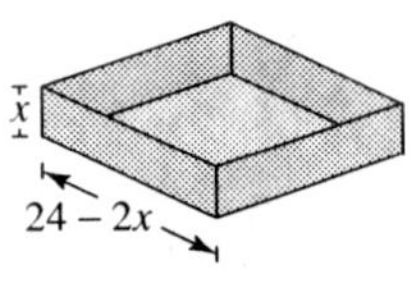

(a) Write the volume V as a function of x, the length of the corner squares. What is the domain of the function?

(b) Use a graphing utility to graph the volume function and approximate the dimensions of the box that yield a maximum volume.

(c) Use the *table* feature of a graphing utility to verify your answer in part (b). (The first two rows of the table are shown.)

Height, x	Length and Width	Volume, V
1	$24 - 2(1)$	$1[24 - 2(1)]^2 = 484$
2	$24 - 2(2)$	$2[24 - 2(2)]^2 = 800$

98. *Length* A right triangle is formed in the first quadrant by the x- and y-axes and a line through the point $(3, 2)$ (see figure). Write the length L of the hypotenuse as a function of x.

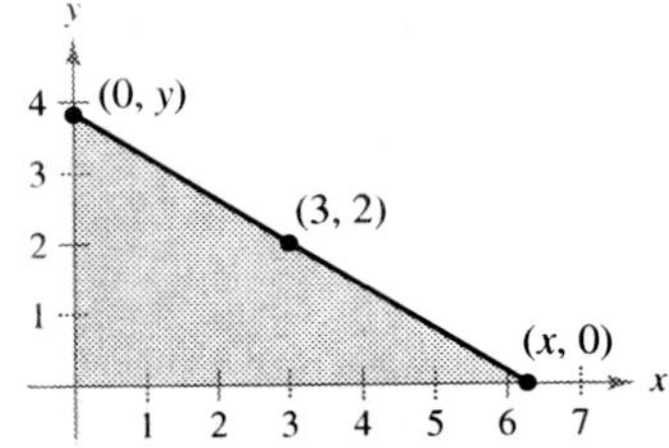

True or False? **In Exercises 99–102, determine whether the statement is true or false. If it is false, explain why or give an example that shows it is false.**

99. If $f(a) = f(b)$, then $a = b$.

100. A vertical line can intersect the graph of a function at most once.

101. If $f(x) = f(-x)$ for all x in the domain of f, then the graph of f is symmetric with respect to the y-axis.

102. If f is a function, then $f(ax) = af(x)$.

PUTNAM EXAM CHALLENGE

103. Let R be the region consisting of the points (x, y) of the Cartesian plane satisfying both $|x| - |y| \le 1$ and $|y| \le 1$. Sketch the region R and find its area.

104. Consider a polynomial $f(x)$ with real coefficients having the property $f(g(x)) = g(f(x))$ for every polynomial $g(x)$ with real coefficients. Determine and prove the nature of $f(x)$.

These problems were composed by the Committee on the Putnam Prize Competition. © The Mathematical Association of America. All rights reserved.

1.4 Fitting Models to Data

- ■ Fit a linear model to a real-life data set.
- ■ Fit a quadratic model to a real-life data set.
- ■ Fit a trigonometric model to a real-life data set.

Fitting a Linear Model to Data

A basic premise of science is that much of the physical world can be described mathematically and that many physical phenomena are predictable. This scientific outlook was part of the scientific revolution that took place in Europe during the late 1500s. Two early publications connected with this revolution were *On the Revolutions of the Heavenly Spheres* by the Polish astronomer Nicolaus Copernicus and *On the Structure of the Human Body* by the Belgian anatomist Andreas Vesalius. Each of these books was published in 1543, and each broke with prior tradition by suggesting the use of a scientific method rather than unquestioned reliance on authority.

One method of modern science is gathering data and then describing the data with a mathematical model. For instance, the data given in Example 1 are inspired by Leonardo da Vinci's famous drawing that indicates that a person's height and arm span are equal.

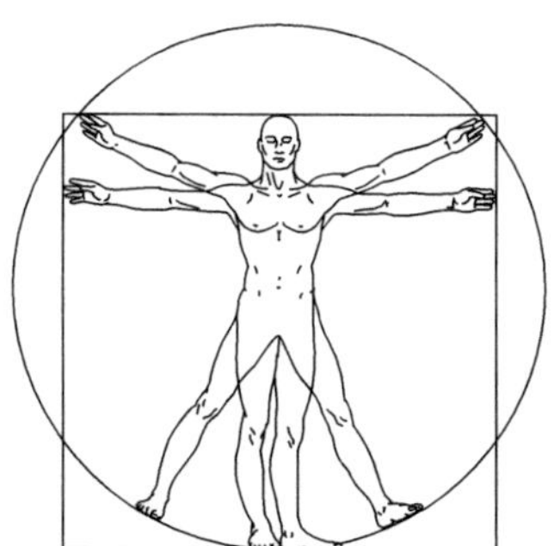

A computer graphics drawing based on the pen and ink drawing of Leonardo da Vinci's famous study of human proportions, called *Vitruvian Man*

EXAMPLE 1 Fitting a Linear Model to Data

A class of 28 people collected the following data, which represent their heights x and arm spans y (rounded to the nearest inch).

(60, 61), (65, 65), (68, 67), (72, 73), (61, 62), (63, 63), (70, 71),
(75, 74), (71, 72), (62, 60), (65, 65), (66, 68), (62, 62), (72, 73),
(70, 70), (69, 68), (69, 70), (60, 61), (63, 63), (64, 64), (71, 71),
(68, 67), (69, 70), (70, 72), (65, 65), (64, 63), (71, 70), (67, 67)

Find a linear model to represent these data.

Solution   There are different ways to model these data with an equation. The simplest would be to observe that x and y are about the same and list the model as simply $y = x$. A more careful analysis would be to use a procedure from statistics called linear regression. (You will study this procedure in Section 13.9.) The least squares regression line for these data is

$$y = 1.006x - 0.23. \qquad \text{Least squares regression line}$$

The graph of the model and the data are shown in Figure 1.32. From this model, you can see that a person's arm span tends to be about the same as his or her height.

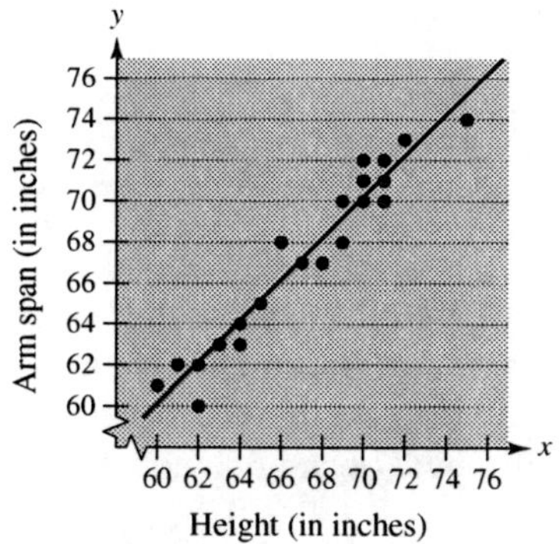

Linear model and data
Figure 1.32

TECHNOLOGY Many scientific and graphing calculators have built-in least squares regression programs. Typically, you enter the data into the calculator and then run the linear regression program. The program usually displays the slope and y-intercept of the best-fitting line and the *correlation coefficient r*. The correlation coefficient gives a measure of how well the model fits the data. The closer $|r|$ is to 1, the better the model fits the data. For instance, the correlation coefficient for the model in Example 1 is $r \approx 0.97$, which indicates that the model is a good fit for the data. If the r-value is positive, the variables have a positive correlation, as in Example 1. If the r-value is negative, the variables have a negative correlation.

Fitting a Quadratic Model to Data

A function that gives the height s of a falling object in terms of the time t is called a *position function*. If air resistance is not considered, the position of a falling object can be modeled by $s(t) = \frac{1}{2}gt^2 + v_0 t + s_0$, where g is the acceleration due to gravity, v_0 is the initial velocity, and s_0 is the initial height. The value of g depends on where the object is dropped. On Earth, g is approximately -32 feet per second per second, or -9.8 meters per second per second.

To discover the value of g experimentally, you could record the heights of a falling object at several increments, as shown in Example 2.

EXAMPLE 2 Fitting a Quadratic Model to Data

A basketball is dropped from a height of about $5\frac{1}{4}$ feet. The height of the basketball is recorded 23 times at intervals of about 0.02 second.* The results are shown in the table.

Time	0.0	0.02	0.04	0.06	0.08	0.099996
Height	5.23594	5.20353	5.16031	5.0991	5.02707	4.95146

Time	0.119996	0.139992	0.159988	0.179988	0.199984	0.219984
Height	4.85062	4.74979	4.63096	4.50132	4.35728	4.19523

Time	0.23998	0.25993	0.27998	0.299976	0.319972	0.339961
Height	4.02958	3.84593	3.65507	3.44981	3.23375	3.01048

Time	0.359961	0.379951	0.399941	0.419941	0.439941
Height	2.76921	2.52074	2.25786	1.98058	1.63488

Find a model to fit these data. Then use the model to predict the time when the basketball hits the ground.

Solution Draw a scatter plot of the data, as shown in Figure 1.33. From the scatter plot, you can see that the data do not appear to be linear. It does appear, however, that they might be quadratic. To find a quadratic model, enter the data into a calculator or computer that has a quadratic regression program. You should obtain the model

$$s = -15.45t^2 - 1.302t + 5.2340. \qquad \text{Least squares regression quadratic}$$

Using this model, you can predict the time when the basketball hits the ground by substituting 0 for s and solving the resulting equation for t.

$$0 = -15.45t^2 - 1.302t + 5.2340 \qquad \text{Let } s = 0.$$

$$t = \frac{1.302 \pm \sqrt{(-1.302)^2 - 4(-15.45)(5.2340)}}{2(-15.45)} \qquad \text{Quadratic Formula}$$

$$t \approx 0.54 \qquad \text{Choose positive solution.}$$

So, the basketball hits the ground about 0.54 second after it is dropped. In other words, the basketball continues to fall for about 0.1 second more before hitting the ground.

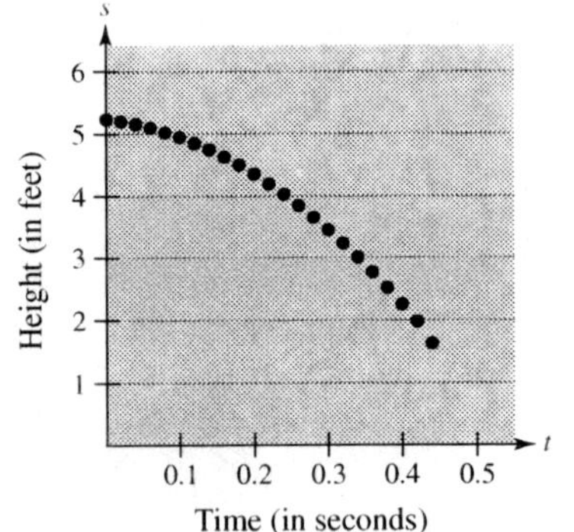

Scatter plot of data

Figure 1.33

** Data were collected with a Texas Instruments CBL (Calculator-Based Laboratory) System.*

The plane of Earth's orbit about the sun and its axis of rotation are not perpendicular. Instead, Earth's axis is tilted with respect to its orbit. The result is that the amount of daylight received by locations on Earth varies with the time of year. That is, it varies with the position of Earth in its orbit.

Fitting a Trigonometric Model to Data

What is mathematical modeling? This is one of the questions that is asked in the book *Guide to Mathematical Modelling*. Here is part of the answer.*

1. Mathematical modeling consists of applying your mathematical skills to obtain useful answers to real problems.

2. Learning to apply mathematical skills is very different from learning mathematics itself.

3. Models are used in a very wide range of applications, some of which do not appear initially to be mathematical in nature.

4. Models often allow quick and cheap evaluation of alternatives, leading to optimal solutions that are not otherwise obvious.

5. There are no precise rules in mathematical modeling and no "correct" answers.

6. Modeling can be learned only by *doing*.

EXAMPLE 3 Fitting a Trigonometric Model to Data

The number of hours of daylight on Earth depends on the latitude and the time of year. Here are the numbers of minutes of daylight at a location of 20°N latitude on the longest and shortest days of the year: June 21, 801 minutes; December 22, 655 minutes. Use these data to write a model for the amount of daylight d (in minutes) on each day of the year at a location of 20°N latitude. How could you check the accuracy of your model?

Solution   Here is one way to create a model. You can hypothesize that the model is a sine function whose period is 365 days. Using the given data, you can conclude that the amplitude of the graph is $(801 - 655)/2$, or 73. So, one possible model is

$$d = 728 - 73 \sin\left(\frac{2\pi t}{365} + \frac{\pi}{2}\right).$$

In this model, t represents the number of the day of the year, with December 22 represented by $t = 0$. A graph of this model is shown in Figure 1.34. To check the accuracy of this model, a weather almanac was used to find the numbers of minutes of daylight on different days of the year at the location of 20°N latitude.

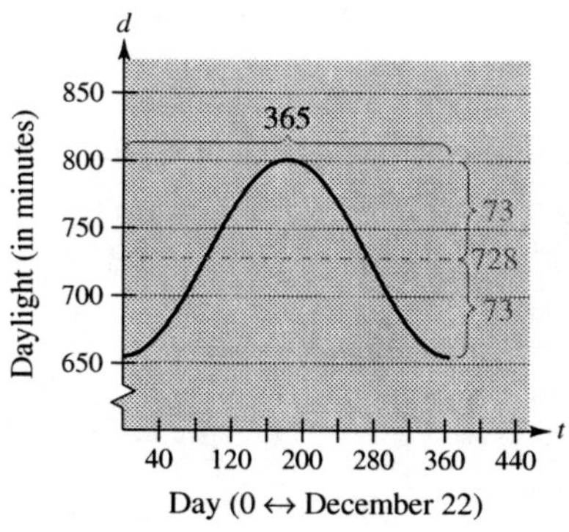

Graph of model
Figure 1.34

NOTE For a review of trigonometric functions, see Appendix C.

Date	Value of t	Actual Daylight	Daylight Given by Model
Dec 22	0	655 min	655 min
Jan 1	10	657 min	656 min
Feb 1	41	676 min	672 min
Mar 1	69	705 min	701 min
Apr 1	100	740 min	739 min
May 1	130	772 min	773 min
Jun 1	161	796 min	796 min
Jun 21	181	801 min	801 min
Jul 1	191	799 min	800 min
Aug 1	222	782 min	785 min
Sep 1	253	752 min	754 min
Oct 1	283	718 min	716 min
Nov 1	314	685 min	681 min
Dec 1	344	661 min	660 min

You can see that the model is fairly accurate.   ■

---

* *Text from Dilwyn Edwards and Mike Hamson,* Guide to Mathematical Modelling *(Boca Raton: CRC Press, 1990), p. 4. Used by permission of the authors.*

1.4 Exercises

See www.CalcChat.com for worked-out solutions to odd-numbered exercises.

In Exercises 1–4, a scatter plot of data is given. Determine whether the data can be modeled by a linear function, a quadratic function, or a trigonometric function, or that there appears to be no relationship between x and y. To print an enlarged copy of the graph, go to the website *www.mathgraphs.com*.

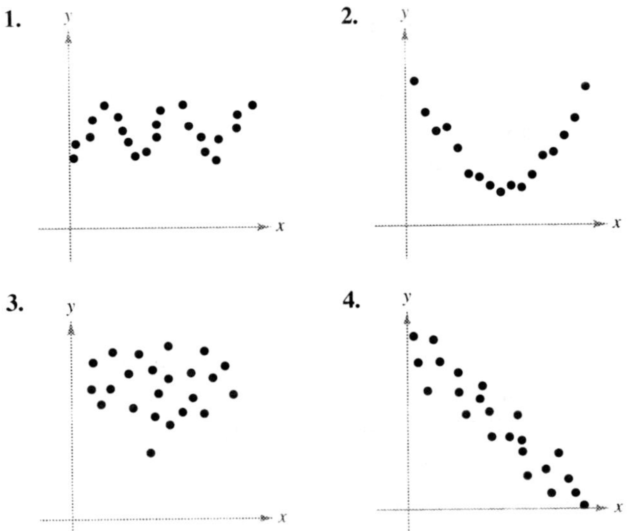

1.   **2.**

3.   **4.**

5. *Carcinogens* Each ordered pair gives the exposure index x of a carcinogenic substance and the cancer mortality y per 100,000 people in the population.

(3.50, 150.1), (3.58, 133.1), (4.42, 132.9),
(2.26, 116.7), (2.63, 140.7), (4.85, 165.5),
(12.65, 210.7), (7.42, 181.0), (9.35, 213.4)

(a) Plot the data. From the graph, do the data appear to be approximately linear?

(b) Visually find a linear model for the data. Graph the model.

(c) Use the model to approximate y if $x = 3$.

6. *Quiz Scores* The ordered pairs represent the scores on two consecutive 15-point quizzes for a class of 18 students.

(7, 13), (9, 7), (14, 14), (15, 15), (10, 15), (9, 7),
(14, 11), (14, 15), (8, 10), (15, 9), (10, 11), (9, 10),
(11, 14), (7, 14), (11, 10), (14, 11), (10, 15), (9, 6)

(a) Plot the data. From the graph, does the relationship between consecutive scores appear to be approximately linear?

(b) If the data appear to be approximately linear, find a linear model for the data. If not, give some possible explanations.

7. *Hooke's Law* Hooke's Law states that the force F required to compress or stretch a spring (within its elastic limits) is proportional to the distance d that the spring is compressed or stretched from its original length. That is, $F = kd$, where k is a measure of the stiffness of the spring and is called the *spring constant*. The table shows the elongation d in centimeters of a spring when a force of F newtons is applied.

F	20	40	60	80	100
d	1.4	2.5	4.0	5.3	6.6

(a) Use the regression capabilities of a graphing utility to find a linear model for the data.

(b) Use a graphing utility to plot the data and graph the model. How well does the model fit the data? Explain your reasoning.

(c) Use the model to estimate the elongation of the spring when a force of 55 newtons is applied.

8. *Falling Object* In an experiment, students measured the speed s (in meters per second) of a falling object t seconds after it was released. The results are shown in the table.

t	0	1	2	3	4
s	0	11.0	19.4	29.2	39.4

(a) Use the regression capabilities of a graphing utility to find a linear model for the data.

(b) Use a graphing utility to plot the data and graph the model. How well does the model fit the data? Explain your reasoning.

(c) Use the model to estimate the speed of the object after 2.5 seconds.

9. *Energy Consumption and Gross National Product* The data show the per capita energy consumptions (in millions of Btu) and the per capita gross national products (in thousands of U.S. dollars) for several countries in 2004. (*Source: U.S. Census Bureau*)

Argentina	(71, 12.53)	Bangladesh	(5, 1.97)
Chile	(75, 10.61)	Ecuador	(29, 3.77)
Greece	(136, 22.23)	Hong Kong	(159, 31.56)
Hungary	(106, 15.8)	India	(15, 3.12)
Mexico	(63, 9.64)	Poland	(95, 12.73)
Portugal	(106, 19.24)	South Korea	(186, 20.53)
Spain	(159, 24.75)	Turkey	(51, 7.72)
United Kingdom	(167, 31.43)	Venezuela	(115, 5.83)

(a) Use the regression capabilities of a graphing utility to find a linear model for the data. What is the correlation coefficient?

(b) Use a graphing utility to plot the data and graph the model.

(c) Interpret the graph in part (b). Use the graph to identify the four countries that differ most from the linear model.

(d) Delete the data for the four countries identified in part (c). Fit a linear model to the remaining data and give the correlation coefficient.

10. ***Brinell Hardness***   The data in the table show the Brinell hardness H of 0.35 carbon steel when hardened and tempered at temperature t (degrees Fahrenheit).   *(Source: Standard Handbook for Mechanical Engineers)*

t	200	400	600	800	1000	1200
H	534	495	415	352	269	217

(a) Use the regression capabilities of a graphing utility to find a linear model for the data.

(b) Use a graphing utility to plot the data and graph the model. How well does the model fit the data? Explain your reasoning.

(c) Use the model to estimate the hardness when t is 500°F.

11. ***Automobile Costs***   The data in the table show the variable costs of operating an automobile in the United States for several recent years. The functions y_1, y_2, and y_3 represent the costs in cents per mile for gas, maintenance, and tires, respectively. *(Source: Bureau of Transportation Statistics)*

Year	y_1	y_2	y_3
0	5.60	3.30	1.70
1	6.90	3.60	1.70
2	7.90	3.90	1.80
3	5.90	4.10	1.80
4	7.20	4.10	1.80
5	6.50	5.40	0.70
6	9.50	4.90	0.70
7	8.90	4.90	0.70

(a) Use the regression capabilities of a graphing utility to find cubic models for y_1 and y_3 and a linear model for y_2.

(b) Use a graphing utility to graph y_1, y_2, y_3, and $y_1 + y_2 + y_3$ in the same viewing window. Use the model to estimate the total variable cost per mile in year 12.

12. ***Beam Strength***   Students in a lab measured the breaking strength S (in pounds) of wood 2 inches thick, x inches high, and 12 inches long. The results are shown in the table.

x	4	6	8	10	12
S	2370	5460	10,310	16,250	23,860

(a) Use the regression capabilities of a graphing utility to fit a quadratic model to the data.

(b) Use a graphing utility to plot the data and graph the model.

(c) Use the model to approximate the breaking strength when $x = 2$.

13. ***Car Performance***   The times t (in seconds) required to attain speeds of s miles per hour from a standing start for a Honda Accord Hybrid are shown in the table.   *(Source: Car & Driver)*

s	30	40	50	60	70	80	90
t	2.5	3.5	5.0	6.7	8.7	11.5	14.4

(a) Use the regression capabilities of a graphing utility to find a quadratic model for the data.

(b) Use a graphing utility to plot the data and graph the model.

(c) Use the graph in part (b) to state why the model is not appropriate for determining the times required to attain speeds of less than 20 miles per hour.

(d) Because the test began from a standing start, add the point $(0, 0)$ to the data. Fit a quadratic model to the revised data and graph the new model.

(e) Does the quadratic model in part (d) more accurately model the behavior of the car? Explain.

CAPSTONE

14. ***Health Maintenance Organizations***   The bar graph shows the numbers of people N (in millions) receiving care in HMOs for the years 1990 through 2004.   *(Source: HealthLeaders-InterStudy)*

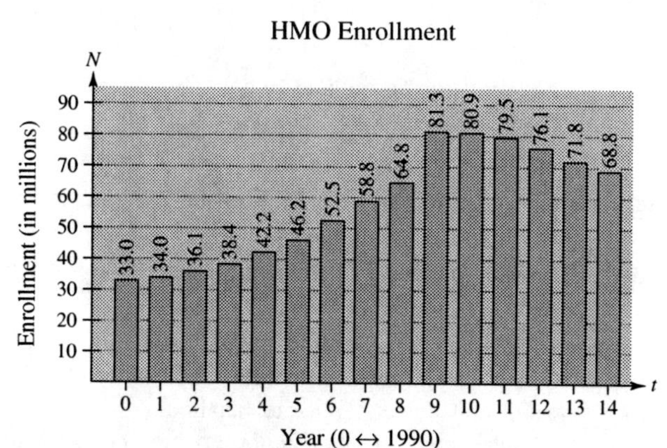

(a) Let t be the time in years, with $t = 0$ corresponding to 1990. Use the regression capabilities of a graphing utility to find linear and cubic models for the data.

(b) Use a graphing utility to graph the data and the linear and cubic models.

(c) Use the graphs in part (b) to determine which is the better model.

(d) Use a graphing utility to find and graph a quadratic model for the data. How well does the model fit the data? Explain your reasoning.

(e) Use the linear and cubic models to estimate the number of people receiving care in HMOs in the year 2007. What do you notice?

(f) Use a graphing utility to find other models for the data. Which models do you think best represent the data? Explain.

15. *Car Performance* A V8 car engine is coupled to a dynamometer, and the horsepower y is measured at different engine speeds x (in thousands of revolutions per minute). The results are shown in the table.

x	1	2	3	4	5	6
y	40	85	140	200	225	245

(a) Use the regression capabilities of a graphing utility to find a cubic model for the data.

(b) Use a graphing utility to plot the data and graph the model.

(c) Use the model to approximate the horsepower when the engine is running at 4500 revolutions per minute.

16. *Boiling Temperature* The table shows the temperatures T (°F) at which water boils at selected pressures p (pounds per square inch). (*Source: Standard Handbook for Mechanical Engineers*)

p	5	10	14.696 (1 atmosphere)	20
T	162.24°	193.21°	212.00°	227.96°

p	30	40	60	80	100
T	250.33°	267.25°	292.71°	312.03°	327.81°

(a) Use the regression capabilities of a graphing utility to find a cubic model for the data.

(b) Use a graphing utility to plot the data and graph the model.

(c) Use the graph to estimate the pressure required for the boiling point of water to exceed 300°F.

(d) Explain why the model would not be accurate for pressures exceeding 100 pounds per square inch.

17. *Harmonic Motion* The motion of an oscillating weight suspended by a spring was measured by a motion detector. The data collected and the approximate maximum (positive and negative) displacements from equilibrium are shown in the figure. The displacement y is measured in centimeters and the time t is measured in seconds.

(a) Is y a function of t? Explain.

(b) Approximate the amplitude and period of the oscillations.

(c) Find a model for the data.

(d) Use a graphing utility to graph the model in part (c). Compare the result with the data in the figure.

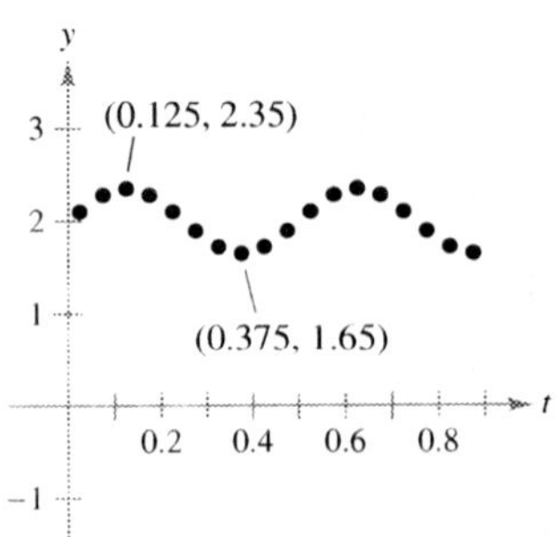

18. *Temperature* The table shows the normal daily high temperatures for Miami M and Syracuse S (in degrees Fahrenheit) for month t, with $t = 1$ corresponding to January. (*Source: NOAA*)

t	1	2	3	4	5	6
M	76.5	77.7	80.7	83.8	87.2	89.5
S	31.4	33.5	43.1	55.7	68.5	77.0

t	7	8	9	10	11	12
M	90.9	90.6	89.0	85.4	81.2	77.5
S	81.7	79.6	71.4	59.8	47.4	36.3

(a) A model for Miami is

$$M(t) = 83.70 + 7.46 \sin(0.4912t - 1.95).$$

Find a model for Syracuse.

(b) Use a graphing utility to graph the data and the model for the temperatures in Miami. How well does the model fit?

(c) Use a graphing utility to graph the data and the model for the temperatures in Syracuse. How well does the model fit?

(d) Use the models to estimate the average annual temperature in each city. Which term of the model did you use? Explain.

(e) What is the period of each model? Is it what you expected? Explain.

(f) Which city has a greater variability in temperature throughout the year? Which factor of the models determines this variability? Explain.

19. 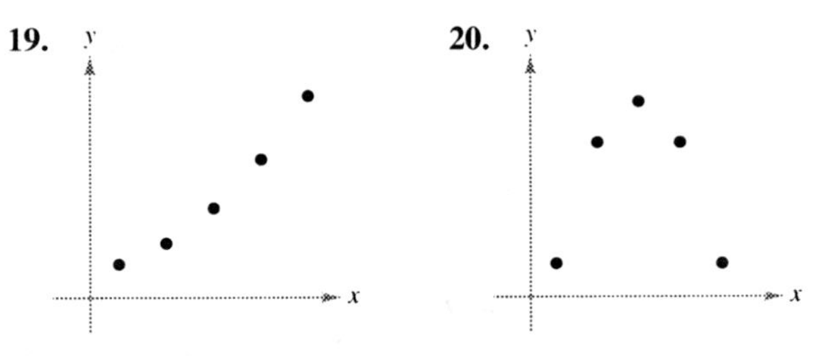

20.

1.5 Inverse Functions

- **Verify that one function is the inverse function of another function.**
- **Determine whether a function has an inverse function.**
- **Develop properties of the six inverse trigonometric functions.**

Inverse Functions

Recall from Section 1.3 that a function can be represented by a set of ordered pairs. For instance, the function $f(x) = x + 3$ from $A = \{1, 2, 3, 4\}$ to $B = \{4, 5, 6, 7\}$ can be written as

$$f: \{(1, 4), (2, 5), (3, 6), (4, 7)\}.$$

By interchanging the first and second coordinates of each ordered pair, you can form the **inverse function** of f. This function is denoted by f^{-1}. It is a function from B to A, and can be written as

$$f^{-1}: \{(4, 1), (5, 2), (6, 3), (7, 4)\}.$$

Note that the domain of f is equal to the range of f^{-1}, and vice versa, as shown in Figure 1.35. The functions f and f^{-1} have the effect of "undoing" each other. That is, when you form the composition of f with f^{-1} or the composition of f^{-1} with f, you obtain the identity function.

$$f(f^{-1}(x)) = x \qquad \text{and} \qquad f^{-1}(f(x)) = x$$

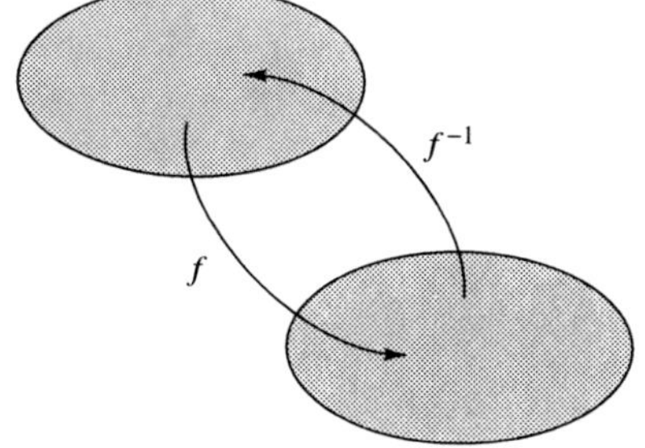

Domain of f = range of f^{-1}
Domain of f^{-1} = range of f
Figure 1.35

Finding Inverse Functions
Explain how to "undo" each of the following functions. Then use your explanation to write the inverse function of f.

a. $f(x) = x - 5$

b. $f(x) = 6x$

c. $f(x) = \dfrac{x}{2}$

d. $f(x) = 3x + 2$

e. $f(x) = x^3$

f. $f(x) = 4(x - 2)$

Use a graphing utility to graph each function and its inverse function in the same "square" viewing window. What observation can you make about each pair of graphs?

DEFINITION OF INVERSE FUNCTION

A function g is the **inverse function** of the function f if

$$f(g(x)) = x \quad \text{for each } x \text{ in the domain of } g$$

and

$$g(f(x)) = x \quad \text{for each } x \text{ in the domain of } f.$$

The function g is denoted by f^{-1} (read "f inverse").

NOTE Although the notation used to denote an inverse function resembles *exponential notation*, it is a different use of -1 as a superscript. That is, in general, $f^{-1}(x) \neq 1/f(x)$. ■

Here are some important observations about inverse functions.

1. If g is the inverse function of f, then f is the inverse function of g.

2. The domain of f^{-1} is equal to the range of f, and the range of f^{-1} is equal to the domain of f.

3. A function need not have an inverse function, but if it does, the inverse function is unique (see Exercise 159).

You can think of f^{-1} as undoing what has been done by f. For example, subtraction can be used to undo addition, and division can be used to undo multiplication. Use the definition of an inverse function to check the following.

$$f(x) = x + c \qquad \text{and} \qquad f^{-1}(x) = x - c \quad \text{are inverse functions of each other.}$$

$$f(x) = cx \qquad \text{and} \qquad f^{-1}(x) = \frac{x}{c}, \; c \neq 0, \quad \text{are inverse functions of each other.}$$

EXAMPLE 1   Verifying Inverse Functions

Show that the functions are inverse functions of each other.

$$f(x) = 2x^3 - 1 \qquad \text{and} \qquad g(x) = \sqrt[3]{\frac{x+1}{2}}$$

Solution   Because the domains and ranges of both f and g consist of all real numbers, you can conclude that both composite functions exist for all x. The composition of f with g is given by

$$f(g(x)) = 2\left(\sqrt[3]{\frac{x+1}{2}}\right)^3 - 1$$
$$= 2\left(\frac{x+1}{2}\right) - 1$$
$$= x + 1 - 1$$
$$= x.$$

The composition of g with f is given by

$$g(f(x)) = \sqrt[3]{\frac{(2x^3 - 1) + 1}{2}}$$
$$= \sqrt[3]{\frac{2x^3}{2}}$$
$$= \sqrt[3]{x^3}$$
$$= x.$$

Because $f(g(x)) = x$ and $g(f(x)) = x$, you can conclude that f and g are inverse functions of each other (see Figure 1.36).

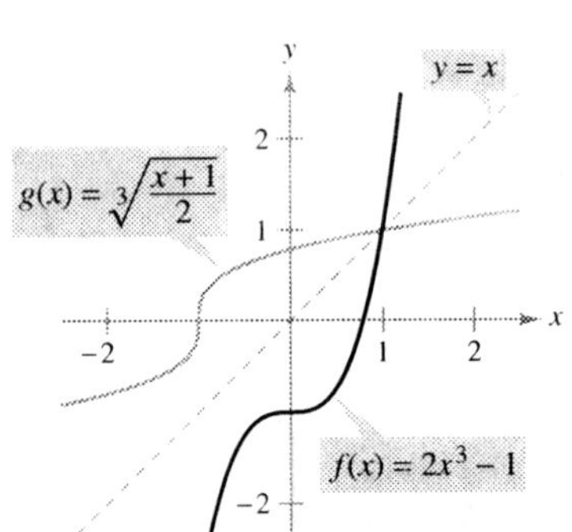

f and g are inverse functions of each other.
Figure 1.36

STUDY TIP   In Example 1, try comparing the functions f and g verbally.

For f: First cube x, then multiply by 2, then subtract 1.

For g: First add 1, then divide by 2, then take the cube root.

Do you see the "undoing pattern"?

In Figure 1.36, the graphs of f and $g = f^{-1}$ appear to be mirror images of each other with respect to the line $y = x$. The graph of f^{-1} is a **reflection** of the graph of f in the line $y = x$. This idea is generalized as follows.

REFLECTIVE PROPERTY OF INVERSE FUNCTIONS

The graph of f contains the point (a, b) if and only if the graph of f^{-1} contains the point (b, a).

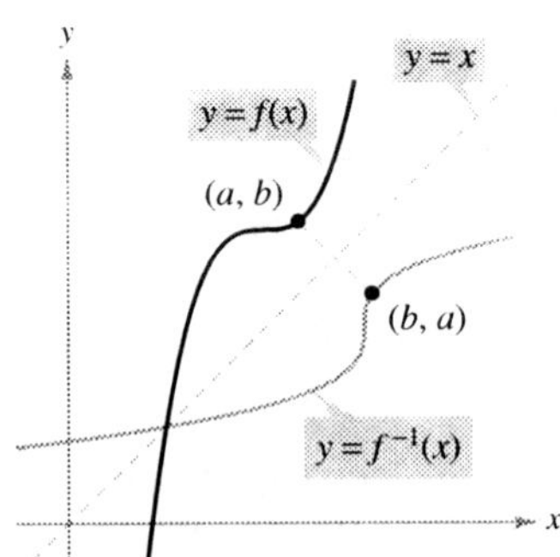

The graph of f^{-1} is a reflection of the graph of f in the line $y = x$.
Figure 1.37

To see this, suppose (a, b) is on the graph of f. Then $f(a) = b$ and you can write

$$f^{-1}(b) = f^{-1}(f(a)) = a.$$

So, (b, a) is on the graph of f^{-1}, as shown in Figure 1.37. A similar argument will verify this result in the other direction.

Existence of an Inverse Function

Not every function has an inverse, and the Reflective Property of Inverse Functions suggests a graphical test for those that do—the **Horizontal Line Test** for an inverse function. This test states that a function f has an inverse function if and only if every horizontal line intersects the graph of f at most once (see Figure 1.38). The following formally states why the Horizontal Line Test is valid.

THE EXISTENCE OF AN INVERSE FUNCTION

A function has an inverse function if and only if it is one-to-one.

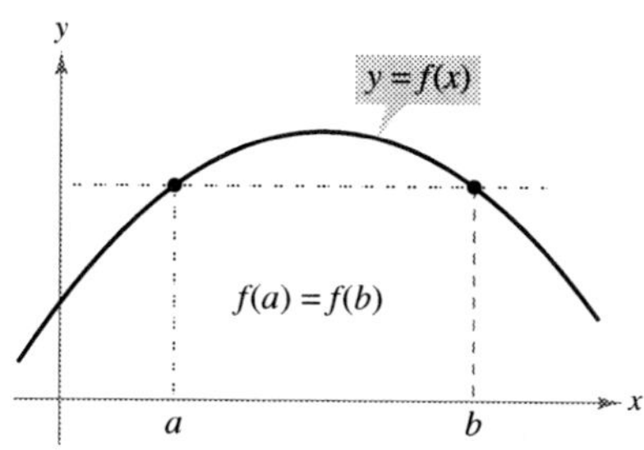

If a horizontal line intersects the graph of f twice, then f is not one-to-one.
Figure 1.38

EXAMPLE 2 The Existence of an Inverse Function

Which of the functions has an inverse function?

a. $f(x) = x^3 - 1$ **b.** $f(x) = x^3 - x + 1$

Solution

a. From the graph of f shown in Figure 1.39(a), it appears that f is one-to-one over its entire domain. To verify this, suppose that there exist x_1 and x_2 such that $f(x_1) = f(x_2)$. By showing that $x_1 = x_2$, it follows that f is one-to-one.

$$f(x_1) = f(x_2)$$
$$x_1^3 - 1 = x_2^3 - 1$$
$$x_1^3 = x_2^3$$
$$\sqrt[3]{x_1^3} = \sqrt[3]{x_2^3}$$
$$x_1 = x_2$$

Because f is one-to-one, you can conclude that f must have an inverse function.

b. From the graph in Figure 1.39(b), you can see that the function does not pass the Horizontal Line Test. In other words, it is not one-to-one. For instance, f has the same value when $x = -1, 0,$ and 1.

$$f(-1) = f(1) = f(0) = 1 \qquad \text{Not one-to-one}$$

Therefore, f does not have an inverse function. ■

NOTE Often it is easier to prove that a function has an inverse function than to find the inverse function. For instance, by sketching the graph of $f(x) = x^3 + x - 1$, you can see that it is one-to-one. Yet it would be difficult to determine the inverse of this function algebraically. ■

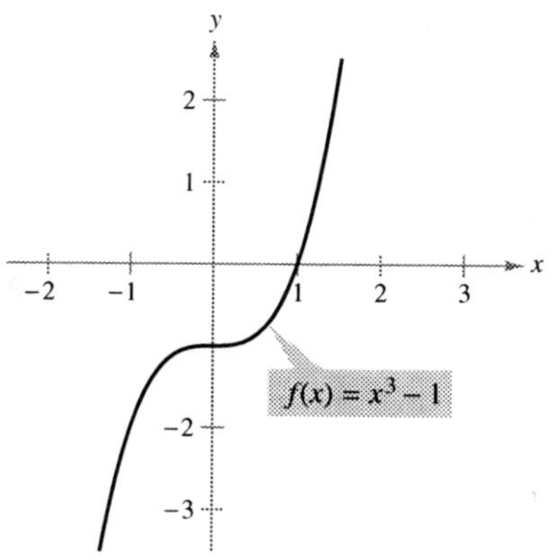

(a) Because f is one-to-one over its entire domain, it has an inverse function.

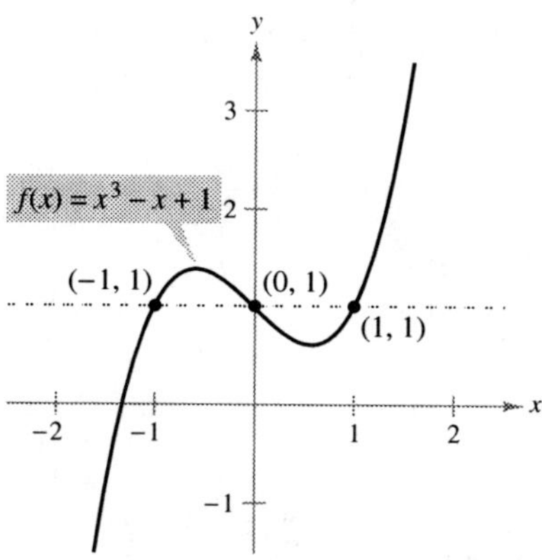

(b) Because f is not one-to-one, it does not have an inverse function.
Figure 1.39

GUIDELINES FOR FINDING AN INVERSE OF A FUNCTION

1. Determine whether the function given by $y = f(x)$ has an inverse function.
2. Solve for x as a function of y: $x = g(y) = f^{-1}(y)$.
3. Interchange x and y. The resulting equation is $y = f^{-1}(x)$.
4. Define the domain of f^{-1} as the range of f.
5. Verify that $f(f^{-1}(x)) = x$ and $f^{-1}(f(x)) = x$.

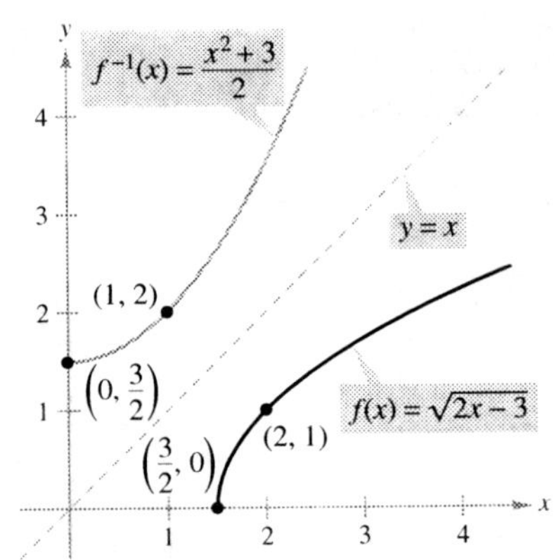

The domain of f^{-1}, $[0, \infty)$, is the range of f.

Figure 1.40

EXAMPLE 3 Finding an Inverse Function

Find the inverse function of

$$f(x) = \sqrt{2x - 3}.$$

Solution   The function has an inverse function because it is one-to-one on its entire domain, $\left[\frac{3}{2}, \infty\right)$, as shown in Figure 1.40. To find an equation for the inverse function, let $y = f(x)$ and solve for x in terms of y.

$$\sqrt{2x - 3} = y \qquad \text{Let } y = f(x).$$

$$2x - 3 = y^2 \qquad \text{Square each side.}$$

$$x = \frac{y^2 + 3}{2} \qquad \text{Solve for } x.$$

$$y = \frac{x^2 + 3}{2} \qquad \text{Interchange } x \text{ and } y.$$

$$f^{-1}(x) = \frac{x^2 + 3}{2} \qquad \text{Replace } y \text{ by } f^{-1}(x).$$

The domain of f^{-1} is the range of f, which is $[0, \infty)$. You can verify this result as shown.

$$f(f^{-1}(x)) = \sqrt{2\left(\frac{x^2 + 3}{2}\right) - 3} = \sqrt{x^2} = x, \quad x \geq 0$$

$$f^{-1}(f(x)) = \frac{\left(\sqrt{2x - 3}\right)^2 + 3}{2} = \frac{2x - 3 + 3}{2} = x, \quad x \geq \frac{3}{2}$$

NOTE   Remember that any letter can be used to represent the independent variable. So,

$$f^{-1}(y) = \frac{y^2 + 3}{2}, \quad f^{-1}(x) = \frac{x^2 + 3}{2}, \quad \text{and} \quad f^{-1}(s) = \frac{s^2 + 3}{2}$$

all represent the same function.

Suppose you are given a function that is *not* one-to-one on its entire domain. By restricting the domain to an interval on which the function *is* one-to-one, you can conclude that the new function has an inverse function on the restricted domain.

EXAMPLE 4 Testing Whether a Function Is One-to-One

Show that the sine function

$$f(x) = \sin x$$

is not one-to-one on the entire real line. Then show that f is one-to-one on the closed interval $[-\pi/2, \pi/2]$.

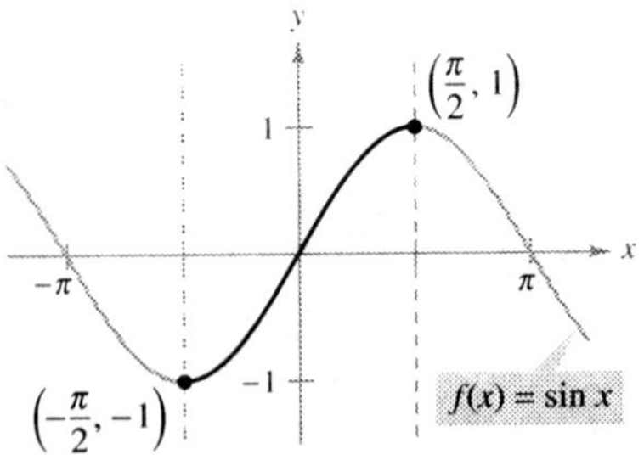

Solution   It is clear that f is not one-to-one, because many different x-values yield the same y-value. For instance,

$$\sin(0) = 0 = \sin(\pi).$$

Moreover, from the graph of $f(x) = \sin x$ in Figure 1.41, you can see that when f is restricted to the interval $[-\pi/2, \pi/2]$, then the restricted function *is* one-to-one.

f is one-to-one on the interval $[-\pi/2, \pi/2]$.

Figure 1.41

Inverse Trigonometric Functions

From the graphs of the six basic trigonometric functions, you can see that they do not have inverse functions. (Graphs of the six basic trigonometric functions are shown in Appendix C.) The functions that are called "inverse trigonometric functions" are actually inverses of trigonometric functions whose domains have been restricted.

For instance, in Example 4, you saw that the sine function is one-to-one on the interval $[-\pi/2, \pi/2]$ (see Figure 1.42). On this interval, you can define the inverse of the *restricted* sine function to be

$$y = \arcsin x \qquad \text{if and only if} \qquad \sin y = x$$

where $-1 \le x \le 1$ and $-\pi/2 \le \arcsin x \le \pi/2$. From Figures 1.42 (a) and (b), you can see that you can obtain the graph of $y = \arcsin x$ by reflecting the graph of $y = \sin x$ in the line $y = x$ on the interval $[-\pi/2, \pi/2]$.

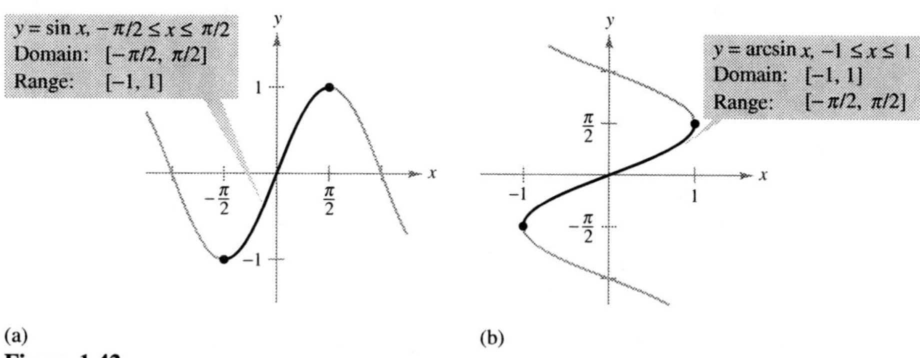

(a)   (b)

Figure 1.42

Under suitable restrictions, each of the six trigonometric functions is one-to-one and so has an inverse function, as indicated in the following definition. (The term "iff" is used to represent the phrase "if and only if.")

DEFINITION OF INVERSE TRIGONOMETRIC FUNCTION

Function	Domain	Range
$y = \arcsin x$ iff $\sin y = x$	$-1 \le x \le 1$	$-\dfrac{\pi}{2} \le y \le \dfrac{\pi}{2}$
$y = \arccos x$ iff $\cos y = x$	$-1 \le x \le 1$	$0 \le y \le \pi$
$y = \arctan x$ iff $\tan y = x$	$-\infty < x < \infty$	$-\dfrac{\pi}{2} < y < \dfrac{\pi}{2}$
$y = \operatorname{arccot} x$ iff $\cot y = x$	$-\infty < x < \infty$	$0 < y < \pi$
$y = \operatorname{arcsec} x$ iff $\sec y = x$	$\lvert x \rvert \ge 1$	$0 \le y \le \pi,\ \ y \ne \dfrac{\pi}{2}$
$y = \operatorname{arccsc} x$ iff $\csc y = x$	$\lvert x \rvert \ge 1$	$-\dfrac{\pi}{2} \le y \le \dfrac{\pi}{2},\ \ y \ne 0$

NOTE   The term $\arcsin x$ is read as "the arcsine of x" or sometimes "the angle whose sine is x." An alternative notation for the inverse sine function is $\sin^{-1} x$.

The graphs of the six inverse trigonometric functions are shown in Figure 1.43.

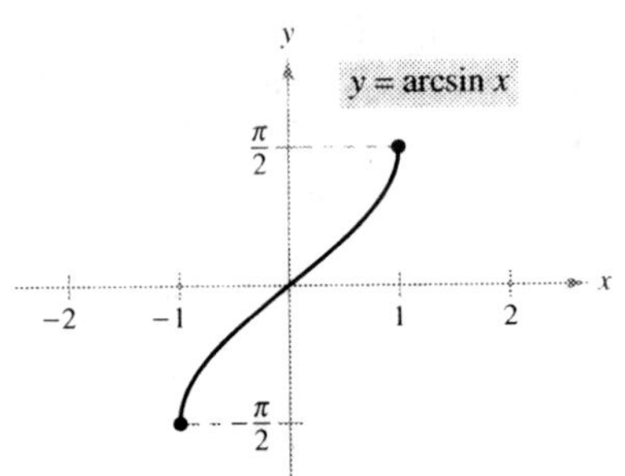

Domain: $[-1, 1]$
Range: $[-\pi/2, \pi/2]$

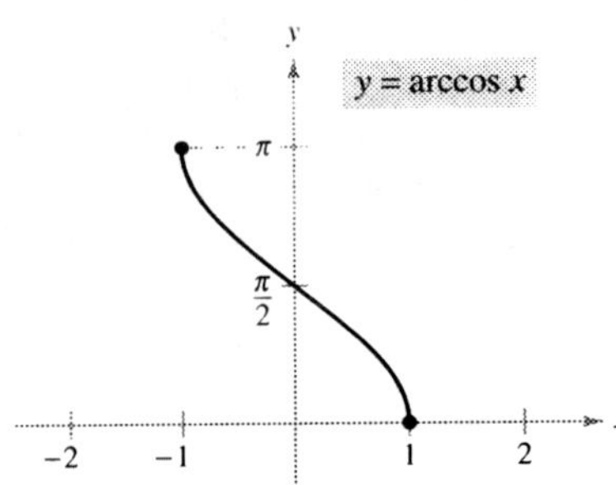

Domain: $[-1, 1]$
Range: $[0, \pi]$

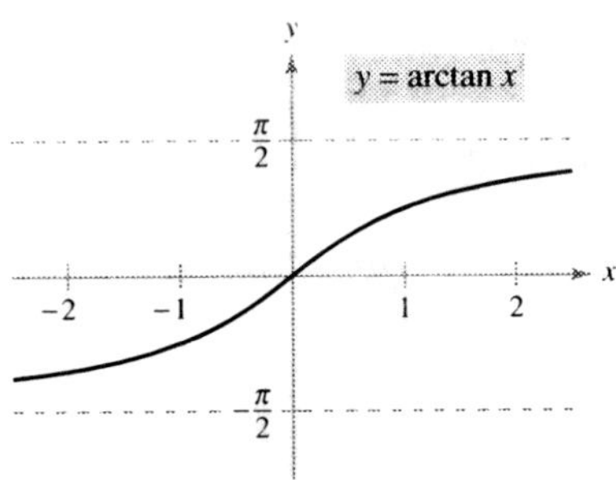

Domain: $(-\infty, \infty)$
Range: $(-\pi/2, \pi/2)$

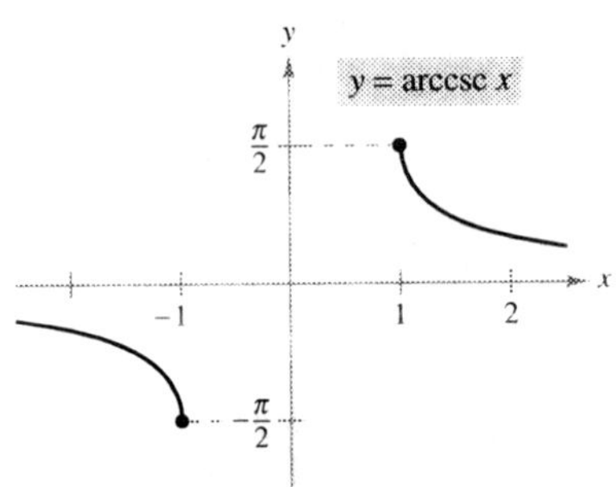

Domain: $(-\infty, -1] \cup [1, \infty)$
Range: $[-\pi/2, 0) \cup (0, \pi/2]$
Figure 1.43

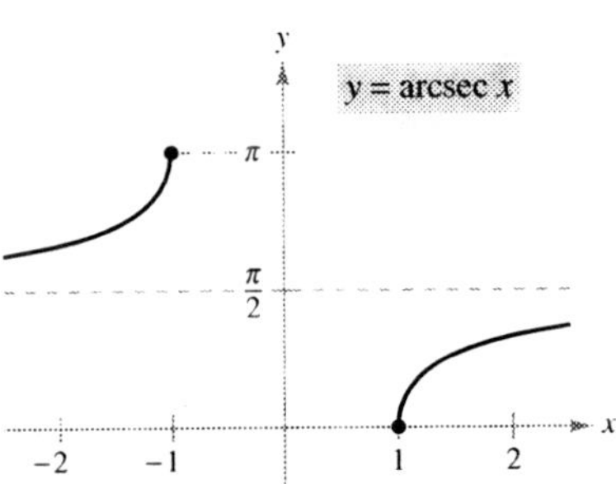

Domain: $(-\infty, -1] \cup [1, \infty)$
Range: $[0, \pi/2) \cup (\pi/2, \pi]$

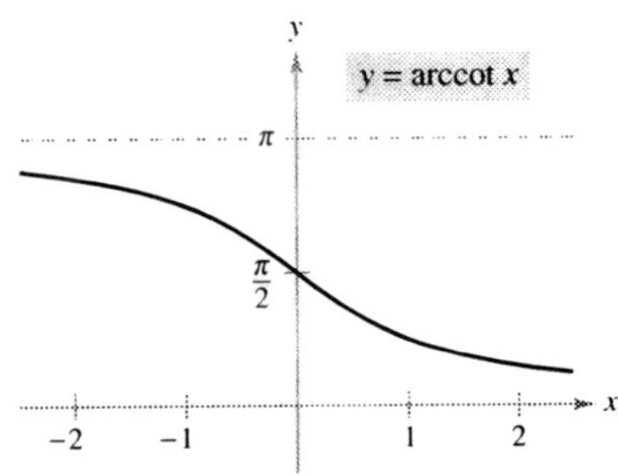

Domain: $(-\infty, \infty)$
Range: $(0, \pi)$

EXAMPLE 5   Evaluating Inverse Trigonometric Functions

Evaluate each of the following.

a. $\arcsin\left(-\dfrac{1}{2}\right)$ **b.** $\arccos 0$ **c.** $\arctan \sqrt{3}$ **d.** $\arcsin(0.3)$

NOTE When evaluating inverse trigonometric functions, remember that they denote *angles in radian measure*.

Solution

a. By definition, $y = \arcsin\left(-\frac{1}{2}\right)$ implies that $\sin y = -\frac{1}{2}$. In the interval $[-\pi/2, \pi/2]$, the correct value of y is $-\pi/6$.

$$\arcsin\left(-\frac{1}{2}\right) = -\frac{\pi}{6}$$

b. By definition, $y = \arccos 0$ implies that $\cos y = 0$. In the interval $[0, \pi]$, you have $y = \pi/2$.

$$\arccos 0 = \frac{\pi}{2}$$

c. By definition, $y = \arctan \sqrt{3}$ implies that $\tan y = \sqrt{3}$. In the interval $(-\pi/2, \pi/2)$, you have $y = \pi/3$.

$$\arctan \sqrt{3} = \frac{\pi}{3}$$

d. Using a calculator set in *radian* mode produces

$$\arcsin(0.3) \approx 0.3047.$$

Inverse functions have the properties

$$f(f^{-1}(x)) = x \quad \text{and} \quad f^{-1}(f(x)) = x.$$

When applying these properties to inverse trigonometric functions, remember that the trigonometric functions have inverse functions only in restricted domains. For x-values outside these domains, these two properties do not hold. For example, $\arcsin(\sin \pi)$ is equal to 0, not π.

PROPERTIES OF INVERSE TRIGONOMETRIC FUNCTIONS

1. If $-1 \le x \le 1$ and $-\pi/2 \le y \le \pi/2$, then

$$\sin(\arcsin x) = x \quad \text{and} \quad \arcsin(\sin y) = y.$$

2. If $-\pi/2 < y < \pi/2$, then

$$\tan(\arctan x) = x \quad \text{and} \quad \arctan(\tan y) = y.$$

3. If $|x| \ge 1$ and $0 \le y < \pi/2$ or $\pi/2 < y \le \pi$, then

$$\sec(\text{arcsec } x) = x \quad \text{and} \quad \text{arcsec}(\sec y) = y.$$

Similar properties hold for the other inverse trigonometric functions.

EXAMPLE 6   Solving an Equation

$$\arctan(2x - 3) = \frac{\pi}{4} \qquad \text{Write original equation.}$$

$$\tan[\arctan(2x - 3)] = \tan\frac{\pi}{4} \qquad \text{Take tangent of both sides.}$$

$$2x - 3 = 1 \qquad \tan(\arctan x) = x$$

$$x = 2 \qquad \text{Solve for } x. \qquad \blacksquare$$

Some problems in calculus require that you evaluate expressions such as $\cos(\arcsin x)$, as shown in Example 7.

EXAMPLE 7   Using Right Triangles

a. Given $y = \arcsin x$, where $0 < y < \pi/2$, find $\cos y$.

b. Given $y = \text{arcsec}\left(\sqrt{5}/2\right)$, find $\tan y$.

Solution

a. Because $y = \arcsin x$, you know that $\sin y = x$. This relationship between x and y can be represented by a right triangle, as shown in Figure 1.44.

$$\cos y = \cos(\arcsin x) = \frac{\text{adj.}}{\text{hyp.}} = \sqrt{1 - x^2}$$

(This result is also valid for $-\pi/2 < y < 0$.)

b. Use the right triangle shown in Figure 1.45.

$$\tan y = \tan\left[\text{arcsec}\left(\frac{\sqrt{5}}{2}\right)\right] = \frac{\text{opp.}}{\text{adj.}} = \frac{1}{2} \qquad \blacksquare$$

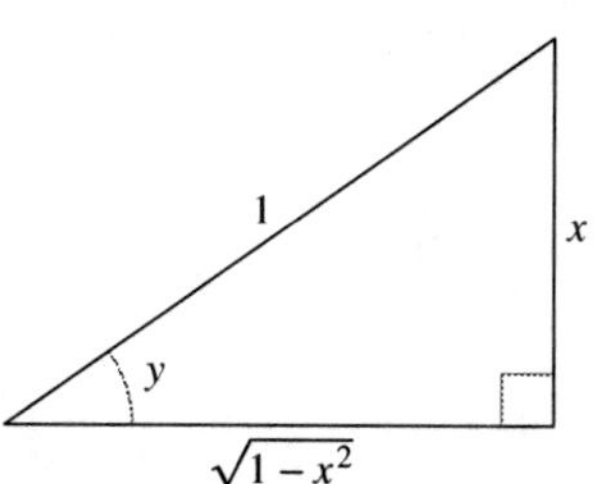

$y = \arcsin x$
Figure 1.44

$y = \text{arcsec } \dfrac{\sqrt{5}}{2}$
Figure 1.45

1.5 Exercises

See www.CalcChat.com for worked-out solutions to odd-numbered exercises.

In Exercises 1–8, show that f and g are inverse functions (a) analytically and (b) graphically.

1. $f(x) = 5x + 1,$ $\qquad g(x) = \dfrac{x - 1}{5}$

2. $f(x) = 3 - 4x,$ $\qquad g(x) = \dfrac{3 - x}{4}$

3. $f(x) = x^3,$ $\qquad g(x) = \sqrt[3]{x}$

4. $f(x) = 1 - x^3,$ $\qquad g(x) = \sqrt[3]{1 - x}$

5. $f(x) = \sqrt{x - 4},$ $\qquad g(x) = x^2 + 4, \quad x \geq 0$

6. $f(x) = 16 - x^2, \quad x \geq 0,$ $\qquad g(x) = \sqrt{16 - x}$

7. $f(x) = \dfrac{1}{x},$ $\qquad g(x) = \dfrac{1}{x}$

8. $f(x) = \dfrac{1}{1 + x}, \quad x \geq 0,$ $\qquad g(x) = \dfrac{1 - x}{x}, \quad 0 < x \leq 1$

In Exercises 9–12, match the graph of the function with the graph of its inverse function. [The graphs of the inverse functions are labeled (a), (b), (c), and (d).]

(a) 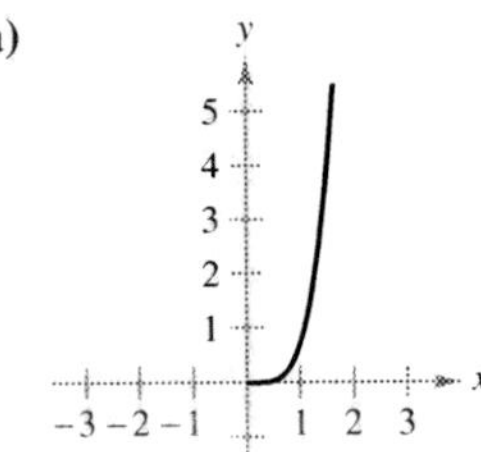

(b) 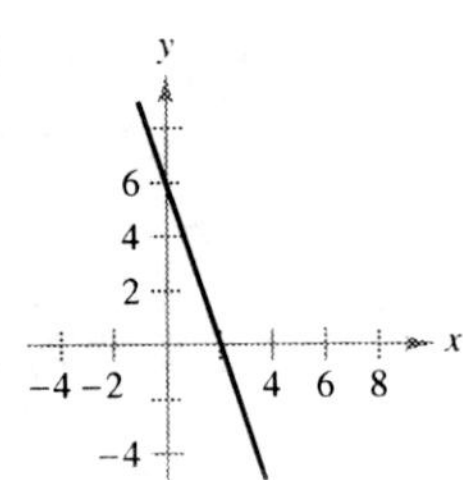

(c) 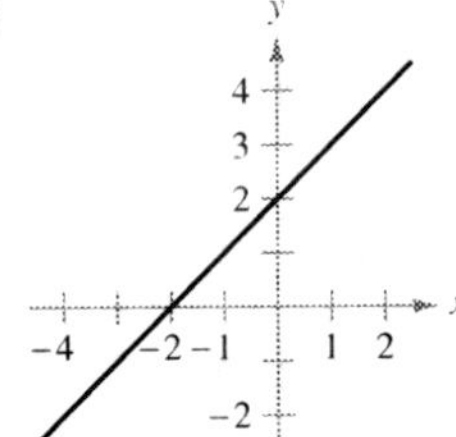

(d) 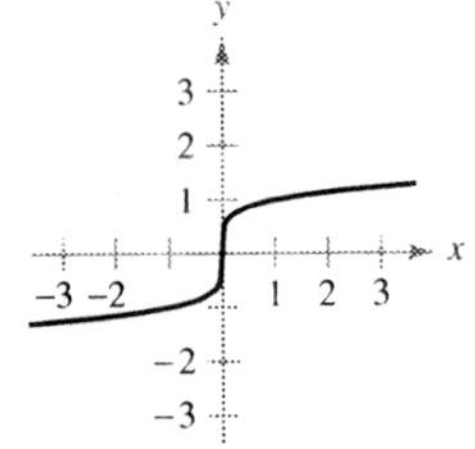

9. 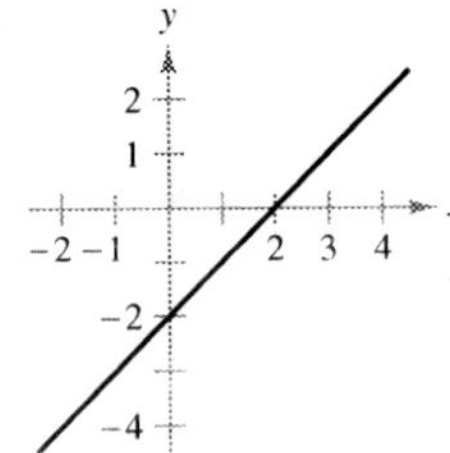

10. 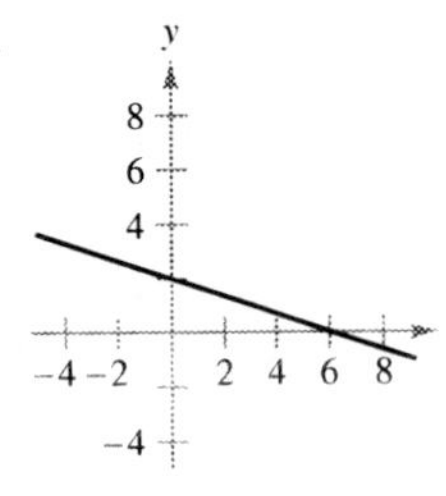

11. 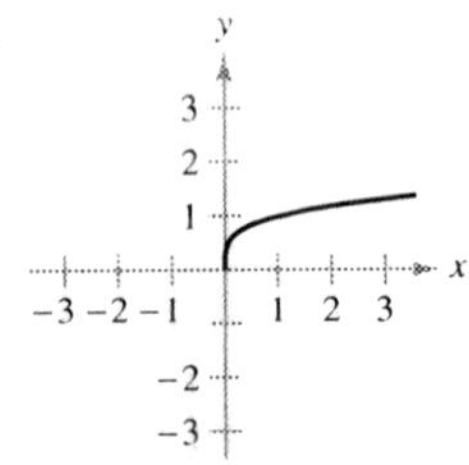

12. 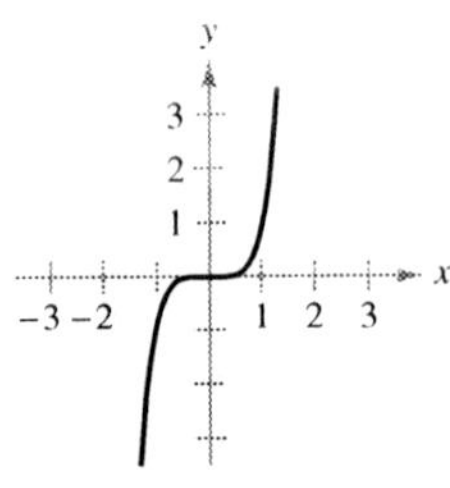

In Exercises 13–16, use the Horizontal Line Test to determine whether the function is one-to-one on its entire domain and therefore has an inverse function. To print an enlarged copy of the graph, go to the website *www.mathgraphs.com*.

13. $f(x) = \frac{3}{4}x + 6$

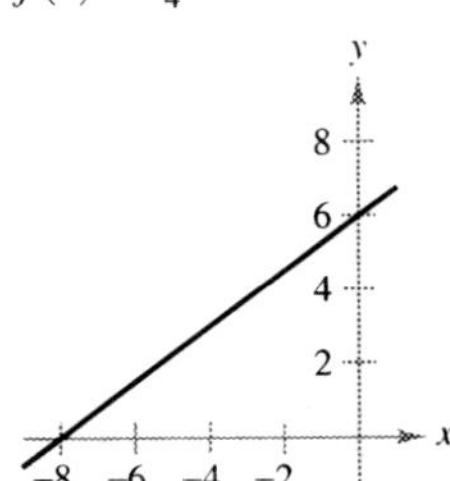

14. $f(x) = 5x - 3$

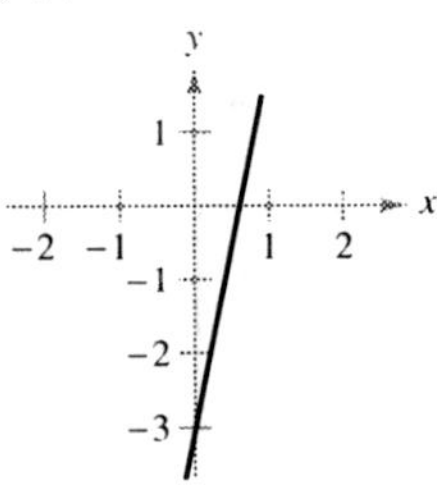

15. $f(\theta) = \sin \theta$

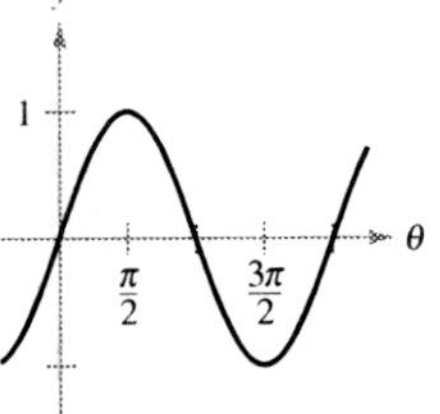

16. $f(x) = \dfrac{x^2}{x^2 + 4}$

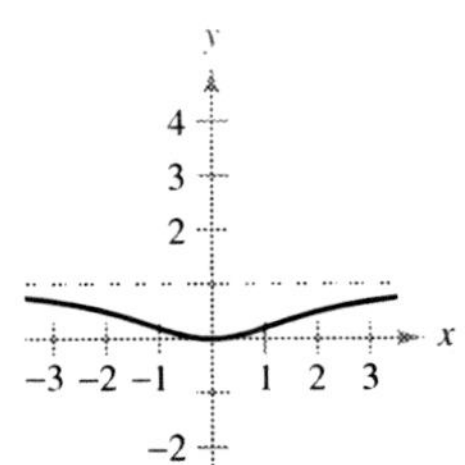

In Exercises 17–22, use a graphing utility to graph the function. Determine whether the function is one-to-one on its entire domain and therefore has an inverse function.

17. $h(s) = \dfrac{1}{s - 2} - 3$

18. $f(x) = \dfrac{1}{x^2 + 1}$

19. $g(t) = \dfrac{1}{\sqrt{t^2 + 1}}$

20. $f(x) = 5x\sqrt{x - 1}$

21. $g(x) = (x + 5)^3$

22. $h(x) = |x + 4| - |x - 4|$

In Exercises 23–28, determine whether the function is one-to-one on its entire domain and therefore has an inverse function.

23. $f(x) = (x + a)^3 + b$

24. $f(x) = \sin \dfrac{3x}{2}$

25. $f(x) = \dfrac{x^4}{4} - 2x^2$

26. $f(x) = x^3 - 6x^2 + 12x$

27. $f(x) = 2 - x - x^3$

28. $f(x) = \sqrt[3]{x + 1}$

In Exercises 29–36, (a) find the inverse function of f, (b) graph f and f^{-1} on the same set of coordinate axes, (c) describe the relationship between the graphs, and (d) state the domains and ranges of f and f^{-1}.

29. $f(x) = 2x - 3$

30. $f(x) = 3x$

31. $f(x) = x^5$

32. $f(x) = x^3 - 1$

33. $f(x) = \sqrt{x}$

34. $f(x) = x^2, \quad x \geq 0$

35. $f(x) = \sqrt{4 - x^2}, \quad 0 \leq x \leq 2$

36. $f(x) = \sqrt{x^2 - 4}, \quad x \geq 2$

In Exercises 37–42, (a) find the inverse function of f, (b) use a graphing utility to graph f and f^{-1} in the same viewing window, (c) describe the relationship between the graphs, and (d) state the domains and ranges of f and f^{-1}.

37. $f(x) = \sqrt[3]{x - 1}$

38. $f(x) = 3\sqrt[5]{2x - 1}$

39. $f(x) = x^{2/3}, \quad x \geq 0$

40. $f(x) = x^{3/5}$

41. $f(x) = \dfrac{x}{\sqrt{x^2 + 7}}$

42. $f(x) = \dfrac{x + 2}{x}$

In Exercises 43 and 44, use the graph of the function f to make a table of values for the given points. Then make a second table that can be used to find f^{-1}, and sketch the graph of f^{-1}. To print an enlarged copy of the graph, go to the website *www.mathgraphs.com.*

43. 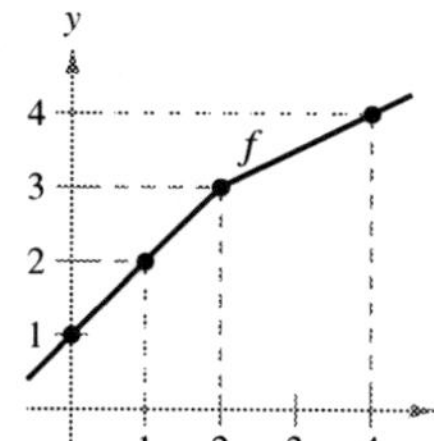

44. 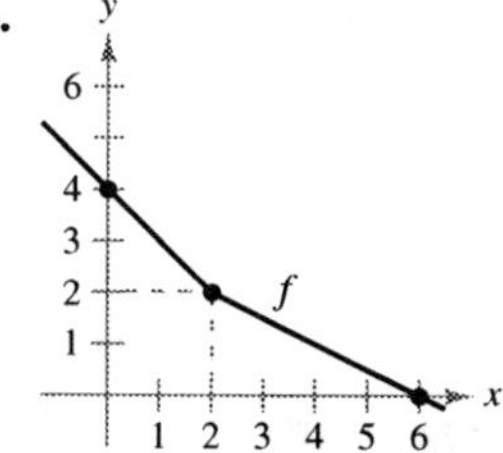

45. *Cost* You need 50 pounds of two commodities costing $1.25 and $1.60 per pound.

(a) Verify that the total cost is $y = 1.25x + 1.60(50 - x)$, where x is the number of pounds of the less expensive commodity.

(b) Find the inverse function of the cost function. What does each variable represent in the inverse function?

(c) What is the domain of the inverse function? Validate or explain your answer using the context of the problem.

(d) Determine the number of pounds of the less expensive commodity purchased if the total cost is $73.

46. *Temperature* The formula $C = \frac{5}{9}(F - 32)$, where $F \geq -459.6$, represents the Celsius temperature C as a function of the Fahrenheit temperature F.

(a) Find the inverse function of C.

(b) What does the inverse function represent?

(c) What is the domain of the inverse function? Validate or explain your answer using the context of the problem.

(d) The temperature is 22°C. What is the corresponding temperature in degrees Fahrenheit?

In Exercises 47 and 48, find f^{-1} over the given interval. Use a graphing utility to graph f and f^{-1} in the same viewing window. Describe the relationship between the graphs.

47. $f(x) = \dfrac{x}{x^2 - 4}; \quad (-2, 2)$

48. $f(x) = 2 - \dfrac{3}{x^2}; \quad (0, 10)$

Graphical Reasoning In Exercises 49–52, (a) use a graphing utility to graph the function, (b) use the *drawing* feature of the graphing utility to draw the inverse of the function, and (c) determine whether the graph of the inverse relation is an inverse function. Explain your reasoning.

49. $f(x) = x^3 + x + 4$

50. $h(x) = x\sqrt{4 - x^2}$

51. $g(x) = \dfrac{3x^2}{x^2 + 1}$

52. $f(x) = \dfrac{4x}{\sqrt{x^2 + 15}}$

In Exercises 53–58, show that f is one-to-one on the given interval and therefore has an inverse function on that interval.

Function	*Interval*
53. $f(x) = (x - 4)^2$	$[4, \infty)$
54. $f(x) = \|x + 2\|$	$[-2, \infty)$
55. $f(x) = \dfrac{4}{x^2}$	$(0, \infty)$
56. $f(x) = \cot x$	$(0, \pi)$
57. $f(x) = \cos x$	$[0, \pi]$
58. $f(x) = \sec x$	$\left[0, \dfrac{\pi}{2}\right)$

In Exercises 59–62, determine whether the function is one-to-one. If it is, find its inverse function.

59. $f(x) = \sqrt{x - 2}$

60. $f(x) = -3$

61. $f(x) = \|x - 2\|, \quad x \leq 2$

62. $f(x) = ax + b, \quad a \neq 0$

In Exercises 63–66, delete part of the domain so that the function that remains is one-to-one. Find the inverse function of the remaining function and give the domain of the inverse function. (*Note:* There is more than one correct answer.)

63. $f(x) = (x - 3)^2$

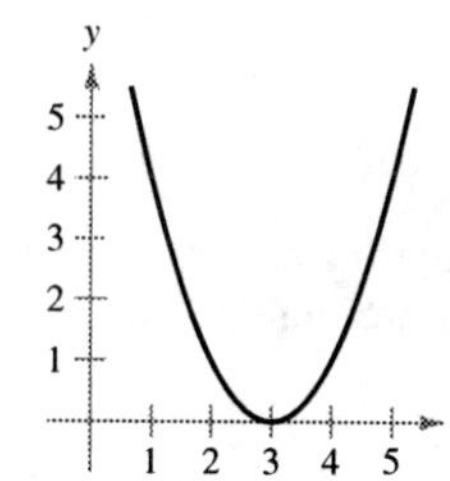

64. $f(x) = 16 - x^4$

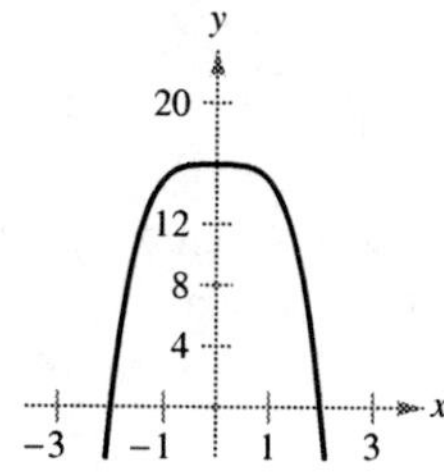

65. $f(x) = \|x + 3\|$

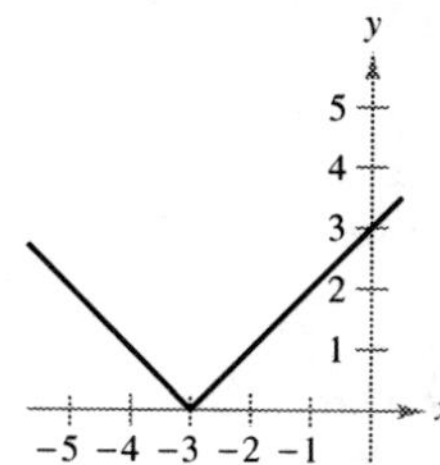

66. $f(x) = \|x - 3\|$

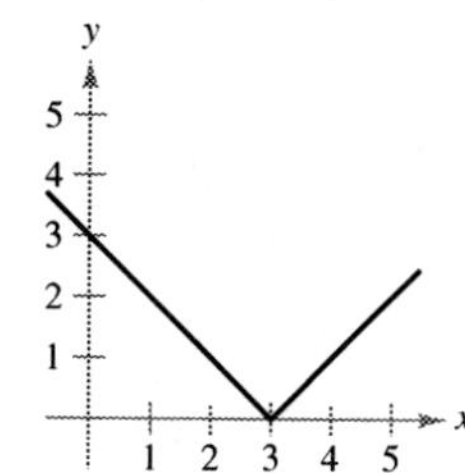

In Exercises 67–72, (a) sketch a graph of the function f, (b) determine an interval on which f is one-to-one, (c) find the inverse function of f on the interval found in part (b), and (d) give the domain of the inverse function. (*Note:* There is more than one correct answer.)

67. $f(x) = (x + 5)^2$

68. $f(x) = (7 - x)^2$

69. $f(x) = \sqrt{x^2 - 4x}$

70. $f(x) = -\sqrt{25 - x^2}$

71. $f(x) = 3 \cos x$

72. $f(x) = 2 \sin x$

In Exercises 73–78, find $f^{-1}(a)$ for the function f and real number a.

Function	Real Number
73. $f(x) = x^3 + 2x - 1$	$a = 2$
74. $f(x) = 2x^5 + x^3 + 1$	$a = -2$
75. $f(x) = \sin x, \quad -\dfrac{\pi}{2} \le x \le \dfrac{\pi}{2}$	$a = \dfrac{1}{2}$
76. $f(x) = \cos 2x, \quad 0 \le x \le \dfrac{\pi}{2}$	$a = 1$
77. $f(x) = x^3 - \dfrac{4}{x}, \quad x > 0$	$a = 6$
78. $f(x) = \sqrt{x - 4}$	$a = 2$

In Exercises 79–82, use the functions $f(x) = \frac{1}{8}x - 3$ and $g(x) = x^3$ to find the indicated value.

79. $(f^{-1} \circ g^{-1})(1)$

80. $(g^{-1} \circ f^{-1})(-3)$

81. $(f^{-1} \circ f^{-1})(6)$

82. $(g^{-1} \circ g^{-1})(-4)$

In Exercises 83–86, use the functions $f(x) = x + 4$ and $g(x) = 2x - 5$ to find the indicated function.

83. $g^{-1} \circ f^{-1}$

84. $f^{-1} \circ g^{-1}$

85. $(f \circ g)^{-1}$

86. $(g \circ f)^{-1}$

In Exercises 87 and 88, (a) use the graph of the function f to determine whether f is one-to-one, (b) state the domain of f^{-1}, and (c) estimate the value of $f^{-1}(2)$.

87. 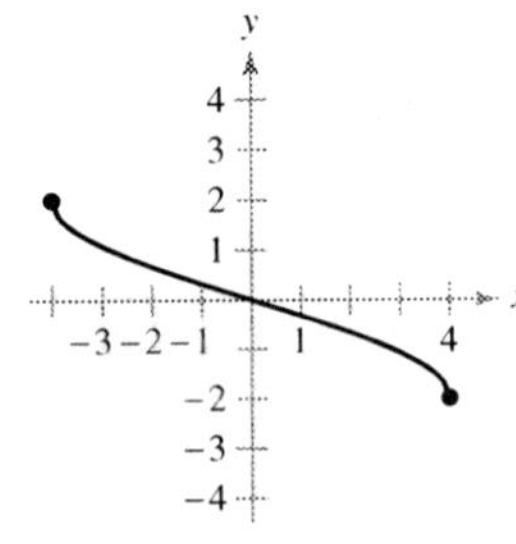

88. 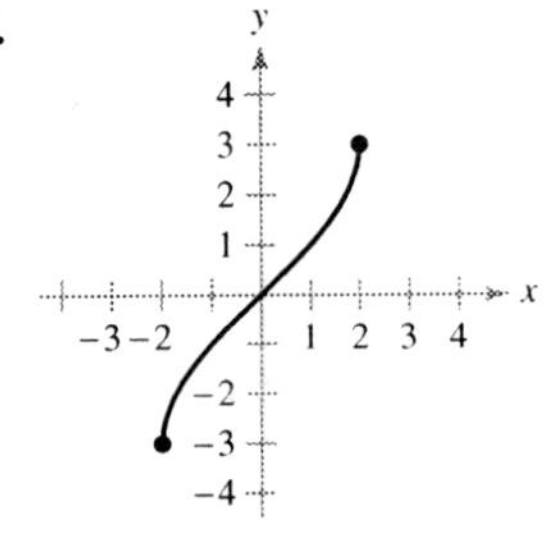

In Exercises 89 and 90, use the graph of the function f to sketch the graph of f^{-1}. To print an enlarged copy of the graph, go to the website *www.mathgraphs.com*.

89.   **90.**

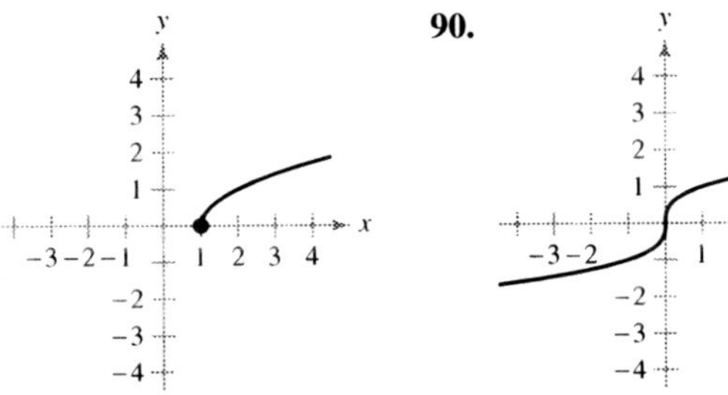

Numerical and Graphical Analysis   In Exercises 91 and 92, (a) use a graphing utility to complete the table, (b) plot the points in the table and graph the function by hand, (c) use a graphing utility to graph the function and compare the result with your hand-drawn graph in part (b), and (d) determine any intercepts and symmetry of the graph.

x	-1	-0.8	-0.6	-0.4	-0.2	0	0.2	0.4	0.6	0.8	1
y											

91. $y = \arcsin x$

92. $y = \arccos x$

93. Determine the missing coordinates of the points on the graph of the function.

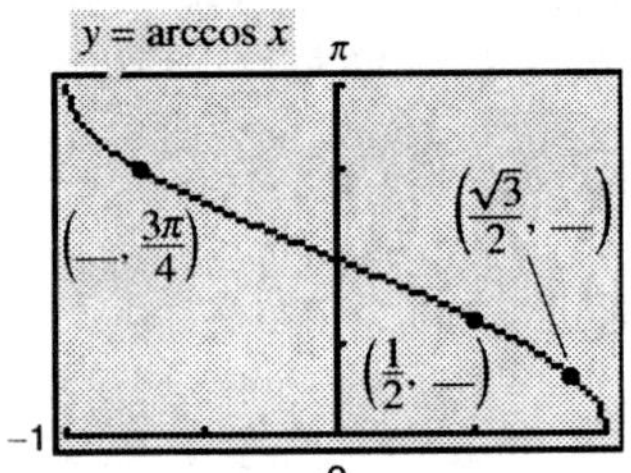

94. Determine the missing coordinates of the points on the graph of the function.

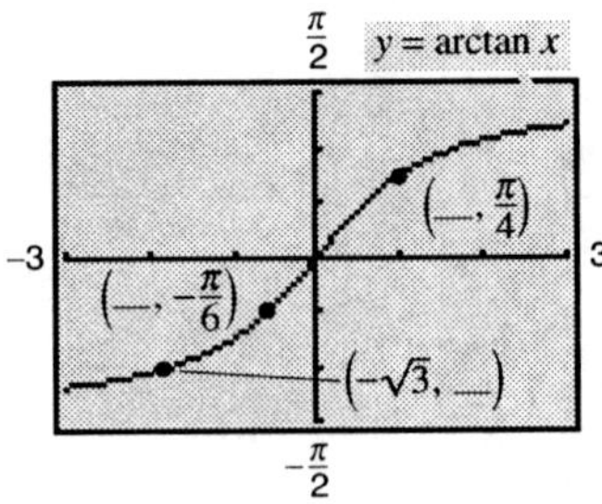

In Exercises 95–102, evaluate the expression without using a calculator.

95. $\arcsin \frac{1}{2}$

96. $\arcsin 0$

97. $\arccos \frac{1}{2}$

98. $\arccos 0$

99. $\arctan \dfrac{\sqrt{3}}{3}$

100. $\operatorname{arccot}\left(-\sqrt{3}\right)$

101. $\operatorname{arccsc}\left(-\sqrt{2}\right)$

102. $\arccos\left(-\dfrac{\sqrt{3}}{2}\right)$

In Exercises 103–106, use a calculator to approximate the value. Round your answer to two decimal places.

103. $\arccos(-0.8)$

104. $\arcsin(-0.39)$

105. $\operatorname{arcsec} 1.269$

106. $\arctan(-3)$

WRITING ABOUT CONCEPTS

107. Describe how to find the inverse function of a one-to-one function given by an equation in x and y. Give an example.

108. Describe the relationship between the graph of a function and the graph of its inverse function.

109. Give an example of a function that does *not* have an inverse function.

110. Explain why $\tan \pi = 0$ does not imply that $\arctan 0 = \pi$.

111. Explain why the domains of the trigonometric functions are restricted when finding the inverse trigonometric functions.

112. Explain how to graph $y = \operatorname{arccot} x$ on a graphing utility that does not have the arccotangent function.

In Exercises 113 and 114, use a graphing utility to confirm that f and g are inverse functions. (Remember to restrict the domain of f properly.)

113. $f(x) = \tan x$

  $g(x) = \arctan x$

114. $f(x) = \sin x$

  $g(x) = \arcsin x$

In Exercises 115 and 116, use the properties of inverse trigonometric functions to evaluate the expression.

115. $\cos[\arccos(-0.1)]$

116. $\arcsin(\sin 3\pi)$

In Exercises 117–122, evaluate the expression without using a calculator. [*Hint:* Sketch a right triangle, as demonstrated in Example 7(b).]

117. (a) $\sin\left(\arcsin \dfrac{1}{2}\right)$

  (b) $\cos\left(\arcsin \dfrac{1}{2}\right)$

118. (a) $\tan\left(\arccos \dfrac{\sqrt{2}}{2}\right)$

  (b) $\cos\left(\arcsin \dfrac{5}{13}\right)$

119. (a) $\sin\left(\arctan \dfrac{3}{4}\right)$

  (b) $\sec\left(\arcsin \dfrac{4}{5}\right)$

120. (a) $\tan(\operatorname{arccot} 2)$

  (b) $\cos(\operatorname{arcsec} \sqrt{5})$

121. (a) $\cot\left[\arcsin\left(-\dfrac{1}{2}\right)\right]$

  (b) $\csc\left[\arctan\left(-\dfrac{5}{12}\right)\right]$

122. (a) $\sec\left[\arctan\left(-\dfrac{3}{5}\right)\right]$

  (b) $\tan\left[\arcsin\left(-\dfrac{5}{6}\right)\right]$

In Exercises 123–128, use the figure to write the expression in algebraic form given $y = \arccos x$, where $0 < y < \pi/2$.

123. $\cos y$

124. $\sin y$

125. $\tan y$

126. $\cot y$

127. $\sec y$

128. $\csc y$

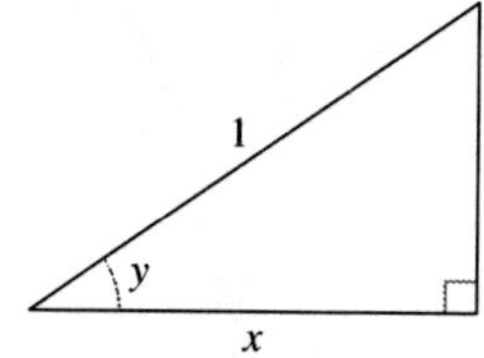

In Exercises 129–138, write the expression in algebraic form. [*Hint:* Sketch a right triangle, as demonstrated in Example 7(a).]

129. $\tan(\arctan x)$

130. $\sin(\arccos x)$

131. $\cos(\arcsin 2x)$

132. $\sec(\arctan 4x)$

133. $\sin(\operatorname{arcsec} x)$

134. $\cos(\operatorname{arccot} x)$

135. $\tan\left(\operatorname{arcsec} \dfrac{x}{3}\right)$

136. $\sec[\arcsin(x - 1)]$

137. $\csc\left(\arctan \dfrac{x}{\sqrt{2}}\right)$

138. $\cos\left(\arcsin \dfrac{x - h}{r}\right)$

In Exercises 139–142, solve the equation for x.

139. $\arcsin(3x - \pi) = \frac{1}{2}$

140. $\arctan(2x - 5) = -1$

141. $\arcsin \sqrt{2x} = \arccos \sqrt{x}$

142. $\arccos x = \text{arcsec } x$

In Exercises 143 and 144, find the point of intersection of the graphs of the functions.

143. $y = \arccos x$

$\quad y = \arctan x$

144. $y = \arcsin x$

$\quad y = \arccos x$

In Exercises 145 and 146, fill in the blank.

145. $\arctan \dfrac{9}{x} = \arcsin (\,\rule{1.2em}{0.8em}\,), \quad x > 0$

146. $\arcsin \dfrac{\sqrt{36 - x^2}}{6} = \arccos (\,\rule{1.2em}{0.8em}\,)$

In Exercises 147 and 148, verify each identity.

147. (a) $\text{arccsc } x = \arcsin \dfrac{1}{x}, \quad |x| \geq 1$

  (b) $\arctan x + \arctan \dfrac{1}{x} = \dfrac{\pi}{2}, \quad x > 0$

148. (a) $\arcsin(-x) = -\arcsin x, \quad |x| \leq 1$

  (b) $\arccos(-x) = \pi - \arccos x, \quad |x| \leq 1$

In Exercises 149–152, sketch the graph of the function. Use a graphing utility to verify your graph.

149. $f(x) = \arcsin(x - 1)$

150. $f(x) = \arctan x + \dfrac{\pi}{2}$

151. $f(x) = \text{arcsec } 2x$

152. $f(x) = \arccos \dfrac{x}{4}$

153. ***Think About It***   Given that f is a one-to-one function and $f(-3) = 8$, find $f^{-1}(8)$.

154. ***Think About It***   Given $f(x) = 5 + \arccos x$, find $f^{-1}\left(5 + \dfrac{\pi}{2}\right)$.

155. Prove that if f and g are one-to-one functions, then $(f \circ g)^{-1}(x) = (g^{-1} \circ f^{-1})(x)$.

156. Prove that if f has an inverse function, then $(f^{-1})^{-1} = f$.

157. Prove that $\cos(\sin^{-1} x) = \sqrt{1 - x^2}$.

158. Prove that if f is a one-to-one function and $f(x) \neq 0$, then

$$g(x) = \dfrac{1}{f(x)}$$

is a one-to-one function.

159. Prove that if a function has an inverse function, then the inverse function is unique.

160. Prove that a function has an inverse function if and only if it is one-to-one.

True or False?   **In Exercises 161–166, determine whether the statement is true or false. If it is false, explain why or give an example that shows it is false.**

161. If f is an even function, then f^{-1} exists.

162. If the inverse function of f exists, then the y-intercept of f is an x-intercept of f^{-1}.

163. $\arcsin^2 x + \arccos^2 x = 1$

164. The range of $y = \arcsin x$ is $[0, \pi]$.

165. If $f(x) = x^n$ where n is odd, then f^{-1} exists.

166. There exists no function f such that $f = f^{-1}$.

167. Verify each identity.

  (a) $\text{arccot } x = \begin{cases} \pi + \arctan(1/x), & x < 0 \\ \pi/2, & x = 0 \\ \arctan(1/x), & x > 0 \end{cases}$

  (b) $\text{arcsec } x = \arccos(1/x), \quad |x| \geq 1$

  (c) $\text{arccsc } x = \arcsin(1/x), \quad |x| \geq 1$

168. Use the results of Exercise 167 and a graphing utility to evaluate the following.

  (a) $\text{arccot } 0.5$

  (b) $\text{arcsec } 2.7$

  (c) $\text{arccsc}(-3.9)$

  (d) $\text{arccot}(-1.4)$

169. Prove that

$$\arctan x + \arctan y = \arctan \dfrac{x + y}{1 - xy}, \quad xy \neq 1.$$

Use this formula to show that

$$\arctan \dfrac{1}{2} + \arctan \dfrac{1}{3} = \dfrac{\pi}{4}.$$

170. ***Think About It***   Use a graphing utility to graph

$$f(x) = \sin x \quad \text{and} \quad g(x) = \arcsin(\sin x).$$

Why isn't the graph of g the line $y = x$?

171. Let $f(x) = ax^2 + bx + c$, where $a > 0$ and the domain is all real numbers such that $x \leq -\dfrac{b}{2a}$. Find f^{-1}.

172. Determine conditions on the constants a, b, and c such that the graph of $f(x) = \dfrac{ax + b}{cx - a}$ is symmetric about the line $y = x$.

173. Determine conditions on the constants a, b, c, and d such that $f(x) = \dfrac{ax + b}{cx + d}$ has an inverse function. Then find f^{-1}.

CAPSTONE

174. The point $\left(\dfrac{3\pi}{2}, 0\right)$ is on the graph of $y = \cos x$. Does $\left(0, \dfrac{3\pi}{2}\right)$ lie on the graph of $y = \arccos x$? If not, does this contradict the definition of inverse function?

1.6 Exponential and Logarithmic Functions

- **Develop and use properties of exponential functions.**
- **Understand the definition of the number *e*.**
- **Understand the definition of the natural logarithmic function.**
- **Develop and use properties of the natural logarithmic function.**

Exponential Functions

An **exponential function** involves a constant raised to a power, such as $f(x) = 2^x$. You already know how to evaluate 2^x for *rational* values of x. For instance,

$$2^0 = 1, \quad 2^2 = 4, \quad 2^{-1} = \frac{1}{2}, \quad \text{and} \quad 2^{1/2} = \sqrt{2} \approx 1.4142136.$$

For *irrational* values of x, you can define 2^x by considering a sequence of rational numbers that approach x. A full discussion of this process would not be appropriate here, but the general idea is as follows. Suppose you want to define the number $2^{\sqrt{2}}$. Because $\sqrt{2} = 1.414213 \ldots$, you consider the following numbers (which are of the form 2^r, where r is rational).

$$2^1 = 2 < 2^{\sqrt{2}} < 4 = 2^2$$
$$2^{1.4} = 2.639015 \ldots < 2^{\sqrt{2}} < 2.828427 \ldots = 2^{1.5}$$
$$2^{1.41} = 2.657371 \ldots < 2^{\sqrt{2}} < 2.675855 \ldots = 2^{1.42}$$
$$2^{1.414} = 2.664749 \ldots < 2^{\sqrt{2}} < 2.666597 \ldots = 2^{1.415}$$
$$2^{1.4142} = 2.665119 \ldots < 2^{\sqrt{2}} < 2.665303 \ldots = 2^{1.4143}$$
$$2^{1.41421} = 2.665137 \ldots < 2^{\sqrt{2}} < 2.665156 \ldots = 2^{1.41422}$$
$$2^{1.414213} = 2.665143 \ldots < 2^{\sqrt{2}} < 2.665144 \ldots = 2^{1.414214}$$

From these calculations, it seems reasonable to conclude that

$$2^{\sqrt{2}} \approx 2.66514.$$

In practice, you can use a calculator to approximate numbers such as $2^{\sqrt{2}}$.

In general, you can use any positive base a, $a \neq 1$, to define an exponential function. So, the exponential function with base a is written as $f(x) = a^x$. Exponential functions, even those with irrational values of x, obey the familiar properties of exponents.

PROPERTIES OF EXPONENTS

Let a and b be positive real numbers, and let x and y be any real numbers.

1. $a^0 = 1$ **2.** $a^x a^y = a^{x+y}$ **3.** $(a^x)^y = a^{xy}$ **4.** $(ab)^x = a^x b^x$

5. $\dfrac{a^x}{a^y} = a^{x-y}$ **6.** $\left(\dfrac{a}{b}\right)^x = \dfrac{a^x}{b^x}$ **7.** $a^{-x} = \dfrac{1}{a^x}$

EXAMPLE 1 Using Properties of Exponents

a. $(2^2)(2^3) = 2^{2+3} = 2^5$ **b.** $\dfrac{2^2}{2^3} = 2^{2-3} = 2^{-1} = \dfrac{1}{2}$

c. $(3^x)^3 = 3^{3x}$ **d.** $\left(\dfrac{1}{3}\right)^{-x} = (3^{-1})^{-x} = 3^x$

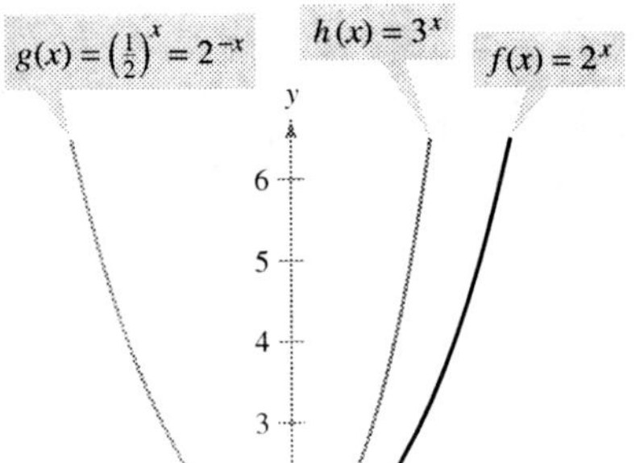

Figure 1.46

EXAMPLE 2 Sketching Graphs of Exponential Functions

Sketch the graphs of the functions

$$f(x) = 2^x, \quad g(x) = \left(\tfrac{1}{2}\right)^x = 2^{-x}, \quad \text{and} \quad h(x) = 3^x.$$

Solution To sketch the graphs of these functions by hand, you can complete a table of values, plot the corresponding points, and connect the points with smooth curves.

x	-3	-2	-1	0	1	2	3	4
2^x	$\frac{1}{8}$	$\frac{1}{4}$	$\frac{1}{2}$	1	2	4	8	16
2^{-x}	8	4	2	1	$\frac{1}{2}$	$\frac{1}{4}$	$\frac{1}{8}$	$\frac{1}{16}$
3^x	$\frac{1}{27}$	$\frac{1}{9}$	$\frac{1}{3}$	1	3	9	27	81

Another way to graph these functions is to use a graphing utility. In either case, you should obtain graphs similar to those shown in Figure 1.46. ■

The shapes of the graphs in Figure 1.46 are typical of the exponential functions $y = a^x$ and $y = a^{-x}$ where $a > 1$, as shown in Figure 1.47.

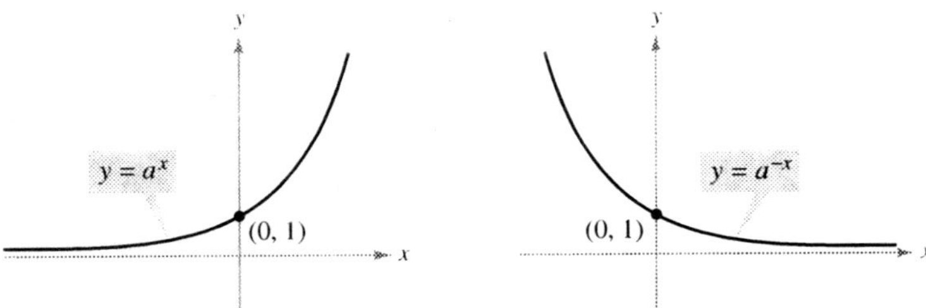

Figure 1.47

PROPERTIES OF EXPONENTIAL FUNCTIONS

Let a be a real number that is greater than 1.

1. The domain of $f(x) = a^x$ and $g(x) = a^{-x}$ is $(-\infty, \infty)$.
2. The range of $f(x) = a^x$ and $g(x) = a^{-x}$ is $(0, \infty)$.
3. The y-intercept of $f(x) = a^x$ and $g(x) = a^{-x}$ is $(0, 1)$.
4. The functions $f(x) = a^x$ and $g(x) = a^{-x}$ are one-to-one.

TECHNOLOGY Functions of the form $h(x) = b^{cx}$ have the same types of properties and graphs as functions of the form $f(x) = a^x$ and $g(x) = a^{-x}$. To see why this is true, notice that

$$b^{cx} = (b^c)^x.$$

For instance, $f(x) = 2^{3x}$ can be written as $f(x) = (2^3)^x$ or $f(x) = 8^x$. Try confirming this by graphing $f(x) = 2^{3x}$ and $g(x) = 8^x$ in the same viewing window.

The Number *e*

In calculus, the natural (or convenient) choice for a base of an exponential number is the irrational number *e*, whose decimal approximation is

$$e \approx 2.71828182846.$$

This choice may seem anything but natural. However, the convenience of this particular base will become apparent as you continue in this course.

EXAMPLE 3   Investigating the Number *e*

Use a graphing utility to graph the function

$$f(x) = (1 + x)^{1/x}.$$

Describe the behavior of the function at values of x that are close to 0.

Solution   One way to examine the values of $f(x)$ near 0 is to construct a table.

x	-0.01	-0.001	-0.0001	0.0001	0.001	0.01
$(1 + x)^{1/x}$	2.7320	2.7196	2.7184	2.7181	2.7169	2.7048

From the table, it appears that the closer x gets to 0, the closer $(1 + x)^{1/x}$ gets to e. You can confirm this by graphing the function f, as shown in Figure 1.48. Try using a graphing calculator to obtain this graph. Then zoom in closer and closer to $x = 0$. Although f is not defined when $x = 0$, it is defined for x-values that are arbitrarily close to zero. By zooming in, you can see that the value of $f(x)$ gets closer and closer to $e \approx 2.71828182846$ as x gets closer and closer to 0. Later, when you study limits, you will learn that this result can be written as

$$\lim_{x \to 0} (1 + x)^{1/x} = e$$

which is read as "the limit of $(1 + x)^{1/x}$ as x approaches 0 is e."

EXAMPLE 4   The Graph of the Natural Exponential Function

Sketch the graph of $f(x) = e^x$.

Solution   To sketch the graph by hand, you can complete a table of values.

x	-2	-1	0	1	2
e^x	0.135	0.368	1	2.718	7.389

You can also use a graphing utility to graph the function. From the values in the table, you can see that a good viewing window for the graph is $-3 \le x \le 3$ and $-1 \le y \le 3$, as shown in Figure 1.49. ■

The Natural Logarithmic Function

Because the natural exponential function $f(x) = e^x$ is one-to-one, it must have an inverse function. Its inverse is called the **natural logarithmic function.** The domain of the natural logarithmic function is the set of positive real numbers.

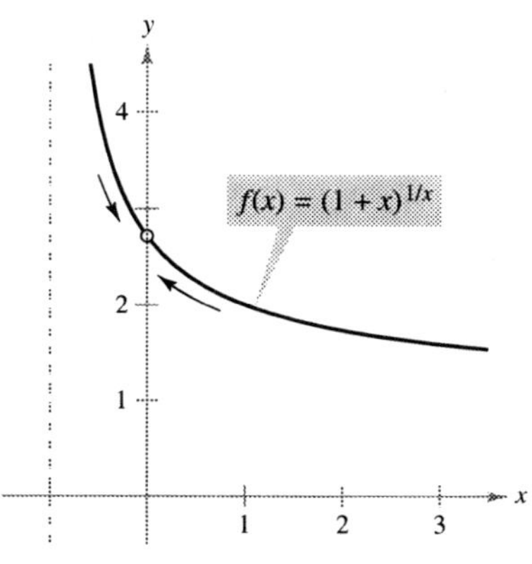

Figure 1.48

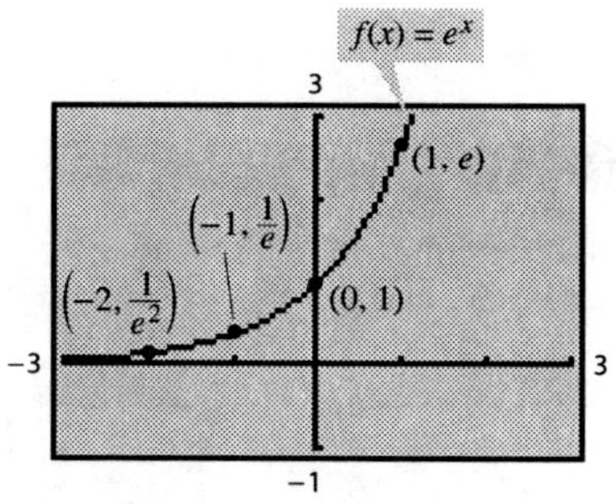

Figure 1.49

> ### DEFINITION OF THE NATURAL LOGARITHMIC FUNCTION
>
> Let x be a positive real number. The **natural logarithmic function,** denoted by $\ln x$, is defined as follows. ($\ln x$ is read as "el en of x" or "the natural log of x.")
>
> $$\ln x = b \quad \text{if and only if} \quad e^b = x.$$

This definition tells you that a logarithmic equation can be written in an equivalent exponential form, and vice versa. Here are some examples.

Logarithmic Form	*Exponential Form*
$\ln 1 = 0$	$e^0 = 1$
$\ln e = 1$	$e^1 = e$
$\ln e^{-1} = -1$	$e^{-1} = \dfrac{1}{e}$

Because the function $g(x) = \ln x$ is defined to be the inverse of $f(x) = e^x$, it follows that the graph of the natural logarithmic function is a reflection of the graph of the natural exponential function in the line $y = x$, as shown in Figure 1.50. Several other properties of the natural logarithmic function also follow directly from its definition as the inverse of the natural exponential function.

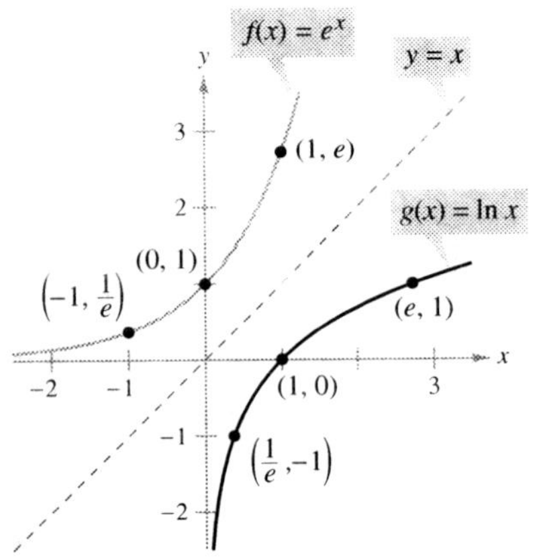

Figure 1.50

> ### PROPERTIES OF THE NATURAL LOGARITHMIC FUNCTION
>
> 1. The domain of $g(x) = \ln x$ is $(0, \infty)$.
> 2. The range of $g(x) = \ln x$ is $(-\infty, \infty)$.
> 3. The x-intercept of $g(x) = \ln x$ is $(1, 0)$.
> 4. The function $g(x) = \ln x$ is one-to-one.

Because $f(x) = e^x$ and $g(x) = \ln x$ are inverses of each other, you can conclude that

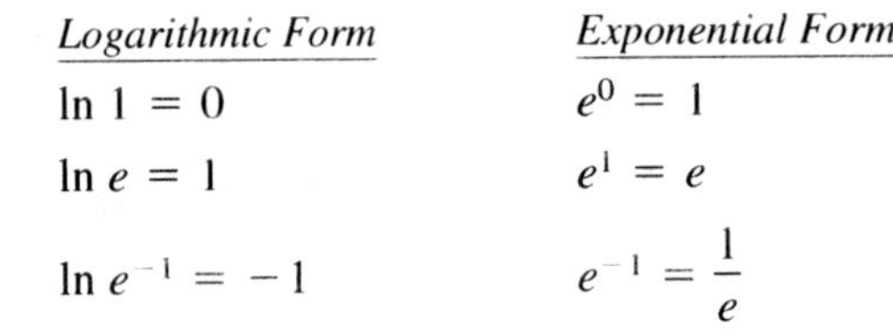

$$\ln e^x = x \quad \text{and} \quad e^{\ln x} = x.$$

EXPLORATION

The graphing utility screen in Figure 1.51 shows the graph of $y_1 = \ln e^x$ or $y_2 = e^{\ln x}$. Which graph is it? What are the domains of y_1 and y_2? Does $\ln e^x = e^{\ln x}$ for all real values of x? Explain.

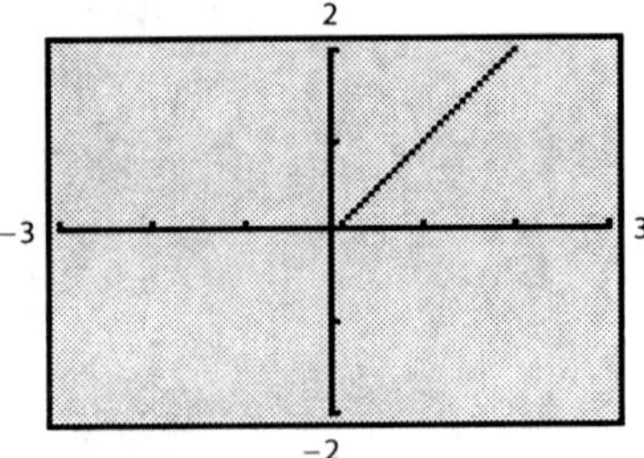

Figure 1.51

Properties of Logarithms

One of the properties of exponents states that when you multiply two exponential functions (having the same base), you add their exponents. For instance,

$$e^x e^y = e^{x+y}.$$

The logarithmic version of this property states that the natural logarithm of the product of two numbers is equal to the sum of the natural logs of the numbers. That is,

$$\ln xy = \ln x + \ln y.$$

This property and the properties dealing with the natural log of a quotient and the natural log of a power are listed here.

PROPERTIES OF LOGARITHMS

Let x, y, and z be real numbers such that $x > 0$ and $y > 0$.

1. $\ln xy = \ln x + \ln y$

2. $\ln \dfrac{x}{y} = \ln x - \ln y$

3. $\ln x^z = z \ln x$

EXAMPLE 5   Expanding Logarithmic Expressions

a. $\ln \dfrac{10}{9} = \ln 10 - \ln 9$ Property 2

b. $\ln \sqrt{3x + 2} = \ln(3x + 2)^{1/2}$ Rewrite with rational exponent.

$\qquad\qquad\quad = \dfrac{1}{2} \ln(3x + 2)$ Property 3

c. $\ln \dfrac{6x}{5} = \ln(6x) - \ln 5$ Property 2

$\qquad\quad = \ln 6 + \ln x - \ln 5$ Property 1

d. $\ln \dfrac{(x^2 + 3)^2}{x\sqrt[3]{x^2 + 1}} = \ln(x^2 + 3)^2 - \ln\!\left(x\sqrt[3]{x^2 + 1}\right)$

$\qquad\qquad\qquad = 2\ln(x^2 + 3) - \left[\ln x + \ln(x^2 + 1)^{1/3}\right]$

$\qquad\qquad\qquad = 2\ln(x^2 + 3) - \ln x - \ln(x^2 + 1)^{1/3}$

$\qquad\qquad\qquad = 2\ln(x^2 + 3) - \ln x - \dfrac{1}{3}\ln(x^2 + 1)$

    When using the properties of logarithms to rewrite logarithmic functions, you must check to see whether the domain of the rewritten function is the same as the domain of the original function. For instance, the domain of $f(x) = \ln x^2$ is all real numbers except $x = 0$, and the domain of $g(x) = 2\ln x$ is all positive real numbers.

TECHNOLOGY Try using a graphing utility to compare the graphs of

$$f(x) = \ln x^2 \quad \text{and} \quad g(x) = 2\ln x.$$

Which of the graphs in Figure 1.52 is the graph of f? Which is the graph of g?

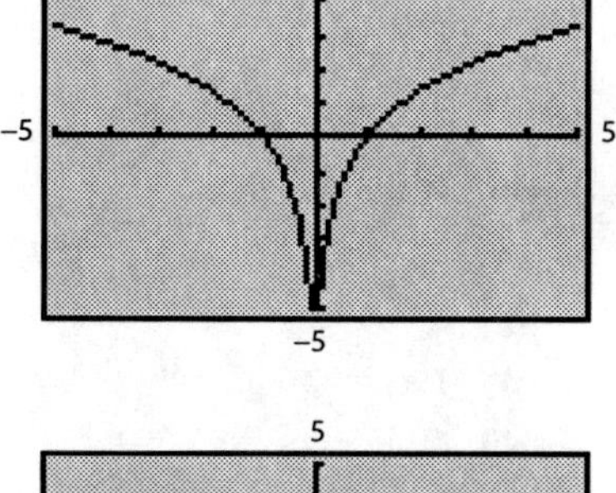
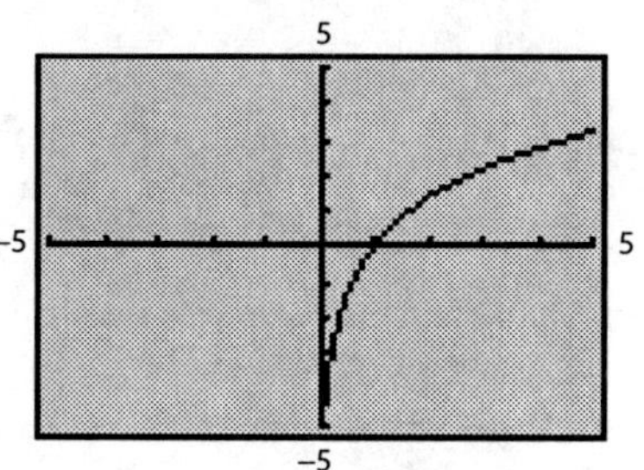

Figure 1.52

EXAMPLE 6 Solving Exponential and Logarithmic Equations

Solve (a) $7 = e^{x+1}$ and (b) $\ln(2x - 3) = 5$.

Solution

a.
$$7 = e^{x+1}$$ Write original equation.
$$\ln 7 = \ln(e^{x+1})$$ Take natural log of each side.
$$\ln 7 = x + 1$$ Apply inverse property.
$$-1 + \ln 7 = x$$ Solve for x.
$$0.946 \approx x$$ Use a calculator.

b.
$$\ln(2x - 3) = 5$$ Write original equation.
$$e^{\ln(2x-3)} = e^5$$ Exponentiate each side.
$$2x - 3 = e^5$$ Apply inverse property.
$$x = \frac{1}{2}(e^5 + 3)$$ Solve for x.
$$x \approx 75.707$$ Use a calculator.

1.6 Exercises

See www.CalcChat.com for worked-out solutions to odd-numbered exercises.

In Exercises 1 and 2, evaluate the expressions.

1. (a) $25^{3/2}$ (b) $81^{1/2}$ (c) 3^{-2} (d) $27^{-1/3}$

2. (a) $64^{1/3}$ (b) 5^{-4} (c) $\left(\frac{1}{8}\right)^{1/3}$ (d) $\left(\frac{1}{4}\right)^3$

In Exercises 3–6, use the properties of exponents to simplify the expressions.

3. (a) $(5^2)(5^3)$ (b) $(5^2)(5^{-3})$

(c) $\dfrac{5^3}{25^2}$ (d) $\left(\dfrac{1}{4}\right)^2 2^6$

4. (a) $(2^2)^3$ (b) $(5^4)^{1/2}$

(c) $[(27^{-1})(27^{2/3})]^3$ (d) $(25^{3/2})(3^2)$

5. (a) $e^2(e^4)$ (b) $(e^3)^4$

(c) $(e^3)^{-2}$ (d) $\dfrac{e^5}{e^3}$

6. (a) $\left(\dfrac{1}{e}\right)^{-2}$ (b) $\left(\dfrac{e^5}{e^2}\right)^{-1}$

(c) e^0 (d) $\dfrac{1}{e^{-3}}$

In Exercises 7–22, solve for x.

7. $3^x = 81$

8. $4^x = 64$

9. $6^{x-2} = 36$

10. $5^{x+1} = 125$

11. $\left(\frac{1}{2}\right)^x = 32$

12. $\left(\frac{1}{4}\right)^x = 16$

13. $\left(\frac{1}{3}\right)^{x-1} = 27$

14. $\left(\frac{1}{5}\right)^{2x} = 625$

15. $4^3 = (x + 2)^3$

16. $18^2 = (5x - 7)^2$

17. $x^{3/4} = 8$

18. $(x + 3)^{4/3} = 16$

19. $e^x = 5$

20. $e^x = 1$

21. $e^{-2x} = e^5$

22. $e^{3x} = e^{-4}$

In Exercises 23 and 24, compare the given number with the number e. Is the number less than or greater than e?

23. $\left(1 + \dfrac{1}{1{,}000{,}000}\right)^{1{,}000{,}000}$

24. $1 + 1 + \frac{1}{2} + \frac{1}{6} + \frac{1}{24} + \frac{1}{120} + \frac{1}{720} + \frac{1}{5040}$

In Exercises 25–34, sketch the graph of the function.

25. $y = 3^x$

26. $y = 3^{x-1}$

27. $y = \left(\frac{1}{3}\right)^x$

28. $y = 2^{-x^2}$

29. $f(x) = 3^{-x^2}$

30. $f(x) = 3^{|x|}$

31. $h(x) = e^{x-2}$

32. $g(x) = -e^{x/2}$

33. $y = e^{-x^2}$

34. $y = e^{-x/4}$

In Exercises 35–40, find the domain of the function.

35. $f(x) = \dfrac{1}{3 + e^x}$

36. $f(x) = \dfrac{1}{2 - e^x}$

37. $f(x) = \sqrt{1 - 4^x}$

38. $f(x) = \sqrt{1 + 3^{-x}}$

39. $f(x) = \sin e^{-x}$

40. $f(x) = \cos e^{-x}$

41. Use a graphing utility to graph $f(x) = e^x$ and the given function in the same viewing window. How are the two graphs related?

(a) $g(x) = e^{x-2}$ (b) $h(x) = -\frac{1}{2}e^x$ (c) $q(x) = e^{-x} + 3$

42. Use a graphing utility to graph the function. Describe the shape of the graph for very large and very small values of x.

(a) $f(x) = \dfrac{8}{1 + e^{-0.5x}}$

(b) $g(x) = \dfrac{8}{1 + e^{-0.5/x}}$

In Exercises 43–46, match the equation with the correct graph. Assume that a and C are positive real numbers. [The graphs are labeled (a), (b), (c), and (d).]

(a)

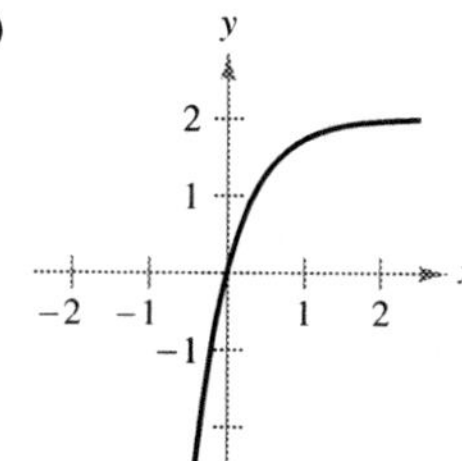

(b)

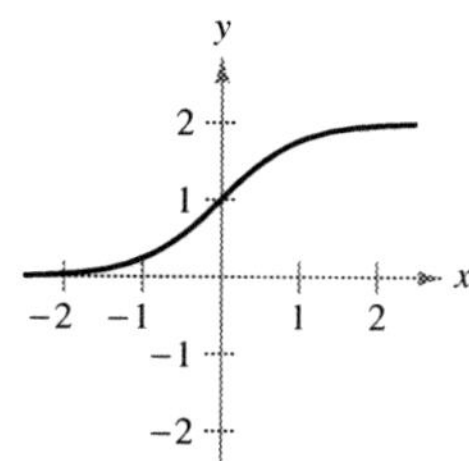

(c)

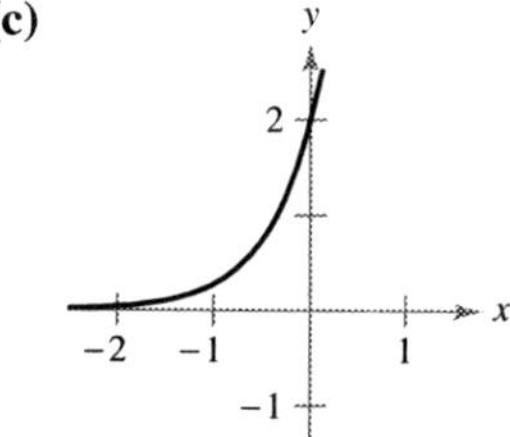

(d)

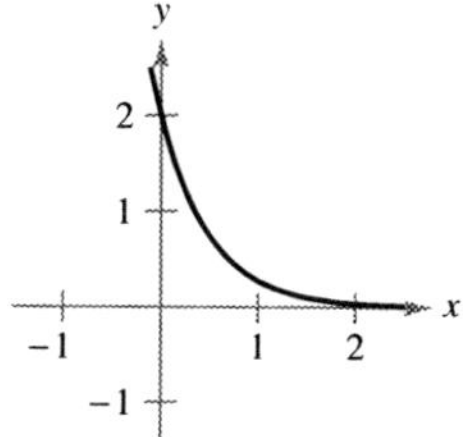

43. $y = Ce^{ax}$

44. $y = Ce^{-ax}$

45. $y = C(1 - e^{-ax})$

46. $y = \dfrac{C}{1 + e^{-ax}}$

In Exercises 47–50, match the function with its graph. [The graphs are labeled (a), (b), (c), and (d).]

(a)

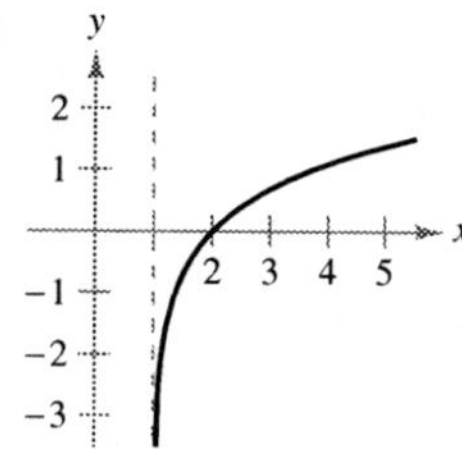

(b)

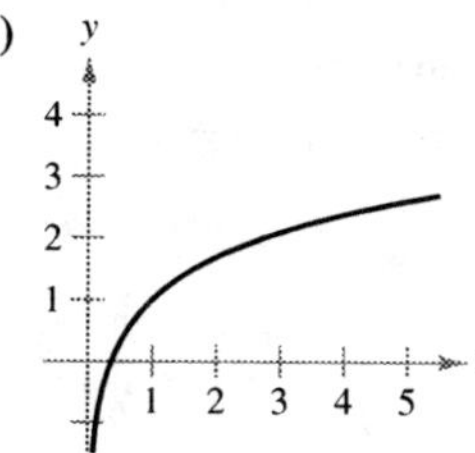

(c)

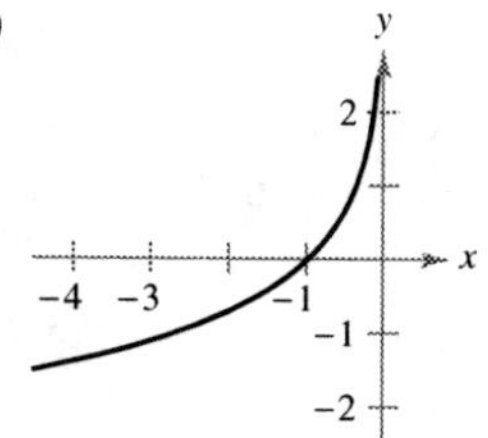

(d)

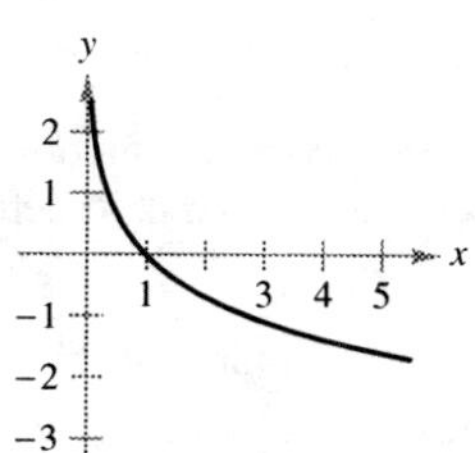

47. $f(x) = \ln x + 1$

48. $f(x) = -\ln x$

49. $f(x) = \ln(x - 1)$

50. $f(x) = -\ln(-x)$

In Exercises 51 and 52, find the exponential function $y = Ca^x$ that fits the graph.

51.

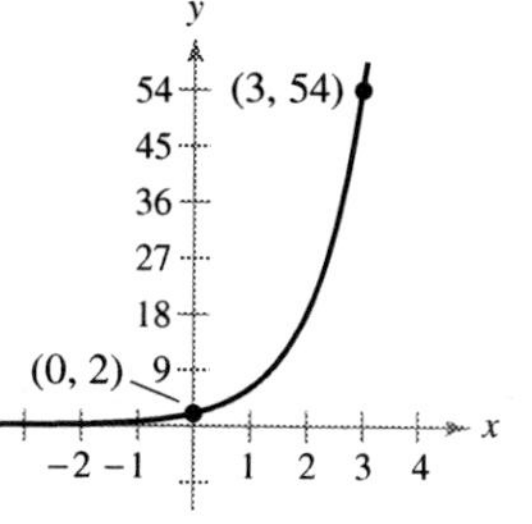

52.

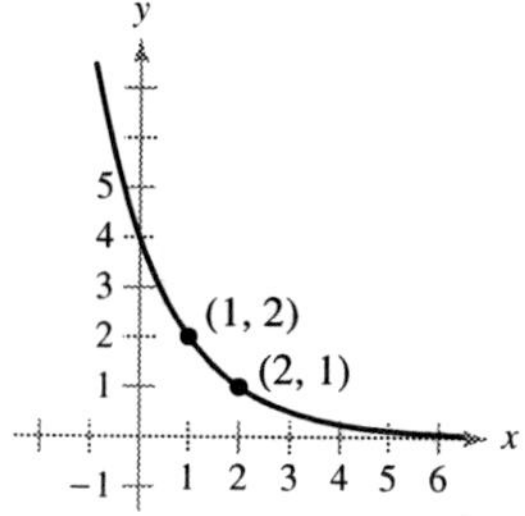

In Exercises 53–56, write the exponential equation as a logarithmic equation, or vice versa.

53. $e^0 = 1$

54. $e^{-2} = 0.1353 \ldots$

55. $\ln 2 = 0.6931 \ldots$

56. $\ln 0.5 = -0.6931 \ldots$

In Exercises 57–62, sketch the graph of the function and state its domain.

57. $f(x) = 3 \ln x$

58. $f(x) = -2 \ln x$

59. $f(x) = \ln 2x$

60. $f(x) = \ln|x|$

61. $f(x) = \ln(x - 1)$

62. $f(x) = 2 + \ln x$

In Exercises 63–66, write an equation for the function having the given characteristics.

63. The shape of $f(x) = e^x$, but shifted eight units upward and reflected in the x-axis

64. The shape of $f(x) = e^x$, but shifted two units to the left and six units downward

65. The shape of $f(x) = \ln x$, but shifted five units to the right and one unit downward

66. The shape of $f(x) = \ln x$, but shifted three units upward and reflected in the y-axis

In Exercises 67–70, show that the functions f and g are inverses of each other by graphing them in the same viewing window.

67. $f(x) = e^{2x}, g(x) = \ln \sqrt{x}$

68. $f(x) = e^{x/3}, g(x) = \ln x^3$

69. $f(x) = e^x - 1, g(x) = \ln(x + 1)$

70. $f(x) = e^{x-1}, g(x) = 1 + \ln x$

In Exercises 71–74, (a) find the inverse of the function, (b) use a graphing utility to graph f and f^{-1} in the same viewing window, and (c) verify that $f^{-1}(f(x)) = x$ and $f(f^{-1}(x)) = x$.

71. $f(x) = e^{4x-1}$

72. $f(x) = 3e^{-x}$

73. $f(x) = 2 \ln(x - 1)$

74. $f(x) = 3 + \ln(2x)$

In Exercises 75–80, apply the inverse properties of $\ln x$ and e^x to simplify the given expression.

75. $\ln e^{x^2}$

76. $\ln e^{2x-1}$

77. $e^{\ln(5x+2)}$

78. $e^{\ln \sqrt{x}}$

79. $-1 + \ln e^{2x}$

80. $-8 + e^{\ln x^3}$

In Exercises 81 and 82, use the properties of logarithms to approximate the indicated logarithms, given that $\ln 2 \approx 0.6931$ and $\ln 3 \approx 1.0986$.

81. (a) $\ln 6$ (b) $\ln \frac{2}{3}$ (c) $\ln 81$ (d) $\ln \sqrt{3}$

82. (a) $\ln 0.25$ (b) $\ln 24$ (c) $\ln \sqrt[3]{12}$ (d) $\ln \frac{1}{72}$

83. In your own words, state the properties of the natural logarithmic function.

84. Explain why $\ln e^x = x$.

85. In your own words, state the properties of the natural exponential function.

86. The table of values below was obtained by evaluating a function. Determine which of the statements may be true and which must be false, and explain why.

  (a) y is an exponential function of x.

  (b) y is a logarithmic function of x.

  (c) x is an exponential function of y.

  (d) y is a linear function of x.

x	1	2	8
y	0	1	3

In Exercises 87–96, use the properties of logarithms to expand the logarithmic expression.

87. $\ln \dfrac{x}{4}$

88. $\ln \sqrt{x^5}$

89. $\ln \dfrac{xy}{z}$

90. $\ln(xyz)$

91. $\ln\left(x\sqrt{x^2 + 5}\right)$

92. $\ln \sqrt[3]{z + 1}$

93. $\ln \sqrt{\dfrac{x - 1}{x}}$

94. $\ln z(z - 1)^2$

95. $\ln(3e^2)$

96. $\ln \dfrac{1}{e}$

In Exercises 97–104, write the expression as the logarithm of a single quantity.

97. $\ln x + \ln 7$

98. $\ln y + \ln x^2$

99. $\ln(x - 2) - \ln(x + 2)$

100. $3 \ln x + 2 \ln y - 4 \ln z$

101. $\frac{1}{3}[2 \ln(x + 3) + \ln x - \ln(x^2 - 1)]$

102. $2[\ln x - \ln(x + 1) - \ln(x - 1)]$

103. $2 \ln 3 - \frac{1}{2} \ln(x^2 + 1)$

104. $\frac{3}{2}[\ln(x^2 + 1) - \ln(x + 1) - \ln(x - 1)]$

In Exercises 105–108, solve for x accurate to three decimal places.

105. (a) $e^{\ln x} = 4$

  (b) $\ln e^{2x} = 3$

106. (a) $e^{\ln 2x} = 12$

  (b) $\ln e^{-x} = 0$

107. (a) $\ln x = 2$

  (b) $e^x = 4$

108. (a) $\ln x^2 = 8$

  (b) $e^{-2x} = 5$

In Exercises 109–112, solve the inequality for x.

109. $e^x > 5$

110. $e^{1-x} < 6$

111. $-2 < \ln x < 0$

112. $1 < \ln x < 100$

 **In Exercises 113 and 114, show that $f = g$ by using a graphing utility to graph f and g in the same viewing window. (Assume $x > 0$.)**

113. $f(x) = \ln(x^2/4)$

  $g(x) = 2 \ln x - \ln 4$

114. $f(x) = \ln \sqrt{x(x^2 + 1)}$

  $g(x) = \frac{1}{2}[\ln x + \ln(x^2 + 1)]$

115. Prove that $\ln (x/y) = \ln x - \ln y, \quad x > 0, y > 0$.

116. Prove that $\ln x^y = y \ln x$.

 **117.** Graph the functions

  $f(x) = 6^x$ and $g(x) = x^6$

  in the same viewing window. Where do these graphs intersect? As x increases, which function grows more rapidly?

118. Graph the functions

  $f(x) = \ln x$ and $g(x) = x^{1/4}$

  in the same viewing window. Where do these graphs intersect? As x increases, which function grows more rapidly?

119. Let $f(x) = \ln\left(x + \sqrt{x^2 + 1}\right)$.

  (a) Use a graphing utility to graph f and determine its domain.

  (b) Show that f is an odd function.

  (c) Find the inverse function of f.

120. Describe the relationship between the graphs of $f(x) = \ln x$ and $g(x) = e^x$.

1 REVIEW EXERCISES

See www.CalcChat.com for worked-out solutions to odd-numbered exercises.

In Exercises 1–4, find the intercepts (if any).

1. $y = 5x - 8$

2. $y = (x - 2)(x - 6)$

3. $y = \dfrac{x - 3}{x - 4}$

4. $xy = 4$

In Exercises 5 and 6, check for symmetry with respect to both axes and to the origin.

5. $x^2y - x^2 + 4y = 0$

6. $y = x^4 - x^2 + 3$

In Exercises 7–14, sketch the graph of the equation.

7. $y = \frac{1}{2}(-x + 3)$

8. $6x - 3y = 12$

9. $-\frac{1}{3}x + \frac{5}{6}y = 1$

10. $0.02x + 0.15y = 0.25$

11. $y = 9 - 8x - x^2$

12. $y = 6x - x^2$

13. $y = 2\sqrt{4 - x}$

14. $y = |x - 4| - 4$

In Exercises 15 and 16, use a graphing utility to find the point(s) of intersection of the graphs of the equations.

15. $5x + 3y = -1$
    $\quad\; x - \;\; y = -5$

16. $x - y + 1 = 0$
    $\quad\; y - x^2 = 7$

In Exercises 17 and 18, plot the points and find the slope of the line passing through the points.

17. $\left(\frac{3}{2}, 1\right), \left(5, \frac{5}{2}\right)$

18. $(-7, 8), (-1, 8)$

In Exercises 19 and 20, use the concept of slope to find t such that the three points are collinear.

19. $(-8, 5), (0, t), (2, -1)$

20. $(-3, 3), (t, -1), (8, 6)$

In Exercises 21–24, find an equation of the line that passes through the point with the given slope. Sketch the line.

21. $(3, -5), \quad m = \frac{7}{4}$

22. $(-8, 1), \quad m$ is undefined.

23. $(-3, 0), \quad m = -\frac{2}{3}$

24. $(5, 4), \quad m = 0$

25. Find equations of the lines passing through $(-3, 5)$ and having the following characteristics.
    (a) Slope of $\frac{7}{16}$
    (b) Parallel to the line $5x - 3y = 3$
    (c) Passing through the origin
    (d) Parallel to the y-axis

26. Find equations of the lines passing through $(2, 4)$ and having the following characteristics.
    (a) Slope of $-\frac{2}{3}$
    (b) Perpendicular to the line $x + y = 0$
    (c) Passing through the point $(6, 1)$
    (d) Parallel to the x-axis

27. **Rate of Change** The purchase price of a new machine is $12,500, and its value will decrease by $850 per year. Use this information to write a linear equation that gives the value V of the machine t years after it is purchased. Find its value at the end of 3 years.

28. **Break-Even Analysis** A contractor purchases a piece of equipment for $36,500 that costs an average of $9.25 per hour for fuel and maintenance. The equipment operator is paid $13.50 per hour, and customers are charged $30 per hour.
    (a) Write an equation for the cost C of operating this equipment for t hours.
    (b) Write an equation for the revenue R derived from t hours of use.
    (c) Find the break-even point for this equipment by finding the time at which $R = C$.

In Exercises 29–32, sketch the graph of the equation and use the Vertical Line Test to determine whether the equation expresses y as a function of x.

29. $x - y^2 = 6$

30. $x^2 - y = 0$

31. $y = |x - 2|/(x - 2)$

32. $x = 9 - y^2$

33. Evaluate (if possible) the function $f(x) = 1/x$ at the specified values of the independent variable, and simplify the results.
    (a) $f(0)$
    (b) $\dfrac{f(1 + \Delta x) - f(1)}{\Delta x}$

34. Evaluate (if possible) the function at each value of the independent variable.
    $$f(x) = \begin{cases} x^2 + 2, & x < 0 \\ |x - 2|, & x \geq 0 \end{cases}$$
    (a) $f(-4)$
    (b) $f(0)$
    (c) $f(1)$

35. Find the domain and range of each function.
    (a) $y = \sqrt{36 - x^2}$
    (b) $y = \dfrac{7}{2x - 10}$
    (c) $y = \begin{cases} x^2, & x < 0 \\ 2 - x, & x \geq 0 \end{cases}$

36. Given $f(x) = 1 - x^2$ and $g(x) = 2x + 1$, find the following.
    (a) $f(x) - g(x)$
    (b) $f(x)g(x)$
    (c) $g(f(x))$

37. Sketch (on the same set of coordinate axes) graphs of f for $c = -2, 0,$ and 2.
    (a) $f(x) = x^3 + c$
    (b) $f(x) = (x - c)^3$
    (c) $f(x) = (x - 2)^3 + c$
    (d) $f(x) = cx^3$

38. Use a graphing utility to graph $f(x) = x^3 - 3x^2$. Use the graph to write an equation for the function g shown in the figure. To print an enlarged copy of the graph, go to the website *www.mathgraphs.com*.

(a)

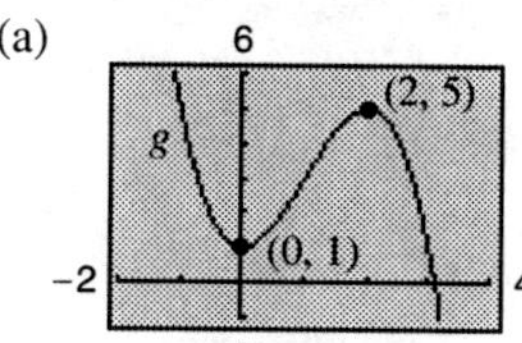

(b)

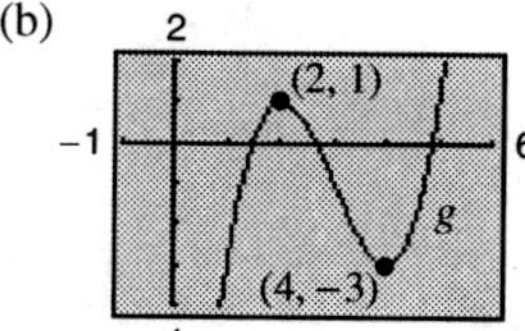

39. *Think About It* What is the minimum degree of the polynomial function whose graph approximates the given graph? What sign must the leading coefficient have?

(a)

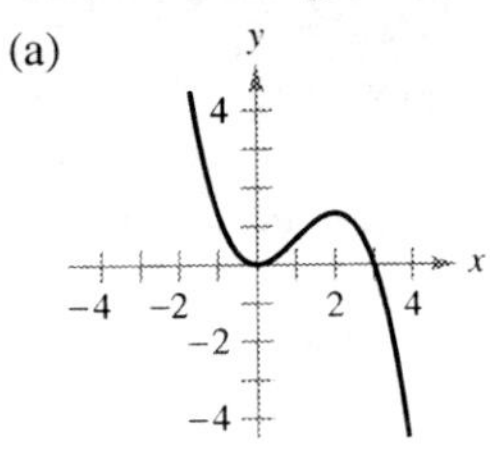

(b)

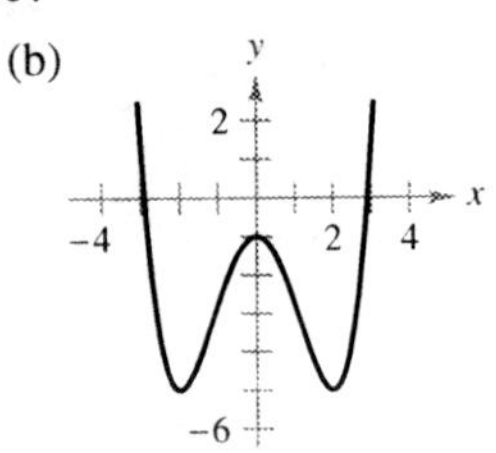

(c)

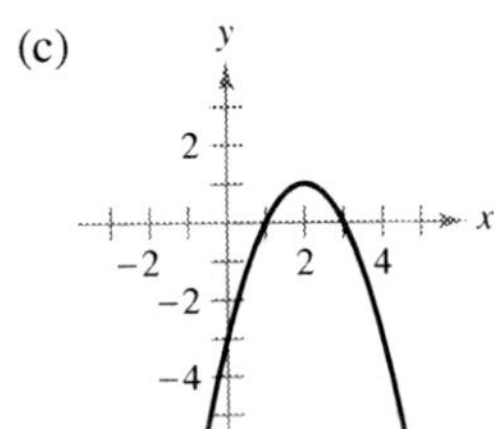

(d)

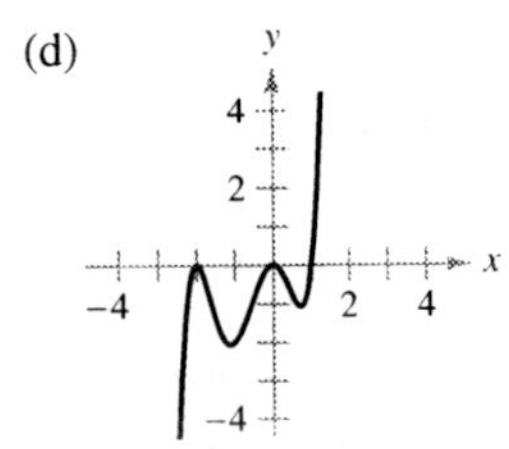

40. *Writing* The following graphs give the profits P for two small companies over a period p of 2 years. Create a story to describe the behavior of each profit function for some hypothetical product the company produces.

(a)

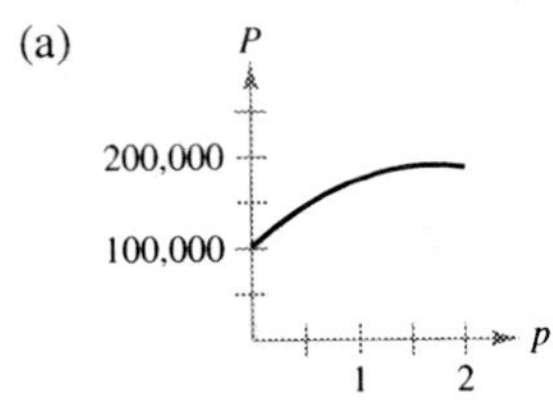

(b)

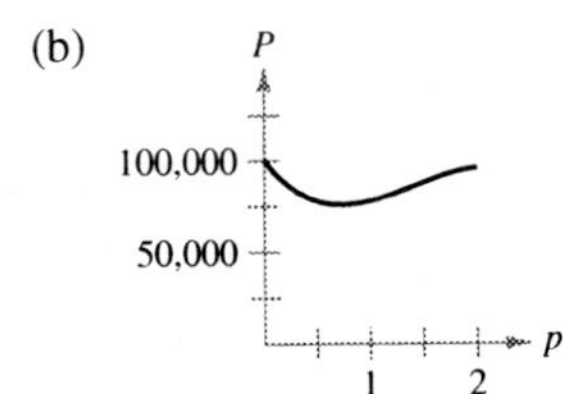

41. *Harmonic Motion* The motion of an oscillating weight suspended by a spring was measured by a motion detector. The data collected and the approximate maximum (positive and negative) displacements from equilibrium are shown in the figure. The displacement y is measured in feet and the time t is measured in seconds.

(a) Is y a function of t? Explain.

(b) Approximate the amplitude and period of the oscillations.

(c) Find a model for the data.

(d) Use a graphing utility to graph the model in part (c). Compare the result with the data in the figure.

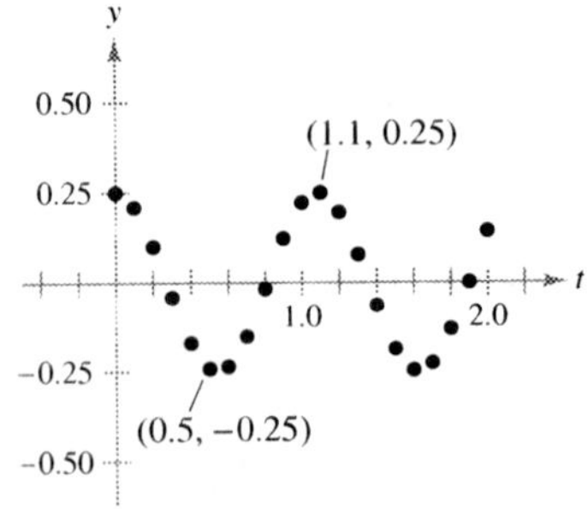

42. *Stress Test* A machine part was tested by bending it x centimeters 10 times per minute until the time y (in hours) of failure. The results are recorded in the table.

x	3	6	9	12	15	18	21	24	27	30
y	61	56	53	55	48	35	36	33	44	23

Table for 42

(a) Use the regression capabilities of a graphing utility to find a linear model for the data.

(b) Use a graphing utility to plot the data and graph the model.

(c) Use the graph to determine whether there may have been an error made in conducting one of the tests or in recording the results. If so, eliminate the erroneous point and find the model for the remaining data.

In Exercises 43–48, (a) find the inverse of the function, (b) use a graphing utility to graph f and f^{-1} in the same viewing window, and (c) verify that $f^{-1}(f(x)) = x$ and $f(f^{-1}(x)) = x$.

43. $f(x) = \frac{1}{2}x - 3$ **44.** $f(x) = 5x - 7$

45. $f(x) = \sqrt{x + 1}$ **46.** $f(x) = x^3 + 2$

47. $f(x) = \sqrt[3]{x + 1}$ **48.** $f(x) = x^2 - 5, \quad x \geq 0$

In Exercises 49 and 50, sketch the graph of the function by hand.

49. $f(x) = 2 \arctan(x + 3)$ **50.** $h(x) = -3 \arcsin 2x$

In Exercises 51 and 52, evaluate the expression without using a calculator. (*Hint:* Make a sketch of a right triangle.)

51. $\sin\left(\arcsin \frac{1}{2}\right)$ **52.** $\tan(\operatorname{arccot} 2)$

In Exercises 53 and 54, sketch the graph of the function by hand.

53. $f(x) = \ln x + 3$ **54.** $f(x) = \ln(x - 3)$

In Exercises 55 and 56, use the properties of logarithms to expand the logarithmic function.

55. $\ln \sqrt[5]{\dfrac{4x^2 - 1}{4x^2 + 1}}$ **56.** $\ln[(x^2 + 1)(x - 1)]$

In Exercises 57 and 58, write the expression as the logarithm of a single quantity.

57. $\ln 3 + \frac{1}{3}\ln(4 - x^2) - \ln x$

58. $3[\ln x - 2\ln(x^2 + 1)] + 2\ln 5$

In Exercises 59 and 60, solve the equation for x.

59. $\ln \sqrt{x + 1} = 2$ **60.** $\ln x + \ln(x - 3) = 0$

In Exercises 61 and 62, (a) find the inverse function of f, (b) use a graphing utility to graph f and f^{-1} in the same viewing window, and (c) verify that $f^{-1}(f(x)) = x$ and $f(f^{-1}(x)) = x$.

61. $f(x) = \ln \sqrt{x}$ **62.** $f(x) = e^{1-x}$

In Exercises 63 and 64, sketch the graph of the function by hand.

63. $y = e^{-x/2}$ **64.** $y = 4e^{-x^2}$

P.S. PROBLEM SOLVING

1. Consider the circle $x^2 + y^2 - 6x - 8y = 0$, as shown in the figure.

  (a) Find the center and radius of the circle.

  (b) Find an equation of the tangent line to the circle at the point $(0, 0)$.

  (c) Find an equation of the tangent line to the circle at the point $(6, 0)$.

  (d) Where do the two tangent lines intersect?

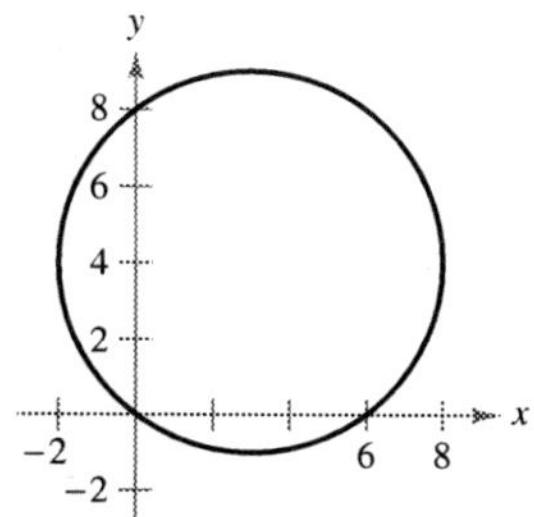 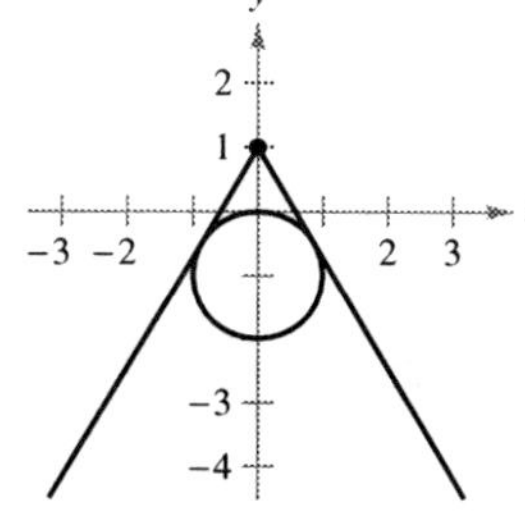

Figure for 1   **Figure for 2**

2. There are two tangent lines from the point $(0, 1)$ to the circle $x^2 + (y + 1)^2 = 1$ (see figure). Find equations of these two lines by using the fact that each tangent line intersects the circle at *exactly* one point.

3. The Heaviside function $H(x)$ is widely used in engineering applications.

$$H(x) = \begin{cases} 1, & x \ge 0 \\ 0, & x < 0 \end{cases}$$

Sketch the graph of the Heaviside function and the graphs of the following functions by hand.

  (a) $H(x) - 2$ (b) $H(x - 2)$ (c) $-H(x)$

  (d) $H(-x)$ (e) $\frac{1}{2}H(x)$ (f) $-H(x - 2) + 2$

OLIVER HEAVISIDE (1850–1925)

Heaviside was a British mathematician and physicist who contributed to the field of applied mathematics, especially applications of mathematics to electrical engineering. The *Heaviside function* is a classic type of "on-off" function that has applications to electricity and computer science.

4. Consider the graph of the function f shown below. Use this graph to sketch the graphs of the following functions. To print an enlarged copy of the graph, go to the website *www.mathgraphs.com*.

  (a) $f(x + 1)$ (b) $f(x) + 1$

  (c) $2f(x)$ (d) $f(-x)$

  (e) $-f(x)$ (f) $|f(x)|$

  (g) $f(|x|)$

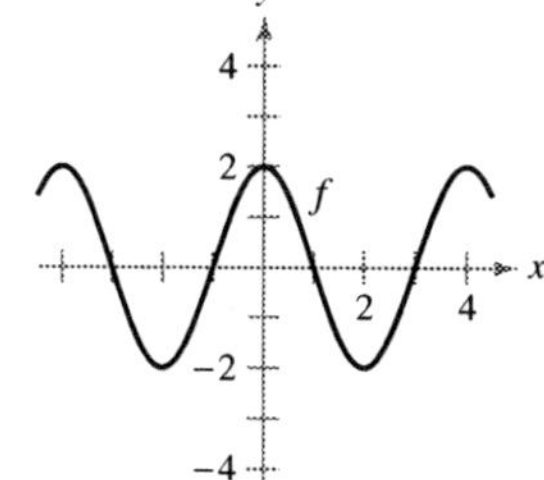

5. A rancher plans to fence a rectangular pasture adjacent to a river. The rancher has 100 meters of fencing, and no fencing is needed along the river (see figure).

  (a) Write the area A of the pasture as a function of x, the length of the side parallel to the river. What is the domain of A?

  (b) Graph the area function $A(x)$ and estimate the dimensions that yield the maximum amount of area for the pasture.

  (c) Find the dimensions that yield the maximum amount of area for the pasture by completing the square.

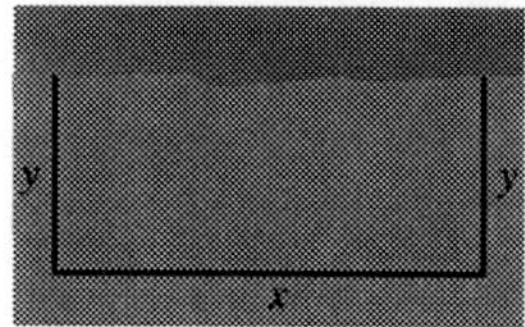 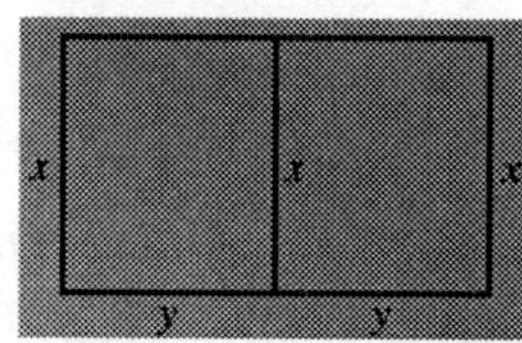

Figure for 5   **Figure for 6**

6. A rancher has 300 feet of fencing to enclose two adjacent pastures.

  (a) Write the total area A of the two pastures as a function of x (see figure). What is the domain of A?

  (b) Graph the area function and estimate the dimensions that yield the maximum amount of area for the pastures.

  (c) Find the dimensions that yield the maximum amount of area for the pastures by completing the square.

7. You are in a boat 2 miles from the nearest point on the coast. You are to go to a point Q located 3 miles down the coast and 1 mile inland (see figure). You can row at 2 miles per hour and walk at 4 miles per hour. Write the total time T of the trip as a function of x.

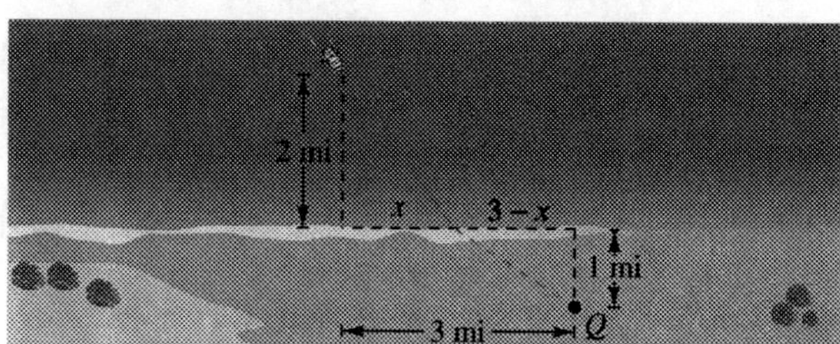

8. Graph the function $f(x) = e^x - e^{-x}$. From the graph, the function appears to be one-to-one. Assuming that the function has an inverse, find $f^{-1}(x)$.

9. One of the fundamental themes of calculus is to find the slope of the tangent line to a curve at a point. To see how this can be done, consider the point $(2, 4)$ on the graph of $f(x) = x^2$.

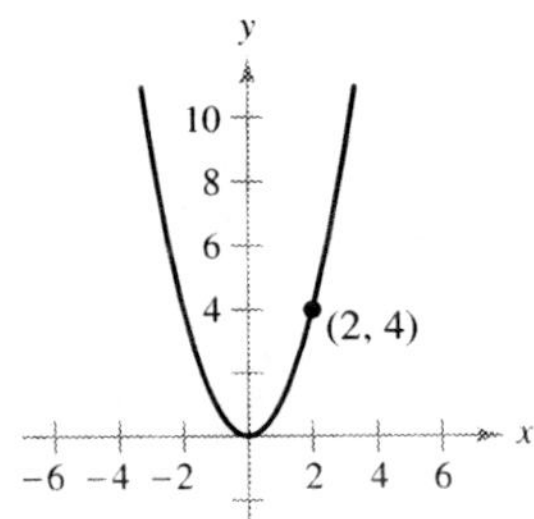

(a) Find the slope of the line joining $(2, 4)$ and $(3, 9)$. Is the slope of the tangent line at $(2, 4)$ greater than or less than this number?

(b) Find the slope of the line joining $(2, 4)$ and $(1, 1)$. Is the slope of the tangent line at $(2, 4)$ greater than or less than this number?

(c) Find the slope of the line joining $(2, 4)$ and $(2.1, 4.41)$. Is the slope of the tangent line at $(2, 4)$ greater than or less than this number?

(d) Find the slope of the line joining $(2, 4)$ and $(2 + h, f(2 + h))$ in terms of the nonzero number h. Verify that $h = 1, -1$, and 0.1 yield the solutions to parts (a)–(c) above.

(e) What is the slope of the tangent line at $(2, 4)$? Explain how you arrived at your answer.

10. Sketch the graph of the function $f(x) = \sqrt{x}$ and label the point $(4, 2)$ on the graph.

(a) Find the slope of the line joining $(4, 2)$ and $(9, 3)$. Is the slope of the tangent line at $(4, 2)$ greater than or less than this number?

(b) Find the slope of the line joining $(4, 2)$ and $(1, 1)$. Is the slope of the tangent line at $(4, 2)$ greater than or less than this number?

(c) Find the slope of the line joining $(4, 2)$ and $(4.41, 2.1)$. Is the slope of the tangent line at $(4, 2)$ greater than or less than this number?

(d) Find the slope of the line joining $(4, 2)$ and $(4 + h, f(4 + h))$ in terms of the nonzero number h.

(e) What is the slope of the tangent line at the point $(4, 2)$? Explain how you arrived at your answer.

11. Explain how you would graph the equation

$$y + |y| = x + |x|.$$

Then sketch the graph.

12. A large room contains two speakers that are 3 meters apart. The sound intensity I of one speaker is twice that of the other, as shown in the figure. (To print an enlarged copy of the graph, go to the website *www.mathgraphs.com*.) Suppose the listener is free to move about the room to find those positions that receive equal amounts of sound from both speakers. Such a location satisfies two conditions: (1) the sound intensity at the listener's position is directly proportional to the sound level of the source, and (2) the sound intensity is inversely proportional to the square of the distance from the source.

(a) Find the points on the x-axis that receive equal amounts of sound from both speakers.

(b) Find and graph the equation of all locations (x, y) where one could stand and receive equal amounts of sound from both speakers.

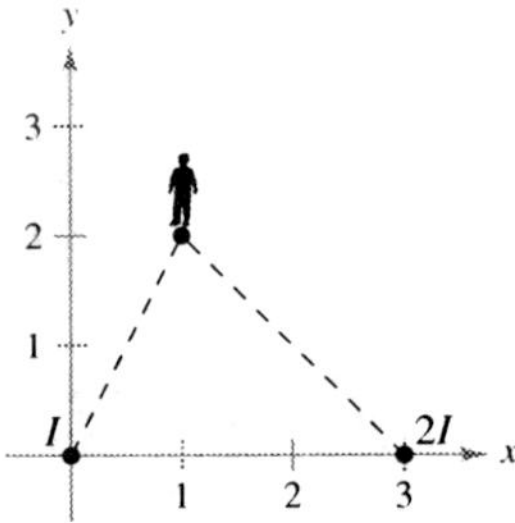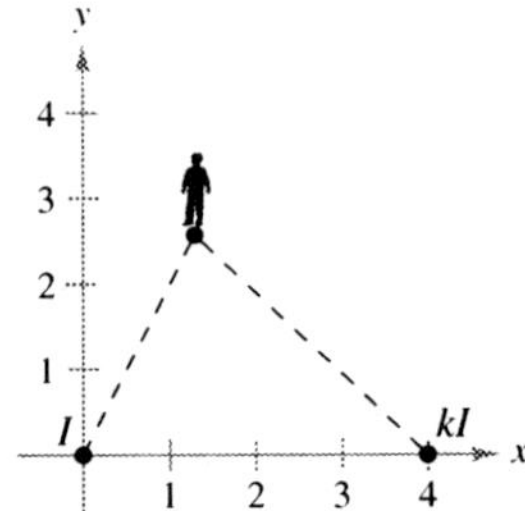

Figure for 12   **Figure for 13**

13. Suppose the speakers in Exercise 12 are 4 meters apart and the sound intensity of one speaker is k times that of the other, as shown in the figure. To print an enlarged copy of the graph, go to the website *www.mathgraphs.com*.

(a) Find the equation of all locations (x, y) where one could stand and receive equal amounts of sound from both speakers.

(b) Graph the equation for the case $k = 3$.

(c) Describe the set of locations of equal sound as k becomes very large.

14. Let d_1 and d_2 be the distances from the point (x, y) to the points $(-1, 0)$ and $(1, 0)$, respectively, as shown in the figure. Show that the equation of the graph of all points (x, y) satisfying $d_1 d_2 = 1$ is $(x^2 + y^2)^2 = 2(x^2 - y^2)$. This curve is called a **lemniscate.** Graph the lemniscate and identify three points on the graph.

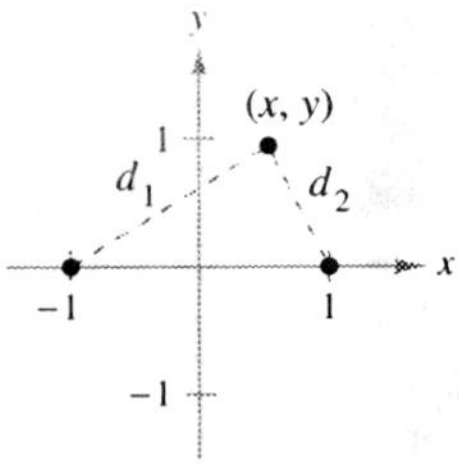

15. Let $f(x) = \dfrac{1}{1 - x}$.

(a) What are the domain and range of f?

(b) Find the composition $f(f(x))$. What is the domain of this function?

(c) Find $f(f(f(x)))$. What is the domain of this function?

(d) Graph $f(f(f(x)))$. Is the graph a line? Why or why not?

2 Limits and Their Properties

The limit of a function is the primary concept that distinguishes calculus from algebra and analytic geometry. The notion of a limit is fundamental to the study of calculus. Thus, it is important to acquire a good working knowledge of limits before moving on to other topics in calculus.

In this chapter, you should learn the following.

- How calculus compares with precalculus. (2.1)
- How to find limits graphically and numerically. (2.2)
- How to evaluate limits analytically. (2.3)
- How to determine continuity at a point and on an open interval, and how to determine one-sided limits. (2.4)
- How to determine infinite limits and find vertical asymptotes. (2.5)

European Space Agency/NASA

According to NASA, the coldest place in the known universe is the Boomerang nebula. The nebula is five thousand light years from Earth and has a temperature of −272°C. That is only 1° warmer than absolute zero, the coldest possible temperature. How did scientists determine that absolute zero is the "lower limit" of the temperature of matter? (See Section 2.4, Example 5.)

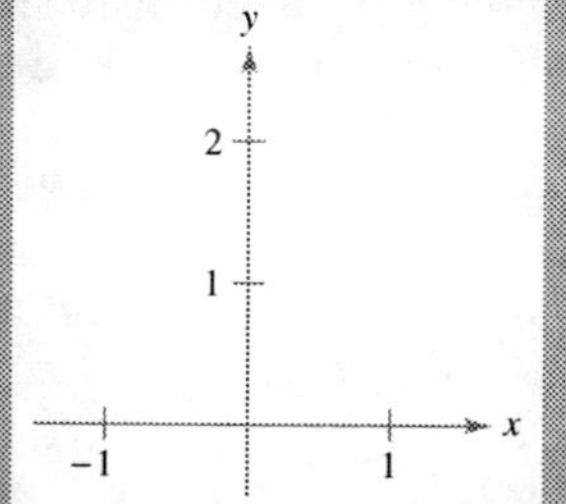

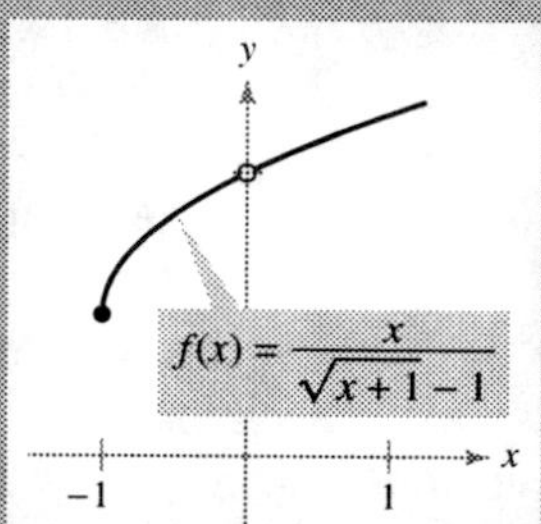

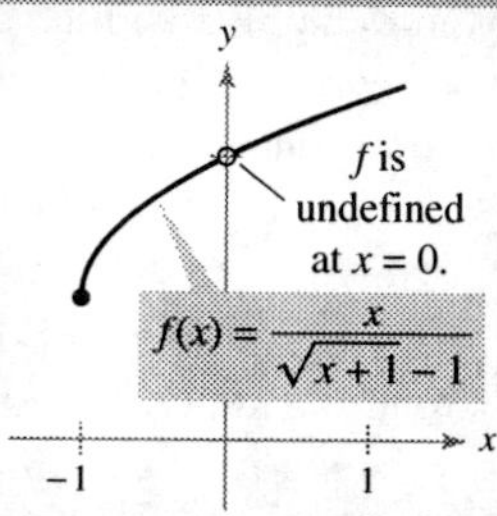

The limit process is a fundamental concept of calculus. One technique you can use to estimate a limit is to graph the function and then determine the behavior of the graph as the independent variable approaches a specific value. (See Section 2.2.)

2.1 A Preview of Calculus

- Understand what calculus is and how it compares with precalculus.
- Understand that the tangent line problem is basic to calculus.
- Understand that the area problem is also basic to calculus.

What Is Calculus?

 As you progress through this course, remember that learning calculus is just one of your goals. Your most important goal is to learn how to use calculus to model and solve real-life problems. Here are a few problem-solving strategies that may help you.

- Be sure you understand the question. What is given? What are you asked to find?
- Outline a plan. There are many approaches you could use: look for a pattern, solve a simpler problem, work backwards, draw a diagram, use technology, or any of many other approaches.
- Complete your plan. Be sure to answer the question. Verbalize your answer. For example, rather than writing the answer as $x = 4.6$, it would be better to write the answer as "The area of the region is 4.6 square meters."
- Look back at your work. Does your answer make sense? Is there a way you can check the reasonableness of your answer?

Calculus is the mathematics of change. For instance, calculus is the mathematics of velocities, accelerations, tangent lines, slopes, areas, volumes, arc lengths, centroids, curvatures, and a variety of other concepts that have enabled scientists, engineers, and economists to model real-life situations.

Although precalculus mathematics deals with velocities, accelerations, tangent lines, slopes, and so on, there is a fundamental difference between precalculus mathematics and calculus. Precalculus mathematics is more static, whereas calculus is more dynamic. Here are some examples.

- An object traveling at a constant velocity can be analyzed with precalculus mathematics. To analyze the velocity of an accelerating object, you need calculus.
- The slope of a line can be analyzed with precalculus mathematics. To analyze the slope of a curve, you need calculus.
- The curvature of a circle is constant and can be analyzed with precalculus mathematics. To analyze the variable curvature of a general curve, you need calculus.
- The area of a rectangle can be analyzed with precalculus mathematics. To analyze the area under a general curve, you need calculus.

Each of these situations involves the same general strategy—the reformulation of precalculus mathematics through the use of a limit process. So, one way to answer the question "What is calculus?" is to say that calculus is a "limit machine" that involves three stages. The first stage is precalculus mathematics, such as finding the slope of a line or the area of a rectangle. The second stage is the limit process, and the third stage is a new calculus formulation, such as a derivative or an integral.

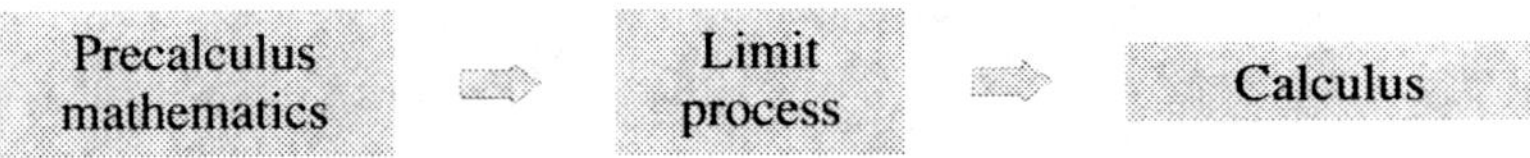

Some students try to learn calculus as if it were simply a collection of new formulas. This is unfortunate. If you reduce calculus to the memorization of differentiation and integration formulas, you will miss a great deal of understanding, self-confidence, and satisfaction.

On the following two pages, some familiar precalculus concepts coupled with their calculus counterparts are listed. Throughout the text, your goal should be to learn how precalculus formulas and techniques are used as building blocks to produce the more general calculus formulas and techniques. Don't worry if you are unfamiliar with some of the "old formulas" listed on the following two pages—you will be reviewing all of them.

As you proceed through this text, we suggest that you come back to this discussion repeatedly. Try to keep track of where you are relative to the three stages involved in the study of calculus. For example, the first three chapters break down as shown.

Chapter 1: Preparation for Calculus   Precalculus

Chapter 2: Limits and Their Properties   Limit process

Chapter 3: Differentiation   Calculus

Without Calculus	**With Differential Calculus**
Value of $f(x)$ when $x = c$	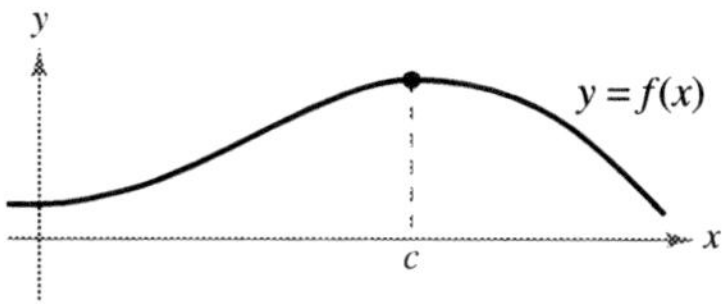 Limit of $f(x)$ as x approaches c 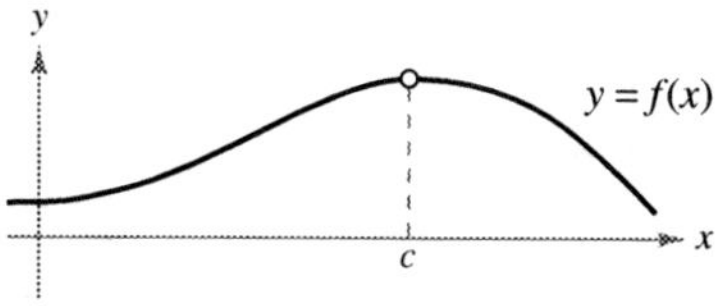
Slope of a line	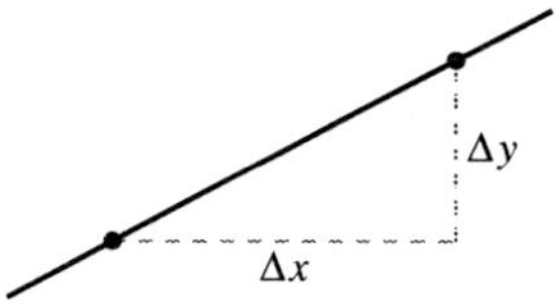 Slope of a curve 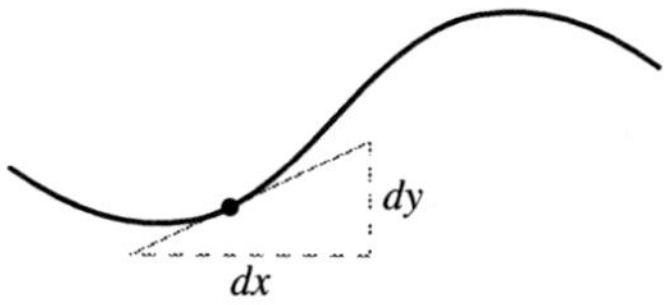
Secant line to a curve	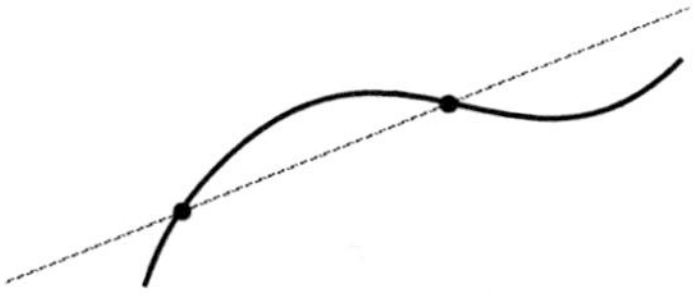 Tangent line to a curve 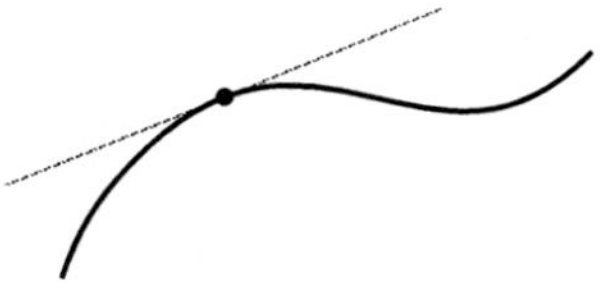
Average rate of change between $t = a$ and $t = b$	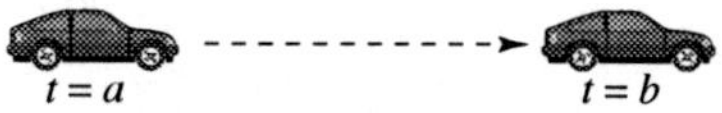 Instantaneous rate of change at $t = c$ 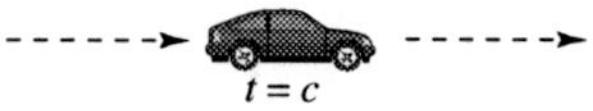
Curvature of a circle	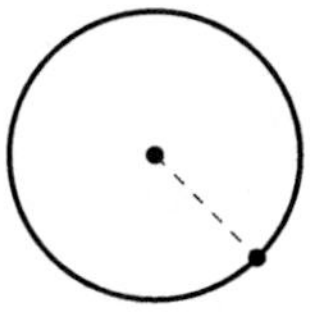 Curvature of a curve 
Height of a curve when $x = c$	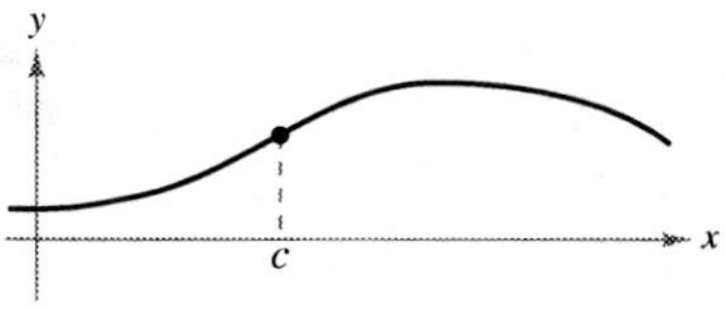 Maximum height of a curve on an interval 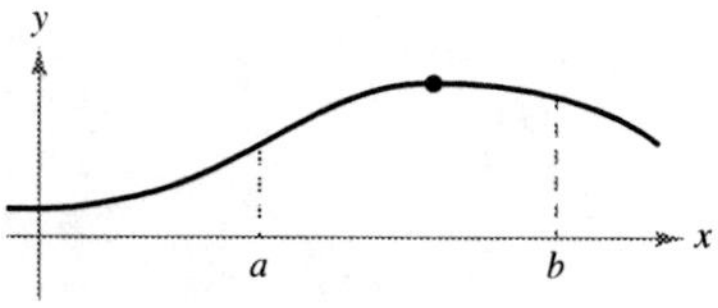
Tangent plane to a sphere	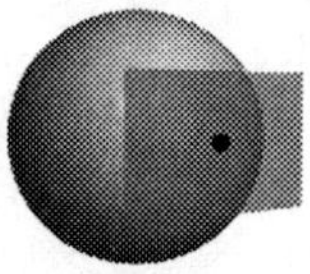 Tangent plane to a surface 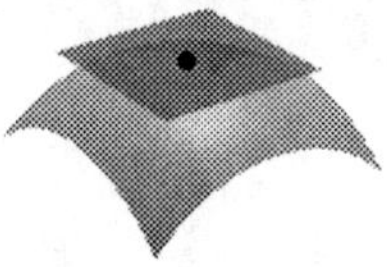
Direction of motion along a straight line	 Direction of motion along a curved line 

Without Calculus	With Integral Calculus
Area of a rectangle	Area under a curve 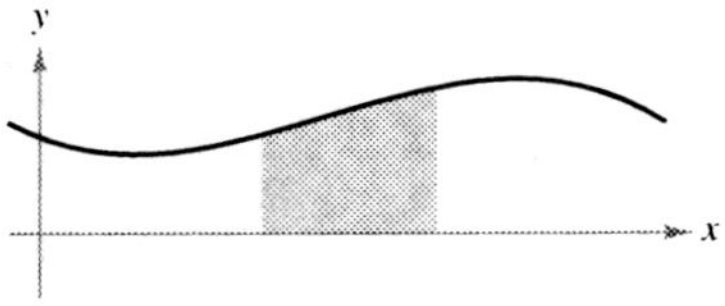
Work done by a constant force	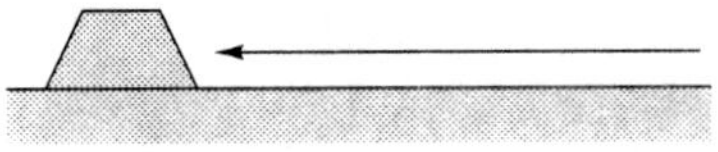 Work done by a variable force 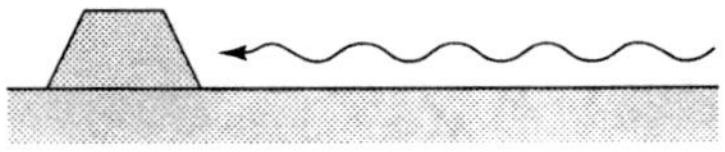
Center of a rectangle	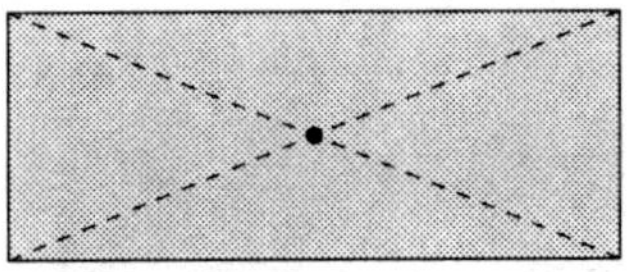 Centroid of a region 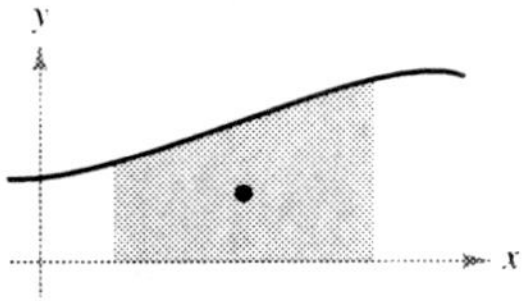
Length of a line segment	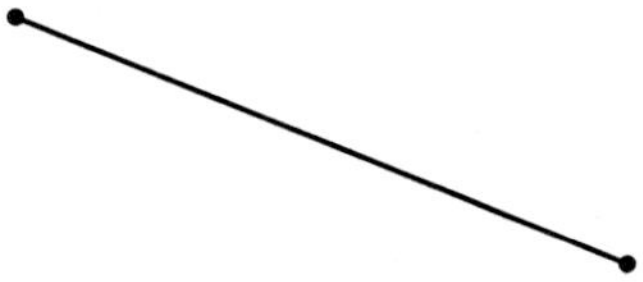 Length of an arc 
Surface area of a cylinder	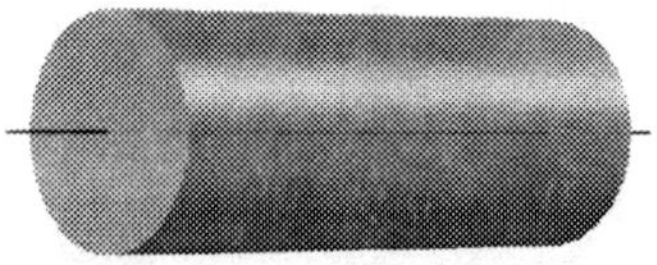 Surface area of a solid of revolution 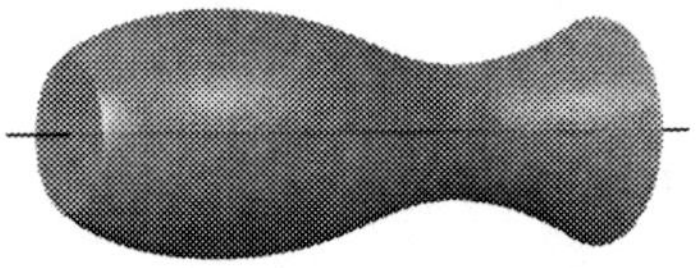
Mass of a solid of constant density	 Mass of a solid of variable density 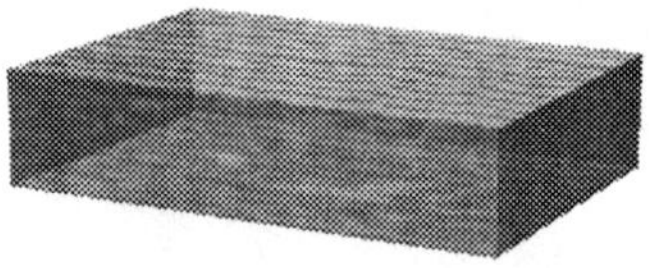
Volume of a rectangular solid	Volume of a region under a surface 
Sum of a finite number of terms $a_1 + a_2 + \cdots + a_n = S$	Sum of an infinite number of terms $a_1 + a_2 + a_3 + \cdots = S$

The Tangent Line Problem

The notion of a limit is fundamental to the study of calculus. The following brief descriptions of two classic problems in calculus—*the tangent line problem* and *the area problem*—should give you some idea of the way limits are used in calculus.

In the tangent line problem, you are given a function f and a point P on its graph and are asked to find an equation of the tangent line to the graph at point P, as shown in Figure 2.1.

Except for cases involving a vertical tangent line, the problem of finding the **tangent line** at a point P is equivalent to finding the *slope* of the tangent line at P. You can approximate this slope by using a line through the point of tangency and a second point on the curve, as shown in Figure 2.2(a). Such a line is called a **secant line.** If $P(c, f(c))$ is the point of tangency and

$$Q(c + \Delta x, f(c + \Delta x))$$

is a second point on the graph of f, then the slope of the secant line through these two points can be found using precalculus and is given by

$$m_{sec} = \frac{f(c + \Delta x) - f(c)}{c + \Delta x - c} = \frac{f(c + \Delta x) - f(c)}{\Delta x}.$$

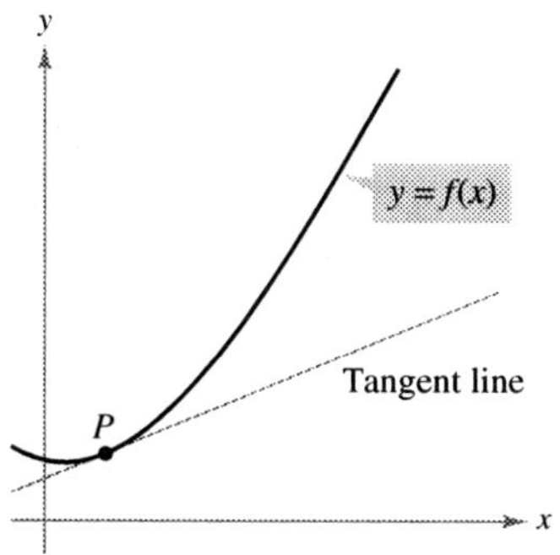

The tangent line to the graph of f at P
Figure 2.1

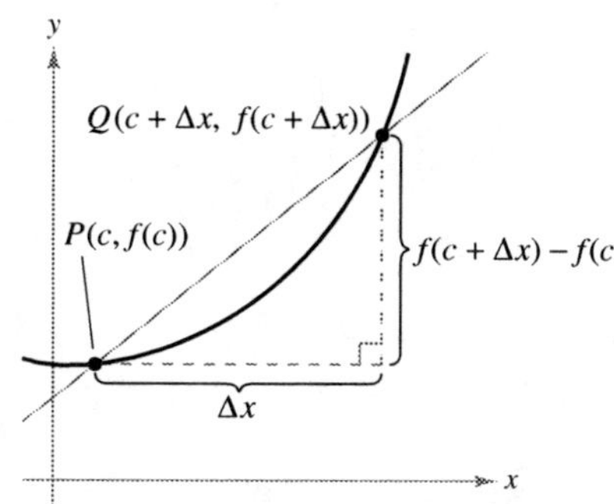

(a) The secant line through $(c, f(c))$ and $(c + \Delta x, f(c + \Delta x))$

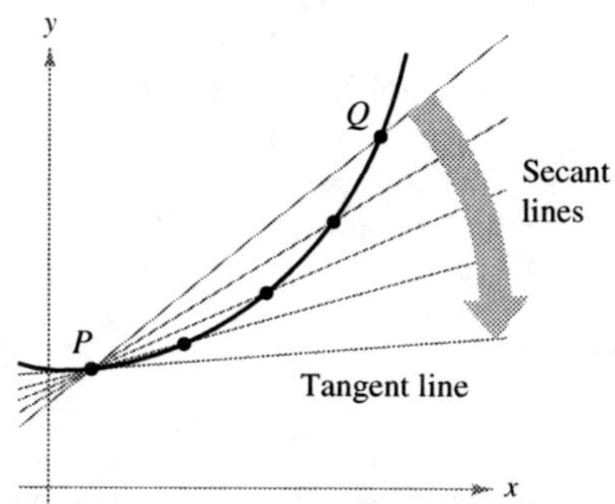

(b) As Q approaches P, the secant lines approach the tangent line.

Figure 2.2

As point Q approaches point P, the slope of the secant line approaches the slope of the tangent line, as shown in Figure 2.2(b). When such a "limiting position" exists, the slope of the tangent line is said to be the **limit** of the slope of the secant line. (Much more will be said about this important calculus concept in Chapter 3.)

GRACE CHISHOLM YOUNG (1868–1944)

Grace Chisholm Young received her degree in mathematics from Girton College in Cambridge, England. Her early work was published under the name of William Young, her husband. Between 1914 and 1916, Grace Young published work on the foundations of calculus that won her the Gamble Prize from Girton College.

EXPLORATION

The following points lie on the graph of $f(x) = x^2$.

$$Q_1(1.5, f(1.5)), \quad Q_2(1.1, f(1.1)), \quad Q_3(1.01, f(1.01)),$$

$$Q_4(1.001, f(1.001)), \quad Q_5(1.0001, f(1.0001))$$

Each successive point gets closer to the point $P(1, 1)$. Find the slopes of the secant lines through Q_1 and P, Q_2 and P, and so on. Graph these secant lines on a graphing utility. Then use your results to estimate the slope of the tangent line to the graph of f at the point P.

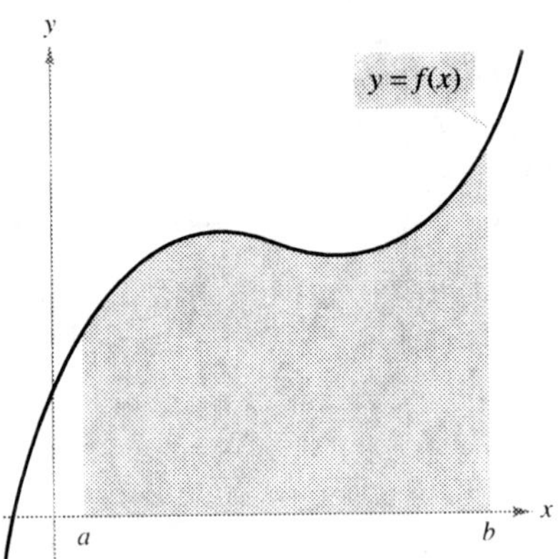

Area under a curve
Figure 2.3

In one of the most astounding events ever to occur in mathematics, it was discovered that the tangent line problem and the area problem are closely related. This discovery led to the birth of calculus. You will learn about the relationship between these two problems when you study the Fundamental Theorem of Calculus in Chapter 5.

The Area Problem

In the tangent line problem, you saw how the limit process can be applied to the slope of a line to find the slope of a general curve. A second classic problem in calculus is finding the area of a plane region that is bounded by the graphs of functions. This problem can also be solved with a limit process. In this case, the limit process is applied to the area of a rectangle to find the area of a general region.

As a simple example, consider the region bounded by the graph of the function $y = f(x)$, the x-axis, and the vertical lines $x = a$ and $x = b$, as shown in Figure 2.3. You can approximate the area of the region with several rectangular regions, as shown in Figure 2.4. As you increase the number of rectangles, the approximation tends to become better and better because the amount of area missed by the rectangles decreases. Your goal is to determine the limit of the sum of the areas of the rectangles as the number of rectangles increases without bound.

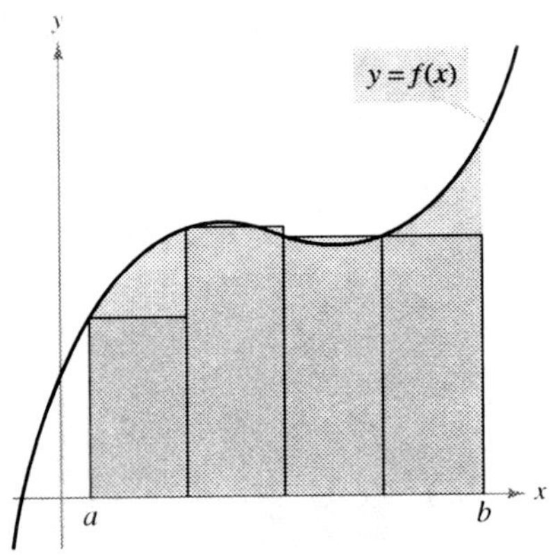

Approximation using four rectangles
Figure 2.4

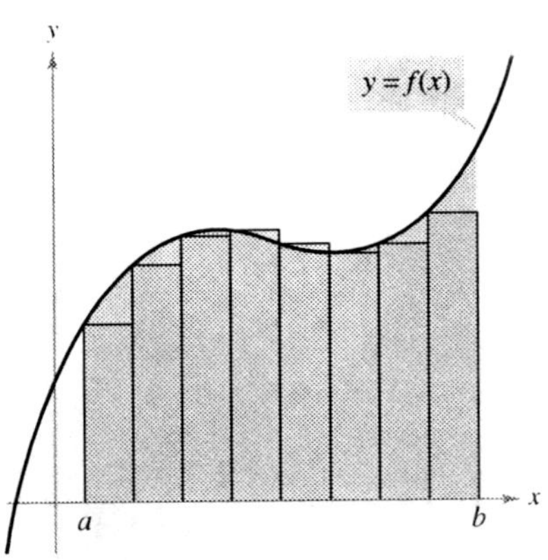

Approximation using eight rectangles

EXPLORATION

Consider the region bounded by the graphs of $f(x) = x^2$, $y = 0$, and $x = 1$, as shown in part (a) of the figure. The area of the region can be approximated by two sets of rectangles—one set inscribed within the region and the other set circumscribed over the region, as shown in parts (b) and (c). Find the sum of the areas of each set of rectangles. Then use your results to approximate the area of the region.

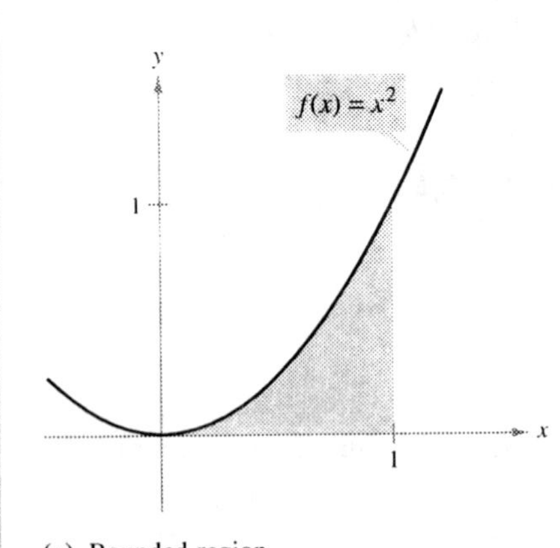

(a) Bounded region

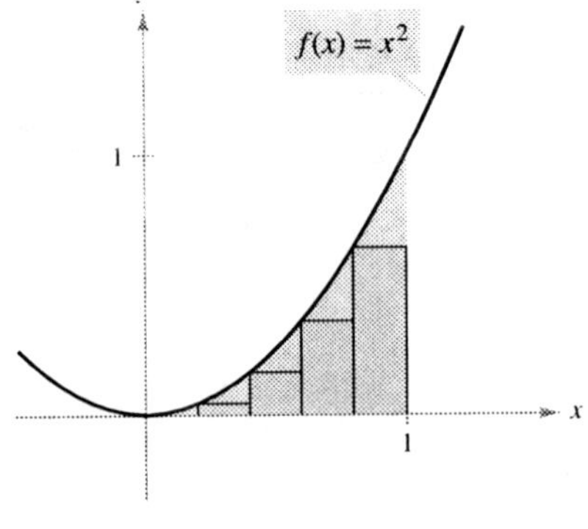
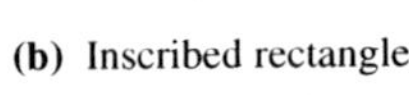

(b) Inscribed rectangles

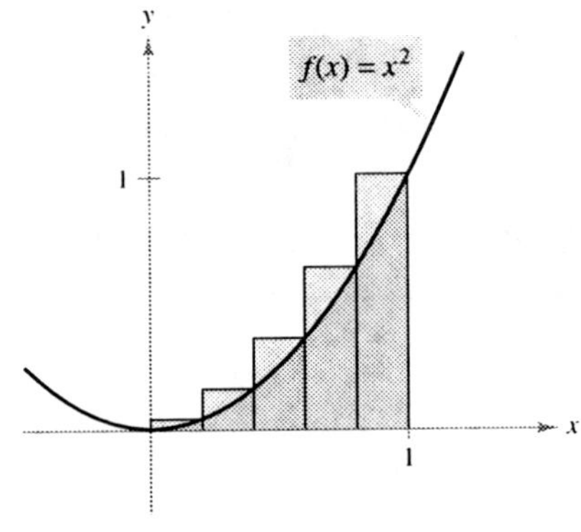

(c) Circumscribed rectangles

2.1 Exercises

See www.CalcChat.com for worked-out solutions to odd-numbered exercises.

In Exercises 1–5, decide whether the problem can be solved using precalculus or whether calculus is required. If the problem can be solved using precalculus, solve it. If the problem seems to require calculus, explain your reasoning and use a graphical or numerical approach to estimate the solution.

1. Find the distance traveled in 15 seconds by an object traveling at a constant velocity of 20 feet per second.

2. Find the distance traveled in 15 seconds by an object moving with a velocity of $v(t) = 20 + 7 \cos t$ feet per second.

3. A bicyclist is riding on a path modeled by the function $f(x) = 0.04(8x - x^2)$, where x and $f(x)$ are measured in miles. Find the rate of change of elevation at $x = 2$.

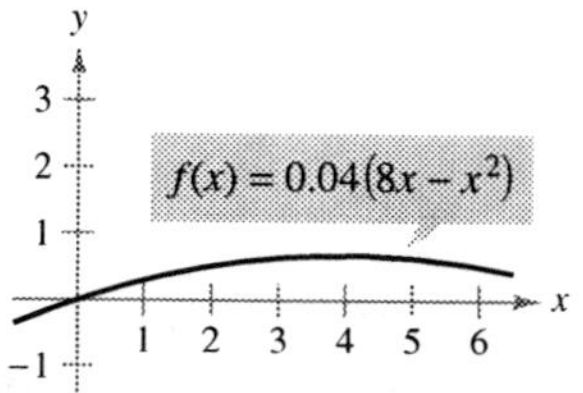

Figure for 3

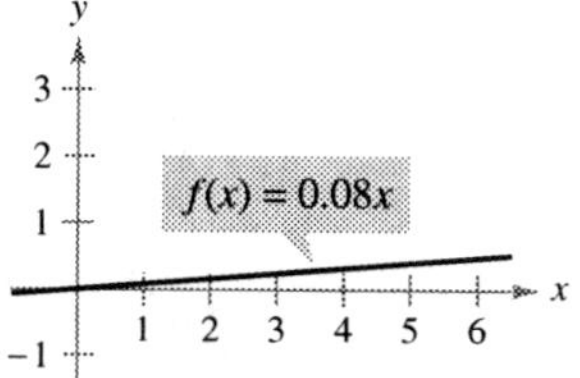

Figure for 4

4. A bicyclist is riding on a path modeled by the function $f(x) = 0.08x$, where x and $f(x)$ are measured in miles. Find the rate of change of elevation at $x = 2$.

5. Find the area of the shaded region.

(a) 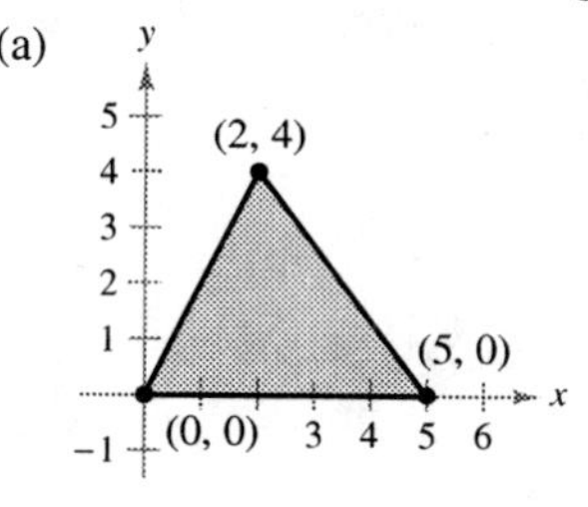

(b) 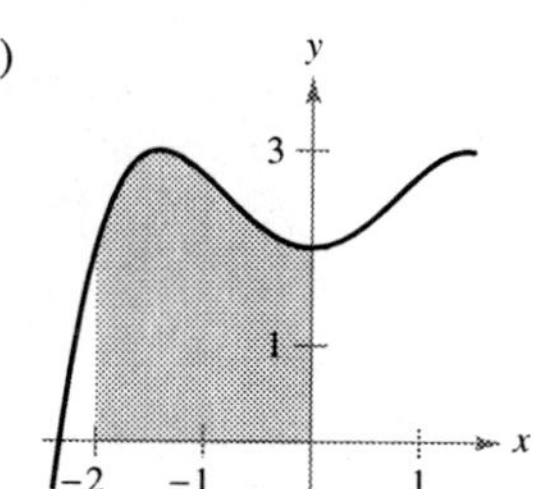

6. **Secant Lines** Consider the function $f(x) = \sqrt{x}$ and the point $P(4, 2)$ on the graph of f.

   (a) Graph f and the secant lines passing through $P(4, 2)$ and $Q(x, f(x))$ for x-values of 1, 3, and 5.

   (b) Find the slope of each secant line.

   (c) Use the results of part (b) to estimate the slope of the tangent line to the graph of f at $P(4, 2)$. Describe how to improve your approximation of the slope.

7. **Secant Lines** Consider the function $f(x) = 6x - x^2$ and the point $P(2, 8)$ on the graph of f.

   (a) Graph f and the secant lines passing through $P(2, 8)$ and $Q(x, f(x))$ for x-values of 3, 2.5, and 1.5.

   (b) Find the slope of each secant line.

   (c) Use the results of part (b) to estimate the slope of the tangent line to the graph of f at $P(2, 8)$. Describe how to improve your approximation of the slope.

8. (a) Use the rectangles in each graph to approximate the area of the region bounded by $y = \sin x$, $y = 0$, $x = 0$, and $x = \pi$.

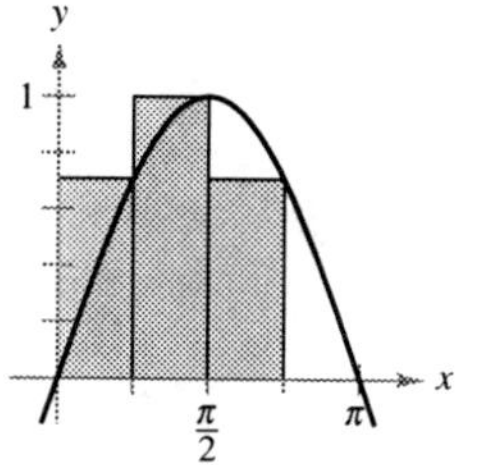
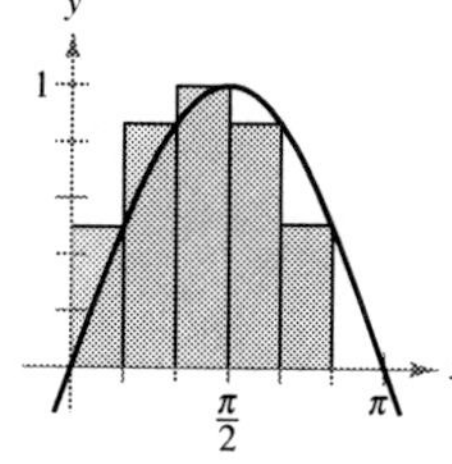

   (b) Describe how you could continue this process to obtain a more accurate approximation of the area.

9. (a) Use the rectangles in each graph to approximate the area of the region bounded by $y = 5/x$, $y = 0$, $x = 1$, and $x = 5$.

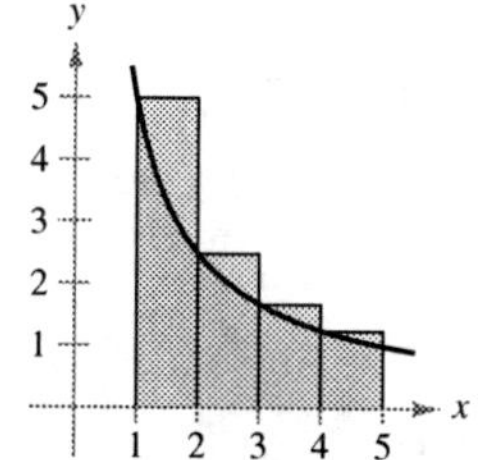
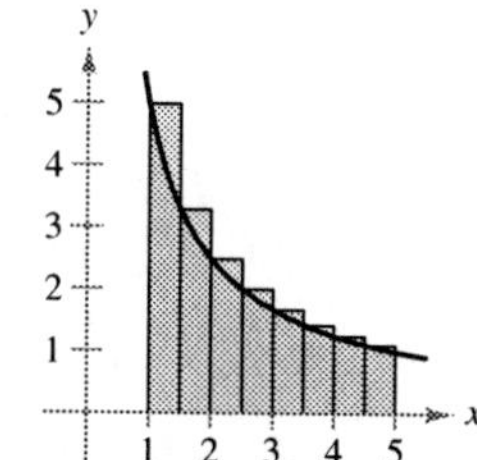

   (b) Describe how you could continue this process to obtain a more accurate approximation of the area.

CAPSTONE

10. How would you describe the instantaneous rate of change of an automobile's position on a highway?

WRITING ABOUT CONCEPTS

11. Consider the length of the graph of $f(x) = 5/x$ from $(1, 5)$ to $(5, 1)$.

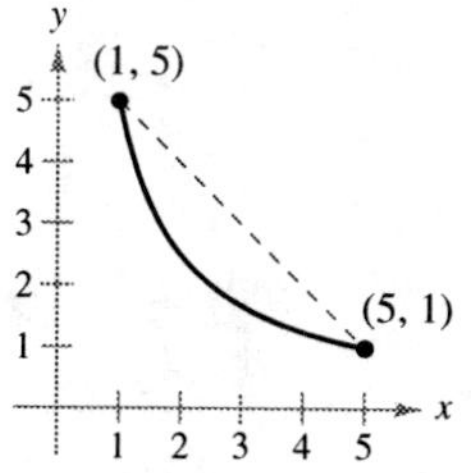

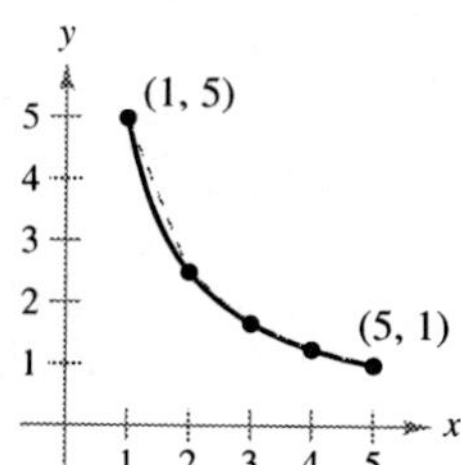

   (a) Approximate the length of the curve by finding the distance between its two endpoints, as shown in the first figure.

   (b) Approximate the length of the curve by finding the sum of the lengths of four line segments, as shown in the second figure.

   (c) Describe how you could continue this process to obtain a more accurate approximation of the length of the curve.

2.2 Finding Limits Graphically and Numerically

- Estimate a limit using a numerical or graphical approach.
- Learn different ways that a limit can fail to exist.
- Study and use a formal definition of limit.

An Introduction to Limits

Suppose you are asked to sketch the graph of the function f given by

$$f(x) = \frac{x^3 - 1}{x - 1}, \quad x \neq 1.$$

For all values other than $x = 1$, you can use standard curve-sketching techniques. However, at $x = 1$, it is not clear what to expect. To get an idea of the behavior of the graph of f near $x = 1$, you can use two sets of x-values—one set that approaches 1 from the left and one that approaches 1 from the right, as shown in the table.

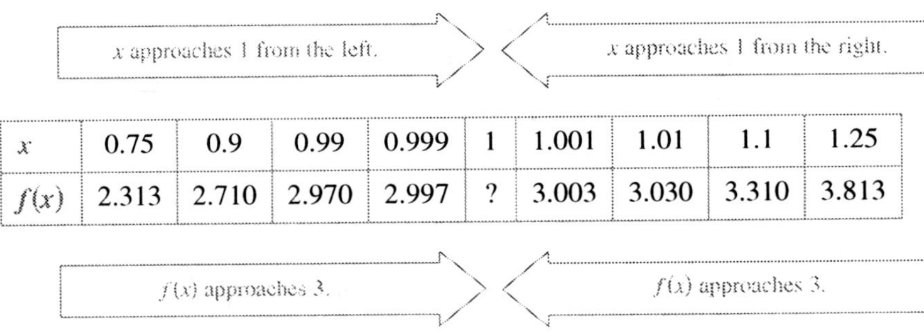

x approaches 1 from the left.					x approaches 1 from the right.			
x	0.75	0.9	0.99	0.999	1	1.001	1.01	1.1	1.25
$f(x)$	2.313	2.710	2.970	2.997	?	3.003	3.030	3.310	3.813

| $f(x)$ approaches 3. | | | | | $f(x)$ approaches 3. | | | |

The graph of f is a parabola that has a gap at the point $(1, 3)$, as shown in Figure 2.5. Although x cannot equal 1, you can move arbitrarily close to 1, and as a result $f(x)$ moves arbitrarily close to 3. Using limit notation, you can write

$$\lim_{x \to 1} f(x) = 3. \qquad \text{This is read as "the limit of } f(x) \text{ as } x \text{ approaches 1 is 3."}$$

This discussion leads to an informal definition of limit. If $f(x)$ becomes arbitrarily close to a single number L as x approaches c from either side, the **limit** of $f(x)$, as x approaches c, is L. This limit is written as

$$\lim_{x \to c} f(x) = L.$$

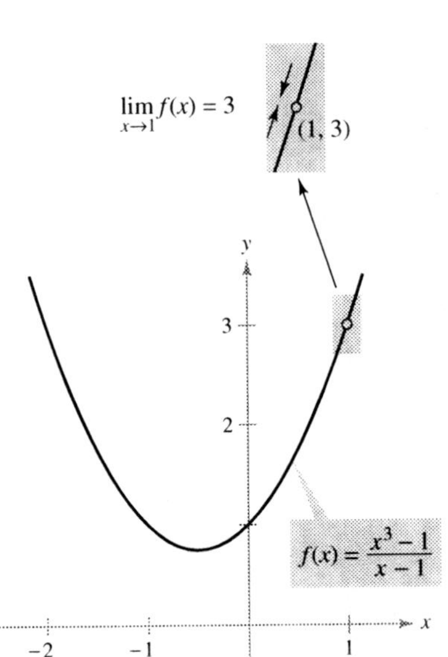

The limit of $f(x)$ as x approaches 1 is 3.
Figure 2.5

EXPLORATION

The discussion above gives an example of how you can estimate a limit *numerically* by constructing a table and *graphically* by drawing a graph. Estimate the following limit numerically by completing the table.

$$\lim_{x \to 2} \frac{x^2 - 3x + 2}{x - 2}$$

x	1.75	1.9	1.99	1.999	2	2.001	2.01	2.1	2.25
$f(x)$	?	?	?	?	?	?	?	?	?

Then use a graphing utility to estimate the limit graphically.

EXAMPLE 1 Estimating a Limit Numerically

Evaluate the function $f(x) = x/(\sqrt{x + 1} - 1)$ at several points near $x = 0$ and use the results to estimate the limit

$$\lim_{x \to 0} \frac{x}{\sqrt{x + 1} - 1}.$$

Solution   The table lists the values of $f(x)$ for several x-values near 0.

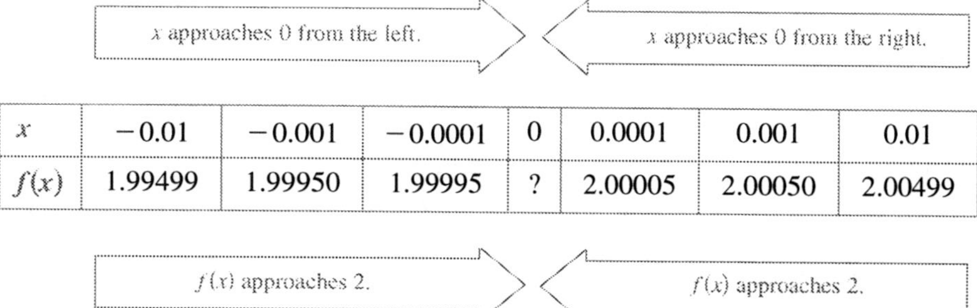

x approaches 0 from the left.				x approaches 0 from the right.		
x	-0.01	-0.001	-0.0001	0	0.0001	0.001	0.01

x	-0.01	-0.001	-0.0001	0	0.0001	0.001	0.01
$f(x)$	1.99499	1.99950	1.99995	?	2.00005	2.00050	2.00499

$f(x)$ approaches 2. | *$f(x)$ approaches 2.*

From the results shown in the table, you can estimate the limit to be 2. This limit is reinforced by the graph of f (see Figure 2.6).

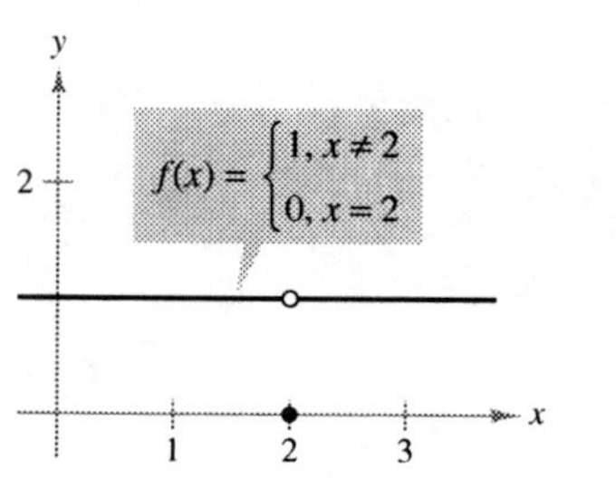

The limit of $f(x)$ as x approaches 0 is 2.
Figure 2.6

In Example 1, note that the function is undefined at $x = 0$ and yet $f(x)$ appears to be approaching a limit as x approaches 0. This often happens, and it is important to realize that *the existence or nonexistence of $f(x)$ at $x = c$ has no bearing on the existence of the limit of $f(x)$ as x approaches c.*

EXAMPLE 2 Finding a Limit

Find the limit of $f(x)$ as x approaches 2, where f is defined as

$$f(x) = \begin{cases} 1, & x \neq 2 \\ 0, & x = 2 \end{cases}.$$

Solution   Because $f(x) = 1$ for all x other than $x = 2$, you can conclude that the limit is 1, as shown in Figure 2.7. So, you can write

$$\lim_{x \to 2} f(x) = 1.$$

The fact that $f(2) = 0$ has no bearing on the existence or value of the limit as x approaches 2. For instance, if the function were defined as

$$f(x) = \begin{cases} 1, & x \neq 2 \\ 2, & x = 2 \end{cases}$$

the limit would be the same.

The limit of $f(x)$ as x approaches 2 is 1.
Figure 2.7

So far in this section, you have been estimating limits numerically and graphically. Each of these approaches produces an estimate of the limit. In Section 2.3, you will study analytic techniques for evaluating limits. Throughout the course, try to develop a habit of using this three-pronged approach to problem solving.

1. Numerical approach   Construct a table of values.

2. Graphical approach   Draw a graph by hand or using technology.

3. Analytic approach   Use algebra or calculus.

Limits That Fail to Exist

In the next three examples you will examine some limits that fail to exist.

EXAMPLE **3** Behavior That Differs from the Right and from the Left

Show that the limit $\lim\limits_{x \to 0} \dfrac{|x|}{x}$ does not exist.

Solution   Consider the graph of the function $f(x) = |x|/x$. From Figure 2.8 and the definition of absolute value

$$|x| = \begin{cases} x, & \text{if } x \geq 0 \\ -x, & \text{if } x < 0 \end{cases} \qquad \text{Definition of absolute value}$$

you can see that

$$\frac{|x|}{x} = \begin{cases} 1, & \text{if } x > 0 \\ -1, & \text{if } x < 0. \end{cases}$$

This means that no matter how close x gets to 0, there will be both positive and negative x-values that yield $f(x) = 1$ or $f(x) = -1$. Specifically, if δ (the lowercase Greek letter *delta*) is a positive number, then for x-values satisfying the inequality $0 < |x| < \delta$, you can classify the values of $|x|/x$ as follows.

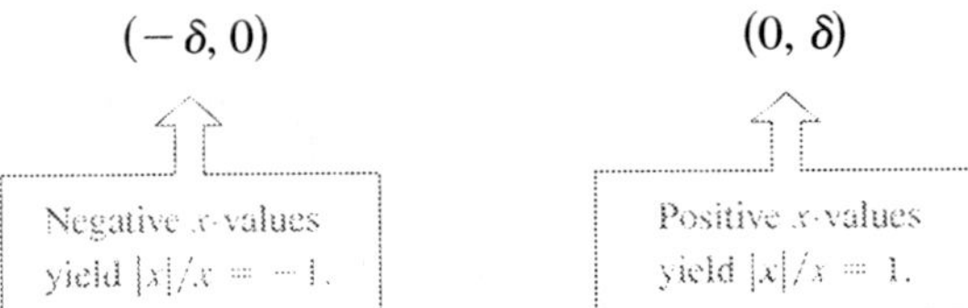

Because $|x|/x$ approaches a different number from the right side of 0 than it approaches from the left side, the limit $\lim\limits_{x \to 0} (|x|/x)$ does not exist.

EXAMPLE **4** Unbounded Behavior

Discuss the existence of the limit $\lim\limits_{x \to 0} \dfrac{1}{x^2}$.

Solution   Let $f(x) = 1/x^2$. In Figure 2.9, you can see that as x approaches 0 from either the right or the left, $f(x)$ increases without bound. This means that by choosing x close enough to 0, you can force $f(x)$ to be as large as you want. For instance, $f(x)$ will be larger than 100 if you choose x that is within $\frac{1}{10}$ of 0. That is,

$$0 < |x| < \frac{1}{10} \quad \Longrightarrow \quad f(x) = \frac{1}{x^2} > 100.$$

Similarly, you can force $f(x)$ to be larger than 1,000,000, as follows.

$$0 < |x| < \frac{1}{1000} \quad \Longrightarrow \quad f(x) = \frac{1}{x^2} > 1,000,000$$

Because $f(x)$ is not approaching a real number L as x approaches 0, you can conclude that the limit does not exist.

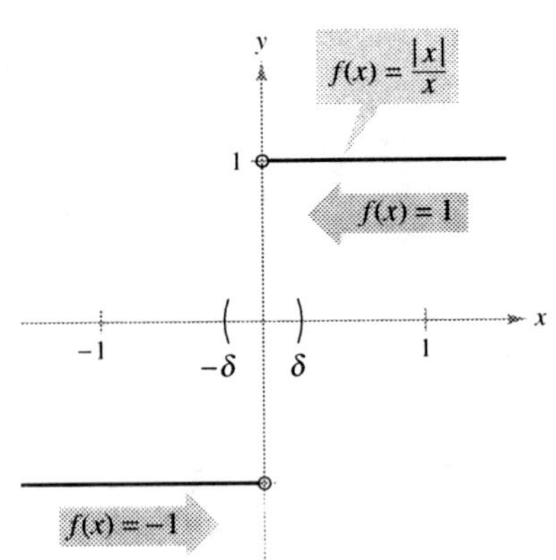

$\lim\limits_{x \to 0} f(x)$ does not exist.

Figure 2.8

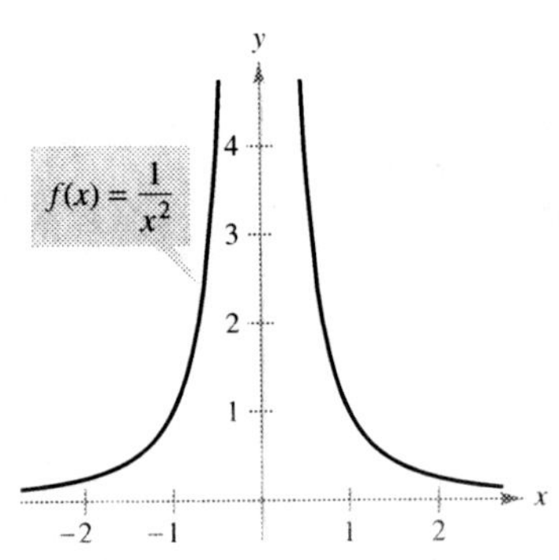

$\lim\limits_{x \to 0} f(x)$ does not exist.

Figure 2.9

EXAMPLE 5 Oscillating Behavior

Discuss the existence of the limit $\displaystyle\lim_{x \to 0} \sin \frac{1}{x}$.

Solution   Let $f(x) = \sin(1/x)$. In Figure 2.10, you can see that as x approaches 0, $f(x)$ oscillates between -1 and 1. So, the limit does not exist because no matter how small you choose δ, it is possible to choose x_1 and x_2 within δ units of 0 such that $\sin(1/x_1) = 1$ and $\sin(1/x_2) = -1$, as shown in the table.

x	$2/\pi$	$2/3\pi$	$2/5\pi$	$2/7\pi$	$2/9\pi$	$2/11\pi$	$x \to 0$
$\sin(1/x)$	1	-1	1	-1	1	-1	Limit does not exist.

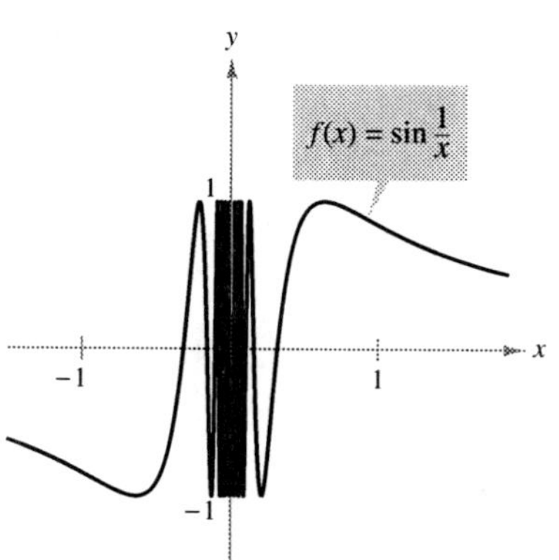

$\displaystyle\lim_{x \to 0} f(x)$ does not exist.

Figure 2.10

> **COMMON TYPES OF BEHAVIOR ASSOCIATED WITH NONEXISTENCE OF A LIMIT**
>
> **1.** $f(x)$ approaches a different number from the right side of c than it approaches from the left side.
>
> **2.** $f(x)$ increases or decreases without bound as x approaches c.
>
> **3.** $f(x)$ oscillates between two fixed values as x approaches c.

There are many other interesting functions that have unusual limit behavior. An often cited one is the *Dirichlet function*.

$$f(x) = \begin{cases} 0, & \text{if } x \text{ is rational.} \\ 1, & \text{if } x \text{ is irrational.} \end{cases}$$

Because this function has *no limit* at any real number c, it is *not continuous* at any real number c. You will study continuity more closely in Section 2.4.

TECHNOLOGY PITFALL   When you use a graphing utility to investigate the behavior of a function near the x-value at which you are trying to evaluate a limit, remember that you can't always trust the pictures that graphing utilities draw. For instance, if you use a graphing utility to graph the function in Example 5 over an interval containing 0, you will most likely obtain an incorrect graph such as that shown in Figure 2.11. The reason that a graphing utility can't show the correct graph is that the graph has infinitely many oscillations over any interval that contains 0.

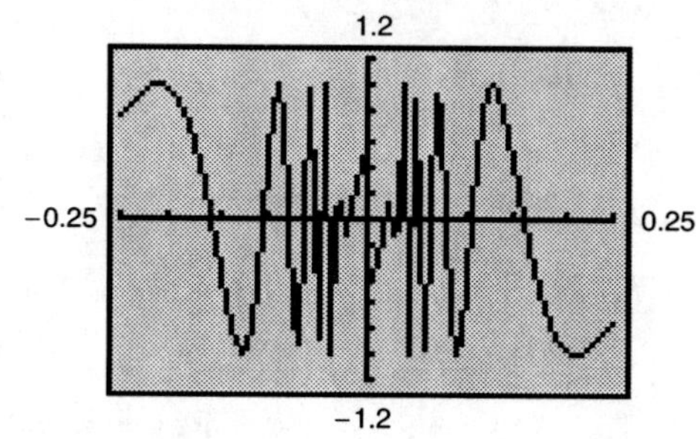

Incorrect graph of $f(x) = \sin(1/x)$.

Figure 2.11

PETER GUSTAV DIRICHLET (1805–1859)

In the early development of calculus, the definition of a function was much more restricted than it is today, and "functions" such as the Dirichlet function would not have been considered. The modern definition of function is attributed to the German mathematician Peter Gustav Dirichlet.

The icon ○ *indicates that you will find a CAS Investigation on the book's website. The CAS Investigation is a collaborative exploration of this example using the computer algebra systems* Maple *and* Mathematica.

A Formal Definition of Limit

Let's take another look at the informal definition of limit. If $f(x)$ becomes arbitrarily close to a single number L as x approaches c from either side, then the limit of $f(x)$ as x approaches c is L, written as

$$\lim_{x \to c} f(x) = L.$$

At first glance, this definition looks fairly technical. Even so, it is informal because exact meanings have not yet been given to the two phrases

"$f(x)$ becomes arbitrarily close to L"

and

"x approaches c."

The first person to assign mathematically rigorous meanings to these two phrases was Augustin-Louis Cauchy. His ε-δ **definition of limit** is the standard used today.

In Figure 2.12, let ε (the lowercase Greek letter *epsilon*) represent a (small) positive number. Then the phrase "$f(x)$ becomes arbitrarily close to L" means that $f(x)$ lies in the interval $(L - \varepsilon, L + \varepsilon)$. Using absolute value, you can write this as

$$|f(x) - L| < \varepsilon.$$

Similarly, the phrase "x approaches c" means that there exists a positive number δ such that x lies in either the interval $(c - \delta, c)$ or the interval $(c, c + \delta)$. This fact can be concisely expressed by the double inequality

$$0 < |x - c| < \delta.$$

The first inequality

$$0 < |x - c| \qquad \text{The distance between } x \text{ and } c \text{ is more than } 0.$$

expresses the fact that $x \neq c$. The second inequality

$$|x - c| < \delta \qquad x \text{ is within } \delta \text{ units of } c.$$

says that x is within a distance δ of c.

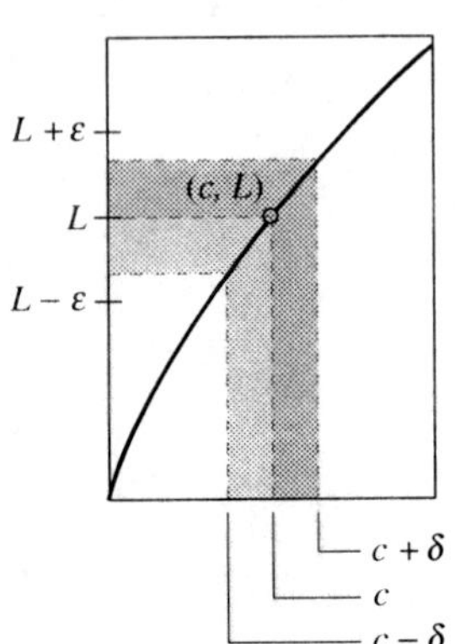

The ε-δ definition of the limit of $f(x)$ as x approaches c

Figure 2.12

DEFINITION OF LIMIT

Let f be a function defined on an open interval containing c (except possibly at c) and let L be a real number. The statement

$$\lim_{x \to c} f(x) = L$$

means that for each $\varepsilon > 0$ there exists a $\delta > 0$ such that if

$$0 < |x - c| < \delta, \quad \text{then} \quad |f(x) - L| < \varepsilon.$$

NOTE Throughout this text, the expression

$$\lim_{x \to c} f(x) = L$$

implies two statements—the limit exists *and* the limit is L. ∎

Some functions do not have limits as $x \to c$, but those that do cannot have two different limits as $x \to c$. That is, *if the limit of a function exists, it is unique* (see Exercise 81).

▨ **FOR FURTHER INFORMATION** For more on the introduction of rigor to calculus, see "Who Gave You the Epsilon? Cauchy and the Origins of Rigorous Calculus" by Judith V. Grabiner in *The American Mathematical Monthly*. To view this article, go to the website *www.matharticles.com*.

The next three examples should help you develop a better understanding of the ε-δ definition of limit.

EXAMPLE 6 Finding a δ for a Given ε

Given the limit

$$\lim_{x \to 3} (2x - 5) = 1$$

find δ such that $|(2x - 5) - 1| < 0.01$ whenever $0 < |x - 3| < \delta$.

Solution In this problem, you are working with a given value of ε—namely, $\varepsilon = 0.01$. To find an appropriate δ, notice that

$$|(2x - 5) - 1| = |2x - 6| = 2|x - 3|.$$

Because the inequality $|(2x - 5) - 1| < 0.01$ is equivalent to $2|x - 3| < 0.01$, you can choose $\delta = \frac{1}{2}(0.01) = 0.005$. This choice works because

$$0 < |x - 3| < 0.005$$

implies that

$$|(2x - 5) - 1| = 2|x - 3| < 2(0.005) = 0.01$$

as shown in Figure 2.13. ▪

NOTE In Example 6, note that 0.005 is the *largest* value of δ that will guarantee $|(2x - 5) - 1| < 0.01$ whenever $0 < |x - 3| < \delta$. Any *smaller* positive value of δ would, of course, also work. ▪

In Example 6, you found a δ-value for a *given* ε. This does not prove the existence of the limit. To do that, you must prove that you can find a δ for *any* ε, as shown in the next example.

EXAMPLE 7 Using the ε-δ Definition of Limit

Use the ε-δ definition of limit to prove that

$$\lim_{x \to 2} (3x - 2) = 4.$$

Solution You must show that for each $\varepsilon > 0$, there exists a $\delta > 0$ such that $|(3x - 2) - 4| < \varepsilon$ whenever $0 < |x - 2| < \delta$. Because your choice of δ depends on ε, you need to establish a connection between the absolute values $|(3x - 2) - 4|$ and $|x - 2|$.

$$|(3x - 2) - 4| = |3x - 6| = 3|x - 2|$$

So, for a given $\varepsilon > 0$ you can choose $\delta = \varepsilon/3$. This choice works because

$$0 < |x - 2| < \delta = \frac{\varepsilon}{3}$$

implies that

$$|(3x - 2) - 4| = 3|x - 2| < 3\left(\frac{\varepsilon}{3}\right) = \varepsilon$$

as shown in Figure 2.14. ▪

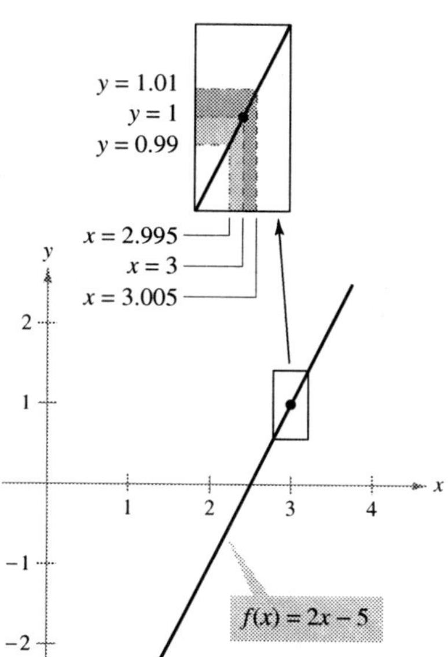

The limit of $f(x)$ as x approaches 3 is 1.
Figure 2.13

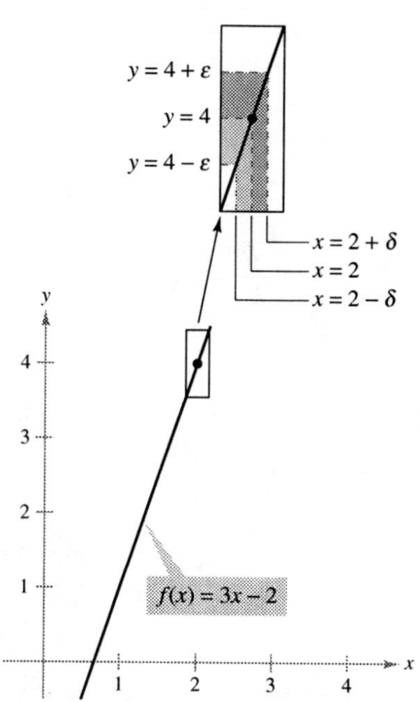

The limit of $f(x)$ as x approaches 2 is 4.
Figure 2.14

EXAMPLE 8 Using the ε-δ Definition of Limit

Use the ε-δ definition of limit to prove that

$$\lim_{x \to 2} x^2 = 4.$$

Solution You must show that for each $\varepsilon > 0$, there exists a $\delta > 0$ such that

$$|x^2 - 4| < \varepsilon \quad \text{whenever} \quad 0 < |x - 2| < \delta.$$

To find an appropriate δ, begin by writing $|x^2 - 4| = |x - 2||x + 2|$. For all x in the interval $(1, 3)$, $x + 2 < 5$ and thus $|x + 2| < 5$. So, letting δ be the minimum of $\varepsilon/5$ and 1, it follows that, whenever $0 < |x - 2| < \delta$, you have

$$|x^2 - 4| = |x - 2||x + 2| < \left(\frac{\varepsilon}{5}\right)(5) = \varepsilon$$

as shown in Figure 2.15.

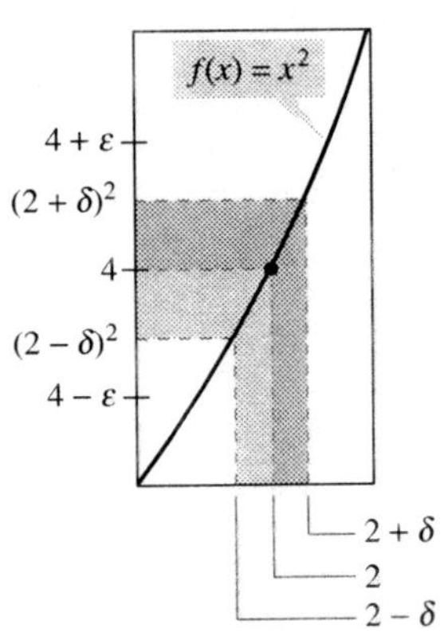

The limit of $f(x)$ as x approaches 2 is 4.
Figure 2.15

Throughout this chapter you will use the ε-δ definition of limit primarily to prove theorems about limits and to establish the existence or nonexistence of particular types of limits. For *finding* limits, you will learn techniques that are easier to use than the ε-δ definition of limit.

2.2 Exercises

In Exercises 1–10, complete the table and use the result to estimate the limit. Use a graphing utility to graph the function to confirm your result.

1. $\lim\limits_{x \to 4} \dfrac{x - 4}{x^2 - 3x - 4}$

x	3.9	3.99	3.999	4.001	4.01	4.1
$f(x)$						

2. $\lim\limits_{x \to 2} \dfrac{x - 2}{x^2 - 4}$

x	1.9	1.99	1.999	2.001	2.01	2.1
$f(x)$						

3. $\lim\limits_{x \to 3} \dfrac{[1/(x + 1)] - (1/4)}{x - 3}$

x	2.9	2.99	2.999	3.001	3.01	3.1
$f(x)$						

4. $\lim\limits_{x \to -5} \dfrac{\sqrt{4 - x} - 3}{x + 5}$

x	-5.1	-5.01	-5.001	-4.999	-4.99	-4.9
$f(x)$						

5. $\lim\limits_{x \to 0} \dfrac{\sin x}{x}$

x	-0.1	-0.01	-0.001	0.001	0.01	0.1
$f(x)$						

6. $\lim\limits_{x \to 0} \dfrac{\cos x - 1}{x}$

x	-0.1	-0.01	-0.001	0.001	0.01	0.1
$f(x)$						

7. $\lim\limits_{x \to 0} \dfrac{e^x - 1}{x}$

x	-0.1	-0.01	-0.001	0.001	0.01	0.1
$f(x)$						

8. $\lim\limits_{x \to 0} \dfrac{4}{1 + e^{1/x}}$

x	-0.1	-0.01	-0.001	0.001	0.01	0.1
$f(x)$						

9. $\displaystyle \lim_{x \to 0} \frac{\ln(x + 1)}{x}$

x	-0.1	-0.01	-0.001	0.001	0.01	0.1
$f(x)$						

10. $\displaystyle \lim_{x \to 2} \frac{\ln x - \ln 2}{x - 2}$

x	1.9	1.99	1.999	2.001	2.01	2.1
$f(x)$						

In Exercises 11–16, create a table of values for the function and use the result to estimate the limit. Use a graphing utility to graph the function to confirm your result.

11. $\displaystyle \lim_{x \to 1} \frac{x - 2}{x^2 + x - 6}$

12. $\displaystyle \lim_{x \to -3} \frac{x + 3}{x^2 + 7x + 12}$

13. $\displaystyle \lim_{x \to 1} \frac{x^4 - 1}{x^6 - 1}$

14. $\displaystyle \lim_{x \to -2} \frac{x^3 + 8}{x + 2}$

15. $\displaystyle \lim_{x \to 0} \frac{\sin 2x}{x}$

16. $\displaystyle \lim_{x \to 0} \frac{\tan x}{\tan 2x}$

In Exercises 17–26, use the graph to find the limit (if it exists). If the limit does not exist, explain why.

17. $\displaystyle \lim_{x \to 3} (4 - x)$

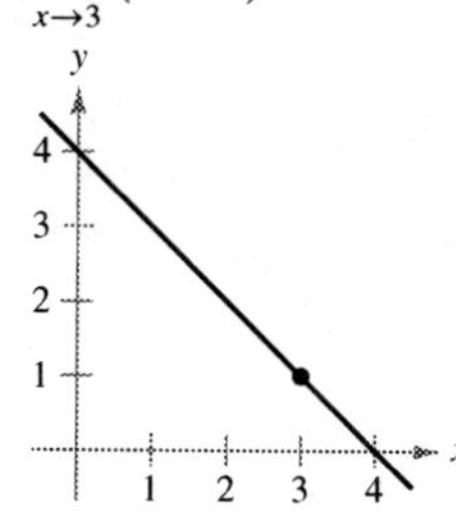

18. $\displaystyle \lim_{x \to 1} (x^2 + 3)$

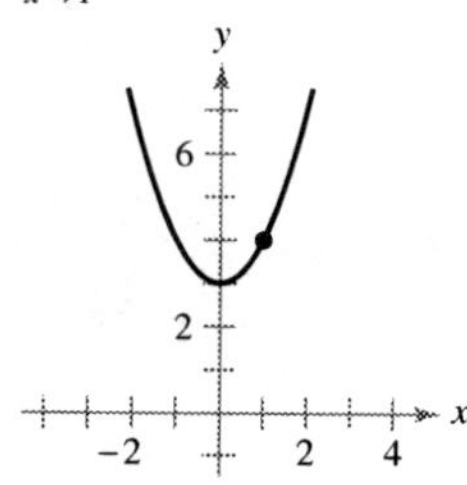

19. $\displaystyle \lim_{x \to 2} \frac{|x - 2|}{x - 2}$

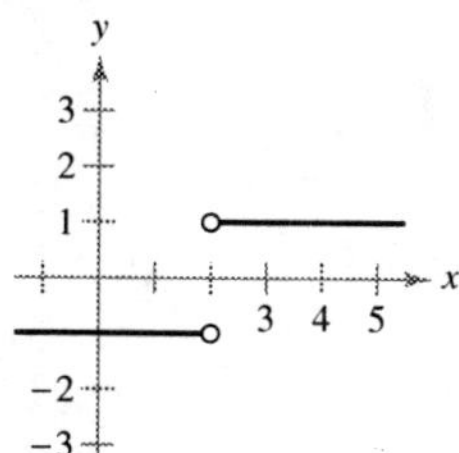

20. $\displaystyle \lim_{x \to 1} f(x)$

$$f(x) = \begin{cases} x^2 + 3, & x \neq 1 \\ 2, & x = 1 \end{cases}$$

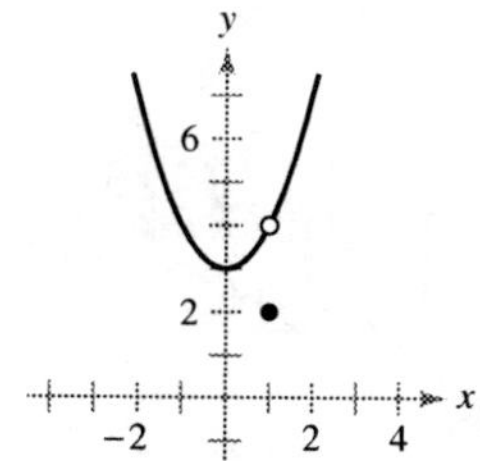

21. $\displaystyle \lim_{x \to 1} \sin \pi x$

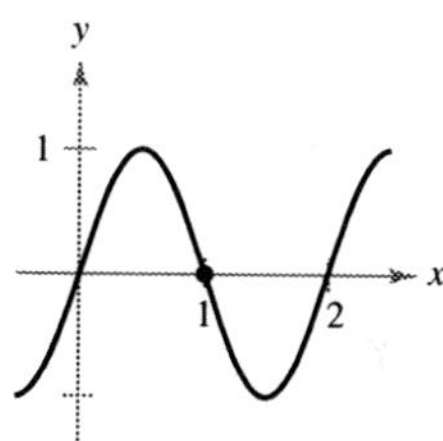

22. $\displaystyle \lim_{x \to 5} \frac{2}{x - 5}$

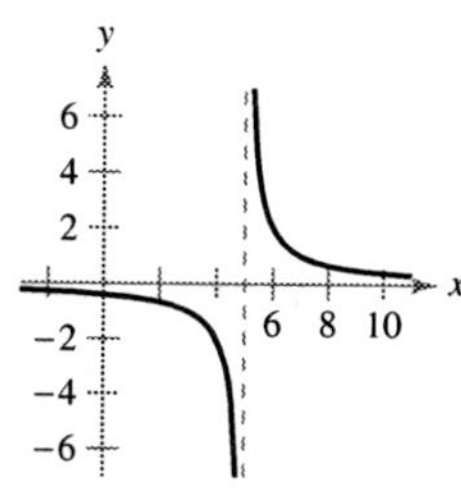

23. $\displaystyle \lim_{x \to 1} \sqrt[3]{x} \ln|x - 2|$

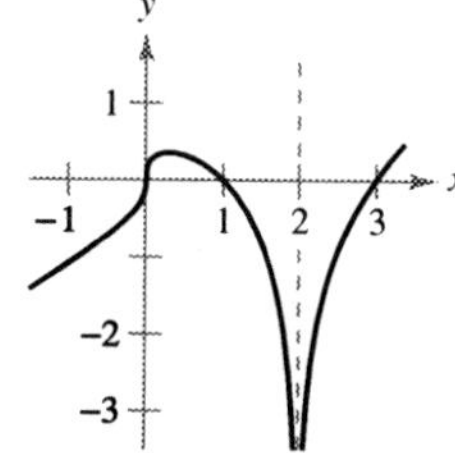

24. $\displaystyle \lim_{x \to 0} \frac{4}{2 + e^{1/x}}$

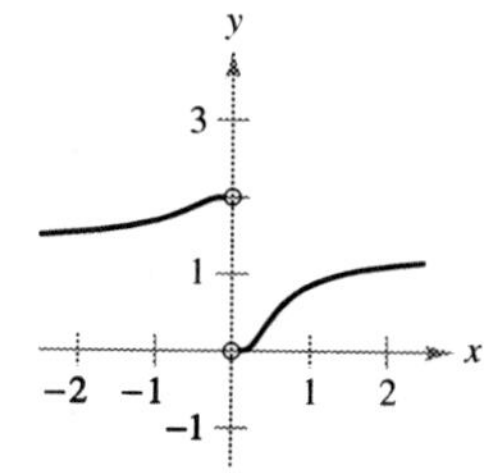

25. $\displaystyle \lim_{x \to 0} \cos \frac{1}{x}$

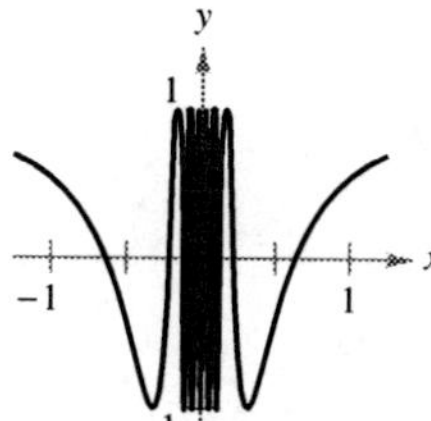

26. $\displaystyle \lim_{x \to \pi/2} \tan x$

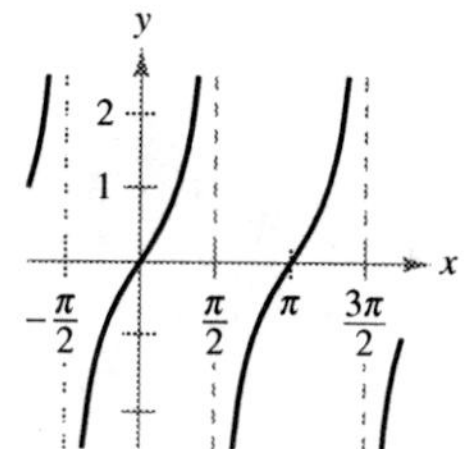

In Exercises 27 and 28, use the graph of the function f to decide whether the value of the given quantity exists. If it does, find it. If not, explain why.

27. (a) $f(1)$

(b) $\displaystyle \lim_{x \to 1} f(x)$

(c) $f(4)$

(d) $\displaystyle \lim_{x \to 4} f(x)$

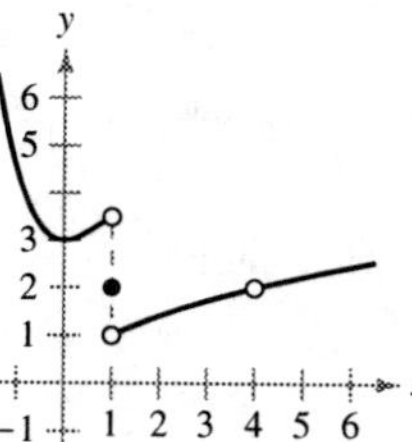

28. (a) $f(-2)$

(b) $\displaystyle \lim_{x \to -2} f(x)$

(c) $f(0)$

(d) $\displaystyle \lim_{x \to 0} f(x)$

(e) $f(2)$

(f) $\displaystyle \lim_{x \to 2} f(x)$

(g) $f(4)$

(h) $\displaystyle \lim_{x \to 4} f(x)$

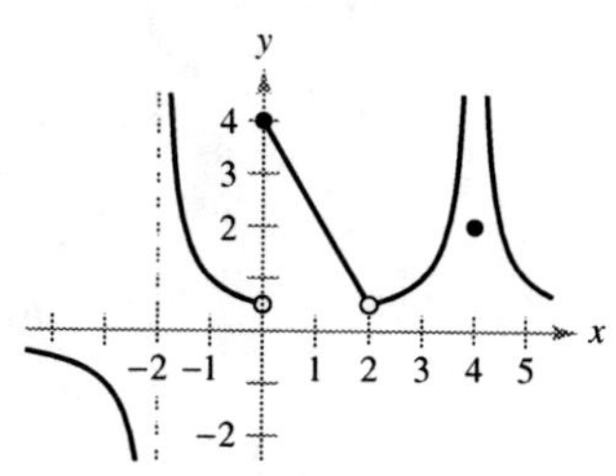

In Exercises 29 and 30, use the graph of f to identify the values of c for which $\lim\limits_{x \to c} f(x)$ exists.

29. 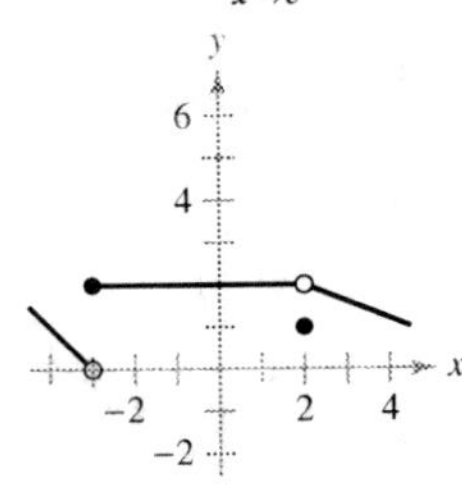

30. 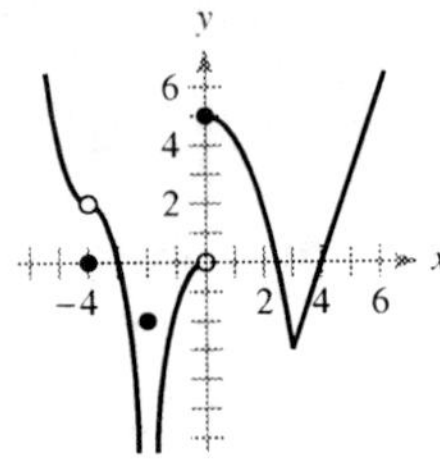

In Exercises 31 and 32, sketch the graph of f. Then identify the values of c for which $\lim\limits_{x \to c} f(x)$ exists.

31. $f(x) = \begin{cases} x^2, & x \le 2 \\ 8 - 2x, & 2 < x < 4 \\ 4, & x \ge 4 \end{cases}$

32. $f(x) = \begin{cases} \sin x, & x < 0 \\ 1 - \cos x, & 0 \le x \le \pi \\ \cos x, & x > \pi \end{cases}$

In Exercises 33 and 34, sketch a graph of a function f that satisfies the given values. (There are many correct answers.)

33. $f(0)$ is undefined.

$\lim\limits_{x \to 0} f(x) = 4$

$f(2) = 6$

$\lim\limits_{x \to 2} f(x) = 3$

34. $f(-2) = 0$

$f(2) = 0$

$\lim\limits_{x \to -2} f(x) = 0$

$\lim\limits_{x \to 2} f(x)$ does not exist.

35. **Modeling Data**   For a long distance phone call, a hotel charges $9.99 for the first minute and $0.79 for each additional minute or fraction thereof. A formula for the cost is given by

$$C(t) = 9.99 - 0.79[\![-(t - 1)]\!]$$

where t is the time in minutes.

(*Note:* $[\![x]\!] = $ greatest integer n such that $n \le x$. For example, $[\![3.2]\!] = 3$ and $[\![-1.6]\!] = -2$.)

(a) Use a graphing utility to graph the cost function for $0 < t \le 6$.

(b) Use the graph to complete the table and observe the behavior of the function as t approaches 3.5. Use the graph and the table to find

$$\lim\limits_{t \to 3.5} C(t).$$

t	3	3.3	3.4	3.5	3.6	3.7	4
C				?			

(c) Use the graph to complete the table and observe the behavior of the function as t approaches 3.

t	2	2.5	2.9	3	3.1	3.5	4
C				?			

Does the limit of $C(t)$ as t approaches 3 exist? Explain.

36. Repeat Exercise 35 for

$$C(t) = 5.79 - 0.99[\![-(t - 1)]\!].$$

37. The graph of $f(x) = x + 1$ is shown in the figure. Find δ such that if $0 < |x - 2| < \delta$, then $|f(x) - 3| < 0.4$.

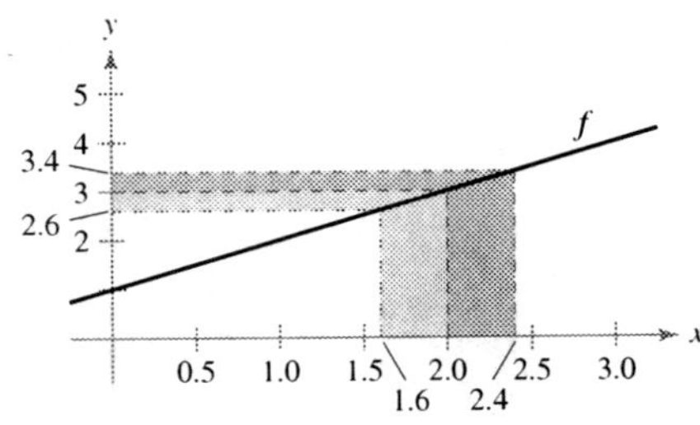

38. The graph of

$$f(x) = \frac{1}{x - 1}$$

is shown in the figure. Find δ such that if $0 < |x - 2| < \delta$, then $|f(x) - 1| < 0.01$.

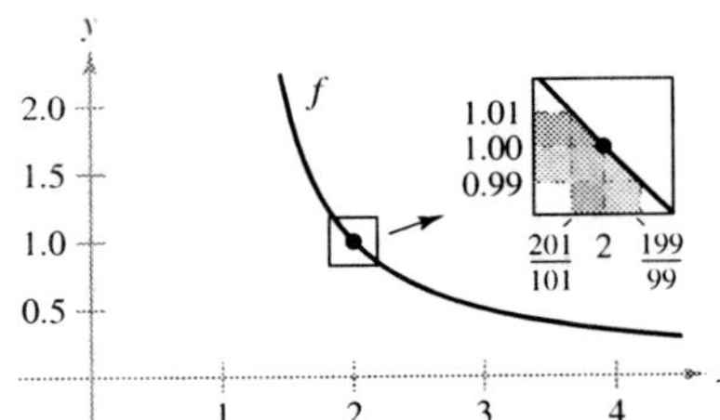

39. The graph of

$$f(x) = 2 - \frac{1}{x}$$

is shown in the figure. Find δ such that if $0 < |x - 1| < \delta$, then $|f(x) - 1| < 0.1$.

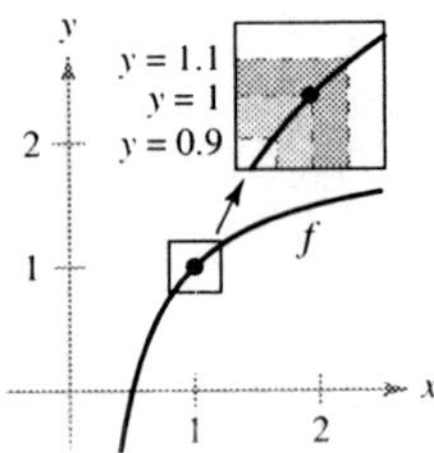

---

40. The graph of $f(x) = x^2 - 1$ is shown in the figure. Find δ such that if $0 < |x - 2| < \delta$, then $|f(x) - 3| < 0.2$.

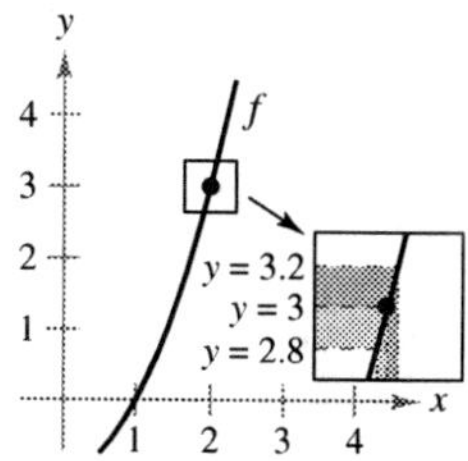

In Exercises 41–44, find the limit L. Then find $\delta > 0$ such that $|f(x) - L| < 0.01$ whenever $0 < |x - c| < \delta$.

41. $\lim\limits_{x \to 2} (3x + 2)$

42. $\lim\limits_{x \to 4} \left(4 - \dfrac{x}{2}\right)$

43. $\lim\limits_{x \to 2} (x^2 - 3)$

44. $\lim\limits_{x \to 5} (x^2 + 4)$

In Exercises 45–56, find the limit L. Then use the ε-δ definition to prove that the limit is L.

45. $\lim\limits_{x \to 2} (x + 3)$

46. $\lim\limits_{x \to -3} (2x + 5)$

47. $\lim\limits_{x \to -4} \left(\tfrac{1}{2}x - 1\right)$

48. $\lim\limits_{x \to 1} \left(\tfrac{2}{3}x + 9\right)$

49. $\lim\limits_{x \to 6} 3$

50. $\lim\limits_{x \to 2} (-1)$

51. $\lim\limits_{x \to 0} \sqrt[3]{x}$

52. $\lim\limits_{x \to 4} \sqrt{x}$

53. $\lim\limits_{x \to -2} |x - 2|$

54. $\lim\limits_{x \to 3} |x - 3|$

55. $\lim\limits_{x \to 1} (x^2 + 1)$

56. $\lim\limits_{x \to -3} (x^2 + 3x)$

57. What is the limit of $f(x) = 4$ as x approaches π?

58. What is the limit of $g(x) = x$ as x approaches π?

Writing **In Exercises 59–62, use a graphing utility to graph the function and estimate the limit (if it exists). What is the domain of the function? Can you detect a possible error in determining the domain of a function solely by analyzing the graph generated by a graphing utility? Write a short paragraph about the importance of examining a function analytically as well as graphically.**

59. $f(x) = \dfrac{\sqrt{x + 5} - 3}{x - 4}$

$\lim\limits_{x \to 4} f(x)$

60. $f(x) = \dfrac{x - 3}{x^2 - 4x + 3}$

$\lim\limits_{x \to 3} f(x)$

61. $f(x) = \dfrac{x - 9}{\sqrt{x} - 3}$

$\lim\limits_{x \to 9} f(x)$

62. $f(x) = \dfrac{e^{x/2} - 1}{x}$

$\lim\limits_{x \to 0} f(x)$

63. Write a brief description of the meaning of the notation

$$\lim\limits_{x \to 8} f(x) = 25.$$

64. The definition of limit on page 72 requires that f is a function defined on an open interval containing c, except possibly at c. Why is this requirement necessary?

65. Identify three types of behavior associated with the nonexistence of a limit. Illustrate each type with a graph of a function.

66. (a) If $f(2) = 4$, can you conclude anything about the limit of $f(x)$ as x approaches 2? Explain your reasoning.

(b) If the limit of $f(x)$ as x approaches 2 is 4, can you conclude anything about $f(2)$? Explain your reasoning.

67. *Jewelry* A jeweler resizes a ring so that its inner circumference is 6 centimeters.

(a) What is the radius of the ring?

(b) If the ring's inner circumference can vary between 5.5 centimeters and 6.5 centimeters, how can the radius vary?

(c) Use the ε-δ definition of limit to describe this situation. Identify ε and δ.

68. *Sports* A sporting goods manufacturer designs a golf ball with a volume of 2.48 cubic inches.

(a) What is the radius of the golf ball?

(b) If the ball's volume can vary between 2.45 cubic inches and 2.51 cubic inches, how can the radius vary?

(c) Use the ε-δ definition of limit to describe this situation. Identify ε and δ.

69. Consider the function $f(x) = (1 + x)^{1/x}$. Estimate the limit

$$\lim\limits_{x \to 0} (1 + x)^{1/x}$$

by evaluating f at x-values near 0. Sketch the graph of f.

70. Consider the function

$$f(x) = \dfrac{|x + 1| - |x - 1|}{x}.$$

Estimate

$$\lim\limits_{x \to 0} \dfrac{|x + 1| - |x - 1|}{x}$$

by evaluating f at x-values near 0. Sketch the graph of f.

71. *Graphical Analysis* The statement

$$\lim_{x \to 2} \frac{x^2 - 4}{x - 2} = 4$$

means that for each $\varepsilon > 0$ there corresponds a $\delta > 0$ such that if $0 < |x - 2| < \delta$, then

$$\left| \frac{x^2 - 4}{x - 2} - 4 \right| < \varepsilon.$$

If $\varepsilon = 0.001$, then

$$\left| \frac{x^2 - 4}{x - 2} - 4 \right| < 0.001.$$

Use a graphing utility to graph each side of this inequality. Use the *zoom* feature to find an interval $(2 - \delta, 2 + \delta)$ such that the graph of the left side is below the graph of the right side of the inequality.

72. *Graphical Analysis* The statement

$$\lim_{x \to 3} \frac{x^2 - 3x}{x - 3} = 3$$

means that for each $\varepsilon > 0$ there corresponds a $\delta > 0$ such that if $0 < |x - 3| < \delta$, then

$$\left| \frac{x^2 - 3x}{x - 3} - 3 \right| < \varepsilon.$$

If $\varepsilon = 0.001$, then

$$\left| \frac{x^2 - 3x}{x - 3} - 3 \right| < 0.001.$$

Use a graphing utility to graph each side of this inequality. Use the *zoom* feature to find an interval $(3 - \delta, 3 + \delta)$ such that the graph of the left side is below the graph of the right side of the inequality.

True or False? **In Exercises 73–76, determine whether the statement is true or false. If it is false, explain why or give an example that shows it is false.**

73. If f is undefined at $x = c$, then the limit of $f(x)$ as x approaches c does not exist.

74. If the limit of $f(x)$ as x approaches c is 0, then there must exist a number k such that $f(k) < 0.001$.

75. If $f(c) = L$, then $\lim_{x \to c} f(x) = L$.

76. If $\lim_{x \to c} f(x) = L$, then $f(c) = L$.

In Exercises 77 and 78, consider the function $f(x) = \sqrt{x}$.

77. Is $\lim_{x \to 0.25} \sqrt{x} = 0.5$ a true statement? Explain.

78. Is $\lim_{x \to 0} \sqrt{x} = 0$ a true statement? Explain.

79. Use a graphing utility to evaluate the limit $\lim_{x \to 0} \dfrac{\sin nx}{x}$ for several values of n. What do you notice?

80. Use a graphing utility to evaluate the limit $\lim_{x \to 0} \dfrac{\tan nx}{x}$ for several values of n. What do you notice?

81. Prove that if the limit of $f(x)$ as $x \to c$ exists, then the limit must be unique. [*Hint:* Let

$$\lim_{x \to c} f(x) = L_1 \quad \text{and} \quad \lim_{x \to c} f(x) = L_2$$

and prove that $L_1 = L_2$.]

82. Consider the line $f(x) = mx + b$, where $m \neq 0$. Use the ε-δ definition of limit to prove that $\lim_{x \to c} f(x) = mc + b$.

83. Prove that $\lim_{x \to c} f(x) = L$ is equivalent to $\lim_{x \to c} [\,f(x) - L\,] = 0$.

84. (a) Given that

$$\lim_{x \to 0} (3x + 1)(3x - 1)x^2 + 0.01 = 0.01$$

prove that there exists an open interval (a, b) containing 0 such that $(3x + 1)(3x - 1)x^2 + 0.01 > 0$ for all $x \neq 0$ in (a, b).

(b) Given that $\lim_{x \to c} g(x) = L$, where $L > 0$, prove that there exists an open interval (a, b) containing c such that $g(x) > 0$ for all $x \neq c$ in (a, b).

85. *Programming* Use the programming capabilities of a graphing utility to write a program for approximating $\lim_{x \to c} f(x)$.

Assume the program will be applied only to functions whose limits exist as x approaches c. Let $y_1 = f(x)$ and generate two lists whose entries form the ordered pairs

$$(c \pm [0.1]^n, f(c \pm [0.1]^n))$$

for $n = 0, 1, 2, 3,$ and 4.

86. *Programming* Use the program you created in Exercise 85 to approximate the limit

$$\lim_{x \to 4} \frac{x^2 - x - 12}{x - 4}.$$

<hr>

PUTNAM EXAM CHALLENGE

87. Inscribe a rectangle of base b and height h and an isosceles triangle of base b in a circle of radius one as shown. For what value of h do the rectangle and triangle have the same area?

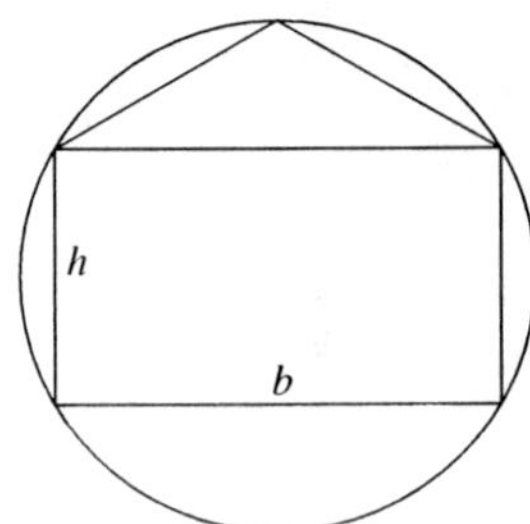

88. A right circular cone has base of radius 1 and height 3. A cube is inscribed in the cone so that one face of the cube is contained in the base of the cone. What is the side-length of the cube?

2.3 Evaluating Limits Analytically

- Evaluate a limit using properties of limits.
- Develop and use a strategy for finding limits.
- Evaluate a limit using dividing out and rationalizing techniques.
- Evaluate a limit using the Squeeze Theorem.

Properties of Limits

In Section 2.2, you learned that the limit of $f(x)$ as x approaches c does not depend on the value of f at $x = c$. It may happen, however, that the limit is precisely $f(c)$. In such cases, the limit can be evaluated by **direct substitution.** That is,

$$\lim_{x \to c} f(x) = f(c). \qquad \text{Substitute } c \text{ for } x.$$

Such *well-behaved* functions are **continuous at c.** You will examine this concept more closely in Section 2.4.

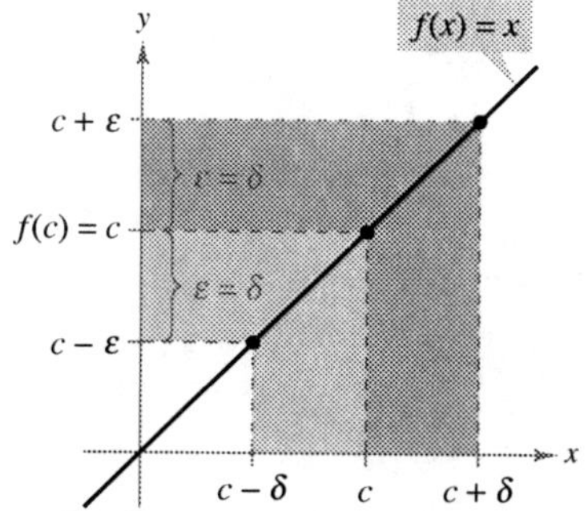

Figure 2.16

NOTE When you encounter new notations or symbols in mathematics, be sure you know how the notations are read. For instance, the limit in Example 1(c) is read as "the limit of x^2 as x approaches 2 is 4."

THEOREM 2.1 SOME BASIC LIMITS

Let b and c be real numbers and let n be a positive integer.

1. $\displaystyle\lim_{x \to c} b = b$ **2.** $\displaystyle\lim_{x \to c} x = c$ **3.** $\displaystyle\lim_{x \to c} x^n = c^n$

PROOF To prove Property 2 of Theorem 2.1, you need to show that for each $\varepsilon > 0$ there exists a $\delta > 0$ such that $|x - c| < \varepsilon$ whenever $0 < |x - c| < \delta$. Because the second inequality is a stricter version of the first, you can simply choose $\delta = \varepsilon$, as shown in Figure 2.16. This completes the proof. (Proofs of the other properties of limits in this section are listed in Appendix A or are discussed in the exercises.) ∎

EXAMPLE 1 Evaluating Basic Limits

a. $\displaystyle\lim_{x \to 2} 3 = 3$ **b.** $\displaystyle\lim_{x \to -4} x = -4$ **c.** $\displaystyle\lim_{x \to 2} x^2 = 2^2 = 4$

THEOREM 2.2 PROPERTIES OF LIMITS

Let b and c be real numbers, let n be a positive integer, and let f and g be functions with the following limits.

$$\lim_{x \to c} f(x) = L \qquad \text{and} \qquad \lim_{x \to c} g(x) = K$$

1. Scalar multiple: $\displaystyle\lim_{x \to c} [b f(x)] = bL$

2. Sum or difference: $\displaystyle\lim_{x \to c} [f(x) \pm g(x)] = L \pm K$

3. Product: $\displaystyle\lim_{x \to c} [f(x)g(x)] = LK$

4. Quotient: $\displaystyle\lim_{x \to c} \frac{f(x)}{g(x)} = \frac{L}{K}, \quad$ provided $K \neq 0$

5. Power: $\displaystyle\lim_{x \to c} [f(x)]^n = L^n$

EXAMPLE 2 The Limit of a Polynomial

$$\lim_{x \to 2} (4x^2 + 3) = \lim_{x \to 2} 4x^2 + \lim_{x \to 2} 3 \qquad \text{Property 2}$$

$$= 4\left(\lim_{x \to 2} x^2\right) + \lim_{x \to 2} 3 \qquad \text{Property 1}$$

$$= 4(2^2) + 3 \qquad \text{See Example 1(c), page 79.}$$

$$= 19 \qquad \text{Simplify.} \qquad \blacksquare$$

In Example 2, note that the limit (as $x \to 2$) of the *polynomial function* $p(x) = 4x^2 + 3$ is simply the value of p at $x = 2$.

$$\lim_{x \to 2} p(x) = p(2) = 4(2^2) + 3 = 19$$

This *direct substitution* property is valid for all polynomial and rational functions with nonzero denominators.

THEOREM 2.3 LIMITS OF POLYNOMIAL AND RATIONAL FUNCTIONS

If p is a polynomial function and c is a real number, then

$$\lim_{x \to c} p(x) = p(c).$$

If r is a rational function given by $r(x) = p(x)/q(x)$ and c is a real number such that $q(c) \neq 0$, then

$$\lim_{x \to c} r(x) = r(c) = \frac{p(c)}{q(c)}.$$

EXAMPLE 3 The Limit of a Rational Function

Find the limit: $\displaystyle \lim_{x \to 1} \frac{x^2 + x + 2}{x + 1}$.

Solution Because the denominator is not 0 when $x = 1$, you can apply Theorem 2.3 to obtain

$$\lim_{x \to 1} \frac{x^2 + x + 2}{x + 1} = \frac{1^2 + 1 + 2}{1 + 1} = \frac{4}{2} = 2. \qquad \blacksquare$$

Polynomial functions and rational functions are two of the three basic types of algebraic functions. The following theorem deals with the limit of the third type of algebraic function—one that involves a radical. See Appendix A for a proof of this theorem.

THE SQUARE ROOT SYMBOL

The first use of a symbol to denote the square root can be traced to the sixteenth century. Mathematicians first used the symbol $\sqrt{\ }$, which had only two strokes. This symbol was chosen because it resembled a lowercase *r*, to stand for the Latin word *radix*, meaning root.

THEOREM 2.4 THE LIMIT OF A FUNCTION INVOLVING A RADICAL

Let n be a positive integer. The following limit is valid for all c if n is odd, and is valid for $c > 0$ if n is even.

$$\lim_{x \to c} \sqrt[n]{x} = \sqrt[n]{c}$$

The following theorem greatly expands your ability to evaluate limits because it shows how to analyze the limit of a composite function. See Appendix A for a proof of this theorem.

THEOREM 2.5 THE LIMIT OF A COMPOSITE FUNCTION

If f and g are functions such that $\lim_{x \to c} g(x) = L$ and $\lim_{x \to L} f(x) = f(L)$, then

$$\lim_{x \to c} f(g(x)) = f\left(\lim_{x \to c} g(x)\right) = f(L).$$

EXAMPLE 4 The Limit of a Composite Function

Because

$$\lim_{x \to 0} (x^2 + 4) = 0^2 + 4 = 4 \quad \text{and} \quad \lim_{x \to 4} \sqrt{x} = \sqrt{4} = 2$$

it follows that

$$\lim_{x \to 0} \sqrt{x^2 + 4} = \sqrt{4} = 2.$$

■

You have seen that the limits of many algebraic functions can be evaluated by direct substitution. The basic transcendental functions (trigonometric, exponential, and logarithmic) also possess this desirable quality, as shown in the next theorem (presented without proof).

NOTE Your goal in this section is to become familiar with limits that can be evaluated by direct substitution. In the following library of elementary functions, what are the values of c for which

$$\lim_{x \to c} f(x) = f(c)?$$

Polynomial function:

$$f(x) = a_n x^n + \cdots + a_1 x + a_0$$

Rational function: (p and q are polynomials):

$$f(x) = \frac{p(x)}{q(x)}$$

Trigonometric functions:

$$f(x) = \sin x, \quad f(x) = \cos x$$
$$f(x) = \tan x, \quad f(x) = \cot x$$
$$f(x) = \sec x, \quad f(x) = \csc x$$

Exponential functions:

$$f(x) = a^x, \quad f(x) = e^x$$

Natural logarithmic function:

$$f(x) = \ln x$$

THEOREM 2.6 LIMITS OF TRANSCENDENTAL FUNCTIONS

Let c be a real number in the domain of the given trigonometric function.

1. $\lim_{x \to c} \sin x = \sin c$ 2. $\lim_{x \to c} \cos x = \cos c$

3. $\lim_{x \to c} \tan x = \tan c$ 4. $\lim_{x \to c} \cot x = \cot c$

5. $\lim_{x \to c} \sec x = \sec c$ 6. $\lim_{x \to c} \csc x = \csc c$

7. $\lim_{x \to c} a^x = a^c, \; a > 0$ 8. $\lim_{x \to c} \ln x = \ln c$

EXAMPLE 5 Limits of Transcendental Functions

a. $\lim_{x \to 0} \sin x = \sin(0) = 0$

b. $\lim_{x \to 2} (2 + \ln x) = 2 + \ln 2$

c. $\lim_{x \to \pi} (x \cos x) = \left(\lim_{x \to \pi} x\right)\left(\lim_{x \to \pi} \cos x\right) = \pi \cos(\pi) = -\pi$

d. $\lim_{x \to 0} \dfrac{\tan x}{x^2 + 1} = \dfrac{\lim_{x \to 0} \tan x}{\lim_{x \to 0} x^2 + 1} = \dfrac{\tan(0)}{0^2 + 1} = \dfrac{0}{1} = 0$

e. $\lim_{x \to -1} x e^x = \left(\lim_{x \to -1} x\right)\left(\lim_{x \to -1} e^x\right) = (-1)(e^{-1}) = -e^{-1}$

f. $\lim_{x \to e} \ln x^3 = \lim_{x \to e} 3 \ln x = 3 \ln(e) = 3(1) = 3$

■

A Strategy for Finding Limits

On the previous three pages, you studied several types of functions whose limits can be evaluated by direct substitution. This knowledge, together with the following theorem, can be used to develop a strategy for finding limits. A proof of this theorem is given in Appendix A.

THEOREM 2.7 FUNCTIONS THAT AGREE AT ALL BUT ONE POINT

Let c be a real number and let $f(x) = g(x)$ for all $x \neq c$ in an open interval containing c. If the limit of $g(x)$ as x approaches c exists, then the limit of $f(x)$ also exists and

$$\lim_{x \to c} f(x) = \lim_{x \to c} g(x).$$

EXAMPLE 6 Finding the Limit of a Function

Find the limit: $\displaystyle \lim_{x \to 1} \frac{x^3 - 1}{x - 1}$.

Solution Let $f(x) = (x^3 - 1)/(x - 1)$. By factoring and dividing out like factors, you can rewrite f as

$$f(x) = \frac{(x - 1)(x^2 + x + 1)}{(x - 1)} = x^2 + x + 1 = g(x), \quad x \neq 1.$$

So, for all x-values other than $x = 1$, the functions f and g agree, as shown in Figure 2.17. Because $\lim_{x \to 1} g(x)$ exists, you can apply Theorem 2.7 to conclude that f and g have the same limit at $x = 1$.

$$\lim_{x \to 1} \frac{x^3 - 1}{x - 1} = \lim_{x \to 1} \frac{(x - 1)(x^2 + x + 1)}{x - 1} \qquad \text{Factor.}$$

$$= \lim_{x \to 1} \frac{(x - 1)(x^2 + x + 1)}{x - 1} \qquad \text{Divide out like factors.}$$

$$= \lim_{x \to 1} (x^2 + x + 1) \qquad \text{Apply Theorem 2.7.}$$

$$= 1^2 + 1 + 1 \qquad \text{Use direct substitution.}$$

$$= 3 \qquad \text{Simplify.}$$

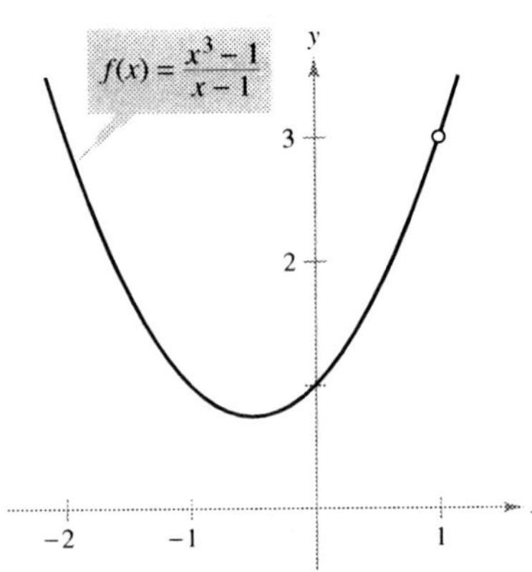

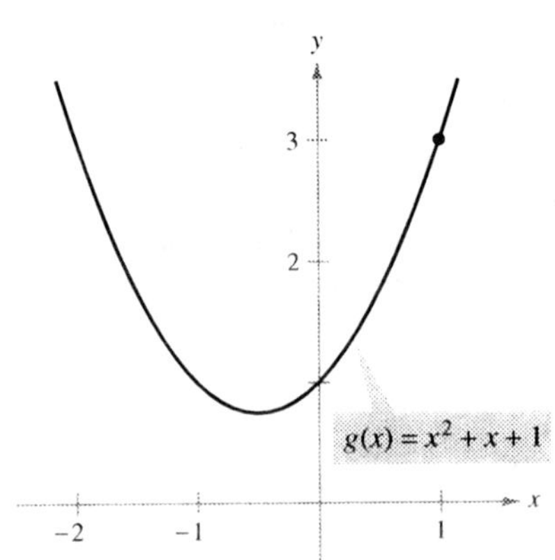

f and g agree at all but one point.
Figure 2.17

STUDY TIP When applying this strategy for finding a limit, remember that some functions do not have a limit (as x approaches c). For instance, the following limit does not exist.

$$\lim_{x \to 1} \frac{x^3 + 1}{x - 1}$$

A STRATEGY FOR FINDING LIMITS

1. Learn to recognize which limits can be evaluated by direct substitution. (These limits are listed in Theorems 2.1 through 2.6.)

2. If the limit of $f(x)$ as x approaches c *cannot* be evaluated by direct substitution, try to find a function g that agrees with f for all x other than $x = c$. [Choose g such that the limit of $g(x)$ *can* be evaluated by direct substitution.]

3. Apply Theorem 2.7 to conclude *analytically* that

$$\lim_{x \to c} f(x) = \lim_{x \to c} g(x) = g(c).$$

4. Use a *graph* or *table* to reinforce your conclusion.

Dividing Out and Rationalizing Techniques

Two techniques for finding limits analytically are shown in Examples 7 and 8. The dividing out technique involves dividing out common factors, and the rationalizing technique involves rationalizing the numerator of a fractional expression.

EXAMPLE 7 Dividing Out Technique

Find the limit: $\displaystyle\lim_{x \to -3} \frac{x^2 + x - 6}{x + 3}$.

Solution   Although you are taking the limit of a rational function, you *cannot* apply Theorem 2.3 because the limit of the denominator is 0.

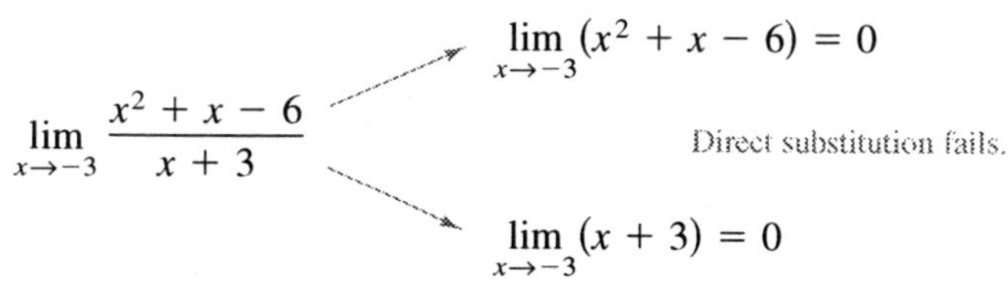

Because the limit of the numerator is also 0, the numerator and denominator have a *common factor* of $(x + 3)$. So, for all $x \neq -3$, you can divide out this factor to obtain

$$f(x) = \frac{x^2 + x - 6}{x + 3} = \frac{(x + 3)(x - 2)}{x + 3} = x - 2 = g(x), \quad x \neq -3.$$

Using Theorem 2.7, it follows that

$$\lim_{x \to -3} \frac{x^2 + x - 6}{x + 3} = \lim_{x \to -3} (x - 2) \qquad \text{Apply Theorem 2.7.}$$

$$= -5. \qquad \text{Use direct substitution.}$$

This result is shown graphically in Figure 2.18. Note that the graph of the function f coincides with the graph of the function $g(x) = x - 2$, except that the graph of f has a gap at the point $(-3, -5)$. ∎

In Example 7, direct substitution produced the meaningless fractional form $0/0$. An expression such as $0/0$ is called an **indeterminate form** because you cannot (from the form alone) determine the limit. When you try to evaluate a limit and encounter this form, remember that you must rewrite the fraction so that the new denominator does not have 0 as its limit. One way to do this is to *divide out common factors*, as shown in Example 7. A second way is to *rationalize the numerator*, as shown in Example 8.

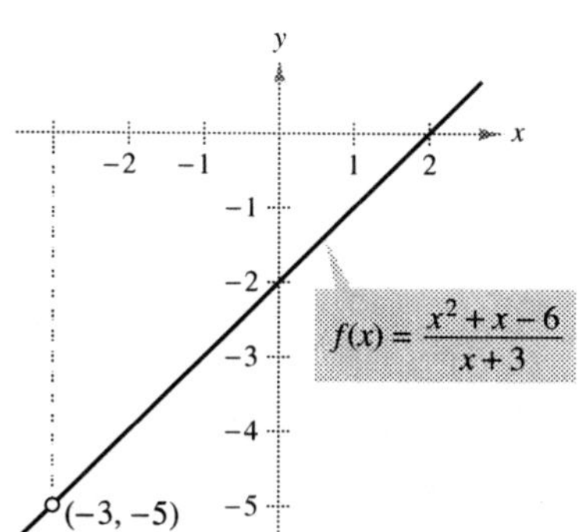

f is undefined when $x = -3$.
Figure 2.18

NOTE In the solution of Example 7, be sure you see the usefulness of the Factor Theorem of Algebra. This theorem states that if c is a zero of a polynomial function, $(x - c)$ is a factor of the polynomial. So, if you apply direct substitution to a rational function and obtain

$$r(c) = \frac{p(c)}{q(c)} = \frac{0}{0}$$

you can conclude that $(x - c)$ must be a common factor of both $p(x)$ and $q(x)$.

TECHNOLOGY PITFALL   Because the graphs of

$$f(x) = \frac{x^2 + x - 6}{x + 3} \qquad \text{and} \qquad g(x) = x - 2$$

differ only at the point $(-3, -5)$, a standard graphing utility setting may not distinguish clearly between these graphs. However, because of the pixel configuration and rounding error of a graphing utility, it may be possible to find screen settings that distinguish between the graphs. Specifically, by repeatedly zooming in near the point $(-3, -5)$ on the graph of f, your graphing utility may show glitches or irregularities that do not exist on the actual graph. (See Figure 2.19.) By changing the screen settings on your graphing utility, you may obtain the correct graph of f.

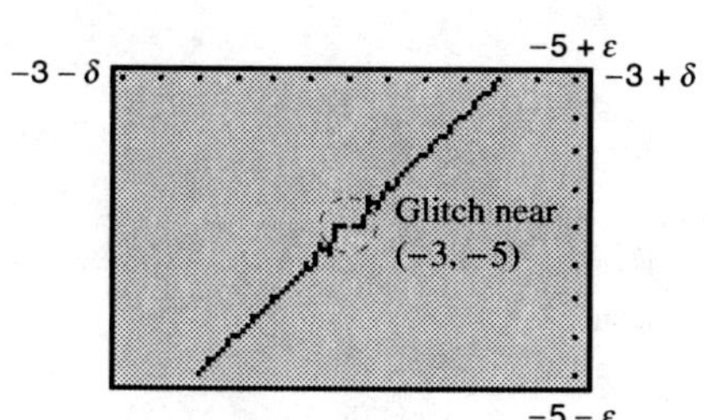

Incorrect graph of f
Figure 2.19

EXAMPLE 8 Rationalizing Technique

Find the limit: $\displaystyle\lim_{x \to 0} \frac{\sqrt{x+1}-1}{x}$.

Solution   By direct substitution, you obtain the indeterminate form $0/0$.

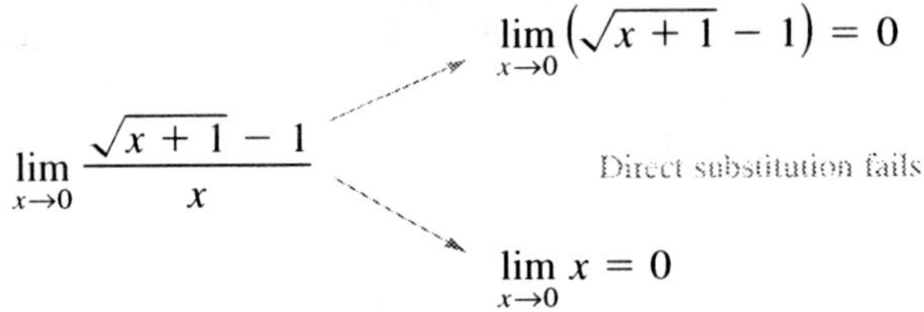

In this case, you can rewrite the fraction by rationalizing the numerator.

$$\frac{\sqrt{x+1}-1}{x} = \left(\frac{\sqrt{x+1}-1}{x}\right)\left(\frac{\sqrt{x+1}+1}{\sqrt{x+1}+1}\right)$$

$$= \frac{(x+1)-1}{x\left(\sqrt{x+1}+1\right)}$$

$$= \frac{\cancel{x}}{\cancel{x}\left(\sqrt{x+1}+1\right)}$$

$$= \frac{1}{\sqrt{x+1}+1}, \quad x \neq 0$$

Now, using Theorem 2.7, you can evaluate the limit as shown.

$$\lim_{x \to 0} \frac{\sqrt{x+1}-1}{x} = \lim_{x \to 0} \frac{1}{\sqrt{x+1}+1}$$

$$= \frac{1}{1+1}$$

$$= \frac{1}{2}$$

A table or a graph can reinforce your conclusion that the limit is $\frac{1}{2}$. (See Figure 2.20.)

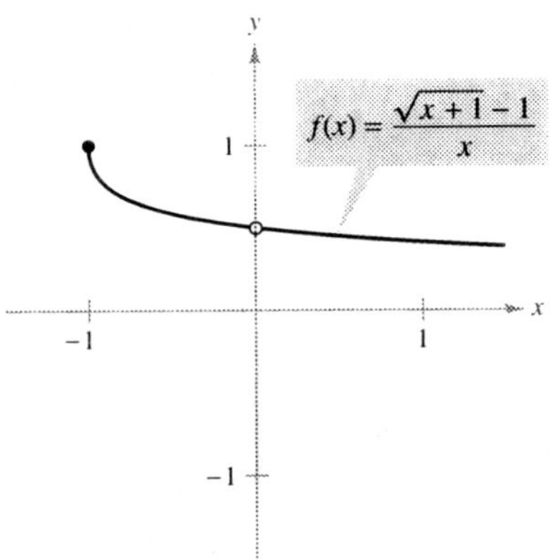

The limit of $f(x)$ as x approaches 0 is $\frac{1}{2}$.
Figure 2.20

			x approaches 0 from the left.			x approaches 0 from the right.			
x	-0.25	-0.1	-0.01	-0.001	0	0.001	0.01	0.1	0.25
$f(x)$	0.5359	0.5132	0.5013	0.5001	?	0.4999	0.4988	0.4881	0.4721

$f(x)$ approaches 0.5. $\qquad\qquad$ $f(x)$ approaches 0.5.

NOTE   The rationalizing technique for evaluating limits is based on multiplication by a convenient form of 1. In Example 8, the convenient form is

$$1 = \frac{\sqrt{x+1}+1}{\sqrt{x+1}+1}.$$

The Squeeze Theorem

The next theorem concerns the limit of a function that is squeezed between two other functions, each of which has the same limit at a given x-value, as shown in Figure 2.21. (The proof of this theorem is given in Appendix A.)

$h(x) \le f(x) \le g(x)$

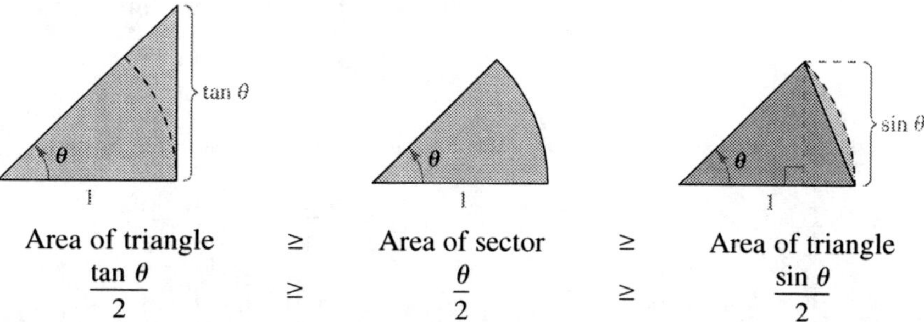

The Squeeze Theorem
Figure 2.21

THEOREM 2.8 THE SQUEEZE THEOREM

If $h(x) \le f(x) \le g(x)$ for all x in an open interval containing c, except possibly at c itself, and if

$$\lim_{x \to c} h(x) = L = \lim_{x \to c} g(x)$$

then $\lim_{x \to c} f(x)$ exists and is equal to L.

You can see the usefulness of the Squeeze Theorem (also called the Sandwich Theorem or the Pinching Theorem) in the proof of Theorem 2.9.

THEOREM 2.9 THREE SPECIAL LIMITS

1. $\displaystyle\lim_{x \to 0} \frac{\sin x}{x} = 1$ **2.** $\displaystyle\lim_{x \to 0} \frac{1 - \cos x}{x} = 0$ **3.** $\displaystyle\lim_{x \to 0} (1 + x)^{1/x} = e$

PROOF To avoid the confusion of two different uses of x, the proof of the first limit is presented using the variable θ, where θ is an acute positive angle *measured in radians*. Figure 2.22 shows a circular sector that is squeezed between two triangles.

A circular sector is used to prove Theorem 2.9.
Figure 2.22

Area of triangle	$\ge$	Area of sector	$\ge$	Area of triangle
$\dfrac{\tan \theta}{2}$	$\ge$	$\dfrac{\theta}{2}$	$\ge$	$\dfrac{\sin \theta}{2}$

Multiplying each expression by $2/\sin \theta$ produces

$$\frac{1}{\cos \theta} \ge \frac{\theta}{\sin \theta} \ge 1$$

and taking reciprocals and reversing the inequalities yields

$$\cos \theta \le \frac{\sin \theta}{\theta} \le 1.$$

Because $\cos \theta = \cos(-\theta)$ and $(\sin \theta)/\theta = [\sin(-\theta)]/(-\theta)$, you can conclude that this inequality is valid for *all* nonzero θ in the open interval $(-\pi/2, \pi/2)$. Finally, because $\displaystyle\lim_{\theta \to 0} \cos \theta = 1$ and $\displaystyle\lim_{\theta \to 0} 1 = 1$, you can apply the Squeeze Theorem to conclude that $\displaystyle\lim_{\theta \to 0} (\sin \theta)/\theta = 1$. The proof of the second limit is left as an exercise (see Exercise 129). Recall from Section 1.6 that the third limit is actually the definition of the number e.

NOTE The third limit of Theorem 2.9 will be used in Chapter 3 in the development of the formula for the derivative of the exponential function $f(x) = e^x$.

EXAMPLE 9 A Limit Involving a Trigonometric Function

Find the limit: $\displaystyle\lim_{x \to 0} \frac{\tan x}{x}$.

Solution Direct substitution yields the indeterminate form $0/0$. To solve this problem, you can write $\tan x$ as $(\sin x)/(\cos x)$ and obtain

$$\lim_{x \to 0} \frac{\tan x}{x} = \lim_{x \to 0} \left(\frac{\sin x}{x}\right)\left(\frac{1}{\cos x}\right).$$

Now, because

$$\lim_{x \to 0} \frac{\sin x}{x} = 1 \qquad \text{and} \qquad \lim_{x \to 0} \frac{1}{\cos x} = 1$$

you can obtain

$$\lim_{x \to 0} \frac{\tan x}{x} = \left(\lim_{x \to 0} \frac{\sin x}{x}\right)\left(\lim_{x \to 0} \frac{1}{\cos x}\right)$$
$$= (1)(1)$$
$$= 1.$$

(See Figure 2.23.)

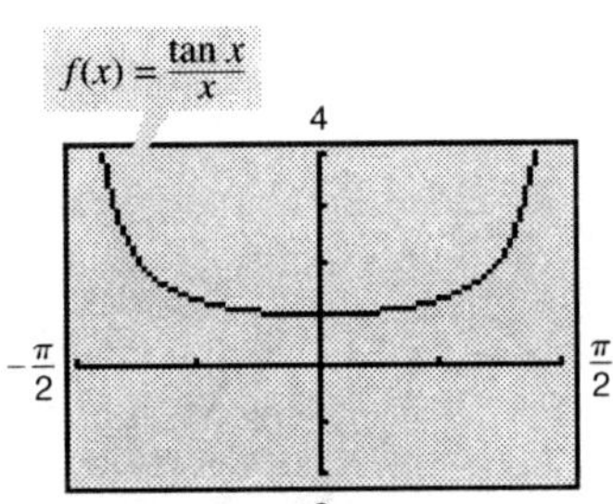

The limit of $f(x)$ as x approaches 0 is 1.
Figure 2.23

EXAMPLE 10 A Limit Involving a Trigonometric Function

Find the limit: $\displaystyle\lim_{x \to 0} \frac{\sin 4x}{x}$.

Solution Direct substitution yields the indeterminate form $0/0$. To solve this problem, you can rewrite the limit as

$$\lim_{x \to 0} \frac{\sin 4x}{x} = 4\left(\lim_{x \to 0} \frac{\sin 4x}{4x}\right). \qquad \text{Multiply and divide by 4}$$

Now, by letting $y = 4x$ and observing that $x \to 0$ if and only if $y \to 0$, you can write

$$\lim_{x \to 0} \frac{\sin 4x}{x} = 4\left(\lim_{x \to 0} \frac{\sin 4x}{4x}\right)$$
$$= 4\left(\lim_{y \to 0} \frac{\sin y}{y}\right)$$
$$= 4(1) \qquad \text{Apply Theorem 2.9(1).}$$
$$= 4.$$

(See Figure 2.24.)

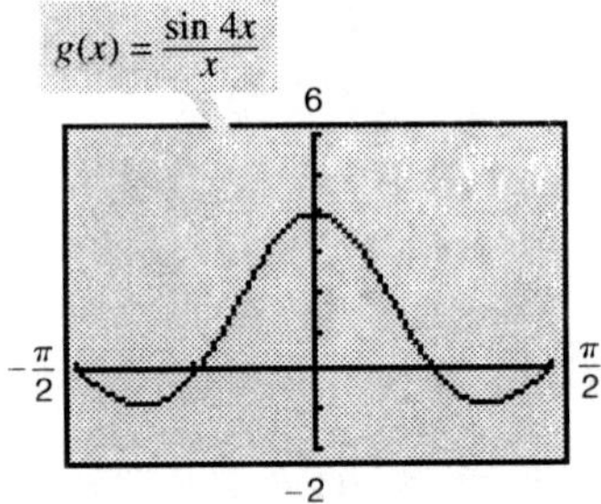

The limit of $g(x)$ as x approaches 0 is 4.
Figure 2.24

TECHNOLOGY Try using a graphing utility to confirm the limits in the examples and in the exercise set. For instance, Figures 2.23 and 2.24 show the graphs of

$$f(x) = \frac{\tan x}{x} \qquad \text{and} \qquad g(x) = \frac{\sin 4x}{x}.$$

Note that the first graph appears to contain the point $(0, 1)$ and the second graph appears to contain the point $(0, 4)$, which lends support to the conclusions obtained in Examples 9 and 10.

2.3 Exercises

See www.CalcChat.com for worked-out solutions to odd-numbered exercises.

In Exercises 1–4, use a graphing utility to graph the function and visually estimate the limits.

1. $h(x) = -x^2 + 4x$

  (a) $\lim\limits_{x \to 4} h(x)$

  (b) $\lim\limits_{x \to -1} h(x)$

2. $g(x) = \dfrac{12(\sqrt{x} - 3)}{x - 9}$

  (a) $\lim\limits_{x \to 4} g(x)$

  (b) $\lim\limits_{x \to 0} g(x)$

3. $f(x) = x \cos x$

  (a) $\lim\limits_{x \to 0} f(x)$

  (b) $\lim\limits_{x \to \pi/3} f(x)$

4. $f(t) = t|t - 4|$

  (a) $\lim\limits_{t \to 4} f(t)$

  (b) $\lim\limits_{t \to -1} f(t)$

In Exercises 5–32, find the limit.

5. $\lim\limits_{x \to 2} x^3$

6. $\lim\limits_{x \to -2} x^4$

7. $\lim\limits_{x \to 0} (2x - 1)$

8. $\lim\limits_{x \to -3} (3x + 2)$

9. $\lim\limits_{x \to -3} (2x^2 + 4x + 1)$

10. $\lim\limits_{x \to 1} (3x^3 - 4x^2 + 3)$

11. $\lim\limits_{x \to 3} \sqrt{x + 1}$

12. $\lim\limits_{x \to 4} \sqrt[3]{x + 4}$

13. $\lim\limits_{x \to 2} \dfrac{1}{x}$

14. $\lim\limits_{x \to -3} \dfrac{2}{x + 2}$

15. $\lim\limits_{x \to 1} \dfrac{x}{x^2 + 4}$

16. $\lim\limits_{x \to 1} \dfrac{2x - 3}{x + 5}$

17. $\lim\limits_{x \to 7} \dfrac{3x}{\sqrt{x + 2}}$

18. $\lim\limits_{x \to 2} \dfrac{\sqrt{x + 2}}{x - 4}$

19. $\lim\limits_{x \to \pi/2} \sin x$

20. $\lim\limits_{x \to \pi} \tan x$

21. $\lim\limits_{x \to 1} \cos\left(\dfrac{\pi x}{3}\right)$

22. $\lim\limits_{x \to 2} \sin\left(\dfrac{\pi x}{2}\right)$

23. $\lim\limits_{x \to 0} \sec 2x$

24. $\lim\limits_{x \to \pi} \cos 3x$

25. $\lim\limits_{x \to 5\pi/6} \sin x$

26. $\lim\limits_{x \to 5\pi/3} \cos x$

27. $\lim\limits_{x \to 3} \tan\left(\dfrac{\pi x}{4}\right)$

28. $\lim\limits_{x \to 7} \sec\left(\dfrac{\pi x}{6}\right)$

29. $\lim\limits_{x \to 0} e^x \cos 2x$

30. $\lim\limits_{x \to 0} e^{-x} \sin \pi x$

31. $\lim\limits_{x \to 1} (\ln 3x + e^x)$

32. $\lim\limits_{x \to 1} \ln\left(\dfrac{x}{e^x}\right)$

In Exercises 33–36, find the limits.

33. $f(x) = 5 - x$, $g(x) = x^3$

  (a) $\lim\limits_{x \to 1} f(x)$ (b) $\lim\limits_{x \to 4} g(x)$ (c) $\lim\limits_{x \to 1} g(f(x))$

34. $f(x) = x + 7$, $g(x) = x^2$

  (a) $\lim\limits_{x \to -3} f(x)$ (b) $\lim\limits_{x \to 4} g(x)$ (c) $\lim\limits_{x \to -3} g(f(x))$

35. $f(x) = 4 - x^2$, $g(x) = \sqrt{x + 1}$

  (a) $\lim\limits_{x \to 1} f(x)$ (b) $\lim\limits_{x \to 3} g(x)$ (c) $\lim\limits_{x \to 1} g(f(x))$

36. $f(x) = 2x^2 - 3x + 1$, $g(x) = \sqrt[3]{x + 6}$

  (a) $\lim\limits_{x \to 4} f(x)$ (b) $\lim\limits_{x \to 21} g(x)$ (c) $\lim\limits_{x \to 4} g(f(x))$

In Exercises 37–40, use the information to evaluate the limits.

37. $\lim\limits_{x \to c} f(x) = 3$

  $\lim\limits_{x \to c} g(x) = 2$

  (a) $\lim\limits_{x \to c} [5g(x)]$

  (b) $\lim\limits_{x \to c} [f(x) + g(x)]$

  (c) $\lim\limits_{x \to c} [f(x)g(x)]$

  (d) $\lim\limits_{x \to c} \dfrac{f(x)}{g(x)}$

38. $\lim\limits_{x \to c} f(x) = \frac{3}{2}$

  $\lim\limits_{x \to c} g(x) = \frac{1}{2}$

  (a) $\lim\limits_{x \to c} [4f(x)]$

  (b) $\lim\limits_{x \to c} [f(x) + g(x)]$

  (c) $\lim\limits_{x \to c} [f(x)g(x)]$

  (d) $\lim\limits_{x \to c} \dfrac{f(x)}{g(x)}$

39. $\lim\limits_{x \to c} f(x) = 4$

  (a) $\lim\limits_{x \to c} [f(x)]^3$

  (b) $\lim\limits_{x \to c} \sqrt{f(x)}$

  (c) $\lim\limits_{x \to c} [3f(x)]$

  (d) $\lim\limits_{x \to c} [f(x)]^{3/2}$

40. $\lim\limits_{x \to c} f(x) = 27$

  (a) $\lim\limits_{x \to c} \sqrt[3]{f(x)}$

  (b) $\lim\limits_{x \to c} \dfrac{f(x)}{18}$

  (c) $\lim\limits_{x \to c} [f(x)]^2$

  (d) $\lim\limits_{x \to c} [f(x)]^{2/3}$

In Exercises 41–44, use the graph to determine the limit visually (if it exists). Write a simpler function that agrees with the given function at all but one point.

41. $g(x) = \dfrac{x^2 - x}{x}$

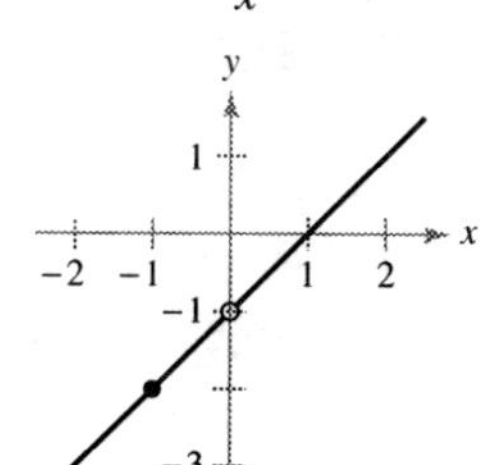

  (a) $\lim\limits_{x \to 0} g(x)$

  (b) $\lim\limits_{x \to -1} g(x)$

42. $h(x) = \dfrac{-x^2 + 3x}{x}$

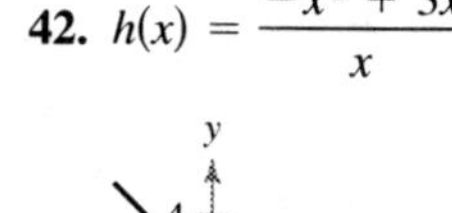

  (a) $\lim\limits_{x \to 2} h(x)$

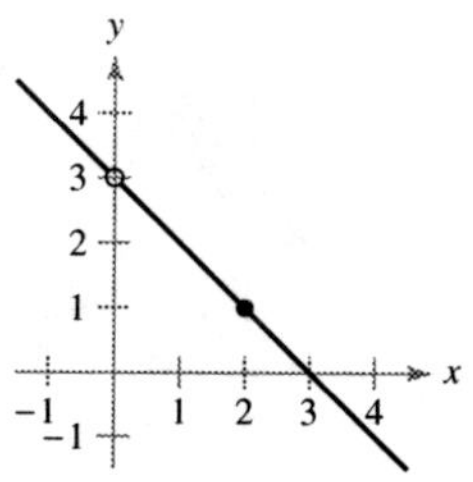

  (b) $\lim\limits_{x \to 0} h(x)$

43. $g(x) = \dfrac{x^3 - x}{x - 1}$

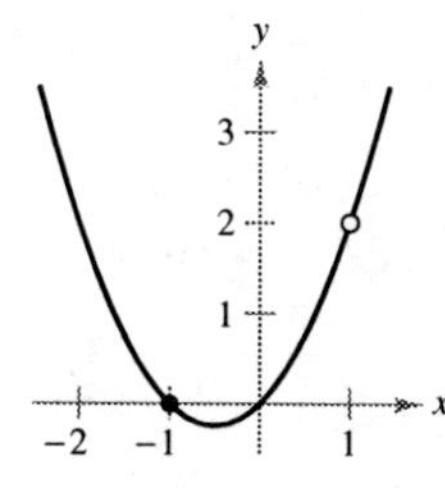

  (a) $\lim\limits_{x \to 1} g(x)$

  (b) $\lim\limits_{x \to -1} g(x)$

44. $f(x) = \dfrac{x}{x^2 - x}$

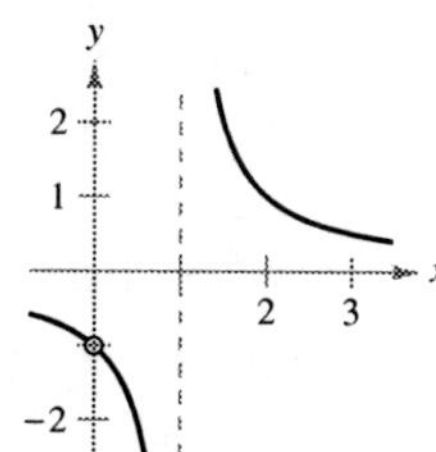

  (a) $\lim\limits_{x \to 1} f(x)$

  (b) $\lim\limits_{x \to 0} f(x)$

In Exercises 45–50, find the limit of the function (if it exists). Write a simpler function that agrees with the given function at all but one point. Use a graphing utility to confirm your result.

45. $\displaystyle\lim_{x \to -1} \frac{x^2 - 1}{x + 1}$

46. $\displaystyle\lim_{x \to -1} \frac{2x^2 - x - 3}{x + 1}$

47. $\displaystyle\lim_{x \to 2} \frac{x^3 - 8}{x - 2}$

48. $\displaystyle\lim_{x \to -1} \frac{x^3 + 1}{x + 1}$

49. $\displaystyle\lim_{x \to -4} \frac{(x + 4)\ln(x + 6)}{x^2 - 16}$

50. $\displaystyle\lim_{x \to 0} \frac{e^{2x} - 1}{e^x - 1}$

In Exercises 51–66, find the limit (if it exists).

51. $\displaystyle\lim_{x \to 0} \frac{x}{x^2 - x}$

52. $\displaystyle\lim_{x \to 0} \frac{3x}{x^2 + 2x}$

53. $\displaystyle\lim_{x \to 4} \frac{x - 4}{x^2 - 16}$

54. $\displaystyle\lim_{x \to 3} \frac{3 - x}{x^2 - 9}$

55. $\displaystyle\lim_{x \to -3} \frac{x^2 + x - 6}{x^2 - 9}$

56. $\displaystyle\lim_{x \to 3} \frac{x^2 - x - 6}{x^2 - 5x + 6}$

57. $\displaystyle\lim_{x \to 4} \frac{\sqrt{x + 5} - 3}{x - 4}$

58. $\displaystyle\lim_{x \to 3} \frac{\sqrt{x + 1} - 2}{x - 3}$

59. $\displaystyle\lim_{x \to 0} \frac{\sqrt{x + 5} - \sqrt{5}}{x}$

60. $\displaystyle\lim_{x \to 0} \frac{\sqrt{3 + x} - \sqrt{3}}{x}$

61. $\displaystyle\lim_{x \to 0} \frac{[1/(3 + x)] - (1/3)}{x}$

62. $\displaystyle\lim_{x \to 0} \frac{[1/(x + 4)] - (1/4)}{x}$

63. $\displaystyle\lim_{\Delta x \to 0} \frac{2(x + \Delta x) - 2x}{\Delta x}$

64. $\displaystyle\lim_{\Delta x \to 0} \frac{(x + \Delta x)^2 - x^2}{\Delta x}$

65. $\displaystyle\lim_{\Delta x \to 0} \frac{(x + \Delta x)^2 - 2(x + \Delta x) + 1 - (x^2 - 2x + 1)}{\Delta x}$

66. $\displaystyle\lim_{\Delta x \to 0} \frac{(x + \Delta x)^3 - x^3}{\Delta x}$

In Exercises 67–80, determine the limit of the transcendental function (if it exists).

67. $\displaystyle\lim_{x \to 0} \frac{\sin x}{5x}$

68. $\displaystyle\lim_{x \to 0} \frac{5(1 - \cos x)}{x}$

69. $\displaystyle\lim_{x \to 0} \frac{\sin x(1 - \cos x)}{x^2}$

70. $\displaystyle\lim_{\theta \to 0} \frac{\cos \theta \tan \theta}{\theta}$

71. $\displaystyle\lim_{x \to 0} \frac{\sin^2 x}{x}$

72. $\displaystyle\lim_{x \to 0} \frac{2 \tan^2 x}{x}$

73. $\displaystyle\lim_{h \to 0} \frac{(1 - \cos h)^2}{h}$

74. $\displaystyle\lim_{\phi \to \pi} \phi \sec \phi$

75. $\displaystyle\lim_{x \to \pi/2} \frac{\cos x}{\cot x}$

76. $\displaystyle\lim_{x \to \pi/4} \frac{1 - \tan x}{\sin x - \cos x}$

77. $\displaystyle\lim_{x \to 0} \frac{1 - e^{-x}}{e^x - 1}$

78. $\displaystyle\lim_{x \to 0} \frac{4(e^{2x} - 1)}{e^x - 1}$

79. $\displaystyle\lim_{t \to 0} \frac{\sin 3t}{2t}$

80. $\displaystyle\lim_{x \to 0} \frac{\sin 2x}{\sin 3x}$ $\left[\textit{Hint:} \text{ Find } \displaystyle\lim_{x \to 0} \left(\frac{2 \sin 2x}{2x}\right)\left(\frac{3x}{3 \sin 3x}\right).\right]$

Graphical, Numerical, and Analytic Analysis **In Exercises 81–90, use a graphing utility to graph the function and estimate the limit. Use a table to reinforce your conclusion. Then find the limit by analytic methods.**

81. $\displaystyle\lim_{x \to 0} \frac{\sqrt{x + 2} - \sqrt{2}}{x}$

82. $\displaystyle\lim_{x \to 16} \frac{4 - \sqrt{x}}{x - 16}$

83. $\displaystyle\lim_{x \to 0} \frac{[1/(2 + x)] - (1/2)}{x}$

84. $\displaystyle\lim_{x \to 2} \frac{x^5 - 32}{x - 2}$

85. $\displaystyle\lim_{t \to 0} \frac{\sin 3t}{t}$

86. $\displaystyle\lim_{x \to 0} \frac{\cos x - 1}{2x^2}$

87. $\displaystyle\lim_{x \to 0} \frac{\sin x^2}{x}$

88. $\displaystyle\lim_{x \to 0} \frac{\sin x}{\sqrt[3]{x}}$

89. $\displaystyle\lim_{x \to 1} \frac{\ln x}{x - 1}$

90. $\displaystyle\lim_{x \to \ln 2} \frac{e^{3x} - 8}{e^{2x} - 4}$

In Exercises 91–94, find $\displaystyle\lim_{\Delta x \to 0} \frac{f(x + \Delta x) - f(x)}{\Delta x}$**.**

91. $f(x) = 3x - 2$

92. $f(x) = \sqrt{x}$

93. $f(x) = \dfrac{1}{x + 3}$

94. $f(x) = x^2 - 4x$

In Exercises 95 and 96, use the Squeeze Theorem to find $\displaystyle\lim_{x \to c} f(x)$**.**

95. $c = 0$

$$4 - x^2 \le f(x) \le 4 + x^2$$

96. $c = a$

$$b - |x - a| \le f(x) \le b + |x - a|$$

In Exercises 97–102, use a graphing utility to graph the given function and the equations $y = |x|$ **and** $y = -|x|$ **in the same viewing window. Using the graphs to observe the Squeeze Theorem visually, find** $\displaystyle\lim_{x \to 0} f(x)$**.**

97. $f(x) = x \cos x$

98. $f(x) = |x \sin x|$

99. $f(x) = |x| \sin x$

100. $f(x) = |x| \cos x$

101. $f(x) = x \sin \dfrac{1}{x}$

102. $h(x) = x \cos \dfrac{1}{x}$

WRITING ABOUT CONCEPTS

103. In the context of finding limits, discuss what is meant by two functions that agree at all but one point.

104. Give an example of two functions that agree at all but one point.

105. What is meant by an indeterminate form?

106. In your own words, explain the Squeeze Theorem.

107. *Writing* Use a graphing utility to graph

$$f(x) = x, \quad g(x) = \sin x, \quad \text{and} \quad h(x) = \frac{\sin x}{x}$$

in the same viewing window. Compare the magnitudes of $f(x)$ and $g(x)$ when x is "close to" 0. Use the comparison to write a short paragraph explaining why $\lim\limits_{x \to 0} h(x) = 1$.

108. *Writing* Use a graphing utility to graph

$$f(x) = x, \quad g(x) = \sin^2 x, \quad \text{and} \quad h(x) = \frac{\sin^2 x}{x}$$

in the same viewing window. Compare the magnitudes of $f(x)$ and $g(x)$ when x is "close to" 0. Use the comparison to write a short paragraph explaining why $\lim\limits_{x \to 0} h(x) = 0$.

Free-Falling Object **In Exercises 109 and 110, use the position function $s(t) = -16t^2 + 500$, which gives the height (in feet) of an object that has fallen for t seconds from a height of 500 feet. The velocity at time $t = a$ seconds is given by**

$$\lim_{t \to a} \frac{s(a) - s(t)}{a - t}.$$

109. If a construction worker drops a wrench from a height of 500 feet, how fast will the wrench be falling after 2 seconds?

110. If a construction worker drops a wrench from a height of 500 feet, when will the wrench hit the ground? At what velocity will the wrench impact the ground?

Free-Falling Object **In Exercises 111 and 112, use the position function $s(t) = -4.9t^2 + 150$, which gives the height (in meters) of an object that has fallen from a height of 150 meters. The velocity at time $t = a$ seconds is given by**

$$\lim_{t \to a} \frac{s(a) - s(t)}{a - t}.$$

111. Find the velocity of the object when $t = 3$.

112. At what velocity will the object impact the ground?

113. Find two functions f and g such that $\lim\limits_{x \to 0} f(x)$ and $\lim\limits_{x \to 0} g(x)$ do not exist, but $\lim\limits_{x \to 0} [f(x) + g(x)]$ does exist.

114. Prove that if $\lim\limits_{x \to c} f(x)$ exists and $\lim\limits_{x \to c} [f(x) + g(x)]$ does not exist, then $\lim\limits_{x \to c} g(x)$ does not exist.

115. Prove Property 1 of Theorem 2.1.

116. Prove Property 3 of Theorem 2.1. (You may use Property 3 of Theorem 2.2.)

117. Prove Property 1 of Theorem 2.2.

118. Prove that if $\lim\limits_{x \to c} f(x) = 0$, then $\lim\limits_{x \to c} |f(x)| = 0$.

119. Prove that if $\lim\limits_{x \to c} f(x) = 0$ and $|g(x)| \le M$ for a fixed number M and all $x \ne c$, then $\lim\limits_{x \to c} f(x)g(x) = 0$.

120. (a) Prove that if $\lim\limits_{x \to c} |f(x)| = 0$, then $\lim\limits_{x \to c} f(x) = 0$.
   (*Note:* This is the converse of Exercise 118.)

   (b) Prove that if $\lim\limits_{x \to c} f(x) = L$, then $\lim\limits_{x \to c} |f(x)| = |L|$.

   [*Hint:* Use the inequality $\big| |f(x)| - |L| \big| \le |f(x) - L|$.]

121. *Think About It* Find a function f to show that the converse of Exercise 120(b) is not true. [*Hint:* Find a function f such that $\lim\limits_{x \to c} |f(x)| = |L|$ but $\lim\limits_{x \to c} f(x)$ does not exist.]

CAPSTONE

122. Let $f(x) = \begin{cases} 3, & x \ne 2 \\ 5, & x = 2 \end{cases}$. Find $\lim\limits_{x \to 2} f(x)$.

True or False? **In Exercises 123–128, determine whether the statement is true or false. If it is false, explain why or give an example that shows it is false.**

123. $\lim\limits_{x \to 0} \dfrac{|x|}{x} = 1$

124. $\lim\limits_{x \to \pi} \dfrac{\sin x}{x} = 1$

125. If $f(x) = g(x)$ for all real numbers other than $x = 0$, and $\lim\limits_{x \to 0} f(x) = L$, then $\lim\limits_{x \to 0} g(x) = L$.

126. If $\lim\limits_{x \to c} f(x) = L$, then $f(c) = L$.

127. $\lim\limits_{x \to 2} f(x) = 3$, where $f(x) = \begin{cases} 3, & x \le 2 \\ 0, & x > 2 \end{cases}$

128. If $f(x) < g(x)$ for all $x \ne a$, then $\lim\limits_{x \to a} f(x) < \lim\limits_{x \to a} g(x)$.

129. Prove the second part of Theorem 2.9 by proving that

$$\lim_{x \to 0} \frac{1 - \cos x}{x} = 0.$$

130. Let $f(x) = \begin{cases} 0, & \text{if } x \text{ is rational} \\ 1, & \text{if } x \text{ is irrational} \end{cases}$

and

$$g(x) = \begin{cases} 0, & \text{if } x \text{ is rational} \\ x, & \text{if } x \text{ is irrational.} \end{cases}$$

Find (if possible) $\lim\limits_{x \to 0} f(x)$ and $\lim\limits_{x \to 0} g(x)$.

131. *Graphical Reasoning* Consider $f(x) = \dfrac{\sec x - 1}{x^2}$.

   (a) Find the domain of f.

   (b) Use a graphing utility to graph f. Is the domain of f obvious from the graph? If not, explain.

   (c) Use the graph of f to approximate $\lim\limits_{x \to 0} f(x)$.

   (d) Confirm your answer to part (c) analytically.

132. *Approximation*

   (a) Find $\lim\limits_{x \to 0} \dfrac{1 - \cos x}{x^2}$.

   (b) Use the result in part (a) to derive the approximation $\cos x \approx 1 - \frac{1}{2}x^2$ for x near 0.

   (c) Use the result in part (b) to approximate $\cos(0.1)$.

   (d) Use a calculator to approximate $\cos(0.1)$ to four decimal places. Compare the result with part (c).

133. *Think About It* When using a graphing utility to generate a table to approximate $\lim\limits_{x \to 0} [(\sin x)/x]$, a student concluded that the limit was 0.01745 rather than 1. Determine the probable cause of the error.

2.4 Continuity and One-Sided Limits

- ■ Determine continuity at a point and continuity on an open interval.
- ■ Determine one-sided limits and continuity on a closed interval.
- ■ Use properties of continuity.
- ■ Understand and use the Intermediate Value Theorem.

Continuity at a Point and on an Open Interval

In mathematics, the term *continuous* has much the same meaning as it has in everyday usage. To say that a function f is continuous at $x = c$ means that there is no interruption in the graph of f at c. That is, its graph is unbroken at c and there are no holes, jumps, or gaps. Figure 2.25 identifies three values of x at which the graph of f is *not* continuous. At all other points in the interval (a, b), the graph of f is uninterrupted and **continuous.**

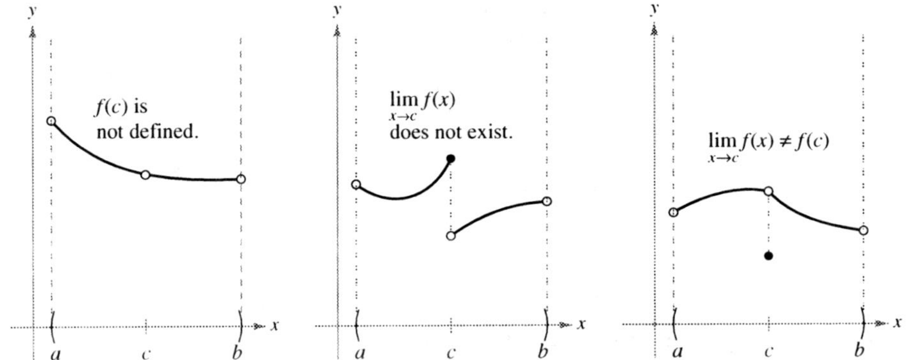

Three conditions exist for which the graph of f is not continuous at $x = c$.
Figure 2.25

In Figure 2.25, it appears that continuity at $x = c$ can be destroyed by any one of the following conditions.

1. The function is not defined at $x = c$.
2. The limit of $f(x)$ does not exist at $x = c$.
3. The limit of $f(x)$ exists at $x = c$, but it is not equal to $f(c)$.

If *none* of the three conditions above is true, the function f is called **continuous at c,** as indicated in the following important definition.

DEFINITION OF CONTINUITY

Continuity at a Point: A function f is **continuous at c** if the following three conditions are met.

1. $f(c)$ is defined.
2. $\lim_{x \to c} f(x)$ exists.
3. $\lim_{x \to c} f(x) = f(c)$

Continuity on an Open Interval: A function is **continuous on an open interval (a, b)** if it is continuous at each point in the interval. A function that is continuous on the entire real line $(-\infty, \infty)$ is **everywhere continuous.**

■ **FOR FURTHER INFORMATION** For more information on the concept of continuity, see the article "Leibniz and the Spell of the Continuous" by Hardy Grant in *The College Mathematics Journal.* To view this article, go to the website *www.matharticles.com.*

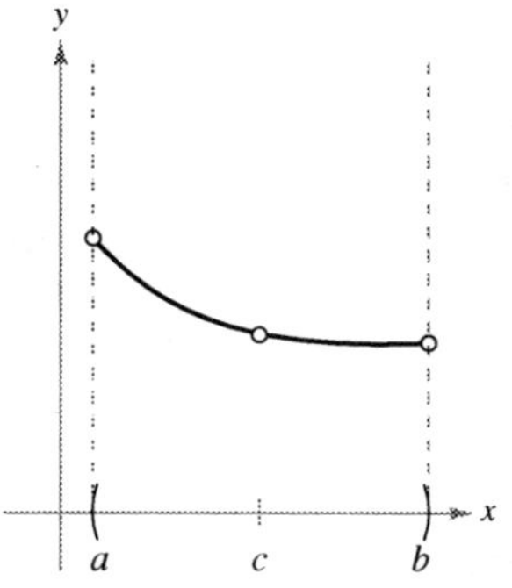

(a) Removable discontinuity

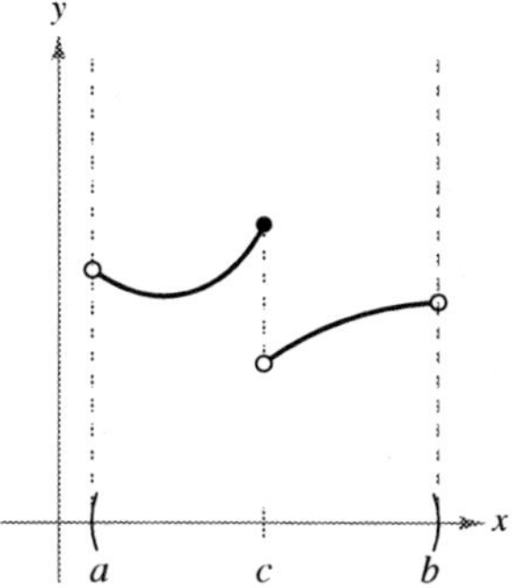

(b) Nonremovable discontinuity

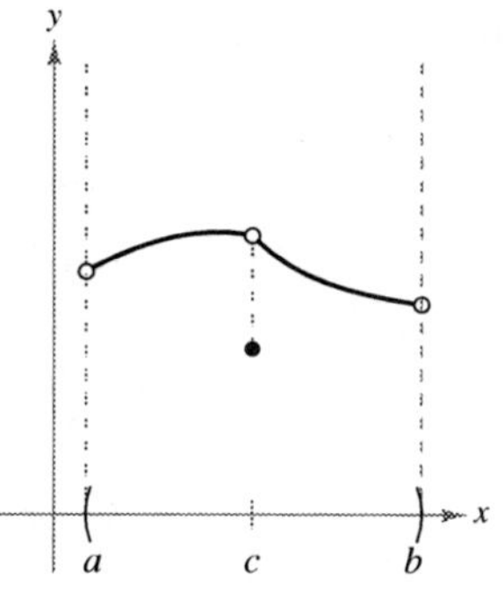

(c) Removable discontinuity
Figure 2.26

Consider an open interval I that contains a real number c. If a function f is defined on I (except possibly at c), and f is not continuous at c, then f is said to have a **discontinuity** at c. Discontinuities fall into two categories: **removable** and **nonremovable.** A discontinuity at c is called removable if f can be made continuous by appropriately defining (or redefining) $f(c)$. For instance, the functions shown in Figures 2.26(a) and (c) have removable discontinuities at c, and the function shown in Figure 2.26(b) has a nonremovable discontinuity at c.

EXAMPLE 1 Continuity of a Function

Discuss the continuity of each function.

a. $f(x) = \dfrac{1}{x}$ **b.** $g(x) = \dfrac{x^2 - 1}{x - 1}$ **c.** $h(x) = \begin{cases} x + 1, & x \le 0 \\ e^x, & x > 0 \end{cases}$ **d.** $y = \sin x$

Solution

a. The domain of f is all nonzero real numbers. From Theorem 2.3, you can conclude that f is continuous at every x-value in its domain. At $x = 0$, f has a nonremovable discontinuity, as shown in Figure 2.27(a). In other words, there is no way to define $f(0)$ so as to make the function continuous at $x = 0$.

b. The domain of g is all real numbers except $x = 1$. From Theorem 2.3, you can conclude that g is continuous at every x-value in its domain. At $x = 1$, the function has a removable discontinuity, as shown in Figure 2.27(b). If $g(1)$ is defined as 2, the "newly defined" function is continuous for all real numbers.

c. The domain of h is all real numbers. The function h is continuous on $(-\infty, 0)$ and $(0, \infty)$, and, because $\lim\limits_{x \to 0} h(x) = 1$, h is continuous on the entire real number line, as shown in Figure 2.27(c).

d. The domain of y is all real numbers. From Theorem 2.6, you can conclude that the function is continuous on its entire domain, $(-\infty, \infty)$, as shown in Figure 2.27(d).

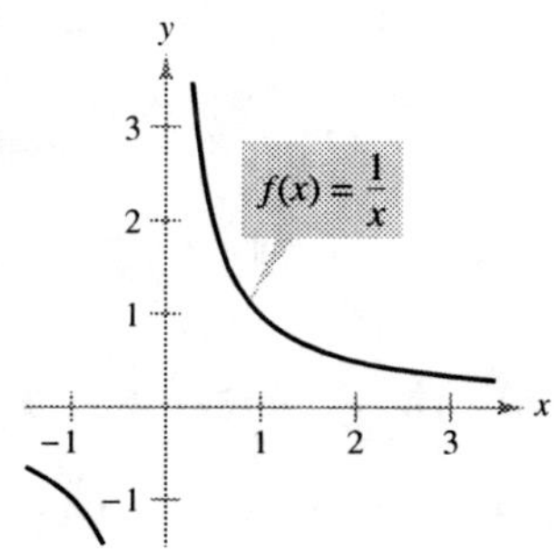

(a) Nonremovable discontinuity at $x = 0$

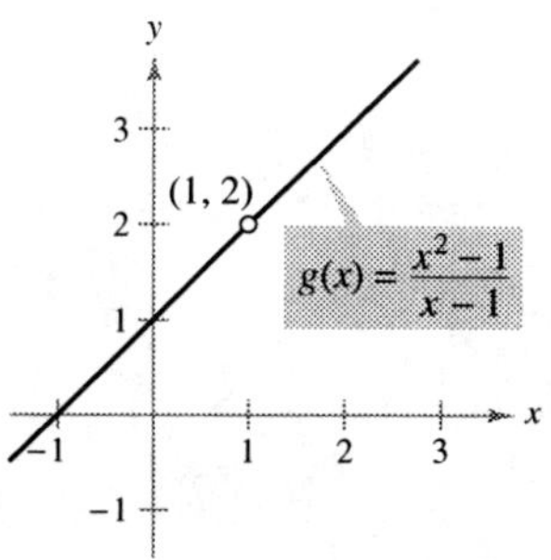

(b) Removable discontinuity at $x = 1$

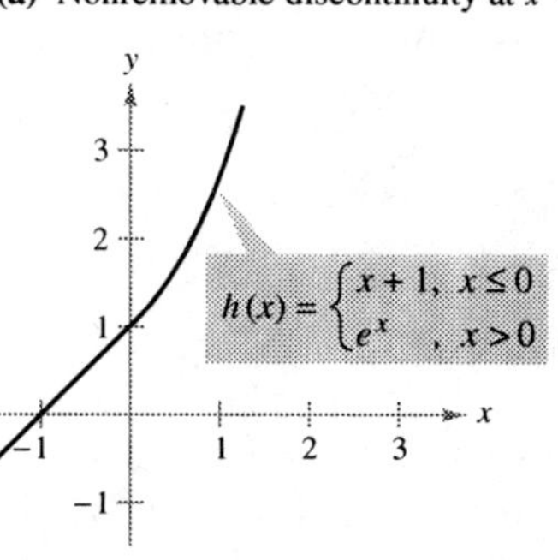

(c) Continuous on entire real line
Figure 2.27

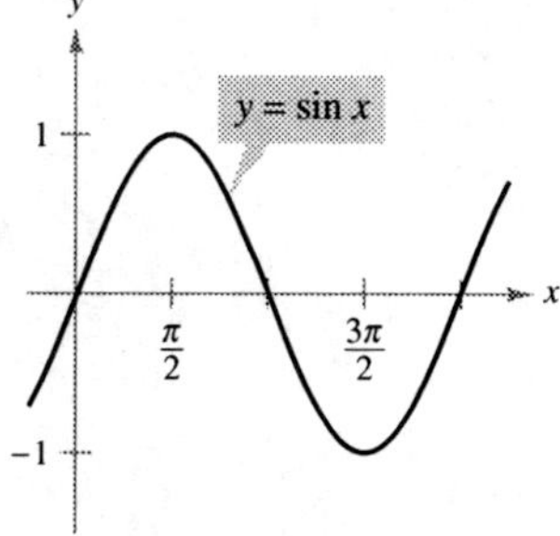

(d) Continuous on entire real line

STUDY TIP Some people may refer to the function in Example 1(a) as "discontinuous." We have found that this terminology can be confusing. Rather than saying the function is discontinuous, we prefer to say that it has a discontinuity at $x = 0$.

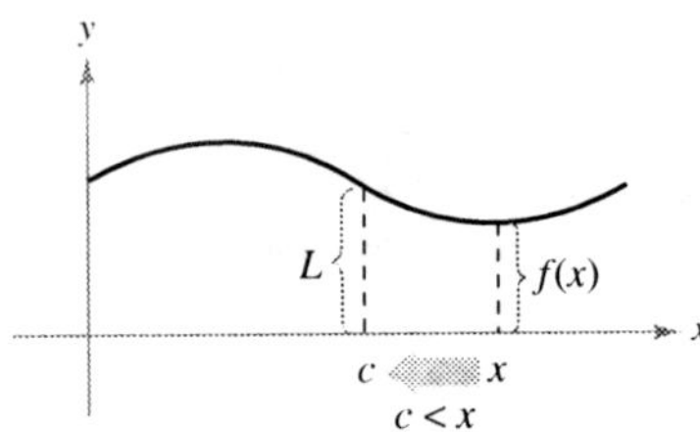

(a) Limit as x approaches c from the right.

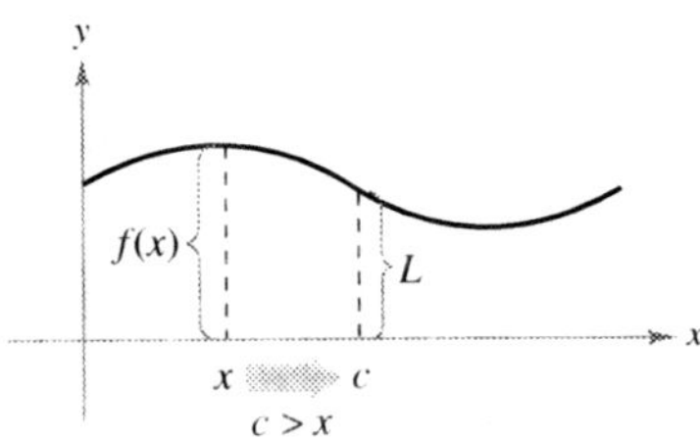

(b) Limit as x approaches c from the left.
Figure 2.28

One-Sided Limits and Continuity on a Closed Interval

To understand continuity on a closed interval, you first need to look at a different type of limit called a **one-sided limit.** For example, the **limit from the right** (or right-hand limit) means that x approaches c from values greater than c [see Figure 2.28(a)]. This limit is denoted as

$$\lim_{x \to c^+} f(x) = L. \qquad \text{Limit from the right}$$

Similarly, the **limit from the left** (or left-hand limit) means that x approaches c from values less than c [see Figure 2.28(b)]. This limit is denoted as

$$\lim_{x \to c^-} f(x) = L. \qquad \text{Limit from the left}$$

One-sided limits are useful in taking limits of functions involving radicals. For instance, if n is an even integer,

$$\lim_{x \to 0^+} \sqrt[n]{x} = 0.$$

EXAMPLE 2 A One-Sided Limit

Find the limit of $f(x) = \sqrt{4 - x^2}$ as x approaches -2 from the right.

Solution As shown in Figure 2.29, the limit as x approaches -2 from the right is

$$\lim_{x \to -2^+} \sqrt{4 - x^2} = 0. \qquad \blacksquare$$

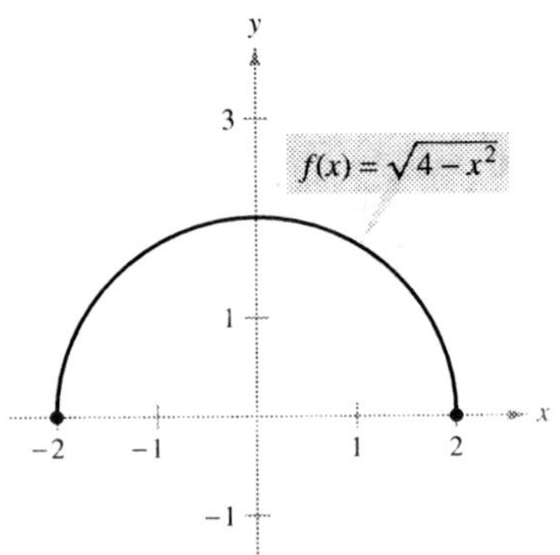

The limit of $f(x)$ as x approaches -2 from the right is 0.
Figure 2.29

One-sided limits can be used to investigate the behavior of **step functions.** One common type of step function is the **greatest integer function** $[\![x]\!]$, defined by

$$[\![x]\!] = \text{greatest integer } n \text{ such that } n \le x. \qquad \text{Greatest integer function}$$

For instance, $[\![2.5]\!] = 2$ and $[\![-2.5]\!] = -3$.

EXAMPLE 3 The Greatest Integer Function

Find the limit of the greatest integer function $f(x) = [\![x]\!]$ as x approaches 0 from the left and from the right.

Solution As shown in Figure 2.30, the limit as x approaches 0 *from the left* is given by

$$\lim_{x \to 0^-} [\![x]\!] = -1$$

and the limit as x approaches 0 *from the right* is given by

$$\lim_{x \to 0^+} [\![x]\!] = 0.$$

The greatest integer function has a discontinuity at zero because the left- and right-hand limits at zero are different. By similar reasoning, you can see that the greatest integer function has a discontinuity at any integer n. $\blacksquare$

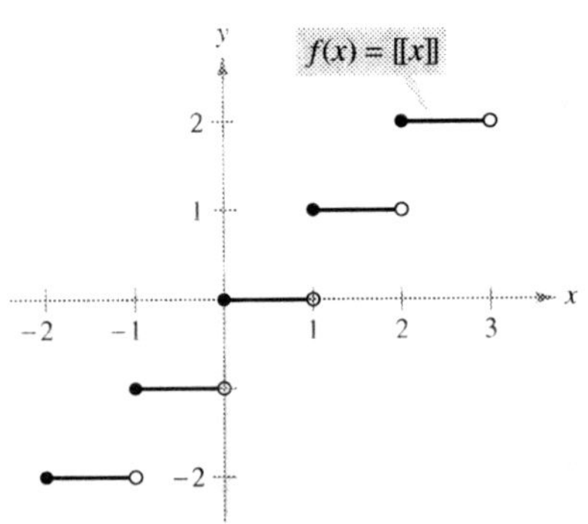

Greatest integer function
Figure 2.30

When the limit from the left is not equal to the limit from the right, the (two-sided) limit *does not exist*. The next theorem makes this more explicit. The proof of this theorem follows directly from the definition of a one-sided limit.

THEOREM 2.10 THE EXISTENCE OF A LIMIT

Let f be a function and let c and L be real numbers. The limit of $f(x)$ as x approaches c is L if and only if

$$\lim_{x \to c^-} f(x) = L \quad \text{and} \quad \lim_{x \to c^+} f(x) = L.$$

The concept of a one-sided limit allows you to extend the definition of continuity to closed intervals. Basically, a function is continuous on a closed interval if it is continuous in the interior of the interval and exhibits one-sided continuity at the endpoints. This is stated formally as follows.

DEFINITION OF CONTINUITY ON A CLOSED INTERVAL

A function f is **continuous on the closed interval** $[a, b]$ if it is continuous on the open interval (a, b) and

$$\lim_{x \to a^+} f(x) = f(a) \quad \text{and} \quad \lim_{x \to b^-} f(x) = f(b).$$

The function f is **continuous from the right** at a and **continuous from the left** at b (see Figure 2.31).

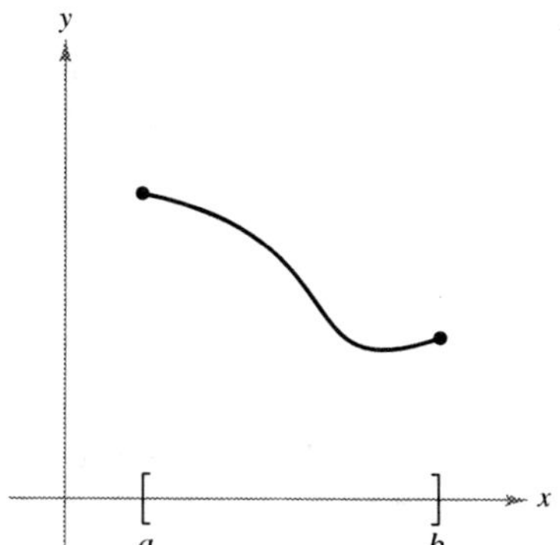

Continuous function on a closed interval
Figure 2.31

Similar definitions can be made to cover continuity on intervals of the form $(a, b]$ and $[a, b)$ that are neither open nor closed, or on infinite intervals. For example, the function

$$f(x) = \sqrt{x}$$

is continuous on the infinite interval $[0, \infty)$, and the function

$$g(x) = \sqrt{2 - x}$$

is continuous on the infinite interval $(-\infty, 2]$.

EXAMPLE 4 Continuity on a Closed Interval

Discuss the continuity of $f(x) = \sqrt{1 - x^2}$.

Solution   The domain of f is the closed interval $[-1, 1]$. At all points in the open interval $(-1, 1)$, the continuity of f follows from Theorems 2.4 and 2.5. Moreover, because

$$\lim_{x \to -1^+} \sqrt{1 - x^2} = 0 = f(-1) \qquad \text{Continuous from the right}$$

and

$$\lim_{x \to 1^-} \sqrt{1 - x^2} = 0 = f(1) \qquad \text{Continuous from the left}$$

you can conclude that f is continuous on the closed interval $[-1, 1]$, as shown in Figure 2.32.

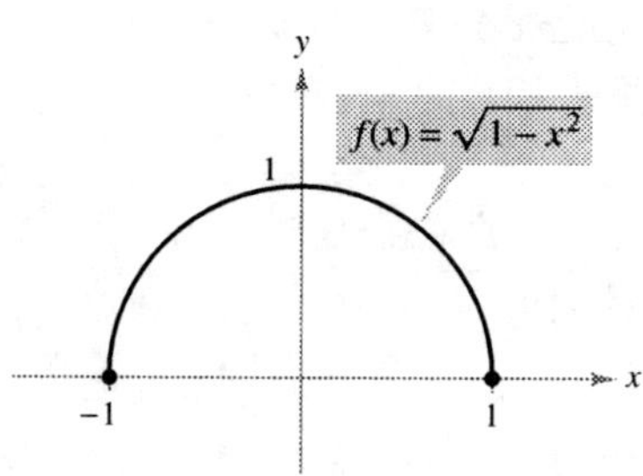

f is continuous on $[-1, 1]$.
Figure 2.32

The next example shows how a one-sided limit can be used to determine the value of absolute zero on the Kelvin scale.

EXAMPLE 5   Charles's Law and Absolute Zero

On the Kelvin scale, *absolute zero* is the temperature 0 K. Although temperatures very close to 0 K have been produced in laboratories, absolute zero has never been attained. In fact, evidence suggests that absolute zero *cannot* be attained. How did scientists determine that 0 K is the "lower limit" of the temperature of matter? What is absolute zero on the Celsius scale?

Solution   The determination of absolute zero stems from the work of the French physicist Jacques Charles (1746–1823). Charles discovered that the volume of gas at a constant pressure increases linearly with the temperature of the gas. The table illustrates this relationship between volume and temperature. To generate the values in the table, one mole of hydrogen is held at a constant pressure of one atmosphere. The volume V is approximated and is measured in liters, and the temperature T is measured in degrees Celsius.

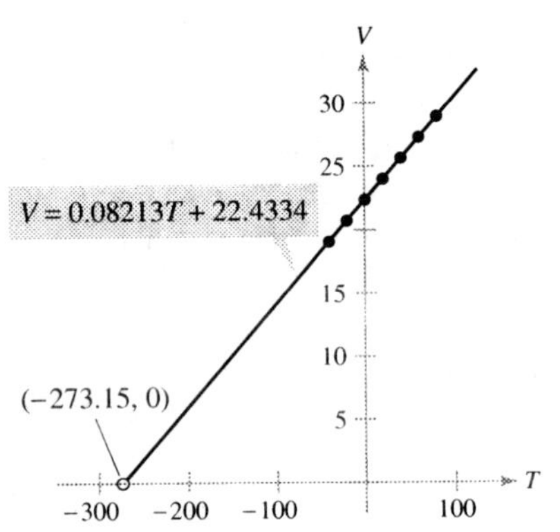

The volume of hydrogen gas depends on its temperature.
Figure 2.33

T	-40	-20	0	20	40	60	80
V	19.1482	20.7908	22.4334	24.0760	25.7186	27.3612	29.0038

The points represented by the table are shown in Figure 2.33. Moreover, by using the points in the table, you can determine that T and V are related by the linear equation

$$V = 0.08213T + 22.4334 \qquad \text{or} \qquad T = \frac{V - 22.4334}{0.08213}.$$

By reasoning that the volume of the gas can approach 0 (but can never equal or go below 0), you can determine that the "least possible temperature" is given by

$$\lim_{V \to 0^+} T = \lim_{V \to 0^+} \frac{V - 22.4334}{0.08213}$$

$$= \frac{0 - 22.4334}{0.08213} \qquad \text{Use direct substitution.}$$

$$\approx -273.15.$$

So, absolute zero on the Kelvin scale (0 K) is approximately $-273.15°$ on the Celsius scale. ∎

The following table shows the temperatures in Example 5 converted to the Fahrenheit scale. Try repeating the solution shown in Example 5 using these temperatures and volumes. Use the result to find the value of absolute zero on the Fahrenheit scale.

T	-40	-4	32	68	104	140	176
V	19.1482	20.7908	22.4334	24.0760	25.7186	27.3612	29.0038

NOTE   Charles's Law for gases (assuming constant pressure) can be stated as

$$V = RT \qquad \text{Charles's Law}$$

where V is volume, R is a constant, and T is temperature. In the statement of this law, what property must the temperature scale have? ∎

In 2003, researchers at the Massachusetts Institute of Technology used lasers and evaporation to produce a supercold gas in which atoms overlap. This gas is called a Bose-Einstein condensate. They measured a temperature of about 450 pK (picokelvin), or approximately $-273.14999999955°$C. (*Source: Science magazine, September 12, 2003*)

Properties of Continuity

In Section 2.3, you studied several properties of limits. Each of those properties yields a corresponding property pertaining to the continuity of a function. For instance, Theorem 2.11 follows directly from Theorem 2.2. (A proof of Theorem 2.11 is given in Appendix A.)

THEOREM 2.11 PROPERTIES OF CONTINUITY

If b is a real number and f and g are continuous at $x = c$, then the following functions are also continuous at c.

1. Scalar multiple: bf

2. Sum or difference: $f \pm g$

3. Product: fg

4. Quotient: $\dfrac{f}{g}$, if $g(c) \neq 0$

The following types of functions are continuous at every point in their domains.

1. Polynomial: $\qquad p(x) = a_n x^n + a_{n-1} x^{n-1} + \cdots + a_1 x + a_0$

2. Rational: $\qquad r(x) = \dfrac{p(x)}{q(x)}, \quad q(x) \neq 0$

3. Radical: $\qquad f(x) = \sqrt[n]{x}$

4. Trigonometric: $\qquad \sin x,\ \cos x,\ \tan x,\ \cot x,\ \sec x,\ \csc x$

5. Exponential and logarithmic: $f(x) = a^x,\ f(x) = e^x,\ f(x) = \ln x$

By combining Theorem 2.11 with this summary, you can conclude that a wide variety of elementary functions are continuous at every point in their domains.

 **EXAMPLE 6   Applying Properties of Continuity**

By Theorem 2.11, it follows that each of the functions below is continuous at every point in its domain.

$$f(x) = x + e^x, \quad f(x) = 3\tan x, \quad f(x) = \frac{x^2 + 1}{\cos x}$$

The next theorem, which is a consequence of Theorem 2.5, allows you to determine the continuity of *composite* functions such as

$$f(x) = \sin 3x, \quad f(x) = \sqrt{x^2 + 1}, \quad f(x) = \tan \frac{1}{x}.$$

THEOREM 2.12 CONTINUITY OF A COMPOSITE FUNCTION

If g is continuous at c and f is continuous at $g(c)$, then the composite function given by $(f \circ g)(x) = f(g(x))$ is continuous at c.

NOTE One consequence of Theorem 2.12 is that if f and g satisfy the given conditions, you can determine the limit of $f(g(x))$ as x approaches c to be

$$\lim_{x \to c} f(g(x)) = f(g(c)).$$

PROOF By the definition of continuity, $\displaystyle\lim_{x \to c} g(x) = g(c)$ and $\displaystyle\lim_{x \to g(c)} f(x) = f(g(c))$.

Apply Theorem 2.5 with $L = g(c)$ to obtain $\displaystyle\lim_{x \to c} f(g(x)) = f\!\left(\lim_{x \to c} g(x)\right) = f(g(c))$. So, $(f \circ g) = f(g(x))$ is continuous at c.

AUGUSTIN-LOUIS CAUCHY (1789–1857)

The concept of a continuous function was first introduced by Augustin-Louis Cauchy in 1821. The definition given in his text *Cours d'Analyse* stated that indefinite small changes in *y* were the result of indefinite small changes in *x*. "...$f(x)$ will be called a *continuous* function if ... the numerical values of the difference $f(x + \alpha) - f(x)$ decrease indefinitely with those of α ...."

EXAMPLE 7 Testing for Continuity

Describe the interval(s) on which each function is continuous.

a. $f(x) = \tan x$ **b.** $g(x) = \begin{cases} \sin\dfrac{1}{x}, & x \neq 0 \\ 0, & x = 0 \end{cases}$ **c.** $h(x) = \begin{cases} x\sin\dfrac{1}{x}, & x \neq 0 \\ 0, & x = 0 \end{cases}$

Solution

a. The tangent function $f(x) = \tan x$ is undefined at

$$x = \frac{\pi}{2} + n\pi, \qquad n \text{ is an integer.}$$

At all other points it is continuous. So, $f(x) = \tan x$ is continuous on the open intervals

$$\ldots, \left(-\frac{3\pi}{2}, -\frac{\pi}{2}\right), \left(-\frac{\pi}{2}, \frac{\pi}{2}\right), \left(\frac{\pi}{2}, \frac{3\pi}{2}\right), \ldots$$

as shown in Figure 2.34(a).

b. Because $y = 1/x$ is continuous except at $x = 0$ and the sine function is continuous for all real values of x, it follows that $y = \sin(1/x)$ is continuous at all real values except $x = 0$. At $x = 0$, the limit of $g(x)$ does not exist (see Example 5, Section 2.2). So, g is continuous on the intervals $(-\infty, 0)$ and $(0, \infty)$, as shown in Figure 2.34(b).

c. This function is similar to the function in part (b) except that the oscillations are damped by the factor x. Using the Squeeze Theorem, you obtain

$$-|x| \leq x\sin\frac{1}{x} \leq |x|, \quad x \neq 0$$

and you can conclude that

$$\lim_{x \to 0} h(x) = 0.$$

So, h is continuous on the entire real line, as shown in Figure 2.34(c).

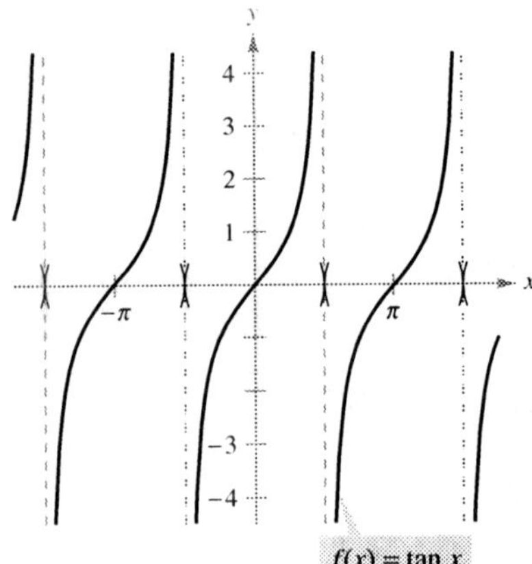

(a) f is continuous on each open interval in its domain.

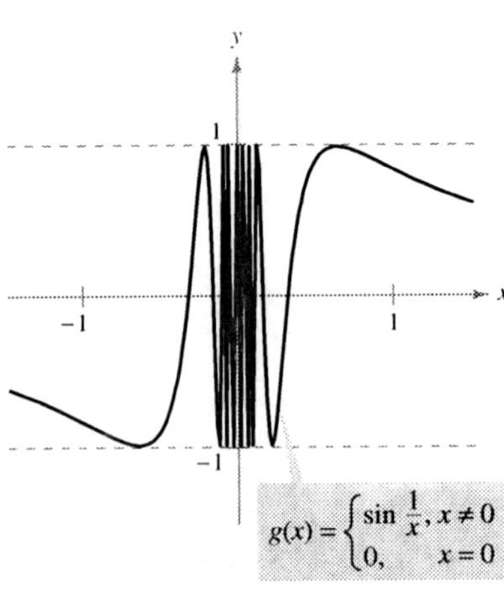

(b) g is continuous on $(-\infty, 0)$ and $(0, \infty)$.

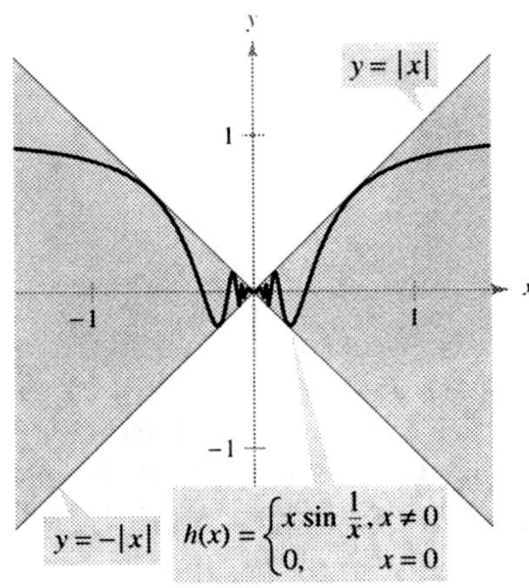

(c) h is continuous on the entire real line.

Figure 2.34

The Intermediate Value Theorem

Theorem 2.13 is an important theorem concerning the behavior of functions that are continuous on a closed interval.

THEOREM 2.13 INTERMEDIATE VALUE THEOREM

If f is continuous on the closed interval $[a, b]$, $f(a) \neq f(b)$, and k is any number between $f(a)$ and $f(b)$, then there is at least one number c in $[a, b]$ such that

$$f(c) = k.$$

NOTE The Intermediate Value Theorem tells you that at least one number c exists, but it does not provide a method for finding c. Such theorems are called **existence theorems.** By referring to a text on advanced calculus, you will find that a proof of this theorem is based on a property of real numbers called *completeness*. The Intermediate Value Theorem states that for a continuous function f, if x takes on all values between a and b, $f(x)$ must take on all values between $f(a)$ and $f(b)$. ■

As a simple example of the application of this theorem, consider a person's height. Suppose that a girl is 5 feet tall on her thirteenth birthday and 5 feet 7 inches tall on her fourteenth birthday. Then, for any height h between 5 feet and 5 feet 7 inches, there must have been a time t when her height was exactly h. This seems reasonable because human growth is continuous and a person's height does not abruptly change from one value to another.

The Intermediate Value Theorem guarantees the existence of *at least one* number c in the closed interval $[a, b]$. There may, of course, be more than one number c such that $f(c) = k$, as shown in Figure 2.35. A function that is not continuous does not necessarily possess the intermediate value property. For example, the graph of the function shown in Figure 2.36 jumps over the horizontal line given by $y = k$, and for this function there is no value of c in $[a, b]$ such that $f(c) = k$.

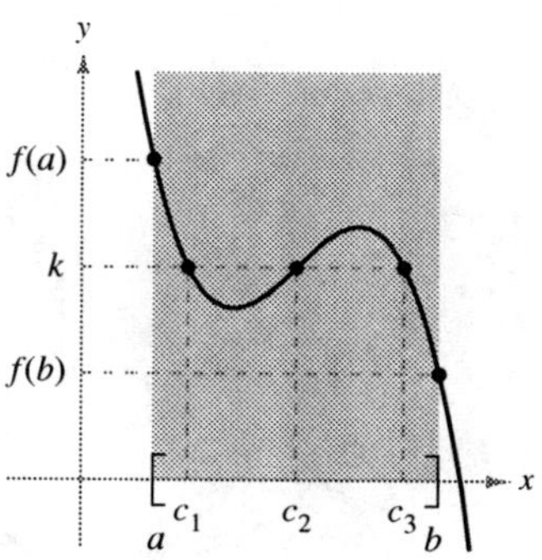

f is continuous on $[a, b]$.
[There exist three c's such that $f(c) = k$.]
Figure 2.35

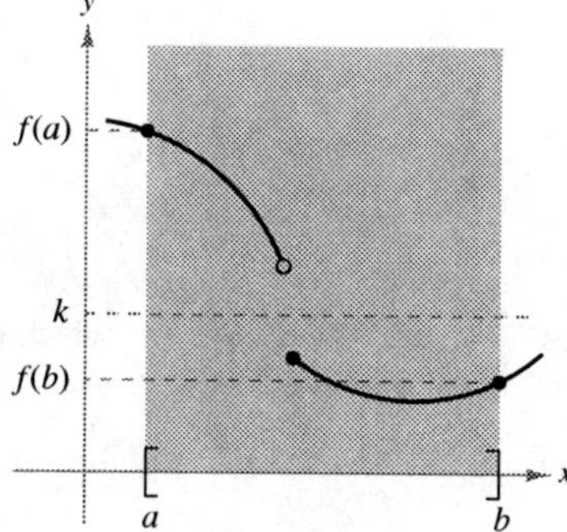

f is not continuous on $[a, b]$.
[There are no c's such that $f(c) = k$.]
Figure 2.36

The Intermediate Value Theorem often can be used to locate the zeros of a function that is continuous on a closed interval. Specifically, if f is continuous on $[a, b]$ and $f(a)$ and $f(b)$ differ in sign, the Intermediate Value Theorem guarantees the existence of at least one zero of f in the closed interval $[a, b]$.

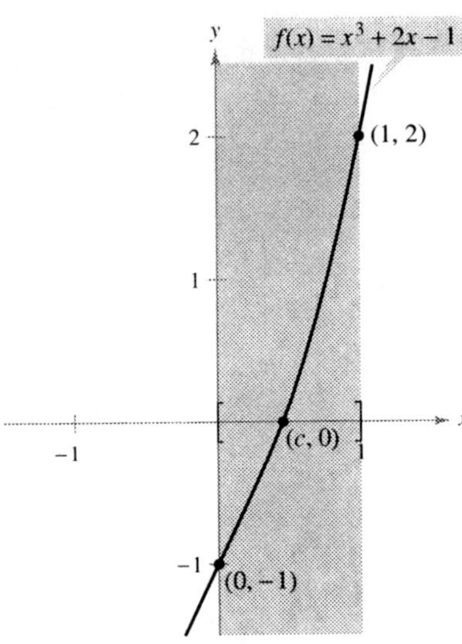
f is continuous on $[0, 1]$ with $f(0) < 0$ and $f(1) > 0$.

Figure 2.37

EXAMPLE 8 An Application of the Intermediate Value Theorem

Use the Intermediate Value Theorem to show that the polynomial function $f(x) = x^3 + 2x - 1$ has a zero in the interval $[0, 1]$.

Solution   Note that f is continuous on the closed interval $[0, 1]$. Because

$$f(0) = 0^3 + 2(0) - 1 = -1 \quad \text{and} \quad f(1) = 1^3 + 2(1) - 1 = 2$$

it follows that $f(0) < 0$ and $f(1) > 0$. You can therefore apply the Intermediate Value Theorem to conclude that there must be some c in $[0, 1]$ such that

$$f(c) = 0 \qquad \textit{f has a zero in the closed interval } [0, 1].$$

as shown in Figure 2.37.

The **bisection method** for approximating the real zeros of a continuous function is similar to the method used in Example 8. If you know that a zero exists in the closed interval $[a, b]$, the zero must lie in the interval $[a, (a + b)/2]$ or $[(a + b)/2, b]$. From the sign of $f([a + b]/2)$, you can determine which interval contains the zero. By repeatedly bisecting the interval, you can "close in" on the zero of the function.

TECHNOLOGY   You can also use the *zoom* feature of a graphing utility to approximate the real zeros of a continuous function. By repeatedly zooming in on the point where the graph crosses the x-axis, and adjusting the x-axis scale, you can approximate the zero of the function to any desired accuracy. The zero of $x^3 + 2x - 1$ is approximately 0.453, as shown in Figure 2.38.

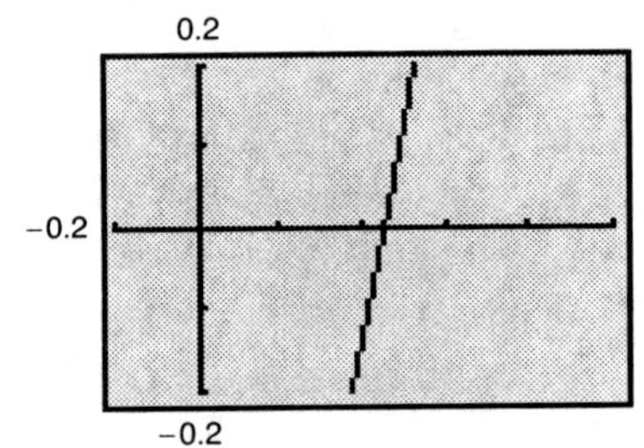
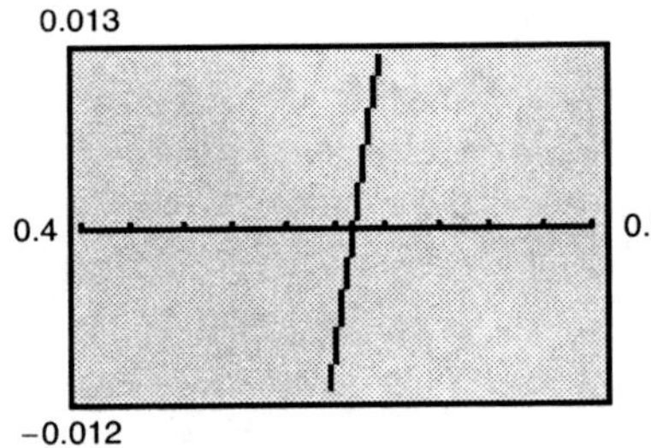

Figure 2.38   Zooming in on the zero of $f(x) = x^3 + 2x - 1$

2.4 Exercises

In Exercises 1–6, use the graph to determine the limit, and discuss the continuity of the function.

(a) $\lim\limits_{x \to c^+} f(x)$ (b) $\lim\limits_{x \to c^-} f(x)$ (c) $\lim\limits_{x \to c} f(x)$

1.
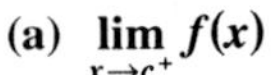
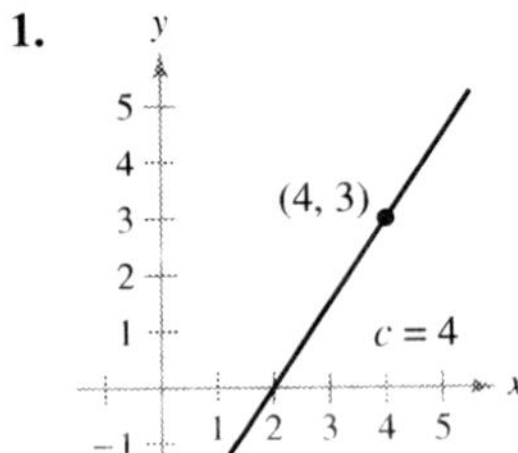

2.
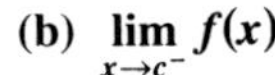
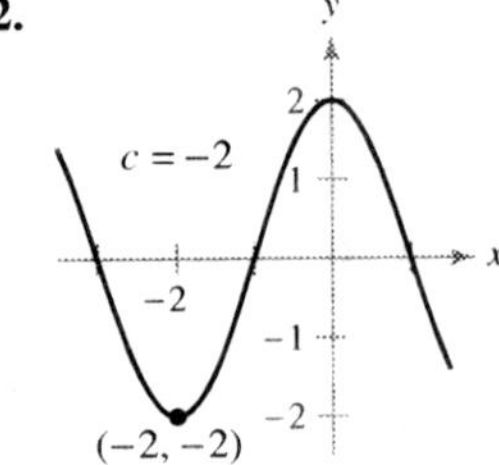

3.
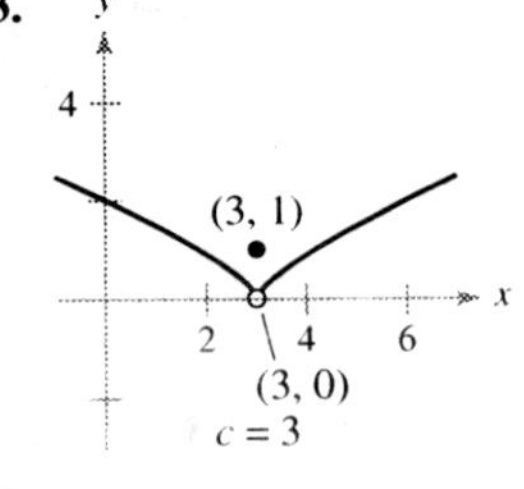

4.
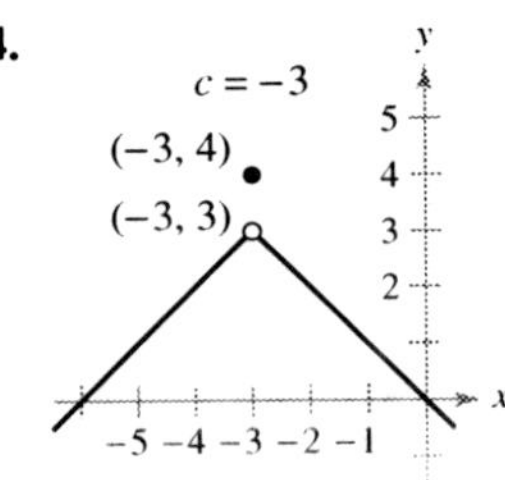

5.
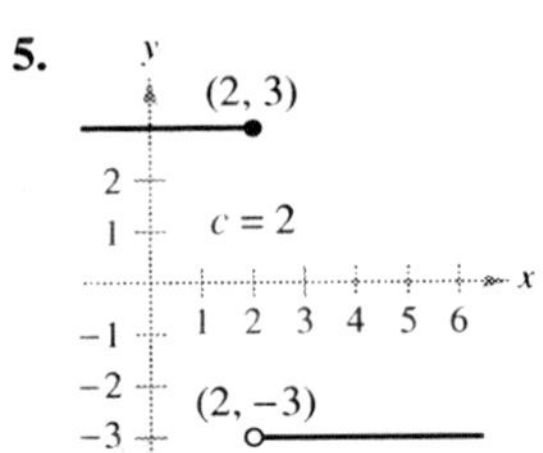

6.
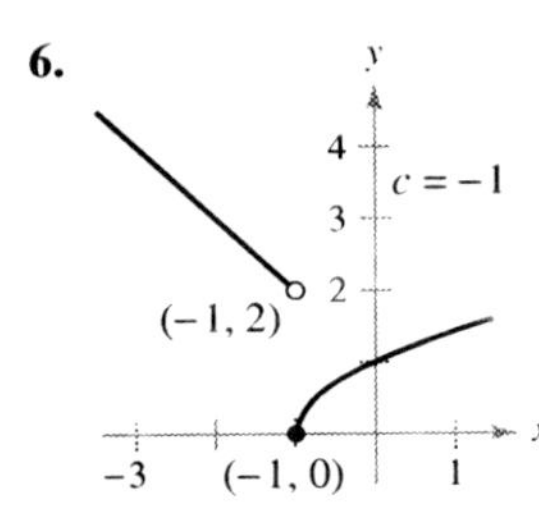

In Exercises 7–30, find the limit (if it exists). If it does not exist, explain why.

7. $\displaystyle \lim_{x \to 8^+} \frac{1}{x + 8}$

8. $\displaystyle \lim_{x \to 5^-} \frac{-3}{x + 5}$

9. $\displaystyle \lim_{x \to 5^+} \frac{x - 5}{x^2 - 25}$

10. $\displaystyle \lim_{x \to 2^+} \frac{2 - x}{x^2 - 4}$

11. $\displaystyle \lim_{x \to -3^-} \frac{x}{\sqrt{x^2 - 9}}$

12. $\displaystyle \lim_{x \to 9^-} \frac{\sqrt{x} - 3}{x - 9}$

13. $\displaystyle \lim_{x \to 0^-} \frac{|x|}{x}$

14. $\displaystyle \lim_{x \to 10^+} \frac{|x - 10|}{x - 10}$

15. $\displaystyle \lim_{\Delta x \to 0^-} \frac{\dfrac{1}{x + \Delta x} - \dfrac{1}{x}}{\Delta x}$

16. $\displaystyle \lim_{\Delta x \to 0^+} \frac{(x + \Delta x)^2 + x + \Delta x - (x^2 + x)}{\Delta x}$

17. $\displaystyle \lim_{x \to 3^-} f(x)$, where $f(x) = \begin{cases} \dfrac{x + 2}{2}, & x \le 3 \\[2mm] \dfrac{12 - 2x}{3}, & x > 3 \end{cases}$

18. $\displaystyle \lim_{x \to 2} f(x)$, where $f(x) = \begin{cases} x^2 - 4x + 6, & x < 2 \\ -x^2 + 4x - 2, & x \ge 2 \end{cases}$

19. $\displaystyle \lim_{x \to 1} f(x)$, where $f(x) = \begin{cases} x^3 + 1, & x < 1 \\ x + 1, & x \ge 1 \end{cases}$

20. $\displaystyle \lim_{x \to 1^+} f(x)$, where $f(x) = \begin{cases} x, & x \le 1 \\ 1 - x, & x > 1 \end{cases}$

21. $\displaystyle \lim_{x \to \pi} \cot x$

22. $\displaystyle \lim_{x \to \pi/2} \sec x$

23. $\displaystyle \lim_{x \to 4^-} (5[\![x]\!] - 7)$

24. $\displaystyle \lim_{x \to 3^+} (3x - [\![x]\!])$

25. $\displaystyle \lim_{x \to 3} (2 - [\![-x]\!])$

26. $\displaystyle \lim_{x \to 1} \left(1 - \left[\!\!\left[-\frac{x}{2} \right]\!\!\right] \right)$

27. $\displaystyle \lim_{x \to 3^+} \ln(x - 3)$

28. $\displaystyle \lim_{x \to 6^-} \ln(6 - x)$

29. $\displaystyle \lim_{x \to 2^-} \ln[x^2(3 - x)]$

30. $\displaystyle \lim_{x \to 5^+} \ln \frac{x}{\sqrt{x - 4}}$

In Exercises 31–34, discuss the continuity of each function.

31. $f(x) = \dfrac{1}{x^2 - 4}$

32. $f(x) = \dfrac{x^2 - 1}{x + 1}$

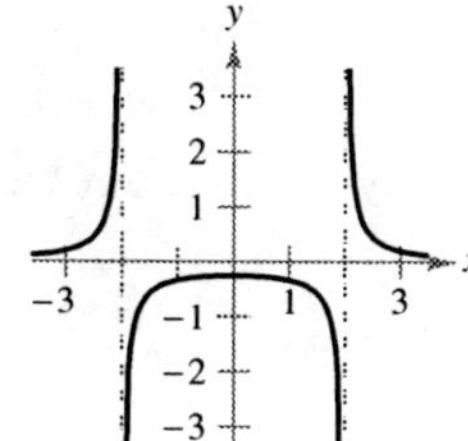

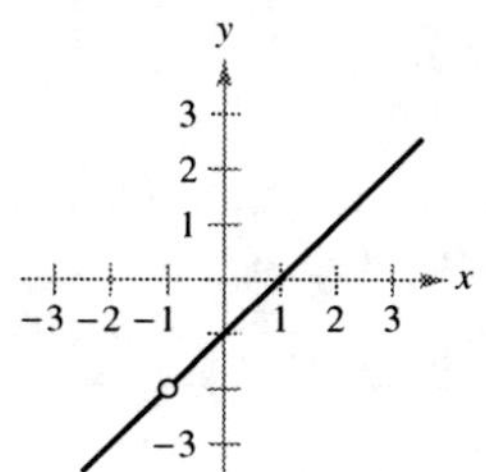

33. $f(x) = \frac{1}{2}[\![x]\!] + x$

34. $f(x) = \begin{cases} x, & x < 1 \\ 2, & x = 1 \\ 2x - 1, & x > 1 \end{cases}$

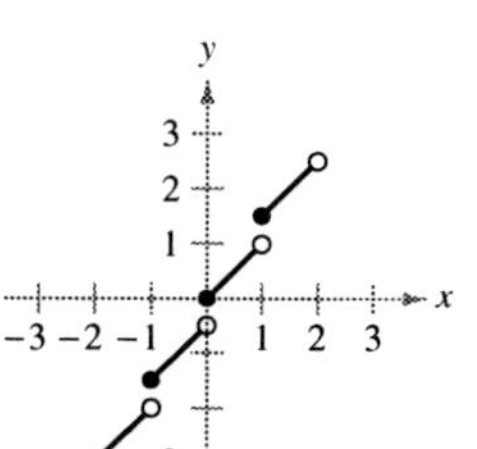

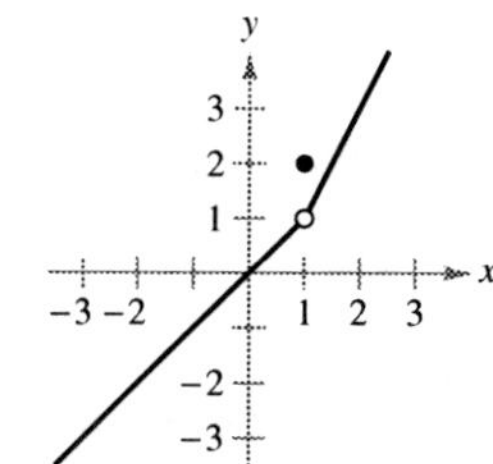

In Exercises 35–38, discuss the continuity of the function on the closed interval.

Function	*Interval*
35. $g(x) = \sqrt{49 - x^2}$	$[-7, 7]$
36. $f(t) = 2 - \sqrt{9 - t^2}$	$[-2, 2]$
37. $f(x) = \begin{cases} 3 - x, & x \le 0 \\ 3 + \frac{1}{2}x, & x > 0 \end{cases}$	$[-1, 4]$
38. $g(x) = \dfrac{1}{x^2 - 4}$	$[-1, 2]$

In Exercises 39–66, find the x-values (if any) at which f is not continuous. Which of the discontinuities are removable?

39. $f(x) = \dfrac{6}{x}$

40. $f(x) = \dfrac{3}{x - 2}$

41. $f(x) = x^2 - 9$

42. $f(x) = x^2 - 2x + 1$

43. $f(x) = \dfrac{1}{4 - x^2}$

44. $f(x) = \dfrac{1}{x^2 + 1}$

45. $f(x) = 3x - \cos x$

46. $f(x) = \cos \dfrac{\pi x}{4}$

47. $f(x) = \dfrac{x}{x^2 - x}$

48. $f(x) = \dfrac{x}{x^2 - 1}$

49. $f(x) = \dfrac{x}{x^2 + 1}$

50. $f(x) = \dfrac{x - 3}{x^2 - 9}$

51. $f(x) = \dfrac{x + 2}{x^2 - 3x - 10}$

52. $f(x) = \dfrac{x - 1}{x^2 + x - 2}$

53. $f(x) = \dfrac{|x + 2|}{x + 2}$

54. $f(x) = \dfrac{|x - 3|}{x - 3}$

55. $f(x) = \begin{cases} x, & x \le 1 \\ x^2, & x > 1 \end{cases}$

56. $f(x) = \begin{cases} -2x + 3, & x < 1 \\ x^2, & x \ge 1 \end{cases}$

57. $f(x) = \begin{cases} \frac{1}{2}x + 1, & x \le 2 \\ 3 - x, & x > 2 \end{cases}$

58. $f(x) = \begin{cases} -2x, & x \le 2 \\ x^2 - 4x + 1, & x > 2 \end{cases}$

59. $f(x) = \begin{cases} \tan \dfrac{\pi x}{4}, & |x| < 1 \\ x, & |x| \ge 1 \end{cases}$

60. $f(x) = \begin{cases} \csc \dfrac{\pi x}{6}, & |x - 3| \le 2 \\ 2, & |x - 3| > 2 \end{cases}$

61. $f(x) = \begin{cases} \ln(x+1), & x \geq 0 \\ 1 - x^2, & x < 0 \end{cases}$

62. $f(x) = \begin{cases} 10 - 3e^{5-x}, & x > 5 \\ 10 - \frac{3}{5}x, & x \leq 5 \end{cases}$

63. $f(x) = \csc 2x$

64. $f(x) = \tan \dfrac{\pi x}{4}$

65. $f(x) = [\![x - 8]\!]$

66. $f(x) = 5 - [\![x]\!]$

In Exercises 67 and 68, use a graphing utility to graph the function. From the graph, estimate

$$\lim_{x \to 0^+} f(x) \quad \text{and} \quad \lim_{x \to 0^-} f(x).$$

Is the function continuous on the entire real number line? Explain.

67. $f(x) = \dfrac{|x^2 - 4|x}{x + 2}$

68. $f(x) = \dfrac{|x^2 + 4x|(x + 2)}{x + 4}$

In Exercises 69–76, find the constant a, or the constants a and b, such that the function is continuous on the entire real number line.

69. $f(x) = \begin{cases} 3x^2, & x \geq 1 \\ ax - 4, & x < 1 \end{cases}$

70. $f(x) = \begin{cases} 3x^3, & x \leq 1 \\ ax + 5, & x > 1 \end{cases}$

71. $f(x) = \begin{cases} x^3, & x \leq 2 \\ ax^2, & x > 2 \end{cases}$

72. $g(x) = \begin{cases} \dfrac{4 \sin x}{x}, & x < 0 \\ a - 2x, & x \geq 0 \end{cases}$

73. $f(x) = \begin{cases} 2, & x \leq -1 \\ ax + b, & -1 < x < 3 \\ -2, & x \geq 3 \end{cases}$

74. $g(x) = \begin{cases} \dfrac{x^2 - a^2}{x - a}, & x \neq a \\ 8, & x = a \end{cases}$

75. $f(x) = \begin{cases} ae^{x-1} + 3, & x < 1 \\ \arctan(x - 1) + 2, & x \geq 1 \end{cases}$

76. $f(x) = \begin{cases} 2e^{ax} - 2, & x \leq 4 \\ \ln(x - 3) + x^2, & x > 4 \end{cases}$

In Exercises 77–80, discuss the continuity of the composite function $h(x) = f(g(x))$.

77. $f(x) = x^2$

$g(x) = x - 1$

78. $f(x) = \dfrac{1}{\sqrt{x}}$

$g(x) = x - 1$

79. $f(x) = \dfrac{1}{x - 6}$

$g(x) = x^2 + 5$

80. $f(x) = \sin x$

$g(x) = x^2$

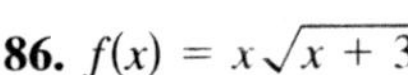 **In Exercises 81–84, use a graphing utility to graph the function. Use the graph to determine any x-values at which the function is not continuous.**

81. $f(x) = [\![x]\!] - x$

82. $h(x) = \dfrac{1}{x^2 - x - 2}$

83. $g(x) = \begin{cases} x^2 - 3x, & x > 4 \\ 2x - 5, & x \leq 4 \end{cases}$

84. $f(x) = \begin{cases} \dfrac{\cos x - 1}{x}, & x < 0 \\ 5x, & x \geq 0 \end{cases}$

In Exercises 85–88, describe the interval(s) on which the function is continuous.

85. $f(x) = \dfrac{x}{x^2 + x + 2}$

86. $f(x) = x\sqrt{x + 3}$

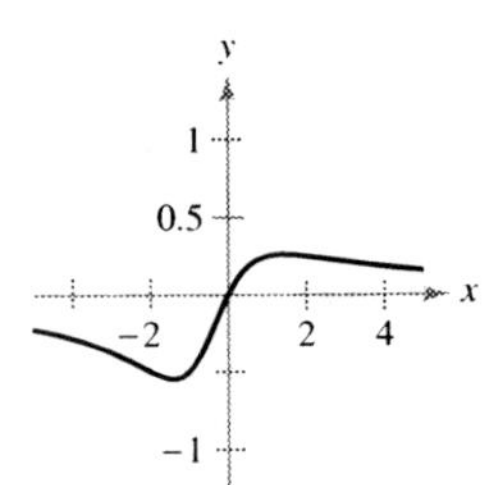

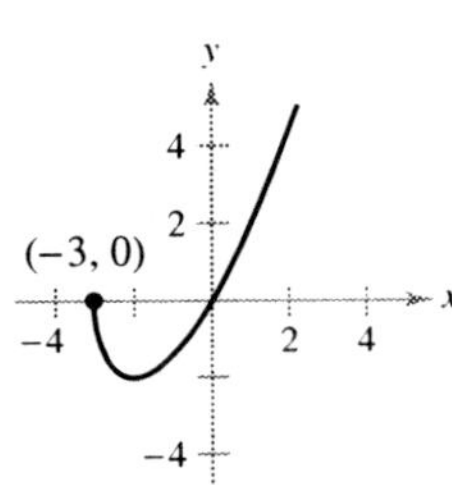

87. $f(x) = \sec \dfrac{\pi x}{4}$

88. $f(x) = \dfrac{x + 1}{\sqrt{x}}$

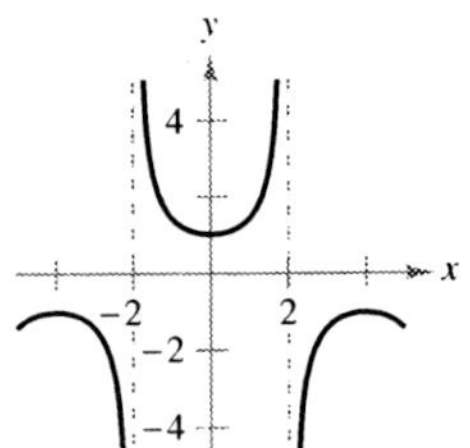

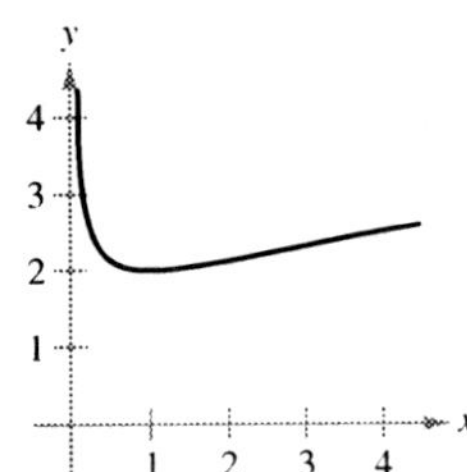

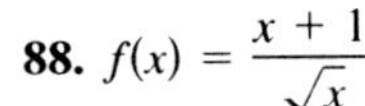 **Writing   In Exercises 89–92, use a graphing utility to graph the function on the interval $[-4, 4]$. Does the graph of the function appear to be continuous on this interval? Is the function continuous on $[-4, 4]$? Write a short paragraph about the importance of examining a function analytically as well as graphically.**

89. $f(x) = \dfrac{\sin x}{x}$

90. $f(x) = \dfrac{x^3 - 8}{x - 2}$

91. $f(x) = \dfrac{\ln(x^2 + 1)}{x}$

92. $f(x) = \dfrac{e^{-x} + 1}{e^x - 1}$

Writing   In Exercises 93–96, explain why the function has a zero in the given interval.

Function	Interval
93. $f(x) = \frac{1}{12}x^4 - x^3 + 4$	$[1, 2]$
94. $f(x) = x^3 + 5x - 3$	$[0, 1]$
95. $h(x) = -2e^{-x/2} \cos 2x$	$\left[0, \dfrac{\pi}{2}\right]$
96. $g(t) = (t^3 + 2t - 2) \ln(t^2 + 4)$	$[0, 1]$

In Exercises 97–102, use the Intermediate Value Theorem and a graphing utility to approximate the zero of the function in the interval $[0, 1]$. Repeatedly "zoom in" on the graph of the function to approximate the zero accurate to two decimal places. Use the *zero* or *root* feature of the graphing utility to approximate the zero accurate to four decimal places.

97. $f(x) = x^3 + x - 1$

98. $f(x) = x^3 + 3x - 3$

99. $g(t) = 2 \cos t - 3t$

100. $h(\theta) = 1 + \theta - 3 \tan \theta$

101. $f(x) = x + e^x - 3$

102. $g(x) = 5 \ln(x + 1) - 2$

In Exercises 103–106, verify that the Intermediate Value Theorem applies to the given interval and find the value of c guaranteed by the theorem.

103. $f(x) = x^2 + x - 1$, $[0, 5]$, $f(c) = 11$

104. $f(x) = x^2 - 6x + 8$, $[0, 3]$, $f(c) = 0$

105. $f(x) = x^3 - x^2 + x - 2$, $[0, 3]$, $f(c) = 4$

106. $f(x) = \dfrac{x^2 + x}{x - 1}$, $\left[\dfrac{5}{2}, 4\right]$, $f(c) = 6$

WRITING ABOUT CONCEPTS

107. State how continuity is destroyed at $x = c$ for each of the following graphs.

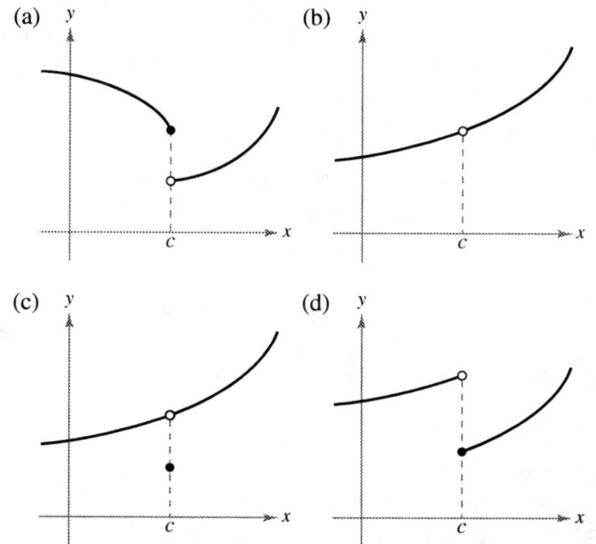

108. Sketch the graph of any function f such that

$$\lim_{x \to 3^+} f(x) = 1 \quad \text{and} \quad \lim_{x \to 3^-} f(x) = 0.$$

Is the function continuous at $x = 3$? Explain.

109. If the functions f and g are continuous for all real x, is $f + g$ always continuous for all real x? Is f/g always continuous for all real x? If either is not continuous, give an example to verify your conclusion.

CAPSTONE

110. Describe the difference between a discontinuity that is removable and one that is nonremovable. In your explanation, give examples of the following descriptions.

(a) A function with a nonremovable discontinuity at $x = 4$

(b) A function with a removable discontinuity at $x = -4$

(c) A function that has both of the characteristics described in parts (a) and (b)

True or False? **In Exercises 111–114, determine whether the statement is true or false. If it is false, explain why or give an example that shows it is false.**

111. If $\lim\limits_{x \to c} f(x) = L$ and $f(c) = L$, then f is continuous at c.

112. If $f(x) = g(x)$ for $x \neq c$ and $f(c) \neq g(c)$, then either f or g is not continuous at c.

113. A rational function can have infinitely many x-values at which it is not continuous.

114. The function $f(x) = |x - 1|/(x - 1)$ is continuous on $(-\infty, \infty)$.

115. *Swimming Pool* Every day you dissolve 28 ounces of chlorine in a swimming pool. The graph shows the amount of chlorine $f(t)$ in the pool after t days.

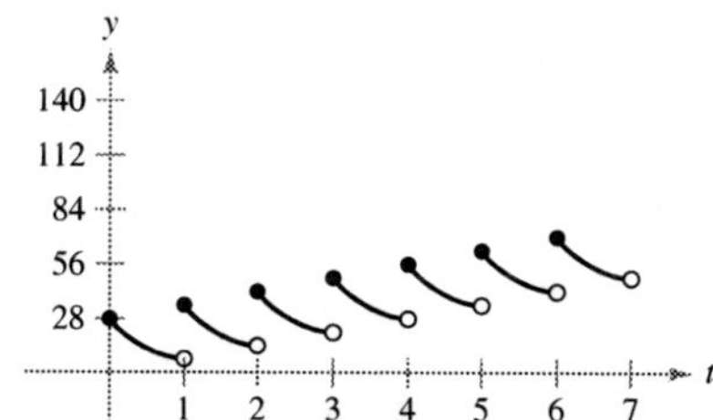

Estimate and interpret $\lim\limits_{t \to 4^-} f(t)$ and $\lim\limits_{t \to 4^+} f(t)$.

116. *Think About It* Describe how $f(x) = 3 + [\![x]\!]$ and $g(x) = 3 - [\![-x]\!]$ differ.

117. *Telephone Charges* A long distance phone service charges $0.40 for the first 10 minutes and $0.05 for each additional minute or fraction thereof. Use the greatest integer function to write the cost C of a call in terms of time t (in minutes). Sketch the graph of this function and discuss its continuity.

118. *Inventory Management* The number of units in inventory in a small company is given by

$$N(t) = 25\left(2\left[\![\frac{t + 2}{2}\right]\!] - t\right)$$

where t is the time in months. Sketch the graph of this function and discuss its continuity. How often must this company replenish its inventory?

119. ***Déjà Vu*** At 8:00 A.M. on Saturday, a man begins running up the side of a mountain to his weekend campsite (see figure). On Sunday morning at 8:00 A.M., he runs back down the mountain. It takes him 20 minutes to run up, but only 10 minutes to run down. At some point on the way down, he realizes that he passed the same place at exactly the same time on Saturday. Prove that he is correct. [*Hint:* Let $s(t)$ and $r(t)$ be the position functions for the runs up and down, and apply the Intermediate Value Theorem to the function $f(t) = s(t) - r(t)$.]

Saturday 8:00 A.M.   Sunday 8:00 A.M.
Not drawn to scale

120. ***Volume*** Use the Intermediate Value Theorem to show that for all spheres with radii in the interval $[5, 8]$, there is one with a volume of 1500 cubic centimeters.

121. Prove that if f is continuous and has no zeros on $[a, b]$, then either

$$f(x) > 0 \text{ for all } x \text{ in } [a, b] \text{ or } f(x) < 0 \text{ for all } x \text{ in } [a, b].$$

122. Show that the Dirichlet function

$$f(x) = \begin{cases} 0, & \text{if } x \text{ is rational} \\ 1, & \text{if } x \text{ is irrational} \end{cases}$$

is not continuous at any real number.

123. Show that the function

$$f(x) = \begin{cases} 0, & \text{if } x \text{ is rational} \\ kx, & \text{if } x \text{ is irrational} \end{cases}$$

is continuous only at $x = 0$. (Assume that k is any nonzero real number.)

124. The **signum function** is defined by

$$\operatorname{sgn}(x) = \begin{cases} -1, & x < 0 \\ 0, & x = 0 \\ 1, & x > 0. \end{cases}$$

Sketch a graph of $\operatorname{sgn}(x)$ and find the following (if possible).

(a) $\displaystyle\lim_{x \to 0^-} \operatorname{sgn}(x)$ (b) $\displaystyle\lim_{x \to 0^+} \operatorname{sgn}(x)$ (c) $\displaystyle\lim_{x \to 0} \operatorname{sgn}(x)$

125. ***Modeling Data*** The table shows the speeds S (in feet per second) of a falling object at various times t (in seconds).

t	0	5	10	15	20	25	30
S	0	48.2	53.5	55.2	55.9	56.2	56.3

(a) Create a line graph of the data.

(b) Does there appear to be a limiting speed of the object? If there is a limiting speed, identify a possible cause.

126. ***Creating Models*** A swimmer crosses a pool of width b by swimming in a straight line from $(0, 0)$ to $(2b, b)$. (See figure.)

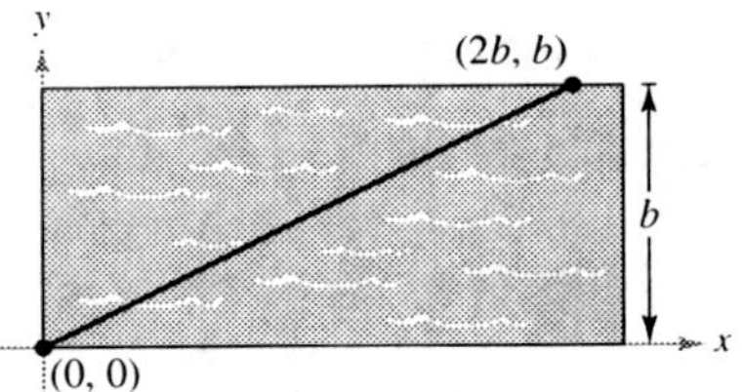

(a) Let f be a function defined as the y-coordinate of the point on the long side of the pool that is nearest the swimmer at any given time during the swimmer's crossing of the pool. Determine the function f and sketch its graph. Is it continuous? Explain.

(b) Let g be the minimum distance between the swimmer and the long sides of the pool. Determine the function g and sketch its graph. Is it continuous? Explain.

127. Find all values of c such that f is continuous on $(-\infty, \infty)$.

$$f(x) = \begin{cases} 1 - x^2, & x \le c \\ x, & x > c \end{cases}$$

128. Prove that for any real number y there exists x in $(-\pi/2, \pi/2)$ such that $\tan x = y$.

129. Let $f(x) = \left(\sqrt{x + c^2} - c\right)/x$, $c > 0$. What is the domain of f? How can you define f at $x = 0$ in order for f to be continuous there?

130. Prove that if $\displaystyle\lim_{\Delta x \to 0} f(c + \Delta x) = f(c)$, then f is continuous at c.

131. Discuss the continuity of the function $h(x) = x[\![x]\!]$.

132. (a) Let $f_1(x)$ and $f_2(x)$ be continuous on the closed interval $[a, b]$. If $f_1(a) < f_2(a)$ and $f_1(b) > f_2(b)$, prove that there exists c between a and b such that $f_1(c) = f_2(c)$.

(b) Show that there exists c in $\left[0, \frac{\pi}{2}\right]$ such that $\cos x = x$. Use a graphing utility to approximate c to three decimal places.

133. ***Think About It*** Consider the function

$$f(x) = \frac{4}{1 + 2^{4/x}}.$$

(a) What is the domain of the function?

(b) Use a graphing utility to graph the function.

(c) Determine $\displaystyle\lim_{x \to 0^-} f(x)$ and $\displaystyle\lim_{x \to 0^+} f(x)$.

(d) Use your knowledge of the exponential function to explain the behavior of f near $x = 0$.

PUTNAM EXAM CHALLENGE

134. Prove or disprove: if x and y are real numbers with $y \ge 0$ and $y(y + 1) \le (x + 1)^2$, then $y(y - 1) \le x^2$.

135. Determine all polynomials $P(x)$ such that

$$P(x^2 + 1) = (P(x))^2 + 1 \text{ and } P(0) = 0.$$

2.5 Infinite Limits

- **Determine infinite limits from the left and from the right.**
- **Find and sketch the vertical asymptotes of the graph of a function.**

Infinite Limits

Let f be the function given by

$$f(x) = \frac{3}{x - 2}.$$

From Figure 2.39 and the table, you can see that $f(x)$ *decreases without bound* as x approaches 2 from the left, and $f(x)$ *increases without bound* as x approaches 2 from the right. This behavior is denoted as

$$\lim_{x \to 2^-} \frac{3}{x-2} = -\infty \qquad \text{$f(x)$ decreases without bound as x approaches 2 from the left.}$$

and

$$\lim_{x \to 2^+} \frac{3}{x-2} = \infty \qquad \text{$f(x)$ increases without bound as x approaches 2 from the right.}$$

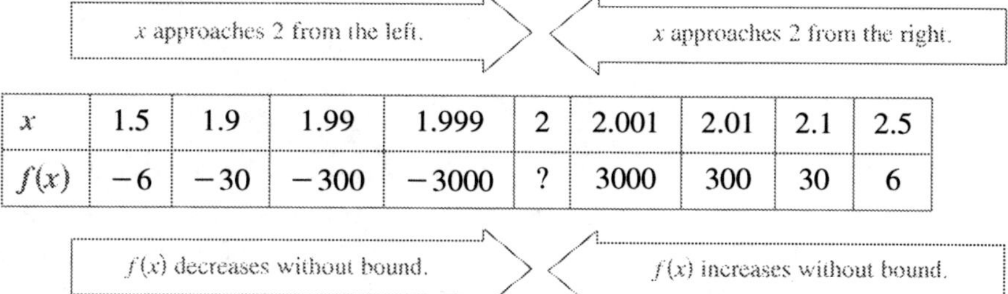

x	1.5	1.9	1.99	1.999	2	2.001	2.01	2.1	2.5
$f(x)$	-6	-30	-300	-3000	?	3000	300	30	6

A limit in which $f(x)$ increases or decreases without bound as x approaches c is called an **infinite limit.**

To the left, Figure 2.39:

$f(x)$ increases and decreases without bound as x approaches 2.

Figure 2.39

DEFINITION OF INFINITE LIMITS

Let f be a function that is defined at every real number in some open interval containing c (except possibly at c itself). The statement

$$\lim_{x \to c} f(x) = \infty$$

means that for each $M > 0$ there exists a $\delta > 0$ such that $f(x) > M$ whenever $0 < |x - c| < \delta$ (see Figure 2.40). Similarly, the statement

$$\lim_{x \to c} f(x) = -\infty$$

means that for each $N < 0$ there exists a $\delta > 0$ such that $f(x) < N$ whenever $0 < |x - c| < \delta$.

To define the **infinite limit from the left,** replace $0 < |x - c| < \delta$ by $c - \delta < x < c$. To define the **infinite limit from the right,** replace $0 < |x - c| < \delta$ by $c < x < c + \delta$.

Infinite limits

Figure 2.40

Be sure you see that the equal sign in the statement $\lim f(x) = \infty$ does not mean that the limit exists. On the contrary, it tells you how the limit **fails to exist** by denoting the unbounded behavior of $f(x)$ as x approaches c.

Use a graphing utility to graph each function. For each function, analytically find the single real number c that is not in the domain. Then graphically find the limit (if it exists) of $f(x)$ as x approaches c from the left and from the right.

a. $f(x) = \dfrac{3}{x - 4}$

b. $f(x) = \dfrac{1}{2 - x}$

c. $f(x) = \dfrac{2}{(x - 3)^2}$

d. $f(x) = \dfrac{-3}{(x + 2)^2}$

EXAMPLE 1 Determining Infinite Limits from a Graph

Determine the limit of each function shown in Figure 2.41 as x approaches 1 from the left and from the right.

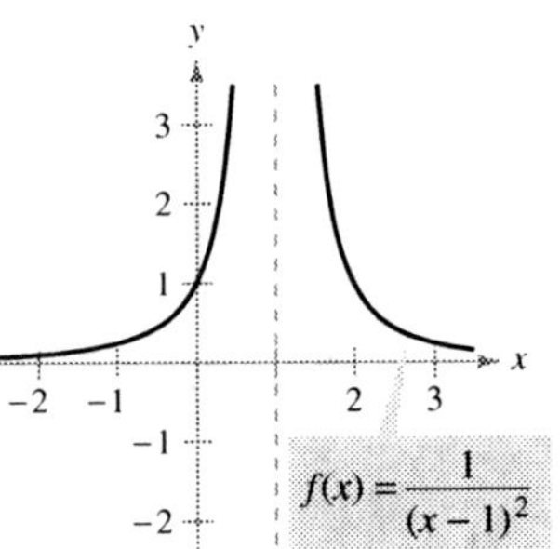

(a)

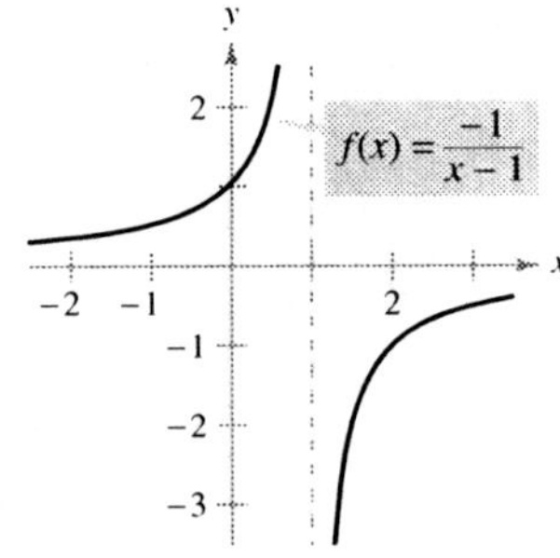

(b)

Each graph has an asymptote at $x = 1$.
Figure 2.41

Solution

a. When x approaches 1 from the left or the right, $(x - 1)^2$ is a small positive number. Thus, the quotient $1/(x - 1)^2$ is a large positive number and $f(x)$ approaches infinity from each side of $x = 1$. So, you can conclude that

$$\lim_{x \to 1} \frac{1}{(x - 1)^2} = \infty. \qquad \text{Limit from each side is infinity.}$$

Figure 2.41(a) confirms this analysis.

b. When x approaches 1 from the left, $x - 1$ is a small negative number. Thus, the quotient $-1/(x - 1)$ is a large positive number and $f(x)$ approaches infinity from the left of $x = 1$. So, you can conclude that

$$\lim_{x \to 1^-} \frac{-1}{x - 1} = \infty. \qquad \text{Limit from the left side is infinity.}$$

When x approaches 1 from the right, $x - 1$ is a small positive number. Thus, the quotient $-1/(x - 1)$ is a large negative number and $f(x)$ approaches negative infinity from the right of $x = 1$. So, you can conclude that

$$\lim_{x \to 1^+} \frac{-1}{x - 1} = -\infty. \qquad \text{Limit from the right side is negative infinity.}$$

Figure 2.41(b) confirms this analysis.

Vertical Asymptotes

If it were possible to extend the graphs in Figure 2.41 toward positive and negative infinity, you would see that each graph becomes arbitrarily close to the vertical line $x = 1$. This line is a **vertical asymptote** of the graph of f. (You will study other types of asymptotes in Sections 4.5 and 4.6.)

NOTE If a function f has a vertical asymptote at $x = c$, then f is *not continuous* at c.

DEFINITION OF VERTICAL ASYMPTOTE

If $f(x)$ approaches infinity (or negative infinity) as x approaches c from the right or the left, then the line $x = c$ is a **vertical asymptote** of the graph of f.

In Example 1, note that each of the functions is a *quotient* and that the vertical asymptote occurs at a number at which the denominator is 0 (and the numerator is not 0). The next theorem generalizes this observation. (A proof of this theorem is given in Appendix A.)

THEOREM 2.14 VERTICAL ASYMPTOTES

Let f and g be continuous on an open interval containing c. If $f(c) \neq 0$, $g(c) = 0$, and there exists an open interval containing c such that $g(x) \neq 0$ for all $x \neq c$ in the interval, then the graph of the function given by

$$h(x) = \frac{f(x)}{g(x)}$$

has a vertical asymptote at $x = c$.

EXAMPLE 2 Finding Vertical Asymptotes

Determine all vertical asymptotes of the graph of each function.

a. $f(x) = \dfrac{1}{2(x + 1)}$ **b.** $f(x) = \dfrac{x^2 + 1}{x^2 - 1}$ **c.** $f(x) = \cot x$

Solution

a. When $x = -1$, the denominator of

$$f(x) = \frac{1}{2(x + 1)}$$

is 0 and the numerator is not 0. So, by Theorem 2.14, you can conclude that $x = -1$ is a vertical asymptote, as shown in Figure 2.42(a).

b. By factoring the denominator as

$$f(x) = \frac{x^2 + 1}{x^2 - 1} = \frac{x^2 + 1}{(x - 1)(x + 1)}$$

you can see that the denominator is 0 at $x = -1$ and $x = 1$. Moreover, because the numerator is not 0 at these two points, you can apply Theorem 2.14 to conclude that the graph of f has two vertical asymptotes, as shown in Figure 2.42(b).

c. By writing the cotangent function in the form

$$f(x) = \cot x = \frac{\cos x}{\sin x}$$

you can apply Theorem 2.14 to conclude that vertical asymptotes occur at all values of x such that $\sin x = 0$ and $\cos x \neq 0$, as shown in Figure 2.42(c). So, the graph of this function has infinitely many vertical asymptotes. These asymptotes occur when $x = n\pi$, where n is an integer.

Theorem 2.14 requires that the value of the numerator at $x = c$ be nonzero. If both the numerator and the denominator are 0 at $x = c$, you obtain the *indeterminate form* 0/0, and you cannot determine the limit behavior at $x = c$ without further investigation, as illustrated in Example 3.

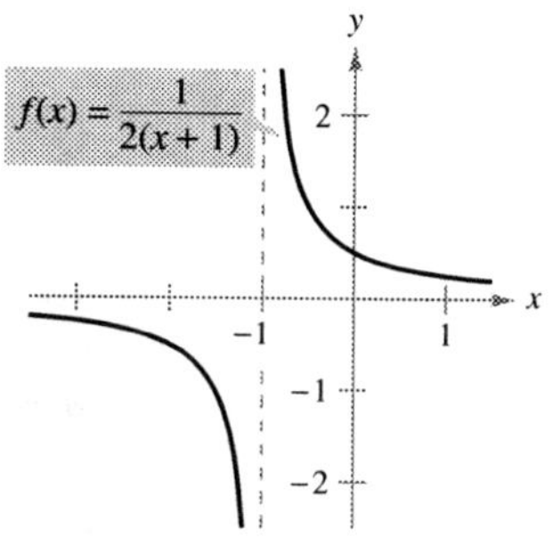

(a)

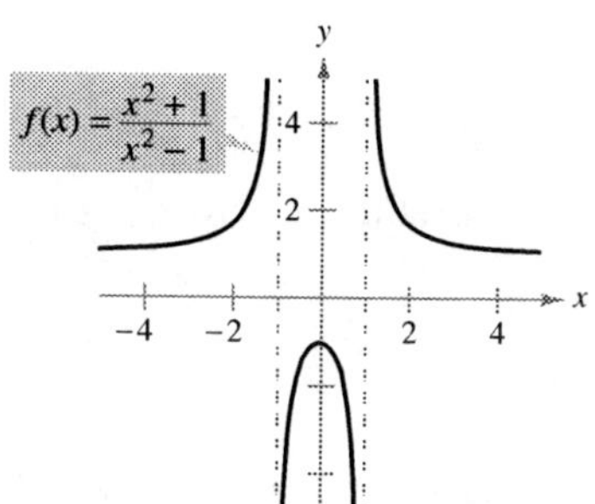

(b)

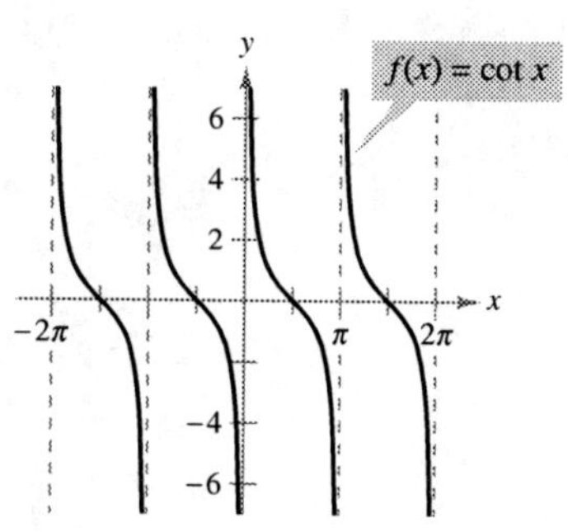

(c)

Functions with vertical asymptotes
Figure 2.42

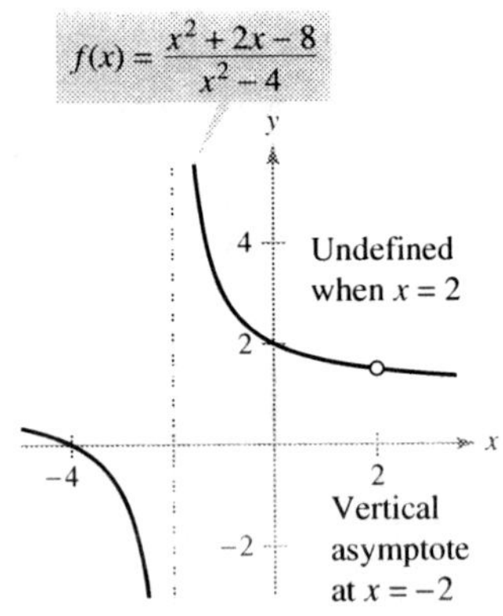

$f(x) = \dfrac{x^2 + 2x - 8}{x^2 - 4}$

Undefined when $x = 2$

Vertical asymptote at $x = -2$

$f(x)$ increases and decreases without bound as x approaches -2.

Figure 2.43

EXAMPLE 3 A Rational Function with Common Factors

Determine all vertical asymptotes of the graph of

$$f(x) = \frac{x^2 + 2x - 8}{x^2 - 4}.$$

Solution Begin by simplifying the expression, as shown.

$$f(x) = \frac{x^2 + 2x - 8}{x^2 - 4}$$

$$= \frac{(x + 4)(x - 2)}{(x + 2)(x - 2)}$$

$$= \frac{x + 4}{x + 2}, \quad x \neq 2$$

At all x-values other than $x = 2$, the graph of f coincides with the graph of $g(x) = (x + 4)/(x + 2)$. So, you can apply Theorem 2.14 to g to conclude that there is a vertical asymptote at $x = -2$, as shown in Figure 2.43. From the graph, you can see that

$$\lim_{x \to -2^-} \frac{x^2 + 2x - 8}{x^2 - 4} = -\infty \quad \text{and} \quad \lim_{x \to -2^+} \frac{x^2 + 2x - 8}{x^2 - 4} = \infty.$$

Note that $x = 2$ is *not* a vertical asymptote. Rather, $x = 2$ is a removable discontinuity.

EXAMPLE 4 Determining Infinite Limits

Find each limit.

$$\lim_{x \to 1^-} \frac{x^2 - 3x}{x - 1} \quad \text{and} \quad \lim_{x \to 1^+} \frac{x^2 - 3x}{x - 1}$$

Solution Because the denominator is 0 when $x = 1$ (and the numerator is not zero), you know that the graph of

$$f(x) = \frac{x^2 - 3x}{x - 1}$$

has a vertical asymptote at $x = 1$. This means that each of the given limits is either ∞ or $-\infty$. You can determine the result by analyzing f at values of x close to 1, or by using a graphing utility. From the graph of f shown in Figure 2.44, you can see that the graph approaches ∞ from the left of $x = 1$ and approaches $-\infty$ from the right of $x = 1$. So, you can conclude that

$$\lim_{x \to 1^-} \frac{x^2 - 3x}{x - 1} = \infty \qquad \text{The limit from the left is infinity.}$$

and

$$\lim_{x \to 1^+} \frac{x^2 - 3x}{x - 1} = -\infty. \qquad \text{The limit from the right is negative infinity.} \qquad \blacksquare$$

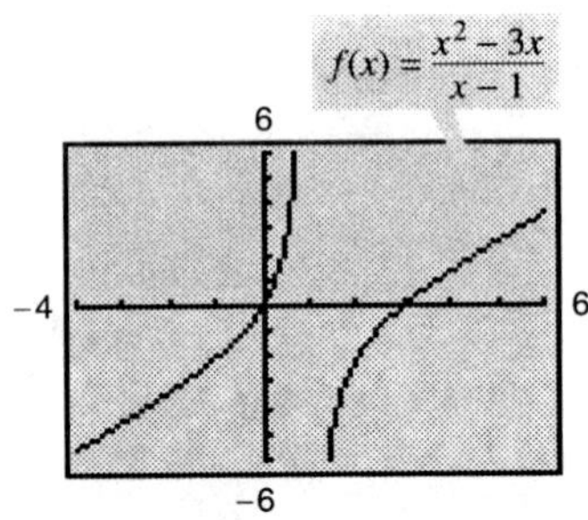

f has a vertical asymptote at $x = 1$.

Figure 2.44

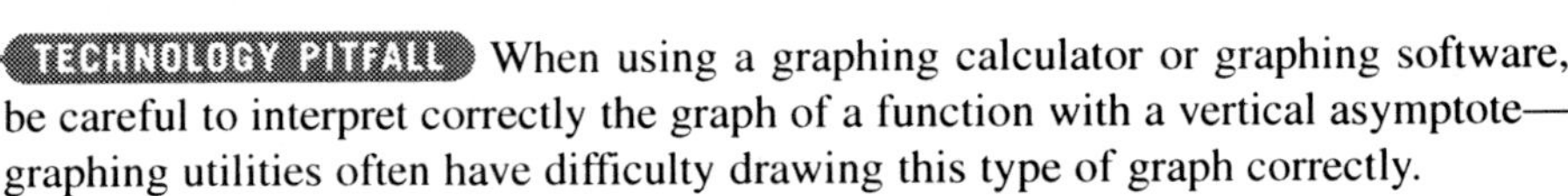

TECHNOLOGY PITFALL When using a graphing calculator or graphing software, be careful to interpret correctly the graph of a function with a vertical asymptote—graphing utilities often have difficulty drawing this type of graph correctly.

THEOREM 2.15 PROPERTIES OF INFINITE LIMITS

Let c and L be real numbers and let f and g be functions such that

$$\lim_{x \to c} f(x) = \infty \quad \text{and} \quad \lim_{x \to c} g(x) = L.$$

1. Sum or difference: $\displaystyle \lim_{x \to c} [f(x) \pm g(x)] = \infty$

2. Product: $\displaystyle \lim_{x \to c} [f(x)g(x)] = \infty, \quad L > 0$

$\displaystyle \lim_{x \to c} [f(x)g(x)] = -\infty, \quad L < 0$

3. Quotient: $\displaystyle \lim_{x \to c} \frac{g(x)}{f(x)} = 0$

Similar properties hold for one-sided limits and for functions for which the limit of $f(x)$ as x approaches c is $-\infty$.

NOTE With a graphing utility, you can confirm that the natural logarithmic function has a vertical asymptote at $x = 0$. (See Figure 2.45.) This implies that

$$\lim_{x \to 0^+} \ln x = -\infty.$$

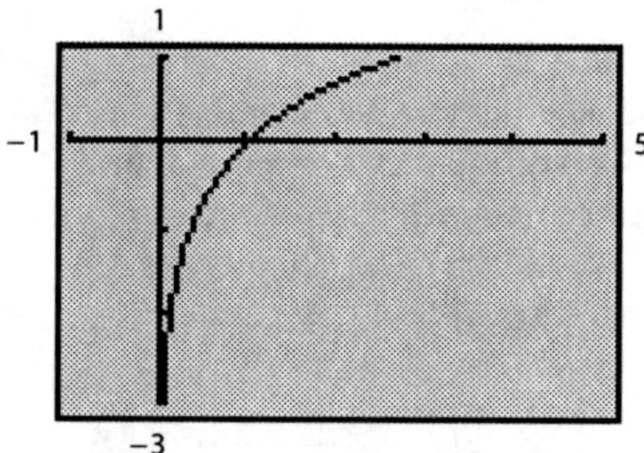

Figure 2.45

PROOF To show that the limit of $f(x) + g(x)$ is infinite, choose $M > 0$. You then need to find $\delta > 0$ such that

$$[f(x) + g(x)] > M$$

whenever $0 < |x - c| < \delta$. For simplicity's sake, you can assume L is positive and let $M_1 = M + 1$. Because the limit of $f(x)$ is infinite, there exists δ_1 such that $f(x) > M_1$ whenever $0 < |x - c| < \delta_1$. Also, because the limit of $g(x)$ is L, there exists δ_2 such that $|g(x) - L| < 1$ whenever $0 < |x - c| < \delta_2$. By letting δ be the smaller of δ_1 and δ_2, you can conclude that $0 < |x - c| < \delta$ implies $f(x) > M + 1$ and $|g(x) - L| < 1$. The second of these two inequalities implies that $g(x) > L - 1$, and, adding this to the first inequality, you can write

$$f(x) + g(x) > (M + 1) + (L - 1) = M + L > M.$$

So, you can conclude that

$$\lim_{x \to c} [f(x) + g(x)] = \infty.$$

The proofs of the remaining properties are left as an exercise (see Exercise 86).

EXAMPLE 5 Determining Limits

a. Because $\displaystyle \lim_{x \to 0} 1 = 1$ and $\displaystyle \lim_{x \to 0} \frac{1}{x^2} = \infty$, you can write

$$\lim_{x \to 0} \left(1 + \frac{1}{x^2} \right) = \infty. \qquad \text{Property 1, Theorem 2.15}$$

b. Because $\displaystyle \lim_{x \to 1^-} (x^2 + 1) = 2$ and $\displaystyle \lim_{x \to 1^-} (\cot \pi x) = -\infty$, you can write

$$\lim_{x \to 1^-} \frac{x^2 + 1}{\cot \pi x} = 0. \qquad \text{Property 3, Theorem 2.15}$$

c. Because $\displaystyle \lim_{x \to 0^+} 3 = 3$ and $\displaystyle \lim_{x \to 0^+} \ln x = -\infty$, you can write

$$\lim_{x \to 0^+} 3 \ln x = -\infty. \qquad \text{Property 2, Theorem 2.15}$$

2.5 Exercises   See www.CalcChat.com for worked-out solutions to odd-numbered exercises.

In Exercises 1–4, determine whether $f(x)$ approaches ∞ or $-\infty$ as x approaches 4 from the left and from the right.

1. $f(x) = \dfrac{1}{x - 4}$

2. $f(x) = \dfrac{-1}{x - 4}$

3. $f(x) = \dfrac{1}{(x - 4)^2}$

4. $f(x) = \dfrac{-1}{(x - 4)^2}$

In Exercises 5–8, determine whether $f(x)$ approaches ∞ or $-\infty$ as x approaches -2 from the left and from the right.

5. $f(x) = 2\left|\dfrac{x}{x^2 - 4}\right|$

6. $f(x) = \dfrac{-1}{x + 2}$

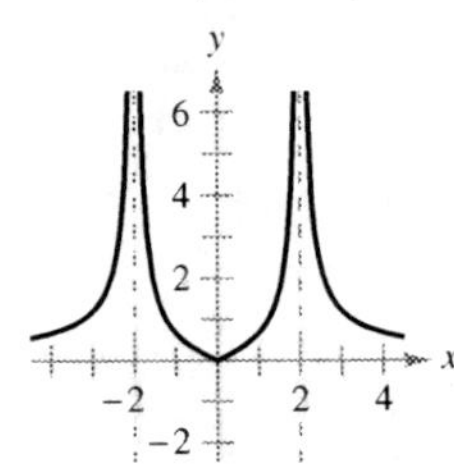

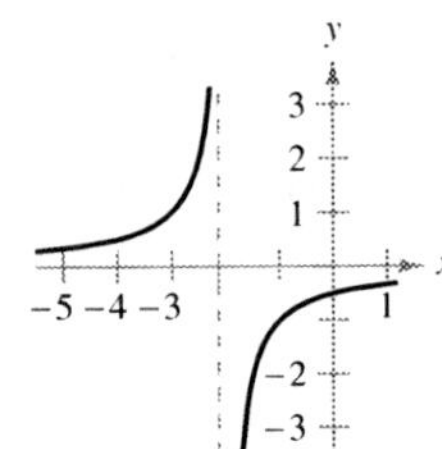

7. $f(x) = \tan\dfrac{\pi x}{4}$

8. $f(x) = \sec\dfrac{\pi x}{4}$

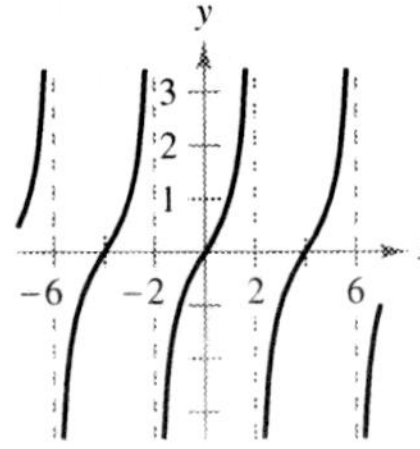

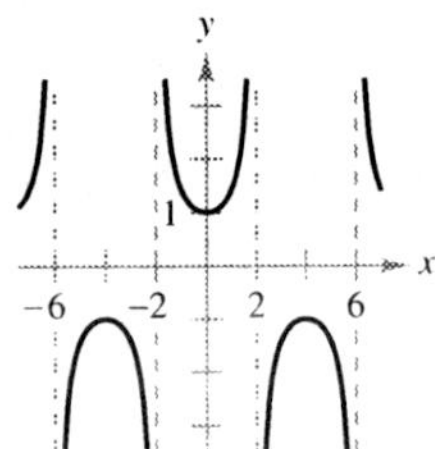

Numerical and Graphical Analysis **In Exercises 9–12, determine whether $f(x)$ approaches ∞ or $-\infty$ as x approaches -3 from the left and from the right by completing the table. Use a graphing utility to graph the function to confirm your answer.**

x	-3.5	-3.1	-3.01	-3.001
$f(x)$				

x	-2.999	-2.99	-2.9	-2.5
$f(x)$				

9. $f(x) = \dfrac{1}{x^2 - 9}$

10. $f(x) = \dfrac{x}{x^2 - 9}$

11. $f(x) = \dfrac{x^2}{x^2 - 9}$

12. $f(x) = \sec\dfrac{\pi x}{6}$

In Exercises 13–36, find the vertical asymptotes (if any) of the graph of the function.

13. $f(x) = \dfrac{1}{x^2}$

14. $f(x) = \dfrac{4}{(x - 2)^3}$

15. $f(x) = \dfrac{x^2}{x^2 - 4}$

16. $f(x) = \dfrac{-4x}{x^2 + 4}$

17. $g(t) = \dfrac{t - 1}{t^2 + 1}$

18. $h(s) = \dfrac{2s - 3}{s^2 - 25}$

19. $h(x) = \dfrac{x^2 - 2}{x^2 - x - 2}$

20. $g(x) = \dfrac{2 + x}{x^2(1 - x)}$

21. $T(t) = 1 - \dfrac{4}{t^2}$

22. $g(x) = \dfrac{\frac{1}{2}x^3 - x^2 - 4x}{3x^2 - 6x - 24}$

23. $f(x) = \dfrac{3}{x^2 + x - 2}$

24. $f(x) = \dfrac{-3x^2 + 12x - 9}{x^4 - 3x^3 - x + 3}$

25. $g(x) = \dfrac{x^3 + 1}{x + 1}$

26. $h(x) = \dfrac{x^2 - 4}{x^3 - 2x^2 + x - 2}$

27. $f(x) = \dfrac{e^{-2x}}{x - 1}$

28. $g(x) = xe^{-2x}$

29. $h(t) = \dfrac{\ln(t^2 + 1)}{t + 2}$

30. $f(z) = \ln(z^2 - 4)$

31. $f(x) = \dfrac{1}{e^x - 1}$

32. $f(x) = \ln(x + 3)$

33. $f(x) = \tan \pi x$

34. $f(x) = \sec \pi x$

35. $s(t) = \dfrac{t}{\sin t}$

36. $g(\theta) = \dfrac{\tan \theta}{\theta}$

In Exercises 37–42, determine whether the function has a vertical asymptote or a removable discontinuity at $x = -1$. Graph the function using a graphing utility to confirm your answer.

37. $f(x) = \dfrac{x^2 - 1}{x + 1}$

38. $f(x) = \dfrac{x^2 - 6x - 7}{x + 1}$

39. $f(x) = \dfrac{x^2 + 1}{x + 1}$

40. $f(x) = \dfrac{\sin(x + 1)}{x + 1}$

41. $f(x) = \dfrac{e^{2(x + 1)} - 1}{e^{x+1} - 1}$

42. $f(x) = \dfrac{\ln(x^2 + 1)}{x + 1}$

In Exercises 43–62, find the limit (if it exists).

43. $\displaystyle\lim_{x \to -1^+} \dfrac{1}{x + 1}$

44. $\displaystyle\lim_{x \to 1^-} \dfrac{-1}{(x - 1)^2}$

45. $\displaystyle\lim_{x \to 2^+} \dfrac{x}{x - 2}$

46. $\displaystyle\lim_{x \to 1^+} \dfrac{x(2 + x)}{1 - x}$

47. $\displaystyle\lim_{x \to 1^+} \dfrac{x^2}{(x - 1)^2}$

48. $\displaystyle\lim_{x \to 4^-} \dfrac{x^2}{x^2 + 16}$

49. $\displaystyle\lim_{x \to -3} \dfrac{x + 3}{x^2 + x - 6}$

50. $\displaystyle\lim_{x \to (-1/2)^+} \dfrac{6x^2 + x - 1}{4x^2 - 4x - 3}$

51. $\displaystyle\lim_{x \to 1} \dfrac{x - 1}{(x^2 + 1)(x - 1)}$

52. $\displaystyle\lim_{x \to 3} \dfrac{x - 2}{x^2}$

53. $\displaystyle\lim_{x \to 0} \left(1 + \dfrac{1}{x}\right)$

54. $\displaystyle\lim_{x \to 0} \left(x^2 - \dfrac{2}{x}\right)$

55. $\displaystyle\lim_{x \to 0^+} \dfrac{2}{\sin x}$

56. $\displaystyle\lim_{x \to (\pi/2)^+} \dfrac{-2}{\cos x}$

57. $\lim\limits_{x \to 8^-} \dfrac{e^x}{(x-8)^3}$

58. $\lim\limits_{x \to 4^+} \ln(x^2 - 16)$

59. $\lim\limits_{x \to (\pi/2)^-} \ln|\cos x|$

60. $\lim\limits_{x \to 0^+} e^{-0.5x} \sin x$

61. $\lim\limits_{x \to 1/2} x \sec \pi x$

62. $\lim\limits_{x \to 1/2} x^2 \tan \pi x$

In Exercises 63–66, use a graphing utility to graph the function and determine the one-sided limit.

63. $f(x) = \dfrac{x^2 + x + 1}{x^3 - 1}$

$\lim\limits_{x \to 1^+} f(x)$

64. $f(x) = \dfrac{x^3 - 1}{x^2 + x + 1}$

$\lim\limits_{x \to 1^-} f(x)$

65. $f(x) = \dfrac{1}{x^2 - 25}$

$\lim\limits_{x \to 5^-} f(x)$

66. $f(x) = \sec \dfrac{\pi x}{8}$

$\lim\limits_{x \to 4^+} f(x)$

WRITING ABOUT CONCEPTS

67. In your own words, describe the meaning of an infinite limit. Is ∞ a real number?

68. In your own words, describe what is meant by an asymptote of a graph.

69. Write a rational function with vertical asymptotes at $x = 6$ and $x = -2$, and with a zero at $x = 3$.

70. Does every rational function have a vertical asymptote? Explain.

71. Use the graph of the function f (see figure) to sketch the graph of $g(x) = 1/f(x)$ on the interval $[-2, 3]$. To print an enlarged copy of the graph, go to the website *www.mathgraphs.com*.

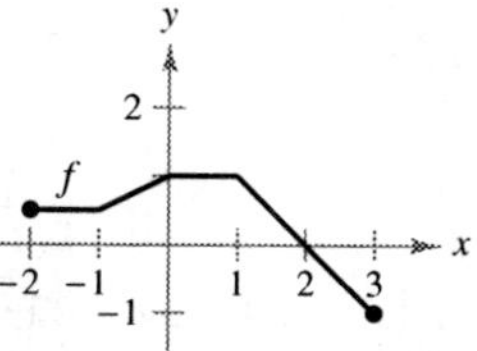

CAPSTONE

72. Given a polynomial $p(x)$, is it true that the graph of the function given by $f(x) = \dfrac{p(x)}{x - 1}$ has a vertical asymptote at $x = 1$? Why or why not?

73. *Relativity* According to the theory of relativity, the mass m of a particle depends on its velocity v. That is,

$$m = \frac{m_0}{\sqrt{1 - (v^2/c^2)}}$$

where m_0 is the mass when the particle is at rest and c is the speed of light. Find the limit of the mass as v approaches c^-.

74. *Boyle's Law* For a quantity of gas at a constant temperature, the pressure P is inversely proportional to the volume V. Find the limit of P as $V \to 0^+$.

75. *Rate of Change* A patrol car is parked 50 feet from a long warehouse (see figure). The revolving light on top of the car turns at a rate of $\frac{1}{2}$ revolution per second. The rate r at which the light beam moves along the wall is $r = 50\pi \sec^2 \theta$ ft/sec.

(a) Find r when θ is $\pi/6$.

(b) Find r when θ is $\pi/3$.

(c) Find the limit of r as $\theta \to (\pi/2)^-$.

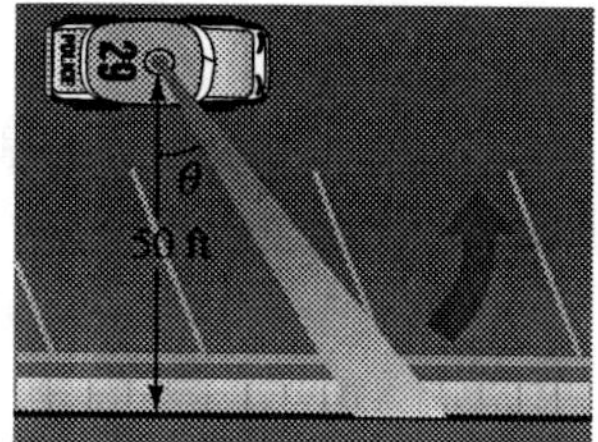

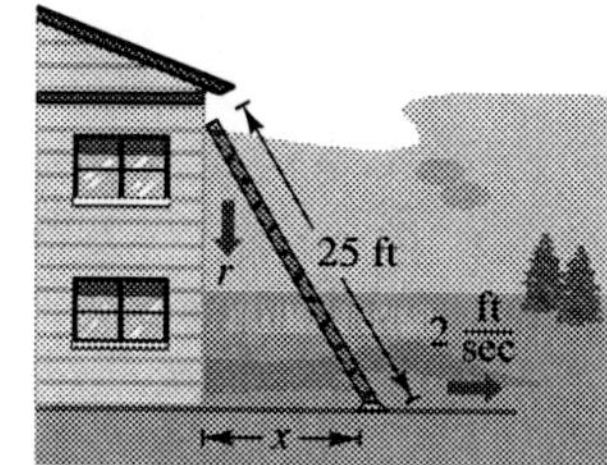

Figure for 75   **Figure for 76**

76. *Rate of Change* A 25-foot ladder is leaning against a house (see figure). If the base of the ladder is pulled away from the house at a rate of 2 feet per second, the top will move down the wall at a rate of

$$r = \frac{2x}{\sqrt{625 - x^2}} \text{ ft/sec}$$

where x is the distance between the ladder base and the house.

(a) Find r when x is 7 feet.

(b) Find r when x is 15 feet.

(c) Find the limit of r as $x \to 25^-$.

77. *Average Speed* On a trip of d miles to another city, a truck driver's average speed was x miles per hour. On the return trip the average speed was y miles per hour. The average speed for the round trip was 50 miles per hour.

(a) Verify that $y = \dfrac{25x}{x - 25}$. What is the domain?

(b) Complete the table.

x	30	40	50	60
y				

Are the values of y different than you expected? Explain.

(c) Find the limit of y as $x \to 25^+$ and interpret its meaning.

78. *Numerical and Graphical Analysis* Use a graphing utility to complete the table for each function and graph each function to estimate the limit. What is the value of the limit when the power of x in the denominator is greater than 3?

x	1	0.5	0.2	0.1	0.01	0.001	0.0001
$f(x)$							

(a) $\lim\limits_{x \to 0^+} \dfrac{x - \sin x}{x}$

(b) $\lim\limits_{x \to 0^+} \dfrac{x - \sin x}{x^2}$

(c) $\lim\limits_{x \to 0^+} \dfrac{x - \sin x}{x^3}$

(d) $\lim\limits_{x \to 0^+} \dfrac{x - \sin x}{x^4}$

79. *Numerical and Graphical Analysis* Consider the shaded region outside the sector of a circle of radius 10 meters and inside a right triangle (see figure).

(a) Write the area $A = f(\theta)$ of the region as a function of θ. Determine the domain of the function.

(b) Use a graphing utility to complete the table and graph the function over the appropriate domain.

θ	0.3	0.6	0.9	1.2	1.5
$f(\theta)$					

(c) Find the limit of A as $\theta \to (\pi/2)^-$.

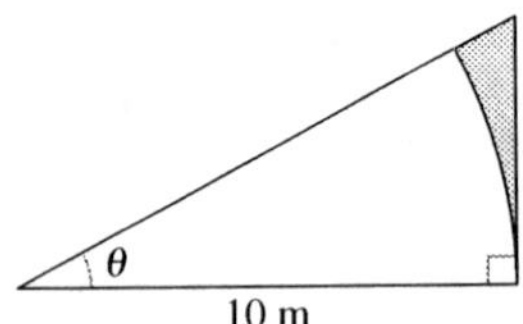

80. *Numerical and Graphical Reasoning* A crossed belt connects a 20-centimeter pulley (10-cm radius) on an electric motor with a 40-centimeter pulley (20-cm radius) on a saw arbor (see figure). The electric motor runs at 1700 revolutions per minute.

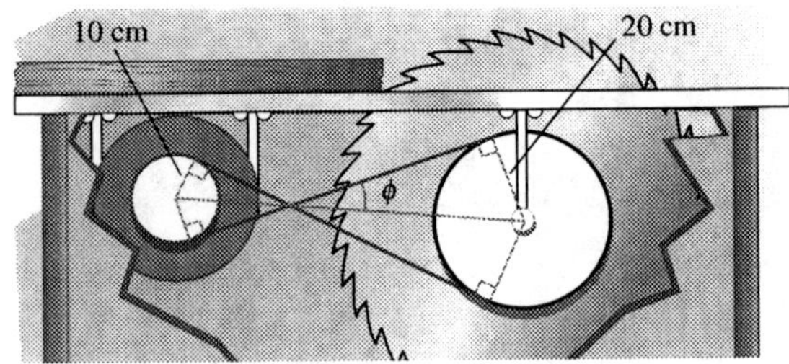

(a) Determine the number of revolutions per minute of the saw.

(b) How does crossing the belt affect the saw in relation to the motor?

(c) Let L be the total length of the belt. Write L as a function of ϕ, where ϕ is measured in radians. What is the domain of the function? (*Hint:* Add the lengths of the straight sections of the belt and the length of the belt around each pulley.)

(d) Use a graphing utility to complete the table.

ϕ	0.3	0.6	0.9	1.2	1.5
L					

(e) Use a graphing utility to graph the function over the appropriate domain.

(f) Find $\displaystyle\lim_{\phi \to (\pi/2)^-} L$. Use a geometric argument as the basis of a second method of finding this limit.

(g) Find $\displaystyle\lim_{\phi \to 0^+} L$.

True or False? **In Exercises 81–84, determine whether the statement is true or false. If it is false, explain why or give an example that shows it is false.**

81. The graph of a rational function has at least one vertical asymptote.

82. The graphs of polynomial functions have no vertical asymptotes.

83. The graphs of trigonometric functions have no vertical asymptotes.

84. If f has a vertical asymptote at $x = 0$, then f is undefined at $x = 0$.

85. Find functions f and g such that $\displaystyle\lim_{x \to c} f(x) = \infty$ and $\displaystyle\lim_{x \to c} g(x) = \infty$ but $\displaystyle\lim_{x \to c} [f(x) - g(x)] \neq 0$.

86. Prove the difference, product, and quotient properties in Theorem 2.15.

87. Prove that if $\displaystyle\lim_{x \to c} f(x) = \infty$, then $\displaystyle\lim_{x \to c} \frac{1}{f(x)} = 0$.

88. Prove that if $\displaystyle\lim_{x \to c} \frac{1}{f(x)} = 0$, then $\displaystyle\lim_{x \to c} f(x)$ does not exist.

Infinite Limits **In Exercises 89 and 90, use the ε-δ definition of infinite limits to prove the statement.**

89. $\displaystyle\lim_{x \to 3^+} \frac{1}{x - 3} = \infty$

90. $\displaystyle\lim_{x \to 5^-} \frac{1}{x - 5} = -\infty$

SECTION PROJECT

Graphs and Limits of Trigonometric Functions

Recall from Theorem 2.9 that the limit of $f(x) = (\sin x)/x$ as x approaches 0 is 1.

(a) Use a graphing utility to graph the function f on the interval $-\pi \le x \le \pi$. Explain how the graph helps confirm that $\displaystyle\lim_{x \to 0} \frac{\sin x}{x} = 1$.

(b) Explain how you could use a table of values to confirm the value of this limit numerically.

(c) Graph $g(x) = \sin x$ by hand. Sketch a tangent line at the point $(0, 0)$ and visually estimate the slope of this tangent line.

(d) Let $(x, \sin x)$ be a point on the graph of g near $(0, 0)$, and write a formula for the slope of the secant line joining $(x, \sin x)$ and $(0, 0)$. Evaluate this formula at $x = 0.1$ and $x = 0.01$. Then find the exact slope of the tangent line to g at the point $(0, 0)$.

(e) Sketch the graph of the cosine function $h(x) = \cos x$. What is the slope of the tangent line at the point $(0, 1)$? Use limits to find this slope analytically.

(f) Find the slope of the tangent line to $k(x) = \tan x$ at $(0, 0)$.

2 REVIEW EXERCISES

In Exercises 1 and 2, determine whether the problem can be solved using precalculus or if calculus is required. If the problem can be solved using precalculus, solve it. If the problem seems to require calculus, explain your reasoning. Use a graphical or numerical approach to estimate the solution.

1. Find the distance between the points $(1, 1)$ and $(3, 9)$ along the curve $y = x^2$.

2. Find the distance between the points $(1, 1)$ and $(3, 9)$ along the line $y = 4x - 3$.

In Exercises 3–6, complete the table and use the result to estimate the limit. Use a graphing utility to graph the function to confirm your result.

x	-0.1	-0.01	-0.001	0.001	0.01	0.1
$f(x)$						

3. $\displaystyle\lim_{x \to 0} \frac{[4/(x + 2)] - 2}{x}$

4. $\displaystyle\lim_{x \to 0} \frac{4\left(\sqrt{x + 2} - \sqrt{2}\right)}{x}$

5. $\displaystyle\lim_{x \to 0} \frac{20(e^{x/2} - 1)}{x - 1}$

6. $\displaystyle\lim_{x \to 0} \frac{\ln(x + 5) - \ln 5}{x}$

In Exercises 7–10, use the graph to determine each limit.

7. $h(x) = \dfrac{4x - x^2}{x}$

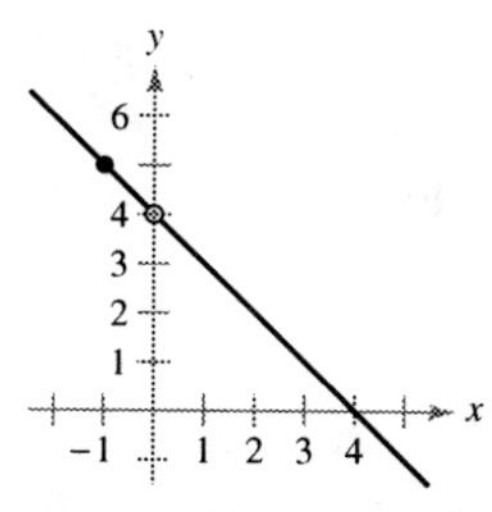

8. $g(x) = \dfrac{-2x}{x - 3}$

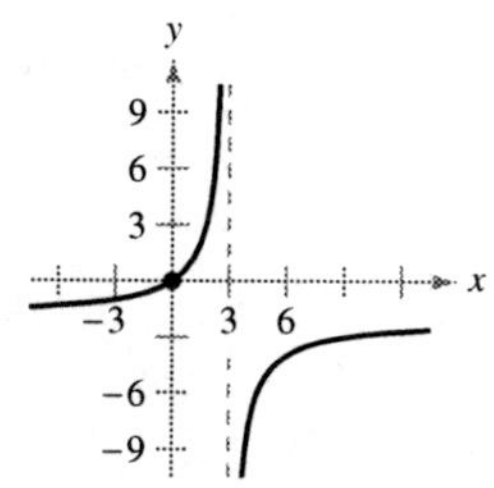

(a) $\displaystyle\lim_{x \to 0} h(x)$ (b) $\displaystyle\lim_{x \to -1} h(x)$

(a) $\displaystyle\lim_{x \to 3} g(x)$ (b) $\displaystyle\lim_{x \to 0} g(x)$

9. $f(t) = \dfrac{\ln(t + 2)}{t}$

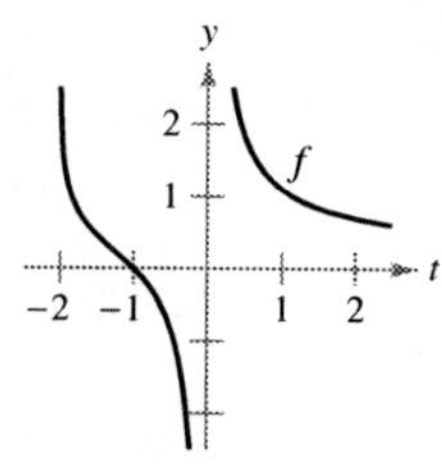

10. $g(x) = e^{-x/2} \sin \pi x$

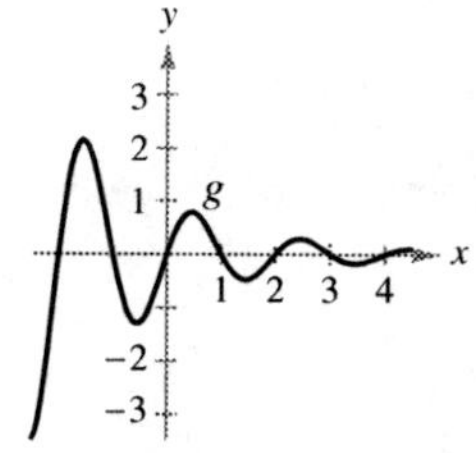

(a) $\displaystyle\lim_{t \to 0} f(t)$ (b) $\displaystyle\lim_{t \to -1} f(t)$

(a) $\displaystyle\lim_{x \to 0} g(x)$ (b) $\displaystyle\lim_{x \to 2} g(x)$

In Exercises 11–14, find the limit L. Then use the ε-δ definition to prove that the limit is L.

11. $\displaystyle\lim_{x \to 1} (x + 4)$

12. $\displaystyle\lim_{x \to 9} \sqrt{x}$

13. $\displaystyle\lim_{x \to 2} (1 - x^2)$

14. $\displaystyle\lim_{x \to 5} 9$

In Exercises 15–32, find the limit (if it exists).

15. $\displaystyle\lim_{x \to 6} (x - 2)^2$

16. $\displaystyle\lim_{x \to 7} (10 - x)^4$

17. $\displaystyle\lim_{t \to 4} \sqrt{t + 2}$

18. $\displaystyle\lim_{y \to 4} 3|y - 1|$

19. $\displaystyle\lim_{t \to -2} \frac{t + 2}{t^2 - 4}$

20. $\displaystyle\lim_{t \to 3} \frac{t^2 - 9}{t - 3}$

21. $\displaystyle\lim_{x \to 4} \frac{\sqrt{x - 3} - 1}{x - 4}$

22. $\displaystyle\lim_{x \to 0} \frac{\sqrt{4 + x} - 2}{x}$

23. $\displaystyle\lim_{x \to 0} \frac{[1/(x + 1)] - 1}{x}$

24. $\displaystyle\lim_{s \to 0} \frac{\left(1/\sqrt{1 + s}\right) - 1}{s}$

25. $\displaystyle\lim_{x \to -5} \frac{x^3 + 125}{x + 5}$

26. $\displaystyle\lim_{x \to -2} \frac{x^2 - 4}{x^3 + 8}$

27. $\displaystyle\lim_{x \to 0} \frac{1 - \cos x}{\sin x}$

28. $\displaystyle\lim_{x \to \pi/4} \frac{4x}{\tan x}$

29. $\displaystyle\lim_{x \to 1} e^{x-1} \sin \frac{\pi x}{2}$

30. $\displaystyle\lim_{x \to 2} \frac{\ln(x - 1)^2}{\ln(x - 1)}$

31. $\displaystyle\lim_{\Delta x \to 0} \frac{\sin[(\pi/6) + \Delta x] - (1/2)}{\Delta x}$

[*Hint:* $\sin(\theta + \phi) = \sin \theta \cos \phi + \cos \theta \sin \phi$]

32. $\displaystyle\lim_{\Delta x \to 0} \frac{\cos(\pi + \Delta x) + 1}{\Delta x}$

[*Hint:* $\cos(\theta + \phi) = \cos \theta \cos \phi - \sin \theta \sin \phi$]

In Exercises 33–36, evaluate the limit given $\displaystyle\lim_{x \to c} f(x) = -\frac{3}{4}$ and $\displaystyle\lim_{x \to c} g(x) = \frac{2}{3}$.

33. $\displaystyle\lim_{x \to c} [f(x)g(x)]$

34. $\displaystyle\lim_{x \to c} \frac{f(x)}{g(x)}$

35. $\displaystyle\lim_{x \to c} [f(x) + 2g(x)]$

36. $\displaystyle\lim_{x \to c} [f(x)]^2$

Numerical, Graphical, and Analytic Analysis **In Exercises 37 and 38, consider $\displaystyle\lim_{x \to 1^+} f(x)$.**

(a) Complete the table to estimate the limit.

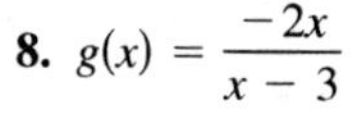

(b) Use a graphing utility to graph the function and use the graph to estimate the limit.

(c) Rationalize the numerator to find the exact value of the limit analytically.

x	1.1	1.01	1.001	1.0001
$f(x)$				

37. $f(x) = \dfrac{\sqrt{2x + 1} - \sqrt{3}}{x - 1}$

38. $f(x) = \dfrac{1 - \sqrt[3]{x}}{x - 1}$

[*Hint:* $a^3 - b^3 = (a - b)(a^2 + ab + b^2)$]

Free-Falling Object   In Exercises 39 and 40, use the position function $s(t) = -4.9t^2 + 250$, which gives the height (in meters) of an object that has fallen from a height of 250 meters. The velocity at time $t = a$ seconds is given by

$$\lim_{t \to a} \frac{s(a) - s(t)}{a - t}.$$

39. Find the velocity of the object when $t = 4$.

40. At what velocity will the object impact the ground?

In Exercises 41–46, find the limit (if it exists). If the limit does not exist, explain why.

41. $\displaystyle \lim_{x \to 3^-} \frac{|x - 3|}{x - 3}$

42. $\displaystyle \lim_{x \to 4} [\![x - 1]\!]$

43. $\displaystyle \lim_{x \to 2} f(x)$, where $f(x) = \begin{cases} (x - 2)^2, & x \le 2 \\ 2 - x, & x > 2 \end{cases}$

44. $\displaystyle \lim_{x \to 1^+} g(x)$, where $g(x) = \begin{cases} \sqrt{1 - x}, & x \le 1 \\ x + 1, & x > 1 \end{cases}$

45. $\displaystyle \lim_{t \to 1} h(t)$, where $h(t) = \begin{cases} t^3 + 1, & t < 1 \\ \frac{1}{2}(t + 1), & t \ge 1 \end{cases}$

46. $\displaystyle \lim_{s \to -2} f(s)$, where $f(s) = \begin{cases} -s^2 - 4s - 2, & s \le -2 \\ s^2 + 4s + 6, & s > -2 \end{cases}$

In Exercises 47–60, determine the intervals on which the function is continuous.

47. $f(x) = -3x^2 + 7$

48. $f(x) = x^2 - \dfrac{2}{x}$

49. $f(x) = [\![x + 3]\!]$

50. $f(x) = \dfrac{3x^2 - x - 2}{x - 1}$

51. $f(x) = \begin{cases} \dfrac{3x^2 - x - 2}{x - 1}, & x \ne 1 \\ 0, & x = 1 \end{cases}$

52. $f(x) = \begin{cases} 5 - x, & x \le 2 \\ 2x - 3, & x > 2 \end{cases}$

53. $f(x) = \dfrac{1}{(x - 2)^2}$

54. $f(x) = \sqrt{\dfrac{x + 1}{x}}$

55. $f(x) = \dfrac{3}{x + 1}$

56. $f(x) = \dfrac{x + 1}{2x + 2}$

57. $f(x) = \csc \dfrac{\pi x}{2}$

58. $f(x) = \tan 2x$

59. $g(x) = 2e^{[\![x]\!]/4}$

60. $h(x) = -2 \ln|5 - x|$

61. Determine the value of c such that the function is continuous on the entire real number line.

$$f(x) = \begin{cases} x + 3, & x \le 2 \\ cx + 6, & x > 2 \end{cases}$$

62. Determine the values of b and c such that the function is continuous on the entire real number line.

$$f(x) = \begin{cases} x + 1, & 1 < x < 3 \\ x^2 + bx + c, & |x - 2| \ge 1 \end{cases}$$

63. Use the Intermediate Value Theorem to show that

$$f(x) = 2x^3 - 3$$

has a zero in the interval $[1, 2]$.

64. *Delivery Charges*   The cost of sending an overnight package from New York to Atlanta is \$12.80 for the first pound and \$2.50 for each additional pound or fraction thereof. Use the greatest integer function to create a model for the cost C of overnight delivery of a package weighing x pounds. Use a graphing utility to graph the function, and discuss its continuity.

65. *Compound Interest*   A sum of \$5000 is deposited in a savings plan that pays 12% interest compounded semiannually. The account balance after t years is given by $A = 5000(1.06)^{[\![2t]\!]}$. Use a graphing utility to graph the function, and discuss its continuity.

66. Let $f(x) = \sqrt{x(x - 1)}$.

(a) Find the domain of f.

(b) Find $\displaystyle \lim_{x \to 0^-} f(x)$.

(c) Find $\displaystyle \lim_{x \to 1^+} f(x)$.

In Exercises 67–72, find the vertical asymptotes (if any) of the graph of the function.

67. $g(x) = 1 + \dfrac{2}{x}$

68. $h(x) = \dfrac{4x}{4 - x^2}$

69. $f(x) = \dfrac{8}{(x - 10)^2}$

70. $f(x) = \csc \pi x$

71. $g(x) = \ln(25 - x^2)$

72. $f(x) = 7e^{-3/x}$

In Exercises 73–84, find the one-sided limit (if it exists).

73. $\displaystyle \lim_{x \to -2^-} \frac{2x^2 + x + 1}{x + 2}$

74. $\displaystyle \lim_{x \to (1/2)^+} \frac{x}{2x - 1}$

75. $\displaystyle \lim_{x \to -1^+} \frac{x + 1}{x^3 + 1}$

76. $\displaystyle \lim_{x \to -1^-} \frac{x + 1}{x^4 - 1}$

77. $\displaystyle \lim_{x \to 1^-} \frac{x^2 + 2x + 1}{x - 1}$

78. $\displaystyle \lim_{x \to -1^+} \frac{x^2 - 2x + 1}{x + 1}$

79. $\displaystyle \lim_{x \to 0^+} \frac{\sin 4x}{5x}$

80. $\displaystyle \lim_{x \to 0^+} \frac{\sec x}{x}$

81. $\displaystyle \lim_{x \to 0^+} \frac{\csc 2x}{x}$

82. $\displaystyle \lim_{x \to 0^-} \frac{\cos^2 x}{x}$

83. $\displaystyle \lim_{x \to 0^+} \ln(\sin x)$

84. $\displaystyle \lim_{x \to 0^-} 12e^{-2/x}$

85. The function f is defined as shown.

$$f(x) = \frac{\tan 2x}{x}, \quad x \ne 0$$

(a) Find $\displaystyle \lim_{x \to 0} \frac{\tan 2x}{x}$ (if it exists).

(b) Can the function f be defined at $x = 0$ such that it is continuous at $x = 0$?

P.S. PROBLEM SOLVING

1. Let $P(x, y)$ be a point on the parabola $y = x^2$ in the first quadrant. Consider the triangle $\triangle PAO$ formed by P, $A(0, 1)$, and the origin $O(0, 0)$, and the triangle $\triangle PBO$ formed by P, $B(1, 0)$, and the origin.

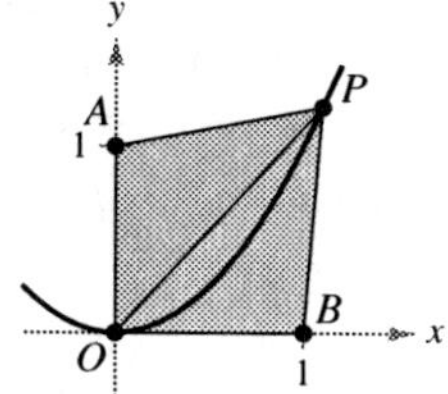

(a) Write the perimeter of each triangle in terms of x.

(b) Let $r(x)$ be the ratio of the perimeters of the two triangles,

$$r(x) = \frac{\text{Perimeter } \triangle PAO}{\text{Perimeter } \triangle PBO}.$$

Complete the table.

x	4	2	1	0.1	0.01
Perimeter $\triangle PAO$					
Perimeter $\triangle PBO$					
$r(x)$					

(c) Calculate $\lim\limits_{x \to 0^+} r(x)$.

2. Let $P(x, y)$ be a point on the parabola $y = x^2$ in the first quadrant. Consider the triangle $\triangle PAO$ formed by P, $A(0, 1)$, and the origin $O(0, 0)$, and the triangle $\triangle PBO$ formed by P, $B(1, 0)$, and the origin.

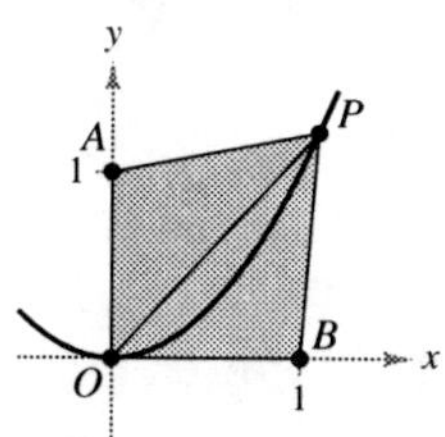

(a) Write the area of each triangle in terms of x.

(b) Let $a(x)$ be the ratio of the areas of the two triangles,

$$a(x) = \frac{\text{Area } \triangle PBO}{\text{Area } \triangle PAO}.$$

Complete the table.

x	4	2	1	0.1	0.01
Area $\triangle PAO$					
Area $\triangle PBO$					
$a(x)$					

(c) Calculate $\lim\limits_{x \to 0^+} a(x)$.

3. (a) Find the area of a regular hexagon inscribed in a circle of radius 1. How close is this area to that of the circle?

(b) Find the area A_n of an n-sided regular polygon inscribed in a circle of radius 1. Write your answer as a function of n.

(c) Complete the table.

n	6	12	24	48	96
A_n					

(d) What number does A_n approach as n gets larger and larger?

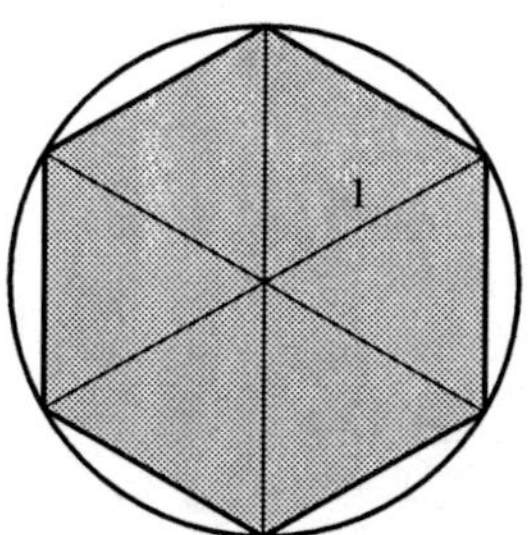

Figure for 3

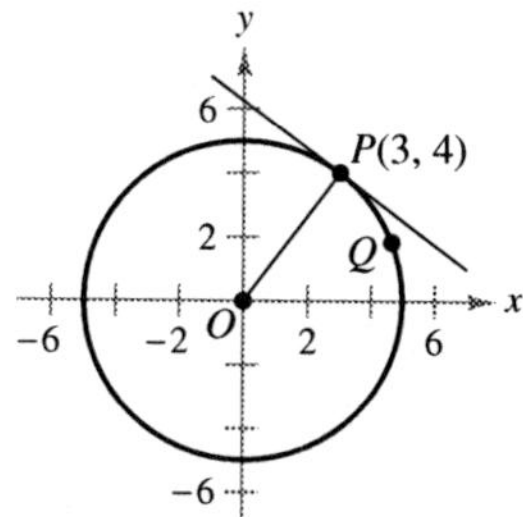

Figure for 4

4. Let $P(3, 4)$ be a point on the circle $x^2 + y^2 = 25$.

(a) What is the slope of the line joining P and $O(0, 0)$?

(b) Find an equation of the tangent line to the circle at P.

(c) Let $Q(x, y)$ be another point on the circle in the first quadrant. Find the slope m_x of the line joining P and Q in terms of x.

(d) Calculate $\lim\limits_{x \to 3} m_x$. How does this number relate to your answer in part (b)?

5. Let $P(5, -12)$ be a point on the circle $x^2 + y^2 = 169$.

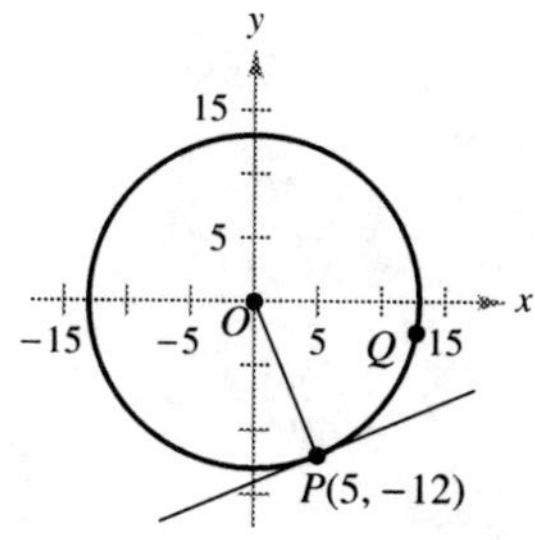

(a) What is the slope of the line joining P and $O(0, 0)$?

(b) Find an equation of the tangent line to the circle at P.

(c) Let $Q(x, y)$ be another point on the circle in the fourth quadrant. Find the slope m_x of the line joining P and Q in terms of x.

(d) Calculate $\lim\limits_{x \to 5} m_x$. How does this number relate to your answer in part (b)?

6. Find the values of the constants a and b such that

$$\lim_{x \to 0} \frac{\sqrt{a + bx} - \sqrt{3}}{x} = \sqrt{3}.$$

7. Consider the function $f(x) = \dfrac{\sqrt{3 + x^{1/3}} - 2}{x - 1}$.

  (a) Find the domain of f.

  (b) Use a graphing utility to graph the function.

  (c) Calculate $\displaystyle\lim_{x \to -27^+} f(x)$.

  (d) Calculate $\displaystyle\lim_{x \to 1} f(x)$.

8. Determine all values of the constant a such that the following function is continuous for all real numbers.

$$f(x) = \begin{cases} \dfrac{ax}{\tan x}, & x \ge 0 \\ a^2 - 2, & x < 0 \end{cases}$$

9. Consider the graphs of the four functions g_1, g_2, g_3, and g_4.

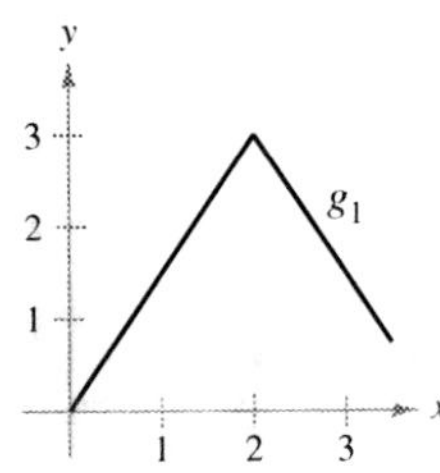 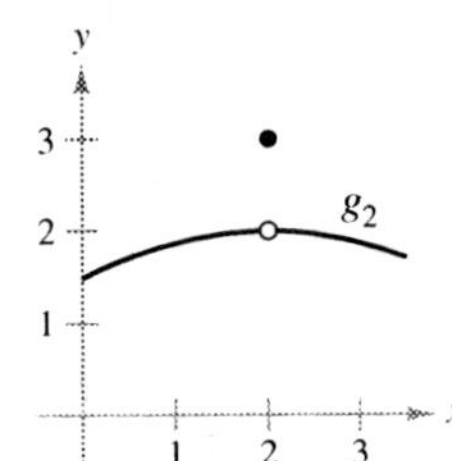

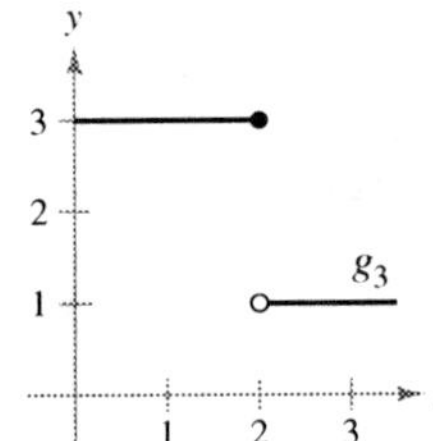 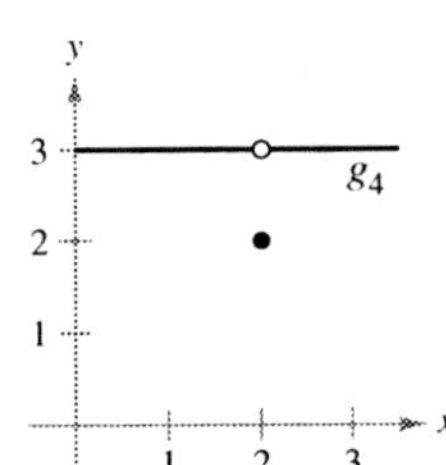

For each given condition of the function f, which of the graphs could be the graph of f?

  (a) $\displaystyle\lim_{x \to 2} f(x) = 3$

  (b) f is continuous at 2.

  (c) $\displaystyle\lim_{x \to 2} f(x) = 3$

10. Sketch the graph of the function $f(x) = \left[\!\left[\dfrac{1}{x} \right]\!\right]$.

  (a) Evaluate $f\left(\tfrac{1}{4}\right)$, $f(3)$, and $f(1)$.

  (b) Evaluate the limits $\displaystyle\lim_{x \to 1^-} f(x)$, $\displaystyle\lim_{x \to 1^+} f(x)$, $\displaystyle\lim_{x \to 0} f(x)$, and $\displaystyle\lim_{x \to 0^+} f(x)$.

  (c) Discuss the continuity of the function.

11. Sketch the graph of the function $f(x) = [\![x]\!] + [\![-x]\!]$.

  (a) Evaluate $f(1)$, $f(0)$, $f\left(\tfrac{1}{2}\right)$, and $f(-2.7)$.

  (b) Evaluate the limits $\displaystyle\lim_{x \to 1^-} f(x)$, $\displaystyle\lim_{x \to 1^+} f(x)$, and $\displaystyle\lim_{x \to \frac{1}{2}} f(x)$.

  (c) Discuss the continuity of the function.

12. To escape Earth's gravitational field, a rocket must be launched with an initial velocity called the **escape velocity**. A rocket launched from the surface of Earth has velocity v (in miles per second) given by

$$v = \sqrt{\frac{2GM}{r} + v_0^2 - \frac{2GM}{R}} \approx \sqrt{\frac{192{,}000}{r} + v_0^2 - 48}$$

where v_0 is the initial velocity, r is the distance from the rocket to the center of Earth, G is the gravitational constant, M is the mass of Earth, and R is the radius of Earth (approximately 4000 miles).

  (a) Find the value of v_0 for which you obtain an infinite limit for r as v approaches zero. This value of v_0 is the escape velocity for Earth.

  (b) A rocket launched from the surface of the moon has velocity v (in miles per second) given by

$$v = \sqrt{\frac{1920}{r} + v_0^2 - 2.17}.$$

  Find the escape velocity for the moon.

  (c) A rocket launched from the surface of a planet has velocity v (in miles per second) given by

$$v = \sqrt{\frac{10{,}600}{r} + v_0^2 - 6.99}.$$

  Find the escape velocity for this planet. Is the mass of this planet larger or smaller than that of Earth? (Assume that the mean density of this planet is the same as that of Earth.)

13. For positive numbers $a < b$, the **pulse function** is defined as

$$P_{a,b}(x) = H(x - a) - H(x - b) = \begin{cases} 0, & x < a \\ 1, & a \le x < b \\ 0, & x \ge b \end{cases}$$

where $H(x) = \begin{cases} 1, & x \ge 0 \\ 0, & x < 0 \end{cases}$ is the Heaviside function.

  (a) Sketch the graph of the pulse function.

  (b) Find the following limits:

    (i) $\displaystyle\lim_{x \to a^+} P_{a,b}(x)$ (ii) $\displaystyle\lim_{x \to a^-} P_{a,b}(x)$

    (iii) $\displaystyle\lim_{x \to b^+} P_{a,b}(x)$ (iv) $\displaystyle\lim_{x \to b^-} P_{a,b}(x)$

  (c) Discuss the continuity of the pulse function.

  (d) Why is

$$U(x) = \frac{1}{b - a} P_{a,b}(x)$$

  called the **unit** pulse function?

14. Let a be a nonzero constant. Prove that if $\displaystyle\lim_{x \to 0} f(x) = L$, then $\displaystyle\lim_{x \to 0} f(ax) = L$. Show by means of an example that a must be nonzero.

3 Differentiation

In this chapter you will study one of the most important processes of calculus—*differentiation*. In each section, you will learn new methods and rules for finding derivatives of functions. Then you will apply these rules to find such things as velocity, acceleration, and the rates of change of two or more related variables.

In this chapter, you should learn the following.

- How to find the derivative of a function using the limit definition and understand the relationship between differentiability and continuity. (3.1)
- How to find the derivative of a function using basic differentiation rules. (3.2)
- How to find the derivative of a function using the Product Rule and the Quotient Rule. (3.3)
- How to find the derivative of a function using the Chain Rule and the General Power Rule. (3.4)
- How to find the derivative of a function using implicit differentiation. (3.5)
- How to find the derivative of an inverse function. (3.6)
- How to find a related rate. (3.7)
- How to approximate a zero of a function using Newton's Method. (3.8)

Al Bello/Getty Images

When jumping from a platform, a diver's velocity is briefly positive because of the upward movement, but then becomes negative when falling. How can you use calculus to determine the velocity of a diver at impact? (See Section 3.2, Example 11.)

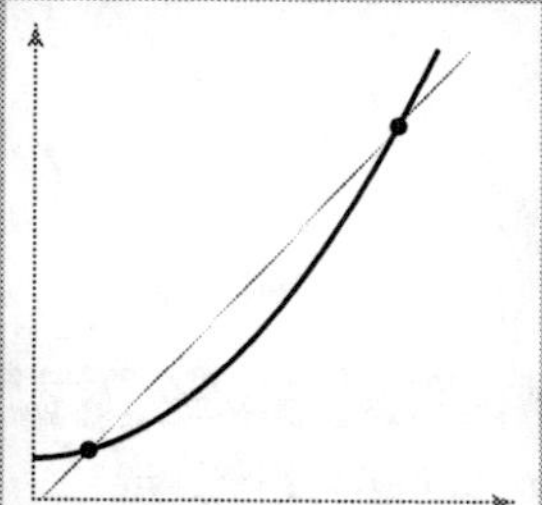
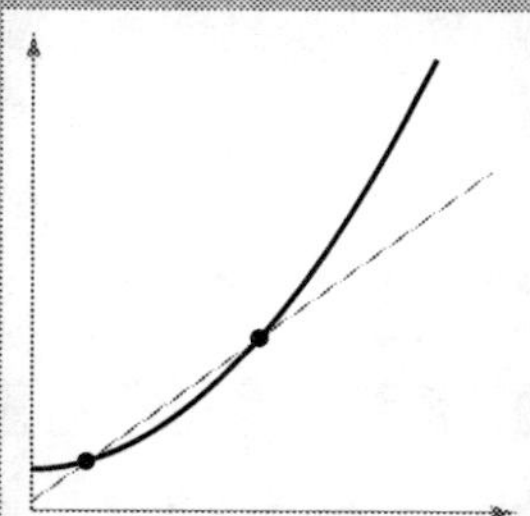
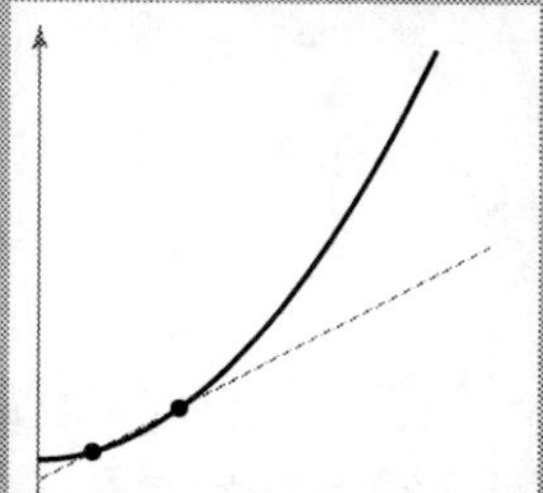
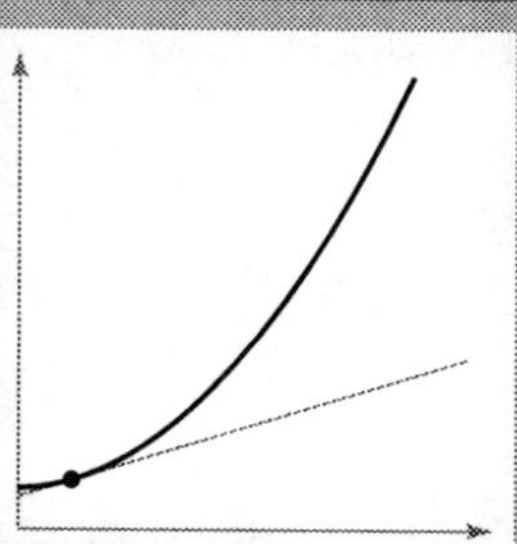

To approximate the slope of a tangent line to a graph at a given point, find the slope of the secant line through the given point and a second point on the graph. As the second point approaches the given point, the approximation tends to become more accurate. (See Section 3.1.)

3.1   The Derivative and the Tangent Line Problem

- Find the slope of the tangent line to a curve at a point.
- Use the limit definition to find the derivative of a function.
- Understand the relationship between differentiability and continuity.

The Tangent Line Problem

Calculus grew out of four major problems that European mathematicians were working on during the seventeenth century.

1. The tangent line problem (Section 2.1 and this section)
2. The velocity and acceleration problem (Sections 3.2 and 3.3)
3. The minimum and maximum problem (Section 4.1)
4. The area problem (Sections 2.1 and 5.2)

Each problem involves the notion of a limit, and calculus can be introduced with any of the four problems.

A brief introduction to the tangent line problem is given in Section 2.1. Although partial solutions to this problem were given by Pierre de Fermat (1601–1665), René Descartes (1596–1650), Christian Huygens (1629–1695), and Isaac Barrow (1630–1677), credit for the first general solution is usually given to Isaac Newton (1642–1727) and Gottfried Leibniz (1646–1716). Newton's work on this problem stemmed from his interest in optics and light refraction.

What does it mean to say that a line is tangent to a curve at a point? For a circle, the tangent line at a point P is the line that is perpendicular to the radial line at point P, as shown in Figure 3.1.

For a general curve, however, the problem is more difficult. For example, how would you define the tangent lines shown in Figure 3.2? You might say that a line is tangent to a curve at a point P if it touches, but does not cross, the curve at point P. This definition would work for the first curve shown in Figure 3.2, but not for the second. *Or* you might say that a line is tangent to a curve if the line touches or intersects the curve at exactly one point. This definition would work for a circle but not for more general curves, as the third curve in Figure 3.2 shows.

ISAAC NEWTON (1642–1727)

In addition to his work in calculus, Newton made revolutionary contributions to physics, including the Law of Universal Gravitation and his three laws of motion.

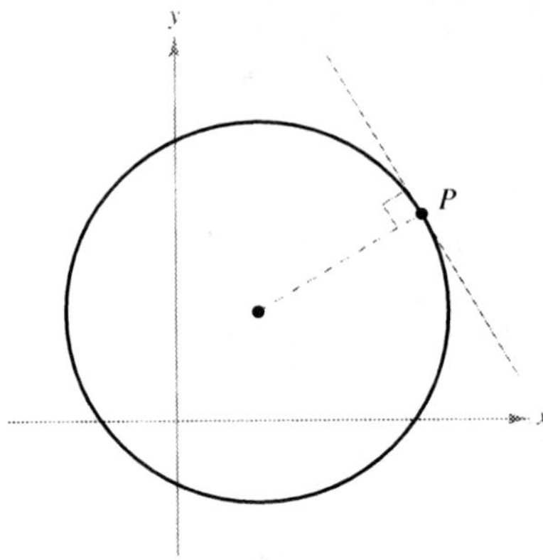

Tangent line to a circle
Figure 3.1

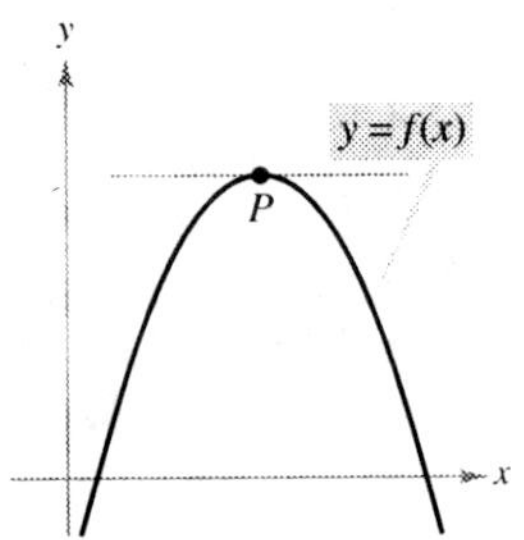

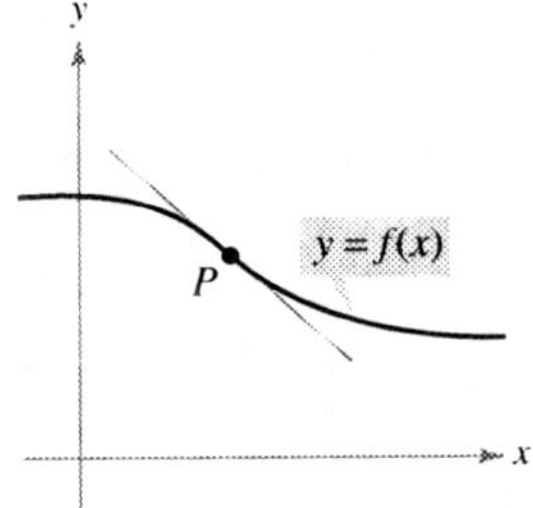

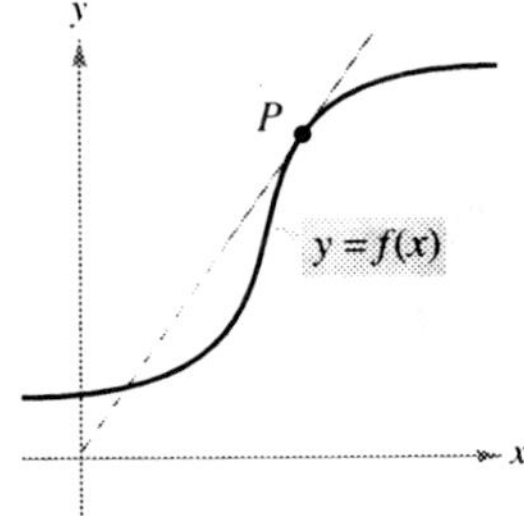

Tangent line to a curve at a point
Figure 3.2

EXPLORATION

Identifying a Tangent Line   Use a graphing utility to graph the function $f(x) = 2x^3 - 4x^2 + 3x - 5$. On the same screen, graph $y = x - 5$, $y = 2x - 5$, and $y = 3x - 5$. Which of these lines, if any, appears to be tangent to the graph of f at the point $(0, -5)$? Explain your reasoning.

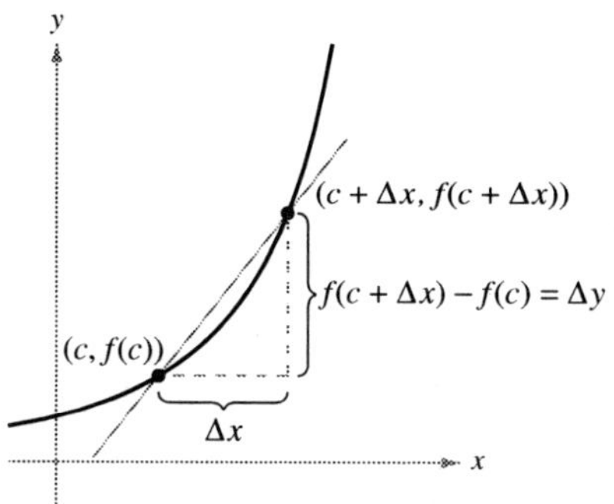

The secant line through $(c, f(c))$ and $(c + \Delta x, f(c + \Delta x))$

Figure 3.3

Essentially, the problem of finding the tangent line at a point P boils down to the problem of finding the *slope* of the tangent line at point P. You can approximate this slope using a **secant line*** through the point of tangency and a second point on the curve, as shown in Figure 3.3. If $(c, f(c))$ is the point of tangency and $(c + \Delta x, f(c + \Delta x))$ is a second point on the graph of f, the slope of the secant line through the two points is given by substitution into the slope formula

$$m = \frac{y_2 - y_1}{x_2 - x_1}$$

$$m_{\text{sec}} = \frac{f(c + \Delta x) - f(c)}{(c + \Delta x) - c} \qquad \text{Change in } y \atop \text{Change in } x$$

$$m_{\text{sec}} = \frac{f(c + \Delta x) - f(c)}{\Delta x}. \qquad \text{Slope of secant line}$$

The right-hand side of this equation is a **difference quotient.** The denominator Δx is the **change in x,** and the numerator $\Delta y = f(c + \Delta x) - f(c)$ is the **change in y.**

The beauty of this procedure is that you can obtain more and more accurate approximations of the slope of the tangent line by choosing points closer and closer to the point of tangency, as shown in Figure 3.4.

<table>
<tr><td>

THE TANGENT LINE PROBLEM

In 1637, mathematician René Descartes stated this about the tangent line problem:

"And I dare say that this is not only the most useful and general problem in geometry that I know, but even that I ever desire to know."

</td></tr>
</table>

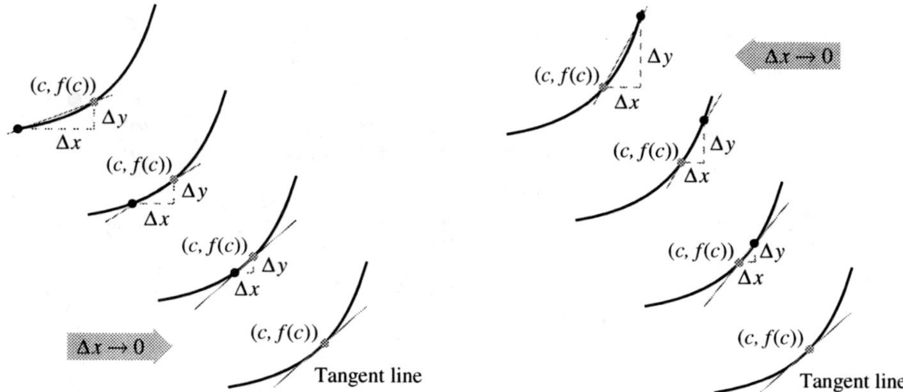

Tangent line approximations

Figure 3.4

DEFINITION OF TANGENT LINE WITH SLOPE m

If f is defined on an open interval containing c, and if the limit

$$\lim_{\Delta x \to 0} \frac{\Delta y}{\Delta x} = \lim_{\Delta x \to 0} \frac{f(c + \Delta x) - f(c)}{\Delta x} = m$$

exists, then the line passing through $(c, f(c))$ with slope m is the **tangent line** to the graph of f at the point $(c, f(c))$.

The slope of the tangent line to the graph of f at the point $(c, f(c))$ is also called the **slope of the graph of f at $x = c$.**

---

* *This use of the word* secant *comes from the Latin* secare, *meaning to cut, and is not a reference to the trigonometric function of the same name.*

EXAMPLE 1 The Slope of the Graph of a Linear Function

Find the slope of the graph of

$$f(x) = 2x - 3$$

at the point $(2, 1)$.

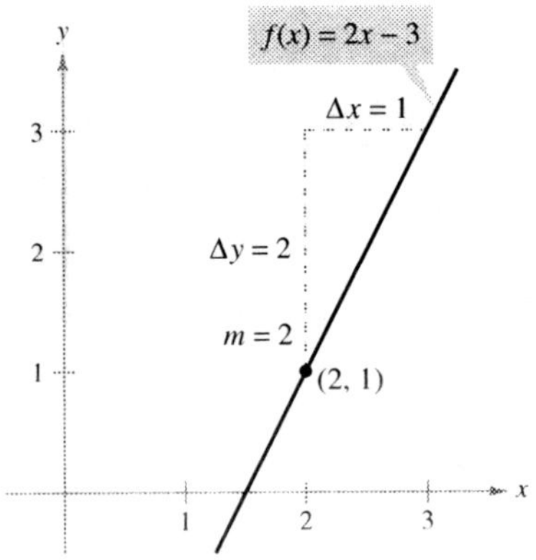

Solution To find the slope of the graph of f when $c = 2$, you can apply the definition of the slope of a tangent line, as shown.

$$\lim_{\Delta x \to 0} \frac{f(2 + \Delta x) - f(2)}{\Delta x} = \lim_{\Delta x \to 0} \frac{[2(2 + \Delta x) - 3] - [2(2) - 3]}{\Delta x}$$

$$= \lim_{\Delta x \to 0} \frac{4 + 2\Delta x - 3 - 4 + 3}{\Delta x}$$

$$= \lim_{\Delta x \to 0} \frac{2\Delta x}{\Delta x}$$

$$= \lim_{\Delta x \to 0} 2$$

$$= 2$$

The slope of f at $(c, f(c)) = (2, 1)$ is $m = 2$, as shown in Figure 3.5.

NOTE In Example 1, the limit definition of the slope of f agrees with the definition of the slope of a line as discussed in Section 1.2.

The graph of a linear function has the same slope at any point. This is not true of nonlinear functions, as shown in the following example.

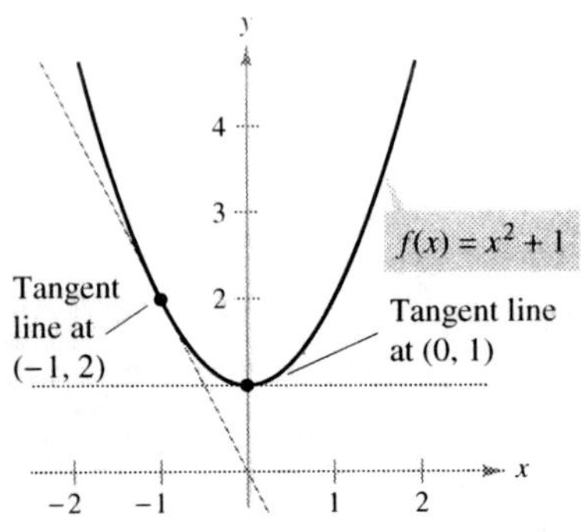

The slope of f at $(2, 1)$ is $m = 2$.
Figure 3.5

EXAMPLE 2 Tangent Lines to the Graph of a Nonlinear Function

Find the slopes of the tangent lines to the graph of

$$f(x) = x^2 + 1$$

at the points $(0, 1)$ and $(-1, 2)$, as shown in Figure 3.6.

Solution Let $(c, f(c))$ represent an arbitrary point on the graph of f. Then the slope of the tangent line at $(c, f(c))$ is given by

$$\lim_{\Delta x \to 0} \frac{f(c + \Delta x) - f(c)}{\Delta x} = \lim_{\Delta x \to 0} \frac{[(c + \Delta x)^2 + 1] - (c^2 + 1)}{\Delta x}$$

$$= \lim_{\Delta x \to 0} \frac{c^2 + 2c(\Delta x) + (\Delta x)^2 + 1 - c^2 - 1}{\Delta x}$$

$$= \lim_{\Delta x \to 0} \frac{2c(\Delta x) + (\Delta x)^2}{\Delta x}$$

$$= \lim_{\Delta x \to 0} (2c + \Delta x)$$

$$= 2c.$$

So, the slope at *any* point $(c, f(c))$ on the graph of f is $m = 2c$. At the point $(0, 1)$, the slope is $m = 2(0) = 0$, and at $(-1, 2)$, the slope is $m = 2(-1) = -2$.

The slope of f at any point $(c, f(c))$ is $m = 2c$.
Figure 3.6

NOTE In Example 2, note that c is held constant in the limit process (as $\Delta x \to 0$).

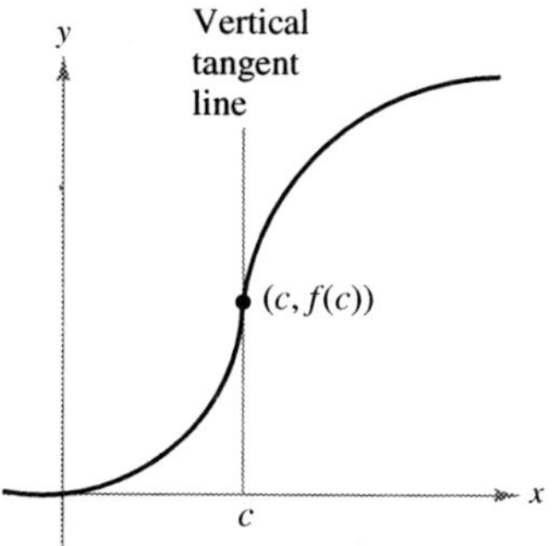

The graph of f has a vertical tangent line at $(c, f(c))$.

Figure 3.7

The definition of a tangent line to a curve does not cover the possibility of a vertical tangent line. For vertical tangent lines, you can use the following definition. If f is continuous at c and

$$\lim_{\Delta x \to 0} \frac{f(c + \Delta x) - f(c)}{\Delta x} = \infty \quad \text{or} \quad \lim_{\Delta x \to 0} \frac{f(c + \Delta x) - f(c)}{\Delta x} = -\infty$$

the vertical line $x = c$ passing through $(c, f(c))$ is a **vertical tangent line** to the graph of f. For example, the function shown in Figure 3.7 has a vertical tangent line at $(c, f(c))$. If the domain of f is the closed interval $[a, b]$, you can extend the definition of a vertical tangent line to include the endpoints by considering continuity and limits from the right (for $x = a$) and from the left (for $x = b$).

The Derivative of a Function

You have now arrived at a crucial point in the study of calculus. The limit used to define the slope of a tangent line is also used to define one of the two fundamental operations of calculus—**differentiation.**

DEFINITION OF THE DERIVATIVE OF A FUNCTION

The **derivative** of f at x is given by

$$f'(x) = \lim_{\Delta x \to 0} \frac{f(x + \Delta x) - f(x)}{\Delta x}$$

provided the limit exists. For all x for which this limit exists, f' is a function of x.

Be sure you see that the derivative of a function of x is also a function of x. This "new" function gives the slope of the tangent line to the graph of f at the point $(x, f(x))$, provided that the graph has a tangent line at this point.

The process of finding the derivative of a function is called **differentiation.** A function is **differentiable** at x if its derivative exists at x and is **differentiable on an open interval (a, b)** if it is differentiable at every point in the interval.

In addition to $f'(x)$, which is read as "f prime of x," other notations are used to denote the derivative of $y = f(x)$. The most common are

$$f'(x), \quad \frac{dy}{dx}, \quad y', \quad \frac{d}{dx}[f(x)], \quad D_x[y].$$

Notation for derivatives

The notation dy/dx is read as "the derivative of y *with respect to* x" or simply "dy, dx." Using limit notation, you can write

$$\frac{dy}{dx} = \lim_{\Delta x \to 0} \frac{\Delta y}{\Delta x}$$

$$= \lim_{\Delta x \to 0} \frac{f(x + \Delta x) - f(x)}{\Delta x}$$

$$= f'(x).$$

▓ FOR FURTHER INFORMATION
For more information on the crediting of mathematical discoveries to the first "discoverers," see the article "Mathematical Firsts—Who Done It?" by Richard H. Williams and Roy D. Mazzagatti in *Mathematics Teacher*. To view this article, go to the website *www.matharticles.com.*

EXAMPLE 3 Finding the Derivative by the Limit Process

Find the derivative of $f(x) = x^3 + 2x$.

Solution

$$
\begin{aligned}
f'(x) &= \lim_{\Delta x \to 0} \frac{f(x + \Delta x) - f(x)}{\Delta x} \qquad \text{\small Definition of derivative} \\[2mm]
&= \lim_{\Delta x \to 0} \frac{(x + \Delta x)^3 + 2(x + \Delta x) - (x^3 + 2x)}{\Delta x} \\[2mm]
&= \lim_{\Delta x \to 0} \frac{x^3 + 3x^2\Delta x + 3x(\Delta x)^2 + (\Delta x)^3 + 2x + 2\Delta x - x^3 - 2x}{\Delta x} \\[2mm]
&= \lim_{\Delta x \to 0} \frac{3x^2\Delta x + 3x(\Delta x)^2 + (\Delta x)^3 + 2\Delta x}{\Delta x} \\[2mm]
&= \lim_{\Delta x \to 0} \frac{\Delta x\left[3x^2 + 3x\Delta x + (\Delta x)^2 + 2\right]}{\Delta x} \\[2mm]
&= \lim_{\Delta x \to 0} \left[3x^2 + 3x\Delta x + (\Delta x)^2 + 2\right] \\[2mm]
&= 3x^2 + 2
\end{aligned}
$$

STUDY TIP When using the definition to find a derivative of a function, the key is to rewrite the difference quotient so that Δx does not occur as a factor of the denominator.

Remember that the derivative of a function f is itself a function, which can be used to find the slope of the tangent line at the point $(x, f(x))$ on the graph of f.

EXAMPLE 4 Using the Derivative to Find the Slope at a Point

Find $f'(x)$ for $f(x) = \sqrt{x}$. Then find the slopes of the graph of f at the points $(1, 1)$ and $(4, 2)$. Discuss the behavior of f at $(0, 0)$.

Solution   Use the procedure for rationalizing numerators, as discussed in Section 2.3.

$$
\begin{aligned}
f'(x) &= \lim_{\Delta x \to 0} \frac{f(x + \Delta x) - f(x)}{\Delta x} \qquad \text{\small Definition of derivative} \\[2mm]
&= \lim_{\Delta x \to 0} \frac{\sqrt{x + \Delta x} - \sqrt{x}}{\Delta x} \\[2mm]
&= \lim_{\Delta x \to 0} \left(\frac{\sqrt{x + \Delta x} - \sqrt{x}}{\Delta x} \right)\left(\frac{\sqrt{x + \Delta x} + \sqrt{x}}{\sqrt{x + \Delta x} + \sqrt{x}} \right) \\[2mm]
&= \lim_{\Delta x \to 0} \frac{(x + \Delta x) - x}{\Delta x\left(\sqrt{x + \Delta x} + \sqrt{x}\right)} \\[2mm]
&= \lim_{\Delta x \to 0} \frac{\Delta x}{\Delta x\left(\sqrt{x + \Delta x} + \sqrt{x}\right)} \\[2mm]
&= \lim_{\Delta x \to 0} \frac{1}{\sqrt{x + \Delta x} + \sqrt{x}} \\[2mm]
&= \frac{1}{2\sqrt{x}}, \quad x > 0
\end{aligned}
$$

At the point $(1, 1)$, the slope is $f'(1) = \frac{1}{2}$. At the point $(4, 2)$, the slope is $f'(4) = \frac{1}{4}$. See Figure 3.8. At the point $(0, 0)$, the slope is undefined. Moreover, the graph of f has a vertical tangent line at $(0, 0)$.

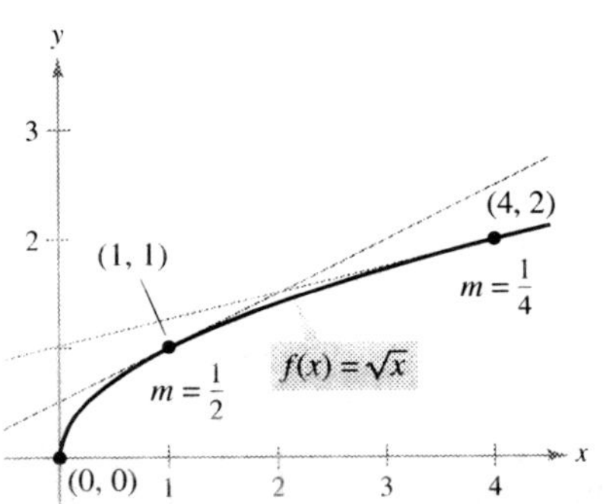

The slope of f at $(x, f(x))$, $x > 0$, is $m = 1/(2\sqrt{x})$.

Figure 3.8

The icon ○ indicates that you will find a CAS Investigation on the book's website. The CAS Investigation is a collaborative exploration of this example using the computer algebra systems Maple and Mathematica.

In many applications, it is convenient to use a variable other than x as the independent variable, as shown in Example 5.

EXAMPLE 5 Finding the Derivative of a Function

Find the derivative with respect to t for the function $y = 2/t$.

Solution Considering $y = f(t)$, you obtain

$$\frac{dy}{dt} = \lim_{\Delta t \to 0} \frac{f(t + \Delta t) - f(t)}{\Delta t} \qquad \text{Definition of derivative}$$

$$= \lim_{\Delta t \to 0} \frac{\dfrac{2}{t + \Delta t} - \dfrac{2}{t}}{\Delta t} \qquad f(t + \Delta t) = 2/(t + \Delta t) \text{ and } f(t) = 2/t$$

$$= \lim_{\Delta t \to 0} \frac{\dfrac{2t - 2(t + \Delta t)}{t(t + \Delta t)}}{\Delta t} \qquad \text{Combine fractions in numerator.}$$

$$= \lim_{\Delta t \to 0} \frac{-2\Delta t}{\Delta t(t)(t + \Delta t)} \qquad \text{Divide out common factor of } \Delta t.$$

$$= \lim_{\Delta t \to 0} \frac{-2}{t(t + \Delta t)} \qquad \text{Simplify.}$$

$$= -\frac{2}{t^2}. \qquad \text{Evaluate limit as } \Delta t \to 0.$$

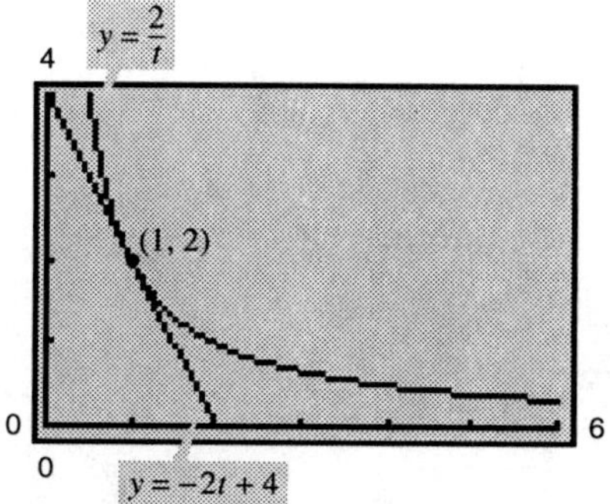

At the point $(1, 2)$, the line $y = -2t + 4$ is tangent to the graph of $y = 2/t$.

Figure 3.9

TECHNOLOGY A graphing utility can be used to reinforce the result given in Example 5. For instance, using the formula $dy/dt = -2/t^2$, you know that the slope of the graph of $y = 2/t$ at the point $(1, 2)$ is $m = -2$. Using the point-slope form, you can find that the equation of the tangent line to the graph at $(1, 2)$ is

$$y - 2 = -2(t - 1) \quad \text{or} \quad y = -2t + 4$$

as shown in Figure 3.9.

Differentiability and Continuity

The following alternative limit form of the derivative is useful in investigating the relationship between differentiability and continuity. The derivative of f at c is

$$f'(c) = \lim_{x \to c} \frac{f(x) - f(c)}{x - c} \qquad \text{Alternative form of derivative}$$

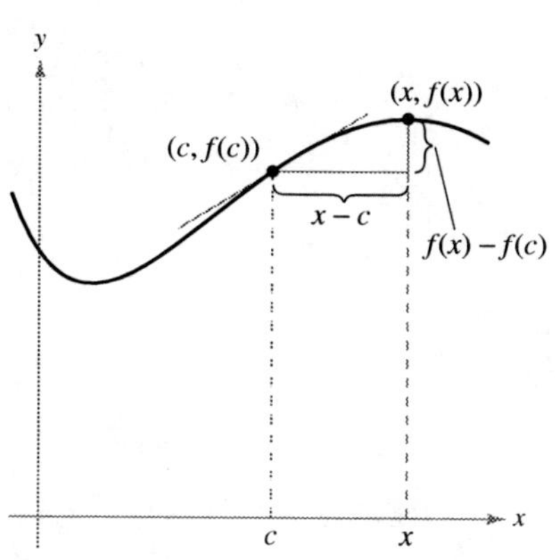

As x approaches c, the secant line approaches the tangent line.

Figure 3.10

provided this limit exists (see Figure 3.10). (A proof of the equivalence of this form is given in Appendix A.) Note that the existence of the limit in this alternative form requires that the one-sided limits

$$\lim_{x \to c^-} \frac{f(x) - f(c)}{x - c} \quad \text{and} \quad \lim_{x \to c^+} \frac{f(x) - f(c)}{x - c}$$

exist and are equal. These one-sided limits are called the **derivatives from the left and from the right,** respectively. It follows that f is **differentiable on the closed interval** $[a, b]$ if it is differentiable on (a, b) and if the derivative from the right at a and the derivative from the left at b both exist.

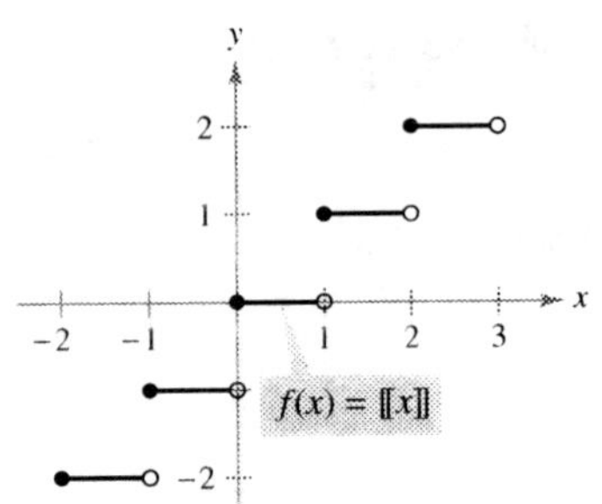

The greatest integer function is not differentiable at $x = 0$, because it is not continuous at $x = 0$.
Figure 3.11

If a function is not continuous at $x = c$, it is also not differentiable at $x = c$. For instance, the greatest integer function

$$f(x) = [\![x]\!]$$

is not continuous at $x = 0$, and so it is not differentiable at $x = 0$ (see Figure 3.11). You can verify this by observing that

$$\lim_{x \to 0^-} \frac{f(x) - f(0)}{x - 0} = \lim_{x \to 0^-} \frac{[\![x]\!] - 0}{x} = \infty \qquad \text{Derivative from the left}$$

and

$$\lim_{x \to 0^+} \frac{f(x) - f(0)}{x - 0} = \lim_{x \to 0^+} \frac{[\![x]\!] - 0}{x} = 0. \qquad \text{Derivative from the right}$$

(See Exercise 99.) Although it is true that differentiability implies continuity (as we will show in Theorem 3.1), the converse is not true. That is, it is possible for a function to be continuous at $x = c$ and *not* differentiable at $x = c$. Examples 6 and 7 illustrate this possibility.

EXAMPLE 6 A Graph with a Sharp Turn

The function

$$f(x) = |x - 2|$$

shown in Figure 3.12 is continuous at $x = 2$. However, the one-sided limits

$$\lim_{x \to 2^-} \frac{f(x) - f(2)}{x - 2} = \lim_{x \to 2^-} \frac{|x - 2| - 0}{x - 2} = -1 \qquad \text{Derivative from the left}$$

and

$$\lim_{x \to 2^+} \frac{f(x) - f(2)}{x - 2} = \lim_{x \to 2^+} \frac{|x - 2| - 0}{x - 2} = 1 \qquad \text{Derivative from the right}$$

are not equal. So, f is not differentiable at $x = 2$ and the graph of f does not have a tangent line at the point $(2, 0)$.

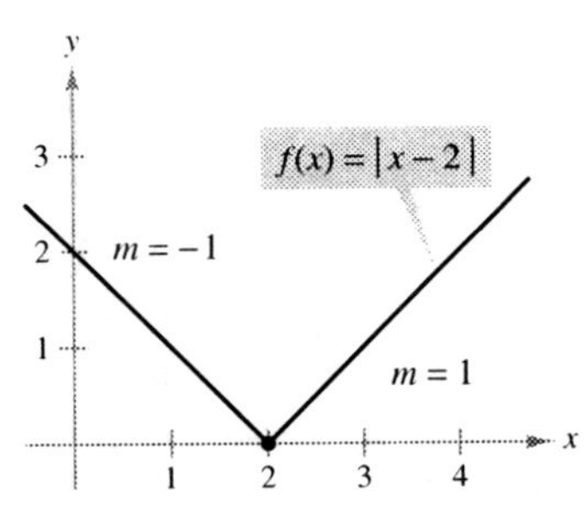

f is not differentiable at $x = 2$, because the derivatives from the left and from the right are not equal.
Figure 3.12

EXAMPLE 7 A Graph with a Vertical Tangent Line

The function

$$f(x) = x^{1/3}$$

is continuous at $x = 0$, as shown in Figure 3.13. However, because the limit

$$\lim_{x \to 0} \frac{f(x) - f(0)}{x - 0} = \lim_{x \to 0} \frac{x^{1/3} - 0}{x}$$

$$= \lim_{x \to 0} \frac{1}{x^{2/3}}$$

$$= \infty$$

is infinite, you can conclude that the tangent line is vertical at $x = 0$. So, f is not differentiable at $x = 0$.

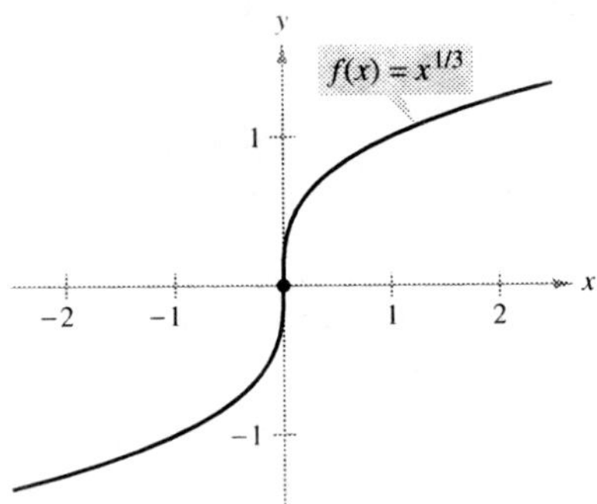

f is not differentiable at $x = 0$, because f has a vertical tangent line at $x = 0$.
Figure 3.13

From Examples 6 and 7, you can see that a function is not differentiable at a point at which its graph has a sharp turn *or* a vertical tangent line.

TECHNOLOGY Some graphing utilities, such as *Maple, Mathematica,* and the *TI-89,* perform symbolic differentiation. Others perform *numerical differentiation* by finding values of derivatives using the formula

$$f'(x) \approx \frac{f(x + \Delta x) - f(x - \Delta x)}{2\Delta x}$$

where Δx is a small number such as 0.001. Can you see any problems with this definition? For instance, using this definition, what is the value of the derivative of $f(x) = |x|$ when $x = 0$?

THEOREM 3.1 DIFFERENTIABILITY IMPLIES CONTINUITY

If f is differentiable at $x = c$, then f is continuous at $x = c$.

PROOF You can prove that f is continuous at $x = c$ by showing that $f(x)$ approaches $f(c)$ as $x \to c$. To do this, use the differentiability of f at $x = c$ and consider the following limit.

$$\lim_{x \to c} [f(x) - f(c)] = \lim_{x \to c} \left[(x - c)\left(\frac{f(x) - f(c)}{x - c} \right) \right]$$

$$= \left[\lim_{x \to c} (x - c) \right]\left[\lim_{x \to c} \frac{f(x) - f(c)}{x - c} \right]$$

$$= (0)[f'(c)]$$

$$= 0$$

Because the difference $f(x) - f(c)$ approaches zero as $x \to c$, you can conclude that $\lim_{x \to c} f(x) = f(c)$. So, f is continuous at $x = c$. ∎

You can summarize the relationship between continuity and differentiability as follows.

1. If a function is differentiable at $x = c$, then it is continuous at $x = c$. So, differentiability implies continuity.

2. It is possible for a function to be continuous at $x = c$ and not be differentiable at $x = c$. So, continuity does not imply differentiability (see Example 6).

3.1 Exercises

See www.CalcChat.com for worked-out solutions to odd-numbered exercises.

In Exercises 1 and 2, estimate the slope of the graph at the points (x_1, y_1) and (x_2, y_2).

1. (a) 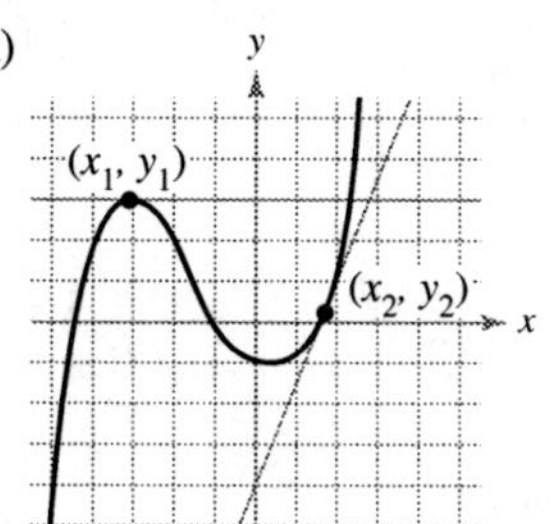   (b) 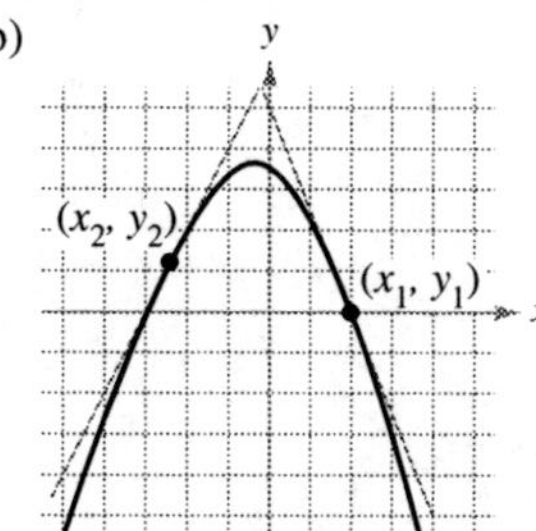

2. (a) 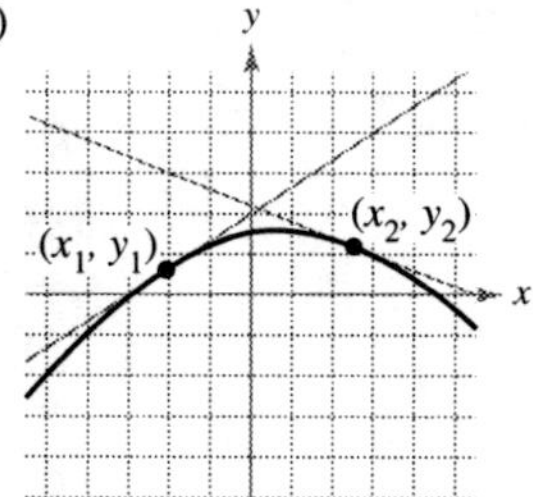   (b) 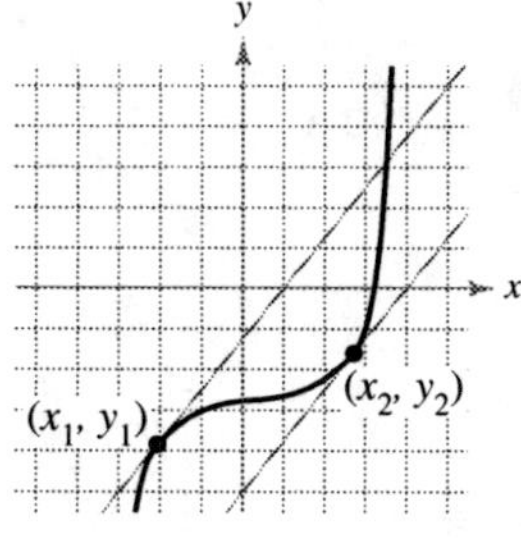

In Exercises 3 and 4, use the graph shown in the figure. To print an enlarged copy of the graph, go to the website *www.mathgraphs.com*.

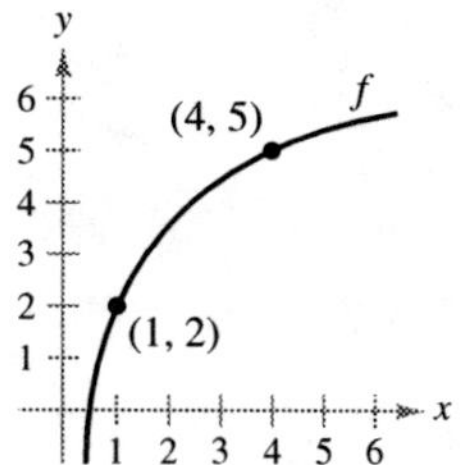

3. Identify or sketch each of the quantities on the figure.

   (a) $f(1)$ and $f(4)$ (b) $f(4) - f(1)$

   (c) $y = \dfrac{f(4) - f(1)}{4 - 1}(x - 1) + f(1)$

4. Insert the proper inequality symbol ($<$ or $>$) between the given quantities.

   (a) $\dfrac{f(4) - f(1)}{4 - 1}$ ▆ $\dfrac{f(4) - f(3)}{4 - 3}$

   (b) $\dfrac{f(4) - f(1)}{4 - 1}$ ▆ $f'(1)$

In Exercises 5–10, find the slope of the tangent line to the graph of the function at the given point.

5. $f(x) = 3 - 5x$, $(-1, 8)$ **6.** $g(x) = \frac{3}{2}x + 1$, $(-2, -2)$

7. $g(x) = x^2 - 9$, $(2, -5)$ **8.** $g(x) = 6 - x^2$, $(1, 5)$

9. $f(t) = 3t - t^2$, $(0, 0)$ **10.** $h(t) = t^2 + 3$, $(-2, 7)$

In Exercises 11–24, find the derivative by the limit process.

11. $f(x) = 7$ **12.** $g(x) = -3$

13. $f(x) = -5x$ **14.** $f(x) = 3x + 2$

15. $h(s) = 3 + \frac{2}{3}s$ **16.** $f(x) = 8 - \frac{1}{5}x$

17. $f(x) = x^2 + x - 3$ **18.** $f(x) = 2 - x^2$

19. $f(x) = x^3 - 12x$ **20.** $f(x) = x^3 + x^2$

21. $f(x) = \dfrac{1}{x - 1}$ **22.** $f(x) = \dfrac{1}{x^2}$

23. $f(x) = \sqrt{x + 4}$ **24.** $f(x) = \dfrac{4}{\sqrt{x}}$

In Exercises 25–32, (a) find an equation of the tangent line to the graph of f at the given point, (b) use a graphing utility to graph the function and its tangent line at the point, and (c) use the _derivative_ feature of a graphing utility to confirm your results.

25. $f(x) = x^2 + 3$, $(1, 4)$

26. $f(x) = x^2 + 3x + 4$, $(-2, 2)$

27. $f(x) = x^3$, $(2, 8)$ **28.** $f(x) = x^3 + 1$, $(1, 2)$

29. $f(x) = \sqrt{x}$, $(1, 1)$ **30.** $f(x) = \sqrt{x - 1}$, $(5, 2)$

31. $f(x) = x + \dfrac{4}{x}$, $(4, 5)$ **32.** $f(x) = \dfrac{1}{x + 1}$, $(0, 1)$

In Exercises 33–38, find an equation of the line that is tangent to the graph of f _and_ parallel to the given line.

Function	_Line_
33. $f(x) = x^2$	$2x - y + 1 = 0$
34. $f(x) = 2x^2$	$4x + y + 3 = 0$
35. $f(x) = x^3$	$3x - y + 1 = 0$
36. $f(x) = x^3 + 2$	$3x - y - 4 = 0$
37. $f(x) = \dfrac{1}{\sqrt{x}}$	$x + 2y - 6 = 0$
38. $f(x) = \dfrac{1}{\sqrt{x - 1}}$	$x + 2y + 7 = 0$

In Exercises 39–42, the graph of f is given. Select the graph of f'.

39.

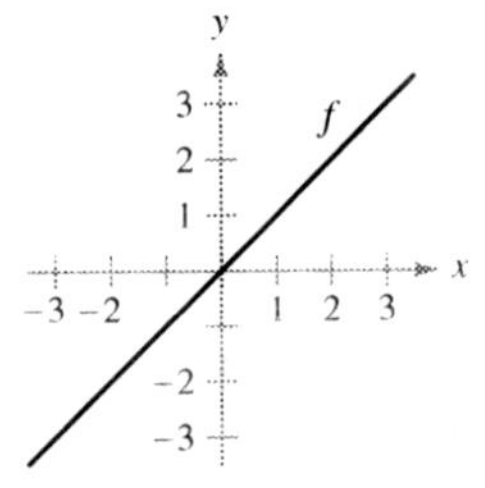

40.

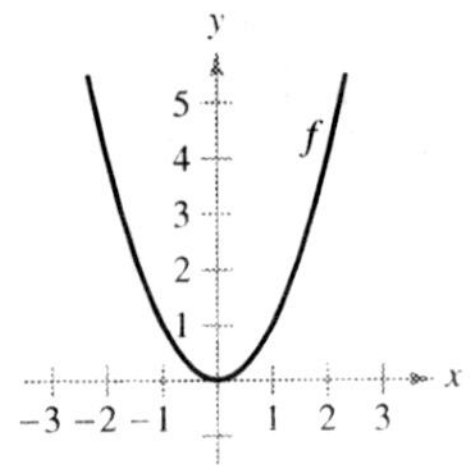

41.

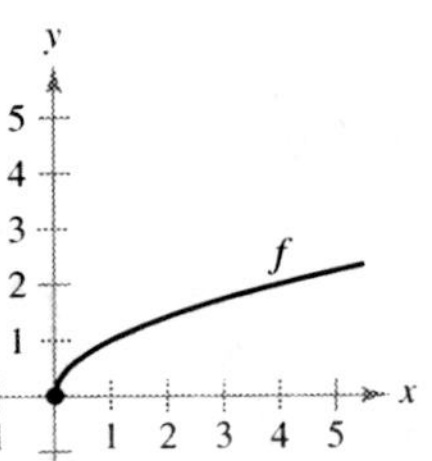

42.

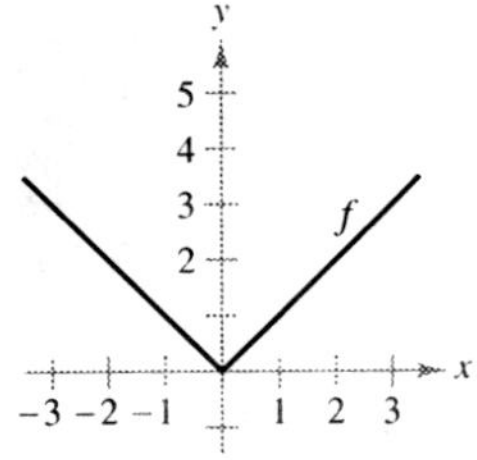

(a)

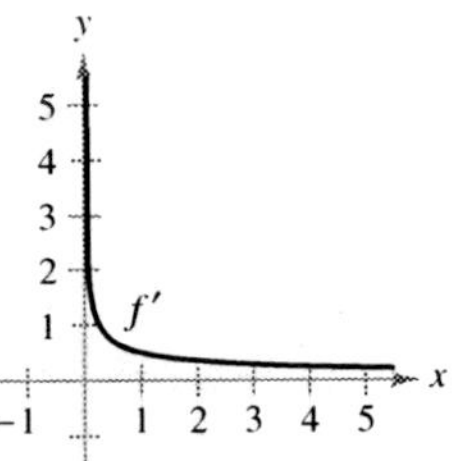

(b)

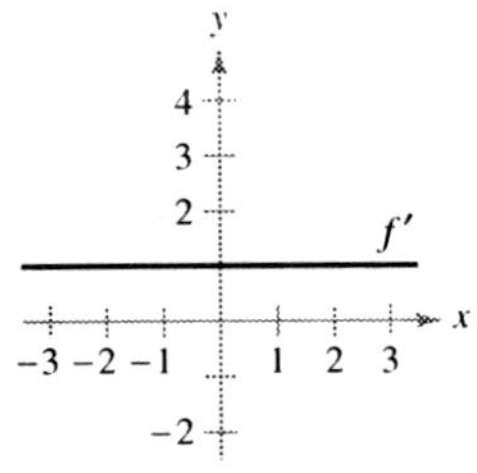

(c)

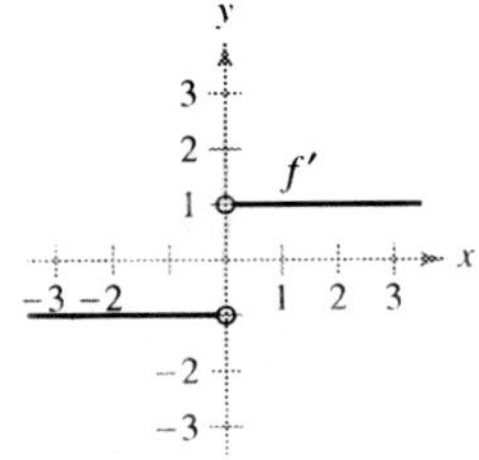

(d)

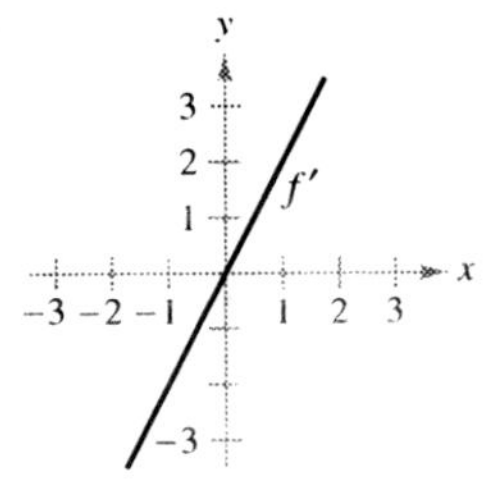

43. The tangent line to the graph of $y = g(x)$ at the point $(4, 5)$ passes through the point $(7, 0)$. Find $g(4)$ and $g'(4)$.

44. The tangent line to the graph of $y = h(x)$ at the point $(-1, 4)$ passes through the point $(3, 6)$. Find $h(-1)$ and $h'(-1)$.

WRITING ABOUT CONCEPTS

In Exercises 45–50, sketch the graph of f'. Explain how you found your answer.

45.

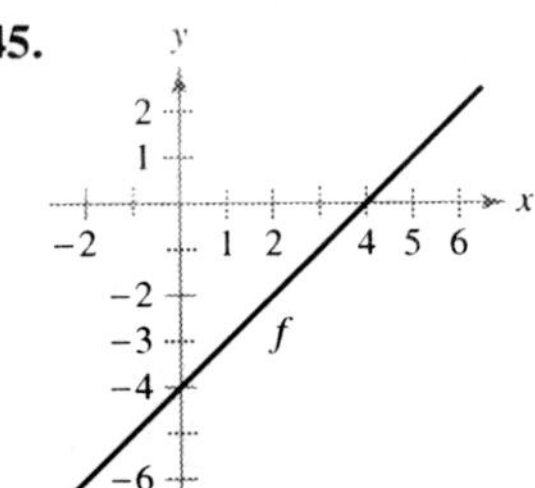

46.

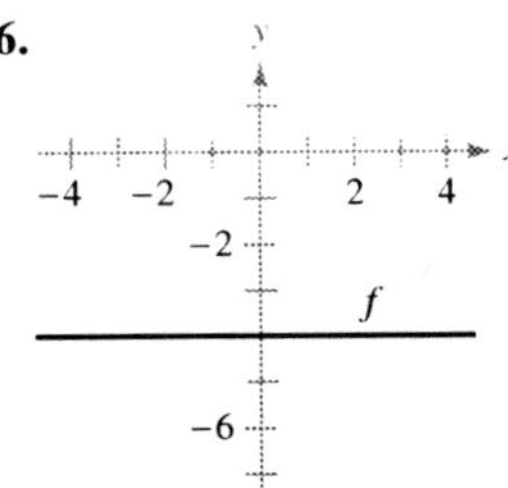

47.

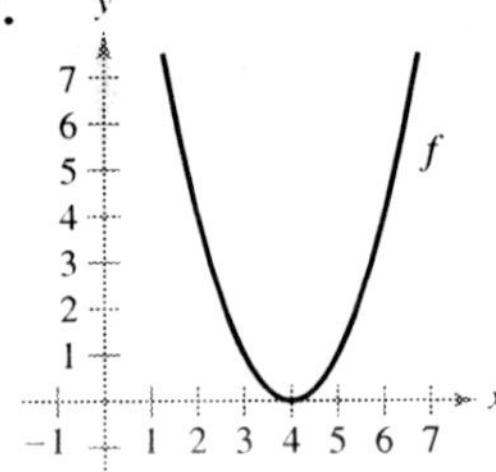

48.

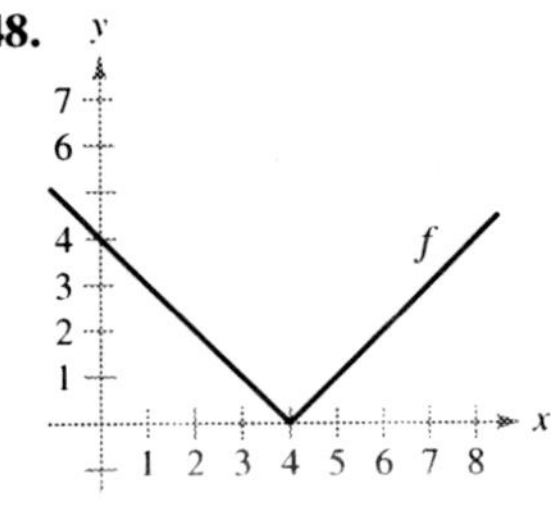

49.

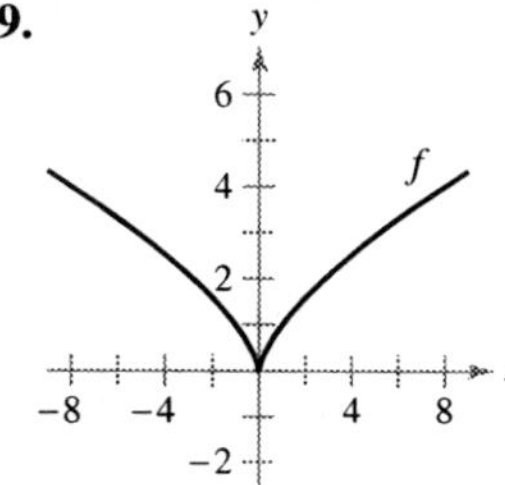

50.

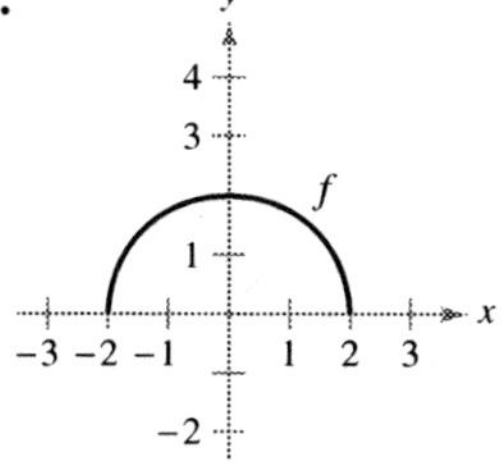

51. Sketch a graph of a function whose derivative is always negative.

52. Sketch a graph of a function whose derivative is always positive.

In Exercises 53–56, the limit represents $f'(c)$ for a function f and a number c. Find f and c.

53. $\displaystyle \lim_{\Delta x \to 0} \frac{[5 - 3(1 + \Delta x)] - 2}{\Delta x}$

54. $\displaystyle \lim_{\Delta x \to 0} \frac{(-2 + \Delta x)^3 + 8}{\Delta x}$

55. $\displaystyle \lim_{x \to 6} \frac{-x^2 + 36}{x - 6}$

56. $\displaystyle \lim_{x \to 9} \frac{2\sqrt{x} - 6}{x - 9}$

In Exercises 57–59, identify a function f that has the given characteristics. Then sketch the function.

57. $f(0) = 2;$
$f'(x) = -3, \ -\infty < x < \infty$

58. $f(0) = 4; f'(0) = 0;$
$f'(x) < 0$ for $x < 0;$
$f'(x) > 0$ for $x > 0$

59. $f(0) = 0; f'(0) = 0; f'(x) > 0$ for $x \neq 0$

60. Assume that $f'(c) = 3$. Find $f'(-c)$ if (a) f is an odd function and (b) f is an even function.

In Exercises 61 and 62, find equations of the two tangent lines to the graph of f that pass through the given point.

61. $f(x) = 4x - x^2$

62. $f(x) = x^2$

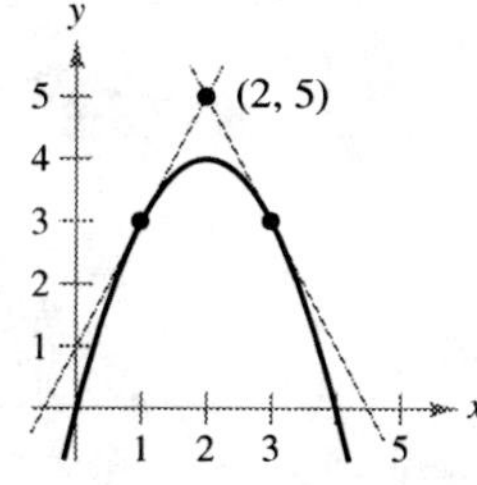

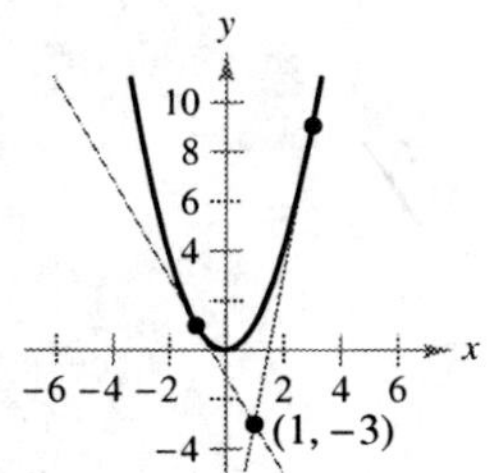

63. *Graphical Reasoning* Use a graphing utility to graph each function and its tangent lines at $x = -1$, $x = 0$, and $x = 1$. Based on the results, determine whether the slopes of tangent lines to the graph of a function at different values of x are always distinct.

(a) $f(x) = x^2$

(b) $g(x) = x^3$

64. The figure shows the graph of g'.

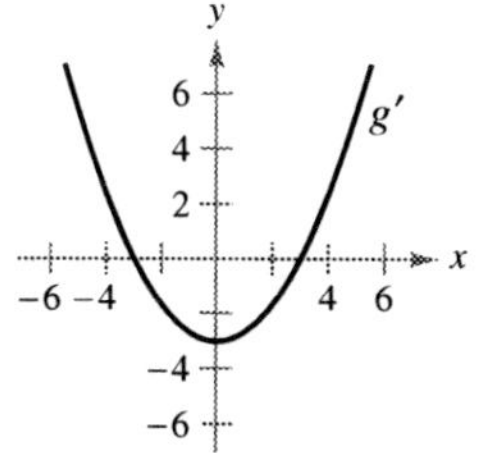

(a) $g'(0) = $ ▨

(b) $g'(3) = $ ▨

(c) What can you conclude about the graph of g knowing that $g'(1) = -\frac{8}{3}$?

(d) What can you conclude about the graph of g knowing that $g'(-4) = \frac{7}{3}$?

(e) Is $g(6) - g(4)$ positive or negative? Explain.

(f) Is it possible to find $g(2)$ from the graph? Explain.

65. *Graphical Analysis* Consider the function $f(x) = \frac{1}{2}x^2$.

(a) Use a graphing utility to graph the function and estimate the values of $f'(0), f'\left(\frac{1}{2}\right), f'(1)$, and $f'(2)$.

(b) Use your results from part (a) to determine the values of $f'\left(-\frac{1}{2}\right), f'(-1)$, and $f'(-2)$.

(c) Sketch a possible graph of f'.

(d) Use the definition of derivative to find $f'(x)$.

66. *Graphical Analysis* Consider the function $f(x) = \frac{1}{3}x^3$.

(a) Use a graphing utility to graph the function and estimate the values of $f'(0), f'\left(\frac{1}{2}\right), f'(1), f'(2)$, and $f'(3)$.

(b) Use your results from part (a) to determine the values of $f'\left(-\frac{1}{2}\right), f'(-1), f'(-2)$, and $f'(-3)$.

(c) Sketch a possible graph of f'.

(d) Use the definition of derivative to find $f'(x)$.

Graphical Reasoning **In Exercises 67 and 68, use a graphing utility to graph the functions f and g in the same viewing window where**
$$g(x) = \frac{f(x + 0.01) - f(x)}{0.01}.$$
Label the graphs and describe the relationship between them.

67. $f(x) = 2x - x^2$

68. $f(x) = 3\sqrt{x}$

In Exercises 69 and 70, evaluate $f(2)$ and $f(2.1)$ and use the results to approximate $f'(2)$.

69. $f(x) = x(4 - x)$

70. $f(x) = \frac{1}{4}x^3$

Graphical Reasoning **In Exercises 71 and 72, use a graphing utility to graph the function and its derivative in the same viewing window. Label the graphs and describe the relationship between them.**

71. $f(x) = \dfrac{1}{\sqrt{x}}$

72. $f(x) = \dfrac{x^3}{4} - 3x$

In Exercises 73–82, use the alternative form of the derivative to find the derivative at $x = c$ (if it exists).

73. $f(x) = x^2 - 5$, $c = 3$ **74.** $g(x) = x(x - 1)$, $c = 1$

75. $f(x) = x^3 + 2x^2 + 1$, $c = -2$

76. $f(x) = x^3 + 6x$, $c = 2$

77. $g(x) = \sqrt{|x|}$, $c = 0$ **78.** $f(x) = 2/x$, $c = 5$

79. $f(x) = (x - 6)^{2/3}$, $c = 6$

80. $g(x) = (x + 3)^{1/3}$, $c = -3$

81. $h(x) = |x + 7|$, $c = -7$ **82.** $f(x) = |x - 6|$, $c = 6$

In Exercises 83–88, describe the x-values at which f is differentiable.

83. $f(x) = \dfrac{2}{x - 3}$ **84.** $f(x) = |x^2 - 9|$

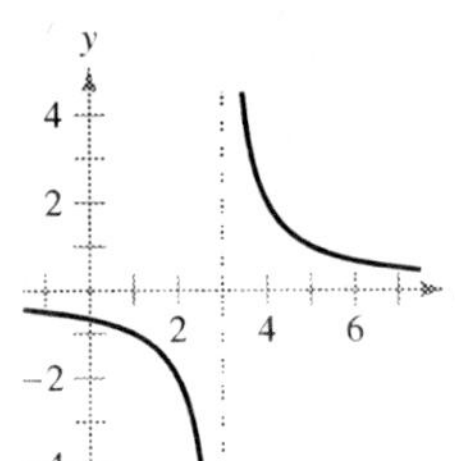   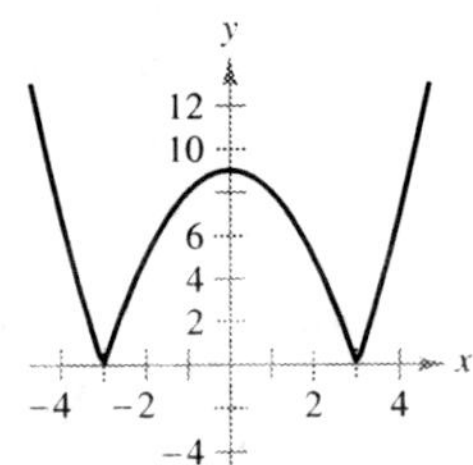

85. $f(x) = (x + 4)^{2/3}$ **86.** $f(x) = \dfrac{x^2}{x^2 - 4}$

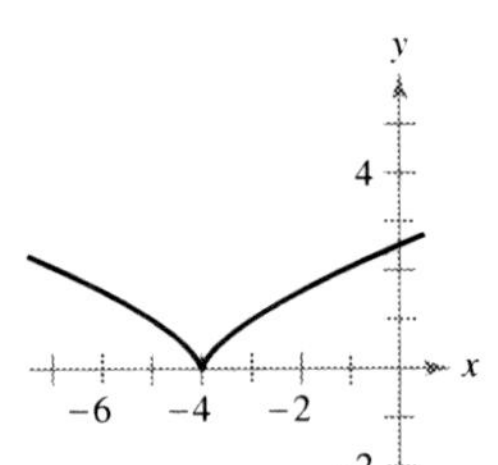   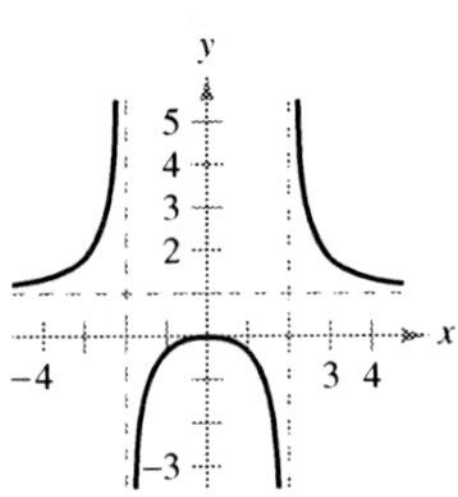

87. $f(x) = \sqrt{x - 1}$ **88.** $f(x) = \begin{cases} x^2 - 4, & x \le 0 \\ 4 - x^2, & x > 0 \end{cases}$

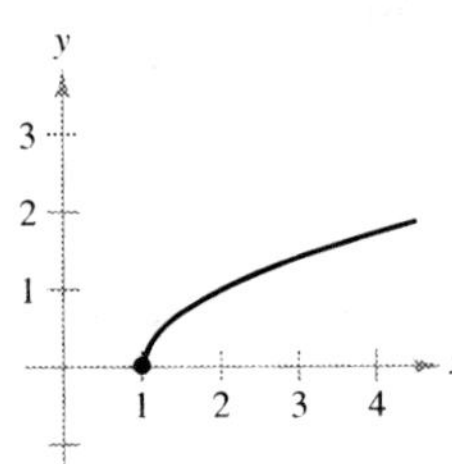   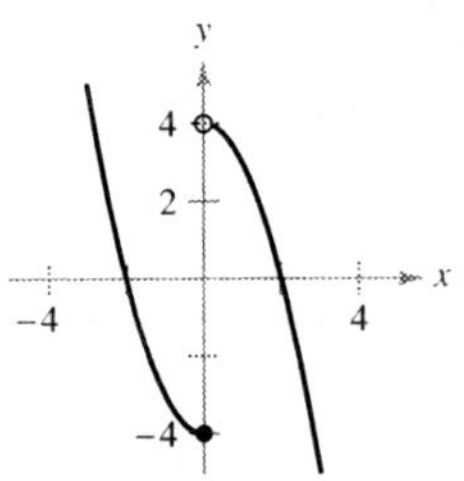

Graphical Analysis **In Exercises 89–92, use a graphing utility to graph the function and find the x-values at which f is differentiable.**

89. $f(x) = |x - 5|$ **90.** $f(x) = \dfrac{4x}{x - 3}$

91. $f(x) = x^{2/5}$

92. $f(x) = \begin{cases} x^3 - 3x^2 + 3x, & x \le 1 \\ x^2 - 2x, & x > 1 \end{cases}$

In Exercises 93–96, find the derivatives from the left and from the right at $x = 1$ (if they exist). Is the function differentiable at $x = 1$?

93. $f(x) = |x - 1|$ **94.** $f(x) = \sqrt{1 - x^2}$

95. $f(x) = \begin{cases} (x - 1)^3, & x \le 1 \\ (x - 1)^2, & x > 1 \end{cases}$ **96.** $f(x) = \begin{cases} x, & x \le 1 \\ x^2, & x > 1 \end{cases}$

In Exercises 97 and 98, determine whether the function is differentiable at $x = 2$.

97. $f(x) = \begin{cases} x^2 + 1, & x \le 2 \\ 4x - 3, & x > 2 \end{cases}$ **98.** $f(x) = \begin{cases} \frac{1}{2}x + 1, & x < 2 \\ \sqrt{2x}, & x \ge 2 \end{cases}$

99. Use a graphing utility to graph $g(x) = [\![x]\!]/x$. Then let $f(x) = [\![x]\!]$ and show that

$$\lim_{x \to 0^-} \frac{f(x) - f(0)}{x - 0} = \infty \text{ and } \lim_{x \to 0^+} \frac{f(x) - f(0)}{x - 0} = 0.$$

Is f differentiable? Explain.

100. *Conjecture*   Consider the functions $f(x) = x^2$ and $g(x) = x^3$.

(a) Graph f and f' on the same set of axes.

(b) Graph g and g' on the same set of axes.

(c) Identify a pattern between f and g and their respective derivatives. Use the pattern to make a conjecture about $h'(x)$ if $h(x) = x^n$, where n is an integer and $n \ge 2$.

(d) Find $f'(x)$ if $f(x) = x^4$. Compare the result with the conjecture in part (c). Is this a proof of your conjecture? Explain.

True or False?   **In Exercises 101–104, determine whether the statement is true or false. If it is false, explain why or give an example that shows it is false.**

101. The slope of the tangent line to the differentiable function f at the point $(2, f(2))$ is $\dfrac{f(2 + \Delta x) - f(2)}{\Delta x}$.

102. If a function is continuous at a point, then it is differentiable at that point.

103. If a function has derivatives from both the right and the left at a point, then it is differentiable at that point.

104. If a function is differentiable at a point, then it is continuous at that point.

105. Let $f(x) = \begin{cases} x \sin \frac{1}{x}, & x \ne 0 \\ 0, & x = 0 \end{cases}$ and $g(x) = \begin{cases} x^2 \sin \frac{1}{x}, & x \ne 0 \\ 0, & x = 0 \end{cases}$.

Show that f is continuous, but not differentiable, at $x = 0$. Show that g is differentiable at 0, and find $g'(0)$.

106. *Writing*   Use a graphing utility to graph the two functions $f(x) = x^2 + 1$ and $g(x) = |x| + 1$ in the same viewing window. Use the *zoom* and *trace* features to analyze the graphs near the point $(0, 1)$. What do you observe? Which function is differentiable at this point? Write a short paragraph describing the geometric significance of differentiability at a point.

3.2 Basic Differentiation Rules and Rates of Change

- **Find the derivative of a function using the Constant Rule.**
- **Find the derivative of a function using the Power Rule.**
- **Find the derivative of a function using the Constant Multiple Rule.**
- **Find the derivative of a function using the Sum and Difference Rules.**
- **Find the derivatives of the sine, cosine, and exponential functions.**
- **Use derivatives to find rates of change.**

The Constant Rule

In Section 3.1 you used the limit definition to find derivatives. In this and the next two sections you will be introduced to several "differentiation rules" that allow you to find derivatives without the *direct* use of the limit definition.

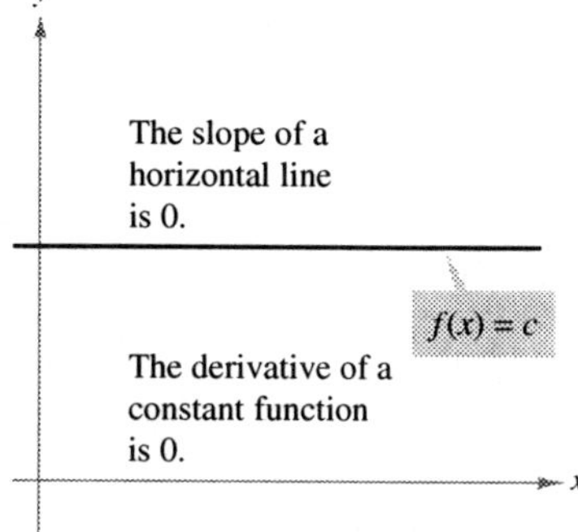

Notice that the Constant Rule is equivalent to saying that the slope of a horizontal line is 0. This demonstrates the relationship between slope and derivative.
Figure 3.14

THEOREM 3.2 THE CONSTANT RULE

The derivative of a constant function is 0. That is, if c is a real number, then

$$\frac{d}{dx}[c] = 0.$$

(See Figure 3.14.)

PROOF Let $f(x) = c$. Then, by the limit definition of the derivative,

$$\frac{d}{dx}[c] = f'(x)$$

$$= \lim_{\Delta x \to 0} \frac{f(x + \Delta x) - f(x)}{\Delta x}$$

$$= \lim_{\Delta x \to 0} \frac{c - c}{\Delta x}$$

$$= \lim_{\Delta x \to 0} 0$$

$$= 0. \qquad \blacksquare$$

EXAMPLE 1 Using the Constant Rule

Function	Derivative
a. $y = 7$	$dy/dx = 0$
b. $f(x) = 0$	$f'(x) = 0$
c. $s(t) = -3$	$s'(t) = 0$
d. $y = k\pi^2$, k is constant	$y' = 0$

$\blacksquare$

EXPLORATION

Writing a Conjecture   Use the definition of the derivative given in Section 3.1 to find the derivative of each of the following. What patterns do you see? Use your results to write a conjecture about the derivative of $f(x) = x^n$.

a. $f(x) = x^1$ **b.** $f(x) = x^2$ **c.** $f(x) = x^3$

d. $f(x) = x^4$ **e.** $f(x) = x^{1/2}$ **f.** $f(x) = x^{-1}$

The Power Rule

Before proving the next rule, review the procedure for expanding a binomial.

$$(x + \Delta x)^2 = x^2 + 2x\Delta x + (\Delta x)^2$$
$$(x + \Delta x)^3 = x^3 + 3x^2\Delta x + 3x(\Delta x)^2 + (\Delta x)^3$$
$$(x + \Delta x)^4 = x^4 + 4x^3\Delta x + 6x^2(\Delta x)^2 + 4x(\Delta x)^3 + (\Delta x)^4$$

The general binomial expansion for a positive integer n is

$$(x + \Delta x)^n = x^n + nx^{n-1}(\Delta x) + \underbrace{\frac{n(n-1)x^{n-2}}{2}(\Delta x)^2 + \cdots + (\Delta x)^n}_{(\Delta x)^2 \text{ is a factor of these terms.}}$$

This binomial expansion is used in proving a special case of the Power Rule.

THEOREM 3.3 THE POWER RULE

If n is a real number, then the function $f(x) = x^n$ is differentiable and

$$\frac{d}{dx}[x^n] = nx^{n-1}.$$

For f to be differentiable at $x = 0$, n must be a number such that x^{n-1} is defined on an interval containing 0.

 From Example 7 in Section 3.1, you know that the function $f(x) = x^{1/3}$ is defined at $x = 0$, but is not differentiable at $x = 0$. This is because $x^{-2/3}$ is not defined on an interval containing 0.

PROOF If n is a positive integer greater than 1, then the binomial expansion produces the following.

$$\frac{d}{dx}[x^n] = \lim_{\Delta x \to 0} \frac{(x + \Delta x)^n - x^n}{\Delta x}$$

$$= \lim_{\Delta x \to 0} \frac{x^n + nx^{n-1}(\Delta x) + \dfrac{n(n-1)x^{n-2}}{2}(\Delta x)^2 + \cdots + (\Delta x)^n - x^n}{\Delta x}$$

$$= \lim_{\Delta x \to 0} \left[nx^{n-1} + \frac{n(n-1)x^{n-2}}{2}(\Delta x) + \cdots + (\Delta x)^{n-1} \right]$$

$$= nx^{n-1} + 0 + \cdots + 0$$

$$= nx^{n-1}.$$

This proves the case for which n is a positive integer greater than 1. It is left to you to prove the case for $n = 1$. Example 7 in Section 3.3 proves the case for which n is a negative integer. The cases for which n is rational and n is irrational are left as an exercise (see Section 3.5, Exercise 100). ■

When using the Power Rule, the case for which $n = 1$ is best thought of as a separate differentiation rule. That is,

$$\frac{d}{dx}[x] = 1. \qquad \text{Power Rule when } n = 1$$

This rule is consistent with the fact that the slope of the line $y = x$ is 1, as shown in Figure 3.15.

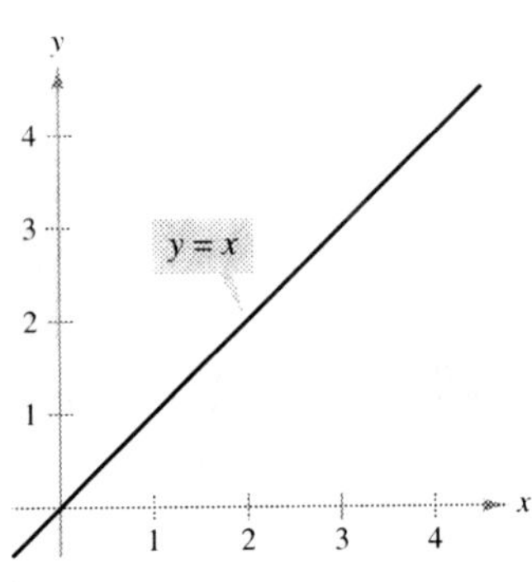

The slope of the line $y = x$ is 1.

Figure 3.15

EXAMPLE 2   Using the Power Rule

Function	*Derivative*
a. $f(x) = x^3$	$f'(x) = 3x^2$
b. $g(x) = \sqrt[3]{x}$	$g'(x) = \dfrac{d}{dx}[x^{1/3}] = \dfrac{1}{3}x^{-2/3} = \dfrac{1}{3x^{2/3}}$
c. $y = \dfrac{1}{x^2}$	$\dfrac{dy}{dx} = \dfrac{d}{dx}[x^{-2}] = (-2)x^{-3} = -\dfrac{2}{x^3}$

In Example 2(c), note that *before* differentiating, $1/x^2$ was rewritten as x^{-2}. Rewriting is the first step in *many* differentiation problems.

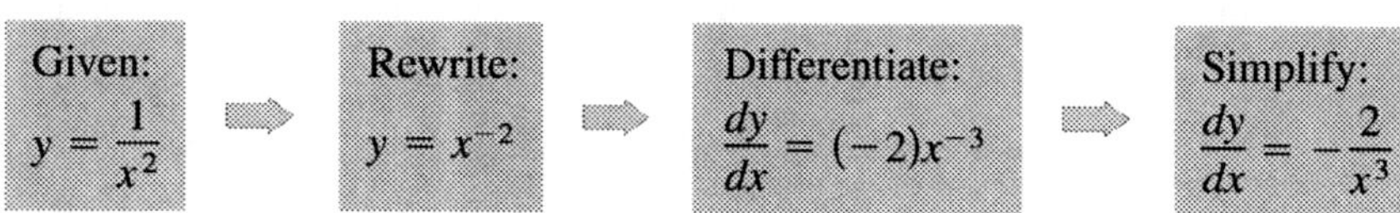

EXAMPLE 3   Finding the Slope of a Graph

Find the slope of the graph of $f(x) = x^4$ when

a. $x = -1$ **b.** $x = 0$ **c.** $x = 1$.

Solution   The derivative of f is $f'(x) = 4x^3$.

a. When $x = -1$, the slope is $f'(-1) = 4(-1)^3 = -4$.   Slope is negative.
b. When $x = 0$, the slope is $f'(0) = 4(0)^3 = 0$.   Slope is zero.
c. When $x = 1$, the slope is $f'(1) = 4(1)^3 = 4$.   Slope is positive.

In Figure 3.16, note that the slope of the graph is negative at the point $(-1, 1)$, the slope is zero at the point $(0, 0)$, and the slope is positive at the point $(1, 1)$.

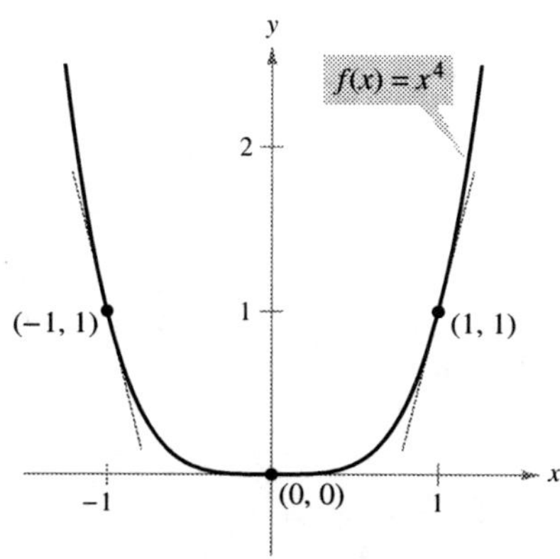

The slope of a graph at a point is the value of the derivative at that point.
Figure 3.16

EXAMPLE 4   Finding an Equation of a Tangent Line

Find an equation of the tangent line to the graph of $f(x) = x^2$ when $x = -2$.

Solution   To find the *point* on the graph of f, evaluate the original function at $x = -2$.

$$(-2, f(-2)) = (-2, 4) \qquad \text{Point on graph}$$

To find the *slope* of the graph when $x = -2$, evaluate the derivative, $f'(x) = 2x$, at $x = -2$.

$$m = f'(-2) = -4 \qquad \text{Slope of graph at } (-2, 4)$$

Now, using the point-slope form of the equation of a line, you can write

$$\begin{aligned} y - y_1 &= m(x - x_1) &&\text{Point-slope form} \\ y - 4 &= -4[x - (-2)] &&\text{Substitute for } y_1, m, \text{ and } x_1. \\ y &= -4x - 4. &&\text{Simplify.} \end{aligned}$$

(See Figure 3.17.)

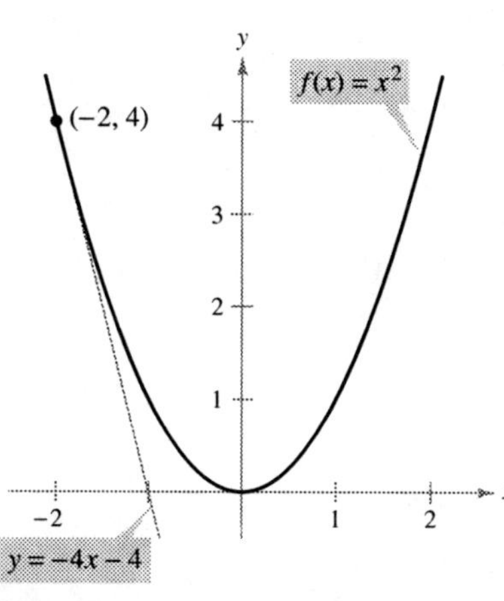

The line $y = -4x - 4$ is tangent to the graph of $f(x) = x^2$ at the point $(-2, 4)$.
Figure 3.17

The Constant Multiple Rule

> **THEOREM 3.4 THE CONSTANT MULTIPLE RULE**
>
> If f is a differentiable function and c is a real number, then cf is also differentiable and $\dfrac{d}{dx}[cf(x)] = cf'(x)$.

PROOF

$$\frac{d}{dx}[cf(x)] = \lim_{\Delta x \to 0} \frac{cf(x + \Delta x) - cf(x)}{\Delta x} \qquad \text{Definition of derivative}$$

$$= \lim_{\Delta x \to 0} c\left[\frac{f(x + \Delta x) - f(x)}{\Delta x}\right]$$

$$= c\left[\lim_{\Delta x \to 0} \frac{f(x + \Delta x) - f(x)}{\Delta x}\right] \qquad \text{Apply Theorem 2.2.}$$

$$= cf'(x) \qquad\qquad\qquad\qquad\qquad\qquad\quad \blacksquare$$

Informally, the Constant Multiple Rule states that constants can be factored out of the differentiation process, even if the constants appear in the denominator.

$$\frac{d}{dx}[cf(x)] = c\frac{d}{dx}[\;f(x)] = cf'(x)$$

$$\frac{d}{dx}\left[\frac{f(x)}{c}\right] = \frac{d}{dx}\left[\left(\frac{1}{c}\right)f(x)\right]$$

$$= \left(\frac{1}{c}\right)\frac{d}{dx}[\;f(x)] = \left(\frac{1}{c}\right)f'(x)$$

EXAMPLE 5   Using the Constant Multiple Rule

Function	*Derivative*
a. $y = \dfrac{2}{x}$	$\dfrac{dy}{dx} = \dfrac{d}{dx}[2x^{-1}] = 2\dfrac{d}{dx}[x^{-1}] = 2(-1)x^{-2} = -\dfrac{2}{x^2}$
b. $f(t) = \dfrac{4t^2}{5}$	$f'(t) = \dfrac{d}{dt}\left[\dfrac{4}{5}t^2\right] = \dfrac{4}{5}\dfrac{d}{dt}[t^2] = \dfrac{4}{5}(2t) = \dfrac{8}{5}t$
c. $y = 2\sqrt{x}$	$\dfrac{dy}{dx} = \dfrac{d}{dx}[2x^{1/2}] = 2\left(\dfrac{1}{2}x^{-1/2}\right) = x^{-1/2} = \dfrac{1}{\sqrt{x}}$
d. $y = \dfrac{1}{2\sqrt[3]{x^2}}$	$\dfrac{dy}{dx} = \dfrac{d}{dx}\left[\dfrac{1}{2}x^{-2/3}\right] = \dfrac{1}{2}\left(-\dfrac{2}{3}\right)x^{-5/3} = -\dfrac{1}{3x^{5/3}}$
e. $y = -\dfrac{3x}{2}$	$y' = \dfrac{d}{dx}\left[-\dfrac{3}{2}x\right] = -\dfrac{3}{2}(1) = -\dfrac{3}{2}$ $\blacksquare$

The Constant Multiple Rule and the Power Rule can be combined into one rule. The combination rule is

$$\frac{d}{dx}[cx^n] = cnx^{n-1}.$$

EXAMPLE 6 Using Parentheses When Differentiating

Original Function	Rewrite	Differentiate	Simplify
a. $y = \dfrac{5}{2x^3}$	$y = \dfrac{5}{2}(x^{-3})$	$y' = \dfrac{5}{2}(-3x^{-4})$	$y' = -\dfrac{15}{2x^4}$
b. $y = \dfrac{5}{(2x)^3}$	$y = \dfrac{5}{8}(x^{-3})$	$y' = \dfrac{5}{8}(-3x^{-4})$	$y' = -\dfrac{15}{8x^4}$
c. $y = \dfrac{7}{3x^{-2}}$	$y = \dfrac{7}{3}(x^2)$	$y' = \dfrac{7}{3}(2x)$	$y' = \dfrac{14x}{3}$
d. $y = \dfrac{7}{(3x)^{-2}}$	$y = 63(x^2)$	$y' = 63(2x)$	$y' = 126x$

The Sum and Difference Rules

> **THEOREM 3.5 THE SUM AND DIFFERENCE RULES**
>
> The sum (or difference) of two differentiable functions f and g is itself differentiable. Moreover, the derivative of $f + g$ (or $f - g$) is the sum (or difference) of the derivatives of f and g.
>
> $$\frac{d}{dx}[f(x) + g(x)] = f'(x) + g'(x) \qquad \text{Sum Rule}$$
>
> $$\frac{d}{dx}[f(x) - g(x)] = f'(x) - g'(x) \qquad \text{Difference Rule}$$

PROOF A proof of the Sum Rule follows from Theorem 2.2. (The Difference Rule can be proved in a similar way.)

$$\frac{d}{dx}[f(x) + g(x)] = \lim_{\Delta x \to 0} \frac{[f(x + \Delta x) + g(x + \Delta x)] - [f(x) + g(x)]}{\Delta x}$$

$$= \lim_{\Delta x \to 0} \frac{f(x + \Delta x) + g(x + \Delta x) - f(x) - g(x)}{\Delta x}$$

$$= \lim_{\Delta x \to 0} \left[\frac{f(x + \Delta x) - f(x)}{\Delta x} + \frac{g(x + \Delta x) - g(x)}{\Delta x} \right]$$

$$= \lim_{\Delta x \to 0} \frac{f(x + \Delta x) - f(x)}{\Delta x} + \lim_{\Delta x \to 0} \frac{g(x + \Delta x) - g(x)}{\Delta x}$$

$$= f'(x) + g'(x)$$

The Sum and Difference Rules can be extended to any finite number of functions. For instance, if $F(x) = f(x) + g(x) - h(x)$, then $F'(x) = f'(x) + g'(x) - h'(x)$.

EXAMPLE 7 Using the Sum and Difference Rules

Function	Derivative
a. $f(x) = x^3 - 4x + 5$	$f'(x) = 3x^2 - 4$
b. $g(x) = -\dfrac{x^4}{2} + 3x^3 - 2x$	$g'(x) = -2x^3 + 9x^2 - 2$

EXPLORATION

Use a graphing utility to graph the function

$$f(x) = \frac{\sin(x + \Delta x) - \sin x}{\Delta x}$$

for $\Delta x = 0.01$. What does this function represent? Compare this graph with that of the cosine function. What do you think the derivative of the sine function equals?

Derivatives of Sine and Cosine Functions

In Section 2.3, you studied the following limits.

$$\lim_{\Delta x \to 0} \frac{\sin \Delta x}{\Delta x} = 1 \quad \text{and} \quad \lim_{\Delta x \to 0} \frac{1 - \cos \Delta x}{\Delta x} = 0$$

These two limits can be used to prove differentiation rules for the sine and cosine functions. (The derivatives of the other four trigonometric functions are discussed in Section 3.3.)

THEOREM 3.6 DERIVATIVES OF SINE AND COSINE FUNCTIONS

$$\frac{d}{dx}[\sin x] = \cos x \qquad \frac{d}{dx}[\cos x] = -\sin x$$

PROOF

$$\frac{d}{dx}[\sin x] = \lim_{\Delta x \to 0} \frac{\sin(x + \Delta x) - \sin x}{\Delta x} \qquad \text{Definition of derivative}$$

$$= \lim_{\Delta x \to 0} \frac{\sin x \cos \Delta x + \cos x \sin \Delta x - \sin x}{\Delta x}$$

$$= \lim_{\Delta x \to 0} \frac{\cos x \sin \Delta x - (\sin x)(1 - \cos \Delta x)}{\Delta x}$$

$$= \lim_{\Delta x \to 0} \left[(\cos x)\left(\frac{\sin \Delta x}{\Delta x}\right) - (\sin x)\left(\frac{1 - \cos \Delta x}{\Delta x}\right) \right]$$

$$= \cos x \left(\lim_{\Delta x \to 0} \frac{\sin \Delta x}{\Delta x}\right) - \sin x \left(\lim_{\Delta x \to 0} \frac{1 - \cos \Delta x}{\Delta x}\right)$$

$$= (\cos x)(1) - (\sin x)(0)$$

$$= \cos x$$

This differentiation rule is shown graphically in Figure 3.18. Note that for each x, the *slope* of the sine curve is equal to the value of the cosine. The proof of the second rule is left as an exercise (see Exercise 124). ■

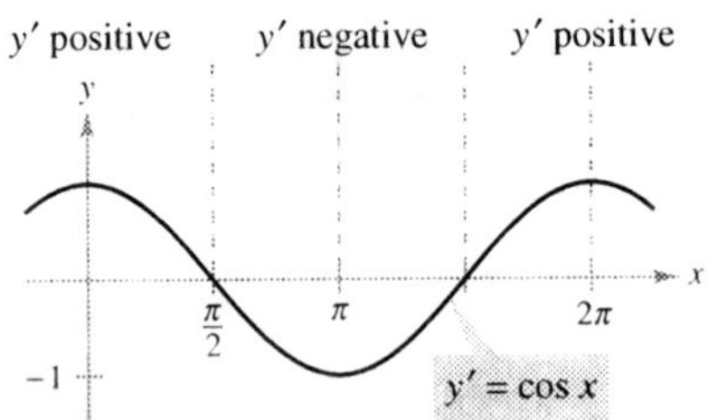

The derivative of the sine function is the cosine function.

Figure 3.18

EXAMPLE 8 Derivatives Involving Sines and Cosines

	Function	Derivative
a.	$y = 2 \sin x$	$y' = 2 \cos x$
b.	$y = \dfrac{\sin x}{2} = \dfrac{1}{2} \sin x$	$y' = \dfrac{1}{2} \cos x = \dfrac{\cos x}{2}$
c.	$y = x + \cos x$	$y' = 1 - \sin x$

■

TECHNOLOGY A graphing utility can provide insight into the interpretation of a derivative. For instance, Figure 3.19 shows the graphs of

$$y = a \sin x$$

for $a = \frac{1}{2}, 1, \frac{3}{2},$ and 2. Estimate the slope of each graph at the point $(0, 0)$. Then verify your estimates analytically by evaluating the derivative of each function when $x = 0$.

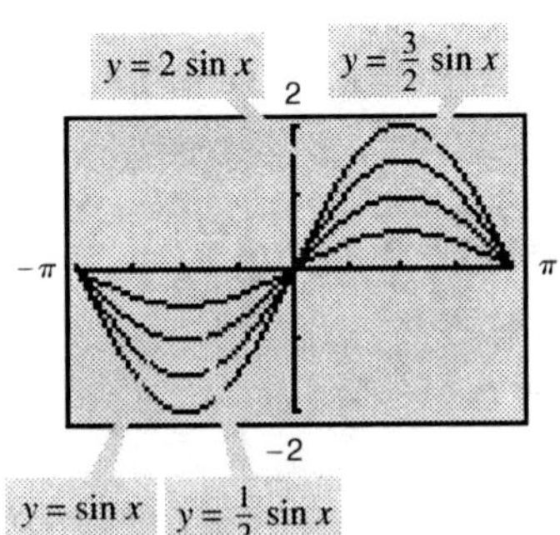

$$\frac{d}{dx}[a \sin x] = a \cos x$$

Figure 3.19

STUDY TIP   The key to the formula for the derivative of $f(x) = e^x$ is the limit

$$\lim_{x \to 0} (1 + x)^{1/x} = e.$$

This important limit was introduced on page 51 and formalized later on page 85. It is used to conclude that for $\Delta x \approx 0$,

$$(1 + \Delta x)^{1/\Delta x} \approx e.$$

Derivatives of Exponential Functions

One of the most intriguing (and useful) characteristics of the natural exponential function is that *it is its own derivative*. Consider the following.

Let $f(x) = e^x$.

$$f'(x) = \lim_{\Delta x \to 0} \frac{f(x + \Delta x) - f(x)}{\Delta x}$$

$$= \lim_{\Delta x \to 0} \frac{e^{x+\Delta x} - e^x}{\Delta x}$$

$$= \lim_{\Delta x \to 0} \frac{e^x(e^{\Delta x} - 1)}{\Delta x}$$

The definition of e

$$\lim_{\Delta x \to 0} (1 + \Delta x)^{1/\Delta x} = e$$

tells you that for small values of Δx, you have $e \approx (1 + \Delta x)^{1/\Delta x}$, which implies that $e^{\Delta x} \approx 1 + \Delta x$. Replacing $e^{\Delta x}$ by this approximation produces the following.

$$f'(x) = \lim_{\Delta x \to 0} \frac{e^x[e^{\Delta x} - 1]}{\Delta x}$$

$$= \lim_{\Delta x \to 0} \frac{e^x[(1 + \Delta x) - 1]}{\Delta x}$$

$$= \lim_{\Delta x \to 0} \frac{e^x \Delta x}{\Delta x}$$

$$= e^x$$

This result is stated in the next theorem.

THEOREM 3.7 DERIVATIVE OF THE NATURAL EXPONENTIAL FUNCTION

$$\frac{d}{dx}[e^x] = e^x$$

You can interpret Theorem 3.7 graphically by saying that the slope of the graph of $f(x) = e^x$ at any point (x, e^x) is equal to the y-coordinate of the point, as shown in Figure 3.20.

EXAMPLE 9 Derivatives of Exponential Functions

Find the derivative of each function.

a. $f(x) = 3e^x$ **b.** $f(x) = x^2 + e^x$ **c.** $f(x) = \sin x - e^x$

Solution

a. $f'(x) = 3 \dfrac{d}{dx}[e^x] = 3e^x$

b. $f'(x) = \dfrac{d}{dx}[x^2] + \dfrac{d}{dx}[e^x] = 2x + e^x$

c. $f'(x) = \dfrac{d}{dx}[\sin x] - \dfrac{d}{dx}[e^x] = \cos x - e^x$

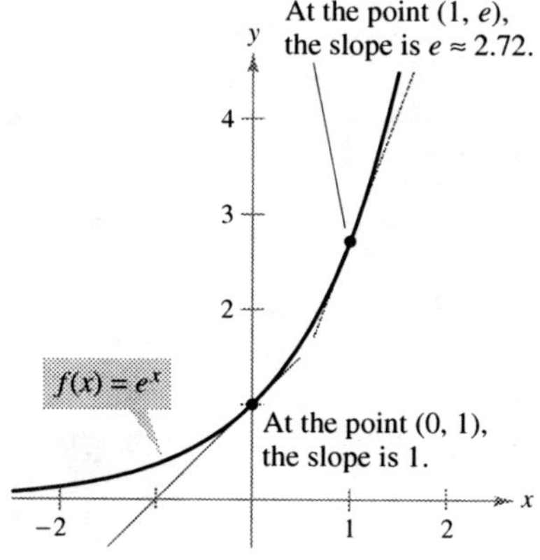

Figure 3.20

Rates of Change

You have seen how the derivative is used to determine slope. The derivative can also be used to determine the rate of change of one variable with respect to another. Applications involving rates of change occur in a wide variety of fields. A few examples are population growth rates, production rates, water flow rates, velocity, and acceleration.

A common use for rate of change is to describe the motion of an object moving in a straight line. In such problems, it is customary to use either a horizontal or a vertical line with a designated origin to represent the line of motion. On such lines, movement to the right (or upward) is considered to be in the positive direction, and movement to the left (or downward) is considered to be in the negative direction.

The function s that gives the position (relative to the origin) of an object as a function of time t is called a **position function.** If, over a period of time Δt, the object changes its position by the amount $\Delta s = s(t + \Delta t) - s(t)$, then, by the familiar formula

$$\text{Rate} = \frac{\text{distance}}{\text{time}}$$

the **average velocity** is

$$\frac{\text{Change in distance}}{\text{Change in time}} = \frac{\Delta s}{\Delta t}. \qquad \text{Average velocity}$$

EXAMPLE 10 Finding Average Velocity of a Falling Object

If a billiard ball is dropped from a height of 100 feet, its height s at time t is given by the position function

$$s = -16t^2 + 100 \qquad \text{Position function}$$

where s is measured in feet and t is measured in seconds. Find the average velocity over each of the following time intervals.

a. $[1, 2]$ **b.** $[1, 1.5]$ **c.** $[1, 1.1]$

Solution

a. For the interval $[1, 2]$, the object falls from a height of $s(1) = -16(1)^2 + 100 = 84$ feet to a height of $s(2) = -16(2)^2 + 100 = 36$ feet. The average velocity is

$$\frac{\Delta s}{\Delta t} = \frac{36 - 84}{2 - 1} = \frac{-48}{1} = -48 \text{ feet per second.}$$

b. For the interval $[1, 1.5]$, the object falls from a height of 84 feet to a height of 64 feet. The average velocity is

$$\frac{\Delta s}{\Delta t} = \frac{64 - 84}{1.5 - 1} = \frac{-20}{0.5} = -40 \text{ feet per second.}$$

c. For the interval $[1, 1.1]$, the object falls from a height of 84 feet to a height of 80.64 feet. The average velocity is

$$\frac{\Delta s}{\Delta t} = \frac{80.64 - 84}{1.1 - 1} = \frac{-3.36}{0.1} = -33.6 \text{ feet per second.}$$

Note that the average velocities are *negative*, indicating that the object is moving downward. ∎

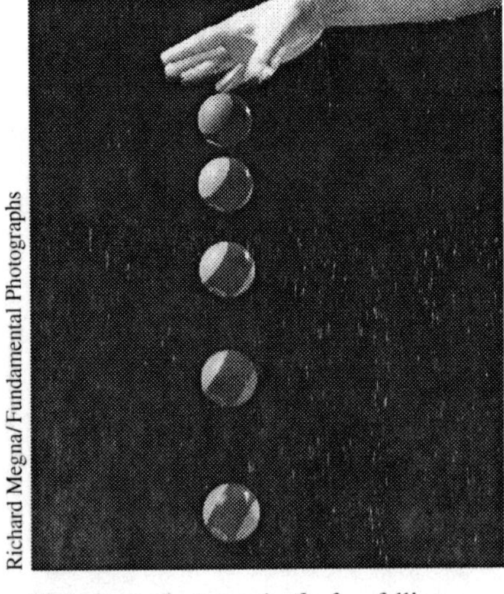

Time-lapse photograph of a free-falling billiard ball

Richard Megna/Fundamental Photographs

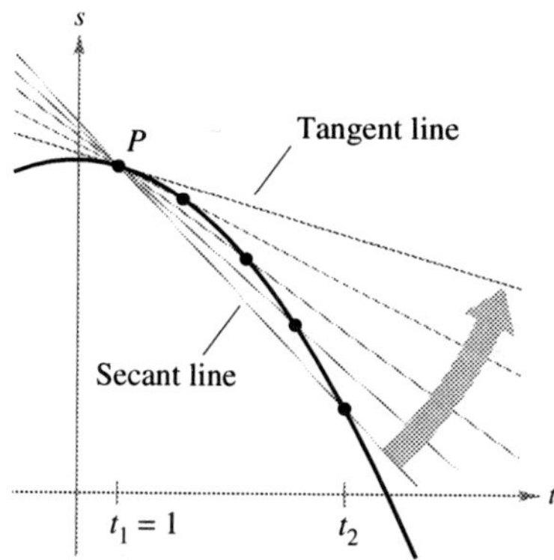

The average velocity between t_1 and t_2 is the slope of the secant line, and the instantaneous velocity at t_1 is the slope of the tangent line.
Figure 3.21

Suppose that in Example 10 you wanted to find the *instantaneous* velocity (or simply the velocity) of the object when $t = 1$. Just as you can approximate the slope of the tangent line by calculating the slope of the secant line, you can approximate the velocity at $t = 1$ by calculating the average velocity over a small interval $[1, 1 + \Delta t]$ (see Figure 3.21). By taking the limit as Δt approaches zero, you obtain the velocity when $t = 1$. Try doing this—you will find that the velocity when $t = 1$ is -32 feet per second.

In general, if $s = s(t)$ is the position function for an object moving along a straight line, the **velocity** of the object at time t is

$$v(t) = \lim_{\Delta t \to 0} \frac{s(t + \Delta t) - s(t)}{\Delta t} = s'(t).$$

Velocity function

In other words, the velocity function is the derivative of the position function. Velocity can be negative, zero, or positive. The **speed** of an object is the absolute value of its velocity. Speed cannot be negative.

The position of a free-falling object (neglecting air resistance) under the influence of gravity can be represented by the equation

$$s(t) = \frac{1}{2} g t^2 + v_0 t + s_0$$

Position function

where s_0 is the initial height of the object, v_0 is the initial velocity of the object, and g is the acceleration due to gravity. On Earth, the value of g is approximately -32 feet per second per second or -9.8 meters per second per second.

EXAMPLE 11 Using the Derivative to Find Velocity

At time $t = 0$, a diver jumps from a platform diving board that is 32 feet above the water (see Figure 3.22). The position of the diver is given by

$$s(t) = -16t^2 + 16t + 32$$

Position function

where s is measured in feet and t is measured in seconds.

a. When does the diver hit the water?

b. What is the diver's velocity at impact?

Solution

a. To find the time t when the diver hits the water, let $s = 0$ and solve for t.

$$-16t^2 + 16t + 32 = 0 \qquad \text{Set position function equal to 0.}$$
$$-16(t + 1)(t - 2) = 0 \qquad \text{Factor.}$$
$$t = -1 \text{ or } 2 \qquad \text{Solve for } t.$$

Because $t \geq 0$, choose the positive value to conclude that the diver hits the water at $t = 2$ seconds.

b. The velocity at time t is given by the derivative $s'(t) = -32t + 16$. So, the velocity at time $t = 2$ is

$$s'(2) = -32(2) + 16 = -48 \text{ feet per second.}$$

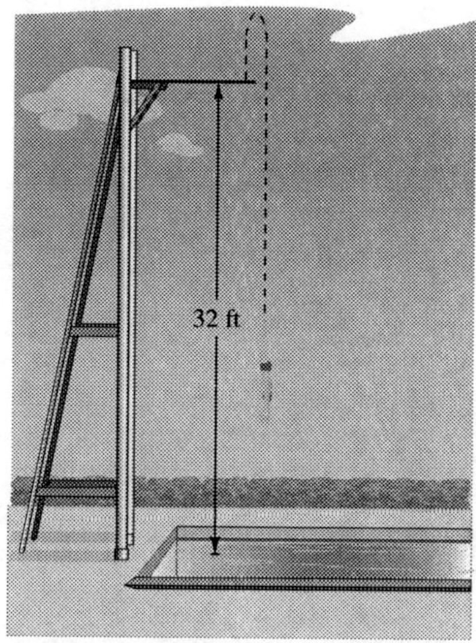

Velocity is positive when an object is rising, and is negative when an object is falling. Notice that the diver moves upward for the first half-second because the velocity is positive for $0 < t < \frac{1}{2}$. When the velocity is 0, the diver has reached the maximum height of the dive.
Figure 3.22

3.2 Exercises

See www.CalcChat.com for worked-out solutions to odd-numbered exercises.

In Exercises 1 and 2, use the graph to estimate the slope of the tangent line to $y = x^n$ at the point $(1, 1)$. Verify your answer analytically. To print an enlarged copy of the graph, go to the website www.mathgraphs.com.

1. (a) $y = x^{1/2}$ (b) $y = x^3$

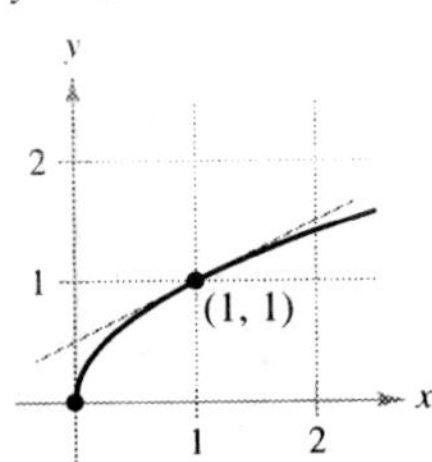 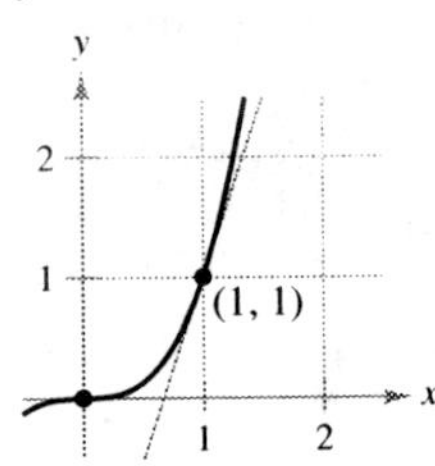

2. (a) $y = x^{-1/2}$ (b) $y = x^{-1}$

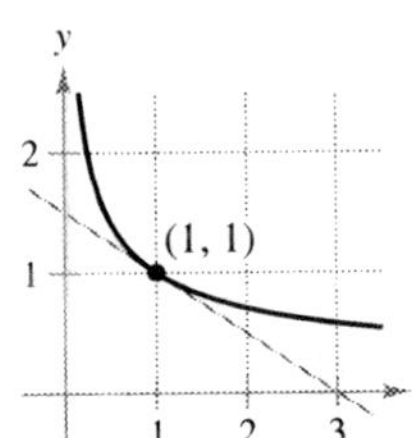 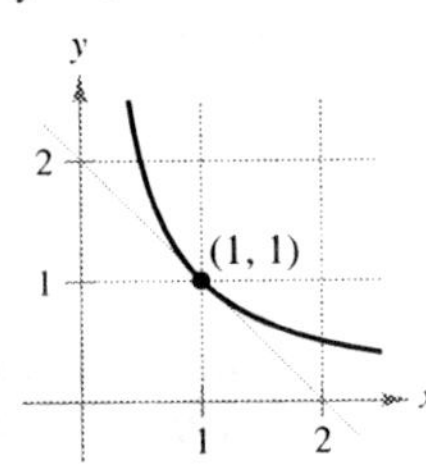

In Exercises 3–26, use the rules of differentiation to find the derivative of the function.

3. $y = 12$ **4.** $f(x) = -9$

5. $y = x^7$ **6.** $y = x^{16}$

7. $y = \dfrac{1}{x^5}$ **8.** $y = \dfrac{1}{x^8}$

9. $f(x) = \sqrt[5]{x}$ **10.** $g(x) = \sqrt[6]{x}$

11. $f(x) = x + 11$ **12.** $g(x) = 3x - 1$

13. $f(t) = -2t^2 + 3t - 6$ **14.** $y = t^2 + 2t - 3$

15. $g(x) = x^2 + 4x^3$ **16.** $y = 8 - x^3$

17. $s(t) = t^3 + 5t^2 - 3t + 8$ **18.** $f(x) = 2x^3 - 4x^2 + 3x$

19. $f(x) = 6x - 5e^x$ **20.** $h(t) = t^3 + 2e^t$

21. $y = \dfrac{\pi}{2}\sin\theta - \cos\theta$ **22.** $g(t) = \pi\cos t$

23. $y = x^2 - \frac{1}{2}\cos x$ **24.** $y = 7 + \sin x$

25. $y = \frac{1}{2}e^x - 3\sin x$ **26.** $y = \frac{3}{4}e^x + 2\cos x$

In Exercises 27–32, complete the table.

Original Function	Rewrite	Differentiate	Simplify
27. $y = \dfrac{5}{2x^2}$			
28. $y = \dfrac{4}{3x^2}$			
29. $y = \dfrac{6}{(5x)^3}$			
30. $y = \dfrac{\pi}{(5x)^2}$			
31. $y = \dfrac{\sqrt{x}}{x}$			
32. $y = \dfrac{4}{x^{-3}}$			

In Exercises 33–40, find the slope of the graph of the function at the given point. Use the *derivative* feature of a graphing utility to confirm your results.

Function	Point
33. $f(x) = \dfrac{8}{x^2}$	$(2, 2)$
34. $f(t) = 3 - \dfrac{3}{5t}$	$\left(\frac{3}{5}, 2\right)$
35. $f(x) = -\frac{1}{2} + \frac{7}{5}x^3$	$\left(0, -\frac{1}{2}\right)$
36. $f(x) = 3(5 - x)^2$	$(5, 0)$
37. $f(\theta) = 4\sin\theta - \theta$	$(0, 0)$
38. $g(t) = -2\cos t + 5$	$(\pi, 7)$
39. $f(t) = \frac{3}{4}e^t$	$\left(0, \frac{3}{4}\right)$
40. $g(x) = -4e^x$	$(1, -4e)$

In Exercises 41–56, find the derivative of the function.

41. $g(t) = t^2 - \dfrac{4}{t^3}$ **42.** $f(x) = x + \dfrac{1}{x^2}$

43. $f(x) = \dfrac{4x^3 + 3x^2}{x}$ **44.** $f(x) = \dfrac{x^3 - 6}{x^2}$

45. $f(x) = \dfrac{x^3 - 3x^2 + 4}{x^2}$ **46.** $h(x) = \dfrac{2x^2 - 3x + 1}{x}$

47. $y = x(x^2 + 1)$ **48.** $y = 3x(6x - 5x^2)$

49. $f(x) = \sqrt{x} - 6\sqrt[3]{x}$ **50.** $f(x) = \sqrt[3]{x} + \sqrt[5]{x}$

51. $h(s) = s^{4/5} - s^{2/3}$ **52.** $f(t) = t^{2/3} - t^{1/3} + 4$

53. $f(x) = 6\sqrt{x} + 5\cos x$ **54.** $f(x) = \dfrac{2}{\sqrt[3]{x}} + 5\cos x$

55. $f(x) = x^{-2} - 2e^x$ **56.** $g(x) = \sqrt{x} - 3e^x$

In Exercises 57–60, (a) find an equation of the tangent line to the graph of f at the given point, (b) use a graphing utility to graph the function and its tangent line at the point, and (c) use the *derivative* feature of a graphing utility to confirm your results.

Function	Point
57. $y = x^4 - x$	$(-1, 2)$
58. $f(x) = \dfrac{2}{\sqrt[4]{x^3}}$	$(1, 2)$
59. $g(x) = x + e^x$	$(0, 1)$
60. $h(t) = \sin t + \frac{1}{2}e^t$	$\left(\pi, \frac{1}{2}e^\pi\right)$

In Exercises 61–68, determine the point(s) (if any) at which the graph of the function has a horizontal tangent line.

61. $y = x^4 - 2x^2 + 3$ **62.** $y = x^3 + x$

63. $y = \dfrac{1}{x^2}$ **64.** $y = x^2 + 9$

65. $y = x + \sin x, \quad 0 \le x < 2\pi$

66. $y = \sqrt{3}x + 2\cos x, \quad 0 \le x < 2\pi$

67. $y = -4x + e^x$

68. $y = x + 4e^x$

In Exercises 69–74, find k such that the line is tangent to the graph of the function.

Function	*Line*
69. $f(x) = x^2 - kx$	$y = 5x - 4$
70. $f(x) = k - x^2$	$y = -6x + 1$
71. $f(x) = \dfrac{k}{x}$	$y = -\dfrac{3}{4}x + 3$
72. $f(x) = k\sqrt{x}$	$y = x + 4$
73. $f(x) = kx^3$	$y = x + 1$
74. $f(x) = kx^4$	$y = 4x - 1$

75. Sketch the graph of a function f such that $f' > 0$ for all x and the rate of change of the function is decreasing.

CAPSTONE

76. Use the graph of f to answer each question. To print an enlarged copy of the graph, go to the website *www.mathgraphs.com*.

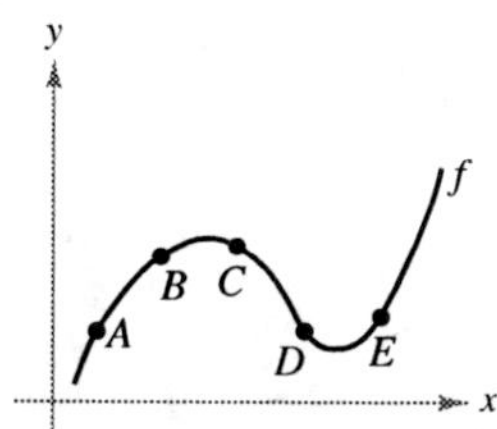

(a) Between which two consecutive points is the average rate of change of the function greatest?

(b) Is the average rate of change of the function between A and B greater than or less than the instantaneous rate of change at B?

(c) Sketch a tangent line to the graph between C and D such that the slope of the tangent line is the same as the average rate of change of the function between C and D.

WRITING ABOUT CONCEPTS

In Exercises 77 and 78, the relationship between f and g is given. Explain the relationship between f' and g'.

77. $g(x) = f(x) + 6$ **78.** $g(x) = -5f(x)$

WRITING ABOUT CONCEPTS (continued)

In Exercises 79 and 80, the graphs of a function f and its derivative f' are shown on the same set of coordinate axes. Label the graphs as f or f' and write a short paragraph stating the criteria you used in making your selection. To print an enlarged copy of the graph, go to the website *www.mathgraphs.com*.

79. 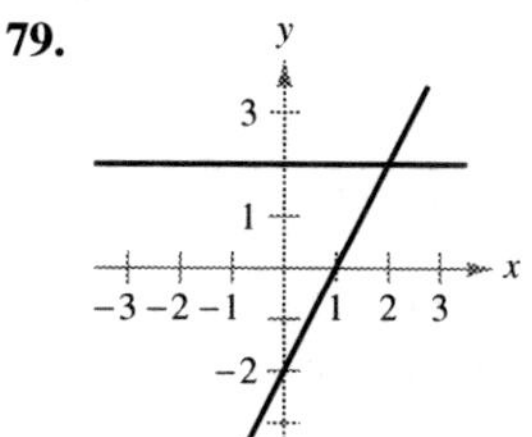   **80.** 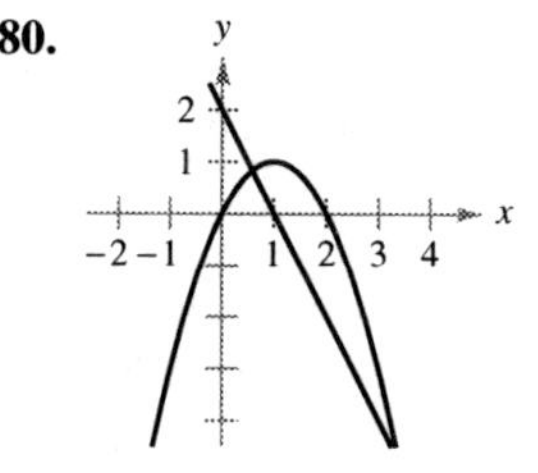

81. Sketch the graphs of $y = x^2$ and $y = -x^2 + 6x - 5$, and sketch the two lines that are tangent to both graphs. Find equations of these lines.

82. Show that the graphs of the two equations $y = x$ and $y = 1/x$ have tangent lines that are perpendicular to each other at their point of intersection.

83. Show that the graph of the function

$$f(x) = 3x + \sin x + 2$$

does not have a horizontal tangent line.

84. Show that the graph of the function

$$f(x) = x^5 + 3x^3 + 5x$$

does not have a tangent line with a slope of 3.

In Exercises 85 and 86, find an equation of the tangent line to the graph of the function f through the point (x_0, y_0) not on the graph. To find the point of tangency (x, y) on the graph of f, solve the equation

$$f'(x) = \frac{y_0 - y}{x_0 - x}.$$

85. $f(x) = \sqrt{x}$ **86.** $f(x) = \dfrac{2}{x}$

$(x_0, y_0) = (-4, 0)$ $(x_0, y_0) = (5, 0)$

87. *Linear Approximation* Use a graphing utility (in square mode) to zoom in on the graph of

$$f(x) = 4 - \tfrac{1}{2}x^2$$

to approximate $f'(1)$. Use the derivative to find $f'(1)$.

88. *Linear Approximation* Use a graphing utility (in square mode) to zoom in on the graph of

$$f(x) = 4\sqrt{x} + 1$$

to approximate $f'(4)$. Use the derivative to find $f'(4)$.

89. *Linear Approximation* Consider the function $f(x) = x^{3/2}$ with the solution point $(4, 8)$.

(a) Use a graphing utility to obtain the graph of f. Use the *zoom* feature to obtain successive magnifications of the graph in the neighborhood of the point $(4, 8)$. After zooming in a few times, the graph should appear nearly linear. Use the *trace* feature to determine the coordinates of a point near $(4, 8)$. Find an equation of the secant line $S(x)$ through the two points.

(b) Find the equation of the line

$$T(x) = f'(4)(x - 4) + f(4)$$

tangent to the graph of f passing through the given point. Why are the linear functions S and T nearly the same?

(c) Use a graphing utility to graph f and T in the same viewing window. Note that T is a good approximation of f when x is close to 4. What happens to the accuracy of the approximation as you move farther away from the point of tangency?

(d) Demonstrate the conclusion in part (c) by completing the table.

Δx	-3	-2	-1	-0.5	-0.1	0
$f(4 + \Delta x)$						
$T(4 + \Delta x)$						

Δx	0.1	0.5	1	2	3
$f(4 + \Delta x)$					
$T(4 + \Delta x)$					

90. *Linear Approximation* Repeat Exercise 89 for the function $f(x) = x^3$, where $T(x)$ is the line tangent to the graph at the point $(1, 1)$. Explain why the accuracy of the linear approximation decreases more rapidly than in Exercise 89.

True or False? In Exercises 91–96, determine whether the statement is true or false. If it is false, explain why or give an example that shows it is false.

91. If $f'(x) = g'(x)$, then $f(x) = g(x)$.

92. If $f(x) = g(x) + c$, then $f'(x) = g'(x)$.

93. If $y = \pi^2$, then $dy/dx = 2\pi$.

94. If $y = x/\pi$, then $dy/dx = 1/\pi$.

95. If $g(x) = 3f(x)$, then $g'(x) = 3f'(x)$.

96. If $f(x) = 1/x^n$, then $f'(x) = 1/(nx^{n-1})$.

In Exercises 97–100, find the average rate of change of the function over the given interval. Compare this average rate of change with the instantaneous rates of change at the endpoints of the interval.

97. $f(x) = \dfrac{-1}{x}$, $[1, 2]$

98. $f(x) = \cos x$, $\left[0, \dfrac{\pi}{3}\right]$

99. $g(x) = x^2 + e^x$, $[0, 1]$

100. $h(x) = x^3 - \frac{1}{2}e^x$, $[0, 2]$

Vertical Motion In Exercises 101 and 102, use the position function $s(t) = -16t^2 + v_0 t + s_0$ for free-falling objects.

101. A silver dollar is dropped from the top of a building that is 1362 feet tall.

(a) Determine the position and velocity functions for the coin.

(b) Determine the average velocity on the interval $[1, 2]$.

(c) Find the instantaneous velocities when $t = 1$ and $t = 2$.

(d) Find the time required for the coin to reach ground level.

(e) Find the velocity of the coin at impact.

102. A ball is thrown straight down from the top of a 220-foot building with an initial velocity of -22 feet per second. What is its velocity after 3 seconds? What is its velocity after falling 108 feet?

Vertical Motion In Exercises 103 and 104, use the position function $s(t) = -4.9t^2 + v_0 t + s_0$ for free-falling objects.

103. A projectile is shot upward from the surface of Earth with an initial velocity of 120 meters per second. What is its velocity after 5 seconds? After 10 seconds?

104. To estimate the height of a building, a stone is dropped from the top of the building into a pool of water at ground level. How high is the building if the splash is seen 5.6 seconds after the stone is dropped?

Think About It In Exercises 105 and 106, the graph of a position function is shown. It represents the distance in miles that a person drives during a 10-minute trip to work. Make a sketch of the corresponding velocity function.

105.
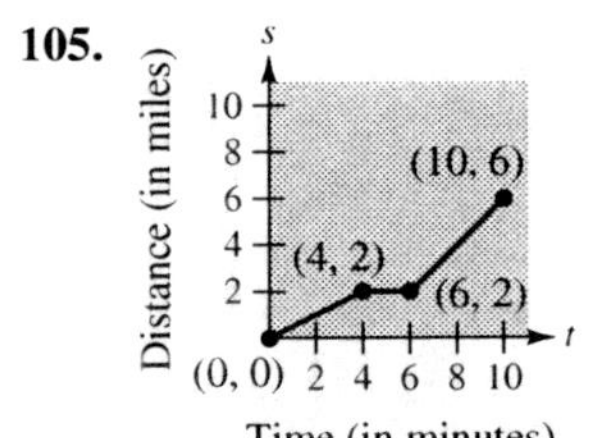

106.
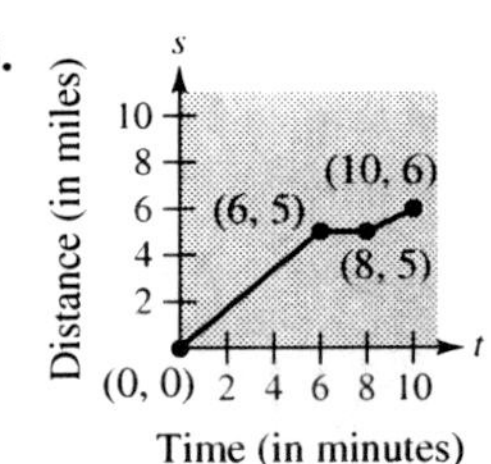

Think About It In Exercises 107 and 108, the graph of a velocity function is shown. It represents the velocity in miles per hour during a 10-minute drive to work. Make a sketch of the corresponding position function.

107.
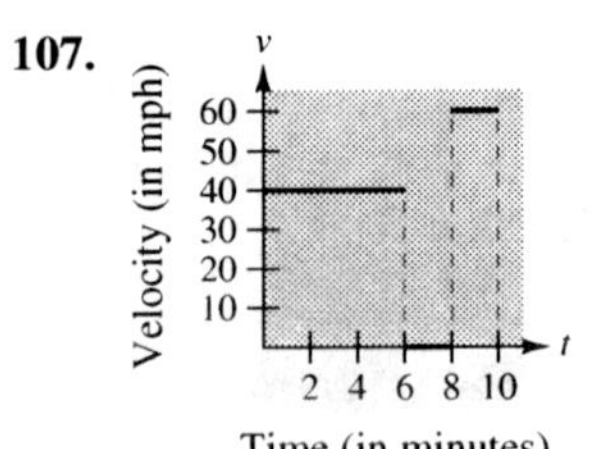

108.
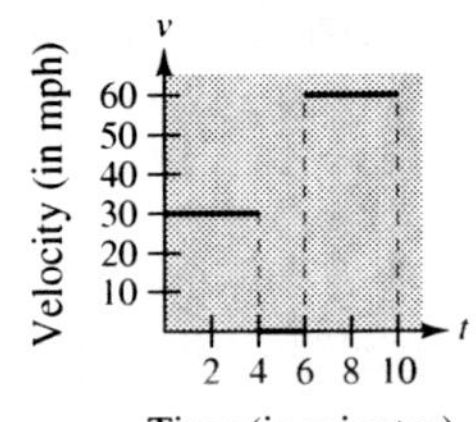

109. *Modeling Data* The stopping distance of an automobile, on dry, level pavement, traveling at a speed v (kilometers per hour) is the distance R (meters) the car travels during the reaction time of the driver plus the distance B (meters) the car travels after the brakes are applied (see figure). The table shows the results of an experiment.

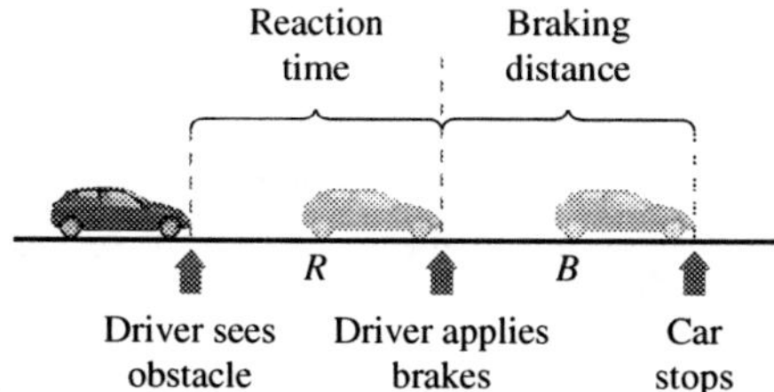

Speed, v	20	40	60	80	100
Reaction Time Distance, R	8.3	16.7	25.0	33.3	41.7
Braking Time Distance, B	2.3	9.0	20.2	35.8	55.9

(a) Use the regression capabilities of a graphing utility to find a linear model for reaction time distance.

(b) Use the regression capabilities of a graphing utility to find a quadratic model for braking distance.

(c) Determine the polynomial giving the total stopping distance T.

(d) Use a graphing utility to graph the functions R, B, and T in the same viewing window.

(e) Find the derivative of T and the rates of change of the total stopping distance for $v = 40$, $v = 80$, and $v = 100$.

(f) Use the results of this exercise to draw conclusions about the total stopping distance as speed increases.

110. *Fuel Cost* A car is driven 15,000 miles a year and gets x miles per gallon. Assume that the average fuel cost is \$2.76 per gallon. Find the annual cost of fuel C as a function of x, and use this function to complete the table.

x	10	15	20	25	30	35	40
C							
dC/dx							

Who would benefit more from a one-mile-per-gallon increase in fuel efficiency—the driver of a car that gets 15 miles per gallon or the driver of a car that gets 35 miles per gallon? Explain.

111. *Volume* The volume of a cube with sides of length s is given by $V = s^3$. Find the rate of change of the volume with respect to s when $s = 6$ centimeters.

112. *Area* The area of a square with sides of length s is given by $A = s^2$. Find the rate of change of the area with respect to s when $s = 6$ meters.

113. *Velocity* Verify that the average velocity over the time interval $[t_0 - \Delta t, t_0 + \Delta t]$ is the same as the instantaneous velocity at $t = t_0$ for the position function

$$s(t) = -\tfrac{1}{2}at^2 + c.$$

114. *Inventory Management* The annual inventory cost C for a manufacturer is

$$C = \frac{1{,}008{,}000}{Q} + 6.3Q$$

where Q is the order size when the inventory is replenished. Find the change in annual cost when Q is increased from 350 to 351, and compare this with the instantaneous rate of change when $Q = 350$.

115. *Writing* The number of gallons N of regular unleaded gasoline sold by a gasoline station at a price of p dollars per gallon is given by $N = f(p)$.

(a) Describe the meaning of $f'(2.979)$.

(b) Is $f'(2.979)$ usually positive or negative? Explain.

116. *Newton's Law of Cooling* This law states that the rate of change of the temperature of an object is proportional to the difference between the object's temperature T and the temperature T_a of the surrounding medium. Write an equation for this law.

117. Find an equation of the parabola $y = ax^2 + bx + c$ that passes through $(0, 1)$ and is tangent to the line $y = x - 1$ at $(1, 0)$.

118. Let (a, b) be an arbitrary point on the graph of $y = 1/x$, $x > 0$. Prove that the area of the triangle formed by the tangent line through (a, b) and the coordinate axes is 2.

119. Find the tangent line(s) to the curve $y = x^3 - 9x$ through the point $(1, -9)$.

120. Find the equation(s) of the tangent line(s) to the parabola $y = x^2$ through the given point.

(a) $(0, a)$ (b) $(a, 0)$

Are there any restrictions on the constant a?

In Exercises 121 and 122, find a and b such that f is differentiable everywhere.

121. $f(x) = \begin{cases} ax^3, & x \le 2 \\ x^2 + b, & x > 2 \end{cases}$

122. $f(x) = \begin{cases} \cos x, & x < 0 \\ ax + b, & x \ge 0 \end{cases}$

123. Where are the functions $f_1(x) = |\sin x|$ and $f_2(x) = \sin |x|$ differentiable?

124. Prove that $\dfrac{d}{dx}[\cos x] = -\sin x$.

FOR FURTHER INFORMATION For a geometric interpretation of the derivatives of trigonometric functions, see the article "Sines and Cosines of the Times" by Victor J. Katz in *Math Horizons*. To view this article, go to the website *www.matharticles.com*.

3.3 Product and Quotient Rules and Higher-Order Derivatives

- Find the derivative of a function using the Product Rule.
- Find the derivative of a function using the Quotient Rule.
- Find the derivative of a trigonometric function.
- Find a higher-order derivative of a function.

The Product Rule

In Section 3.2 you learned that the derivative of the sum of two functions is simply the sum of their derivatives. The rules for the derivatives of the product and quotient of two functions are not as simple.

THEOREM 3.8 THE PRODUCT RULE

The product of two differentiable functions f and g is itself differentiable. Moreover, the derivative of fg is the first function times the derivative of the second, plus the second function times the derivative of the first.

$$\frac{d}{dx}[f(x)g(x)] = f(x)g'(x) + g(x)f'(x)$$

NOTE A version of the Product Rule that some people prefer is

$$\frac{d}{dx}[f(x)g(x)] = f'(x)g(x) + f(x)g'(x).$$

The advantage of this form is that it generalizes easily to products of three or more factors.

PROOF Some mathematical proofs, such as the proof of the Sum Rule, are straightforward. Others involve clever steps that may appear unmotivated to a reader. This proof involves such a step—subtracting and adding the same quantity—which is shown in color.

$$\frac{d}{dx}[f(x)g(x)] = \lim_{\Delta x \to 0} \frac{f(x + \Delta x)g(x + \Delta x) - f(x)g(x)}{\Delta x}$$

$$= \lim_{\Delta x \to 0} \frac{f(x + \Delta x)g(x + \Delta x) - f(x + \Delta x)g(x) + f(x + \Delta x)g(x) - f(x)g(x)}{\Delta x}$$

$$= \lim_{\Delta x \to 0} \left[f(x + \Delta x)\frac{g(x + \Delta x) - g(x)}{\Delta x} + g(x)\frac{f(x + \Delta x) - f(x)}{\Delta x} \right]$$

$$= \lim_{\Delta x \to 0} \left[f(x + \Delta x)\frac{g(x + \Delta x) - g(x)}{\Delta x} \right] + \lim_{\Delta x \to 0} \left[g(x)\frac{f(x + \Delta x) - f(x)}{\Delta x} \right]$$

$$= \lim_{\Delta x \to 0} f(x + \Delta x) \cdot \lim_{\Delta x \to 0} \frac{g(x + \Delta x) - g(x)}{\Delta x} + \lim_{\Delta x \to 0} g(x) \cdot \lim_{\Delta x \to 0} \frac{f(x + \Delta x) - f(x)}{\Delta x}$$

$$= f(x)g'(x) + g(x)f'(x) \qquad \blacksquare$$

Note that $\lim\limits_{\Delta x \to 0} f(x + \Delta x) = f(x)$ because f is given to be differentiable and therefore is continuous.

The Product Rule can be extended to cover products involving more than two factors. For example, if f, g, and h are differentiable functions of x, then

$$\frac{d}{dx}[f(x)g(x)h(x)] = f'(x)g(x)h(x) + f(x)g'(x)h(x) + f(x)g(x)h'(x).$$

For instance, the derivative of $y = x^2 \sin x \cos x$ is

$$\frac{dy}{dx} = 2x \sin x \cos x + x^2 \cos x \cos x + x^2 \sin x(-\sin x)$$

$$= 2x \sin x \cos x + x^2(\cos^2 x - \sin^2 x).$$

NOTE The proof of the Product Rule for products of more than two factors is left as an exercise (see Exercise 141).

THE PRODUCT RULE

When Leibniz originally wrote a formula for the Product Rule, he was motivated by the expression

$$(x + dx)(y + dy) - xy$$

from which he subtracted $dx\, dy$ (as being negligible) and obtained the differential form $x\, dy + y\, dx$. This derivation resulted in the traditional form of the Product Rule. (*Source: The History of Mathematics by David M. Burton*)

The derivative of a product of two functions is not (in general) given by the product of the derivatives of the two functions. To see this, try comparing the product of the derivatives of $f(x) = 3x - 2x^2$ and $g(x) = 5 + 4x$ with the derivative in Example 1.

EXAMPLE 1 Using the Product Rule

Find the derivative of $h(x) = (3x - 2x^2)(5 + 4x)$.

Solution

$$h'(x) = \underbrace{(3x - 2x^2)}_{\text{First}} \underbrace{\frac{d}{dx}[5 + 4x]}_{\substack{\text{Derivative} \\ \text{of second}}} + \underbrace{(5 + 4x)}_{\text{Second}} \underbrace{\frac{d}{dx}[3x - 2x^2]}_{\substack{\text{Derivative} \\ \text{of first}}} \qquad \text{Apply Product Rule.}$$

$$= (3x - 2x^2)(4) + (5 + 4x)(3 - 4x)$$
$$= (12x - 8x^2) + (15 - 8x - 16x^2)$$
$$= -24x^2 + 4x + 15$$

In Example 1, you have the option of finding the derivative with or without the Product Rule. To find the derivative without the Product Rule, you can write

$$D_x[(3x - 2x^2)(5 + 4x)] = D_x[-8x^3 + 2x^2 + 15x]$$
$$= -24x^2 + 4x + 15.$$

In the next example, you must use the Product Rule.

EXAMPLE 2 Using the Product Rule

Find the derivative of $y = xe^x$.

Solution

$$\frac{d}{dx}[xe^x] = x\frac{d}{dx}[e^x] + e^x\frac{d}{dx}[x] \qquad \text{Apply Product Rule.}$$

$$= xe^x + e^x(1)$$
$$= e^x(x + 1)$$

EXAMPLE 3 Using the Product Rule

Find the derivative of $y = 2x \cos x - 2 \sin x$.

Solution

$$\frac{dy}{dx} = \underbrace{(2x)\left(\frac{d}{dx}[\cos x]\right) + (\cos x)\left(\frac{d}{dx}[2x]\right)}_{\text{Product Rule}} - \underbrace{2\frac{d}{dx}[\sin x]}_{\text{Constant Multiple Rule}}$$

$$= (2x)(-\sin x) + (\cos x)(2) - 2(\cos x)$$
$$= -2x \sin x$$

The Quotient Rule

THEOREM 3.9 THE QUOTIENT RULE

The quotient f/g of two differentiable functions f and g is itself differentiable at all values of x for which $g(x) \neq 0$. Moreover, the derivative of f/g is given by the denominator times the derivative of the numerator minus the numerator times the derivative of the denominator, all divided by the square of the denominator.

$$\frac{d}{dx}\left[\frac{f(x)}{g(x)}\right] = \frac{g(x)f'(x) - f(x)g'(x)}{[g(x)]^2}, \quad g(x) \neq 0$$

PROOF As with the proof of Theorem 3.8, the key to this proof is subtracting and adding the same quantity.

$$\frac{d}{dx}\left[\frac{f(x)}{g(x)}\right] = \lim_{\Delta x \to 0} \frac{\dfrac{f(x + \Delta x)}{g(x + \Delta x)} - \dfrac{f(x)}{g(x)}}{\Delta x} \qquad \text{Definition of derivative}$$

$$= \lim_{\Delta x \to 0} \frac{g(x)f(x + \Delta x) - f(x)g(x + \Delta x)}{\Delta x g(x)g(x + \Delta x)}$$

$$= \lim_{\Delta x \to 0} \frac{g(x)f(x + \Delta x) - f(x)g(x) + f(x)g(x) - f(x)g(x + \Delta x)}{\Delta x g(x)g(x + \Delta x)}$$

$$= \frac{\displaystyle\lim_{\Delta x \to 0} \frac{g(x)[f(x + \Delta x) - f(x)]}{\Delta x} - \lim_{\Delta x \to 0} \frac{f(x)[g(x + \Delta x) - g(x)]}{\Delta x}}{\displaystyle\lim_{\Delta x \to 0} [g(x)g(x + \Delta x)]}$$

$$= \frac{g(x)\left[\displaystyle\lim_{\Delta x \to 0} \frac{f(x + \Delta x) - f(x)}{\Delta x}\right] - f(x)\left[\displaystyle\lim_{\Delta x \to 0} \frac{g(x + \Delta x) - g(x)}{\Delta x}\right]}{\displaystyle\lim_{\Delta x \to 0} [g(x)g(x + \Delta x)]}$$

$$= \frac{g(x)f'(x) - f(x)g'(x)}{[g(x)]^2} \qquad \blacksquare$$

Note that $\displaystyle\lim_{\Delta x \to 0} g(x + \Delta x) = g(x)$ because g is given to be differentiable and therefore is continuous.

EXAMPLE 4 Using the Quotient Rule

Find the derivative of $y = \dfrac{5x - 2}{x^2 + 1}$.

Solution

$$\frac{d}{dx}\left[\frac{5x - 2}{x^2 + 1}\right] = \frac{(x^2 + 1)\dfrac{d}{dx}[5x - 2] - (5x - 2)\dfrac{d}{dx}[x^2 + 1]}{(x^2 + 1)^2} \qquad \text{Apply Quotient Rule.}$$

$$= \frac{(x^2 + 1)(5) - (5x - 2)(2x)}{(x^2 + 1)^2}$$

$$= \frac{(5x^2 + 5) - (10x^2 - 4x)}{(x^2 + 1)^2}$$

$$= \frac{-5x^2 + 4x + 5}{(x^2 + 1)^2} \qquad \blacksquare$$

TECHNOLOGY Graphing utilities can be used to compare the graph of a function with the graph of its derivative. For instance, in Figure 3.23, the graph of the function in Example 4 appears to have two points that have horizontal tangent lines. What are the values of y' at these two points?

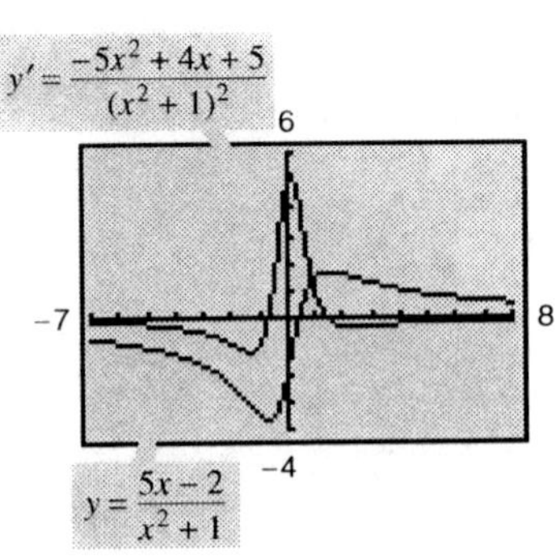

Graphical comparison of a function and its derivative
Figure 3.23

Note the use of parentheses in Example 4. A liberal use of parentheses is recommended for *all* types of differentiation problems. For instance, with the Quotient Rule, it is a good idea to enclose all factors and derivatives in parentheses, and to pay special attention to the subtraction required in the numerator.

When differentiation rules were introduced in the preceding section, the need for rewriting *before* differentiating was emphasized. The next example illustrates this point with the Quotient Rule.

EXAMPLE 5   Rewriting Before Differentiating

Find an equation of the tangent line to the graph of $f(x) = \dfrac{3 - (1/x)}{x + 5}$ at $(-1, 1)$.

Solution   Begin by rewriting the function.

$$
\begin{aligned}
f(x) &= \frac{3 - (1/x)}{x + 5} && \text{Write original function.}\\[2mm]
&= \frac{x\left(3 - \dfrac{1}{x}\right)}{x(x + 5)} && \text{Multiply numerator and denominator by } x.\\[2mm]
&= \frac{3x - 1}{x^2 + 5x} && \text{Rewrite.}\\[2mm]
f'(x) &= \frac{(x^2 + 5x)(3) - (3x - 1)(2x + 5)}{(x^2 + 5x)^2} && \text{Apply Quotient Rule.}\\[2mm]
&= \frac{(3x^2 + 15x) - (6x^2 + 13x - 5)}{(x^2 + 5x)^2}\\[2mm]
&= \frac{-3x^2 + 2x + 5}{(x^2 + 5x)^2} && \text{Simplify.}
\end{aligned}
$$

To find the slope at $(-1, 1)$, evaluate $f'(-1)$.

$$f'(-1) = 0 \qquad \text{Slope of graph at } (-1, 1)$$

Then, using the point-slope form of the equation of a line, you can determine that the equation of the tangent line at $(-1, 1)$ is $y = 1$. See Figure 3.24.

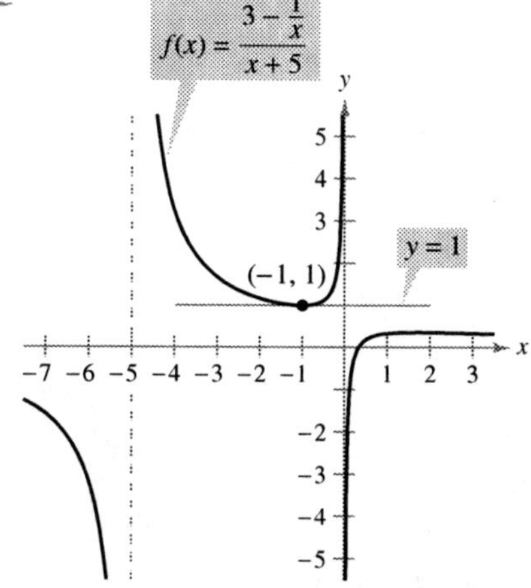

The line $y = 1$ is tangent to the graph of $f(x)$ at the point $(-1, 1)$.
Figure 3.24

Not every quotient needs to be differentiated by the Quotient Rule. For example, each quotient in the next example can be considered as the product of a constant times a function of x. In such cases it is more convenient to use the Constant Multiple Rule.

EXAMPLE 6   Using the Constant Multiple Rule

Original Function	*Rewrite*	*Differentiate*	*Simplify*
a. $y = \dfrac{x^2 + 3x}{6}$	$y = \dfrac{1}{6}(x^2 + 3x)$	$y' = \dfrac{1}{6}(2x + 3)$	$y' = \dfrac{2x + 3}{6}$
b. $y = \dfrac{5x^4}{8}$	$y = \dfrac{5}{8}x^4$	$y' = \dfrac{5}{8}(4x^3)$	$y' = \dfrac{5}{2}x^3$
c. $y = \dfrac{-3(3x - 2x^2)}{7x}$	$y = -\dfrac{3}{7}(3 - 2x)$	$y' = -\dfrac{3}{7}(-2)$	$y' = \dfrac{6}{7}$
d. $y = \dfrac{9}{5x^2}$	$y = \dfrac{9}{5}(x^{-2})$	$y' = \dfrac{9}{5}(-2x^{-3})$	$y' = -\dfrac{18}{5x^3}$

NOTE   To see the benefit of using the Constant Multiple Rule for some quotients, try using the Quotient Rule to differentiate the functions in Example 6—you should obtain the same results, but with more work.

In Section 3.2, the Power Rule was proved only for the case in which the exponent n is a positive integer greater than 1. The next example extends the proof to include negative integer exponents.

EXAMPLE 7 Proof of the Power Rule (Negative Integer Exponents)

If n is a negative integer, there exists a positive integer k such that $n = -k$. So, by the Quotient Rule, you can write

$$
\begin{aligned}
\frac{d}{dx}[x^n] &= \frac{d}{dx}\left[\frac{1}{x^k}\right] \\
&= \frac{x^k(0) - (1)(kx^{k-1})}{(x^k)^2} \qquad \text{Quotient Rule and Power Rule} \\
&= \frac{0 - kx^{k-1}}{x^{2k}} \\
&= -kx^{-k-1} \\
&= nx^{n-1}. \qquad n = -k
\end{aligned}
$$

So, the Power Rule

$$
\frac{d}{dx}[x^n] = nx^{n-1} \qquad \text{Power Rule}
$$

is valid for any integer. The cases for which n is rational and n is irrational are left as an exercise (see Section 3.5, Exercise 100). ∎

Derivatives of Trigonometric Functions

Knowing the derivatives of the sine and cosine functions, you can use the Quotient Rule to find the derivatives of the four remaining trigonometric functions.

THEOREM 3.10 DERIVATIVES OF TRIGONOMETRIC FUNCTIONS

$$
\frac{d}{dx}[\tan x] = \sec^2 x \qquad\qquad \frac{d}{dx}[\cot x] = -\csc^2 x
$$

$$
\frac{d}{dx}[\sec x] = \sec x \tan x \qquad\qquad \frac{d}{dx}[\csc x] = -\csc x \cot x
$$

(PROOF) Considering $\tan x = (\sin x)/(\cos x)$ and applying the Quotient Rule, you obtain

$$
\begin{aligned}
\frac{d}{dx}[\tan x] &= \frac{(\cos x)(\cos x) - (\sin x)(-\sin x)}{\cos^2 x} \qquad \text{Apply Quotient Rule.} \\
&= \frac{\cos^2 x + \sin^2 x}{\cos^2 x} \\
&= \frac{1}{\cos^2 x} \\
&= \sec^2 x.
\end{aligned}
$$

The proofs of the other three parts of the theorem are left as an exercise (see Exercise 91). ∎

EXAMPLE 8 Differentiating Trigonometric Functions

Function	*Derivative*
a. $y = x - \tan x$	$\dfrac{dy}{dx} = 1 - \sec^2 x$
b. $y = x \sec x$	$y' = x(\sec x \tan x) + (\sec x)(1)$
	$\quad = (\sec x)(1 + x \tan x)$

EXAMPLE 9 Different Forms of a Derivative

Differentiate both forms of $y = \dfrac{1 - \cos x}{\sin x} = \csc x - \cot x.$

Solution

First form: $\quad y = \dfrac{1 - \cos x}{\sin x}$

$$y' = \frac{(\sin x)(\sin x) - (1 - \cos x)(\cos x)}{\sin^2 x}$$

$$= \frac{\sin^2 x + \cos^2 x - \cos x}{\sin^2 x}$$

$$= \frac{1 - \cos x}{\sin^2 x}$$

Second form: $\quad y = \csc x - \cot x$

$$y' = -\csc x \cot x + \csc^2 x$$

To verify that the two derivatives are equal, you can write

$$\frac{1 - \cos x}{\sin^2 x} = \frac{1}{\sin^2 x} - \left(\frac{1}{\sin x}\right)\left(\frac{\cos x}{\sin x}\right)$$

$$= \csc^2 x - \csc x \cot x.$$

The summary below shows that much of the work in obtaining a simplified form of a derivative occurs *after* differentiating. Note that two characteristics of a simplified form are the absence of negative exponents and the combining of like terms.

	$f'(x)$ **After Differentiating**	$f'(x)$ **After Simplifying**
Example 1	$(3x - 2x^2)(4) + (5 + 4x)(3 - 4x)$	$-24x^2 + 4x + 15$
Example 3	$(2x)(-\sin x) + (\cos x)(2) - 2(\cos x)$	$-2x \sin x$
Example 4	$\dfrac{(x^2 + 1)(5) - (5x - 2)(2x)}{(x^2 + 1)^2}$	$\dfrac{-5x^2 + 4x + 5}{(x^2 + 1)^2}$
Example 5	$\dfrac{(x^2 + 5x)(3) - (3x - 1)(2x + 5)}{(x^2 + 5x)^2}$	$\dfrac{-3x^2 + 2x + 5}{(x^2 + 5x)^2}$
Example 9	$\dfrac{(\sin x)(\sin x) - (1 - \cos x)(\cos x)}{\sin^2 x}$	$\dfrac{1 - \cos x}{\sin^2 x}$

THE MOON

Higher-Order Derivatives

Just as you can obtain a velocity function by differentiating a position function, you can obtain an **acceleration** function by differentiating a velocity function. Another way of looking at this is that you can obtain an acceleration function by differentiating a position function *twice*.

$$s(t) \qquad \text{Position function}$$
$$v(t) = s'(t) \qquad \text{Velocity function}$$
$$a(t) = v'(t) = s''(t) \qquad \text{Acceleration function}$$

The function given by $a(t)$ is the **second derivative** of $s(t)$ and is denoted by $s''(t)$.

The second derivative is an example of a **higher-order derivative.** You can define derivatives of any positive integer order. For instance, the **third derivative** is the derivative of the second derivative. Higher-order derivatives are denoted as follows.

First derivative: y', $f'(x)$, $\dfrac{dy}{dx}$, $\dfrac{d}{dx}[f(x)]$, $D_x[y]$

Second derivative: y'', $f''(x)$, $\dfrac{d^2 y}{dx^2}$, $\dfrac{d^2}{dx^2}[f(x)]$, $D_x^2[y]$

Third derivative: y''', $f'''(x)$, $\dfrac{d^3 y}{dx^3}$, $\dfrac{d^3}{dx^3}[f(x)]$, $D_x^3[y]$

Fourth derivative: $y^{(4)}$, $f^{(4)}(x)$, $\dfrac{d^4 y}{dx^4}$, $\dfrac{d^4}{dx^4}[f(x)]$, $D_x^4[y]$

$$\vdots$$

nth derivative: $y^{(n)}$, $f^{(n)}(x)$, $\dfrac{d^n y}{dx^n}$, $\dfrac{d^n}{dx^n}[f(x)]$, $D_x^n[y]$

EXAMPLE 10   Finding the Acceleration Due to Gravity

Because the moon has no atmosphere, a falling object on the moon encounters no air resistance. In 1971, astronaut David Scott demonstrated that a feather and a hammer fall at the same rate on the moon. The position function for each of these falling objects is given by

$$s(t) = -0.81t^2 + 2$$

where $s(t)$ is the height in meters and t is the time in seconds. What is the ratio of Earth's gravitational force to the moon's?

Solution   To find the acceleration, differentiate the position function twice.

$$s(t) = -0.81t^2 + 2 \qquad \text{Position function}$$
$$s'(t) = -1.62t \qquad \text{Velocity function}$$
$$s''(t) = -1.62 \qquad \text{Acceleration function}$$

So, the acceleration due to gravity on the moon is -1.62 meters per second per second. Because the acceleration due to gravity on Earth is -9.8 meters per second per second, the ratio of Earth's gravitational force to the moon's is

$$\frac{\text{Earth's gravitational force}}{\text{Moon's gravitational force}} = \frac{-9.8}{-1.62}$$
$$\approx 6.0.$$

3.3 Exercises

In Exercises 1–6, use the Product Rule to differentiate the function.

1. $g(x) = (x^2 + 3)(x^2 - 4x)$

2. $f(x) = (6x + 5)(x^3 - 2)$

3. $h(t) = \sqrt{t}(1 - t^2)$

4. $g(s) = \sqrt{s}(s^2 + 8)$

5. $f(x) = e^x \cos x$

6. $g(x) = \sqrt{x} \sin x$

In Exercises 7–12, use the Quotient Rule to differentiate the function.

7. $f(x) = \dfrac{x}{x^2 + 1}$

8. $g(t) = \dfrac{t^2 + 4}{5t - 3}$

9. $h(x) = \dfrac{\sqrt{x}}{x^3 + 1}$

10. $h(s) = \dfrac{s}{\sqrt{s} - 1}$

11. $g(x) = \dfrac{\sin x}{e^x}$

12. $f(t) = \dfrac{\cos t}{t^3}$

In Exercises 13–18, find $f'(x)$ and $f'(c)$.

Function	Value of c
13. $f(x) = (x^3 + 4x)(3x^2 + 2x - 5)$	$c = 0$
14. $f(x) = \dfrac{x + 5}{x - 5}$	$c = 4$
15. $f(x) = x \cos x$	$c = \dfrac{\pi}{4}$
16. $f(x) = \dfrac{\sin x}{x}$	$c = \dfrac{\pi}{6}$
17. $f(x) = e^x \sin x$	$c = 0$
18. $f(x) = \dfrac{\cos x}{e^x}$	$c = 0$

In Exercises 19–24, complete the table without using the Quotient Rule.

Function	Rewrite	Differentiate	Simplify
19. $y = \dfrac{x^2 + 3x}{7}$			
20. $y = \dfrac{5x^2 - 3}{4}$			
21. $y = \dfrac{6}{7x^2}$			
22. $y = \dfrac{10}{3x^3}$			
23. $y = \dfrac{4x^{3/2}}{x}$			
24. $y = \dfrac{5x^2 - 8}{11}$			

In Exercises 25–38, find the derivative of the algebraic function.

25. $f(x) = \dfrac{4 - 3x - x^2}{x^2 - 1}$

26. $f(x) = \dfrac{x^3 + 5x + 3}{x^2 - 1}$

27. $f(x) = x\left(1 - \dfrac{4}{x + 3}\right)$

28. $f(x) = x^4\left(1 - \dfrac{2}{x + 1}\right)$

29. $f(x) = \dfrac{3x - 1}{\sqrt{x}}$

30. $f(x) = \sqrt[3]{x}(\sqrt{x} + 3)$

31. $h(s) = (s^3 - 2)^2$

32. $h(x) = (x^2 - 1)^2$

33. $f(x) = \dfrac{2 - \dfrac{1}{x}}{x - 3}$

34. $g(x) = x^2\left(\dfrac{2}{x} - \dfrac{1}{x + 1}\right)$

35. $f(x) = (2x^3 + 5x)(x - 3)(x + 2)$

36. $f(x) = (x^3 - x)(x^2 + 2)(x^2 + x - 1)$

37. $f(x) = \dfrac{x^2 + c^2}{x^2 - c^2}$, c is a constant

38. $f(x) = \dfrac{c^2 - x^2}{c^2 + x^2}$, c is a constant

In Exercises 39–56, find the derivative of the transcendental function.

39. $f(t) = t^2 \sin t$

40. $f(\theta) = (\theta + 1) \cos \theta$

41. $f(t) = \dfrac{\cos t}{t}$

42. $f(x) = \dfrac{\sin x}{x^3}$

43. $f(x) = -e^x + \tan x$

44. $y = e^x - \cot x$

45. $g(t) = \sqrt[4]{t} + 6 \csc t$

46. $h(x) = \dfrac{1}{x} - 12 \sec x$

47. $y = \dfrac{3(1 - \sin x)}{2 \cos x}$

48. $y = \dfrac{\sec x}{x}$

49. $y = -\csc x - \sin x$

50. $y = x \cos x + \sin x$

51. $f(x) = x^2 \tan x$

52. $f(x) = 2 \sin x \cos x$

53. $y = 2x \sin x + x^2 e^x$

54. $h(x) = 2e^x \cos x$

55. $y = \dfrac{e^x}{4\sqrt{x}}$

56. $y = \dfrac{2e^x}{x^2 + 1}$

CAS **In Exercises 57–60, use a computer algebra system to differentiate the function.**

57. $g(x) = \left(\dfrac{x + 1}{x + 2}\right)(2x - 5)$

58. $f(x) = \left(\dfrac{x^2 - x - 3}{x^2 + 1}\right)(x^2 + x + 1)$

59. $g(\theta) = \dfrac{\theta}{1 - \sin \theta}$

60. $f(\theta) = \dfrac{\sin \theta}{1 - \cos \theta}$

The symbol **CAS** *indicates an exercise in which you are instructed to specifically use a computer algebra system.*

In Exercises 61–64, evaluate the derivative of the function at the given point. Use a graphing utility to verify your result.

Function	Point
61. $y = \dfrac{1 + \csc x}{1 - \csc x}$	$\left(\dfrac{\pi}{6}, -3\right)$
62. $f(x) = \tan x \cot x$	$(1, 1)$
63. $h(t) = \dfrac{\sec t}{t}$	$\left(\pi, -\dfrac{1}{\pi}\right)$
64. $f(x) = \sin x(\sin x + \cos x)$	$\left(\dfrac{\pi}{4}, 1\right)$

In Exercises 65–70, (a) find an equation of the tangent line to the graph of f at the given point, (b) use a graphing utility to graph the function and its tangent line at the point, and (c) use the *derivative* feature of a graphing utility to confirm your results.

Function	Point
65. $f(x) = (x^3 + 4x - 1)(x - 2)$	$(1, -4)$
66. $f(x) = \dfrac{x - 1}{x + 1}$	$\left(2, \dfrac{1}{3}\right)$
67. $f(x) = \tan x$	$\left(\dfrac{\pi}{4}, 1\right)$
68. $f(x) = \sec x$	$\left(\dfrac{\pi}{3}, 2\right)$
69. $f(x) = (x - 1)e^x$	$(1, 0)$
70. $f(x) = \dfrac{e^x}{x + 4}$	$\left(0, \dfrac{1}{4}\right)$

Famous Curves **In Exercises 71–74, find an equation of the tangent line to the graph at the given point. (The graphs in Exercises 71 and 72 are called *witches of Agnesi*. The graphs in Exercises 73 and 74 are called *serpentines*.)**

71.

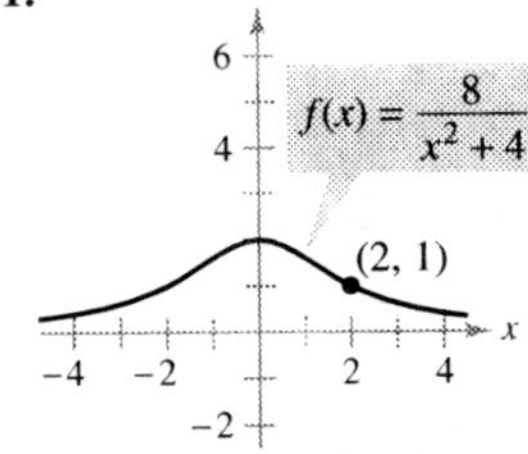

72.

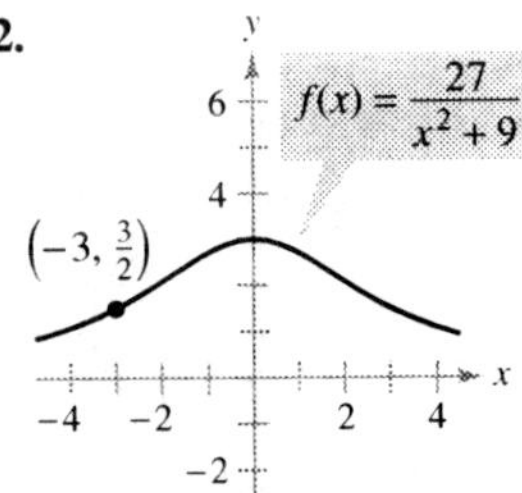

73.

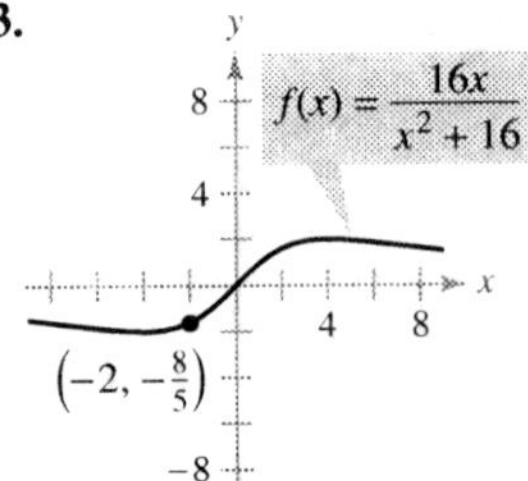

74.

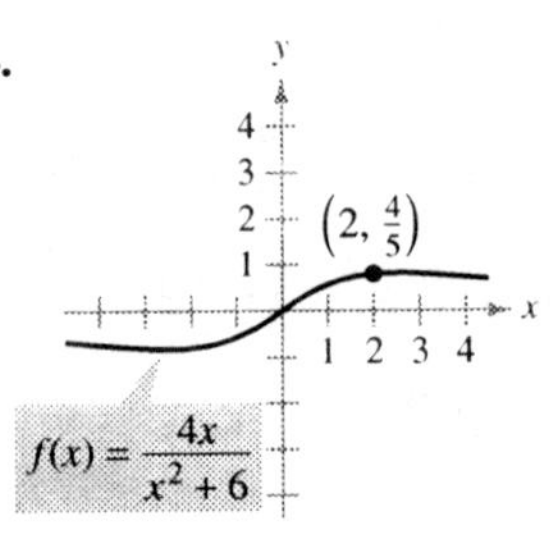

In Exercises 75–78, determine the point(s) at which the graph of the function has a horizontal tangent line.

75. $f(x) = \dfrac{2x - 1}{x^2}$

76. $f(x) = \dfrac{x^2}{x^2 + 1}$

77. $g(x) = \dfrac{8(x - 2)}{e^x}$

78. $f(x) = e^x \sin x, \quad [0, \pi]$

79. ***Tangent Lines*** Find equations of the tangent lines to the graph of $f(x) = (x + 1)/(x - 1)$ that are parallel to the line $2y + x = 6$. Then graph the function and the tangent lines.

80. ***Tangent Lines*** Find equations of the tangent lines to the graph of $f(x) = x/(x - 1)$ that pass through the point $(-1, 5)$. Then graph the function and the tangent lines.

In Exercises 81 and 82, verify that $f'(x) = g'(x)$, and explain the relationship between f and g.

81. $f(x) = \dfrac{3x}{x + 2}, \quad g(x) = \dfrac{5x + 4}{x + 2}$

82. $f(x) = \dfrac{\sin x - 3x}{x}, \quad g(x) = \dfrac{\sin x + 2x}{x}$

In Exercises 83 and 84, use the graphs of f and g. Let $p(x) = f(x)g(x)$ and $q(x) = f(x)/g(x)$.

83. (a) Find $p'(1)$.

    (b) Find $q'(4)$.

84. (a) Find $p'(4)$.

    (b) Find $q'(7)$.

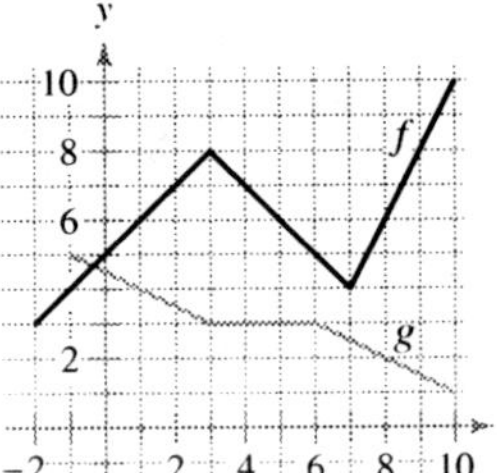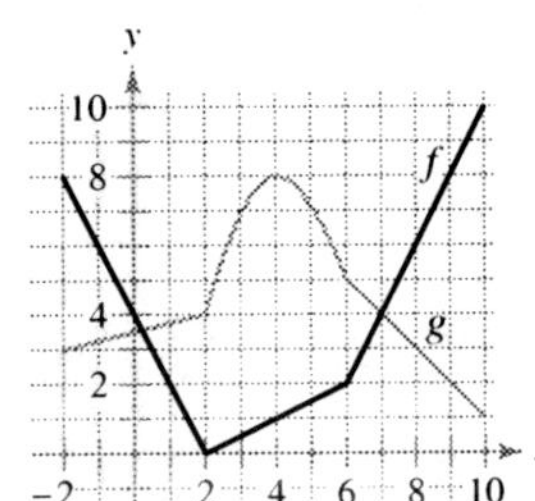

85. ***Area*** The length of a rectangle is given by $6t + 5$ and its height is $\sqrt{t}$, where t is time in seconds and the dimensions are in centimeters. Find the rate of change of the area with respect to time.

86. ***Volume*** The radius of a right circular cylinder is given by $\sqrt{t + 2}$ and its height is $\frac{1}{2}\sqrt{t}$, where t is time in seconds and the dimensions are in inches. Find the rate of change of the volume with respect to time.

87. ***Inventory Replenishment*** The ordering and transportation cost C for the components used in manufacturing a product is

$$C = \frac{375{,}000 + 6x^2}{x}, \quad x \geq 1$$

where C is measured in dollars and x is the order size. Find the rate of change of C with respect to x when (a) $x = 200$, (b) $x = 250$, and (c) $x = 300$. Interpret the meanings of these values.

88. ***Boyle's Law*** This law states that if the temperature of a gas remains constant, its pressure is inversely proportional to its volume. Use the derivative to show that the rate of change of the pressure is inversely proportional to the square of the volume.

89. Population Growth   A population of 500 bacteria is introduced into a culture and grows in number according to the equation

$$P(t) = 500\left(1 + \frac{4t}{50 + t^2}\right)$$

where t is measured in hours. Find the rate at which the population is growing when $t = 2$.

90. Gravitational Force   Newton's Law of Universal Gravitation states that the force F between two masses, m_1 and m_2, is

$$F = \frac{Gm_1m_2}{d^2}$$

where G is a constant and d is the distance between the masses. Find an equation that gives an instantaneous rate of change of F with respect to d. (Assume that m_1 and m_2 represent moving points.)

91. Prove the following differentiation rules.

(a) $\dfrac{d}{dx}[\sec x] = \sec x \tan x$ (b) $\dfrac{d}{dx}[\csc x] = -\csc x \cot x$

(c) $\dfrac{d}{dx}[\cot x] = -\csc^2 x$

92. Rate of Change   Determine whether there exist any values of x in the interval $[0, 2\pi)$ such that the rate of change of $f(x) = \sec x$ and the rate of change of $g(x) = \csc x$ are equal.

93. Modeling Data   The table shows the quantities q (in millions) of personal computers shipped in the United States and the values v (in billions of dollars) of these shipments for the years 1999 through 2004. The year is represented by t, with $t = 9$ corresponding to 1999.   (*Source: U.S. Census Bureau*)

Year, t	9	10	11	12	13	14
q	19.6	15.9	14.6	12.9	15.0	15.8
v	26.8	22.6	18.9	16.2	14.7	15.3

(a) Use a graphing utility to find cubic models for the quantity of personal computers shipped $q(t)$ and the value $v(t)$ of the personal computers.

(b) Graph each model found in part (a).

(c) Find $A = v(t)/q(t)$, then graph A. What does this function represent?

(d) Interpret $A'(t)$ in the context of these data.

94. Satellites   When satellites observe Earth, they can scan only part of Earth's surface. Some satellites have sensors that can measure the angle θ shown in the figure. Let h represent the satellite's distance from Earth's surface and let r represent Earth's radius.

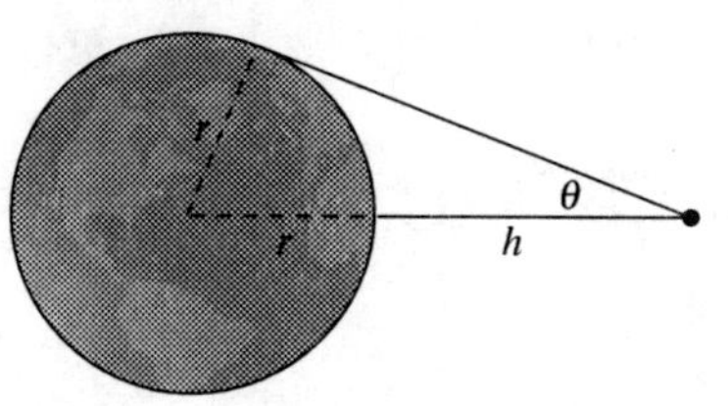

(a) Show that $h = r(\csc \theta - 1)$.

(b) Find the rate at which h is changing with respect to θ when $\theta = 30°$. (Assume $r = 3960$ miles.)

In Exercises 95–102, find the second derivative of the function.

95. $f(x) = 4x^{3/2}$

96. $f(x) = x + 32x^{-2}$

97. $f(x) = \dfrac{x}{x - 1}$

98. $f(x) = \dfrac{x^2 + 2x - 1}{x}$

99. $f(x) = x \sin x$

100. $f(x) = \sec x$

101. $g(x) = \dfrac{e^x}{x}$

102. $h(t) = e^t \sin t$

In Exercises 103–106, find the given higher-order derivative.

Given	*Find*
103. $f'(x) = x^2$	$f''(x)$
104. $f''(x) = 2 - \dfrac{2}{x}$	$f'''(x)$
105. $f'''(x) = 2\sqrt{x}$	$f^{(4)}(x)$
106. $f^{(4)}(x) = 2x + 1$	$f^{(6)}(x)$

WRITING ABOUT CONCEPTS

107. Sketch the graph of a differentiable function f such that $f(2) = 0$, $f' < 0$ for $-\infty < x < 2$, and $f' > 0$ for $2 < x < \infty$.

108. Sketch the graph of a differentiable function f such that $f > 0$ and $f' < 0$ for all real numbers x.

In Exercises 109–112, use the given information to find $f'(2)$.

$g(2) = 3$ and $g'(2) = -2$

$h(2) = -1$ and $h'(2) = 4$

109. $f(x) = 2g(x) + h(x)$

110. $f(x) = 4 - h(x)$

111. $f(x) = \dfrac{g(x)}{h(x)}$

112. $f(x) = g(x)h(x)$

In Exercises 113 and 114, the graphs of f, f', and f'' are shown on the same set of coordinate axes. Identify each graph. Explain your reasoning. To print an enlarged copy of the graph, go to the website *www.mathgraphs.com*.

113.

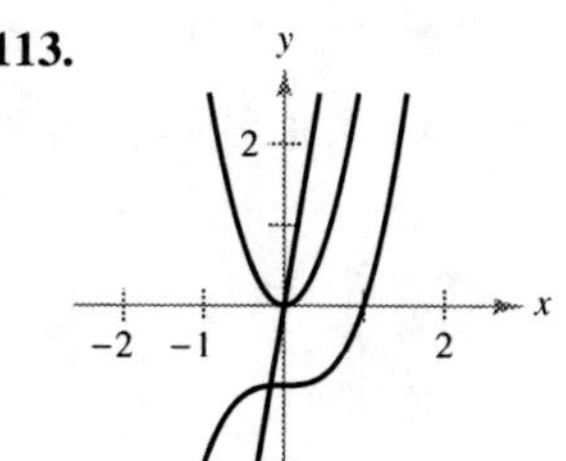

114.

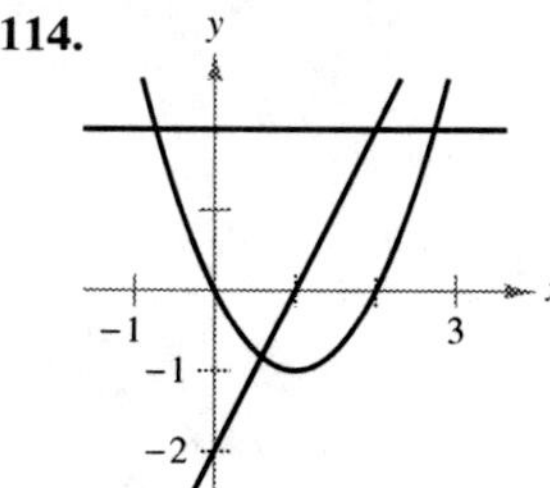

In Exercises 115–118, the graph of f is shown. Sketch the graphs of f' and f''. To print an enlarged copy of the graph, go to the website *www.mathgraphs.com.*

115.

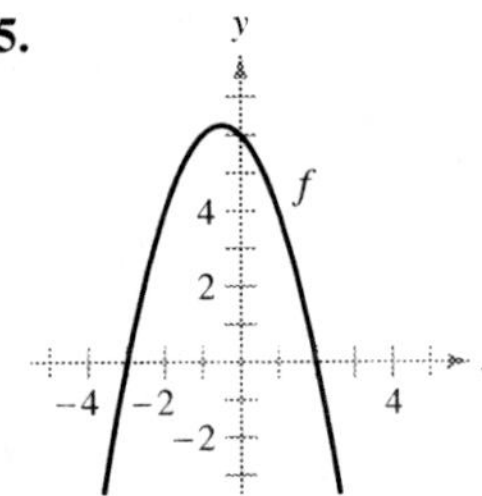

116.

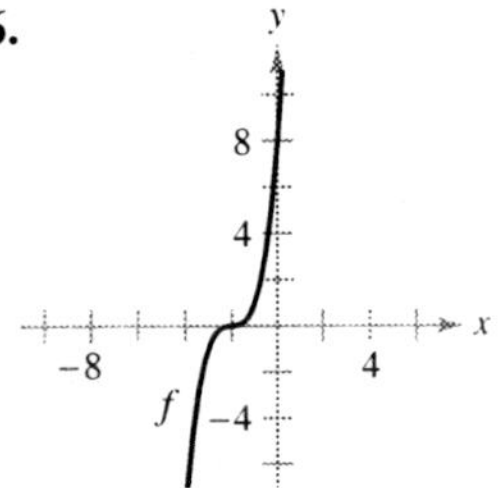

117.

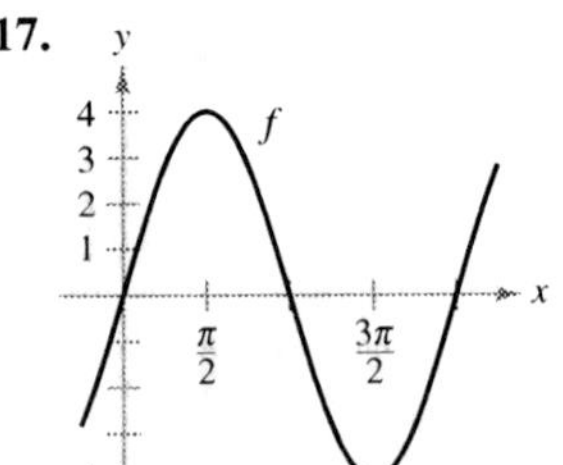

118.

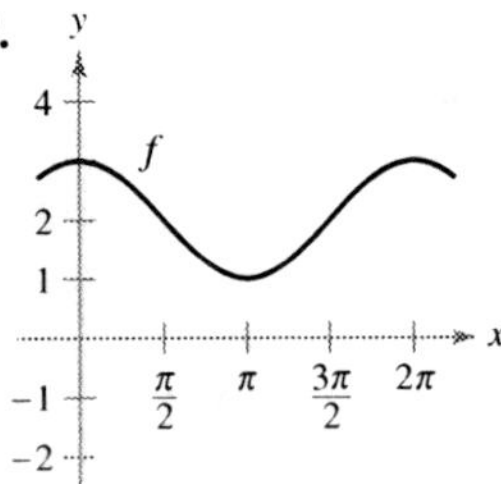

119. *Acceleration* The velocity of an object in meters per second is $v(t) = 36 - t^2, 0 \le t \le 6$. Find the velocity and acceleration of the object when $t = 3$. What can be said about the speed of the object when the velocity and acceleration have opposite signs?

CAPSTONE

120. *Particle Motion* The figure shows the graphs of the position, velocity, and acceleration functions of a particle.

(a) Copy the graphs of the functions shown. Identify each graph. Explain your reasoning. To print an enlarged copy of the graph, go to the website *www.mathgraphs.com.*

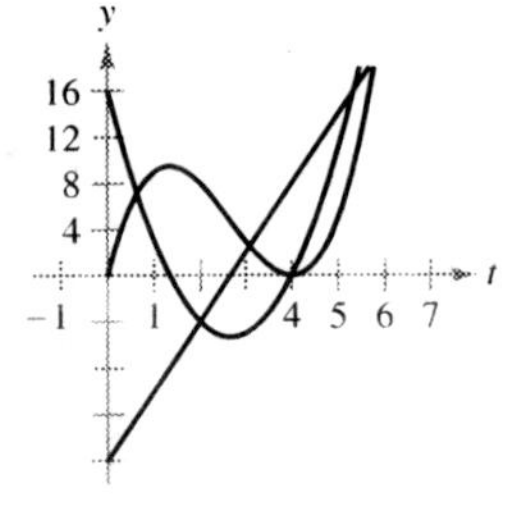

(b) On your sketch, identify when the particle speeds up and when it slows down. Explain your reasoning.

Finding a Pattern In Exercises 121 and 122, develop a general rule for $f^{(n)}(x)$ given $f(x)$.

121. $f(x) = x^n$

122. $f(x) = \dfrac{1}{x}$

123. *Finding a Pattern* Consider the function $f(x) = g(x)h(x)$.

(a) Use the Product Rule to generate rules for finding $f''(x)$, $f'''(x)$, and $f^{(4)}(x)$.

(b) Use the results of part (a) to write a general rule for $f^{(n)}(x)$.

124. *Finding a Pattern* Develop a general rule for $[xf(x)]^{(n)}$, where f is a differentiable function of x.

In Exercises 125 and 126, find the derivatives of the function f for $n = 1, 2, 3,$ and 4. Use the results to write a general rule for $f'(x)$ in terms of n.

125. $f(x) = x^n \sin x$

126. $f(x) = \dfrac{\cos x}{x^n}$

Differential Equations In Exercises 127–130, verify that the function satisfies the differential equation.

Function	Differential Equation
127. $y = \dfrac{1}{x}, x > 0$	$x^3 y'' + 2x^2 y' = 0$
128. $y = 2x^3 - 6x + 10$	$-y''' - xy'' - 2y' = -24x^2$
129. $y = 2 \sin x + 3$	$y'' + y = 3$
130. $y = 3 \cos x + \sin x$	$y'' + y = 0$

True or False? In Exercises 131–136, determine whether the statement is true or false. If it is false, explain why or give an example that shows it is false.

131. If $y = f(x)g(x)$, then $dy/dx = f'(x)g'(x)$.

132. If $y = (x + 1)(x + 2)(x + 3)(x + 4)$, then $d^5y/dx^5 = 0$.

133. If $f'(c)$ and $g'(c)$ are zero and $h(x) = f(x)g(x)$, then $h'(c) = 0$.

134. If $f(x)$ is an nth-degree polynomial, then $f^{(n+1)}(x) = 0$.

135. The second derivative represents the rate of change of the first derivative.

136. If the velocity of an object is constant, then its acceleration is zero.

137. Find a second-degree polynomial $f(x) = ax^2 + bx + c$ such that its graph has a tangent line with slope 10 at the point $(2, 7)$ and an x-intercept at $(1, 0)$.

138. Consider the third-degree polynomial

$$f(x) = ax^3 + bx^2 + cx + d, \quad a \ne 0.$$

Determine conditions for a, b, c, and d if the graph of f has (a) no horizontal tangent lines, (b) exactly one horizontal tangent line, and (c) exactly two horizontal tangent lines. Give an example for each case.

139. Find the derivative of $f(x) = x|x|$. Does $f''(0)$ exist?

140. *Think About It* Let f and g be functions whose first and second derivatives exist on an interval I. Which of the following formulas is (are) true?

(a) $fg'' - f''g = (fg' - f'g)'$

(b) $fg'' + f''g = (fg)''$

141. Use the Product Rule twice to prove that if f, g, and h are differentiable functions of x, then

$$\frac{d}{dx}[f(x)g(x)h(x)] = f'(x)g(x)h(x) + f(x)g'(x)h(x) + f(x)g(x)h'(x).$$

3.4   The Chain Rule

- ■ Find the derivative of a composite function using the Chain Rule.
- ■ Find the derivative of a function using the General Power Rule.
- ■ Simplify the derivative of a function using algebra.
- ■ Find the derivative of a transcendental function using the Chain Rule.
- ■ Find the derivative of a function involving the natural logarithmic function.
- ■ Define and differentiate exponential functions that have bases other than e.

The Chain Rule

This text has yet to discuss one of the most powerful differentiation rules—the **Chain Rule.** This rule deals with composite functions and adds a surprising versatility to the rules discussed in the two previous sections. For example, compare the following functions. Those on the left can be differentiated without the Chain Rule, and those on the right are best differentiated with the Chain Rule.

Without the Chain Rule	*With the Chain Rule*
$y = x^2 + 1$	$y = \sqrt{x^2 + 1}$
$y = \sin x$	$y = \sin 6x$
$y = 3x + 2$	$y = (3x + 2)^5$
$y = e^x + \tan x$	$y = e^{5x} + \tan x^2$

Basically, the Chain Rule states that if y changes dy/du times as fast as u, and u changes du/dx times as fast as x, then y changes $(dy/du)(du/dx)$ times as fast as x.

EXAMPLE 1   The Derivative of a Composite Function

A set of gears is constructed, as shown in Figure 3.25, such that the second and third gears are on the same axle. As the first axle revolves, it drives the second axle, which in turn drives the third axle. Let y, u, and x represent the numbers of revolutions per minute of the first, second, and third axles, respectively. Find dy/du, du/dx, and dy/dx, and show that

$$\frac{dy}{dx} = \frac{dy}{du} \cdot \frac{du}{dx}.$$

Solution   Because the circumference of the second gear is three times that of the first, the first axle must make three revolutions to turn the second axle once. Similarly, the second axle must make two revolutions to turn the third axle once, and you can write

$$\frac{dy}{du} = 3 \quad \text{and} \quad \frac{du}{dx} = 2.$$

Combining these two results, you know that the first axle must make six revolutions to turn the third axle once. So, you can write

$$\frac{dy}{dx} = \begin{array}{c} \text{Rate of change of first axle} \\ \text{with respect to second axle} \end{array} \cdot \begin{array}{c} \text{Rate of change of second axle} \\ \text{with respect to third axle} \end{array}$$

$$= \frac{dy}{du} \cdot \frac{du}{dx} = 3 \cdot 2 = 6 = \begin{array}{c} \text{Rate of change of first axle} \\ \text{with respect to third axle} \end{array}$$

In other words, the rate of change of y with respect to x is the product of the rate of change of y with respect to u and the rate of change of u with respect to x.   ■

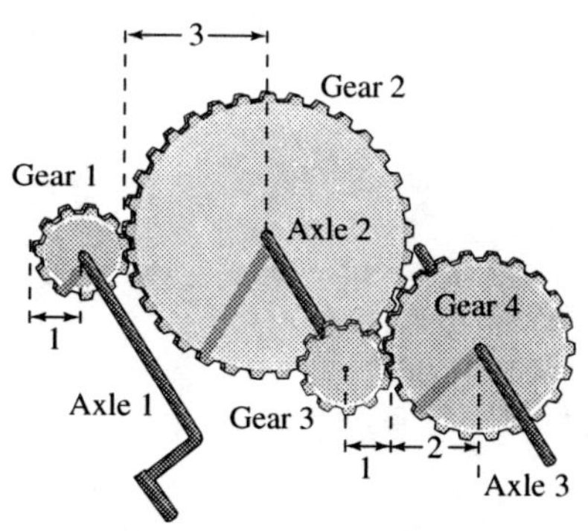

Axle 1: y revolutions per minute
Axle 2: u revolutions per minute
Axle 3: x revolutions per minute
Figure 3.25

Example 1 illustrates a simple case of the Chain Rule. The general rule is stated below.

THEOREM 3.11   THE CHAIN RULE

If $y = f(u)$ is a differentiable function of u and $u = g(x)$ is a differentiable function of x, then $y = f(g(x))$ is a differentiable function of x and

$$\frac{dy}{dx} = \frac{dy}{du} \cdot \frac{du}{dx}$$

or, equivalently,

$$\frac{d}{dx}[f(g(x))] = f'(g(x))g'(x).$$

PROOF   Let $h(x) = f(g(x))$. Then, using the alternative form of the derivative, you need to show that, for $x = c$,

$$h'(c) = f'(g(c))g'(c).$$

An important consideration in this proof is the behavior of g as x approaches c. A problem occurs if there are values of x, other than c, such that $g(x) = g(c)$. Appendix A shows how to use the differentiability of f and g to overcome this problem. For now, assume that $g(x) \neq g(c)$ for values of x other than c. In the proofs of the Product Rule and the Quotient Rule, the same quantity was added and subtracted to obtain the desired form. This proof uses a similar technique—multiplying and dividing by the same (nonzero) quantity. Note that because g is differentiable, it is also continuous, and it follows that $g(x) \to g(c)$ as $x \to c$.

$$h'(c) = \lim_{x \to c} \frac{f(g(x)) - f(g(c))}{x - c}$$

$$= \lim_{x \to c} \left[\frac{f(g(x)) - f(g(c))}{g(x) - g(c)} \cdot \frac{g(x) - g(c)}{x - c} \right], \quad g(x) \neq g(c)$$

$$= \left[\lim_{x \to c} \frac{f(g(x)) - f(g(c))}{g(x) - g(c)} \right] \left[\lim_{x \to c} \frac{g(x) - g(c)}{x - c} \right]$$

$$= f'(g(c))g'(c) \qquad \blacksquare$$

When applying the Chain Rule, it is helpful to think of the composite function $f \circ g$ as having two parts—an inner part and an outer part.

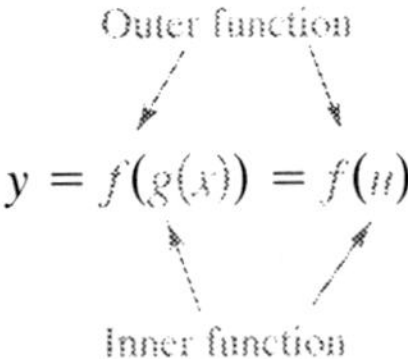

The derivative of $y = f(u)$ is the derivative of the outer function (at the inner function u) *times* the derivative of the inner function.

$$y' = f'(u) \cdot u'$$

EXAMPLE 2   Decomposition of a Composite Function

$y = f(g(x))$	$u = g(x)$	$y = f(u)$
a. $y = \dfrac{1}{x + 1}$	$u = x + 1$	$y = \dfrac{1}{u}$
b. $y = \sin 2x$	$u = 2x$	$y = \sin u$
c. $y = \sqrt{3x^2 - x + 1}$	$u = 3x^2 - x + 1$	$y = \sqrt{u}$
d. $y = \tan^2 x$	$u = \tan x$	$y = u^2$

EXAMPLE 3   Using the Chain Rule

Find dy/dx for $y = (x^2 + 1)^3$.

Solution   For this function, you can consider the inside function to be $u = x^2 + 1$. By the Chain Rule, you obtain

$$\frac{dy}{dx} = \underbrace{3(x^2 + 1)^2}_{\frac{dy}{du}}\underbrace{(2x)}_{\frac{du}{dx}} = 6x(x^2 + 1)^2.$$

STUDY TIP   You could also solve the problem in Example 3 without using the Chain Rule by observing that

$$y = x^6 + 3x^4 + 3x^2 + 1$$

and

$$y' = 6x^5 + 12x^3 + 6x.$$

Verify that this is the same result as the derivative in Example 3. Which method would you use to find

$$\frac{d}{dx}(x^2 + 1)^{50}?$$

The General Power Rule

The function in Example 3 is an example of one of the most common types of composite functions, $y = [u(x)]^n$. The rule for differentiating such functions is called the **General Power Rule,** and it is a special case of the Chain Rule.

THEOREM 3.12 THE GENERAL POWER RULE

If $y = [u(x)]^n$, where u is a differentiable function of x and n is a real number, then

$$\frac{dy}{dx} = n[u(x)]^{n-1}\frac{du}{dx}$$

or, equivalently,

$$\frac{d}{dx}[u^n] = nu^{n-1}u'.$$

PROOF   Because $y = u^n$, you apply the Chain Rule to obtain

$$\frac{dy}{dx} = \left(\frac{dy}{du}\right)\left(\frac{du}{dx}\right)$$

$$= \frac{d}{du}[u^n]\frac{du}{dx}.$$

By the (Simple) Power Rule in Section 3.2, you have $D_u[u^n] = nu^{n-1}$, and it follows that

$$\frac{dy}{dx} = n[u(x)]^{n-1}\frac{du}{dx}.$$

EXAMPLE 4 Applying the General Power Rule

Find the derivative of $f(x) = (3x - 2x^2)^3$.

Solution   Let $u = 3x - 2x^2$. Then

$$f(x) = (3x - 2x^2)^3 = u^3$$

and, by the General Power Rule, the derivative is

$$f'(x) = 3(3x - 2x^2)^2 \frac{d}{dx}[3x - 2x^2] \qquad \text{Apply General Power Rule.}$$

$$= 3(3x - 2x^2)^2(3 - 4x). \qquad \text{Differentiate } 3x - 2x^2.$$

EXAMPLE 5 Differentiating Functions Involving Radicals

Find all points on the graph of $f(x) = \sqrt[3]{(x^2 - 1)^2}$ for which $f'(x) = 0$ and those for which $f'(x)$ does not exist.

Solution   Begin by rewriting the function as

$$f(x) = (x^2 - 1)^{2/3}.$$

Then, applying the General Power Rule (with $u = x^2 - 1$) produces

$$f'(x) = \frac{2}{3}(x^2 - 1)^{-1/3}(2x) \qquad \text{Apply General Power Rule.}$$

$$= \frac{4x}{3\sqrt[3]{x^2 - 1}}. \qquad \text{Write in radical form.}$$

So, $f'(x) = 0$ when $x = 0$ and $f'(x)$ does not exist when $x = \pm 1$, as shown in Figure 3.26.

EXAMPLE 6 Differentiating Quotients with Constant Numerators

Differentiate $g(t) = \dfrac{-7}{(2t - 3)^2}$.

Solution   Begin by rewriting the function as

$$g(t) = -7(2t - 3)^{-2}.$$

Then, applying the General Power Rule produces

$$g'(t) = (-7)(-2)(2t - 3)^{-3}(2) \qquad \text{Apply General Power Rule.}$$

Constant Multiple Rule

$$= 28(2t - 3)^{-3} \qquad \text{Simplify.}$$

$$= \frac{28}{(2t - 3)^3}. \qquad \text{Write with positive exponent.}$$

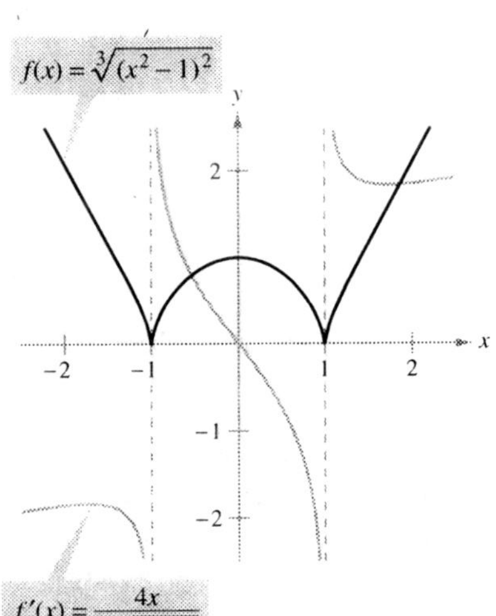

$f(x) = \sqrt[3]{(x^2 - 1)^2}$

$f'(x) = \dfrac{4x}{3\sqrt[3]{x^2 - 1}}$

The derivative of f is 0 at $x = 0$ and is undefined at $x = \pm 1$.
Figure 3.26

NOTE   Try differentiating the function in Example 6 using the Quotient Rule. You should obtain the same result, but using the Quotient Rule is less efficient than using the General Power Rule.

Simplifying Derivatives

The next three examples illustrate some techniques for simplifying the "raw derivatives" of functions involving products, quotients, and composites.

EXAMPLE 7 Simplifying by Factoring Out the Least Powers

$$f(x) = x^2\sqrt{1 - x^2}$$ Original function

$$= x^2(1 - x^2)^{1/2}$$ Rewrite.

$$f'(x) = x^2\frac{d}{dx}\left[(1 - x^2)^{1/2}\right] + (1 - x^2)^{1/2}\frac{d}{dx}[x^2]$$ Product Rule

$$= x^2\left[\frac{1}{2}(1 - x^2)^{-1/2}(-2x)\right] + (1 - x^2)^{1/2}(2x)$$ General Power Rule

$$= -x^3(1 - x^2)^{-1/2} + 2x(1 - x^2)^{1/2}$$ Simplify.

$$= x(1 - x^2)^{-1/2}[-x^2(1) + 2(1 - x^2)]$$ Factor.

$$= \frac{x(2 - 3x^2)}{\sqrt{1 - x^2}}$$ Simplify.

EXAMPLE 8 Simplifying the Derivative of a Quotient

$$f(x) = \frac{x}{\sqrt[3]{x^2 + 4}}$$ Original function

$$= \frac{x}{(x^2 + 4)^{1/3}}$$ Rewrite.

$$f'(x) = \frac{(x^2 + 4)^{1/3}(1) - x(1/3)(x^2 + 4)^{-2/3}(2x)}{(x^2 + 4)^{2/3}}$$ Quotient Rule

$$= \frac{1}{3}(x^2 + 4)^{-2/3}\left[\frac{3(x^2 + 4) - (2x^2)(1)}{(x^2 + 4)^{2/3}}\right]$$ Factor.

$$= \frac{x^2 + 12}{3(x^2 + 4)^{4/3}}$$ Simplify.

EXAMPLE 9 Simplifying the Derivative of a Power

$$y = \left(\frac{3x - 1}{x^2 + 3}\right)^2$$ Original function

$$y' = 2\left(\frac{3x - 1}{x^2 + 3}\right)\frac{d}{dx}\left[\frac{3x - 1}{x^2 + 3}\right]$$ General Power Rule

$$= \left[\frac{2(3x - 1)}{x^2 + 3}\right]\left[\frac{(x^2 + 3)(3) - (3x - 1)(2x)}{(x^2 + 3)^2}\right]$$ Quotient Rule

$$= \frac{2(3x - 1)(3x^2 + 9 - 6x^2 + 2x)}{(x^2 + 3)^3}$$ Multiply.

$$= \frac{2(3x - 1)(-3x^2 + 2x + 9)}{(x^2 + 3)^3}$$ Simplify.

TECHNOLOGY Symbolic differentiation utilities are capable of differentiating very complicated functions. Often, however, the result is given in unsimplified form. If you have access to such a utility, use it to find the derivatives of the functions given in Examples 7, 8, and 9. Then compare the results with those given on this page.

Transcendental Functions and the Chain Rule

The "Chain Rule versions" of the derivatives of the six trigonometric functions and the natural exponential function are as follows.

$$\frac{d}{dx}[\sin u] = (\cos u)\, u' \qquad\qquad \frac{d}{dx}[\cos u] = -(\sin u)\, u'$$

$$\frac{d}{dx}[\tan u] = (\sec^2 u)\, u' \qquad\qquad \frac{d}{dx}[\cot u] = -(\csc^2 u)\, u'$$

$$\frac{d}{dx}[\sec u] = (\sec u \tan u)\, u' \qquad\qquad \frac{d}{dx}[\csc u] = -(\csc u \cot u)\, u'$$

$$\frac{d}{dx}[e^u] = e^u u'$$

EXAMPLE 10 Applying the Chain Rule to Transcendental Functions

NOTE Be sure that you understand the mathematical conventions regarding parentheses and trigonometric functions. For instance, in Example 10(a), $\sin 2x$ is written to mean $\sin(2x)$.

a. $y = \sin 2x$

$$y' = \cos 2x \frac{d}{dx}[2x] = (\cos 2x)(2) = 2\cos 2x$$

b. $y = \cos(x - 1)$

$$y' = -\sin(x - 1)\frac{d}{dx}[x - 1] = -\sin(x - 1)$$

c. $y = e^{3x}$

$$y' = e^{3x}\frac{d}{dx}[3x] = 3e^{3x}$$

EXAMPLE 11 Parentheses and Trigonometric Functions

a. $y = \cos 3x^2 = \cos(3x^2)$ $\qquad y' = (-\sin 3x^2)(6x) = -6x \sin 3x^2$

b. $y = (\cos 3)x^2$ $\qquad\qquad\qquad y' = (\cos 3)(2x) = 2x \cos 3$

c. $y = \cos(3x)^2 = \cos(9x^2)$ $\qquad y' = (-\sin 9x^2)(18x) = -18x \sin 9x^2$

d. $y = \cos^2 x = (\cos x)^2$ $\qquad\quad y' = 2(\cos x)(-\sin x) = -2\cos x \sin x$　■

To find the derivative of a function of the form $k(x) = f(g(h(x)))$, you need to apply the Chain Rule twice, as shown in Example 12.

EXAMPLE 12 Repeated Application of the Chain Rule

$$f(t) = \sin^3 4t \qquad\qquad \text{Original function}$$

$$= (\sin 4t)^3 \qquad\qquad \text{Rewrite.}$$

$$f'(t) = 3(\sin 4t)^2 \frac{d}{dt}[\sin 4t] \qquad \text{Apply Chain Rule once.}$$

$$= 3(\sin 4t)^2(\cos 4t)\frac{d}{dt}[4t] \qquad \text{Apply Chain Rule a second time.}$$

$$= 3(\sin 4t)^2(\cos 4t)(4)$$

$$= 12 \sin^2 4t \cos 4t \qquad\qquad \text{Simplify.}$$　■

The Derivative of the Natural Logarithmic Function

Up to this point in the text, derivatives of algebraic functions have been algebraic and derivatives of transcendental functions have been transcendental. The next theorem looks at an unusual situation in which the derivative of a transcendental function is algebraic. Specifically, the derivative of the natural logarithmic function is the algebraic function $1/x$.

THEOREM 3.13 DERIVATIVE OF THE NATURAL LOGARITHMIC FUNCTION

Let u be a differentiable function of x.

1. $\dfrac{d}{dx}[\ln x] = \dfrac{1}{x}, \quad x > 0$

2. $\dfrac{d}{dx}[\ln u] = \dfrac{1}{u}\dfrac{du}{dx} = \dfrac{u'}{u}, \quad u > 0$

Use the *table* feature of a graphing utility to display the values of $f(x) = \ln x$ and its derivative for $x = 0, 1, 2, 3, \ldots$. What do these values tell you about the derivative of the natural logarithmic function?

PROOF To prove the first part, let $y = \ln x$, which implies that $e^y = x$. Differentiating both sides of this equation produces the following.

$$y = \ln x$$

$$e^y = x$$

$$\frac{d}{dx}[e^y] = \frac{d}{dx}[x]$$

$$e^y \frac{dy}{dx} = 1 \qquad \text{Chain Rule}$$

$$\frac{dy}{dx} = \frac{1}{e^y}$$

$$\frac{dy}{dx} = \frac{1}{x}$$

The second part of the theorem can be obtained by applying the Chain Rule to the first part. ∎

EXAMPLE 13 Differentiation of Logarithmic Functions

a. $\dfrac{d}{dx}[\ln(2x)] = \dfrac{u'}{u} = \dfrac{2}{2x} = \dfrac{1}{x} \qquad u = 2x$

b. $\dfrac{d}{dx}[\ln(x^2 + 1)] = \dfrac{u'}{u} = \dfrac{2x}{x^2 + 1} \qquad u = x^2 + 1$

c. $\dfrac{d}{dx}[x \ln x] = x\left(\dfrac{d}{dx}[\ln x]\right) + (\ln x)\left(\dfrac{d}{dx}[x]\right) \qquad \text{Product Rule}$

$$= x\left(\frac{1}{x}\right) + (\ln x)(1)$$

$$= 1 + \ln x$$

d. $\dfrac{d}{dx}[(\ln x)^3] = 3(\ln x)^2 \dfrac{d}{dx}[\ln x] \qquad \text{Chain Rule}$

$$= 3(\ln x)^2 \frac{1}{x}$$

JOHN NAPIER (1550–1617)

Logarithms were invented by the Scottish mathematician John Napier. Although he did not introduce the *natural* logarithmic function, it is sometimes called the *Napierian* logarithm.

John Napier used logarithmic properties to simplify *calculations* involving products, quotients, and powers. Of course, given the availability of calculators, there is now little need for this particular application of logarithms. However, there is great value in using logarithmic properties to simplify *differentiation* involving products, quotients, and powers.

EXAMPLE 14 Logarithmic Properties as Aids to Differentiation

Differentiate $f(x) = \ln \sqrt{x + 1}$.

Solution Because

$$f(x) = \ln \sqrt{x + 1} = \ln(x + 1)^{1/2} = \frac{1}{2} \ln(x + 1) \qquad \text{Rewrite before differentiating.}$$

you can write

$$f'(x) = \frac{1}{2}\left(\frac{1}{x + 1}\right) = \frac{1}{2(x + 1)}. \qquad \text{Differentiate.}$$

EXAMPLE 15 Logarithmic Properties as Aids to Differentiation

Differentiate $f(x) = \ln \dfrac{x(x^2 + 1)^2}{\sqrt{2x^3 - 1}}$.

Solution

$$f(x) = \ln \frac{x(x^2 + 1)^2}{\sqrt{2x^3 - 1}} \qquad \text{Write original function.}$$

$$= \ln x + 2 \ln(x^2 + 1) - \frac{1}{2} \ln(2x^3 - 1) \qquad \text{Rewrite before differentiating.}$$

$$f'(x) = \frac{1}{x} + 2\left(\frac{2x}{x^2 + 1}\right) - \frac{1}{2}\left(\frac{6x^2}{2x^3 - 1}\right) \qquad \text{Differentiate.}$$

$$= \frac{1}{x} + \frac{4x}{x^2 + 1} - \frac{3x^2}{2x^3 - 1} \qquad \text{Simplify.} \qquad \blacksquare$$

NOTE In Examples 14 and 15, be sure that you see the benefit of applying logarithmic properties *before* differentiation. Consider, for instance, the difficulty of direct differentiation of the function given in Example 15.

Because the natural logarithm is undefined for negative numbers, you will often encounter expressions of the form $\ln|u|$. Theorem 3.14 states that you can differentiate functions of the form $y = \ln|u|$ as if the absolute value notation was not present.

THEOREM 3.14 DERIVATIVE INVOLVING ABSOLUTE VALUE

If u is a differentiable function of x such that $u \neq 0$, then

$$\frac{d}{dx}[\ln|u|] = \frac{u'}{u}.$$

PROOF If $u > 0$, then $|u| = u$, and the result follows from Theorem 3.13. If $u < 0$, then $|u| = -u$, and you have

$$\frac{d}{dx}[\ln|u|] = \frac{d}{dx}[\ln(-u)] = \frac{-u'}{-u} = \frac{u'}{u}. \qquad \blacksquare$$

Bases Other than e

The **base** of the natural exponential function is e. This "natural" base can be used to assign a meaning to a general base a.

DEFINITION OF EXPONENTIAL FUNCTION TO BASE a

If a is a positive real number ($a \neq 1$) and x is any real number, then the **exponential function to the base a** is denoted by a^x and is defined by

$$a^x = e^{(\ln a)x}.$$

If $a = 1$, then $y = 1^x = 1$ is a constant function.

Logarithmic functions to bases other than e can be defined in much the same way as exponential functions to other bases are defined.

DEFINITION OF LOGARITHMIC FUNCTION TO BASE a

If a is a positive real number ($a \neq 1$) and x is any positive real number, then the **logarithmic function to the base a** is denoted by $\log_a x$ and is defined as

$$\log_a x = \frac{1}{\ln a} \ln x.$$

To differentiate exponential and logarithmic functions to other bases, you have two options: (1) use the definitions of a^x and $\log_a x$ and differentiate using the rules for the natural exponential and logarithmic functions, or (2) use the following differentiation rules for bases other than e.

NOTE These differentiation rules are similar to those for the natural exponential function and the natural logarithmic function. In fact, they differ only by the constant factors $\ln a$ and $1/\ln a$. This points out one reason why, for calculus, e is the most convenient base.

THEOREM 3.15 DERIVATIVES FOR BASES OTHER THAN e

Let a be a positive real number ($a \neq 1$) and let u be a differentiable function of x.

1. $\dfrac{d}{dx}[a^x] = (\ln a)a^x$ **2.** $\dfrac{d}{dx}[a^u] = (\ln a)a^u \dfrac{du}{dx}$

3. $\dfrac{d}{dx}[\log_a x] = \dfrac{1}{(\ln a)x}$ **4.** $\dfrac{d}{dx}[\log_a u] = \dfrac{1}{(\ln a)u}\dfrac{du}{dx}$

PROOF By definition, $a^x = e^{(\ln a)x}$. Therefore, you can prove the first rule by letting $u = (\ln a)x$ and differentiating with base e to obtain

$$\frac{d}{dx}[a^x] = \frac{d}{dx}[e^{(\ln a)x}] = e^u \frac{du}{dx} = e^{(\ln a)x}(\ln a) = (\ln a)a^x.$$

To prove the third rule, you can write

$$\frac{d}{dx}[\log_a x] = \frac{d}{dx}\left[\frac{1}{\ln a} \ln x\right] = \frac{1}{\ln a}\left(\frac{1}{x}\right) = \frac{1}{(\ln a)x}.$$

The second and fourth rules are simply the Chain Rule versions of the first and third rules. ∎

EXAMPLE 16 Differentiating Functions to Other Bases

Find the derivative of each function.

a. $y = 2^x$ **b.** $y = 2^{3x}$ **c.** $y = \log_{10} \cos x$

Solution

a. $y' = \dfrac{d}{dx}[2^x] = (\ln 2)2^x$

b. $y' = \dfrac{d}{dx}[2^{3x}] = (\ln 2)2^{3x}(3) = (3 \ln 2)2^{3x}$

Try writing 2^{3x} as 8^x and differentiating to see that you obtain the same result.

c. $y' = \dfrac{d}{dx}[\log_{10} \cos x] = \dfrac{-\sin x}{(\ln 10) \cos x} = -\dfrac{1}{\ln 10} \tan x$

STUDY TIP   To become skilled at differentiation, you should memorize each rule in words, not symbols. As an aid to memorization, note that the cofunctions (cosine, cotangent, and cosecant) require a negative sign as part of their derivatives.

This section concludes with a summary of the differentiation rules studied so far.

SUMMARY OF DIFFERENTIATION RULES

General Differentiation Rules

Let u and v be differentiable functions of x.

Constant Rule:

$$\frac{d}{dx}[c] = 0$$

(Simple) Power Rule:

$$\frac{d}{dx}[x^n] = nx^{n-1}, \quad \frac{d}{dx}[x] = 1$$

Constant Multiple Rule:

$$\frac{d}{dx}[cu] = cu'$$

Sum or Difference Rule:

$$\frac{d}{dx}[u \pm v] = u' \pm v'$$

Product Rule:

$$\frac{d}{dx}[uv] = uv' + vu'$$

Quotient Rule:

$$\frac{d}{dx}\left[\frac{u}{v}\right] = \frac{vu' - uv'}{v^2}$$

Chain Rule:

$$\frac{d}{dx}[f(u)] = f'(u)\,u'$$

General Power Rule:

$$\frac{d}{dx}[u^n] = nu^{n-1}\,u'$$

Derivatives of Trigonometric Functions

$$\frac{d}{dx}[\sin x] = \cos x \qquad \frac{d}{dx}[\tan x] = \sec^2 x \qquad \frac{d}{dx}[\sec x] = \sec x \tan x$$

$$\frac{d}{dx}[\cos x] = -\sin x \qquad \frac{d}{dx}[\cot x] = -\csc^2 x \qquad \frac{d}{dx}[\csc x] = -\csc x \cot x$$

Derivatives of Exponential and Logarithmic Functions

$$\frac{d}{dx}[e^x] = e^x \qquad \frac{d}{dx}[\ln x] = \frac{1}{x}$$

$$\frac{d}{dx}[a^x] = (\ln a)a^x \qquad \frac{d}{dx}[\log_a x] = \frac{1}{(\ln a)x}$$

3.4 Exercises   See www.CalcChat.com for worked-out solutions to odd-numbered exercises.

In Exercises 1–8, complete the table.

$y = f(g(x))$	$u = g(x)$	$y = f(u)$
1. $y = (5x - 8)^4$		
2. $y = \dfrac{1}{\sqrt{x+1}}$		
3. $y = \sqrt{x^3 - 7}$		
4. $y = 3\tan(\pi x^2)$		
5. $y = \csc^3 x$		
6. $y = \sin\dfrac{5x}{2}$		
7. $y = e^{-2x}$		
8. $y = (\ln x)^3$		

In Exercises 9–38, find the derivative of the function.

9. $y = (4x - 1)^3$

10. $y = 2(6 - x^2)^5$

11. $g(x) = 3(4 - 9x)^4$

12. $y = 3(5 - x^2)^5$

13. $f(x) = (9 - x^2)^{2/3}$

14. $f(t) = (9t + 7)^{2/3}$

15. $f(t) = \sqrt{5 - t}$

16. $g(x) = \sqrt{9 - 4x}$

17. $y = \sqrt[3]{6x^2 + 1}$

18. $g(x) = \sqrt{x^2 - 2x + 1}$

19. $y = 2\sqrt[4]{9 - x^2}$

20. $f(x) = -3\sqrt[4]{2 - 9x}$

21. $y = \dfrac{1}{x - 2}$

22. $s(t) = \dfrac{1}{t^2 + 3t - 1}$

23. $f(t) = \left(\dfrac{1}{t - 3}\right)^2$

24. $y = -\dfrac{8}{(t + 3)^3}$

25. $y = \dfrac{1}{\sqrt{x + 2}}$

26. $g(t) = \sqrt{\dfrac{1}{t^2 - 2}}$

27. $f(x) = x^2(x - 2)^4$

28. $f(x) = x(3x - 7)^3$

29. $y = x\sqrt{1 - x^2}$

30. $y = \tfrac{1}{2}x^2\sqrt{16 - x^2}$

31. $y = \dfrac{x}{\sqrt{x^2 + 1}}$

32. $y = \dfrac{x}{\sqrt{x^4 + 2}}$

33. $g(x) = \left(\dfrac{x + 5}{x^2 + 2}\right)^2$

34. $g(x) = \left(\dfrac{3x^2 - 1}{2x + 5}\right)^3$

35. $f(x) = ((x^2 + 3)^5 + x)^2$

36. $g(x) = (2 + (x^2 + 1)^4)^3$

37. $f(x) = \sqrt{2 + \sqrt{2 + \sqrt{x}}}$

38. $g(t) = \sqrt{\sqrt{t + 1} + 1}$

CAS **In Exercises 39–48, use a computer algebra system to find the derivative of the function. Then use the utility to graph the function and its derivative on the same set of coordinate axes. Describe the behavior of the function that corresponds to any zeros of the graph of the derivative.**

39. $y = \dfrac{\sqrt{x + 1}}{x^2 + 1}$

40. $y = \sqrt{\dfrac{2x}{x + 1}}$

41. $g(t) = \dfrac{3t^2}{\sqrt{t^2 + 2t - 1}}$

42. $f(x) = \sqrt{x}(2 - x)^2$

43. $y = \sqrt{\dfrac{x + 1}{x}}$

44. $y = (t^2 - 9)\sqrt{t + 2}$

45. $s(t) = \dfrac{-2(2 - t)\sqrt{1 + t}}{3}$

46. $g(x) = \sqrt{x - 1} + \sqrt{x + 1}$

47. $y = \dfrac{\cos \pi x + 1}{x}$

48. $y = x^2\tan\dfrac{1}{x}$

In Exercises 49 and 50, find the slope of the tangent line to the sine function at the origin. Compare this value with the number of complete cycles in the interval $[0, 2\pi]$. What can you conclude about the slope of the sine function $\sin ax$ at the origin?

49. (a) 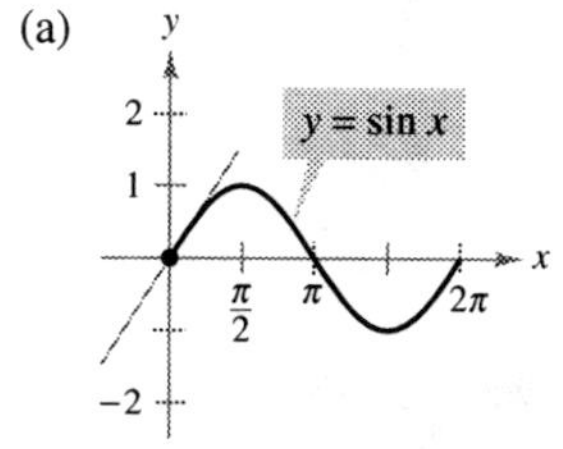

(b) 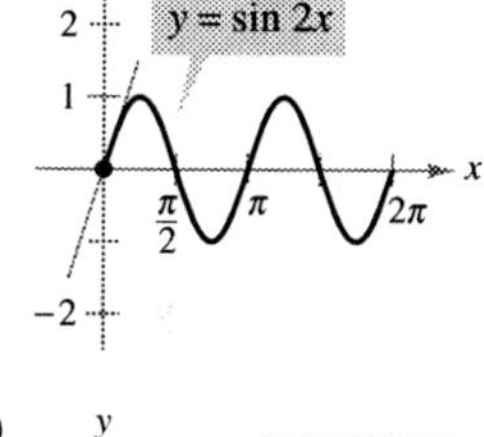

50. (a) 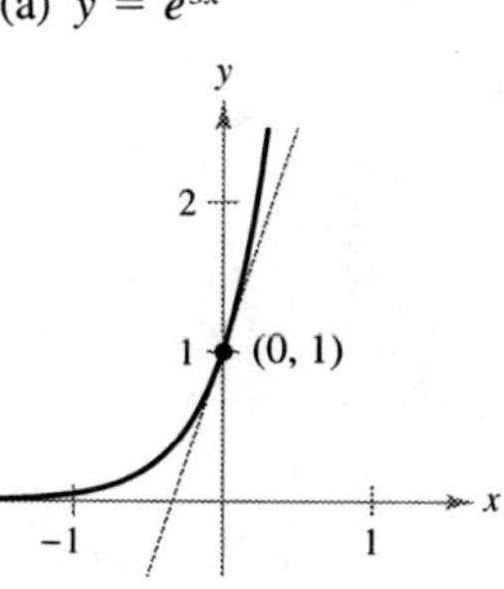

(b) 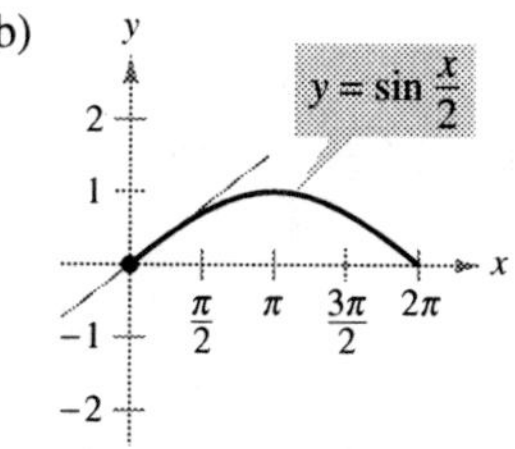

In Exercises 51 and 52, find the slope of the tangent line to the graph of the function at the point $(0, 1)$.

51. (a) $y = e^{3x}$

(b) $y = e^{-3x}$

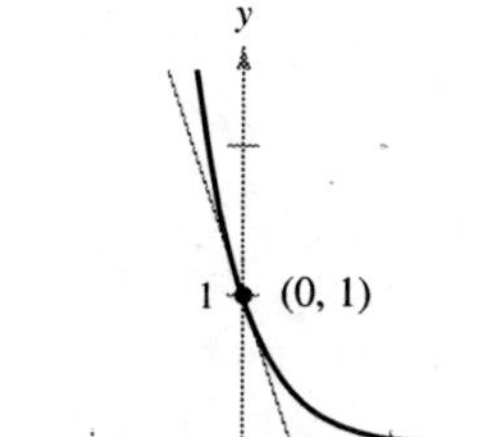

52. (a) $y = e^{2x}$

(b) $y = e^{-2x}$

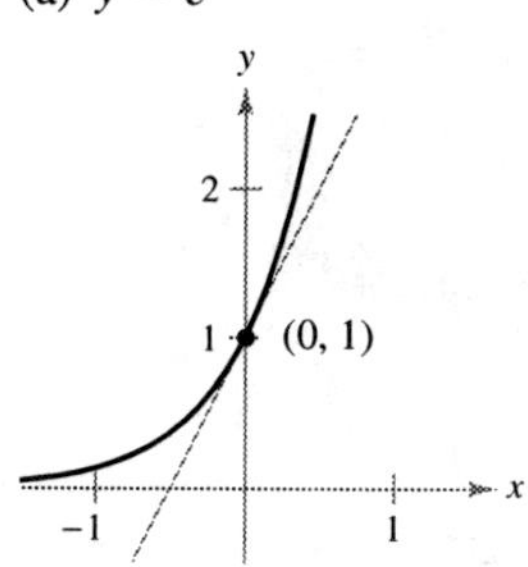

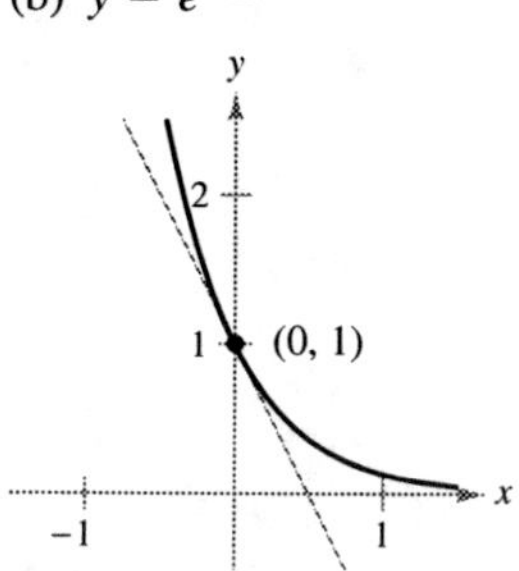

In Exercises 53–56, find the slope of the tangent line to the graph of the logarithmic function at the point (1, 0).

53. $y = \ln x^3$

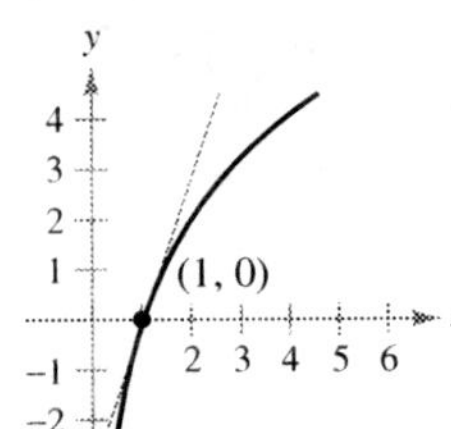

54. $y = \ln x^{3/2}$

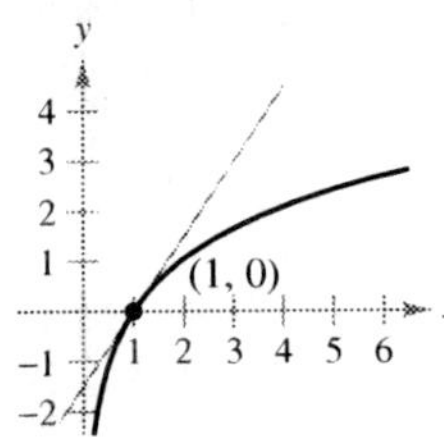

55. $y = \ln x^2$

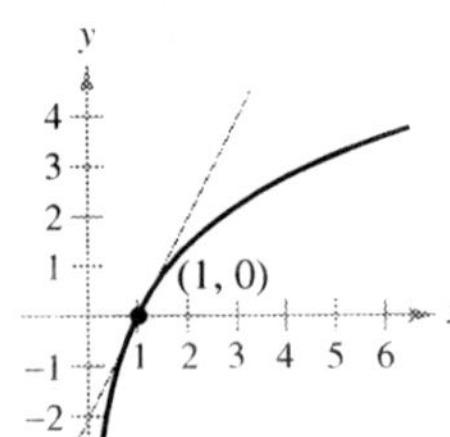

56. $y = \ln x^{1/2}$

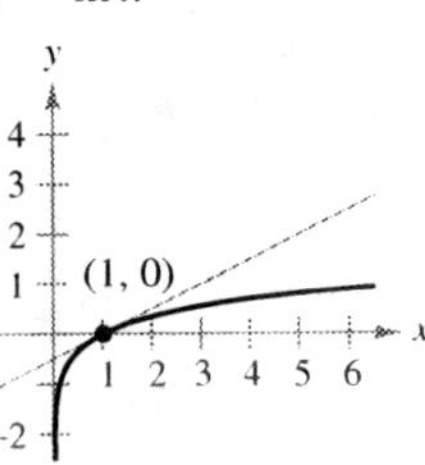

In Exercises 57–106, find the derivative of the function.

57. $y = \cos 4x$

58. $y = \sin \pi x$

59. $g(x) = 5 \tan 3x$

60. $h(x) = \sec x^3$

61. $f(\theta) = \tan^2 5\theta$

62. $g(\theta) = \cos^2 8\theta$

63. $f(\theta) = \frac{1}{4} \sin^2 2\theta$

64. $g(t) = 5 \cos^3 \pi t$

65. $y = \sqrt{x} + \frac{1}{4} \sin(2x)^2$

66. $y = 3x - 5 \cos(2x)^2$

67. $y = \sin(\cos x)$

68. $y = \sin \sqrt[3]{x} + \sqrt[3]{\sin x}$

69. $y = \sin(\tan 2x)$

70. $y = \cos \sqrt{\sin(\tan \pi x)}$

71. $f(x) = e^{2x}$

72. $y = e^{-x^2}$

73. $y = e^{\sqrt{x}}$

74. $y = x^2 e^{-x}$

75. $g(t) = (e^{-t} + e^t)^3$

76. $g(t) = e^{-3/t^2}$

77. $y = \ln(e^{x^2})$

78. $y = \ln\left(\frac{1 + e^x}{1 - e^x}\right)$

79. $y = \frac{2}{e^x + e^{-x}}$

80. $y = \frac{e^x - e^{-x}}{2}$

81. $y = x^2 e^x - 2x e^x + 2e^x$

82. $y = x e^x - e^x$

83. $f(x) = e^{-x} \ln x$

84. $f(x) = e^3 \ln x$

85. $y = e^x(\sin x + \cos x)$

86. $y = \ln e^x$

87. $g(x) = \ln x^2$

88. $h(x) = \ln(2x^2 + 3)$

89. $y = (\ln x)^4$

90. $y = x \ln x$

91. $y = \ln\left(x \sqrt{x^2 - 1}\right)$

92. $y = \ln \sqrt{x^2 - 9}$

93. $f(x) = \ln\left(\frac{x}{x^2 + 1}\right)$

94. $f(x) = \ln\left(\frac{2x}{x + 3}\right)$

95. $g(t) = \frac{\ln t}{t^2}$

96. $h(t) = \frac{\ln t}{t}$

97. $y = \ln \sqrt{\frac{x + 1}{x - 1}}$

98. $y = \ln \sqrt[3]{\frac{x - 2}{x + 2}}$

99. $y = \frac{-\sqrt{x^2 + 1}}{x} + \ln\left(x + \sqrt{x^2 + 1}\right)$

100. $y = \frac{-\sqrt{x^2 + 4}}{2x^2} - \frac{1}{4} \ln\left(\frac{2 + \sqrt{x^2 + 4}}{x}\right)$

101. $y = \ln|\sin x|$

102. $y = \ln|\csc x|$

103. $y = \ln\left|\frac{\cos x}{\cos x - 1}\right|$

104. $y = \ln|\sec x + \tan x|$

105. $y = \ln\left|\frac{-1 + \sin x}{2 + \sin x}\right|$

106. $y = \ln \sqrt{1 + \sin^2 x}$

In Exercises 107–114, find the second derivative of the function.

107. $f(x) = 5(2 - 7x)^4$

108. $f(x) = 4(x^2 - 2)^3$

109. $f(x) = \frac{1}{x - 6}$

110. $f(x) = \frac{4}{(x + 2)^3}$

111. $f(x) = \sin x^2$

112. $f(x) = \sec^2 \pi x$

113. $f(x) = (3 + 2x)e^{-3x}$

114. $g(x) = \sqrt{x} + e^x \ln x$

In Exercises 115–122, evaluate the derivative of the function at the given point. Use a graphing utility to verify your result.

Function	*Point*
115. $s(t) = \sqrt{t^2 + 6t - 2}$	$(3, 5)$
116. $y = \sqrt[5]{3x^3 + 4x}$	$(2, 2)$
117. $f(x) = \dfrac{5}{x^3 - 2}$	$\left(-2, -\dfrac{1}{2}\right)$
118. $f(x) = \dfrac{1}{(x^2 - 3x)^2}$	$\left(4, \dfrac{1}{16}\right)$
119. $f(t) = \dfrac{3t + 2}{t - 1}$	$(0, -2)$
120. $f(x) = \dfrac{x + 1}{2x - 3}$	$(2, 3)$
121. $y = 26 - \sec^3 4x$	$(0, 25)$
122. $y = \dfrac{1}{x} + \sqrt{\cos x}$	$\left(\dfrac{\pi}{2}, \dfrac{2}{\pi}\right)$

In Exercises 123–130, (a) find an equation of the tangent line to the graph of f at the given point, (b) use a graphing utility to graph the function and its tangent line at the point, and (c) use the *derivative* feature of a graphing utility to confirm your results.

Function	*Point*
123. $f(x) = \sqrt{2x^2 - 7}$	$(4, 5)$
124. $f(x) = \frac{1}{3}x \sqrt{x^2 + 5}$	$(2, 2)$
125. $f(x) = \sin 2x$	$(\pi, 0)$
126. $y = \cos 3x$	$\left(\dfrac{\pi}{4}, -\dfrac{\sqrt{2}}{2}\right)$
127. $y = 2 \tan^3 x$	$\left(\dfrac{\pi}{4}, 2\right)$
128. $f(x) = \tan^2 x$	$\left(\dfrac{\pi}{4}, 1\right)$
129. $y = 4 - x^2 - \ln(\frac{1}{2}x + 1)$	$(0, 4)$
130. $y = 2e^{1 - x^2}$	$(1, 2)$

In Exercises 131–146, find the derivative of the function.

131. $f(x) = 4^x$

132. $g(x) = 5^{-x}$

133. $y = 5^{x-2}$

134. $y = x(6^{-2x})$

135. $g(t) = t^2 2^t$

136. $f(t) = \dfrac{3^{2t}}{t}$

137. $h(\theta) = 2^{-\theta} \cos \pi\theta$

138. $g(\alpha) = 5^{-\alpha/2} \sin 2\alpha$

139. $y = \log_3 x$

140. $y = \log_{10} 2x$

141. $f(x) = \log_2 \dfrac{x^2}{x-1}$

142. $h(x) = \log_3 \dfrac{x\sqrt{x-1}}{2}$

143. $y = \log_5 \sqrt{x^2 - 1}$

144. $y = \log_{10} \dfrac{x^2 - 1}{x}$

145. $g(t) = \dfrac{10 \log_4 t}{t}$

146. $f(t) = t^{3/2} \log_2 \sqrt{t+1}$

WRITING ABOUT CONCEPTS

In Exercises 147–150, the graphs of a function f and its derivative f' are shown. Label the graphs as f or f' and write a short paragraph stating the criteria you used in making your selection. To print an enlarged copy of the graph, go to the website *www.mathgraphs.com*.

147.

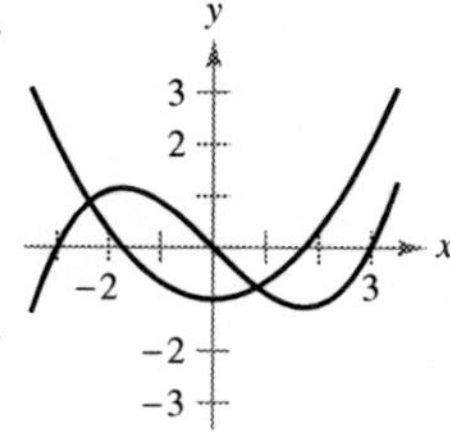

148.

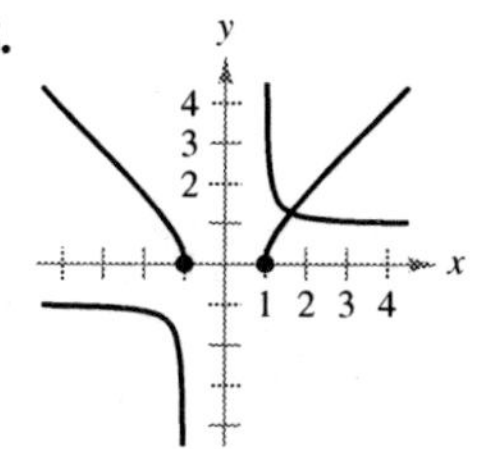

149.

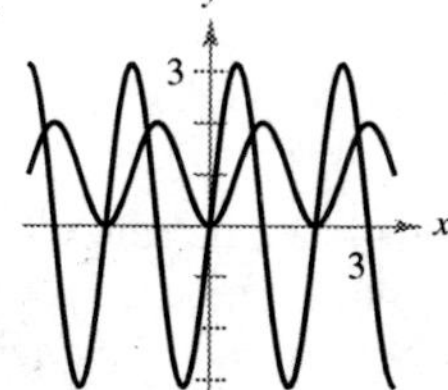

150.

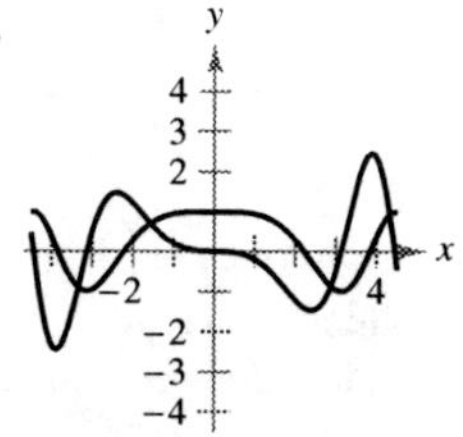

In Exercises 151 and 152, the relationship between f and g is given. Explain the relationship between f' and g'.

151. $g(x) = f(3x)$

152. $g(x) = f(x^2)$

---

153. (a) Find the derivative of the function $g(x) = \sin^2 x + \cos^2 x$ in two ways.

(b) For $f(x) = \sec^2 x$ and $g(x) = \tan^2 x$, show that

$$f'(x) = g'(x).$$

CAPSTONE

154. Given that $g(5) = -3$, $g'(5) = 6$, $h(5) = 3$, and $h'(5) = -2$, find $f'(5)$ (if possible) for each of the following. If it is not possible, state what additional information is required.

(a) $f(x) = g(x)h(x)$

(b) $f(x) = g(h(x))$

(c) $f(x) = \dfrac{g(x)}{h(x)}$

(d) $f(x) = [g(x)]^3$

In Exercises 155–158, (a) use a graphing utility to find the derivative of the function at the given point, (b) find an equation of the tangent line to the graph of the function at the given point, and (c) use the utility to graph the function and its tangent line in the same viewing window.

155. $g(t) = \dfrac{3t^2}{\sqrt{t^2 + 2t - 1}}, \quad \left(\dfrac{1}{2}, \dfrac{3}{2}\right)$

156. $f(x) = \sqrt{x}(2-x)^2, \quad (4, 8)$

157. $s(t) = \dfrac{(4 - 2t)\sqrt{1+t}}{3}, \quad \left(0, \dfrac{4}{3}\right)$

158. $y = (t^2 - 9)\sqrt{t+2}, \quad (2, -10)$

Famous Curves **In Exercises 159 and 160, find an equation of the tangent line to the graph at the given point. Then use a graphing utility to graph the function and its tangent line in the same viewing window.**

159. Top half of circle

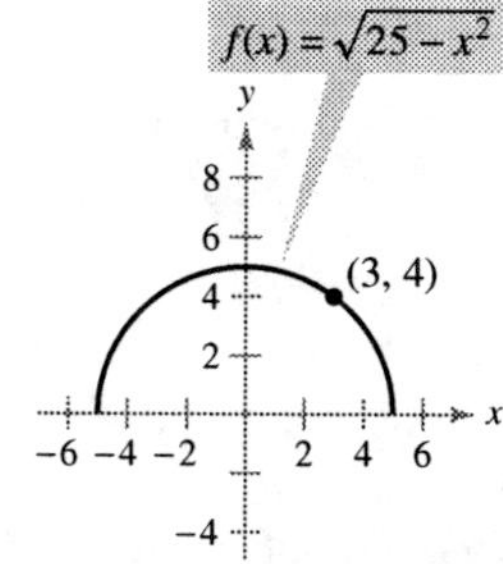

160. Bullet-nose curve

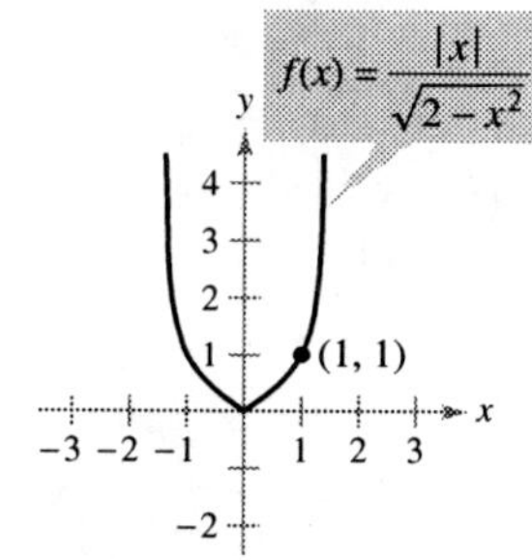

161. ***Horizontal Tangent Line*** Determine the point(s) in the interval $(0, 2\pi)$ at which the graph of $f(x) = 2\cos x + \sin 2x$ has a horizontal tangent line.

162. ***Horizontal Tangent Line*** Determine the point(s) at which the graph of $f(x) = \dfrac{x}{\sqrt{2x-1}}$ has a horizontal tangent line.

In Exercises 163–166, evaluate the second derivative of the function at the given point. Use a computer algebra system to verify your result.

163. $h(x) = \frac{1}{9}(3x + 1)^3, \quad \left(1, \frac{64}{9}\right)$

164. $f(x) = \dfrac{1}{\sqrt{x+4}}, \quad \left(0, \dfrac{1}{2}\right)$

165. $f(x) = \cos x^2, \quad (0, 1)$

166. $g(t) = \tan 2t, \quad \left(\dfrac{\pi}{6}, \sqrt{3}\right)$

167. *Doppler Effect* The frequency F of a fire truck siren heard by a stationary observer is

$$F = \frac{132{,}400}{331 \pm v}$$

where $\pm v$ represents the velocity of the accelerating fire truck in meters per second. Find the rate of change of F with respect to v when

(a) the fire truck is approaching at a velocity of 30 meters per second (use $-v$).

(b) the fire truck is moving away at a velocity of 30 meters per second (use $+v$).

168. *Harmonic Motion* The displacement from equilibrium of an object in harmonic motion on the end of a spring is

$$y = \tfrac{1}{3} \cos 12t - \tfrac{1}{4} \sin 12t$$

where y is measured in feet and t is the time in seconds. Determine the position and velocity of the object when $t = \pi/8$.

169. *Pendulum* A 15-centimeter pendulum moves according to the equation $\theta = 0.2 \cos 8t$, where θ is the angular displacement from the vertical in radians and t is the time in seconds. Determine the maximum angular displacement and the rate of change of θ when $t = 3$ seconds.

170. *Wave Motion* A buoy oscillates in simple harmonic motion $y = A \cos \omega t$ as waves move past it. The buoy moves a total of 3.5 feet (vertically) from its low point to its high point. It returns to its high point every 10 seconds.

(a) Write an equation describing the motion of the buoy if it is at its high point at $t = 0$.

(b) Determine the velocity of the buoy as a function of t.

171. *Circulatory System* The speed S of blood that is r centimeters from the center of an artery is

$$S = C(R^2 - r^2)$$

where C is a constant, R is the radius of the artery, and S is measured in centimeters per second. Suppose a drug is administered and the artery begins to dilate at a rate of dR/dt. At a constant distance r, find the rate at which S changes with respect to t for $C = 1.76 \times 10^5$, $R = 1.2 \times 10^{-2}$, and $dR/dt = 10^{-5}$.

172. *Modeling Data* The normal daily maximum temperatures T (in degrees Fahrenheit) for Chicago, Illinois are shown in the table. (*Source: National Oceanic and Atmospheric Administration*)

Month	Jan	Feb	Mar	Apr	May	Jun
Temperature	29.6	34.7	46.1	58.0	69.9	79.2

Month	Jul	Aug	Sep	Oct	Nov	Dec
Temperature	83.5	81.2	73.9	62.1	47.1	34.4

(a) Use a graphing utility to plot the data and find a model for the data of the form

$$T(t) = a + b \sin(ct - d)$$

where T is the temperature and t is the time in months, with $t = 1$ corresponding to January.

(b) Use a graphing utility to graph the model. How well does the model fit the data?

(c) Find T' and use a graphing utility to graph the derivative.

(d) Based on the graph of the derivative, during what times does the temperature change most rapidly? Most slowly? Do your answers agree with your observations of the temperature changes? Explain.

173. *Volume* Air is being pumped into a spherical balloon so that the radius is increasing at the rate of $dr/dt = 3$ inches per second. What is the rate of change of the volume of the balloon, in cubic inches per second, when $r = 8$ inches? $\left[\text{Hint: } V = \tfrac{4}{3}\pi r^3 \right]$

174. *Think About It* The table shows some values of the derivative of an unknown function f. Complete the table by finding (if possible) the derivative of each transformation of f.

(a) $g(x) = f(x) - 2$ (b) $h(x) = 2 f(x)$

(c) $r(x) = f(-3x)$ (d) $s(x) = f(x + 2)$

x	-2	-1	0	1	2	3
$f'(x)$	4	$\tfrac{2}{3}$	$-\tfrac{1}{3}$	-1	-2	-4
$g'(x)$						
$h'(x)$						
$r'(x)$						
$s'(x)$						

175. *Modeling Data* The table shows the temperatures T (°F) at which water boils at selected pressures p (pounds per square inch). (*Source: Standard Handbook of Mechanical Engineers*)

p	5	10	14.696 (1 atm)	20
T	$162.24°$	$193.21°$	$212.00°$	$227.96°$

p	30	40	60	80	100
T	$250.33°$	$267.25°$	$292.71°$	$312.03°$	$327.81°$

A model that approximates the data is

$$T = 87.97 + 34.96 \ln p + 7.91 \sqrt{p}.$$

(a) Use a graphing utility to plot the data and graph the model.

(b) Find the rates of change of T with respect to p when $p = 10$ and $p = 70$.

176. Depreciation After t years, the value of a car purchased for $25,000 is

$$V(t) = 25,000\left(\tfrac{3}{4}\right)^t.$$

(a) Use a graphing utility to graph the function and determine the value of the car 2 years after it was purchased.

(b) Find the rates of change of V with respect to t when $t = 1$ and $t = 4$.

177. Inflation If the annual rate of inflation averages 5% over the next 10 years, the approximate cost C of goods or services during any year in that decade is $C(t) = P(1.05)^t$, where t is the time in years and P is the present cost.

(a) If the price of an oil change for your car is presently $29.95, estimate the price 10 years from now.

(b) Find the rates of change of C with respect to t when $t = 1$ and $t = 8$.

(c) Verify that the rate of change of C is proportional to C. What is the constant of proportionality?

178. Finding a Pattern Consider the function $f(x) = \sin \beta x$, where β is a constant.

(a) Find the first-, second-, third-, and fourth-order derivatives of the function.

(b) Verify that the function and its second derivative satisfy the equation $f''(x) + \beta^2 f(x) = 0$.

(c) Use the results of part (a) to write general rules for the even- and odd-order derivatives

$$f^{(2k)}(x) \quad \text{and} \quad f^{(2k-1)}(x).$$

[*Hint:* $(-1)^k$ is positive if k is even and negative if k is odd.]

179. Conjecture Let f be a differentiable function of period p.

(a) Is the function f' periodic? Verify your answer.

(b) Consider the function $g(x) = f(2x)$. Is the function $g'(x)$ periodic? Verify your answer.

180. Think About It Let $r(x) = f(g(x))$ and $s(x) = g(f(x))$, where f and g are shown in the figure. Find (a) $r'(1)$ and (b) $s'(4)$.

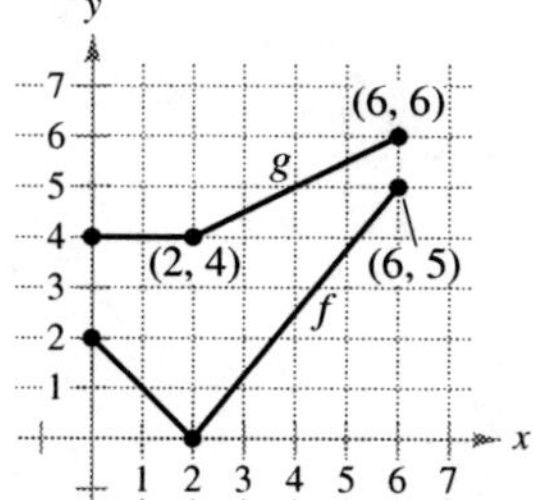

181. (a) Show that the derivative of an odd function is even. That is, if $f(-x) = -f(x)$, then $f'(-x) = f'(x)$.

(b) Show that the derivative of an even function is odd. That is, if $f(-x) = f(x)$, then $f'(-x) = -f'(x)$.

182. Let u be a differentiable function of x. Use the fact that $|u| = \sqrt{u^2}$ to prove that

$$\frac{d}{dx}[|u|] = u'\frac{u}{|u|}, \quad u \neq 0.$$

In Exercises 183–186, use the result of Exercise 182 to find the derivative of the function.

183. $g(x) = |3x - 5|$

184. $f(x) = |x^2 - 9|$

185. $h(x) = |x| \cos x$

186. $f(x) = |\sin x|$

Linear and Quadratic Approximations The linear and quadratic approximations of a function f at $x = a$ are

$$P_1(x) = f'(a)(x - a) + f(a) \text{ and}$$

$$P_2(x) = \tfrac{1}{2}f''(a)(x - a)^2 + f'(a)(x - a) + f(a).$$

In Exercises 187–190, (a) find the specified linear and quadratic approximations of f, (b) use a graphing utility to graph f and the approximations, (c) determine whether P_1 or P_2 is the better approximation, and (d) state how the accuracy changes as you move farther from $x = a$.

187. $f(x) = \tan x$

$$a = \frac{\pi}{4}$$

188. $f(x) = \sec x$

$$a = \frac{\pi}{6}$$

189. $f(x) = e^x$

$$a = 0$$

190. $f(x) = \ln x$

$$a = 1$$

True or False? In Exercises 191–194, determine whether the statement is true or false. If it is false, explain why or give an example that shows it is false.

191. If $y = (1 - x)^{1/2}$, then $y' = \tfrac{1}{2}(1 - x)^{-1/2}$.

192. If $f(x) = \sin^2(2x)$, then $f'(x) = 2(\sin 2x)(\cos 2x)$.

193. If y is a differentiable function of u, u is a differentiable function of v, and v is a differentiable function of x, then

$$\frac{dy}{dx} = \frac{dy}{du}\frac{du}{dv}\frac{dv}{dx}.$$

194. If f and g are differentiable functions of x and $h(x) = f(g(x))$, then $h'(x) = f'(g(x))g'(x)$.

PUTNAM EXAM CHALLENGE

195. Let $f(x) = a_1 \sin x + a_2 \sin 2x + \cdots + a_n \sin nx$, where $a_1, a_2, \ldots, a_n$ are real numbers and where n is a positive integer. Given that $|f(x)| \leq |\sin x|$ for all real x, prove that $|a_1 + 2a_2 + \cdots + na_n| \leq 1$.

196. Let k be a fixed positive integer. The nth derivative of $\dfrac{1}{x^k - 1}$ has the form

$$\frac{P_n(x)}{(x^k - 1)^{n+1}}$$

where $P_n(x)$ is a polynomial. Find $P_n(1)$.

These problems were composed by the Committee on the Putnam Prize Competition. © The Mathematical Association of America. All rights reserved.

3.5 Implicit Differentiation

- ■ **Distinguish between functions written in implicit form and explicit form.**
- ■ **Use implicit differentiation to find the derivative of a function.**
- ■ **Find derivatives of functions using logarithmic differentiation.**

Implicit and Explicit Functions

Up to this point in the text, most functions have been expressed in **explicit form.** For example, in the equation

$$y = 3x^2 - 5 \qquad \text{Explicit form}$$

the variable y is explicitly written as a function of x. Some functions, however, are only *implied* by an equation. For instance, the function $y = 1/x$ is defined **implicitly** by the equation $xy = 1$. Suppose you were asked to find dy/dx for this equation. You could begin by writing y explicitly as a function of x and then differentiating.

Implicit Form	*Explicit Form*	*Derivative*
$xy = 1$	$y = \dfrac{1}{x} = x^{-1}$	$\dfrac{dy}{dx} = -x^{-2} = -\dfrac{1}{x^2}$

This strategy works whenever you can solve for the function explicitly. You cannot, however, use this procedure when you are unable to solve for y as a function of x. For instance, how would you find dy/dx for the equation $x^2 - 2y^3 + 4y = 2$, where it is very difficult to express y as a function of x explicitly? To do this, you can use **implicit differentiation.**

To understand how to find dy/dx implicitly, you must realize that the differentiation is taking place *with respect to x*. This means that when you differentiate terms involving x alone, you can differentiate as usual. However, when you differentiate terms involving y, you must apply the Chain Rule, because you are assuming that y is defined implicitly as a differentiable function of x.

EXAMPLE 1 Differentiating with Respect to *x*

a. $\dfrac{d}{dx}[x^3] = 3x^2$

Variables agree: use Simple Power Rule.

Variables agree

b. $\dfrac{d}{dx}[y^3] = 3y^2\dfrac{dy}{dx}$

Variables disagree: use Chain Rule.

Variables disagree

c. $\dfrac{d}{dx}[x + 3y] = 1 + 3\dfrac{dy}{dx}$

Chain Rule: $\dfrac{d}{dx}[3y] = 3y'$

d. $\dfrac{d}{dx}[xy^2] = x\dfrac{d}{dx}[y^2] + y^2\dfrac{d}{dx}[x]$

Product Rule

$\qquad = x\left(2y\dfrac{dy}{dx}\right) + y^2(1)$

Chain Rule

$\qquad = 2xy\dfrac{dy}{dx} + y^2$

Simplify.

Implicit Differentiation

> **GUIDELINES FOR IMPLICIT DIFFERENTIATION**
>
> 1. Differentiate both sides of the equation *with respect to x*.
> 2. Collect all terms involving dy/dx on the left side of the equation and move all other terms to the right side of the equation.
> 3. Factor dy/dx out of the left side of the equation.
> 4. Solve for dy/dx by dividing both sides of the equation by the left-hand factor that does not contain dy/dx.

In Example 2, note that implicit differentiation can produce an expression for dy/dx that contains both x and y.

EXAMPLE 2 Implicit Differentiation

Find dy/dx given that $y^3 + y^2 - 5y - x^2 = -4$.

Solution

1. Differentiate both sides of the equation with respect to x.

$$\frac{d}{dx}[y^3 + y^2 - 5y - x^2] = \frac{d}{dx}[-4]$$

$$\frac{d}{dx}[y^3] + \frac{d}{dx}[y^2] - \frac{d}{dx}[5y] - \frac{d}{dx}[x^2] = \frac{d}{dx}[-4]$$

$$3y^2\frac{dy}{dx} + 2y\frac{dy}{dx} - 5\frac{dy}{dx} - 2x = 0$$

2. Collect the dy/dx terms on the left side of the equation.

$$3y^2\frac{dy}{dx} + 2y\frac{dy}{dx} - 5\frac{dy}{dx} = 2x$$

3. Factor dy/dx out of the left side of the equation.

$$\frac{dy}{dx}(3y^2 + 2y - 5) = 2x$$

4. Solve for dy/dx by dividing by $(3y^2 + 2y - 5)$.

$$\frac{dy}{dx} = \frac{2x}{3y^2 + 2y - 5}$$

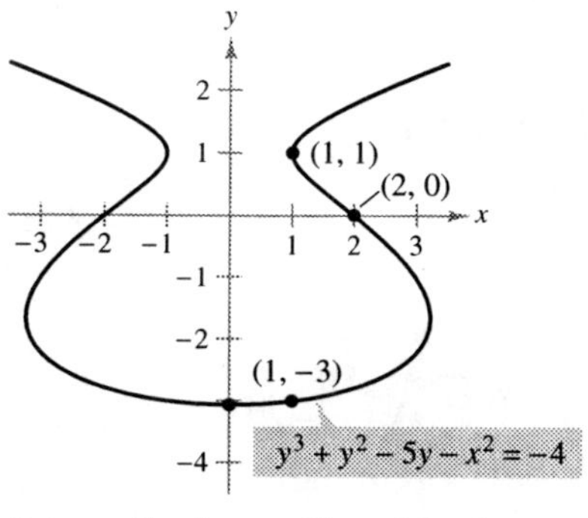

To see how you can use an *implicit derivative*, consider the graph shown in Figure 3.27. From the graph, you can see that y is not a function of x. Even so, the derivative found in Example 2 gives a formula for the slope of the tangent line at a point on this graph. The slopes at several points on the graph are shown below the graph.

TECHNOLOGY With most graphing utilities, it is easy to graph an equation that explicitly represents y as a function of x. Graphing other equations, however, can require some ingenuity. For instance, to graph the equation given in Example 2, use a graphing utility, set in *parametric* mode, to graph the parametric representations $x = \sqrt{t^3 + t^2 - 5t + 4}$, $y = t$, and $x = -\sqrt{t^3 + t^2 - 5t + 4}$, $y = t$, for $-5 \le t \le 5$. How does the result compare with the graph shown in Figure 3.27?

Point on Graph	Slope of Graph
$(2, 0)$	$-\frac{4}{5}$
$(1, -3)$	$\frac{1}{8}$
$x = 0$	0
$(1, 1)$	Undefined

The implicit equation

$$y^3 + y^2 - 5y - x^2 = -4$$

has the derivative

$$\frac{dy}{dx} = \frac{2x}{3y^2 + 2y - 5}.$$

Figure 3.27

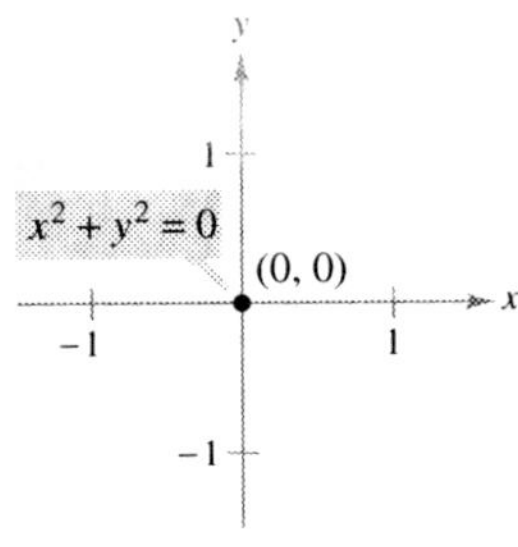

(a)

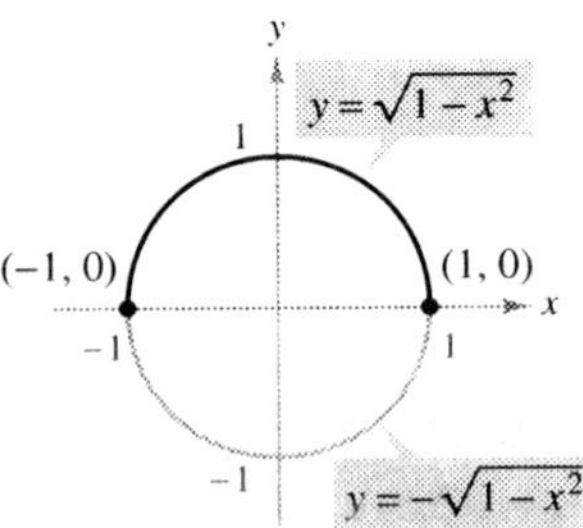

(b)

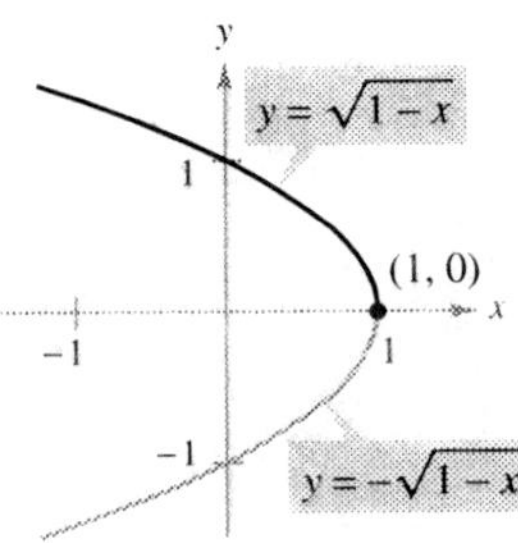

(c)

Some graph segments can be represented by differentiable functions.

Figure 3.28

It is meaningless to solve for dy/dx in an equation that has no solution points. (For example, $x^2 + y^2 = -4$ has no solution points.) If, however, a segment of a graph can be represented by a differentiable function, dy/dx will have meaning as the slope at each point on the segment. Recall that a function is not differentiable at (1) points with vertical tangents and (2) points at which the function is not continuous.

EXAMPLE 3 Representing a Graph by Differentiable Functions

If possible, represent y as a differentiable function of x (see Figure 3.28).

a. $x^2 + y^2 = 0$ **b.** $x^2 + y^2 = 1$ **c.** $x + y^2 = 1$

Solution

a. The graph of this equation is a single point. So, the equation does not define y as a differentiable function of x.

b. The graph of this equation is the unit circle, centered at $(0, 0)$. The upper semicircle is given by the differentiable function

$$y = \sqrt{1 - x^2}, \quad -1 < x < 1$$

and the lower semicircle is given by the differentiable function

$$y = -\sqrt{1 - x^2}, \quad -1 < x < 1.$$

At the points $(-1, 0)$ and $(1, 0)$, the slope of the graph is undefined.

c. The upper half of this parabola is given by the differentiable function

$$y = \sqrt{1 - x}, \quad x < 1$$

and the lower half of this parabola is given by the differentiable function

$$y = -\sqrt{1 - x}, \quad x < 1.$$

At the point $(1, 0)$, the slope of the graph is undefined.

EXAMPLE 4 Finding the Slope of a Graph Implicitly

Determine the slope of the tangent line to the graph of

$$x^2 + 4y^2 = 4$$

at the point $\left(\sqrt{2}, -1/\sqrt{2} \right)$. See Figure 3.29.

Solution

$$x^2 + 4y^2 = 4 \qquad \text{Write original equation.}$$

$$2x + 8y\frac{dy}{dx} = 0 \qquad \text{Differentiate with respect to } x.$$

$$\frac{dy}{dx} = \frac{-2x}{8y} = \frac{-x}{4y} \qquad \text{Solve for } \frac{dy}{dx}.$$

So, at $\left(\sqrt{2}, -1/\sqrt{2} \right)$, the slope is

$$\frac{dy}{dx} = \frac{-\sqrt{2}}{-4/\sqrt{2}} = \frac{1}{2}. \qquad \text{Evaluate } \frac{dy}{dx} \text{ when } x = \sqrt{2} \text{ and } y = -\frac{1}{\sqrt{2}}.$$

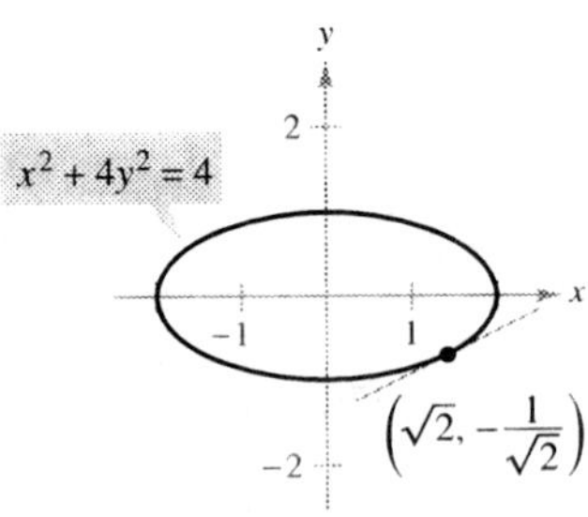

Figure 3.29

NOTE To see the benefit of implicit differentiation, try doing Example 4 using the explicit function $y = -\frac{1}{2}\sqrt{4 - x^2}$.

EXAMPLE ▣5 Finding the Slope of a Graph Implicitly

Determine the slope of the graph of $3(x^2 + y^2)^2 = 100xy$ at the point $(3, 1)$.

Solution

$$\frac{d}{dx}[3(x^2 + y^2)^2] = \frac{d}{dx}[100xy]$$

$$3(2)(x^2 + y^2)\left(2x + 2y\frac{dy}{dx}\right) = 100\left[x\frac{dy}{dx} + y(1)\right]$$

$$12y(x^2 + y^2)\frac{dy}{dx} - 100x\frac{dy}{dx} = 100y - 12x(x^2 + y^2)$$

$$[12y(x^2 + y^2) - 100x]\frac{dy}{dx} = 100y - 12x(x^2 + y^2)$$

$$\frac{dy}{dx} = \frac{100y - 12x(x^2 + y^2)}{-100x + 12y(x^2 + y^2)}$$

$$= \frac{25y - 3x(x^2 + y^2)}{-25x + 3y(x^2 + y^2)}$$

At the point $(3, 1)$, the slope of the graph is

$$\frac{dy}{dx} = \frac{25(1) - 3(3)(3^2 + 1^2)}{-25(3) + 3(1)(3^2 + 1^2)} = \frac{25 - 90}{-75 + 30} = \frac{-65}{-45} = \frac{13}{9}$$

as shown in Figure 3.30. This graph is called a **lemniscate.**

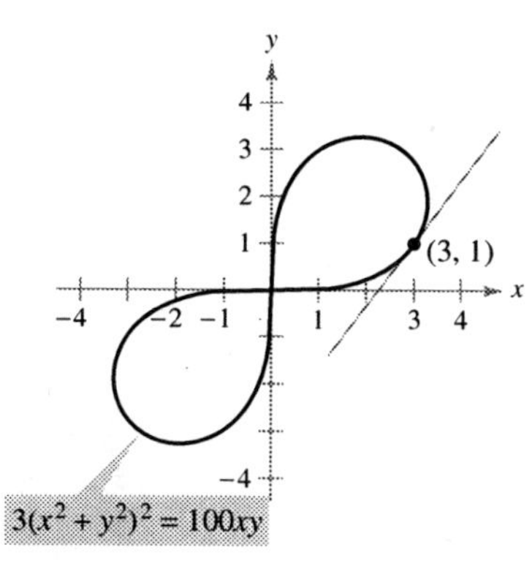

Lemniscate
Figure 3.30

EXAMPLE ▣6 Determining a Differentiable Function

Find dy/dx implicitly for the equation $\sin y = x$. Then find the largest interval of the form $-a < y < a$ on which y is a differentiable function of x (see Figure 3.31).

Solution

$$\frac{d}{dx}[\sin y] = \frac{d}{dx}[x]$$

$$\cos y\frac{dy}{dx} = 1$$

$$\frac{dy}{dx} = \frac{1}{\cos y}$$

The largest interval about the origin for which y is a differentiable function of x is $-\pi/2 < y < \pi/2$. To see this, note that $\cos y$ is positive for all y in this interval and is 0 at the endpoints. If you restrict y to the interval $-\pi/2 < y < \pi/2$, you should be able to write dy/dx explicitly as a function of x. To do this, you can use

$$\cos y = \sqrt{1 - \sin^2 y}$$

$$= \sqrt{1 - x^2}, \quad -\frac{\pi}{2} < y < \frac{\pi}{2}$$

and conclude that

$$\frac{dy}{dx} = \frac{1}{\sqrt{1 - x^2}}.$$

You will study this example further when derivatives of inverse trigonometric functions are defined in Section 3.6.

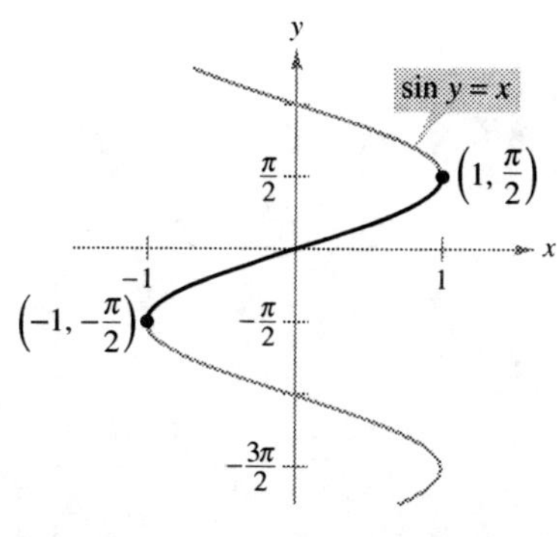

The derivative is $\dfrac{dy}{dx} = \dfrac{1}{\sqrt{1 - x^2}}.$

Figure 3.31

ISAAC BARROW (1630–1677)

The graph in Example 8 is called the **kappa curve** because it resembles the Greek letter kappa, κ. The general solution for the tangent line to this curve was discovered by the English mathematician Isaac Barrow. Newton was Barrow's student, and they corresponded frequently regarding their work in the early development of calculus.

With implicit differentiation, the form of the derivative often can be simplified (as in Example 6) by an appropriate use of the *original* equation. A similar technique can be used to find and simplify higher-order derivatives obtained implicitly.

EXAMPLE 7 Finding the Second Derivative Implicitly

Given $x^2 + y^2 = 25$, find $\dfrac{d^2y}{dx^2}$.

Solution Differentiating each term with respect to x produces

$$2x + 2y\frac{dy}{dx} = 0$$

$$2y\frac{dy}{dx} = -2x$$

$$\frac{dy}{dx} = \frac{-2x}{2y} = -\frac{x}{y}.$$

Differentiating a second time with respect to x yields

$$\frac{d^2y}{dx^2} = -\frac{(y)(1) - (x)(dy/dx)}{y^2} \qquad \text{Quotient Rule}$$

$$= -\frac{y - (x)(-x/y)}{y^2} \qquad \text{Substitute } -x/y \text{ for } \frac{dy}{dx}$$

$$= -\frac{y^2 + x^2}{y^3} \qquad \text{Simplify.}$$

$$= -\frac{25}{y^3}. \qquad \text{Substitute 25 for } x^2 + y^2.$$

EXAMPLE 8 Finding a Tangent Line to a Graph

Find the tangent line to the graph given by $x^2(x^2 + y^2) = y^2$ at the point $\left(\sqrt{2}/2, \sqrt{2}/2\right)$, as shown in Figure 3.32.

Solution By rewriting and differentiating implicitly, you obtain

$$x^4 + x^2y^2 - y^2 = 0$$

$$4x^3 + x^2\left(2y\frac{dy}{dx}\right) + 2xy^2 - 2y\frac{dy}{dx} = 0$$

$$2y(x^2 - 1)\frac{dy}{dx} = -2x(2x^2 + y^2)$$

$$\frac{dy}{dx} = \frac{x(2x^2 + y^2)}{y(1 - x^2)}.$$

At the point $\left(\sqrt{2}/2, \sqrt{2}/2\right)$, the slope is

$$\frac{dy}{dx} = \frac{\left(\sqrt{2}/2\right)[2(1/2) + (1/2)]}{\left(\sqrt{2}/2\right)[1 - (1/2)]} = \frac{3/2}{1/2} = 3$$

and the equation of the tangent line at this point is

$$y - \frac{\sqrt{2}}{2} = 3\left(x - \frac{\sqrt{2}}{2}\right)$$

$$y = 3x - \sqrt{2}.$$

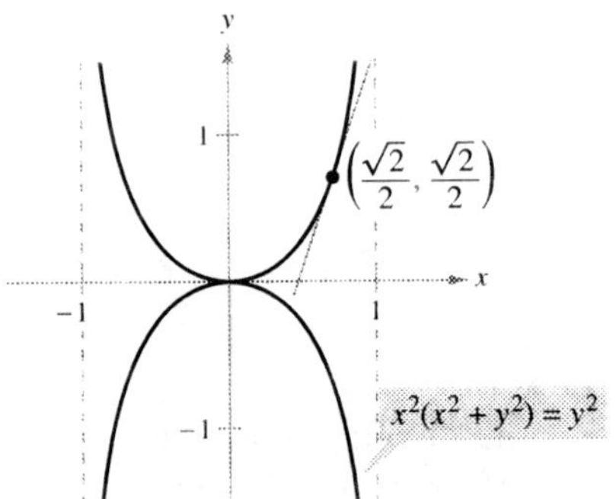

Kappa curve
Figure 3.32

Logarithmic Differentiation

On occasion, it is convenient to use logarithms as aids in differentiating nonlogarithmic functions. This procedure is called **logarithmic differentiation.**

EXAMPLE 9 Logarithmic Differentiation

Find the derivative of $y = \dfrac{(x - 2)^2}{\sqrt{x^2 + 1}}, \; x \neq 2.$

Solution Note that $y > 0$ and so $\ln y$ is defined. Begin by taking the natural logarithms of both sides of the equation. Then apply logarithmic properties and differentiate implicitly. Finally, solve for y'.

$$\ln y = \ln \frac{(x - 2)^2}{\sqrt{x^2 + 1}} \qquad \text{Take ln of both sides.}$$

$$\ln y = 2 \ln(x - 2) - \frac{1}{2} \ln(x^2 + 1) \qquad \text{Logarithmic properties}$$

$$\frac{y'}{y} = 2\left(\frac{1}{x - 2}\right) - \frac{1}{2}\left(\frac{2x}{x^2 + 1}\right) \qquad \text{Differentiate.}$$

$$= \frac{x^2 + 2x + 2}{(x - 2)(x^2 + 1)} \qquad \text{Simplify.}$$

$$y' = y\left[\frac{x^2 + 2x + 2}{(x - 2)(x^2 + 1)}\right] \qquad \text{Solve for } y'.$$

$$= \frac{(x - 2)^2}{\sqrt{x^2 + 1}}\left[\frac{x^2 + 2x + 2}{(x - 2)(x^2 + 1)}\right] \qquad \text{Substitute for } y.$$

$$= \frac{(x - 2)(x^2 + 2x + 2)}{(x^2 + 1)^{3/2}} \qquad \text{Simplify.} \qquad \blacksquare$$

3.5 Exercises

See www.CalcChat.com for worked-out solutions to odd-numbered exercises.

In Exercises 1–22, find dy/dx by implicit differentiation.

1. $x^2 + y^2 = 9$

2. $x^2 - y^2 = 25$

3. $x^{1/2} + y^{1/2} = 16$

4. $x^3 + y^3 = 64$

5. $x^3 - xy + y^2 = 7$

6. $x^2 y + y^2 x = -3$

7. $xe^y - 10x + 3y = 0$

8. $e^{xy} + x^2 - y^2 = 10$

9. $x^3 y^3 - y = x$

10. $\sqrt{xy} = x^2 y + 1$

11. $x^3 - 3x^2 y + 2xy^2 = 12$

12. $4 \cos x \sin y = 1$

13. $\sin x + 2 \cos 2y = 1$

14. $(\sin \pi x + \cos \pi y)^2 = 2$

15. $\sin x = x(1 + \tan y)$

16. $\cot y = x - y$

17. $y = \sin(xy)$

18. $x = \sec \dfrac{1}{y}$

19. $x^2 - 3 \ln y + y^2 = 10$

20. $\ln xy + 5x = 30$

21. $4x^3 + \ln y^2 + 2y = 2x$

22. $4xy + \ln x^2 y = 7$

In Exercises 23–26, (a) find two explicit functions by solving the equation for y in terms of x, (b) sketch the graph of the equation and label the parts given by the corresponding explicit functions, (c) differentiate the explicit functions, and (d) find dy/dx implicitly and show that the result is equivalent to that of part (c).

23. $x^2 + y^2 = 64$

24. $x^2 + y^2 - 4x + 6y + 9 = 0$

25. $16x^2 + 25y^2 = 400$

26. $16y^2 - x^2 = 16$

In Exercises 27–36, find dy/dx **by implicit differentiation and evaluate the derivative at the given point.**

Equation	Point
27. $xy = 6$	$(-6, -1)$
28. $x^3 - y^2 = 0$	$(1, 1)$
29. $y^2 = \dfrac{x^2 - 49}{x^2 + 49}$	$(7, 0)$
30. $(x + y)^3 = x^3 + y^3$	$(-1, 1)$
31. $x^{2/3} + y^{2/3} = 5$	$(8, 1)$
32. $x^3 + y^3 = 6xy - 1$	$(2, 3)$
33. $\tan(x + y) = x$	$(0, 0)$
34. $x \cos y = 1$	$\left(2, \dfrac{\pi}{3}\right)$
35. $3e^{xy} - x = 0$	$(3, 0)$
36. $y^2 = \ln x$	$(e, 1)$

Famous Curves **In Exercises 37–40, find the slope of the tangent line to the graph at the given point.**

37. Witch of Agnesi:

$(x^2 + 4)y = 8$

Point: $(2, 1)$

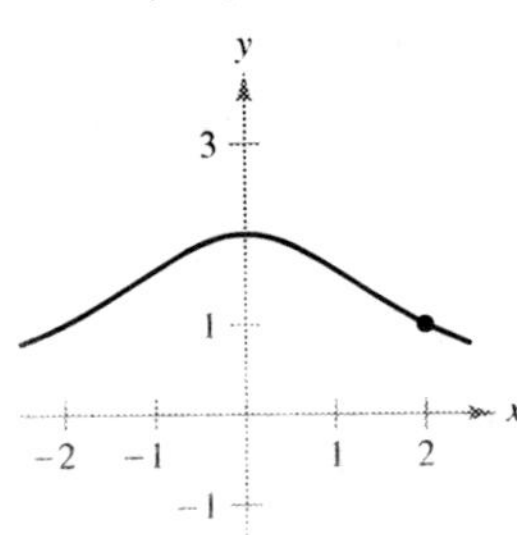

38. Cissoid:

$(4 - x)y^2 = x^3$

Point: $(2, 2)$

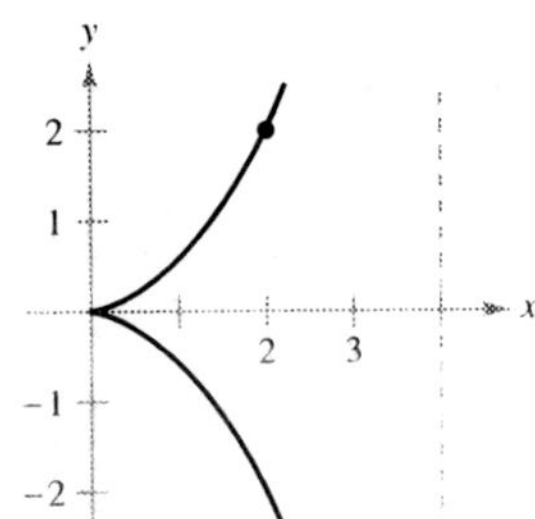

39. Bifolium:

$(x^2 + y^2)^2 = 4x^2 y$

Point: $(1, 1)$

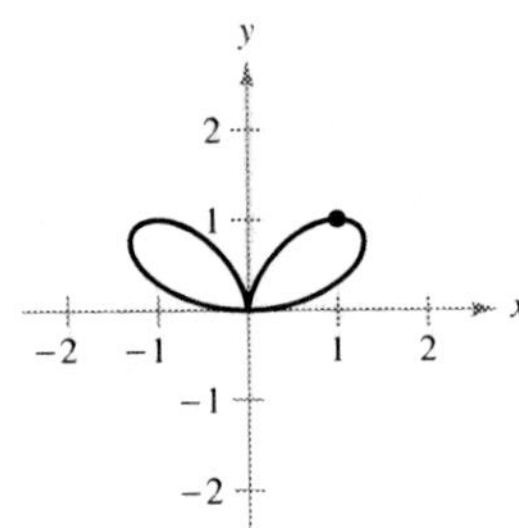

40. Folium of Descartes:

$x^3 + y^3 - 6xy = 0$

Point: $\left(\frac{4}{3}, \frac{8}{3}\right)$

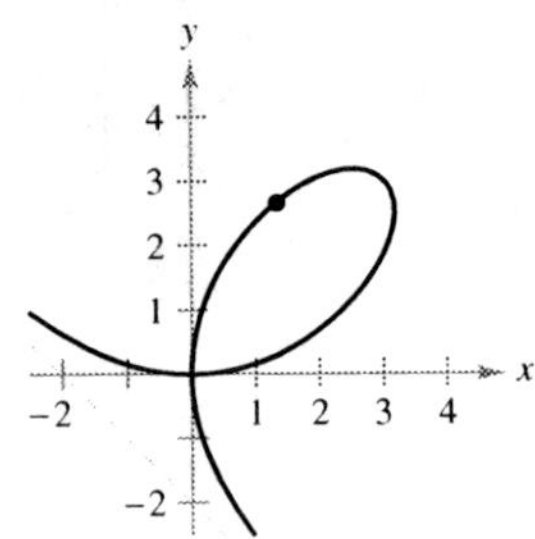

In Exercises 41–44, use implicit differentiation to find an equation of the tangent line to the graph at the given point.

41. $4xy = 9$, $\left(1, \frac{9}{4}\right)$

42. $x^2 + xy + y^2 = 4$, $(2, 0)$

43. $x + y - 1 = \ln(x^2 + y^2)$, $(1, 0)$

44. $y^2 + \ln xy = 2$, $(e, 1)$

Famous Curves **In Exercises 45–52, find an equation of the tangent line to the graph at the given point. To print an enlarged copy of the graph, go to the website *www.mathgraphs.com*.**

45. Parabola

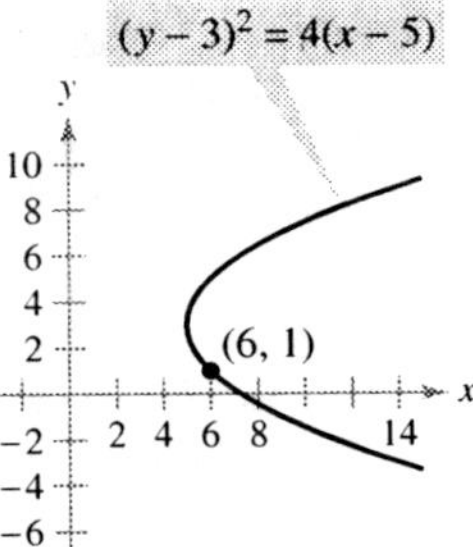

46. Circle

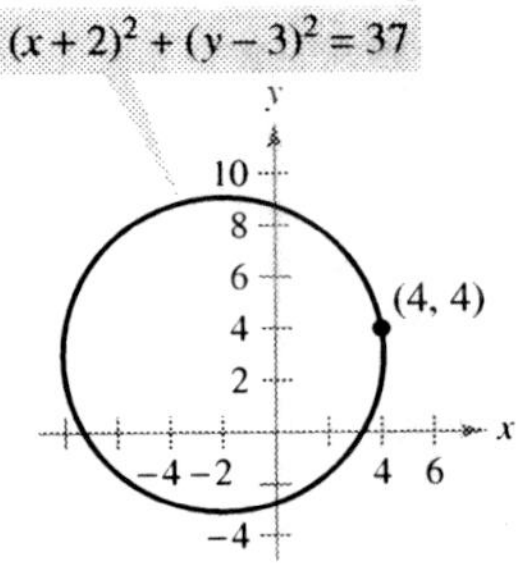

47. Rotated hyperbola

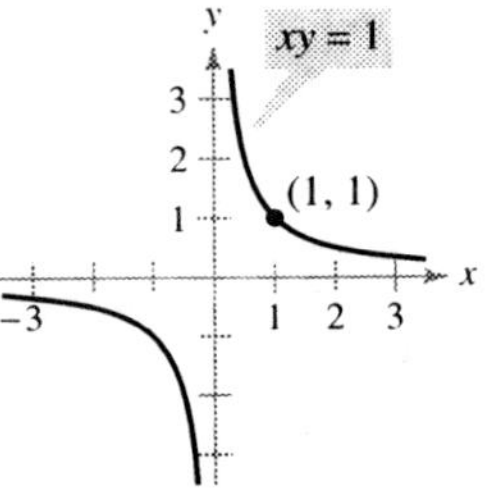

48. Rotated ellipse

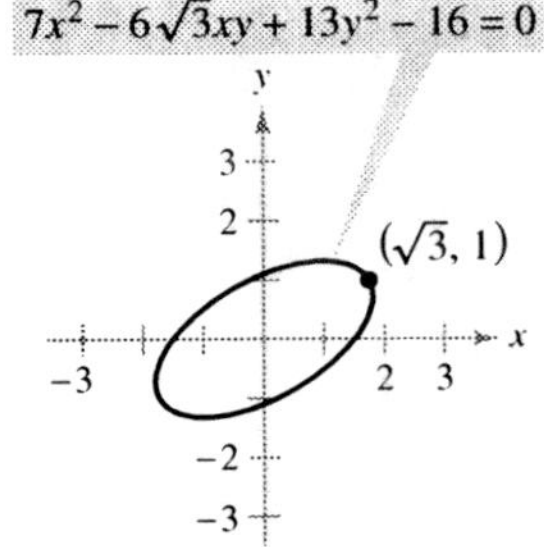

49. Cruciform

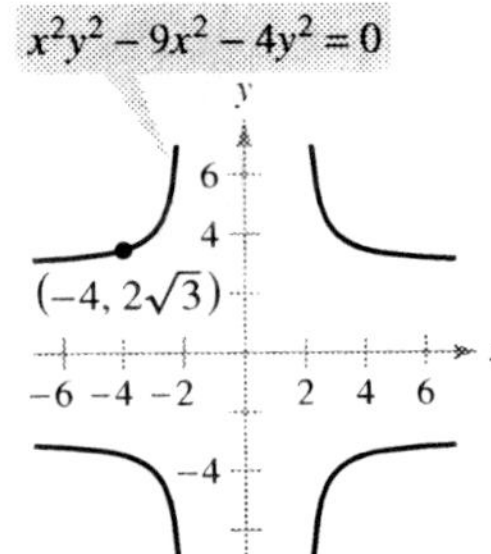

50. Astroid

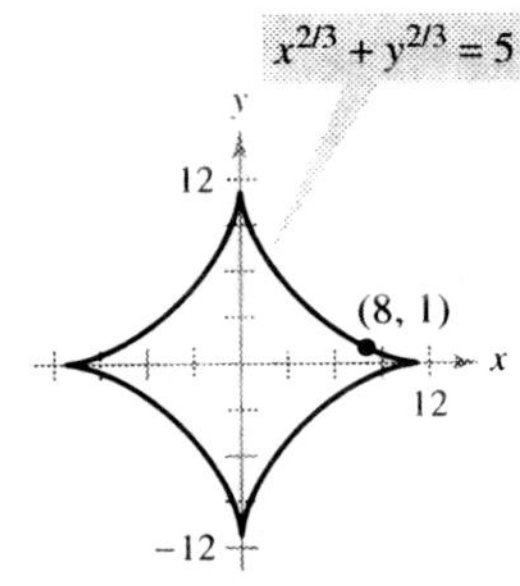

51. Lemniscate

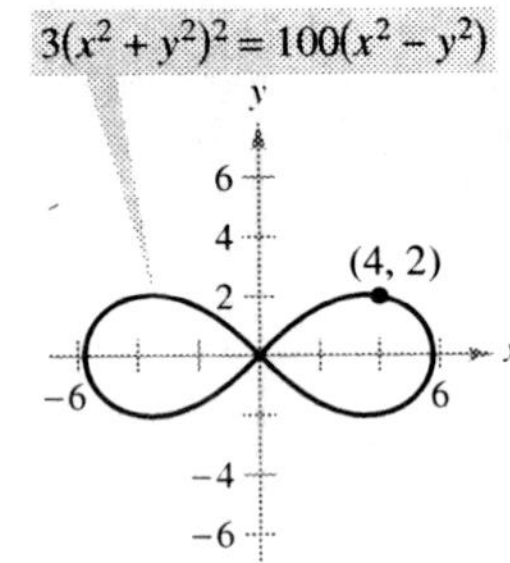

52. Kappa curve

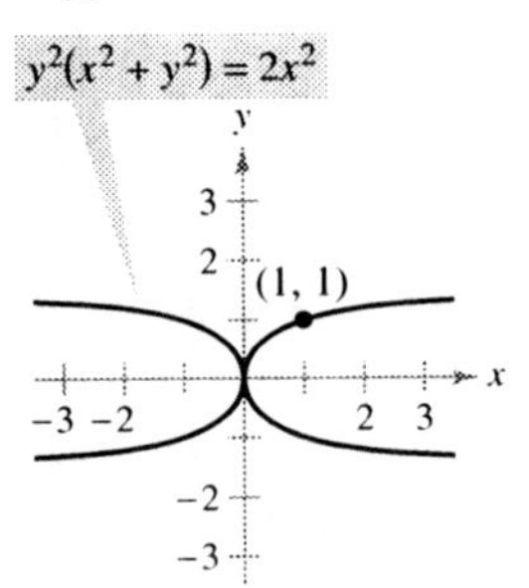

53. (a) Use implicit differentiation to find an equation of the tangent line to the ellipse $\dfrac{x^2}{2} + \dfrac{y^2}{8} = 1$ at $(1, 2)$.

(b) Show that the equation of the tangent line to the ellipse $\dfrac{x^2}{a^2} + \dfrac{y^2}{b^2} = 1$ at (x_0, y_0) is $\dfrac{x_0 x}{a^2} + \dfrac{y_0 y}{b^2} = 1$.

54. (a) Use implicit differentiation to find an equation of the tangent line to the hyperbola $\dfrac{x^2}{6} - \dfrac{y^2}{8} = 1$ at $(3, -2)$.

(b) Show that the equation of the tangent line to the hyperbola $\dfrac{x^2}{a^2} - \dfrac{y^2}{b^2} = 1$ at (x_0, y_0) is $\dfrac{x_0 x}{a^2} - \dfrac{y_0 y}{b^2} = 1$.

In Exercises 55 and 56, find dy/dx implicitly and find the largest interval of the form $-a < y < a$ or $0 < y < a$ such that y is a differentiable function of x. Write dy/dx as a function of x.

55. $\tan y = x$ **56.** $\cos y = x$

In Exercises 57–62, find d^2y/dx^2 in terms of x and y.

57. $x^2 + y^2 = 4$ **58.** $x^2 y^2 - 2x = 3$

59. $x^2 - y^2 = 36$ **60.** $1 - xy = x - y$

61. $y^2 = x^3$ **62.** $y^2 = 10x$

In Exercises 63 and 64, use a graphing utility to graph the equation. Find an equation of the tangent line to the graph at the given point and graph the tangent line in the same viewing window.

63. $\sqrt{x} + \sqrt{y} = 5$, $(9, 4)$ **64.** $y^2 = \dfrac{x-1}{x^2+1}$, $\left(2, \dfrac{\sqrt{5}}{5}\right)$

In Exercises 65 and 66, find equations of the tangent line and normal line to the circle at each given point. (The *normal line* at a point is perpendicular to the tangent line at the point.) Use a graphing utility to graph the equation, tangent line, and normal line.

65. $x^2 + y^2 = 25$ **66.** $x^2 + y^2 = 36$

     $(4, 3), (-3, 4)$ $(6, 0), \left(5, \sqrt{11}\right)$

67. Show that the normal line at any point on the circle $x^2 + y^2 = r^2$ passes through the origin.

68. Two circles of radius 4 are tangent to the graph of $y^2 = 4x$ at the point $(1, 2)$. Find equations of these two circles.

In Exercises 69 and 70, find the points at which the graph of the equation has a vertical or horizontal tangent line.

69. $25x^2 + 16y^2 + 200x - 160y + 400 = 0$

70. $4x^2 + y^2 - 8x + 4y + 4 = 0$

In Exercises 71–82, find dy/dx using logarithmic differentiation.

71. $y = x\sqrt{x^2 + 1}$, $x > 0$

72. $y = \sqrt{x^2(x + 1)(x + 2)}$, $x > 0$

73. $y = \dfrac{x^2\sqrt{3x - 2}}{(x + 1)^2}$, $x > \dfrac{2}{3}$ **74.** $y = \sqrt{\dfrac{x^2 - 1}{x^2 + 1}}$, $x > 1$

75. $y = \dfrac{x(x - 1)^{3/2}}{\sqrt{x + 1}}$, $x > 1$ **76.** $y = \dfrac{(x + 1)(x - 2)}{(x - 1)(x + 2)}$, $x > 2$

77. $y = x^{2/x}$, $x > 0$ **78.** $y = x^{x - 1}$, $x > 0$

79. $y = (x - 2)^{x + 1}$, $x > 2$ **80.** $y = (1 + x)^{1/x}$, $x > 0$

81. $y = x^{\ln x}$, $x > 0$ **82.** $y = (\ln x)^{\ln x}$, $x > 1$

Orthogonal Trajectories **In Exercises 83–86, use a graphing utility to graph the intersecting graphs of the equations and show that they are orthogonal. [Two graphs are *orthogonal* if at their point(s) of intersection their tangent lines are perpendicular to each other.]**

83. $2x^2 + y^2 = 6$ **84.** $y^2 = x^3$

     $y^2 = 4x$ $2x^2 + 3y^2 = 5$

85. $x + y = 0$ **86.** $x^3 = 3(y - 1)$

     $x = \sin y$ $x(3y - 29) = 3$

Orthogonal Trajectories **In Exercises 87 and 88, verify that the two families of curves are orthogonal, where C and K are real numbers. Use a graphing utility to graph the two families for two values of C and two values of K.**

87. $xy = C$, $x^2 - y^2 = K$

88. $x^2 + y^2 = C^2$, $y = Kx$

In Exercises 89–92, differentiate (a) with respect to x (y is a function of x) and (b) with respect to t (x and y are functions of t).

89. $2y^2 - 3x^4 = 0$

90. $x^2 - 3xy^2 + y^3 = 10$

91. $\cos \pi y - 3 \sin \pi x = 1$

92. $4 \sin x \cos y = 1$

WRITING ABOUT CONCEPTS

93. Describe the difference between the explicit form of a function and an implicit equation. Give an example of each.

94. In your own words, state the guidelines for implicit differentiation.

---

95. *Orthogonal Trajectories* The figure below shows the topographic map carried by a group of hikers. The hikers are in a wooded area on top of the hill shown on the map and they decide to follow a path of steepest descent (orthogonal trajectories to the contours on the map). Draw their routes if they start from point A and if they start from point B. If their goal is to reach the road along the top of the map, which starting point should they use? To print an enlarged copy of the map, go to the website *www.mathgraphs.com*.

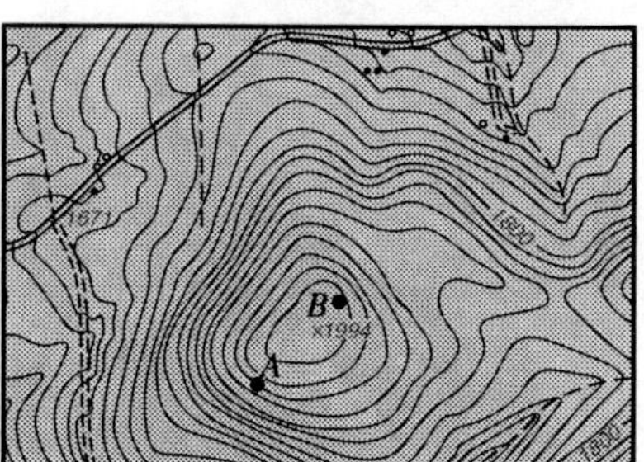

96. *Weather Map* The weather map shows several *isobars*—curves that represent areas of constant air pressure. Three high pressures H and one low pressure L are shown on the map. Given that wind speed is greatest along the orthogonal trajectories of the isobars, use the map to determine the areas having high wind speed.

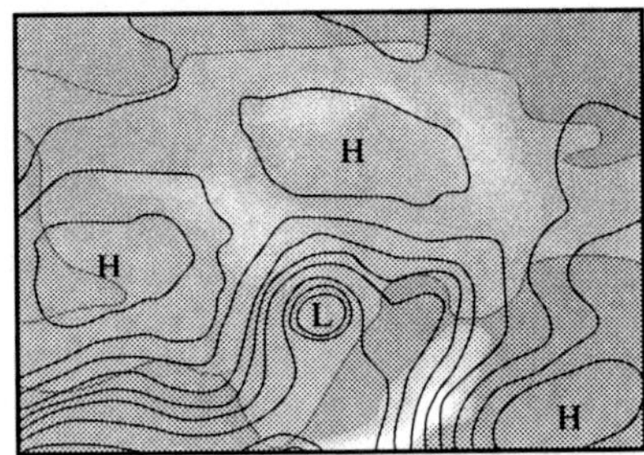

97. Consider the equation $x^4 = 4(4x^2 - y^2)$.

(a) Use a graphing utility to graph the equation.

(b) Find and graph the four tangent lines to the curve for $y = 3$.

(c) Find the exact coordinates of the point of intersection of the two tangent lines in the first quadrant.

CAPSTONE

98. Determine if the statement is true. If it is false, explain why and correct it. For each statement, assume y is a function of x.

(a) $\dfrac{d}{dx}\cos(x^2) = -2x \sin(x^2)$ (b) $\dfrac{d}{dy}\cos(y^2) = 2y \sin(y^2)$

(c) $\dfrac{d}{dx}\cos(y^2) = -2y \sin(y^2)$

99. Let L be any tangent line to the curve $\sqrt{x} + \sqrt{y} = \sqrt{c}$. Show that the sum of the x- and y-intercepts of L is c.

100. (a) Prove (Theorem 3.3) that $d/dx\,[x^n] = nx^{n-1}$ for the case in which n is a rational number. (*Hint:* Write $y = x^{p/q}$ in the form $y^q = x^p$ and differentiate implicitly. Assume that p and q are integers, where $q > 0$.)

(b) Prove part (a) for the case in which n is an irrational number. (*Hint:* Let $y = x^r$, where r is a real number, and use logarithmic differentiation.)

101. *Slope* Find all points on the circle $x^2 + y^2 = 100$ where the slope is $\frac{3}{4}$.

102. *Horizontal Tangent Line* Determine the point(s) at which the graph of $y^4 = y^2 - x^2$ has a horizontal tangent line.

103. *Tangent Lines* Find equations of both tangent lines to the ellipse $\dfrac{x^2}{4} + \dfrac{y^2}{9} = 1$ that passes through the point $(4, 0)$.

104. *Normals to a Parabola* The graph shows the normal lines from the point $(2, 0)$ to the graph of the parabola $x = y^2$. How many normal lines are there from the point $(x_0, 0)$ to the graph of the parabola if (a) $x_0 = \frac{1}{4}$, (b) $x_0 = \frac{1}{2}$, and (c) $x_0 = 1$? For what value of x_0 are two of the normal lines perpendicular to each other?

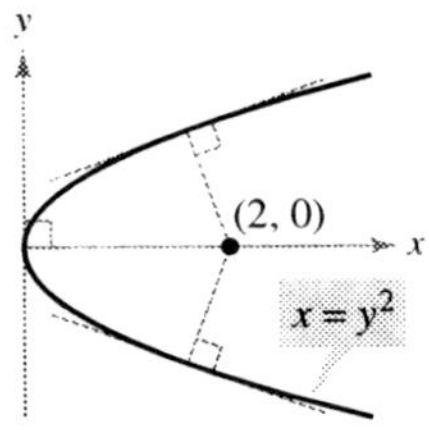

105. *Normal Lines* (a) Find an equation of the normal line to the ellipse $\dfrac{x^2}{32} + \dfrac{y^2}{8} = 1$ at the point $(4, 2)$. (b) Use a graphing utility to graph the ellipse and the normal line. (c) At what other point does the normal line intersect the ellipse?

SECTION PROJECT

Optical Illusions

In each graph below, an optical illusion is created by having lines intersect a family of curves. In each case, the lines appear to be curved. Find the value of dy/dx for the given values of x and y.

(a) Circles: $x^2 + y^2 = C^2$
$x = 3, y = 4, C = 5$

(b) Hyperbolas: $xy = C$
$x = 1, y = 4, C = 4$

(c) Lines: $ax = by$
$x = \sqrt{3}, y = 3,$
$a = \sqrt{3}, b = 1$

(d) Cosine curves: $y = C \cos x$
$x = \dfrac{\pi}{3}, y = \dfrac{1}{3}, C = \dfrac{2}{3}$

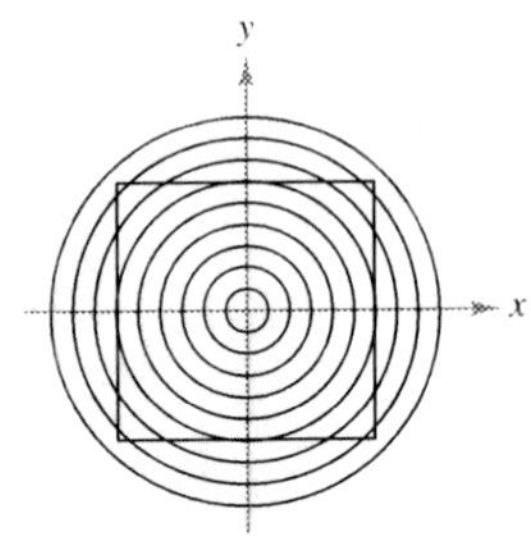
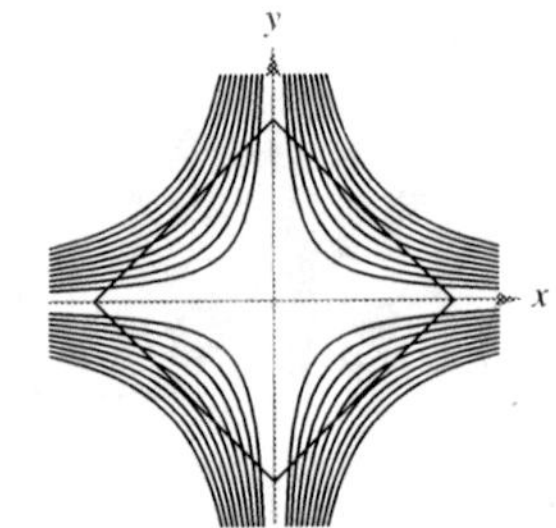
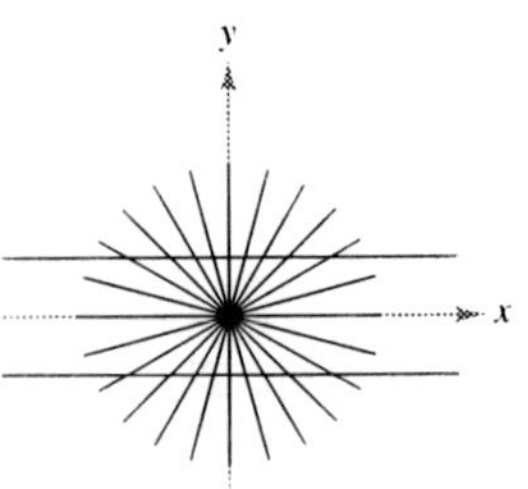
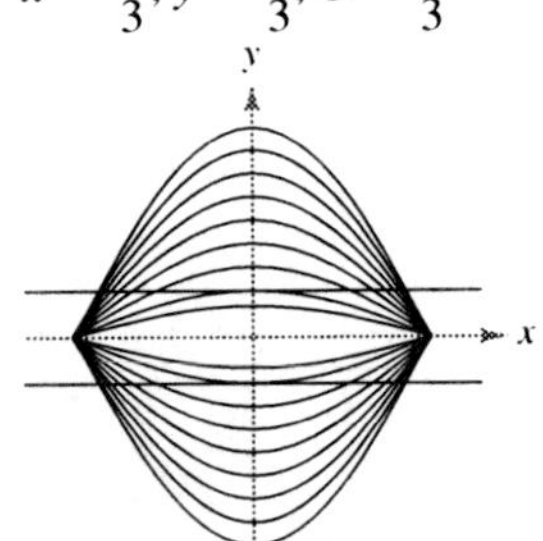

FOR FURTHER INFORMATION For more information on the mathematics of optical illusions, see the article "Descriptive Models for Perception of Optical Illusions" by David A. Smith in *The UMAP Journal*.

3.6 Derivatives of Inverse Functions

- **Find the derivative of an inverse function.**
- **Differentiate an inverse trigonometric function.**
- **Review the basic differentiation rules for elementary functions.**

Derivative of an Inverse Function

The next two theorems discuss the derivative of an inverse function. The reasonableness of Theorem 3.16 follows from the reflective property of inverse functions, as shown in Figure 3.33. Proofs of the two theorems are given in Appendix A.

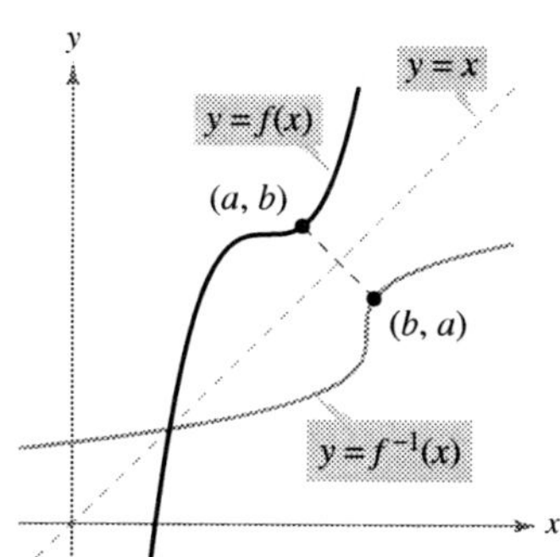

The graph of f^{-1} is a reflection of the graph of f in the line $y = x$.

Figure 3.33

THEOREM 3.16 CONTINUITY AND DIFFERENTIABILITY OF INVERSE FUNCTIONS

Let f be a function whose domain is an interval I. If f has an inverse function, then the following statements are true.

1. If f is continuous on its domain, then f^{-1} is continuous on its domain.
2. If f is differentiable on an interval containing c and $f'(c) \neq 0$, then f^{-1} is differentiable at $f(c)$.

THEOREM 3.17 THE DERIVATIVE OF AN INVERSE FUNCTION

Let f be a function that is differentiable on an interval I. If f has an inverse function g, then g is differentiable at any x for which $f'(g(x)) \neq 0$. Moreover,

$$g'(x) = \frac{1}{f'(g(x))}, \quad f'(g(x)) \neq 0.$$

EXAMPLE 1 Evaluating the Derivative of an Inverse Function

Let $f(x) = \frac{1}{4}x^3 + x - 1$.

a. What is the value of $f^{-1}(x)$ when $x = 3$?

b. What is the value of $(f^{-1})'(x)$ when $x = 3$?

Solution Notice that f is one-to-one and therefore has an inverse function.

a. Because $f(2) = 3$, you know that $f^{-1}(3) = 2$.

b. Because the function f is differentiable and has an inverse function, you can apply Theorem 3.17 (with $g = f^{-1}$) to write

$$(f^{-1})'(3) = \frac{1}{f'(f^{-1}(3))} = \frac{1}{f'(2)}.$$

Moreover, using $f'(x) = \frac{3}{4}x^2 + 1$, you can conclude that

$$(f^{-1})'(3) = \frac{1}{f'(2)}$$
$$= \frac{1}{\frac{3}{4}(2^2) + 1}$$
$$= \frac{1}{4}. \qquad \text{(See Figure 3.34.)}$$

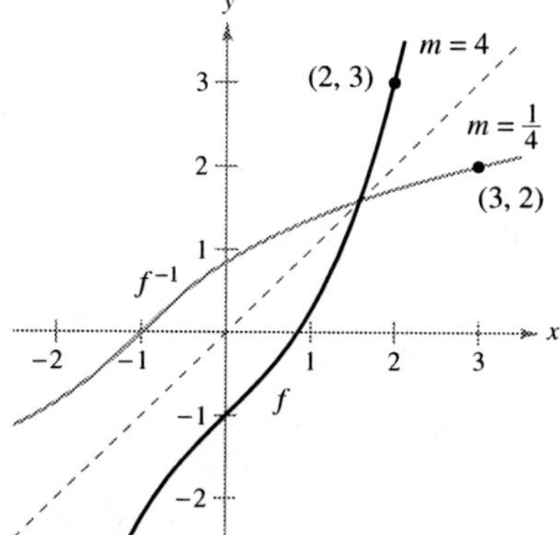

The graphs of the inverse functions f and f^{-1} have reciprocal slopes at points (a, b) and (b, a).

Figure 3.34

In Example 1, note that at the point $(2, 3)$ the slope of the graph of f is 4 and at the point $(3, 2)$ the slope of the graph of f^{-1} is $\frac{1}{4}$ (see Figure 3.34). This reciprocal relationship (which follows from Theorem 3.17) is sometimes written as

$$\frac{dy}{dx} = \frac{1}{dx/dy}.$$

EXAMPLE 2 Graphs of Inverse Functions Have Reciprocal Slopes

Let $f(x) = x^2$ (for $x \geq 0$) and let $f^{-1}(x) = \sqrt{x}$. Show that the slopes of the graphs of f and f^{-1} are reciprocals at each of the following points.

a. $(2, 4)$ and $(4, 2)$ **b.** $(3, 9)$ and $(9, 3)$

Solution   The derivatives of f and f^{-1} are $f'(x) = 2x$ and $(f^{-1})'(x) = \dfrac{1}{2\sqrt{x}}$.

a. At $(2, 4)$, the slope of the graph of f is $f'(2) = 2(2) = 4$. At $(4, 2)$, the slope of the graph of f^{-1} is

$$(f^{-1})'(4) = \frac{1}{2\sqrt{4}} = \frac{1}{2(2)} = \frac{1}{4}.$$

b. At $(3, 9)$, the slope of the graph of f is $f'(3) = 2(3) = 6$. At $(9, 3)$, the slope of the graph of f^{-1} is

$$(f^{-1})'(9) = \frac{1}{2\sqrt{9}} = \frac{1}{2(3)} = \frac{1}{6}.$$

So, in both cases, the slopes are reciprocals, as shown in Figure 3.35. ■

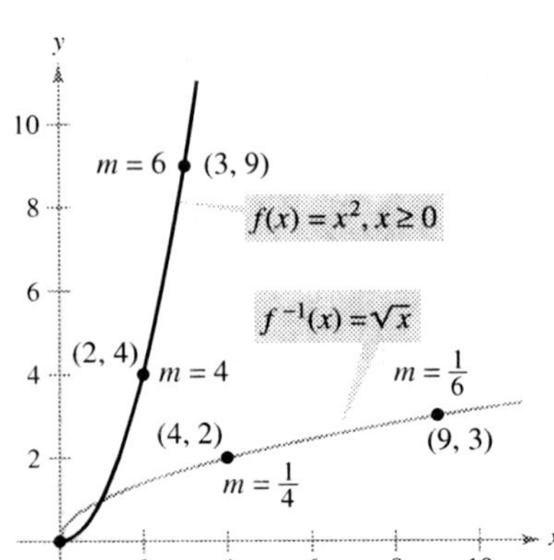

At $(0, 0)$, the derivative of f is 0 and the derivative of f^{-1} does not exist.
Figure 3.35

When determining the derivative of an inverse function, you have two options: (1) you can apply Theorem 3.17, or (2) you can use implicit differentiation. The first approach is illustrated in Example 3, and the second in the proof of Theorem 3.18.

EXAMPLE 3 Finding the Derivative of an Inverse Function

Find the derivative of the inverse tangent function.

Solution   Let $f(x) = \tan x$, $-\pi/2 < x < \pi/2$. Then let $g(x) = \arctan x$ be the inverse tangent function. To find the derivative of $g(x)$, use the fact that $f'(x) = \sec^2 x = \tan^2 x + 1$, and apply Theorem 3.17 as follows.

$$g'(x) = \frac{1}{f'(g(x))} = \frac{1}{f'(\arctan x)} = \frac{1}{[\tan(\arctan x)]^2 + 1} = \frac{1}{x^2 + 1}$$ ■

Derivatives of Inverse Trigonometric Functions

In Section 3.4, you saw that the derivative of the *transcendental* function $f(x) = \ln x$ is the *algebraic* function $f'(x) = 1/x$. You will now see that the derivatives of the inverse trigonometric functions also are algebraic (even though the inverse trigonometric functions are themselves transcendental).

The following theorem lists the derivatives of the six inverse trigonometric functions. Note that the derivatives of arccos u, arccot u, and arccsc u are the *negatives* of the derivatives of arcsin u, arctan u, and arcsec u, respectively.

> ### THEOREM 3.18 DERIVATIVES OF INVERSE TRIGONOMETRIC FUNCTIONS
>
> Let u be a differentiable function of x.
>
> $$\frac{d}{dx}[\arcsin u] = \frac{u'}{\sqrt{1-u^2}} \qquad \frac{d}{dx}[\arccos u] = \frac{-u'}{\sqrt{1-u^2}}$$
>
> $$\frac{d}{dx}[\arctan u] = \frac{u'}{1+u^2} \qquad \frac{d}{dx}[\operatorname{arccot} u] = \frac{-u'}{1+u^2}$$
>
> $$\frac{d}{dx}[\operatorname{arcsec} u] = \frac{u'}{|u|\sqrt{u^2-1}} \qquad \frac{d}{dx}[\operatorname{arccsc} u] = \frac{-u'}{|u|\sqrt{u^2-1}}$$

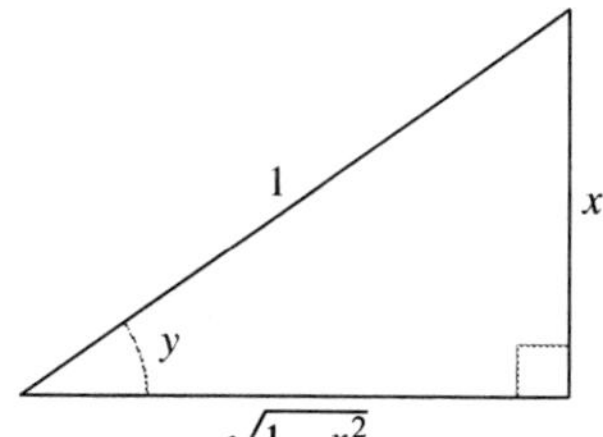

$y = \arcsin x$
Figure 3.36

(**PROOF**) Let $y = \arcsin x$, $-\pi/2 \le y \le \pi/2$ (see Figure 3.36). So, $\sin y = x$, and you can use implicit differentiation as follows.

$$\sin y = x$$

$$(\cos y)\!\left(\frac{dy}{dx}\right) = 1$$

$$\frac{dy}{dx} = \frac{1}{\cos y} = \frac{1}{\sqrt{1 - \sin^2 y}} = \frac{1}{\sqrt{1 - x^2}}$$

Because u is a differentiable function of x, you can use the Chain Rule to write

$$\frac{d}{dx}[\arcsin u] = \frac{u'}{\sqrt{1-u^2}}, \quad \text{where} \quad u' = \frac{du}{dx}.$$

Proofs of the other differentiation rules are left as an exercise (see Exercise 73). ∎

There is no common agreement on the definition of arcsec x (or arccsc x) for negative values of x. When we defined the range of the arcsecant, we chose to preserve the reciprocal identity $\operatorname{arcsec} x = \arccos(1/x)$. For example, to evaluate $\operatorname{arcsec}(-2)$, you can write

$$\operatorname{arcsec}(-2) = \arccos(-0.5)$$
$$\approx 2.09.$$

One of the consequences of the definition of the inverse secant function given in this text is that its graph has a positive slope at every x-value in its domain. This accounts for the absolute value sign in the formula for the derivative of arcsec x.

EXAMPLE 4 A Derivative That Can Be Simplified

Differentiate $y = \arcsin x + x\sqrt{1-x^2}$.

Solution

$$y' = \frac{1}{\sqrt{1-x^2}} + x\!\left(\frac{1}{2}\right)(-2x)(1-x^2)^{-1/2} + \sqrt{1-x^2}$$

$$= \frac{1}{\sqrt{1-x^2}} - \frac{x^2}{\sqrt{1-x^2}} + \sqrt{1-x^2}$$

$$= \sqrt{1-x^2} + \sqrt{1-x^2}$$

$$= 2\sqrt{1-x^2}$$

■

> ### EXPLORATION
>
> Suppose that you want to find a linear approximation to the graph of the function in Example 4. You decide to use the tangent line at the origin, as shown below. Use a graphing utility to describe an interval about the origin where the tangent line is within 0.01 unit of the graph of the function. What might a person mean by saying that the original function is "locally linear"?
>
> 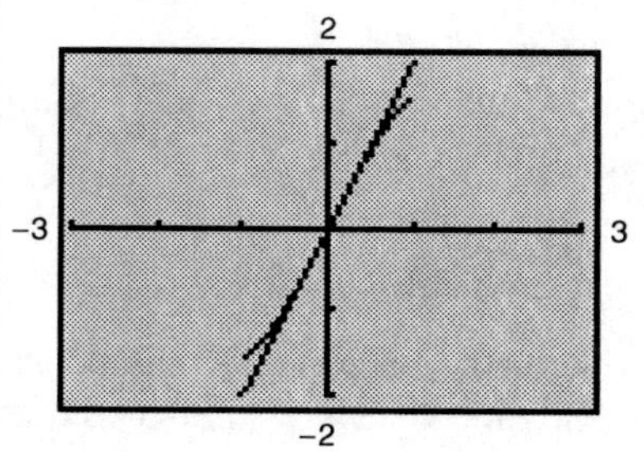
>

EXAMPLE 5 Differentiating Inverse Trigonometric Functions

a. $\dfrac{d}{dx}[\arcsin(2x)] = \dfrac{2}{\sqrt{1-(2x)^2}}$ $\qquad u = 2x$

$\qquad\qquad\quad = \dfrac{2}{\sqrt{1-4x^2}}$

b. $\dfrac{d}{dx}[\arctan(3x)] = \dfrac{3}{1+(3x)^2}$ $\qquad u = 3x$

$\qquad\qquad\quad = \dfrac{3}{1+9x^2}$

c. $\dfrac{d}{dx}\big[\arcsin\sqrt{x}\,\big] = \dfrac{(1/2)x^{-1/2}}{\sqrt{1-x}}$ $\qquad u = \sqrt{x}$

$\qquad\qquad\quad = \dfrac{1}{2\sqrt{x}\sqrt{1-x}}$

$\qquad\qquad\quad = \dfrac{1}{2\sqrt{x-x^2}}$

d. $\dfrac{d}{dx}[\operatorname{arcsec} e^{2x}] = \dfrac{2e^{2x}}{e^{2x}\sqrt{(e^{2x})^2-1}}$ $\qquad u = e^{2x}$

$\qquad\qquad\quad = \dfrac{2e^{2x}}{e^{2x}\sqrt{e^{4x}-1}}$

$\qquad\qquad\quad = \dfrac{2}{\sqrt{e^{4x}-1}}$

In part (d), the absolute value sign is not necessary because $e^{2x} > 0$. ∎

Review of Basic Differentiation Rules

In the 1600s, Europe was ushered into the scientific age by such great thinkers as Descartes, Galileo, Huygens, Newton, and Kepler. These men believed that nature is governed by basic laws—laws that can, for the most part, be written in terms of mathematical equations. One of the most influential publications of this period—*Dialogue on the Great World Systems*, by Galileo Galilei—has become a classic description of modern scientific thought.

As mathematics has developed during the past few hundred years, a small number of elementary functions has proven sufficient for modeling most* phenomena in physics, chemistry, biology, engineering, economics, and a variety of other fields. An **elementary function** is a function from the following list or one that can be formed as the sum, product, quotient, or composition of functions in the list.

Algebraic Functions	*Transcendental Functions*
Polynomial functions	Logarithmic functions
Rational functions	Exponential functions
Functions involving radicals	Trigonometric functions
	Inverse trigonometric functions

With the differentiation rules introduced so far in the text, you can differentiate any elementary function. For convenience, these differentiation rules are summarized on the next page.

GALILEO GALILEI (1564–1642)

Galileo's approach to science departed from the accepted Aristotelian view that nature had describable *qualities*, such as "fluidity" and "potentiality." He chose to describe the physical world in terms of measurable *quantities*, such as time, distance, force, and mass.

---

* *Some important functions used in engineering and science (such as Bessel functions and gamma functions) are not elementary functions.*

BASIC DIFFERENTIATION RULES FOR ELEMENTARY FUNCTIONS

1. $\dfrac{d}{dx}[cu] = cu'$

2. $\dfrac{d}{dx}[u \pm v] = u' \pm v'$

3. $\dfrac{d}{dx}[uv] = uv' + vu'$

4. $\dfrac{d}{dx}\left[\dfrac{u}{v}\right] = \dfrac{vu' - uv'}{v^2}$

5. $\dfrac{d}{dx}[c] = 0$

6. $\dfrac{d}{dx}[u^n] = nu^{n-1}u'$

7. $\dfrac{d}{dx}[x] = 1$

8. $\dfrac{d}{dx}\big[|u|\big] = \dfrac{u}{|u|}(u'), \quad u \neq 0$

9. $\dfrac{d}{dx}[\ln u] = \dfrac{u'}{u}$

10. $\dfrac{d}{dx}[e^u] = e^u u'$

11. $\dfrac{d}{dx}[\log_a u] = \dfrac{u'}{(\ln a)u}$

12. $\dfrac{d}{dx}[a^u] = (\ln a)a^u u'$

13. $\dfrac{d}{dx}[\sin u] = (\cos u)u'$

14. $\dfrac{d}{dx}[\cos u] = -(\sin u)u'$

15. $\dfrac{d}{dx}[\tan u] = (\sec^2 u)u'$

16. $\dfrac{d}{dx}[\cot u] = -(\csc^2 u)u'$

17. $\dfrac{d}{dx}[\sec u] = (\sec u \tan u)u'$

18. $\dfrac{d}{dx}[\csc u] = -(\csc u \cot u)u'$

19. $\dfrac{d}{dx}[\arcsin u] = \dfrac{u'}{\sqrt{1-u^2}}$

20. $\dfrac{d}{dx}[\arccos u] = \dfrac{-u'}{\sqrt{1-u^2}}$

21. $\dfrac{d}{dx}[\arctan u] = \dfrac{u'}{1+u^2}$

22. $\dfrac{d}{dx}[\text{arccot } u] = \dfrac{-u'}{1+u^2}$

23. $\dfrac{d}{dx}[\text{arcsec } u] = \dfrac{u'}{|u|\sqrt{u^2-1}}$

24. $\dfrac{d}{dx}[\text{arccsc } u] = \dfrac{-u'}{|u|\sqrt{u^2-1}}$

3.6 Exercises

See www.CalcChat.com for worked-out solutions to odd-numbered exercises.

In Exercises 1–8, verify that f has an inverse. Then use the function f and the given real number a to find $(f^{-1})'(a)$. (Hint: See Example 1.)

Function	Real Number
1. $f(x) = x^3 - 1$	$a = 26$
2. $f(x) = 5 - 2x^3$	$a = 7$
3. $f(x) = x^3 + 2x - 1$	$a = 2$
4. $f(x) = \frac{1}{27}(x^5 + 2x^3)$	$a = -11$
5. $f(x) = \sin x, \ -\frac{\pi}{2} \le x \le \frac{\pi}{2}$	$a = \frac{1}{2}$
6. $f(x) = \cos 2x, \ 0 \le x \le \frac{\pi}{2}$	$a = 1$
7. $f(x) = \dfrac{x+6}{x-2}, \ x > 2$	$a = 3$
8. $f(x) = \sqrt{x-4}$	$a = 2$

In Exercises 9–12, show that the slopes of the graphs of f and f^{-1} are reciprocals at the given points.

Function	Point
9. $f(x) = x^3$	$\left(\frac{1}{2}, \frac{1}{8}\right)$
$f^{-1}(x) = \sqrt[3]{x}$	$\left(\frac{1}{8}, \frac{1}{2}\right)$
10. $f(x) = 3 - 4x$	$(1, -1)$
$f^{-1}(x) = \dfrac{3-x}{4}$	$(-1, 1)$
11. $f(x) = \sqrt{x-4}$	$(5, 1)$
$f^{-1}(x) = x^2 + 4, \ x \ge 0$	$(1, 5)$

Function	Point
12. $f(x) = \dfrac{4}{1+x^2}, \ x \ge 0$	$(1, 2)$
$f^{-1}(x) = \sqrt{\dfrac{4-x}{x}}$	$(2, 1)$

In Exercises 13–16, (a) find an equation of the tangent line to the graph of f at the given point and (b) use a graphing utility to graph the function and its tangent line at the point.

Function	Point
13. $f(x) = \arccos x^2$	$\left(0, \frac{\pi}{2}\right)$
14. $f(x) = \arctan x$	$\left(-1, -\frac{\pi}{4}\right)$
15. $f(x) = \arcsin 3x$	$\left(\frac{\sqrt{2}}{6}, \frac{\pi}{4}\right)$
16. $f(x) = \text{arcsec } x$	$\left(\sqrt{2}, \frac{\pi}{4}\right)$

In Exercises 17–20, find dy/dx at the given point for the equation.

17. $x = y^3 - 7y^2 + 2, \ (-4, 1)$

18. $x = 2\ln(y^2 - 3), \ (0, 2)$

19. $x \arctan x = e^y, \ \left(1, \ln\frac{\pi}{4}\right)$

20. $\arcsin xy = \frac{2}{3}\arctan 2x, \ \left(\frac{1}{2}, 1\right)$

In Exercises 21–46, find the derivative of the function.

21. $f(x) = \arcsin(x + 1)$

22. $f(t) = \arcsin t^2$

23. $g(x) = 3 \arccos \dfrac{x}{2}$

24. $f(x) = \text{arcsec } 4x$

25. $f(x) = \arctan e^x$

26. $f(x) = \text{arccot} \sqrt{2x}$

27. $g(x) = \dfrac{\arcsin 3x}{x}$

28. $h(x) = x^2 \arctan 5x$

29. $g(x) = \dfrac{\arccos x}{x + 1}$

30. $g(x) = e^{2x} \arcsin x$

31. $h(x) = \text{arccot } 6x$

32. $f(x) = \text{arccsc } 3x$

33. $h(t) = \sin(\arccos t)$

34. $f(x) = \arcsin x + \arccos x$

35. $y = 2x \arccos x - 2\sqrt{1 - x^2}$

36. $y = \ln(t^2 + 4) - \dfrac{1}{2} \arctan \dfrac{t}{2}$

37. $y = \dfrac{1}{2}\left(\dfrac{1}{2} \ln \dfrac{x + 1}{x - 1} + \arctan x \right)$

38. $y = \dfrac{1}{2}\left[x\sqrt{4 - x^2} + 4 \arcsin\left(\dfrac{x}{2}\right) \right]$

39. $g(t) = \tan(\arcsin t)$

40. $f(x) = \text{arcsec } x + \text{arccsc } x$

41. $y = x \arcsin x + \sqrt{1 - x^2}$

42. $y = x \arctan 2x - \dfrac{1}{4} \ln(1 + 4x^2)$

43. $y = 8 \arcsin \dfrac{x}{4} - \dfrac{x\sqrt{16 - x^2}}{2}$

44. $y = 25 \arcsin \dfrac{x}{5} - x\sqrt{25 - x^2}$

45. $y = \arctan x + \dfrac{x}{1 + x^2}$

46. $y = \arctan \dfrac{x}{2} - \dfrac{1}{2(x^2 + 4)}$

In Exercises 47–52, find an equation of the tangent line to the graph of the function at the given point.

47. $y = 2 \arcsin x$

48. $y = \dfrac{1}{2} \arccos x$

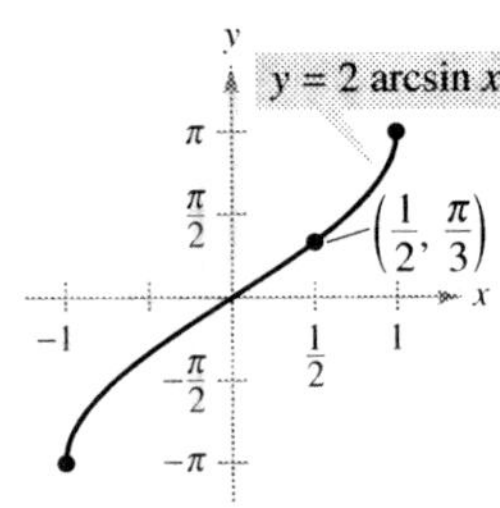

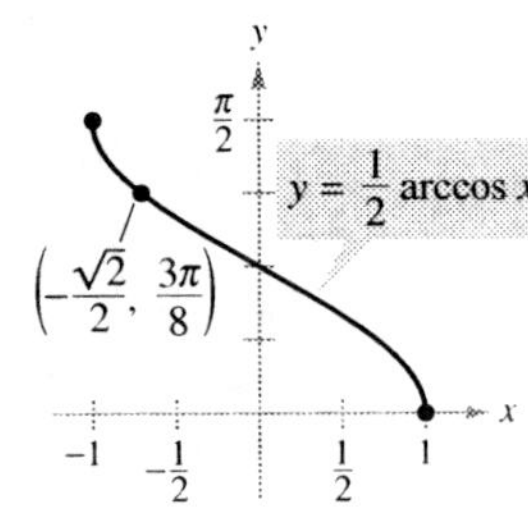

49. $y = \arctan \dfrac{x}{2}$

50. $y = \text{arcsec } 4x$

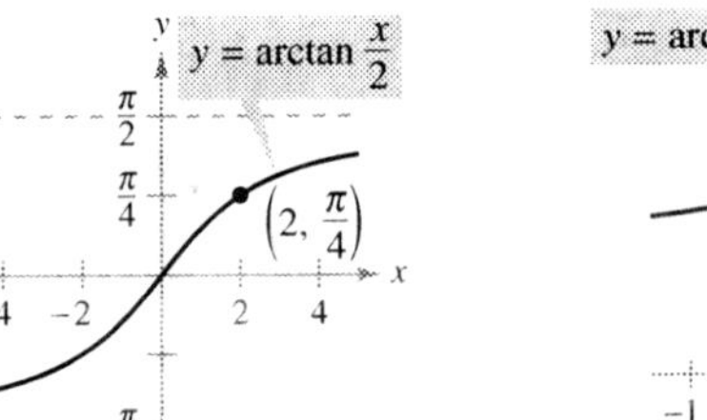

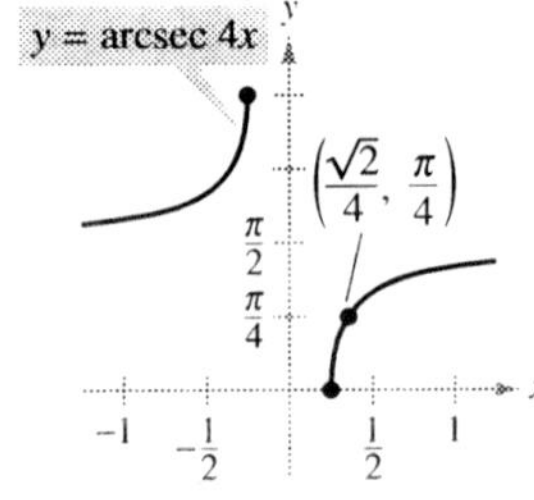

51. $y = 4x \arccos(x - 1)$

52. $y = 3x \arcsin x$

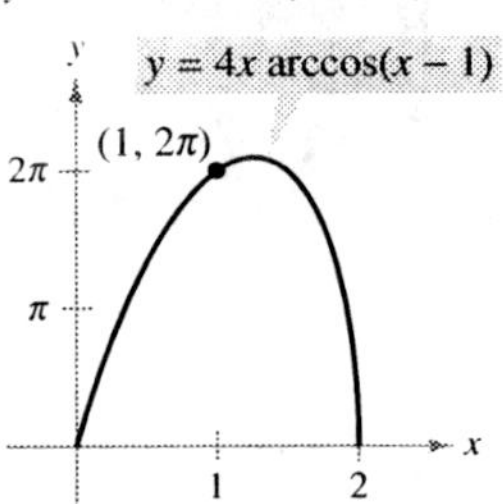

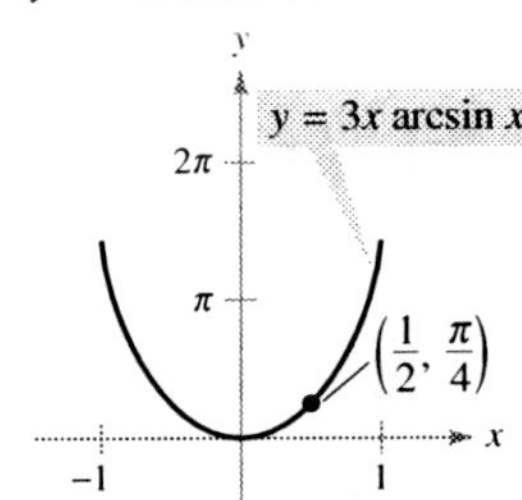

53. Find equations of all tangent lines to the graph of $f(x) = \arccos x$ that have slope -2.

54. Find an equation of the tangent line to the graph of $g(x) = \arctan x$ when $x = 1$.

CAS *Linear and Quadratic Approximations* **In Exercises 55–58, use a computer algebra system to find the linear approximation**

$$P_1(x) = f(a) + f'(a)(x - a)$$

and the quadratic approximation

$$P_2(x) = f(a) + f'(a)(x - a) + \tfrac{1}{2}f''(a)(x - a)^2$$

to the function f at $x = a$. Sketch the graph of the function and its linear and quadratic approximations.

55. $f(x) = \arcsin x, \quad a = \dfrac{1}{2}$

56. $f(x) = \arctan x, \quad a = 1$

57. $f(x) = \arctan x, \quad a = 0$

58. $f(x) = \arccos x, \quad a = 0$

Implicit Differentiation **In Exercises 59–62, find an equation of the tangent line to the graph of the equation at the given point.**

59. $x^2 + x \arctan y = y - 1, \quad \left(-\dfrac{\pi}{4}, 1\right)$

60. $\arctan(xy) = \arcsin(x + y), \quad (0, 0)$

61. $\arcsin x + \arcsin y = \dfrac{\pi}{2}, \quad \left(\dfrac{\sqrt{2}}{2}, \dfrac{\sqrt{2}}{2}\right)$

62. $\arctan(x + y) = y^2 + \dfrac{\pi}{4}, \quad (1, 0)$

WRITING ABOUT CONCEPTS

In Exercises 63 and 64, the derivative of the function has the same sign for all x in its domain, but the function is not one-to-one. Explain.

63. $f(x) = \tan x$

64. $f(x) = \dfrac{x}{x^2 - 4}$

65. State the theorem that gives the method for finding the derivative of an inverse function.

66. Are the derivatives of the inverse trigonometric functions algebraic or transcendental functions? List the derivatives of the inverse trigonometric functions.

67. *Angular Rate of Change* An airplane flies at an altitude of 5 miles toward a point directly over an observer. Consider θ and x as shown in the figure.

(a) Write θ as a function of x.

(b) The speed of the plane is 400 miles per hour. Find $d\theta/dt$ when $x = 10$ miles and $x = 3$ miles.

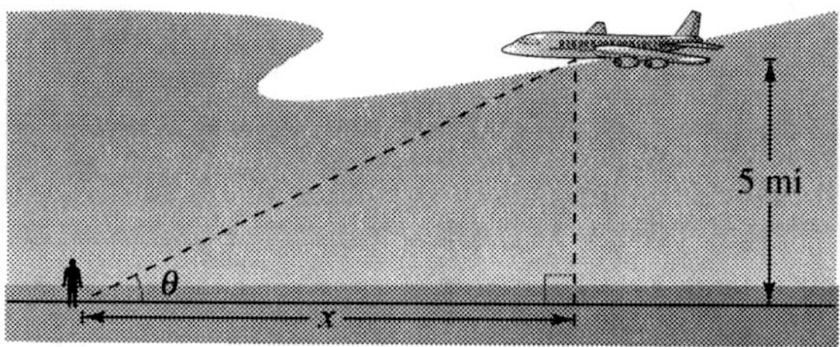

Not drawn to scale

68. *Writing* Repeat Exercise 67 if the altitude of the plane is 3 miles and describe how the altitude affects the rate of change of θ.

69. *Angular Rate of Change* In a free-fall experiment, an object is dropped from a height of 256 feet. A camera on the ground 500 feet from the point of impact records the fall of the object (see figure).

(a) Find the position function giving the height of the object at time t, assuming the object is released at time $t = 0$. At what time will the object reach ground level?

(b) Find the rates of change of the angle of elevation of the camera when $t = 1$ and $t = 2$.

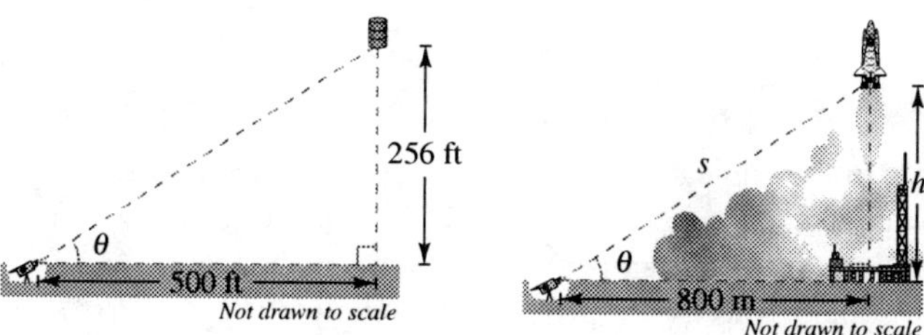

Figure for 69   **Figure for 70**

70. *Angular Rate of Change* A television camera at ground level is filming the lift-off of a space shuttle at a point 800 meters from the launch pad. Let θ be the angle of elevation of the shuttle and let s be the distance between the camera and the shuttle (see figure). Write θ as a function of s for the period of time when the shuttle is moving vertically. Differentiate the result to find $d\theta/dt$ in terms of s and ds/dt.

71. *Angular Rate of Change* An observer is standing 300 feet from the point at which a balloon is released. The balloon rises at a rate of 5 feet per second. How fast is the angle of elevation of the observer's line of sight increasing when the balloon is 100 feet high?

72. *Angular Speed* A patrol car is parked 50 feet from a long warehouse (see figure). The revolving light on top of the car turns at a rate of 30 revolutions per minute. Write θ as a function of x. How fast is the light beam moving along the wall when the beam makes an angle of $\theta = 45°$ with the line perpendicular from the light to the wall?

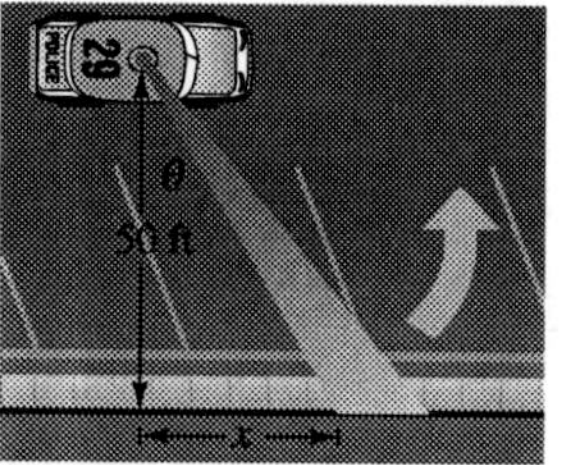

Figure for 72

73. Verify each differentiation formula.

(a) $\dfrac{d}{dx}[\arccos u] = \dfrac{-u'}{\sqrt{1 - u^2}}$

(b) $\dfrac{d}{dx}[\arctan u] = \dfrac{u'}{1 + u^2}$

(c) $\dfrac{d}{dx}[\operatorname{arcsec} u] = \dfrac{u'}{|u|\sqrt{u^2 - 1}}$

(d) $\dfrac{d}{dx}[\operatorname{arccot} u] = \dfrac{-u'}{1 + u^2}$

(e) $\dfrac{d}{dx}[\operatorname{arccsc} u] = \dfrac{-u'}{|u|\sqrt{u^2 - 1}}$

74. *Existence of an Inverse* Determine the values of k such that the function $f(x) = kx + \sin x$ has an inverse function.

True or False? **In Exercises 75 and 76, determine whether the statement is true or false. If it is false, explain why or give an example that shows it is false.**

75. The slope of the graph of the inverse tangent function is positive for all x.

76. $\dfrac{d}{dx}[\arctan(\tan x)] = 1$ for all x in the domain.

77. Prove that $\arcsin x = \arctan\!\left(\dfrac{x}{\sqrt{1 - x^2}}\right)$, $|x| < 1$.

78. Prove that $\arccos x = \dfrac{\pi}{2} - \arctan\!\left(\dfrac{x}{\sqrt{1 - x^2}}\right)$, $|x| < 1$.

79. Some calculus textbooks define the inverse secant function using the range $[0, \pi/2) \cup [\pi, 3\pi/2)$.

(a) Sketch the graph of $y = \operatorname{arcsec} x$ using this range.

(b) Show that $y' = \dfrac{1}{x\sqrt{x^2 - 1}}$.

80. Compare the graphs of $y_1 = \sin(\arcsin x)$ and $y_2 = \arcsin(\sin x)$. What are the domains and ranges of y_1 and y_2?

81. Show that the function $f(x) = \arcsin\dfrac{x - 2}{2} - 2\arcsin\!\left(\dfrac{\sqrt{x}}{2}\right)$ is constant for $0 \le x \le 4$.

CAPSTONE

82. *Think About It* The point $(1, 3)$ lies on the graph of f, and the slope of the tangent line through this point is $m = 2$. Assume f^{-1} exists. What is the slope of the tangent line to the graph of f^{-1} at the point $(3, 1)$?

3.7 Related Rates

- Find a related rate.
- Use related rates to solve real-life problems.

Finding Related Rates

You have seen how the Chain Rule can be used to find dy/dx implicitly. Another important use of the Chain Rule is to find the rates of change of two or more related variables that are changing with respect to *time*.

For example, when water is drained out of a conical tank (see Figure 3.37), the volume V, the radius r, and the height h of the water level are all functions of time t. Knowing that these variables are related by the equation

$$V = \frac{\pi}{3} r^2 h \qquad \text{Original equation}$$

you can differentiate implicitly with respect to t to obtain the **related-rate** equation

$$\frac{d}{dt}(V) = \frac{d}{dt}\left(\frac{\pi}{3} r^2 h\right)$$

$$\frac{dV}{dt} = \frac{\pi}{3}\left[r^2 \frac{dh}{dt} + h\left(2r \frac{dr}{dt}\right)\right] \qquad \text{Differentiate with respect to } t.$$

$$= \frac{\pi}{3}\left(r^2 \frac{dh}{dt} + 2rh \frac{dr}{dt}\right).$$

From this equation you can see that the rate of change of V is related to the rates of change of both h and r.

EXPLORATION

Finding a Related Rate   In the conical tank shown in Figure 3.37, suppose that the height of the water level is changing at a rate of -0.2 foot per minute and the radius is changing at a rate of -0.1 foot per minute. What is the rate of change in the volume when the radius is $r = 1$ foot and the height is $h = 2$ feet? Does the rate of change of the volume depend on the values of r and h? Explain.

EXAMPLE 1   Two Rates That Are Related

Suppose x and y are both differentiable functions of t and are related by the equation $y = x^2 + 3$. Find dy/dt when $x = 1$, given that $dx/dt = 2$ when $x = 1$.

Solution   Using the Chain Rule, you can differentiate both sides of the equation *with respect to t.*

$$y = x^2 + 3 \qquad \text{Write original equation.}$$

$$\frac{d}{dt}[y] = \frac{d}{dt}[x^2 + 3] \qquad \text{Differentiate with respect to } t.$$

$$\frac{dy}{dt} = 2x \frac{dx}{dt} \qquad \text{Chain Rule}$$

When $x = 1$ and $dx/dt = 2$, you have

$$\frac{dy}{dt} = 2(1)(2) = 4.$$

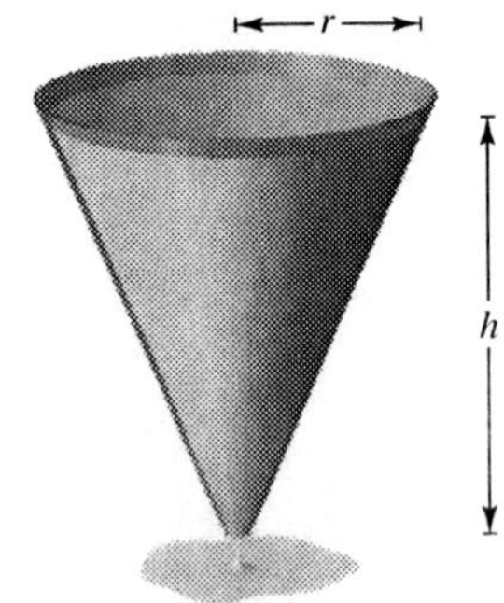
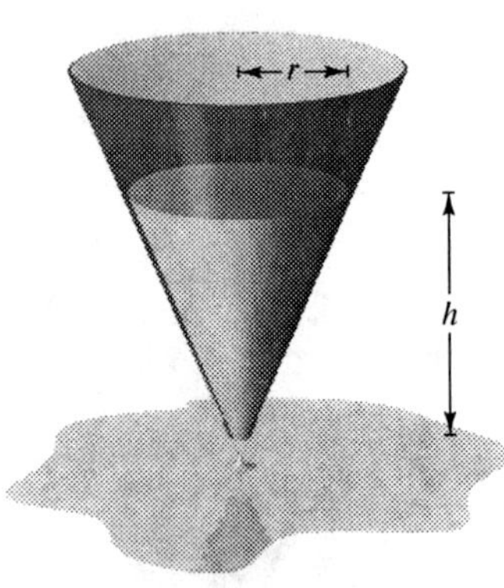
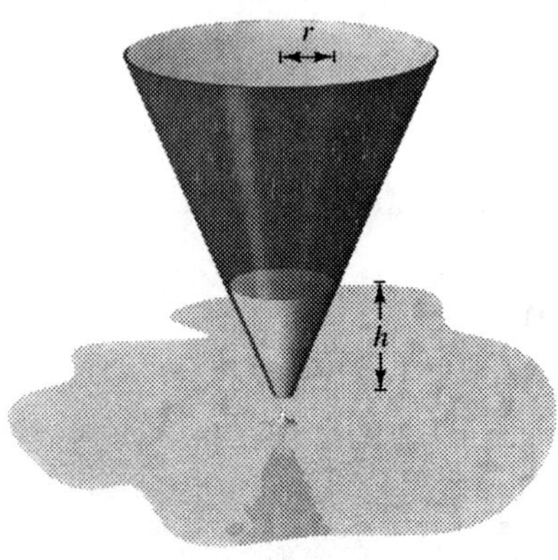

Volume is related to radius and height.
Figure 3.37

■ **FOR FURTHER INFORMATION** To learn more about the history of related-rate problems, see the article "The Lengthening Shadow: The Story of Related Rates" by Bill Austin, Don Barry, and David Berman in *Mathematics Magazine*. To view this article, go to the website *www.matharticles.com*.

Problem Solving with Related Rates

In Example 1, you were *given* an equation that related the variables x and y and were asked to find the rate of change of y when $x = 1$.

Equation: $\quad y = x^2 + 3$

Given rate: $\quad \dfrac{dx}{dt} = 2 \quad$ when $\quad x = 1$

Find: $\quad \dfrac{dy}{dt} \quad$ when $\quad x = 1$

In each of the remaining examples in this section, you must *create* a mathematical model from a verbal description.

EXAMPLE 2 Ripples in a Pond

A pebble is dropped into a calm pond, causing ripples in the form of concentric circles, as shown in Figure 3.38. The radius r of the outer ripple is increasing at a constant rate of 1 foot per second. When the radius is 4 feet, at what rate is the total area A of the disturbed water changing?

Solution The variables r and A are related by $A = \pi r^2$. The rate of change of the radius r is $dr/dt = 1$.

Equation: $\quad A = \pi r^2$

Given rate: $\quad \dfrac{dr}{dt} = 1$

Find: $\quad \dfrac{dA}{dt} \quad$ when $\quad r = 4$

With this information, you can proceed as in Example 1.

$$\frac{d}{dt}[A] = \frac{d}{dt}[\pi r^2] \qquad \text{Differentiate with respect to } t.$$

$$\frac{dA}{dt} = 2\pi r \frac{dr}{dt} \qquad \text{Chain Rule}$$

$$\frac{dA}{dt} = 2\pi(4)(1) = 8\pi \qquad \text{Substitute 4 for } r \text{ and 1 for } dr/dt.$$

When the radius is 4 feet, the area is changing at a rate of 8π square feet per second.

Total area increases as the outer radius increases.
Figure 3.38

GUIDELINES FOR SOLVING RELATED-RATE PROBLEMS

1. Identify all *given* quantities and quantities *to be determined*. Make a sketch and label the quantities.
2. Write an equation involving the variables whose rates of change either are given or are to be determined.
3. Using the Chain Rule, implicitly differentiate both sides of the equation *with respect to time t.*
4. *After* completing Step 3, substitute into the resulting equation all known values for the variables and their rates of change. Then solve for the required rate of change.

NOTE When using these guidelines, be sure you perform Step 3 before Step 4. Substituting the known values of the variables before differentiating will produce an inappropriate derivative.

The table below lists examples of mathematical models involving rates of change. For instance, the rate of change in the first example is the velocity of a car.

Verbal Statement	Mathematical Model
The velocity of a car after traveling for 1 hour is 50 miles per hour.	$x=$ distance traveled $\dfrac{dx}{dt}=50$ when $t=1$
Water is being pumped into a swimming pool at a rate of 10 cubic meters per hour.	$V=$ volume of water in pool $\dfrac{dV}{dt}=10$ m^3/hr
A gear is revolving at a rate of 25 revolutions per minute (1 revolution $=2\pi$ radians).	$\theta=$ angle of revolution $\dfrac{d\theta}{dt}=25(2\pi)$ rad/min

EXAMPLE 3   An Inflating Balloon

Air is being pumped into a spherical balloon (see Figure 3.39) at a rate of 4.5 cubic feet per minute. Find the rate of change of the radius when the radius is 2 feet.

Solution   Let V be the volume of the balloon and let r be its radius. Because the volume is increasing at a rate of 4.5 cubic feet per minute, you know that at time t the rate of change of the volume is $dV/dt = \frac{9}{2}$. So, the problem can be stated as shown.

Given rate: $\dfrac{dV}{dt} = \dfrac{9}{2}$ (constant rate)

Find: $\dfrac{dr}{dt}$ when $r = 2$

To find the rate of change of the radius, you must find an equation that relates the radius r to the volume V.

Equation: $V = \dfrac{4}{3}\pi r^3$ Volume of a sphere

Differentiating both sides of the equation with respect to t produces

$$\frac{dV}{dt} = 4\pi r^2 \frac{dr}{dt} \qquad \text{Differentiate with respect to } t.$$

$$\frac{dr}{dt} = \frac{1}{4\pi r^2}\left(\frac{dV}{dt}\right). \qquad \text{Solve for } dr/dt.$$

Finally, when $r = 2$, the rate of change of the radius is

$$\frac{dr}{dt} = \frac{1}{16\pi}\left(\frac{9}{2}\right) \approx 0.09 \text{ foot per minute.} \qquad \blacksquare$$

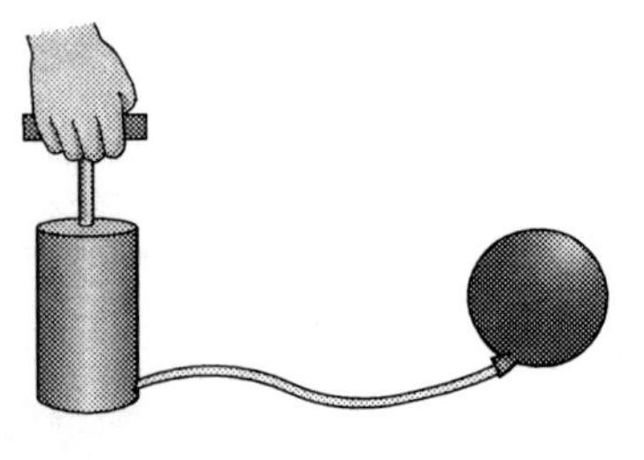
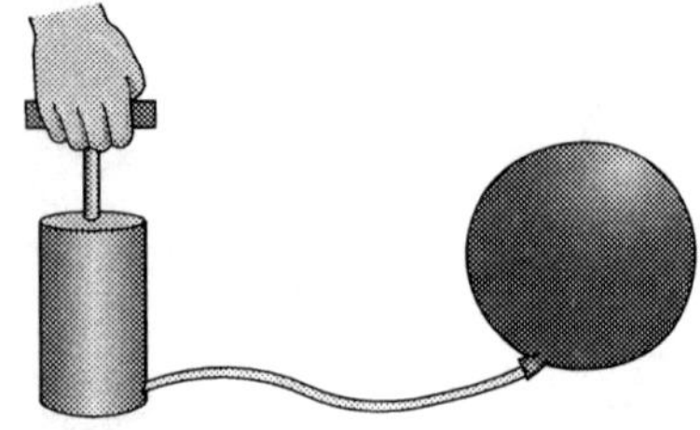
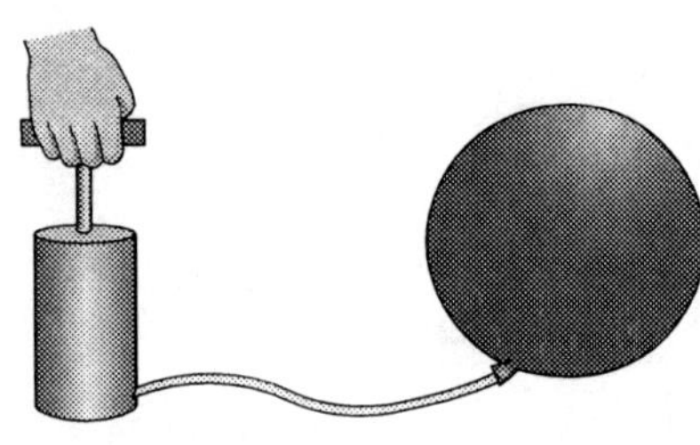

Inflating a balloon
Figure 3.39

In Example 3, note that the volume is increasing at a *constant* rate but the radius is increasing at a *variable* rate. Just because two rates are related does not mean that they are proportional. In this particular case, the radius is growing more and more slowly as t increases. Do you see why?

EXAMPLE 4 The Speed of an Airplane Tracked by Ra...

An airplane is flying on a flight path that will take it directly o...
station, as shown in Figure 3.40. If s is decreasing at a rate of 400 miles pe...
$s = 10$ miles, what is the speed of the plane?

Solution Let x be the horizontal distance from the station, as shown in Figure 3.40.
Notice that when $s = 10$, $x = \sqrt{10^2 - 36} = 8$.

Given rate: $ds/dt = -400$ when $s = 10$

Find: dx/dt when $s = 10$ and $x = 8$

You can find the velocity of the plane as shown.

$$\textbf{Equation:} \quad x^2 + 6^2 = s^2 \qquad \text{Pythagorean Theorem}$$

$$2x\frac{dx}{dt} = 2s\frac{ds}{dt} \qquad \text{Differentiate with respect to } t.$$

$$\frac{dx}{dt} = \frac{s}{x}\left(\frac{ds}{dt}\right) \qquad \text{Solve for } dx/dt.$$

$$\frac{dx}{dt} = \frac{10}{8}(-400) \qquad \text{Substitute for } s, x, \text{ and } ds/dt.$$

$$= -500 \text{ miles per hour} \qquad \text{Simplify.}$$

Because the velocity is -500 miles per hour, the *speed* is 500 miles per hour.

NOTE Note that the velocity in Example 4 is negative because x represents a distance that is decreasing.

An airplane is flying at an altitude of 6 miles,
s miles from the station.
Figure 3.40

EXAMPLE 5 A Changing Angle of Elevation

Find the rate of change in the angle of elevation of the camera shown in Figure 3.41
at 10 seconds after lift-off.

Solution Let θ be the angle of elevation, as shown in Figure 3.41. When $t = 10$, the
height s of the rocket is $s = 50t^2 = 50(10)^2 = 5000$ feet.

Given rate: $ds/dt = 100t = $ velocity of rocket

Find: $d\theta/dt$ when $t = 10$ and $s = 5000$

Using Figure 3.41, you can relate s and θ by the equation $\tan\theta = s/2000$.

$$\textbf{Equation:} \quad \tan\theta = \frac{s}{2000} \qquad \text{See Figure 3.41.}$$

$$(\sec^2\theta)\frac{d\theta}{dt} = \frac{1}{2000}\left(\frac{ds}{dt}\right) \qquad \text{Differentiate with respect to } t.$$

$$\frac{d\theta}{dt} = \cos^2\theta\,\frac{100t}{2000} \qquad \text{Substitute } 100t \text{ for } ds/dt.$$

$$= \left(\frac{2000}{\sqrt{s^2 + 2000^2}}\right)^2 \frac{100t}{2000} \qquad \cos\theta = 2000/\sqrt{s^2 + 2000^2}$$

When $t = 10$ and $s = 5000$, you have

$$\frac{d\theta}{dt} = \frac{2000(100)(10)}{5000^2 + 2000^2} = \frac{2}{29} \text{ radian per second.}$$

So, when $t = 10$, θ is changing at a rate of $\frac{2}{29}$ radian per second.

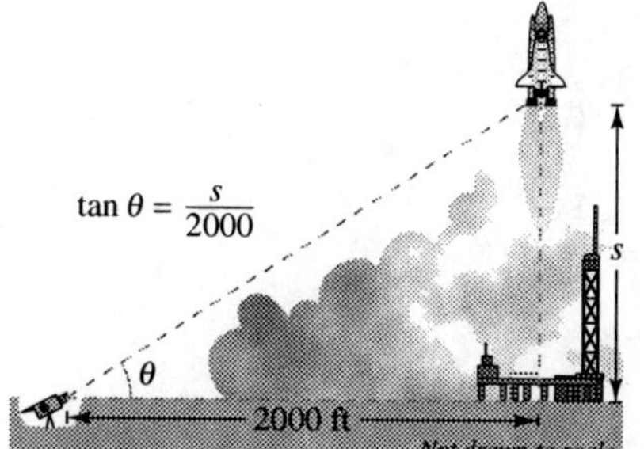

A television camera at ground level is filming
the lift-off of a space shuttle that is rising
vertically according to the position equation
$s = 50t^2$, where s is measured in feet and t is
measured in seconds. The camera is 2000 feet
from the launch pad.
Figure 3.41

EXAMPLE 6 The Velocity of a Piston

In the engine shown in Figure 3.42, a 7-inch connecting rod is fastened to a crank of radius 3 inches. The crankshaft rotates counterclockwise at a constant rate of 200 revolutions per minute. Find the velocity of the piston when $\theta = \pi/3$.

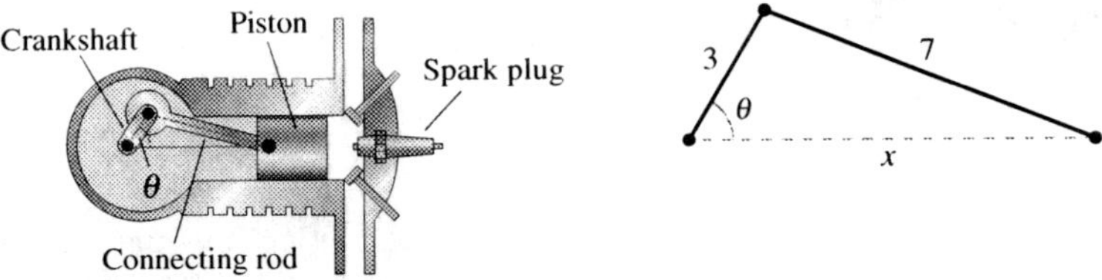

The velocity of a piston is related to the angle of the crankshaft.
Figure 3.42

Solution Label the distances as shown in Figure 3.42. Because a complete revolution corresponds to 2π radians, it follows that $d\theta/dt = 200(2\pi) = 400\pi$ radians per minute.

Given rate: $\dfrac{d\theta}{dt} = 400\pi$ (constant rate)

Find: $\dfrac{dx}{dt}$ when $\theta = \dfrac{\pi}{3}$

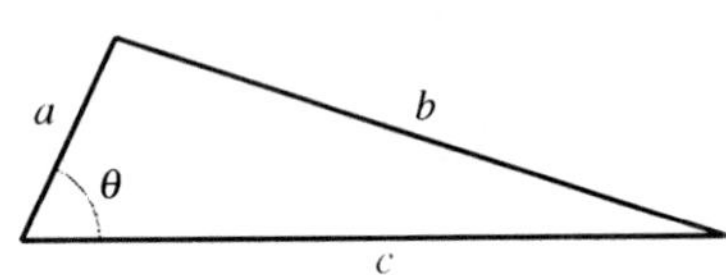

Law of Cosines:
$b^2 = a^2 + c^2 - 2ac \cos \theta$
Figure 3.43

You can use the Law of Cosines (Figure 3.43) to find an equation that relates x and θ.

Equation:
$$7^2 = 3^2 + x^2 - 2(3)(x) \cos \theta$$

$$0 = 2x \frac{dx}{dt} - 6\left(-x \sin \theta \frac{d\theta}{dt} + \cos \theta \frac{dx}{dt}\right)$$

$$(6 \cos \theta - 2x)\frac{dx}{dt} = 6x \sin \theta \frac{d\theta}{dt}$$

$$\frac{dx}{dt} = \frac{6x \sin \theta}{6 \cos \theta - 2x}\left(\frac{d\theta}{dt}\right)$$

When $\theta = \pi/3$, you can solve for x as shown.

$$7^2 = 3^2 + x^2 - 2(3)(x) \cos \frac{\pi}{3}$$

$$49 = 9 + x^2 - 6x\left(\frac{1}{2}\right)$$

$$0 = x^2 - 3x - 40$$
$$0 = (x - 8)(x + 5)$$
$$x = 8 \qquad\qquad \text{Choose positive solution.}$$

So, when $x = 8$ and $\theta = \pi/3$, the velocity of the piston is

$$\frac{dx}{dt} = \frac{6(8)\left(\sqrt{3}/2\right)}{6(1/2) - 16}(400\pi)$$

$$= \frac{9600\pi\sqrt{3}}{-13}$$

$$\approx -4018 \text{ inches per minute.}$$

NOTE Note that the velocity in Example 6 is negative because x represents a distance that is decreasing.

3.7 Exercises

In Exercises 1–4, assume that x and y are both differentiable functions of t and find the required values of dy/dt and dx/dt.

Equation	Find	Given
1. $y = \sqrt{x}$	(a) $\dfrac{dy}{dt}$ when $x = 4$	$\dfrac{dx}{dt} = 3$
	(b) $\dfrac{dx}{dt}$ when $x = 25$	$\dfrac{dy}{dt} = 2$
2. $y = 4(x^2 - 5x)$	(a) $\dfrac{dy}{dt}$ when $x = 3$	$\dfrac{dx}{dt} = 2$
	(b) $\dfrac{dx}{dt}$ when $x = 1$	$\dfrac{dy}{dt} = 5$
3. $xy = 4$	(a) $\dfrac{dy}{dt}$ when $x = 8$	$\dfrac{dx}{dt} = 10$
	(b) $\dfrac{dx}{dt}$ when $x = 1$	$\dfrac{dy}{dt} = -6$
4. $x^2 + y^2 = 25$	(a) $\dfrac{dy}{dt}$ when $x = 3$, $y = 4$	$\dfrac{dx}{dt} = 8$
	(b) $\dfrac{dx}{dt}$ when $x = 4$, $y = 3$	$\dfrac{dy}{dt} = -2$

In Exercises 5–8, a point is moving along the graph of the given function such that dx/dt is 2 centimeters per second. Find dy/dt for the given values of x.

5. $y = 2x^2 + 1$ (a) $x = -1$ (b) $x = 0$ (c) $x = 1$

6. $y = \dfrac{1}{1 + x^2}$ (a) $x = -2$ (b) $x = 0$ (c) $x = 2$

7. $y = \tan x$ (a) $x = -\dfrac{\pi}{3}$ (b) $x = -\dfrac{\pi}{4}$ (c) $x = 0$

8. $y = \cos x$ (a) $x = \dfrac{\pi}{6}$ (b) $x = \dfrac{\pi}{4}$ (c) $x = \dfrac{\pi}{3}$

WRITING ABOUT CONCEPTS

9. Consider the linear function $y = ax + b$. If x changes at a constant rate, does y change at a constant rate? If so, does it change at the same rate as x? Explain.

10. In your own words, state the guidelines for solving related-rate problems.

11. Find the rate of change of the distance between the origin and a moving point on the graph of $y = x^2 + 1$ if $dx/dt = 2$ centimeters per second.

12. Find the rate of change of the distance between the origin and a moving point on the graph of $y = \sin x$ if $dx/dt = 2$ centimeters per second.

13. *Area* The radius r of a circle is increasing at a rate of 4 centimeters per minute. Find the rates of change of the area when (a) $r = 8$ centimeters and (b) $r = 32$ centimeters.

14. *Area* Let A be the area of a circle of radius r that is changing with respect to time. If dr/dt is constant, is dA/dt constant? Explain.

15. *Area* The included angle of the two sides of constant equal length s of an isosceles triangle is θ.

(a) Show that the area of the triangle is given by $A = \frac{1}{2}s^2 \sin \theta$.

(b) If θ is increasing at the rate of $\frac{1}{2}$ radian per minute, find the rates of change of the area when $\theta = \pi/6$ and $\theta = \pi/3$.

(c) Explain why the rate of change of the area of the triangle is not constant even though $d\theta/dt$ is constant.

16. *Volume* The radius r of a sphere is increasing at a rate of 3 inches per minute.

(a) Find the rates of change of the volume when $r = 9$ inches and $r = 36$ inches.

(b) Explain why the rate of change of the volume of the sphere is not constant even though dr/dt is constant.

17. *Volume* A hemispherical water tank with radius 6 meters is filled to a depth of h meters. The volume of water in the tank is given by $V = \frac{1}{3}\pi h(108 - h^2)$, $0 < h < 6$. If water is being pumped into the tank at the rate of 3 cubic meters per minute, find the rate of change of the depth of the water when $h = 2$ meters.

18. *Volume* All edges of a cube are expanding at a rate of 6 centimeters per second. How fast is the volume changing when each edge is (a) 2 centimeters and (b) 10 centimeters?

19. *Surface Area* The conditions are the same as in Exercise 18. Determine how fast the *surface area* is changing when each edge is (a) 2 centimeters and (b) 10 centimeters.

20. *Volume* The formula for the volume of a cone is $V = \frac{1}{3}\pi r^2 h$. Find the rates of change of the volume if dr/dt is 2 inches per minute and $h = 3r$ when (a) $r = 6$ inches and (b) $r = 24$ inches.

21. *Volume* At a sand and gravel plant, sand is falling off a conveyor and onto a conical pile at a rate of 10 cubic feet per minute. The diameter of the base of the cone is approximately three times the altitude. At what rate is the height of the pile changing when the pile is 15 feet high?

22. *Depth* A conical tank (with vertex down) is 10 feet across the top and 12 feet deep. If water is flowing into the tank at a rate of 10 cubic feet per minute, find the rate of change of the depth of the water when the water is 8 feet deep.

23. *Depth* A swimming pool is 12 meters long, 6 meters wide, 1 meter deep at the shallow end, and 3 meters deep at the deep end (see figure on next page). Water is being pumped into the pool at $\frac{1}{4}$ cubic meter per minute, and there is 1 meter of water at the deep end.

(a) What percent of the pool is filled?

(b) At what rate is the water level rising?

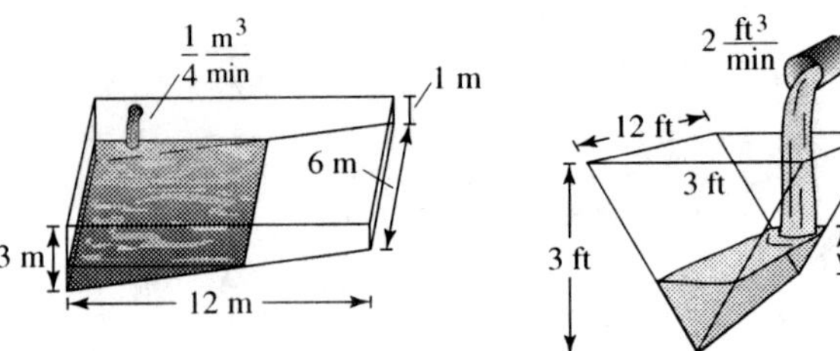

Figure for 23

Figure for 24

24. Depth A trough is 12 feet long and 3 feet across the top (see figure). Its ends are isosceles triangles with altitudes of 3 feet.

(a) If water is being pumped into the trough at 2 cubic feet per minute, how fast is the water level rising when the depth h is 1 foot?

(b) If the water is rising at a rate of $\frac{3}{8}$ inch per minute when $h = 2$, determine the rate at which water is being pumped into the trough.

25. Moving Ladder A ladder 25 feet long is leaning against the wall of a house (see figure). The base of the ladder is pulled away from the wall at a rate of 2 feet per second.

(a) How fast is the top of the ladder moving down the wall when its base is 7 feet, 15 feet, and 24 feet from the wall?

(b) Consider the triangle formed by the side of the house, the ladder, and the ground. Find the rate at which the area of the triangle is changing when the base of the ladder is 7 feet from the wall.

(c) Find the rate at which the angle between the ladder and the wall of the house is changing when the base of the ladder is 7 feet from the wall.

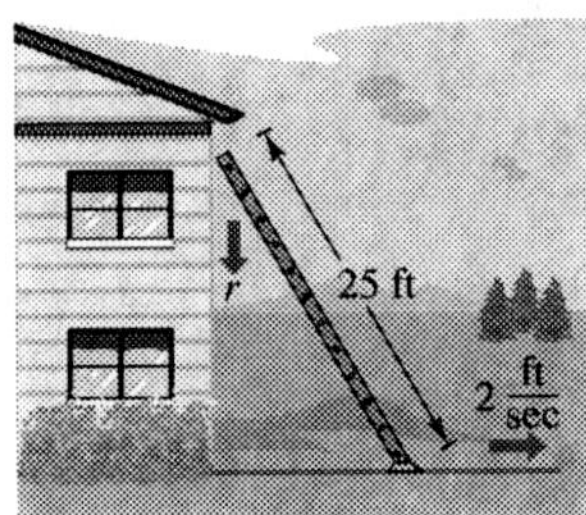

Figure for 25

Figure for 26

▓ **FOR FURTHER INFORMATION** For more information on the mathematics of moving ladders, see the article "The Falling Ladder Paradox" by Paul Scholten and Andrew Simoson in *The College Mathematics Journal*. To view this article, go to the website *www.matharticles.com*.

26. Construction A construction worker pulls a five-meter plank up the side of a building under construction by means of a rope tied to one end of the plank (see figure). Assume the opposite end of the plank follows a path perpendicular to the wall of the building and the worker pulls the rope at a rate of 0.15 meter per second. How fast is the end of the plank sliding along the ground when it is 2.5 meters from the wall of the building?

27. Construction A winch at the top of a 12-meter building pulls a pipe of the same length to a vertical position, as shown in the figure. The winch pulls in rope at a rate of -0.2 meter per second. Find the rate of vertical change and the rate of horizontal change at the end of the pipe when $y = 6$.

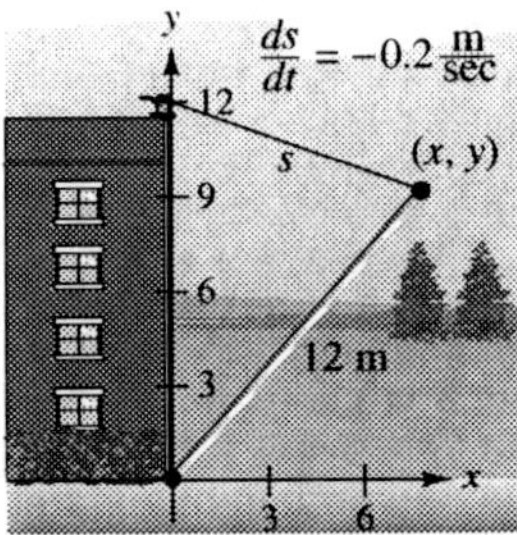

Figure for 27

Figure for 28

28. Boating A boat is pulled into a dock by means of a winch 12 feet above the deck of the boat (see figure).

(a) The winch pulls in rope at a rate of 4 feet per second. Determine the speed of the boat when there is 13 feet of rope out. What happens to the speed of the boat as it gets closer to the dock?

(b) Suppose the boat is moving at a constant rate of 4 feet per second. Determine the speed at which the winch pulls in rope when there is a total of 13 feet of rope out. What happens to the speed at which the winch pulls in rope as the boat gets closer to the dock?

29. Air Traffic Control An air traffic controller spots two planes at the same altitude converging on a point as they fly at right angles to each other (see figure). One plane is 225 miles from the point moving at 450 miles per hour. The other plane is 300 miles from the point moving at 600 miles per hour.

(a) At what rate is the distance between the planes decreasing?

(b) How much time does the air traffic controller have to get one of the planes on a different flight path?

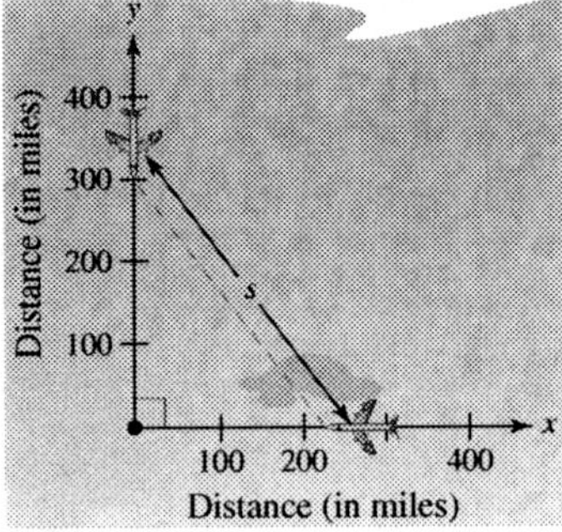

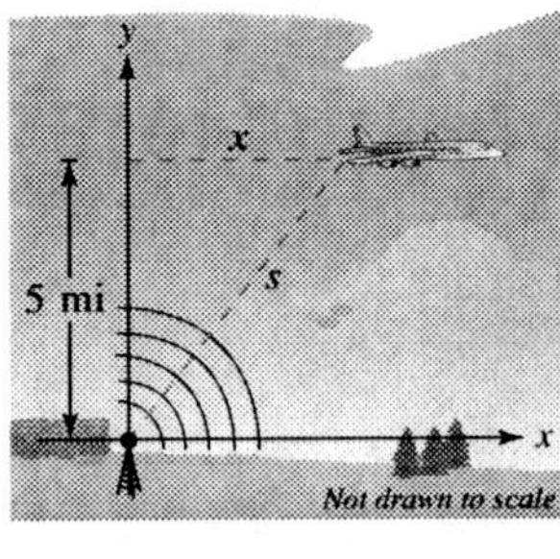

Figure for 29

Figure for 30

30. Air Traffic Control An airplane is flying at an altitude of 5 miles and passes directly over a radar antenna (see figure). When the plane is 10 miles away ($s = 10$), the radar detects that the distance s is changing at a rate of 240 miles per hour. What is the speed of the plane?

31. *Sports* A baseball diamond has the shape of a square with sides 90 feet long (see figure). A player running from second base to third base at a speed of 25 feet per second is 20 feet from third base. At what rate is the player's distance s from home plate changing?

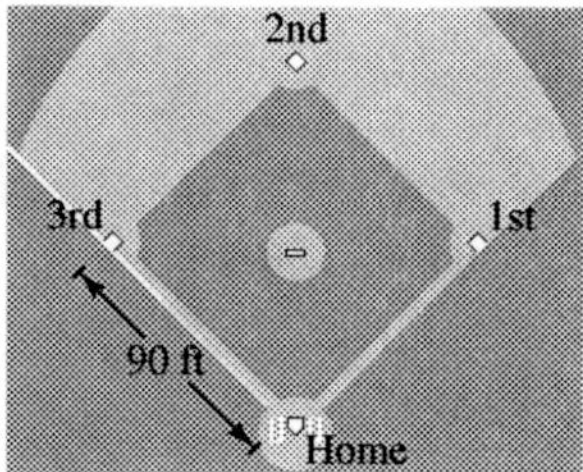

Figure for 31 and 32   **Figure for 33**

32. *Sports* For the baseball diamond in Exercise 31, suppose the player is running from first base to second base at a speed of 25 feet per second. Find the rate at which the distance from home plate is changing when the player is 20 feet from second base.

33. *Shadow Length* A man 6 feet tall walks at a rate of 5 feet per second away from a light that is 15 feet above the ground (see figure). When he is 10 feet from the base of the light,

(a) at what rate is the tip of his shadow moving?

(b) at what rate is the length of his shadow changing?

34. *Shadow Length* Repeat Exercise 33 for a man 6 feet tall walking at a rate of 5 feet per second *toward* a light that is 20 feet above the ground (see figure).

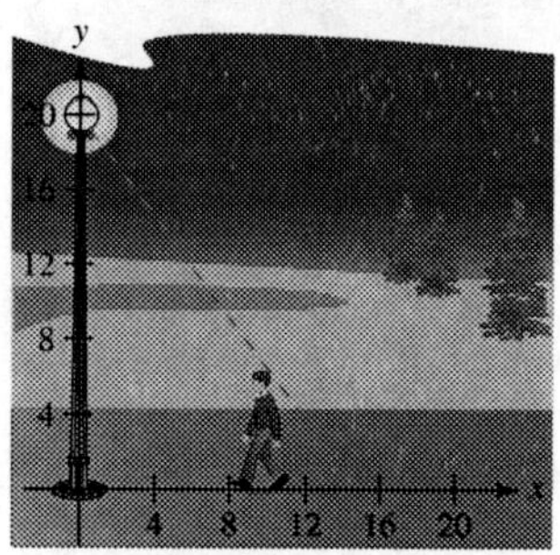

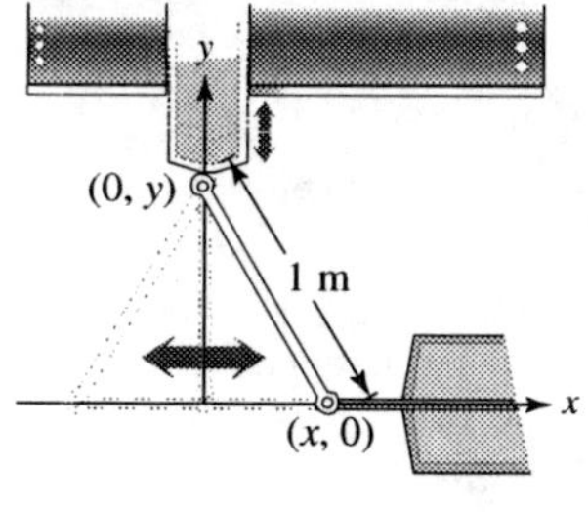

Figure for 34   **Figure for 35**

35. *Machine Design* The endpoints of a movable rod of length 1 meter have coordinates $(x, 0)$ and $(0, y)$ (see figure). The position of the end on the x-axis is

$$x(t) = \frac{1}{2} \sin \frac{\pi t}{6}$$

where t is the time in seconds.

(a) Find the time of one complete cycle of the rod.

(b) What is the lowest point reached by the end of the rod on the y-axis?

(c) Find the speed of the y-axis endpoint when the x-axis endpoint is $\left(\frac{1}{4}, 0\right)$.

36. *Machine Design* Repeat Exercise 35 for a position function of $x(t) = \frac{3}{5} \sin \pi t$. Use the point $\left(\frac{3}{10}, 0\right)$ for part (c).

37. *Evaporation* As a spherical raindrop falls, it reaches a layer of dry air and begins to evaporate at a rate that is proportional to its surface area $(S = 4\pi r^2)$. Show that the radius of the raindrop decreases at a constant rate.

38. *Electricity* The combined electrical resistance R of R_1 and R_2, connected in parallel, is given by

$$\frac{1}{R} = \frac{1}{R_1} + \frac{1}{R_2}$$

where R, R_1, and R_2 are measured in ohms. R_1 and R_2 are increasing at rates of 1 and 1.5 ohms per second, respectively. At what rate is R changing when $R_1 = 50$ ohms and $R_2 = 75$ ohms?

39. *Adiabatic Expansion* When a certain polyatomic gas undergoes adiabatic expansion, its pressure p and volume V satisfy the equation $pV^{1.3} = k$, where k is a constant. Find the relationship between the related rates dp/dt and dV/dt.

40. *Roadway Design* Cars on a certain roadway travel on a circular arc of radius r. In order not to rely on friction alone to overcome the centrifugal force, the road is banked at an angle of magnitude θ from the horizontal (see figure). The banking angle must satisfy the equation $rg \tan \theta = v^2$, where v is the velocity of the cars and $g = 32$ feet per second per second is the acceleration due to gravity. Find the relationship between the related rates dv/dt and $d\theta/dt$.

41. *Angle of Elevation* A balloon rises at a rate of 4 meters per second from a point on the ground 50 meters from an observer. Find the rate of change of the angle of elevation of the balloon from the observer when the balloon is 50 meters above the ground.

42. *Angle of Elevation* A fish is reeled in at a rate of 1 foot per second from a point 10 feet above the water (see figure). At what rate is the angle θ between the line and the water changing when there is a total of 25 feet of line from the end of the rod to the water?

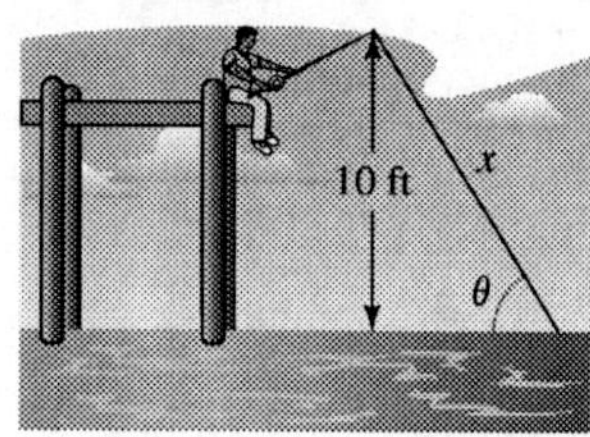

43. *Relative Humidity* When the dewpoint is 65° Fahrenheit, the relative humidity H is

$$H = \frac{4347}{400,000,000} e^{369,444/(50t + 19,793)}$$

where t is the temperature in degrees Fahrenheit.

(a) Determine the relative humidity when $t = 65°$ and $t = 80°$.

(b) At 10 A.M., the temperature is 75° and increasing at the rate of 2° per hour. Find the rate at which the relative humidity is changing.

44. *Linear vs. Angular Speed* A patrol car is parked 50 feet from a long warehouse (see figure). The revolving light on top of the car turns at a rate of 30 revolutions per minute. How fast is the light beam moving along the wall when the beam makes angles of (a) $\theta = 30°$, (b) $\theta = 60°$, and (c) $\theta = 70°$ with the line perpendicular from the light to the wall?

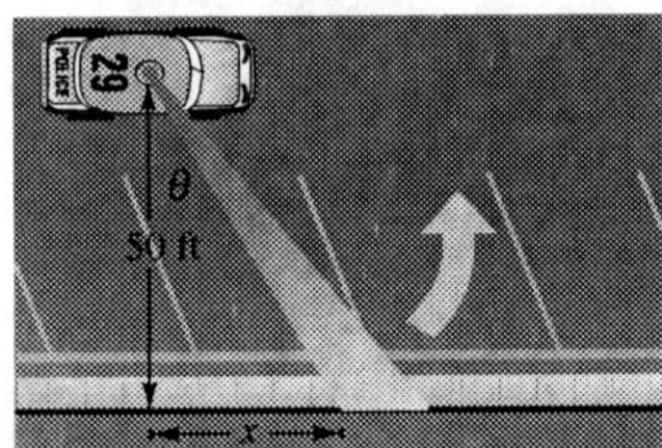
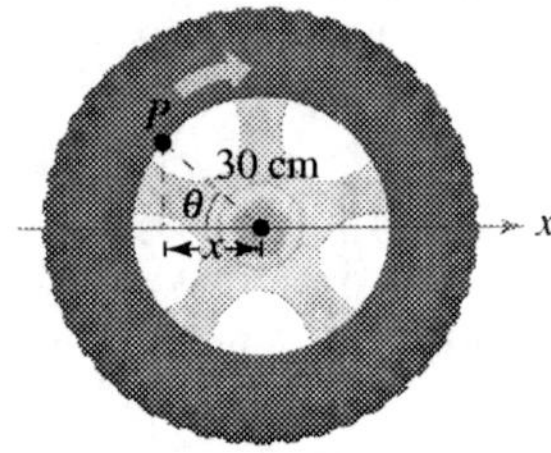

Figure for 44   **Figure for 45**

45. *Linear vs. Angular Speed* A wheel of radius 30 centimeters revolves at a rate of 10 revolutions per second. A dot is painted at a point P on the rim of the wheel (see figure).

(a) Find dx/dt as a function of θ.

(b) Use a graphing utility to graph the function in part (a).

(c) When is the absolute value of the rate of change of x greatest? When is it least?

(d) Find dx/dt when $\theta = 30°$ and $\theta = 60°$.

46. *Flight Control* An airplane is flying in still air with an airspeed of 275 miles per hour. If it is climbing at an angle of 18°, find the rate at which it is gaining altitude.

47. *Security Camera* A security camera is centered 50 feet above a 100-foot hallway (see figure). It is easiest to design the camera with a constant angular rate of rotation, but this results in a variable rate at which the images of the surveillance area are recorded. So, it is desirable to design a system with a variable rate of rotation and a constant rate of movement of the scanning beam along the hallway. Find a model for the variable rate of rotation if $|dx/dt| = 2$ feet per second.

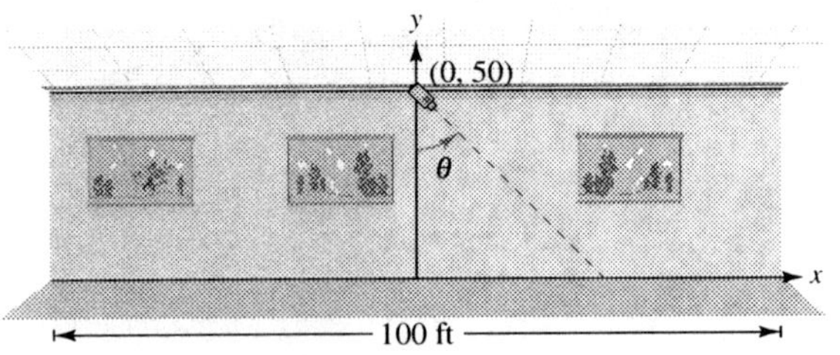

48. Using the graph of f, (a) determine whether dy/dt is positive or negative given that dx/dt is negative, and (b) determine whether dx/dt is positive or negative given that dy/dt is positive.

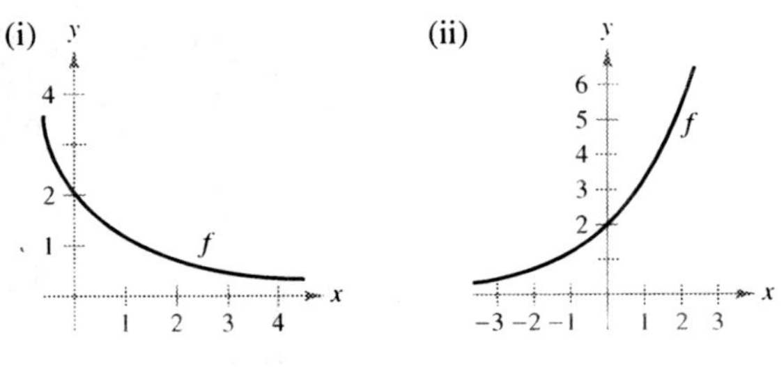

49. *Angle of Elevation* An airplane flies at an altitude of 5 miles toward a point directly over an observer (see figure). The speed of the plane is 600 miles per hour. Find the rates at which the angle of elevation θ is changing when the angle is (a) $\theta = 30°$, (b) $\theta = 60°$, and (c) $\theta = 75°$.

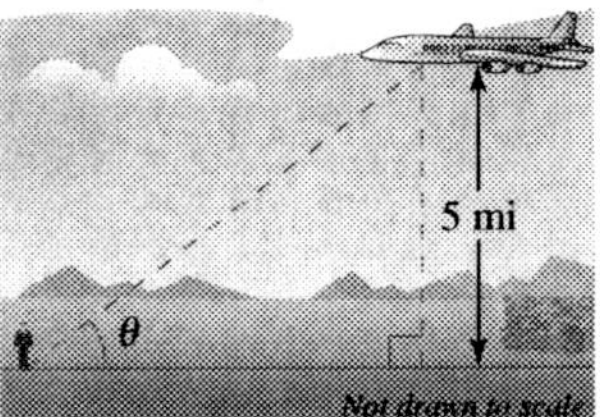

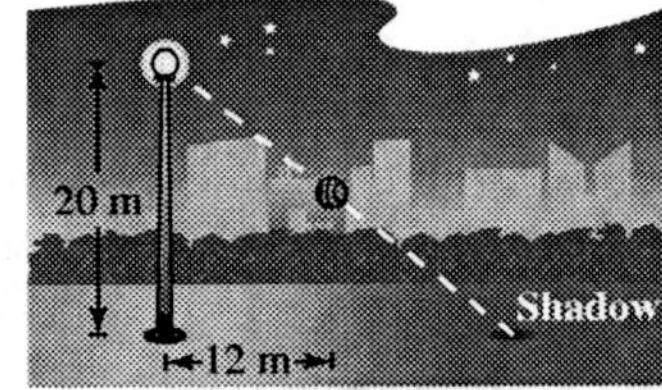

Figure for 49   **Figure for 50**

50. *Moving Shadow* A ball is dropped from a height of 20 meters, 12 meters away from the top of a 20-meter lamppost (see figure). The ball's shadow, caused by the light at the top of the lamppost, is moving along the level ground. How fast is the shadow moving 1 second after the ball is released? (*Submitted by Dennis Gittinger, St. Philips College, San Antonio, TX*)

Acceleration **In Exercises 51 and 52, find the acceleration of the specified object. (*Hint:* Recall that if a variable is changing at a constant rate, its acceleration is zero.)**

51. Find the acceleration of the top of the ladder described in Exercise 25 when the base of the ladder is 7 feet from the wall.

52. Find the acceleration of the boat in Exercise 28(a) when there is a total of 13 feet of rope out.

53. *Think About It* Describe the relationship between the rate of change of y and the rate of change of x in each expression. Assume all variables and derivatives are positive.

(a) $\dfrac{dy}{dt} = 3\dfrac{dx}{dt}$

(b) $\dfrac{dy}{dt} = x(L - x)\dfrac{dx}{dt}, \quad 0 \le x \le L$

3.8 Newton's Method

■ **Approximate a zero of a function using Newton's Method.**

Newton's Method

In this section you will study a technique for approximating the real zeros of a function. The technique is called **Newton's Method,** and it uses tangent lines to approximate the graph of the function near its x-intercepts.

To see how Newton's Method works, consider a function f that is continuous on the interval $[a, b]$ and differentiable on the interval (a, b). If $f(a)$ and $f(b)$ differ in sign, then, by the Intermediate Value Theorem, f must have at least one zero in the interval (a, b). Suppose you estimate this zero to occur at

$$x = x_1 \qquad \text{First estimate}$$

as shown in Figure 3.44(a). Newton's Method is based on the assumption that the graph of f and the tangent line at $(x_1, f(x_1))$ both cross the x-axis at *about* the same point. Because you can easily calculate the x-intercept for this tangent line, you can use it as a second (and, usually, better) estimate of the zero of f. The tangent line passes through the point $(x_1, f(x_1))$ with a slope of $f'(x_1)$. In point-slope form, the equation of the tangent line is therefore

$$y - f(x_1) = f'(x_1)(x - x_1)$$
$$y = f'(x_1)(x - x_1) + f(x_1).$$

Letting $y = 0$ and solving for x produces

$$x = x_1 - \frac{f(x_1)}{f'(x_1)}.$$

So, from the initial estimate x_1 you obtain a new estimate

$$x_2 = x_1 - \frac{f(x_1)}{f'(x_1)}. \qquad \text{Second estimate [see Figure 3.44(b)]}$$

You can improve on x_2 and calculate yet a third estimate

$$x_3 = x_2 - \frac{f(x_2)}{f'(x_2)}. \qquad \text{Third estimate}$$

Repeated application of this process is called Newton's Method.

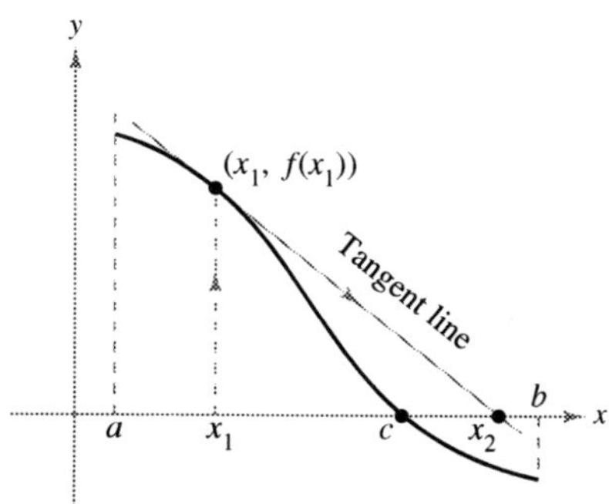

(a)

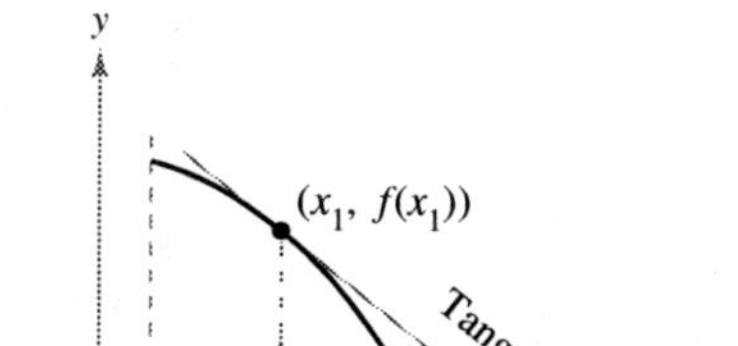

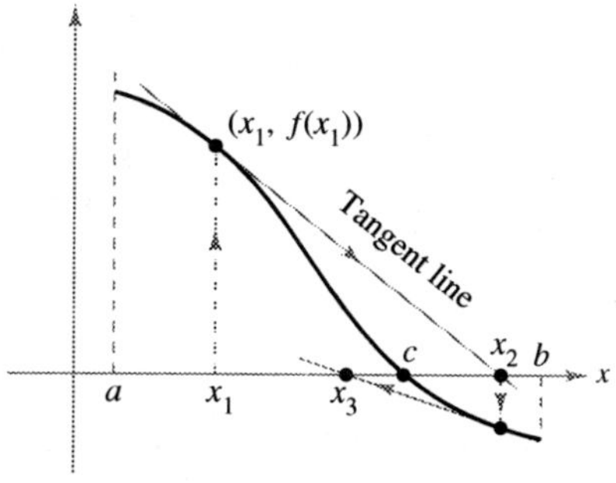

(b)
The x-intercept of the tangent line approximates the zero of f.
Figure 3.44

NEWTON'S METHOD FOR APPROXIMATING THE ZEROS OF A FUNCTION

Let $f(c) = 0$, where f is differentiable on an open interval containing c. Then, to approximate c, use the following steps.

1. Make an initial estimate x_1 that is close to c. (A graph is helpful.)
2. Determine a new approximation

$$x_{n+1} = x_n - \frac{f(x_n)}{f'(x_n)}.$$

3. If $|x_n - x_{n+1}|$ is within the desired accuracy, let x_{n+1} serve as the final approximation. Otherwise, return to Step 2 and calculate a new approximation.

Each successive application of this procedure is called an **iteration.**

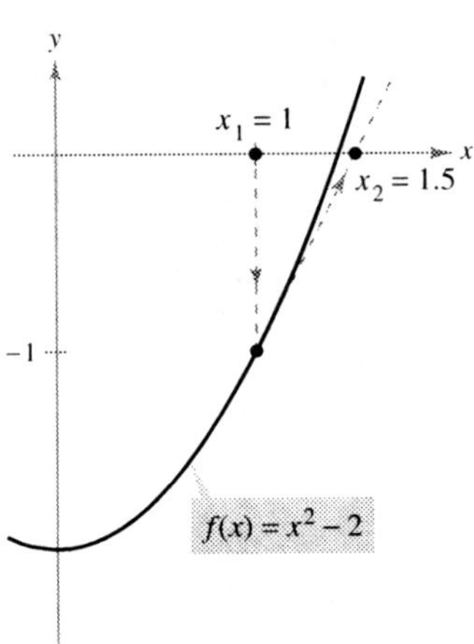

The first iteration of Newton's Method
Figure 3.45

EXAMPLE 1 Using Newton's Method

Calculate three iterations of Newton's Method to approximate a zero of $f(x) = x^2 - 2$. Use $x_1 = 1$ as the initial guess.

Solution Because $f(x) = x^2 - 2$, you have $f'(x) = 2x$, and the iterative process is given by the formula

$$x_{n+1} = x_n - \frac{f(x_n)}{f'(x_n)} = x_n - \frac{x_n^2 - 2}{2x_n}.$$

The calculations for three iterations are shown in the table.

n	x_n	$f(x_n)$	$f'(x_n)$	$\dfrac{f(x_n)}{f'(x_n)}$	$x_n - \dfrac{f(x_n)}{f'(x_n)}$
1	1.000000	-1.000000	2.000000	-0.500000	1.500000
2	1.500000	0.250000	3.000000	0.083333	1.416667
3	1.416667	0.006945	2.833334	0.002451	1.414216
4	1.414216				

Of course, in this case you know that the two zeros of the function are $\pm\sqrt{2}$. To six decimal places, $\sqrt{2} = 1.414214$. So, after only three iterations of Newton's Method, you have obtained an approximation that is within 0.000002 of an actual root. The first iteration of this process is shown in Figure 3.45.

EXAMPLE 2 Using Newton's Method

Use Newton's Method to approximate the zero(s) of

$$f(x) = e^x + x.$$

Continue the iterations until two successive approximations differ by less than 0.0001.

Solution Begin by sketching a graph of f, as shown in Figure 3.46. From the graph, you can observe that the function has only one zero, which occurs near $x = -0.6$. Next, differentiate f and form the iterative formula

$$x_{n+1} = x_n - \frac{f(x_n)}{f'(x_n)} = x_n - \frac{e^{x_n} + x_n}{e^{x_n} + 1}.$$

The calculations are shown in the table.

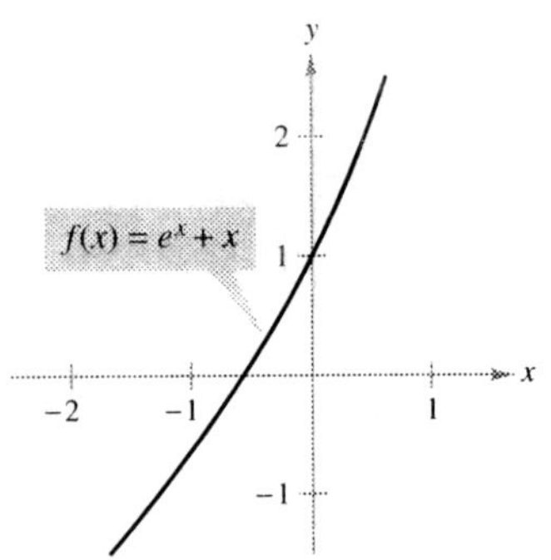

After three iterations of Newton's Method, the zero of f is approximated to the desired accuracy.
Figure 3.46

n	x_n	$f(x_n)$	$f'(x_n)$	$\dfrac{f(x_n)}{f'(x_n)}$	$x_n - \dfrac{f(x_n)}{f'(x_n)}$
1	-0.60000	-0.05119	1.54881	-0.03305	-0.56695
2	-0.56695	0.00030	1.56725	0.00019	-0.56714
3	-0.56714	0.00000	1.56714	0.00000	-0.56714
4	-0.56714				

Because two successive approximations differ by less than the required 0.0001, you can estimate the zero of f to be -0.56714.

When, as in Examples 1 and 2, the approximations approach a limit, the sequence $x_1, x_2, x_3, \ldots, x_n, \ldots$ is said to **converge.** Moreover, if the limit is c, it can be shown that c must be a zero of f.

Newton's Method does not always yield a convergent sequence. One way it can fail to do so is shown in Figure 3.47. Because Newton's Method involves division by $f'(x_n)$, it is clear that the method will fail if the derivative is zero for any x_n in the sequence. When you encounter this problem, you can usually overcome it by choosing a different value for x_1. Another way Newton's Method can fail is shown in the next example.

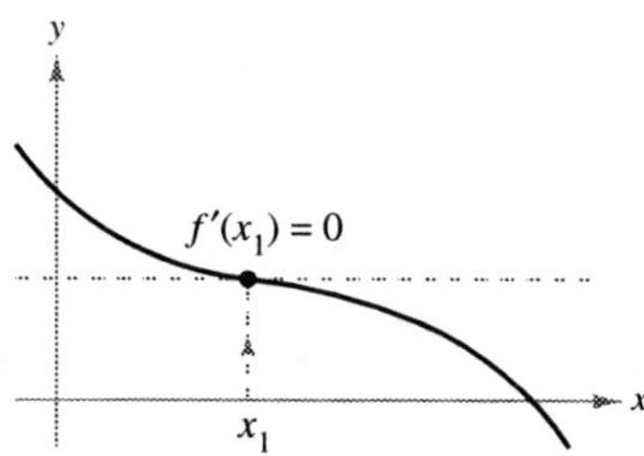

Newton's Method fails to converge if $f'(x_n) = 0$.
Figure 3.47

EXAMPLE 3   An Example in Which Newton's Method Fails

The function $f(x) = x^{1/3}$ is not differentiable at $x = 0$. Show that Newton's Method fails to converge using $x_1 = 0.1$.

Solution   Because $f'(x) = \frac{1}{3}x^{-2/3}$, the iterative formula is

$$x_{n+1} = x_n - \frac{f(x_n)}{f'(x_n)}$$

$$= x_n - \frac{x_n^{1/3}}{\frac{1}{3}x_n^{-2/3}}$$

$$= x_n - 3x_n$$

$$= -2x_n.$$

The calculations are shown in the table. This table and Figure 3.48 indicate that x_n continues to increase in magnitude as $n \to \infty$, and so the limit of the sequence does not exist.

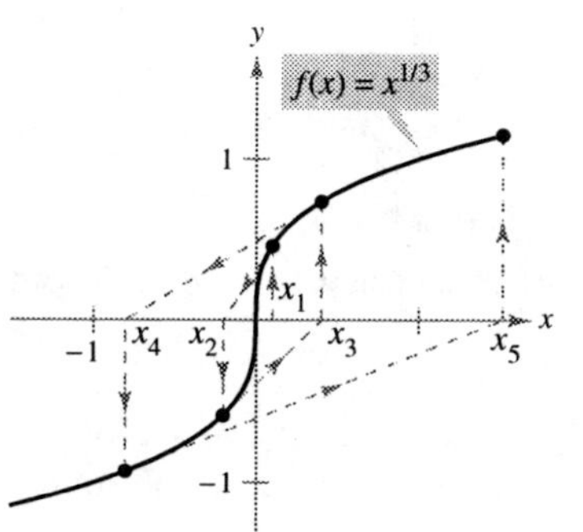

Newton's Method fails to converge for every x-value other than the actual zero of f.
Figure 3.48

n	x_n	$f(x_n)$	$f'(x_n)$	$\dfrac{f(x_n)}{f'(x_n)}$	$x_n - \dfrac{f(x_n)}{f'(x_n)}$
1	0.10000	0.46416	1.54720	0.30000	−0.20000
2	−0.20000	−0.58480	0.97467	−0.60000	0.40000
3	0.40000	0.73681	0.61401	1.20000	−0.80000
4	−0.80000	−0.92832	0.38680	−2.40000	1.60000

NOTE   In Example 3, the initial estimate $x_1 = 0.1$ fails to produce a convergent sequence. Try showing that Newton's Method also fails for every other choice of x_1 (other than the actual zero).

It can be shown that a condition sufficient to produce convergence of Newton's Method to a zero of f is that

$$\left|\frac{f(x)\,f''(x)}{[f'(x)]^2}\right| < 1 \qquad \text{Condition for convergence}$$

on an open interval containing the zero. For instance, in Example 1 this test would yield $f(x) = x^2 - 2, f'(x) = 2x, f''(x) = 2$, and

$$\left|\frac{f(x)\,f''(x)}{[f'(x)]^2}\right| = \left|\frac{(x^2 - 2)(2)}{4x^2}\right| = \left|\frac{1}{2} - \frac{1}{x^2}\right|. \qquad \text{Example 1}$$

On the interval $(1, 3)$, this quantity is less than 1 and therefore the convergence of Newton's Method is guaranteed. On the other hand, in Example 3, you have $f(x) = x^{1/3}$, $f'(x) = \frac{1}{3}x^{-2/3}, f''(x) = -\frac{2}{9}x^{-5/3}$, and

$$\left|\frac{f(x)\,f''(x)}{[f'(x)]^2}\right| = \left|\frac{x^{1/3}(-2/9)(x^{-5/3})}{(1/9)(x^{-4/3})}\right| = 2 \qquad \text{Example 3}$$

which is not less than 1 for any value of x, so you cannot conclude that Newton's Method will converge.

Algebraic Solutions of Polynomial Equations

The zeros of some functions, such as

$$f(x) = x^3 - 2x^2 - x + 2$$

can be found by simple algebraic techniques, such as factoring. The zeros of other functions, such as

$$f(x) = x^3 - x + 1$$

cannot be found by *elementary* algebraic methods. This particular function has only one real zero, and by using more advanced algebraic techniques you can determine the zero to be

$$x = -\sqrt[3]{\frac{3 - \sqrt{23/3}}{6}} - \sqrt[3]{\frac{3 + \sqrt{23/3}}{6}}.$$

Because the *exact* solution is written in terms of square roots and cube roots, it is called a **solution by radicals.**

NOTE Try approximating the real zero of $f(x) = x^3 - x + 1$ and compare your result with the exact solution shown above. ∎

The determination of radical solutions of a polynomial equation is one of the fundamental problems of algebra. The earliest such result is the Quadratic Formula, which dates back at least to Babylonian times. The general formula for the zeros of a cubic function was developed much later. In the sixteenth century an Italian mathematician, Jerome Cardan, published a method for finding radical solutions to cubic and quartic equations. Then, for 300 years, the problem of finding a general quintic formula remained open. Finally, in the nineteenth century, the problem was answered independently by two young mathematicians. Niels Henrik Abel, a Norwegian mathematician, and Evariste Galois, a French mathematician, proved that it is not possible to solve a *general* fifth- (or higher-) degree polynomial equation by radicals. Of course, you can solve particular fifth-degree equations such as $x^5 - 1 = 0$, but Abel and Galois were able to show that no general *radical* solution exists.

NIELS HENRIK ABEL (1802–1829)

EVARISTE GALOIS (1811–1832)

Although the lives of both Abel and Galois were brief, their work in the fields of analysis and abstract algebra was far-reaching.

3.8 Exercises

In Exercises 1–4, complete two iterations of Newton's Method for the function using the given initial guess.

1. $f(x) = x^2 - 5$, $x_1 = 2.2$ **2.** $f(x) = x^3 - 3$, $x_1 = 1.4$

3. $f(x) = \cos x$, $x_1 = 1.6$ **4.** $f(x) = \tan x$, $x_1 = 0.1$

In Exercises 5–16, approximate the zero(s) of the function. Use Newton's Method and continue the process until two successive approximations differ by less than 0.001. Then find the zero(s) using a graphing utility and compare the results.

5. $f(x) = x^3 + 4$ **6.** $f(x) = 2 - x^3$

7. $f(x) = x^3 + x - 1$ **8.** $f(x) = x^5 + x - 1$

9. $f(x) = 5\sqrt{x - 1} - 2x$ **10.** $f(x) = x - 2\sqrt{x + 1}$

11. $f(x) = x - e^{-x}$ **12.** $f(x) = x - 3 + \ln x$

13. $f(x) = x^3 - 3.9x^2 + 4.79x - 1.881$

14. $f(x) = x^4 + x^3 - 1$ **15.** $f(x) = 1 - x + \sin x$

16. $f(x) = x^3 - \cos x$

In Exercises 17–24, apply Newton's Method to approximate the x-value(s) of the given point(s) of intersection of the two graphs. Continue the process until two successive approximations differ by less than 0.001. [Hint: Let $h(x) = f(x) - g(x)$.]

17. $f(x) = 2x + 1$
   $g(x) = \sqrt{x + 4}$

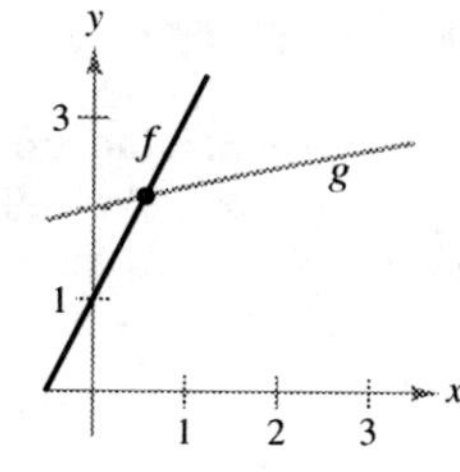

18. $f(x) = 3 - x$
   $g(x) = 1/(x^2 + 1)$

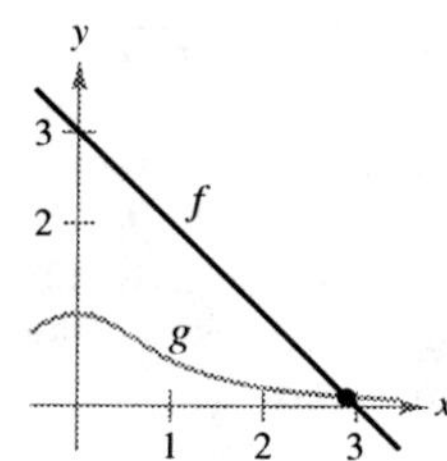

19. $f(x) = x$
   $g(x) = \tan x$

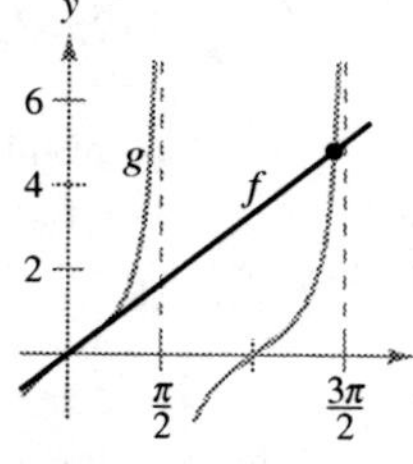

20. $f(x) = x^2$
   $g(x) = \cos x$

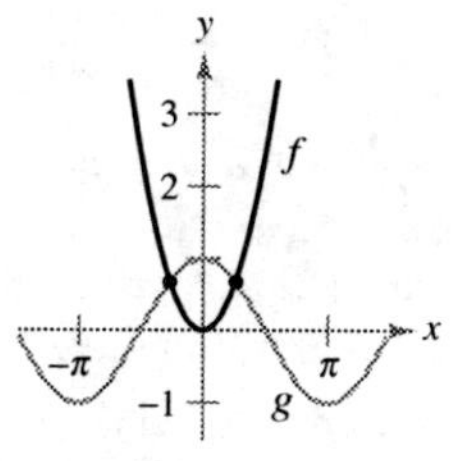

21. $f(x) = -x$
   $g(x) = \ln x$

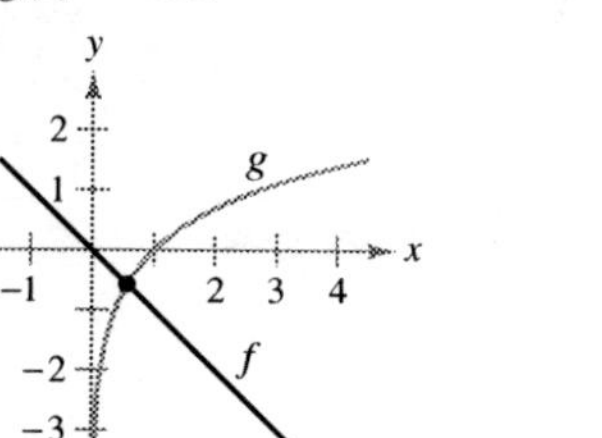

22. $f(x) = 2 - x^2$
   $g(x) = e^{x/2}$

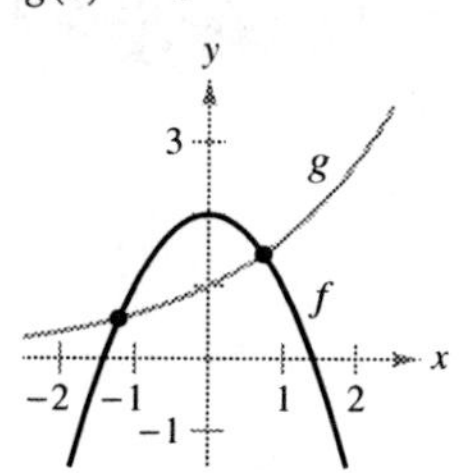

23. $f(x) = \arccos x$
   $g(x) = \arctan x$

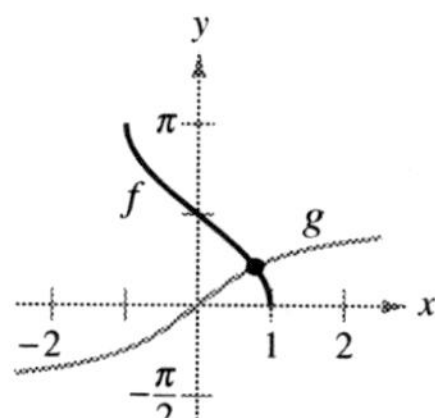

24. $f(x) = 1 - x$
   $g(x) = \arcsin x$

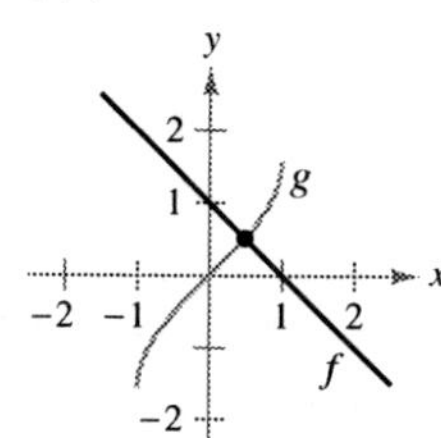

25. *Mechanic's Rule*   The Mechanic's Rule for approximating $\sqrt{a}$, $a > 0$, is

$$x_{n+1} = \frac{1}{2}\left(x_n + \frac{a}{x_n}\right), \quad n = 1, 2, 3 \ldots$$

where x_1 is an approximation of $\sqrt{a}$.

(a) Use Newton's Method and the function $f(x) = x^2 - a$ to derive the Mechanic's Rule.

(b) Use the Mechanic's Rule to approximate $\sqrt{5}$ and $\sqrt{7}$ to three decimal places.

26. (a) Use Newton's Method and the function $f(x) = x^n - a$ to obtain a general rule for approximating $x = \sqrt[n]{a}$.

(b) Use the general rule found in part (a) to approximate $\sqrt[4]{6}$ and $\sqrt[3]{15}$ to three decimal places.

In Exercises 27–30, apply Newton's Method using the given initial guess, and explain why the method fails.

27. $y = 2x^3 - 6x^2 + 6x - 1$, $x_1 = 1$

28. $y = x^3 - 2x - 2$, $x_1 = 0$

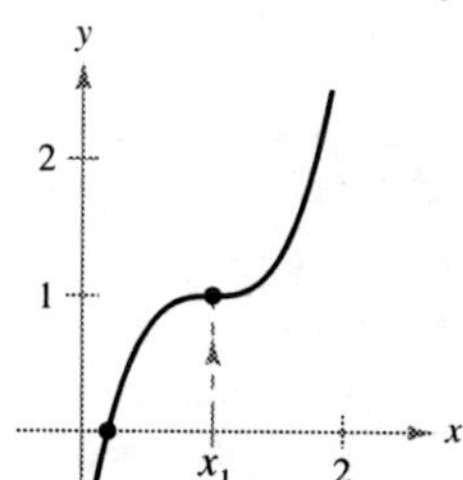

Figure for 27

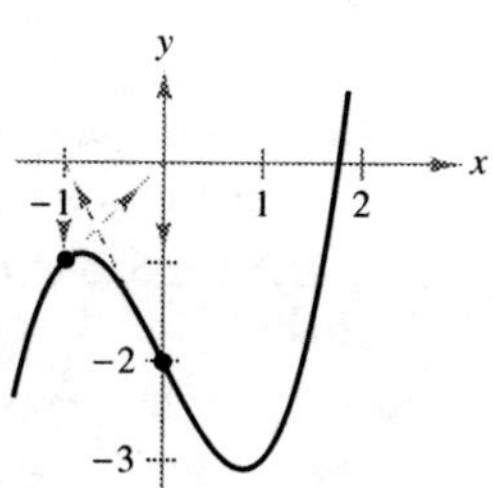

Figure for 28

29. $f(x) = -x^3 + 6x^2 - 10x + 6$, $x_1 = 2$

30. $f(x) = 2 \sin x + \cos 2x$, $x_1 = \dfrac{3\pi}{2}$

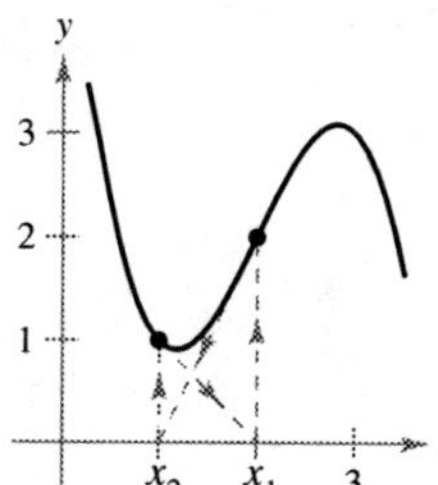

Figure for 29

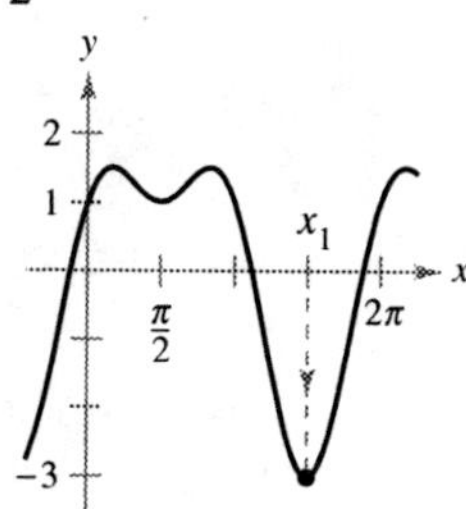

Figure for 30

Fixed Point In Exercises 31–34, approximate the fixed point of the function to two decimal places. [A *fixed point* x_0 of a function f is a value of x such that $f(x_0) = x_0$.]

31. $f(x) = \cos x$

32. $f(x) = \cot x, \quad 0 < x < \pi$

33. $f(x) = e^{x/10}$

34. $f(x) = -\ln x$

WRITING ABOUT CONCEPTS

35. Consider the function $f(x) = x^3 - 3x^2 + 3$.

(a) Use a graphing utility to graph f.

(b) Use Newton's Method with $x_1 = 1$ as an initial guess.

(c) Repeat part (b) using $x_1 = \frac{1}{4}$ as an initial guess and observe that the result is different.

(d) To understand why the results in parts (b) and (c) are different, sketch the tangent lines to the graph of f at the points $(1, f(1))$ and $\left(\frac{1}{4}, f\left(\frac{1}{4}\right)\right)$. Find the x-intercept of each tangent line and compare the intercepts with the first iteration of Newton's Method using the respective initial guesses.

(e) Write a short paragraph summarizing how Newton's Method works. Use the results of this exercise to describe why it is important to select the initial guess carefully.

36. Repeat the steps in Exercise 35 for the function $f(x) = \sin x$ with initial guesses of $x_1 = 1.8$ and $x_1 = 3$.

37. In your own words and using a sketch, describe Newton's Method for approximating the zeros of a function.

CAPSTONE

38. Under what conditions will Newton's Method fail?

39. Use Newton's Method to show that the equation $x_{n+1} = x_n(2 - ax_n)$ can be used to approximate $1/a$ if x_1 is an initial guess of the reciprocal of a. Note that this method of approximating reciprocals uses only the operations of multiplication and subtraction. [*Hint:* Consider $f(x) = (1/x) - a$.]

40. Use the result of Exercise 39 to approximate (a) $\frac{1}{3}$ and (b) $\frac{1}{11}$ to three decimal places.

41. ***Crime*** The total number of arrests T (in thousands) for all males ages 14 to 27 in 2006 is approximated by the model

$$T = 0.602x^3 - 41.44x^2 + 922.8x - 6330, \quad 14 \le x \le 27$$

where x is the age in years (see figure). Approximate the two ages that had total arrests of 225 thousand. (*Source: U.S. Department of Justice*)

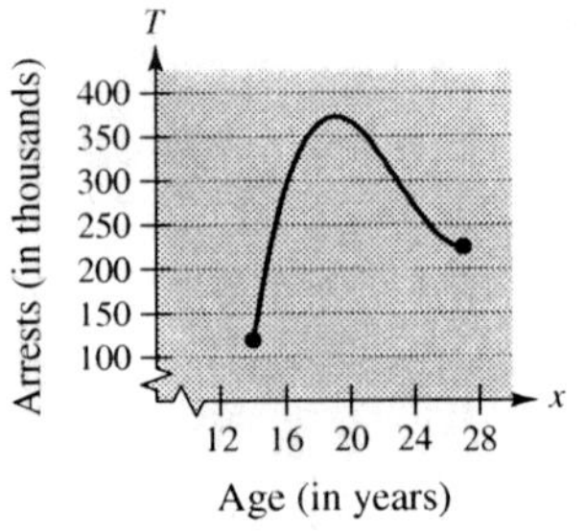

42. ***Advertising Costs*** A company that produces digital audio players estimates that the profit for selling a particular model is

$$P = -76x^3 + 4830x^2 - 320,000, \quad 0 \le x \le 60$$

where P is the profit in dollars and x is the advertising expense in 10,000s of dollars (see figure). According to this model, find the smaller of two advertising amounts that yield a profit P of \$2,500,000.

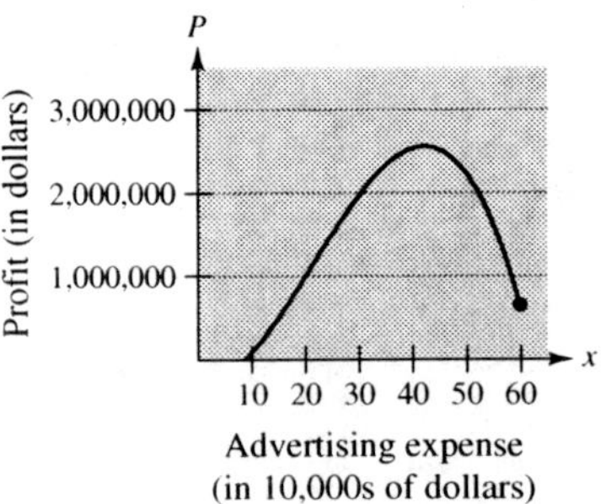

True or False? In Exercises 43–46, determine whether the statement is true or false. If it is false, explain why or give an example that shows it is false.

43. The zeros of $f(x) = p(x)/q(x)$ coincide with the zeros of $p(x)$.

44. If the coefficients of a polynomial function are all positive, then the polynomial has no positive zeros.

45. If $f(x)$ is a cubic polynomial such that $f'(x)$ is never zero, then any initial guess will force Newton's Method to converge to the zero of f.

46. The roots of $\sqrt{f(x)} = 0$ coincide with the roots of $f(x) = 0$.

In Exercises 47 and 48, *write a computer program or use a spreadsheet to find the zeros of a function using Newton's Method. Approximate the zeros of the function accurate to three decimal places. The output should be a table with the following headings.*

$$n, \quad x_n, \quad f(x_n), \quad f'(x_n), \quad \frac{f(x_n)}{f'(x_n)}, \quad x_n - \frac{f(x_n)}{f'(x_n)}$$

47. $f(x) = \frac{1}{4}x^3 - 3x^2 + \frac{3}{4}x - 2$

48. $f(x) = \sqrt{4 - x^2} \sin(x - 2)$

49. ***Tangent Lines*** The graph of $f(x) = -\sin x$ has infinitely many tangent lines that pass through the origin. Use Newton's Method to approximate the slope of the tangent line having the greatest slope to three decimal places.

50. ***Point of Tangency*** The graph of $f(x) = \cos x$ and a tangent line to f through the origin are shown. Find the coordinates of the point of tangency to three decimal places.

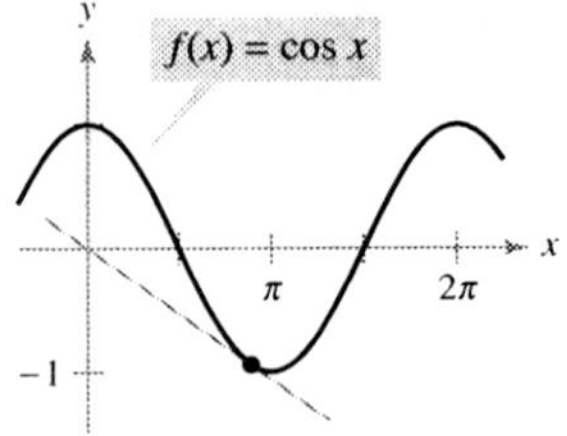

3 REVIEW EXERCISES

See www.CalcChat.com for worked-out solutions to odd-numbered exercises.

In Exercises 1–4, find the derivative of the function by using the definition of the derivative.

1. $f(x) = x^2 - 4x + 5$

2. $f(x) = \dfrac{x+1}{x-1}$

3. $f(x) = \sqrt{x} + 1$

4. $f(x) = \dfrac{6}{x}$

In Exercises 5 and 6, describe the x-values at which f is differentiable.

5. $f(x) = (x-3)^{2/5}$

6. $f(x) = \dfrac{3x}{x+1}$

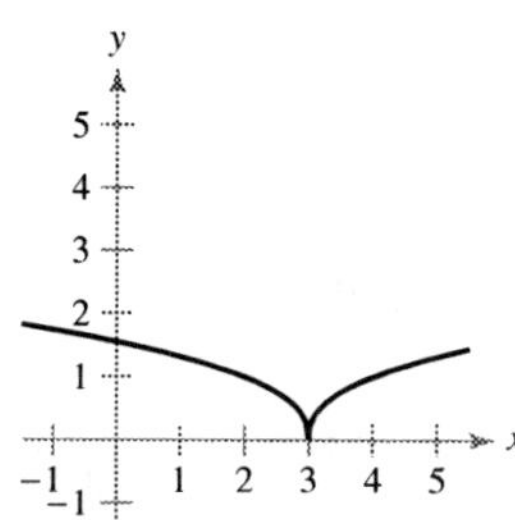

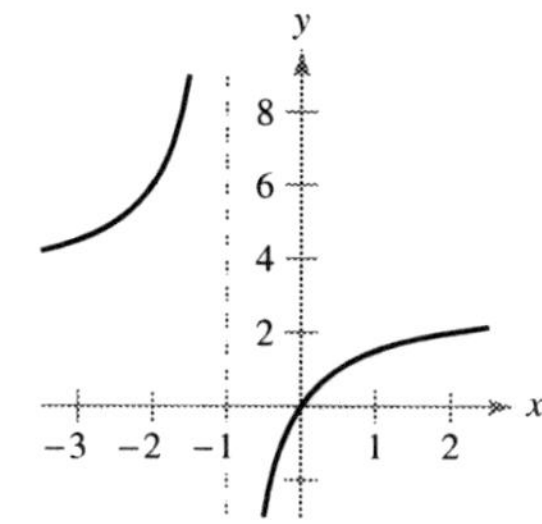

7. Sketch the graph of $f(x) = 4 - |x - 2|$.

  (a) Is f continuous at $x = 2$?

  (b) Is f differentiable at $x = 2$? Explain.

8. Sketch the graph of $f(x) = \begin{cases} x^2 + 4x + 2, & x < -2 \\ 1 - 4x - x^2, & x \geq -2. \end{cases}$

  (a) Is f continuous at $x = -2$?

  (b) Is f differentiable at $x = -2$? Explain.

In Exercises 9 and 10, find the slope of the tangent line to the graph of the function at the given point.

9. $g(x) = \dfrac{2}{3}x^2 - \dfrac{x}{6}, \quad \left(-1, \dfrac{5}{6}\right)$

10. $h(x) = \dfrac{3x}{8} - 2x^2, \quad \left(-2, -\dfrac{35}{4}\right)$

In Exercises 11 and 12, (a) find an equation of the tangent line to the graph of f at the given point, (b) use a graphing utility to graph the function and its tangent line at the point, and (c) use the *derivative* feature of the graphing utility to confirm your results.

11. $f(x) = x^3 - 1, \quad (-1, -2)$

12. $f(x) = \dfrac{2}{x+1}, \quad (0, 2)$

In Exercises 13 and 14, use the alternative form of the derivative to find the derivative at $x = c$ (if it exists).

13. $g(x) = x^2(x - 1), \quad c = 2$

14. $f(x) = \dfrac{1}{x+4}, \quad c = 3$

In Exercises 15–30, use the rules of differentiation to find the derivative of the function.

15. $y = 25$

16. $y = -30$

17. $f(x) = x^8$

18. $g(x) = x^{20}$

19. $h(t) = 3t^4$

20. $f(t) = -8t^5$

21. $f(x) = x^3 - 3x^2$

22. $g(s) = 4s^4 - 5s^2$

23. $h(x) = 6\sqrt{x} + 3\sqrt[3]{x}$

24. $f(x) = x^{1/2} - x^{-1/2}$

25. $g(t) = \dfrac{2}{3t^2}$

26. $h(x) = \dfrac{10}{(7x)^2}$

27. $f(\theta) = 4\theta - 5\sin\theta$

28. $g(\alpha) = 4\cos\alpha + 6$

29. $f(t) = 3\cos t - 4e^t$

30. $g(s) = \dfrac{5}{3}\sin s - 2e^s$

Writing **In Exercises 31 and 32, the figure shows the graphs of a function and its derivative. Label the graphs as f or f' and write a short paragraph stating the criteria you used in making your selection. To print an enlarged copy of the graph, go to the website *www.mathgraphs.com*.**

31.

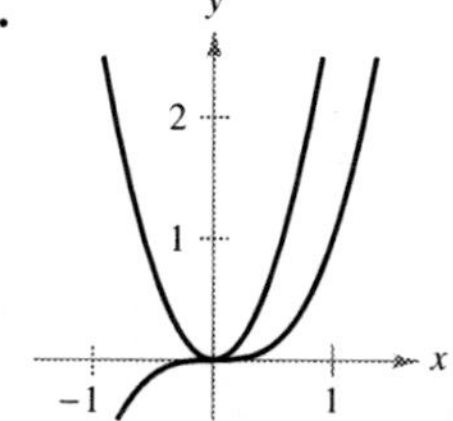

32.

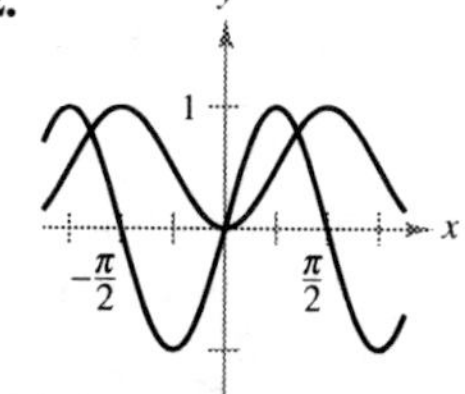

33. ***Vibrating String*** When a guitar string is plucked, it vibrates with a frequency of $F = 200\sqrt{T}$, where F is measured in vibrations per second and the tension T is measured in pounds. Find the rates of change of F when (a) $T = 4$ and (b) $T = 9$.

34. ***Vertical Motion*** A ball is dropped from a height of 100 feet. One second later, another ball is dropped from a height of 75 feet. Which ball hits the ground first?

35. ***Vertical Motion*** To estimate the height of a building, a weight is dropped from the top of the building into a pool at ground level. How high (in feet) is the building if the splash is seen 9.2 seconds after the weight is dropped?

36. ***Vertical Motion*** A bomb is dropped from an airplane at an altitude of 14,400 feet. How long will it take for the bomb to reach the ground? (Because of the motion of the plane, the fall will not be vertical, but the time will be the same as that for a vertical fall.) The plane is moving at 600 miles per hour. How far will the bomb move horizontally after it is released from the plane?

37. ***Projectile Motion*** A thrown ball follows a path described by $y = x - 0.02x^2$.

  (a) Sketch a graph of the path.

  (b) Find the total horizontal distance the ball is thrown.

  (c) At what x-value does the ball reach its maximum height? (Use the symmetry of the path.)

  (d) Find an equation that gives the instantaneous rate of change of the height of the ball with respect to the horizontal change. Evaluate the equation at $x = 0, 10, 25, 30,$ and 50.

  (e) What is the instantaneous rate of change of the height when the ball reaches its maximum height?

38. *Projectile Motion*   The path of a projectile thrown at an angle of 45° with level ground is

$$y = x - \frac{32}{v_0^2}(x^2)$$

where the initial velocity is v_0 feet per second.

(a) Find the x-coordinate of the point where the projectile strikes the ground. Use the symmetry of the path of the projectile to locate the x-coordinate of the point where the projectile reaches its maximum height.

(b) What is the instantaneous rate of change of the height when the projectile is at its maximum height?

(c) Show that doubling the initial velocity of the projectile multiplies both the maximum height and the range by a factor of 4.

(d) Find the maximum height and range of a projectile thrown with an initial velocity of 70 feet per second. Use a graphing utility to graph the path of the projectile.

39. *Horizontal Motion*   The position function of a particle moving along the x-axis is

$$x(t) = t^2 - 3t + 2 \quad \text{for} \quad -\infty < t < \infty.$$

(a) Find the velocity of the particle.

(b) Find the open t-interval(s) in which the particle is moving to the left.

(c) Find the position of the particle when the velocity is 0.

(d) Find the speed of the particle when the position is 0.

40. *Modeling Data*   The speed of a car in miles per hour and the stopping distance in feet are recorded in the table.

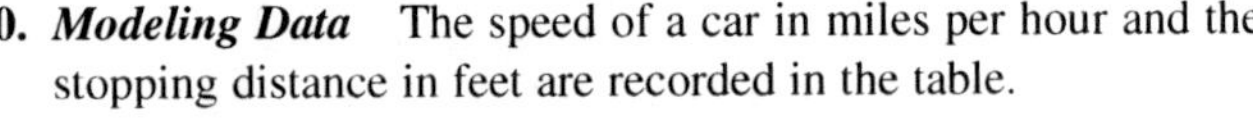

Speed, x	20	30	40	50	60
Stopping Distance, y	25	55	105	188	300

(a) Use the regression capabilities of a graphing utility to find a quadratic model for the data.

(b) Use a graphing utility to plot the data and graph the model.

(c) Use a graphing utility to graph dy/dx.

(d) Use the model to approximate the stopping distance at a speed of 65 miles per hour.

(e) Use the graphs in parts (b) and (c) to explain the change in stopping distance as the speed increases.

In Exercises 41–56, find the derivative of the function.

41. $f(x) = (5x^2 + 8)(x^2 - 4x - 6)$

42. $g(x) = (x^3 + 7x)(x + 3)$

43. $h(x) = \sqrt{x} \sin x$

44. $f(t) = 2t^5 \cos t$

45. $f(x) = \dfrac{x^2 + x - 1}{x^2 - 1}$

46. $f(x) = \dfrac{6x - 5}{x^2 + 1}$

47. $f(x) = \dfrac{1}{9 - 4x^2}$

48. $f(x) = \dfrac{9}{3x^2 - 2x}$

49. $y = \dfrac{x^4}{\cos x}$

50. $y = \dfrac{\sin x}{x^4}$

51. $y = 3x^2 \sec x$

52. $y = 2x - x^2 \tan x$

53. $y = x \cos x - \sin x$

54. $g(x) = 3x \sin x + x^2 \cos x$

55. $y = 4xe^x$

56. $y = \dfrac{1 + \sin x}{1 - \sin x}$

In Exercises 57–60, find an equation of the tangent line to the graph of f at the given point.

57. $f(x) = \dfrac{2x^3 - 1}{x^2}, \quad (1, 1)$

58. $f(x) = \dfrac{x + 1}{x - 1}, \quad \left(\dfrac{1}{2}, -3\right)$

59. $f(x) = -x \tan x, \quad (0, 0)$

60. $f(x) = \dfrac{1 + \cos x}{1 - \cos x}, \quad \left(\dfrac{\pi}{2}, 1\right)$

61. *Acceleration*   The velocity of an object in meters per second is $v(t) = 36 - t^2$, $0 \le t \le 6$. Find the velocity and acceleration of the object when $t = 4$.

62. *Acceleration*   The velocity of an automobile starting from rest is

$$v(t) = \frac{90t}{4t + 10}$$

where v is measured in feet per second. Find the vehicle's velocity and acceleration at each of the following times.

(a) 1 second   (b) 5 seconds   (c) 10 seconds

In Exercises 63–68, find the second derivative of the function.

63. $g(t) = -8t^3 - 5t + 12$

64. $h(x) = 21x^{-3} + 3x$

65. $f(x) = 15x^{5/2}$

66. $f(x) = 20\sqrt[5]{x}$

67. $f(\theta) = 3 \tan \theta$

68. $h(t) = 10 \cos t - 15 \sin t$

In Exercises 69 and 70, show that the function satisfies the equation.

Function	*Equation*
69. $y = 2 \sin x + 3 \cos x$	$y'' + y = 0$
70. $y = \dfrac{10 - \cos x}{x}$	$xy' + y = \sin x$

71. *Rate of Change*   Determine whether there exist any values of x in the interval $[0, 2\pi)$ such that the rate of change of $f(x) = \sec x$ and the rate of change of $g(x) = \csc x$ are equal.

72. *Volume*   The radius of a right circular cylinder is given by $\sqrt{t + 2}$ and its height is $\frac{1}{2}\sqrt{t}$, where t is time in seconds and the dimensions are in inches. Find the rate of change of the volume with respect to time.

In Exercises 73–98, find the derivative of the function.

73. $h(x) = \left(\dfrac{x + 5}{x^2 + 3}\right)^2$

74. $f(x) = \left(x^2 + \dfrac{1}{x}\right)^5$

75. $f(s) = (s^2 - 1)^{5/2}(s^3 + 5)$

76. $h(\theta) = \dfrac{\theta}{(1 - \theta)^3}$

77. $y = 5\cos(9x + 1)$

78. $y = 1 - \cos 2x + 2\cos^2 x$

79. $y = \dfrac{x}{2} - \dfrac{\sin 2x}{4}$

80. $y = \dfrac{\sec^7 x}{7} - \dfrac{\sec^5 x}{5}$

81. $y = \dfrac{2}{3}\sin^{3/2} x - \dfrac{2}{7}\sin^{7/2} x$

82. $f(x) = \dfrac{3x}{\sqrt{x^2 + 1}}$

83. $y = \dfrac{\sin \pi x}{x + 2}$

84. $y = \dfrac{\cos(x - 1)}{x - 1}$

85. $g(t) = t^2 e^{t/4}$

86. $h(z) = e^{-z^2/2}$

87. $y = \sqrt{e^{2x} + e^{-2x}}$

88. $y = 3e^{-3/t}$

89. $g(x) = \dfrac{x^2}{e^x}$

90. $f(\theta) = \dfrac{1}{2}e^{\sin 2\theta}$

91. $g(x) = \ln \sqrt{x}$

92. $h(x) = \ln \dfrac{x(x - 1)}{x - 2}$

93. $f(x) = x\sqrt{\ln x}$

94. $f(x) = \ln[x(x^2 - 2)^{2/3}]$

95. $y = \dfrac{1}{b^2}\left[\ln(a + bx) + \dfrac{a}{a + bx}\right]$

96. $y = \dfrac{1}{b^2}[a + bx - a\ln(a + bx)]$

97. $y = -\dfrac{1}{a}\ln\dfrac{a + bx}{x}$

98. $y = -\dfrac{1}{ax} + \dfrac{b}{a^2}\ln\dfrac{a + bx}{x}$

In Exercises 99–102, find the derivative of the function at the given point.

99. $f(x) = \sqrt{1 - x^3}$, $(-2, 3)$

100. $f(x) = \sqrt[3]{x^2 - 1}$, $(3, 2)$

101. $y = \dfrac{1}{2}\csc 2x$, $\left(\dfrac{\pi}{4}, \dfrac{1}{2}\right)$

102. $y = \csc 3x + \cot 3x$, $\left(\dfrac{\pi}{6}, 1\right)$

CAS **In Exercises 103–110, use a computer algebra system to find the derivative of the function. Use the utility to graph the function and its derivative on the same set of coordinate axes. Describe the behavior of the function that corresponds to any zeros of the graph of the derivative.**

103. $f(t) = t^2(t - 1)^5$

104. $f(x) = [(x - 2)(x + 4)]^2$

105. $g(x) = \dfrac{2x}{\sqrt{x + 1}}$

106. $g(x) = x\sqrt{x^2 + 1}$

107. $f(t) = \sqrt{t + 1}\,\sqrt[3]{t + 1}$

108. $y = \sqrt{3x}\,(x + 2)^3$

109. $y = \tan\sqrt{1 - x}$

110. $y = 2\csc^3(\sqrt{x})$

In Exercises 111–114, find the second derivative of the function.

111. $y = 7x^2 + \cos 2x$

112. $y = \dfrac{1}{x} + \tan x$

113. $f(x) = \cot x$

114. $y = \sin^2 x$

CAS **In Exercises 115–120, use a computer algebra system to find the second derivative of the function.**

115. $f(t) = \dfrac{4t^2}{(1 - t)^2}$

116. $g(x) = \dfrac{6x - 5}{x^2 + 1}$

117. $g(\theta) = \tan 3\theta - \sin(\theta - 1)$

118. $h(x) = 5x\sqrt{x^2 - 16}$

119. $g(x) = x^3 \ln x$

120. $f(x) = 6x^2 e^{-x/3}$

121. Refrigeration The temperature T of food put in a freezer is

$$T = \dfrac{700}{t^2 + 4t + 10}$$

where t is the time in hours. Find the rate of change of T with respect to t at each of the following times.

(a) $t = 1$ (b) $t = 3$ (c) $t = 5$ (d) $t = 10$

122. Fluid Flow The emergent velocity v of a liquid flowing from a hole in the bottom of a tank is given by $v = \sqrt{2gh}$, where g is the acceleration due to gravity (32 feet per second per second) and h is the depth of the liquid in the tank. Find the rate of change of v with respect to h when (a) $h = 9$ and (b) $h = 4$. (Note that $g = +32$ feet per second per second. The sign of g depends on how a problem is modeled. In this case, letting g be negative would produce an imaginary value for v.)

123. Modeling Data The atmospheric pressure decreases with increasing altitude. At sea level, the average air pressure is one atmosphere (1.033227 kilograms per square centimeter). The table gives the pressures p (in atmospheres) at various altitudes h (in kilometers).

h	0	5	10	15	20	25
p	1	0.55	0.25	0.12	0.06	0.02

(a) Use a graphing utility to find a model of the form $p = a + b \ln h$ for the data. Explain why the result is an error message.

(b) Use a graphing utility to find the logarithmic model $h = a + b \ln p$ for the data.

(c) Use a graphing utility to plot the data and graph the logarithmic model.

(d) Use the model to estimate the altitude at which the pressure is 0.75 atmosphere.

(e) Use the model to estimate the pressure at an altitude of 13 kilometers.

(f) Find the rates of change of pressure when $h = 5$ and $h = 20$. Interpret the results in the context of the problem.

124. Tractrix A person walking along a dock drags a boat by a 10-meter rope. The boat travels along a path known as a *tractrix* (see figure). The equation of this path is

$$y = 10\ln\left(\dfrac{10 + \sqrt{100 - x^2}}{x}\right) - \sqrt{100 - x^2}.$$

(a) Use a graphing utility to graph the function.

(b) What is the slope of the path when $x = 5$ and $x = 9$?

(c) What does the slope of the path approach as $x \to 10$?

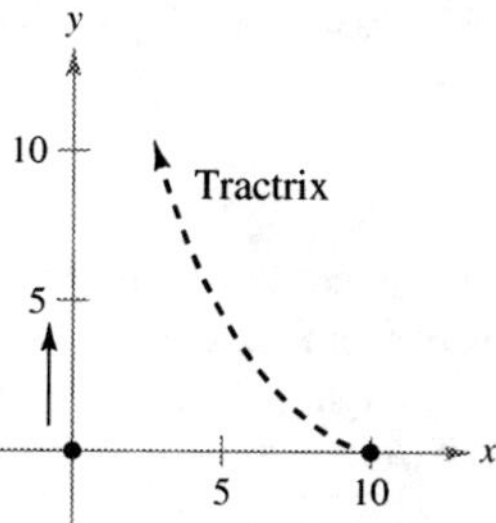

In Exercises 125–132, find *dy/dx* by implicit differentiation.

125. $x^2 + 3xy + y^3 = 10$

126. $y^2 = (x - y)(x^2 + y)$

127. $\cos x^2 = xe^y$

128. $ye^x + xe^y = xy$

129. $\sqrt{xy} = x - 4y$

130. $y\sqrt{x} - x\sqrt{y} = 25$

131. $x \sin y = y \cos x$

132. $\cos(x + y) = x$

In Exercises 133–136, find the equations of the tangent line and the normal line to the graph of the equation at the given point. Use a graphing utility to graph the equation, the tangent line, and the normal line.

133. $x^2 + y^2 = 10$, $(3, 1)$

134. $x^2 - y^2 = 20$, $(6, 4)$

135. $y \ln x + y^2 = 0$, $(e, -1)$

136. $\ln(x + y) = x$, $(0, 1)$

In Exercises 137 and 138, use logarithmic differentiation to find *dy/dx*.

137. $y = \dfrac{x\sqrt{x^2 + 1}}{x + 4}$

138. $y = \dfrac{(2x + 1)^3(x^2 - 1)^2}{x + 3}$

In Exercises 139–142, verify that *f* has an inverse. Then use the function *f* and the given real number *a* to find $(f^{-1})'(a)$. (*Hint: Use Theorem 3.17.*)

Function	*Real number*
139. $f(x) = x^3 + 2$	$a = -1$
140. $f(x) = x\sqrt{x - 3}$	$a = 4$
141. $f(x) = \tan x,\ -\dfrac{\pi}{4} \le x \le \dfrac{\pi}{4}$	$a = \dfrac{\sqrt{3}}{3}$
142. $f(x) = \cos x,\ 0 \le x \le \pi$	$a = 0$

In Exercises 143–148, find the derivative of the function.

143. $y = \tan(\arcsin x)$

144. $y = \arctan(x^2 - 1)$

145. $y = x \operatorname{arcsec} x$

146. $y = \frac{1}{2}\arctan e^{2x}$

147. $y = x(\arcsin x)^2 - 2x + 2\sqrt{1 - x^2}\,\arcsin x$

148. $y = \sqrt{x^2 - 4} - 2\operatorname{arcsec}\dfrac{x}{2},\quad 2 < x < 4$

149. A point moves along the curve $y = \sqrt{x}$ in such a way that the y-value is increasing at a rate of 2 units per second. At what rate is x changing for each of the following values?

(a) $x = \frac{1}{2}$ (b) $x = 1$ (c) $x = 4$

150. *Surface Area*   The edges of a cube are expanding at a rate of 8 centimeters per second. How fast is the surface area changing when each edge is 6.5 centimeters?

151. *Changing Depth*   The cross section of a 5-meter trough is an isosceles trapezoid with a 2-meter lower base, a 3-meter upper base, and an altitude of 2 meters. Water is running into the trough at a rate of 1 cubic meter per minute. How fast is the water level rising when the water is 1 meter deep?

152. *Linear and Angular Velocity*   A rotating beacon is located 1 kilometer off a straight shoreline (see figure). If the beacon rotates at a rate of 3 revolutions per minute, how fast (in kilometers per hour) does the beam of light appear to be moving to a viewer who is $\frac{1}{2}$ kilometer down the shoreline?

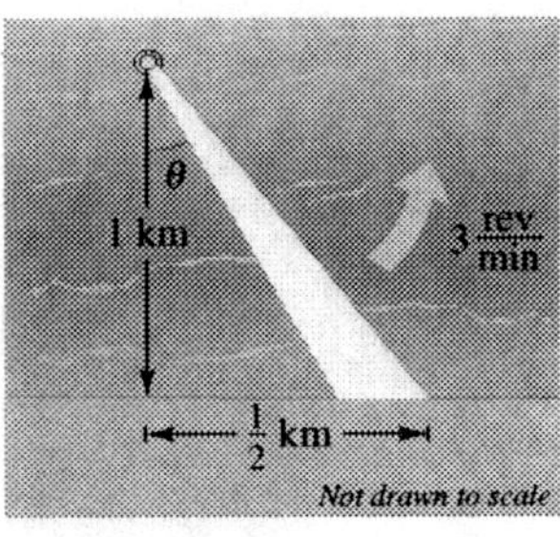

Figure for 152

153. *Moving Shadow*   A sandbag is dropped from a balloon at a height of 60 meters when the angle of elevation to the sun is $30°$ (see figure). Find the rate at which the shadow of the sandbag is traveling along the ground when the sandbag is at a height of 35 meters. [*Hint:* The position of the sandbag is given by $s(t) = 60 - 4.9t^2$.]

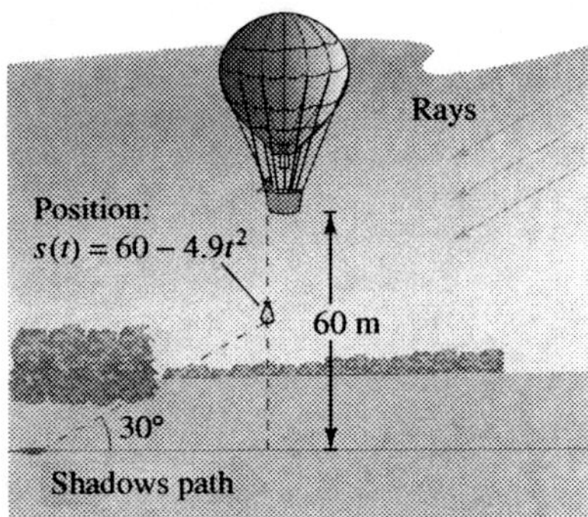

154. *Geometry*   Consider the rectangle shown in the figure.

(a) Find the area of the rectangle as a function of x.

(b) Find the rate of change of the area when $x = 4$ centimeters if $dx/dt = 4$ centimeters per minute.

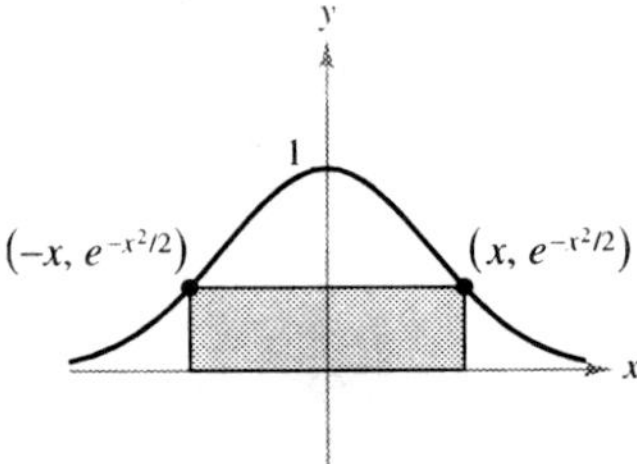

In Exercises 155–158, use Newton's Method to approximate any real zeros of the function accurate to three decimal places. Use the root-finding capabilities of a graphing utility to verify your results.

155. $f(x) = x^3 - 3x - 1$

156. $f(x) = x^3 + 2x + 1$

157. $g(x) = xe^x - 4$

158. $f(x) = 3 - x \ln x$

In Exercises 159 and 160, use Newton's Method to approximate, to three decimal places, the *x*-values of any points of intersection of the graphs of the equations. Use a graphing utility to verify your results.

159. $y = x^4$

$y = x + 3$

160. $y = \sin \pi x$

$y = 1 - x$

 **P.S. PROBLEM SOLVING**

1. Consider the graph of the parabola $y = x^2$.

(a) Find the radius r of the largest possible circle centered on the y-axis that is tangent to the parabola at the origin, as indicated in the figure. This circle is called the **circle of curvature** (see Section 12.5). Use a graphing utility to graph the circle and parabola in the same viewing window.

(b) Find the center $(0, b)$ of the circle of radius 1 centered on the y-axis that is tangent to the parabola at two points, as indicated in the figure. Use a graphing utility to graph the circle and parabola in the same viewing window.

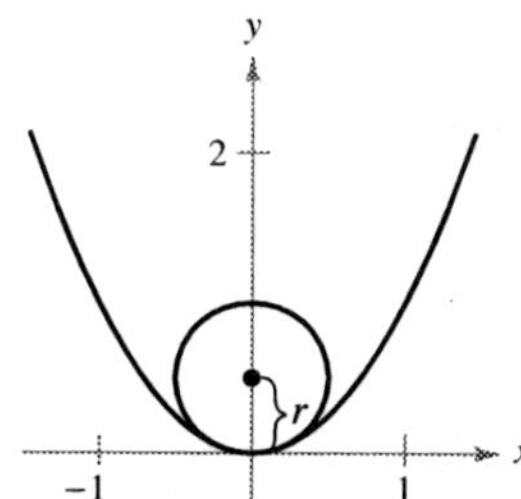

Figure for 1(a)

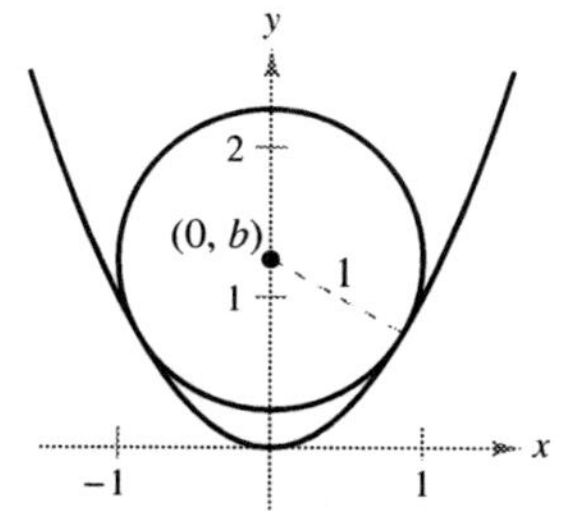

Figure for 1(b)

2. Graph the two parabolas $y = x^2$ and $y = -x^2 + 2x - 5$ in the same coordinate plane. Find equations of the two lines simultaneously tangent to both parabolas.

3. (a) Find the polynomial $P_1(x) = a_0 + a_1 x$ whose value and slope agree with the value and slope of $f(x) = \cos x$ at the point $x = 0$.

(b) Find the polynomial $P_2(x) = a_0 + a_1 x + a_2 x^2$ whose value and first two derivatives agree with the value and first two derivatives of $f(x) = \cos x$ at the point $x = 0$. This polynomial is called the second-degree **Taylor polynomial** of $f(x) = \cos x$ at $x = 0$.

(c) Complete the table comparing the values of $f(x) = \cos x$ and $P_2(x)$. What do you observe?

x	-1.0	-0.1	-0.001	0	0.001	0.1	1.0
$\cos x$							
$P_2(x)$							

(d) Find the third-degree Taylor polynomial of $f(x) = \sin x$ at $x = 0$.

4. (a) Find an equation of the tangent line to the parabola $y = x^2$ at the point $(2, 4)$.

(b) Find an equation of the normal line to $y = x^2$ at the point $(2, 4)$. (The normal line is perpendicular to the tangent line.) Where does this line intersect the parabola a second time?

(c) Find equations of the tangent line and normal line to $y = x^2$ at the point $(0, 0)$.

(d) Prove that for any point $(a, b) \neq (0, 0)$ on the parabola $y = x^2$, the normal line intersects the graph a second time.

5. Find a third-degree polynomial $p(x)$ that is tangent to the line $y = 14x - 13$ at the point $(1, 1)$, and tangent to the line $y = -2x - 5$ at the point $(-1, -3)$.

6. Find a function of the form $f(x) = a + b \cos cx$ that is tangent to the line $y = 1$ at the point $(0, 1)$, and tangent to the line

$$y = x + \frac{3}{2} - \frac{\pi}{4}$$

at the point $\left(\dfrac{\pi}{4}, \dfrac{3}{2}\right)$.

7. The graph of the **eight curve**

$$x^4 = a^2(x^2 - y^2), \quad a \neq 0$$

is shown below.

(a) Explain how you could use a graphing utility to graph this curve.

(b) Use a graphing utility to graph the curve for various values of the constant a. Describe how a affects the shape of the curve.

(c) Determine the points on the curve at which the tangent line is horizontal.

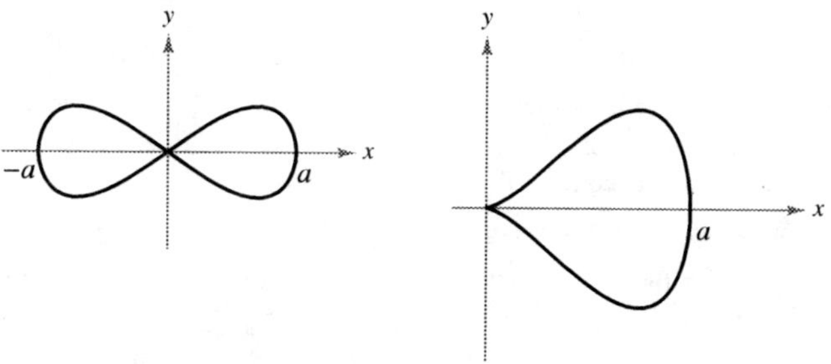

Figure for 7   **Figure for 8**

8. The graph of the **pear-shaped quartic**

$$b^2 y^2 = x^3(a - x), \quad a, b > 0$$

is shown above.

(a) Explain how you could use a graphing utility to graph this curve.

(b) Use a graphing utility to graph the curve for various values of the constants a and b. Describe how a and b affect the shape of the curve.

(c) Determine the points on the curve at which the tangent line is horizontal.

9. To approximate e^x, you can use a function of the form $f(x) = \dfrac{a + bx}{1 + cx}$. (This function is known as a **Padé approximation**.) The values of $f(0)$, $f'(0)$, and $f''(0)$ are equal to the corresponding values of e^x. Show that these values are equal to 1 and find the values of a, b, and c such that $f(0) = f'(0) = f''(0) = 1$. Then use a graphing utility to compare the graphs of f and e^x.

10. A man 6 feet tall walks at a rate of 5 feet per second toward a streetlight that is 30 feet high (see figure). The man's 3-foot-tall child follows at the same speed, but 10 feet behind the man. At times, the shadow behind the child is caused by the man, and at other times, by the child.

(a) Suppose the man is 90 feet from the streetlight. Show that the man's shadow extends beyond the child's shadow.

(b) Suppose the man is 60 feet from the streetlight. Show that the child's shadow extends beyond the man's shadow.

(c) Determine the distance d from the man to the streetlight at which the tips of the two shadows are exactly the same distance from the streetlight.

(d) Determine how fast the tip of the shadow is moving as a function of x, the distance between the man and the streetlight. Discuss the continuity of this shadow speed function.

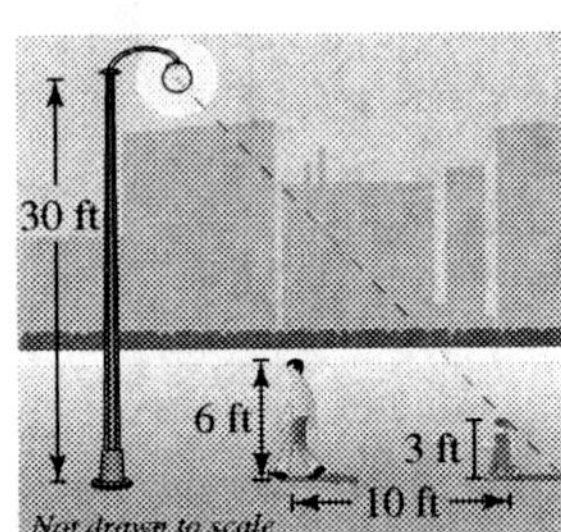

Figure for 10   **Figure for 11**

11. A particle is moving along the graph of $y = \sqrt[3]{x}$ (see figure). When $x = 8$, the y-component of the position of the particle is increasing at the rate of 1 centimeter per second.

(a) How fast is the x-component changing at this moment?

(b) How fast is the distance from the origin changing at this moment?

(c) How fast is the angle of inclination θ changing at this moment?

12. The figure shows the graph of the function $y = \ln x$ and its tangent line L at the point (a, b). Show that the distance between b and c is always equal to 1.

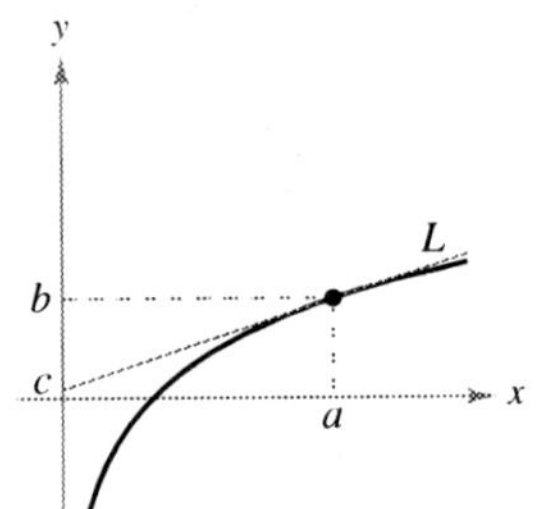

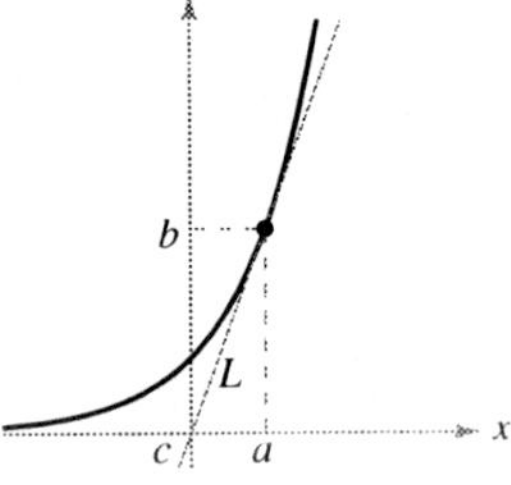

Figure for 12   **Figure for 13**

13. The figure shows the graph of the function $y = e^x$ and its tangent line L at the point (a, b). Show that the distance between a and c is always equal to 1.

 **14.** The fundamental limit

$$\lim_{x \to 0} \frac{\sin x}{x} = 1$$

assumes that x is measured in radians. What happens if we assume that x is measured in degrees instead of radians?

(a) Set your calculator to degree mode and complete the table.

z (in degrees)	0.1	0.01	0.0001
$\dfrac{\sin z}{z}$			

(b) Use the table to estimate $\displaystyle\lim_{z \to 0} \frac{\sin z}{z}$ for z in degrees. What is the exact value of this limit? (*Hint:* $180° = \pi$ radians)

(c) Use the limit definition of the derivative to find $\dfrac{d}{dz} \sin z$ for z in degrees.

(d) Define the new functions $S(z) = \sin(cz)$ and $C(z) = \cos(cz)$, where $c = \pi/180$. Find $S(90)$ and $C(180)$. Use the Chain Rule to calculate $\dfrac{d}{dz} S(z)$.

(e) Explain why calculus is made easier by using radians instead of degrees.

15. An astronaut standing on the moon throws a rock upward. The height of the rock is $s = -\frac{27}{10}t^2 + 27t + 6$, where s is measured in feet and t is measured in seconds.

(a) Find expressions for the velocity and acceleration of the rock.

(b) Find the time when the rock is at its highest point by finding the time when the velocity is zero. What is the rock's height at this time?

(c) How does the acceleration of the rock compare with the acceleration due to gravity on Earth?

16. If a is the acceleration of an object, the *jerk* j is defined by $j = a'(t)$.

(a) Use this definition to give a physical interpretation of j.

(b) The figure shows the graphs of the position, velocity, acceleration, and jerk functions of a vehicle. Identify each graph and explain your reasoning.

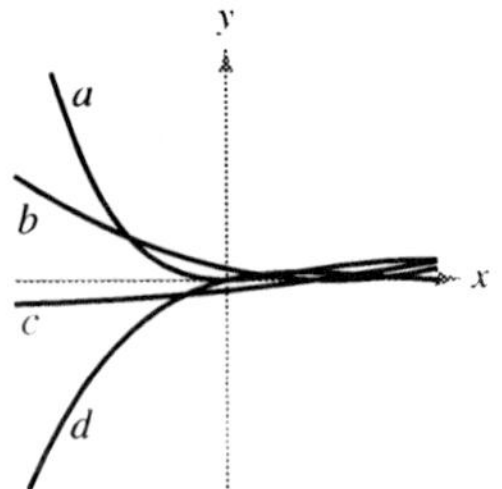

Applications of Differentiation

This chapter discusses several applications of the derivative of a function. These applications fall into three basic categories—curve sketching, optimization, and approximation techniques.

In this chapter, you should learn the following.

- How to use a derivative to locate the minimum and maximum values of a function on a closed interval. (4.1)
- How numerous results in this chapter depend on two important theorems called *Rolle's Theorem* and the *Mean Value Theorem*. (4.2)
- How to use the first derivative to determine whether a function is increasing or decreasing. (4.3)
- How to use the second derivative to determine whether the graph of a function is concave upward or concave downward. (4.4)
- How to find horizontal asymptotes of the graph of a function. (4.5)
- How to graph a function using the techniques from Chapters 1–4. (4.6)
- How to solve optimization problems. (4.7)
- How to use approximation techniques to solve problems. (4.8)

© E.J. Baumeister Jr./Alamy

A small aircraft starts its descent from an altitude of 1 mile, 4 miles west of the runway. Given a function that models the glide path of the plane, when would the plane be descending at the greatest rate? (See Section 4.4, Exercise 91.)

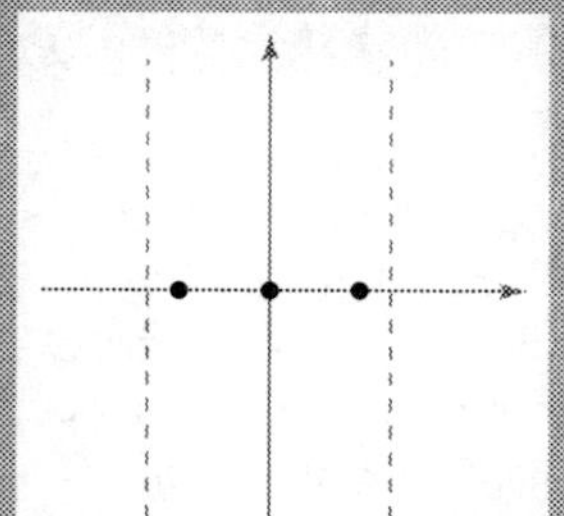
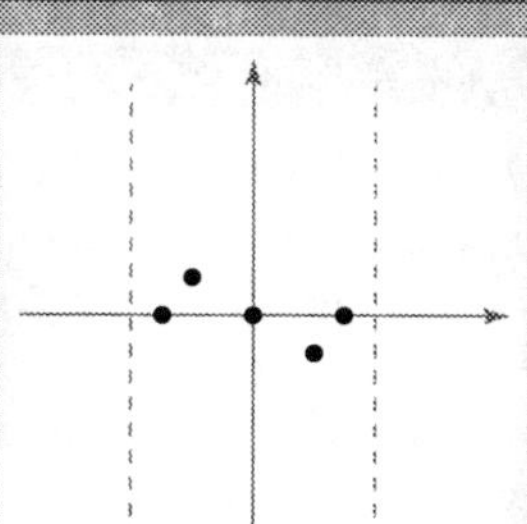
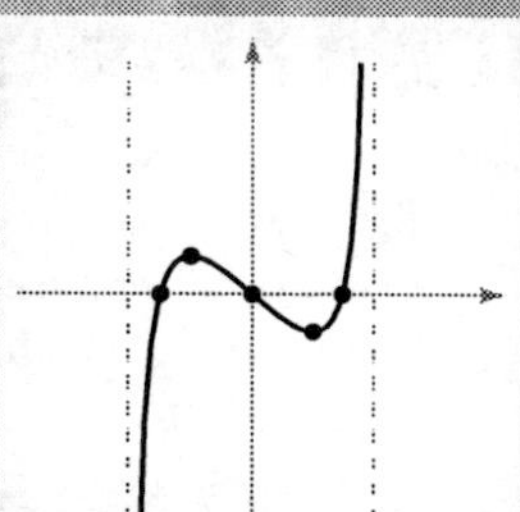

In Chapter 4, you will use calculus to analyze graphs of functions. For example, you can use the derivative of a function to determine the function's maximum and minimum values. You can use limits to identify any asymptotes of the function's graph. In Section 4.6, you will combine these techniques to sketch the graph of a function.

4.1 Extrema on an Interval

- Understand the definition of extrema of a function on an interval.
- Understand the definition of relative extrema of a function on an open interval.
- Find extrema on a closed interval.

Extrema of a Function

In calculus, much effort is devoted to determining the behavior of a function f on an interval I. Does f have a maximum value on I? Does it have a minimum value? Where is the function increasing? Where is it decreasing? In this chapter you will learn how derivatives can be used to answer these questions. You will also see why these questions are important in real-life applications.

DEFINITION OF EXTREMA

Let f be defined on an interval I containing c.

1. $f(c)$ is the **minimum of f on I** if $f(c) \le f(x)$ for all x in I.
2. $f(c)$ is the **maximum of f on I** if $f(c) \ge f(x)$ for all x in I.

The minimum and maximum of a function on an interval are the **extreme values**, or **extrema** (the singular form of extrema is extremum), of the function on the interval. The minimum and maximum of a function on an interval are also called the **absolute minimum** and **absolute maximum**, or the **global minimum** and **global maximum,** on the interval.

A function need not have a minimum or a maximum on an interval. For instance, in Figure 4.1(a) and (b), you can see that the function $f(x) = x^2 + 1$ has both a minimum and a maximum on the closed interval $[-1, 2]$, but does not have a maximum on the open interval $(-1, 2)$. Moreover, in Figure 4.1(c), you can see that continuity (or the lack of it) can affect the existence of an extremum on the interval. This suggests the following theorem. (Although the Extreme Value Theorem is intuitively plausible, a proof of this theorem is not within the scope of this text.)

THEOREM 4.1 THE EXTREME VALUE THEOREM

If f is continuous on a closed interval $[a, b]$, then f has both a minimum and a maximum on the interval.

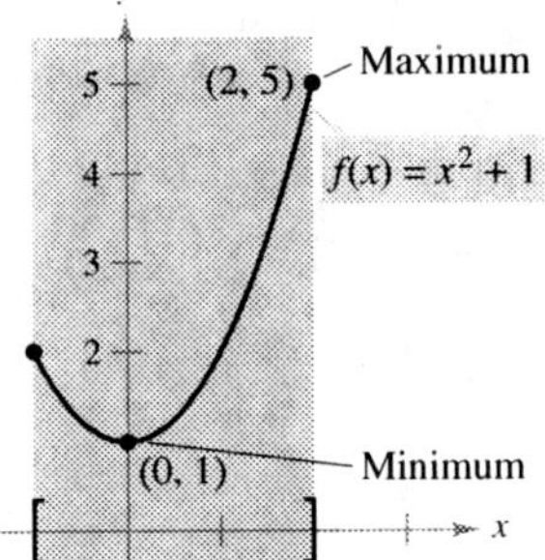

(a) f is continuous, $[-1, 2]$ is closed.

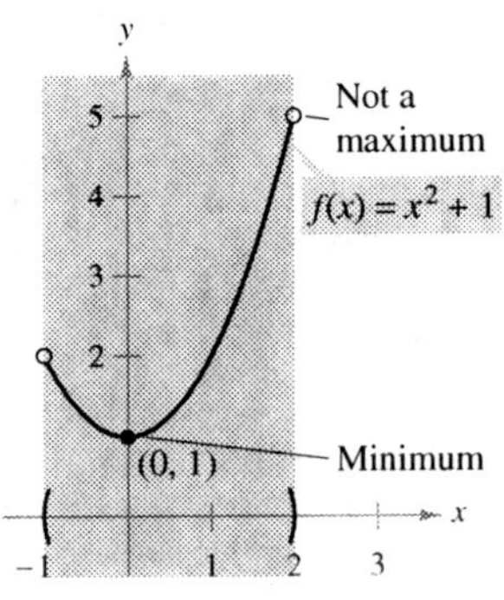

(b) f is continuous, $(-1, 2)$ is open.

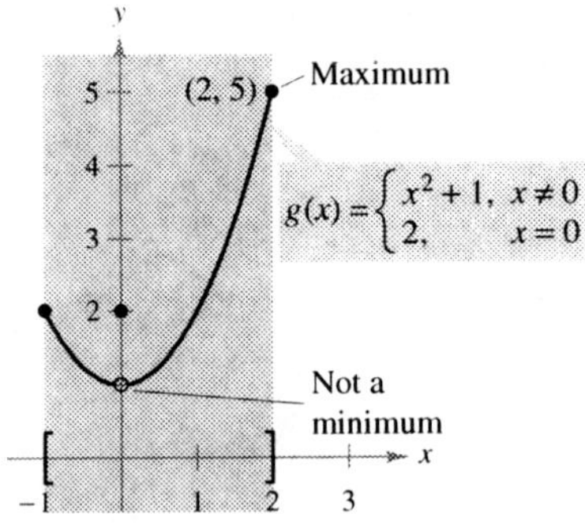

(c) g is not continuous, $[-1, 2]$ is closed.

Extrema can occur at interior points or endpoints of an interval. Extrema that occur at the endpoints are called **endpoint extrema.**

Figure 4.1

EXPLORATION

Finding Minimum and Maximum Values   The Extreme Value Theorem (like the Intermediate Value Theorem) is an *existence theorem* because it tells of the existence of minimum and maximum values but does not show how to find these values. Use the extreme-value capability of a graphing utility to find the minimum and maximum values of each of the following functions. In each case, do you think the x-values are exact or approximate? Explain your reasoning.

a. $f(x) = x^2 - 4x + 5$ on the closed interval $[-1, 3]$

b. $f(x) = x^3 - 2x^2 - 3x - 2$ on the closed interval $[-1, 3]$

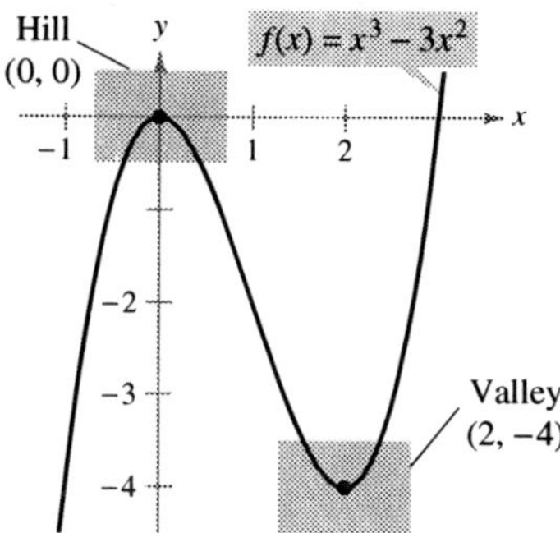

f has a relative maximum at $(0, 0)$ and a relative minimum at $(2, -4)$.

Figure 4.2

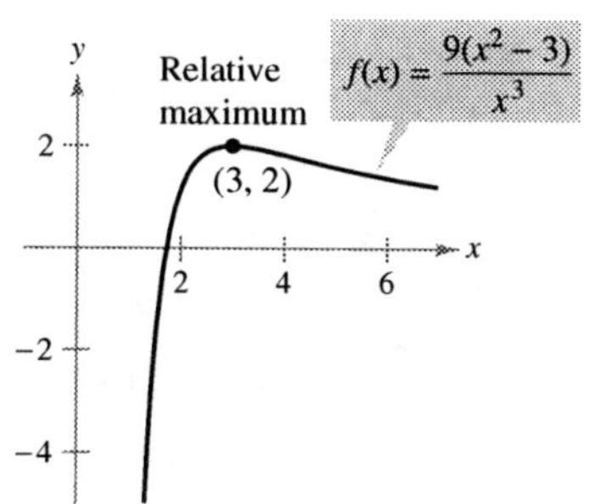

(a) $f'(3) = 0$

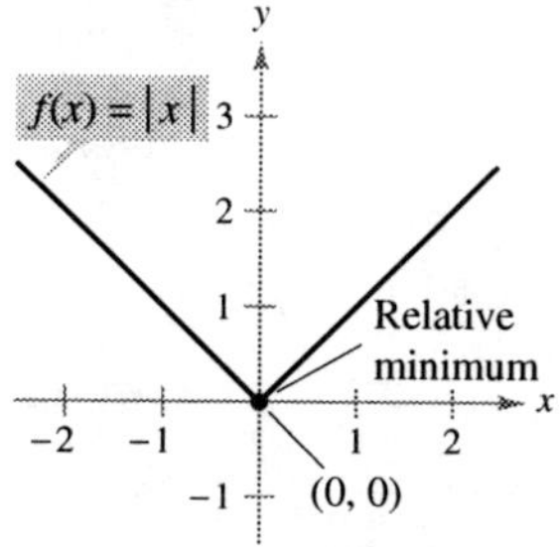

(b) $f'(0)$ does not exist.

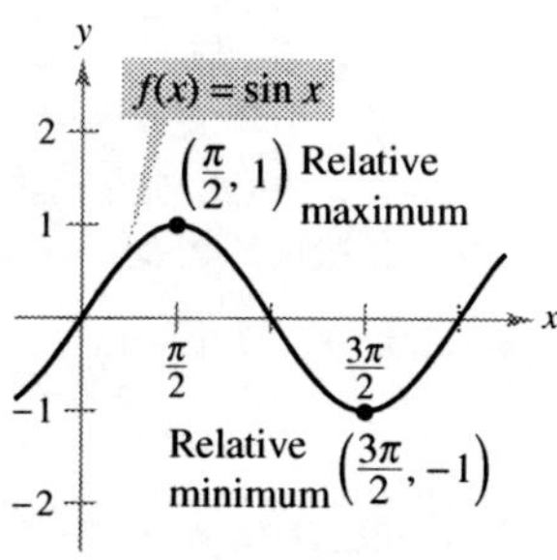

(c) $f'\left(\dfrac{\pi}{2}\right) = 0; f'\left(\dfrac{3\pi}{2}\right) = 0$

Figure 4.3

Relative Extrema and Critical Numbers

In Figure 4.2, the graph of $f(x) = x^3 - 3x^2$ has a **relative maximum** at the point $(0, 0)$ and a **relative minimum** at the point $(2, -4)$. Informally, for a continuous function, you can think of a relative maximum as occurring on a "hill" on the graph, and a relative minimum as occurring in a "valley" on the graph. Such a hill and valley can occur in two ways. If the hill (or valley) is smooth and rounded, the graph has a horizontal tangent line at the high point (or low point). If the hill (or valley) is sharp and peaked, the graph represents a function that is not differentiable at the high point (or low point).

DEFINITION OF RELATIVE EXTREMA

1. If there is an open interval containing c on which $f(c)$ is a maximum, then $f(c)$ is called a **relative maximum** of f, or you can say that f has a **relative maximum at $(c, f(c))$**.

2. If there is an open interval containing c on which $f(c)$ is a minimum, then $f(c)$ is called a **relative minimum** of f, or you can say that f has a **relative minimum at $(c, f(c))$**.

The plural of relative maximum is relative maxima, and the plural of relative minimum is relative minima. Relative maximum and relative minimum are sometimes called **local maximum** and **local minimum,** respectively.

Example 1 examines the derivatives of functions at *given* relative extrema. (Much more is said about *finding* the relative extrema of a function in Section 4.3.)

EXAMPLE 1 The Value of the Derivative at Relative Extrema

Find the value of the derivative at each relative extremum shown in Figure 4.3.

Solution

a. The derivative of $f(x) = \dfrac{9(x^2 - 3)}{x^3}$ is

$$f'(x) = \frac{x^3(18x) - (9)(x^2 - 3)(3x^2)}{(x^3)^2} \qquad \text{Differentiate using Quotient Rule.}$$

$$= \frac{9(9 - x^2)}{x^4}. \qquad \text{Simplify.}$$

At the point $(3, 2)$, the value of the derivative is $f'(3) = 0$ [see Figure 4.3(a)].

b. At $x = 0$, the derivative of $f(x) = |x|$ *does not exist* because the following one-sided limits differ [see Figure 4.3(b)].

$$\lim_{x \to 0^-} \frac{f(x) - f(0)}{x - 0} = \lim_{x \to 0^-} \frac{|x|}{x} = -1 \qquad \text{Limit from the left}$$

$$\lim_{x \to 0^+} \frac{f(x) - f(0)}{x - 0} = \lim_{x \to 0^+} \frac{|x|}{x} = 1 \qquad \text{Limit from the right}$$

c. The derivative of $f(x) = \sin x$ is

$$f'(x) = \cos x.$$

At the point $(\pi/2, 1)$, the value of the derivative is $f'(\pi/2) = \cos(\pi/2) = 0$. At the point $(3\pi/2, -1)$, the value of the derivative is $f'(3\pi/2) = \cos(3\pi/2) = 0$ [see Figure 4.3(c)]. ∎

Note in Example 1 that at each relative extremum, the derivative either is zero or does not exist. The x-values at these special points are called **critical numbers.** Figure 4.4 illustrates the two types of critical numbers. Notice in the definition that the critical number c has to be in the domain of f, but c does not have to be in the domain of f'.

 Use a graphing utility to examine the graphs of the following four functions. Only one of the functions has critical numbers. Which is it?

$$f(x) = e^x$$
$$f(x) = \ln x$$
$$f(x) = \sin x$$
$$f(x) = \tan x$$

DEFINITION OF CRITICAL NUMBER

Let f be defined at c. If $f'(c) = 0$ or if f is not differentiable at c, then c is a **critical number** of f.

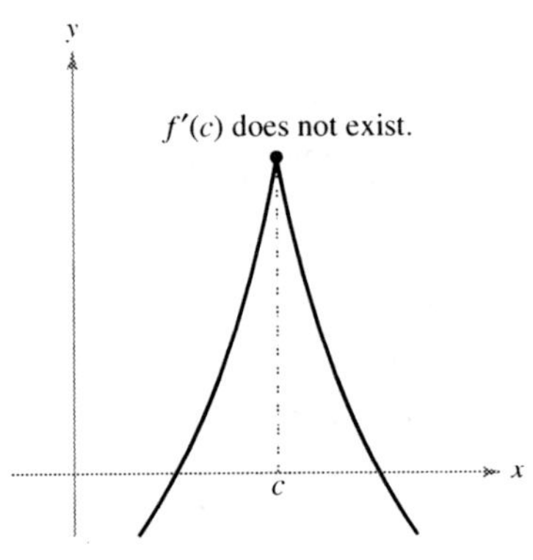

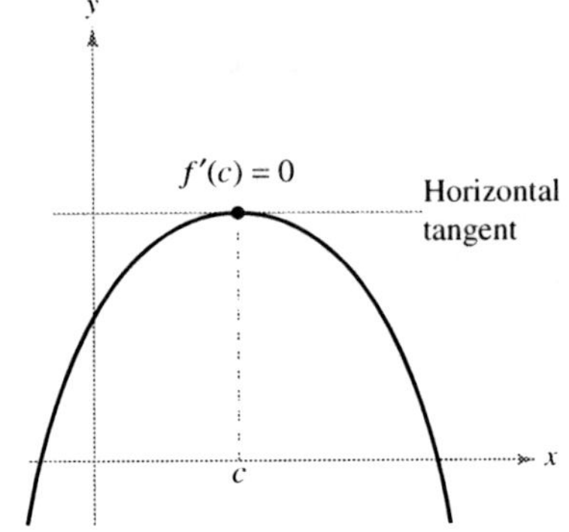

c is a critical number of f.
Figure 4.4

THEOREM 4.2 RELATIVE EXTREMA OCCUR ONLY AT CRITICAL NUMBERS

If f has a relative minimum or relative maximum at $x = c$, then c is a critical number of f.

PIERRE DE FERMAT (1601–1665)

For Fermat, who was trained as a lawyer, mathematics was more of a hobby than a profession. Nevertheless, Fermat made many contributions to analytic geometry, number theory, calculus, and probability. In letters to friends, he wrote of many of the fundamental ideas of calculus, long before Newton or Leibniz. For instance, Theorem 4.2 is sometimes attributed to Fermat.

Case 1: If f is *not* differentiable at $x = c$, then, by definition, c is a critical number of f and the theorem is valid.

Case 2: If f is differentiable at $x = c$, then $f'(c)$ must be positive, negative, or 0. Suppose $f'(c)$ is positive. Then

$$f'(c) = \lim_{x \to c} \frac{f(x) - f(c)}{x - c} > 0$$

which implies that there exists an interval (a, b) containing c such that

$$\frac{f(x) - f(c)}{x - c} > 0, \text{ for all } x \neq c \text{ in } (a, b). \qquad \text{[See Exercise 84(b), Section 2.2.]}$$

Because this quotient is positive, the signs of the denominator and numerator must agree. This produces the following inequalities for x-values in the interval (a, b).

Left of c: $\quad x < c$ and $f(x) < f(c) \implies f(c)$ is not a relative minimum.

Right of c: $\quad x > c$ and $f(x) > f(c) \implies f(c)$ is not a relative maximum.

So, the assumption that $f'(c) > 0$ contradicts the hypothesis that $f(c)$ is a relative extremum. Assuming that $f'(c) < 0$ produces a similar contradiction, you are left with only one possibility—namely, $f'(c) = 0$. So, by definition, c is a critical number of f and the theorem is valid. ∎

Finding Extrema on a Closed Interval

Theorem 4.2 states that the relative extrema of a function can occur *only* at the critical numbers of the function. Knowing this, you can use the following guidelines to find extrema on a closed interval.

GUIDELINES FOR FINDING EXTREMA ON A CLOSED INTERVAL

To find the extrema of a continuous function f on a closed interval $[a, b]$, use the following steps.

1. Find the critical numbers of f in (a, b).
2. Evaluate f at each critical number in (a, b).
3. Evaluate f at each endpoint of $[a, b]$.
4. The least of these values is the minimum. The greatest is the maximum.

The next three examples show how to apply these guidelines. Be sure you see that finding the critical numbers of the function is only part of the procedure. Evaluating the function at the critical numbers *and* the endpoints is the other part.

EXAMPLE 2 Finding Extrema on a Closed Interval

Find the extrema of $f(x) = 3x^4 - 4x^3$ on the interval $[-1, 2]$.

Solution   Begin by differentiating the function.

$$f(x) = 3x^4 - 4x^3 \qquad \text{Write original function.}$$
$$f'(x) = 12x^3 - 12x^2 \qquad \text{Differentiate.}$$

To find the critical numbers of f in the interval $(-1, 2)$, you must find all x-values for which $f'(x) = 0$ and all x-values for which $f'(x)$ does not exist.

$$f'(x) = 12x^3 - 12x^2 = 0 \qquad \text{Set } f'(x) \text{ equal to 0.}$$
$$12x^2(x - 1) = 0 \qquad \text{Factor.}$$
$$x = 0, 1 \qquad \text{Critical numbers}$$

Because f' is defined for all x, you can conclude that these are the only critical numbers of f. By evaluating f at these two critical numbers and at the endpoints of $[-1, 2]$, you can determine that the maximum is $f(2) = 16$ and the minimum is $f(1) = -1$, as shown in the table. The graph of f is shown in Figure 4.5.

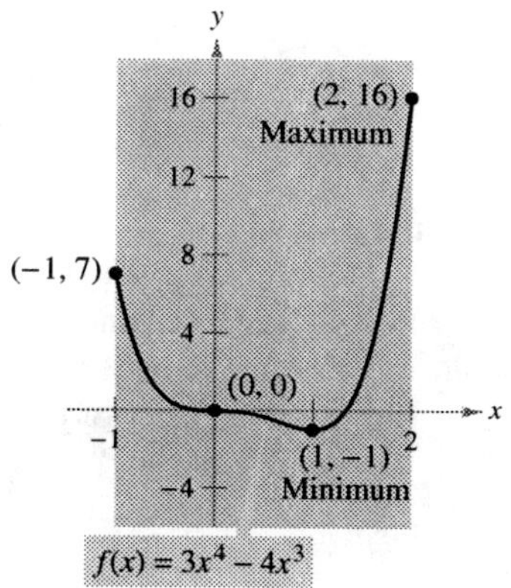

On the closed interval $[-1, 2]$, f has a minimum at $(1, -1)$ and a maximum at $(2, 16)$.

Figure 4.5

Left Endpoint	Critical Number	Critical Number	Right Endpoint
$f(-1) = 7$	$f(0) = 0$	$f(1) = -1$ Minimum	$f(2) = 16$ Maximum

In Figure 4.5, note that the critical number $x = 0$ does not yield a relative minimum or a relative maximum. This tells you that the converse of Theorem 4.2 is not true. In other words, *the critical numbers of a function need not produce relative extrema.*

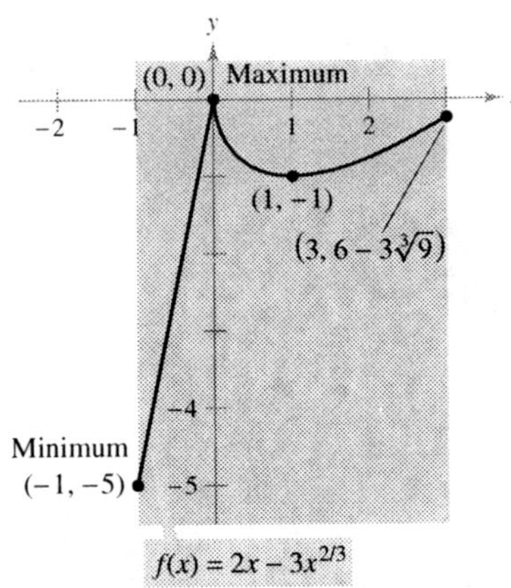

On the closed interval $[-1, 3]$, f has a minimum at $(-1, -5)$ and a maximum at $(0, 0)$.

Figure 4.6

EXAMPLE 3 Finding Extrema on a Closed Interval

Find the extrema of $f(x) = 2x - 3x^{2/3}$ on the interval $[-1, 3]$.

Solution   Begin by differentiating the function.

$$f(x) = 2x - 3x^{2/3} \qquad \text{Write original function.}$$

$$f'(x) = 2 - \frac{2}{x^{1/3}} = 2\left(\frac{x^{1/3} - 1}{x^{1/3}}\right) \qquad \text{Differentiate.}$$

From this derivative, you can see that the function has two critical numbers in the interval $[-1, 3]$. The number 1 is a critical number because $f'(1) = 0$, and the number 0 is a critical number because $f'(0)$ does not exist. By evaluating f at these two numbers and at the endpoints of the interval, you can conclude that the minimum is $f(-1) = -5$ and the maximum is $f(0) = 0$, as shown in the table. The graph of f is shown in Figure 4.6.

Left Endpoint	Critical Number	Critical Number	Right Endpoint
$f(-1) = -5$	$f(0) = 0$	$f(1) = -1$	$f(3) = 6 - 3\sqrt[3]{9} \approx -0.24$
Minimum	Maximum		

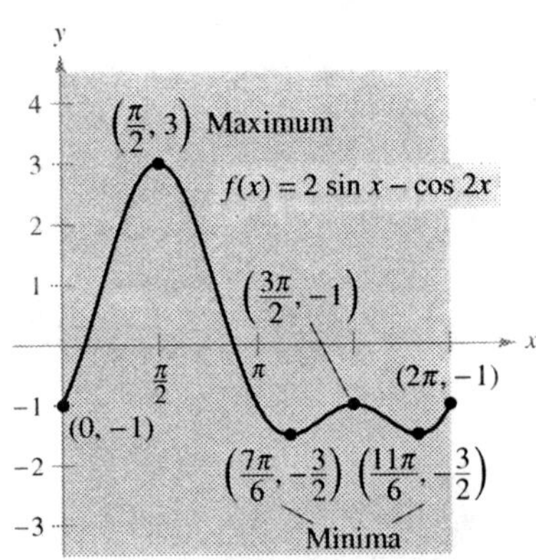

On the closed interval $[0, 2\pi]$, f has minima at $(7\pi/6, -3/2)$ and $(11\pi/6, -3/2)$ and a maximum at $(\pi/2, 3)$.

Figure 4.7

EXAMPLE 4 Finding Extrema on a Closed Interval

Find the extrema of $f(x) = 2\sin x - \cos 2x$ on the interval $[0, 2\pi]$.

Solution   This function is differentiable for all real x, so you can find all critical numbers by differentiating the function and setting $f'(x)$ equal to zero, as shown.

$$f(x) = 2\sin x - \cos 2x \qquad \text{Write original function.}$$

$$f'(x) = 2\cos x + 2\sin 2x = 0 \qquad \text{Set } f'(x) \text{ equal to 0.}$$

$$2\cos x + 4\cos x \sin x = 0 \qquad \sin 2x = 2\cos x \sin x$$

$$2(\cos x)(1 + 2\sin x) = 0 \qquad \text{Factor.}$$

In the interval $[0, 2\pi]$, the factor $\cos x$ is zero when $x = \pi/2$ and when $x = 3\pi/2$. The factor $(1 + 2\sin x)$ is zero when $x = 7\pi/6$ and when $x = 11\pi/6$. By evaluating f at these four critical numbers and at the endpoints of the interval, you can conclude that the maximum is $f(\pi/2) = 3$ and the minimum occurs at *two* points, $f(7\pi/6) = -3/2$ and $f(11\pi/6) = -3/2$, as shown in the table. The graph of f is shown in Figure 4.7.

Left Endpoint	Critical Number	Critical Number	Critical Number	Critical Number	Right Endpoint
$f(0) = -1$	$f\left(\dfrac{\pi}{2}\right) = 3$	$f\left(\dfrac{7\pi}{6}\right) = -\dfrac{3}{2}$	$f\left(\dfrac{3\pi}{2}\right) = -1$	$f\left(\dfrac{11\pi}{6}\right) = -\dfrac{3}{2}$	$f(2\pi) = -1$
	Maximum	Minimum		Minimum	

The icon ⟳ *indicates that you will find a CAS Investigation on the book's website. The CAS Investigation is a collaborative exploration of this example using the computer algebra systems* Maple *and* Mathematica.

4.1 Exercises

In Exercises 1–6, find the value of the derivative (if it exists) at each given extremum.

1. $f(x) = \dfrac{x^2}{x^2 + 4}$

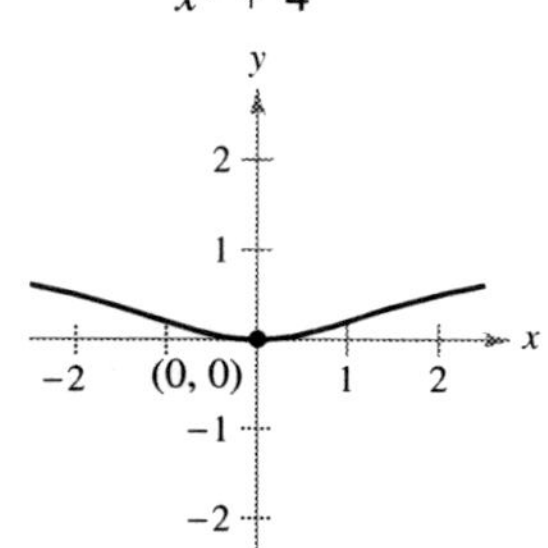

2. $f(x) = \cos \dfrac{\pi x}{2}$

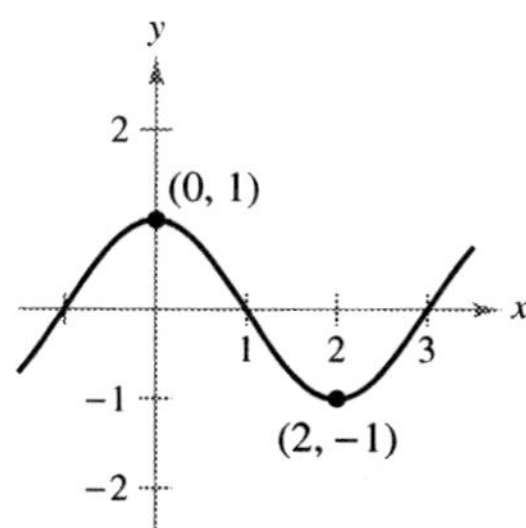

3. $g(x) = x + \dfrac{4}{x^2}$

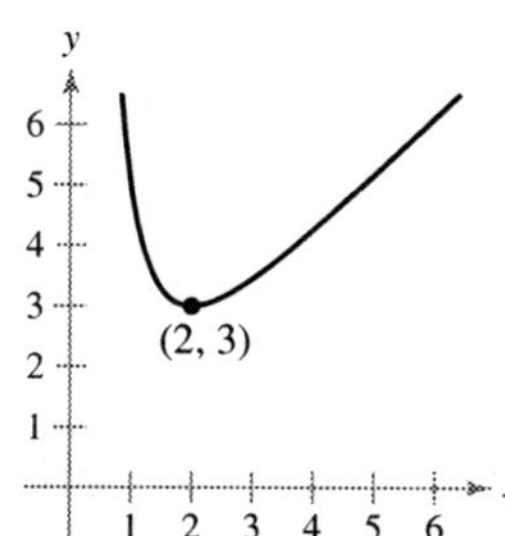

4. $f(x) = -3x\sqrt{x + 1}$

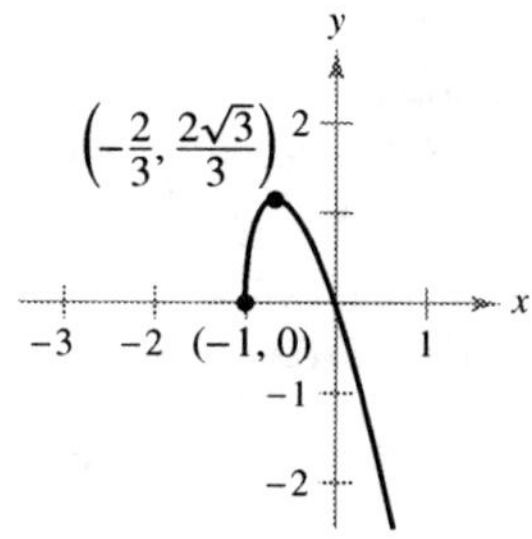

5. $f(x) = (x + 2)^{2/3}$

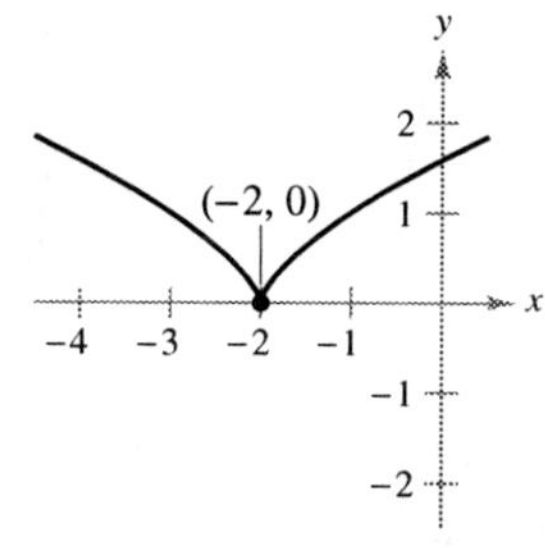

6. $f(x) = 4 - |x|$

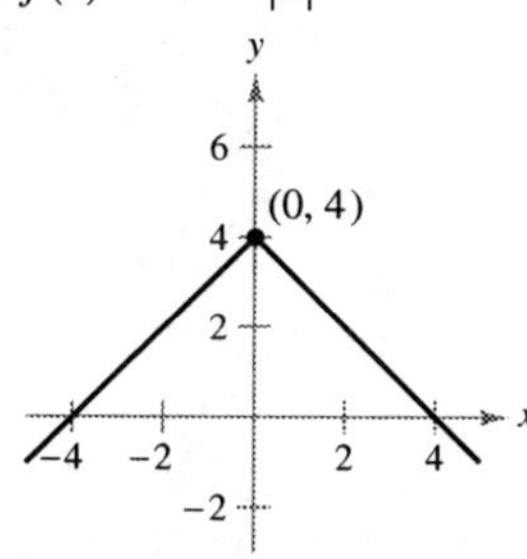

In Exercises 7–10, approximate the critical numbers of the function shown in the graph. Determine whether the function has a relative maximum, a relative minimum, an absolute maximum, an absolute minimum, or none of these at each critical number on the interval shown.

7.

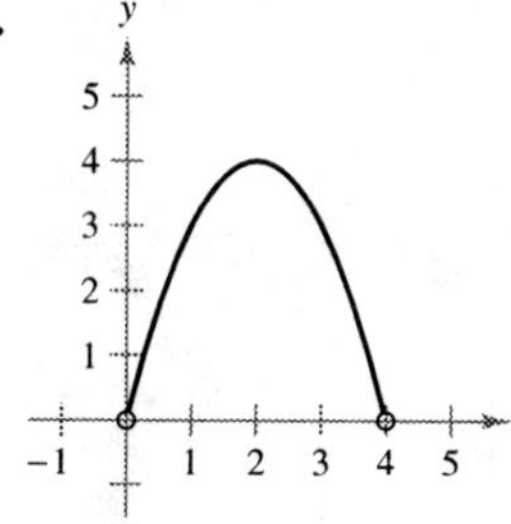

8.

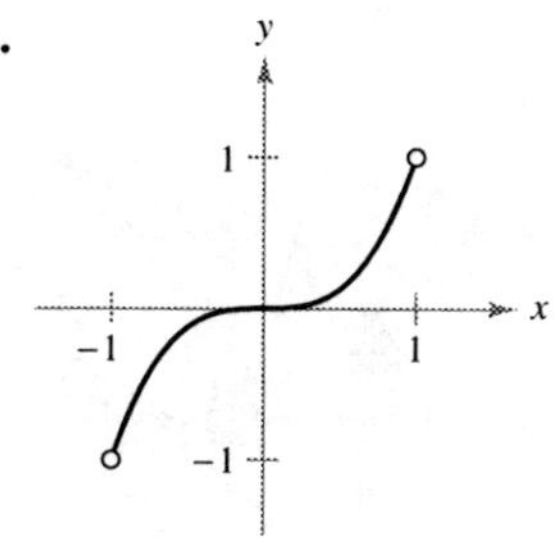

9.

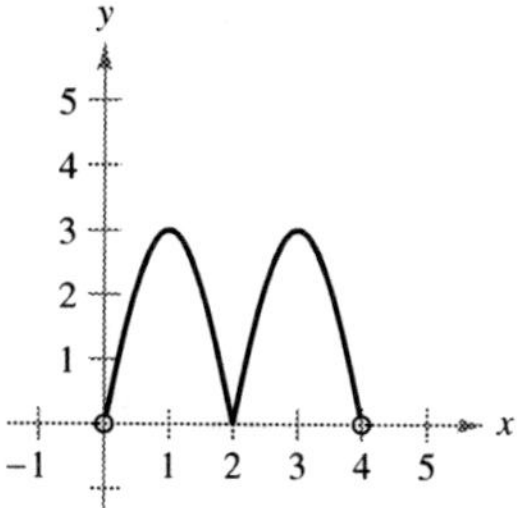

10.

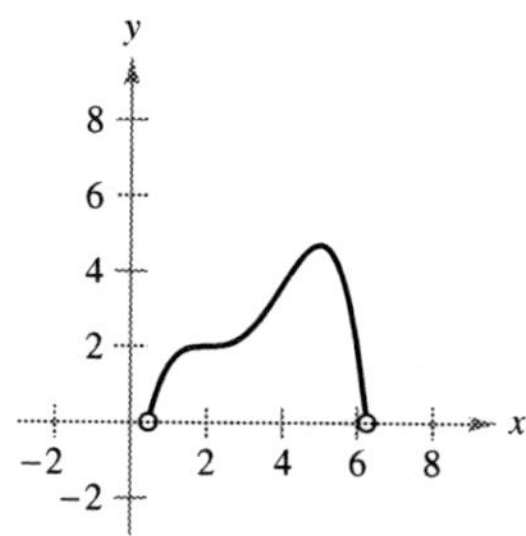

In Exercises 11–20, find any critical numbers of the function.

11. $f(x) = x^3 - 3x^2$

12. $g(x) = x^4 - 4x^2$

13. $g(t) = t\sqrt{4 - t}$, $t < 3$

14. $f(x) = \dfrac{4x}{x^2 + 1}$

15. $h(x) = \sin^2 x + \cos x$
$0 < x < 2\pi$

16. $f(\theta) = 2 \sec \theta + \tan \theta$
$0 < \theta < 2\pi$

17. $f(t) = te^{-2t}$

18. $g(t) = 2t \ln t$

19. $f(x) = x^2 \log_2(x^2 + 1)$

20. $g(x) = 4x^2(3^x)$

In Exercises 21–42, locate the absolute extrema of the function on the closed interval.

21. $f(x) = 3 - x$, $[-1, 2]$

22. $f(x) = \dfrac{2x + 5}{3}$, $[0, 5]$

23. $g(x) = x^2 - 2x$, $[0, 4]$

24. $h(x) = -x^2 + 3x - 5$, $[-2, 1]$

25. $f(x) = x^3 - \dfrac{3}{2}x^2$, $[-1, 2]$

26. $f(x) = x^3 - 12x$, $[0, 4]$

27. $y = 3x^{2/3} - 2x$, $[-1, 1]$

28. $g(x) = \sqrt[3]{x}$, $[-1, 1]$

29. $h(s) = \dfrac{1}{s - 2}$, $[0, 1]$

30. $h(t) = \dfrac{t}{t - 2}$, $[3, 5]$

31. $g(t) = \dfrac{t^2}{t^2 + 3}$, $[-1, 1]$

32. $y = 3 - |t - 3|$, $[-1, 5]$

33. $f(x) = \cos \pi x$, $\left[0, \dfrac{1}{6}\right]$

34. $g(x) = \sec x$, $\left[-\dfrac{\pi}{6}, \dfrac{\pi}{3}\right]$

35. $y = 3 \cos x$, $[0, 2\pi]$

36. $y = \tan\left(\dfrac{\pi x}{8}\right)$, $[0, 2]$

37. $f(x) = \arctan x^2$, $[-2, 1]$

38. $g(x) = \dfrac{\ln x}{x}$, $[1, 4]$

39. $h(x) = 5e^x - e^{2x}$, $[-1, 2]$

40. $y = x^2 - 8 \ln x$, $[1, 5]$

41. $y = e^x \sin x$, $[0, \pi]$

42. $y = x \ln(x + 3)$, $[0, 3]$

In Exercises 43 and 44, locate the absolute extrema of the function (if any exist) over each interval.

43. $f(x) = 2x - 3$
(a) $[0, 2]$ (b) $[0, 2)$
(c) $(0, 2]$ (d) $(0, 2)$

44. $f(x) = \sqrt{9 - x^2}$
(a) $[-3, 3]$ (b) $[-3, 0)$
(c) $(-3, 3)$ (d) $[1, 3)$

In Exercises 45–50, use a graphing utility to graph the function. Locate the absolute extrema of the function on the given interval.

Function	Interval
45. $f(x) = \begin{cases} 2x + 2, & 0 \le x \le 1 \\ 4x^2, & 1 < x \le 3 \end{cases}$	$[0, 3]$
46. $f(x) = \begin{cases} 2 - x^2, & 1 \le x < 3 \\ 2 - 3x, & 3 \le x \le 5 \end{cases}$	$[1, 5]$
47. $f(x) = \dfrac{3}{x - 1}$	$(1, 4]$
48. $f(x) = \dfrac{2}{2 - x}$	$[0, 2)$
49. $f(x) = x^4 - 2x^3 + x + 1$	$[-1, 3]$
50. $f(x) = \sqrt{x} + \cos\dfrac{x}{2}$	$[0, 2\pi]$

CAS In Exercises 51–56, (a) use a computer algebra system to graph the function and approximate any absolute extrema on the given interval. (b) Use the utility to find any critical numbers, and use them to find any absolute extrema not located at the endpoints. Compare the results with those in part (a).

Function	Interval
51. $f(x) = 3.2x^5 + 5x^3 - 3.5x$	$[0, 1]$
52. $f(x) = \dfrac{4}{3}x\sqrt{3 - x}$	$[0, 3]$
53. $f(x) = (x^2 - 2x)\ln(x + 3)$	$[0, 3]$
54. $f(x) = \sqrt{x + 4}\, e^{x^2/10}$	$[-2, 2]$
55. $f(x) = 2x \arctan(x - 1)$	$[0, 2]$
56. $f(x) = (x - 4)\arcsin\dfrac{x}{4}$	$[-2, 4]$

CAS In Exercises 57–60, use a computer algebra system to find the maximum value of $|f''(x)|$ on the closed interval. (This value is used in the error estimate for the Trapezoidal Rule, as discussed in Section 5.6.)

57. $f(x) = \sqrt{1 + x^3}$, $[0, 2]$

58. $f(x) = \dfrac{1}{x^2 + 1}$, $\left[\dfrac{1}{2}, 3\right]$

59. $f(x) = e^{-x^2/2}$, $[0, 1]$

60. $f(x) = x \ln(x + 1)$, $[0, 2]$

CAS In Exercises 61 and 62, use a computer algebra system to find the maximum value of $|f^{(4)}(x)|$ on the closed interval. (This value is used in the error estimate for Simpson's Rule, as discussed in Section 5.6.)

61. $f(x) = (x + 1)^{2/3}$, $[0, 2]$ **62.** $f(x) = \dfrac{1}{x^2 + 1}$, $[-1, 1]$

63. *Writing*   Write a short paragraph explaining why a continuous function on an open interval may not have a maximum or minimum. Illustrate your explanation with a sketch of the graph of such a function.

64. Decide whether each labeled point is an absolute maximum or minimum, a relative maximum or minimum, or neither.

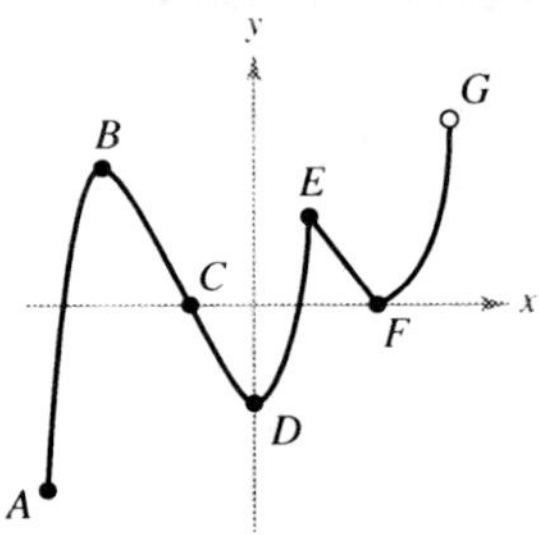

In Exercises 65 and 66, graph a function on the interval $[-2, 5]$ having the given characteristics.

65. Absolute maximum at $x = -2$, absolute minimum at $x = 1$, relative maximum at $x = 3$.

66. Relative minimum at $x = -1$, critical number (but no extremum) at $x = 0$, absolute maximum at $x = 2$, absolute minimum at $x = 5$.

In Exercises 67–69, determine from the graph whether f has a minimum in the open interval (a, b).

67. (a)   (b)

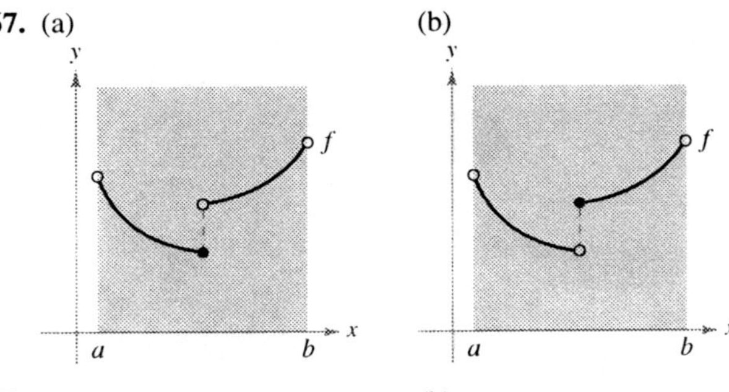

68. (a)   (b)

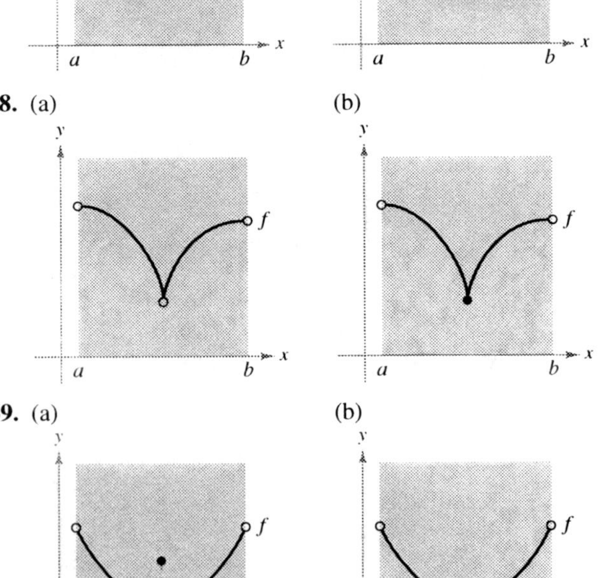

69. (a)   (b)

70. Explain why the function $f(x) = \tan x$ has a maximum on $[0, \pi/4]$ but not on $[0, \pi]$.

71. *Lawn Sprinkler* A lawn sprinkler is constructed in such a way that $d\theta/dt$ is constant, where θ ranges between $45°$ and $135°$ (see figure). The distance the water travels horizontally is

$$x = \frac{v^2 \sin 2\theta}{32}, \quad 45° \le \theta \le 135°$$

where v is the speed of the water. Find dx/dt and explain why this lawn sprinkler does not water evenly. What part of the lawn receives the most water?

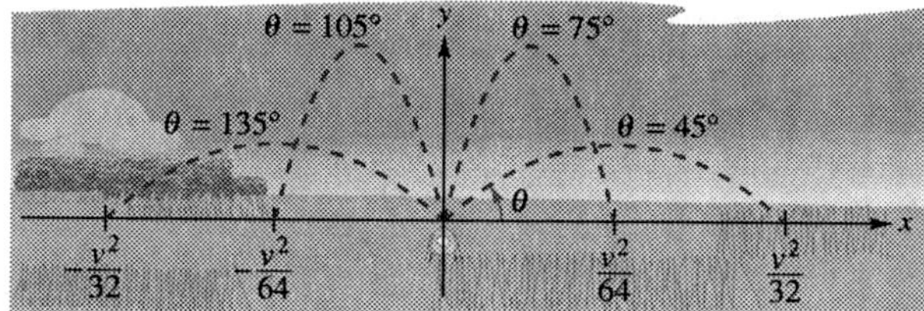

Water sprinkler: $45° \le \theta \le 135°$

■ **FOR FURTHER INFORMATION** For more information on the "calculus of lawn sprinklers," see the article "Design of an Oscillating Sprinkler" by Bart Braden in *Mathematics Magazine*. To view this article, go to the website *www.matharticles.com*.

72. *Honeycomb* The surface area of a cell in a honeycomb is

$$S = 6hs + \frac{3s^2}{2}\left(\frac{\sqrt{3} - \cos \theta}{\sin \theta}\right)$$

where h and s are positive constants and θ is the angle at which the upper faces meet the altitude of the cell (see figure). Find the angle $\theta \ (\pi/6 \le \theta \le \pi/2)$ that minimizes the surface area S.

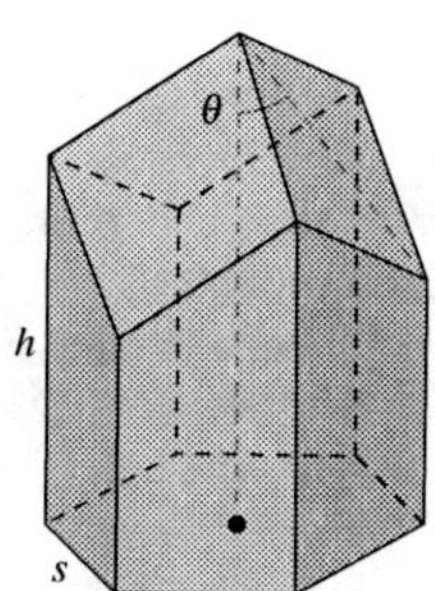

■ **FOR FURTHER INFORMATION** For more information on the geometric structure of a honeycomb cell, see the article "The Design of Honeycombs" by Anthony L. Peressini in UMAP Module 502, published by COMAP, Inc., Suite 210, 57 Bedford Street, Lexington, MA.

True or False? **In Exercises 73–76, determine whether the statement is true or false. If it is false, explain why or give an example that shows it is false.**

73. The maximum of a function that is continuous on a closed interval can occur at two different values in the interval.

74. If a function is continuous on a closed interval, then it must have a minimum on the interval.

75. If $x = c$ is a critical number of the function f, then it is also a critical number of the function $g(x) = f(x) + k$, where k is a constant.

76. If $x = c$ is a critical number of the function f, then it is also a critical number of the function $g(x) = f(x - k)$, where k is a constant.

77. Let the function f be differentiable on an interval I containing c. If f has a maximum value at $x = c$, show that $-f$ has a minimum value at $x = c$.

78. Consider the cubic function $f(x) = ax^3 + bx^2 + cx + d$, where $a \ne 0$. Show that f can have zero, one, or two critical numbers and give an example of each case.

79. *Highway Design* In order to build a highway, it is necessary to fill a section of a valley where the grades (slopes) of the sides are 9% and 6% (see figure). The top of the filled region will have the shape of a parabolic arc that is tangent to the two slopes at the points A and B. The horizontal distances from A to the y-axis and from B to the y-axis are both 500 feet.

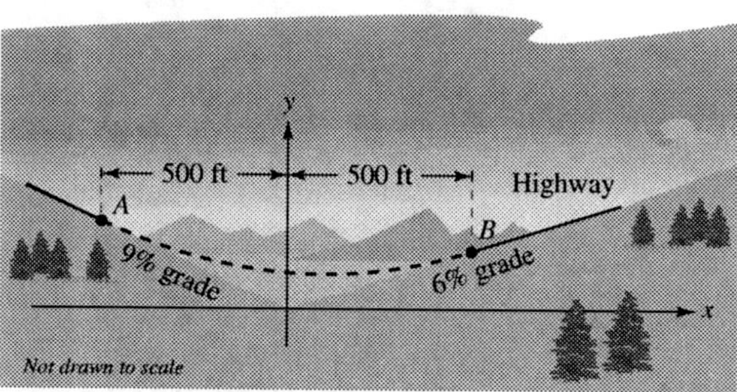

(a) Find the coordinates of A and B.

(b) Find a quadratic function $y = ax^2 + bx + c$, $-500 \le x \le 500$, that describes the top of the filled region.

(c) Construct a table giving the depths d of the fill for $x = -500, -400, -300, -200, -100, 0, 100, 200, 300, 400,$ and 500.

(d) What will be the lowest point on the completed highway? Will it be directly over the point where the two hillsides come together?

80. Determine all real numbers $a > 0$ for which there exists a nonnegative continuous function $f(x)$ defined on $[0, a]$ with the property that the region $R = \{(x, y); \ 0 \le x \le a, \ 0 \le y \le f(x)\}$ has perimeter k units and area k square units for some real number k.

4.2 Rolle's Theorem and the Mean Value Theorem

- Understand and use Rolle's Theorem.
- Understand and use the Mean Value Theorem.

Rolle's Theorem

> **ROLLE'S THEOREM**
>
> French mathematician Michel Rolle first published the theorem that bears his name in 1691. Before this time, however, Rolle was one of the most vocal critics of calculus, stating that it gave erroneous results and was based on unsound reasoning. Later in life, Rolle came to see the usefulness of calculus.

The Extreme Value Theorem (Section 4.1) states that a continuous function on a closed interval $[a, b]$ must have both a minimum and a maximum on the interval. Both of these values, however, can occur at the endpoints. **Rolle's Theorem,** named after the French mathematician Michel Rolle (1652–1719), gives conditions that guarantee the existence of an extreme value in the *interior* of a closed interval.

EXPLORATION

Extreme Values in a Closed Interval   Sketch a rectangular coordinate plane on a piece of paper. Label the points $(1, 3)$ and $(5, 3)$. Using a pencil or pen, draw the graph of a differentiable function f that starts at $(1, 3)$ and ends at $(5, 3)$. Is there at least one point on the graph for which the derivative is zero? Would it be possible to draw the graph so that there *isn't* a point for which the derivative is zero? Explain your reasoning.

THEOREM 4.3 ROLLE'S THEOREM

Let f be continuous on the closed interval $[a, b]$ and differentiable on the open interval (a, b). If

$$f(a) = f(b)$$

then there is at least one number c in (a, b) such that $f'(c) = 0$.

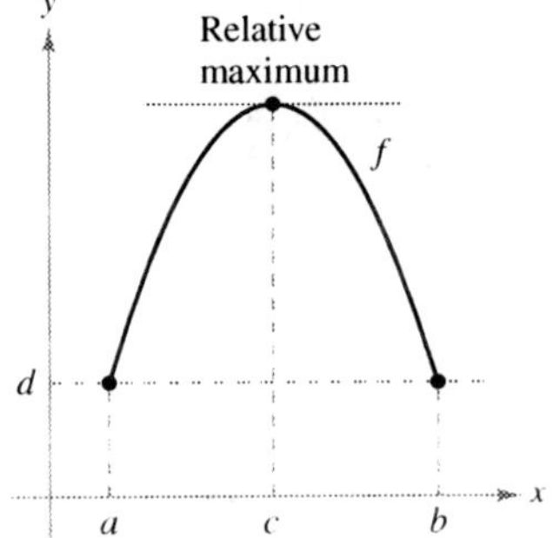

(a) f is continuous on $[a, b]$ and differentiable on (a, b).

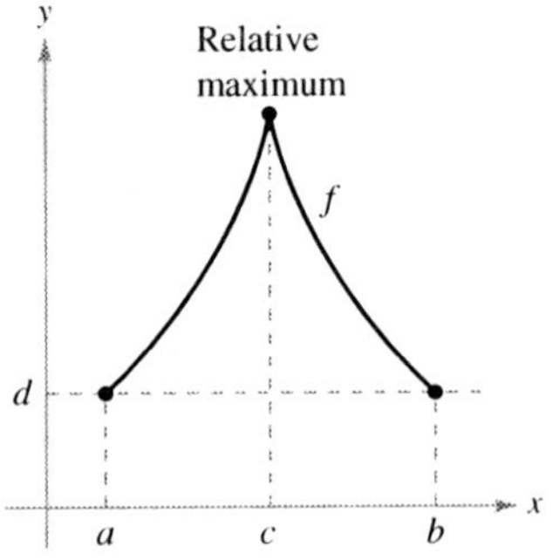

(b) f is continuous on $[a, b]$.

Figure 4.8

PROOF Let $f(a) = d = f(b)$.

Case 1: If $f(x) = d$ for all x in $[a, b]$, then f is constant on the interval and, by Theorem 3.2, $f'(x) = 0$ for all x in (a, b).

Case 2: Suppose $f(x) > d$ for some x in (a, b). By the Extreme Value Theorem, you know that f has a maximum at some c in the interval. Moreover, because $f(c) > d$, this maximum does not occur at either endpoint. So, f has a maximum in the *open* interval (a, b). This implies that $f(c)$ is a *relative* maximum and, by Theorem 4.2, c is a critical number of f. Finally, because f is differentiable at c, you can conclude that $f'(c) = 0$.

Case 3: If $f(x) < d$ for some x in (a, b), you can use an argument similar to that in Case 2, but involving the minimum instead of the maximum. ∎

From Rolle's Theorem, you can see that if a function f is continuous on $[a, b]$ and differentiable on (a, b), and if $f(a) = f(b)$, then there must be at least one x-value between a and b at which the graph of f has a horizontal tangent, as shown in Figure 4.8(a). If the differentiability requirement is dropped from Rolle's Theorem, f will still have a critical number in (a, b), but it may not yield a horizontal tangent. Such a case is shown in Figure 4.8(b).

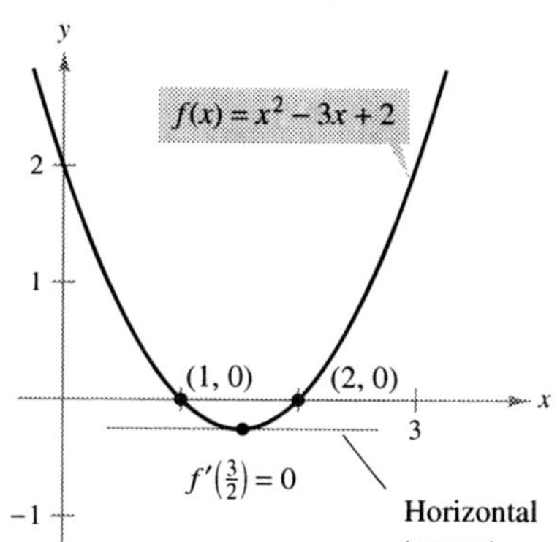

The x-value for which $f'(x) = 0$ is between the two x-intercepts.
Figure 4.9

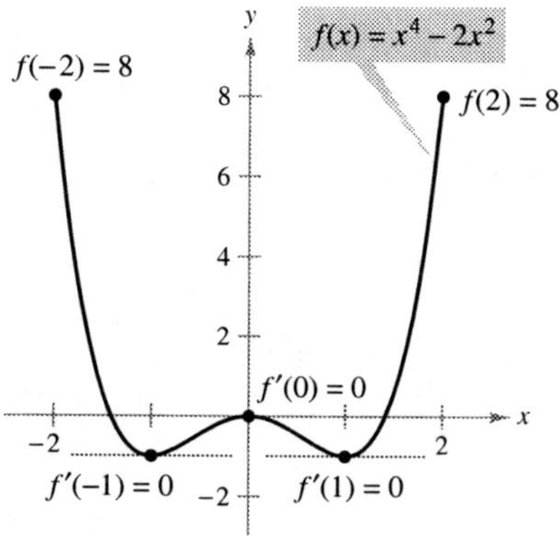

$f'(x) = 0$ for more than one x-value in the interval $(-2, 2)$.
Figure 4.10

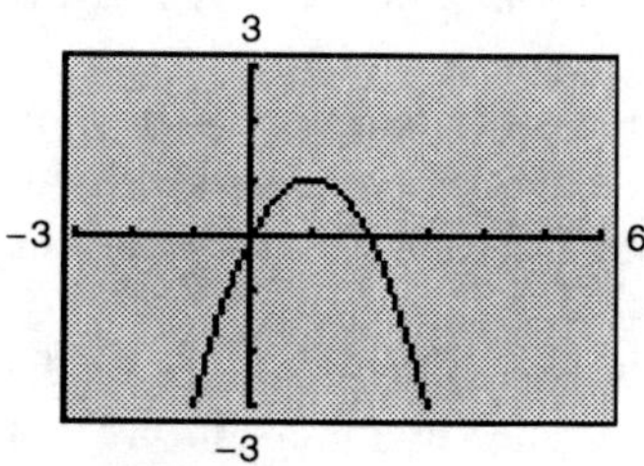

Figure 4.11

EXAMPLE 1 Illustrating Rolle's Theorem

Find the two x-intercepts of

$$f(x) = x^2 - 3x + 2$$

and show that $f'(x) = 0$ at some point between the two x-intercepts.

Solution   Note that f is differentiable on the entire real number line. Setting $f(x)$ equal to 0 produces

$$x^2 - 3x + 2 = 0 \qquad \text{Set } f(x) \text{ equal to 0.}$$
$$(x - 1)(x - 2) = 0. \qquad \text{Factor.}$$

So, $f(1) = f(2) = 0$, and from Rolle's Theorem you know that there *exists* at least one c in the interval $(1, 2)$ such that $f'(c) = 0$. To *find* such a c, you can solve the equation

$$f'(x) = 2x - 3 = 0 \qquad \text{Set } f'(x) \text{ equal to 0.}$$

and determine that $f'(x) = 0$ when $x = \frac{3}{2}$. Note that this x-value lies in the open interval $(1, 2)$, as shown in Figure 4.9. ∎

Rolle's Theorem states that if f satisfies the conditions of the theorem, there must be *at least* one point between a and b at which the derivative is 0. There may of course be more than one such point, as shown in the next example.

EXAMPLE 2 Illustrating Rolle's Theorem

Let $f(x) = x^4 - 2x^2$. Find all values of c in the interval $(-2, 2)$ such that $f'(c) = 0$.

Solution   To begin, note that the function satisfies the conditions of Rolle's Theorem. That is, f is continuous on the interval $[-2, 2]$ and differentiable on the interval $(-2, 2)$. Moreover, because $f(-2) = f(2) = 8$, you can conclude that there exists at least one c in $(-2, 2)$ such that $f'(c) = 0$. Setting the derivative equal to 0 produces

$$f'(x) = 4x^3 - 4x = 0 \qquad \text{Set } f'(x) \text{ equal to 0.}$$
$$4x(x - 1)(x + 1) = 0 \qquad \text{Factor.}$$
$$x = 0, 1, -1. \qquad x\text{-values for which } f'(x) = 0$$

So, in the interval $(-2, 2)$, the derivative is zero at three different values of x, as shown in Figure 4.10. ∎

TECHNOLOGY PITFALL   A graphing utility can be used to indicate whether the points on the graphs in Examples 1 and 2 are relative minima or relative maxima of the functions. When using a graphing utility, however, you should keep in mind that it can give misleading pictures of graphs. For example, use a graphing utility to graph

$$f(x) = 1 - (x - 1)^2 - \frac{1}{1000(x - 1)^{1/7} + 1}.$$

With most viewing windows, it appears that the function has a maximum of 1 when $x = 1$ (see Figure 4.11). By evaluating the function at $x = 1$, however, you can see that $f(1) = 0$. To determine the behavior of this function near $x = 1$, you need to examine the graph analytically to get the complete picture.

The Mean Value Theorem

Rolle's Theorem can be used to prove another theorem—the **Mean Value Theorem.**

THEOREM 4.4 THE MEAN VALUE THEOREM

If f is continuous on the closed interval $[a, b]$ and differentiable on the open interval (a, b), then there exists a number c in (a, b) such that

$$f'(c) = \frac{f(b) - f(a)}{b - a}.$$

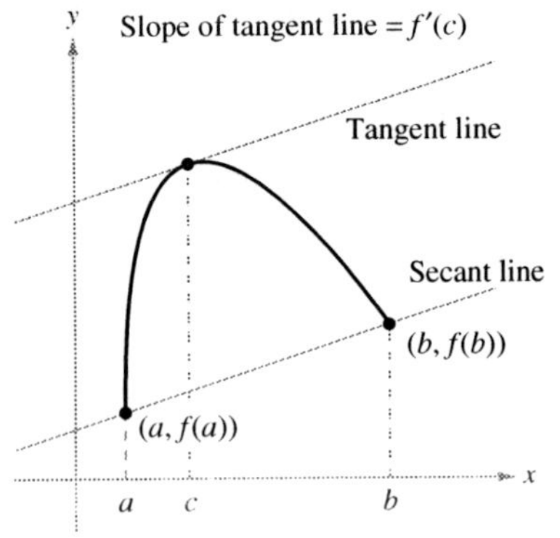

Figure 4.12

PROOF Refer to Figure 4.12. The equation of the secant line that passes through the points $(a, f(a))$ and $(b, f(b))$ is

$$y = \left[\frac{f(b) - f(a)}{b - a} \right](x - a) + f(a).$$

Let $g(x)$ be the difference between $f(x)$ and y. Then

$$g(x) = f(x) - y$$
$$= f(x) - \left[\frac{f(b) - f(a)}{b - a} \right](x - a) - f(a).$$

By evaluating g at a and b, you can see that $g(a) = 0 = g(b)$. Because f is continuous on $[a, b]$, it follows that g is also continuous on $[a, b]$. Furthermore, because f is differentiable, g is also differentiable, and you can apply Rolle's Theorem to the function g. So, there exists a number c in (a, b) such that $g'(c) = 0$, which implies that

$$0 = g'(c)$$
$$= f'(c) - \frac{f(b) - f(a)}{b - a}.$$

So, there exists a number c in (a, b) such that

$$f'(c) = \frac{f(b) - f(a)}{b - a}. \qquad \blacksquare$$

NOTE The "mean" in the Mean Value Theorem refers to the mean (or average) rate of change of f in the interval $[a, b]$. $\qquad \blacksquare$

Although the Mean Value Theorem can be used directly in problem solving, it is used more often to prove other theorems. In fact, some people consider this to be the most important theorem in calculus—it is closely related to the Fundamental Theorem of Calculus discussed in Section 5.4. For now, you can get an idea of the versatility of the Mean Value Theorem by looking at the results stated in Exercises 89–97 in this section.

The Mean Value Theorem has implications for both basic interpretations of the derivative. Geometrically, the theorem guarantees the existence of a tangent line that is parallel to the secant line through the points $(a, f(a))$ and $(b, f(b))$, as shown in Figure 4.12. Example 3 illustrates this geometric interpretation of the Mean Value Theorem. In terms of rates of change, the Mean Value Theorem implies that there must be a point in the open interval (a, b) at which the instantaneous rate of change is equal to the average rate of change over the interval $[a, b]$. This is illustrated in Example 4.

JOSEPH-LOUIS LAGRANGE (1736–1813)

The Mean Value Theorem was first proved by the famous mathematician Joseph-Louis Lagrange. Born in Italy, Lagrange held a position in the court of Frederick the Great in Berlin for 20 years. Afterward, he moved to France, where he met emperor Napoleon Bonaparte, who is quoted as saying, "Lagrange is the lofty pyramid of the mathematical sciences."

EXAMPLE 3   Finding a Tangent Line

Given $f(x) = 5 - (4/x)$, find all values of c in the open interval $(1, 4)$ such that

$$f'(c) = \frac{f(4) - f(1)}{4 - 1}.$$

Solution   The slope of the secant line through $(1, f(1))$ and $(4, f(4))$ is

$$\frac{f(4) - f(1)}{4 - 1} = \frac{4 - 1}{4 - 1} = 1.$$

Note that the function satisfies the conditions of the Mean Value Theorem. That is, f is continuous on the interval $[1, 4]$ and differentiable on the interval $(1, 4)$. So, there exists at least one number c in $(1, 4)$ such that $f'(c) = 1$. Solving the equation $f'(x) = 1$ yields

$$f'(x) = \frac{4}{x^2} = 1$$

which implies that $x = \pm 2$. So, in the interval $(1, 4)$, you can conclude that $c = 2$, as shown in Figure 4.13.

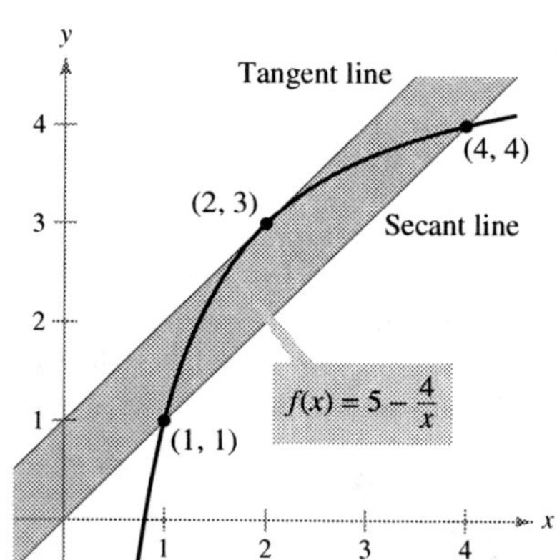

The tangent line at $(2, 3)$ is parallel to the secant line through $(1, 1)$ and $(4, 4)$.

Figure 4.13

EXAMPLE 4   Finding an Instantaneous Rate of Change

Two stationary patrol cars equipped with radar are 5 miles apart on a highway, as shown in Figure 4.14. As a truck passes the first patrol car, its speed is clocked at 55 miles per hour. Four minutes later, when the truck passes the second patrol car, its speed is clocked at 50 miles per hour. Prove that the truck must have exceeded the speed limit (of 55 miles per hour) at some time during the 4 minutes.

Solution   Let $t = 0$ be the time (in hours) when the truck passes the first patrol car. The time when the truck passes the second patrol car is

$$t = \frac{4}{60} = \frac{1}{15} \text{ hour.}$$

By letting $s(t)$ represent the distance (in miles) traveled by the truck, you have $s(0) = 0$ and $s\left(\frac{1}{15}\right) = 5$. So, the average velocity of the truck over the five-mile stretch of highway is

$$\text{Average velocity} = \frac{s(1/15) - s(0)}{(1/15) - 0} = \frac{5}{1/15} = 75 \text{ miles per hour.}$$

Assuming that the position function is differentiable, you can apply the Mean Value Theorem to conclude that the truck must have been traveling at a rate of 75 miles per hour sometime during the 4 minutes.   ■

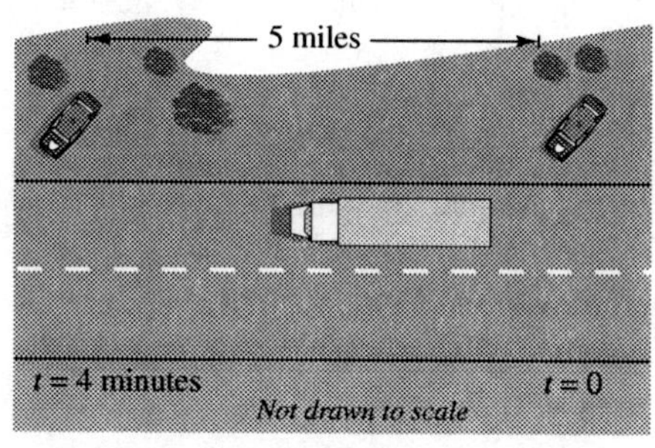

At some time t, the instantaneous velocity is equal to the average velocity over 4 minutes.

Figure 4.14

A useful alternative form of the Mean Value Theorem is as follows: If f is continuous on $[a, b]$ and differentiable on (a, b), then there exists a number c in (a, b) such that

$$f(b) = f(a) + (b - a)f'(c).$$ Alternative form of Mean Value Theorem

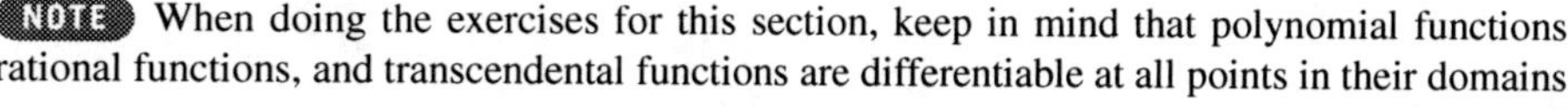

NOTE   When doing the exercises for this section, keep in mind that polynomial functions, rational functions, and transcendental functions are differentiable at all points in their domains.   ■

4.2 Exercises

In Exercises 1–4, explain why Rolle's Theorem does not apply to the function even though there exist a and b such that $f(a) = f(b)$.

1. $f(x) = \left| \dfrac{1}{x} \right|$, $[-1, 1]$

2. $f(x) = \cot \dfrac{x}{2}$, $[\pi, 3\pi]$

3. $f(x) = 1 - |x - 1|$, $[0, 2]$

4. $f(x) = \sqrt{(2 - x^{2/3})^3}$, $[-1, 1]$

In Exercises 5–8, find the two x-intercepts of the function f and show that $f'(x) = 0$ at some point between the two x-intercepts.

5. $f(x) = x^2 - x - 2$

6. $f(x) = x(x - 3)$

7. $f(x) = x\sqrt{x + 4}$

8. $f(x) = -3x\sqrt{x + 1}$

Rolle's Theorem   **In Exercises 9 and 10, the graph of f is shown. Apply Rolle's Theorem and find all values of c such that $f'(c) = 0$ at some point between the labeled intercepts.**

9.

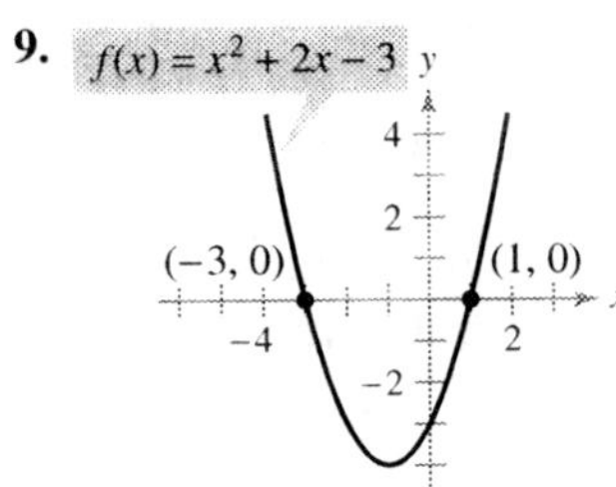

10.

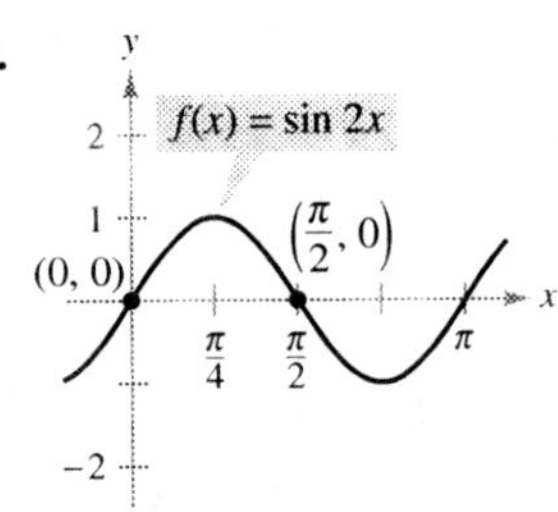

In Exercises 11–26, determine whether Rolle's Theorem can be applied to f on the closed interval $[a, b]$. If Rolle's Theorem can be applied, find all values of c in the open interval (a, b) such that $f'(c) = 0$. If Rolle's Theorem cannot be applied, explain why not.

11. $f(x) = -x^2 + 3x$, $[0, 3]$

12. $f(x) = x^2 - 5x + 4$, $[1, 4]$

13. $f(x) = (x - 1)(x - 2)(x - 3)$, $[1, 3]$

14. $f(x) = (x - 3)(x + 1)^2$, $[-1, 3]$

15. $f(x) = x^{2/3} - 1$, $[-8, 8]$

16. $f(x) = 3 - |x - 3|$, $[0, 6]$

17. $f(x) = \dfrac{x^2 - 2x - 3}{x + 2}$, $[-1, 3]$

18. $f(x) = \dfrac{x^2 - 1}{x}$, $[-1, 1]$

19. $f(x) = (x^2 - 2x)e^x$, $[0, 2]$

20. $f(x) = x - 2 \ln x$, $[1, 3]$

21. $f(x) = \sin x$, $[0, 2\pi]$

22. $f(x) = \cos x$, $[0, 2\pi]$

23. $f(x) = \dfrac{6x}{\pi} - 4 \sin^2 x$, $\left[0, \dfrac{\pi}{6}\right]$

24. $f(x) = \cos 2x$, $[-\pi, \pi]$

25. $f(x) = \tan x$, $[0, \pi]$

26. $f(x) = \sec x$, $[\pi, 2\pi]$

In Exercises 27–32, use a graphing utility to graph the function on the closed interval $[a, b]$. Determine whether Rolle's Theorem can be applied to f on the interval and, if so, find all values of c in the open interval (a, b) such that $f'(c) = 0$.

27. $f(x) = |x| - 1$, $[-1, 1]$

28. $f(x) = x - x^{1/3}$, $[0, 1]$

29. $f(x) = x - \tan \pi x$, $\left[-\frac{1}{4}, \frac{1}{4}\right]$

30. $f(x) = \dfrac{x}{2} - \sin \dfrac{\pi x}{6}$, $[-1, 0]$

31. $f(x) = 2 + \arcsin(x^2 - 1)$, $[-1, 1]$

32. $f(x) = 2 + (x^2 - 4x)(2^{-x/4})$, $[0, 4]$

33. *Vertical Motion*   The height of a ball t seconds after it is thrown upward from a height of 6 feet and with an initial velocity of 48 feet per second is $f(t) = -16t^2 + 48t + 6$.

(a) Verify that $f(1) = f(2)$.

(b) According to Rolle's Theorem, what must the velocity be at some time in the interval $(1, 2)$? Find that time.

34. *Reorder Costs*   The ordering and transportation cost C of components used in a manufacturing process is approximated by

$$C(x) = 10\left(\frac{1}{x} + \frac{x}{x + 3} \right)$$

where C is measured in thousands of dollars and x is the order size in hundreds.

(a) Verify that $C(3) = C(6)$.

(b) According to Rolle's Theorem, the rate of change of cost must be 0 for some order size in the interval $(3, 6)$. Find that order size.

In Exercises 35 and 36, copy the graph and sketch the secant line to the graph through the points $(a, f(a))$ and $(b, f(b))$. Then sketch any tangent lines to the graph for each value of c guaranteed by the Mean Value Theorem. To print an enlarged copy of the graph, go to the website *www.mathgraphs.com*.

35.

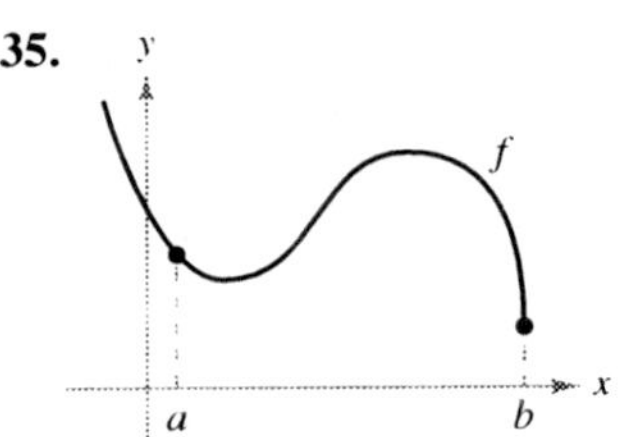

36.

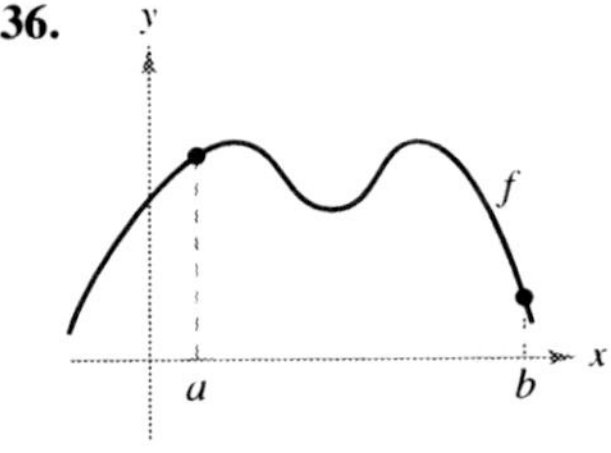

Writing In Exercises 37–40, explain why the Mean Value Theorem does not apply to the function f on the interval $[0, 6]$.

37.
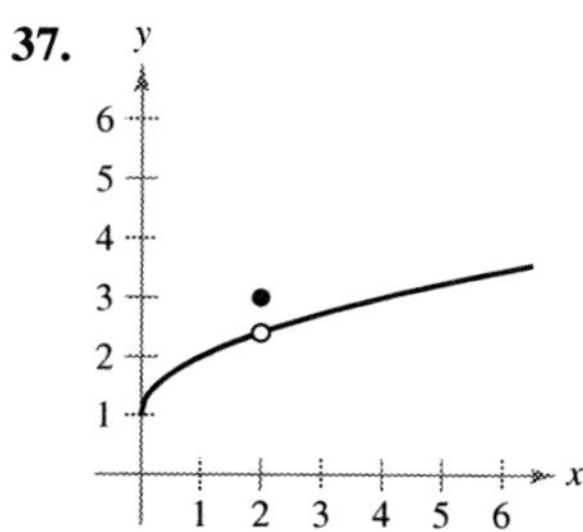

38.
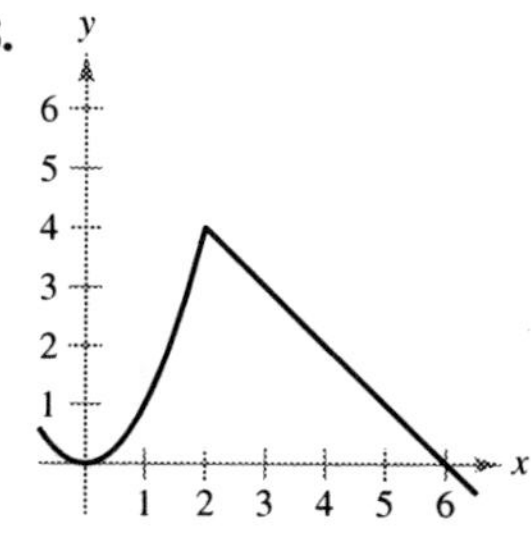

39. $f(x) = \dfrac{1}{x - 3}$

40. $f(x) = |x - 3|$

41. *Mean Value Theorem* Consider the graph of the function $f(x) = -x^2 + 5$. (a) Find the equation of the secant line joining the points $(-1, 4)$ and $(2, 1)$. (b) Use the Mean Value Theorem to determine a point c in the interval $(-1, 2)$ such that the tangent line at c is parallel to the secant line. (c) Find the equation of the tangent line through c. (d) Then use a graphing utility to graph f, the secant line, and the tangent line.

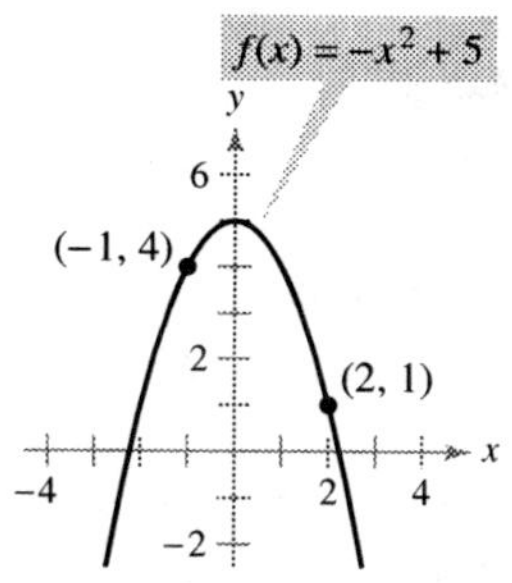

Figure for 41

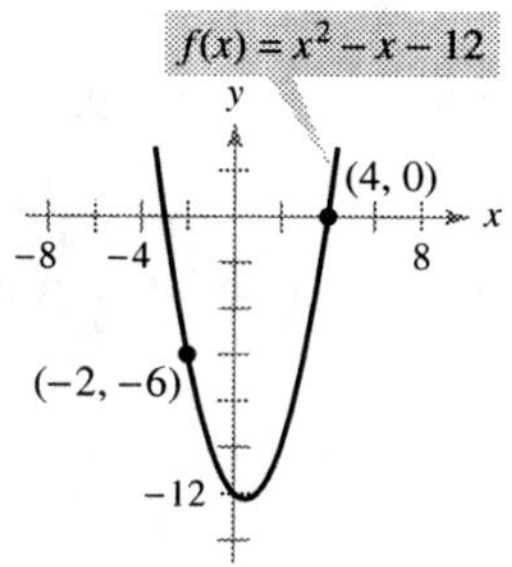

Figure for 42

42. *Mean Value Theorem* Consider the graph of the function $f(x) = x^2 - x - 12$. (a) Find the equation of the secant line joining the points $(-2, -6)$ and $(4, 0)$. (b) Use the Mean Value Theorem to determine a point c in the interval $(-2, 4)$ such that the tangent line at c is parallel to the secant line. (c) Find the equation of the tangent line through c. (d) Then use a graphing utility to graph f, the secant line, and the tangent line.

In Exercises 43–56, determine whether the Mean Value Theorem can be applied to f on the closed interval $[a, b]$. If the Mean Value Theorem can be applied, find all values of c in the open interval (a, b) such that $f'(c) = \dfrac{f(b) - f(a)}{b - a}$. If the Mean Value Theorem cannot be applied, explain why not.

43. $f(x) = x^2$, $[-2, 1]$

44. $f(x) = x^3$, $[0, 1]$

45. $f(x) = x^3 + 2x$, $[-1, 1]$

46. $f(x) = x^4 - 8x$, $[0, 2]$

47. $f(x) = x^{2/3}$, $[0, 1]$

48. $f(x) = \dfrac{x + 1}{x}$, $[-1, 2]$

49. $f(x) = |2x + 1|$, $[-1, 3]$

50. $f(x) = \sqrt{2 - x}$, $[-7, 2]$

51. $f(x) = \sin x$, $[0, \pi]$

52. $f(x) = \cos x + \tan x$, $[0, \pi]$

53. $f(x) = e^{-3x}$, $[0, 2]$

54. $f(x) = (x + 3) \ln(x + 3)$, $[-2, -1]$

55. $f(x) = x \log_2 x$, $[1, 2]$

56. $f(x) = \arctan(1 - x)$, $[0, 1]$

In Exercises 57–62, use a graphing utility to (a) graph the function f on the given interval, (b) find and graph the secant line through points on the graph of f at the endpoints of the given interval, and (c) find and graph any tangent lines to the graph of f that are parallel to the secant line.

57. $f(x) = \dfrac{x}{x + 1}$, $\left[-\dfrac{1}{2}, 2\right]$

58. $f(x) = x - 2 \sin x$, $[-\pi, \pi]$

59. $f(x) = \sqrt{x}$, $[1, 9]$

60. $f(x) = x^4 - 2x^3 + x^2$, $[0, 6]$

61. $f(x) = 2e^{x/4} \cos \dfrac{\pi x}{4}$, $[0, 2]$

62. $f(x) = \ln|\sec \pi x|$, $\left[0, \tfrac{1}{4}\right]$

WRITING ABOUT CONCEPTS

63. Let f be continuous on $[a, b]$ and differentiable on (a, b). If there exists c in (a, b) such that $f'(c) = 0$, does it follow that $f(a) = f(b)$? Explain.

64. Let f be continuous on the closed interval $[a, b]$ and differentiable on the open interval (a, b). Also, suppose that $f(a) = f(b)$ and that c is a real number in the interval such that $f'(c) = 0$. Find an interval for the function g over which Rolle's Theorem can be applied, and find the corresponding critical number of g (k is a constant).

(a) $g(x) = f(x) + k$ (b) $g(x) = f(x - k)$

(c) $g(x) = f(kx)$

65. The function
$$f(x) = \begin{cases} 0, & x = 0 \\ 1 - x, & 0 < x \le 1 \end{cases}$$
is differentiable on $(0, 1)$ and satisfies $f(0) = f(1)$. However, its derivative is never zero on $(0, 1)$. Does this contradict Rolle's Theorem? Explain.

66. Can you find a function f such that $f(-2) = -2$, $f(2) = 6$, and $f'(x) < 1$ for all x? Why or why not?

67. *Speed* A plane begins its takeoff at 2:00 P.M. on a 2500-mile flight. After 5.5 hours, the plane arrives at its destination. Explain why there are at least two times during the flight when the speed of the plane is 400 miles per hour.

68. *Temperature* When an object is removed from a furnace and placed in an environment with a constant temperature of 90°F, its core temperature is 1500°F. Five hours later the core temperature is 390°F. Explain why there must exist a time in the interval when the temperature is decreasing at a rate of 222°F per hour.

69. *Velocity* Two bicyclists begin a race at 8:00 A.M. They both finish the race 2 hours and 15 minutes later. Prove that at some time during the race, the bicyclists are traveling at the same velocity.

70. *Acceleration* At 9:13 A.M., a sports car is traveling 35 miles per hour. Two minutes later, the car is traveling 85 miles per hour. Prove that at some time during this two-minute interval, the car's acceleration is exactly 1500 miles per hour squared.

71. Consider the function $f(x) = 3 \cos^2\left(\dfrac{\pi x}{2}\right)$.

(a) Use a graphing utility to graph f and f'.

(b) Is f a continuous function? Is f' a continuous function?

(c) Does Rolle's Theorem apply on the interval $[-1, 1]$? Does it apply on the interval $[1, 2]$? Explain.

(d) Evaluate, if possible, $\displaystyle\lim_{x \to 3^-} f'(x)$ and $\displaystyle\lim_{x \to 3^+} f'(x)$.

CAPSTONE

72. *Graphical Reasoning* The figure shows two parts of the graph of a continuous differentiable function f on $[-10, 4]$. The derivative f' is also continuous. To print an enlarged copy of the graph, go to the website *www.mathgraphs.com*.

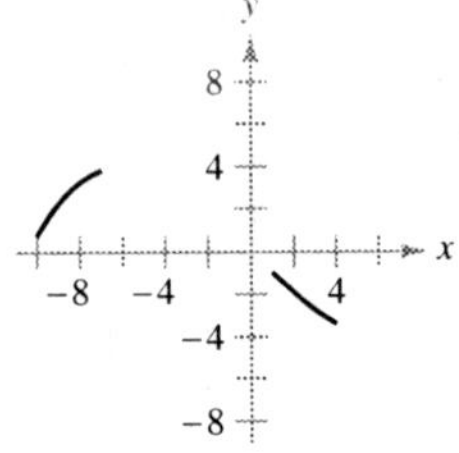

(a) Explain why f must have at least one zero in $[-10, 4]$.

(b) Explain why f' must also have at least one zero in the interval $[-10, 4]$. What are these zeros called?

(c) Make a possible sketch of the function with one zero of f' on the interval $[-10, 4]$.

(d) Make a possible sketch of the function with two zeros of f' on the interval $[-10, 4]$.

(e) Were the conditions of continuity of f and f' necessary to do parts (a) through (d)? Explain.

Think About It In Exercises 73 and 74, sketch the graph of an arbitrary function f that satisfies the given condition but does not satisfy the conditions of the Mean Value Theorem on the interval $[-5, 5]$.

73. f is continuous on $[-5, 5]$.

74. f is not continuous on $[-5, 5]$.

In Exercises 75–78, use the Intermediate Value Theorem and Rolle's Theorem to prove that the equation has exactly one real solution.

75. $x^5 + x^3 + x + 1 = 0$ **76.** $2x^5 + 7x - 1 = 0$

77. $3x + 1 - \sin x = 0$ **78.** $2x - 2 - \cos x = 0$

79. Determine the values of a, b, and c such that the function f satisfies the hypotheses of the Mean Value Theorem on the interval $[0, 3]$.

$$f(x) = \begin{cases} 1, & x = 0 \\ ax + b, & 0 < x \le 1 \\ x^2 + 4x + c, & 1 < x \le 3 \end{cases}$$

80. Determine the values a, b, c, and d such that the function f satisfies the hypotheses of the Mean Value Theorem on the interval $[-1, 2]$.

$$f(x) = \begin{cases} a, & x = -1 \\ 2, & -1 < x \le 0 \\ bx^2 + c, & 0 < x \le 1 \\ dx + 4, & 1 < x \le 2 \end{cases}$$

Differential Equations In Exercises 81–84, find a function f that has the derivative $f'(x)$ and whose graph passes through the given point. Explain your reasoning.

81. $f'(x) = 0$, $(2, 5)$ **82.** $f'(x) = 4$, $(0, 1)$

83. $f'(x) = 2x$, $(1, 0)$ **84.** $f'(x) = 2x + 3$, $(1, 0)$

True or False? In Exercises 85–88, determine whether the statement is true or false. If it is false, explain why or give an example that shows it is false.

85. The Mean Value Theorem can be applied to $f(x) = 1/x$ on the interval $[-1, 1]$.

86. If the graph of a function has three x-intercepts, then it must have at least two points at which its tangent line is horizontal.

87. If the graph of a polynomial function has three x-intercepts, then it must have at least two points at which its tangent line is horizontal.

88. If $f'(x) = 0$ for all x in the domain of f, then f is a constant function.

89. Prove that if $a > 0$ and n is any positive integer, then the polynomial function $p(x) = x^{2n+1} + ax + b$ cannot have two real roots.

90. Prove that if $f'(x) = 0$ for all x in an interval (a, b), then f is constant on (a, b).

91. Let $p(x) = Ax^2 + Bx + C$. Prove that for any interval $[a, b]$, the value c guaranteed by the Mean Value Theorem is the midpoint of the interval.

92. (a) Let $f(x) = x^2$ and $g(x) = -x^3 + x^2 + 3x + 2$. Then $f(-1) = g(-1)$ and $f(2) = g(2)$. Show that there is at least one value c in the interval $(-1, 2)$ where the tangent line to f at $(c, f(c))$ is parallel to the tangent line to g at $(c, g(c))$. Identify c.

(b) Let f and g be differentiable functions on $[a, b]$ where $f(a) = g(a)$ and $f(b) = g(b)$. Show that there is at least one value c in the interval (a, b) where the tangent line to f at $(c, f(c))$ is parallel to the tangent line to g at $(c, g(c))$.

93. Prove that if f is differentiable on $(-\infty, \infty)$ and $f'(x) < 1$ for all real numbers, then f has at most one fixed point. A fixed point of a function f is a real number c such that $f(c) = c$.

94. Use the result of Exercise 93 to show that $f(x) = \frac{1}{2} \cos x$ has at most one fixed point.

95. Prove that $|\cos a - \cos b| \le |a - b|$ for all a and b.

96. Prove that $|\sin a - \sin b| \le |a - b|$ for all a and b.

97. Let $0 < a < b$. Use the Mean Value Theorem to show that

$$\sqrt{b} - \sqrt{a} < \frac{b - a}{2\sqrt{a}}.$$

4.3 Increasing and Decreasing Functions and the First Derivative Test

- **Determine intervals on which a function is increasing or decreasing.**
- **Apply the First Derivative Test to find relative extrema of a function.**

Increasing and Decreasing Functions

In this section you will learn how derivatives can be used to *classify* relative extrema as either relative minima or relative maxima. First, it is important to define increasing and decreasing functions.

DEFINITIONS OF INCREASING AND DECREASING FUNCTIONS

A function f is **increasing** on an interval if for any two numbers x_1 and x_2 in the interval, $x_1 < x_2$ implies $f(x_1) < f(x_2)$.

A function f is **decreasing** on an interval if for any two numbers x_1 and x_2 in the interval, $x_1 < x_2$ implies $f(x_1) > f(x_2)$.

A function is increasing if, *as x moves to the right*, its graph moves up, and is decreasing if its graph moves down. For example, the function in Figure 4.15 is decreasing on the interval $(-\infty, a)$, is constant on the interval (a, b), and is increasing on the interval (b, ∞). As shown in Theorem 4.5 below, a positive derivative implies that the function is increasing; a negative derivative implies that the function is decreasing; and a zero derivative on an entire interval implies that the function is constant on that interval.

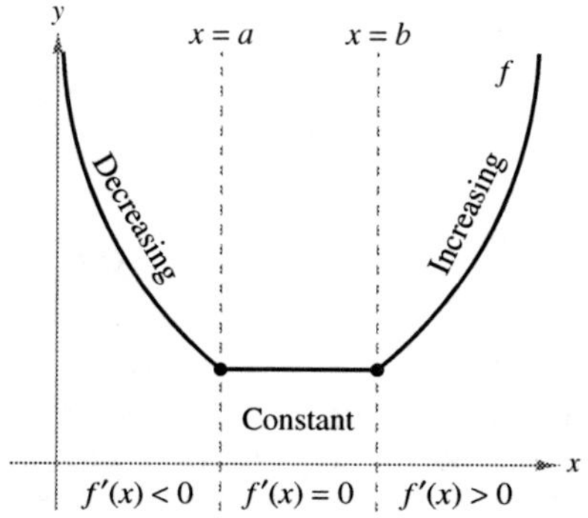

The derivative is related to the slope of a function.

Figure 4.15

THEOREM 4.5 TEST FOR INCREASING AND DECREASING FUNCTIONS

Let f be a function that is continuous on the closed interval $[a, b]$ and differentiable on the open interval (a, b).

1. If $f'(x) > 0$ for all x in (a, b), then f is increasing on $[a, b]$.
2. If $f'(x) < 0$ for all x in (a, b), then f is decreasing on $[a, b]$.
3. If $f'(x) = 0$ for all x in (a, b), then f is constant on $[a, b]$.

PROOF To prove the first case, assume that $f'(x) > 0$ for all x in the interval (a, b) and let $x_1 < x_2$ be any two points in the interval. By the Mean Value Theorem, you know that there exists a number c such that $x_1 < c < x_2$, and

$$f'(c) = \frac{f(x_2) - f(x_1)}{x_2 - x_1}.$$

Because $f'(c) > 0$ and $x_2 - x_1 > 0$, you know that

$$f(x_2) - f(x_1) > 0$$

which implies that $f(x_1) < f(x_2)$. So, f is increasing on the interval. The second case has a similar proof (see Exercise 119), and the third case was given as Exercise 90 in Section 4.2. ∎

NOTE The conclusions in the first two cases of Theorem 4.5 are valid even if $f'(x) = 0$ at a finite number of x-values in (a, b). ∎

EXAMPLE 1   Intervals on Which f Is Increasing or Decreasing

Find the open intervals on which $f(x) = x^3 - \frac{3}{2}x^2$ is increasing or decreasing.

Solution   Note that f is differentiable on the entire real number line. To determine the critical numbers of f, set $f'(x)$ equal to zero.

$$f(x) = x^3 - \frac{3}{2}x^2 \qquad \text{Write original function.}$$

$$f'(x) = 3x^2 - 3x = 0 \qquad \text{Differentiate and set } f'(x) \text{ equal to 0.}$$

$$3(x)(x - 1) = 0 \qquad \text{Factor.}$$

$$x = 0, 1 \qquad \text{Critical numbers}$$

Because there are no points for which f' does not exist, you can conclude that $x = 0$ and $x = 1$ are the only critical numbers. The table summarizes the testing of the three intervals determined by these two critical numbers.

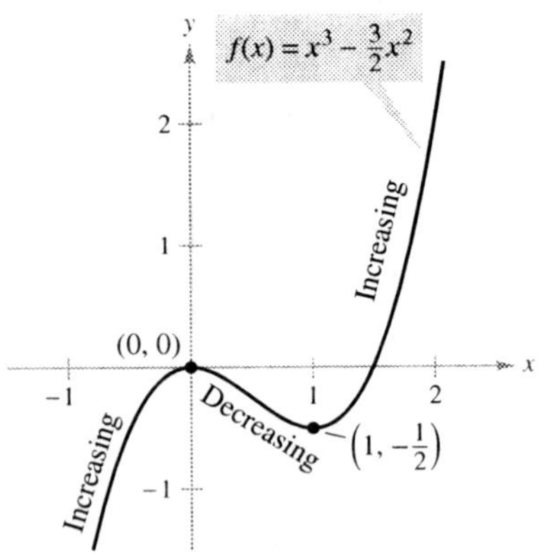
Figure 4.16

Interval	$-\infty < x < 0$	$0 < x < 1$	$1 < x < \infty$
Test Value	$x = -1$	$x = \frac{1}{2}$	$x = 2$
Sign of $f'(x)$	$f'(-1) = 6 > 0$	$f'\left(\frac{1}{2}\right) = -\frac{3}{4} < 0$	$f'(2) = 6 > 0$
Conclusion	Increasing	Decreasing	Increasing

So, f is increasing on the intervals $(-\infty, 0)$ and $(1, \infty)$ and decreasing on the interval $(0, 1)$, as shown in Figure 4.16.

Example 1 gives you one example of how to find intervals on which a function is increasing or decreasing. The guidelines below summarize the steps followed in that example.

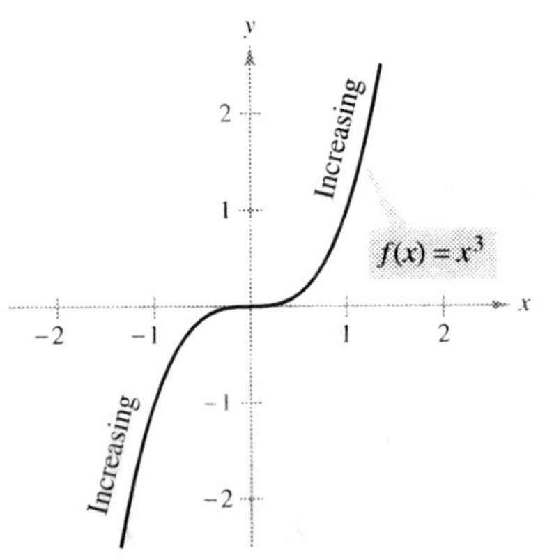
(a) Strictly monotonic function

GUIDELINES FOR FINDING INTERVALS ON WHICH A FUNCTION IS INCREASING OR DECREASING

Let f be continuous on the interval (a, b). To find the open intervals on which f is increasing or decreasing, use the following steps.

1. Locate the critical numbers of f in (a, b), and use these numbers to determine test intervals.
2. Determine the sign of $f'(x)$ at one test value in each of the intervals.
3. Use Theorem 4.5 to determine whether f is increasing or decreasing on each interval.

These guidelines are also valid if the interval (a, b) is replaced by an interval of the form $(-\infty, b)$, (a, ∞), or $(-\infty, \infty)$.

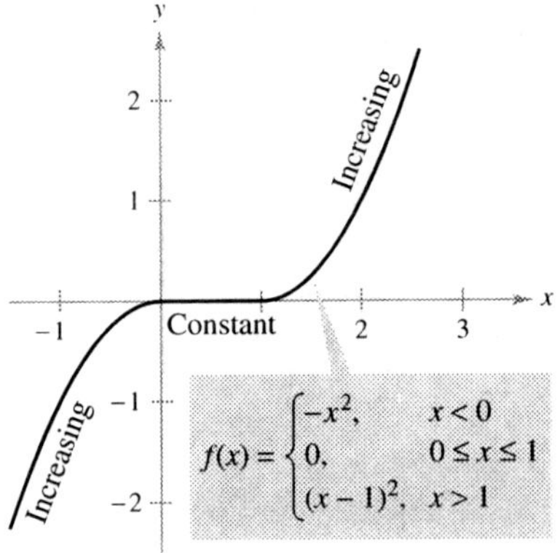
(b) Not strictly monotonic

Figure 4.17

A function is **strictly monotonic** on an interval if it is either increasing on the entire interval or decreasing on the entire interval. For instance, the function $f(x) = x^3$ is strictly monotonic on the entire real number line because it is increasing on the entire real number line, as shown in Figure 4.17(a). The function shown in Figure 4.17(b) is not strictly monotonic on the entire real number line because it is constant on the interval $[0, 1]$.

The First Derivative Test

After you have determined the intervals on which a function is increasing or decreasing, it is not difficult to locate the relative extrema of the function. For instance, in Figure 4.18 (from Example 1), the function

$$f(x) = x^3 - \frac{3}{2}x^2$$

has a relative maximum at the point $(0, 0)$ because f is increasing immediately to the left of $x = 0$ and decreasing immediately to the right of $x = 0$. Similarly, f has a relative minimum at the point $\left(1, -\frac{1}{2}\right)$ because f is decreasing immediately to the left of $x = 1$ and increasing immediately to the right of $x = 1$. The following theorem, called the First Derivative Test, makes this more explicit.

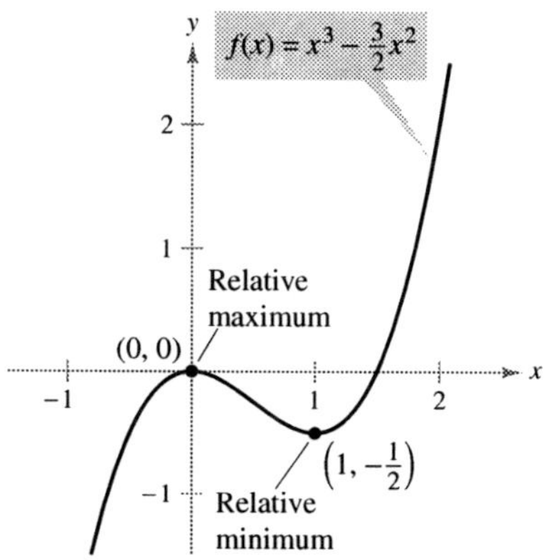

Relative extrema of f
Figure 4.18

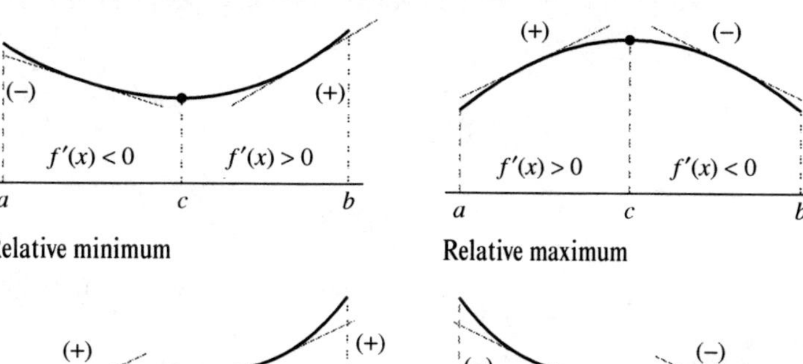

THEOREM 4.6 THE FIRST DERIVATIVE TEST

Let c be a critical number of a function f that is continuous on an open interval I containing c. If f is differentiable on the interval, except possibly at c, then $f(c)$ can be classified as follows.

1. If $f'(x)$ changes from negative to positive at c, then f has a *relative minimum* at $(c, f(c))$.

2. If $f'(x)$ changes from positive to negative at c, then f has a *relative maximum* at $(c, f(c))$.

3. If $f'(x)$ is positive on both sides of c or negative on both sides of c, then $f(c)$ is neither a relative minimum nor a relative maximum.

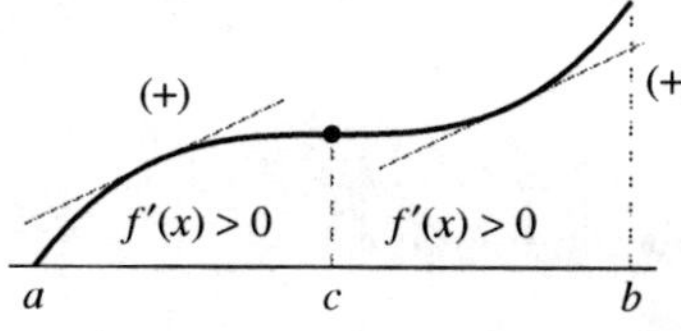

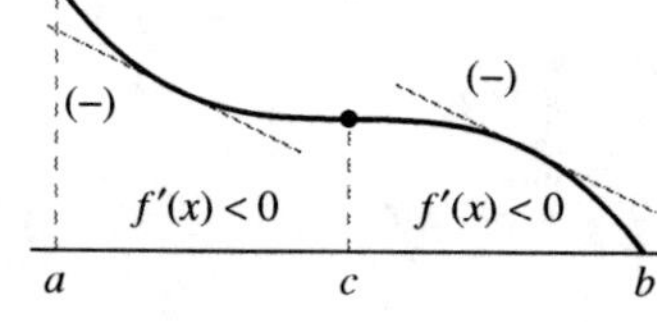

Neither relative minimum nor relative maximum

(**PROOF**) Assume that $f'(x)$ changes from negative to positive at c. Then there exist a and b in I such that

$$f'(x) < 0 \text{ for all } x \text{ in } (a, c)$$

and

$$f'(x) > 0 \text{ for all } x \text{ in } (c, b).$$

By Theorem 4.5, f is decreasing on $[a, c]$ and increasing on $[c, b]$. So, $f(c)$ is a minimum of f on the open interval (a, b) and, consequently, a relative minimum of f. This proves the first case of the theorem. The second case can be proved in a similar way (see Exercise 120). ∎

EXAMPLE 2 Applying the First Derivative Test

Find the relative extrema of the function $f(x) = \frac{1}{2}x - \sin x$ in the interval $(0, 2\pi)$.

Solution   Note that f is continuous on the interval $(0, 2\pi)$. To determine the critical numbers of f in this interval, set $f'(x)$ equal to 0.

$$f'(x) = \frac{1}{2} - \cos x = 0 \qquad \text{Set } f'(x) \text{ equal to 0.}$$

$$\cos x = \frac{1}{2}$$

$$x = \frac{\pi}{3}, \frac{5\pi}{3} \qquad \text{Critical numbers}$$

Because there are no points for which f' does not exist, you can conclude that $x = \pi/3$ and $x = 5\pi/3$ are the only critical numbers. The table summarizes the testing of the three intervals determined by these two critical numbers.

Interval	$0 < x < \dfrac{\pi}{3}$	$\dfrac{\pi}{3} < x < \dfrac{5\pi}{3}$	$\dfrac{5\pi}{3} < x < 2\pi$
Test Value	$x = \dfrac{\pi}{4}$	$x = \pi$	$x = \dfrac{7\pi}{4}$
Sign of $f'(x)$	$f'\left(\dfrac{\pi}{4}\right) < 0$	$f'(\pi) > 0$	$f'\left(\dfrac{7\pi}{4}\right) < 0$
Conclusion	Decreasing	Increasing	Decreasing

By applying the First Derivative Test, you can conclude that f has a relative minimum at the point where

$$x = \frac{\pi}{3} \qquad \text{\textit{x}-value where relative minimum occurs}$$

and a relative maximum at the point where

$$x = \frac{5\pi}{3} \qquad \text{\textit{x}-value where relative maximum occurs}$$

as shown in Figure 4.19.

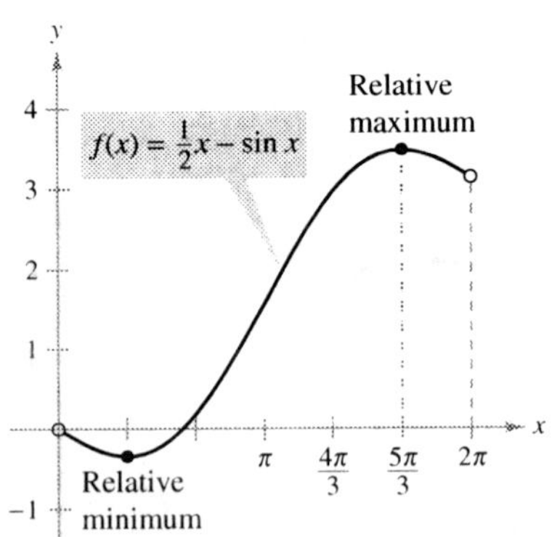

A relative minimum occurs where f changes from decreasing to increasing, and a relative maximum occurs where f changes from increasing to decreasing.
Figure 4.19

EXPLORATION

Comparing Graphical and Analytic Approaches   From Section 4.2, you know that, *by itself*, a graphing utility can give misleading information about the relative extrema of a graph. *Used in conjunction with an analytic approach*, however, a graphing utility can provide a good way to reinforce your conclusions. Try using a graphing utility to graph the function in Example 2. Then use the *zoom* and *trace* features to estimate the relative extrema. How close are your graphical approximations?

Note that in Examples 1 and 2 the given functions are differentiable on the entire real number line. For such functions, the only critical numbers are those for which $f'(x) = 0$. Example 3 concerns a function that has two types of critical numbers—those for which $f'(x) = 0$ and those for which f is not differentiable.

EXAMPLE 3   Applying the First Derivative Test

Find the relative extrema of

$$f(x) = (x^2 - 4)^{2/3}.$$

Solution   Begin by noting that f is continuous on the entire real number line. The derivative of f

$$f'(x) = \frac{2}{3}(x^2 - 4)^{-1/3}(2x) \qquad \text{General Power Rule}$$

$$= \frac{4x}{3(x^2 - 4)^{1/3}} \qquad \text{Simplify.}$$

is 0 when $x = 0$ and does not exist when $x = \pm 2$. So, the critical numbers are $x = -2$, $x = 0$, and $x = 2$. The table summarizes the testing of the four intervals determined by these three critical numbers.

Interval	$-\infty < x < -2$	$-2 < x < 0$	$0 < x < 2$	$2 < x < \infty$
Test Value	$x = -3$	$x = -1$	$x = 1$	$x = 3$
Sign of $f'(x)$	$f'(-3) < 0$	$f'(-1) > 0$	$f'(1) < 0$	$f'(3) > 0$
Conclusion	Decreasing	Increasing	Decreasing	Increasing

By applying the First Derivative Test, you can conclude that f has a relative minimum at the point $(-2, 0)$, a relative maximum at the point $\left(0, \sqrt[3]{16}\right)$, and another relative minimum at the point $(2, 0)$, as shown in Figure 4.20.

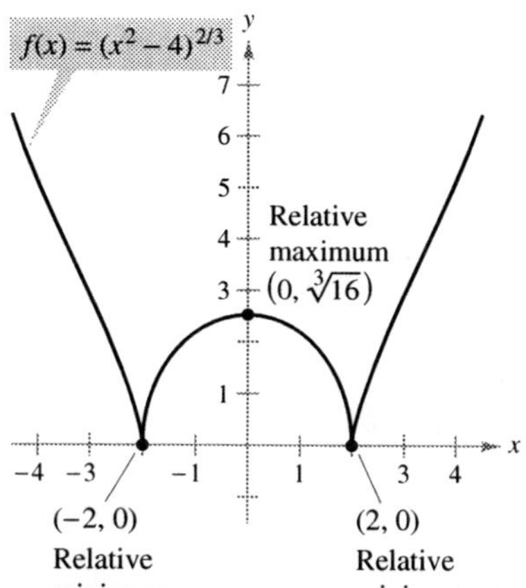

You can apply the First Derivative Test to find relative extrema.
Figure 4.20

TECHNOLOGY PITFALL   When using a graphing utility to graph a function involving radicals or rational exponents, be sure you understand the way the utility evaluates radical expressions. For instance, even though

$$f(x) = (x^2 - 4)^{2/3}$$

and

$$g(x) = [(x^2 - 4)^2]^{1/3}$$

are the same algebraically, some graphing utilities distinguish between these two functions. Which of the graphs shown in Figure 4.21 is incorrect? Why did the graphing utility produce an incorrect graph?

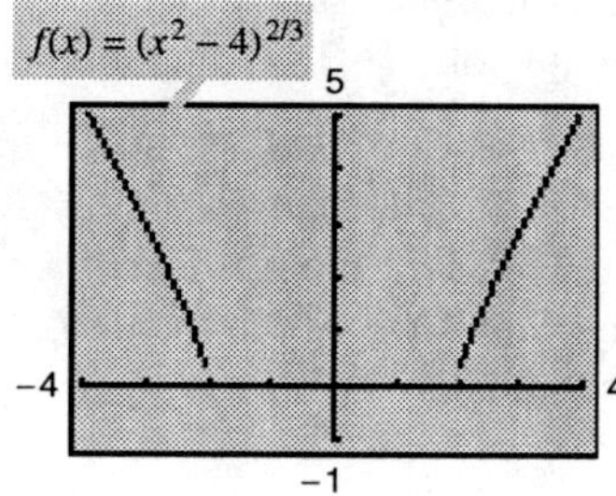

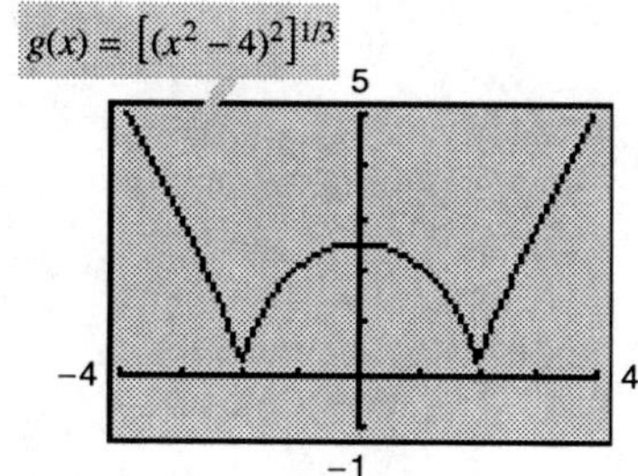

Which graph is incorrect?
Figure 4.21

When using the First Derivative Test, be sure to consider the domain of the function. For instance, in the next example, the function

$$f(x) = \frac{x^4 + 1}{x^2}$$

is not defined when $x = 0$. This x-value must be used with the critical numbers to determine the test intervals.

EXAMPLE 4 Applying the First Derivative Test

Find the relative extrema of $f(x) = \dfrac{x^4 + 1}{x^2}$.

Solution

$$\begin{aligned}
f(x) &= x^2 + x^{-2} &&\text{Rewrite original function.} \\
f'(x) &= 2x - 2x^{-3} &&\text{Differentiate.} \\
&= 2x - \frac{2}{x^3} &&\text{Rewrite with positive exponent.} \\
&= \frac{2(x^4 - 1)}{x^3} &&\text{Simplify.} \\
&= \frac{2(x^2 + 1)(x - 1)(x + 1)}{x^3} &&\text{Factor.}
\end{aligned}$$

So, $f'(x)$ is zero at $x = \pm 1$. Moreover, because $x = 0$ is not in the domain of f, you should use this x-value along with the critical numbers to determine the test intervals.

$$\begin{aligned}
x &= \pm 1 &&\text{Critical numbers, } f'(\pm 1) = 0 \\
x &= 0 &&0 \text{ is not in the domain of } f
\end{aligned}$$

The table summarizes the testing of the four intervals determined by these three x-values.

Interval	$-\infty < x < -1$	$-1 < x < 0$	$0 < x < 1$	$1 < x < \infty$
Test Value	$x = -2$	$x = -\frac{1}{2}$	$x = \frac{1}{2}$	$x = 2$
Sign of $f'(x)$	$f'(-2) < 0$	$f'(-\frac{1}{2}) > 0$	$f'(\frac{1}{2}) < 0$	$f'(2) > 0$
Conclusion	Decreasing	Increasing	Decreasing	Increasing

By applying the First Derivative Test, you can conclude that f has one relative minimum at the point $(-1, 2)$ and another at the point $(1, 2)$, as shown in Figure 4.22.

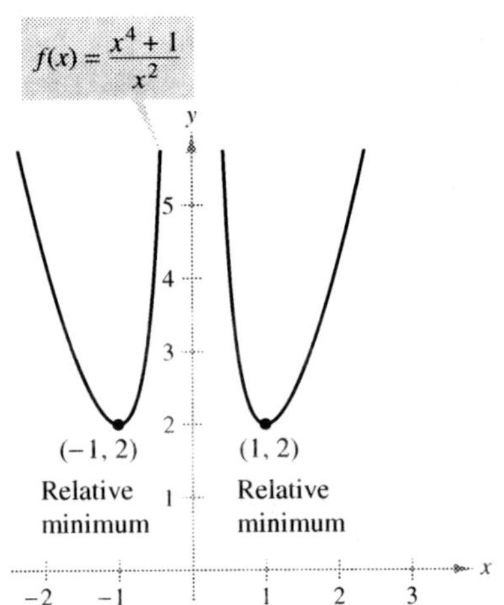

x-values that are not in the domain of f, as well as critical numbers, determine test intervals for f'.

Figure 4.22

TECHNOLOGY The most difficult step in applying the First Derivative Test is finding the values for which the derivative is equal to 0. For instance, the values of x for which the derivative of

$$f(x) = \frac{x^4 + 1}{x^2 + 1}$$

is equal to zero are $x = 0$ and $x = \pm\sqrt{\sqrt{2} - 1}$. If you have access to technology that can perform symbolic differentiation and solve equations, use it to apply the First Derivative Test to this function.

EXAMPLE 5 The Path of a Projectile

Neglecting air resistance, the path of a projectile that is propelled at an angle θ is

$$y = \frac{g \sec^2 \theta}{2v_0^2}x^2 + (\tan \theta)x + h, \quad 0 \le \theta \le \frac{\pi}{2}$$

where y is the height, x is the horizontal distance, g is the acceleration due to gravity, v_0 is the initial velocity, and h is the initial height. (This equation is derived in Section 12.3.) Let $g = -32$ feet per second per second, $v_0 = 24$ feet per second, and $h = 9$ feet. What value of θ will produce a maximum horizontal distance?

Solution To find the distance the projectile travels, let $y = 0$ and use the Quadratic Formula to solve for x.

$$\frac{g \sec^2 \theta}{2v_0^2}x^2 + (\tan \theta)x + h = 0$$

$$\frac{-32 \sec^2 \theta}{2(24^2)}x^2 + (\tan \theta)x + 9 = 0$$

$$-\frac{\sec^2 \theta}{36}x^2 + (\tan \theta)x + 9 = 0$$

$$x = \frac{-\tan \theta \pm \sqrt{\tan^2 \theta + \sec^2 \theta}}{-\sec^2 \theta/18}$$

$$x = 18 \cos \theta\left(\sin \theta + \sqrt{\sin^2 \theta + 1}\right), \quad x \ge 0$$

At this point, you need to find the value of θ that produces a maximum value of x. Applying the First Derivative Test by hand would be very tedious. Using technology to solve the equation $dx/d\theta = 0$, however, eliminates most of the messy computations. The result is that the maximum value of x occurs when

$$\theta \approx 0.61548 \text{ radian, or } 35.3°.$$

This conclusion is reinforced by sketching the path of the projectile for different values of θ, as shown in Figure 4.23. Of the three paths shown, note that the distance traveled is greatest for $\theta = 35°$.

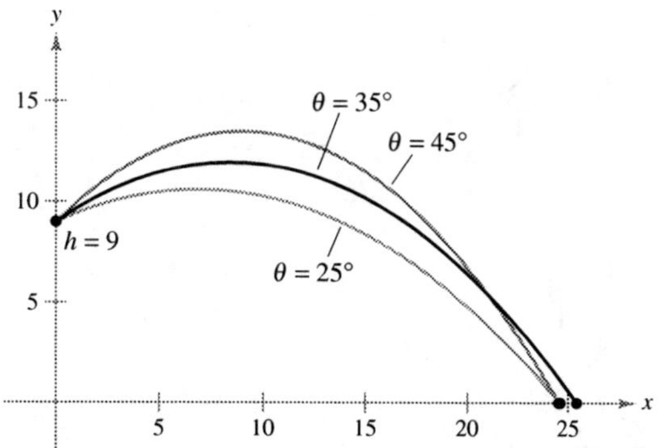

The path of a projectile with initial angle θ
Figure 4.23

NOTE A computer simulation of this example is given in the premium eBook for this text. Using that simulation, you can experimentally discover that the maximum value of x occurs when $\theta \approx 35.3°$.

 **Exercises** See www.CalcChat.com for worked-out solutions to odd-numbered exercises.

In Exercises 1 and 2, use the graph of f to find (a) the largest open interval on which f is increasing, and (b) the largest open interval on which f is decreasing.

1.
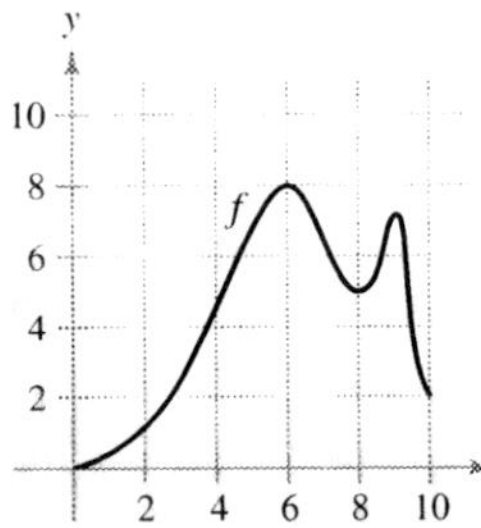

2.
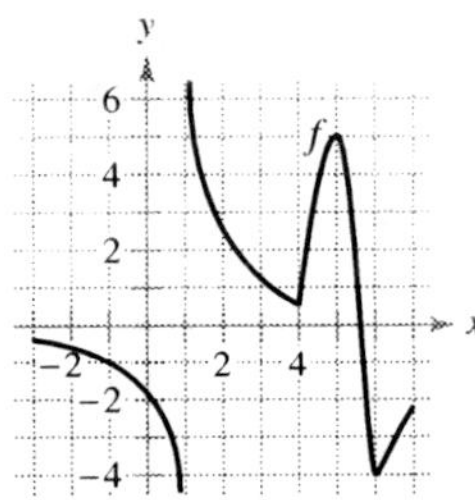

In Exercises 3–8, use the graph to estimate the open intervals on which the function is increasing or decreasing. Then find the open intervals analytically.

3. $f(x) = x^2 - 6x + 8$

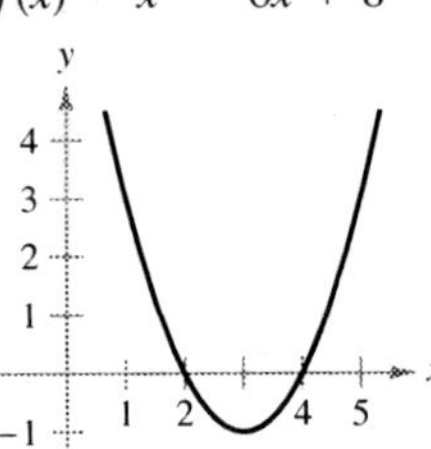

4. $y = -(x + 1)^2$

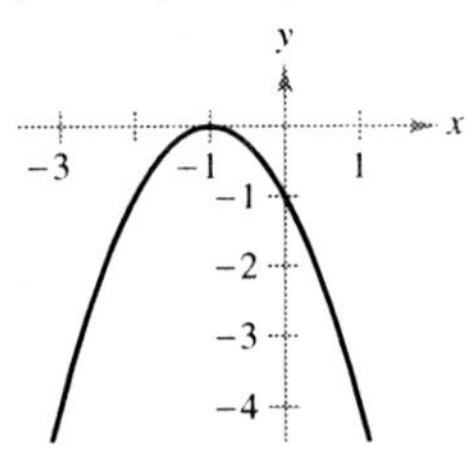

5. $y = \dfrac{x^3}{4} - 3x$

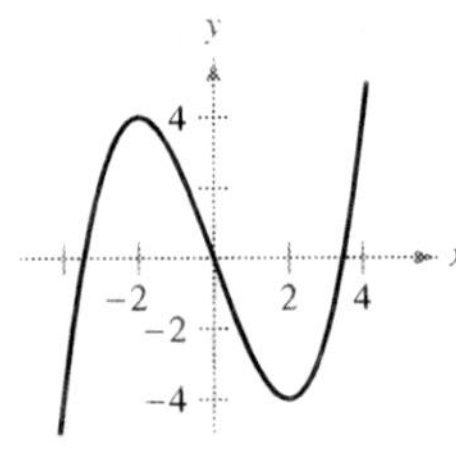

6. $f(x) = x^4 - 2x^2$

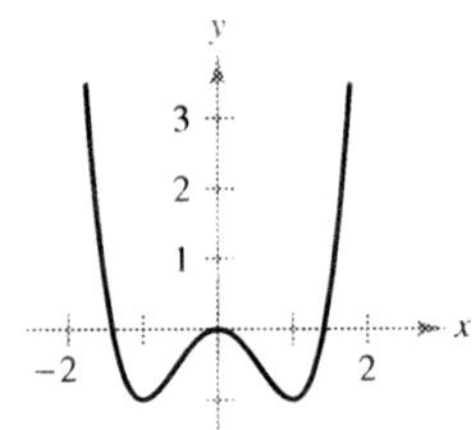

7. $f(x) = \dfrac{1}{(x + 1)^2}$

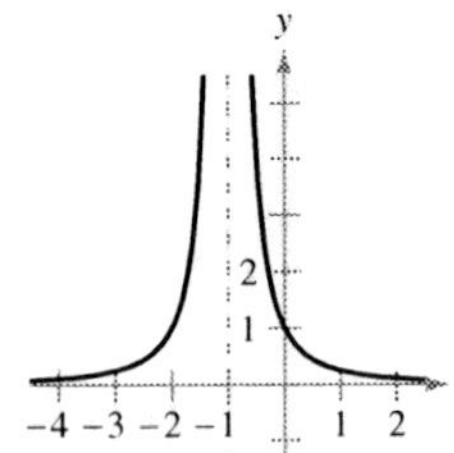

8. $y = \dfrac{x^2}{2x - 1}$

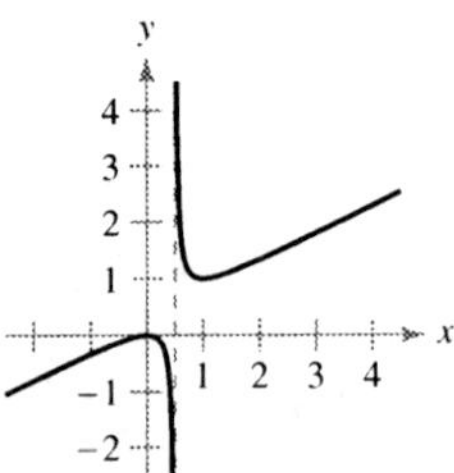

In Exercises 9–20, identify the open intervals on which the function is increasing or decreasing.

9. $g(x) = x^2 - 2x - 8$

10. $h(x) = 27x - x^3$

11. $y = x\sqrt{16 - x^2}$

12. $y = x + \dfrac{4}{x}$

13. $f(x) = \sin x - 1, \quad 0 < x < 2\pi$

14. $h(x) = \cos \dfrac{x}{2}, \quad 0 < x < 2\pi$

15. $y = x - 2\cos x, \quad 0 < x < 2\pi$

16. $f(x) = \cos^2 x - \cos x, \quad 0 < x < 2\pi$

17. $g(x) = e^{-x} + e^{3x}$

18. $h(x) = \sqrt{x}\, e^{-x}$

19. $f(x) = x^2 \ln\left(\dfrac{x}{2}\right)$

20. $f(x) = \dfrac{\ln x}{\sqrt{x}}$

In Exercises 21–58, find the critical numbers of f (if any). Find the open intervals on which the function is increasing or decreasing and locate all relative extrema. Use a graphing utility to confirm your results.

21. $f(x) = x^2 - 4x$

22. $f(x) = x^2 + 6x + 10$

23. $f(x) = -2x^2 + 4x + 3$

24. $f(x) = -(x^2 + 8x + 12)$

25. $f(x) = 2x^3 + 3x^2 - 12x$

26. $f(x) = x^3 - 6x^2 + 15$

27. $f(x) = (x - 1)^2(x + 3)$

28. $f(x) = (x + 2)^2(x - 1)$

29. $f(x) = \dfrac{x^5 - 5x}{5}$

30. $f(x) = x^4 - 32x + 4$

31. $f(x) = x^{1/3} + 1$

32. $f(x) = x^{2/3} - 4$

33. $f(x) = (x + 2)^{2/3}$

34. $f(x) = (x - 3)^{1/3}$

35. $f(x) = 5 - |x - 5|$

36. $f(x) = |x + 3| - 1$

37. $f(x) = 2x + \dfrac{1}{x}$

38. $f(x) = \dfrac{x}{x + 3}$

39. $f(x) = \dfrac{x^2}{x^2 - 9}$

40. $f(x) = \dfrac{x + 4}{x^2}$

41. $f(x) = \dfrac{x^2 - 2x + 1}{x + 1}$

42. $f(x) = \dfrac{x^2 - 3x - 4}{x - 2}$

43. $f(x) = \begin{cases} 4 - x^2, & x \leq 0 \\ -2x, & x > 0 \end{cases}$

44. $f(x) = \begin{cases} 2x + 1, & x \leq -1 \\ x^2 - 2, & x > -1 \end{cases}$

45. $f(x) = \begin{cases} 3x + 1, & x \leq 1 \\ 5 - x^2, & x > 1 \end{cases}$

46. $f(x) = \begin{cases} -x^3 + 1, & x \leq 0 \\ -x^2 + 2x, & x > 0 \end{cases}$

47. $f(x) = (3 - x)e^{x-3}$

48. $f(x) = (x - 1)e^x$

49. $f(x) = 4(x - \arcsin x)$

50. $f(x) = x \arctan x$

51. $f(x) = (x)3^{-x}$

52. $f(x) = 2^{x^2 - 3}$

53. $f(x) = x - \log_4 x$

54. $f(x) = \dfrac{x^3}{3} - \ln x$

55. $f(x) = \dfrac{e^{2x}}{e^{2x} + 1}$

56. $f(x) = \ln(2 - \ln x)$

57. $f(x) = e^{-1/(x-2)}$

58. $f(x) = e^{\arctan x}$

In Exercises 59–66, consider the function on the interval $(0, 2\pi)$. For each function, (a) find the open interval(s) on which the function is increasing or decreasing, (b) apply the First Derivative Test to identify all relative extrema, and (c) use a graphing utility to confirm your results.

59. $f(x) = \dfrac{x}{2} + \cos x$

60. $f(x) = \sin x \cos x + 5$

61. $f(x) = \sin x + \cos x$

62. $f(x) = x + 2 \sin x$

63. $f(x) = \cos^2(2x)$

64. $f(x) = \sqrt{3} \sin x + \cos x$

65. $f(x) = \sin^2 x + \sin x$

66. $f(x) = \dfrac{\sin x}{1 + \cos^2 x}$

CAS In Exercises 67–72, (a) use a computer algebra system to differentiate the function, (b) sketch the graphs of f and f' on the same set of coordinate axes over the given interval, (c) find the critical numbers of f in the open interval, (d) find the interval(s) on which f' is positive and the interval(s) on which it is negative, and (e) compare the behavior of f and the sign of f'.

67. $f(x) = 2x\sqrt{9 - x^2}, \quad [-3, 3]$

68. $f(x) = 10\!\left(5 - \sqrt{x^2 - 3x + 16}\right), \quad [0, 5]$

69. $f(t) = t^2 \sin t, \quad [0, 2\pi]$

70. $f(x) = \dfrac{x}{2} + \cos \dfrac{x}{2}, \quad [0, 4\pi]$

71. $f(x) = \dfrac{1}{2}(x^2 - \ln x), \quad (0, 3]$

72. $f(x) = (4 - x^2)e^x, \quad [0, 2]$

In Exercises 73 and 74, use symmetry, extrema, and zeros to sketch the graph of f. How do the functions f and g differ? Explain.

73. $f(x) = \dfrac{x^5 - 4x^3 + 3x}{x^2 - 1}, \quad g(x) = x(x^2 - 3)$

74. $f(t) = \cos^2 t - \sin^2 t, \quad g(t) = 1 - 2 \sin^2 t \quad (-2, 2)$

Think About It In Exercises 75–80, the graph of f is shown in the figure. Sketch a graph of the derivative of f. To print an enlarged copy of the graph, go to the website *www.mathgraphs.com*.

75.
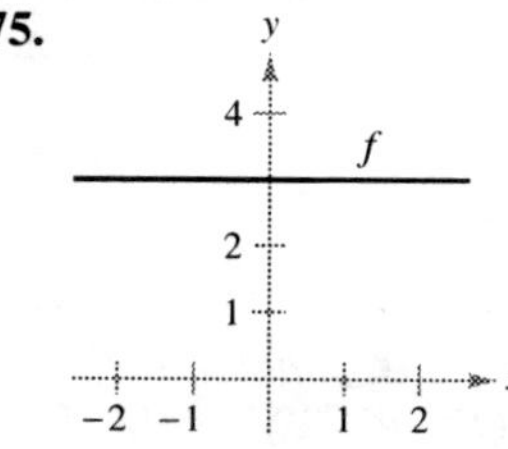

76.
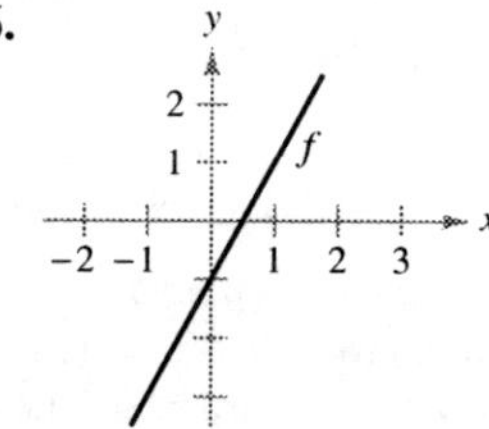

77.
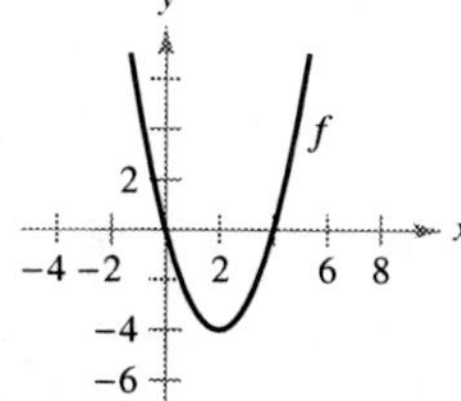

78.
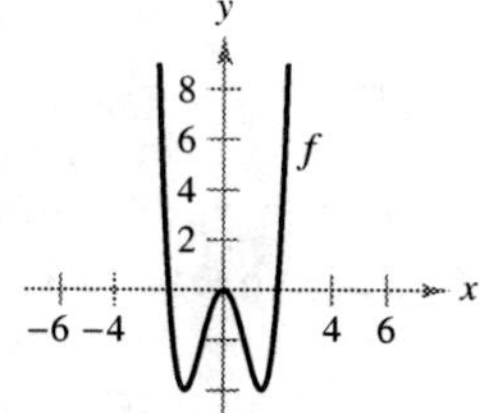

79.
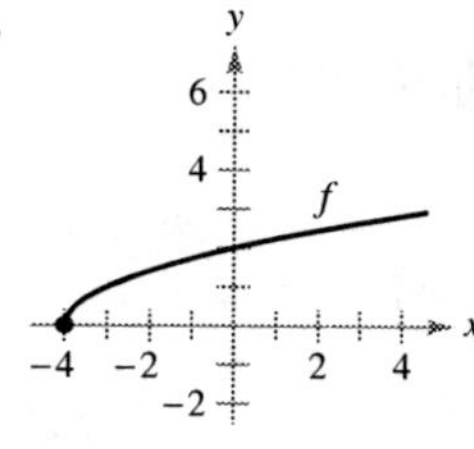

80.
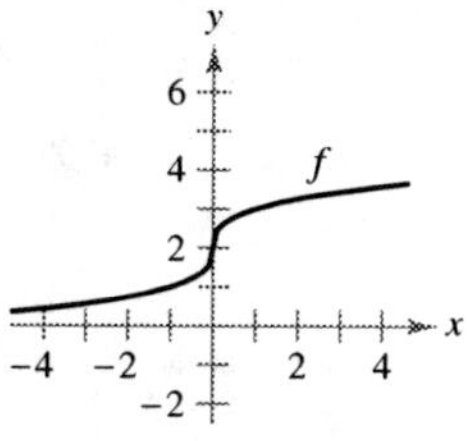

In Exercises 81–84, use the graph of f' to (a) identify the interval(s) on which f is increasing or decreasing, and (b) estimate the value(s) of x at which f has a relative maximum or minimum.

81.
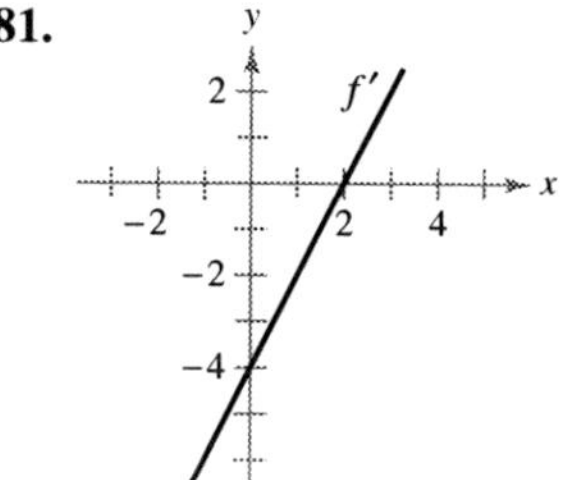

82.
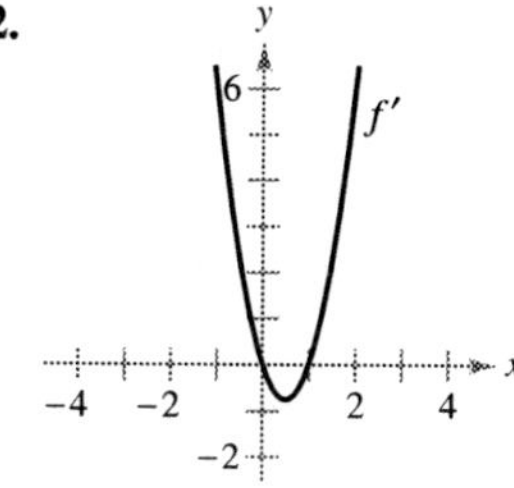

83.
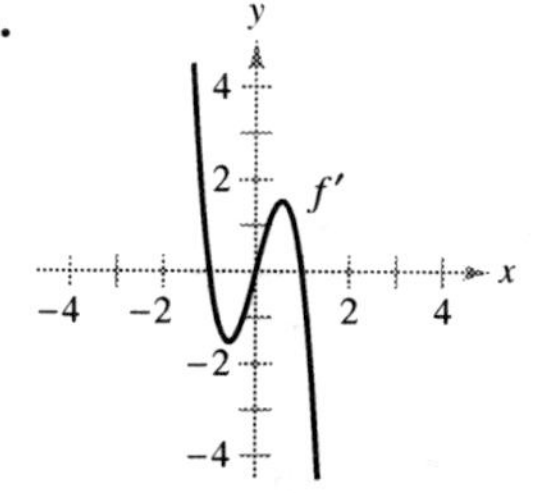

84.
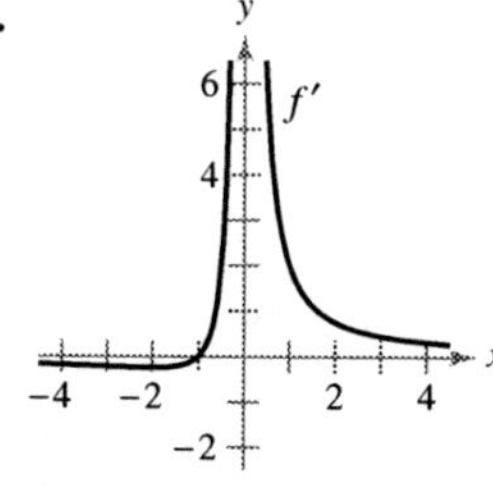

In Exercises 85 and 86, use the graph of f' to (a) identify the critical numbers of f, and (b) determine whether f has a relative maximum, a relative minimum, or neither at each critical number.

85.
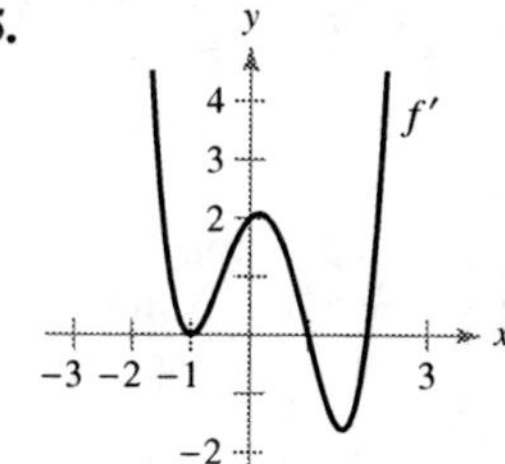

86.
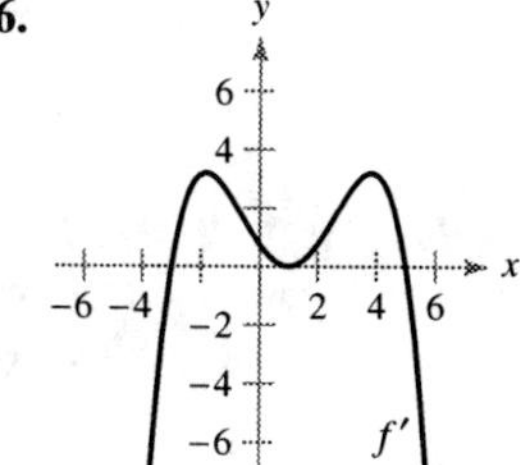

WRITING ABOUT CONCEPTS

In Exercises 87–92, assume that f is differentiable for all x, where $f'(x) > 0$ on $(-\infty, -4)$, $f'(x) < 0$ on $(-4, 6)$, and $f'(x) > 0$ on $(6, \infty)$. Supply the appropriate inequality symbol for the given value of c.

Function	Sign of $g'(c)$
87. $g(x) = f(x) + 5$	$g'(0) \ \blacksquare\ 0$
88. $g(x) = 3f(x) - 3$	$g'(-5) \ \blacksquare\ 0$
89. $g(x) = -f(x)$	$g'(-6) \ \blacksquare\ 0$
90. $g(x) = -f(x)$	$g'(0) \ \blacksquare\ 0$
91. $g(x) = f(x - 10)$	$g'(0) \ \blacksquare\ 0$
92. $g(x) = f(x - 10)$	$g'(8) \ \blacksquare\ 0$

93. Sketch the graph of the arbitrary function f such that

$$f'(x) \begin{cases} > 0, & x < 4 \\ \text{undefined}, & x = 4. \\ < 0, & x > 4 \end{cases}$$

CAPSTONE

94. A differentiable function f has one critical number at $x = 5$. Identify the relative extrema of f at the critical number if $f'(4) = -2.5$ and $f'(6) = 3$.

Think About It **In Exercises 95 and 96, the function f is differentiable on the given interval. The table shows $f'(x)$ for selected values of x. (a) Sketch the graph of f, (b) approximate the critical numbers, and (c) identify the relative extrema.**

95. f is differentiable on $[-1, 1]$.

x	-1	-0.75	-0.50	-0.25
$f'(x)$	-10	-3.2	-0.5	0.8

x	0	0.25	0.50	0.75	1
$f'(x)$	5.6	3.6	-0.2	-6.7	-20.1

96. f is differentiable on $[0, \pi]$.

x	0	$\pi/6$	$\pi/4$	$\pi/3$	$\pi/2$
$f'(x)$	3.14	-0.23	-2.45	-3.11	0.69

x	$2\pi/3$	$3\pi/4$	$5\pi/6$	π
$f'(x)$	3.00	1.37	-1.14	-2.84

97. *Rolling a Ball Bearing* A ball bearing is placed on an inclined plane and begins to roll. The angle of elevation of the plane is θ. The distance (in meters) the ball bearing rolls in t seconds is $s(t) = 4.9(\sin \theta)t^2$.

(a) Determine the speed of the ball bearing after t seconds.

(b) Complete the table and use it to determine the value of θ that produces the maximum speed at a particular time.

θ	0	$\pi/4$	$\pi/3$	$\pi/2$	$2\pi/3$	$3\pi/4$	π
$s'(t)$							

98. *Numerical, Graphical, and Analytic Analysis* The concentration C of a chemical in the bloodstream t hours after injection into muscle tissue is

$$C(t) = \frac{3t}{27 + t^3}, \quad t \geq 0.$$

(a) Complete the table and use it to approximate the time when the concentration is greatest.

t	0	0.5	1	1.5	2	2.5	3
$C(t)$							

(b) Use a graphing utility to graph the concentration function and use the graph to approximate the time when the concentration is greatest.

(c) Use calculus to determine analytically the time when the concentration is greatest.

99. *Numerical, Graphical, and Analytic Analysis* Consider the functions $f(x) = x$ and $g(x) = \sin x$ on the interval $(0, \pi)$.

(a) Complete the table and make a conjecture about which is the greater function on the interval $(0, \pi)$.

x	0.5	1	1.5	2	2.5	3
$f(x)$						
$g(x)$						

(b) Use a graphing utility to graph the functions and use the graphs to make a conjecture about which is the greater function on the interval $(0, \pi)$.

(c) Prove that $f(x) > g(x)$ on the interval $(0, \pi)$. [*Hint:* Show that $h'(x) > 0$, where $h = f - g$.]

100. *Numerical, Graphical, and Analytic Analysis* Consider the functions $f(x) = x$ and $g(x) = \tan x$ on the interval $(0, \pi/2)$.

(a) Complete the table and make a conjecture about which is the greater function on the interval $(0, \pi/2)$.

x	0.25	0.5	0.75	1	1.25	1.5
$f(x)$						
$g(x)$						

(b) Use a graphing utility to graph the functions and use the graphs to make a conjecture about which is the greater function on the interval $(0, \pi/2)$.

(c) Prove that $f(x) < g(x)$ on the interval $(0, \pi/2)$. [*Hint:* Show that $h'(x) > 0$, where $h = g - f$.]

101. *Trachea Contraction* Coughing forces the trachea (windpipe) to contract, which affects the velocity v of the air passing through the trachea. The velocity of the air during coughing is

$$v = k(R - r)r^2, \quad 0 \leq r < R$$

where k is a constant, R is the normal radius of the trachea, and r is the radius during coughing. What radius will produce the maximum air velocity?

102. *Modeling Data* The end-of-year assets of the Medicare Hospital Insurance Trust Fund (in billions of dollars) for the years 1995 through 2006 are shown.

1995: 130.3; 1996: 124.9; 1997: 115.6; 1998: 120.4;

1999: 141.4; 2000: 177.5; 2001: 208.7; 2002: 234.8;

2003: 256.0; 2004: 269.3; 2005: 285.8; 2006: 305.4

(*Source: U.S. Centers for Medicare and Medicaid Services*)

(a) Use the regression capabilities of a graphing utility to find a model of the form $M = at^4 + bt^3 + ct^2 + dt + e$ for the data. (Let $t = 5$ represent 1995.)

(b) Use a graphing utility to plot the data and graph the model.

(c) Find the minimum value of the model and compare the result with the actual data.

Motion Along a Line In Exercises 103–106, the function $s(t)$ describes the motion of a particle along a line. For each function, **(a) find the velocity function of the particle at any time $t \geq 0$, (b) identify the time interval(s) in which the particle is moving in a positive direction, (c) identify the time interval(s) in which the particle is moving in a negative direction, and (d) identify the time(s) at which the particle changes direction.**

103. $s(t) = 6t - t^2$

104. $s(t) = t^2 - 7t + 10$

105. $s(t) = t^3 - 5t^2 + 4t$

106. $s(t) = t^3 - 20t^2 + 128t - 280$

Motion Along a Line In Exercises 107 and 108, the graph shows the position of a particle moving along a line. Describe how the particle's position changes with respect to time.

107.

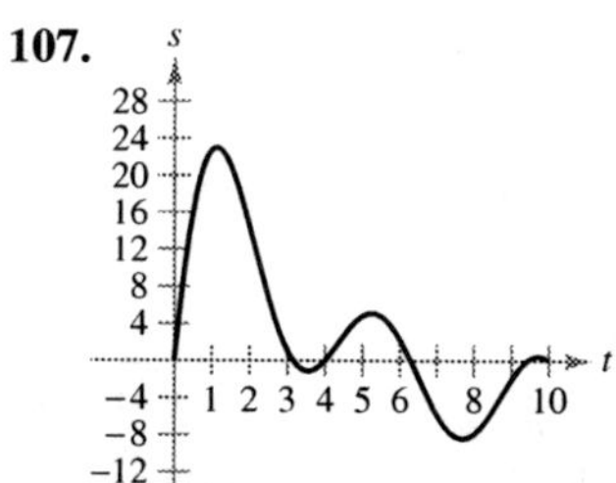

108.

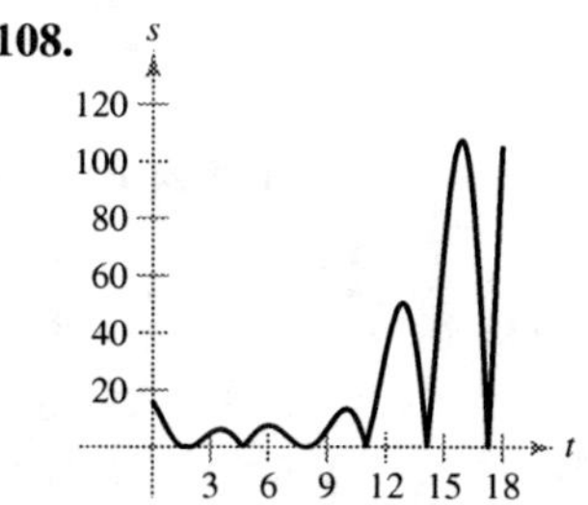

Creating Polynomial Functions In Exercises 109–112, find a polynomial function

$$f(x) = a_n x^n + a_{n-1}x^{n-1} + \cdots + a_2 x^2 + a_1 x + a_0$$

that has only the specified extrema. (a) Determine the minimum degree of the function and give the criteria you used in determining the degree. (b) Using the fact that the coordinates of the extrema are solution points of the function, and that the x-coordinates are critical numbers, determine a system of linear equations whose solution yields the coefficients of the required function. (c) Use a graphing utility to solve the system of equations and determine the function. (d) Use a graphing utility to confirm your result graphically.

109. Relative minimum: $(0, 0)$; Relative maximum: $(2, 2)$

110. Relative minimum: $(0, 0)$; Relative maximum: $(4, 1000)$

111. Relative minima: $(0, 0)$, $(4, 0)$; Relative maximum: $(2, 4)$

112. Relative minimum: $(1, 2)$; Relative maxima: $(-1, 4)$, $(3, 4)$

True or False? In Exercises 113–118, determine whether the statement is true or false. If it is false, explain why or give an example that shows it is false.

113. The sum of two increasing functions is increasing.

114. The product of two increasing functions is increasing.

115. Every nth-degree polynomial has $(n - 1)$ critical numbers.

116. An nth-degree polynomial has at most $(n - 1)$ critical numbers.

117. There is a relative maximum or minimum at each critical number.

118. The relative maxima of the function f are $f(1) = 4$ and $f(3) = 10$. So, f has at least one minimum for some x in the interval $(1, 3)$.

119. Prove the second case of Theorem 4.5.

120. Prove the second case of Theorem 4.6.

121. Let $x > 0$ and $n > 1$ be real numbers. Prove that $(1 + x)^n > 1 + nx$.

122. Use the definitions of increasing and decreasing functions to prove that $f(x) = x^3$ is increasing on $(-\infty, \infty)$.

123. Use the definitions of increasing and decreasing functions to prove that $f(x) = 1/x$ is decreasing on $(0, \infty)$.

124. Consider $f(x) = axe^{bx^2}$. Find a and b such that the relative maximum of f is $f(4) = 2$.

PUTNAM EXAM CHALLENGE

125. Find the minimum value of

$$|\sin x + \cos x + \tan x + \cot x + \sec x + \csc x|$$

for real numbers x.

This problem was composed by the Committee on the Putnam Prize Competition. © The Mathematical Association of America. All rights reserved.

SECTION PROJECT

Rainbows

Rainbows are formed when light strikes raindrops and is reflected and refracted, as shown in the figure. (This figure shows a cross section of a spherical raindrop.) The Law of Refraction states that $(\sin \alpha)/(\sin \beta) = k$, where $k \approx 1.33$ (for water). The angle of deflection is given by $D = \pi + 2\alpha - 4\beta$.

(a) Use a graphing utility to graph

$$D = \pi + 2\alpha - 4 \sin^{-1}(1/k \sin \alpha),$$
$$0 \leq \alpha \leq \pi/2.$$

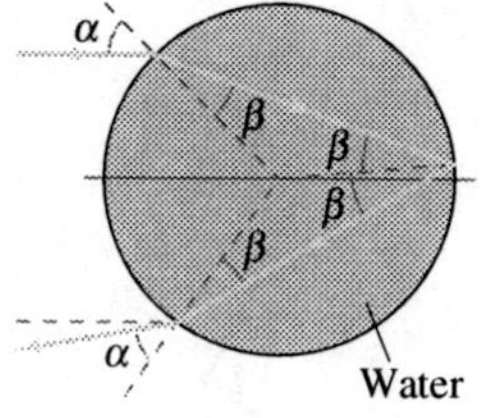

(b) Prove that the minimum angle of deflection occurs when

$$\cos \alpha = \sqrt{\frac{k^2 - 1}{3}}.$$

For water, what is the minimum angle of deflection, D_{min}? (The angle $\pi - D_{min}$ is called the *rainbow angle*.) What value of α produces this minimum angle? (A ray of sunlight that strikes a raindrop at this angle, α, is called a *rainbow ray*.)

FOR FURTHER INFORMATION For more information about the mathematics of rainbows, see the article "Somewhere Within the Rainbow" by Steven Janke in *The UMAP Journal*.

4.4 Concavity and the Second Derivative Test

- Determine intervals on which a function is concave upward or concave downward.
- Find any points of inflection of the graph of a function.
- Apply the Second Derivative Test to find relative extrema of a function.

Concavity

You have already seen that locating the intervals on which a function f increases or decreases helps to describe its graph. In this section, you will see how locating the intervals on which f' increases or decreases can be used to determine where the graph of f is *curving upward* or *curving downward*.

DEFINITION OF CONCAVITY

Let f be differentiable on an open interval I. The graph of f is **concave upward** on I if f' is increasing on the interval and **concave downward** on I if f' is decreasing on the interval.

The following graphical interpretation of concavity is useful. (See Appendix A for a proof of these results.)

1. Let f be differentiable on an open interval I. If the graph of f is concave *upward* on I, then the graph of f lies *above* all of its tangent lines on I. [See Figure 4.24(a).]

2. Let f be differentiable on an open interval I. If the graph of f is concave *downward* on I, then the graph of f lies *below* all of its tangent lines on I. [See Figure 4.24(b).]

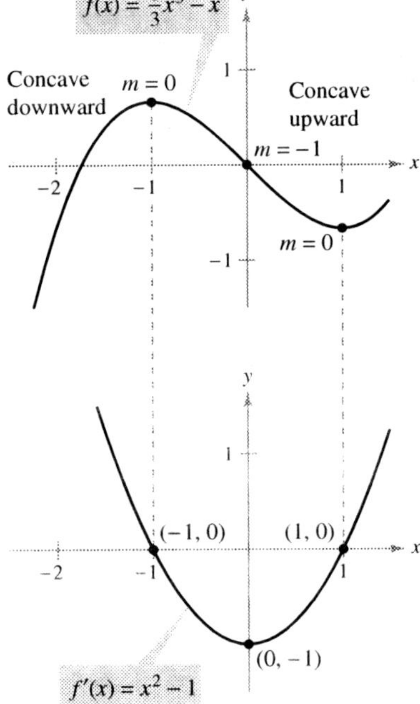

The concavity of f is related to the slope of the derivative.
Figure 4.25

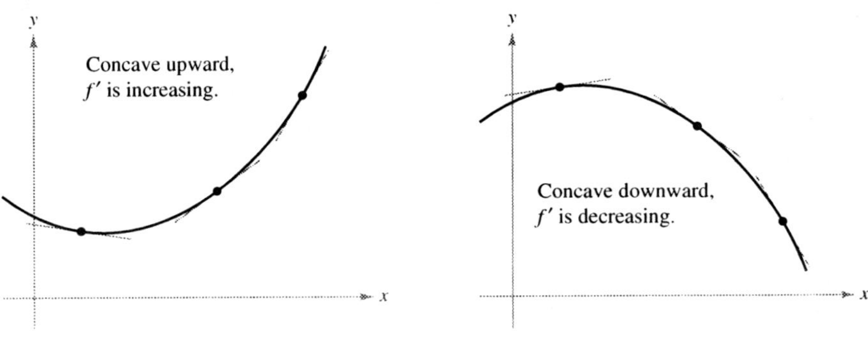

(a) The graph of f lies above its tangent lines. **(b)** The graph of f lies below its tangent lines.

Figure 4.24

To find the open intervals on which the graph of a function f is concave upward or downward, you need to find the intervals on which f' is increasing or decreasing. For instance, the graph of

$$f(x) = \tfrac{1}{3}x^3 - x$$

is concave downward on the open interval $(-\infty, 0)$ because $f'(x) = x^2 - 1$ is decreasing there. (See Figure 4.25.) Similarly, the graph of f is concave upward on the interval $(0, \infty)$ because f' is increasing on $(0, \infty)$.

The following theorem shows how to use the *second* derivative of a function f to determine intervals on which the graph of f is concave upward or concave downward. A proof of this theorem (see Appendix A) follows directly from Theorem 4.5 and the definition of concavity.

THEOREM 4.7 TEST FOR CONCAVITY

Let f be a function whose second derivative exists on an open interval I.

1. If $f''(x) > 0$ for all x in I, then the graph of f is concave upward on I.
2. If $f''(x) < 0$ for all x in I, then the graph of f is concave downward on I.

Note that a third case of Theorem 4.7 could be that if $f''(x) = 0$ for all x in I, then f is linear. Note, however, that concavity is not defined for a line. In other words, a straight line is neither concave upward nor concave downward.

To apply Theorem 4.7, first locate the x-values at which $f''(x) = 0$ or f'' does not exist. Second, use these x-values to determine test intervals. Finally, test the sign of $f''(x)$ in each of the test intervals.

EXAMPLE 1 Determining Concavity

Determine the open intervals on which the graph of

$$f(x) = e^{-x^2/2}$$

is concave upward or concave downward.

Solution Begin by observing that f is continuous on the entire real number line. Next, find the second derivative of f.

$$f'(x) = -xe^{-x^2/2} \qquad \text{First derivative}$$
$$f''(x) = (-x)(-x)e^{-x^2/2} + e^{-x^2/2}(-1) \qquad \text{Differentiate.}$$
$$= e^{-x^2/2}(x^2 - 1) \qquad \text{Second derivative}$$

Because $f''(x) = 0$ when $x = \pm 1$ and f'' is defined on the entire real number line, you should test f'' in the intervals $(-\infty, -1)$, $(-1, 1)$, and $(1, \infty)$. The results are shown in the table and in Figure 4.26.

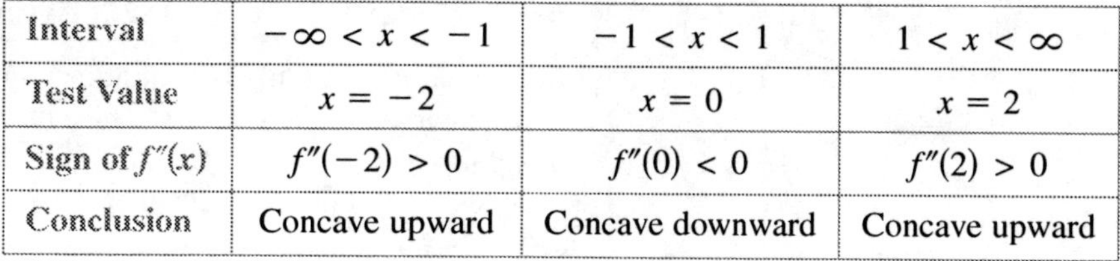

Interval	$-\infty < x < -1$	$-1 < x < 1$	$1 < x < \infty$
Test Value	$x = -2$	$x = 0$	$x = 2$
Sign of $f''(x)$	$f''(-2) > 0$	$f''(0) < 0$	$f''(2) > 0$
Conclusion	Concave upward	Concave downward	Concave upward

The function given in Example 1 is continuous on the entire real number line. If there are x-values at which the function is not continuous, these values should be used, along with the points at which $f''(x) = 0$ or $f''(x)$ does not exist, to form the test intervals.

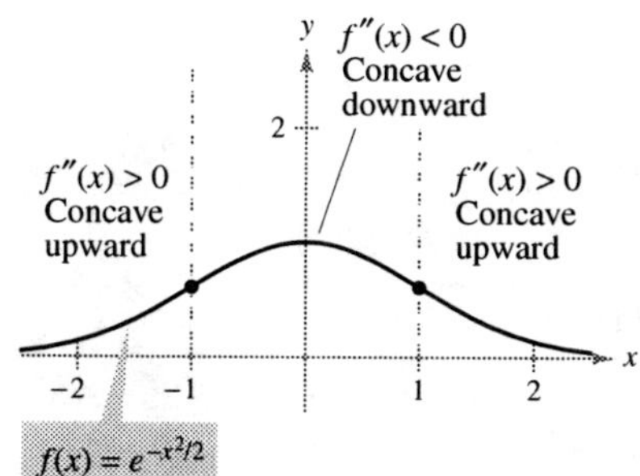

From the sign of f'' you can determine the concavity of the graph of f.
Figure 4.26

NOTE The function in Example 1 is similar to the normal probability density function, whose general form is

$$f(x) = \frac{1}{\sigma\sqrt{2\pi}} e^{-x^2/2\sigma^2}$$

where σ is the standard deviation (σ is the lowercase Greek letter sigma). This "bell-shaped" curve is concave downward on the interval $(-\sigma, \sigma)$.

EXAMPLE 2 Determining Concavity

Determine the open intervals on which the graph of $f(x) = \dfrac{x^2 + 1}{x^2 - 4}$ is concave upward or concave downward.

Solution Differentiating twice produces the following.

$$f(x) = \frac{x^2 + 1}{x^2 - 4} \qquad \text{Write original function.}$$

$$f'(x) = \frac{(x^2 - 4)(2x) - (x^2 + 1)(2x)}{(x^2 - 4)^2} \qquad \text{Differentiate.}$$

$$= \frac{-10x}{(x^2 - 4)^2} \qquad \text{First derivative}$$

$$f''(x) = \frac{(x^2 - 4)^2(-10) - (-10x)(2)(x^2 - 4)(2x)}{(x^2 - 4)^4} \qquad \text{Differentiate.}$$

$$= \frac{10(3x^2 + 4)}{(x^2 - 4)^3} \qquad \text{Second derivative}$$

There are no points at which $f''(x) = 0$, but at $x = \pm 2$ the function f is not continuous, so test for concavity in the intervals $(-\infty, -2)$, $(-2, 2)$, and $(2, \infty)$, as shown in the table. The graph of f is shown in Figure 4.27.

Interval	$-\infty < x < -2$	$-2 < x < 2$	$2 < x < \infty$
Test Value	$x = -3$	$x = 0$	$x = 3$
Sign of $f''(x)$	$f''(-3) > 0$	$f''(0) < 0$	$f''(3) > 0$
Conclusion	Concave upward	Concave downward	Concave upward

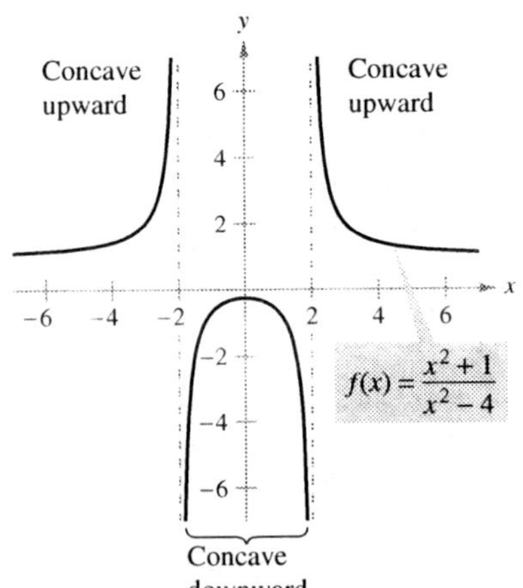

Figure 4.27

Points of Inflection

The graph in Figure 4.26 has two points at which the concavity changes. If the tangent line to the graph exists at such a point, that point is a **point of inflection.** Three types of points of inflection are shown in Figure 4.28.

DEFINITION OF POINT OF INFLECTION

Let f be a function that is continuous on an open interval and let c be a point in the interval. If the graph of f has a tangent line at this point $(c, f(c))$, then this point is a **point of inflection** of the graph of f if the concavity of f changes from upward to downward (or downward to upward) at the point.

NOTE The definition of *point of inflection* given above requires that the tangent line exists at the point of inflection. Some books do not require this. For instance, we do not consider the function

$$f(x) = \begin{cases} x^3, & x < 0 \\ x^2 + 2x, & x \geq 0 \end{cases}$$

to have a point of inflection at the origin, even though the concavity of the graph changes from concave downward to concave upward.

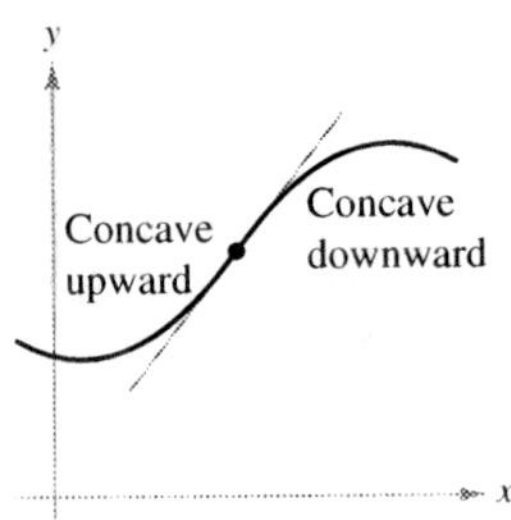

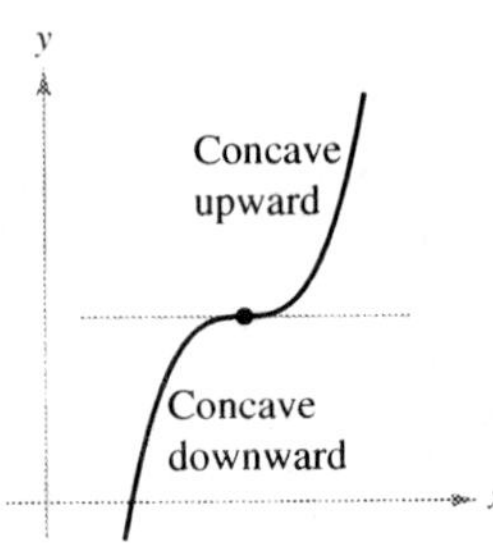

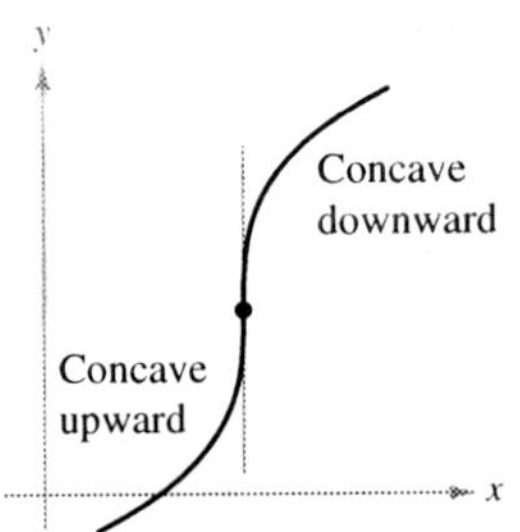

The concavity of f changes at a point of inflection. Note that a graph crosses its tangent line at a point of inflection.

Figure 4.28

To locate *possible* points of inflection, you can determine the values of x for which $f''(x) = 0$ or $f''(x)$ does not exist. This is similar to the procedure for locating relative extrema of f.

THEOREM 4.8 POINTS OF INFLECTION

If $(c, f(c))$ is a point of inflection of the graph of f, then either $f''(c) = 0$ or f'' does not exist at $x = c$.

EXAMPLE 3 Finding Points of Inflection

Determine the points of inflection and discuss the concavity of the graph of

$$f(x) = x^4 - 4x^3.$$

Solution Differentiating twice produces the following.

$$f(x) = x^4 - 4x^3 \qquad \text{Write original function.}$$
$$f'(x) = 4x^3 - 12x^2 \qquad \text{Find first derivative.}$$
$$f''(x) = 12x^2 - 24x = 12x(x - 2) \qquad \text{Find second derivative.}$$

Setting $f''(x) = 0$, you can determine that the possible points of inflection occur at $x = 0$ and $x = 2$. By testing the intervals determined by these x-values, you can conclude that they both yield points of inflection. A summary of this testing is shown in the table, and the graph of f is shown in Figure 4.29.

Interval	$-\infty < x < 0$	$0 < x < 2$	$2 < x < \infty$
Test Value	$x = -1$	$x = 1$	$x = 3$
Sign of $f''(x)$	$f''(-1) > 0$	$f''(1) < 0$	$f''(3) > 0$
Conclusion	Concave upward	Concave downward	Concave upward

The converse of Theorem 4.8 is not generally true. That is, it is possible for the second derivative to be 0 at a point that is *not* a point of inflection. For instance, the graph of $f(x) = x^4$ is shown in Figure 4.30. The second derivative is 0 when $x = 0$, but the point $(0, 0)$ is not a point of inflection because the graph of f is concave upward on both intervals $-\infty < x < 0$ and $0 < x < \infty$.

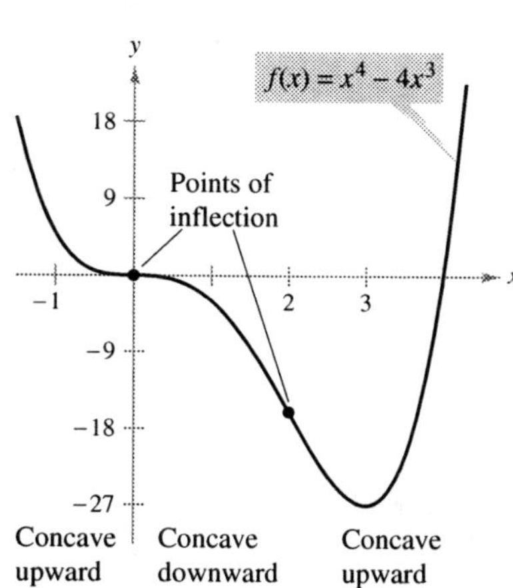

Concave upward   Concave downward   Concave upward

Points of inflection can occur where $f''(x) = 0$ or f'' does not exist.
Figure 4.29

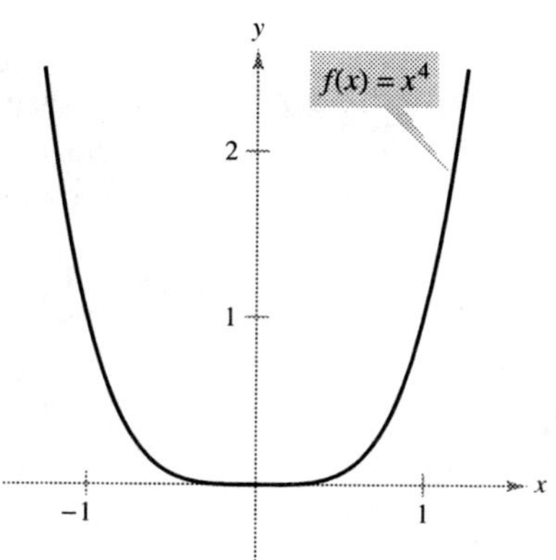

$f''(x) = 0$, but $(0, 0)$ is not a point of inflection.
Figure 4.30

EXPLORATION

Consider a general cubic function of the form

$$f(x) = ax^3 + bx^2 + cx + d.$$

You know that the value of d has a bearing on the location of the graph but has no bearing on the value of the first derivative at given values of x. Graphically, this is true because changes in the value of d shift the graph up or down but do not change its basic shape. Use a graphing utility to graph several cubics with different values of c. Then give a graphical explanation of why changes in c do not affect the values of the second derivative.

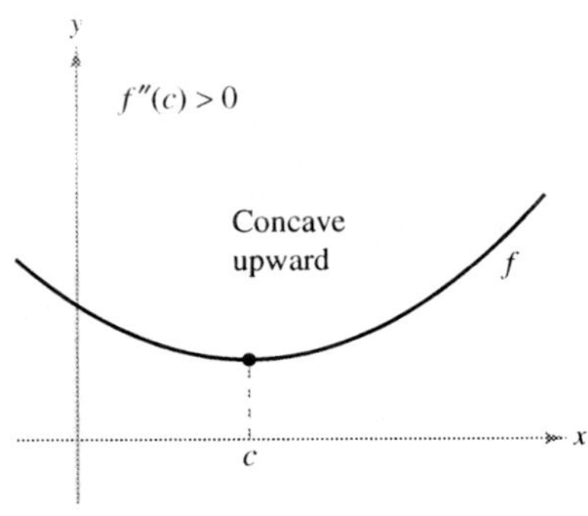

If $f'(c) = 0$ and $f''(c) > 0$, $f(c)$ is a relative minimum.

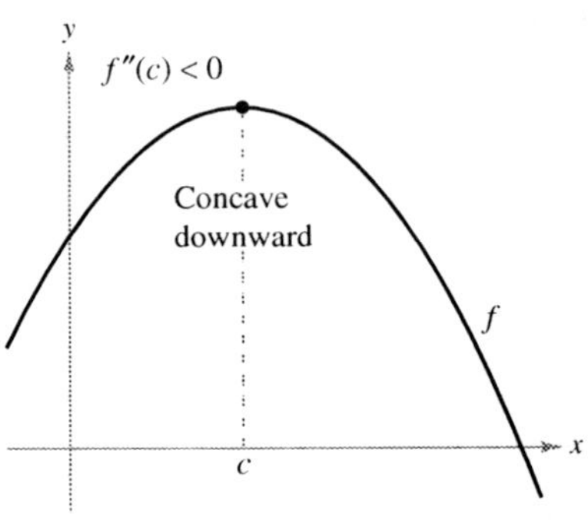

If $f'(c) = 0$ and $f''(c) < 0$, $f(c)$ is a relative maximum.

Figure 4.31

The Second Derivative Test

In addition to testing for concavity, the second derivative can be used to perform a simple test for relative maxima and minima. The test is based on the fact that if the graph of a function f is concave upward on an open interval containing c, and $f'(c) = 0$, $f(c)$ must be a relative minimum of f. Similarly, if the graph of a function f is concave downward on an open interval containing c, and $f'(c) = 0$, $f(c)$ must be a relative maximum of f (see Figure 4.31).

THEOREM 4.9 SECOND DERIVATIVE TEST

Let f be a function such that $f'(c) = 0$ and the second derivative of f exists on an open interval containing c.

1. If $f''(c) > 0$, then $f(c)$ is a relative minimum.

2. If $f''(c) < 0$, then $f(c)$ is a relative maximum.

If $f''(c) = 0$, the test fails. That is, f may have a relative maximum, a relative minimum, or neither. In such cases, you can use the First Derivative Test.

PROOF If $f'(c) = 0$ and $f''(c) > 0$, there exists an open interval I containing c for which

$$\frac{f'(x) - f'(c)}{x - c} = \frac{f'(x)}{x - c} > 0$$

for all $x \neq c$ in I. If $x < c$, then $x - c < 0$ and $f'(x) < 0$. Also, if $x > c$, then $x - c > 0$ and $f'(x) > 0$. So, $f'(x)$ changes from negative to positive at c, and the First Derivative Test implies that $f(c)$ is a relative minimum. A proof of the second case is left to you. ∎

EXAMPLE 4 Using the Second Derivative Test

Find the relative extrema for $f(x) = -3x^5 + 5x^3$.

Solution Begin by finding the critical numbers of f.

$$f'(x) = -15x^4 + 15x^2 = 15x^2(1 - x^2) = 0 \qquad \text{Set } f'(x) \text{ equal to 0.}$$
$$x = -1, 0, 1 \qquad \text{Critical numbers}$$

Using

$$f''(x) = -60x^3 + 30x = 30(-2x^3 + x)$$

you can apply the Second Derivative Test as shown below.

Point	$(-1, -2)$	$(1, 2)$	$(0, 0)$
Sign of $f''(x)$	$f''(-1) > 0$	$f''(1) < 0$	$f''(0) = 0$
Conclusion	Relative minimum	Relative maximum	Test fails

Because the Second Derivative Test fails at $(0, 0)$, you can use the First Derivative Test and observe that f increases to the left and right of $x = 0$. So, $(0, 0)$ is neither a relative minimum nor a relative maximum (even though the graph has a horizontal tangent line at this point). The graph of f is shown in Figure 4.32. ∎

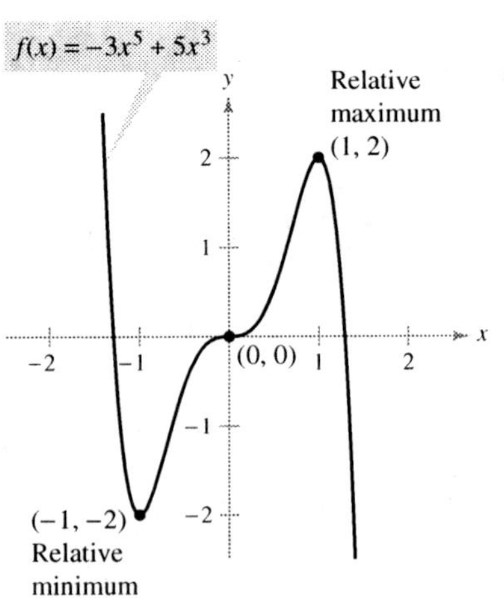

$(0, 0)$ is neither a relative minimum nor a relative maximum.

Figure 4.32

4.4 Exercises

See www.CalcChat.com for worked-out solutions to odd-numbered exercises.

In Exercises 1–4, the graph of f is shown. State the signs of f' and f'' on the interval $(0, 2)$.

1. 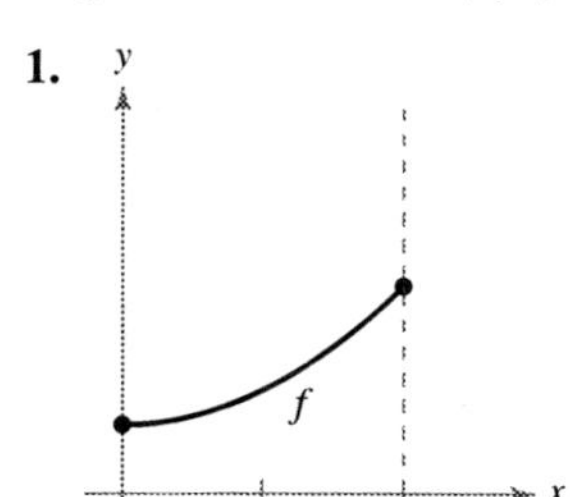

2. 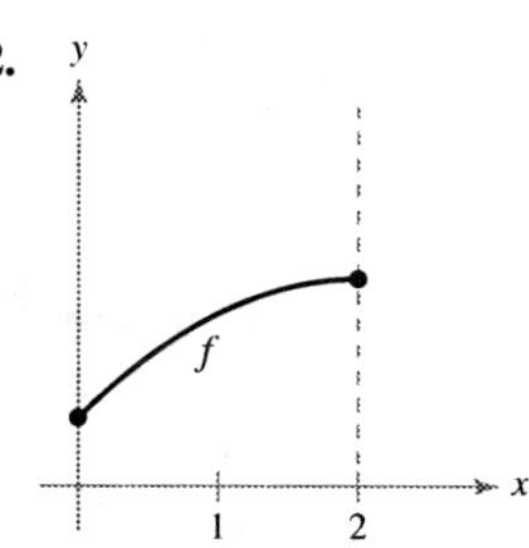

3. 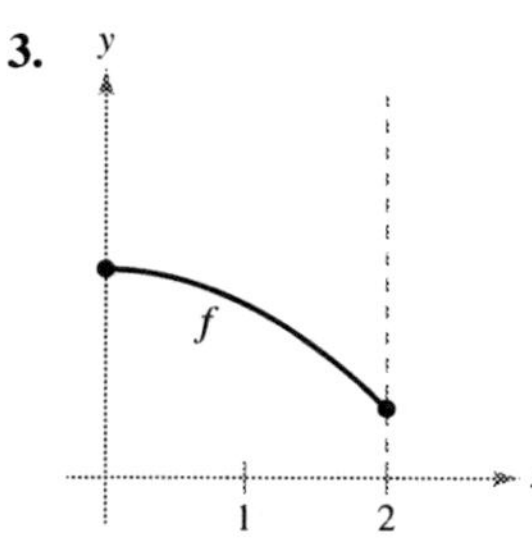

4. 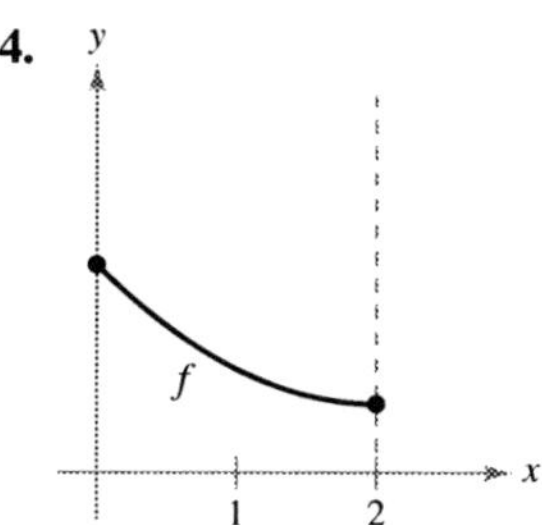

In Exercises 5–18, determine the open intervals on which the graph is concave upward or concave downward.

5. $y = x^2 - x - 2$

6. $y = -x^3 + 3x^2 - 2$

7. $g(x) = 3x^2 - x^3$

8. $h(x) = x^5 - 5x + 2$

9. $f(x) = -x^3 + 6x^2 - 9x - 1$

10. $f(x) = x^5 + 5x^4 - 40x^2$

11. $f(x) = \dfrac{24}{x^2 + 12}$

12. $f(x) = \dfrac{x^2}{x^2 + 1}$

13. $f(x) = \dfrac{x^2 + 1}{x^2 - 1}$

14. $y = \dfrac{-3x^5 + 40x^3 + 135x}{270}$

15. $g(x) = \dfrac{x^2 + 4}{4 - x^2}$

16. $h(x) = \dfrac{x^2 - 1}{2x - 1}$

17. $y = 2x - \tan x, \quad \left(-\dfrac{\pi}{2}, \dfrac{\pi}{2}\right)$

18. $y = x + \dfrac{2}{\sin x}, \quad (-\pi, \pi)$

In Exercises 19–42, find the points of inflection and discuss the concavity of the graph of the function.

19. $f(x) = \frac{1}{2}x^4 + 2x^3$

20. $f(x) = -x^4 + 24x^2$

21. $f(x) = x^3 - 6x^2 + 12x$

22. $f(x) = 2x^3 - 3x^2 - 12x + 5$

23. $f(x) = \frac{1}{4}x^4 - 2x^2$

24. $f(x) = 2x^4 - 8x + 3$

25. $f(x) = x(x - 4)^3$

26. $f(x) = (x - 2)^3(x - 1)$

27. $f(x) = x\sqrt{x + 3}$

28. $f(x) = x\sqrt{9 - x}$

29. $f(x) = \dfrac{4}{x^2 + 1}$

30. $f(x) = \dfrac{x + 1}{\sqrt{x}}$

31. $f(x) = \sin \dfrac{x}{2}, \quad [0, 4\pi]$

32. $f(x) = 2 \csc \dfrac{3x}{2}, \quad (0, 2\pi)$

33. $f(x) = \sec\left(x - \dfrac{\pi}{2}\right), \quad (0, 4\pi)$

34. $f(x) = \sin x + \cos x, \quad [0, 2\pi]$

35. $f(x) = 2 \sin x + \sin 2x, \quad [0, 2\pi]$

36. $f(x) = x + 2 \cos x, \quad [0, 2\pi]$

37. $y = e^{-3/x}$

38. $y = \frac{1}{2}(e^x - e^{-x})$

39. $y = x - \ln x$

40. $y = \ln\sqrt{x^2 + 9}$

41. $f(x) = \arcsin x^{4/5}$

42. $f(x) = \arctan(x^2)$

In Exercises 43–70, find all relative extrema. Use the Second Derivative Test where applicable.

43. $f(x) = (x - 5)^2$

44. $f(x) = -(x - 5)^2$

45. $f(x) = 6x - x^2$

46. $f(x) = x^2 + 3x - 8$

47. $f(x) = x^3 - 3x^2 + 3$

48. $f(x) = x^3 - 5x^2 + 7x$

49. $f(x) = x^4 - 4x^3 + 2$

50. $f(x) = -x^4 + 4x^3 + 8x^2$

51. $g(x) = x^2(6 - x)^3$

52. $g(x) = -\frac{1}{8}(x + 2)^2(x - 4)^2$

53. $f(x) = x^{2/3} - 3$

54. $f(x) = \sqrt{x^2 + 1}$

55. $f(x) = x + \dfrac{4}{x}$

56. $f(x) = \dfrac{x}{x - 1}$

57. $f(x) = \cos x - x, \quad [0, 4\pi]$

58. $f(x) = 2 \sin x + \cos 2x, \quad [0, 2\pi]$

59. $y = \dfrac{1}{2}x^2 - \ln x$

60. $y = x \ln x$

61. $y = \dfrac{x}{\ln x}$

62. $y = x^2 \ln \dfrac{x}{4}$

63. $f(x) = \dfrac{e^x + e^{-x}}{2}$

64. $g(x) = \dfrac{1}{\sqrt{2\pi}} e^{-(x - 3)^2/2}$

65. $f(x) = x^2 e^{-x}$

66. $f(x) = xe^{-x}$

67. $f(x) = 8x(4^{-x})$

68. $y = x^2 \log_3 x$

69. $f(x) = \text{arcsec } x - x$

70. $f(x) = \arcsin x - 2x$

CAS **In Exercises 71–74, use a computer algebra system to analyze the function over the given interval. (a) Find the first and second derivatives of the function. (b) Find any relative extrema and points of inflection. (c) Graph f, f', and f'' on the same set of coordinate axes and state the relationship between the behavior of f and the signs of f' and f''.**

71. $f(x) = 0.2x^2(x - 3)^3, \quad [-1, 4]$

72. $f(x) = x^2\sqrt{6 - x^2}, \quad [-\sqrt{6}, \sqrt{6}]$

73. $f(x) = \sin x - \frac{1}{3} \sin 3x + \frac{1}{5} \sin 5x, \quad [0, \pi]$

74. $f(x) = \sqrt{2x} \sin x, \quad [0, 2\pi]$

75. Consider a function f such that f' is increasing. Sketch graphs of f for (a) $f' < 0$ and (b) $f' > 0$.

76. Consider a function f such that f' is decreasing. Sketch graphs of f for (a) $f' < 0$ and (b) $f' > 0$.

77. Sketch the graph of a function f that does *not* have a point of inflection at $(c, f(c))$ even though $f''(c) = 0$.

78. S represents weekly sales of a product. What can be said of S' and S'' for each of the following statements?

(a) The rate of change of sales is increasing.

(b) Sales are increasing at a slower rate.

(c) The rate of change of sales is constant.

(d) Sales are steady.

(e) Sales are declining, but at a slower rate.

(f) Sales have bottomed out and have started to rise.

In Exercises 79–82, the graph of f is shown. Graph f, f', and f'' on the same set of coordinate axes. To print an enlarged copy of the graph, go to the website www.mathgraphs.com.

79.

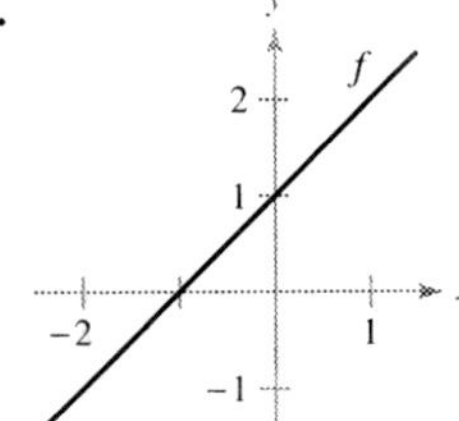

80.

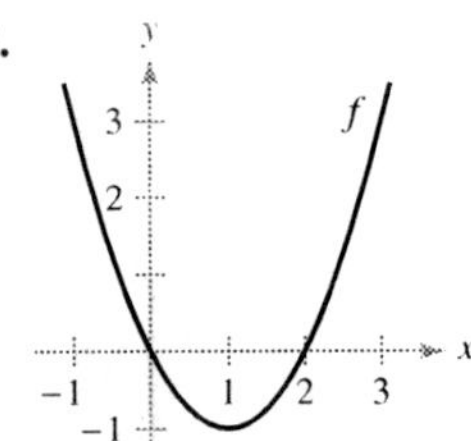

81.

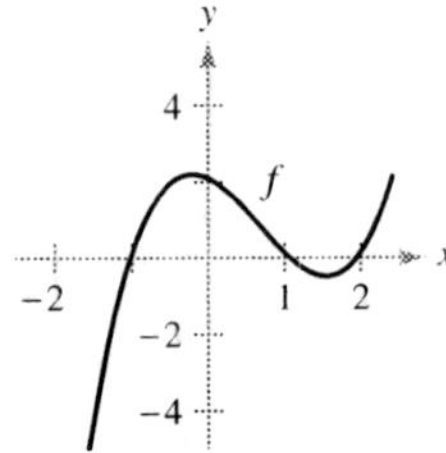

82.

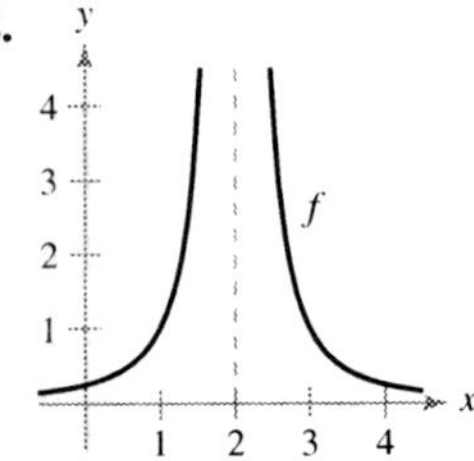

Think About It **In Exercises 83–86, sketch the graph of a function f having the given characteristics.**

83. $f(2) = f(4) = 0$
$f'(x) < 0$ if $x < 3$
$f'(3)$ does not exist.
$f'(x) > 0$ if $x > 3$
$f''(x) < 0$, $x \neq 3$

84. $f(0) = f(2) = 0$
$f'(x) > 0$ if $x < 1$
$f'(1) = 0$
$f'(x) < 0$ if $x > 1$
$f''(x) < 0$

85. $f(2) = f(4) = 0$
$f'(x) > 0$ if $x < 3$
$f'(3)$ does not exist.
$f'(x) < 0$ if $x > 3$
$f''(x) > 0$, $x \neq 3$

86. $f(0) = f(2) = 0$
$f'(x) < 0$ if $x < 1$
$f'(1) = 0$
$f'(x) > 0$ if $x > 1$
$f''(x) > 0$

87. ***Conjecture*** Consider the function $f(x) = (x - 2)^n$.

(a) Use a graphing utility to graph f for $n = 1, 2, 3,$ and 4. Use the graphs to make a conjecture about the relationship between n and any inflection points of the graph of f.

(b) Verify your conjecture in part (a).

88. ***Think About It*** Water is running into the vase shown in the figure at a constant rate.

(a) Graph the depth d of water in the vase as a function of time.

(b) Does the function have any extrema? Explain.

(c) Interpret the inflection points of the graph of d.

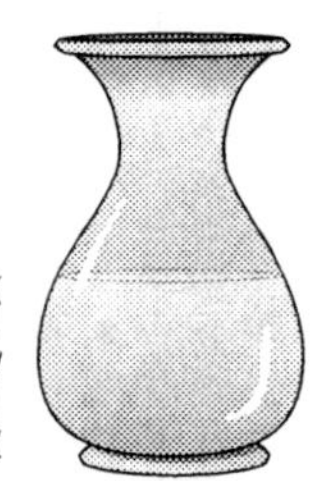

In Exercises 89 and 90, find a, b, c, and d such that the cubic $f(x) = ax^3 + bx^2 + cx + d$ satisfies the given conditions.

89. Relative maximum: $(3, 3)$
Relative minimum: $(5, 1)$
Inflection point: $(4, 2)$

90. Relative maximum: $(2, 4)$
Relative minimum: $(4, 2)$
Inflection point: $(3, 3)$

91. ***Aircraft Glide Path*** A small aircraft starts its descent from an altitude of 1 mile, 4 miles west of the runway (see figure).

(a) Find the cubic $f(x) = ax^3 + bx^2 + cx + d$ on the interval $[-4, 0]$ that describes a smooth glide path for the landing.

(b) The function in part (a) models the glide path of the plane. When would the plane be descending at the greatest rate?

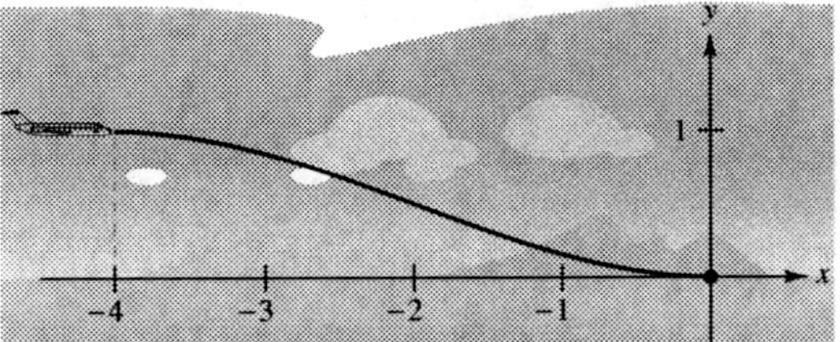

FOR FURTHER INFORMATION For more information on this type of modeling, see the article "How Not to Land at Lake Tahoe!" by Richard Barshinger in *The American Mathematical Monthly*. To view this article, go to the website www.matharticles.com.

92.  ***Highway Design*** A section of highway connecting two hillsides with grades of 6% and 4% is to be built between two points that are separated by a horizontal distance of 2000 feet (see figure on the next page). At the point where the two hillsides come together, there is a 50-foot difference in elevation.

(a) Design a section of highway connecting the hillsides modeled by the function $f(x) = ax^3 + bx^2 + cx + d$ $(-1000 \le x \le 1000)$. At the points A and B, the slope of the model must match the grade of the hillside.

(b) Use a graphing utility to graph the model.

(c) Use a graphing utility to graph the derivative of the model.

(d) Determine the grade at the steepest part of the transitional section of the highway.

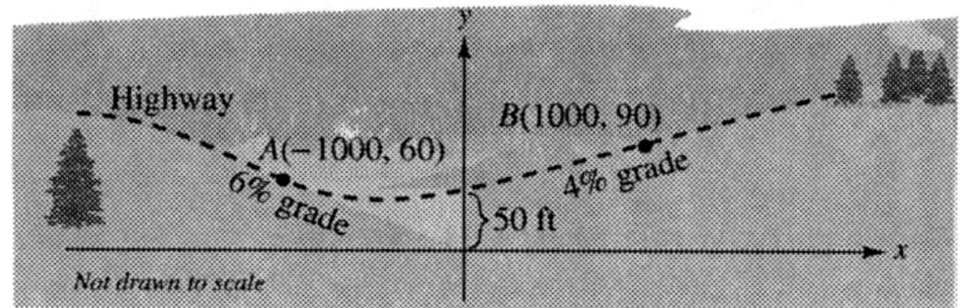

93. Beam Deflection The deflection D of a beam of length L is $D = 2x^4 - 5Lx^3 + 3L^2x^2$, where x is the distance from one end of the beam. Find the value of x that yields the maximum deflection.

94. Specific Gravity A model for the specific gravity of water S is

$$S = \frac{5.755}{10^8}T^3 - \frac{8.521}{10^6}T^2 + \frac{6.540}{10^5}T + 0.99987, \quad 0 < T < 25$$

where T is the water temperature in degrees Celsius.

CAS (a) Use a computer algebra system to find the coordinates of the maximum value of the function.

(b) Sketch a graph of the function over the specified domain. (Use a setting in which $0.996 \le S \le 1.001$.)

(c) Estimate the specific gravity of water when $T = 20°$.

95. Average Cost A manufacturer has determined that the total cost C of operating a factory is $C = 0.5x^2 + 15x + 5000$, where x is the number of units produced. At what level of production will the average cost per unit be minimized? (The average cost per unit is C/x.)

96. Modeling Data The average typing speeds S (in words per minute) of a typing student after t weeks of lessons are shown in the table.

t	5	10	15	20	25	30
S	38	56	79	90	93	94

A model for the data is $S = \dfrac{100t^2}{65 + t^2}$, $t > 0$.

(a) Use a graphing utility to plot the data and graph the model.

(b) Use the second derivative to determine the concavity of S. Compare the result with the graph in part (a).

(c) What is the sign of the first derivative for $t > 0$? By combining this information with the concavity of the model, what inferences can be made about the typing speed as t increases?

Linear and Quadratic Approximations In Exercises 97–100, use a graphing utility to graph the function. Then graph the linear and quadratic approximations

$$P_1(x) = f(a) + f'(a)(x - a)$$

and

$$P_2(x) = f(a) + f'(a)(x - a) + \tfrac{1}{2}f''(a)(x - a)^2$$

in the same viewing window. Compare the values of f, P_1, and P_2 and their first derivatives at $x = a$. How do the approximations change as you move farther away from $x = a$?

Function	Value of a
97. $f(x) = 2(\sin x + \cos x)$	$a = \dfrac{\pi}{4}$
98. $f(x) = 2(\sin x + \cos x)$	$a = 0$
99. $f(x) = \arctan x$	$a = -1$
100. $f(x) = \dfrac{\sqrt{x}}{x - 1}$	$a = 2$

True or False? In Exercises 101–106, determine whether the statement is true or false. If it is false, explain why or give an example that shows it is false.

101. The graph of every cubic polynomial has precisely one point of inflection.

102. The graph of $f(x) = 1/x$ is concave downward for $x < 0$ and concave upward for $x > 0$, and thus it has a point of inflection at $x = 0$.

103. The maximum value of $y = 3 \sin x + 2 \cos x$ is 5.

104. The maximum slope of the graph of $y = \sin(bx)$ is b.

105. If $f'(c) > 0$, then f is concave upward at $x = c$.

106. If $f''(2) = 0$, then the graph of f must have a point of inflection at $x = 2$.

In Exercises 107 and 108, let f and g represent differentiable functions such that $f'' \ne 0$ and $g'' \ne 0$.

107. Show that if f and g are concave upward on the interval (a, b), then $f + g$ is also concave upward on (a, b).

108. Prove that if f and g are positive, increasing, and concave upward on the interval (a, b), then fg is also concave upward on (a, b).

109. Use a graphing utility to graph $y = x \sin(1/x)$. Show that the graph is concave downward to the right of $x = 1/\pi$.

110. (a) Graph $f(x) = \sqrt[3]{x}$ and identify the inflection point.

(b) Does $f''(x)$ exist at the inflection point? Explain.

111. Show that the point of inflection of $f(x) = x(x - 6)^2$ lies midway between the relative extrema of f.

112. Prove that every cubic function with three distinct real zeros has a point of inflection whose x-coordinate is the average of the three zeros.

4.5 Limits at Infinity

- Determine (finite) limits at infinity.
- Determine the horizontal asymptotes, if any, of the graph of a function.
- Determine infinite limits at infinity.

Limits at Infinity

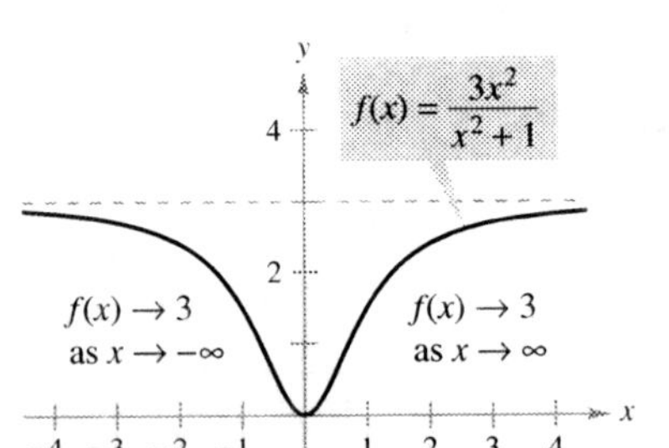

The limit of $f(x)$ as x approaches $-\infty$ or ∞ is 3.

Figure 4.33

This section discusses the "end behavior" of a function on an *infinite* interval. Consider the graph of

$$f(x) = \frac{3x^2}{x^2 + 1}$$

as shown in Figure 4.33. Graphically, you can see that the values of $f(x)$ appear to approach 3 as x increases without bound or decreases without bound. You can come to the same conclusions numerically, as shown in the table.

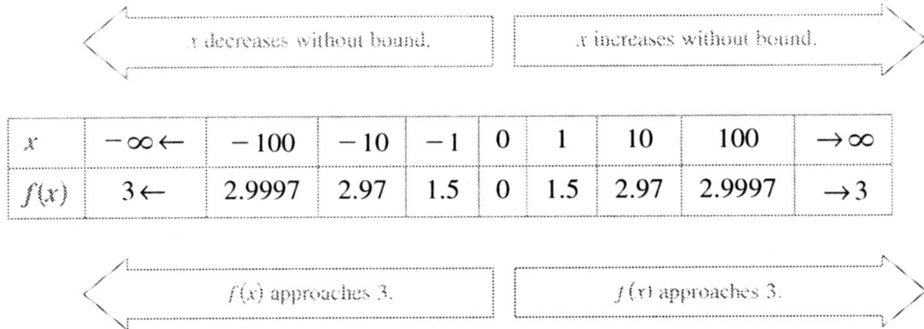

	x decreases without bound.						x increases without bound.		
x	$-\infty \leftarrow$	-100	-10	-1	0	1	10	100	$\to \infty$
$f(x)$	$3 \leftarrow$	2.9997	2.97	1.5	0	1.5	2.97	2.9997	$\to 3$
	$f(x)$ approaches 3.						$f(x)$ approaches 3.		

The table suggests that the value of $f(x)$ approaches 3 as x increases without bound $(x \to \infty)$. Similarly, $f(x)$ approaches 3 as x decreases without bound $(x \to -\infty)$. These **limits at infinity** are denoted by

$$\lim_{x \to -\infty} f(x) = 3 \qquad \text{Limit at negative infinity}$$

and

$$\lim_{x \to \infty} f(x) = 3. \qquad \text{Limit at positive infinity}$$

NOTE The statement $\lim\limits_{x \to -\infty} f(x) = L$ or $\lim\limits_{x \to \infty} f(x) = L$ means that the limit exists *and* the limit is equal to L.

To say that a statement is true as x increases *without bound* means that for some (large) real number M, the statement is true for *all* x in the interval $\{x : x > M\}$. The following definition uses this concept.

DEFINITION OF LIMITS AT INFINITY

Let L be a real number.

1. The statement $\lim\limits_{x \to \infty} f(x) = L$ means that for each $\varepsilon > 0$ there exists an $M > 0$ such that $|f(x) - L| < \varepsilon$ whenever $x > M$.

2. The statement $\lim\limits_{x \to -\infty} f(x) = L$ means that for each $\varepsilon > 0$ there exists an $N < 0$ such that $|f(x) - L| < \varepsilon$ whenever $x < N$.

$f(x)$ is within ε units of L as $x \to \infty$.

Figure 4.34

The definition of a limit at infinity is shown in Figure 4.34. In this figure, note that for a given positive number ε there exists a positive number M such that, for $x > M$, the graph of f will lie between the horizontal lines given by $y = L + \varepsilon$ and $y = L - \varepsilon$.

Horizontal Asymptotes

In Figure 4.34, the graph of f approaches the line $y = L$ as x increases without bound. The line $y = L$ is called a **horizontal asymptote** of the graph of f.

DEFINITION OF HORIZONTAL ASYMPTOTE

The line $y = L$ is a **horizontal asymptote** of the graph of f if

$$\lim_{x \to -\infty} f(x) = L \quad \text{or} \quad \lim_{x \to \infty} f(x) = L.$$

Note that from this definition, it follows that the graph of a *function* of x can have at most two horizontal asymptotes—one to the right and one to the left.

Limits at infinity have many of the same properties of limits discussed in Section 2.3. For example, if $\lim\limits_{x \to \infty} f(x)$ and $\lim\limits_{x \to \infty} g(x)$ both exist, then

$$\lim_{x \to \infty} [f(x) + g(x)] = \lim_{x \to \infty} f(x) + \lim_{x \to \infty} g(x)$$

and

$$\lim_{x \to \infty} [f(x)g(x)] = \left[\lim_{x \to \infty} f(x) \right]\left[\lim_{x \to \infty} g(x) \right].$$

Similar properties hold for limits at $-\infty$.

When evaluating limits at infinity, the following theorem is helpful. (A proof of part 1 of this theorem is given in Appendix A.)

THEOREM 4.10 LIMITS AT INFINITY

1. If r is a positive rational number and c is any real number, then

$$\lim_{x \to \infty} \frac{c}{x^r} = 0 \quad \text{and} \quad \lim_{x \to -\infty} \frac{c}{x^r} = 0.$$

The second limit is valid only if x^r is defined when $x < 0$.

2. $\lim\limits_{x \to -\infty} e^x = 0$ and $\lim\limits_{x \to \infty} e^{-x} = 0$

EXAMPLE 1 Evaluating a Limit at Infinity

a. $\lim\limits_{x \to \infty} \left(5 - \dfrac{2}{x^2} \right) = \lim\limits_{x \to \infty} 5 - \lim\limits_{x \to \infty} \dfrac{2}{x^2}$ Property of limits

$$= 5 - 0$$

$$= 5$$

b. $\lim\limits_{x \to \infty} \dfrac{3}{e^x} = \lim\limits_{x \to \infty} 3e^{-x}$

$$= 3 \lim_{x \to \infty} e^{-x} \qquad \text{Property of limits}$$

$$= 3(0)$$

$$= 0$$

EXAMPLE 2 Evaluating a Limit at Infinity

Find the limit: $\lim\limits_{x \to \infty} \dfrac{2x - 1}{x + 1}$.

Solution Note that both the numerator and the denominator approach infinity as x approaches infinity.

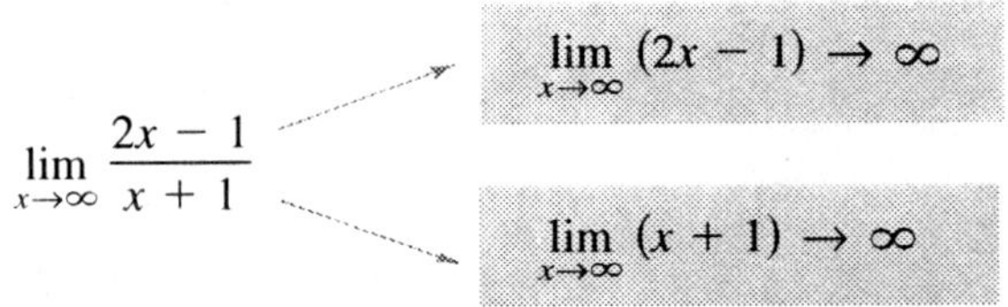

NOTE When you encounter an indeterminate form such as the one in Example 2, you should divide the numerator and denominator by the highest power of x in the *denominator*.

This results in $\dfrac{\infty}{\infty}$, an **indeterminate form.** To resolve this problem, you can divide both the numerator and the denominator by x. After dividing, the limit may be evaluated as follows.

$$\lim_{x \to \infty} \frac{2x - 1}{x + 1} = \lim_{x \to \infty} \frac{\dfrac{2x - 1}{x}}{\dfrac{x + 1}{x}} \qquad \text{Divide numerator and denominator by } x.$$

$$= \lim_{x \to \infty} \frac{2 - \dfrac{1}{x}}{1 + \dfrac{1}{x}} \qquad \text{Simplify.}$$

$$= \frac{\lim\limits_{x \to \infty} 2 - \lim\limits_{x \to \infty} \dfrac{1}{x}}{\lim\limits_{x \to \infty} 1 + \lim\limits_{x \to \infty} \dfrac{1}{x}} \qquad \text{Take limits of numerator and denominator.}$$

$$= \frac{2 - 0}{1 + 0} \qquad \text{Apply Theorem 4.10.}$$

$$= 2$$

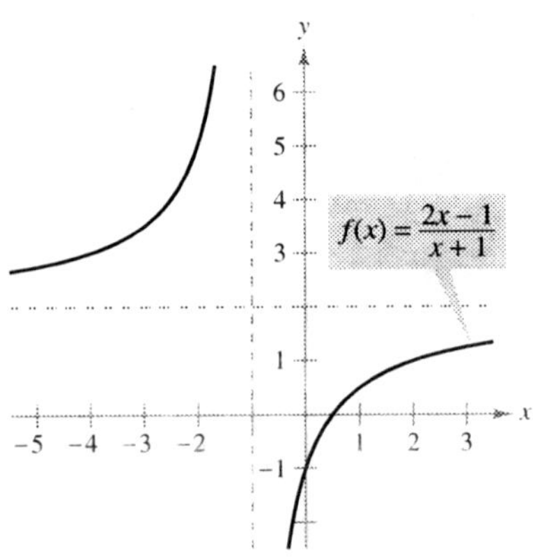

$y = 2$ is a horizontal asymptote.

Figure 4.35

So, the line $y = 2$ is a horizontal asymptote to the right. By taking the limit as $x \to -\infty$, you can see that $y = 2$ is also a horizontal asymptote to the left. The graph of the function is shown in Figure 4.35.

TECHNOLOGY You can test the reasonableness of the limit found in Example 2 by evaluating $f(x)$ for a few large positive values of x. For instance,

$$f(100) \approx 1.9703, \quad f(1000) \approx 1.9970, \quad \text{and} \quad f(10{,}000) \approx 1.9997.$$

Another way to test the reasonableness of the limit is to use a graphing utility. For instance, in Figure 4.36, the graph of

$$f(x) = \frac{2x - 1}{x + 1}$$

As x increases, the graph of f moves closer and closer to the line $y = 2$.

Figure 4.36

is shown with the horizontal line $y = 2$. Note that as x increases, the graph of f moves closer and closer to its horizontal asymptote.

EXAMPLE 3   A Comparison of Three Rational Functions

Find each limit.

a. $\displaystyle\lim_{x\to\infty}\frac{2x+5}{3x^2+1}$ **b.** $\displaystyle\lim_{x\to\infty}\frac{2x^2+5}{3x^2+1}$ **c.** $\displaystyle\lim_{x\to\infty}\frac{2x^3+5}{3x^2+1}$

Solution   In each case, attempting to evaluate the limit produces the indeterminate form ∞/∞.

a. Divide both the numerator and the denominator by x^2.

$$\lim_{x\to\infty}\frac{2x+5}{3x^2+1}=\lim_{x\to\infty}\frac{(2/x)+(5/x^2)}{3+(1/x^2)}=\frac{0+0}{3+0}=\frac{0}{3}=0$$

b. Divide both the numerator and the denominator by x^2.

$$\lim_{x\to\infty}\frac{2x^2+5}{3x^2+1}=\lim_{x\to\infty}\frac{2+(5/x^2)}{3+(1/x^2)}=\frac{2+0}{3+0}=\frac{2}{3}$$

c. Divide both the numerator and the denominator by x^2.

$$\lim_{x\to\infty}\frac{2x^3+5}{3x^2+1}=\lim_{x\to\infty}\frac{2x+(5/x^2)}{3+(1/x^2)}=\frac{\infty}{3}$$

You can conclude that the limit *does not exist* because the numerator increases without bound while the denominator approaches 3. ∎

GUIDELINES FOR FINDING LIMITS AT $\pm\infty$ OF RATIONAL FUNCTIONS

1. If the degree of the numerator is *less than* the degree of the denominator, then the limit of the rational function is 0.
2. If the degree of the numerator is *equal to* the degree of the denominator, then the limit of the rational function is the ratio of the leading coefficients.
3. If the degree of the numerator is *greater than* the degree of the denominator, then the limit of the rational function does not exist.

Use these guidelines to check the results in Example 3. These limits seem reasonable when you consider that for large values of x, the highest-power term of the rational function is the most "influential" in determining the limit. For instance, the limit as x approaches infinity of the function

$$f(x)=\frac{1}{x^2+1}$$

is 0 because the denominator overpowers the numerator as x increases or decreases without bound, as shown in Figure 4.37.

The function shown in Figure 4.37 is a special case of a type of curve studied by the Italian mathematician Maria Gaetana Agnesi. The general form of this function is

$$f(x)=\frac{8a^3}{x^2+4a^2} \qquad \text{Witch of Agnesi}$$

and, through a mistranslation of the Italian word *vertéré*, the curve has come to be known as the Witch of Agnesi. Agnesi's work with this curve first appeared in a comprehensive text on calculus that was published in 1748.

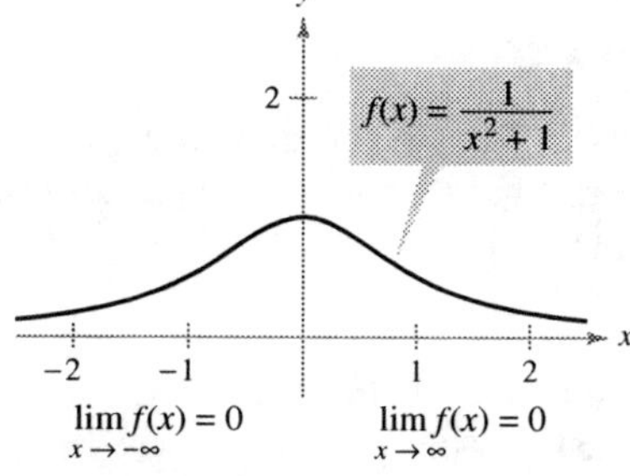

MARIA GAETANA AGNESI (1718–1799)

Agnesi was one of a handful of women to receive credit for significant contributions to mathematics before the twentieth century. In her early twenties, she wrote the first text that included both differential and integral calculus. By age 30, she was an honorary member of the faculty at the University of Bologna.

For more information on the contributions of women to mathematics, see the article "Why Women Succeed in Mathematics" by Mona Fabricant, Sylvia Svitak, and Patricia Clark Kenschaft in *Mathematics Teacher*. To view this article, go to the website *www.matharticles.com*.

f has a horizontal asymptote at $y=0$.
Figure 4.37

In Figure 4.37, you can see that the function

$$f(x) = \frac{1}{x^2 + 1}$$

approaches the same horizontal asymptote to the right and to the left. This is always true of rational functions. Functions that are not rational, however, may approach different horizontal asymptotes to the right and to the left. A common example of such a function is the **logistic function** shown in the next example.

EXAMPLE 4 A Function with Two Horizontal Asymptotes

Show that the *logistic function*

$$f(x) = \frac{1}{1 + e^{-x}}$$

has different horizontal asymptotes to the left and to the right.

Solution To begin, try using a graphing utility to graph the function. From Figure 4.38 it appears that

$$y = 0 \quad \text{and} \quad y = 1$$

are horizontal asymptotes to the left and to the right, respectively. The following table shows the same results numerically.

x	-10	-5	-2	-1	1	2	5	10
$f(x)$	0.000	0.007	0.119	0.269	0.731	0.881	0.9933	1.0000

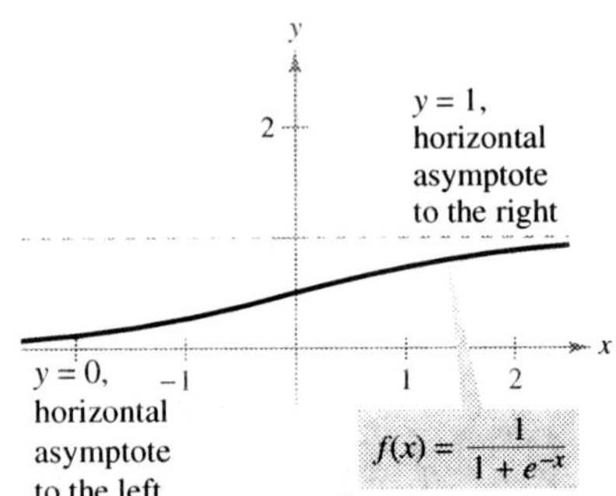

Functions that are not rational may have different right and left horizontal asymptotes.
Figure 4.38

Finally, you can obtain the same results analytically, as follows.

$$\lim_{x \to \infty} \frac{1}{1 + e^{-x}} = \frac{\lim\limits_{x \to \infty} 1}{\lim\limits_{x \to \infty} (1 + e^{-x})}$$

$$= \frac{1}{1 + 0}$$

$$= 1 \qquad \text{$y = 1$ is a horizontal asymptote to the right.}$$

For the horizontal asymptote to the left, note that as $x \to -\infty$ the denominator of $1/(1 + e^{-x})$ approaches infinity. So, the quotient approaches 0 and thus the limit is 0. You can conclude that $y = 0$ is a horizontal asymptote to the left. ∎

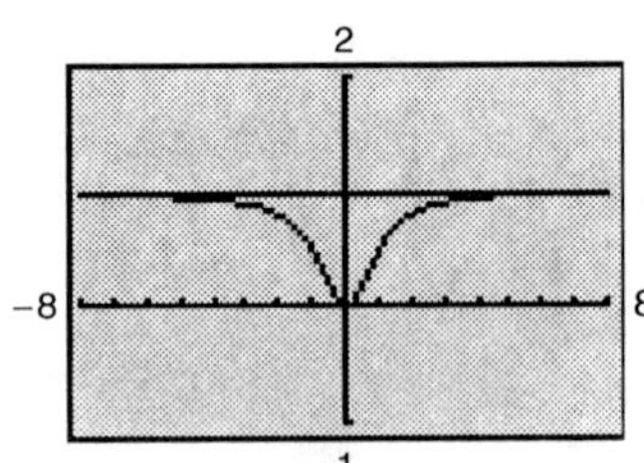

The horizontal asymptote appears to be the line $y = 1$ but it is actually the line $y = 2$.
Figure 4.39

TECHNOLOGY PITFALL If you use a graphing utility to help estimate a limit, be sure that you also confirm the estimate analytically—the pictures shown by a graphing utility can be misleading. For instance, Figure 4.39 shows one view of the graph of

$$y = \frac{2x^3 + 1000x^2 + x}{x^3 + 1000x^2 + x + 1000}.$$

From this view, one could be convinced that the graph has $y = 1$ as a horizontal asymptote. An analytical approach shows that the horizontal asymptote is actually $y = 2$. Confirm this by enlarging the viewing window on the graphing utility.

In Section 2.3 (Example 9), you saw how the Squeeze Theorem can be used to evaluate limits involving trigonometric functions. This theorem is also valid for limits at infinity.

EXAMPLE 5   Limits Involving Trigonometric Functions

Find each limit.

a. $\displaystyle \lim_{x \to \infty} \sin x$ **b.** $\displaystyle \lim_{x \to \infty} \frac{\sin x}{x}$

Solution

a. As x approaches infinity, the sine function oscillates between 1 and -1. So, this limit does not exist.

b. Because $-1 \le \sin x \le 1$, it follows that for $x > 0$,

$$-\frac{1}{x} \le \frac{\sin x}{x} \le \frac{1}{x}$$

where $\displaystyle \lim_{x \to \infty} (-1/x) = 0$ and $\displaystyle \lim_{x \to \infty} (1/x) = 0$. So, by the Squeeze Theorem, you can obtain

$$\lim_{x \to \infty} \frac{\sin x}{x} = 0$$

as shown in Figure 4.40.

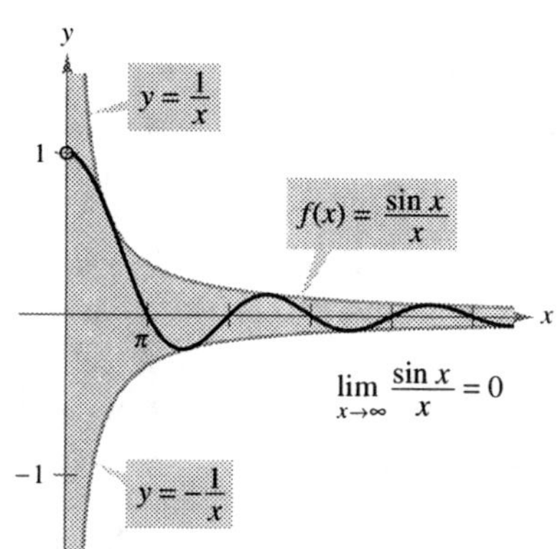

As x increases without bound, $f(x)$ approaches 0.
Figure 4.40

EXAMPLE 6   Oxygen Level in a Pond

Suppose that $f(t)$ measures the level of oxygen in a pond, where $f(t) = 1$ is the normal (unpolluted) level and the time t is measured in weeks. When $t = 0$, organic waste is dumped into the pond, and as the waste material oxidizes, the level of oxygen in the pond is

$$f(t) = \frac{t^2 - t + 1}{t^2 + 1}.$$

What percent of the normal level of oxygen exists in the pond after 1 week? After 2 weeks? After 10 weeks? What is the limit as t approaches infinity?

Solution   When $t = 1, 2,$ and 10, the levels of oxygen are as shown.

$$f(1) = \frac{1^2 - 1 + 1}{1^2 + 1} = \frac{1}{2} = 50\% \qquad \text{1 week}$$

$$f(2) = \frac{2^2 - 2 + 1}{2^2 + 1} = \frac{3}{5} = 60\% \qquad \text{2 weeks}$$

$$f(10) = \frac{10^2 - 10 + 1}{10^2 + 1} = \frac{91}{101} \approx 90.1\% \qquad \text{10 weeks}$$

To find the limit as t approaches infinity, you can use the guidelines on page 241, or divide the numerator and the denominator by t^2 to obtain

$$\lim_{t \to \infty} \frac{t^2 - t + 1}{t^2 + 1} = \lim_{t \to \infty} \frac{1 - (1/t) + (1/t^2)}{1 + (1/t^2)} = \frac{1 - 0 + 0}{1 + 0} = 1 = 100\%.$$

See Figure 4.41.

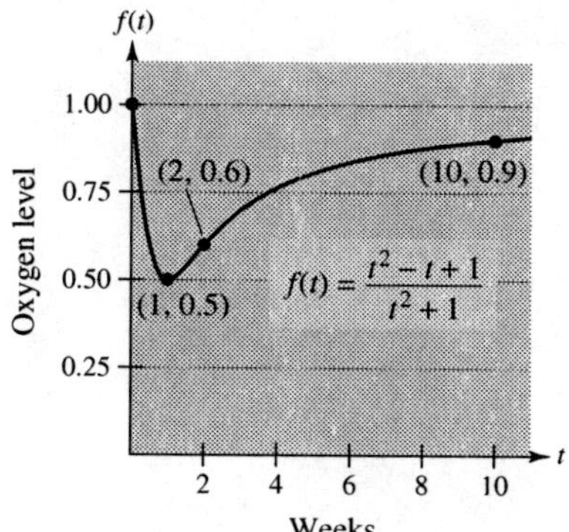

The level of oxygen in a pond approaches the normal level of 1 as t approaches ∞.
Figure 4.41

Infinite Limits at Infinity

Many functions do not approach a finite limit as x increases (or decreases) without bound. For instance, no polynomial function has a finite limit at infinity. The following definition is used to describe the behavior of polynomial and other functions at infinity.

NOTE Determining whether a function has an infinite limit at infinity is useful in analyzing the "end behavior" of its graph. You will see examples of this in Section 4.6 on curve sketching.

DEFINITION OF INFINITE LIMITS AT INFINITY

Let f be a function defined on the interval (a, ∞).

1. The statement $\lim\limits_{x \to \infty} f(x) = \infty$ means that for each positive number M, there is a corresponding number $N > 0$ such that $f(x) > M$ whenever $x > N$.

2. The statement $\lim\limits_{x \to \infty} f(x) = -\infty$ means that for each negative number M, there is a corresponding number $N > 0$ such that $f(x) < M$ whenever $x > N$.

Similar definitions can be given for the statements $\lim\limits_{x \to -\infty} f(x) = \infty$ and $\lim\limits_{x \to -\infty} f(x) = -\infty$.

EXAMPLE 7 Finding Infinite Limits at Infinity

Find each limit.

a. $\lim\limits_{x \to \infty} x^3$ **b.** $\lim\limits_{x \to -\infty} x^3$

Solution

a. As x increases without bound, x^3 also increases without bound. So, you can write
$$\lim_{x \to \infty} x^3 = \infty.$$

b. As x decreases without bound, x^3 also decreases without bound. So, you can write
$$\lim_{x \to -\infty} x^3 = -\infty.$$

The graph of $f(x) = x^3$ in Figure 4.42 illustrates these two results. These results agree with the Leading Coefficient Test for polynomial functions as described in Section 1.3.

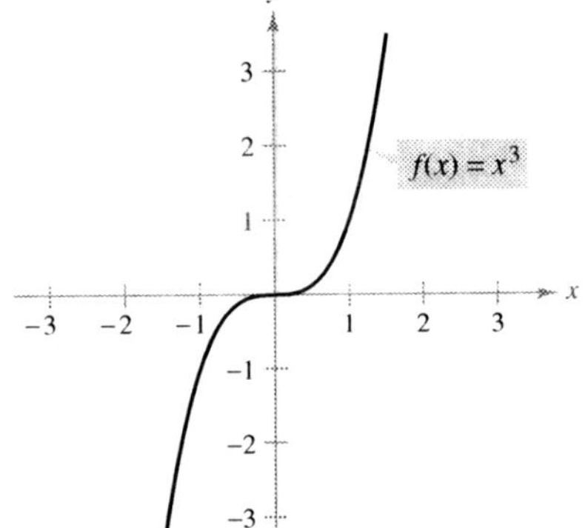

Figure 4.42

EXAMPLE 8 Finding Infinite Limits at Infinity

Find each limit.

a. $\lim\limits_{x \to \infty} \dfrac{2x^2 - 4x}{x + 1}$ **b.** $\lim\limits_{x \to -\infty} \dfrac{2x^2 - 4x}{x + 1}$

Solution One way to evaluate each of these limits is to use long division to rewrite the improper rational function as the sum of a polynomial and a rational function.

a. $\lim\limits_{x \to \infty} \dfrac{2x^2 - 4x}{x + 1} = \lim\limits_{x \to \infty} \left(2x - 6 + \dfrac{6}{x + 1}\right) = \infty$

b. $\lim\limits_{x \to -\infty} \dfrac{2x^2 - 4x}{x + 1} = \lim\limits_{x \to -\infty} \left(2x - 6 + \dfrac{6}{x + 1}\right) = -\infty$

The statements above can be interpreted as saying that as x approaches $\pm\infty$, the function $f(x) = (2x^2 - 4x)/(x + 1)$ behaves like the function $g(x) = 2x - 6$. In Section 4.6, you will see that this is graphically described by saying that the line $y = 2x - 6$ is a slant asymptote of the graph of f, as shown in Figure 4.43. ∎

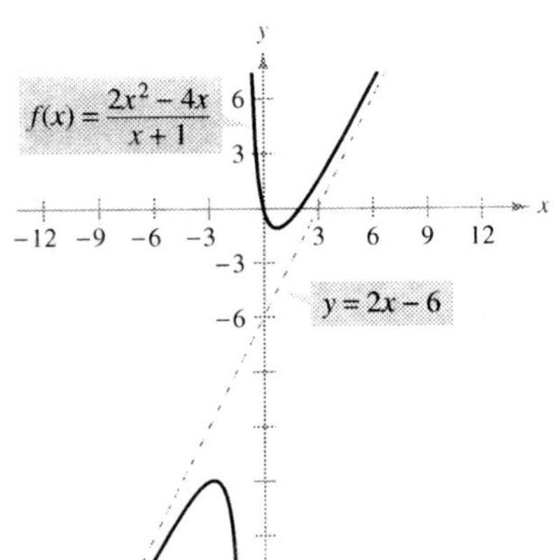

Figure 4.43

4.5 Exercises

See www.CalcChat.com for worked-out solutions to odd-numbered exercises.

In Exercises 1–6, match the function with one of the graphs [(a), (b), (c), (d), (e), or (f)] using horizontal asymptotes as an aid.

(a) 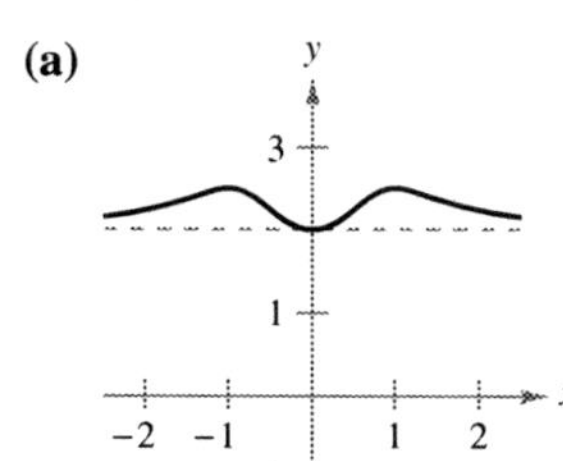

(b) 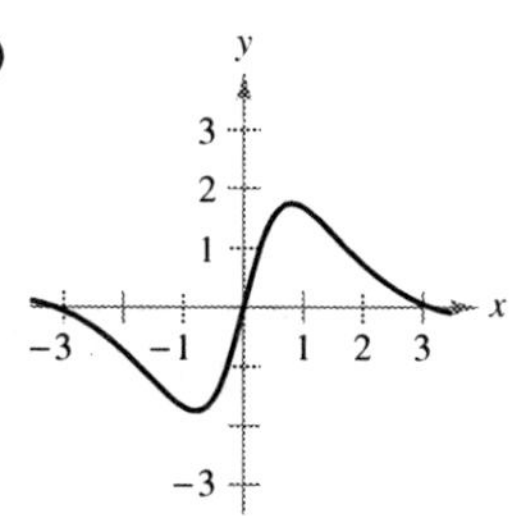

(c) 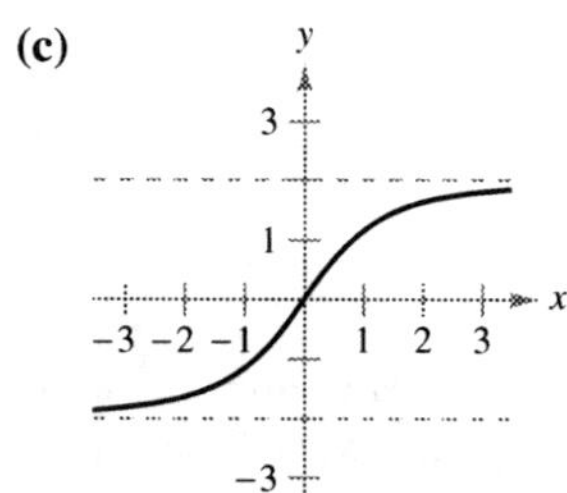

(d) 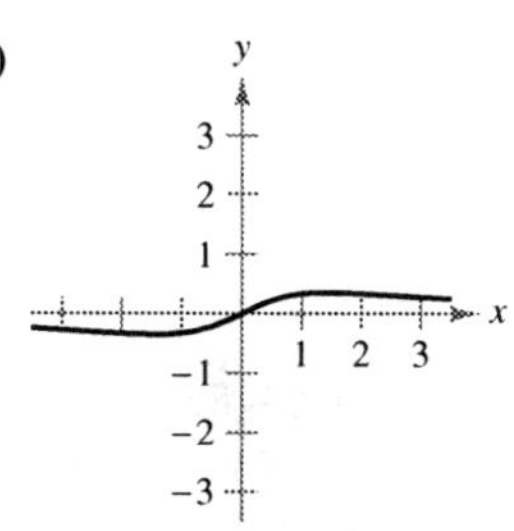

(e) 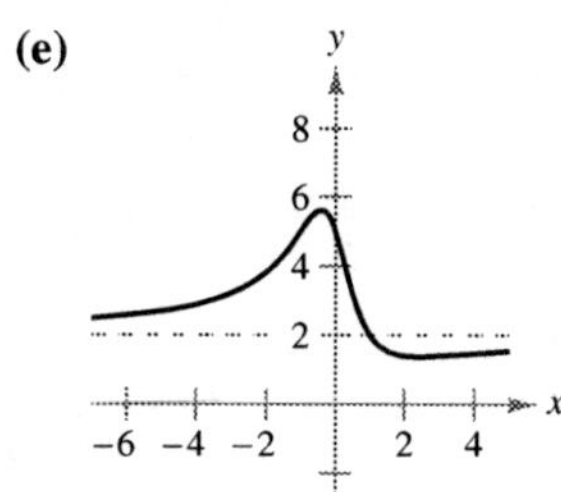

(f) 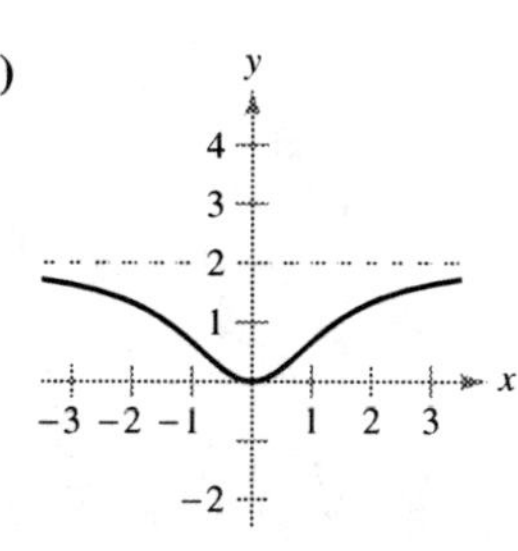

1. $f(x) = \dfrac{2x^2}{x^2 + 2}$

2. $f(x) = \dfrac{2x}{\sqrt{x^2 + 2}}$

3. $f(x) = \dfrac{x}{x^2 + 2}$

4. $f(x) = 2 + \dfrac{x^2}{x^4 + 1}$

5. $f(x) = \dfrac{4 \sin x}{x^2 + 1}$

6. $f(x) = \dfrac{2x^2 - 3x + 5}{x^2 + 1}$

Numerical and Graphical Analysis **In Exercises 7–12, use a graphing utility to complete the table and estimate the limit as x approaches infinity. Then use a graphing utility to graph the function and estimate the limit graphically.**

x	10^0	10^1	10^2	10^3	10^4	10^5	10^6
$f(x)$							

7. $f(x) = \dfrac{4x + 3}{2x - 1}$

8. $f(x) = \dfrac{2x^2}{x + 1}$

9. $f(x) = \dfrac{-6x}{\sqrt{4x^2 + 5}}$

10. $f(x) = \dfrac{20x}{\sqrt{9x^2 - 1}}$

11. $f(x) = 5 - \dfrac{1}{x^2 + 1}$

12. $f(x) = 4 + \dfrac{3}{x^2 + 2}$

In Exercises 13 and 14, find $\lim\limits_{x \to \infty} h(x)$, if possible.

13. $f(x) = 5x^3 - 3x^2 + 10x$

(a) $h(x) = \dfrac{f(x)}{x^2}$

(b) $h(x) = \dfrac{f(x)}{x^3}$

(c) $h(x) = \dfrac{f(x)}{x^4}$

14. $f(x) = -4x^2 + 2x - 5$

(a) $h(x) = \dfrac{f(x)}{x}$

(b) $h(x) = \dfrac{f(x)}{x^2}$

(c) $h(x) = \dfrac{f(x)}{x^3}$

In Exercises 15–18, find each limit, if possible.

15. (a) $\lim\limits_{x \to \infty} \dfrac{x^2 + 2}{x^3 - 1}$

(b) $\lim\limits_{x \to \infty} \dfrac{x^2 + 2}{x^2 - 1}$

(c) $\lim\limits_{x \to \infty} \dfrac{x^2 + 2}{x - 1}$

16. (a) $\lim\limits_{x \to \infty} \dfrac{3 - 2x}{3x^3 - 1}$

(b) $\lim\limits_{x \to \infty} \dfrac{3 - 2x}{3x - 1}$

(c) $\lim\limits_{x \to \infty} \dfrac{3 - 2x^2}{3x - 1}$

17. (a) $\lim\limits_{x \to \infty} \dfrac{5 - 2x^{3/2}}{3x^2 - 4}$

(b) $\lim\limits_{x \to \infty} \dfrac{5 - 2x^{3/2}}{3x^{3/2} - 4}$

(c) $\lim\limits_{x \to \infty} \dfrac{5 - 2x^{3/2}}{3x - 4}$

18. (a) $\lim\limits_{x \to \infty} \dfrac{5x^{3/2}}{4x^2 + 1}$

(b) $\lim\limits_{x \to \infty} \dfrac{5x^{3/2}}{4x^{3/2} + 1}$

(c) $\lim\limits_{x \to \infty} \dfrac{5x^{3/2}}{4\sqrt{x} + 1}$

In Exercises 19–44, find the limit.

19. $\lim\limits_{x \to \infty} \left(4 + \dfrac{3}{x} \right)$

20. $\lim\limits_{x \to -\infty} \left(\dfrac{5}{x} - \dfrac{x}{3} \right)$

21. $\lim\limits_{x \to \infty} \dfrac{2x - 1}{3x + 2}$

22. $\lim\limits_{x \to \infty} \dfrac{x^2 + 3}{2x^2 - 1}$

23. $\lim\limits_{x \to \infty} \dfrac{x}{x^2 - 1}$

24. $\lim\limits_{x \to \infty} \dfrac{5x^3 + 1}{10x^3 - 3x^2 + 7}$

25. $\lim\limits_{x \to -\infty} \dfrac{x}{\sqrt{x^2 - x}}$

26. $\lim\limits_{x \to -\infty} \dfrac{x}{\sqrt{x^2 + 1}}$

27. $\lim\limits_{x \to -\infty} \dfrac{2x + 1}{\sqrt{x^2 - x}}$

28. $\lim\limits_{x \to -\infty} \dfrac{-3x + 1}{\sqrt{x^2 + x}}$

29. $\lim\limits_{x \to \infty} \dfrac{\sqrt{x^2 - 1}}{2x - 1}$

30. $\lim\limits_{x \to -\infty} \dfrac{\sqrt{x^4 - 1}}{x^3 - 1}$

31. $\lim\limits_{x \to \infty} \dfrac{x + 1}{(x^2 + 1)^{1/3}}$

32. $\lim\limits_{x \to -\infty} \dfrac{2x}{(x^6 - 1)^{1/3}}$

33. $\lim\limits_{x \to \infty} \dfrac{1}{2x + \sin x}$

34. $\lim\limits_{x \to \infty} \cos \dfrac{1}{x}$

35. $\lim\limits_{x \to \infty} \dfrac{\sin 2x}{x}$

36. $\lim\limits_{x \to \infty} \dfrac{x - \cos x}{x}$

37. $\lim\limits_{x \to \infty} (2 - 5e^{-x})$

38. $\lim\limits_{x \to -\infty} (2 + 5e^x)$

39. $\lim\limits_{x \to -\infty} \dfrac{3}{1 + 2e^x}$

40. $\lim\limits_{x \to \infty} \dfrac{8}{4 - 10^{-x/2}}$

41. $\lim\limits_{x \to \infty} \log_{10}(1 + 10^{-x})$

42. $\lim\limits_{x \to \infty} \left[\dfrac{5}{2} + \ln\left(\dfrac{x^2 + 1}{x^2} \right) \right]$

43. $\lim\limits_{t \to \infty} (8t^{-1} - \arctan t)$

44. $\lim\limits_{u \to \infty} \operatorname{arcsec}(u + 1)$

In Exercises 45–48, use a graphing utility to graph the function and identify any horizontal asymptotes.

45. $f(x) = \dfrac{|x|}{x + 1}$

46. $f(x) = \dfrac{|3x + 2|}{x - 2}$

47. $f(x) = \dfrac{3x}{\sqrt{x^2 + 2}}$

48. $f(x) = \dfrac{\sqrt{9x^2 - 2}}{2x + 1}$

In Exercises 49 and 50, find the limit. (*Hint:* Let $x = 1/t$ and find the limit as $t \to 0^+$.)

49. $\displaystyle\lim_{x \to \infty} x \sin \dfrac{1}{x}$

50. $\displaystyle\lim_{x \to \infty} x \tan \dfrac{1}{x}$

In Exercises 51–54, find the limit. (*Hint:* Treat the expression as a fraction whose denominator is 1, and rationalize the numerator.) **Use a graphing utility to verify your result.**

51. $\displaystyle\lim_{x \to -\infty} \left(x + \sqrt{x^2 + 3}\right)$

52. $\displaystyle\lim_{x \to \infty} \left(2x - \sqrt{4x^2 + 1}\right)$

53. $\displaystyle\lim_{x \to -\infty} \left(3x + \sqrt{9x^2 - x}\right)$

54. $\displaystyle\lim_{x \to \infty} \left(4x - \sqrt{16x^2 - x}\right)$

Numerical, Graphical, and Analytic Analysis **In Exercises 55–58, use a graphing utility to complete the table and estimate the limit as x approaches infinity. Then use a graphing utility to graph the function and estimate the limit. Finally, find the limit analytically and compare your results with the estimates.**

x	10^0	10^1	10^2	10^3	10^4	10^5	10^6
$f(x)$							

55. $f(x) = x - \sqrt{x(x - 1)}$

56. $f(x) = x^2 - x\sqrt{x(x - 1)}$

57. $f(x) = x \sin \dfrac{1}{2x}$

58. $f(x) = \dfrac{x + 1}{x\sqrt{x}}$

WRITING ABOUT CONCEPTS

In Exercises 59 and 60, describe in your own words what the statement means.

59. $\displaystyle\lim_{x \to \infty} f(x) = 4$

60. $\displaystyle\lim_{x \to -\infty} f(x) = 2$

61. Sketch a graph of a differentiable function f that satisfies the following conditions and has $x = 2$ as its only critical number.

$$f'(x) < 0 \text{ for } x < 2 \qquad f'(x) > 0 \text{ for } x > 2$$

$$\lim_{x \to -\infty} f(x) = \lim_{x \to \infty} f(x) = 6$$

62. Is it possible to sketch a graph of a function that satisfies the conditions of Exercise 61 and has *no* points of inflection? Explain.

63. If f is a continuous function such that $\displaystyle\lim_{x \to \infty} f(x) = 5$, find, if possible, $\displaystyle\lim_{x \to -\infty} f(x)$ for each specified condition.

(a) The graph of f is symmetric with respect to the y-axis.

(b) The graph of f is symmetric with respect to the origin.

CAPSTONE

64. The graph of a function f is shown below. To print an enlarged copy of the graph, go to the website *www.mathgraphs.com*.

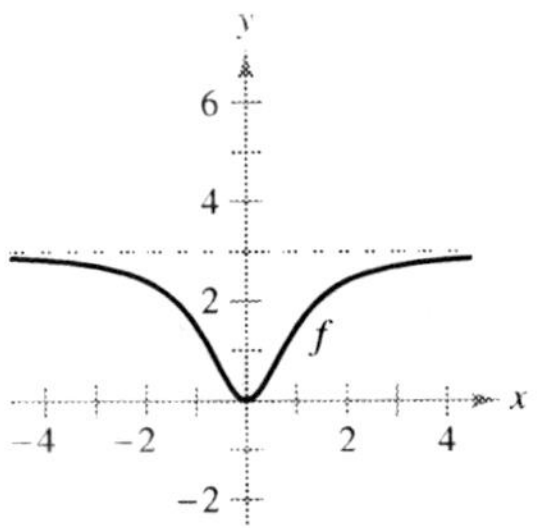

(a) Sketch f'.

(b) Use the graphs to estimate $\displaystyle\lim_{x \to \infty} f(x)$ and $\displaystyle\lim_{x \to \infty} f'(x)$.

(c) Explain the answers you gave in part (b).

In Exercises 65–82, sketch the graph of the equation. Look for extrema, intercepts, symmetry, and asymptotes as necessary. Use a graphing utility to verify your result.

65. $y = \dfrac{x}{1 - x}$

66. $y = \dfrac{x - 4}{x - 3}$

67. $y = \dfrac{x + 1}{x^2 - 4}$

68. $y = \dfrac{2x}{9 - x^2}$

69. $y = \dfrac{x^2}{x^2 + 16}$

70. $y = \dfrac{x^2}{x^2 - 16}$

71. $y = \dfrac{2x^2}{x^2 - 4}$

72. $y = \dfrac{2x^2}{x^2 + 4}$

73. $xy^2 = 9$

74. $x^2y = 9$

75. $y = \dfrac{3x}{1 - x}$

76. $y = \dfrac{3x}{1 - x^2}$

77. $y = 2 - \dfrac{3}{x^2}$

78. $y = 1 + \dfrac{1}{x}$

79. $y = 3 + \dfrac{2}{x}$

80. $y = 4\left(1 - \dfrac{1}{x^2}\right)$

81. $y = \dfrac{x^3}{\sqrt{x^2 - 4}}$

82. $y = \dfrac{x}{\sqrt{x^2 - 4}}$

CAS **In Exercises 83–92, use a computer algebra system to analyze the graph of the function. Label any extrema and/or asymptotes that exist.**

83. $f(x) = 9 - \dfrac{5}{x^2}$

84. $f(x) = \dfrac{1}{x^2 - x - 2}$

85. $f(x) = \dfrac{x - 2}{x^2 - 4x + 3}$

86. $f(x) = \dfrac{x + 1}{x^2 + x + 1}$

87. $f(x) = \dfrac{3x}{\sqrt{4x^2 + 1}}$

88. $g(x) = \dfrac{2x}{\sqrt{3x^2 + 1}}$

89. $g(x) = \sin\left(\dfrac{x}{x - 2}\right), \quad x > 3$

90. $f(x) = \dfrac{2 \sin 2x}{x}$

91. $f(x) = 2 + (x^2 - 3)e^{-x}$ **92.** $f(x) = \dfrac{10 \ln x}{x^2 \sqrt{x}}$

In Exercises 93 and 94, (a) use a graphing utility to graph f and g in the same viewing window, (b) verify algebraically that f and g represent the same function, and (c) zoom out sufficiently far so that the graph appears as a line. What equation does this line appear to have? (Note that the points at which the function is not continuous are not readily seen when you zoom out.)

93. $f(x) = \dfrac{x^3 - 3x^2 + 2}{x(x - 3)}$ **94.** $f(x) = -\dfrac{x^3 - 2x^2 + 2}{2x^2}$

$g(x) = x + \dfrac{2}{x(x - 3)}$ $g(x) = -\dfrac{1}{2}x + 1 - \dfrac{1}{x^2}$

95. Engine Efficiency   The efficiency of an internal combustion engine is

$$\text{Efficiency } (\%) = 100 \left[1 - \dfrac{1}{(v_1/v_2)^c} \right]$$

where v_1/v_2 is the ratio of the uncompressed gas to the compressed gas and c is a positive constant dependent on the engine design. Find the limit of the efficiency as the compression ratio approaches infinity.

96. Average Cost   A business has a cost of $C = 0.5x + 500$ for producing x units. The average cost per unit is

$$\overline{C} = \dfrac{C}{x}.$$

Find the limit of $\overline{C}$ as x approaches infinity.

97. Physics   Newton's First Law of Motion and Einstein's Special Theory of Relativity differ concerning a particle's behavior as its velocity approaches the speed of light c. In the graph, functions N and E represent the velocity v, with respect to time t, of a particle accelerated by a constant force as predicted by Newton and Einstein. Write limit statements that describe these two theories.

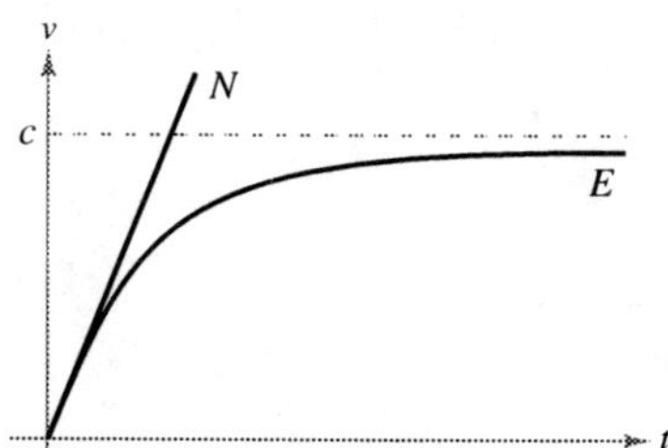

98. Temperature   The graph shows the temperature T, in degrees Fahrenheit, of molten glass t seconds after it is removed from a kiln.

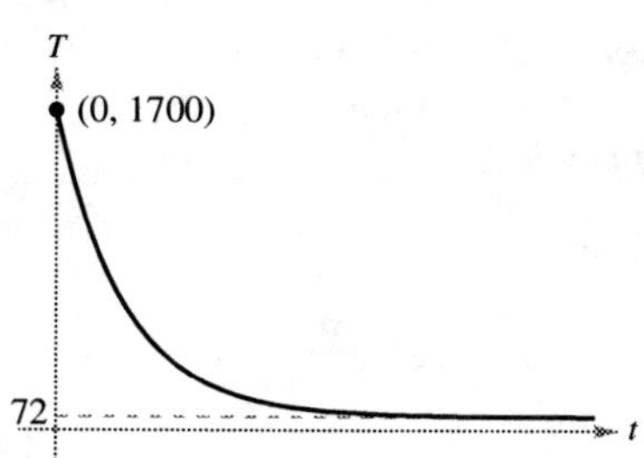

(a) Find $\lim\limits_{t \to 0^+} T$. What does this limit represent?

(b) Find $\lim\limits_{t \to \infty} T$. What does this limit represent?

(c) Will the temperature of the glass ever actually reach room temperature? Why?

99. Modeling Data   A heat probe is attached to the heat exchanger of a heating system. The temperature T (in degrees Celsius) is recorded t seconds after the furnace is started. The results for the first 2 minutes are recorded in the table.

t	0	15	30	45	60
T	25.2°	36.9°	45.5°	51.4°	56.0°

t	75	90	105	120
T	59.6°	62.0°	64.0°	65.2°

(a) Use the regression capabilities of a graphing utility to find a model of the form $T_1 = at^2 + bt + c$ for the data.

(b) Use a graphing utility to graph T_1.

(c) A rational model for the data is $T_2 = \dfrac{1451 + 86t}{58 + t}$. Use a graphing utility to graph T_2.

(d) Find $T_1(0)$ and $T_2(0)$.

(e) Find $\lim\limits_{t \to \infty} T_2$.

(f) Interpret the result in part (e) in the context of the problem. Is it possible to do this type of analysis using T_1? Explain.

100. Modeling Data   A container holds 5 liters of a 25% brine solution. The table shows the concentrations C of the mixture after adding x liters of a 75% brine solution to the container.

x	0	0.5	1	1.5	2
C	0.25	0.295	0.333	0.365	0.393

x	2.5	3	3.5	4
C	0.417	0.438	0.456	0.472

(a) Use the regression features of a graphing utility to find a model of the form $C_1 = ax^2 + bx + c$ for the data.

(b) Use a graphing utility to graph C_1.

(c) A rational model for these data is $C_2 = \dfrac{5 + 3x}{20 + 4x}$. Use a graphing utility to graph C_2.

(d) Find $\lim\limits_{x \to \infty} C_1$ and $\lim\limits_{x \to \infty} C_2$. Which model do you think best represents the concentration of the mixture? Explain.

(e) What is the limiting concentration?

101. Timber Yield   The yield V (in millions of cubic feet per acre) for a stand of timber at age t (in years) is $V = 7.1e^{(-48.1)/t}$.

(a) Find the limiting volume of wood per acre as t approaches infinity.

(b) Find the rates at which the yield is changing when $t = 20$ years and $t = 60$ years.

102. ***Learning Theory***   In a group project in learning theory, a mathematical model for the proportion P of correct responses after n trials was found to be

$$P = \frac{0.83}{1 + e^{-0.2n}}.$$

(a) Find the limiting proportion of correct responses as n approaches infinity.

(b) Find the rates at which P is changing after $n = 3$ trials and $n = 10$ trials.

103. ***Writing***   Consider the function $f(x) = \dfrac{2}{1 + e^{1/x}}$.

(a) Use a graphing utility to graph f.

(b) Write a short paragraph explaining why the graph has a horizontal asymptote at $y = 1$ and why the function has a nonremovable discontinuity at $x = 0$.

104. ***Writing***   In your own words, state the guidelines for finding the limit of a rational function. Give examples.

105. A line with slope m passes through the point $(0, 4)$.

(a) Write the distance d between the line and the point $(3, 1)$ as a function of m.

(b) Use a graphing utility to graph the equation in part (a).

(c) Find $\lim\limits_{m \to \infty} d(m)$ and $\lim\limits_{m \to -\infty} d(m)$. Interpret the results geometrically.

106. A line with slope m passes through the point $(0, -2)$.

(a) Write the distance d between the line and the point $(4, 2)$ as a function of m.

(b) Use a graphing utility to graph the equation in part (a).

(c) Find $\lim\limits_{m \to \infty} d(m)$ and $\lim\limits_{m \to -\infty} d(m)$. Interpret the results geometrically.

107. The graph of $f(x) = \dfrac{2x^2}{x^2 + 2}$ is shown.

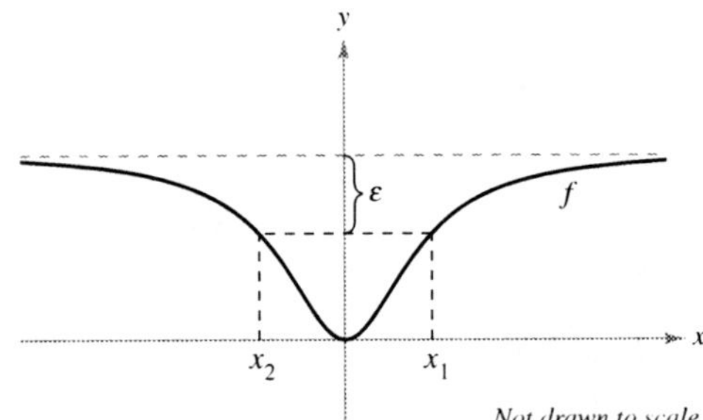

Not drawn to scale

(a) Find $L = \lim\limits_{x \to \infty} f(x)$.

(b) Determine x_1 and x_2 in terms of ε.

(c) Determine M, where $M > 0$, such that $|f(x) - L| < \varepsilon$ for $x > M$.

(d) Determine N, where $N < 0$, such that $|f(x) - L| < \varepsilon$ for $x < N$.

108. The graph of $f(x) = \dfrac{6x}{\sqrt{x^2 + 2}}$ is shown.

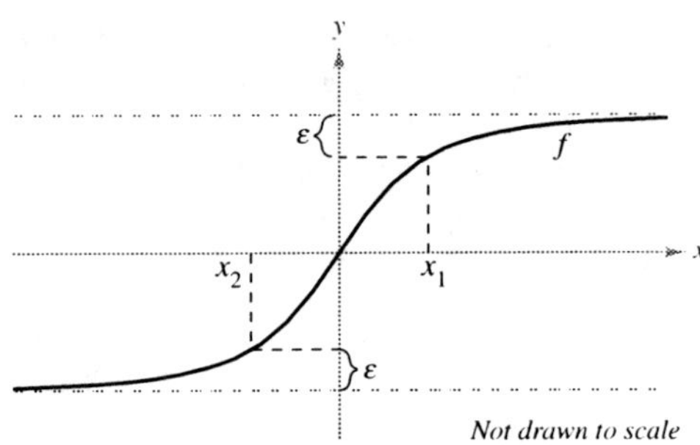

Not drawn to scale

(a) Find $L = \lim\limits_{x \to \infty} f(x)$ and $K = \lim\limits_{x \to -\infty} f(x)$.

(b) Determine x_1 and x_2 in terms of ε.

(c) Determine M, where $M > 0$, such that $|f(x) - L| < \varepsilon$ for $x > M$.

(d) Determine N, where $N < 0$, such that $|f(x) - K| < \varepsilon$ for $x < N$.

109. Consider $\lim\limits_{x \to \infty} \dfrac{3x}{\sqrt{x^2 + 3}}$. Use the definition of limits at infinity to find values of M that correspond to (a) $\varepsilon = 0.5$ and (b) $\varepsilon = 0.1$.

110. Consider $\lim\limits_{x \to -\infty} \dfrac{3x}{\sqrt{x^2 + 3}}$. Use the definition of limits at infinity to find values of N that correspond to (a) $\varepsilon = 0.5$ and (b) $\varepsilon = 0.1$.

In Exercises 111–114, use the definition of limits at infinity to prove the limit.

111. $\lim\limits_{x \to \infty} \dfrac{1}{x^2} = 0$

112. $\lim\limits_{x \to \infty} \dfrac{2}{\sqrt{x}} = 0$

113. $\lim\limits_{x \to -\infty} \dfrac{1}{x^3} = 0$

114. $\lim\limits_{x \to -\infty} \dfrac{1}{x - 2} = 0$

115. Prove that if $p(x) = a_n x^n + \cdots + a_1 x + a_0$ and $q(x) = b_m x^m + \cdots + b_1 x + b_0 \ (a_n \neq 0, b_m \neq 0)$, then

$$\lim_{x \to \infty} \frac{p(x)}{q(x)} = \begin{cases} 0, & n < m \\ \dfrac{a_n}{b_m}, & n = m. \\ \pm\infty, & n > m \end{cases}$$

116. Use the definition of infinite limits at infinity to prove that $\lim\limits_{x \to \infty} x^3 = \infty$.

True or False?   **In Exercises 117 and 118, determine whether the statement is true or false. If it is false, explain why or give an example that shows it is false.**

117. If $f'(x) > 0$ for all real numbers x, then f increases without bound.

118. If $f''(x) < 0$ for all real numbers x, then f decreases without bound.

4.6 | A Summary of Curve Sketching

■ **Analyze and sketch the graph of a function.**

Analyzing the Graph of a Function

It would be difficult to overstate the importance of using graphs in mathematics. Descartes's introduction of analytic geometry contributed significantly to the rapid advances in calculus that began during the mid-seventeenth century. In the words of Lagrange, "As long as algebra and geometry traveled separate paths, their advance was slow and their applications limited. But when these two sciences joined company, they drew from each other fresh vitality and thenceforth marched on at a rapid pace toward perfection."

So far, you have studied several concepts that are useful in analyzing the graph of a function.

• x-intercepts and y-intercepts	(Section 1.1)
• Symmetry	(Section 1.1)
• Domain and range	(Section 1.3)
• Continuity	(Section 2.4)
• Vertical asymptotes	(Section 2.5)
• Differentiability	(Section 3.1)
• Relative extrema	(Section 4.1)
• Concavity	(Section 4.4)
• Points of inflection	(Section 4.4)
• Horizontal asymptotes	(Section 4.5)
• Infinite limits at infinity	(Section 4.5)

When you are sketching the graph of a function, either by hand or with a graphing utility, remember that normally you cannot show the *entire* graph. The decision as to which part of the graph you choose to show is often crucial. For instance, which of the viewing windows in Figure 4.44 better represents the graph of

$$f(x) = x^3 - 25x^2 + 74x - 20?$$

By seeing both views, it is clear that the second viewing window gives a more complete representation of the graph. But would a third viewing window reveal other interesting portions of the graph? To answer this, you need to use calculus to interpret the first and second derivatives. Here are some guidelines for determining a good viewing window for the graph of a function.

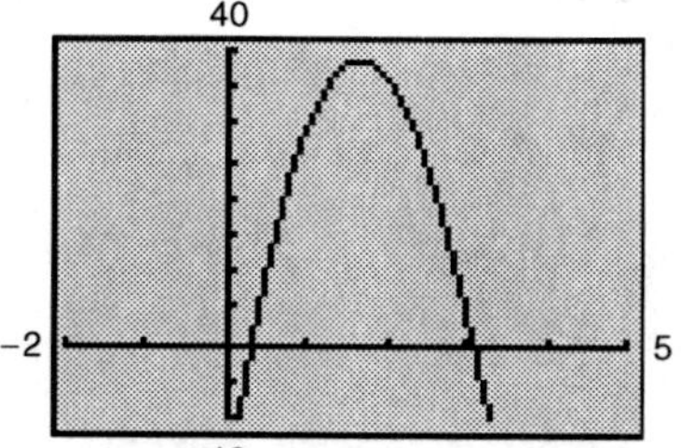

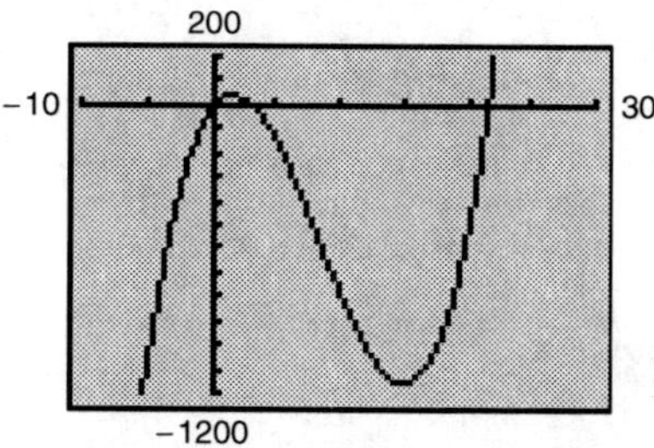

Different viewing windows for the graph of $f(x) = x^3 - 25x^2 + 74x - 20$

Figure 4.44

GUIDELINES FOR ANALYZING THE GRAPH OF A FUNCTION

1. Determine the domain and range of the function.
2. Determine the intercepts, asymptotes, and symmetry of the graph.
3. Locate the x-values for which $f'(x)$ and $f''(x)$ either are zero or do not exist. Use the results to determine relative extrema and points of inflection.

NOTE   In these guidelines, note the importance of *algebra* (as well as calculus) for solving the equations $f(x) = 0$, $f'(x) = 0$, and $f''(x) = 0$.   ■

EXAMPLE 1   Sketching the Graph of a Rational Function

Analyze and sketch the graph of $f(x) = \dfrac{2(x^2 - 9)}{x^2 - 4}$.

Solution

$$\textit{First derivative:} \quad f'(x) = \frac{20x}{(x^2 - 4)^2}$$

$$\textit{Second derivative:} \quad f''(x) = \frac{-20(3x^2 + 4)}{(x^2 - 4)^3}$$

x-intercepts:	$(-3, 0), (3, 0)$
y-intercept:	$\left(0, \frac{9}{2}\right)$
Vertical asymptotes:	$x = -2, x = 2$
Horizontal asymptote:	$y = 2$
Critical number:	$x = 0$
Possible points of inflection:	None
Domain:	All real numbers except $x = \pm 2$
Symmetry:	With respect to y-axis
Test intervals:	$(-\infty, -2), (-2, 0), (0, 2), (2, \infty)$

The table shows how the test intervals are used to determine several characteristics of the graph. The graph of f is shown in Figure 4.45.

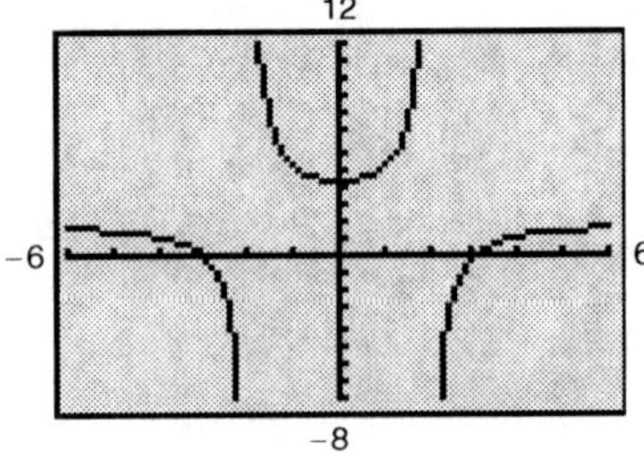

Using calculus, you can be certain that you have determined all characteristics of the graph of f.
Figure 4.45

	$f(x)$	$f'(x)$	$f''(x)$	Characteristic of Graph
$-\infty < x < -2$		$-$	$-$	Decreasing, concave downward
$x = -2$	Undef.	Undef.	Undef.	Vertical asymptote
$-2 < x < 0$		$-$	$+$	Decreasing, concave upward
$x = 0$	$\frac{9}{2}$	0	$+$	Relative minimum
$0 < x < 2$		$+$	$+$	Increasing, concave upward
$x = 2$	Undef.	Undef.	Undef.	Vertical asymptote
$2 < x < \infty$		$+$	$-$	Increasing, concave downward

Be sure you understand all of the implications of creating a table such as that shown in Example 1. By using calculus, you can *be sure* that the graph has no relative extrema or points of inflection other than those shown in Figure 4.45.

TECHNOLOGY PITFALL   Without using the type of analysis outlined in Example 1, it is easy to obtain an incomplete view of a graph's basic characteristics. For instance, Figure 4.46 shows a view of the graph of

$$g(x) = \frac{2(x^2 - 9)(x - 20)}{(x^2 - 4)(x - 21)}.$$

From this view, it appears that the graph of g is about the same as the graph of f shown in Figure 4.45. The graphs of these two functions, however, differ significantly. Try enlarging the viewing window to see the differences.

FOR FURTHER INFORMATION For more information on the use of technology to graph rational functions, see the article "Graphs of Rational Functions for Computer Assisted Calculus" by Stan Byrd and Terry Walters in *The College Mathematics Journal*. To view this article, go to the website *www.matharticles.com*.

By not using calculus you may overlook important characteristics of the graph of g.
Figure 4.46

EXAMPLE 2 Sketching the Graph of a Rational Function

Analyze and sketch the graph of $f(x) = \dfrac{x^2 - 2x + 4}{x - 2}$.

Solution

First derivative:	$f'(x) = \dfrac{x(x - 4)}{(x - 2)^2}$
Second derivative:	$f''(x) = \dfrac{8}{(x - 2)^3}$
x-intercepts:	None
y-intercept:	$(0, -2)$
Vertical asymptote:	$x = 2$
Horizontal asymptotes:	None
End behavior:	$\displaystyle\lim_{x \to -\infty} f(x) = -\infty, \ \lim_{x \to \infty} f(x) = \infty$
Critical numbers:	$x = 0, \ x = 4$
Possible points of inflection:	None
Domain:	All real numbers except $x = 2$
Test intervals:	$(-\infty, 0), (0, 2), (2, 4), (4, \infty)$

The analysis of the graph of f is shown in the table, and the graph is shown in Figure 4.47.

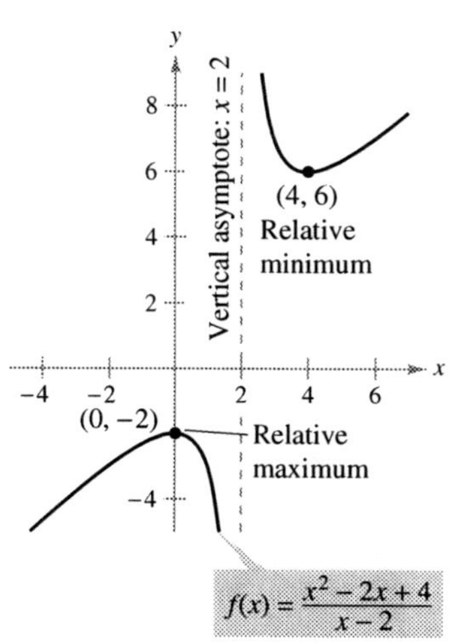

Figure 4.47

	$f(x)$	$f'(x)$	$f''(x)$	Characteristic of Graph
$-\infty < x < 0$		$+$	$-$	Increasing, concave downward
$x = 0$	-2	0	$-$	Relative maximum
$0 < x < 2$		$-$	$-$	Decreasing, concave downward
$x = 2$	Undef.	Undef.	Undef.	Vertical asymptote
$2 < x < 4$		$-$	$+$	Decreasing, concave upward
$x = 4$	6	0	$+$	Relative minimum
$4 < x < \infty$		$+$	$+$	Increasing, concave upward

Although the graph of the function in Example 2 has no horizontal asymptote, it does have a slant asymptote. The graph of a rational function (having no common factors and whose denominator is of degree 1 or greater) has a **slant asymptote** if the degree of the numerator exceeds the degree of the denominator by exactly 1. To find the slant asymptote, use long division to rewrite the rational function as the sum of a first-degree polynomial and another rational function.

$$f(x) = \frac{x^2 - 2x + 4}{x - 2} \qquad \text{Write original equation.}$$

$$= x + \frac{4}{x - 2} \qquad \text{Rewrite using long division.}$$

In Figure 4.48, note that the graph of f approaches the slant asymptote $y = x$ as x approaches $-\infty$ or ∞.

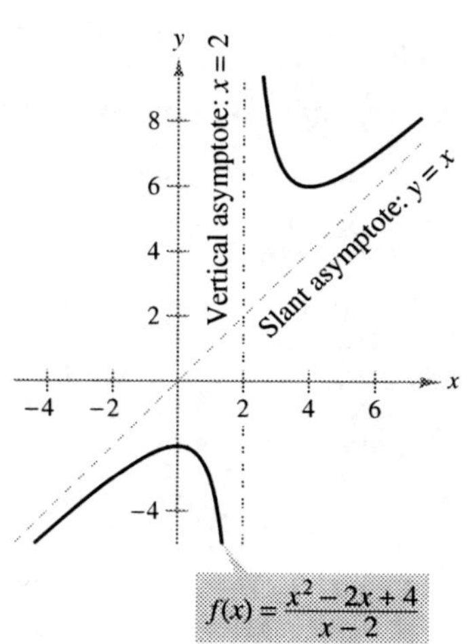

A slant asymptote
Figure 4.48

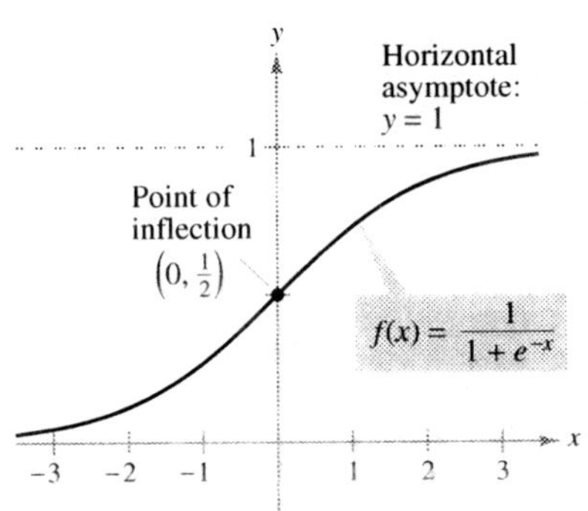

Figure 4.49

EXAMPLE 3 Sketching the Graph of a Logistic Function

Analyze and sketch the graph of the *logistic function* $f(x) = \dfrac{1}{1 + e^{-x}}$.

Solution

$$f'(x) = \frac{e^{-x}}{(1 + e^{-x})^2} \quad \text{and} \quad f''(x) = -\frac{e^{-x}(e^{-x} - 1)}{(1 + e^{-x})^3}$$

The graph has only one intercept, $\left(0, \frac{1}{2}\right)$. It has no vertical asymptotes, but it has two horizontal asymptotes: $y = 1$ (to the right) and $y = 0$ (to the left). The function has no critical numbers and one possible point of inflection (at $x = 0$). The domain of the function is all real numbers. The analysis of the graph of f is shown in the table, and the graph is shown in Figure 4.49.

	$f(x)$	$f'(x)$	$f''(x)$	Characteristic of Graph
$-\infty < x < 0$		$+$	$+$	Increasing, concave upward
$x = 0$	$\dfrac{1}{2}$	$\dfrac{1}{4}$	0	Point of inflection
$0 < x < \infty$		$+$	$-$	Increasing, concave downward

EXAMPLE 4 Sketching the Graph of a Radical Function

Analyze and sketch the graph of $f(x) = 2x^{5/3} - 5x^{4/3}$.

Solution

$$f'(x) = \frac{10}{3}x^{1/3}(x^{1/3} - 2) \qquad f''(x) = \frac{20(x^{1/3} - 1)}{9x^{2/3}}$$

The function has two intercepts: $(0, 0)$ and $\left(\frac{125}{8}, 0\right)$. There are no horizontal or vertical asymptotes. The function has two critical numbers ($x = 0$ and $x = 8$) and two possible points of inflection ($x = 0$ and $x = 1$). The domain is all real numbers. The analysis of the graph of f is shown in the table, and the graph is shown in Figure 4.50.

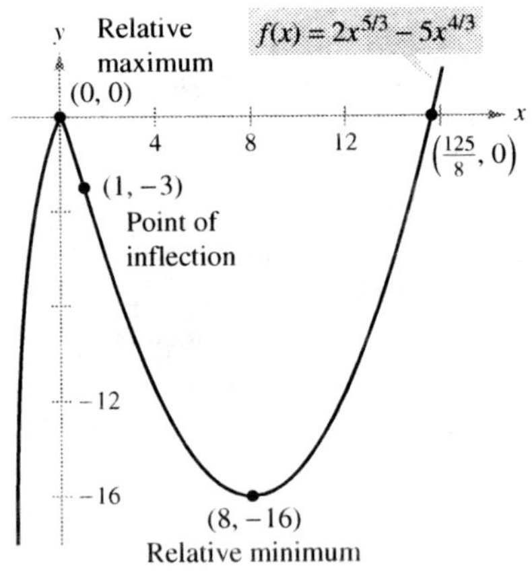

Figure 4.50

	$f(x)$	$f'(x)$	$f''(x)$	Characteristic of Graph
$-\infty < x < 0$		$+$	$-$	Increasing, concave downward
$x = 0$	0	0	Undef.	Relative maximum
$0 < x < 1$		$-$	$-$	Decreasing, concave downward
$x = 1$	-3	$-$	0	Point of inflection
$1 < x < 8$		$-$	$+$	Decreasing, concave upward
$x = 8$	-16	0	$+$	Relative minimum
$8 < x < \infty$		$+$	$+$	Increasing, concave upward

EXAMPLE 5   Sketching the Graph of a Polynomial Function

Analyze and sketch the graph of $f(x) = x^4 - 12x^3 + 48x^2 - 64x$.

Solution   Begin by factoring to obtain

$$f(x) = x^4 - 12x^3 + 48x^2 - 64x$$
$$= x(x - 4)^3.$$

Then, using the factored form of $f(x)$, you can perform the following analysis.

First derivative:	$f'(x) = 4(x - 1)(x - 4)^2$
Second derivative:	$f''(x) = 12(x - 4)(x - 2)$
x-intercepts:	$(0, 0), (4, 0)$
y-intercept:	$(0, 0)$
Vertical asymptotes:	None
Horizontal asymptotes:	None
End behavior:	$\lim\limits_{x \to -\infty} f(x) = \infty, \quad \lim\limits_{x \to \infty} f(x) = \infty$
Critical numbers:	$x = 1, x = 4$
Possible points of inflection:	$x = 2, x = 4$
Domain:	All real numbers
Test intervals:	$(-\infty, 1), (1, 2), (2, 4), (4, \infty)$

The analysis of the graph of f is shown in the table, and the graph is shown in Figure 4.51(a). Using a computer algebra system such as *Maple* [see Figure 4.51(b)] can help you verify your analysis.

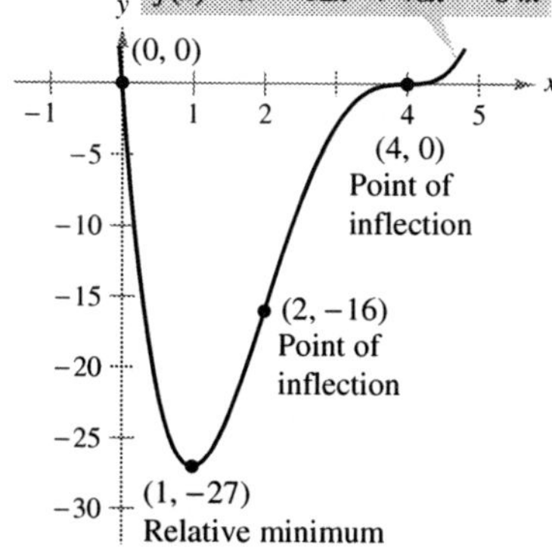

(a)

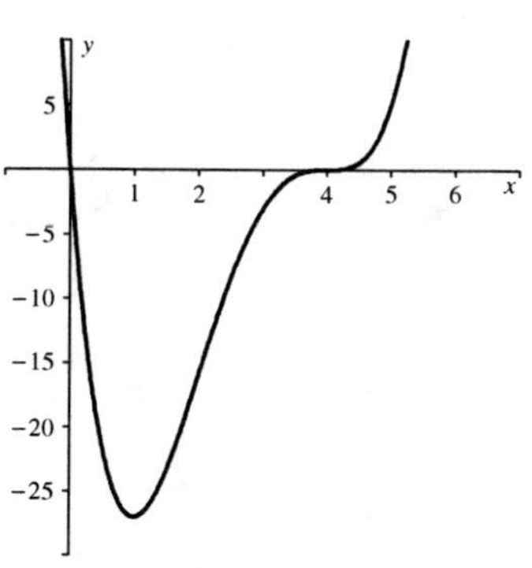

(b)

A polynomial function of even degree must have at least one relative extremum.

Figure 4.51

	$f(x)$	$f'(x)$	$f''(x)$	Characteristic of Graph
$-\infty < x < 1$		$-$	$+$	Decreasing, concave upward
$x = 1$	-27	0	$+$	Relative minimum
$1 < x < 2$		$+$	$+$	Increasing, concave upward
$x = 2$	-16	$+$	0	Point of inflection
$2 < x < 4$		$+$	$-$	Increasing, concave downward
$x = 4$	0	0	0	Point of inflection
$4 < x < \infty$		$+$	$+$	Increasing, concave upward

The fourth-degree polynomial function in Example 5 has one relative minimum and no relative maxima. In general, a polynomial function of degree n can have *at most* $n - 1$ relative extrema, and *at most* $n - 2$ points of inflection. Moreover, polynomial functions of even degree must have *at least* one relative extremum.

Remember from the Leading Coefficient Test described in Section 1.3 that the "end behavior" of the graph of a polynomial function is determined by its leading coefficient and its degree. For instance, because the polynomial in Example 5 has a positive leading coefficient, the graph rises to the right. Moreover, because the degree is even, the graph also rises to the left.

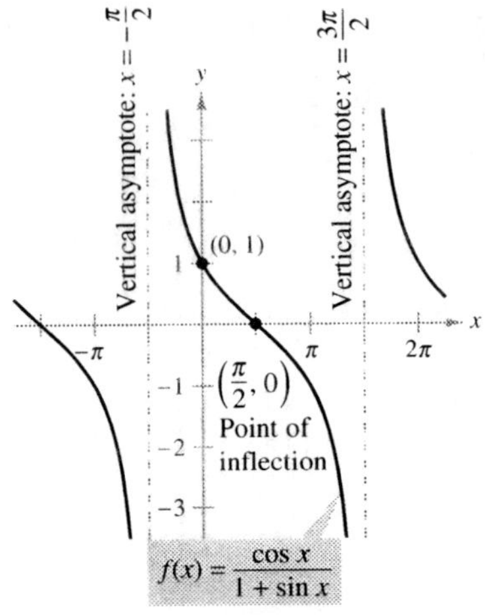

(a)

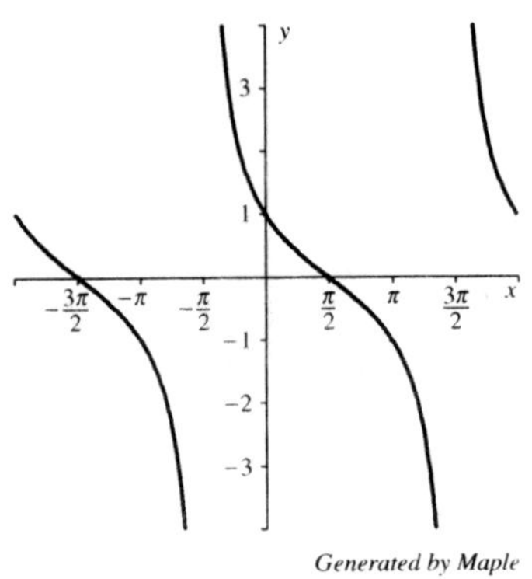

(b)

Figure 4.52

EXAMPLE 6   Sketching the Graph of a Trigonometric Function

Analyze and sketch the graph of $f(x) = \dfrac{\cos x}{1 + \sin x}$.

Solution   Because the function has a period of 2π, you can restrict the analysis of the graph to any interval of length 2π. For convenience, choose $(-\pi/2, 3\pi/2)$.

$$\textit{First derivative:} \quad f'(x) = -\frac{1}{1 + \sin x}$$

$$\textit{Second derivative:} \quad f''(x) = \frac{\cos x}{(1 + \sin x)^2}$$

$$\textit{Period:} \quad 2\pi$$

$$\textit{x-intercept:} \quad \left(\frac{\pi}{2}, 0\right)$$

$$\textit{y-intercept:} \quad (0, 1)$$

$$\textit{Vertical asymptotes:} \quad x = -\frac{\pi}{2}, \; x = \frac{3\pi}{2} \qquad \text{See Note below.}$$

$$\textit{Horizontal asymptotes:} \quad \text{None}$$

$$\textit{Critical numbers:} \quad \text{None}$$

$$\textit{Possible points of inflection:} \quad x = \frac{\pi}{2}$$

$$\textit{Domain:} \quad \text{All real numbers except } x = \frac{3 + 4n}{2}\pi$$

$$\textit{Test intervals:} \quad \left(-\frac{\pi}{2}, \frac{\pi}{2}\right), \left(\frac{\pi}{2}, \frac{3\pi}{2}\right)$$

The analysis of the graph of f on the interval $(-\pi/2, 3\pi/2)$ is shown in the table, and the graph is shown in Figure 4.52(a). Compare this with the graph generated by the computer algebra system *Maple* in Figure 4.52(b).

	$f(x)$	$f'(x)$	$f''(x)$	Characteristic of Graph
$x = -\dfrac{\pi}{2}$	Undef.	Undef.	Undef.	Vertical asymptote
$-\dfrac{\pi}{2} < x < \dfrac{\pi}{2}$		$-$	$+$	Decreasing, concave upward
$x = \dfrac{\pi}{2}$	0	$-\dfrac{1}{2}$	0	Point of inflection
$\dfrac{\pi}{2} < x < \dfrac{3\pi}{2}$		$-$	$-$	Decreasing, concave downward
$x = \dfrac{3\pi}{2}$	Undef.	Undef.	Undef.	Vertical asymptote

NOTE   By substituting $-\pi/2$ or $3\pi/2$ into the function, you obtain the form $0/0$. This is called an indeterminate form, which you will study in Section 8.7. To determine that the function has vertical asymptotes at these two values, you can rewrite the function as follows.

$$f(x) = \frac{\cos x}{1 + \sin x} = \frac{(\cos x)(1 - \sin x)}{(1 + \sin x)(1 - \sin x)} = \frac{(\cos x)(1 - \sin x)}{\cos^2 x} = \frac{1 - \sin x}{\cos x}$$

In this form, it is clear that the graph of f has vertical asymptotes at $x = -\pi/2$ and $3\pi/2$.

EXAMPLE 7 Analyzing an Inverse Trigonometric Graph

Analyze the graph of $y = (\arctan x)^2$.

Solution From the derivative

$$y' = 2(\arctan x)\left(\frac{1}{1 + x^2}\right)$$

$$= \frac{2 \arctan x}{1 + x^2}$$

you can see that the only critical number is $x = 0$. By the First Derivative Test, this value corresponds to a relative minimum. From the second derivative

$$y'' = \frac{(1 + x^2)\left(\dfrac{2}{1 + x^2}\right) - (2 \arctan x)(2x)}{(1 + x^2)^2}$$

$$= \frac{2(1 - 2x \arctan x)}{(1 + x^2)^2}$$

it follows that points of inflection occur when $2x \arctan x = 1$. Using Newton's Method, these points occur when $x \approx \pm 0.765$. Finally, because

$$\lim_{x \to \pm\infty} (\arctan x)^2 = \frac{\pi^2}{4}$$

it follows that the graph has a horizontal asymptote at $y = \pi^2/4$. The graph is shown in Figure 4.53.

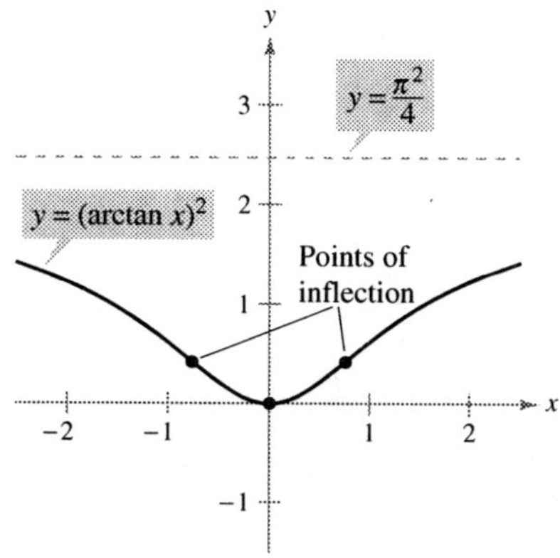

The graph of $y = (\arctan x)^2$ has a horizontal asymptote at $y = \pi^2/4$.
Figure 4.53

4.6 Exercises

See www.CalcChat.com for worked-out solutions to odd-numbered exercises.

In Exercises 1–4, match the graph of f in the left column with that of its derivative in the right column.

Graph of f

1.
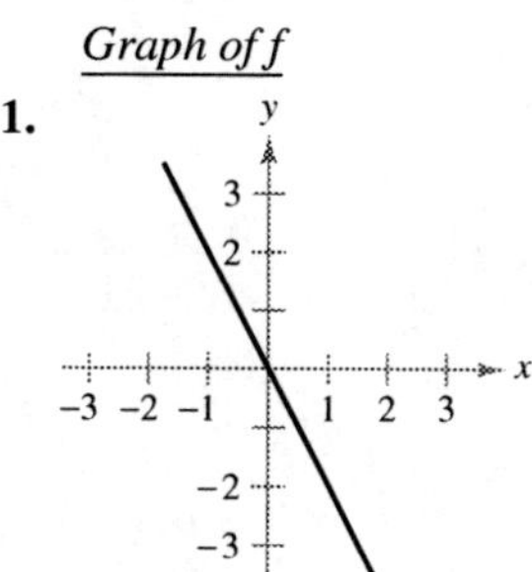

2.
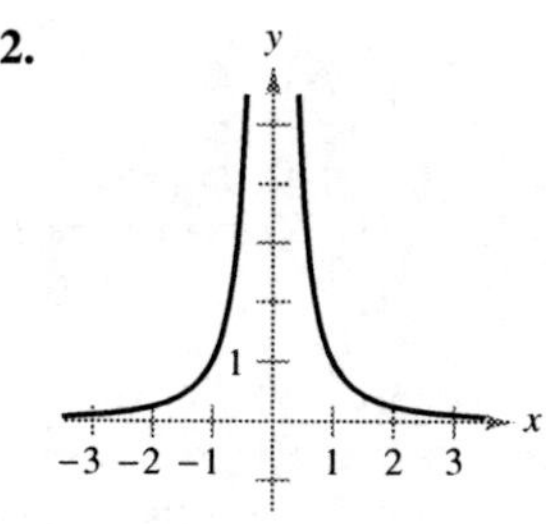

Graph of f'

(a)
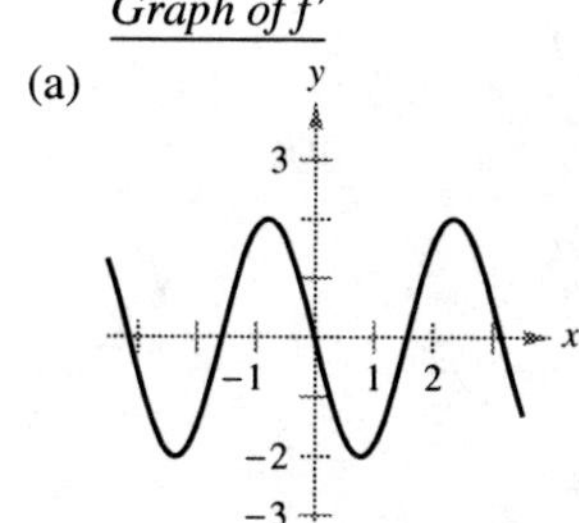

(b)
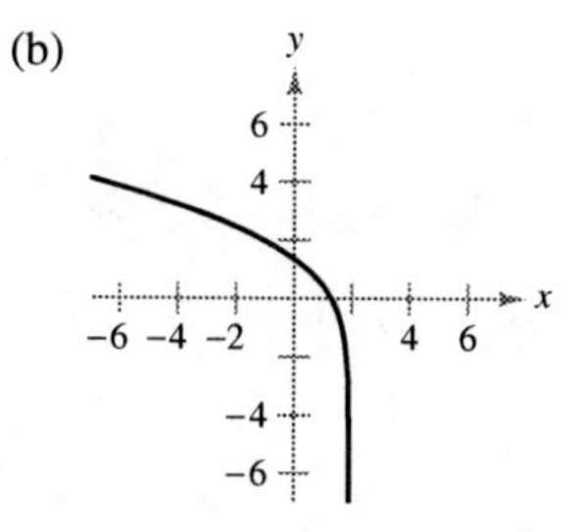

3.
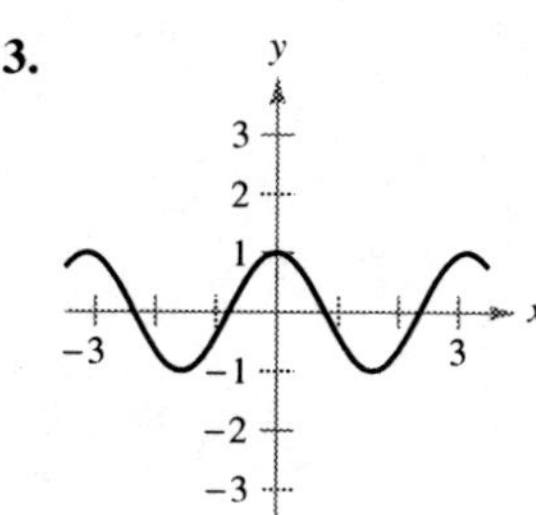

4.
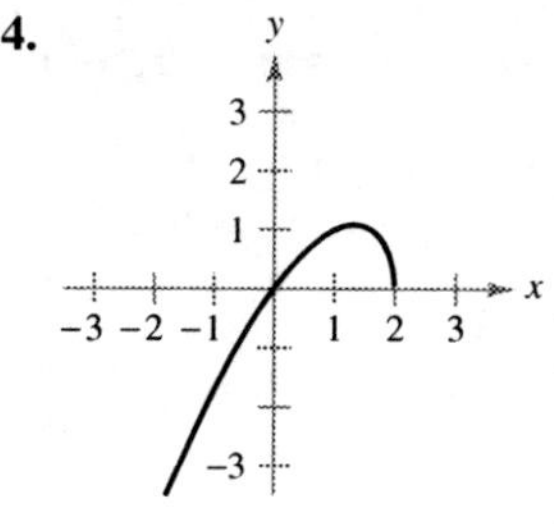

(c)
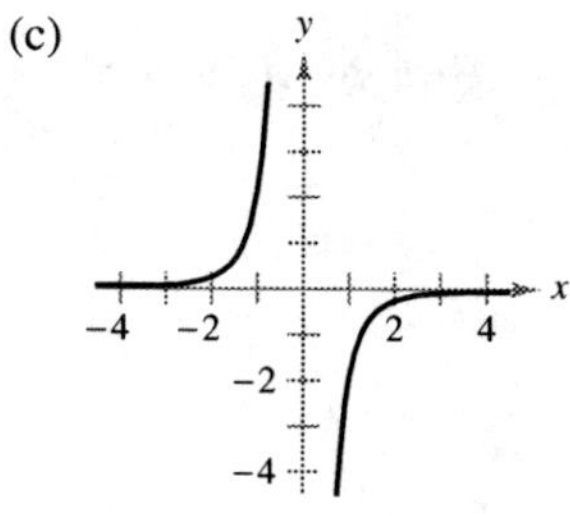

(d)
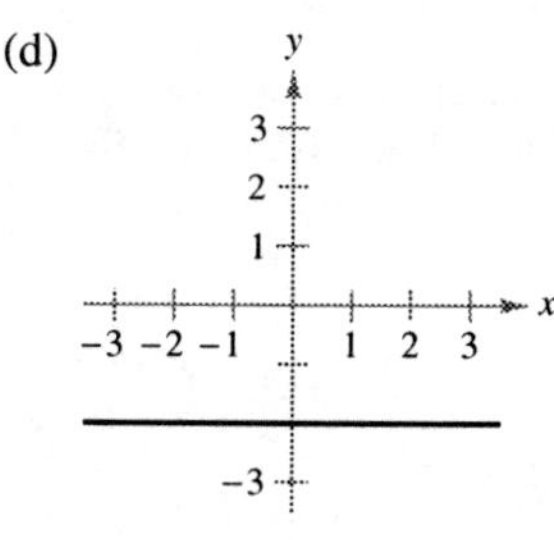

In Exercises 5–48, analyze and sketch a graph of the function. Label any intercepts, relative extrema, points of inflection, and asymptotes. Use a graphing utility to verify your results.

5. $y = \dfrac{1}{x-2} - 3$

6. $y = \dfrac{x}{x^2+1}$

7. $y = \dfrac{x^2}{x^2+3}$

8. $y = \dfrac{x^2+1}{x^2-4}$

9. $y = \dfrac{3x}{x^2-1}$

10. $f(x) = \dfrac{x-3}{x}$

11. $g(x) = x - \dfrac{8}{x^2}$

12. $f(x) = x + \dfrac{32}{x^2}$

13. $f(x) = \dfrac{x^2+1}{x}$

14. $f(x) = \dfrac{x^3}{x^2-9}$

15. $y = \dfrac{x^2-6x+12}{x-4}$

16. $y = \dfrac{2x^2-5x+5}{x-2}$

17. $y = x\sqrt{4-x}$

18. $g(x) = x\sqrt{9-x}$

19. $h(x) = x\sqrt{4-x^2}$

20. $g(x) = x\sqrt{9-x^2}$

21. $y = 3x^{2/3} - 2x$

22. $y = 3(x-1)^{2/3} - (x-1)^2$

23. $y = x^3 - 3x^2 + 3$

24. $y = -\frac{1}{3}(x^3 - 3x + 2)$

25. $y = 2 - x - x^3$

26. $f(x) = \frac{1}{3}(x-1)^3 + 2$

27. $f(x) = 3x^3 - 9x + 1$

28. $f(x) = (x+1)(x-2)(x-5)$

29. $y = 3x^4 + 4x^3$

30. $y = 3x^4 - 6x^2 + \frac{5}{3}$

31. $f(x) = x^4 - 4x^3 + 16x$

32. $f(x) = x^4 - 8x^3 + 18x^2 - 16x + 5$

33. $y = x^5 - 5x$

34. $y = (x-1)^5$

35. $y = |2x-3|$

36. $y = |x^2-6x+5|$

37. $f(x) = e^{3x}(2-x)$

38. $f(x) = -2 + e^{3x}(4-2x)$

39. $g(t) = \dfrac{10}{1+4e^{-t}}$

40. $h(x) = \dfrac{8}{2+3e^{-x/2}}$

41. $y = (x-1)\ln(x-1)$

42. $y = \frac{1}{24}x^3 - \ln x$

43. $g(x) = 6\arcsin\!\left(\dfrac{x-2}{2}\right)^2$

44. $h(x) = 7\arctan(x+1) - \ln(x^2+2x+2)$

45. $f(x) = \dfrac{x}{3^{x-3}}$

46. $g(t) = (5-t)5^t$

47. $g(x) = \log_4(x - x^2)$

48. $f(x) = \log_2|x^2 - 4x|$

CAS **In Exercises 49–54, use a computer algebra system to analyze and graph the function. Identify any relative extrema, points of inflection, and asymptotes.**

49. $f(x) = \dfrac{20x}{x^2+1} - \dfrac{1}{x}$

50. $f(x) = x + \dfrac{4}{x^2+1}$

51. $f(x) = \dfrac{-2x}{\sqrt{x^2+7}}$

52. $f(x) = \dfrac{4x}{\sqrt{x^2+15}}$

53. $y = \dfrac{x}{2} + \ln\!\left(\dfrac{x}{x+3}\right)$

54. $y = \dfrac{3x}{2}(1 + 4e^{-x/3})$

In Exercises 55–64, sketch a graph of the function over the given interval. Use a graphing utility to verify your graph.

Function	*Interval*
55. $f(x) = 2x - 4\sin x$	$0 \le x \le 2\pi$
56. $f(x) = -x + 2\cos x$	$0 \le x \le 2\pi$
57. $y = \sin x - \frac{1}{18}\sin 3x$	$0 \le x \le 2\pi$
58. $y = \cos x - \frac{1}{4}\cos 2x$	$0 \le x \le 2\pi$
59. $y = 2x - \tan x$	$-\dfrac{\pi}{2} < x < \dfrac{\pi}{2}$
60. $y = 2(x-2) + \cot x$	$0 < x < \pi$
61. $y = 2(\csc x + \sec x)$	$0 < x < \dfrac{\pi}{2}$
62. $y = \sec^2\!\left(\dfrac{\pi x}{8}\right) - 2\tan\!\left(\dfrac{\pi x}{8}\right) - 1$	$-3 < x < 3$
63. $g(x) = x\tan x$	$-\dfrac{3\pi}{2} < x < \dfrac{3\pi}{2}$
64. $g(x) = x\cot x$	$-2\pi < x < 2\pi$

WRITING ABOUT CONCEPTS

In Exercises 65 and 66, the graphs of f, f', and f'' are shown on the same set of coordinate axes. Which is which? Explain your reasoning. To print an enlarged copy of the graph, go to the website www.mathgraphs.com.

65.

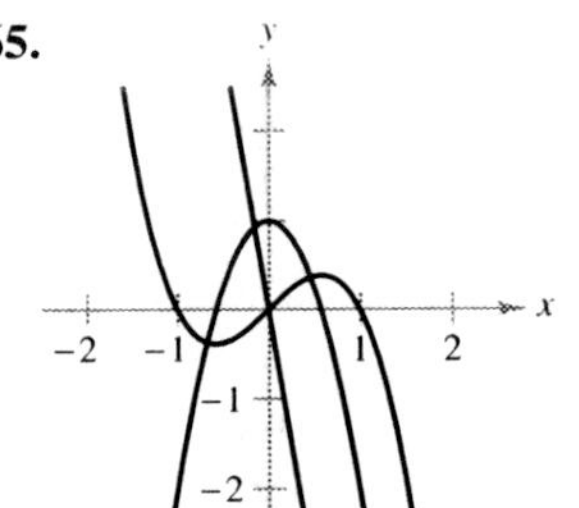

66.

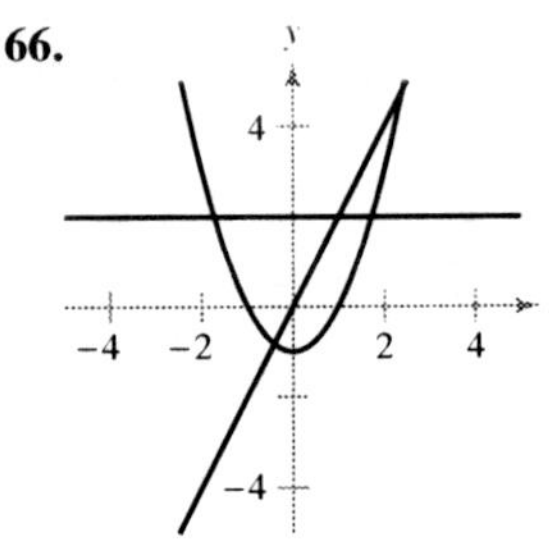

67. Suppose $f'(t) < 0$ for all t in the interval $(2, 8)$. Explain why $f(3) > f(5)$.

68. Suppose $f(0) = 3$ and $2 \le f'(x) \le 4$ for all x in the interval $[-5, 5]$. Determine the greatest and least possible values of $f(2)$.

In Exercises 69–72, use a graphing utility to graph the function. Use the graph to determine whether it is possible for the graph of a function to cross its horizontal asymptote. Do you think it is possible for the graph of a function to cross its vertical asymptote? Why or why not?

69. $f(x) = \dfrac{4(x-1)^2}{x^2-4x+5}$

70. $g(x) = \dfrac{3x^4-5x+3}{x^4+1}$

71. $h(x) = \dfrac{\sin 2x}{x}$

72. $f(x) = \dfrac{\cos 3x}{4x}$

WRITING ABOUT CONCEPTS (continued)

In Exercises 73 and 74, use a graphing utility to graph the function. Explain why there is no vertical asymptote when a superficial examination of the function may indicate that there should be one.

73. $h(x) = \dfrac{6 - 2x}{3 - x}$

74. $g(x) = \dfrac{x^2 + x - 2}{x - 1}$

In Exercises 75–78, use a graphing utility to graph the function and determine the slant asymptote of the graph. Zoom out repeatedly and describe how the graph on the display appears to change. Why does this occur?

75. $f(x) = -\dfrac{x^2 - 3x - 1}{x - 2}$

76. $g(x) = \dfrac{2x^2 - 8x - 15}{x - 5}$

77. $f(x) = \dfrac{x^3}{x^2 + 1}$

78. $h(x) = \dfrac{-x^3 + x^2 + 4}{x^2}$

Graphical Reasoning In Exercises 79–82, use the graph of f' to sketch a graph of f and the graph of f''. To print an enlarged copy of the graph, go to the website *www.mathgraphs.com*.

79.

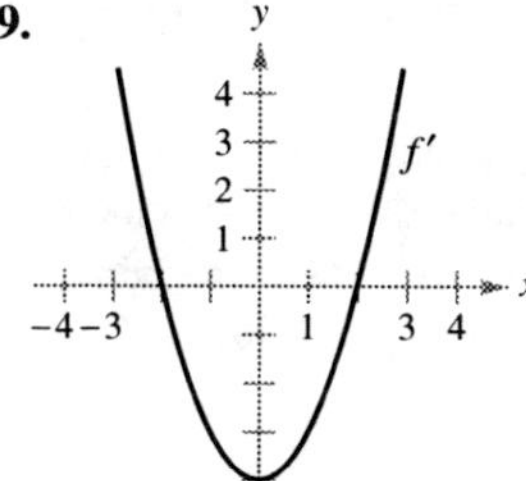

80.

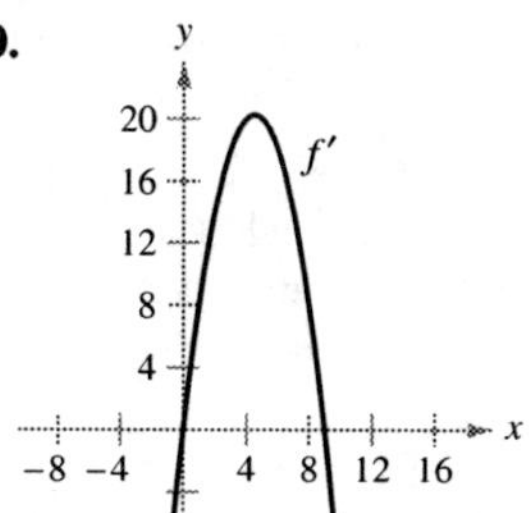

81.

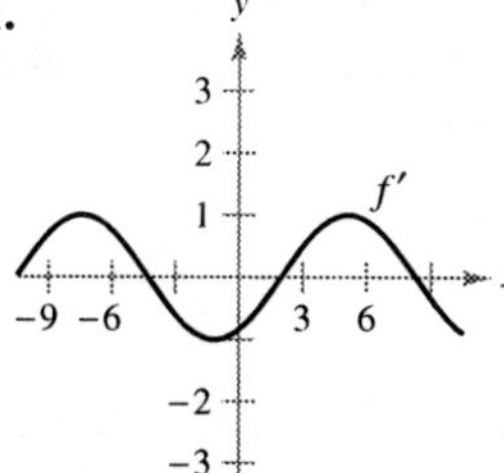

82.

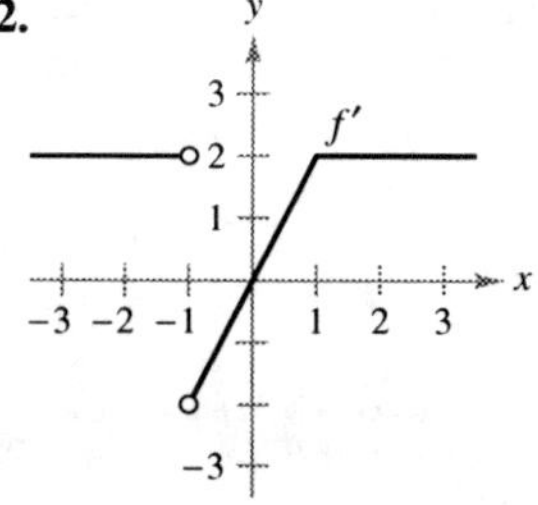

(*Submitted by Bill Fox, Moberly Area Community College, Moberly, MO*)

CAS **83.** *Graphical Reasoning* Consider the function

$$f(x) = \frac{\cos^2 \pi x}{\sqrt{x^2 + 1}}, \quad 0 < x < 4.$$

(a) Use a computer algebra system to graph the function and use the graph to approximate the critical numbers visually.

(b) Use a computer algebra system to find f' and approximate the critical numbers. Are the results the same as the visual approximation in part (a)? Explain.

84. *Graphical Reasoning* Consider the function

$$f(x) = \tan(\sin \pi x).$$

(a) Use a graphing utility to graph the function.

(b) Identify any symmetry of the graph.

(c) Is the function periodic? If so, what is the period?

(d) Identify any extrema on $(-1, 1)$.

(e) Use a graphing utility to determine the concavity of the graph on $(0, 1)$.

Think About It In Exercises 85–88, create a function whose graph has the given characteristics. (There is more than one correct answer.)

85. Vertical asymptote: $x = 3$

Horizontal asymptote: $y = 0$

86. Vertical asymptote: $x = -5$

Horizontal asymptote: None

87. Vertical asymptote: $x = 3$

Slant asymptote: $y = 3x + 2$

88. Vertical asymptote: $x = 2$

Slant asymptote: $y = -x$

89. *Graphical Reasoning* The graph of f is shown in the figure.

(a) For which values of x is $f'(x)$ zero? Positive? Negative?

(b) For which values of x is $f''(x)$ zero? Positive? Negative?

(c) On what interval is f' an increasing function?

(d) For which value of x is $f'(x)$ minimum? For this value of x, how does the rate of change of f compare with the rates of change of f for other values of x? Explain.

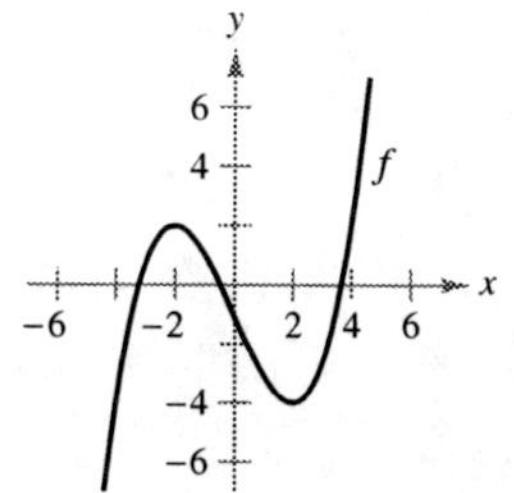

Figure for 89

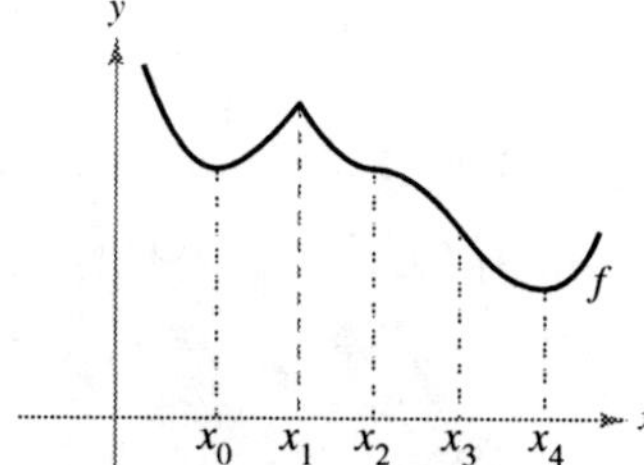

Figure for 90

CAPSTONE

90. *Graphical Reasoning* Identify the real numbers $x_0, x_1, x_2, x_3,$ and x_4 in the figure such that each of the following is true.

(a) $f'(x) = 0$

(b) $f''(x) = 0$

(c) $f'(x)$ does not exist.

(d) f has a relative maximum.

(e) f has a point of inflection.

91. *Graphical Reasoning* Consider the function

$$f(x) = \frac{ax}{(x-b)^2}.$$

Determine the effect on the graph of f as a and b are changed. Consider cases where a and b are both positive or both negative, and cases where a and b have opposite signs.

92. Consider the function $f(x) = \frac{1}{2}(ax)^2 - ax$, $a \neq 0$.

(a) Determine the changes (if any) in the intercepts, extrema, and concavity of the graph of f when a is varied.

(b) In the same viewing window, use a graphing utility to graph the function for four different values of a.

93. *Investigation* Consider the function $f(x) = \dfrac{2x^n}{x^4 + 1}$ for nonnegative integer values of n.

(a) Discuss the relationship between the value of n and the symmetry of the graph.

(b) For which values of n will the x-axis be the horizontal asymptote?

(c) For which value of n will $y = 2$ be the horizontal asymptote?

(d) What is the asymptote of the graph when $n = 5$?

(e) Use a graphing utility to graph f for the given values of n in the table. Use the graph to determine the number of extrema M and the number of inflection points N of the graph.

n	0	1	2	3	4	5
M						
N						

94. *Investigation* Let $P(x_0, y_0)$ be an arbitrary point on the graph of f such that $f'(x_0) \neq 0$, as shown in the figure. Verify each statement.

(a) The x-intercept of the tangent line is $\left(x_0 - \dfrac{f(x_0)}{f'(x_0)}, 0\right)$.

(b) The y-intercept of the tangent line is $(0, f(x_0) - x_0 f'(x_0))$.

(c) The x-intercept of the normal line is $(x_0 + f(x_0)f'(x_0), 0)$.

(d) The y-intercept of the normal line is $\left(0, y_0 + \dfrac{x_0}{f'(x_0)}\right)$.

(e) $|BC| = \left|\dfrac{f(x_0)}{f'(x_0)}\right|$ (f) $|PC| = \left|\dfrac{f(x_0)\sqrt{1 + [f'(x_0)]^2}}{f'(x_0)}\right|$

(g) $|AB| = |f(x_0)f'(x_0)|$

(h) $|AP| = |f(x_0)|\sqrt{1 + [f'(x_0)]^2}$

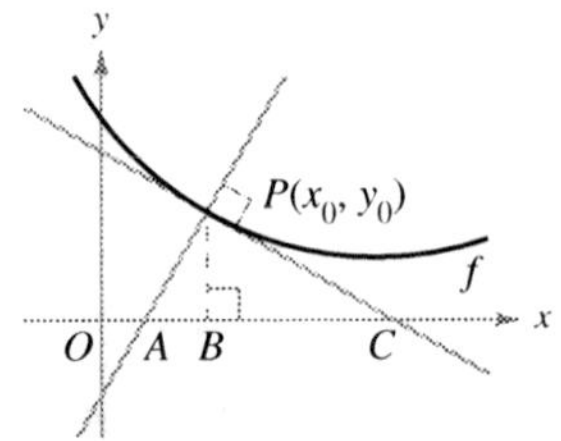

95. *Modeling Data* A meteorologist measures the atmospheric pressure P (in kilograms per square meter) at altitude h (in kilometers). The data are shown below.

h	0	5	10	15	20
P	10,332	5583	2376	1240	517

(a) Use a graphing utility to plot the points $(h, \ln P)$. Use the regression capabilities of the graphing utility to find a linear model for the revised data points.

(b) The line in part (a) has the form

$$\ln P = ah + b.$$

Write the equation in exponential form.

(c) Use a graphing utility to plot the original data and graph the exponential model in part (b).

(d) Find the rate of change of the pressure when $h = 5$ and $h = 18$.

96. Let f be a function that is positive and differentiable on the entire real number line. Let $g(x) = \ln f(x)$.

(a) If g is increasing, must f be increasing? Explain.

(b) If the graph of f is concave upward, must the graph of g be concave upward? Explain.

97. *Conjecture* Use a graphing utility to graph f and g in the same viewing window and determine which is increasing at the faster rate for "large" values of x. What can you conclude about the rate of growth of the natural logarithmic function?

(a) $f(x) = \ln x$, $g(x) = \sqrt{x}$

(b) $f(x) = \ln x$, $g(x) = \sqrt[4]{x}$

Slant Asymptotes **In Exercises 98 and 99, the graph of the function has two slant asymptotes. Identify each slant asymptote. Then graph the function and its asymptotes.**

98. $y = \sqrt{4 + 16x^2}$

99. $y = \sqrt{x^2 + 6x}$

PUTNAM EXAM CHALLENGE

100. Let $f(x)$ be defined for $a \leq x \leq b$. Assuming appropriate properties of continuity and derivability, prove for $a < x < b$ that

$$\frac{\dfrac{f(x) - f(a)}{x - a} - \dfrac{f(b) - f(a)}{b - a}}{x - b} = \frac{1}{2}f''(\beta)$$

where β is some number between a and b.

4.7   Optimization Problems

■ Solve applied minimum and maximum problems.

Applied Minimum and Maximum Problems

One of the most common applications of calculus involves the determination of minimum and maximum values. Consider how frequently you hear or read terms such as greatest profit, least cost, least time, greatest voltage, optimum size, least size, greatest strength, and greatest distance. Before outlining a general problem-solving strategy for such problems, consider the next example.

EXAMPLE 1   Finding Maximum Volume

A manufacturer wants to design an open box having a square base and a surface area of 108 square inches, as shown in Figure 4.54. What dimensions will produce a box with maximum volume?

Solution   Because the box has a square base, its volume is

$$V = x^2 h. \qquad \text{Primary equation}$$

This equation is called the **primary equation** because it gives a formula for the quantity to be optimized. The surface area of the box is

$$S = (\text{area of base}) + (\text{area of four sides})$$
$$S = x^2 + 4xh = 108. \qquad \text{Secondary equation}$$

Because V is to be maximized, you want to write V as a function of just one variable. To do this, you can solve the equation $x^2 + 4xh = 108$ for h in terms of x to obtain $h = (108 - x^2)/(4x)$. Substituting into the primary equation produces

$$V = x^2 h \qquad \text{Function of two variables}$$

$$= x^2 \left(\frac{108 - x^2}{4x} \right) \qquad \text{Substitute for } h.$$

$$= 27x - \frac{x^3}{4}. \qquad \text{Function of one variable}$$

Before finding which x-value will yield a maximum value of V, you should determine the *feasible domain*. That is, what values of x make sense in this problem? You know that $V \geq 0$. You also know that x must be nonnegative and that the area of the base ($A = x^2$) is at most 108. So, the feasible domain is

$$0 \leq x \leq \sqrt{108}. \qquad \text{Feasible domain}$$

To maximize V, find the critical numbers of the volume function on the interval $\left(0, \sqrt{108}\right)$.

$$\frac{dV}{dx} = 27 - \frac{3x^2}{4} = 0 \qquad \text{Set derivative equal to 0.}$$

$$3x^2 = 108 \qquad \text{Simplify.}$$

$$x = \pm 6 \qquad \text{Critical numbers}$$

So, the critical numbers are $x = \pm 6$. You do not need to consider $x = -6$ because it is outside the domain. Evaluating V at the critical number $x = 6$ and at the endpoints of the domain produces $V(0) = 0$, $V(6) = 108$, and $V\left(\sqrt{108}\right) = 0$. So, V is maximum when $x = 6$ and the dimensions of the box are $6 \times 6 \times 3$ inches.   ■

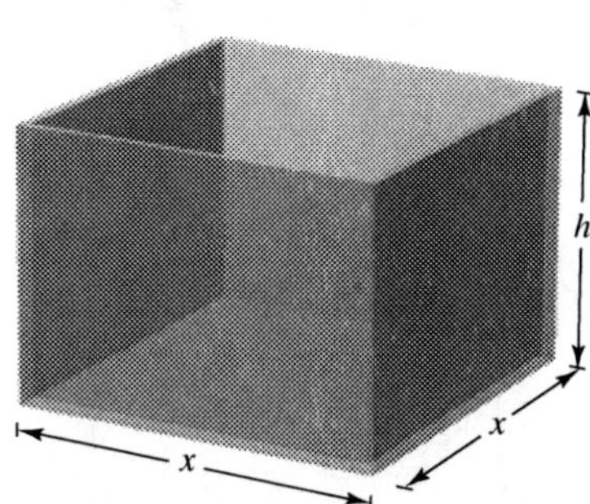

Open box with square base:
$S = x^2 + 4xh = 108$
Figure 4.54

TECHNOLOGY   You can verify your answer in Example 1 by using a graphing utility to graph the volume function

$$V = 27x - \frac{x^3}{4}.$$

Use a viewing window in which $0 \leq x \leq \sqrt{108} \approx 10.4$ and $0 \leq y \leq 120$ and use the *trace* feature to determine the maximum value of V.

In Example 1, you should realize that there are infinitely many open boxes having 108 square inches of surface area. To begin solving the problem, you might ask yourself which basic shape would seem to yield a maximum volume. Should the box be tall, squat, or nearly cubical?

You might even try calculating a few volumes, as shown in Figure 4.55, to see if you can get a better feeling for what the optimum dimensions should be. Remember that you are not ready to begin solving a problem until you have clearly identified what the problem is.

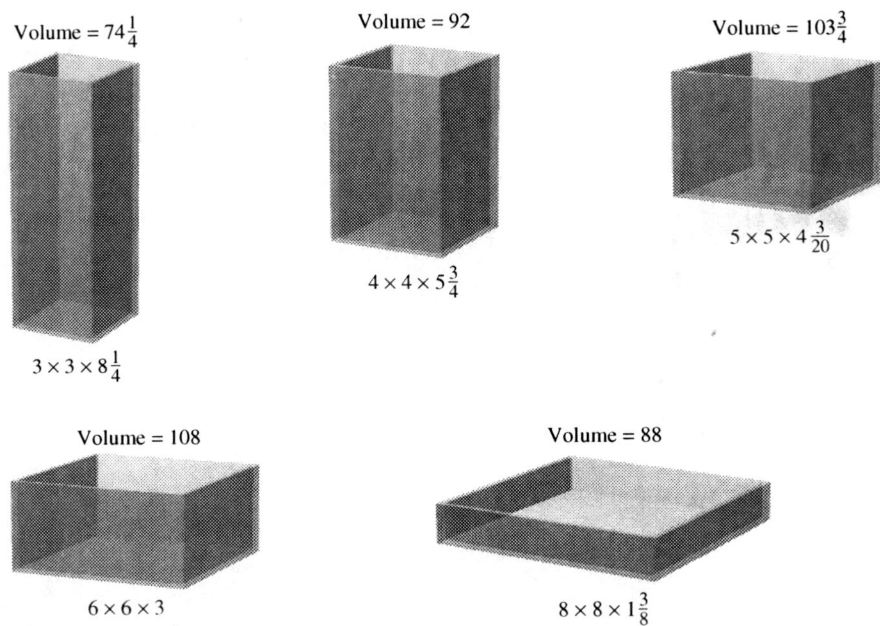

Which box has the greatest volume?
Figure 4.55

Example 1 illustrates the following guidelines for solving applied minimum and maximum problems.

> ### GUIDELINES FOR SOLVING APPLIED MINIMUM AND MAXIMUM PROBLEMS
>
> 1. Identify all *given* quantities and all quantities *to be determined*. If possible, make a sketch.
> 2. Write a **primary equation** for the quantity that is to be maximized or minimized. (A review of several useful formulas from geometry is presented inside the back cover.)
> 3. Reduce the primary equation to one having a *single independent variable*. This may involve the use of **secondary equations** relating the independent variables of the primary equation.
> 4. Determine the feasible domain of the primary equation. That is, determine the values for which the stated problem makes sense.
> 5. Determine the desired maximum or minimum value by the calculus techniques discussed in Sections 4.1 through 4.4.

NOTE When performing Step 5, recall that to determine the maximum or minimum value of a continuous function f on a closed interval, you should compare the values of f at its critical numbers with the values of f at the endpoints of the interval.

EXAMPLE 2 Finding Minimum Distance

Which points on the graph of $y = 4 - x^2$ are closest to the point $(0, 2)$?

Solution Figure 4.56 shows that there are two points at a minimum distance from the point $(0, 2)$. The distance between the point $(0, 2)$ and a point (x, y) on the graph of $y = 4 - x^2$ is given by

$$d = \sqrt{(x - 0)^2 + (y - 2)^2}.$$ Primary equation

Using the secondary equation $y = 4 - x^2$, you can rewrite the primary equation as

$$d = \sqrt{x^2 + (4 - x^2 - 2)^2} = \sqrt{x^4 - 3x^2 + 4}.$$

Because d is smallest when the expression inside the radical is smallest, you need only find the critical numbers of $f(x) = x^4 - 3x^2 + 4$. Note that the domain of f is the entire real number line. So, there are no endpoints of the domain to consider. Moreover, setting $f'(x)$ equal to 0 yields

$$f'(x) = 4x^3 - 6x = 2x(2x^2 - 3) = 0$$

$$x = 0, \sqrt{\frac{3}{2}}, -\sqrt{\frac{3}{2}}.$$

The First Derivative Test verifies that $x = 0$ yields a relative maximum, whereas both $x = \sqrt{3/2}$ and $x = -\sqrt{3/2}$ yield a minimum distance. So, the closest points are $\left(\sqrt{3/2}, 5/2\right)$ and $\left(-\sqrt{3/2}, 5/2\right)$.

EXAMPLE 3 Finding Minimum Area

A rectangular page is to contain 24 square inches of print. The margins at the top and bottom of the page are to be $1\frac{1}{2}$ inches, and the margins on the left and right are to be 1 inch (see Figure 4.57). What should the dimensions of the page be so that the least amount of paper is used?

Solution Let A be the area to be minimized.

$$A = (x + 3)(y + 2)$$ Primary equation

The printed area inside the margins is given by

$$24 = xy.$$ Secondary equation

Solving this equation for y produces $y = 24/x$. Substitution into the primary equation produces

$$A = (x + 3)\left(\frac{24}{x} + 2\right) = 30 + 2x + \frac{72}{x}.$$ Function of one variable

Because x must be positive, you are interested only in values of A for $x > 0$. To find the critical numbers, differentiate with respect to x.

$$\frac{dA}{dx} = 2 - \frac{72}{x^2} = 0 \implies x^2 = 36$$

So, the critical numbers are $x = \pm 6$. You do not have to consider $x = -6$ because it is outside the domain. The First Derivative Test confirms that A is a minimum when $x = 6$. So, $y = \frac{24}{6} = 4$ and the dimensions of the page should be $x + 3 = 9$ inches by $y + 2 = 6$ inches. ∎

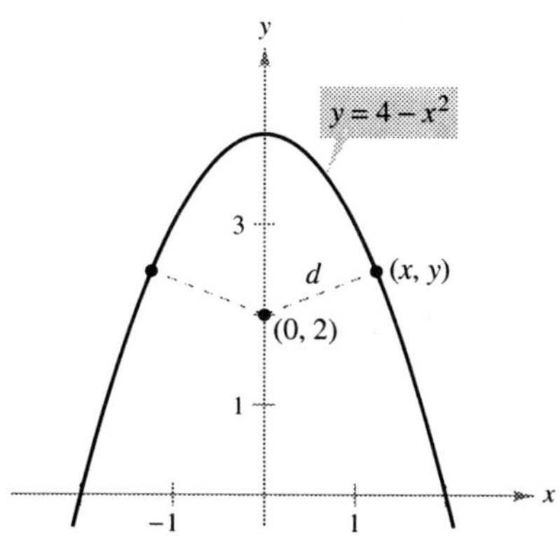

The quantity to be minimized is distance:
$d = \sqrt{(x - 0)^2 + (y - 2)^2}.$
Figure 4.56

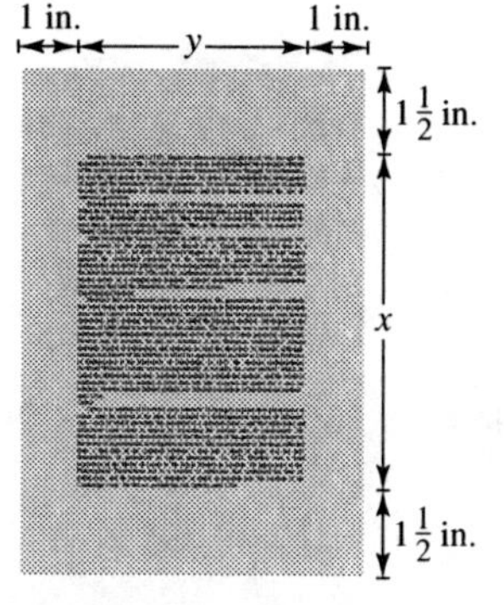

The quantity to be minimized is area:
$A = (x + 3)(y + 2).$
Figure 4.57

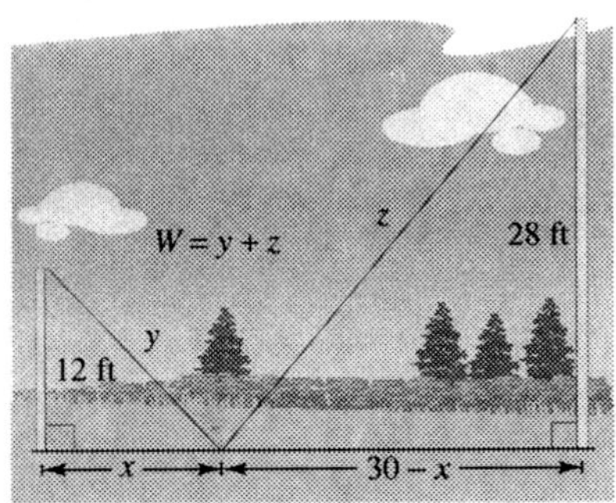

The quantity to be minimized is length. From the diagram, you can see that x varies between 0 and 30.

Figure 4.58

EXAMPLE 4 Finding Minimum Length

Two posts, one 12 feet high and the other 28 feet high, stand 30 feet apart. They are to be stayed by two wires, attached to a single stake, running from ground level to the top of each post. Where should the stake be placed to use the least amount of wire?

Solution   Let W be the wire length to be minimized. Using Figure 4.58, you can write

$$W = y + z. \qquad \text{Primary equation}$$

In this problem, rather than solving for y in terms of z (or vice versa), you can solve for both y and z in terms of a third variable x, as shown in Figure 4.58. From the Pythagorean Theorem, you obtain

$$x^2 + 12^2 = y^2$$
$$(30 - x)^2 + 28^2 = z^2$$

which implies that

$$y = \sqrt{x^2 + 144}$$
$$z = \sqrt{x^2 - 60x + 1684}.$$

So, W is given by

$$W = y + z$$
$$= \sqrt{x^2 + 144} + \sqrt{x^2 - 60x + 1684}, \quad 0 \le x \le 30.$$

Differentiating W with respect to x yields

$$\frac{dW}{dx} = \frac{x}{\sqrt{x^2 + 144}} + \frac{x - 30}{\sqrt{x^2 - 60x + 1684}}.$$

By letting $dW/dx = 0$, you obtain

$$\frac{x}{\sqrt{x^2 + 144}} + \frac{x - 30}{\sqrt{x^2 - 60x + 1684}} = 0$$
$$x\sqrt{x^2 - 60x + 1684} = (30 - x)\sqrt{x^2 + 144}$$
$$x^2(x^2 - 60x + 1684) = (30 - x)^2(x^2 + 144)$$
$$x^4 - 60x^3 + 1684x^2 = x^4 - 60x^3 + 1044x^2 - 8640x + 129{,}600$$
$$640x^2 + 8640x - 129{,}600 = 0$$
$$320(x - 9)(2x + 45) = 0$$
$$x = 9, \, -22.5.$$

Because $x = -22.5$ is not in the domain and

$$W(0) \approx 53.04, \quad W(9) = 50, \quad \text{and} \quad W(30) \approx 60.31$$

you can conclude that the wire should be staked at 9 feet from the 12-foot pole.

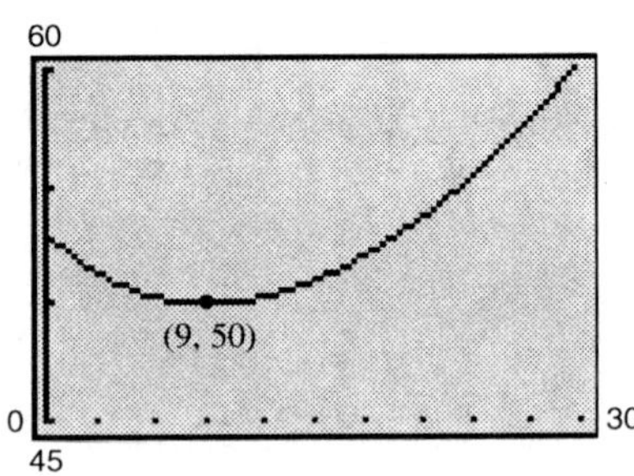

You can confirm the minimum value of W with a graphing utility.

Figure 4.59

TECHNOLOGY From Example 4, you can see that applied optimization problems can involve a lot of algebra. If you have access to a graphing utility, you can confirm that $x = 9$ yields a minimum value of W by graphing

$$W = \sqrt{x^2 + 144} + \sqrt{x^2 - 60x + 1684}$$

as shown in Figure 4.59.

In each of the first four examples, the extreme value occurred at a critical number. Although this happens often, remember that an extreme value can also occur at an endpoint of an interval, as shown in Example 5.

EXAMPLE 5 An Endpoint Maximum

Four feet of wire is to be used to form a square and a circle. How much of the wire should be used for the square and how much should be used for the circle to enclose the maximum total area?

Solution The total area (see Figure 4.60) is given by

$$A = (\text{area of square}) + (\text{area of circle})$$
$$A = x^2 + \pi r^2. \qquad \text{Primary equation}$$

Because the total length of wire is 4 feet, you obtain

$$4 = (\text{perimeter of square}) + (\text{circumference of circle})$$
$$4 = 4x + 2\pi r.$$

So, $r = 2(1 - x)/\pi$, and by substituting into the primary equation you have

$$A = x^2 + \pi \left[\frac{2(1 - x)}{\pi} \right]^2$$
$$= x^2 + \frac{4(1 - x)^2}{\pi}$$
$$= \frac{1}{\pi}[(\pi + 4)x^2 - 8x + 4].$$

You know that x must be nonnegative and that the perimeter of the square ($P = 4x$) is at most 4. So, the feasible domain is $0 \le x \le 1$. Because

$$\frac{dA}{dx} = \frac{2(\pi + 4)x - 8}{\pi}$$

the only critical number in $(0, 1)$ is $x = 4/(\pi + 4) \approx 0.56$. So, using

$$A(0) \approx 1.273, \quad A(0.56) \approx 0.56, \quad \text{and} \quad A(1) = 1$$

you can conclude that the maximum area occurs when $x = 0$. That is, *all* the wire is used for the circle. ∎

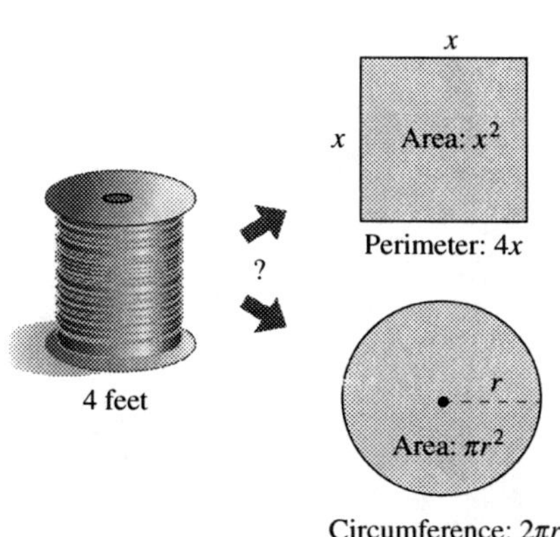

The quantity to be maximized is area:
$A = x^2 + \pi r^2$.

Figure 4.60

Let's review the primary equations developed in the first five examples. As applications go, these five examples are fairly simple, and yet the resulting primary equations are quite complicated.

$$V = 27x - \frac{x^3}{4} \qquad\qquad W = \sqrt{x^2 + 144} + \sqrt{x^2 - 60x + 1684}$$

$$d = \sqrt{x^4 - 3x^2 + 4} \qquad\qquad A = \frac{1}{\pi}[(\pi + 4)x^2 - 8x + 4]$$

$$A = 30 + 2x + \frac{72}{x}$$

You must expect that real-life applications often involve equations that are *at least as complicated* as these five. Remember that one of the main goals of this course is to learn to use calculus to analyze equations that initially seem formidable.

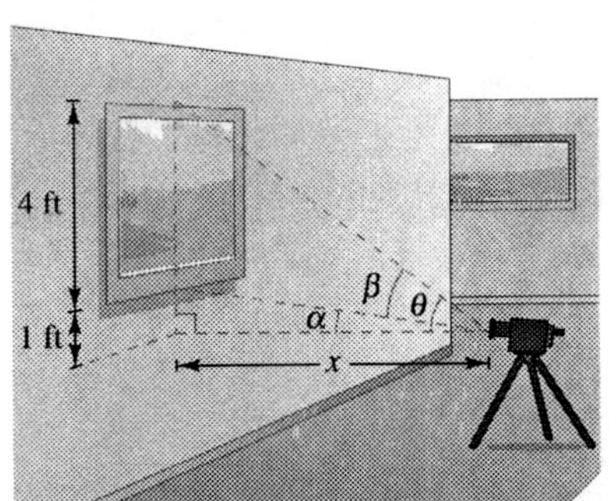

The camera should be 2.236 feet from the painting to maximize the angle β.
Figure 4.61

$\left(\ \right)$**EXAMPLE 6 Maximizing an Angle**

A photographer is taking a picture of a 4-foot painting hung in an art gallery. The camera lens is 1 foot below the lower edge of the painting, as shown in Figure 4.61. How far should the camera be from the painting to maximize the angle subtended by the camera lens?

Solution In Figure 4.61, let β be the angle to be maximized.

$$\beta = \theta - \alpha \qquad \text{Primary equation}$$

From Figure 4.61, you can see that $\cot \theta = \dfrac{x}{5}$ and $\cot \alpha = \dfrac{x}{1}$. Therefore, $\theta = \operatorname{arccot} \dfrac{x}{5}$ and $\alpha = \operatorname{arccot} x$. So,

$$\beta = \operatorname{arccot} \frac{x}{5} - \operatorname{arccot} x.$$

Differentiating β with respect to x produces

$$\frac{d\beta}{dx} = \frac{-1/5}{1 + (x^2/25)} - \frac{-1}{1 + x^2}$$

$$= \frac{-5}{25 + x^2} + \frac{1}{1 + x^2}$$

$$= \frac{4(5 - x^2)}{(25 + x^2)(1 + x^2)}.$$

Because $d\beta/dx = 0$ when $x = \sqrt{5}$, you can conclude from the First Derivative Test that this distance yields a maximum value of β. So, the distance is $x \approx 2.236$ feet and the angle is $\beta \approx 0.7297$ radian $\approx 41.81°$.

EXAMPLE 7 Finding a Maximum Revenue

The demand function for a product is modeled by

$$p = 56e^{-0.000012x} \qquad \text{Demand function}$$

where p is the price per unit (in dollars) and x is the number of units. What price will yield a maximum revenue?

Solution Substituting for p (from the demand function) produces the revenue function

$$R = xp = 56xe^{-0.000012x}. \qquad \text{Primary equation}$$

The rate of change of revenue R with respect to the number of units sold x is called the *marginal revenue*. Setting the marginal revenue equal to zero,

$$\frac{dR}{dx} = 56x(e^{-0.000012x})(-0.000012) + e^{-0.000012x}(56) = 0$$

yields $x \approx 83{,}333$ units. From this, you can conclude that the maximum revenue occurs when the price is

$$p = 56e^{-0.000012(83{,}333)} \approx \$20.60.$$

So, a price of about \$20.60 will yield a maximum revenue (see Figure 4.62). ∎

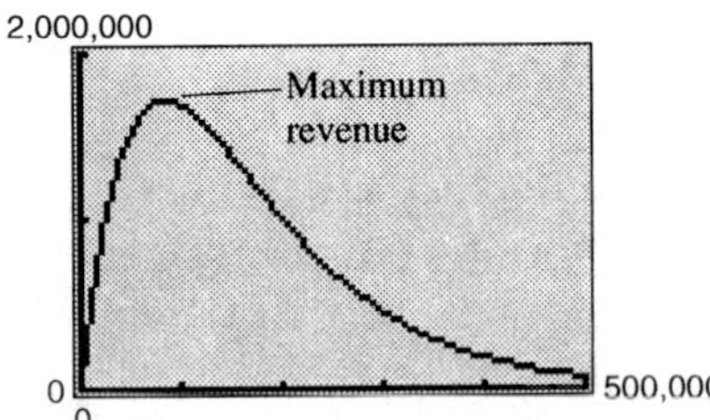

Figure 4.62

4.7 Exercises

1. ***Numerical, Graphical, and Analytic Analysis*** Find two positive numbers whose sum is 110 and whose product is a maximum.

(a) Analytically complete six rows of a table such as the one below. (The first two rows are shown.)

First Number x	Second Number	Product P
10	$110 - 10$	$10(110 - 10) = 1000$
20	$110 - 20$	$20(110 - 20) = 1800$

(b) Use a graphing utility to generate additional rows of the table. Use the table to estimate the solution. (*Hint:* Use the *table* feature of the graphing utility.)

(c) Write the product P as a function of x.

(d) Use a graphing utility to graph the function in part (c) and estimate the solution from the graph.

(e) Use calculus to find the critical number of the function in part (c). Then find the two numbers.

2. ***Numerical, Graphical, and Analytic Analysis*** An open box of maximum volume is to be made from a square piece of material, 24 inches on a side, by cutting equal squares from the corners and turning up the sides (see figure).

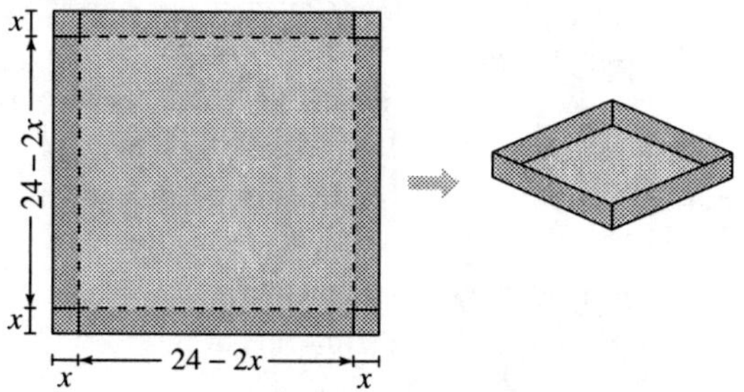

(a) Analytically complete six rows of a table such as the one below. (The first two rows are shown.) Use the table to guess the maximum volume.

Height x	Length and Width	Volume V
1	$24 - 2(1)$	$1[24 - 2(1)]^2 = 484$
2	$24 - 2(2)$	$2[24 - 2(2)]^2 = 800$

(b) Write the volume V as a function of x.

(c) Use calculus to find the critical number of the function in part (b) and find the maximum value.

(d) Use a graphing utility to graph the function in part (b) and verify the maximum volume from the graph.

In Exercises 3–8, find two positive numbers that satisfy the given requirements.

3. The sum is S and the product is a maximum.

4. The product is 185 and the sum is a minimum.

5. The product is 147 and the sum of the first number plus three times the second number is a minimum.

6. The second number is the reciprocal of the first number and the sum is a minimum.

7. The sum of the first number and twice the second number is 108 and the product is a maximum.

8. The sum of the first number squared and the second number is 54 and the product is a maximum.

In Exercises 9 and 10, find the length and width of a rectangle that has the given perimeter and a maximum area.

9. Perimeter: 80 meters

10. Perimeter: P units

In Exercises 11 and 12, find the length and width of a rectangle that has the given area and a minimum perimeter.

11. Area: 32 square feet

12. Area: A square centimeters

In Exercises 13–16, find the point on the graph of the function that is closest to the given point.

Function	*Point*		*Function*	*Point*
13. $f(x) = x^2$	$\left(2, \frac{1}{2}\right)$		**14.** $f(x) = (x - 1)^2$	$(-5, 3)$
15. $f(x) = \sqrt{x}$	$(4, 0)$		**16.** $f(x) = \sqrt{x - 8}$	$(12, 0)$

17. ***Area*** A rectangular page is to contain 30 square inches of print. The margins on each side are 1 inch. Find the dimensions of the page such that the least amount of paper is used.

18. ***Area*** A rectangular page is to contain 36 square inches of print. The margins on each side are $1\frac{1}{2}$ inches. Find the dimensions of the page such that the least amount of paper is used.

19. ***Chemical Reaction*** In an autocatalytic chemical reaction, the product formed is a catalyst for the reaction. If Q_0 is the amount of the original substance and x is the amount of catalyst formed, the rate of chemical reaction is

$$\frac{dQ}{dx} = kx(Q_0 - x).$$

For what value of x will the rate of chemical reaction be greatest?

20. ***Traffic Control*** On a given day, the flow rate F (in cars per hour) on a congested roadway is

$$F = \frac{v}{22 + 0.02v^2}$$

where v is the speed of the traffic in miles per hour. What speed will maximize the flow rate on the road?

21. *Area*   A farmer plans to fence a rectangular pasture adjacent to a river (see figure). The pasture must contain 245,000 square meters in order to provide enough grass for the herd. What dimensions will require the least amount of fencing if no fencing is needed along the river?

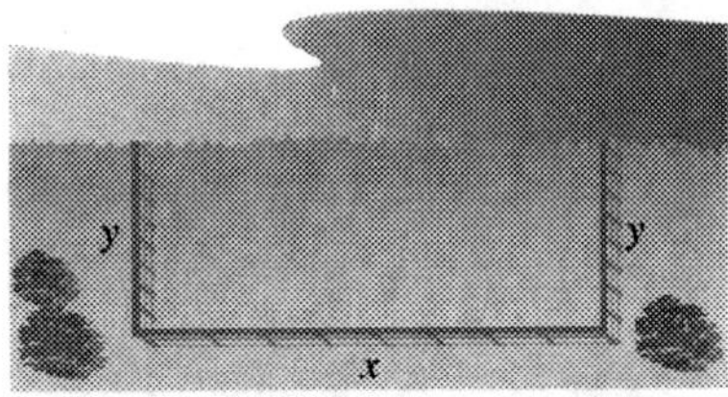

22. *Maximum Area*   A rancher has 400 feet of fencing with which to enclose two adjacent rectangular corrals (see figure). What dimensions should be used so that the enclosed area will be a maximum?

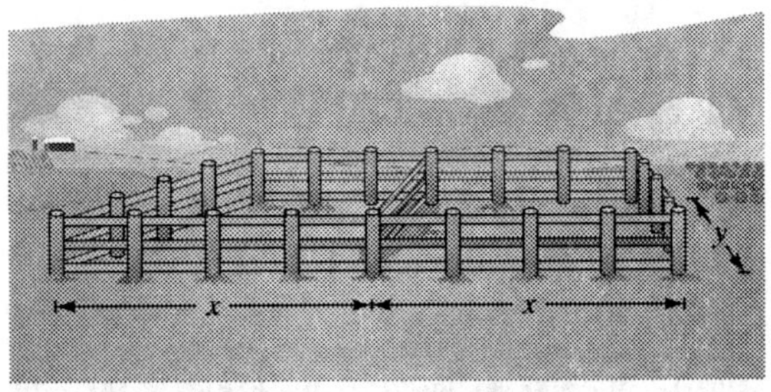

23. *Maximum Volume*

(a) Verify that each of the rectangular solids shown in the figure has a surface area of 150 square inches.

(b) Find the volume of each solid.

(c) Determine the dimensions of a rectangular solid (with a square base) of maximum volume if its surface area is 150 square inches.

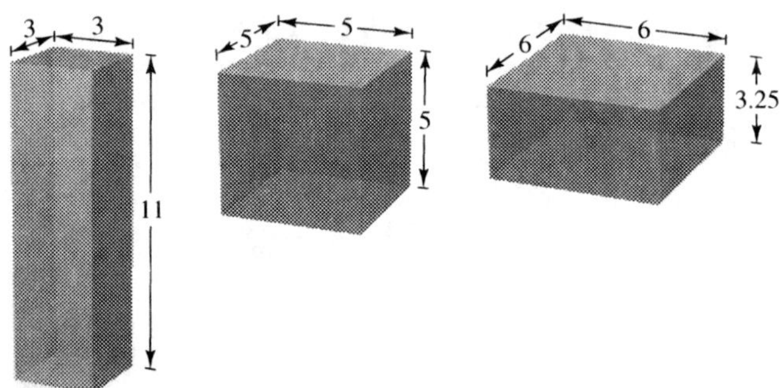

24. *Maximum Volume*   Determine the dimensions of a rectangular solid (with a square base) with maximum volume if its surface area is 337.5 square centimeters.

25. *Maximum Area*   A Norman window is constructed by adjoining a semicircle to the top of an ordinary rectangular window (see figure). Find the dimensions of a Norman window of maximum area if the total perimeter is 16 feet.

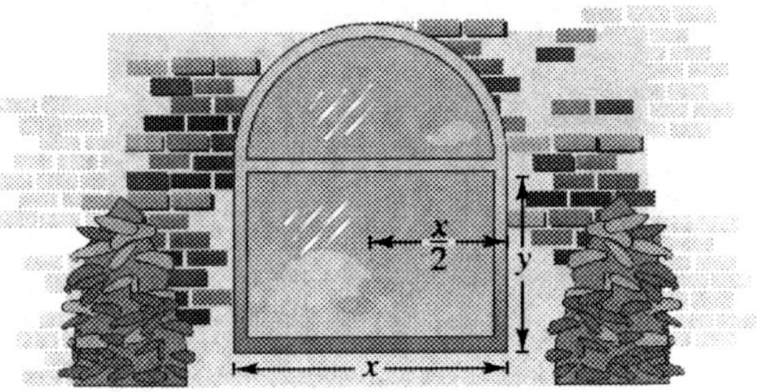

Figure for 25

26. *Maximum Area*   A rectangle is bounded by the x- and y-axes and the graph of $y = (6 - x)/2$ (see figure). What length and width should the rectangle have so that its area is a maximum?

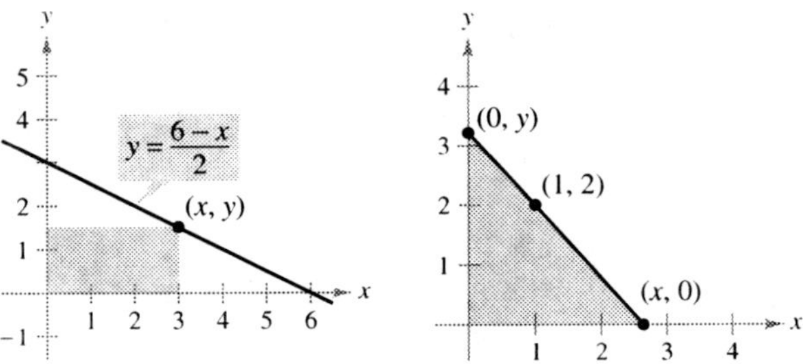

Figure for 26   **Figure for 27**

27. *Minimum Length*   A right triangle is formed in the first quadrant by the x- and y-axes and a line through the point $(1, 2)$ (see figure).

(a) Write the length L of the hypotenuse as a function of x.

(b) Use a graphing utility to approximate x graphically such that the length of the hypotenuse is a minimum.

(c) Find the vertices of the triangle such that its area is a minimum.

28. *Maximum Area*   Find the area of the largest isosceles triangle that can be inscribed in a circle of radius 6 (see figure).

(a) Solve by writing the area as a function of h.

(b) Solve by writing the area as a function of α.

(c) Identify the type of triangle of maximum area.

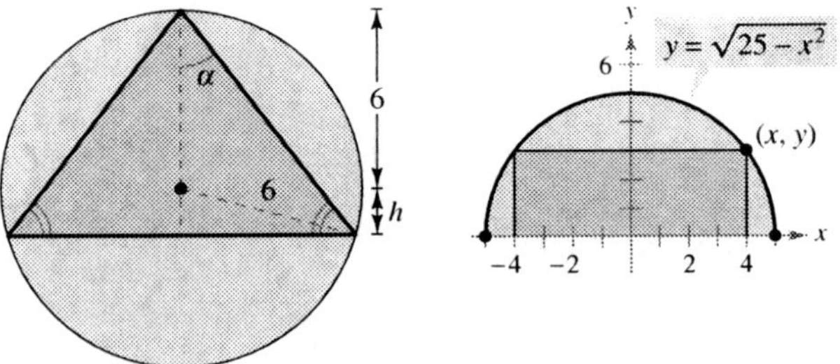

Figure for 28   **Figure for 29**

29. *Maximum Area*   A rectangle is bounded by the x-axis and the semicircle $y = \sqrt{25 - x^2}$ (see figure). What length and width should the rectangle have so that its area is a maximum?

30. *Area*   Find the dimensions of the largest rectangle that can be inscribed in a semicircle of radius r (see Exercise 29).

31. *Numerical, Graphical, and Analytic Analysis* An exercise room consists of a rectangle with a semicircle on each end. A 200-meter running track runs around the outside of the room.

(a) Draw a figure to represent the problem. Let x and y represent the length and width of the rectangle.

(b) Analytically complete six rows of a table such as the one below. (The first two rows are shown.) Use the table to guess the maximum area of the rectangular region.

Length x	Width y	Area A
10	$\dfrac{2}{\pi}(100 - 10)$	$(10)\dfrac{2}{\pi}(100 - 10) \approx 573$
20	$\dfrac{2}{\pi}(100 - 20)$	$(20)\dfrac{2}{\pi}(100 - 20) \approx 1019$

(c) Write the area A as a function of x.

(d) Use calculus to find the critical number of the function in part (c) and find the maximum value.

(e) Use a graphing utility to graph the function in part (c) and verify the maximum area from the graph.

32. *Numerical, Graphical, and Analytic Analysis* A right circular cylinder is to be designed to hold 22 cubic inches of a soft drink (approximately 12 fluid ounces).

(a) Analytically complete six rows of a table such as the one below. (The first two rows are shown.)

Radius r	Height	Surface Area S
0.2	$\dfrac{22}{\pi(0.2)^2}$	$2\pi(0.2)\left[0.2 + \dfrac{22}{\pi(0.2)^2}\right] \approx 220.3$
0.4	$\dfrac{22}{\pi(0.4)^2}$	$2\pi(0.4)\left[0.4 + \dfrac{22}{\pi(0.4)^2}\right] \approx 111.0$

(b) Use a graphing utility to generate additional rows of the table. Use the table to estimate the minimum surface area. (*Hint:* Use the *table* feature of the graphing utility.)

(c) Write the surface area S as a function of r.

(d) Use a graphing utility to graph the function in part (c) and estimate the minimum surface area from the graph.

(e) Use calculus to find the critical number of the function in part (c) and find dimensions that will yield the minimum surface area.

33. *Maximum Volume* A rectangular package to be sent by a postal service can have a maximum combined length and girth (perimeter of a cross section) of 108 inches (see figure). Find the dimensions of the package of maximum volume that can be sent. (Assume the cross section is square.)

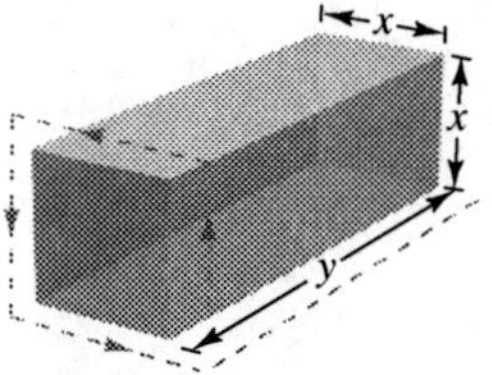

34. *Maximum Volume* Rework Exercise 33 for a cylindrical package. (The cross section is circular.)

35. *Maximum Volume* Find the volume of the largest right circular cone that can be inscribed in a sphere of radius r.

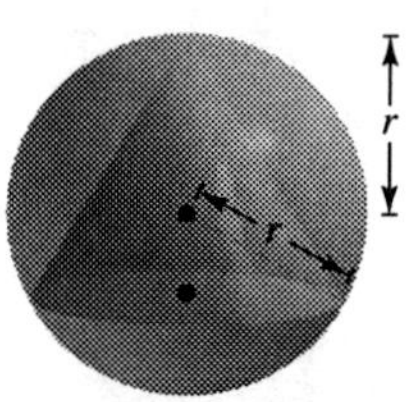

36. *Maximum Volume* Find the volume of the largest right circular cylinder that can be inscribed in a sphere of radius r.

WRITING ABOUT CONCEPTS

37. A shampoo bottle is a right circular cylinder. Because the surface area of the bottle does not change when it is squeezed, is it true that the volume remains the same? Explain.

CAPSTONE

38. The perimeter of a rectangle is 20 feet. Of all possible dimensions, the maximum area is 25 square feet when its length and width are both 5 feet. Are there dimensions that yield a minimum area? Explain.

39. *Minimum Surface Area* A solid is formed by adjoining two hemispheres to the ends of a right circular cylinder. The total volume of the solid is 14 cubic centimeters. Find the radius of the cylinder that produces the minimum surface area.

40. *Minimum Cost* An industrial tank of the shape described in Exercise 39 must have a volume of 4000 cubic feet. The hemispherical ends cost twice as much per square foot of surface area as the sides. Find the dimensions that will minimize cost.

41. *Minimum Area* The sum of the perimeters of an equilateral triangle and a square is 10. Find the dimensions of the triangle and the square that produce a minimum total area.

42. *Maximum Area* Twenty feet of wire is to be used to form two figures. In each of the following cases, how much wire should be used for each figure so that the total enclosed area is a maximum?

(a) Equilateral triangle and square

(b) Square and regular pentagon

(c) Regular pentagon and regular hexagon

(d) Regular hexagon and circle

What can you conclude from this pattern? {*Hint:* The area of a regular polygon with n sides of length x is $A = (n/4)[\cot(\pi/n)]x^2$.}

43. *Beam Strength* A wooden beam has a rectangular cross section of height h and width w (see figure on the next page). The strength S of the beam is directly proportional to the width and the square of the height. What are the dimensions of the strongest beam that can be cut from a round log of diameter 20 inches? (*Hint:* $S = kh^2w$, where k is the proportionality constant.)

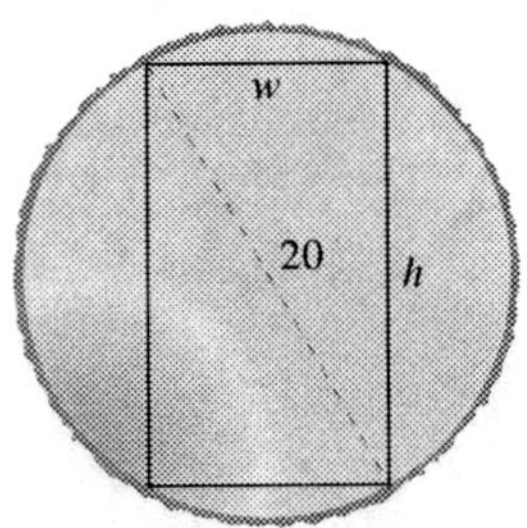

Figure for 43

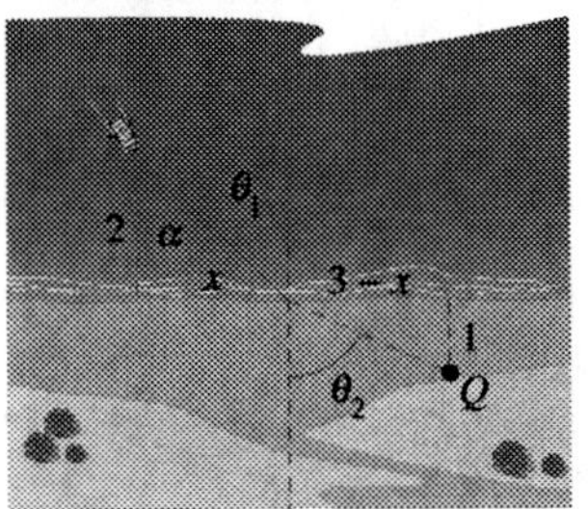

Figure for 44

44. *Minimum Length* Two factories are located at the coordinates $(-x, 0)$ and $(x, 0)$, and their power supply is at $(0, h)$ (see figure). Find y such that the total length of power line from the power supply to the factories is a minimum.

45. *Projectile Range* The range R of a projectile fired with an initial velocity v_0 at an angle θ with the horizontal is

$$R = \frac{v_0^2 \sin 2\theta}{g}$$

where g is the acceleration due to gravity. Find the angle θ such that the range is a maximum.

46. *Conjecture* Consider the functions $f(x) = \frac{1}{2}x^2$ and $g(x) = \frac{1}{16}x^4 - \frac{1}{2}x^2$ on the domain $[0, 4]$.

(a) Use a graphing utility to graph the functions on the specified domain.

(b) Write the vertical distance d between the functions as a function of x and use calculus to find the value of x for which d is maximum.

(c) Find the equations of the tangent lines to the graphs of f and g at the critical number found in part (b). Graph the tangent lines. What is the relationship between the lines?

(d) Make a conjecture about the relationship between tangent lines to the graphs of two functions at the value of x at which the vertical distance between the functions is greatest, and prove your conjecture.

47. *Illumination* A light source is located over the center of a circular table of diameter 4 feet (see figure). Find the height h of the light source such that the illumination I at the perimeter of the table is maximum if $I = k(\sin \alpha)/s^2$, where s is the slant height, α is the angle at which the light strikes the table, and k is a constant.

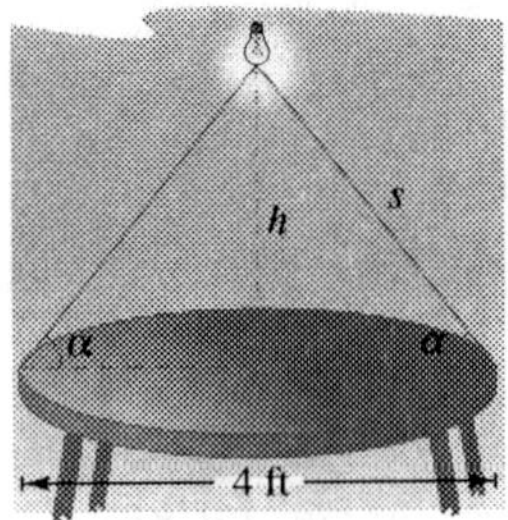

48. *Illumination* The illumination from a light source is directly proportional to the strength of the source and inversely proportional to the square of the distance from the source. Two light sources of intensities I_1 and I_2 are d units apart. What point on the line segment joining the two sources has the least illumination?

49. *Minimum Time* A man is in a boat 2 miles from the nearest point on the coast. He is to go to a point Q, located 3 miles down the coast and 1 mile inland (see figure). He can row at 2 miles per hour and walk at 4 miles per hour. Toward what point on the coast should he row in order to reach point Q in the least time?

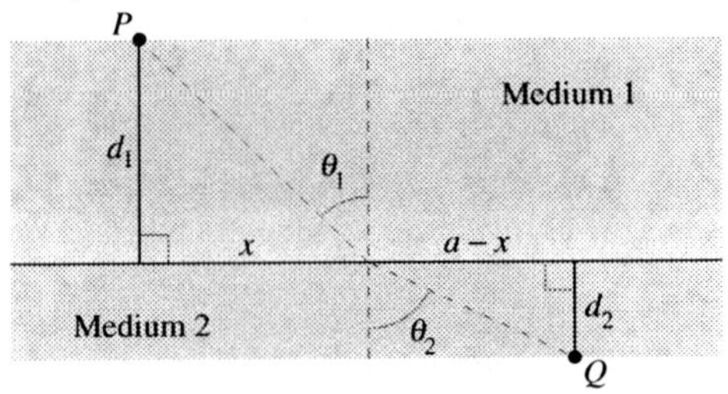

50. *Minimum Time* Consider Exercise 49 if the point Q is on the shoreline rather than 1 mile inland.

(a) Write the travel time T as a function of α.

(b) Use the result of part (a) to find the minimum time to reach Q.

(c) The man can row at v_1 miles per hour and walk at v_2 miles per hour. Write the time T as a function of α. Show that the critical number of T depends only on v_1 and v_2 and not on the distances. Explain how this result would be more beneficial to the man than the result of Exercise 49.

(d) Describe how to apply the result of part (c) to minimizing the cost of constructing a power transmission cable that costs c_1 dollars per mile under water and c_2 dollars per mile over land.

51. *Minimum Time* The conditions are the same as in Exercise 49 except that the man can row at v_1 miles per hour and walk at v_2 miles per hour. If θ_1 and θ_2 are the magnitudes of the angles, show that the man will reach point Q in the least time when

$$\frac{\sin \theta_1}{v_1} = \frac{\sin \theta_2}{v_2}.$$

52. *Minimum Time* When light waves traveling in a transparent medium strike the surface of a second transparent medium, they change direction. This change of direction is called *refraction* and is defined by **Snell's Law of Refraction,**

$$\frac{\sin \theta_1}{v_1} = \frac{\sin \theta_2}{v_2}$$

where θ_1 and θ_2 are the magnitudes of the angles shown in the figure and v_1 and v_2 are the velocities of light in the two media. Show that this problem is equivalent to that in Exercise 51, and that light waves traveling from P to Q follow the path of minimum time.

53. Sketch the graph of $f(x) = 2 - 2\sin x$ on the interval $[0, \pi/2]$.

(a) Find the distance from the origin to the y-intercept and the distance from the origin to the x-intercept.

(b) Write the distance d from the origin to a point on the graph of f as a function of x. Use your graphing utility to graph d and find the minimum distance.

(c) Use calculus and the *zero* or *root* feature of a graphing utility to find the value of x that minimizes the function d on the interval $[0, \pi/2]$. What is the minimum distance?

(Submitted by Tim Chapell, Penn Valley Community College, Kansas City, MO)

54. *Minimum Cost* An offshore oil well is 2 kilometers off the coast. The refinery is 4 kilometers down the coast. Laying pipe in the ocean is twice as expensive as on land. What path should the pipe follow in order to minimize the cost?

55. *Maximum Volume* A sector with central angle θ is cut from a circle of radius 12 inches (see figure), and the edges of the sector are brought together to form a cone. Find the magnitude of θ such that the volume of the cone is a maximum.

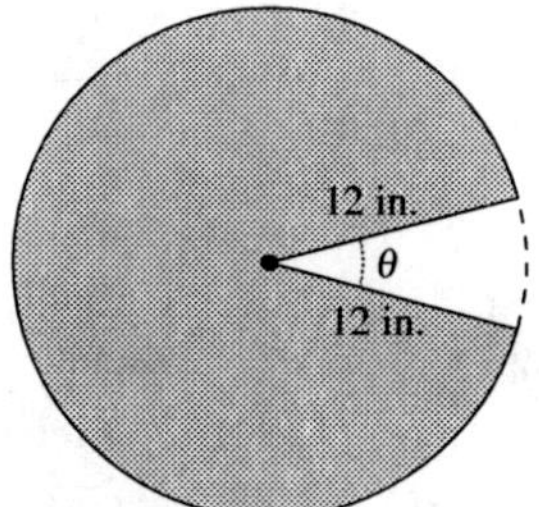

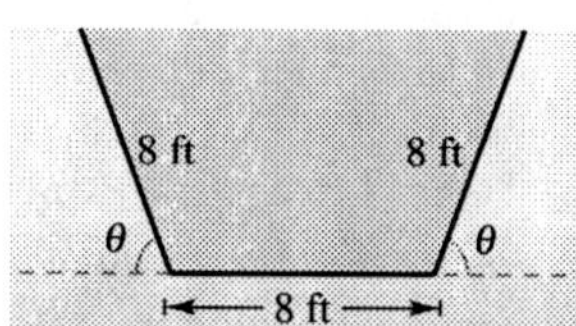

Figure for 55   **Figure for 56**

56. *Numerical, Graphical, and Analytic Analysis* The cross sections of an irrigation canal are isosceles trapezoids of which three sides are 8 feet long (see figure). Determine the angle of elevation θ of the sides such that the area of the cross sections is a maximum by completing the following.

(a) Analytically complete six rows of a table such as the one below. (The first two rows are shown.)

Base 1	Base 2	Altitude	Area
8	$8 + 16\cos 10°$	$8\sin 10°$	≈ 22.1
8	$8 + 16\cos 20°$	$8\sin 20°$	≈ 42.5

(b) Use a graphing utility to generate additional rows of the table and estimate the maximum cross-sectional area. (*Hint:* Use the *table* feature of the graphing utility.)

(c) Write the cross-sectional area A as a function of θ.

(d) Use calculus to find the critical number of the function in part (c) and find the angle that will yield the maximum cross-sectional area.

(e) Use a graphing utility to graph the function in part (c) and verify the maximum cross-sectional area.

57. *Minimum Cost* The ordering and transportation cost C of the components used in manufacturing a product is

$$C = 100\left(\frac{200}{x^2} + \frac{x}{x + 30}\right), \quad x \geq 1$$

where C is measured in thousands of dollars and x is the order size in hundreds. Find the order size that minimizes the cost. (*Hint:* Use the *root* feature of a graphing utility.)

58. *Diminishing Returns* The profit P (in thousands of dollars) for a company spending an amount s (in thousands of dollars) on advertising is

$$P = -\tfrac{1}{10}s^3 + 6s^2 + 400.$$

(a) Find the amount of money the company should spend on advertising in order to yield a maximum profit.

(b) The *point of diminishing returns* is the point at which the rate of growth of the profit function begins to decline. Find the point of diminishing returns.

59. *Area* Find the area of the largest rectangle that can be inscribed under the curve $y = e^{-x^2}$ in the first and second quadrants.

60. *Population Growth* Fifty elk are introduced into a game preserve. It is estimated that their population will increase according to the model $p(t) = 250/(1 + 4e^{-t/3})$, where t is measured in years. At what rate is the population increasing when $t = 2$? After how many years is the population increasing most rapidly?

61. Verify that the function

$$y = \frac{L}{1 + ae^{-x/b}}, \quad a > 0, \ b > 0, \ L > 0$$

increases at the maximum rate when $y = L/2$.

62. *Area* Perform the following steps to find the maximum area of the rectangle shown in the figure.

(a) Solve for c in the equation $f(c) = f(c + x)$.

(b) Use the result in part (a) to write the area A as a function of x. [*Hint:* $A = xf(c)$]

(c) Use a graphing utility to graph the area function. Use the graph to approximate the dimensions of the rectangle of maximum area. Determine the required area.

(d) Use a graphing utility to graph the expression for c found in part (a). Use the graph to approximate

$$\lim_{x \to 0^+} c \quad \text{and} \quad \lim_{x \to \infty} c.$$

Use this result to describe the changes in the dimensions and position of the rectangle for $0 < x < \infty$.

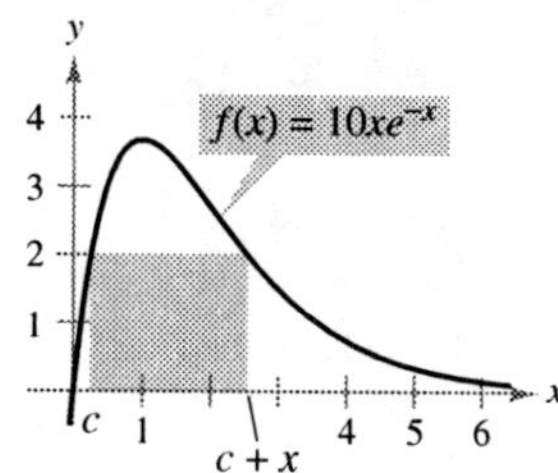

Minimum Distance In Exercises 63–65, consider a fuel distribution center located at the origin of the rectangular coordinate system (units in miles; see figures). The center supplies three factories with coordinates $(4, 1)$, $(5, 6)$, and $(10, 3)$. A trunk line will run from the distribution center along the line $y = mx$, and feeder lines will run to the three factories. The objective is to find m such that the lengths of the feeder lines are minimized.

63. Minimize the sum of the squares of the lengths of the vertical feeder lines given by

$$S_1 = (4m - 1)^2 + (5m - 6)^2 + (10m - 3)^2.$$

Find the equation of the trunk line by this method and then determine the sum of the lengths of the feeder lines.

64. Minimize the sum of the absolute values of the lengths of the vertical feeder lines given by

$$S_2 = |4m - 1| + |5m - 6| + |10m - 3|.$$

Find the equation of the trunk line by this method and then determine the sum of the lengths of the feeder lines. (*Hint:* Use a graphing utility to graph the function S_2 and approximate the required critical number.)

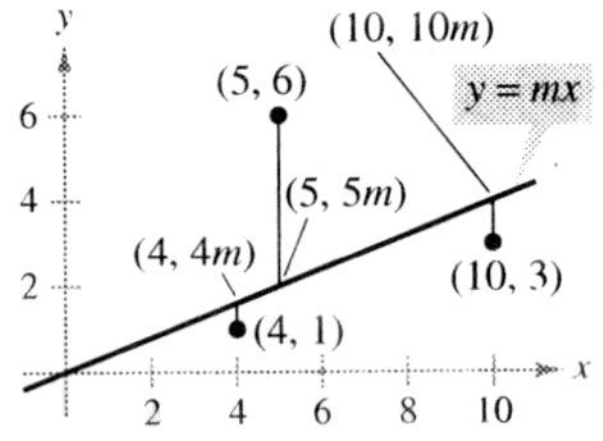
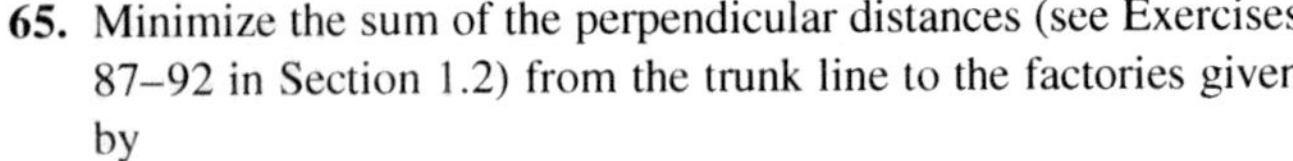

Figure for 63 and 64   **Figure for 65**

65. Minimize the sum of the perpendicular distances (see Exercises 87–92 in Section 1.2) from the trunk line to the factories given by

$$S_3 = \frac{|4m - 1|}{\sqrt{m^2 + 1}} + \frac{|5m - 6|}{\sqrt{m^2 + 1}} + \frac{|10m - 3|}{\sqrt{m^2 + 1}}.$$

Find the equation of the trunk line by this method and then determine the sum of the lengths of the feeder lines. (*Hint:* Use a graphing utility to graph the function S_3 and approximate the required critical number.)

66. ***Maximum Area*** Consider a symmetric cross inscribed in a circle of radius r (see figure).

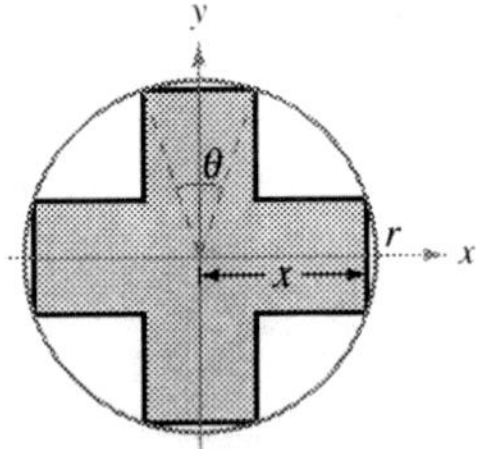

(a) Write the area A of the cross as a function of x and find the value of x that maximizes the area.

(b) Write the area A of the cross as a function of θ and find the value of θ that maximizes the area.

(c) Show that the critical numbers of parts (a) and (b) yield the same maximum area. What is that area?

67. Find the maximum value of $f(x) = x^3 - 3x$ on the set of all real numbers x satisfying $x^4 + 36 \leq 13x^2$. Explain your reasoning.

68. Find the minimum value of

$$\frac{(x + 1/x)^6 - (x^6 + 1/x^6) - 2}{(x + 1/x)^3 + (x^3 + 1/x^3)} \quad \text{for } x > 0.$$

These problems were composed by the Committee on the Putnam Prize Competition. © The Mathematical Association of America. All rights reserved.

Connecticut River

Whenever the Connecticut River reaches a level of 105 feet above sea level, two Northampton, Massachusetts flood control station operators begin a round-the-clock river watch. Every 2 hours, they check the height of the river, using a scale marked off in tenths of a foot, and record the data in a log book. In the spring of 1996, the flood watch lasted from April 4, when the river reached 105 feet and was rising at 0.2 foot per hour, until April 25, when the level subsided again to 105 feet. Between those dates, their log shows that the river rose and fell several times, at one point coming close to the 115-foot mark. If the river had reached 115 feet, the city would have closed down Mount Tom Road (Route 5, south of Northampton).

The graph below shows the rate of change of the level of the river during one portion of the flood watch. Use the graph to answer each question.

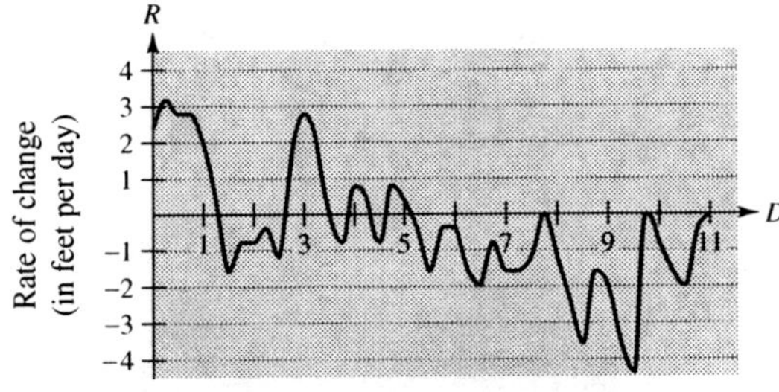

Day (0 ↔ 12:01 A.M. April 14)

(a) On what date was the river rising most rapidly? How do you know?

(b) On what date was the river falling most rapidly? How do you know?

(c) There were two dates in a row on which the river rose, then fell, then rose again during the course of the day. On which days did this occur, and how do you know?

(d) At 1 minute past midnight, April 14, the river level was 111.0 feet. Estimate its height 24 hours later and 48 hours later. Explain how you made your estimates.

(e) The river crested at 114.4 feet. On what date do you think this occurred?

(Submitted by Mary Murphy, Smith College, Northampton, MA)

4.8   Differentials

- Understand the concept of a tangent line approximation.
- Compare the value of the differential, *dy*, with the actual change in *y*, Δ*y*.
- Estimate a propagated error using a differential.
- Find the differential of a function using differentiation formulas.

Tangent Line Approximations

Newton's Method (Section 3.8) is an example of the use of a tangent line to a graph to approximate the graph. In this section, you will study other situations in which the graph of a function can be approximated by a straight line.

To begin, consider a function f that is differentiable at c. The equation for the tangent line at the point $(c, f(c))$ is given by

$$y - f(c) = f'(c)(x - c)$$

$$y = f(c) + f'(c)(x - c)$$

and is called the **tangent line approximation** (or **linear approximation**) of f at c. Because c is a constant, y is a linear function of x. Moreover, by restricting the values of x to those sufficiently close to c, the values of y can be used as approximations (to any desired degree of accuracy) of the values of the function f. In other words, as $x \to c$, the limit of y is $f(c)$.

EXAMPLE 1   Using a Tangent Line Approximation

Find the tangent line approximation of

$$f(x) = 1 + \sin x$$

at the point $(0, 1)$. Then use a table to compare the y-values of the linear function with those of $f(x)$ on an open interval containing $x = 0$.

Solution   The derivative of f is

$$f'(x) = \cos x. \qquad \text{First derivative}$$

So, the equation of the tangent line to the graph of f at the point $(0, 1)$ is

$$y - f(0) = f'(0)(x - 0)$$
$$y - 1 = (1)(x - 0)$$
$$y = 1 + x. \qquad \text{Tangent line approximation}$$

The table compares the values of y given by this linear approximation with the values of $f(x)$ near $x = 0$. Notice that the closer x is to 0, the better the approximation is. This conclusion is reinforced by the graph shown in Figure 4.63.

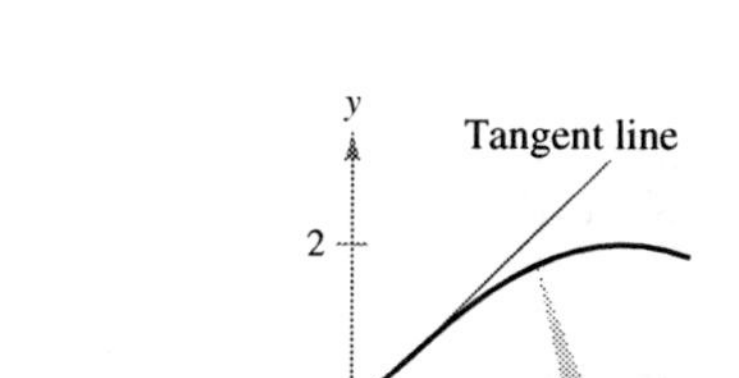

The tangent line approximation of f at the point $(0, 1)$
Figure 4.63

x	-0.5	-0.1	-0.01	0	0.01	0.1	0.5
$f(x) = 1 + \sin x$	0.521	0.9002	0.9900002	1	1.0099998	1.0998	1.479
$y = 1 + x$	0.5	0.9	0.99	1	1.01	1.1	1.5

NOTE   Be sure you see that this linear approximation of $f(x) = 1 + \sin x$ depends on the point of tangency. At a different point on the graph of f, you would obtain a different tangent line approximation.

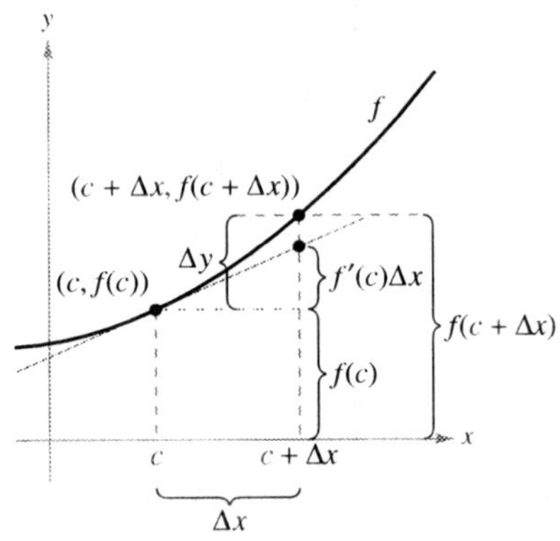

When Δx is small, $\Delta y = f(c + \Delta x) - f(c)$ is approximated by $f'(c)\Delta x$.

Figure 4.64

Differentials

When the tangent line to the graph of f at the point $(c, f(c))$

$$y = f(c) + f'(c)(x - c) \qquad \text{Tangent line at } (c, f(c))$$

is used as an approximation of the graph of f, the quantity $x - c$ is called the change in x, and is denoted by Δx, as shown in Figure 4.64. When Δx is small, the change in y (denoted by Δy) can be approximated as shown.

$$\Delta y = f(c + \Delta x) - f(c) \qquad \text{Actual change in } y$$
$$\approx f'(c)\Delta x \qquad \text{Approximate change in } y$$

For such an approximation, the quantity Δx is traditionally denoted by dx, and is called the **differential of x.** The expression $f'(x)\,dx$ is denoted by dy, and is called the **differential of y.**

DEFINITION OF DIFFERENTIALS

Let $y = f(x)$ represent a function that is differentiable on an open interval containing x. The **differential of x** (denoted by dx) is any nonzero real number. The **differential of y** (denoted by dy) is

$$dy = f'(x)\,dx.$$

In many types of applications, the differential of y can be used as an approximation of the change in y. That is, for small values of $\Delta x = dx$,

$$\Delta y \approx dy \qquad \text{or} \qquad \Delta y \approx f'(x)dx.$$

EXAMPLE 2 Comparing Δy and dy

Let $y = x^2$. Find dy when $x = 1$ and $dx = 0.01$. Compare this value with Δy for $x = 1$ and $\Delta x = 0.01$.

Solution Because $y = f(x) = x^2$, you have $f'(x) = 2x$, and the differential dy is given by

$$dy = f'(x)\,dx = f'(1)(0.01) = 2(0.01) = 0.02. \qquad \text{Differential of } y$$

Now, using $\Delta x = 0.01$, the change in y is

$$\Delta y = f(x + \Delta x) - f(x) = f(1.01) - f(1) = (1.01)^2 - 1^2 = 0.0201.$$

Figure 4.65 shows the geometric comparison of dy and Δy. Try comparing other values of dy and Δy. You will see that the values become closer to each other as dx (or Δx) approaches 0. ∎

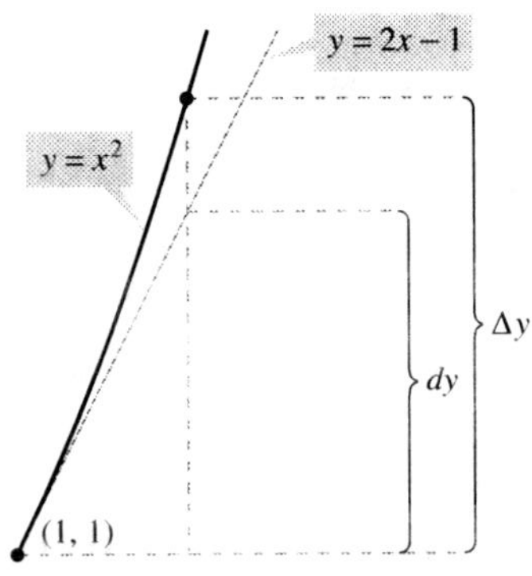

The change in y, Δy, is approximated by the differential of y, dy.

Figure 4.65

In Example 2, the tangent line to the graph of $f(x) = x^2$ at $x = 1$ is

$$y = 2x - 1 \qquad \text{or} \qquad g(x) = 2x - 1. \qquad \text{Tangent line to the graph of } f \text{ at } x = 1$$

For x-values near 1, this line is close to the graph of f, as shown in Figure 4.65. For instance,

$$f(1.01) = 1.01^2 = 1.0201 \qquad \text{and} \qquad g(1.01) = 2(1.01) - 1 = 1.02.$$

Error Propagation

Physicists and engineers tend to make liberal use of the approximation of Δy by dy. One way this occurs in practice is in the estimation of errors propagated by physical measuring devices. For example, if you let x represent the measured value of a variable and let $x + \Delta x$ represent the exact value, then Δx is the *error in measurement*. Finally, if the measured value x is used to compute another value $f(x)$, the difference between $f(x + \Delta x)$ and $f(x)$ is the **propagated error.**

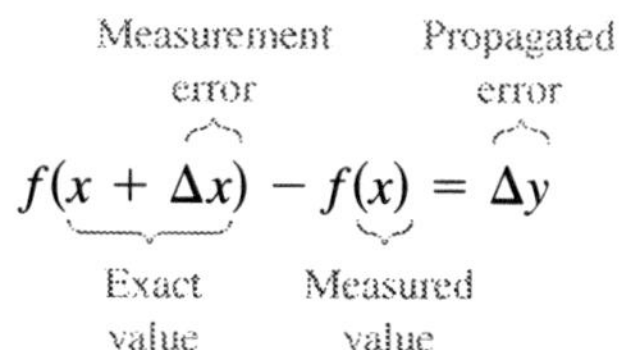

$$\underbrace{f(\overbrace{x + \Delta x}^{\text{Measurement error}}) - f(\overbrace{x}^{})}_{\text{Exact value}} = \underbrace{\Delta y}^{\text{Propagated error}}$$

EXAMPLE 3 Estimation of Error

The measured radius of a ball bearing is 0.7 inch, as shown in Figure 4.66. If the measurement is correct to within 0.01 inch, estimate the propagated error in the volume V of the ball bearing.

Solution The formula for the volume of a sphere is $V = \frac{4}{3}\pi r^3$, where r is the radius of the sphere. So, you can write

$$r = 0.7 \qquad \text{Measured radius}$$

and

$$-0.01 \le \Delta r \le 0.01. \qquad \text{Possible error}$$

To approximate the propagated error in the volume, differentiate V to obtain $dV/dr = 4\pi r^2$ and write

$$\begin{aligned}
\Delta V &\approx dV && \text{Approximate } \Delta V \text{ by } dV. \\
&= 4\pi r^2\, dr \\
&= 4\pi (0.7)^2 (\pm 0.01) && \text{Substitute for } r \text{ and } dr. \\
&\approx \pm 0.06158 \text{ cubic inch.}
\end{aligned}$$

So, the volume has a propagated error of about 0.06 cubic inch. ■

Would you say that the propagated error in Example 3 is large or small? The answer is best given in *relative* terms by comparing dV with V. The ratio

$$\begin{aligned}
\frac{dV}{V} &= \frac{4\pi r^2\, dr}{\frac{4}{3}\pi r^3} && \text{Ratio of } dV \text{ to } V \\
&= \frac{3\, dr}{r} && \text{Simplify.} \\
&\approx \frac{3}{0.7}(\pm 0.01) && \text{Substitute for } dr \text{ and } r. \\
&\approx \pm 0.0429
\end{aligned}$$

is called the **relative error.** The corresponding **percent error** is approximately 4.29%.

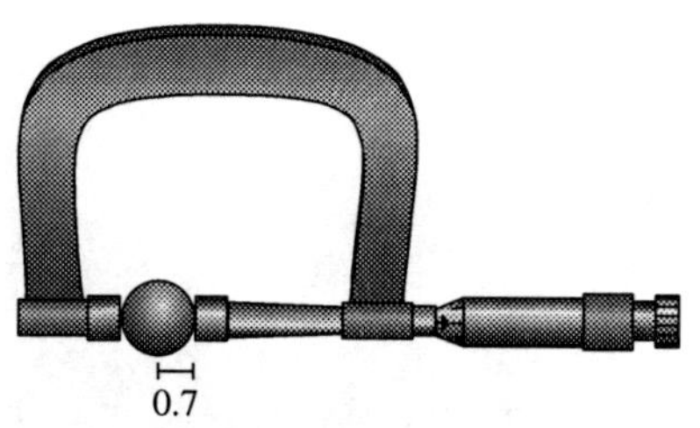

Ball bearing with measured radius that is correct to within 0.01 inch
Figure 4.66

Calculating Differentials

Each of the differentiation rules that you studied in Chapter 3 can be written in **differential form.** For example, suppose u and v are differentiable functions of x. By the definition of differentials, you have

$$du = u'\,dx \quad \text{and} \quad dv = v'\,dx.$$

So, you can write the differential form of the Product Rule as shown below.

$$d[uv] = \frac{d}{dx}[uv]\,dx \qquad \text{Differential of } uv$$
$$= [uv' + vu']\,dx \qquad \text{Product Rule}$$
$$= uv'\,dx + vu'\,dx$$
$$= u\,dv + v\,du$$

DIFFERENTIAL FORMULAS

Let u and v be differentiable functions of x.

Constant multiple: $\quad d[cu] = c\,du$

Sum or difference: $\quad d[u \pm v] = du \pm dv$

Product: $\quad d[uv] = u\,dv + v\,du$

Quotient: $\quad d\left[\dfrac{u}{v}\right] = \dfrac{v\,du - u\,dv}{v^2}$

EXAMPLE 4 Finding Differentials

Function	*Derivative*	*Differential*
a. $y = x^2$	$\dfrac{dy}{dx} = 2x$	$dy = 2x\,dx$
b. $y = 2\sin x$	$\dfrac{dy}{dx} = 2\cos x$	$dy = 2\cos x\,dx$
c. $y = xe^x$	$\dfrac{dy}{dx} = e^x(x + 1)$	$dy = e^x(x + 1)\,dx$
d. $y = \dfrac{1}{x}$	$\dfrac{dy}{dx} = -\dfrac{1}{x^2}$	$dy = -\dfrac{dx}{x^2}$

GOTTFRIED WILHELM LEIBNIZ (1646–1716)

Both Leibniz and Newton are credited with creating calculus. It was Leibniz, however, who tried to broaden calculus by developing rules and formal notation. He often spent days choosing an appropriate notation for a new concept.

The notation in Example 4 is called the **Leibniz notation** for derivatives and differentials, named after the German mathematician Gottfried Wilhelm Leibniz. The beauty of this notation is that it provides an easy way to remember several important calculus formulas by making it seem as though the formulas were derived from algebraic manipulations of differentials. For instance, in Leibniz notation, the *Chain Rule*

$$\frac{dy}{dx} = \frac{dy}{du}\frac{du}{dx}$$

would appear to be true because the du's divide out. Even though this reasoning is *incorrect*, the notation does help one remember the Chain Rule.

EXAMPLE 5   Finding the Differential of a Composite Function

$$y = f(x) = \sin 3x \qquad \text{Original function}$$
$$f'(x) = 3 \cos 3x \qquad \text{Apply Chain Rule.}$$
$$dy = f'(x)\, dx = 3 \cos 3x\, dx \qquad \text{Differential form}$$

EXAMPLE 6   Finding the Differential of a Composite Function

$$y = f(x) = (x^2 + 1)^{1/2} \qquad \text{Original function}$$
$$f'(x) = \frac{1}{2}(x^2 + 1)^{-1/2}(2x) = \frac{x}{\sqrt{x^2 + 1}} \qquad \text{Apply Chain Rule.}$$
$$dy = f'(x)\, dx = \frac{x}{\sqrt{x^2 + 1}}\, dx \qquad \text{Differential form}$$

Differentials can be used to approximate function values. To do this for the function given by $y = f(x)$, use the formula

$$f(x + \Delta x) \approx f(x) + dy = f(x) + f'(x)\, dx$$

which is derived from the approximation $\Delta y = f(x + \Delta x) - f(x) \approx dy$. The key to using this formula is to choose a value for x that makes the calculations easier, as shown in Example 7. (This formula is equivalent to the tangent line approximation given earlier in this section.)

EXAMPLE 7   Approximating Function Values

Use differentials to approximate $\sqrt{16.5}$.

Solution   Using $f(x) = \sqrt{x}$, you can write

$$f(x + \Delta x) \approx f(x) + f'(x)\, dx = \sqrt{x} + \frac{1}{2\sqrt{x}}\, dx.$$

Now, choosing $x = 16$ and $dx = 0.5$, you obtain the following approximation.

$$f(x + \Delta x) = \sqrt{16.5} \approx \sqrt{16} + \frac{1}{2\sqrt{16}}(0.5) = 4 + \left(\frac{1}{8}\right)\left(\frac{1}{2}\right) = 4.0625$$

The tangent line approximation to $f(x) = \sqrt{x}$ at $x = 16$ is the line $g(x) = \frac{1}{8}x + 2$. For x-values near 16, the graphs of f and g are close together, as shown in Figure 4.67. For instance,

$$f(16.5) = \sqrt{16.5} \approx 4.0620 \quad \text{and} \quad g(16.5) = \frac{1}{8}(16.5) + 2 = 4.0625.$$

In fact, if you use a graphing utility to zoom in near the point of tangency $(16, 4)$, you will see that the two graphs appear to coincide. Notice also that as you move farther away from the point of tangency, the linear approximation becomes less accurate.

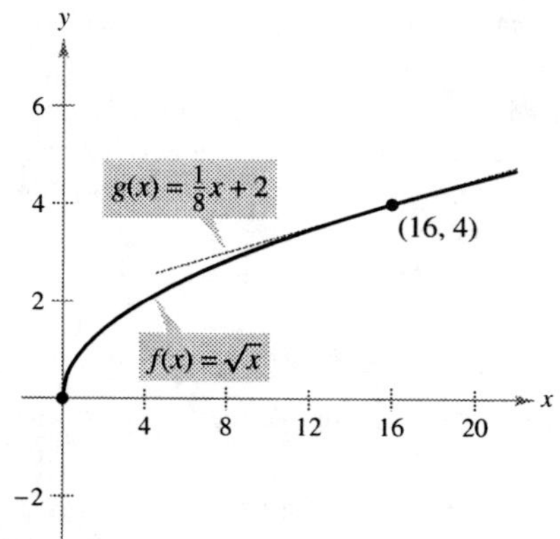

Figure 4.67

4.8 Exercises

See www.CalcChat.com for worked-out solutions to odd-numbered exercises.

In Exercises 1–6, find the equation of the tangent line T to the graph of f at the given point. Use this linear approximation to complete the table.

x	1.9	1.99	2	2.01	2.1
$f(x)$					
$T(x)$					

Function	Point
1. $f(x) = x^2$	$(2, 4)$
2. $f(x) = \dfrac{6}{x^2}$	$\left(2, \dfrac{3}{2}\right)$
3. $f(x) = x^5$	$(2, 32)$
4. $f(x) = \sqrt{x}$	$\left(2, \sqrt{2}\right)$
5. $f(x) = \sin x$	$(2, \sin 2)$
6. $f(x) = \log_2 x$	$(2, 1)$

In Exercises 7–10, use the information to evaluate and compare Δy and dy.

7. $y = x^3$ ⟶ $x = 1$ ⟶ $\Delta x = dx = 0.1$
8. $y = 1 - 2x^2$ ⟶ $x = 0$ ⟶ $\Delta x = dx = -0.1$
9. $y = x^4 + 1$ ⟶ $x = -1$ ⟶ $\Delta x = dx = 0.01$
10. $y = 2 - x^4$ ⟶ $x = 2$ ⟶ $\Delta x = dx = 0.01$

In Exercises 11–24, find the differential dy of the given function.

11. $y = 3x^2 - 4$
12. $y = 3x^{2/3}$
13. $y = \dfrac{x + 1}{2x - 1}$
14. $y = \sqrt{9 - x^2}$
15. $y = x\sqrt{1 - x^2}$
16. $y = \sqrt{x} + \dfrac{1}{\sqrt{x}}$
17. $y = \ln \sqrt{4 - x^2}$
18. $y = e^{-0.5x} \cos 4x$
19. $y = 3x - \sin^2 x$
20. $y = x \cos x$
21. $y = \dfrac{1}{3} \cos\left(\dfrac{6\pi x - 1}{2}\right)$
22. $y = \dfrac{\sec^2 x}{x^2 + 1}$
23. $y = x \arcsin x$
24. $y = \arctan(x - 2)$

In Exercises 25–28, use differentials and the graph of f to approximate (a) $f(1.9)$ and (b) $f(2.04)$. To print an enlarged copy of the graph, go to the website www.mathgraphs.com.

25.
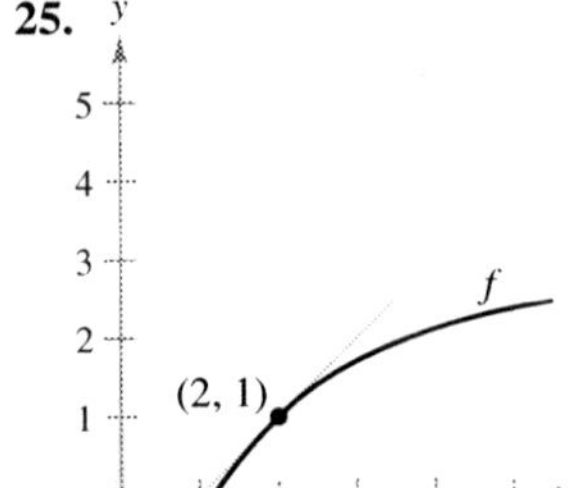

26.
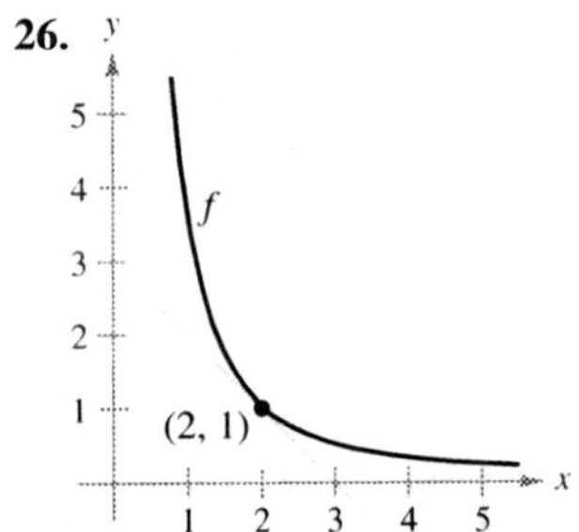

27.
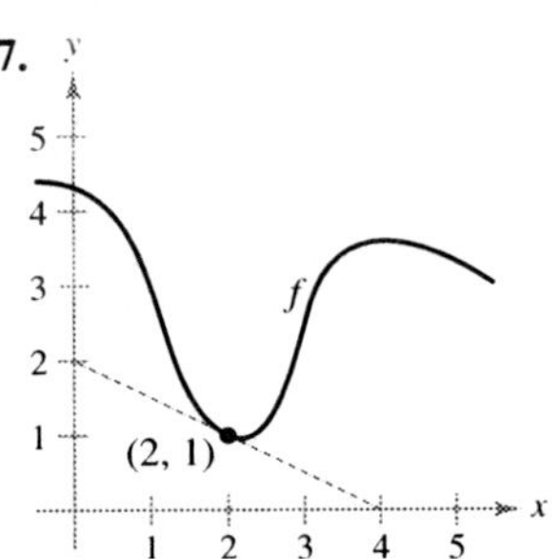

28.
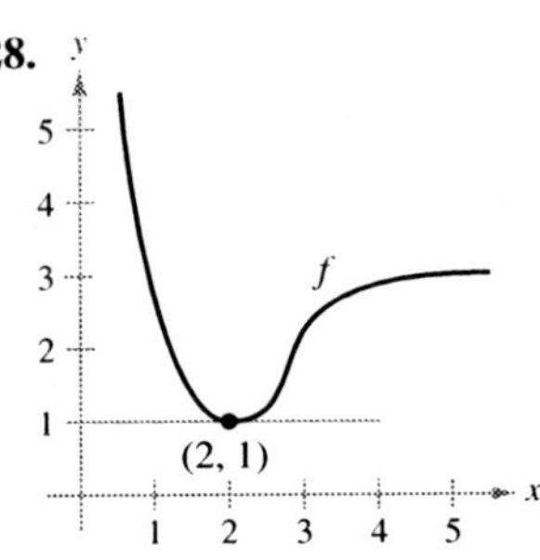

In Exercises 29 and 30, use differentials and the graph of g' to approximate (a) $g(2.93)$ and (b) $g(3.1)$ given that $g(3) = 8$.

29.
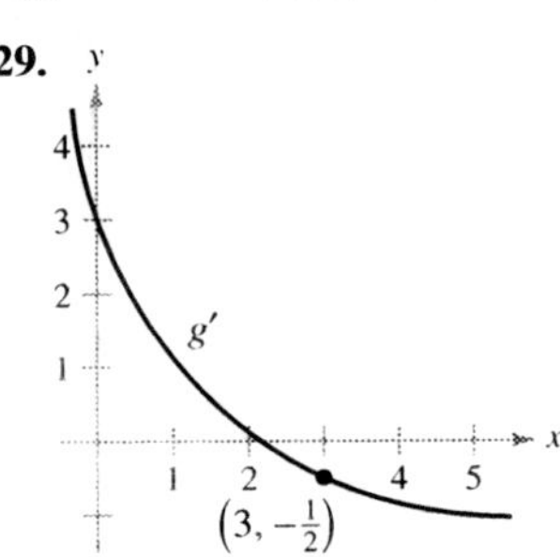

30.
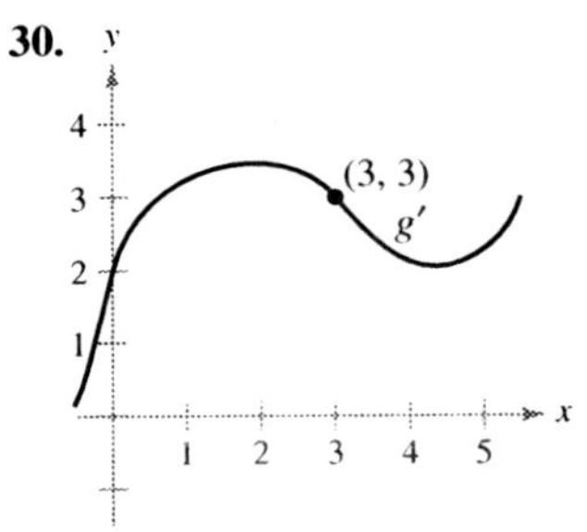

31. **Area** The measurement of the side of a square floor tile is 10 inches, with a possible error of $\frac{1}{32}$ inch. Use differentials to approximate the possible propagated error in computing the area of the square.

32. **Area** The measurements of the base and altitude of a triangle are found to be 36 and 50 centimeters, respectively. The possible error in each measurement is 0.25 centimeter. Use differentials to approximate the possible propagated error in computing the area of the triangle.

33. **Area** The measurement of the radius of the end of a log is found to be 16 inches, with a possible error of $\frac{1}{4}$ inch. Use differentials to approximate the possible propagated error in computing the area of the end of the log.

34. **Volume and Surface Area** The measurement of the edge of a cube is found to be 15 inches, with a possible error of 0.03 inch. Use differentials to approximate the maximum possible propagated error in computing (a) the volume of the cube and (b) the surface area of the cube.

35. **Area** The measurement of a side of a square is found to be 12 centimeters, with a possible error of 0.05 centimeter.

    (a) Approximate the percent error in computing the area of the square.

    (b) Estimate the maximum allowable percent error in measuring the side if the error in computing the area cannot exceed 2.5%.

36. **Circumference** The measurement of the circumference of a circle is found to be 64 centimeters, with a possible error of 0.9 centimeter.

    (a) Approximate the percent error in computing the area of the circle.

(b) Estimate the maximum allowable percent error in measuring the circumference if the error in computing the area cannot exceed 3%.

37. *Volume and Surface Area* The radius of a spherical balloon is measured as 8 inches, with a possible error of 0.02 inch. Use differentials to approximate the maximum possible error in calculating (a) the volume of the sphere, (b) the surface area of the sphere, and (c) the relative errors in parts (a) and (b).

38. *Stopping Distance* The total stopping distance T of a vehicle is

$$T = 2.5x + 0.5x^2$$

where T is in feet and x is the speed in miles per hour. Approximate the change and percent change in total stopping distance as speed increases from $x = 25$ to $x = 26$ miles per hour.

39. *Profit* The profit P for a company is $P = 100xe^{-x/400}$, where x is sales. Approximate the change and percent change in profit as sales increase from $x = 115$ to $x = 120$ units.

40. *Relative Humidity* When the dewpoint is 65° Fahrenheit, the relative humidity H is modeled by

$$H = \frac{4347}{400,000,000} e^{369,444/(50t + 19,793)}$$

where t is the air temperature in degrees Fahrenheit. Use differentials to approximate the change in relative humidity at $t = 72$ for a 1-degree change in the air temperature.

In Exercises 41 and 42, the thickness of the shell is 0.2 centimeter. Use differentials to approximate the volume of the shell.

41. A cylindrical shell with height 40 centimeters and radius 5 centimeters

42. A spherical shell of radius 100 centimeters

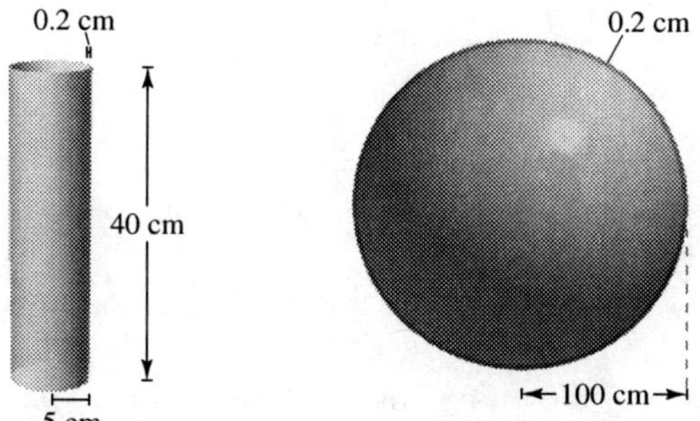

Figure for 41   **Figure for 42**

43. *Triangle Measurements* The measurement of one side of a right triangle is found to be 9.5 inches, and the angle opposite that side is 26°45′ with a possible error of 15′.

(a) Approximate the percent error in computing the length of the hypotenuse.

(b) Estimate the maximum allowable percent error in measuring the angle if the error in computing the length of the hypotenuse cannot exceed 2%.

44. *Ohm's Law* A current of I amperes passes through a resistor of R ohms. **Ohm's Law** states that the voltage E applied to the resistor is $E = IR$. If the voltage is constant, show that the magnitude of the relative error in R caused by a change in I is equal in magnitude to the relative error in I.

45. *Projectile Motion* The range R of a projectile is

$$R = \frac{v_0^2}{32}(\sin 2\theta)$$

where v_0 is the initial velocity in feet per second and θ is the angle of elevation. If $v_0 = 2500$ feet per second and θ is changed from 10° to 11°, use differentials to approximate the change in the range.

46. *Surveying* A surveyor standing 50 feet from the base of a large tree measures the angle of elevation to the top of the tree as 71.5°. How accurately must the angle be measured if the percent error in estimating the height of the tree is to be less than 6%?

In Exercises 47–50, use differentials to approximate the value of the expression. Compare your answer with that of a calculator.

47. $\sqrt{99.4}$ **48.** $\sqrt[3]{26}$ **49.** $\sqrt[4]{624}$ **50.** $(2.99)^3$

In Exercises 51 and 52, verify the tangent line approximation of the function at the given point. Then use a graphing utility to graph the function and its approximation in the same viewing window.

Function	Approximation	Point
51. $f(x) = \sqrt{x + 4}$	$y = 2 + \dfrac{x}{4}$	$(0, 2)$
52. $f(x) = \tan x$	$y = x$	$(0, 0)$

WRITING ABOUT CONCEPTS

53. Describe the change in the accuracy of dy as an approximation for Δy when Δx is decreased.

54. When using differentials, what is meant by the terms *propagated error*, *relative error*, and *percent error*?

55. Give a short explanation of why the approximation is valid.

(a) $\sqrt{4.02} \approx 2 + \frac{1}{4}(0.02)$

(b) $\tan 0.05 \approx 0 + 1(0.05)$

CAPSTONE

56. Would you use $y = x$ to approximate $f(x) = \sin x$ near $x = 0$? Why or why not?

True or False? **In Exercises 57–60, determine whether the statement is true or false. If it is false, explain why or give an example that shows it is false.**

57. If $y = x + c$, then $dy = dx$.

58. If $y = ax + b$, then $\Delta y/\Delta x = dy/dx$.

59. If y is differentiable, then $\lim\limits_{\Delta x \to 0} (\Delta y - dy) = 0$.

60. If $y = f(x)$, f is increasing and differentiable, and $\Delta x > 0$, then $\Delta y \geq dy$.

4   REVIEW EXERCISES

1. Give the definition of a critical number, and graph a function f showing the different types of critical numbers.

2. Consider the odd function f that is continuous and differentiable and has the functional values shown in the table.

x	-5	-4	-1	0	2	3	6
$f(x)$	1	3	2	0	-1	-4	0

   (a) Determine $f(4)$.

   (b) Determine $f(-3)$.

   (c) Plot the points and make a possible sketch of the graph of f on the interval $[-6, 6]$. What is the smallest number of critical points in the interval? Explain.

   (d) Does there exist at least one real number c in the interval $(-6, 6)$ where $f'(c) = -1$? Explain.

   (e) Is it possible that $\lim_{x \to 0} f(x)$ does not exist? Explain.

   (f) Is it necessary that $f'(x)$ exists at $x = 2$? Explain.

In Exercises 3–6, find the absolute extrema of the function on the closed interval. Use a graphing utility to graph the function over the given interval to confirm your results.

3. $f(x) = x^2 + 5x$, $[-4, 0]$

4. $h(x) = 3\sqrt{x} - x$, $[0, 9]$

5. $g(x) = 2x + 5 \cos x$, $[0, 2\pi]$

6. $f(x) = \dfrac{x}{\sqrt{x^2 + 1}}$, $[0, 2]$

In Exercises 7–10, determine whether Rolle's Theorem can be applied to f on the closed interval $[a, b]$. If Rolle's Theorem can be applied, find all values of c in the open interval (a, b) such that $f'(c) = 0$. If Rolle's Theorem cannot be applied, explain why not.

7. $f(x) = 2x^2 - 7$, $[0, 4]$

8. $f(x) = (x - 2)(x + 3)^2$, $[-3, 2]$

9. $f(x) = \dfrac{x^2}{1 - x^2}$, $[-2, 2]$

10. $f(x) = |x - 2| - 2$, $[0, 4]$

11. Consider the function $f(x) = 3 - |x - 4|$.

   (a) Graph the function and verify that $f(1) = f(7)$.

   (b) Note that $f'(x)$ is not equal to zero for any x in $[1, 7]$. Explain why this does not contradict Rolle's Theorem.

12. Can the Mean Value Theorem be applied to the function $f(x) = 1/x^2$ on the interval $[-2, 1]$? Explain.

In Exercises 13–18, determine whether the Mean Value Theorem can be applied to f on the closed interval $[a, b]$. If the Mean Value Theorem can be applied, find all values of c in the open interval (a, b) such that $f'(c) = \dfrac{f(b) - f(a)}{b - a}$. If the Mean Value Theorem cannot be applied, explain why not.

13. $f(x) = x^{2/3}$, $[1, 8]$

14. $f(x) = \dfrac{1}{x}$, $[1, 4]$

15. $f(x) = |5 - x|$, $[2, 6]$

16. $f(x) = 2x - 3\sqrt{x}$, $[-1, 1]$

17. $f(x) = x - \cos x$, $\left[-\dfrac{\pi}{2}, \dfrac{\pi}{2}\right]$

18. $f(x) = x \log_2 x$, $[1, 2]$

19. For the function $f(x) = Ax^2 + Bx + C$, determine the value of c guaranteed by the Mean Value Theorem on the interval $[x_1, x_2]$.

20. Demonstrate the result of Exercise 19 for $f(x) = 2x^2 - 3x + 1$ on the interval $[0, 4]$.

In Exercises 21–28, find the critical numbers (if any) and the open intervals on which the function is increasing or decreasing.

21. $f(x) = x^2 + 3x - 12$

22. $h(x) = (x + 2)^{1/3} + 8$

23. $f(x) = (x - 1)^2(x - 3)$

24. $g(x) = (x + 1)^3$

25. $h(x) = \sqrt{x}(x - 3)$, $x > 0$

26. $f(x) = \sin x + \cos x$, $[0, 2\pi]$

27. $f(t) = (2 - t)2^t$

28. $g(x) = 2x \ln x$

In Exercises 29–32, use the First Derivative Test to find any relative extrema of the function. Use a graphing utility to confirm your results.

29. $f(x) = 4x^3 - 5x$

30. $g(x) = \dfrac{x^3 - 8x}{4}$

31. $h(t) = \dfrac{1}{4}t^4 - 8t$

32. $g(x) = \dfrac{3}{2} \sin\left(\dfrac{\pi x}{2} - 1\right)$, $[0, 4]$

33. **Harmonic Motion** The height of an object attached to a spring is given by the harmonic equation

$$y = \tfrac{1}{3} \cos 12t - \tfrac{1}{4} \sin 12t$$

   where y is measured in inches and t is measured in seconds.

   (a) Calculate the height and velocity of the object when $t = \pi/8$ second.

   (b) Show that the maximum displacement of the object is $\dfrac{5}{12}$ inch.

   (c) Find the period P of y. Also, find the frequency f (number of oscillations per second) if $f = 1/P$.

34. **Writing** The general equation giving the height of an oscillating object attached to a spring is

$$y = A \sin \sqrt{\dfrac{k}{m}}t + B \cos \sqrt{\dfrac{k}{m}}t$$

   where k is the spring constant and m is the mass of the object.

   (a) Show that the maximum displacement of the object is $\sqrt{A^2 + B^2}$.

   (b) Show that the object oscillates with a frequency of

$$f = \dfrac{1}{2\pi} \sqrt{\dfrac{k}{m}}.$$

In Exercises 35–38, determine the points of inflection of the function and discuss the concavity of the graph of the function.

35. $f(x) = x^3 - 9x^2$

36. $g(x) = x\sqrt{x + 5}$

37. $f(x) = x + \cos x, \quad [0, 2\pi]$

38. $f(x) = (x + 2)^2(x - 4)$

In Exercises 39–42, use the Second Derivative Test to find all relative extrema.

39. $f(x) = (x + 9)^2$

40. $h(x) = x - 2\cos x, \quad [0, 4\pi]$

41. $g(x) = 2x^2(1 - x^2)$

42. $h(t) = t - 4\sqrt{t + 1}$

Think About It   **In Exercises 43 and 44, sketch the graph of a function f having the given characteristics.**

43. $f(0) = f(6) = 0$

   $f'(3) = f'(5) = 0$

   $f'(x) > 0$ if $x < 3$

   $f'(x) > 0$ if $3 < x < 5$

   $f'(x) < 0$ if $x > 5$

   $f''(x) < 0$ if $x < 3$ or $x > 4$

   $f''(x) > 0, \ 3 < x < 4$

44. $f(0) = 4, \ f(6) = 0$

   $f'(x) < 0$ if $x < 2$ or $x > 4$

   $f'(2)$ does not exist.

   $f'(4) = 0$

   $f'(x) > 0$ if $2 < x < 4$

   $f''(x) < 0, \ x \neq 2$

45. **Writing**   A newspaper headline states that "The rate of growth of the national deficit is decreasing." What does this mean? What does it imply about the graph of the deficit as a function of time?

46. **Inventory Cost**   The cost of inventory depends on the ordering and storage costs according to the inventory model

$$C = \left(\frac{Q}{x}\right)s + \left(\frac{x}{2}\right)r.$$

Determine the order size that will minimize the cost, assuming that sales occur at a constant rate, Q is the number of units sold per year, r is the cost of storing one unit for 1 year, s is the cost of placing an order, and x is the number of units per order.

47. **Modeling Data**   Outlays for national defense D (in billions of dollars) for selected years from 1970 through 2005 are shown in the table, where t is time in years, with $t = 0$ corresponding to 1970. (*Source: U.S. Office of Management and Budget*)

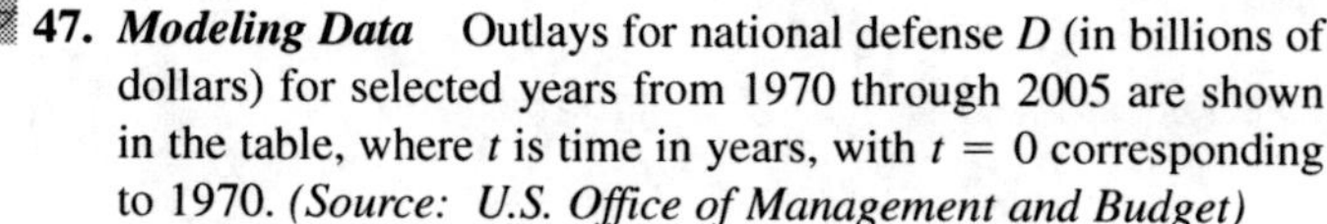

t	0	5	10	15	20
D	81.7	86.5	134.0	252.7	299.3

t	25	30	35
D	272.1	294.5	495.3

(a) Use the regression capabilities of a graphing utility to fit a model of the form $D = at^4 + bt^3 + ct^2 + dt + e$ to the data.

(b) Use a graphing utility to plot the data and graph the model.

(c) For the years shown in the table, when does the model indicate that the outlay for national defense was at a maximum? When was it at a minimum?

(d) For the years shown in the table, when does the model indicate that the outlay for national defense was increasing at the greatest rate?

48. **Climb Rate**   The time t (in minutes) for a small plane to climb to an altitude of h feet is

$$t = 50 \log_{10} \frac{18{,}000}{18{,}000 - h}$$

where 18,000 feet is the plane's absolute ceiling.

(a) Determine the domain of the function appropriate for the context of the problem.

(b) Use a graphing utility to graph the time function and identify any asymptotes.

(c) Find the time when the altitude is increasing at the greatest rate.

In Exercises 49–58, find the limit.

49. $\displaystyle \lim_{x \to \infty} \left(8 + \frac{1}{x}\right)$

50. $\displaystyle \lim_{x \to \infty} \frac{3 - x}{2x + 5}$

51. $\displaystyle \lim_{x \to \infty} \frac{2x^2}{3x^2 + 5}$

52. $\displaystyle \lim_{x \to \infty} \frac{2x}{3x^2 + 5}$

53. $\displaystyle \lim_{x \to -\infty} \frac{3x^2}{x + 5}$

54. $\displaystyle \lim_{x \to -\infty} \frac{\sqrt{x^2 + x}}{-2x}$

55. $\displaystyle \lim_{x \to \infty} \frac{5\cos x}{x}$

56. $\displaystyle \lim_{x \to \infty} \frac{3x}{\sqrt{x^2 + 4}}$

57. $\displaystyle \lim_{x \to -\infty} \frac{6x}{x + \cos x}$

58. $\displaystyle \lim_{x \to -\infty} \frac{x}{2\sin x}$

In Exercises 59–66, find any vertical and horizontal asymptotes of the graph of the function. Use a graphing utility to verify your results.

59. $f(x) = \dfrac{3}{x} - 2$

60. $g(x) = \dfrac{5x^2}{x^2 + 2}$

61. $h(x) = \dfrac{2x + 3}{x - 4}$

62. $f(x) = \dfrac{3x}{\sqrt{x^2 + 2}}$

63. $f(x) = \dfrac{5}{3 + 2e^{-x}}$

64. $g(x) = 30xe^{-2x}$

65. $g(x) = 3\ln(1 + e^{-x/4})$

66. $h(x) = 10\ln\left(\dfrac{x}{x + 1}\right)$

In Exercises 67–70, use a graphing utility to graph the function. Use the graph to approximate any relative extrema or asymptotes.

67. $f(x) = x^3 + \dfrac{243}{x}$

68. $f(x) = |x^3 - 3x^2 + 2x|$

69. $f(x) = \dfrac{x - 1}{1 + 3x^2}$

70. $g(x) = \dfrac{\pi^2}{3} - 4\cos x + \cos 2x$

In Exercises 71–96, analyze and sketch the graph of the function.

71. $f(x) = 4x - x^2$

72. $f(x) = 4x^3 - x^4$

73. $f(x) = x\sqrt{16 - x^2}$

74. $f(x) = (x^2 - 4)^2$

75. $f(x) = (x - 1)^3(x - 3)^2$

76. $f(x) = (x - 3)(x + 2)^3$

77. $f(x) = x^{1/3}(x + 3)^{2/3}$

78. $f(x) = (x - 2)^{1/3}(x + 1)^{2/3}$

79. $f(x) = \dfrac{5 - 3x}{x - 2}$

80. $f(x) = \dfrac{2x}{1 + x^2}$

81. $f(x) = \dfrac{7}{1 + x^2}$

82. $f(x) = \dfrac{x^2}{1 + x^4}$

83. $f(x) = x^3 + x + \dfrac{2}{x}$

84. $f(x) = x^2 + \dfrac{8}{x}$

85. $f(x) = |x^2 - 9|$

86. $f(x) = |x - 1| + |x - 3|$

87. $h(x) = (1 - x)e^x$

88. $g(x) = 5xe^{-x^2}$

89. $g(x) = (x + 3)\ln(x + 3)$

90. $h(t) = \dfrac{\ln t}{t^2}$

91. $f(x) = \dfrac{10 \log_4 x}{x}$

92. $g(x) = 100x(3^{-x})$

93. $f(x) = x + \cos x, \quad 0 \le x \le 2\pi$

94. $f(x) = \dfrac{1}{\pi}(2 \sin \pi x - \sin 2\pi x), \quad -1 \le x \le 1$

95. $y = 4x - 6 \arctan x$

96. $y = \dfrac{1}{2}x^2 - \arcsin \dfrac{x}{2}$

97. Find the maximum and minimum points on the graph of

$$x^2 + 4y^2 - 2x - 16y + 13 = 0$$

  (a) without using calculus.

  (b) using calculus.

98. Consider the function $f(x) = x^n$ for positive integer values of n.

  (a) For what values of n does the function have a relative minimum at the origin?

  (b) For what values of n does the function have a point of inflection at the origin?

99. *Minimum Distance*   At noon, ship A is 100 kilometers due east of ship B. Ship A is sailing west at 12 kilometers per hour, and ship B is sailing south at 10 kilometers per hour. At what time will the ships be nearest to each other, and what will this distance be?

100. *Maximum Area*   Find the dimensions of the rectangle of maximum area, with sides parallel to the coordinate axes, that can be inscribed in the ellipse given by

$$\frac{x^2}{144} + \frac{y^2}{16} = 1.$$

101. *Minimum Length*   A right triangle in the first quadrant has the coordinate axes as sides, and the hypotenuse passes through the point $(1, 8)$. Find the vertices of the triangle such that the length of the hypotenuse is minimum.

102. *Minimum Length*   The wall of a building is to be braced by a beam that must pass over a parallel fence 5 feet high and 4 feet from the building. Find the length of the shortest beam that can be used.

103. *Maximum Area*   Three sides of a trapezoid have the same length s. Of all such possible trapezoids, show that the one of maximum area has a fourth side of length $2s$.

104. *Maximum Area*   Show that the greatest area of any rectangle inscribed in a triangle is one-half the area of the triangle.

105. *Maximum Length*   Find the length of the longest pipe that can be carried level around a right-angle corner at the intersection of two corridors of widths 4 feet and 6 feet. (Do not use trigonometry.)

106. *Maximum Length*   Rework Exercise 105, given corridors of widths a meters and b meters.

107. *Maximum Length*   A hallway of width 6 feet meets a hallway of width 9 feet at right angles. Find the length of the longest pipe that can be carried level around this corner. [*Hint:* If L is the length of the pipe, show that

$$L = 6 \csc \theta + 9 \csc\left(\frac{\pi}{2} - \theta\right)$$

where θ is the angle between the pipe and the wall of the narrower hallway.]

108. *Maximum Length*   Rework Exercise 107, given that one hallway is of width a meters and the other is of width b meters. Show that the result is the same as in Exercise 106.

Minimum Cost   **In Exercises 109 and 110, find the speed v (in miles per hour) that will minimize costs on a 110-mile delivery trip. The cost per hour for fuel is C dollars, and the driver is paid W dollars per hour. (Assume there are no costs other than wages and fuel.)**

109. Fuel cost: $C = \dfrac{v^2}{600}$

  Driver: $W = \$5$

110. Fuel cost: $C = \dfrac{v^2}{500}$

  Driver: $W = \$7.50$

In Exercises 111 and 112, find the differential dy.

111. $y = x(1 - \cos x)$

112. $y = \sqrt{36 - x^2}$

113. *Surface Area and Volume*   The diameter of a sphere is measured as 18 centimeters, with a maximum possible error of 0.05 centimeter. Use differentials to approximate the possible propagated error and percent error in calculating the surface area and the volume of the sphere.

114. *Demand Function*   A company finds that the demand for its commodity is $p = 75 - \frac{1}{4}x$. If x changes from 7 to 8, find and compare the values of Δp and dp.

P.S. PROBLEM SOLVING

1. Graph the fourth-degree polynomial $p(x) = x^4 + ax^2 + 1$ for various values of the constant a.

(a) Determine the values of a for which p has exactly one relative minimum.

(b) Determine the values of a for which p has exactly one relative maximum.

(c) Determine the values of a for which p has exactly two relative minima.

(d) Show that the graph of p cannot have exactly two relative extrema.

2. (a) Graph the fourth-degree polynomial $p(x) = ax^4 - 6x^2$ for $a = -3, -2, -1, 0, 1, 2,$ and 3. For what values of the constant a does p have a relative minimum or relative maximum?

(b) Show that p has a relative maximum for all values of the constant a.

(c) Determine analytically the values of a for which p has a relative minimum.

(d) Let $(x, y) = (x, p(x))$ be a relative extremum of p. Show that (x, y) lies on the graph of $y = -3x^2$. Verify this result graphically by graphing $y = -3x^2$ together with the seven curves from part (a).

3. Let $f(x) = \dfrac{c}{x} + x^2$. Determine all values of the constant c such that f has a relative minimum, but no relative maximum.

4. (a) Let $f(x) = ax^2 + bx + c$, $a \neq 0$, be a quadratic polynomial. How many points of inflection does the graph of f have?

(b) Let $f(x) = ax^3 + bx^2 + cx + d$, $a \neq 0$, be a cubic polynomial. How many points of inflection does the graph of f have?

(c) Suppose the function $y = f(x)$ satisfies the equation $\dfrac{dy}{dx} = ky\left(1 - \dfrac{y}{L}\right)$, where k and L are positive constants. Show that the graph of f has a point of inflection at the point where $y = \dfrac{L}{2}$. (This equation is called the **logistic differential equation**.)

5. Prove Darboux's Theorem: Let f be differentiable on the closed interval $[a, b]$ such that $f'(a) = y_1$ and $f'(b) = y_2$. If d lies between y_1 and y_2, then there exists c in (a, b) such that $f'(c) = d$.

6. Let f and g be functions that are continuous on $[a, b]$ and differentiable on (a, b). Prove that if $f(a) = g(a)$ and $g'(x) > f'(x)$ for all x in (a, b), then $g(b) > f(b)$.

7. Prove the following **Extended Mean Value Theorem.** If f and f' are continuous on the closed interval $[a, b]$, and if f'' exists on the open interval (a, b), then there exists a number c in (a, b) such that

$$f(b) = f(a) + f'(a)(b - a) + \frac{1}{2}f''(c)(b - a)^2.$$

8. (a) Let $V = x^3$. Find dV and ΔV. Show that for small values of x, the difference $\Delta V - dV$ is very small in the sense that there exists ε such that $\Delta V - dV = \varepsilon \Delta x$, where $\varepsilon \to 0$ as $\Delta x \to 0$.

(b) Generalize this result by showing that if $y = f(x)$ is a differentiable function, then $\Delta y - dy = \varepsilon \Delta x$, where $\varepsilon \to 0$ as $\Delta x \to 0$.

9. The amount of illumination of a surface is proportional to the intensity of the light source, inversely proportional to the square of the distance from the light source, and proportional to $\sin \theta$, where θ is the angle at which the light strikes the surface. A rectangular room measures 10 feet by 24 feet, with a 10-foot ceiling. Determine the height at which the light should be placed to allow the corners of the floor to receive as much light as possible.

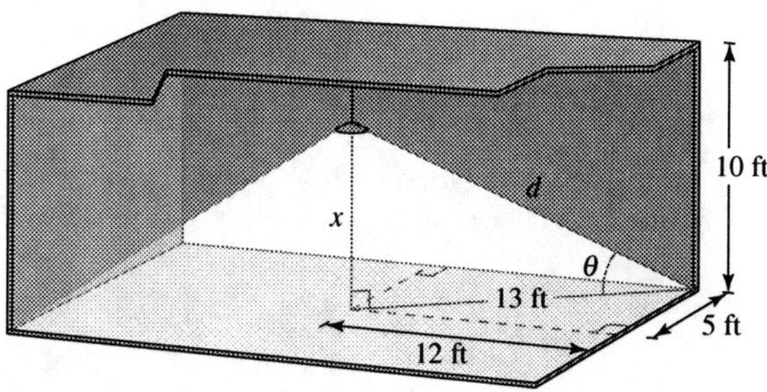

10. Consider a room in the shape of a cube, 4 meters on each side. A bug at point P wants to walk to point Q at the opposite corner, as shown in the figure. Use calculus to determine the shortest path. Can you solve the problem without calculus?

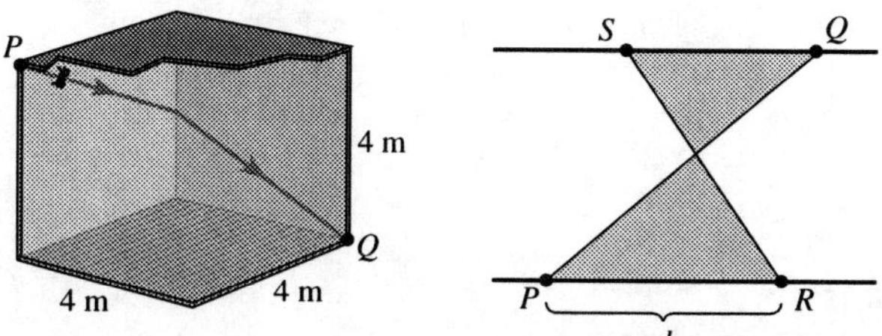

Figure for 10   **Figure for 11**

11. The line joining P and Q crosses two parallel lines, as shown in the figure. The point R is d units from P. How far from Q should the point S be positioned so that the sum of the areas of the two shaded triangles is a minimum? So that the sum is a maximum?

12. (a) Let x be a positive number. Use the *table* feature of a graphing utility to verify that $\sin x < x$.

(b) Use the Mean Value Theorem to prove that $\sin x < x$ for all positive real numbers x.

13. (a) Let x be a positive number. Use the *table* feature of a graphing utility to verify that $\sqrt{1 + x} < \frac{1}{2}x + 1$.

(b) Use the Mean Value Theorem to prove that $\sqrt{1 + x} < \frac{1}{2}x + 1$ for all positive real numbers x.

14. The figures show a rectangle, a circle, and a semicircle inscribed in a triangle bounded by the coordinate axes and the first-quadrant portion of the line with intercepts $(3, 0)$ and $(0, 4)$. Find the dimensions of each inscribed figure such that its area is maximum. State whether calculus was helpful in finding the required dimensions. Explain your reasoning.

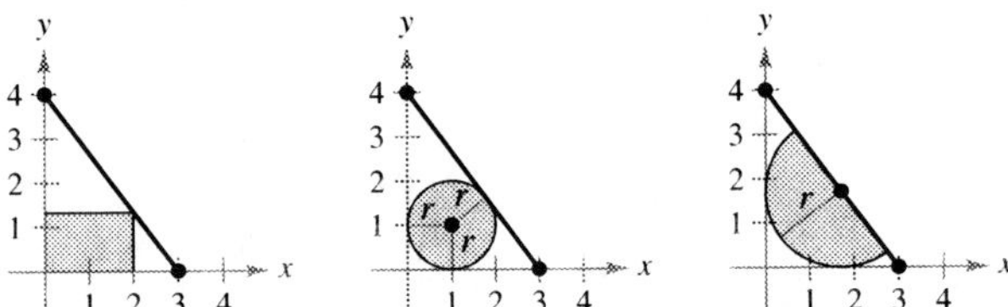

15. (a) Prove that $\lim\limits_{x \to \infty} x^2 = \infty$.

    (b) Prove that $\lim\limits_{x \to \infty} \left(\dfrac{1}{x^2} \right) = 0$.

    (c) Let L be a real number. Prove that if $\lim\limits_{x \to \infty} f(x) = L$, then

$$\lim_{y \to 0^+} f\left(\frac{1}{y} \right) = L.$$

16. Find the point on the graph of $y = \dfrac{1}{1 + x^2}$ (see figure) where the tangent line has the greatest slope, and the point where the tangent line has the least slope.

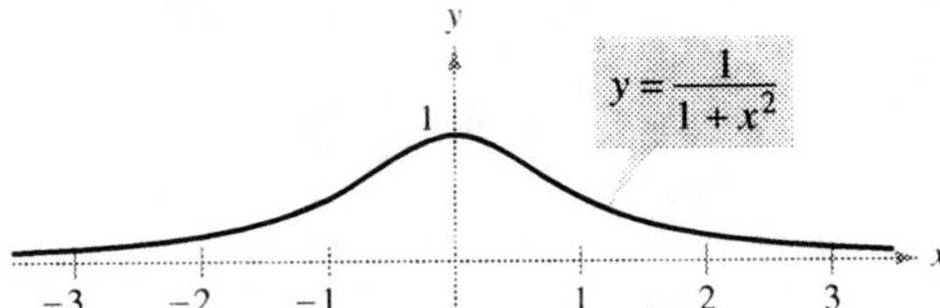

17. The police department must determine the speed limit on a bridge such that the flow rate of cars is maximized per unit time. The greater the speed limit, the farther apart the cars must be in order to keep a safe stopping distance. Experimental data on the stopping distances d (in meters) for various speeds v (in kilometers per hour) are shown in the table.

v	20	40	60	80	100
d	5.1	13.7	27.2	44.2	66.4

    (a) Convert the speeds v in the table to speeds s in meters per second. Use the regression capabilities of a graphing utility to find a model of the form $d(s) = as^2 + bs + c$ for the data.

    (b) Consider two consecutive vehicles of average length 5.5 meters, traveling at a safe speed on the bridge. Let T be the difference between the times (in seconds) when the front bumpers of the two vehicles pass a given point on the bridge. Verify that this difference in times is given by

$$T = \frac{d(s)}{s} + \frac{5.5}{s}.$$

    (c) Use a graphing utility to graph the function T and estimate the speed s that minimizes the time between vehicles.

    (d) Use calculus to determine the speed that minimizes T. What is the minimum value of T? Convert the required speed to kilometers per hour.

    (e) Find the optimal distance between vehicles for the posted speed limit determined in part (d).

18. A legal-sized sheet of paper (8.5 inches by 14 inches) is folded so that corner P touches the opposite 14-inch edge at R (see figure). $\left(\textit{Note: } PQ = \sqrt{C^2 - x^2}. \right)$

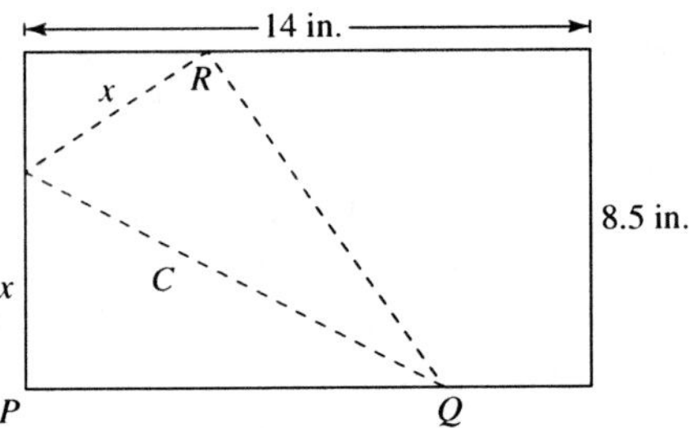

    (a) Show that $C^2 = \dfrac{2x^3}{2x - 8.5}$.

    (b) What is the domain of C?

    (c) Determine the x-value that minimizes C.

    (d) Determine the minimum length of C.

19. Let $f(x) = \sin(\ln x)$.

    (a) Determine the domain of the function f.

    (b) Find two values of x satisfying $f(x) = 1$.

    (c) Find two values of x satisfying $f(x) = -1$.

    (d) What is the range of the function f?

    (e) Calculate $f'(x)$ and use calculus to find the maximum value of f on the interval $[1, 10]$.

    (f) Use a graphing utility to graph f in the viewing window $[0, 5] \times [-2, 2]$ and estimate $\lim\limits_{x \to 0^+} f(x)$, if it exists.

    (g) Determine $\lim\limits_{x \to 0^+} f(x)$ analytically, if it exists.

20. Show that the cubic polynomial $p(x) = ax^3 + bx^2 + cx + d$ has exactly one point of inflection (x_0, y_0), where

$$x_0 = \frac{-b}{3a} \quad \text{and} \quad y_0 = \frac{2b^3}{27a^2} - \frac{bc}{3a} + d.$$

Use this formula to find the point of inflection of

$$p(x) = x^3 - 3x^2 + 2.$$

5.1   Antiderivatives and Indefinite Integration

- Write the general solution of a differential equation.
- Use indefinite integral notation for antiderivatives.
- Use basic integration rules to find antiderivatives.
- Find a particular solution of a differential equation.

Antiderivatives

Suppose you were asked to find a function F whose derivative is $f(x) = 3x^2$. From your knowledge of derivatives, you would probably say that

$$F(x) = x^3 \text{ because } \frac{d}{dx}[x^3] = 3x^2.$$

The function F is an *antiderivative* of f.

DEFINITION OF ANTIDERIVATIVE

A function F is an **antiderivative** of f on an interval I if $F'(x) = f(x)$ for all x in I.

Note that F is called *an* antiderivative of f, rather than *the* antiderivative of f. To see why, observe that

$$F_1(x) = x^3, \quad F_2(x) = x^3 - 5, \quad \text{and} \quad F_3(x) = x^3 + 97$$

are all antiderivatives of $f(x) = 3x^2$. In fact, for any constant C, the function given by $F(x) = x^3 + C$ is an antiderivative of f.

THEOREM 5.1   REPRESENTATION OF ANTIDERIVATIVES

If F is an antiderivative of f on an interval I, then G is an antiderivative of f on the interval I if and only if G is of the form $G(x) = F(x) + C$, for all x in I, where C is a constant.

(PROOF) The proof of Theorem 5.1 in one direction is straightforward. That is, if $G(x) = F(x) + C$, $F'(x) = f(x)$, and C is a constant, then

$$G'(x) = \frac{d}{dx}[F(x) + C] = F'(x) + 0 = f(x).$$

To prove this theorem in the other direction, assume that G is an antiderivative of f. Define a function H such that

$$H(x) = G(x) - F(x).$$

For any two points a and b $(a < b)$ in the interval, H is continuous on $[a, b]$ and differentiable on (a, b). By the Mean Value Theorem,

$$H'(c) = \frac{H(b) - H(a)}{b - a}$$

for some c in (a, b). However, $H'(c) = 0$, so $H(a) = H(b)$. Because a and b are arbitrary points in the interval, you know that H is a constant function C. So, $G(x) - F(x) = C$ and it follows that $G(x) = F(x) + C$.   ∎

Using Theorem 5.1, you can represent the entire family of antiderivatives of a function by adding a constant to a *known* antiderivative. For example, knowing that $D_x[x^2] = 2x$, you can represent the family of *all* antiderivatives of $f(x) = 2x$ by

$$G(x) = x^2 + C \qquad \text{Family of all antiderivatives of } f(x) = 2x$$

where C is a constant. The constant C is called the **constant of integration.** The family of functions represented by G is the **general antiderivative** of f, and $G(x) = x^2 + C$ is the **general solution** of the *differential equation*

$$G'(x) = 2x. \qquad \text{Differential equation}$$

A **differential equation** in x and y is an equation that involves x, y, and derivatives of y. For instance, $y' = 3x$ and $y' = x^2 + 1$ are examples of differential equations.

EXAMPLE 1 Solving a Differential Equation

Find the general solution of the differential equation $y' = 2$.

Solution   To begin, you need to find a function whose derivative is 2. One such function is

$$y = 2x. \qquad 2x \text{ is } an \text{ antiderivative of } 2.$$

Now, you can use Theorem 5.1 to conclude that the general solution of the differential equation is

$$y = 2x + C. \qquad \text{General solution}$$

The graphs of several functions of the form $y = 2x + C$ are shown in Figure 5.1.

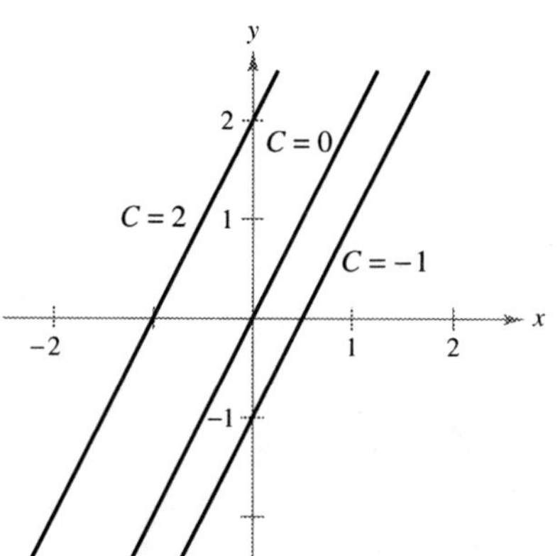

Functions of the form $y = 2x + C$
Figure 5.1

Notation for Antiderivatives

When solving a differential equation of the form

$$\frac{dy}{dx} = f(x)$$

it is convenient to write it in the equivalent differential form

$$dy = f(x)\, dx.$$

The operation of finding all solutions of this equation is called **antidifferentiation** (or **indefinite integration**) and is denoted by an integral sign $\int$. The general solution is denoted by

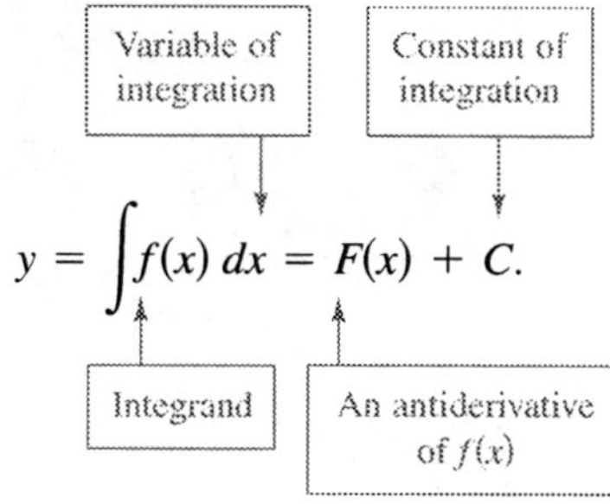

$$y = \int f(x)\, dx = F(x) + C.$$

The expression $\int f(x)\,dx$ is read as the *antiderivative of f with respect to x.* So, the differential dx serves to identify x as the variable of integration. The term **indefinite integral** is a synonym for antiderivative.

Basic Integration Rules

The inverse nature of integration and differentiation can be verified by substituting $F'(x)$ for $f(x)$ in the definition of indefinite integration to obtain

$$\int F'(x)\, dx = F(x) + C.$$

Integration is the "inverse" of differentiation.

Moreover, if $\int f(x)\, dx = F(x) + C$, then

$$\frac{d}{dx}\left[\int f(x)\, dx\right] = f(x).$$

Differentiation is the "inverse" of integration.

NOTE The Power Rule for Integration has the restriction that $n \neq -1$. To evaluate $\int x^{-1}\, dx$, you must use the natural log rule. (See Exercise 106.)

These two equations allow you to obtain integration formulas directly from differentiation formulas, as shown in the following summary.

BASIC INTEGRATION RULES

Differentiation Formula	*Integration Formula*
$\dfrac{d}{dx}[C] = 0$	$\displaystyle\int 0\, dx = C$
$\dfrac{d}{dx}[kx] = k$	$\displaystyle\int k\, dx = kx + C$
$\dfrac{d}{dx}[kf(x)] = kf'(x)$	$\displaystyle\int kf(x)\, dx = k\int f(x)\, dx$
$\dfrac{d}{dx}[f(x) \pm g(x)] = f'(x) \pm g'(x)$	$\displaystyle\int [f(x) \pm g(x)]\, dx = \int f(x)\, dx \pm \int g(x)\, dx$
$\dfrac{d}{dx}[x^n] = nx^{n-1}$	$\displaystyle\int x^n\, dx = \frac{x^{n+1}}{n+1} + C, \quad n \neq -1$ Power Rule
$\dfrac{d}{dx}[\sin x] = \cos x$	$\displaystyle\int \cos x\, dx = \sin x + C$
$\dfrac{d}{dx}[\cos x] = -\sin x$	$\displaystyle\int \sin x\, dx = -\cos x + C$
$\dfrac{d}{dx}[\tan x] = \sec^2 x$	$\displaystyle\int \sec^2 x\, dx = \tan x + C$
$\dfrac{d}{dx}[\sec x] = \sec x \tan x$	$\displaystyle\int \sec x \tan x\, dx = \sec x + C$
$\dfrac{d}{dx}[\cot x] = -\csc^2 x$	$\displaystyle\int \csc^2 x\, dx = -\cot x + C$
$\dfrac{d}{dx}[\csc x] = -\csc x \cot x$	$\displaystyle\int \csc x \cot x\, dx = -\csc x + C$
$\dfrac{d}{dx}[e^x] = e^x$	$\displaystyle\int e^x\, dx = e^x + C$
$\dfrac{d}{dx}[a^x] = (\ln a)a^x$	$\displaystyle\int a^x\, dx = \left(\frac{1}{\ln a}\right)a^x + C$
$\dfrac{d}{dx}[\ln x] = \dfrac{1}{x}, \ x > 0$	$\displaystyle\int \frac{1}{x}\, dx = \ln|x| + C$

EXAMPLE 2 Applying the Basic Integration Rules

Describe the antiderivatives of $3x$.

Solution

$$\int 3x \, dx = 3 \int x \, dx \qquad \text{Constant Multiple Rule}$$

$$= 3 \int x^1 \, dx \qquad \text{Rewrite } x \text{ as } x^1.$$

$$= 3 \left(\frac{x^2}{2} \right) + C \qquad \text{Power Rule } (n = 1)$$

$$= \frac{3}{2} x^2 + C \qquad \text{Simplify.}$$

When indefinite integrals are evaluated, a strict application of the basic integration rules tends to produce complicated constants of integration. For instance, in Example 2, you could have written

$$\int 3x \, dx = 3 \int x \, dx = 3 \left(\frac{x^2}{2} + C \right) = \frac{3}{2} x^2 + 3C.$$

However, because C represents *any* constant, it is both cumbersome and unnecessary to write $3C$ as the constant of integration. So, $\frac{3}{2}x^2 + 3C$ is written in the simpler form $\frac{3}{2}x^2 + C$.

In Example 2, note that the general pattern of integration is similar to that of differentiation.

Original integral $\Rightarrow$ Rewrite $\Rightarrow$ Integrate $\Rightarrow$ Simplify

EXAMPLE 3 Rewriting Before Integrating

TECHNOLOGY Some software programs, such as *Maple*, *Mathematica*, and the *TI-89*, are capable of performing integration symbolically. If you have access to such a symbolic integration utility, try using it to evaluate the indefinite integrals in Example 3.

NOTE The properties of logarithms presented on page 53 can be used to rewrite antiderivatives in different forms. For instance, the antiderivative in Example 3(d) can be rewritten as

$$3 \ln |x| + C = \ln|x|^3 + C.$$

	Original Integral	Rewrite	Integrate	Simplify
a.	$\int \dfrac{1}{x^3} \, dx$	$\int x^{-3} \, dx$	$\dfrac{x^{-2}}{-2} + C$	$-\dfrac{1}{2x^2} + C$
b.	$\int \sqrt{x} \, dx$	$\int x^{1/2} \, dx$	$\dfrac{x^{3/2}}{3/2} + C$	$\dfrac{2}{3} x^{3/2} + C$
c.	$\int 2 \sin x \, dx$	$2 \int \sin x \, dx$	$2(-\cos x) + C$	$-2 \cos x + C$
d.	$\int \dfrac{3}{x} \, dx$	$3 \int \dfrac{1}{x} \, dx$	$3(\ln|x|) + C$	$3 \ln|x| + C$

Remember that you can check your answer to an antidifferentiation problem by differentiating. For instance, in Example 3(b), you can check that $\frac{2}{3}x^{3/2} + C$ is the correct antiderivative by differentiating the answer to obtain

$$D_x \left[\frac{2}{3} x^{3/2} + C \right] = \left(\frac{2}{3} \right) \left(\frac{3}{2} \right) x^{1/2} = \sqrt{x}. \qquad \text{Use differentiation to check antiderivative.}$$

The icon ⟳ indicates that you will find a CAS Investigation on the book's website. The CAS Investigation is a collaborative exploration of this example using the computer algebra systems Maple *and* Mathematica.

The basic integration rules listed earlier in this section allow you to integrate any polynomial function, as shown in Example 4.

EXAMPLE 4 Integrating Polynomial Functions

a. $\displaystyle \int dx = \int 1 \, dx$ — *Integrand is understood to be 1.*

$\qquad = x + C$ — *Integrate.*

b. $\displaystyle \int (x + 2) \, dx = \int x \, dx \; + \int 2 \, dx$

$\qquad\qquad\qquad = \dfrac{x^2}{2} + C_1 + 2x + C_2$ — *Integrate.*

$\qquad\qquad\qquad = \dfrac{x^2}{2} + 2x + C$ — $C = C_1 + C_2$

The second line in the solution is usually omitted.

c. $\displaystyle \int (3x^4 - 5x^2 + x) \, dx = 3\left(\dfrac{x^5}{5}\right) - 5\left(\dfrac{x^3}{3}\right) + \dfrac{x^2}{2} + C$ — *Integrate.*

$\qquad\qquad\qquad\qquad\qquad = \dfrac{3}{5}x^5 - \dfrac{5}{3}x^3 + \dfrac{1}{2}x^2 + C$ — *Simplify.*

EXAMPLE 5 Rewriting Before Integrating

$\displaystyle \int \dfrac{x + 1}{\sqrt{x}} \, dx = \int \left(\dfrac{x}{\sqrt{x}} + \dfrac{1}{\sqrt{x}}\right) dx$ — *Rewrite as two fractions*

$\qquad\qquad\quad = \int (x^{1/2} + x^{-1/2}) \, dx$ — *Rewrite with fractional exponents*

$\qquad\qquad\quad = \dfrac{x^{3/2}}{3/2} + \dfrac{x^{1/2}}{1/2} + C$ — *Integrate.*

$\qquad\qquad\quad = \dfrac{2}{3}x^{3/2} + 2x^{1/2} + C$ — *Simplify.*

$\qquad\qquad\quad = \dfrac{2}{3}\sqrt{x}(x + 3) + C$ — *Factor.*

STUDY TIP Remember that you can check your answer by differentiating.

NOTE When integrating quotients, do not integrate the numerator and denominator separately. This is no more valid in integration than it is in differentiation. For instance, in Example 5, be sure you understand that

$$\int \dfrac{x + 1}{\sqrt{x}} \, dx = \dfrac{2}{3}\sqrt{x}(x + 3) + C \text{ is not the same as } \dfrac{\int (x + 1) \, dx}{\int \sqrt{x} \, dx} = \dfrac{\frac{1}{2}x^2 + x + C_1}{\frac{2}{3}x\sqrt{x} + C_2}.$$

EXAMPLE 6 Rewriting Before Integrating

$\displaystyle \int \dfrac{\sin x}{\cos^2 x} \, dx = \int \left(\dfrac{1}{\cos x}\right)\left(\dfrac{\sin x}{\cos x}\right) dx$ — *Rewrite as a product.*

$\qquad\qquad\quad = \int \sec x \tan x \, dx$ — *Rewrite using trigonometric identities.*

$\qquad\qquad\quad = \sec x + C$ — *Integrate.*

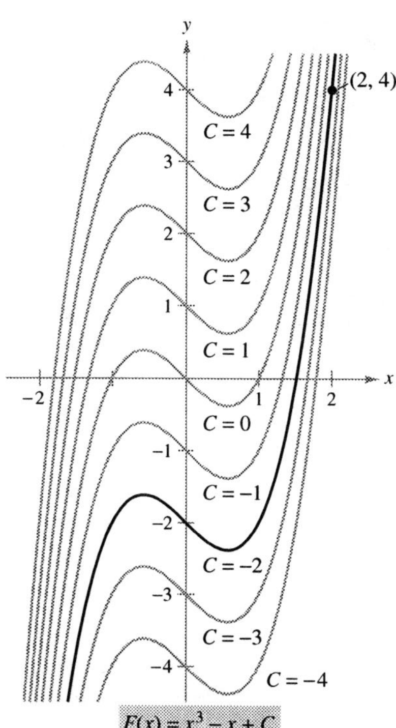

The particular solution that satisfies the initial condition $F(2) = 4$ is $F(x) = x^3 - x - 2$.

Figure 5.2

Initial Conditions and Particular Solutions

You have already seen that the equation $y = \int f(x)\,dx$ has many solutions (each differing from the others by a constant). This means that the graphs of any two antiderivatives of f are vertical translations of each other. For example, Figure 5.2 shows the graphs of several antiderivatives of the form

$$y = \int (3x^2 - 1)\,dx$$

$$= x^3 - x + C \qquad \text{General solution}$$

for various integer values of C. Each of these antiderivatives is a solution of the differential equation

$$\frac{dy}{dx} = 3x^2 - 1.$$

In many applications of integration, you are given enough information to determine a **particular solution.** To do this, you need only know the value of $y = F(x)$ for one value of x. This information is called an **initial condition.** For example, in Figure 5.2, only one curve passes through the point $(2, 4)$. To find this curve, you can use the following information.

$$F(x) = x^3 - x + C \qquad \text{General solution}$$
$$F(2) = 4 \qquad \text{Initial condition}$$

By using the initial condition in the general solution, you can determine that $F(2) = 8 - 2 + C = 4$, which implies that $C = -2$. So, you obtain

$$F(x) = x^3 - x - 2. \qquad \text{Particular solution}$$

EXAMPLE 7 Finding a Particular Solution

Find the general solution of

$$F'(x) = e^x$$

and find the particular solution that satisfies the initial condition $F(0) = 3$.

Solution To find the general solution, integrate to obtain

$$F(x) = \int e^x\,dx$$

$$= e^x + C. \qquad \text{General solution}$$

Using the initial condition $F(0) = 3$, you can solve for C as follows.

$$F(0) = e^0 + C$$
$$3 = 1 + C$$
$$2 = C$$

So, the particular solution, as shown in Figure 5.3, is

$$F(x) = e^x + 2. \qquad \text{Particular solution} \qquad \blacksquare$$

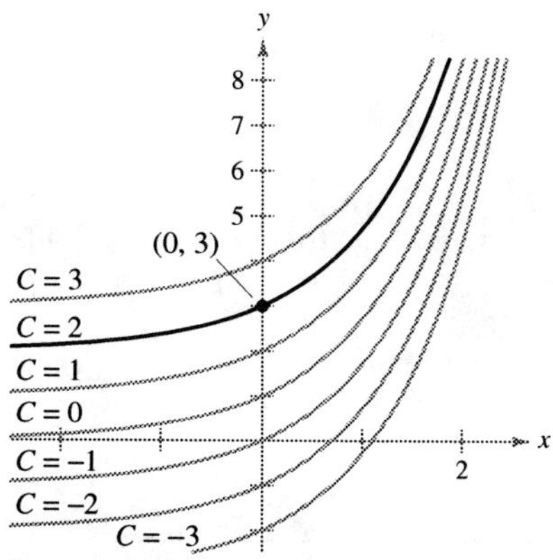

The particular solution that satisfies the initial condition $F(0) = 3$ is $F(x) = e^x + 2$.

Figure 5.3

So far in this section you have been using x as the variable of integration. In applications, it is often convenient to use a different variable. For instance, in the following example involving *time*, the variable of integration is t.

EXAMPLE 8   Solving a Vertical Motion Problem

A ball is thrown upward with an initial velocity of 64 feet per second from an initial height of 80 feet.

a. Find the position function giving the height s as a function of the time t.

b. When does the ball hit the ground?

Solution

a. Let $t = 0$ represent the initial time. The two given initial conditions can be written as follows.

$$s(0) = 80 \qquad \text{Initial height is 80 feet.}$$
$$s'(0) = 64 \qquad \text{Initial velocity is 64 feet per second.}$$

Using -32 feet per second per second as the acceleration due to gravity, you can write

$$s''(t) = -32$$
$$s'(t) = \int s''(t)\, dt$$
$$= \int -32\, dt = -32t + C_1.$$

Using the initial velocity, you obtain $s'(0) = 64 = -32(0) + C_1$, which implies that $C_1 = 64$. Next, by integrating $s'(t)$, you obtain

$$s(t) = \int s'(t)\, dt$$
$$= \int (-32t + 64)\, dt$$
$$= -16t^2 + 64t + C_2.$$

Using the initial height, you obtain

$$s(0) = 80 = -16(0^2) + 64(0) + C_2$$

which implies that $C_2 = 80$. So, the position function is

$$s(t) = -16t^2 + 64t + 80. \qquad \text{See Figure 5.4.}$$

b. Using the position function found in part (a), you can find the time that the ball hits the ground by solving the equation $s(t) = 0$.

$$s(t) = -16t^2 + 64t + 80 = 0$$
$$-16(t + 1)(t - 5) = 0$$
$$t = -1, 5$$

Because t must be positive, you can conclude that the ball hits the ground 5 seconds after it is thrown.

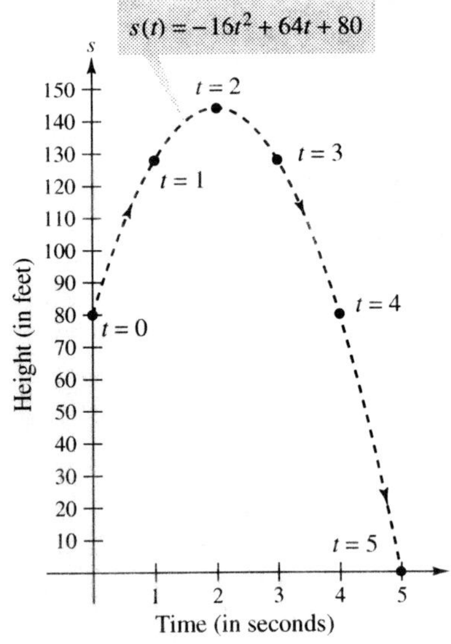

Height of a ball at time t

Figure 5.4

NOTE In Example 8, note that the position function has the form

$$s(t) = \tfrac{1}{2}gt^2 + v_0 t + s_0$$

where $g = -32$, v_0 is the initial velocity, and s_0 is the initial height, as presented in Section 3.2.

Example 8 shows how to use calculus to analyze vertical motion problems in which the acceleration is determined by a gravitational force. You can use a similar strategy to analyze other linear motion problems (vertical or horizontal) in which the acceleration (or deceleration) is the result of some other force, as you will see in Exercises 87–94.

Before you begin the exercise set, be sure you realize that one of the most important steps in integration is *rewriting the integrand* in a form that fits the basic integration rules. To illustrate this point further, here are some additional examples.

Original Integral	Rewrite	Integrate	Simplify
$\displaystyle\int \frac{2}{\sqrt{x}}\,dx$	$\displaystyle 2\int x^{-1/2}\,dx$	$\displaystyle 2\left(\frac{x^{1/2}}{1/2}\right) + C$	$4x^{1/2} + C$
$\displaystyle\int (t^2 + 1)^2\,dt$	$\displaystyle\int (t^4 + 2t^2 + 1)\,dt$	$\displaystyle \frac{t^5}{5} + 2\left(\frac{t^3}{3}\right) + t + C$	$\displaystyle \frac{1}{5}t^5 + \frac{2}{3}t^3 + t + C$
$\displaystyle\int \frac{x^3 + 3}{x^2}\,dx$	$\displaystyle\int (x + 3x^{-2})\,dx$	$\displaystyle \frac{x^2}{2} + 3\left(\frac{x^{-1}}{-1}\right) + C$	$\displaystyle \frac{1}{2}x^2 - \frac{3}{x} + C$
$\displaystyle\int \sqrt[3]{x}\,(x - 4)\,dx$	$\displaystyle\int (x^{4/3} - 4x^{1/3})\,dx$	$\displaystyle \frac{x^{7/3}}{7/3} - 4\left(\frac{x^{4/3}}{4/3}\right) + C$	$\displaystyle \frac{3}{7}x^{4/3}(x - 7) + C$

5.1 Exercises

In Exercises 1–4, verify the statement by showing that the derivative of the right side equals the integrand of the left side.

1. $\displaystyle\int \left(-\frac{6}{x^4}\right) dx = \frac{2}{x^3} + C$

2. $\displaystyle\int \left(8x^3 + \frac{1}{2x^2}\right) dx = 2x^4 - \frac{1}{2x} + C$

3. $\displaystyle\int (x - 4)(x + 4)\,dx = \frac{1}{3}x^3 - 16x + C$

4. $\displaystyle\int \frac{x^2 - 1}{x^{3/2}}\,dx = \frac{2(x^2 + 3)}{3\sqrt{x}} + C$

In Exercises 5–8, find the general solution of the differential equation and check the result by differentiation.

5. $\displaystyle \frac{dy}{dt} = 9t^2$

6. $\displaystyle \frac{dr}{d\theta} = \pi$

7. $\displaystyle \frac{dy}{dx} = x^{3/2}$

8. $\displaystyle \frac{dy}{dx} = 2x^{-3}$

In Exercises 9–14, complete the table using Example 3 and the examples at the top of this page as a model.

Original Integral	Rewrite	Integrate	Simplify
9. $\displaystyle\int \sqrt[3]{x}\,dx$			
10. $\displaystyle\int \frac{1}{4x^2}\,dx$			
11. $\displaystyle\int \frac{1}{x\sqrt{x}}\,dx$			
12. $\displaystyle\int x(x^3 + 1)\,dx$			
13. $\displaystyle\int \frac{1}{2x^3}\,dx$			
14. $\displaystyle\int \frac{1}{(3x)^2}\,dx$			

In Exercises 15–44, find the indefinite integral and check the result by differentiation.

15. $\displaystyle\int (x + 7)\,dx$

16. $\displaystyle\int (13 - x)\,dx$

17. $\displaystyle\int (x^5 + 1)\,dx$

18. $\displaystyle\int (8x^3 - 9x^2 + 4)\,dx$

19. $\displaystyle\int (x^{3/2} + 2x + 1)\,dx$

20. $\displaystyle\int \left(\sqrt[4]{x^3} + 1\right) dx$

21. $\displaystyle\int \frac{1}{x^5}\,dx$

22. $\displaystyle\int \frac{1}{x^6}\,dx$

23. $\displaystyle\int \frac{x + 6}{\sqrt{x}}\,dx$

24. $\displaystyle\int \frac{x^2 + 2x - 3}{x^4}\,dx$

25. $\displaystyle\int (x + 1)(3x - 2)\,dx$

26. $\displaystyle\int (2t^2 - 1)^2\,dt$

27. $\displaystyle\int y^2\sqrt{y}\,dy$

28. $\displaystyle\int (1 + 3t)t^2\,dt$

29. $\displaystyle\int dx$

30. $\displaystyle\int 14\,dt$

31. $\displaystyle\int (5\cos x + 4\sin x)\,dx$

32. $\displaystyle\int (t^2 - \cos t)\,dt$

33. $\displaystyle\int (1 - \csc t \cot t)\,dt$

34. $\displaystyle\int (\theta^2 + \sec^2 \theta)\,d\theta$

35. $\displaystyle\int (2\sin x - 5e^x)\,dx$

36. $\displaystyle\int (3x^2 + 2e^x)\,dx$

37. $\displaystyle\int (\sec^2 \theta - \sin \theta)\,d\theta$

38. $\displaystyle\int \sec y\,(\tan y - \sec y)\,dy$

39. $\displaystyle\int (\tan^2 y + 1)\,dy$

40. $\displaystyle\int \frac{\cos x}{1 - \cos^2 x}\,dx$

41. $\displaystyle\int (2x - 4^x)\,dx$

42. $\displaystyle\int (\cos x + 3^x)\,dx$

43. $\displaystyle\int \left(x - \frac{5}{x}\right) dx$

44. $\displaystyle\int \left(\frac{4}{x} + \sec^2 x\right) dx$

In Exercises 45–48, sketch the graphs of the function $g(x) = f(x) + C$ for $C = -2$, $C = 0$, and $C = 3$ on the same set of coordinate axes.

45. $f(x) = \cos x$

46. $f(x) = \sqrt{x}$

47. $f(x) = \ln x$

48. $f(x) = \frac{1}{2}e^x$

In Exercises 49–52, the graph of the derivative of a function is given. Sketch the graphs of *two* functions that have the given derivative. (There is more than one correct answer.) To print an enlarged copy of the graph, go to the website *www.mathgraphs.com*.

49.

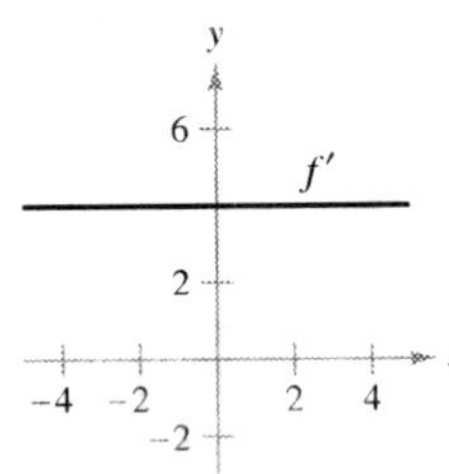

50.

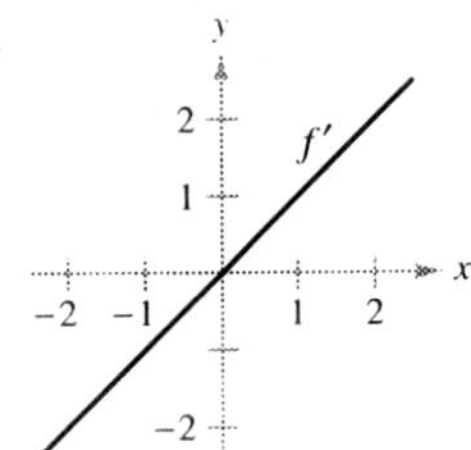

51.

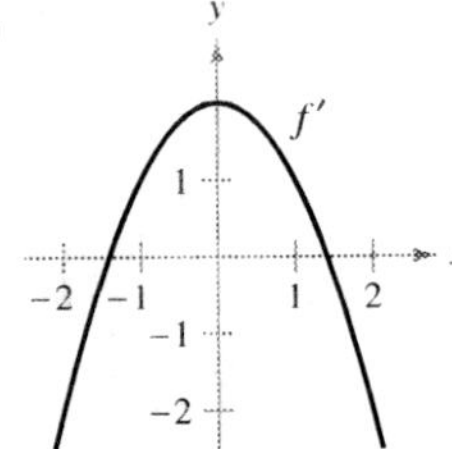

52.

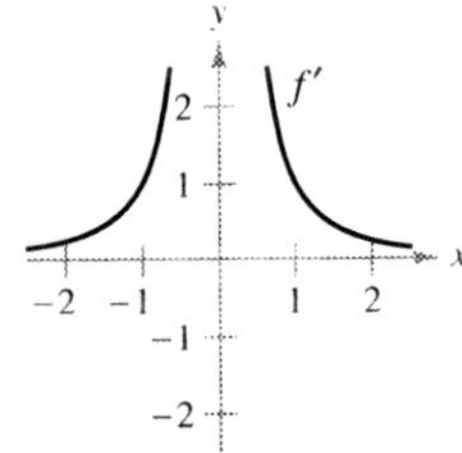

In Exercises 53–56, find the equation of y, given the derivative and the given point on the curve.

53. $\dfrac{dy}{dx} = 2x - 1$

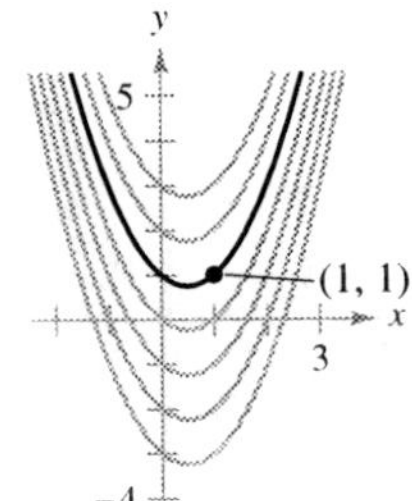

54. $\dfrac{dy}{dx} = 2(x - 1)$

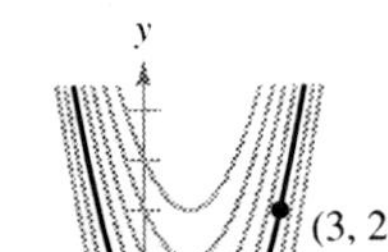

55. $\dfrac{dy}{dx} = \cos x$

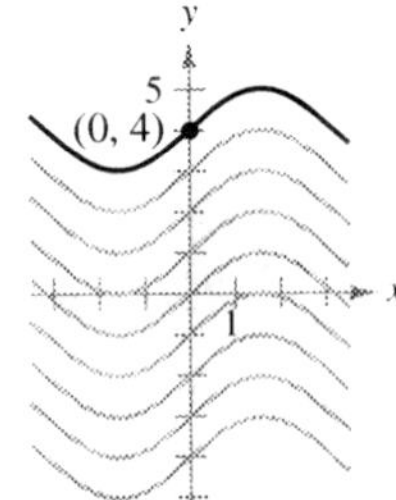

56. $\dfrac{dy}{dx} = \dfrac{3}{x}, \quad x > 0$

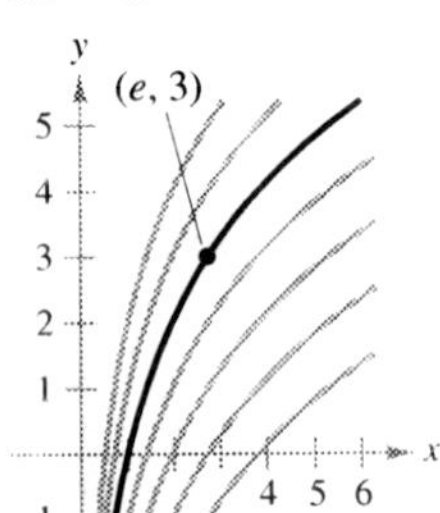

Slope Fields In Exercises 57–60, a differential equation, a point, and a slope field are given. A *slope field* (or *direction field*) consists of line segments with slopes given by the differential equation. These line segments give a visual perspective of the slopes of the solutions of the differential equation. (a) Sketch two approximate solutions of the differential equation on the slope field, one of which passes through the given point. (To print an enlarged copy of the graph, go to the website *www.mathgraphs.com*.) (b) Use integration to find the particular solution of the differential equation and use a graphing utility to graph the solution. Compare the result with the sketches in part (a).

57. $\dfrac{dy}{dx} = \dfrac{1}{2}x - 1, \; (4, 2)$

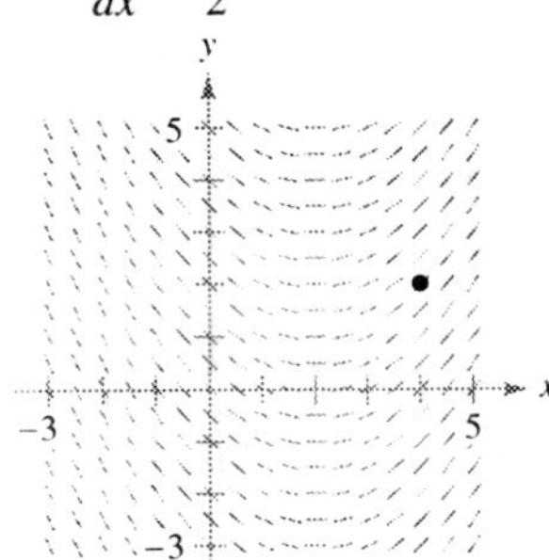

58. $\dfrac{dy}{dx} = x^2 - 1, \; (-1, 3)$

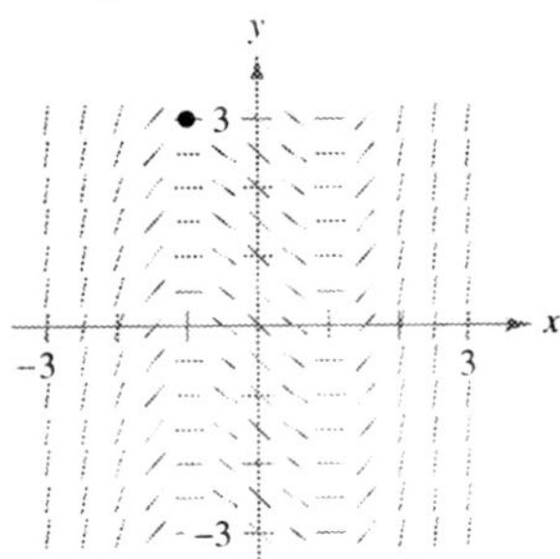

59. $\dfrac{dy}{dx} = \cos x, \; (0, 4)$

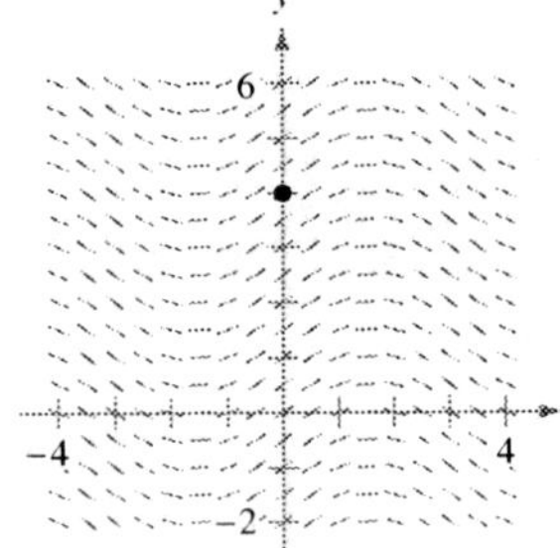

60. $\dfrac{dy}{dx} = -\dfrac{1}{x^2}, \; x > 0, \; (1, 3)$

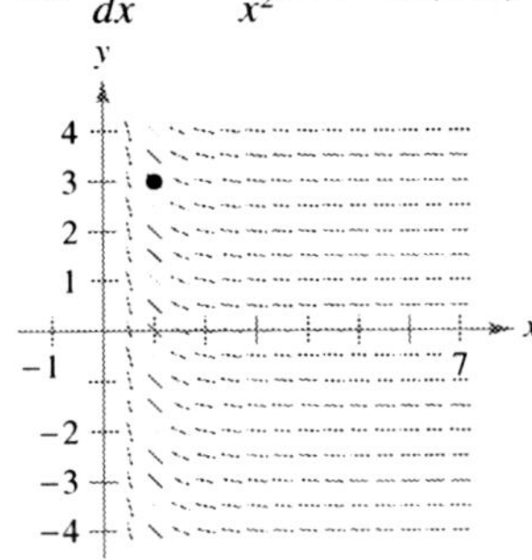

Slope Fields In Exercises 61 and 62, (a) use a graphing utility to graph a slope field for the differential equation, (b) use integration and the given point to find the particular solution of the differential equation, and (c) graph the solution and the slope field in the same viewing window.

61. $\dfrac{dy}{dx} = 2x, \; (-2, -2)$

62. $\dfrac{dy}{dx} = 2\sqrt{x}, \; (4, 12)$

In Exercises 63–72, solve the differential equation.

63. $f'(x) = 6x, \; f(0) = 8$

64. $g'(x) = 6x^2, \; g(0) = -1$

65. $h'(t) = 8t^3 + 5, \; h(1) = -4$

66. $f'(s) = 10s - 12s^3, \; f(3) = 2$

67. $f''(x) = 2, \; f'(2) = 5, \; f(2) = 10$

68. $f''(x) = x^2, \; f'(0) = 8, \; f(0) = 4$

69. $f''(x) = x^{-3/2}, \; f'(4) = 2, \; f(0) = 0$

70. $f''(x) = \sin x, \; f'(0) = 1, \; f(0) = 6$

71. $f''(x) = e^x$, $f'(0) = 2$, $f(0) = 5$

72. $f''(x) = \dfrac{2}{x^2}$, $f'(1) = 4$, $f(1) = 3$

WRITING ABOUT CONCEPTS

73. What is the difference, if any, between finding the antiderivative of $f(x)$ and evaluating the integral $\int f(x)\,dx$?

74. Consider $f(x) = \tan^2 x$ and $g(x) = \sec^2 x$. What do you notice about the derivatives of $f(x)$ and $g(x)$? What can you conclude about the relationship between $f(x)$ and $g(x)$?

75. The graphs of f and f' each pass through the origin. Use the graph of f'' shown in the figure to sketch the graphs of f and f'. To print an enlarged copy of the graph, go to the website *www.mathgraphs.com*.

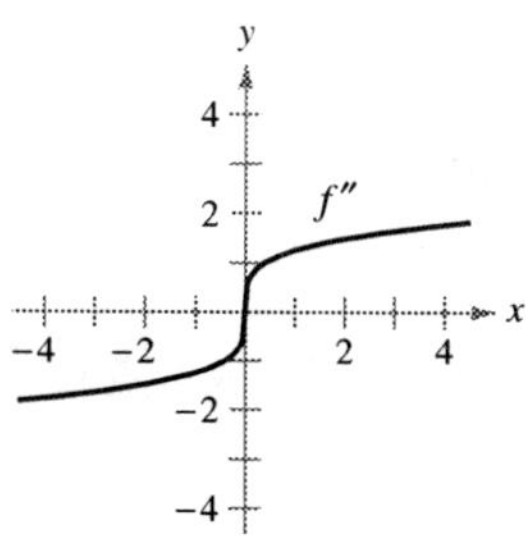

CAPSTONE

76. Use the graph of f' shown in the figure to answer the following, given that $f(0) = -4$.

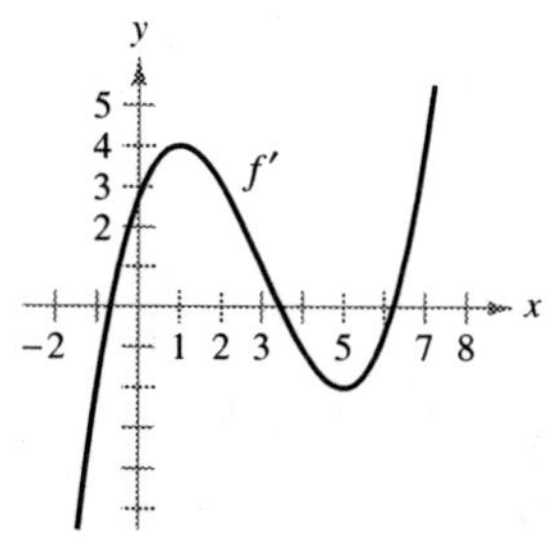

(a) Approximate the slope of f at $x = 4$. Explain.

(b) Is it possible that $f(2) = -1$? Explain.

(c) Is $f(5) - f(4) > 0$? Explain.

(d) Approximate the value of x where f is maximum. Explain.

(e) Approximate any intervals in which the graph of f is concave upward and any intervals in which it is concave downward. Approximate the x-coordinates of any points of inflection.

(f) Approximate the x-coordinate of the minimum of $f''(x)$.

(g) Sketch an approximate graph of f. To print an enlarged copy of the graph, go to the website *www.mathgraphs.com*.

Vertical Motion In Exercises 77–80, use $a(t) = -32$ feet per second per second as the acceleration due to gravity. (Neglect air resistance.)

77. A ball is thrown vertically upward from a height of 6 feet with an initial velocity of 60 feet per second. How high will the ball go?

78. Show that the height above the ground of an object thrown upward from a point s_0 feet above the ground with an initial velocity of v_0 feet per second is given by the function
$$f(t) = -16t^2 + v_0 t + s_0.$$

79. With what initial velocity must an object be thrown upward (from ground level) to reach the top of the Washington Monument (approximately 550 feet)?

80. A balloon, rising vertically at a velocity of 8 feet per second, releases a sandbag at the instant it is 64 feet above the ground.

(a) How many seconds after its release will the bag strike the ground?

(b) At what velocity will it hit the ground?

Vertical Motion In Exercises 81–84, use $a(t) = -9.8$ meters per second per second as the acceleration due to gravity. (Neglect air resistance.)

81. Show that the height above the ground of an object thrown upward from a point s_0 meters above the ground with an initial velocity of v_0 meters per second is given by the function
$$f(t) = -4.9t^2 + v_0 t + s_0.$$

82. The Grand Canyon is 1800 meters deep at its deepest point. A rock is dropped from the rim above this point. Express the height of the rock as a function of the time t in seconds. How long will it take the rock to hit the canyon floor?

83. A baseball is thrown upward from a height of 2 meters with an initial velocity of 10 meters per second. Determine its maximum height.

84. With what initial velocity must an object be thrown upward (from a height of 2 meters) to reach a maximum height of 200 meters?

85. *Lunar Gravity* On the moon, the acceleration due to gravity is -1.6 meters per second per second. A stone is dropped from a cliff on the moon and hits the surface of the moon 20 seconds later. How far did it fall? What was its velocity at impact?

86. *Escape Velocity* The minimum velocity required for an object to escape Earth's gravitational pull is obtained from the solution of the equation
$$\int v\,dv = -GM \int \frac{1}{y^2}\,dy$$

where v is the velocity of the object projected from Earth, y is the distance from the center of Earth, G is the gravitational constant, and M is the mass of Earth. Show that v and y are related by the equation
$$v^2 = v_0^2 + 2GM\left(\frac{1}{y} - \frac{1}{R}\right)$$

where v_0 is the initial velocity of the object and R is the radius of Earth.

Rectilinear Motion **In Exercises 87–90, consider a particle moving along the x-axis where $x(t)$ is the position of the particle at time t, $x'(t)$ is its velocity, and $x''(t)$ is its acceleration.**

87. $x(t) = t^3 - 6t^2 + 9t - 2, \quad 0 \le t \le 5$

   (a) Find the velocity and acceleration of the particle.

   (b) Find the open t-intervals on which the particle is moving to the right.

   (c) Find the velocity of the particle when the acceleration is 0.

88. Repeat Exercise 87 for the position function

$$x(t) = (t - 1)(t - 3)^2, \quad 0 \le t \le 5.$$

89. A particle moves along the x-axis at a velocity of $v(t) = 1/\sqrt{t}$, $t > 0$. At time $t = 1$, its position is $x = 4$. Find the acceleration and position functions for the particle.

90. A particle, initially at rest, moves along the x-axis such that its acceleration at time $t > 0$ is given by $a(t) = \cos t$. At the time $t = 0$, its position is $x = 3$.

   (a) Find the velocity and position functions for the particle.

   (b) Find the values of t for which the particle is at rest.

91. *Acceleration* The maker of an automobile advertises that it takes 13 seconds to accelerate from 25 kilometers per hour to 80 kilometers per hour. Assuming constant acceleration, compute the following.

   (a) The acceleration in meters per second per second

   (b) The distance the car travels during the 13 seconds

92. *Deceleration* A car traveling at 45 miles per hour is brought to a stop, at constant deceleration, 132 feet from where the brakes are applied.

   (a) How far has the car moved when its speed has been reduced to 30 miles per hour?

   (b) How far has the car moved when its speed has been reduced to 15 miles per hour?

   (c) Draw the real number line from 0 to 132, and plot the points found in parts (a) and (b). What can you conclude?

93. *Acceleration* At the instant the traffic light turns green, a car that has been waiting at an intersection starts with a constant acceleration of 6 feet per second per second. At the same instant, a truck traveling with a constant velocity of 30 feet per second passes the car.

   (a) How far beyond its starting point will the car pass the truck?

   (b) How fast will the car be traveling when it passes the truck?

94. *Modeling Data* The table shows the velocities (in miles per hour) of two cars on an entrance ramp to an interstate highway. The time t is in seconds.

t	0	5	10	15	20	25	30
v_1	0	2.5	7	16	29	45	65
v_2	0	21	38	51	60	64	65

   (a) Rewrite the table, converting miles per hour to feet per second.

   (b) Use the regression capabilities of a graphing utility to find quadratic models for the data in part (a).

   (c) Approximate the distance traveled by each car during the 30 seconds. Explain the difference in the distances.

True or False? **In Exercises 95–100, determine whether the statement is true or false. If it is false, explain why or give an example that shows it is false.**

95. Each antiderivative of an nth-degree polynomial function is an $(n + 1)$th-degree polynomial function.

96. If $p(x)$ is a polynomial function, then p has exactly one antiderivative whose graph contains the origin.

97. If $F(x)$ and $G(x)$ are antiderivatives of $f(x)$, then $F(x) = G(x) + C$.

98. If $f'(x) = g(x)$, then $\int g(x)\, dx = f(x) + C$.

99. $\int f(x)g(x)\, dx = \int f(x)\, dx \int g(x)\, dx$

100. The antiderivative of $f(x)$ is unique.

101. Find a function f such that the graph of f has a horizontal tangent at $(2, 0)$ and $f''(x) = 2x$.

102. The graph of f' is shown. Sketch the graph of f given that f is continuous and $f(0) = 1$.

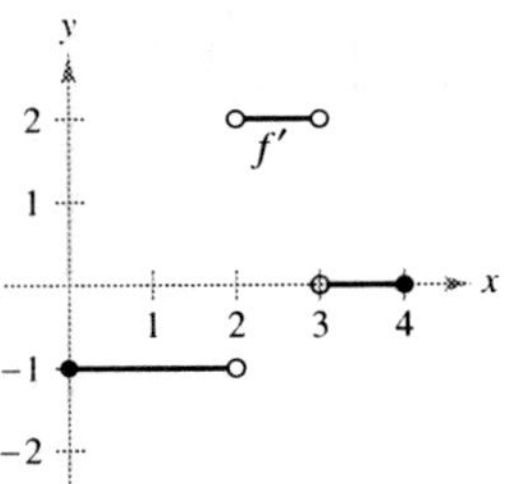

103. If $f'(x) = \begin{cases} 1, & 0 \le x < 2 \\ 3x, & 2 \le x \le 5 \end{cases}$, f is continuous, and $f(1) = 3$, find f. Is f differentiable at $x = 2$?

104. Let $s(x)$ and $c(x)$ be two functions satisfying $s'(x) = c(x)$ and $c'(x) = -s(x)$ for all x. If $s(0) = 0$ and $c(0) = 1$, prove that $[s(x)]^2 + [c(x)]^2 = 1$.

105. *Verification* Verify the natural log rule $\int \dfrac{1}{x}\, dx = \ln|Cx|$, $C \ne 0$, by showing that the derivative of $\ln|Cx|$ is $1/x$.

106. *Verification* Verify the natural log rule $\int \dfrac{1}{x}\, dx = \ln|x| + C$ by showing that the derivative of $\ln|x| + C$ is $1/x$.

PUTNAM EXAM CHALLENGE

107. Suppose f and g are nonconstant, differentiable, real-valued functions on R. Furthermore, suppose that for each pair of real numbers x and y, $f(x + y) = f(x)f(y) - g(x)g(y)$ and $g(x + y) = f(x)g(y) + g(x)f(y)$. If $f'(0) = 0$, prove that $(f(x))^2 + (g(x))^2 = 1$ for all x.

6.2 Differential Equations: Growth and Decay

■ Use separation of variables to solve a simple differential equation.
■ Use exponential functions to model growth and decay in applied problems.

Differential Equations

In the preceding section, you learned to analyze visually the solutions of differential equations using slope fields and to approximate solutions numerically using Euler's Method. Analytically, you have learned to solve only two types of differential equations—those of the forms $y' = f(x)$ and $y'' = f(x)$. In this section, you will learn how to solve a more general type of differential equation. The strategy is to rewrite the equation so that each variable occurs on only one side of the equation. This strategy is called *separation of variables*. (You will study this strategy in detail in Section 6.3.)

EXAMPLE 1 Solving a Differential Equation

$$y' = \frac{2x}{y} \qquad \text{Original equation}$$

$$yy' = 2x \qquad \text{Multiply both sides by } y.$$

$$\int yy'\, dx = \int 2x\, dx \qquad \text{Integrate with respect to } x.$$

$$\int y\, dy = \int 2x\, dx \qquad dy = y'\, dx$$

$$\frac{1}{2}y^2 = x^2 + C_1 \qquad \text{Apply Power Rule.}$$

$$y^2 - 2x^2 = C \qquad \text{Rewrite, letting } C = 2C_1.$$

So, the general solution is given by $y^2 - 2x^2 = C$.   ■

Notice that when you integrate both sides of the equation in Example 1, you don't need to add a constant of integration to both sides. If you did, you would obtain the same result.

$$\int y\, dy = \int 2x\, dx$$
$$\tfrac{1}{2}y^2 + C_2 = x^2 + C_3$$
$$\tfrac{1}{2}y^2 = x^2 + (C_3 - C_2)$$
$$\tfrac{1}{2}y^2 = x^2 + C_1$$

Some people prefer to use Leibniz notation and differentials when applying separation of variables. The solution of Example 1 is shown below using this notation.

$$\frac{dy}{dx} = \frac{2x}{y}$$
$$y\, dy = 2x\, dx$$
$$\int y\, dy = \int 2x\, dx$$
$$\frac{1}{2}y^2 = x^2 + C_1$$
$$y^2 - 2x^2 = C$$

STUDY TIP   You can use implicit differentiation to check the solution in Example 1.

EXPLORATION

In Example 1, the general solution of the differential equation is

$$y^2 - 2x^2 = C.$$

Use a graphing utility to sketch the particular solutions for $C = \pm 2$, $C = \pm 1$, and $C = 0$. Describe the solutions graphically. Is the following statement true of each solution?

The slope of the graph at the point (x, y) is equal to twice the ratio of x and y.

Explain your reasoning. Are all curves for which this statement is true represented by the general solution?

Growth and Decay Models

In many applications, the rate of change of a variable y is proportional to the value of y. If y is a function of time t, the proportion can be written as follows.

Rate of change of y is   proportional to y.

$$\frac{dy}{dt} = ky$$

The general solution of this differential equation is given in the following theorem.

THEOREM 6.1 EXPONENTIAL GROWTH AND DECAY MODEL

If y is a differentiable function of t such that $y > 0$ and $y' = ky$ for some constant k, then

$$y = Ce^{kt}.$$

C is the **initial value** of y, and k is the **proportionality constant. Exponential growth** occurs when $k > 0$, and **exponential decay** occurs when $k < 0$.

(PROOF)

$$y' = ky \qquad \text{Write original equation.}$$

$$\frac{y'}{y} = k \qquad \text{Separate variables.}$$

$$\int \frac{y'}{y}\, dt = \int k\, dt \qquad \text{Integrate with respect to } t.$$

$$\int \frac{1}{y}\, dy = \int k\, dt \qquad dy = y'\, dt$$

$$\ln y = kt + C_1 \qquad \text{Find antiderivative of each side.}$$

$$y = e^{kt}e^{C_1} \qquad \text{Solve for } y.$$

$$y = Ce^{kt} \qquad \text{Let } C = e^{C_1}.$$

So, all solutions of $y' = ky$ are of the form $y = Ce^{kt}$. Remember that you can differentiate the function $y = Ce^{kt}$ with respect to t to verify that $y' = ky$. ∎

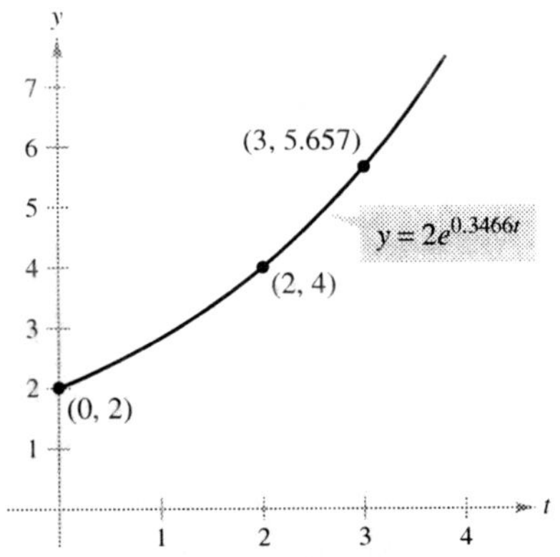

EXAMPLE 2 Using an Exponential Growth Model

The rate of change of y is proportional to y. When $t = 0$, $y = 2$, and when $t = 2$, $y = 4$. What is the value of y when $t = 3$?

Solution Because $y' = ky$, you know that y and t are related by the equation $y = Ce^{kt}$. You can find the values of the constants C and k by applying the initial conditions.

$$2 = Ce^0 \implies C = 2 \qquad \text{When } t = 0, y = 2.$$

$$4 = 2e^{2k} \implies k = \frac{1}{2}\ln 2 \approx 0.3466 \qquad \text{When } t = 2, y = 4.$$

So, the model is $y \approx 2e^{0.3466t}$. When $t = 3$, the value of y is $2e^{0.3466(3)} \approx 5.657$ (see Figure 6.8). ∎

If the rate of change of y is proportional to y, then y follows an exponential model.

Figure 6.8

STUDY TIP Using logarithmic properties, note that the value of k in Example 2 can also be written as $\ln(\sqrt{2})$. So, the model becomes $y = 2e^{(\ln\sqrt{2})t}$, which can then be rewritten as $y = 2(\sqrt{2})^t$.

TECHNOLOGY Most graphing utilities have curve-fitting capabilities that can be used to find models that represent data. Use the *exponential regression* feature of a graphing utility and the information in Example 2 to find a model for the data. How does your model compare with the given model?

Radioactive decay is measured in terms of *half-life*—the number of years required for half of the atoms in a sample of radioactive material to decay. The rate of decay is proportional to the amount present. The half-lives of some common radioactive isotopes are shown below.

Uranium (^{238}U)	4,470,000,000 years
Plutonium (^{239}Pu)	24,100 years
Carbon (^{14}C)	5715 years
Radium (^{226}Ra)	1599 years
Einsteinium (^{254}Es)	276 days
Nobelium (^{257}No)	25 seconds

EXAMPLE 3 Radioactive Decay

Suppose that 10 grams of the plutonium isotope ^{239}Pu was released in the Chernobyl nuclear accident. How long will it take for the 10 grams to decay to 1 gram?

Solution Let y represent the mass (in grams) of the plutonium. Because the rate of decay is proportional to y, you know that

$$y = Ce^{kt}$$

where t is the time in years. To find the values of the constants C and k, apply the initial conditions. Using the fact that $y = 10$ when $t = 0$, you can write

$$10 = Ce^{k(0)} = Ce^0$$

which implies that $C = 10$. Next, using the fact that the half-life of ^{239}Pu is 24,100 years, you have $y = 10/2 = 5$ when $t = 24,100$, so you can write

$$5 = 10e^{k(24,100)}$$

$$\frac{1}{2} = e^{24,100k}$$

$$\frac{1}{24,100} \ln \frac{1}{2} = k$$

$$-0.000028761 \approx k.$$

So, the model is

$$y = 10e^{-0.000028761t}. \qquad \text{Half-life model}$$

NOTE The exponential decay model in Example 3 could also be written as $y = 10\left(\frac{1}{2}\right)^{t/24,100}$. This model is much easier to derive, but for some applications it is not as convenient to use.

To find the time it would take for 10 grams to decay to 1 gram, you can solve for t in the equation

$$1 = 10e^{-0.000028761t}.$$

The solution is approximately 80,059 years. ∎

From Example 3, notice that in an exponential growth or decay problem, it is easy to solve for C when you are given the value of y at $t = 0$. The next example demonstrates a procedure for solving for C and k when you do not know the value of y at $t = 0$.

EXAMPLE 4 Population Growth

Suppose an experimental population of fruit flies increases according to the law of exponential growth. There were 100 flies after the second day of the experiment and 300 flies after the fourth day. Approximately how many flies were in the original population?

Solution Let $y = Ce^{kt}$ be the number of flies at time t, where t is measured in days. Note that y is continuous whereas the number of flies is discrete. Because $y = 100$ when $t = 2$ and $y = 300$ when $t = 4$, you can write

$$100 = Ce^{2k} \quad \text{and} \quad 300 = Ce^{4k}.$$

From the first equation, you know that $C = 100e^{-2k}$. Substituting this value into the second equation produces the following.

$$300 = 100e^{-2k}e^{4k}$$
$$300 = 100e^{2k}$$
$$\ln 3 = 2k$$
$$\frac{1}{2}\ln 3 = k$$
$$0.5493 \approx k$$

So, the exponential growth model is

$$y = Ce^{0.5493t}.$$

To solve for C, reapply the condition $y = 100$ when $t = 2$ and obtain

$$100 = Ce^{0.5493(2)}$$
$$C = 100e^{-1.0986} \approx 33.$$

So, the original population (when $t = 0$) consisted of approximately $y = C = 33$ flies, as shown in Figure 6.9.

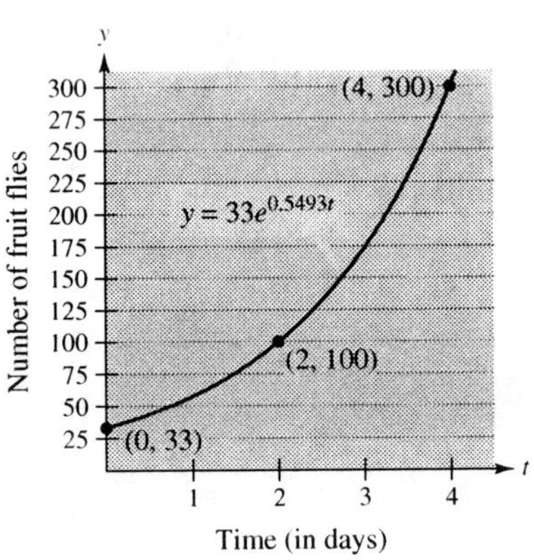

Figure 6.9

EXAMPLE 5 Declining Sales

Four months after it stops advertising, a manufacturing company notices that its sales have dropped from 100,000 units per month to 80,000 units per month. If the sales follow an exponential pattern of decline, what will they be after another 2 months?

Solution Use the exponential decay model $y = Ce^{kt}$, where t is measured in months. From the initial condition ($t = 0$), you know that $C = 100,000$. Moreover, because $y = 80,000$ when $t = 4$, you have

$$80,000 = 100,000e^{4k}$$
$$0.8 = e^{4k}$$
$$\ln(0.8) = 4k$$
$$-0.0558 \approx k.$$

So, after 2 more months ($t = 6$), you can expect the monthly sales rate to be

$$y \approx 100,000e^{-0.0558(6)}$$
$$\approx 71,500 \text{ units.}$$

See Figure 6.10.

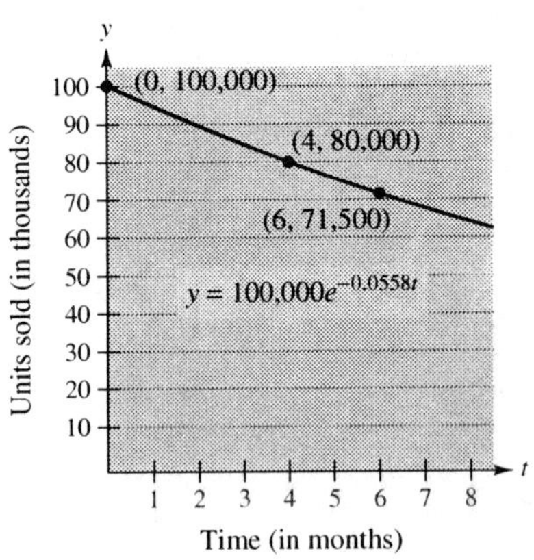

Figure 6.10

In Examples 2 through 5, you did not actually have to solve the differential equation

$$y' = ky.$$

(This was done once in the proof of Theorem 6.1.) The next example demonstrates a problem whose solution involves the separation of variables technique. The example concerns **Newton's Law of Cooling,** which states that the rate of change in the temperature of an object is proportional to the difference between the object's temperature and the temperature of the surrounding medium.

EXAMPLE 6 Newton's Law of Cooling

Let y represent the temperature (in °F) of an object in a room whose temperature is kept at a constant 60°. If the object cools from 100° to 90° in 10 minutes, how much longer will it take for its temperature to decrease to 80°?

Solution From Newton's Law of Cooling, you know that the rate of change in y is proportional to the difference between y and 60. This can be written as

$$y' = k(y - 60), \quad 80 \le y \le 100.$$

To solve this differential equation, use separation of variables, as follows.

$$\frac{dy}{dt} = k(y - 60) \qquad \text{Differential equation}$$

$$\left(\frac{1}{y - 60}\right) dy = k\, dt \qquad \text{Separate variables.}$$

$$\int \frac{1}{y - 60}\, dy = \int k\, dt \qquad \text{Integrate each side.}$$

$$\ln|y - 60| = kt + C_1 \qquad \text{Find antiderivative of each side.}$$

Because $y > 60$, $|y - 60| = y - 60$, and you can omit the absolute value signs. Using exponential notation, you have

$$y - 60 = e^{kt + C_1} \quad \Longrightarrow \quad y = 60 + Ce^{kt}. \qquad C = e^{C_1}$$

Using $y = 100$ when $t = 0$, you obtain $100 = 60 + Ce^{k(0)} = 60 + C$, which implies that $C = 40$. Because $y = 90$ when $t = 10$,

$$90 = 60 + 40e^{k(10)}$$

$$30 = 40e^{10k}$$

$$k = \tfrac{1}{10} \ln \tfrac{3}{4} \approx -0.02877.$$

So, the model is

$$y = 60 + 40e^{-0.02877t} \qquad \text{Cooling model}$$

and finally, when $y = 80$, you obtain

$$80 = 60 + 40e^{-0.02877t}$$

$$20 = 40e^{-0.02877t}$$

$$\tfrac{1}{2} = e^{-0.02877t}$$

$$\ln \tfrac{1}{2} = -0.02877t$$

$$t \approx 24.09 \text{ minutes.}$$

So, it will require about 14.09 *more* minutes for the object to cool to a temperature of 80° (see Figure 6.11).

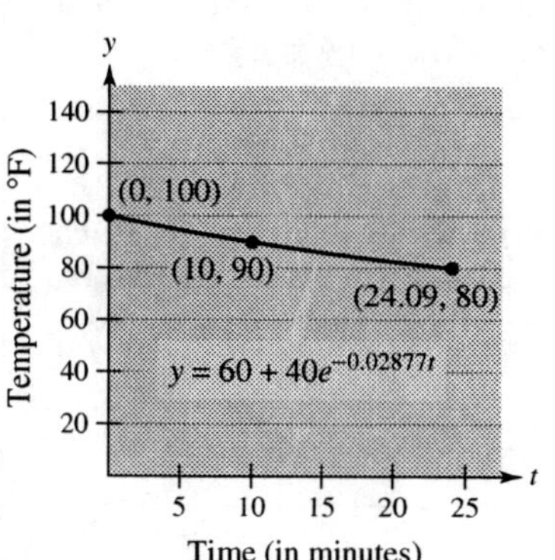

Figure 6.11

6.2 Exercises

In Exercises 1–10, solve the differential equation.

1. $\dfrac{dy}{dx} = x + 3$ 2. $\dfrac{dy}{dx} = 6 - x$

3. $\dfrac{dy}{dx} = y + 3$ 4. $\dfrac{dy}{dx} = 6 - y$

5. $y' = \dfrac{5x}{y}$ 6. $y' = \dfrac{\sqrt{x}}{7y}$

7. $y' = \sqrt{x}\,y$ 8. $y' = x(1 + y)$

9. $(1 + x^2)y' - 2xy = 0$ 10. $xy + y' = 100x$

In Exercises 11–14, write and solve the differential equation that models the verbal statement.

11. The rate of change of Q with respect to t is inversely proportional to the square of t.

12. The rate of change of P with respect to t is proportional to $25 - t$.

13. The rate of change of N with respect to s is proportional to $500 - s$.

14. The rate of change of y with respect to x varies jointly as x and $L - y$.

Slope Fields In Exercises 15 and 16, a differential equation, a point, and a slope field are given. (a) Sketch two approximate solutions of the differential equation on the slope field, one of which passes through the given point. (b) Use integration to find the particular solution of the differential equation and use a graphing utility to graph the solution. Compare the result with the sketch in part (a). To print an enlarged copy of the graph, go to the website *www.mathgraphs.com*.

15. $\dfrac{dy}{dx} = x(6 - y)$, $(0, 0)$ 16. $\dfrac{dy}{dx} = xy$, $\left(0, \tfrac{1}{2}\right)$

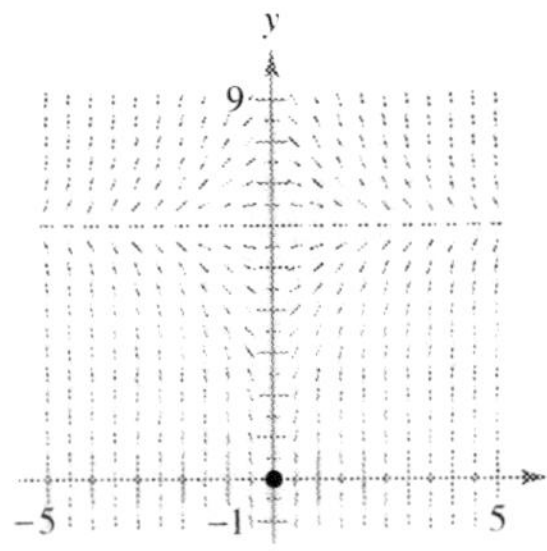

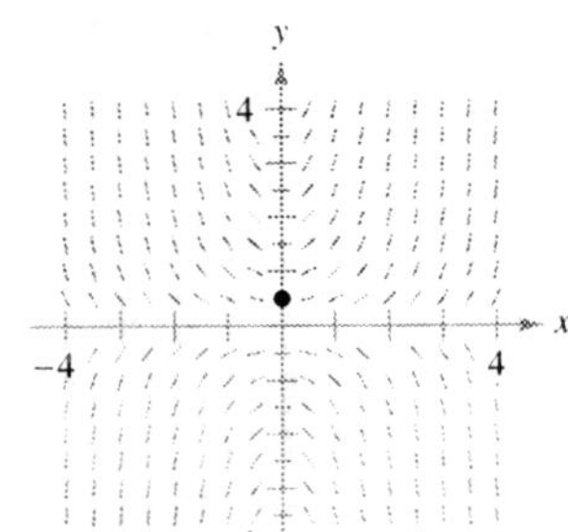

In Exercises 17–20, find the function $y = f(t)$ passing through the point $(0, 10)$ with the given first derivative. Use a graphing utility to graph the solution.

17. $\dfrac{dy}{dt} = \dfrac{1}{2}t$ 18. $\dfrac{dy}{dt} = -\dfrac{3}{4}\sqrt{t}$

19. $\dfrac{dy}{dt} = -\dfrac{1}{2}y$ 20. $\dfrac{dy}{dt} = \dfrac{3}{4}y$

In Exercises 21–24, write and solve the differential equation that models the verbal statement. Evaluate the solution at the specified value of the independent variable.

21. The rate of change of y is proportional to y. When $x = 0$, $y = 6$, and when $x = 4$, $y = 15$. What is the value of y when $x = 8$?

22. The rate of change of N is proportional to N. When $t = 0$, $N = 250$, and when $t = 1$, $N = 400$. What is the value of N when $t = 4$?

23. The rate of change of V is proportional to V. When $t = 0$, $V = 20{,}000$, and when $t = 4$, $V = 12{,}500$. What is the value of V when $t = 6$?

24. The rate of change of P is proportional to P. When $t = 0$, $P = 5000$, and when $t = 1$, $P = 4750$. What is the value of P when $t = 5$?

In Exercises 25–28, find the exponential function $y = Ce^{kt}$ that passes through the two given points.

25. 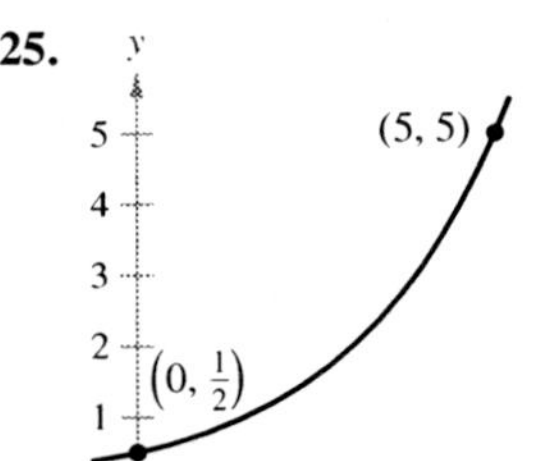

26. 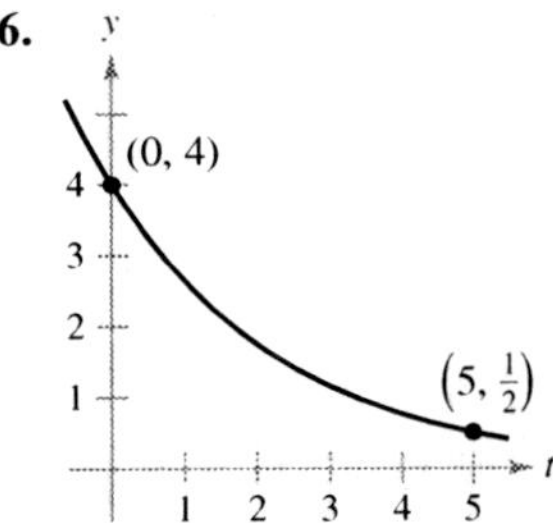

27. 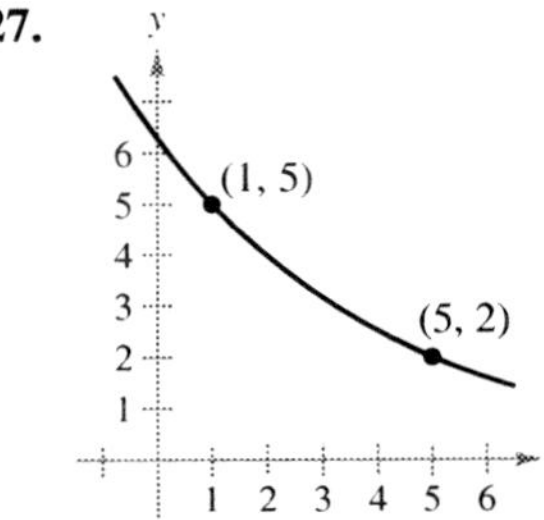

28. 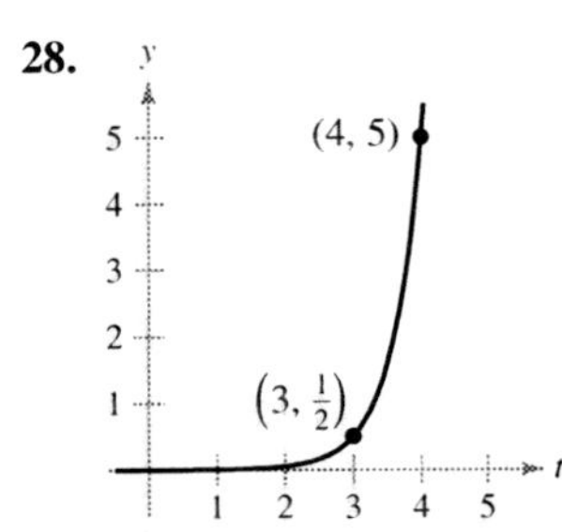

WRITING ABOUT CONCEPTS

29. Describe what the values of C and k represent in the exponential growth and decay model, $y = Ce^{kt}$.

30. Give the differential equation that models exponential growth and decay.

In Exercises 31 and 32, determine the quadrants in which the solution of the differential equation is an increasing function. Explain. (Do not solve the differential equation.)

31. $\dfrac{dy}{dx} = \dfrac{1}{2}xy$ 32. $\dfrac{dy}{dx} = \dfrac{1}{2}x^2 y$

Radioactive Decay **In Exercises 33–40, complete the table for the radioactive isotope.**

Isotope	Half-Life (in years)	Initial Quantity	Amount After 1000 Years	Amount After 10,000 Years
33. ^{226}Ra	1599	20 g		
34. ^{226}Ra	1599		1.5 g	
35. ^{226}Ra	1599			0.1 g
36. ^{14}C	5715			3 g
37. ^{14}C	5715	5 g		
38. ^{14}C	5715		1.6 g	
39. ^{239}Pu	24,100		2.1 g	
40. ^{239}Pu	24,100			0.4 g

41. *Radioactive Decay* Radioactive radium has a half-life of approximately 1599 years. What percent of a given amount remains after 100 years?

42. *Carbon Dating* Carbon-14 dating assumes that the carbon dioxide on Earth today has the same radioactive content as it did centuries ago. If this is true, the amount of ^{14}C absorbed by a tree that grew several centuries ago should be the same as the amount of ^{14}C absorbed by a tree growing today. A piece of ancient charcoal contains only 15% as much of the radioactive carbon as a piece of modern charcoal. How long ago was the tree burned to make the ancient charcoal? (The half-life of ^{14}C is 5715 years.)

Compound Interest **In Exercises 43–48, complete the table for a savings account in which interest is compounded continuously.**

Initial Investment	Annual Rate	Time to Double	Amount After 10 Years
43. $4000	6%		
44. $18,000	$5\frac{1}{2}\%$		
45. $750		$7\frac{3}{4}$ yr	
46. $12,500		5 yr	
47. $500			$1292.85
48. $2000			$5436.56

Compound Interest **In Exercises 49–52, find the principal P that must be invested at rate r, compounded monthly, so that $1,000,000 will be available for retirement in t years.**

49. $r = 7\frac{1}{2}\%, \quad t = 20$ **50.** $r = 6\%, \quad t = 40$

51. $r = 8\%, \quad t = 35$ **52.** $r = 9\%, \quad t = 25$

Compound Interest **In Exercises 53–56, find the time necessary for $1000 to double if it is invested at a rate of r compounded (a) annually, (b) monthly, (c) daily, and (d) continuously.**

53. $r = 7\%$ **54.** $r = 6\%$

55. $r = 8.5\%$ **56.** $r = 5.5\%$

Population **In Exercises 57–61, the population (in millions) of a country in 2007 and the expected continuous annual rate of change k of the population are given.** (*Source: U.S. Census Bureau, International Data Base*)

(a) Find the exponential growth model $P = Ce^{kt}$ for the population by letting $t = 0$ correspond to 2000.

(b) Use the model to predict the population of the country in 2015.

(c) Discuss the relationship between the sign of k and the change in the population of the country.

Country	2007 Population	k
57. Latvia	2.3	-0.006
58. Egypt	80.3	0.017
59. Paraguay	6.7	0.024
60. Hungary	10.0	-0.003
61. Uganda	30.3	0.036

CAPSTONE

62. **(a)** Suppose an insect population increases by a constant number each month. Explain why the number of insects can be represented by a linear function.

    (b) Suppose an insect population increases by a constant percentage each month. Explain why the number of insects can be represented by an exponential function.

63. *Modeling Data* One hundred bacteria are started in a culture and the number N of bacteria is counted each hour for 5 hours. The results are shown in the table, where t is the time in hours.

t	0	1	2	3	4	5
N	100	126	151	198	243	297

(a) Use the regression capabilities of a graphing utility to find an exponential model for the data.

(b) Use the model to estimate the time required for the population to quadruple in size.

64. *Bacteria Growth* The number of bacteria in a culture is increasing according to the law of exponential growth. There are 125 bacteria in the culture after 2 hours and 350 bacteria after 4 hours.

(a) Find the initial population.

(b) Write an exponential growth model for the bacteria population. Let t represent time in hours.

(c) Use the model to determine the number of bacteria after 8 hours.

(d) After how many hours will the bacteria count be 25,000?

65. *Learning Curve* The management at a certain factory has found that a worker can produce at most 30 units in a day. The learning curve for the number of units N produced per day after a new employee has worked t days is $N = 30(1 - e^{kt})$. After 20 days on the job, a particular worker produces 19 units.

(a) Find the learning curve for this worker.

(b) How many days should pass before this worker is producing 25 units per day?

66. Learning Curve If the management in Exercise 65 requires a new employee to produce at least 20 units per day after 30 days on the job, find (a) the learning curve that describes this minimum requirement and (b) the number of days before a minimal achiever is producing 25 units per day.

67. Modeling Data The table shows the populations P (in millions) of the United States from 1960 to 2000. *(Source: U.S. Census Bureau)*

Year	1960	1970	1980	1990	2000
Population, P	181	205	228	250	282

(a) Use the 1960 and 1970 data to find an exponential model P_1 for the data. Let $t = 0$ represent 1960.

(b) Use a graphing utility to find an exponential model P_2 for all the data. Let $t = 0$ represent 1960.

(c) Use a graphing utility to plot the data and graph models P_1 and P_2 in the same viewing window. Compare the actual data with the predictions. Which model better fits the data?

(d) Estimate when the population will be 320 million.

68. Modeling Data The table shows the net receipts and the amounts required to service the national debt (interest on Treasury debt securities) of the United States from 2001 through 2010. The years 2007 through 2010 are estimated, and the monetary amounts are given in billions of dollars. *(Source: U.S. Office of Management and Budget)*

Year	2001	2002	2003	2004	2005
Receipts	1991.4	1853.4	1782.5	1880.3	2153.9
Interest	359.5	332.5	318.1	321.7	352.3

Year	2006	2007	2008	2009	2010
Receipts	2407.3	2540.1	2662.5	2798.3	2954.7
Interest	405.9	433.0	469.9	498.0	523.2

(a) Use the regression capabilities of a graphing utility to find an exponential model R for the receipts and a quartic model I for the amount required to service the debt. Let t represent the time in years, with $t = 1$ corresponding to 2001.

(b) Use a graphing utility to plot the points corresponding to the receipts, and graph the exponential model. Based on the model, what is the continuous rate of growth of the receipts?

(c) Use a graphing utility to plot the points corresponding to the amounts required to service the debt, and graph the quartic model.

(d) Find a function $P(t)$ that approximates the percent of the receipts that is required to service the national debt. Use a graphing utility to graph this function.

69. Sound Intensity The level of sound β (in decibels) with an intensity of I is $\beta(I) = 10 \log_{10}(I/I_0)$, where I_0 is an intensity of 10^{-16} watt per square centimeter, corresponding roughly to the faintest sound that can be heard. Determine $\beta(I)$ for the following.

(a) $I = 10^{-14}$ watt per square centimeter (whisper)

(b) $I = 10^{-9}$ watt per square centimeter (busy street corner)

(c) $I = 10^{-6.5}$ watt per square centimeter (air hammer)

(d) $I = 10^{-4}$ watt per square centimeter (threshold of pain)

70. Noise Level With the installation of noise suppression materials, the noise level in an auditorium was reduced from 93 to 80 decibels. Use the function in Exercise 69 to find the percent decrease in the intensity level of the noise as a result of the installation of these materials.

71. Forestry The value of a tract of timber is $V(t) = 100,000e^{0.8\sqrt{t}}$, where t is the time in years, with $t = 0$ corresponding to 2008. If money earns interest continuously at 10%, the present value of the timber at any time t is $A(t) = V(t)e^{-0.10t}$. Find the year in which the timber should be harvested to maximize the present value function.

72. Earthquake Intensity On the Richter scale, the magnitude R of an earthquake of intensity I is

$$R = \frac{\ln I - \ln I_0}{\ln 10}$$

where I_0 is the minimum intensity used for comparison. Assume that $I_0 = 1$.

(a) Find the intensity of the 1906 San Francisco earthquake ($R = 8.3$).

(b) Find the factor by which the intensity is increased if the Richter scale measurement is doubled.

(c) Find dR/dI.

73. Newton's Law of Cooling When an object is removed from a furnace and placed in an environment with a constant temperature of 80°F, its core temperature is 1500°F. One hour after it is removed, the core temperature is 1120°F. Find the core temperature 5 hours after the object is removed from the furnace.

74. Newton's Law of Cooling A container of hot liquid is placed in a freezer that is kept at a constant temperature of 20°F. The initial temperature of the liquid is 160°F. After 5 minutes, the liquid's temperature is 60°F. How much longer will it take for its temperature to decrease to 30°F?

True or False? In Exercises 75–78, determine whether the statement is true or false. If it is false, explain why or give an example that shows it is false.

75. In exponential growth, the rate of growth is constant.

76. In linear growth, the rate of growth is constant.

77. If prices are rising at a rate of 0.5% per month, then they are rising at a rate of 6% per year.

78. The differential equation modeling exponential growth is $dy/dx = ky$, where k is a constant.

6.3 Differential Equations: Separation of Variables

- **Recognize and solve differential equations that can be solved by separation of variables.**
- **Recognize and solve homogeneous differential equations.**
- **Use differential equations to model and solve applied problems.**

Separation of Variables

Consider a differential equation that can be written in the form

$$M(x) + N(y)\frac{dy}{dx} = 0$$

where M is a continuous function of x alone and N is a continuous function of y alone. As you saw in the preceding section, for this type of equation, all x terms can be collected with dx and all y terms with dy, and a solution can be obtained by integration. Such equations are said to be **separable,** and the solution procedure is called **separation of variables.** Below are some examples of differential equations that are separable.

Original Differential Equation	*Rewritten with Variables Separated*
$x^2 + 3y\dfrac{dy}{dx} = 0$	$3y\,dy = -x^2\,dx$
$(\sin x)y' = \cos x$	$dy = \cot x\,dx$
$\dfrac{xy'}{e^y + 1} = 2$	$\dfrac{1}{e^y + 1}\,dy = \dfrac{2}{x}\,dx$

 **EXAMPLE 1 Separation of Variables**

Find the general solution of $(x^2 + 4)\dfrac{dy}{dx} = xy$.

Solution   To begin, note that $y = 0$ is a solution. To find other solutions, assume that $y \neq 0$ and separate variables as shown.

$$(x^2 + 4)\,dy = xy\,dx \qquad \text{Differential form}$$

$$\frac{dy}{y} = \frac{x}{x^2 + 4}\,dx \qquad \text{Separate variables.}$$

Now, integrate to obtain

$$\int \frac{dy}{y} = \int \frac{x}{x^2 + 4}\,dx \qquad \text{Integrate.}$$

$$\ln|y| = \frac{1}{2}\ln(x^2 + 4) + C_1$$

$$\ln|y| = \ln\sqrt{x^2 + 4} + C_1$$

$$|y| = e^{C_1}\sqrt{x^2 + 4}$$

$$y = \pm e^{C_1}\sqrt{x^2 + 4}.$$

Because $y = 0$ is also a solution, you can write the general solution as

$$y = C\sqrt{x^2 + 4}. \qquad \text{General solution } (C = \pm e^{C_1})$$

NOTE   Be sure to check your solutions throughout this chapter. In Example 1, you can check the solution $y = C\sqrt{x^2 + 4}$ by differentiating and substituting into the original equation.

$$(x^2 + 4)\frac{dy}{dx} = xy$$

$$(x^2 + 4)\frac{Cx}{\sqrt{x^2 + 4}} \stackrel{?}{=} x\left(C\sqrt{x^2 + 4}\right)$$

$$Cx\sqrt{x^2 + 4} = Cx\sqrt{x^2 + 4}$$

So, the solution checks.

In some cases it is not feasible to write the general solution in the explicit form $y = f(x)$. The next example illustrates such a solution. Implicit differentiation can be used to verify this solution.

EXAMPLE 2 Finding a Particular Solution

Given the initial condition $y(0) = 1$, find the particular solution of the equation

$$xy \, dx + e^{-x^2}(y^2 - 1) \, dy = 0.$$

Solution   Note that $y = 0$ is a solution of the differential equation—but this solution does not satisfy the initial condition. So, you can assume that $y \neq 0$. To separate variables, you must rid the first term of y and the second term of e^{-x^2}. So, you should multiply by e^{x^2}/y and obtain the following.

$$xy \, dx + e^{-x^2}(y^2 - 1) \, dy = 0$$

$$e^{-x^2}(y^2 - 1) \, dy = -xy \, dx$$

$$\int \left(y - \frac{1}{y} \right) dy = \int -xe^{x^2} \, dx$$

$$\frac{y^2}{2} - \ln |y| = -\frac{1}{2}e^{x^2} + C$$

From the initial condition $y(0) = 1$, you have $\frac{1}{2} - 0 = -\frac{1}{2} + C$, which implies that $C = 1$. So, the particular solution has the implicit form

$$\frac{y^2}{2} - \ln |y| = -\frac{1}{2}e^{x^2} + 1$$

$$y^2 - \ln y^2 + e^{x^2} = 2.$$

You can check this by differentiating and rewriting to get the original equation.

EXAMPLE 3 Finding a Particular Solution Curve

Find the equation of the curve that passes through the point $(1, 3)$ and has a slope of y/x^2 at any point (x, y).

Solution   Because the slope of the curve is given by y/x^2, you have

$$\frac{dy}{dx} = \frac{y}{x^2}$$

with the initial condition $y(1) = 3$. Separating variables and integrating produces

$$\int \frac{dy}{y} = \int \frac{dx}{x^2}, \quad y \neq 0$$

$$\ln|y| = -\frac{1}{x} + C_1$$

$$y = e^{-(1/x)+C_1} = Ce^{-1/x}.$$

Because $y = 3$ when $x = 1$, it follows that $3 = Ce^{-1}$ and $C = 3e$. So, the equation of the specified curve is

$$y = (3e)e^{-1/x} = 3e^{(x-1)/x}, \quad x > 0.$$

Because the solution is not defined at $x = 0$ and the initial condition is given at $x = 1$, x is restricted to positive values. See Figure 6.12.

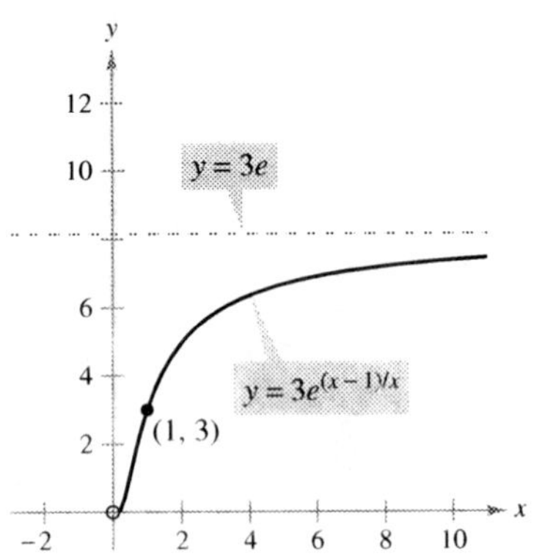

Figure 6.12

Homogeneous Differential Equations

Some differential equations that are not separable in x and y can be made separable by a change of variables. This is true for differential equations of the form $y' = f(x, y)$, where f is a **homogeneous function**. The function given by $f(x, y)$ is **homogeneous of degree n** if

NOTE The notation $f(x, y)$ is used to denote a function of two variables in much the same way as $f(x)$ denotes a function of one variable. You will study functions of two variables in detail in Chapter 13.

$$f(tx, ty) = t^n f(x, y)$$ Homogeneous function of degree n

where n is an integer.

EXAMPLE 4   Verifying Homogeneous Functions

a. $f(x, y) = x^2 y - 4x^3 + 3xy^2$ is a homogeneous function of degree 3 because

$$\begin{aligned}
f(tx, ty) &= (tx)^2(ty) - 4(tx)^3 + 3(tx)(ty)^2 \\
&= t^3(x^2 y) - t^3(4x^3) + t^3(3xy^2) \\
&= t^3(x^2 y - 4x^3 + 3xy^2) \\
&= t^3 f(x, y).
\end{aligned}$$

b. $f(x, y) = xe^{x/y} + y \sin(y/x)$ is a homogeneous function of degree 1 because

$$\begin{aligned}
f(tx, ty) &= txe^{tx/ty} + ty \sin \frac{ty}{tx} \\
&= t\left(xe^{x/y} + y \sin \frac{y}{x} \right) \\
&= tf(x, y).
\end{aligned}$$

c. $f(x, y) = x + y^2$ is *not* a homogeneous function because

$$f(tx, ty) = tx + t^2 y^2 = t(x + ty^2) \neq t^n(x + y^2).$$

d. $f(x, y) = x/y$ is a homogeneous function of degree 0 because

$$f(tx, ty) = \frac{tx}{ty} = t^0 \frac{x}{y}.$$

DEFINITION OF HOMOGENEOUS DIFFERENTIAL EQUATION

A **homogeneous differential equation** is an equation of the form

$$M(x, y)\,dx + N(x, y)dy = 0$$

where M and N are homogeneous functions of the same degree.

EXAMPLE 5   Testing for Homogeneous Differential Equations

a. $(x^2 + xy)\,dx + y^2\,dy = 0$ is homogeneous of degree 2.

b. $x^3\,dx = y^3\,dy$ is homogeneous of degree 3.

c. $(x^2 + 1)\,dx + y^2\,dy = 0$ is *not* a homogeneous differential equation.

To solve a homogeneous differential equation by the method of separation of variables, use the following change of variables theorem.

> **THEOREM 6.2 CHANGE OF VARIABLES FOR HOMOGENEOUS EQUATIONS**
>
> If $M(x, y)\,dx + N(x, y)\,dy = 0$ is homogeneous, then it can be transformed into a differential equation whose variables are separable by the substitution
>
> $$y = vx$$
>
> where v is a differentiable function of x.

EXAMPLE 6 Solving a Homogeneous Differential Equation

Find the general solution of

$$(x^2 - y^2)\,dx + 3xy\,dy = 0.$$

STUDY TIP The substitution $y = vx$ will yield a differential equation that is separable with respect to the variables x and v. You must write your final solution, however, in terms of x and y.

Solution Because $(x^2 - y^2)$ and $3xy$ are both homogeneous of degree 2, let $y = vx$ to obtain $dy = x\,dv + v\,dx$. Then, by substitution, you have

$$
\begin{aligned}
(x^2 - v^2 x^2)\,dx + 3x(vx)(x\,dv + v\,dx) &= 0 \\
(x^2 + 2v^2 x^2)\,dx + 3x^3 v\,dv &= 0 \\
x^2(1 + 2v^2)\,dx + x^2(3vx)\,dv &= 0.
\end{aligned}
$$

Dividing by x^2 and separating variables produces

$$
\begin{aligned}
(1 + 2v^2)\,dx &= -3vx\,dv \\
\int \frac{dx}{x} &= \int \frac{-3v}{1 + 2v^2}\,dv \\
\ln|x| &= -\frac{3}{4}\ln(1 + 2v^2) + C_1 \\
4\ln|x| &= -3\ln(1 + 2v^2) + \ln|C| \\
\ln x^4 &= \ln\left|C(1 + 2v^2)^{-3}\right| \\
x^4 &= C(1 + 2v^2)^{-3}.
\end{aligned}
$$

Substituting for v produces the following general solution.

$$
\begin{aligned}
x^4 &= C\left[1 + 2\left(\frac{y}{x}\right)^2\right]^{-3} \\
\left(1 + \frac{2y^2}{x^2}\right)^3 x^4 &= C \\
(x^2 + 2y^2)^3 &= Cx^2 \qquad \text{General solution}
\end{aligned}
$$

You can check this by differentiating and rewriting to get the original equation. ∎

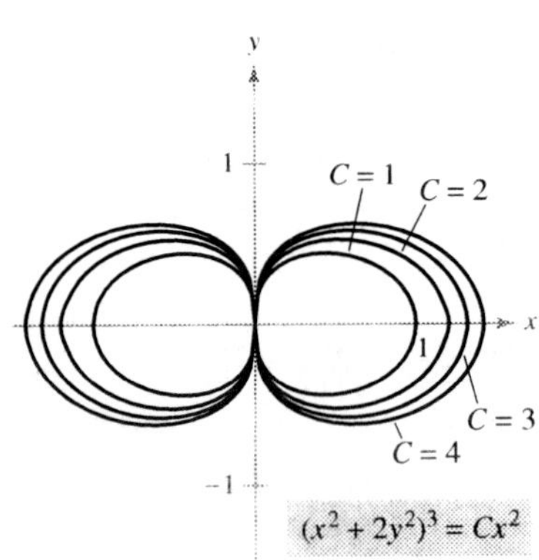

General solution of
$(x^2 - y^2)\,dx + 3xy\,dy = 0$
Figure 6.13

TECHNOLOGY If you have access to a graphing utility, try using it to graph several solutions of the equation in Example 6. For instance, Figure 6.13 shows the graphs of

$$(x^2 + 2y^2)^3 = Cx^2$$

for $C = 1, 2, 3,$ and 4.

Applications

EXAMPLE 7 Wildlife Population

The rate of change of the number of coyotes $N(t)$ in a population is directly proportional to $650 - N(t)$, where t is the time in years. When $t = 0$, the population is 300, and when $t = 2$, the population has increased to 500. Find the population when $t = 3$.

Solution   Because the rate of change of the population is proportional to $650 - N(t)$, you can write the following differential equation.

$$\frac{dN}{dt} = k(650 - N)$$

You can solve this equation using separation of variables.

$$dN = k(650 - N)\,dt \qquad \text{Differential form}$$

$$\frac{dN}{650 - N} = k\,dt \qquad \text{Separate variables.}$$

$$-\ln|650 - N| = kt + C_1 \qquad \text{Integrate.}$$

$$\ln|650 - N| = -kt - C_1 \qquad \text{Multiply each side by } -1.$$

$$650 - N = e^{-kt - C_1} \qquad \text{Assume } N < 650.$$

$$N = 650 - Ce^{-kt} \qquad \text{General solution}$$

Using $N = 300$ when $t = 0$, you can conclude that $C = 350$, which produces

$$N = 650 - 350e^{-kt}.$$

Then, using $N = 500$ when $t = 2$, it follows that

$$500 = 650 - 350e^{-2k} \quad \Longrightarrow \quad e^{-2k} = \tfrac{3}{7} \quad \Longrightarrow \quad k \approx 0.4236.$$

So, the model for the coyote population is

$$N = 650 - 350e^{-0.4236t}. \qquad \text{Model for population}$$

When $t = 3$, you can approximate the population to be

$$N = 650 - 350e^{-0.4236(3)} \approx 552 \text{ coyotes.}$$

The model for the population is shown in Figure 6.14. Note that $N = 650$ is the horizontal asymptote of the graph and is the *carrying capacity* of the model. You will learn more about carrying capacity in Section 6.4.

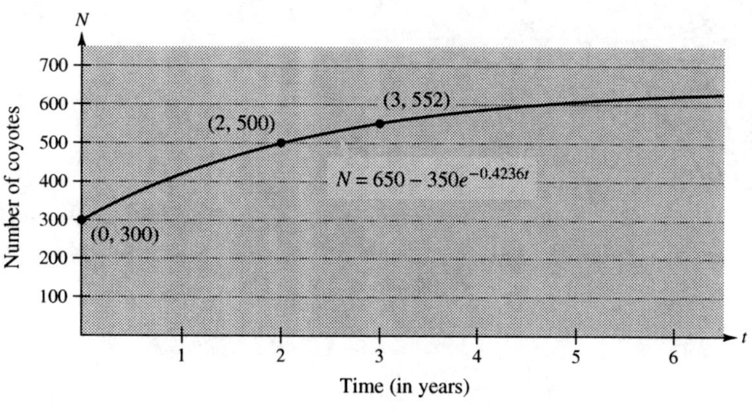

Figure 6.14

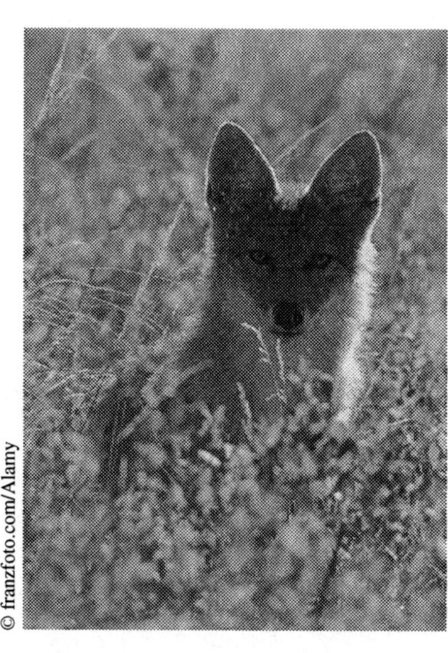

A common problem in electrostatics, thermodynamics, and hydrodynamics involves finding a family of curves, each of which is orthogonal to all members of a given family of curves. For example, Figure 6.15 shows a family of circles

$$x^2 + y^2 = C \qquad \text{Family of circles}$$

each of which intersects the lines in the family

$$y = Kx \qquad \text{Family of lines}$$

at right angles. Two such families of curves are said to be **mutually orthogonal,** and each curve in one of the families is called an **orthogonal trajectory** of the other family. In electrostatics, lines of force are orthogonal to the *equipotential curves.* In thermodynamics, the flow of heat across a plane surface is orthogonal to the *isothermal curves.* In hydrodynamics, the flow (stream) lines are orthogonal trajectories of the *velocity potential curves.*

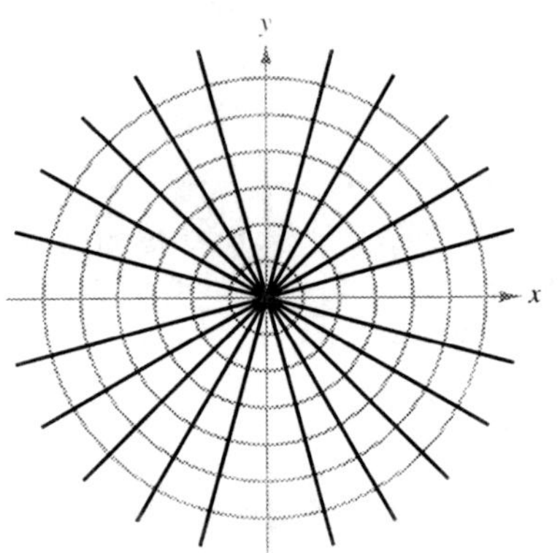

Each line $y = Kx$ is an orthogonal trajectory of the family of circles.

Figure 6.15

EXAMPLE 8 Finding Orthogonal Trajectories

Describe the orthogonal trajectories for the family of curves given by

$$y = \frac{C}{x}$$

for $C \neq 0$. Sketch several members of each family.

Solution   First, solve the given equation for C and write $xy = C$. Then, by differentiating implicitly with respect to x, you obtain the differential equation

$$xy' + y = 0 \qquad \text{Differential equation}$$

$$x\frac{dy}{dx} = -y$$

$$\frac{dy}{dx} = -\frac{y}{x}. \qquad \text{Slope of given family}$$

Because y' represents the slope of the given family of curves at (x, y), it follows that the orthogonal family has the negative reciprocal slope x/y. So,

$$\frac{dy}{dx} = \frac{x}{y}. \qquad \text{Slope of orthogonal family}$$

Now you can find the orthogonal family by separating variables and integrating.

$$\int y \, dy = \int x \, dx$$

$$\frac{y^2}{2} = \frac{x^2}{2} + C_1$$

$$y^2 - x^2 = K$$

The centers are at the origin, and the transverse axes are vertical for $K > 0$ and horizontal for $K < 0$. If $K = 0$, the orthogonal trajectories are the lines $y = \pm x$. If $K \neq 0$, the orthogonal trajectories are hyperbolas. Several trajectories are shown in Figure 6.16.

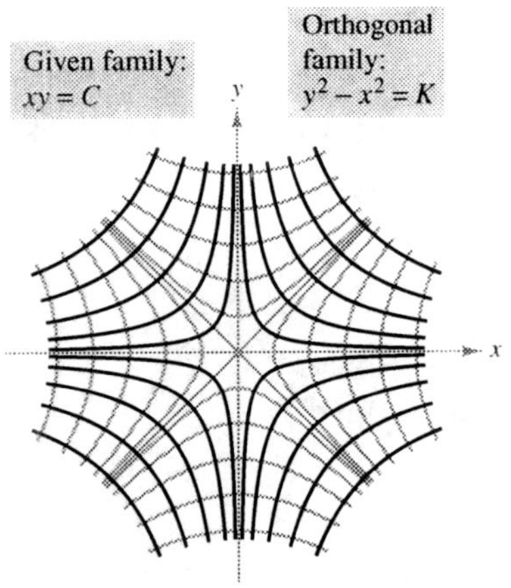

Orthogonal trajectories

Figure 6.16

EXAMPLE 9 Modeling Advertising Awareness

A new cereal product is introduced through an advertising campaign to a population of 1 million potential customers. The rate at which the population hears about the product is assumed to be proportional to the number of people who are not yet aware of the product. By the end of 1 year, half of the population has heard of the product. How many will have heard of it by the end of 2 years?

Solution Let y be the number (in millions) of people at time t who have heard of the product. This means that $(1 - y)$ is the number of people who have not heard of it, and dy/dt is the rate at which the population hears about the product. From the given assumption, you can write the differential equation as shown.

$$\frac{dy}{dt} = k(1 - y)$$

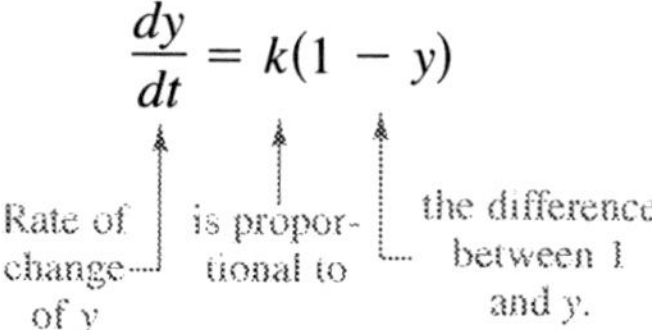

You can solve this equation using separation of variables.

$$dy = k(1 - y)\, dt \qquad \text{Differential form}$$

$$\frac{dy}{1 - y} = k\, dt \qquad \text{Separate variables.}$$

$$-\ln|1 - y| = kt + C_1 \qquad \text{Integrate.}$$

$$\ln|1 - y| = -kt - C_1 \qquad \text{Multiply each side by } -1.$$

$$1 - y = e^{-kt - C_1} \qquad \text{Assume } y < 1.$$

$$y = 1 - Ce^{-kt} \qquad \text{General solution}$$

To solve for the constants C and k, use the initial conditions. That is, because $y = 0$ when $t = 0$, you can determine that $C = 1$. Similarly, because $y = 0.5$ when $t = 1$, it follows that $0.5 = 1 - e^{-k}$, which implies that

$$k = \ln 2 \approx 0.693.$$

So, the particular solution is

$$y = 1 - e^{-0.693t}. \qquad \text{Particular solution}$$

This model is shown in Figure 6.17. Using the model, you can determine that the number of people who have heard of the product after 2 years is

$$y = 1 - e^{-0.693(2)}$$

$$\approx 0.75 \text{ or } 750{,}000 \text{ people.}$$

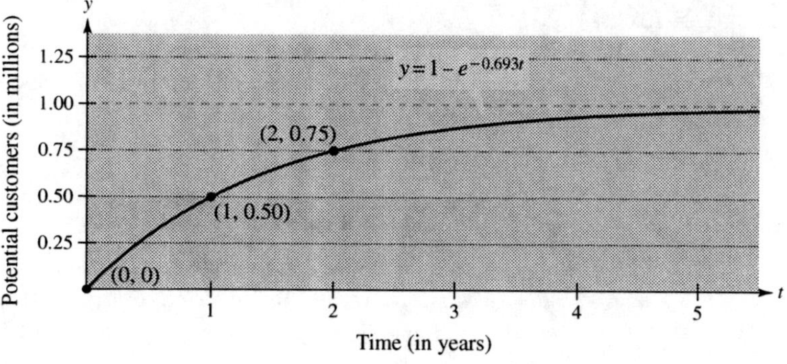

Figure 6.17

EXAMPLE 10 Modeling a Chemical Reaction

During a chemical reaction, substance A is converted into substance B at a rate that is proportional to the square of the amount of A. When $t = 0$, 60 grams of A is present, and after 1 hour ($t = 1$), only 10 grams of A remains unconverted. How much of A is present after 2 hours?

Solution Let y be the unconverted amount of substance A at any time t. From the given assumption about the conversion rate, you can write the differential equation as shown.

$$\frac{dy}{dt} = ky^2$$

Rate of change of y is proportional to   the square of y.

You can solve this equation using separation of variables.

$$dy = ky^2\, dt \qquad \text{Differential form}$$

$$\frac{dy}{y^2} = k\, dt \qquad \text{Separate variables}$$

$$-\frac{1}{y} = kt + C \qquad \text{Integrate.}$$

$$y = \frac{-1}{kt + C} \qquad \text{General solution}$$

To solve for the constants C and k, use the initial conditions. That is, because $y = 60$ when $t = 0$, you can determine that $C = -\frac{1}{60}$. Similarly, because $y = 10$ when $t = 1$, it follows that

$$10 = \frac{-1}{k - (1/60)}$$

which implies that $k = -\frac{1}{12}$. So, the particular solution is

$$y = \frac{-1}{(-1/12)t - (1/60)} \qquad \text{Substitute for } k \text{ and } C.$$

$$= \frac{60}{5t + 1}. \qquad \text{Particular solution}$$

Using the model, you can determine that the unconverted amount of substance A after 2 hours is

$$y = \frac{60}{5(2) + 1}$$

$$\approx 5.45 \text{ grams.}$$

In Figure 6.18, note that the chemical conversion is occurring rapidly during the first hour. Then, as more and more of substance A is converted, the conversion rate slows down.

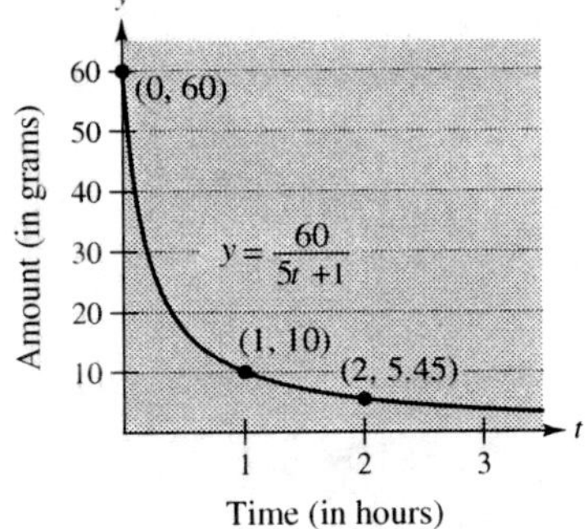

Figure 6.18

EXPLORATION

In Example 10, the rate of conversion was assumed to be proportional to the *square* of the unconverted amount. How would the result change if the rate of conversion were assumed to be proportional to the unconverted amount?

The next example describes a growth model called a **Gompertz growth model.** This model assumes that the rate of change of y is proportional to the product of y and the natural log of L/y, where L is the population limit.

EXAMPLE 11   Modeling Population Growth

A population of 20 wolves has been introduced into a national park. The forest service estimates that the maximum population the park can sustain is 200 wolves. After 3 years, the population is estimated to be 40 wolves. If the population follows a Gompertz growth model, how many wolves will there be 10 years after their introduction?

Solution   Let y be the number of wolves at any time t. From the given assumption about the rate of growth of the population, you can write the differential equation as shown.

$$\frac{dy}{dt} = ky \ln \frac{200}{y}$$

Rate of change of y — is proportional to — the product of y and — the natural log of the ratio of 200 and y.

Using separation of variables *or* a computer algebra system, you can find the general solution to be

$$y = 200e^{-Ce^{-kt}}. \qquad \text{General solution}$$

To solve for the constants C and k, use the initial conditions. That is, because $y = 20$ when $t = 0$, you can determine that

$$C = \ln 10 \approx 2.3026.$$

Similarly, because $y = 40$ when $t = 3$, it follows that

$$40 = 200e^{-2.3026e^{-3k}}$$

which implies that $k \approx 0.1194$. So, the particular solution is

$$y = 200e^{-2.3026e^{-0.1194t}}. \qquad \text{Particular solution}$$

Using the model, you can estimate the wolf population after 10 years to be

$$y = 200e^{-2.3026e^{-0.1194(10)}} \approx 100 \text{ wolves}.$$

In Figure 6.19, note that after 10 years the population has reached about half of the estimated maximum population. Try checking the growth model to see that it yields $y = 20$ when $t = 0$ and $y = 40$ when $t = 3$.

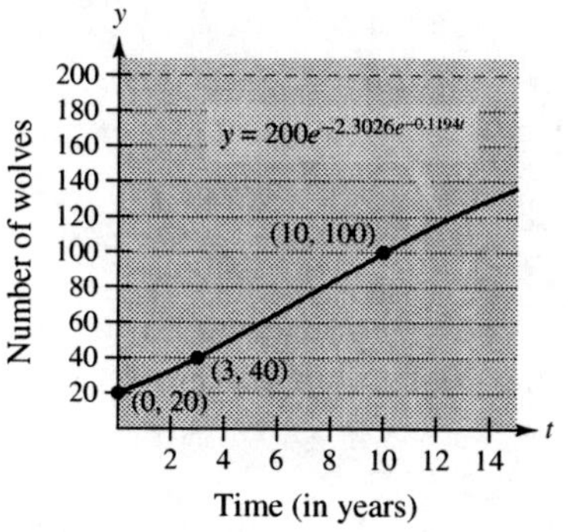

Figure 6.19

In genetics, a commonly used hybrid selection model is based on the differential equation

$$\frac{dy}{dt} = ky(1 - y)(a - by).$$

In this model, y represents the portion of the population that has a certain characteristic and t represents the time (measured in generations). The numbers a, b, and k are constants that depend on the genetic characteristic that is being studied.

EXAMPLE 12 Modeling Hybrid Selection

You are studying a population of beetles to determine how quickly characteristic D will pass from one generation to the next. At the beginning of your study ($t = 0$), you find that half the population has characteristic D. After four generations ($t = 4$), you find that 80% of the population has characteristic D. Use the hybrid selection model above with $a = 2$ and $b = 1$ to find the percent of the population that will have characteristic D after 10 generations.

Solution Using $a = 2$ and $b = 1$, the differential equation for the hybrid selection model is

$$\frac{dy}{dt} = ky(1 - y)(2 - y).$$

Using separation of variables or a computer algebra system, you can find the general solution to be

$$\frac{y(2 - y)}{(1 - y)^2} = Ce^{2kt}. \qquad \text{General solution}$$

To solve for the constants C and k, use the initial conditions. That is, because $y = 0.5$ when $t = 0$, you can determine that $C = 3$. Similarly, because $y = 0.8$ when $t = 4$, it follows that

$$\frac{0.8(1.2)}{(0.2)^2} = 3e^{8k}$$

which implies that

$$k = \frac{1}{8} \ln 8 \approx 0.2599.$$

So, the particular solution is

$$\frac{y(2 - y)}{(1 - y)^2} = 3e^{0.5199t}. \qquad \text{Particular solution}$$

Using the model, you can estimate the percent of the population that will have characteristic D after 10 generations to be given by

$$\frac{y(2 - y)}{(1 - y)^2} = 3e^{0.5199(10)}.$$

Using a computer algebra system, you can solve this equation for y to obtain $y \approx 0.96$ or 96% of the population. The graph of the model is shown in Figure 6.20. ∎

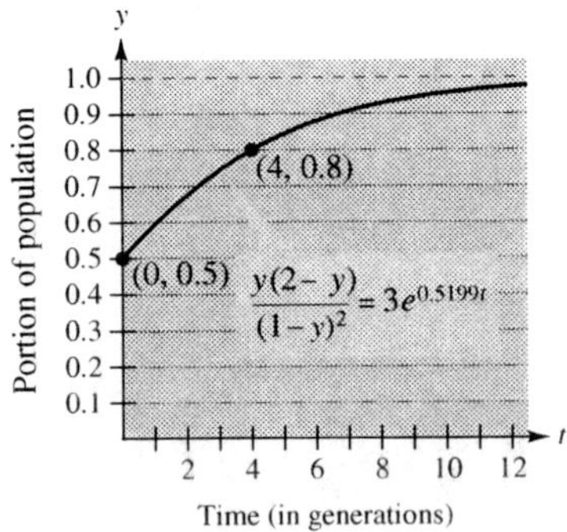

Figure 6.20

6.3 Exercises

In Exercises 1–14, find the general solution of the differential equation.

1. $\dfrac{dy}{dx} = \dfrac{x}{y}$

2. $\dfrac{dy}{dx} = \dfrac{3x^2}{y^2}$

3. $x^2 + 5y\dfrac{dy}{dx} = 0$

4. $\dfrac{dy}{dx} = \dfrac{x^2 - 3}{6y^2}$

5. $\dfrac{dr}{ds} = 0.75r$

6. $\dfrac{dr}{ds} = 0.75s$

7. $(2 + x)y' = 3y$

8. $xy' = y$

9. $yy' = 4\sin x$

10. $yy' = -8\cos(\pi x)$

11. $\sqrt{1 - 4x^2}\, y' = x$

12. $\sqrt{x^2 - 16}\, y' = 11x$

13. $y\ln x - xy' = 0$

14. $12yy' - 7e^x = 0$

In Exercises 15–24, find the particular solution that satisfies the initial condition.

Differential Equation	*Initial Condition*
15. $yy' - 2e^x = 0$	$y(0) = 3$
16. $\sqrt{x} + \sqrt{y}\,y' = 0$	$y(1) = 9$
17. $y(x + 1) + y' = 0$	$y(-2) = 1$
18. $2xy' - \ln x^2 = 0$	$y(1) = 2$
19. $y(1 + x^2)y' - x(1 + y^2) = 0$	$y(0) = \sqrt{3}$
20. $y\sqrt{1 - x^2}\,y' - x\sqrt{1 - y^2} = 0$	$y(0) = 1$
21. $\dfrac{du}{dv} = uv\sin v^2$	$u(0) = 1$
22. $\dfrac{dr}{ds} = e^{r-2s}$	$r(0) = 0$
23. $dP - kP\,dt = 0$	$P(0) = P_0$
24. $dT + k(T - 70)\,dt = 0$	$T(0) = 140$

In Exercises 25–28, find an equation of the graph that passes through the point and has the given slope.

25. $(0, 2), \quad y' = \dfrac{x}{4y}$

26. $(1, 1), \quad y' = -\dfrac{9x}{16y}$

27. $(9, 1), \quad y' = \dfrac{y}{2x}$

28. $(8, 2), \quad y' = \dfrac{2y}{3x}$

In Exercises 29 and 30, find all functions f having the indicated property.

29. The tangent to the graph of f at the point (x, y) intersects the x-axis at $(x + 2, 0)$.

30. All tangents to the graph of f pass through the origin.

In Exercises 31–38, determine whether the function is homogeneous, and if it is, determine its degree.

31. $f(x, y) = x^3 - 4xy^2 + y^3$

32. $f(x, y) = x^3 + 3x^2y^2 - 2y^2$

33. $f(x, y) = \dfrac{x^2y^2}{\sqrt{x^2 + y^2}}$

34. $f(x, y) = \dfrac{xy}{\sqrt{x^2 + y^2}}$

35. $f(x, y) = 2\ln xy$

36. $f(x, y) = \tan(x + y)$

37. $f(x, y) = 2\ln\dfrac{x}{y}$

38. $f(x, y) = \tan\dfrac{y}{x}$

In Exercises 39–44, solve the homogeneous differential equation.

39. $y' = \dfrac{x + y}{2x}$

40. $y' = \dfrac{x^3 + y^3}{xy^2}$

41. $y' = \dfrac{x - y}{x + y}$

42. $y' = \dfrac{x^2 + y^2}{2xy}$

43. $y' = \dfrac{xy}{x^2 - y^2}$

44. $y' = \dfrac{2x + 3y}{x}$

In Exercises 45–48, find the particular solution that satisfies the initial condition.

Differential Equation	*Initial Condition*
45. $x\,dy - (2xe^{-y/x} + y)\,dx = 0$	$y(1) = 0$
46. $-y^2\,dx + x(x + y)\,dy = 0$	$y(1) = 1$
47. $\left(x\sec\dfrac{y}{x} + y\right)dx - x\,dy = 0$	$y(1) = 0$
48. $(2x^2 + y^2)\,dx + xy\,dy = 0$	$y(1) = 0$

Slope Fields **In Exercises 49–52, sketch a few solutions of the differential equation on the slope field and then find the general solution analytically. To print an enlarged copy of the graph, go to the website *www.mathgraphs.com*.**

49. $\dfrac{dy}{dx} = x$

50. $\dfrac{dy}{dx} = -\dfrac{x}{y}$

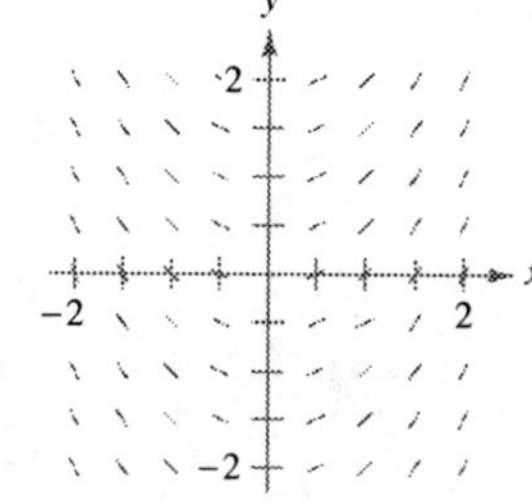

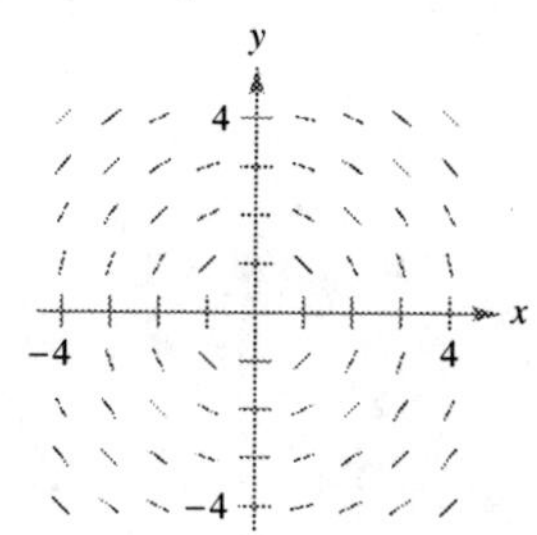

51. $\dfrac{dy}{dx} = 4 - y$

52. $\dfrac{dy}{dx} = 0.25x(4 - y)$

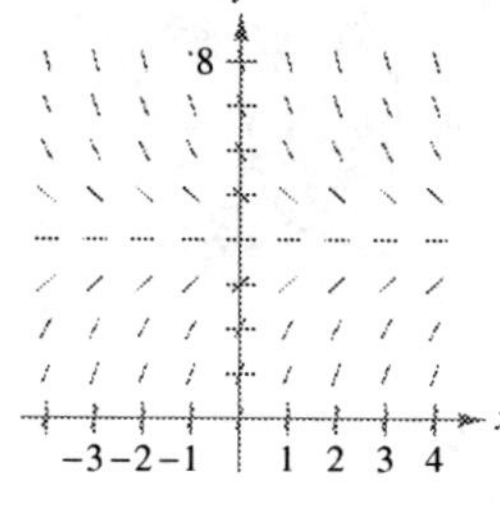

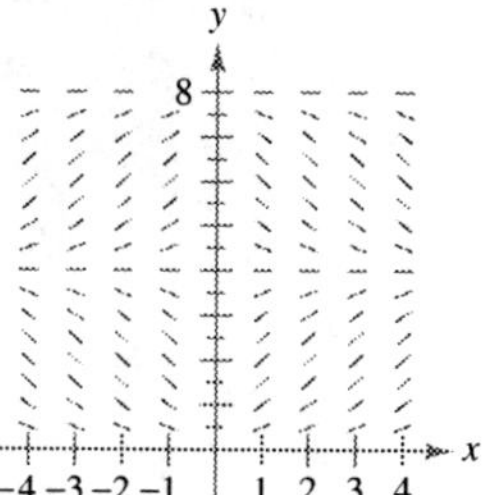

Euler's Method **In Exercises 53–56, (a) use Euler's Method with a step size of $h = 0.1$ to approximate the particular solution of the initial value problem at the given x-value, (b) find the exact solution of the differential equation analytically, and (c) compare the solutions at the given x-value.**

Differential Equation	*Initial Condition*	*x-value*
53. $\dfrac{dy}{dx} = -6xy$	$(0, 5)$	$x = 1$
54. $\dfrac{dy}{dx} + 6xy^2 = 0$	$(0, 3)$	$x = 1$
55. $\dfrac{dy}{dx} = \dfrac{2x + 12}{3y^2 - 4}$	$(1, 2)$	$x = 2$
56. $\dfrac{dy}{dx} = 2x(1 + y^2)$	$(1, 0)$	$x = 1.5$

57. *Radioactive Decay* The rate of decomposition of radioactive radium is proportional to the amount present at any time. The half-life of radioactive radium is 1599 years. What percent of a present amount will remain after 50 years?

58. *Chemical Reaction* In a chemical reaction, a certain compound changes into another compound at a rate proportional to the unchanged amount. If initially there is 40 grams of the original compound, and there is 35 grams after 1 hour, when will 75 percent of the compound be changed?

Slope Fields **In Exercises 59–62, (a) write a differential equation for the statement, (b) match the differential equation with a possible slope field, and (c) verify your result by using a graphing utility to graph a slope field for the differential equation. [The slope fields are labeled (a), (b), (c), and (d).] To print an enlarged copy of the graph, go to the website *www.mathgraphs.com*.**

(a)
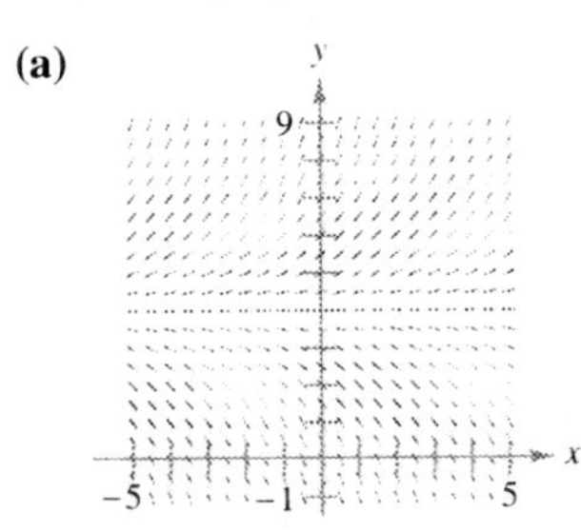

(b)
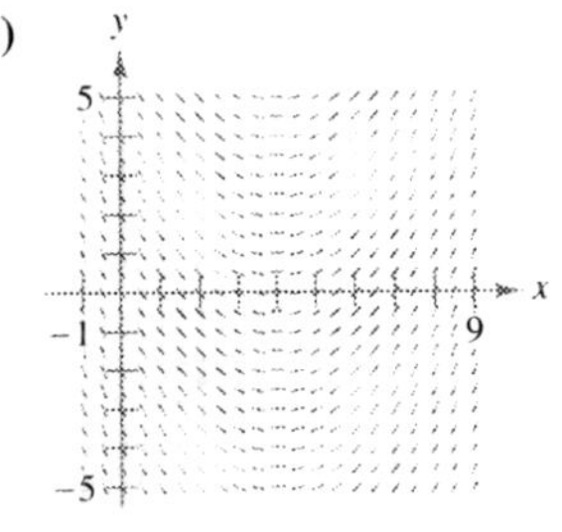

(c)
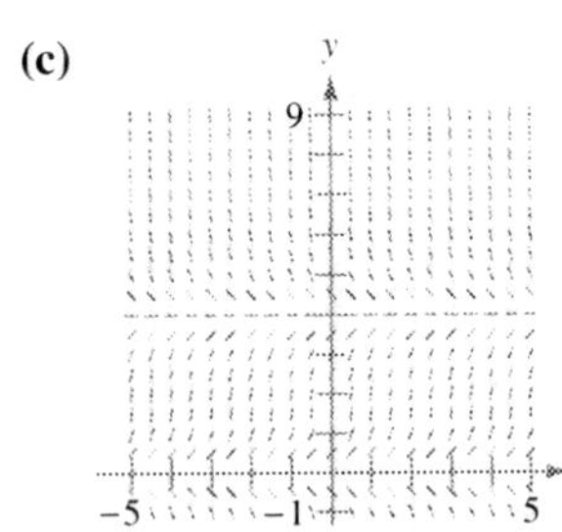

(d)
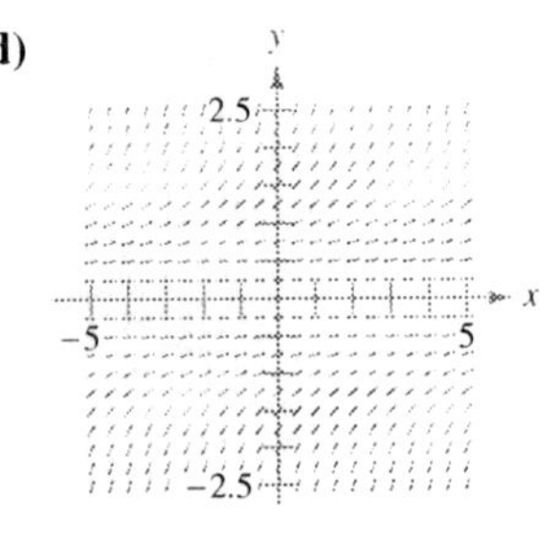

59. The rate of change of y with respect to x is proportional to the difference between y and 4.

60. The rate of change of y with respect to x is proportional to the difference between x and 4.

61. The rate of change of y with respect to x is proportional to the product of y and the difference between y and 4.

62. The rate of change of y with respect to x is proportional to y^2.

63. *Weight Gain* A calf that weighs 60 pounds at birth gains weight at the rate

$$\frac{dw}{dt} = k(1200 - w)$$

where w is weight in pounds and t is time in years.

(a) Solve the differential equation.

(b) Use a graphing utility to graph the solution when $k = 0.8$, 0.9, and 1.

(c) If the animal is sold when its weight reaches 800 pounds, find the time of sale for each of the models in part (b).

(d) What is the maximum weight of the animal for each of the models?

64. *Weight Gain* A calf that weighs w_0 pounds at birth gains weight at the rate

$$\frac{dw}{dt} = 1200 - w$$

where w is weight in pounds and t is time in years. Solve the differential equation.

In Exercises 65–70, find the orthogonal trajectories of the family. Use a graphing utility to graph several members of each family.

65. $x^2 + y^2 = C$

66. $x^2 - 2y^2 = C$

67. $x^2 = Cy$

68. $y^2 = 2Cx$

69. $y^2 = Cx^3$

70. $y = Ce^x$

71. *Biology* At any time t, the rate of growth of the population N of deer in a state park is proportional to the product of N and $L - N$, where $L = 500$ is the maximum number of deer the park can sustain. When $t = 0$, $N = 100$, and when $t = 4$, $N = 200$. Write N as a function of t.

72. *Sales Growth* The rate of change in sales S (in thousands of units) of a new product is proportional to the product of S and $L - S$. L (in thousands of units) is the estimated maximum level of sales, and $S = 10$ when $t = 0$. Write and solve the differential equation for this sales model.

Chemical Reaction **In Exercises 73 and 74, use the chemical reaction model given in Example 10 to find the amount y as a function of t, and use a graphing utility to graph the function.**

73. $y = 45$ grams when $t = 0$; $y = 4$ grams when $t = 2$

74. $y = 75$ grams when $t = 0$; $y = 12$ grams when $t = 1$

In Exercises 75 and 76, use the Gompertz growth model described in Example 11 to find the growth function, and sketch its graph.

75. $L = 500$; $y = 100$ when $t = 0$; $y = 150$ when $t = 2$

76. $L = 5000$; $y = 500$ when $t = 0$; $y = 625$ when $t = 1$

77. *Biology* A population of eight beavers has been introduced into a new wetlands area. Biologists estimate that the maximum population the wetlands can sustain is 60 beavers. After 3 years, the population is 15 beavers. If the population follows a Gompertz growth model, how many beavers will be present in the wetlands after 10 years?

78. *Biology* A population of 30 rabbits has been introduced into a new region. It is estimated that the maximum population the region can sustain is 400 rabbits. After 1 year, the population is estimated to be 90 rabbits. If the population follows a Gompertz growth model, how many rabbits will be present after 3 years?

Biology **In Exercises 79 and 80, use the hybrid selection model described in Example 12 to find the percent of the population that has the given characteristic.**

79. You are studying a population of mayflies to determine how quickly characteristic A will pass from one generation to the next. At the start of the study, half the population has characteristic A. After four generations, 75% of the population has characteristic A. Find the percent of the population that will have characteristic A after 10 generations. (Assume $a = 2$ and $b = 1$.)

80. A research team is studying a population of snails to determine how quickly characteristic B will pass from one generation to the next. At the start of the study, 40% of the snails have characteristic B. After five generations, 80% of the population has characteristic B. Find the percent of the population that will have characteristic B after eight generations. (Assume $a = 2$ and $b = 1$.)

81. *Chemical Mixture* A 100-gallon tank is full of a solution containing 25 pounds of a concentrate. Starting at time $t = 0$, distilled water is admitted to the tank at the rate of 5 gallons per minute, and the well-stirred solution is withdrawn at the same rate, as shown in the figure.

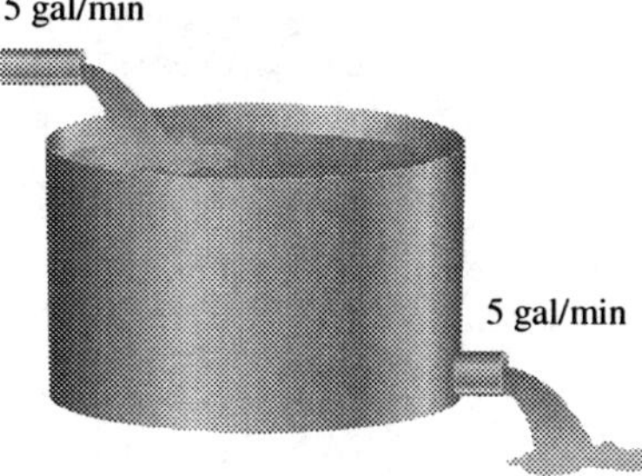

(a) Find the amount Q of the concentrate in the solution as a function of t. (*Hint:* $Q' + Q/20 = 0$)

(b) Find the time when the amount of concentrate in the tank reaches 15 pounds.

82. *Chemical Mixture* A 200-gallon tank is half full of distilled water. At time $t = 0$, a solution containing 0.5 pound of concentrate per gallon enters the tank at the rate of 5 gallons per minute, and the well-stirred mixture is withdrawn at the same rate. Find the amount Q of concentrate in the tank after 30 minutes. $\left(\text{*Hint:* } Q' + Q/20 = \frac{5}{2}\right)$

83. *Chemical Reaction* In a chemical reaction, a compound changes into another compound at a rate proportional to the unchanged amount, according to the model

$$\frac{dy}{dt} = ky.$$

(a) Solve the differential equation.

(b) If the initial amount of the original compound is 20 grams, and the amount remaining after 1 hour is 16 grams, when will 75% of the compound have been changed?

84. *Safety* Assume that the rate of change in the number of miles s of road cleared per hour by a snowplow is inversely proportional to the depth h of snow. That is,

$$\frac{ds}{dh} = \frac{k}{h}.$$

Find s as a function of h if $s = 25$ miles when $h = 2$ inches and $s = 12$ miles when $h = 6$ inches ($2 \leq h \leq 15$).

85. *Chemistry* A wet towel hung from a clothesline to dry loses moisture through evaporation at a rate proportional to its moisture content. If after 1 hour the towel has lost 40% of its original moisture content, after how long will it have lost 80%?

86. *Biology* Let x and y be the sizes of two internal organs of a particular mammal at time t. Empirical data indicate that the relative growth rates of these two organs are equal, and can be modeled by

$$\frac{1}{x}\frac{dx}{dt} = \frac{1}{y}\frac{dy}{dt}.$$

Use this differential equation to write y as a function of x.

87. *Population Growth* When predicting population growth, demographers must consider birth and death rates as well as the net change caused by the difference between the rates of immigration and emigration. Let P be the population at time t and let N be the net increase per unit time due to the difference between immigration and emigration. So, the rate of growth of the population is given by

$$\frac{dP}{dt} = kP + N, \quad N \text{ is constant.}$$

Solve this differential equation to find P as a function of time.

88. *Meteorology* The barometric pressure y (in inches of mercury) at an altitude of x miles above sea level decreases at a rate proportional to the current pressure according to the model

$$\frac{dy}{dx} = -0.2y$$

where $y = 29.92$ inches when $x = 0$. Find the barometric pressure (a) at the top of Mt. St. Helens (8364 feet) and (b) at the top of Mt. McKinley (20,320 feet).

89. *Investment* A large corporation starts at time $t = 0$ to invest part of its receipts at a rate of P dollars per year in a fund for future corporate expansion. Assume that the fund earns r percent interest per year compounded continuously. So, the rate of growth of the amount A in the fund is given by

$$\frac{dA}{dt} = rA + P$$

where $A = 0$ when $t = 0$. Solve this differential equation for A as a function of t.

Investment **In Exercises 90–92, use the result of Exercise 89.**

90. Find A for each situation.

   (a) $P = \$275{,}000$, $r = 8\%$, and $t = 10$ years

   (b) $P = \$550{,}000$, $r = 5.9\%$, and $t = 25$ years

91. Find P if the corporation needs $\$260{,}000{,}000$ in 8 years and the fund earns $7\frac{1}{4}\%$ interest compounded continuously.

92. Find t if the corporation needs $\$1{,}000{,}000$ and it can invest $\$125{,}000$ per year in a fund earning 8% interest compounded continuously.

In Exercises 93 and 94, use the Gompertz growth model described in Example 11.

93. (a) Use a graphing utility to graph the slope field for the growth model when $k = 0.02$ and $L = 5000$.

   (b) Describe the behavior of the graph as $t \to \infty$.

   (c) Solve the growth model for $L = 5000$, $y_0 = 500$, and $k = 0.02$.

   (d) Graph the equation you found in part (c). Determine the concavity of the graph.

94. (a) Use a graphing utility to graph the slope field for the growth model when $k = 0.05$ and $L = 1000$.

   (b) Describe the behavior of the graph as $t \to \infty$.

   (c) Solve the growth model for $L = 1000$, $y_0 = 100$, and $k = 0.05$.

   (d) Graph the equation you found in part (c). Determine the concavity of the graph.

WRITING ABOUT CONCEPTS

95. In your own words, describe how to recognize and solve differential equations that can be solved by separation of variables.

96. State the test for determining if a differential equation is homogeneous. Give an example.

97. In your own words, describe the relationship between two families of curves that are mutually orthogonal.

CAPSTONE

98. Determine whether the differential equation is separable. If the equation is separable, rewrite it in the form $N(y)dy = M(x)dx$. (Do not solve the differential equation.)

   (a) $y(1 + x)\,dx + x\,dy = 0$

   (b) $y' = y^{1/2}$

   (c) $y' + xy = 5$

   (d) $y' = x - xy - y + 1$

True or False? **In Exercises 99–102, determine whether the statement is true or false. If it is false, explain why or give an example that shows it is false.**

99. The function $y = 0$ is always a solution of a differential equation that can be solved by separation of variables.

100. The differential equation

$$y' = xy - 2y + x - 2$$

can be written in separated variables form.

101. The function

$$f(x, y) = x^2 - 4xy + 6y^2 + 1$$

is homogeneous.

102. The families

$$x^2 + y^2 = 2Cy \text{ and } x^2 + y^2 = 2Kx$$

are mutually orthogonal.

PUTNAM EXAM CHALLENGE

103. A not uncommon calculus mistake is to believe that the product rule for derivatives says that $(fg)' = f'g'$. If $f(x) = e^{x^2}$, determine, with proof, whether there exists an open interval (a, b) and a nonzero function g defined on (a, b) such that this wrong product rule is true for x in (a, b).

This problem was composed by the Committee on the Putnam Prize Competition.
© The Mathematical Association of America. All rights reserved.

A Proofs of Selected Theorems

THEOREM 2.2 PROPERTIES OF LIMITS (PROPERTIES 2, 3, 4, AND 5) (PAGE 79)

Let b and c be real numbers, let n be a positive integer, and let f and g be functions with the following limits.

$$\lim_{x \to c} f(x) = L \quad \text{and} \quad \lim_{x \to c} g(x) = K$$

2. Sum or difference: $\quad \lim\limits_{x \to c} [f(x) \pm g(x)] = L \pm K$

3. Product: $\quad \lim\limits_{x \to c} [f(x)g(x)] = LK$

4. Quotient: $\quad \lim\limits_{x \to c} \dfrac{f(x)}{g(x)} = \dfrac{L}{K}, \quad$ provided $K \neq 0$

5. Power: $\quad \lim\limits_{x \to c} [f(x)]^n = L^n$

PROOF To prove Property 2, choose $\varepsilon > 0$. Because $\varepsilon/2 > 0$, you know that there exists $\delta_1 > 0$ such that $0 < |x - c| < \delta_1$ implies $|f(x) - L| < \varepsilon/2$. You also know that there exists $\delta_2 > 0$ such that $0 < |x - c| < \delta_2$ implies $|g(x) - K| < \varepsilon/2$. Let δ be the smaller of δ_1 and δ_2; then $0 < |x - c| < \delta$ implies that

$$|f(x) - L| < \frac{\varepsilon}{2} \quad \text{and} \quad |g(x) - K| < \frac{\varepsilon}{2}.$$

So, you can apply the triangle inequality to conclude that

$$|[f(x) + g(x)] - (L + K)| \leq |f(x) - L| + |g(x) - K| < \frac{\varepsilon}{2} + \frac{\varepsilon}{2} = \varepsilon$$

which implies that

$$\lim_{x \to c} [f(x) + g(x)] = L + K = \lim_{x \to c} f(x) + \lim_{x \to c} g(x).$$

The proof that

$$\lim_{x \to c} [f(x) - g(x)] = L - K$$

is similar.

To prove Property 3, given that

$$\lim_{x \to c} f(x) = L \quad \text{and} \quad \lim_{x \to c} g(x) = K$$

you can write

$$f(x)g(x) = [f(x) - L][g(x) - K] + [Lg(x) + Kf(x)] - LK.$$

Because the limit of $f(x)$ is L, and the limit of $g(x)$ is K, you have

$$\lim_{x \to c} [f(x) - L] = 0 \quad \text{and} \quad \lim_{x \to c} [g(x) - K] = 0.$$

Let $0 < \varepsilon < 1$. Then there exists $\delta > 0$ such that if $0 < |x - c| < \delta$, then

$$|f(x) - L - 0| < \varepsilon \quad \text{and} \quad |g(x) - K - 0| < \varepsilon$$

which implies that

$$|[f(x) - L][g(x) - K] - 0| = |f(x) - L|\,|g(x) - K| < \varepsilon\varepsilon < \varepsilon.$$

So,

$$\lim_{x \to c} [f(x) - L][g(x) - K] = 0.$$

Furthermore, by Property 1, you have

$$\lim_{x \to c} Lg(x) = LK \quad \text{and} \quad \lim_{x \to c} Kf(x) = KL.$$

Finally, by Property 2, you obtain

$$\lim_{x \to c} f(x)g(x) = \lim_{x \to c} [f(x) - L][g(x) - K] + \lim_{x \to c} Lg(x) + \lim_{x \to c} Kf(x) - \lim_{x \to c} LK$$

$$= 0 + LK + KL - LK$$

$$= LK.$$

To prove Property 4, note that it is sufficient to prove that

$$\lim_{x \to c} \frac{1}{g(x)} = \frac{1}{K}.$$

Then you can use Property 3 to write

$$\lim_{x \to c} \frac{f(x)}{g(x)} = \lim_{x \to c} f(x) \frac{1}{g(x)} = \lim_{x \to c} f(x) \cdot \lim_{x \to c} \frac{1}{g(x)} = \frac{L}{K}.$$

Let $\varepsilon > 0$. Because $\lim\limits_{x \to c} g(x) = K$, there exists $\delta_1 > 0$ such that if

$$0 < |x - c| < \delta_1, \text{ then } |g(x) - K| < \frac{|K|}{2}$$

which implies that

$$|K| = |g(x) + [|K| - g(x)]| \le |g(x)| + ||K| - g(x)| < |g(x)| + \frac{|K|}{2}.$$

That is, for $0 < |x - c| < \delta_1$,

$$\frac{|K|}{2} < |g(x)| \quad \text{or} \quad \frac{1}{|g(x)|} < \frac{2}{|K|}.$$

Similarly, there exists a $\delta_2 > 0$ such that if $0 < |x - c| < \delta_2$, then

$$|g(x) - K| < \frac{|K|^2}{2}\,\varepsilon.$$

Let δ be the smaller of δ_1 and δ_2. For $0 < |x - c| < \delta$, you have

$$\left| \frac{1}{g(x)} - \frac{1}{K} \right| = \left| \frac{K - g(x)}{g(x)K} \right| = \frac{1}{|K|} \cdot \frac{1}{|g(x)|} |K - g(x)| < \frac{1}{|K|} \cdot \frac{2}{|K|} \frac{|K|^2}{2} \varepsilon = \varepsilon.$$

So, $\lim\limits_{x \to c} \dfrac{1}{g(x)} = \dfrac{1}{K}.$

Finally, the proof of Property 5 can be obtained by a straightforward application of mathematical induction coupled with Property 3. ∎

THEOREM 2.4 THE LIMIT OF A FUNCTION INVOLVING A RADICAL (PAGE 80)

Let n be a positive integer. The following limit is valid for all c if n is odd, and is valid for $c > 0$ if n is even.

$$\lim_{x \to c} \sqrt[n]{x} = \sqrt[n]{c}$$

(**PROOF**) Consider the case for which $c > 0$ and n is any positive integer. For a given $\varepsilon > 0$, you need to find $\delta > 0$ such that

$$\left| \sqrt[n]{x} - \sqrt[n]{c} \right| < \varepsilon \quad \text{whenever} \quad 0 < |x - c| < \delta$$

which is the same as saying

$$-\varepsilon < \sqrt[n]{x} - \sqrt[n]{c} < \varepsilon \quad \text{whenever} \quad -\delta < x - c < \delta.$$

Assume $\varepsilon < \sqrt[n]{c}$, which implies that $0 < \sqrt[n]{c} - \varepsilon < \sqrt[n]{c}$. Now, let δ be the smaller of the two numbers

$$c - \left(\sqrt[n]{c} - \varepsilon \right)^n \quad \text{and} \quad \left(\sqrt[n]{c} + \varepsilon \right)^n - c.$$

Then you have

$$
\begin{aligned}
-\delta &< x - c & &< \delta \\
-\left[c - \left(\sqrt[n]{c} - \varepsilon \right)^n \right] &< x - c & &< \left(\sqrt[n]{c} + \varepsilon \right)^n - c \\
\left(\sqrt[n]{c} - \varepsilon \right)^n - c &< x - c & &< \left(\sqrt[n]{c} + \varepsilon \right)^n - c \\
\left(\sqrt[n]{c} - \varepsilon \right)^n &< x & &< \left(\sqrt[n]{c} + \varepsilon \right)^n \\
\sqrt[n]{c} - \varepsilon &< \sqrt[n]{x} & &< \sqrt[n]{c} + \varepsilon \\
-\varepsilon &< \sqrt[n]{x} - \sqrt[n]{c} < \varepsilon. &
\end{aligned}
$$

$\blacksquare$

THEOREM 2.5 THE LIMIT OF A COMPOSITE FUNCTION (PAGE 81)

If f and g are functions such that $\displaystyle\lim_{x \to c} g(x) = L$ and $\displaystyle\lim_{x \to L} f(x) = f(L)$, then

$$\lim_{x \to c} f(g(x)) = f\left(\lim_{x \to c} g(x) \right) = f(L).$$

(**PROOF**) For a given $\varepsilon > 0$, you must find $\delta > 0$ such that

$$|f(g(x)) - f(L)| < \varepsilon \quad \text{whenever} \quad 0 < |x - c| < \delta.$$

Because the limit of $f(x)$ as $x \to L$ is $f(L)$, you know there exists $\delta_1 > 0$ such that

$$|f(u) - f(L)| < \varepsilon \quad \text{whenever} \quad |u - L| < \delta_1.$$

Moreover, because the limit of $g(x)$ as $x \to c$ is L, you know there exists $\delta > 0$ such that

$$|g(x) - L| < \delta_1 \quad \text{whenever} \quad 0 < |x - c| < \delta.$$

Finally, letting $u = g(x)$, you have

$$|f(g(x)) - f(L)| < \varepsilon \quad \text{whenever} \quad 0 < |x - c| < \delta.$$

$\blacksquare$

THEOREM 2.7 FUNCTIONS THAT AGREE AT ALL BUT ONE POINT (PAGE 82)

Let c be a real number and let $f(x) = g(x)$ for all $x \neq c$ in an open interval containing c. If the limit of $g(x)$ as x approaches c exists, then the limit of $f(x)$ also exists and

$$\lim_{x \to c} f(x) = \lim_{x \to c} g(x).$$

(**PROOF**) Let L be the limit of $g(x)$ as $x \to c$. Then, for each $\varepsilon > 0$ there exists a $\delta > 0$ such that $f(x) = g(x)$ in the open intervals $(c - \delta, c)$ and $(c, c + \delta)$, and

$$|g(x) - L| < \varepsilon \quad \text{whenever} \quad 0 < |x - c| < \delta.$$

Because $f(x) = g(x)$ for all x in the open interval other than $x = c$, it follows that

$$|f(x) - L| < \varepsilon \quad \text{whenever} \quad 0 < |x - c| < \delta.$$

So, the limit of $f(x)$ as $x \to c$ is also L. ∎

THEOREM 2.8 THE SQUEEZE THEOREM (PAGE 85)

If $h(x) \leq f(x) \leq g(x)$ for all x in an open interval containing c, except possibly at c itself, and if

$$\lim_{x \to c} h(x) = L = \lim_{x \to c} g(x)$$

then $\lim\limits_{x \to c} f(x)$ exists and is equal to L.

(**PROOF**) For $\varepsilon > 0$ there exist $\delta_1 > 0$ and $\delta_2 > 0$ such that

$$|h(x) - L| < \varepsilon \quad \text{whenever} \quad 0 < |x - c| < \delta_1$$

and

$$|g(x) - L| < \varepsilon \quad \text{whenever} \quad 0 < |x - c| < \delta_2.$$

Because $h(x) \leq f(x) \leq g(x)$ for all x in an open interval containing c, except possibly at c itself, there exists $\delta_3 > 0$ such that $h(x) \leq f(x) \leq g(x)$ for $0 < |x - c| < \delta_3$. Let δ be the smallest of δ_1, δ_2, and δ_3. Then, if $0 < |x - c| < \delta$, it follows that $|h(x) - L| < \varepsilon$ and $|g(x) - L| < \varepsilon$, which implies that

$$-\varepsilon < h(x) - L < \varepsilon \quad \text{and} \quad -\varepsilon < g(x) - L < \varepsilon$$
$$L - \varepsilon < h(x) \quad \text{and} \quad g(x) < L + \varepsilon.$$

Now, because $h(x) \leq f(x) \leq g(x)$, it follows that $L - \varepsilon < f(x) < L + \varepsilon$, which implies that $|f(x) - L| < \varepsilon$. Therefore,

$$\lim_{x \to c} f(x) = L. \quad ∎$$

THEOREM 2.11 PROPERTIES OF CONTINUITY (PAGE 95)

If b is a real number and f and g are continuous at $x = c$, then the following functions are also continuous at c.

1. Scalar multiple: bf

2. Sum or difference: $f \pm g$

3. Product: fg

4. Quotient: $\dfrac{f}{g}, \quad$ if $g(c) \neq 0$

PROOF Because f and g are continuous at $x = c$, you can write

$$\lim_{x \to c} f(x) = f(c) \quad \text{and} \quad \lim_{x \to c} g(x) = g(c).$$

For Property 1, when b is a real number, it follows from Theorem 2.2 that

$$\lim_{x \to c} [(bf)(x)] = \lim_{x \to c} [bf(x)] = b \lim_{x \to c} [f(x)] = b f(c) = (bf)(c).$$

Thus, bf is continuous at $x = c$.

For Property 2, it follows from Theorem 2.2 that

$$\begin{aligned}
\lim_{x \to c} (f \pm g)(x) &= \lim_{x \to c} [f(x) \pm g(x)] \\
&= \lim_{x \to c} [f(x)] \pm \lim_{x \to c} [g(x)] \\
&= f(c) \pm g(c) \\
&= (f \pm g)(c).
\end{aligned}$$

Thus, $f \pm g$ is continuous at $x = c$.

For Property 3, it follows from Theorem 2.2 that

$$\begin{aligned}
\lim_{x \to c} (fg)(x) &= \lim_{x \to c} [f(x)g(x)] \\
&= \lim_{x \to c} [f(x)] \lim_{x \to c} [g(x)] \\
&= f(c)g(c) \\
&= (fg)(c).
\end{aligned}$$

Thus, fg is continuous at $x = c$.

For Property 4, when $g(c) \neq 0$, it follows from Theorem 2.2 that

$$\begin{aligned}
\lim_{x \to c} \frac{f}{g}(x) &= \lim_{x \to c} \frac{f(x)}{g(x)} \\
&= \frac{\lim\limits_{x \to c} f(x)}{\lim\limits_{x \to c} g(x)} \\
&= \frac{f(c)}{g(c)} \\
&= \frac{f}{g}(c).
\end{aligned}$$

Thus, $\dfrac{f}{g}$ is continuous at $x = c$. ∎

THEOREM 2.14 VERTICAL ASYMPTOTES (PAGE 105)

Let f and g be continuous on an open interval containing c. If $f(c) \neq 0$, $g(c) = 0$, and there exists an open interval containing c such that $g(x) \neq 0$ for all $x \neq c$ in the interval, then the graph of the function given by

$$h(x) = \frac{f(x)}{g(x)}$$

has a vertical asymptote at $x = c$.

$\boxed{\text{PROOF}}$ Consider the case for which $f(c) > 0$, and there exists $b > c$ such that $c < x < b$ implies $g(x) > 0$. Then for $M > 0$, choose δ_1 such that

$$0 < x - c < \delta_1 \quad \text{implies that} \quad \frac{f(c)}{2} < f(x) < \frac{3f(c)}{2}$$

and δ_2 such that

$$0 < x - c < \delta_2 \quad \text{implies that} \quad 0 < g(x) < \frac{f(c)}{2M}.$$

Now let δ be the smaller of δ_1 and δ_2. Then it follows that

$$0 < x - c < \delta \quad \text{implies that} \quad \frac{f(x)}{g(x)} > \frac{f(c)}{2}\left[\frac{2M}{f(c)}\right] = M.$$

So, it follows that

$$\lim_{x \to c^+} \frac{f(x)}{g(x)} = \infty$$

and the line $x = c$ is a vertical asymptote of the graph of h. ∎

ALTERNATIVE FORM OF THE DERIVATIVE (PAGE 121)

The derivative of f at c is given by

$$f'(c) = \lim_{x \to c} \frac{f(x) - f(c)}{x - c}$$

provided this limit exists.

$\boxed{\text{PROOF}}$ The derivative of f at c is given by

$$f'(c) = \lim_{\Delta x \to 0} \frac{f(c + \Delta x) - f(c)}{\Delta x}.$$

Let $x = c + \Delta x$. Then $x \to c$ as $\Delta x \to 0$. So, replacing $c + \Delta x$ by x, you have

$$f'(c) = \lim_{\Delta x \to 0} \frac{f(c + \Delta x) - f(c)}{\Delta x} = \lim_{x \to c} \frac{f(x) - f(c)}{x - c}.$$ ∎

THEOREM 3.11 THE CHAIN RULE (PAGE 152)

If $y = f(u)$ is a differentiable function of u and $u = g(x)$ is a differentiable function of x, then $y = f(g(x))$ is a differentiable function of x and

$$\frac{dy}{dx} = \frac{dy}{du} \cdot \frac{du}{dx}$$

or, equivalently,

$$\frac{d}{dx}[f(g(x))] = f'(g(x))g'(x).$$

PROOF   In Section 3.4, you let $h(x) = f(g(x))$ and used the alternative form of the derivative to show that $h'(c) = f'(g(c))g'(c)$, provided $g(x) \neq g(c)$ for values of x other than c. Now consider a more general proof. Begin by considering the derivative of f.

$$f'(x) = \lim_{\Delta x \to 0} \frac{f(x + \Delta x) - f(x)}{\Delta x} = \lim_{\Delta x \to 0} \frac{\Delta y}{\Delta x}$$

For a fixed value of x, define a function η such that

$$\eta(\Delta x) = \begin{cases} 0, & \Delta x = 0 \\ \dfrac{\Delta y}{\Delta x} - f'(x), & \Delta x \neq 0. \end{cases}$$

Because the limit of $\eta(\Delta x)$ as $\Delta x \to 0$ doesn't depend on the value of $\eta(0)$, you have

$$\lim_{\Delta x \to 0} \eta(\Delta x) = \lim_{\Delta x \to 0}\left[\frac{\Delta y}{\Delta x} - f'(x)\right] = 0$$

and you can conclude that η is continuous at 0. Moreover, because $\Delta y = 0$ when $\Delta x = 0$, the equation

$$\Delta y = \Delta x \eta(\Delta x) + \Delta x f'(x)$$

is valid whether Δx is zero or not. Now, by letting $\Delta u = g(x + \Delta x) - g(x)$, you can use the continuity of g to conclude that

$$\lim_{\Delta x \to 0} \Delta u = \lim_{\Delta x \to 0} [g(x + \Delta x) - g(x)] = 0$$

which implies that

$$\lim_{\Delta x \to 0} \eta(\Delta u) = 0.$$

Finally,

$$\Delta y = \Delta u \eta(\Delta u) + \Delta u f'(u) \to \frac{\Delta y}{\Delta x} = \frac{\Delta u}{\Delta x} \eta(\Delta u) + \frac{\Delta u}{\Delta x} f'(u), \quad \Delta x \neq 0$$

and taking the limit as $\Delta x \to 0$, you have

$$\frac{dy}{dx} = \frac{du}{dx}\left[\lim_{\Delta x \to 0} \eta(\Delta u)\right] + \frac{du}{dx} f'(u) = \frac{dy}{dx}(0) + \frac{du}{dx} f'(u)$$

$$= \frac{du}{dx} f'(u)$$

$$= \frac{du}{dx} \cdot \frac{dy}{du}.$$

THEOREM 3.16 CONTINUITY AND DIFFERENTIABILITY OF INVERSE FUNCTIONS (PAGE 175)

Let f be a function whose domain is an interval I. If f has an inverse, then the following statements are true.

1. If f is continuous on its domain, then f^{-1} is continuous on its domain.
2. If f is differentiable on an interval containing c and $f'(c) \neq 0$, then f^{-1} is differentiable at $f(c)$.

PROOF To prove Property 1, you first need to define what is meant by a *strictly increasing* function or a *strictly decreasing* function. A function f is **strictly increasing** on an entire interval I if for any two numbers x_1 and x_2 in the interval, $x_1 < x_2$ implies $f(x_1) < f(x_2)$. The function f is **strictly decreasing** on the entire interval I if $x_1 < x_2$ implies $f(x_1) > f(x_2)$. The function f is **strictly monotonic** on the interval I if it is either strictly increasing or strictly decreasing. Now show that if f is continuous on I, and has an inverse, then f is strictly monotonic on I. Suppose that f were not strictly monotonic. Then there would exist numbers x_1, x_2, x_3 in I such that $x_1 < x_2 < x_3$, but $f(x_2)$ is not between $f(x_1)$ and $f(x_3)$. Without loss of generality, assume $f(x_1) < f(x_3) < f(x_2)$. By the Intermediate Value Theorem, there exists a number x_0 between x_1 and x_2 such that $f(x_0) = f(x_3)$. So, f is not one-to-one and cannot have an inverse. So, f must be strictly monotonic.

Because f is continuous, the Intermediate Value Theorem implies that the set of values of f, $\{f(x): x \in I\}$, forms an interval J. Assume that a is an interior point of J. From the previous argument, $f^{-1}(a)$ is an interior point of I. Let $\varepsilon > 0$. There exists $0 < \varepsilon_1 < \varepsilon$ such that

$$I_1 = (f^{-1}(a) - \varepsilon_1, f^{-1}(a) + \varepsilon_1) \subseteq I.$$

Because f is strictly monotonic on I_1, the set of values $\{f(x): x \in I_1\}$ forms an interval $J_1 \subseteq J$. Let $\delta > 0$ such that $(a - \delta, a + \delta) \subseteq J_1$. Finally, if $|y - a| < \delta$, then $|f^{-1}(y) - f^{-1}(a)| < \varepsilon_1 < \varepsilon$. So, f^{-1} is continuous at a. A similar proof can be given if a is an endpoint.

To prove Property 2, consider the limit

$$(f^{-1})'(a) = \lim_{y \to a} \frac{f^{-1}(y) - f^{-1}(a)}{y - a}$$

where a is in the domain of f^{-1} and $f^{-1}(a) = c$. Because f is differentiable on an interval containing c, f is continuous on that interval, and so is f^{-1} at a. So, $y \to a$ implies that $x \to c$, and you have

$$(f^{-1})'(a) = \lim_{x \to c} \frac{x - c}{f(x) - f(c)} = \lim_{x \to c} \frac{1}{\left(\dfrac{f(x) - f(c)}{x - c}\right)} = \frac{1}{\lim\limits_{x \to c} \dfrac{f(x) - f(c)}{x - c}} = \frac{1}{f'(c)}.$$

So, $(f^{-1})'(a)$ exists, and f^{-1} is differentiable at $f(c)$. ∎

THEOREM 3.17 THE DERIVATIVE OF AN INVERSE FUNCTION (PAGE 175)

Let f be a function that is differentiable on an interval I. If f has an inverse function g, then g is differentiable at any x for which $f'(g(x)) \neq 0$. Moreover,

$$g'(x) = \frac{1}{f'(g(x))}, \quad f'(g(x)) \neq 0.$$

(PROOF) From the proof of Theorem 3.16, letting $a = x$, you know that g is differentiable. Using the Chain Rule, differentiate both sides of the equation $x = f(g(x))$ to obtain

$$1 = f'(g(x)) \frac{d}{dx}[g(x)].$$

Because $f'(g(x)) \neq 0$, you can divide by this quantity to obtain

$$\frac{d}{dx}[g(x)] = \frac{1}{f'(g(x))}.$$

CONCAVITY INTERPRETATION (PAGE 230)

1. Let f be differentiable on an open interval I. If the graph of f is concave *upward* on I, then the graph of f lies *above* all of its tangent lines on I.

2. Let f be differentiable on an open interval I. If the graph of f is concave *downward* on I, then the graph of f lies *below* all of its tangent lines on I.

(PROOF) Assume that f is concave upward on $I = (a, b)$. Then, f' is increasing on (a, b). Let c be a point in the interval $I = (a, b)$. The equation of the tangent line to the graph of f at c is given by

$$g(x) = f(c) + f'(c)(x - c).$$

If x is in the open interval (c, b), then the directed distance from point $(x, f(x))$ (on the graph of f) to the point $(x, g(x))$ (on the tangent line) is given by

$$d = f(x) - [f(c) + f'(c)(x - c)]$$
$$= f(x) - f(c) - f'(c)(x - c).$$

Moreover, by the Mean Value Theorem there exists a number z in (c, x) such that

$$f'(z) = \frac{f(x) - f(c)}{x - c}.$$

So, you have

$$d = f(x) - f(c) - f'(c)(x - c)$$
$$= f'(z)(x - c) - f'(c)(x - c)$$
$$= [f'(z) - f'(c)](x - c).$$

The second factor $(x - c)$ is positive because $c < x$. Moreover, because f' is increasing, it follows that the first factor $[f'(z) - f'(c)]$ is also positive. Therefore, $d > 0$ and you can conclude that the graph of f lies above the tangent line at x. If x is in the open interval (a, c), a similar argument can be given. This proves the first statement. The proof of the second statement is similar.

THEOREM 4.7 TEST FOR CONCAVITY (PAGE 231)

Let f be a function whose second derivative exists on an open interval I.

1. If $f''(x) > 0$ for all x in I, then the graph of f is concave upward on I.

2. If $f''(x) < 0$ for all x in I, then the graph of f is concave downward on I.

(**PROOF**) For Property 1, assume $f''(x) > 0$ for all x in (a, b). Then, by Theorem 4.5, f' is increasing on $[a, b]$. Thus, by the definition of concavity, the graph of f is concave upward on (a, b).

For Property 2, assume $f''(x) < 0$ for all x in (a, b). Then, by Theorem 4.5, f' is decreasing on $[a, b]$. Thus, by the definition of concavity, the graph of f is concave downward on (a, b). ∎

THEOREM 4.10 LIMITS AT INFINITY (PAGE 239)

If r is a positive rational number and c is any real number, then

$$\lim_{x \to \infty} \frac{c}{x^r} = 0.$$

Furthermore, if x^r is defined when $x < 0$, then $\displaystyle\lim_{x \to -\infty} \frac{c}{x^r} = 0.$

(**PROOF**) Begin by proving that

$$\lim_{x \to \infty} \frac{1}{x} = 0.$$

For $\varepsilon > 0$, let $M = 1/\varepsilon$. Then, for $x > M$, you have

$$x > M = \frac{1}{\varepsilon} \quad\Longrightarrow\quad \frac{1}{x} < \varepsilon \quad\Longrightarrow\quad \left| \frac{1}{x} - 0 \right| < \varepsilon.$$

So, by the definition of a limit at infinity, you can conclude that the limit of $1/x$ as $x \to \infty$ is 0. Now, using this result, and letting $r = m/n$, you can write the following.

$$\lim_{x \to \infty} \frac{c}{x^r} = \lim_{x \to \infty} \frac{c}{x^{m/n}}$$

$$= c \left[\lim_{x \to \infty} \left(\frac{1}{\sqrt[n]{x}} \right)^m \right]$$

$$= c \left(\lim_{x \to \infty} \sqrt[n]{\frac{1}{x}} \right)^m$$

$$= c \left(\sqrt[n]{\lim_{x \to \infty} \frac{1}{x}} \right)^m$$

$$= c \left(\sqrt[n]{0} \right)^m$$

$$= 0$$

The proof of the second part of the theorem is similar. ∎

THEOREM 5.2 SUMMATION FORMULAS (PAGE 296)

1. $\displaystyle\sum_{i=1}^{n} c = cn$ **2.** $\displaystyle\sum_{i=1}^{n} i = \frac{n(n+1)}{2}$

3. $\displaystyle\sum_{i=1}^{n} i^2 = \frac{n(n+1)(2n+1)}{6}$ **4.** $\displaystyle\sum_{i=1}^{n} i^3 = \frac{n^2(n+1)^2}{4}$

(**PROOF**) The proof of Property 1 is straightforward. By adding c to itself n times, you obtain a sum of cn.

To prove Property 2, write the sum in increasing and decreasing order and add corresponding terms, as follows.

$$\sum_{i=1}^{n} i = \quad 1 \quad + \quad 2 \quad + \quad 3 \quad + \cdots + (n-1) + \quad n$$

$$\sum_{i=1}^{n} i = \quad n \quad + (n-1) + (n-2) + \cdots + \quad 2 \quad + \quad 1$$

$$2\sum_{i=1}^{n} i = (n+1) + (n+1) + (n+1) + \cdots + (n+1) + (n+1)$$

$$n \text{ terms}$$

So,

$$\sum_{i=1}^{n} i = \frac{n(n+1)}{2}.$$

To prove Property 3, use mathematical induction. First, if $n = 1$, the result is true because

$$\sum_{i=1}^{1} i^2 = 1^2 = 1 = \frac{1(1+1)(2+1)}{6}.$$

Now, assuming the result is true for $n = k$, you can show that it is true for $n = k + 1$, as follows.

$$\sum_{i=1}^{k+1} i^2 = \sum_{i=1}^{k} i^2 + (k+1)^2$$

$$= \frac{k(k+1)(2k+1)}{6} + (k+1)^2$$

$$= \frac{k+1}{6}(2k^2 + k + 6k + 6)$$

$$= \frac{k+1}{6}[(2k+3)(k+2)]$$

$$= \frac{(k+1)(k+2)[2(k+1)+1]}{6}$$

Property 4 can be proved using a similar argument with mathematical induction. ∎

THEOREM 5.8 PRESERVATION OF INEQUALITY (PAGE 314)

1. If f is integrable and nonnegative on the closed interval $[a, b]$, then

$$0 \le \int_{a}^{b} f(x)\, dx.$$

2. If f and g are integrable on the closed interval $[a, b]$ and $f(x) \le g(x)$ for every x in $[a, b]$, then

$$\int_{a}^{b} f(x)\, dx \le \int_{a}^{b} g(x)\, dx.$$

(PROOF) To prove Property 1, suppose, on the contrary, that

$$\int_a^b f(x)\, dx = I < 0.$$

Then, let $a = x_0 < x_1 < x_2 < \cdots < x_n = b$ be a partition of $[a, b]$, and let

$$R = \sum_{i=1}^n f(c_i)\, \Delta x_i$$

be a Riemann sum. Because $f(x) \geq 0$, it follows that $R \geq 0$. Now, for $\|\Delta\|$ sufficiently small, you have $|R - I| < -I/2$, which implies that

$$\sum_{i=1}^n f(c_i)\, \Delta x_i = R < I - \frac{I}{2} < 0$$

which is not possible. From this contradiction, you can conclude that

$$0 \leq \int_a^b f(x)\, dx.$$

To prove Property 2 of the theorem, note that $f(x) \leq g(x)$ implies that $g(x) - f(x) \geq 0$. So, you can apply the result of Property 1 to conclude that

$$0 \leq \int_a^b [g(x) - f(x)]\, dx$$

$$0 \leq \int_a^b g(x)\, dx - \int_a^b f(x)\, dx$$

$$\int_a^b f(x)\, dx \leq \int_a^b g(x)\, dx. \qquad \blacksquare$$

THEOREM 8.3 THE EXTENDED MEAN VALUE THEOREM (PAGE 570)

If f and g are differentiable on an open interval (a, b) and continuous on $[a, b]$ such that $g'(x) \neq 0$ for any x in (a, b), then there exists a point c in (a, b) such that $\dfrac{f'(c)}{g'(c)} = \dfrac{f(b) - f(a)}{g(b) - g(a)}$.

(PROOF) You can assume that $g(a) \neq g(b)$, because otherwise, by Rolle's Theorem, it would follow that $g'(x) = 0$ for some x in (a, b). Now, define $h(x)$ to be

$$h(x) = f(x) - \left[\frac{f(b) - f(a)}{g(b) - g(a)}\right] g(x).$$

Then

$$h(a) = f(a) - \left[\frac{f(b) - f(a)}{g(b) - g(a)}\right] g(a) = \frac{f(a)g(b) - f(b)g(a)}{g(b) - g(a)}$$

and

$$h(b) = f(b) - \left[\frac{f(b) - f(a)}{g(b) - g(a)}\right] g(b) = \frac{f(a)g(b) - f(b)g(a)}{g(b) - g(a)}$$

and by Rolle's Theorem there exists a point c in (a, b) such that

$$h'(c) = f'(c) - \frac{f(b) - f(a)}{g(b) - g(a)} g'(c) = 0$$

which implies that $\dfrac{f'(c)}{g'(c)} = \dfrac{f(b) - f(a)}{g(b) - g(a)}.$ \qquad $\blacksquare$

THEOREM 8.4 L'HÔPITAL'S RULE (PAGE 570)

Let f and g be functions that are differentiable on an open interval (a, b) containing c, except possibly at c itself. Assume that $g'(x) \neq 0$ for all x in (a, b), except possibly at c itself. If the limit of $f(x)/g(x)$ as x approaches c produces the indeterminate form $0/0$, then

$$\lim_{x \to c} \frac{f(x)}{g(x)} = \lim_{x \to c} \frac{f'(x)}{g'(x)}$$

provided the limit on the right exists (or is infinite). This result also applies if the limit of $f(x)/g(x)$ as x approaches c produces any one of the indeterminate forms ∞/∞, $(-\infty)/\infty$, $\infty/(-\infty)$, or $(-\infty)/(-\infty)$.

You can use the Extended Mean Value Theorem to prove L'Hôpital's Rule. Of the several different cases of this rule, the proof of only one case is illustrated. The remaining cases, where $x \to c^-$ and $x \to c$, are left for you to prove.

PROOF Consider the case for which

$$\lim_{x \to c^+} f(x) = 0 \quad \text{and} \quad \lim_{x \to c^+} g(x) = 0.$$

Define the following new functions:

$$F(x) = \begin{cases} f(x), & x \neq c \\ 0, & x = c \end{cases} \quad \text{and} \quad G(x) = \begin{cases} g(x), & x \neq c \\ 0, & x = c. \end{cases}$$

For any x, $c < x < b$, F and G are differentiable on $(c, x]$ and continuous on $[c, x]$. You can apply the Extended Mean Value Theorem to conclude that there exists a number z in (c, x) such that

$$\frac{F'(z)}{G'(z)} = \frac{F(x) - F(c)}{G(x) - G(c)}$$

$$= \frac{F(x)}{G(x)}$$

$$= \frac{f'(z)}{g'(z)}$$

$$= \frac{f(x)}{g(x)}.$$

Finally, by letting x approach c from the right, $x \to c^+$, you have $z \to c^+$ because $c < z < x$, and

$$\lim_{x \to c^+} \frac{f(x)}{g(x)} = \lim_{x \to c^+} \frac{f'(z)}{g'(z)}$$

$$= \lim_{z \to c^+} \frac{f'(z)}{g'(z)}$$

$$= \lim_{x \to c^+} \frac{f'(x)}{g'(x)}.$$

THEOREM 9.19 TAYLOR'S THEOREM (PAGE 656)

If a function f is differentiable through order $n + 1$ in an interval I containing c, then, for each x in I, there exists z between x and c such that

$$f(x) = f(c) + f'(c)(x - c) + \frac{f''(c)}{2!}(x - c)^2 + \cdots + \frac{f^{(n)}(c)}{n!}(x - c)^n + R_n(x)$$

where

$$R_n(x) = \frac{f^{(n+1)}(z)}{(n + 1)!}(x - c)^{n+1}.$$

PROOF To find $R_n(x)$, fix x in I ($x \neq c$) and write

$$R_n(x) = f(x) - P_n(x)$$

where $P_n(x)$ is the nth Taylor polynomial for $f(x)$. Then let g be a function of t defined by

$$g(t) = f(x) - f(t) - f'(t)(x - t) - \cdots - \frac{f^{(n)}(t)}{n!}(x - t)^n - R_n(x)\frac{(x - t)^{n+1}}{(x - c)^{n+1}}.$$

The reason for defining g in this way is that differentiation with respect to t has a telescoping effect. For example, you have

$$\frac{d}{dt}\left[-f(t) - f'(t)(x - t)\right] = -f'(t) + f'(t) - f''(t)(x - t)$$

$$= -f''(t)(x - t).$$

The result is that the derivative $g'(t)$ simplifies to

$$g'(t) = -\frac{f^{(n+1)}(t)}{n!}(x - t)^n + (n + 1)R_n(x)\frac{(x - t)^n}{(x - c)^{n+1}}$$

for all t between c and x. Moreover, for a fixed x,

$$g(c) = f(x) - [P_n(x) + R_n(x)] = f(x) - f(x) = 0$$

and

$$g(x) = f(x) - f(x) - 0 - \cdots - 0 = f(x) - f(x) = 0.$$

Therefore, g satisfies the conditions of Rolle's Theorem, and it follows that there is a number z between c and x such that $g'(z) = 0$. Substituting z for t in the equation for $g'(t)$ and then solving for $R_n(x)$, you obtain

$$g'(z) = -\frac{f^{(n+1)}(z)}{n!}(x - z)^n + (n + 1)R_n(x)\frac{(x - z)^n}{(x - c)^{n+1}} = 0$$

$$R_n(x) = \frac{f^{(n+1)}(z)}{(n + 1)!}(x - c)^{n+1}.$$

Finally, because $g(c) = 0$, you have

$$0 = f(x) - f(c) - f'(c)(x - c) - \cdots - \frac{f^{(n)}(c)}{n!}(x - c)^n - R_n(x)$$

$$f(x) = f(c) + f'(c)(x - c) + \cdots + \frac{f^{(n)}(c)}{n!}(x - c)^n + R_n(x).$$ ∎

THEOREM 9.20 CONVERGENCE OF A POWER SERIES (PAGE 662)

For a power series centered at c, precisely one of the following is true.

1. The series converges only at c.

2. There exists a real number $R > 0$ such that the series converges absolutely for $|x - c| < R$, and diverges for $|x - c| > R$.

3. The series converges absolutely for all x.

The number R is the **radius of convergence** of the power series. If the series converges only at c, the radius of convergence is $R = 0$, and if the series converges for all x, the radius of convergence is $R = \infty$. The set of all values of x for which the power series converges is the **interval of convergence** of the power series.

(**PROOF**) In order to simplify the notation, the theorem for the power series $\sum a_n x^n$ centered at $x = 0$ will be proved. The proof for a power series centered at $x = c$ follows easily. A key step in this proof uses the completeness property of the set of real numbers: If a nonempty set S of real numbers has an upper bound, then it must have a least upper bound (see page 603).

It must be shown that if a power series $\sum a_n x^n$ converges at $x = d$, $d \neq 0$, then it converges for all b satisfying $|b| < |d|$. Because $\sum a_n x^n$ converges, $\lim\limits_{x \to \infty} a_n d^n = 0$. So, there exists $N > 0$ such that $a_n d^n < 1$ for all $n \geq N$. Then for $n \geq N$,

$$\left| a_n b^n \right| = \left| a_n b^n \frac{d^n}{d^n} \right| = \left| a_n d^n \right| \left| \frac{b^n}{d^n} \right| < \left| \frac{b^n}{d^n} \right|.$$

So, for $|b| < |d|$, $\left| \dfrac{b}{d} \right| < 1$, which implies that

$$\sum \left| \frac{b^n}{d^n} \right|$$

is a convergent geometric series. By the Comparison Test, the series $\sum a_n b^n$ converges.

Similarly, if the power series $\sum a_n x^n$ diverges at $x = b$, where $b \neq 0$, then it diverges for all d satisfying $|d| > |b|$. If $\sum a_n d^n$ converged, then the argument above would imply that $\sum a_n b^n$ converged as well.

Finally, to prove the theorem, suppose that neither case 1 nor case 3 is true. Then there exist points b and d such that $\sum a_n x^n$ converges at b and diverges at d. Let $S = \{x : \sum a_n x^n \text{ converges}\}$. S is nonempty because $b \in S$. If $x \in S$, then $|x| \leq |d|$, which shows that $|d|$ is an upper bound for the nonempty set S. By the completeness property, S has a least upper bound, R.

Now, if $|x| > R$, then $x \notin S$, so $\sum a_n x^n$ diverges. And if $|x| < R$, then $|x|$ is not an upper bound for S, so there exists b in S satisfying $|b| > |x|$. Since $b \in S$, $\sum a_n b^n$ converges, which implies that $\sum a_n x^n$ converges. ■

> ### THEOREM 10.16 CLASSIFICATION OF CONICS BY ECCENTRICITY (PAGE 750)
>
> Let F be a fixed point (*focus*) and let D be a fixed line (*directrix*) in the plane. Let P be another point in the plane and let e (*eccentricity*) be the ratio of the distance between P and F to the distance between P and D. The collection of all points P with a given eccentricity is a conic.
>
> 1. The conic is an ellipse if $0 < e < 1$.
> 2. The conic is a parabola if $e = 1$.
> 3. The conic is a hyperbola if $e > 1$.

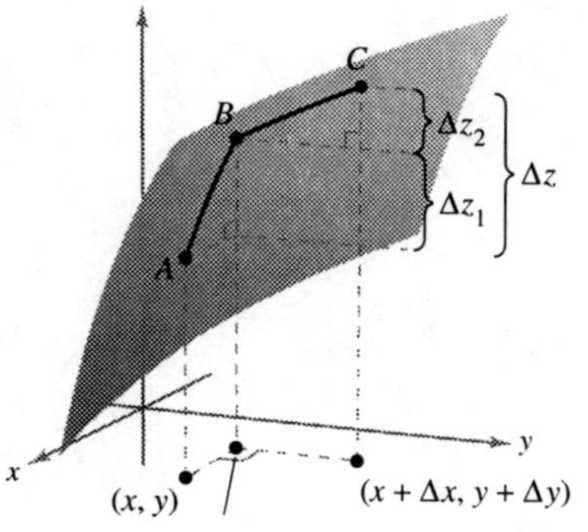

Figure A.1

(**PROOF**) If $e = 1$, then, by definition, the conic must be a parabola. If $e \neq 1$, then you can consider the focus F to lie at the origin and the directrix $x = d$ to lie to the right of the origin, as shown in Figure A.1. For the point $P = (r, \theta) = (x, y)$, you have $|PF| = r$ and $|PQ| = d - r\cos\theta$. Given that $e = |PF|/|PQ|$, it follows that

$$|PF| = |PQ|e \quad \Longrightarrow \quad r = e(d - r\cos\theta).$$

By converting to rectangular coordinates and squaring each side, you obtain

$$x^2 + y^2 = e^2(d - x)^2 = e^2(d^2 - 2dx + x^2).$$

Completing the square produces

$$\left(x + \frac{e^2 d}{1 - e^2}\right)^2 + \frac{y^2}{1 - e^2} = \frac{e^2 d^2}{(1 - e^2)^2}.$$

If $e < 1$, this equation represents an ellipse. If $e > 1$, then $1 - e^2 < 0$, and the equation represents a hyperbola. ∎

> ### THEOREM 13.4 SUFFICIENT CONDITION FOR DIFFERENTIABILITY (PAGE 919)
>
> If f is a function of x and y, where f_x and f_y are continuous in an open region R, then f is differentiable on R.

(**PROOF**) Let S be the surface defined by $z = f(x, y)$, where $f, f_x,$ and f_y are continuous at (x, y). Let A, B, and C be points on surface S, as shown in Figure A.2. From this figure, you can see that the change in f from point A to point C is given by

$$\begin{aligned}
\Delta z &= f(x + \Delta x, y + \Delta y) - f(x, y) \\
&= [f(x + \Delta x, y) - f(x, y)] + [f(x + \Delta x, y + \Delta y) - f(x + \Delta x, y)] \\
&= \Delta z_1 + \Delta z_2.
\end{aligned}$$

Between A and B, y is fixed and x changes. So, by the Mean Value Theorem, there is a value x_1 between x and $x + \Delta x$ such that

$$\Delta z_1 = f(x + \Delta x, y) - f(x, y) = f_x(x_1, y)\,\Delta x.$$

Similarly, between B and C, x is fixed and y changes, and there is a value y_1 between y and $y + \Delta y$ such that

$$\Delta z_2 = f(x + \Delta x, y + \Delta y) - f(x + \Delta x, y) = f_y(x + \Delta x, y_1)\,\Delta y.$$

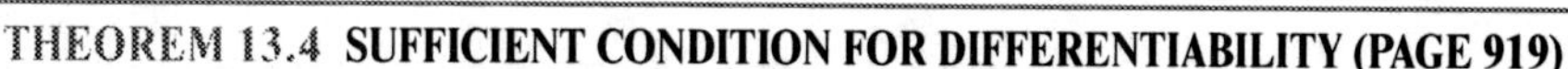

$$\Delta z = f(x + \Delta x, y + \Delta y) - f(x, y)$$

Figure A.2

By combining these two results, you can write

$$\Delta z = \Delta z_1 + \Delta z_2 = f_x(x_1, y)\Delta x + f_y(x + \Delta x, y_1)\,\Delta y.$$

If you define ε_1 and ε_2 as

$$\varepsilon_1 = f_x(x_1, y) - f_x(x, y) \text{ and } \varepsilon_2 = f_y(x + \Delta x, y_1) - f_y(x, y)$$

it follows that

$$\Delta z = \Delta z_1 + \Delta z_2 = [\varepsilon_1 + f_x(x, y)]\,\Delta x + [\varepsilon_2 + f_y(x, y)]\,\Delta y$$
$$= [f_x(x, y)\,\Delta x + f_y(x, y)\,\Delta y] + \varepsilon_1\Delta x + \varepsilon_2\Delta y.$$

By the continuity of f_x and f_y and the fact that $x \le x_1 \le x + \Delta x$ and $y \le y_1 \le y + \Delta y$, it follows that $\varepsilon_1 \to 0$ and $\varepsilon_2 \to 0$ as $\Delta x \to 0$ and $\Delta y \to 0$. Therefore, by definition, f is differentiable. ◼

THEOREM 13.6 CHAIN RULE: ONE INDEPENDENT VARIABLE (PAGE 925)

Let $w = f(x, y)$, where f is a differentiable function of x and y. If $x = g(t)$ and $y = h(t)$, where g and h are differentiable functions of t, then w is a differentiable function of t, and

$$\frac{dw}{dt} = \frac{\partial w}{\partial x}\frac{dx}{dt} + \frac{\partial w}{\partial y}\frac{dy}{dt}.$$

PROOF Because g and h are differentiable functions of t, you know that both Δx and Δy approach zero as Δt approaches zero. Moreover, because f is a differentiable function of x and y, you know that

$$\Delta w = \frac{\partial w}{\partial x}\Delta x + \frac{\partial w}{\partial y}\Delta y + \varepsilon_1\Delta x + \varepsilon_2\Delta y$$

where both ε_1 and $\varepsilon_2 \to 0$ as $(\Delta x, \Delta y) \to (0, 0)$. So, for $\Delta t \ne 0$,

$$\frac{\Delta w}{\Delta t} = \frac{\partial w}{\partial x}\frac{\Delta x}{\Delta t} + \frac{\partial w}{\partial y}\frac{\Delta y}{\Delta t} + \varepsilon_1\frac{\Delta x}{\Delta t} + \varepsilon_2\frac{\Delta y}{\Delta t}$$

from which it follows that

$$\frac{dw}{dt} = \lim_{\Delta t \to 0}\frac{\Delta w}{\Delta t} = \frac{\partial w}{\partial x}\frac{dx}{dt} + \frac{\partial w}{\partial y}\frac{dy}{dt} + 0\left(\frac{dx}{dt}\right) + 0\left(\frac{dy}{dt}\right)$$
$$= \frac{\partial w}{\partial x}\frac{dx}{dt} + \frac{\partial w}{\partial y}\frac{dy}{dt}.$$ ◼

B Integration Tables

Forms Involving u^n

1. $\displaystyle \int u^n\, du = \frac{u^{n+1}}{n+1} + C, \; n \neq -1$

2. $\displaystyle \int \frac{1}{u}\, du = \ln|u| + C$

Forms Involving $a + bu$

3. $\displaystyle \int \frac{u}{a+bu}\, du = \frac{1}{b^2}\left(bu - a\ln|a+bu|\right) + C$

4. $\displaystyle \int \frac{u}{(a+bu)^2}\, du = \frac{1}{b^2}\left(\frac{a}{a+bu} + \ln|a+bu|\right) + C$

5. $\displaystyle \int \frac{u}{(a+bu)^n}\, du = \frac{1}{b^2}\left[\frac{-1}{(n-2)(a+bu)^{n-2}} + \frac{a}{(n-1)(a+bu)^{n-1}}\right] + C, \quad n \neq 1, 2$

6. $\displaystyle \int \frac{u^2}{a+bu}\, du = \frac{1}{b^3}\left[-\frac{bu}{2}(2a - bu) + a^2\ln|a+bu|\right] + C$

7. $\displaystyle \int \frac{u^2}{(a+bu)^2}\, du = \frac{1}{b^3}\left(bu - \frac{a^2}{a+bu} - 2a\ln|a+bu|\right) + C$

8. $\displaystyle \int \frac{u^2}{(a+bu)^3}\, du = \frac{1}{b^3}\left[\frac{2a}{a+bu} - \frac{a^2}{2(a+bu)^2} + \ln|a+bu|\right] + C$

9. $\displaystyle \int \frac{u^2}{(a+bu)^n}\, du = \frac{1}{b^3}\left[\frac{-1}{(n-3)(a+bu)^{n-3}} + \frac{2a}{(n-2)(a+bu)^{n-2}} - \frac{a^2}{(n-1)(a+bu)^{n-1}}\right] + C, \quad n \neq 1, 2, 3$

10. $\displaystyle \int \frac{1}{u(a+bu)}\, du = \frac{1}{a}\ln\left|\frac{u}{a+bu}\right| + C$

11. $\displaystyle \int \frac{1}{u(a+bu)^2}\, du = \frac{1}{a}\left(\frac{1}{a+bu} + \frac{1}{a}\ln\left|\frac{u}{a+bu}\right|\right) + C$

12. $\displaystyle \int \frac{1}{u^2(a+bu)}\, du = -\frac{1}{a}\left(\frac{1}{u} + \frac{b}{a}\ln\left|\frac{u}{a+bu}\right|\right) + C$

13. $\displaystyle \int \frac{1}{u^2(a+bu)^2}\, du = -\frac{1}{a^2}\left[\frac{a+2bu}{u(a+bu)} + \frac{2b}{a}\ln\left|\frac{u}{a+bu}\right|\right] + C$

Forms Involving $a + bu + cu^2$, $b^2 \neq 4ac$

14. $\displaystyle \int \frac{1}{a + bu + cu^2}\, du = \begin{cases} \dfrac{2}{\sqrt{4ac - b^2}} \arctan \dfrac{2cu + b}{\sqrt{4ac - b^2}} + C, & b^2 < 4ac \\[3mm] \dfrac{1}{\sqrt{b^2 - 4ac}} \ln\left| \dfrac{2cu + b - \sqrt{b^2 - 4ac}}{2cu + b + \sqrt{b^2 - 4ac}} \right| + C, & b^2 > 4ac \end{cases}$

15. $\displaystyle \int \frac{u}{a + bu + cu^2}\, du = \frac{1}{2c}\left(\ln\left|a + bu + cu^2\right| - b \int \frac{1}{a + bu + cu^2}\, du \right)$

Forms Involving $\sqrt{a + bu}$

16. $\displaystyle \int u^n \sqrt{a + bu}\, du = \frac{2}{b(2n + 3)}\left[u^n (a + bu)^{3/2} - na \int u^{n-1} \sqrt{a + bu}\, du \right]$

17. $\displaystyle \int \frac{1}{u\sqrt{a + bu}}\, du = \begin{cases} \dfrac{1}{\sqrt{a}} \ln\left| \dfrac{\sqrt{a + bu} - \sqrt{a}}{\sqrt{a + bu} + \sqrt{a}} \right| + C, & a > 0 \\[3mm] \dfrac{2}{\sqrt{-a}} \arctan \sqrt{\dfrac{a + bu}{-a}} + C, & a < 0 \end{cases}$

18. $\displaystyle \int \frac{1}{u^n \sqrt{a + bu}}\, du = \frac{-1}{a(n - 1)}\left[\frac{\sqrt{a + bu}}{u^{n-1}} + \frac{(2n - 3)b}{2} \int \frac{1}{u^{n-1}\sqrt{a + bu}}\, du \right], \quad n \neq 1$

19. $\displaystyle \int \frac{\sqrt{a + bu}}{u}\, du = 2\sqrt{a + bu} + a \int \frac{1}{u\sqrt{a + bu}}\, du$

20. $\displaystyle \int \frac{\sqrt{a + bu}}{u^n}\, du = \frac{-1}{a(n - 1)}\left[\frac{(a + bu)^{3/2}}{u^{n-1}} + \frac{(2n - 5)b}{2} \int \frac{\sqrt{a + bu}}{u^{n-1}}\, du \right], \quad n \neq 1$

21. $\displaystyle \int \frac{u}{\sqrt{a + bu}}\, du = \frac{-2(2a - bu)}{3b^2}\sqrt{a + bu} + C$

22. $\displaystyle \int \frac{u^n}{\sqrt{a + bu}}\, du = \frac{2}{(2n + 1)b}\left(u^n \sqrt{a + bu} - na \int \frac{u^{n-1}}{\sqrt{a + bu}}\, du \right)$

Forms Involving $a^2 \pm u^2$, $a > 0$

23. $\displaystyle \int \frac{1}{a^2 + u^2}\, du = \frac{1}{a} \arctan \frac{u}{a} + C$

24. $\displaystyle \int \frac{1}{u^2 - a^2}\, du = -\int \frac{1}{a^2 - u^2}\, du = \frac{1}{2a} \ln\left| \frac{u - a}{u + a} \right| + C$

25. $\displaystyle \int \frac{1}{(a^2 \pm u^2)^n}\, du = \frac{1}{2a^2(n - 1)}\left[\frac{u}{(a^2 \pm u^2)^{n-1}} + (2n - 3) \int \frac{1}{(a^2 \pm u^2)^{n-1}}\, du \right], \quad n \neq 1$

Forms Involving $\sqrt{u^2 \pm a^2}$, $a > 0$

26. $\displaystyle \int \sqrt{u^2 \pm a^2}\, du = \frac{1}{2}\left(u\sqrt{u^2 \pm a^2} \pm a^2 \ln\left| u + \sqrt{u^2 \pm a^2} \right| \right) + C$

27. $\displaystyle \int u^2 \sqrt{u^2 \pm a^2}\, du = \frac{1}{8}\left[u(2u^2 \pm a^2)\sqrt{u^2 \pm a^2} - a^4 \ln\left| u + \sqrt{u^2 \pm a^2} \right| \right] + C$

28. $\displaystyle \int \frac{\sqrt{u^2 + a^2}}{u}\, du = \sqrt{u^2 + a^2} - a \ln\left| \frac{a + \sqrt{u^2 + a^2}}{u} \right| + C$

29. $\displaystyle \int \frac{\sqrt{u^2 - a^2}}{u}\,du = \sqrt{u^2 - a^2} - a\,\text{arcsec}\,\frac{|u|}{a} + C$

30. $\displaystyle \int \frac{\sqrt{u^2 \pm a^2}}{u^2}\,du = \frac{-\sqrt{u^2 \pm a^2}}{u} + \ln|u + \sqrt{u^2 \pm a^2}| + C$

31. $\displaystyle \int \frac{1}{\sqrt{u^2 \pm a^2}}\,du = \ln|u + \sqrt{u^2 \pm a^2}| + C$

32. $\displaystyle \int \frac{1}{u\sqrt{u^2 + a^2}}\,du = \frac{-1}{a}\ln\left|\frac{a + \sqrt{u^2 + a^2}}{u}\right| + C$

33. $\displaystyle \int \frac{1}{u\sqrt{u^2 - a^2}}\,du = \frac{1}{a}\,\text{arcsec}\,\frac{|u|}{a} + C$

34. $\displaystyle \int \frac{u^2}{\sqrt{u^2 \pm a^2}}\,du = \frac{1}{2}\left(u\sqrt{u^2 \pm a^2} \mp a^2 \ln|u + \sqrt{u^2 \pm a^2}|\right) + C$

35. $\displaystyle \int \frac{1}{u^2\sqrt{u^2 \pm a^2}}\,du = \mp\frac{\sqrt{u^2 \pm a^2}}{a^2 u} + C$

36. $\displaystyle \int \frac{1}{(u^2 \pm a^2)^{3/2}}\,du = \frac{\pm u}{a^2\sqrt{u^2 \pm a^2}} + C$

Forms Involving $\sqrt{a^2 - u^2},\ a > 0$

37. $\displaystyle \int \sqrt{a^2 - u^2}\,du = \frac{1}{2}\left(u\sqrt{a^2 - u^2} + a^2 \arcsin\frac{u}{a}\right) + C$

38. $\displaystyle \int u^2\sqrt{a^2 - u^2}\,du = \frac{1}{8}\left[u(2u^2 - a^2)\sqrt{a^2 - u^2} + a^4 \arcsin\frac{u}{a}\right] + C$

39. $\displaystyle \int \frac{\sqrt{a^2 - u^2}}{u}\,du = \sqrt{a^2 - u^2} - a\ln\left|\frac{a + \sqrt{a^2 - u^2}}{u}\right| + C$

40. $\displaystyle \int \frac{\sqrt{a^2 - u^2}}{u^2}\,du = \frac{-\sqrt{a^2 - u^2}}{u} - \arcsin\frac{u}{a} + C$

41. $\displaystyle \int \frac{1}{\sqrt{a^2 - u^2}}\,du = \arcsin\frac{u}{a} + C$

42. $\displaystyle \int \frac{1}{u\sqrt{a^2 - u^2}}\,du = \frac{-1}{a}\ln\left|\frac{a + \sqrt{a^2 - u^2}}{u}\right| + C$

43. $\displaystyle \int \frac{u^2}{\sqrt{a^2 - u^2}}\,du = \frac{1}{2}\left(-u\sqrt{a^2 - u^2} + a^2 \arcsin\frac{u}{a}\right) + C$

44. $\displaystyle \int \frac{1}{u^2\sqrt{a^2 - u^2}}\,du = \frac{-\sqrt{a^2 - u^2}}{a^2 u} + C$

45. $\displaystyle \int \frac{1}{(a^2 - u^2)^{3/2}}\,du = \frac{u}{a^2\sqrt{a^2 - u^2}} + C$

Forms Involving $\sin u$ or $\cos u$

46. $\displaystyle \int \sin u \, du = -\cos u + C$

47. $\displaystyle \int \cos u \, du = \sin u + C$

48. $\displaystyle \int \sin^2 u \, du = \frac{1}{2}(u - \sin u \cos u) + C$

49. $\displaystyle \int \cos^2 u \, du = \frac{1}{2}(u + \sin u \cos u) + C$

50. $\displaystyle \int \sin^n u \, du = -\frac{\sin^{n-1} u \cos u}{n} + \frac{n-1}{n} \int \sin^{n-2} u \, du$

51. $\displaystyle \int \cos^n u \, du = \frac{\cos^{n-1} u \sin u}{n} + \frac{n-1}{n} \int \cos^{n-2} u \, du$

52. $\displaystyle \int u \sin u \, du = \sin u - u \cos u + C$

53. $\displaystyle \int u \cos u \, du = \cos u + u \sin u + C$

54. $\displaystyle \int u^n \sin u \, du = -u^n \cos u + n \int u^{n-1} \cos u \, du$

55. $\displaystyle \int u^n \cos u \, du = u^n \sin u - n \int u^{n-1} \sin u \, du$

56. $\displaystyle \int \frac{1}{1 \pm \sin u} \, du = \tan u \mp \sec u + C$

57. $\displaystyle \int \frac{1}{1 \pm \cos u} \, du = -\cot u \pm \csc u + C$

58. $\displaystyle \int \frac{1}{\sin u \cos u} \, du = \ln|\tan u| + C$

Forms Involving $\tan u$, $\cot u$, $\sec u$, $\csc u$

59. $\displaystyle \int \tan u \, du = -\ln|\cos u| + C$

60. $\displaystyle \int \cot u \, du = \ln|\sin u| + C$

61. $\displaystyle \int \sec u \, du = \ln|\sec u + \tan u| + C$

62. $\displaystyle \int \csc u \, du = \ln|\csc u - \cot u| + C$ or $\displaystyle \int \csc u \, du = -\ln|\csc u + \cot u| + C$

63. $\displaystyle \int \tan^2 u \, du = -u + \tan u + C$

64. $\displaystyle \int \cot^2 u \, du = -u - \cot u + C$

65. $\displaystyle \int \sec^2 u \, du = \tan u + C$

66. $\displaystyle \int \csc^2 u \, du = -\cot u + C$

67. $\displaystyle \int \tan^n u \, du = \frac{\tan^{n-1} u}{n-1} - \int \tan^{n-2} u \, du, \ n \neq 1$

68. $\displaystyle \int \cot^n u \, du = -\frac{\cot^{n-1} u}{n-1} - \int (\cot^{n-2} u) \, du, \ n \neq 1$

69. $\displaystyle \int \sec^n u \, du = \frac{\sec^{n-2} u \tan u}{n-1} + \frac{n-2}{n-1} \int \sec^{n-2} u \, du, \ n \neq 1$

70. $\displaystyle \int \csc^n u \, du = -\frac{\csc^{n-2} u \cot u}{n-1} + \frac{n-2}{n-1} \int \csc^{n-2} u \, du, \ n \neq 1$

71. $\displaystyle \int \frac{1}{1 \pm \tan u} \, du = \frac{1}{2}(u \pm \ln|\cos u \pm \sin u|) + C$

72. $\displaystyle \int \frac{1}{1 \pm \cot u} \, du = \frac{1}{2}(u \mp \ln|\sin u \pm \cos u|) + C$

73. $\displaystyle \int \frac{1}{1 \pm \sec u} \, du = u + \cot u \mp \csc u + C$

74. $\displaystyle \int \frac{1}{1 \pm \csc u} \, du = u - \tan u \pm \sec u + C$

Forms Involving Inverse Trigonometric Functions

75. $\displaystyle\int \arcsin u \, du = u \arcsin u + \sqrt{1 - u^2} + C$

76. $\displaystyle\int \arccos u \, du = u \arccos u - \sqrt{1 - u^2} + C$

77. $\displaystyle\int \arctan u \, du = u \arctan u - \ln\sqrt{1 + u^2} + C$

78. $\displaystyle\int \text{arccot } u \, du = u \text{ arccot } u + \ln\sqrt{1 + u^2} + C$

79. $\displaystyle\int \text{arcsec } u \, du = u \text{ arcsec } u - \ln\left|u + \sqrt{u^2 - 1}\right| + C$

80. $\displaystyle\int \text{arccsc } u \, du = u \text{ arccsc } u + \ln\left|u + \sqrt{u^2 - 1}\right| + C$

Forms Involving e^u

81. $\displaystyle\int e^u \, du = e^u + C$

82. $\displaystyle\int u e^u \, du = (u - 1)e^u + C$

83. $\displaystyle\int u^n e^u \, du = u^n e^u - n \int u^{n-1} e^u \, du$

84. $\displaystyle\int \frac{1}{1 + e^u} \, du = u - \ln(1 + e^u) + C$

85. $\displaystyle\int e^{au} \sin bu \, du = \frac{e^{au}}{a^2 + b^2}(a \sin bu - b \cos bu) + C$

86. $\displaystyle\int e^{au} \cos bu \, du = \frac{e^{au}}{a^2 + b^2}(a \cos bu + b \sin bu) + C$

Forms Involving $\ln u$

87. $\displaystyle\int \ln u \, du = u(-1 + \ln u) + C$

88. $\displaystyle\int u \ln u \, du = \frac{u^2}{4}(-1 + 2 \ln u) + C$

89. $\displaystyle\int u^n \ln u \, du = \frac{u^{n+1}}{(n + 1)^2}[-1 + (n + 1) \ln u] + C, \ n \neq -1$

90. $\displaystyle\int (\ln u)^2 \, du = u\left[2 - 2 \ln u + (\ln u)^2\right] + C$

91. $\displaystyle\int (\ln u)^n \, du = u(\ln u)^n - n \int (\ln u)^{n-1} \, du$

Forms Involving Hyperbolic Functions

92. $\displaystyle\int \cosh u \, du = \sinh u + C$

93. $\displaystyle\int \sinh u \, du = \cosh u + C$

94. $\displaystyle\int \text{sech}^2 u \, du = \tanh u + C$

95. $\displaystyle\int \text{csch}^2 u \, du = -\coth u + C$

96. $\displaystyle\int \text{sech } u \tanh u \, du = -\text{sech } u + C$

97. $\displaystyle\int \text{csch } u \coth u \, du = -\text{csch } u + C$

Forms Involving Inverse Hyperbolic Functions (in logarithmic form)

98. $\displaystyle\int \frac{du}{\sqrt{u^2 \pm a^2}} = \ln\left(u + \sqrt{u^2 \pm a^2}\right) + C$

99. $\displaystyle\int \frac{du}{a^2 - u^2} = \frac{1}{2a} \ln\left|\frac{a + u}{a - u}\right| + C$

100. $\displaystyle\int \frac{du}{u\sqrt{a^2 \pm u^2}} = -\frac{1}{a} \ln \frac{a + \sqrt{a^2 \pm u^2}}{|u|} + C$

C Precalculus Review

Real Numbers and the Real Number Line

- **Represent and classify real numbers.**
- **Order real numbers and use inequalities.**
- **Find the absolute values of real numbers and find the distance between two real numbers.**

Real Numbers and the Real Number Line

Real numbers can be represented by a coordinate system called the **real number line** or x-axis (see Figure C.1). The real number corresponding to a point on the real number line is the **coordinate** of the point. As Figure C.1 shows, it is customary to identify those points whose coordinates are integers.

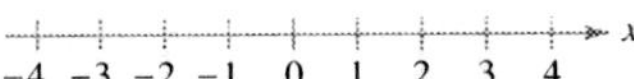

The real number line
Figure C.1

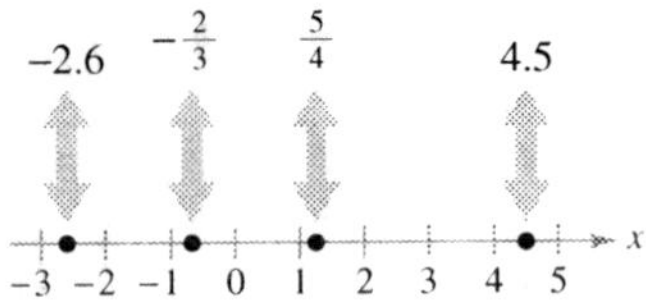

Rational numbers
Figure C.2

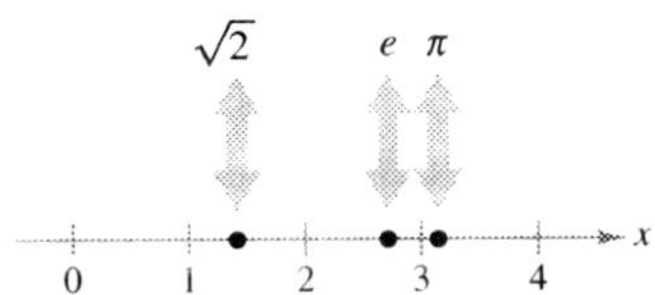

Irrational numbers
Figure C.3

The point on the real number line corresponding to zero is the **origin** and is denoted by 0. The **positive direction** (to the right) is denoted by an arrowhead and is the direction of increasing values of x. Numbers to the right of the origin are **positive.** Numbers to the left of the origin are **negative.** The term **nonnegative** describes a number that is either positive or zero. The term **nonpositive** describes a number that is either negative or zero.

Each point on the real number line corresponds to one and only one real number, and each real number corresponds to one and only one point on the real number line. This type of relationship is called a **one-to-one correspondence.**

Each of the four points in Figure C.2 corresponds to a **rational number**—one that can be written as the ratio of two integers. $\left(\text{Note that } 4.5 = \frac{9}{2} \text{ and } -2.6 = -\frac{13}{5}.\right)$ Rational numbers can be represented either by *terminating decimals* such as $\frac{2}{5} = 0.4$, or by *repeating decimals* such as $\frac{1}{3} = 0.333 \ldots = 0.\overline{3}$.

Real numbers that are not rational are **irrational.** Irrational numbers cannot be represented as terminating or repeating decimals. In computations, irrational numbers are represented by decimal approximations. Here are three familiar examples.

$$\sqrt{2} \approx 1.414213562$$

$$\pi \approx 3.141592654$$

$$e \approx 2.718281828$$

(See Figure C.3.)

Order and Inequalities

One important property of real numbers is that they are **ordered.** If a and b are real numbers, a is **less than** b if $b - a$ is positive. This order is denoted by the **inequality**

$$a < b.$$

This relationship can also be described by saying that b is **greater than** a and writing $b > a$. When three real numbers a, b, and c are ordered such that $a < b$ and $b < c$, you say that b is **between** a and c and $a < b < c$.

Geometrically, $a < b$ if and only if a lies to the *left* of b on the real number line (see Figure C.4). For example, $1 < 2$ because 1 lies to the left of 2 on the real number line.

The following properties are used in working with inequalities. Similar properties are obtained if $<$ is replaced by $\leq$ and $>$ is replaced by $\geq$. (The symbols $\leq$ and $\geq$ mean **less than or equal to** and **greater than or equal to,** respectively.)

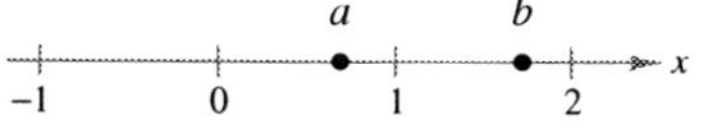

$a < b$ if and only if a lies to the left of b.
Figure C.4

PROPERTIES OF INEQUALITIES

Let a, b, c, d, and k be real numbers.

1. If $a < b$ and $b < c$, then $a < c$.   Transitive Property

2. If $a < b$ and $c < d$, then $a + c < b + d$.   Add inequalities.

3. If $a < b$, then $a + k < b + k$.   Add a constant.

4. If $a < b$ and $k > 0$, then $ak < bk$.   Multiply by a positive constant.

5. If $a < b$ and $k < 0$, then $ak > bk$.   Multiply by a negative constant.

NOTE   Note that you *reverse the inequality* when you multiply the inequality by a negative number. For example, if $x < 3$, then $-4x > -12$. This also applies to division by a negative number. So, if $-2x > 4$, then $x < -2$.   ■

A **set** is a collection of elements. Two common sets are the set of real numbers and the set of points on the real number line. Many problems in calculus involve **subsets** of one of these two sets. In such cases, it is convenient to use **set notation** of the form $\{x:\text{ condition on }x\}$, which is read as follows.

$$\underbrace{\text{The set of}}\ \underbrace{\text{all } x}\ \underbrace{\text{such that}}\ \underbrace{\text{a certain condition is true.}}$$
$$\{\qquad x\qquad :\qquad\qquad \text{condition on } x\}$$

For example, you can describe the set of positive real numbers as

$$\{x:\ x > 0\}.\qquad \text{Set of positive real numbers}$$

Similarly, you can describe the set of nonnegative real numbers as

$$\{x:\ x \geq 0\}.\qquad \text{Set of nonnegative real numbers}$$

The **union** of two sets A and B, denoted by $A \cup B$, is the set of elements that are members of A *or* B or both. The **intersection** of two sets A and B, denoted by $A \cap B$, is the set of elements that are members of A *and* B. Two sets are **disjoint** if they have no elements in common.

The most commonly used subsets are **intervals** on the real number line. For example, the **open** interval

$$(a, b) = \{x : a < x < b\} \qquad \text{Open interval}$$

is the set of all real numbers greater than a and less than b, where a and b are the **endpoints** of the interval. Note that the endpoints are not included in an open interval. Intervals that include their endpoints are **closed** and are denoted by

$$[a, b] = \{x : a \le x \le b\}. \qquad \text{Closed interval}$$

The nine basic types of intervals on the real number line are shown in the table below. The first four are **bounded intervals** and the remaining five are **unbounded intervals.** Unbounded intervals are also classified as open or closed. The intervals $(-\infty, b)$ and (a, ∞) are open, the intervals $(-\infty, b]$ and $[a, \infty)$ are closed, and the interval $(-\infty, \infty)$ is considered to be both open *and* closed.

Intervals on the Real Number Line

	Interval Notation	Set Notation	Graph
Bounded open interval	(a, b)	$\{x : a < x < b\}$	
Bounded closed interval	$[a, b]$	$\{x : a \le x \le b\}$	
Bounded intervals (neither open nor closed)	$[a, b)$	$\{x : a \le x < b\}$	
	$(a, b]$	$\{x : a < x \le b\}$	
Unbounded open intervals	$(-\infty, b)$	$\{x : x < b\}$	
	(a, ∞)	$\{x : x > a\}$	
Unbounded closed intervals	$(-\infty, b]$	$\{x : x \le b\}$	
	$[a, \infty)$	$\{x : x \ge a\}$	
Entire real line	$(-\infty, \infty)$	$\{x : x \text{ is a real number}\}$	

NOTE The symbols ∞ and $-\infty$ refer to positive and negative infinity, respectively. These symbols do not denote real numbers. They simply enable you to describe unbounded conditions more concisely. For instance, the interval $[a, \infty)$ is unbounded to the right because it includes *all* real numbers that are greater than or equal to a.

EXAMPLE 1 Liquid and Gaseous States of Water

Describe the intervals on the real number line that correspond to the temperatures x (in degrees Celsius) of water in

a. a liquid state.   **b.** a gaseous state.

Solution

a. Water is in a liquid state at temperatures greater than 0°C and less than 100°C, as shown in Figure C.5(a).

$$(0, 100) = \{x : 0 < x < 100\}$$

b. Water is in a gaseous state (steam) at temperatures greater than or equal to 100°C, as shown in Figure C.5(b).

$$[100, \infty) = \{x : x \geq 100\}$$

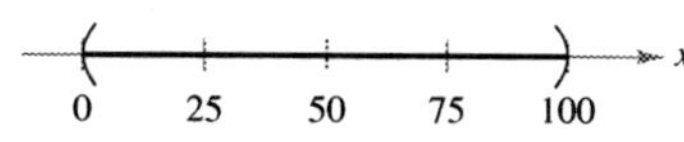

(a) Temperature range of water
    (in degrees Celsius)

(b) Temperature range of steam
    (in degrees Celsius)

Figure C.5

A real number a is a **solution** of an inequality if the inequality is **satisfied** (is true) when a is substituted for x. The set of all solutions is the **solution set** of the inequality.

EXAMPLE 2 Solving an Inequality

Solve $2x - 5 < 7$.

Solution

$$
\begin{aligned}
2x - 5 &< 7 && \text{Write original inequality.} \\
2x - 5 + 5 &< 7 + 5 && \text{Add 5 to each side.} \\
2x &< 12 && \text{Simplify.} \\
\frac{2x}{2} &< \frac{12}{2} && \text{Divide each side by 2.} \\
x &< 6 && \text{Simplify.}
\end{aligned}
$$

The solution set is $(-\infty, 6)$.

NOTE In Example 2, all five inequalities listed as steps in the solution are called **equivalent** because they have the same solution set.

Once you have solved an inequality, check some x-values in your solution set to verify that they satisfy the original inequality. You should also check some values outside your solution set to verify that they *do not* satisfy the inequality. For example, Figure C.6 shows that when $x = 0$ or $x = 5$ the inequality $2x - 5 < 7$ is satisfied, but when $x = 7$ the inequality $2x - 5 < 7$ is not satisfied.

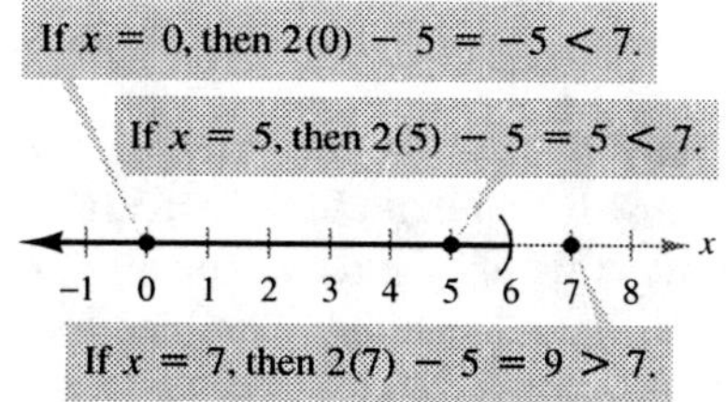

If $x = 0$, then $2(0) - 5 = -5 < 7$.

If $x = 5$, then $2(5) - 5 = 5 < 7$.

If $x = 7$, then $2(7) - 5 = 9 > 7$.

Checking solutions of $2x - 5 < 7$
Figure C.6

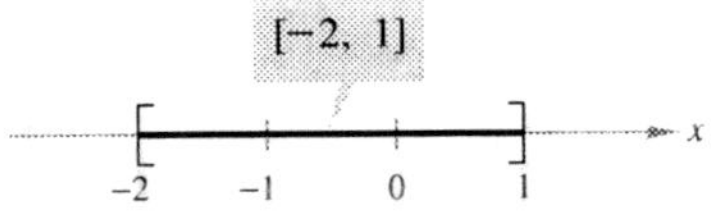

$[-2, 1]$

Solution set of $-3 \leq 2 - 5x \leq 12$
Figure C.7

EXAMPLE 3 Solving a Double Inequality

Solve $-3 \leq 2 - 5x \leq 12$.

Solution

$$
\begin{array}{lll}
-3 \leq \quad 2 - 5x \quad \leq 12 & \text{Write original inequality.} \\[4pt]
-3 - 2 \leq 2 - 5x - 2 \leq 12 - 2 & \text{Subtract 2 from each part.} \\[4pt]
-5 \leq \quad -5x \quad \leq 10 & \text{Simplify.} \\[4pt]
\dfrac{-5}{-5} \geq \dfrac{-5x}{-5} \geq \dfrac{10}{-5} & \text{Divide each part by } -5 \text{ and} \\
& \text{reverse both inequalities.} \\[4pt]
1 \geq \quad x \quad \geq -2 & \text{Simplify.}
\end{array}
$$

The solution set is $[-2, 1]$, as shown in Figure C.7. ∎

The inequalities in Examples 2 and 3 are **linear inequalities**—that is, they involve first-degree polynomials. To solve inequalities involving polynomials of higher degree, use the fact that a polynomial can change signs *only* at its real **zeros** (the x-values that make the polynomial equal to zero). Between two consecutive real zeros, a polynomial must be either entirely positive or entirely negative. This means that when the real zeros of a polynomial are put in order, they divide the real number line into **test intervals** in which the polynomial has no sign changes. So, if a polynomial has the factored form

$$(x - r_1)(x - r_2) \cdots (x - r_n), \qquad r_1 < r_2 < r_3 < \cdots < r_n$$

the test intervals are

$$(-\infty, r_1), \quad (r_1, r_2), \ldots, \quad (r_{n-1}, r_n), \quad \text{and} \quad (r_n, \infty).$$

To determine the sign of the polynomial in each test interval, you need to test only *one value* from the interval.

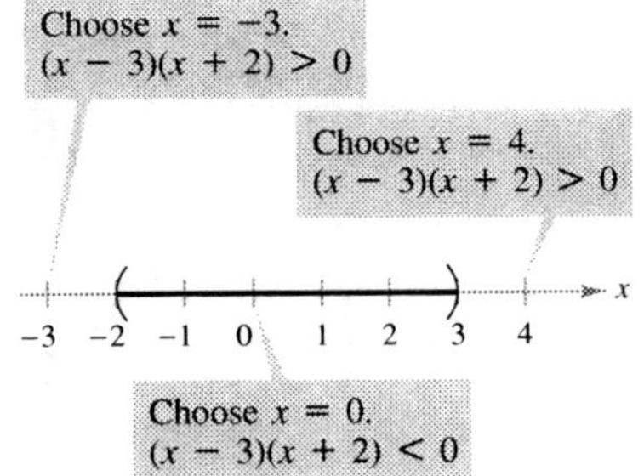

Testing an interval
Figure C.8

EXAMPLE 4 Solving a Quadratic Inequality

Solve $x^2 < x + 6$.

Solution

$$
\begin{array}{ll}
x^2 < x + 6 & \text{Write original inequality.} \\[4pt]
x^2 - x - 6 < 0 & \text{Write in general form.} \\[4pt]
(x - 3)(x + 2) < 0 & \text{Factor.}
\end{array}
$$

The polynomial $x^2 - x - 6$ has $x = -2$ and $x = 3$ as its zeros. So, you can solve the inequality by testing the sign of $x^2 - x - 6$ in each of the test intervals $(-\infty, -2)$, $(-2, 3)$, and $(3, \infty)$. To test an interval, choose any number in the interval and compute the sign of $x^2 - x - 6$. After doing this, you will find that the polynomial is positive for all real numbers in the first and third intervals and negative for all real numbers in the second interval. The solution of the original inequality is therefore $(-2, 3)$, as shown in Figure C.8. ∎

Absolute Value and Distance

If a is a real number, the **absolute value** of a is

$$|a| = \begin{cases} a, & \text{if } a \geq 0 \\ -a, & \text{if } a < 0. \end{cases}$$

The absolute value of a number cannot be negative. For example, let $a = -4$. Then, because $-4 < 0$, you have

$$|a| = |-4| = -(-4) = 4.$$

Remember that the symbol $-a$ does not necessarily mean that $-a$ is negative.

OPERATIONS WITH ABSOLUTE VALUE

Let a and b be real numbers and let n be a positive integer.

1. $|ab| = |a|\,|b|$ **2.** $\left|\dfrac{a}{b}\right| = \dfrac{|a|}{|b|}, \quad b \neq 0$

3. $|a| = \sqrt{a^2}$ **4.** $|a^n| = |a|^n$

NOTE You are asked to prove these properties in Exercises 73, 75, 76, and 77. ■

PROPERTIES OF INEQUALITIES AND ABSOLUTE VALUE

Let a and b be real numbers and let k be a positive real number.

1. $-|a| \leq a \leq |a|$

2. $|a| \leq k$ if and only if $-k \leq a \leq k$.

3. $|a| \geq k$ if and only if $a \leq -k$ or $a \geq k$.

4. *Triangle Inequality:* $|a + b| \leq |a| + |b|$

Properties 2 and 3 are also true if $\leq$ is replaced by $<$ and $\geq$ is replaced by $>$.

EXAMPLE 5   Solving an Absolute Value Inequality

Solve $|x - 3| \leq 2$.

Solution Using the second property of inequalities and absolute value, you can rewrite the original inequality as a double inequality.

$$\begin{aligned} -2 \leq \quad x - 3 \quad &\leq 2 \qquad &&\text{Write as double inequality.} \\ -2 + 3 \leq x - 3 + 3 &\leq 2 + 3 \qquad &&\text{Add 3 to each part.} \\ 1 \leq \quad x \quad &\leq 5 \qquad &&\text{Simplify.} \end{aligned}$$

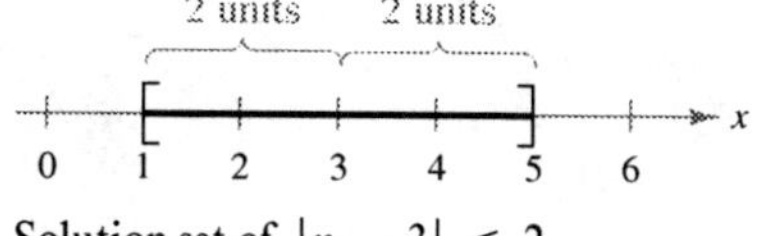

Solution set of $|x - 3| \leq 2$
Figure C.9

The solution set is $[1, 5]$, as shown in Figure C.9. ■

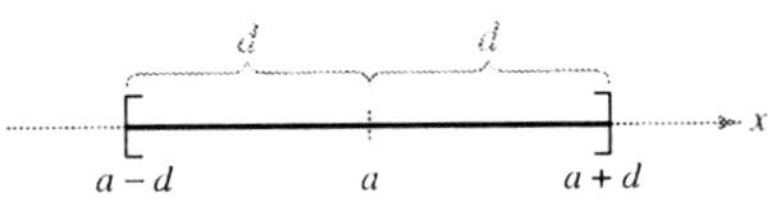

Solution set of $|x + 2| > 3$
Figure C.10

Solution set of $|x - a| \leq d$

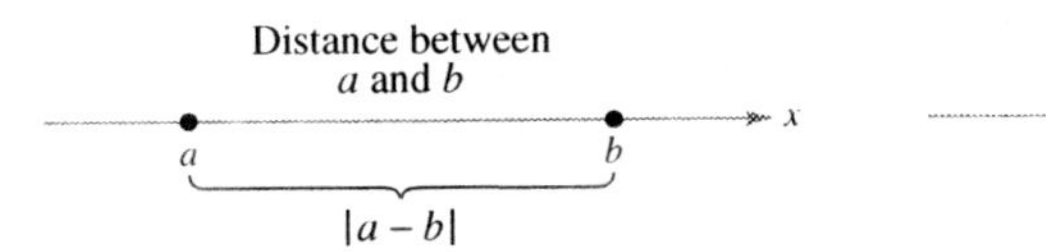

Solution set of $|x - a| \geq d$
Figure C.11

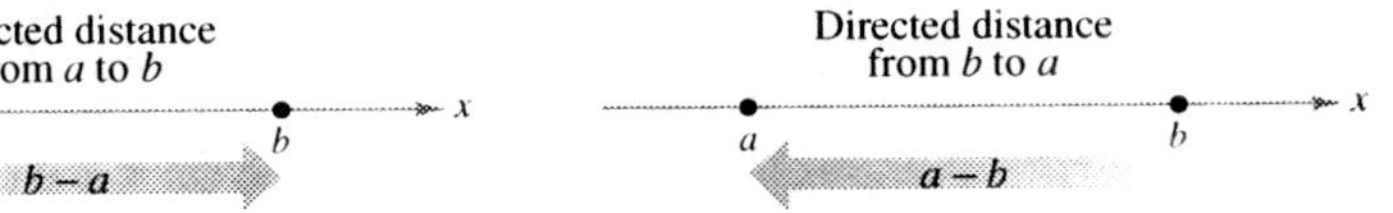

Figure C.12

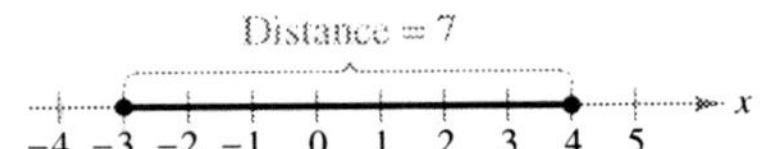

Figure C.13

EXAMPLE 6 A Two-Interval Solution Set

Solve $|x + 2| > 3$.

Solution   Using the third property of inequalities and absolute value, you can rewrite the original inequality as two linear inequalities.

$$x + 2 < -3 \quad \text{or} \quad x + 2 > 3$$
$$x < -5 \qquad\qquad x > 1$$

The solution set is the union of the disjoint intervals $(-\infty, -5)$ and $(1, \infty)$, as shown in Figure C.10.

Examples 5 and 6 illustrate the general results shown in Figure C.11. Note that if $d > 0$, the solution set for the inequality $|x - a| \leq d$ is a *single* interval, whereas the solution set for the inequality $|x - a| \geq d$ is the union of *two* disjoint intervals.

The **distance between two points** a and b on the real number line is given by

$$d = |a - b| = |b - a|.$$

The **directed distance from a to b** is $b - a$ and the **directed distance from b to a** is $a - b$, as shown in Figure C.12.

EXAMPLE 7 Distance on the Real Number Line

a. The distance between -3 and 4 is

$$|4 - (-3)| = |7| = 7 \quad \text{or} \quad |-3 - 4| = |-7| = 7.$$

(See Figure C.13.)

b. The directed distance from -3 to 4 is $4 - (-3) = 7$.

c. The directed distance from 4 to -3 is $-3 - 4 = -7$.

The **midpoint** of an interval with endpoints a and b is the average value of a and b. That is,

$$\text{Midpoint of interval } (a, b) = \frac{a + b}{2}.$$

To show that this is the midpoint, you need only show that $(a + b)/2$ is equidistant from a and b.

C.1 Exercises

See www.CalcChat.com for worked-out solutions to odd-numbered exercises.

In Exercises 1–10, determine whether the real number is rational or irrational.

1. 0.7
2. -3678
3. $\dfrac{3\pi}{2}$
4. $3\sqrt{2} - 1$
5. $4.3\overline{451}$
6. $\frac{22}{7}$
7. $\sqrt[3]{64}$
8. $0.\overline{8177}$
9. $4\frac{5}{8}$
10. $(\sqrt{2})^3$

In Exercises 11–14, write the repeating decimal as a ratio of two integers using the following procedure. If $x = 0.6363\ldots$, then $100x = 63.6363\ldots$. Subtracting the first equation from the second produces $99x = 63$ or $x = \frac{63}{99} = \frac{7}{11}$.

11. $0.\overline{36}$
12. $0.3\overline{18}$
13. $0.\overline{297}$
14. $0.\overline{9900}$

15. Given $a < b$, determine which of the following are true.

(a) $a + 2 < b + 2$ (b) $5b < 5a$

(c) $5 - a > 5 - b$ (d) $\dfrac{1}{a} < \dfrac{1}{b}$

(e) $(a - b)(b - a) > 0$ (f) $a^2 < b^2$

16. Complete the table with the appropriate interval notation, set notation, and graph on the real number line.

Interval Notation	Set Notation	Graph
		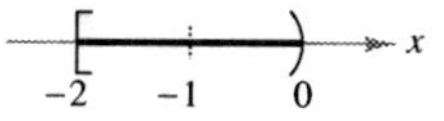
$(-\infty, -4]$		
	$\{x: 3 \le x \le \frac{11}{2}\}$	
$(-1, 7)$		

In Exercises 17–20, verbally describe the subset of real numbers represented by the inequality. Sketch the subset on the real number line, and state whether the interval is bounded or unbounded.

17. $-3 < x < 3$
18. $x \ge 4$
19. $x \le 5$
20. $0 \le x < 8$

In Exercises 21–24, use inequality and interval notation to describe the set.

21. y is at least 4.

22. q is nonnegative.

23. The interest rate r on loans is expected to be greater than 3% and no more than 7%.

24. The temperature T is forecast to be above 90°F today.

In Exercises 25–44, solve the inequality and graph the solution on the real number line.

25. $2x - 1 \ge 0$
26. $3x + 1 \ge 2x + 2$
27. $-4 < 2x - 3 < 4$
28. $0 \le x + 3 < 5$
29. $\dfrac{x}{2} + \dfrac{x}{3} > 5$
30. $x > \dfrac{1}{x}$
31. $|x| < 1$
32. $\dfrac{x}{2} - \dfrac{x}{3} > 5$
33. $\left|\dfrac{x - 3}{2}\right| \ge 5$
34. $\left|\dfrac{x}{2}\right| > 3$
35. $|x - a| < b,\ b > 0$
36. $|x + 2| < 5$
37. $|2x + 1| < 5$
38. $|3x + 1| \ge 4$
39. $\left|1 - \dfrac{2}{3}x\right| < 1$
40. $|9 - 2x| < 1$
41. $x^2 \le 3 - 2x$
42. $x^4 - x \le 0$
43. $x^2 + x - 1 \le 5$
44. $2x^2 + 1 < 9x - 3$

In Exercises 45–48, find the directed distance from a to b, the directed distance from b to a, and the distance between a and b.

45. 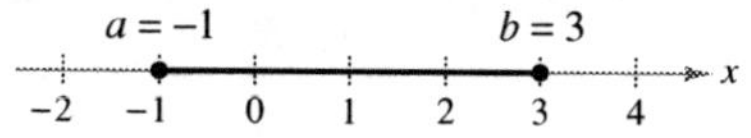

46. 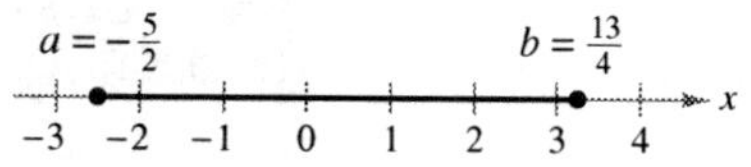

47. (a) $a = 126, b = 75$

(b) $a = -126, b = -75$

48. (a) $a = 9.34, b = -5.65$

(b) $a = \frac{16}{5}, b = \frac{112}{75}$

In Exercises 49–54, use absolute value notation to define the interval or pair of intervals on the real number line.

49. 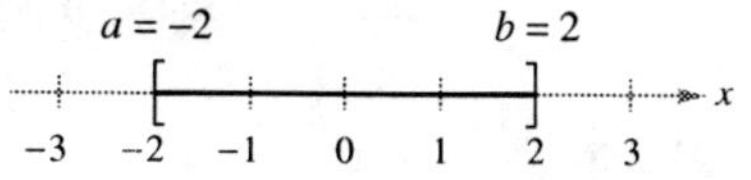

50. 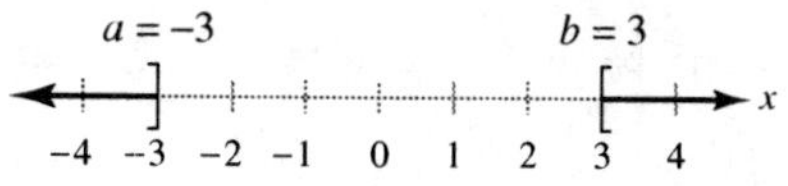

51. 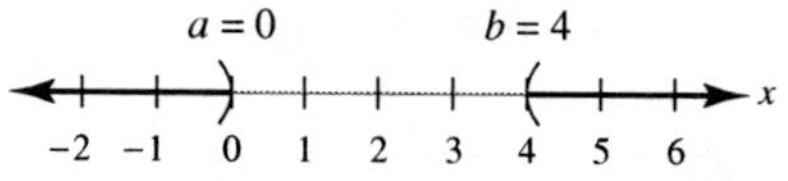

52. 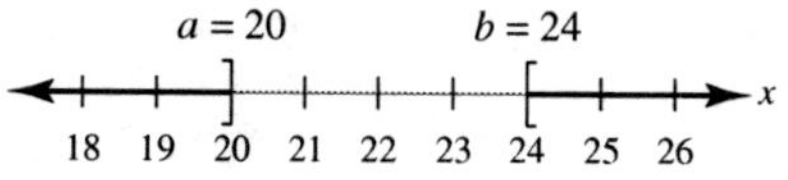

53. (a) All numbers that are at most 10 units from 12

(b) All numbers that are at least 10 units from 12

54. (a) y is at most two units from a.

(b) y is less than δ units from c.

In Exercises 55–58, find the midpoint of the interval.

55.

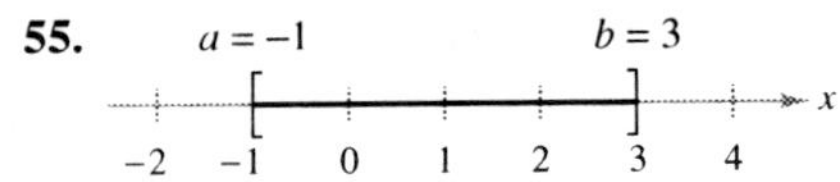

56.

57. (a) $[7, 21]$

(b) $[8.6, 11.4]$

58. (a) $[-6.85, 9.35]$

(b) $[-4.6, -1.3]$

59. Profit   The revenue R from selling x units of a product is

$$R = 115.95x$$

and the cost C of producing x units is

$$C = 95x + 750.$$

To make a (positive) profit, R must be greater than C. For what values of x will the product return a profit?

60. Fleet Costs   A utility company has a fleet of vans. The annual operating cost C (in dollars) of each van is estimated to be

$$C = 0.32m + 2300$$

where m is measured in miles. The company wants the annual operating cost of each van to be less than \$10,000. To do this, m must be less than what value?

61. Fair Coin   To determine whether a coin is fair (has an equal probability of landing tails up or heads up), you toss the coin 100 times and record the number of heads x. The coin is declared unfair if

$$\left| \frac{x - 50}{5} \right| \geq 1.645.$$

For what values of x will the coin be declared unfair?

62. Daily Production   The estimated daily oil production p at a refinery is

$$|p - 2,250,000| < 125,000$$

where p is measured in barrels. Determine the high and low production levels.

In Exercises 63 and 64, determine which of the two real numbers is greater.

63. (a) π or $\frac{355}{113}$

(b) π or $\frac{22}{7}$

64. (a) $\frac{224}{151}$ or $\frac{144}{97}$

(b) $\frac{73}{81}$ or $\frac{6427}{7132}$

65. Approximation—Powers of 10   Light travels at the speed of 2.998×10^8 meters per second. Which best estimates the distance in meters that light travels in a year?

(a) 9.5×10^5

(b) 9.5×10^{15}

(c) 9.5×10^{12}

(d) 9.6×10^{16}

66. Writing   The accuracy of an approximation of a number is related to how many significant digits there are in the approximation. Write a definition of significant digits and illustrate the concept with examples.

True or False?   **In Exercises 67–72, determine whether the statement is true or false. If it is false, explain why or give an example that shows it is false.**

67. The reciprocal of a nonzero integer is an integer.

68. The reciprocal of a nonzero rational number is a rational number.

69. Each real number is either rational or irrational.

70. The absolute value of each real number is positive.

71. If $x < 0$, then $\sqrt{x^2} = -x$.

72. If a and b are any two distinct real numbers, then $a < b$ or $a > b$.

In Exercises 73–80, prove the property.

73. $|ab| = |a||b|$

74. $|a - b| = |b - a|$

$$\left[Hint:\ (a - b) = (-1)(b - a) \right]$$

75. $\left| \dfrac{a}{b} \right| = \dfrac{|a|}{|b|}, \quad b \neq 0$

76. $|a| = \sqrt{a^2}$

77. $|a^n| = |a|^n, \quad n = 1, 2, 3, \ldots$

78. $-|a| \leq a \leq |a|$

79. $|a| \leq k$ if and only if $-k \leq a \leq k, \quad k > 0$.

80. $|a| \geq k$ if and only if $a \leq -k$ or $a \geq k, \quad k > 0$.

81. Find an example for which $|a - b| > |a| - |b|$, and an example for which $|a - b| = |a| - |b|$. Then prove that $|a - b| \geq |a| - |b|$ for all a, b.

82. Show that the maximum of two numbers a and b is given by the formula

$$\max(a, b) = \tfrac{1}{2}(a + b + |a - b|).$$

Derive a similar formula for $\min(a, b)$.

C.2 The Cartesian Plane

- Understand the Cartesian plane.
- Use the Distance Formula to find the distance between two points and use the Midpoint Formula to find the midpoint of a line segment.
- Find equations of circles and sketch the graphs of circles.

The Cartesian Plane

Just as you can represent real numbers by points on a real number line, you can represent ordered pairs of real numbers by points in a plane called the **rectangular coordinate system,** or the **Cartesian plane,** after the French mathematician René Descartes.

The Cartesian plane is formed by using two real number lines intersecting at right angles, as shown in Figure C.14. The horizontal real number line is usually called the **x-axis,** and the vertical real number line is usually called the **y-axis.** The point of intersection of these two axes is the **origin.** The two axes divide the plane into four parts called **quadrants.**

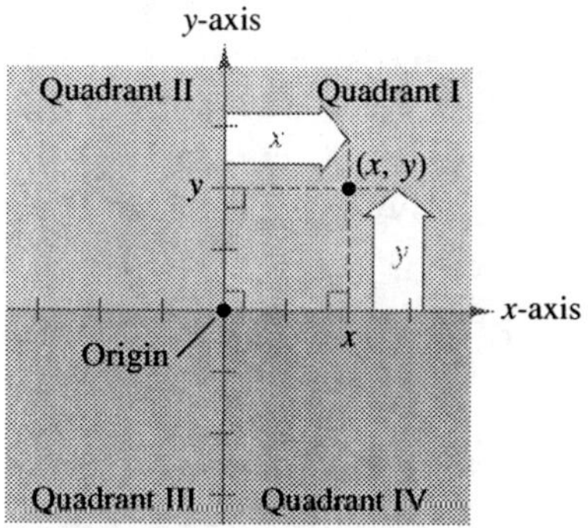

The Cartesian plane
Figure C.14

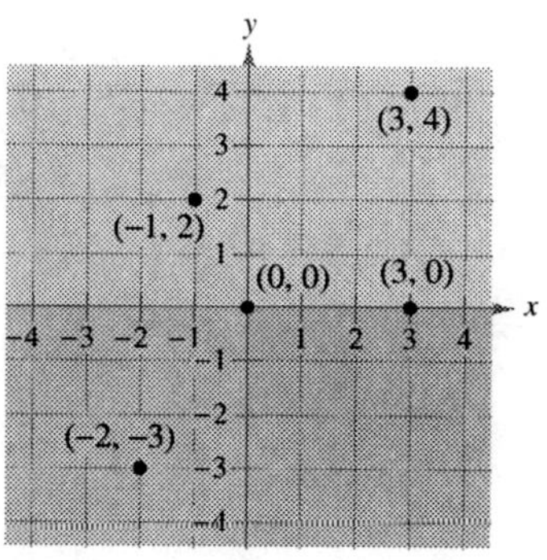

Points represented by orderd pairs
Figure C.15

Each point in the plane is identified by an **ordered pair** (x, y) of real numbers x and y, called the **coordinates** of the point. The number x represents the directed distance from the y-axis to the point, and the number y represents the directed distance from the x-axis to the point (see Figure C.14). For the point (x, y), the first coordinate is the **x-coordinate** or **abscissa,** and the second coordinate is the **y-coordinate** or **ordinate.** For example, Figure C.15 shows the locations of the points $(-1, 2)$, $(3, 4)$, $(0, 0)$, $(3, 0)$, and $(-2, -3)$ in the Cartesian plane.

NOTE The signs of the coordinates of a point determine the quadrant in which the point lies. For instance, if $x > 0$ and $y < 0$, then the point (x, y) lies in Quadrant IV.

Note that an ordered pair (a, b) is used to denote either a point in the plane *or* an open interval on the real number line. This, however, should not be confusing—the nature of the problem should clarify whether a point in the plane or an open interval is being discussed.

The Distance and Midpoint Formulas

Recall from the Pythagorean Theorem that, in a right triangle, the hypotenuse c and sides a and b are related by $a^2 + b^2 = c^2$. Conversely, if $a^2 + b^2 = c^2$, then the triangle is a right triangle (see Figure C.16).

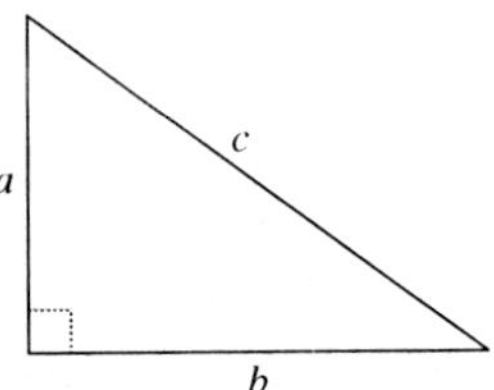

The Pythagorean Theorem:
$a^2 + b^2 = c^2$
Figure C.16

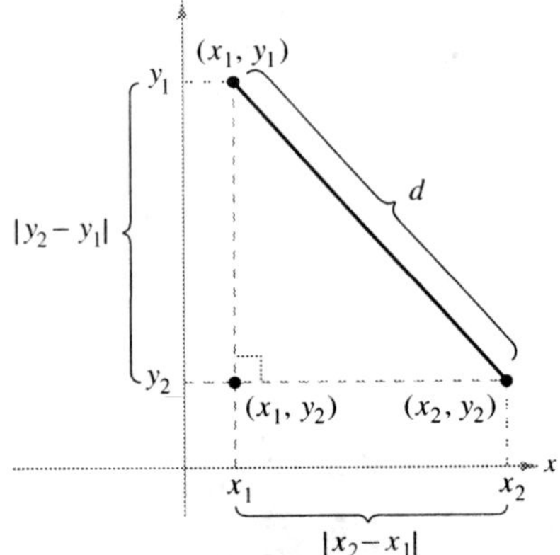

The distance between two points
Figure C.17

Suppose you want to determine the distance d between the two points (x_1, y_1) and (x_2, y_2) in the plane. If the points lie on a horizontal line, then $y_1 = y_2$ and the distance between the points is $|x_2 - x_1|$. If the points lie on a vertical line, then $x_1 = x_2$ and the distance between the points is $|y_2 - y_1|$. If the two points do not lie on a horizontal or vertical line, they can be used to form a right triangle, as shown in Figure C.17. The length of the vertical side of the triangle is $|y_2 - y_1|$, and the length of the horizontal side is $|x_2 - x_1|$. By the Pythagorean Theorem, it follows that

$$d^2 = |x_2 - x_1|^2 + |y_2 - y_1|^2$$
$$d = \sqrt{|x_2 - x_1|^2 + |y_2 - y_1|^2}.$$

Replacing $|x_2 - x_1|^2$ and $|y_2 - y_1|^2$ by the equivalent expressions $(x_2 - x_1)^2$ and $(y_2 - y_1)^2$ produces the following result.

DISTANCE FORMULA

The distance d between the points (x_1, y_1) and (x_2, y_2) in the plane is given by

$$d = \sqrt{(x_2 - x_1)^2 + (y_2 - y_1)^2}.$$

EXAMPLE 1 Finding the Distance Between Two Points

Find the distance between the points $(-2, 1)$ and $(3, 4)$.

Solution

$$
\begin{aligned}
d &= \sqrt{(x_2 - x_1)^2 + (y_2 - y_1)^2} && \text{Distance Formula} \\
&= \sqrt{[3 - (-2)]^2 + (4 - 1)^2} && \text{Substitute for } x_1, y_1, x_2, \text{ and } y_2. \\
&= \sqrt{5^2 + 3^2} \\
&= \sqrt{25 + 9} \\
&= \sqrt{34} \\
&\approx 5.83
\end{aligned}
$$

EXAMPLE 2   Verifying a Right Triangle

Verify that the points $(2, 1)$, $(4, 0)$, and $(5, 7)$ form the vertices of a right triangle.

Solution   Figure C.18 shows the triangle formed by the three points. The lengths of the three sides are as follows.

$$d_1 = \sqrt{(5 - 2)^2 + (7 - 1)^2} = \sqrt{9 + 36} = \sqrt{45}$$
$$d_2 = \sqrt{(4 - 2)^2 + (0 - 1)^2} = \sqrt{4 + 1} = \sqrt{5}$$
$$d_3 = \sqrt{(5 - 4)^2 + (7 - 0)^2} = \sqrt{1 + 49} = \sqrt{50}$$

Because

$$d_1^2 + d_2^2 = 45 + 5 = 50 \qquad \text{Sum of squares of sides}$$

and

$$d_3^2 = 50 \qquad \text{Square of hypotenuse}$$

you can apply the Pythagorean Theorem to conclude that the triangle is a right triangle.

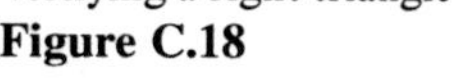

Verifying a right triangle
Figure C.18

EXAMPLE 3   Using the Distance Formula

Find x such that the distance between $(x, 3)$ and $(2, -1)$ is 5.

Solution   Using the Distance Formula, you can write the following.

$$5 = \sqrt{(x - 2)^2 + [3 - (-1)]^2} \qquad \text{Distance Formula}$$
$$25 = (x^2 - 4x + 4) + 16 \qquad \text{Square each side.}$$
$$0 = x^2 - 4x - 5 \qquad \text{Write in general form.}$$
$$0 = (x - 5)(x + 1) \qquad \text{Factor.}$$

Therefore, $x = 5$ or $x = -1$, and you can conclude that there are two solutions. That is, each of the points $(5, 3)$ and $(-1, 3)$ lies five units from the point $(2, -1)$, as shown in Figure C.19.

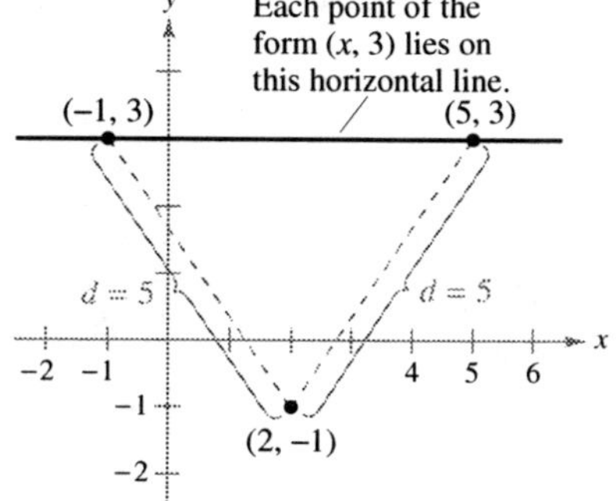

Given a distance, find a point.
Figure C.19

The coordinates of the **midpoint** of the line segment joining two points can be found by "averaging" the x-coordinates of the two points and "averaging" the y-coordinates of the two points. That is, the midpoint of the line segment joining the points (x_1, y_1) and (x_2, y_2) in the plane is

$$\left(\frac{x_1 + x_2}{2}, \frac{y_1 + y_2}{2} \right). \qquad \text{Midpoint Formula}$$

For instance, the midpoint of the line segment joining the points $(-5, -3)$ and $(9, 3)$ is

$$\left(\frac{-5 + 9}{2}, \frac{-3 + 3}{2} \right) = (2, 0)$$

as shown in Figure C.20.

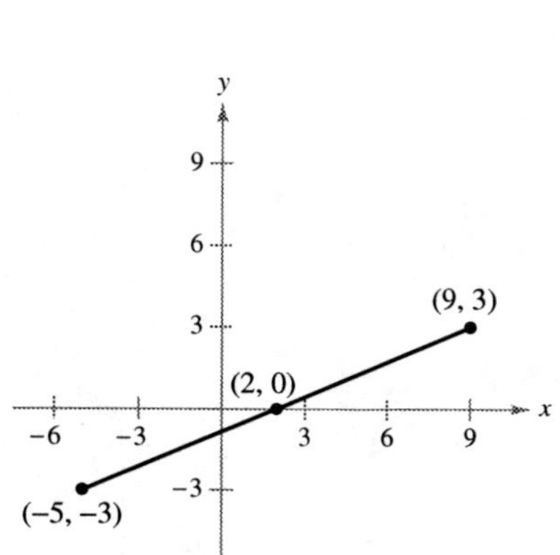

Midpoint of a line segment
Figure C.20

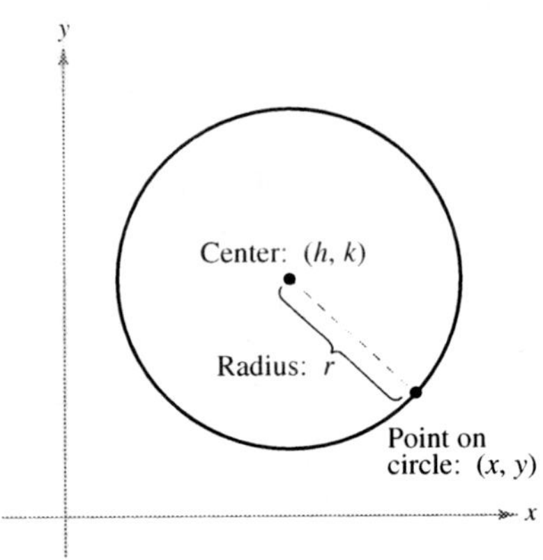

Definition of a circle
Figure C.21

Equations of Circles

A **circle** can be defined as the set of all points in a plane that are equidistant from a fixed point. The fixed point is the **center** of the circle, and the distance between the center and a point on the circle is the **radius** (see Figure C.21).

You can use the Distance Formula to write an equation for the circle with center (h, k) and radius r. Let (x, y) be any point on the circle. Then the distance between (x, y) and the center (h, k) is given by

$$\sqrt{(x - h)^2 + (y - k)^2} = r.$$

By squaring each side of this equation, you obtain the **standard form of the equation of a circle.**

STANDARD FORM OF THE EQUATION OF A CIRCLE

The point (x, y) lies on the circle of radius r and center (h, k) if and only if

$$(x - h)^2 + (y - k)^2 = r^2.$$

The standard form of the equation of a circle with center at the origin, $(h, k) = (0, 0)$, is

$$x^2 + y^2 = r^2.$$

If $r = 1$, the circle is called the **unit circle.**

EXAMPLE 4 Writing the Equation of a Circle

The point $(3, 4)$ lies on a circle whose center is at $(-1, 2)$, as shown in Figure C.22. Write the standard form of the equation of this circle.

Solution   The radius of the circle is the distance between $(-1, 2)$ and $(3, 4)$.

$$r = \sqrt{[3 - (-1)]^2 + (4 - 2)^2} = \sqrt{16 + 4} = \sqrt{20}$$

You can write the standard form of the equation of this circle as

$$[x - (-1)]^2 + (y - 2)^2 = \left(\sqrt{20}\right)^2$$
$$(x + 1)^2 + (y - 2)^2 = 20. \qquad \text{Write in standard form.} \qquad \blacksquare$$

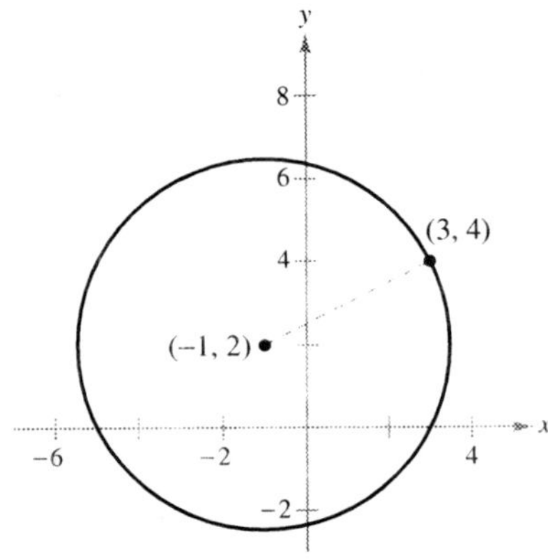

Figure C.22

By squaring and simplifying, the equation $(x - h)^2 + (y - k)^2 = r^2$ can be written in the following **general form of the equation of a circle.**

$$Ax^2 + Ay^2 + Dx + Ey + F = 0, \quad A \neq 0$$

To convert such an equation to the standard form

$$(x - h)^2 + (y - k)^2 = p$$

you can use a process called **completing the square.** If $p > 0$, the graph of the equation is a circle. If $p = 0$, the graph is the single point (h, k). If $p < 0$, the equation has no graph.

EXAMPLE 5 Completing the Square

Sketch the graph of the circle whose general equation is

$$4x^2 + 4y^2 + 20x - 16y + 37 = 0.$$

Solution   To complete the square, first divide by 4 so that the coefficients of x^2 and y^2 are both 1.

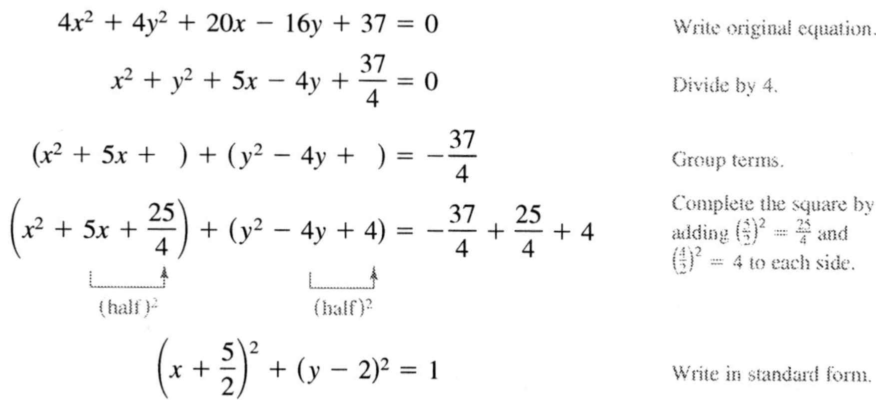

$$4x^2 + 4y^2 + 20x - 16y + 37 = 0 \qquad \text{Write original equation.}$$

$$x^2 + y^2 + 5x - 4y + \frac{37}{4} = 0 \qquad \text{Divide by 4.}$$

$$(x^2 + 5x + \quad) + (y^2 - 4y + \quad) = -\frac{37}{4} \qquad \text{Group terms.}$$

$$\left(x^2 + 5x + \frac{25}{4}\right) + (y^2 - 4y + 4) = -\frac{37}{4} + \frac{25}{4} + 4 \qquad \text{Complete the square by adding } \left(\tfrac{5}{2}\right)^2 = \tfrac{25}{4} \text{ and } \left(\tfrac{4}{2}\right)^2 = 4 \text{ to each side.}$$

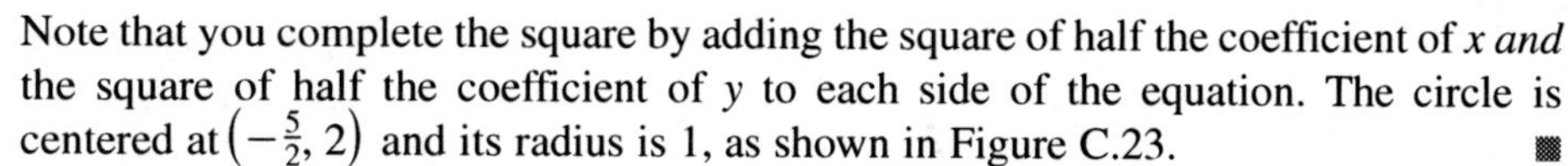

$$\underbrace{\qquad}_{(\text{half})^2} \qquad \underbrace{\qquad}_{(\text{half})^2}$$

$$\left(x + \frac{5}{2}\right)^2 + (y - 2)^2 = 1 \qquad \text{Write in standard form.}$$

Note that you complete the square by adding the square of half the coefficient of x *and* the square of half the coefficient of y to each side of the equation. The circle is centered at $\left(-\frac{5}{2}, 2\right)$ and its radius is 1, as shown in Figure C.23.

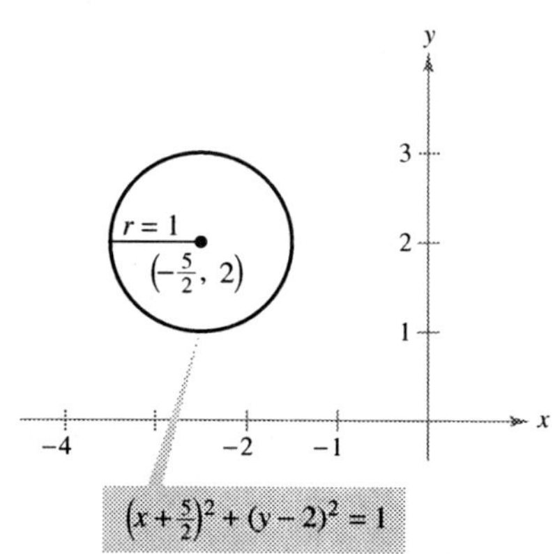

A circle with a radius of 1 and center at $\left(-\frac{5}{2}, 2\right)$

Figure C.23

You have now reviewed some fundamental concepts of *analytic geometry*. Because these concepts are in common use today, it is easy to overlook their revolutionary nature. At the time analytic geometry was being developed by Pierre de Fermat and René Descartes, the two major branches of mathematics—geometry and algebra—were largely independent of each other. Circles belonged to geometry and equations belonged to algebra. The coordination of the points on a circle and the solutions of an equation belongs to what is now called analytic geometry.

It is important to become skilled in analytic geometry so that you can move easily between geometry and algebra. For instance, in Example 4, you were given a geometric description of a circle and were asked to find an algebraic equation for the circle. So, you were moving from geometry to algebra. Similarly, in Example 5 you were given an algebraic equation and asked to sketch a geometric picture. In this case, you were moving from algebra to geometry. These two examples illustrate the two most common problems in analytic geometry.

1. Given a graph, find its equation.

2. Given an equation, find its graph.

In the next section, you will review other examples of these two types of problems.

C.2 Exercises

See www.CalcChat.com for worked-out solutions to odd-numbered exercises.

In Exercises 1–6, (a) plot the points, (b) find the distance between the points, and (c) find the midpoint of the line segment joining the points.

1. $(2, 1), (4, 5)$

2. $(-3, 2), (3, -2)$

3. $\left(\frac{1}{2}, 1\right), \left(-\frac{3}{2}, -5\right)$

4. $\left(\frac{2}{3}, -\frac{1}{3}\right), \left(\frac{5}{6}, 1\right)$

5. $\left(1, \sqrt{3}\right), (-1, 1)$

6. $(-2, 0), \left(0, \sqrt{2}\right)$

In Exercises 7–10, determine the quadrant(s) in which (x, y) is located so that the condition(s) is (are) satisfied.

7. $x = -2$ and $y > 0$

8. $y < -2$

9. $xy > 0$

10. $(x, -y)$ is in Quadrant II.

In Exercises 11–14, show that the points are the vertices of the polygon. (A rhombus is a quadrilateral whose sides are all of the same length.)

Vertices	Polygon
11. $(4, 0), (2, 1), (-1, -5)$	Right triangle
12. $(1, -3), (3, 2), (-2, 4)$	Isosceles triangle
13. $(0, 0), (1, 2), (2, 1), (3, 3)$	Rhombus
14. $(0, 1), (3, 7), (4, 4), (1, -2)$	Parallelogram

15. *Number of Stores* The table shows the number y of Target stores for each year x from 1998 through 2007. *(Source: Target Corp.)*

Year, x	1998	1999	2000	2001	2002
Number, y	851	912	977	1053	1147

Year, x	2003	2004	2005	2006	2007
Number, y	1225	1308	1397	1488	1591

Select reasonable scales on the coordinate axes and plot the points (x, y).

16. *Conjecture* Plot the points $(2, 1), (-3, 5)$, and $(7, -3)$ on a rectangular coordinate system. Then change the sign of the x-coordinate of each point and plot the three new points on the same rectangular coordinate system. What conjecture can you make about the location of a point when the sign of the x-coordinate is changed? Repeat the exercise for the case in which the signs of the y-coordinates are changed.

In Exercises 17–20, use the Distance Formula to determine whether the points lie on the same line.

17. $(0, -4), (2, 0), (3, 2)$

18. $(0, 4), (7, -6), (-5, 11)$

19. $(-2, 1), (-1, 0), (2, -2)$

20. $(-1, 1), (3, 3), (5, 5)$

In Exercises 21 and 22, find x such that the distance between the points is 5.

21. $(0, 0), (x, -4)$

22. $(2, -1), (x, 2)$

In Exercises 23 and 24, find y such that the distance between the points is 8.

23. $(0, 0), (3, y)$

24. $(5, 1), (5, y)$

25. Use the Midpoint Formula to find the three points that divide the line segment joining (x_1, y_1) and (x_2, y_2) into four equal parts.

26. Use the result of Exercise 25 to find the points that divide the line segment joining the given points into four equal parts.

(a) $(1, -2), (4, -1)$ (b) $(-2, -3), (0, 0)$

In Exercises 27–30, match the equation with its graph. [The graphs are labeled (a), (b), (c), and (d).]

(a)
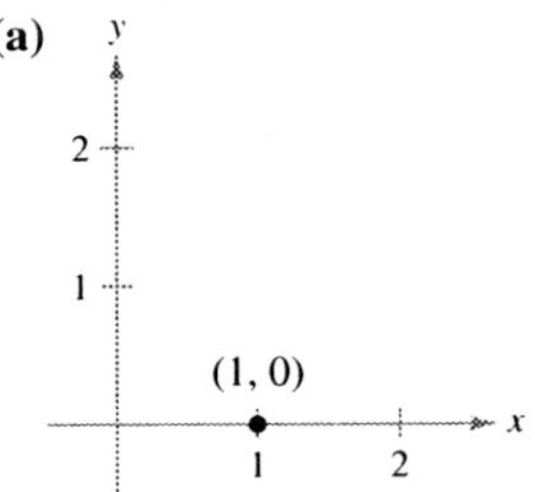

(b)
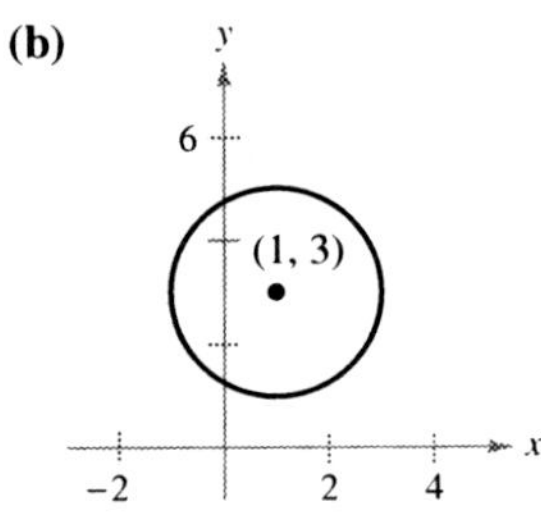

(c)
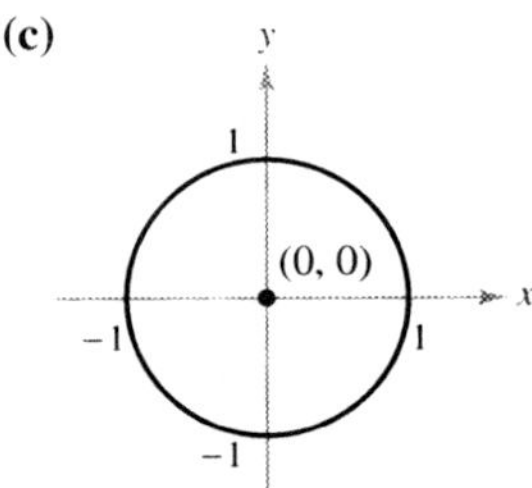

(d)
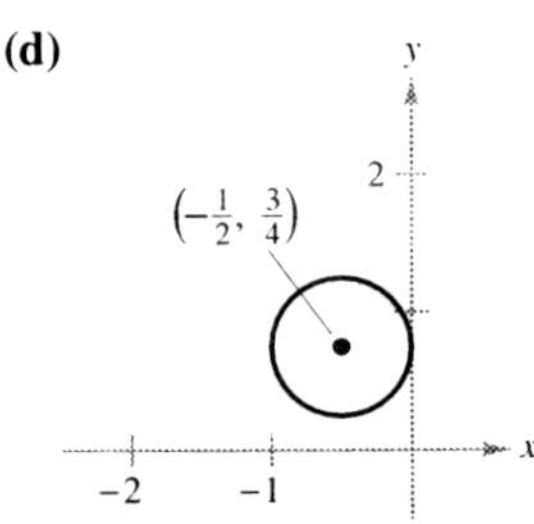

27. $x^2 + y^2 = 1$

28. $(x - 1)^2 + (y - 3)^2 = 4$

29. $(x - 1)^2 + y^2 = 0$

30. $\left(x + \frac{1}{2}\right)^2 + \left(y - \frac{3}{4}\right)^2 = \frac{1}{4}$

In Exercises 31–38, write the general form of the equation of the circle.

31. Center: $(0, 0)$
Radius: 3

32. Center: $(0, 0)$
Radius: 5

33. Center: $(2, -1)$
Radius: 4

34. Center: $(-4, 3)$
Radius: $\frac{5}{8}$

35. Center: $(-1, 2)$

Point on circle: $(0, 0)$

36. Center: $(3, -2)$

Point on circle: $(-1, 1)$

37. Endpoints of a diameter: $(2, 5), (4, -1)$

38. Endpoints of a diameter: $(1, 1), (-1, -1)$

39. *Satellite Communication* Write the standard form of the equation for the path of a communications satellite in a circular orbit 22,000 miles above Earth. (Assume that the radius of Earth is 4000 miles.)

40. *Building Design* A circular air duct of diameter D is fit firmly into the right-angle corner where a basement wall meets the floor (see figure). Find the diameter of the largest water pipe that can be run in the right-angle corner behind the air duct.

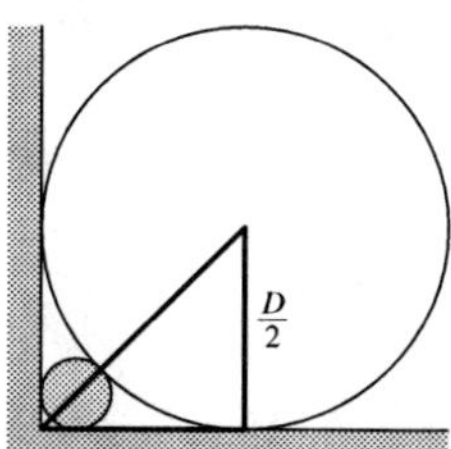

In Exercises 41–48, write the standard form of the equation of the circle and sketch its graph.

41. $x^2 + y^2 - 2x + 6y + 6 = 0$

42. $x^2 + y^2 - 2x + 6y - 15 = 0$

43. $x^2 + y^2 - 2x + 6y + 10 = 0$

44. $3x^2 + 3y^2 - 6y - 1 = 0$

45. $2x^2 + 2y^2 - 2x - 2y - 3 = 0$

46. $4x^2 + 4y^2 - 4x + 2y - 1 = 0$

47. $16x^2 + 16y^2 + 16x + 40y - 7 = 0$

48. $x^2 + y^2 - 4x + 2y + 3 = 0$

In Exercises 49 and 50, use a graphing utility to graph the equation. Use a square setting. (*Hint:* It may be necessary to solve the equation for y and graph the resulting two equations.)

49. $4x^2 + 4y^2 - 4x + 24y - 63 = 0$

50. $x^2 + y^2 - 8x - 6y - 11 = 0$

In Exercises 51 and 52, sketch the set of all points satisfying the inequality. Use a graphing utility to verify your result.

51. $x^2 + y^2 - 4x + 2y + 1 \le 0$

52. $(x - 1)^2 + \left(y - \tfrac{1}{2}\right)^2 > 1$

53. Prove that

$$\left(\frac{2x_1 + x_2}{3}, \frac{2y_1 + y_2}{3}\right)$$

is one of the points of trisection of the line segment joining (x_1, y_1) and (x_2, y_2). Find the midpoint of the line segment joining

$$\left(\frac{2x_1 + x_2}{3}, \frac{2y_1 + y_2}{3}\right)$$

and (x_2, y_2) to find the second point of trisection.

54. Use the results of Exercise 53 to find the points of trisection of the line segment joining the following points.

(a) $(1, -2), (4, 1)$ (b) $(-2, -3), (0, 0)$

True or False? In Exercises 55–58, determine whether the statement is true or false. If it is false, explain why or give an example that shows it is false.

55. If $ab < 0$, the point (a, b) lies in either Quadrant II or Quadrant IV.

56. The distance between the points $(a + b, a)$ and $(a - b, a)$ is $2b$.

57. If the distance between two points is zero, the two points must coincide.

58. If $ab = 0$, the point (a, b) lies on the x-axis or on the y-axis.

In Exercises 59–62, prove the statement.

59. The line segments joining the midpoints of the opposite sides of a quadrilateral bisect each other.

60. The perpendicular bisector of a chord of a circle passes through the center of the circle.

61. An angle inscribed in a semicircle is a right angle.

62. The midpoint of the line segment joining the points (x_1, y_1) and (x_2, y_2) is

$$\left(\frac{x_1 + x_2}{2}, \frac{y_1 + y_2}{2}\right).$$

C.3   Review of Trigonometric Functions

- Describe angles and use degree measure.
- Use radian measure.
- Understand the definitions of the six trigonometric functions.
- Evaluate trigonometric functions.
- Solve trigonometric equations.
- Graph trigonometric functions.

Angles and Degree Measure

An **angle** has three parts: an **initial ray,** a **terminal ray,** and a **vertex** (the point of intersection of the two rays), as shown in Figure C.24. An angle is in **standard position** if its initial ray coincides with the positive x-axis and its vertex is at the origin. It is assumed that you are familiar with the degree measure of an angle.* It is common practice to use θ (the Greek lowercase *theta*) to represent both an angle and its measure. Angles between $0°$ and $90°$ are **acute,** and angles between $90°$ and $180°$ are **obtuse.**

Positive angles are measured *counterclockwise*, and negative angles are measured *clockwise*. For instance, Figure C.25 shows an angle whose measure is $-45°$. You cannot assign a measure to an angle by simply knowing where its initial and terminal rays are located. To measure an angle, you must also know how the terminal ray was revolved. For example, Figure C.25 shows that the angle measuring $-45°$ has the same terminal ray as the angle measuring $315°$. Such angles are **coterminal.** In general, if θ is any angle, then

$$\theta + n(360), \quad n \text{ is a nonzero integer}$$

is coterminal with θ.

An angle that is larger than $360°$ is one whose terminal ray has been revolved more than one full revolution counterclockwise, as shown in Figure C.26. You can form an angle whose measure is less than $-360°$ by revolving a terminal ray more than one full revolution clockwise.

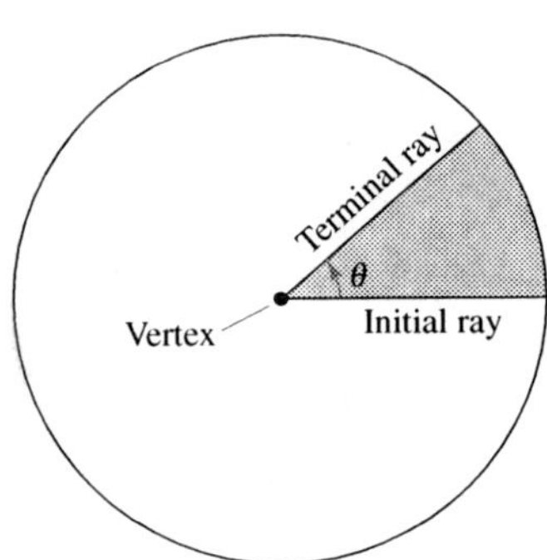

Standard position of an angle
Figure C.24

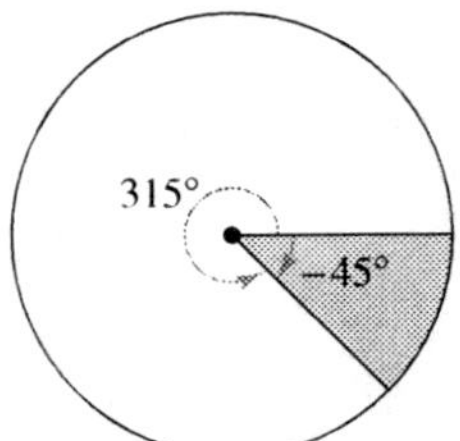

Coterminal angles
Figure C.25

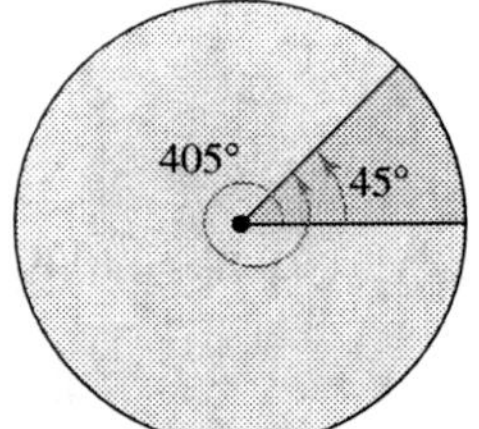

Coterminal angles
Figure C.26

NOTE It is common to use the symbol θ to refer to both an *angle* and its *measure*. For instance, in Figure C.26, you can write the measure of the smaller angle as $\theta = 45°$.

For a more complete review of trigonometry, see Precalculus, 8th edition, by Larson (Brooks/Cole, Cengage Learning, 2011).

Radian Measure

To assign a radian measure to an angle θ, consider θ to be a central angle of a circle of radius 1, as shown in Figure C.27. The **radian measure** of θ is then defined to be the length of the arc of the sector. Because the circumference of a circle is $2\pi r$, the circumference of a **unit circle** (of radius 1) is 2π. This implies that the radian measure of an angle measuring $360°$ is 2π. In other words, $360° = 2\pi$ radians.

Using radian measure for θ, the length s of a circular arc of radius r is $s = r\theta$, as shown in Figure C.28.

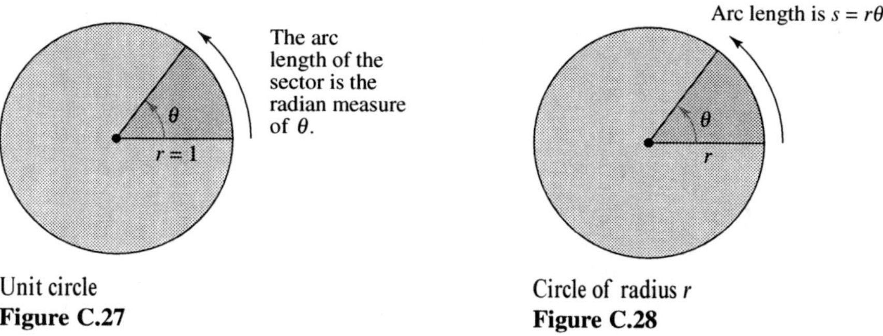

Unit circle
Figure C.27

Circle of radius r
Figure C.28

You should know the conversions of the common angles shown in Figure C.29. For other angles, use the fact that $180°$ is equal to π radians.

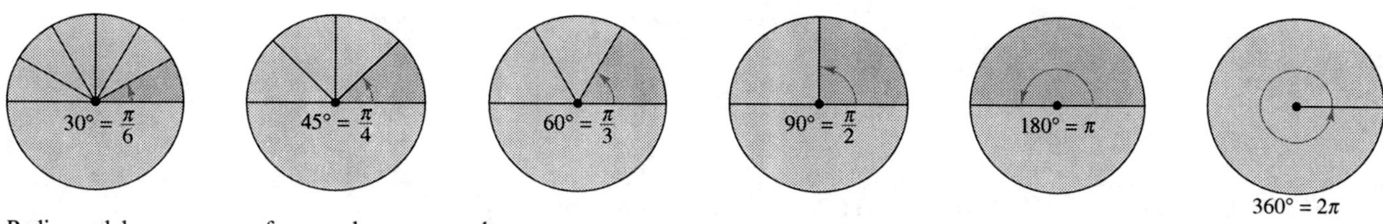

Radian and degree measures for several common angles
Figure C.29

EXAMPLE 1   Conversions Between Degrees and Radians

a. $40° = (40 \text{ deg})\left(\dfrac{\pi \text{ rad}}{180 \text{ deg}}\right) = \dfrac{2\pi}{9} \text{ radian}$

b. $540° = (540 \text{ deg})\left(\dfrac{\pi \text{ rad}}{180 \text{ deg}}\right) = 3\pi \text{ radians}$

c. $-270° = (-270 \text{ deg})\left(\dfrac{\pi \text{ rad}}{180 \text{ deg}}\right) = -\dfrac{3\pi}{2} \text{ radians}$

d. $-\dfrac{\pi}{2} \text{ radians} = \left(-\dfrac{\pi}{2} \text{ rad}\right)\left(\dfrac{180 \text{ deg}}{\pi \text{ rad}}\right) = -90°$

e. $2 \text{ radians} = (2 \text{ rad})\left(\dfrac{180 \text{ deg}}{\pi \text{ rad}}\right) = \dfrac{360}{\pi} \approx 114.59°$

f. $\dfrac{9\pi}{2} \text{ radians} = \left(\dfrac{9\pi}{2} \text{ rad}\right)\left(\dfrac{180 \text{ deg}}{\pi \text{ rad}}\right) = 810°$

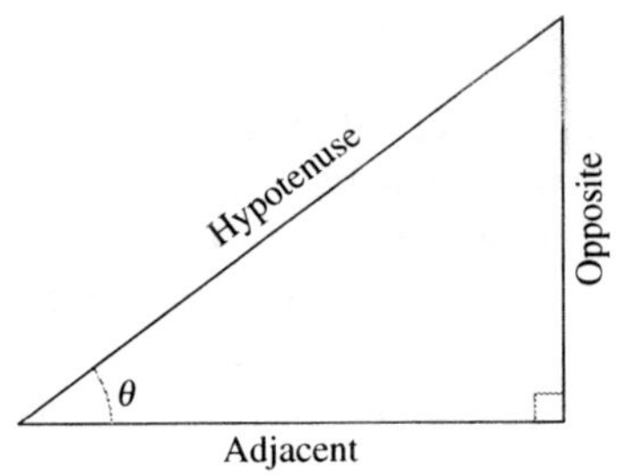

Sides of a right triangle
Figure C.30

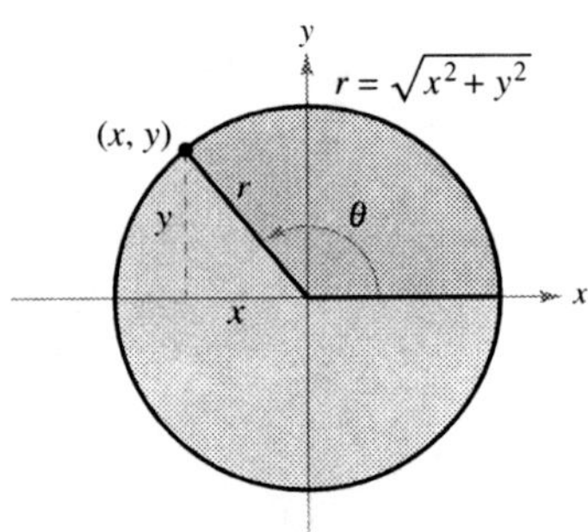

An angle in standard position
Figure C.31

The Trigonometric Functions

There are two common approaches to the study of trigonometry. In one, the trigonometric functions are defined as ratios of two sides of a right triangle. In the other, these functions are defined in terms of a point on the terminal side of an angle in standard position. The six trigonometric functions, **sine, cosine, tangent, cotangent, secant,** and **cosecant** (abbreviated as sin, cos, etc.), are defined below from both viewpoints.

DEFINITION OF THE SIX TRIGONOMETRIC FUNCTIONS

Right triangle definitions, where $0 < \theta < \dfrac{\pi}{2}$ *(see Figure C.30).*

$$\sin \theta = \frac{\text{opposite}}{\text{hypotenuse}} \qquad \cos \theta = \frac{\text{adjacent}}{\text{hypotenuse}} \qquad \tan \theta = \frac{\text{opposite}}{\text{adjacent}}$$

$$\csc \theta = \frac{\text{hypotenuse}}{\text{opposite}} \qquad \sec \theta = \frac{\text{hypotenuse}}{\text{adjacent}} \qquad \cot \theta = \frac{\text{adjacent}}{\text{opposite}}$$

Circular function definitions, where θ *is any angle (see Figure C.31).*

$$\sin \theta = \frac{y}{r} \qquad \cos \theta = \frac{x}{r} \qquad \tan \theta = \frac{y}{x},\ x \neq 0$$

$$\csc \theta = \frac{r}{y},\ y \neq 0 \qquad \sec \theta = \frac{r}{x},\ x \neq 0 \qquad \cot \theta = \frac{x}{y},\ y \neq 0$$

The following trigonometric identities are direct consequences of the definitions. (ϕ is the Greek letter *phi*.)

TRIGONOMETRIC IDENTITIES   [*Note that* $\sin^2 \theta$ *is used to represent* $(\sin \theta)^2$.]

Pythagorean Identities:

$$\sin^2 \theta + \cos^2 \theta = 1$$
$$1 + \tan^2 \theta = \sec^2 \theta$$
$$1 + \cot^2 \theta = \csc^2 \theta$$

Sum or Difference of Two Angles:

$$\sin(\theta \pm \phi) = \sin \theta \cos \phi \pm \cos \theta \sin \phi$$

$$\cos(\theta \pm \phi) = \cos \theta \cos \phi \mp \sin \theta \sin \phi$$

$$\tan(\theta \pm \phi) = \frac{\tan \theta \pm \tan \phi}{1 \mp \tan \theta \tan \phi}$$

Law of Cosines:

$$a^2 = b^2 + c^2 - 2bc \cos A$$

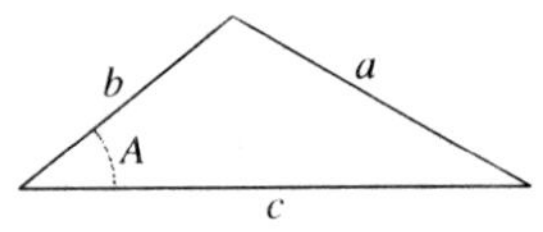

Reduction Formulas:

$$\sin(-\theta) = -\sin \theta \qquad \sin \theta = -\sin(\theta - \pi)$$
$$\cos(-\theta) = \cos \theta \qquad \cos \theta = -\cos(\theta - \pi)$$
$$\tan(-\theta) = -\tan \theta \qquad \tan \theta = \tan(\theta - \pi)$$

Half–Angle Formulas:

$$\sin^2 \theta = \frac{1 - \cos 2\theta}{2}$$

$$\cos^2 \theta = \frac{1 + \cos 2\theta}{2}$$

Double–Angle Formulas:

$$\sin 2\theta = 2 \sin \theta \cos \theta$$

$$\cos 2\theta = 2 \cos^2 \theta - 1$$
$$= 1 - 2 \sin^2 \theta$$
$$= \cos^2 \theta - \sin^2 \theta$$

Reciprocal Identities:

$$\csc \theta = \frac{1}{\sin \theta}$$

$$\sec \theta = \frac{1}{\cos \theta}$$

$$\cot \theta = \frac{1}{\tan \theta}$$

Quotient Identities:

$$\tan \theta = \frac{\sin \theta}{\cos \theta}$$

$$\cot \theta = \frac{\cos \theta}{\sin \theta}$$

Evaluating Trigonometric Functions

There are two ways to evaluate trigonometric functions: (1) decimal approximations with a calculator and (2) exact evaluations using trigonometric identities and formulas from geometry. When using a calculator to evaluate a trigonometric function, remember to set the calculator to the appropriate mode—*degree* mode or *radian* mode.

EXAMPLE 2 Exact Evaluation of Trigonometric Functions

Evaluate the sine, cosine, and tangent of $\dfrac{\pi}{3}$.

Solution   Begin by drawing the angle $\theta = \pi/3$ in standard position, as shown in Figure C.32. Then, because $60° = \pi/3$ radians, you can draw an equilateral triangle with sides of length 1 and θ as one of its angles. Because the altitude of this triangle bisects its base, you know that $x = \frac{1}{2}$. Using the Pythagorean Theorem, you obtain

$$y = \sqrt{r^2 - x^2} = \sqrt{1 - \left(\frac{1}{2}\right)^2} = \sqrt{\frac{3}{4}} = \frac{\sqrt{3}}{2}.$$

Now, knowing the values of x, y, and r, you can write the following.

$$\sin \frac{\pi}{3} = \frac{y}{r} = \frac{\sqrt{3}/2}{1} = \frac{\sqrt{3}}{2}$$

$$\cos \frac{\pi}{3} = \frac{x}{r} = \frac{1/2}{1} = \frac{1}{2}$$

$$\tan \frac{\pi}{3} = \frac{y}{x} = \frac{\sqrt{3}/2}{1/2} = \sqrt{3}$$

NOTE All angles in this text are measured in radians unless stated otherwise. For example, when sin 3 is written, the sine of 3 radians is meant, and when sin 3° is written, the sine of 3 degrees is meant.

The degree and radian measures of several common angles are shown in the table below, along with the corresponding values of the sine, cosine, and tangent (see Figure C.33).

Common First-Quadrant Angles

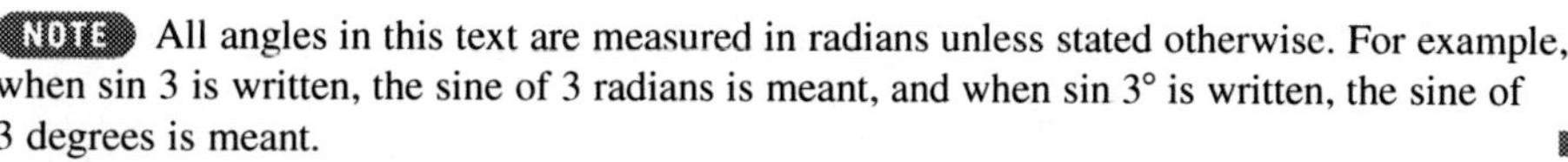

Degrees	0	30°	45°	60°	90°
Radians	0	$\dfrac{\pi}{6}$	$\dfrac{\pi}{4}$	$\dfrac{\pi}{3}$	$\dfrac{\pi}{2}$
$\sin \theta$	0	$\dfrac{1}{2}$	$\dfrac{\sqrt{2}}{2}$	$\dfrac{\sqrt{3}}{2}$	1
$\cos \theta$	1	$\dfrac{\sqrt{3}}{2}$	$\dfrac{\sqrt{2}}{2}$	$\dfrac{1}{2}$	0
$\tan \theta$	0	$\dfrac{\sqrt{3}}{3}$	1	$\sqrt{3}$	Undefined

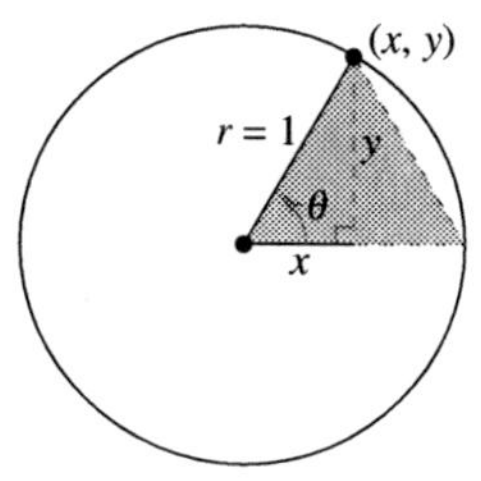

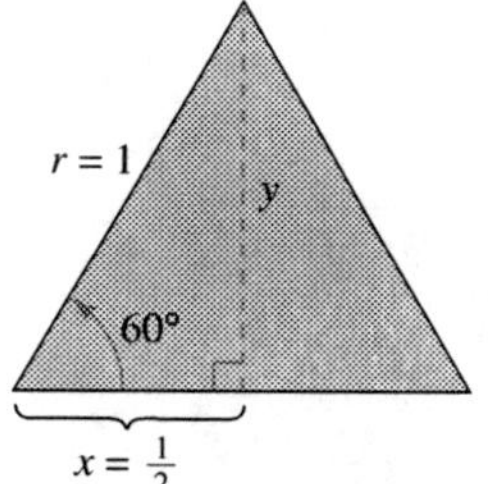

The angle $\pi/3$ in standard position
Figure C.32

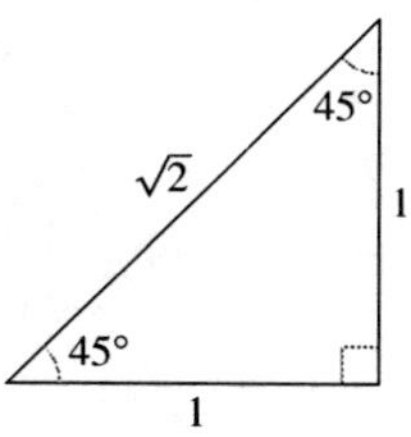

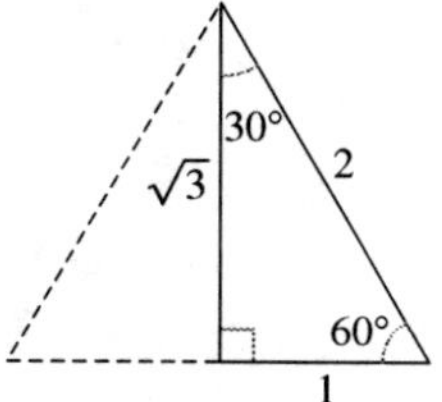

Common angles
Figure C.33

Quadrant II	Quadrant I
sin θ: +	sin θ: +
cos θ: −	cos θ: +
tan θ: −	tan θ: +
Quadrant III	Quadrant IV
sin θ: −	sin θ: −
cos θ: −	cos θ: +
tan θ: +	tan θ: −

Quadrant signs for trigonometric functions
Figure C.34

The quadrant signs for the sine, cosine, and tangent functions are shown in Figure C.34. To extend the use of the table on the preceding page to angles in quadrants other than the first quadrant, you can use the concept of a **reference angle** (see Figure C.35), with the appropriate quadrant sign. For instance, the reference angle for $3\pi/4$ is $\pi/4$, and because the sine is positive in Quadrant II, you can write

$$\sin\frac{3\pi}{4} = +\sin\frac{\pi}{4} = \frac{\sqrt{2}}{2}.$$

Similarly, because the reference angle for $330°$ is $30°$, and the tangent is negative in Quadrant IV, you can write

$$\tan 330° = -\tan 30° = -\frac{\sqrt{3}}{3}.$$

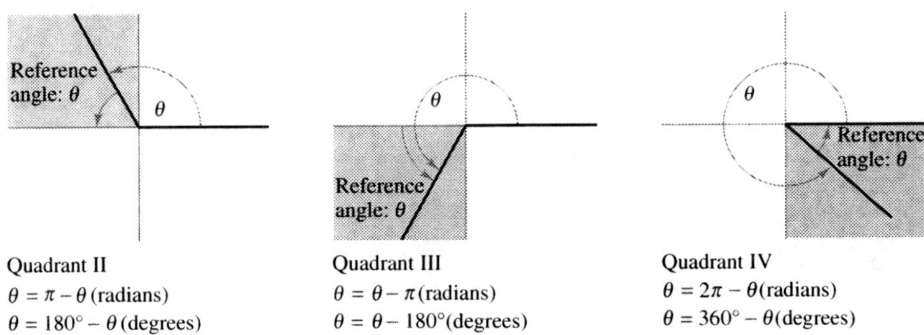

Figure C.35

EXAMPLE 3 Trigonometric Identities and Calculators

Evaluate each trigonometric expression.

a. $\sin\left(-\dfrac{\pi}{3}\right)$ **b.** $\sec 60°$ **c.** $\cos(1.2)$

Solution

a. Using the reduction formula $\sin(-\theta) = -\sin\theta$, you can write

$$\sin\left(-\frac{\pi}{3}\right) = -\sin\frac{\pi}{3} = -\frac{\sqrt{3}}{2}.$$

b. Using the reciprocal identity $\sec\theta = 1/\cos\theta$, you can write

$$\sec 60° = \frac{1}{\cos 60°} = \frac{1}{1/2} = 2.$$

c. Using a calculator, you obtain

$$\cos(1.2) \approx 0.3624.$$

Remember that 1.2 is given in *radian* measure. Consequently, your calculator must be set in *radian* mode.

Solving Trigonometric Equations

How would you solve the equation $\sin \theta = 0$? You know that $\theta = 0$ is one solution, but this is not the only solution. Any one of the following values of θ is also a solution.

$$\ldots,\, -3\pi,\, -2\pi,\, -\pi,\, 0,\, \pi,\, 2\pi,\, 3\pi,\, \ldots$$

You can write this infinite solution set as $\{n\pi \colon n \text{ is an integer}\}$.

EXAMPLE 4   Solving a Trigonometric Equation

Solve the equation

$$\sin \theta = -\frac{\sqrt{3}}{2}.$$

Solution   To solve the equation, you should consider that the sine is negative in Quadrants III and IV and that

$$\sin \frac{\pi}{3} = \frac{\sqrt{3}}{2}.$$

So, you are seeking values of θ in the third and fourth quadrants that have a reference angle of $\pi/3$. In the interval $[0, 2\pi]$, the two angles fitting these criteria are

$$\theta = \pi + \frac{\pi}{3} = \frac{4\pi}{3} \quad \text{and} \quad \theta = 2\pi - \frac{\pi}{3} = \frac{5\pi}{3}.$$

By adding integer multiples of 2π to each of these solutions, you obtain the following general solution.

$$\theta = \frac{4\pi}{3} + 2n\pi \quad \text{or} \quad \theta = \frac{5\pi}{3} + 2n\pi, \quad \text{where } n \text{ is an integer.}$$

See Figure C.36.

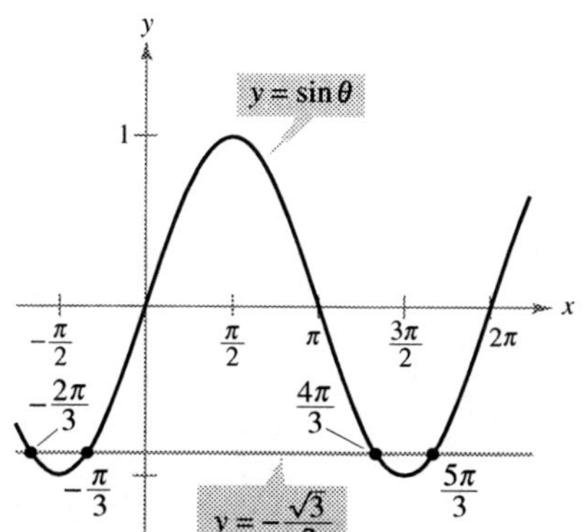

Solution points of $\sin \theta = -\dfrac{\sqrt{3}}{2}$

Figure C.36

EXAMPLE 5   Solving a Trigonometric Equation

Solve $\cos 2\theta = 2 - 3 \sin \theta$, where $0 \le \theta \le 2\pi$.

Solution   Using the double-angle identity $\cos 2\theta = 1 - 2 \sin^2 \theta$, you can rewrite the equation as follows.

$$
\begin{aligned}
\cos 2\theta &= 2 - 3 \sin \theta & &\text{Write original equation.}\\
1 - 2 \sin^2 \theta &= 2 - 3 \sin \theta & &\text{Trigonometric identity}\\
0 &= 2 \sin^2 \theta - 3 \sin \theta + 1 & &\text{Quadratic form}\\
0 &= (2 \sin \theta - 1)(\sin \theta - 1) & &\text{Factor.}
\end{aligned}
$$

If $2 \sin \theta - 1 = 0$, then $\sin \theta = 1/2$ and $\theta = \pi/6$ or $\theta = 5\pi/6$. If $\sin \theta - 1 = 0$, then $\sin \theta = 1$ and $\theta = \pi/2$. So, for $0 \le \theta \le 2\pi$, there are three solutions.

$$\theta = \frac{\pi}{6},\, \frac{5\pi}{6}, \quad \text{or} \quad \frac{\pi}{2}$$

Graphs of Trigonometric Functions

A function f is **periodic** if there exists a nonzero number p such that $f(x + p) = f(x)$ for all x in the domain of f. The smallest such positive value of p (if it exists) is the **period** of f. The sine, cosine, secant, and cosecant functions each have a period of 2π, and the other two trigonometric functions, tangent and cotangent, have a period of π, as shown in Figure C.37.

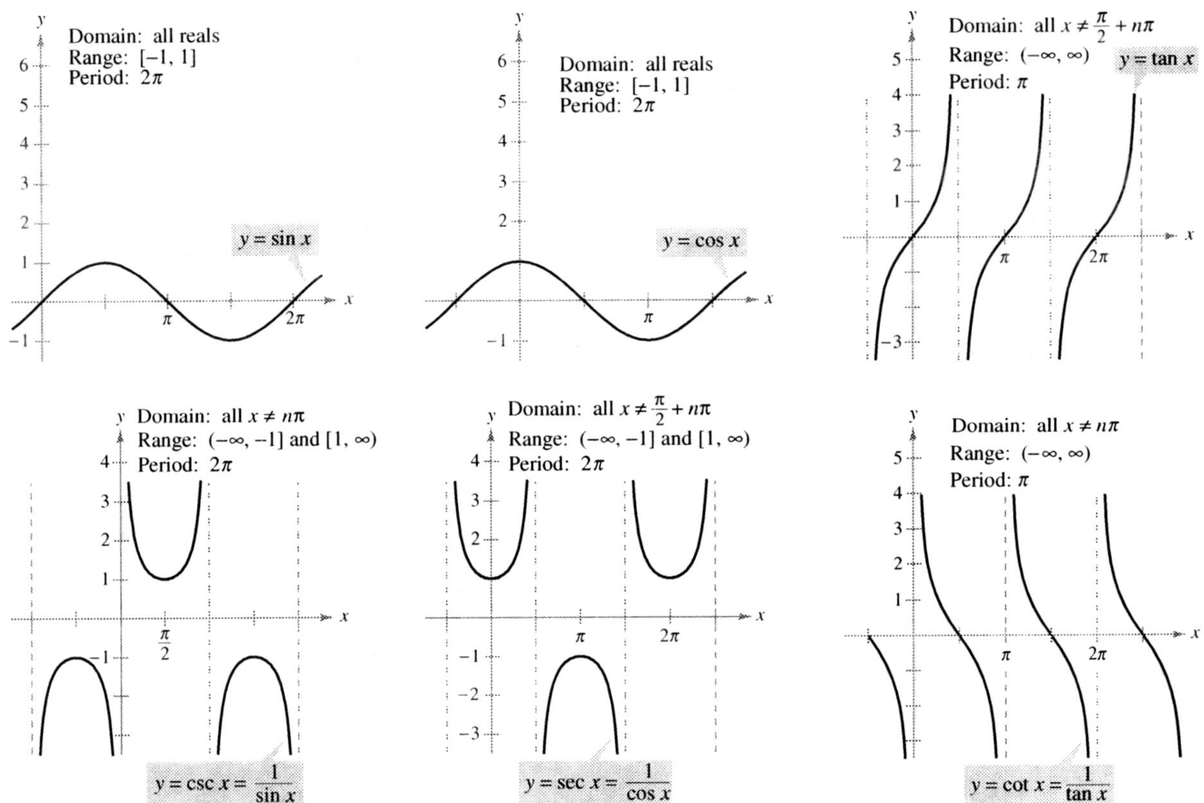

The graphs of the six trigonometric functions
Figure C.37

Note in Figure C.37 that the maximum value of $\sin x$ and $\cos x$ is 1 and the minimum value is -1. The graphs of the functions $y = a \sin bx$ and $y = a \cos bx$ oscillate between $-a$ and a, and so have an **amplitude** of $|a|$. Furthermore, because $bx = 0$ when $x = 0$ and $bx = 2\pi$ when $x = 2\pi/b$, it follows that the functions $y = a \sin bx$ and $y = a \cos bx$ each have a period of $2\pi/|b|$. The table below summarizes the amplitudes and periods of some types of trigonometric functions.

Function	Period	Amplitude
$y = a \sin bx$ or $y = a \cos bx$	$\dfrac{2\pi}{\|b\|}$	$\|a\|$
$y = a \tan bx$ or $y = a \cot bx$	$\dfrac{\pi}{\|b\|}$	Not applicable
$y = a \sec bx$ or $y = a \csc bx$	$\dfrac{2\pi}{\|b\|}$	Not applicable

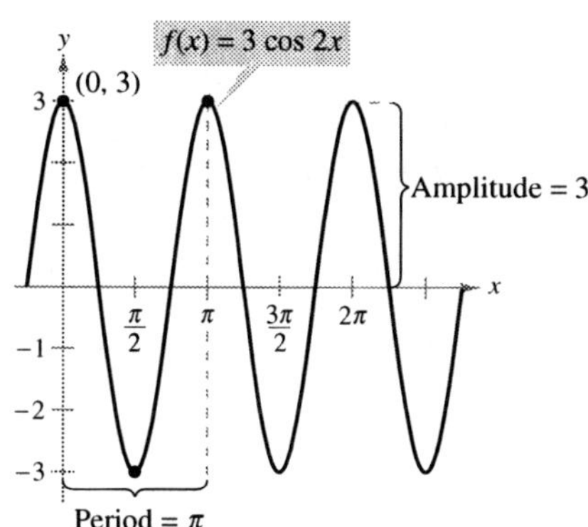

Figure C.38

EXAMPLE 6   Sketching the Graph of a Trigonometric Function

Sketch the graph of $f(x) = 3 \cos 2x$.

Solution   The graph of $f(x) = 3 \cos 2x$ has an amplitude of 3 and a period of $2\pi/2 = \pi$. Using the basic shape of the graph of the cosine function, sketch one period of the function on the interval $[0, \pi]$, using the following pattern.

Maximum: $(0, 3)$

Minimum: $\left(\dfrac{\pi}{2}, -3\right)$

Maximum: $(\pi, 3)$

By continuing this pattern, you can sketch several cycles of the graph, as shown in Figure C.38.

Horizontal shifts, vertical shifts, and reflections can be applied to the graphs of trigonometric functions, as illustrated in Example 7.

EXAMPLE 7   Shifts of Graphs of Trigonometric Functions

Sketch the graph of each function.

a. $f(x) = \sin\left(x + \dfrac{\pi}{2}\right)$ **b.** $f(x) = 2 + \sin x$ **c.** $f(x) = 2 + \sin\left(x - \dfrac{\pi}{4}\right)$

Solution

a. To sketch the graph of $f(x) = \sin(x + \pi/2)$, shift the graph of $y = \sin x$ to the left $\pi/2$ units, as shown in Figure C.39(a).

b. To sketch the graph of $f(x) = 2 + \sin x$, shift the graph of $y = \sin x$ upward two units, as shown in Figure C.39(b).

c. To sketch the graph of $f(x) = 2 + \sin(x - \pi/4)$, shift the graph of $y = \sin x$ upward two units and to the right $\pi/4$ units, as shown in Figure C.39(c).

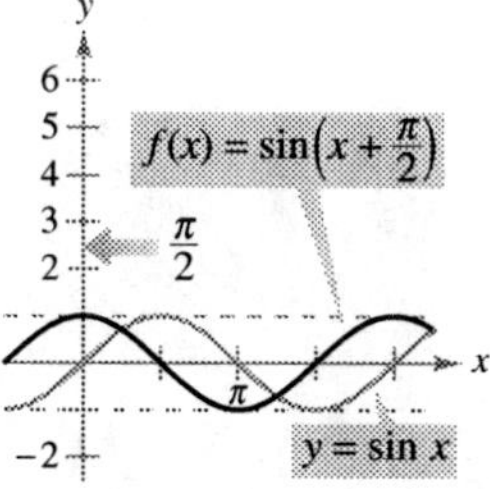

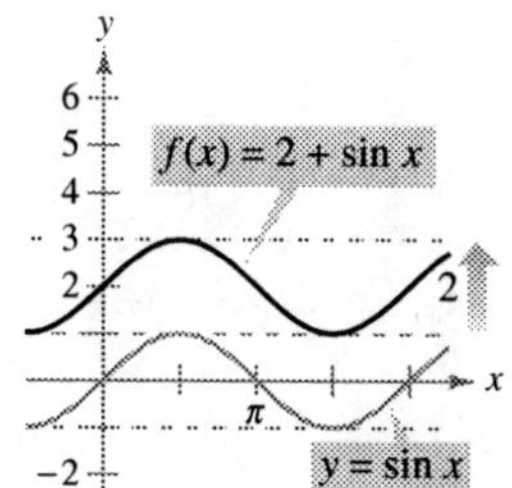

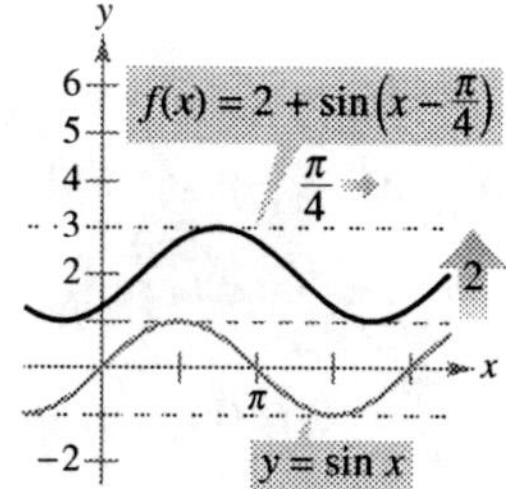

(a) Horizontal shift to the left   (b) Vertical shift upward   (c) Horizontal and vertical shifts

Transformations of the graph of $y = \sin x$
Figure C.39

C.3 Exercises

See www.CalcChat.com for worked-out solutions to odd-numbered exercises.

In Exercises 1 and 2, determine two coterminal angles in degree measure (one positive and one negative) for each angle.

1. (a) 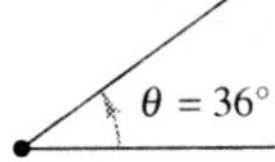

(b) $\theta = 36°$

(b) $\theta = -120°$

2. (a) $\theta = 300°$ 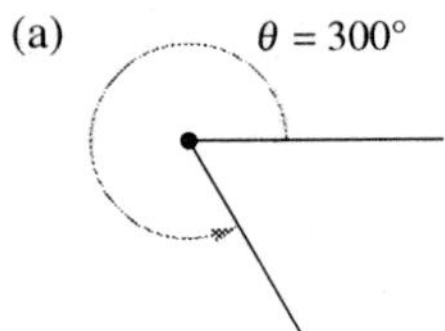

(b) $\theta = -420°$ 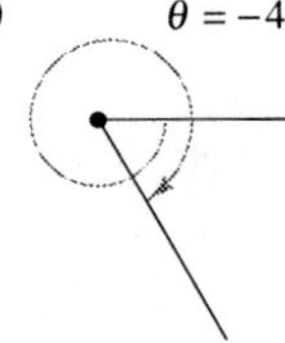

In Exercises 3 and 4, determine two coterminal angles in radian measure (one positive and one negative) for each angle.

3. (a) $\theta = \dfrac{\pi}{9}$ 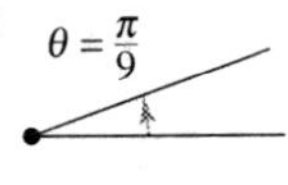

(b) $\theta = \dfrac{4\pi}{3}$ 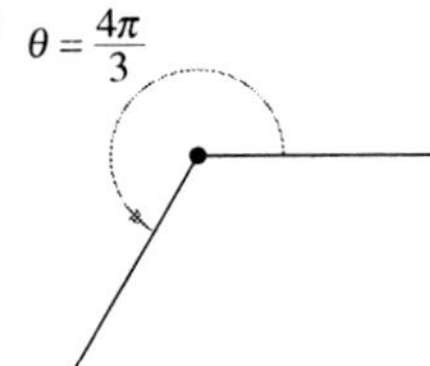

4. (a) $\theta = -\dfrac{9\pi}{4}$ 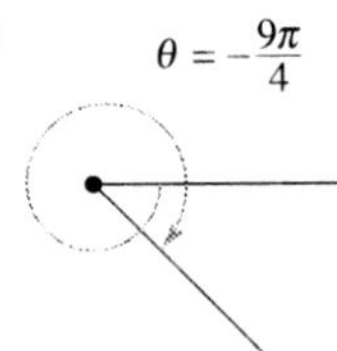

(b) $\theta = \dfrac{8\pi}{9}$ 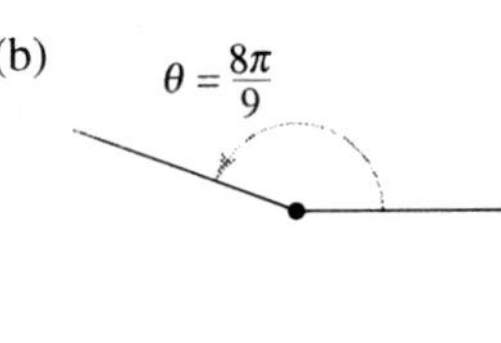

In Exercises 5 and 6, rewrite each angle in radian measure as a multiple of π and as a decimal accurate to three decimal places.

5. (a) $30°$ (b) $150°$ (c) $315°$ (d) $120°$

6. (a) $-20°$ (b) $-240°$ (c) $-270°$ (d) $144°$

In Exercises 7 and 8, rewrite each angle in degree measure.

7. (a) $\dfrac{3\pi}{2}$ (b) $\dfrac{7\pi}{6}$ (c) $-\dfrac{7\pi}{12}$ (d) -2.367

8. (a) $\dfrac{7\pi}{3}$ (b) $-\dfrac{11\pi}{30}$ (c) $\dfrac{11\pi}{6}$ (d) 0.438

9. Let r represent the radius of a circle, θ the central angle (measured in radians), and s the length of the arc subtended by the angle. Use the relationship $s = r\theta$ to complete the table.

r	8 ft	15 in.	85 cm		
s	12 ft			96 in.	8642 mi
θ		1.6	$\dfrac{3\pi}{4}$	4	$\dfrac{2\pi}{3}$

10. *Angular Speed* A car is moving at the rate of 50 miles per hour, and the diameter of its wheels is 2.5 feet.

(a) Find the number of revolutions per minute that the wheels are rotating.

(b) Find the angular speed of the wheels in radians per minute.

In Exercises 11 and 12, determine all six trigonometric functions for the angle θ.

11. (a) 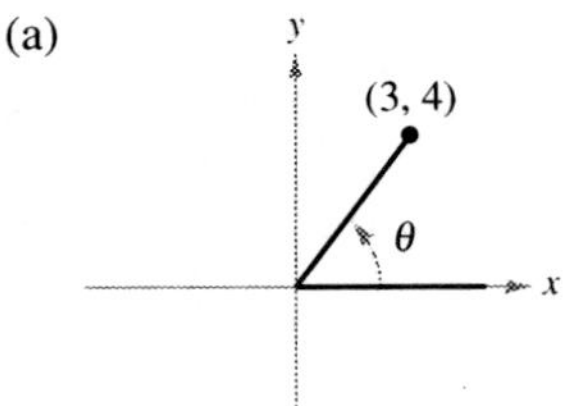

(b) 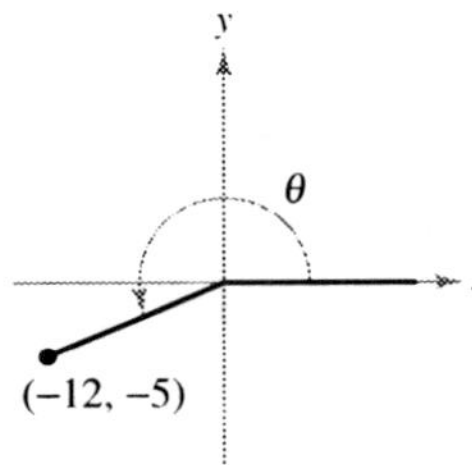

12. (a) 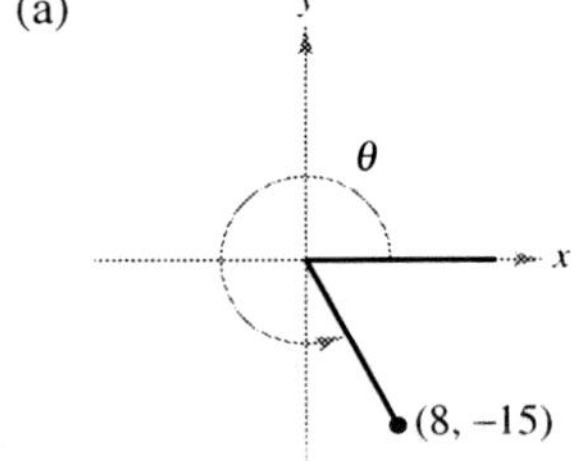

(b) 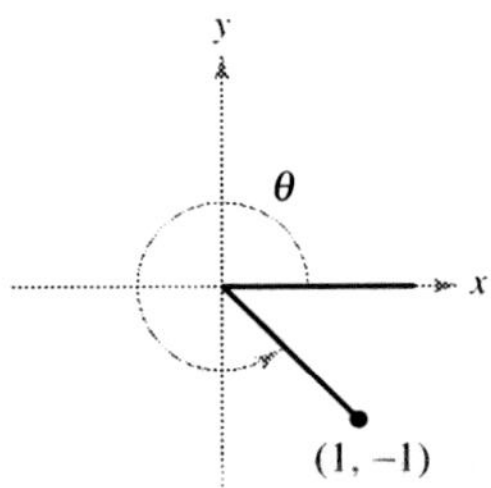

In Exercises 13 and 14, determine the quadrant in which θ lies.

13. (a) $\sin \theta < 0$ and $\cos \theta < 0$

(b) $\sec \theta > 0$ and $\cot \theta < 0$

14. (a) $\sin \theta > 0$ and $\cos \theta < 0$

(b) $\csc \theta < 0$ and $\tan \theta > 0$

In Exercises 15–18, evaluate the trigonometric function.

15. $\sin \theta = \dfrac{1}{2}$

$\cos \theta = $ ▨

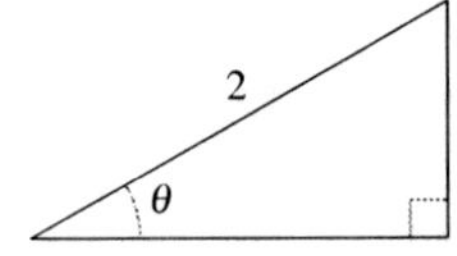

16. $\sin \theta = \dfrac{1}{3}$

$\tan \theta = $ ▨

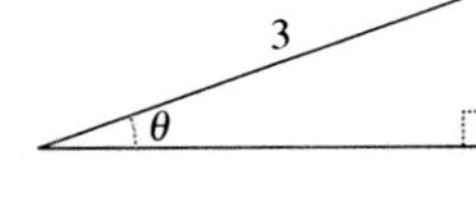

17. $\cos \theta = \dfrac{4}{5}$

$\cot \theta = $ ▨

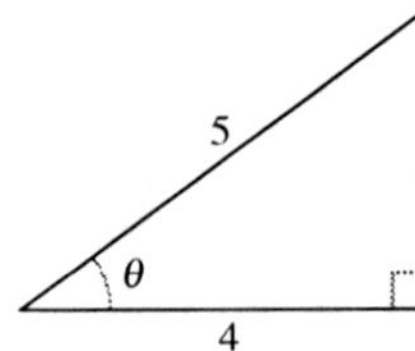

18. $\sec \theta = \dfrac{13}{5}$

$\csc \theta = $ ▨

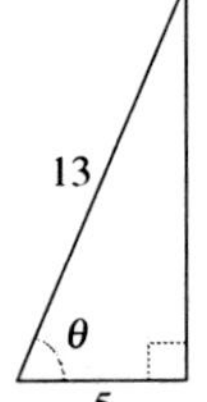

In Exercises 19–22, evaluate the sine, cosine, and tangent of each angle *without* using a calculator.

19. (a) $60°$

(b) $120°$

(c) $\dfrac{\pi}{4}$

(d) $\dfrac{5\pi}{4}$

20. (a) $-30°$

(b) $150°$

(c) $-\dfrac{\pi}{6}$

(d) $\dfrac{\pi}{2}$

21. (a) $225°$

(b) $-225°$

(c) $\dfrac{5\pi}{3}$

(d) $\dfrac{11\pi}{6}$

22. (a) $750°$

(b) $510°$

(c) $\dfrac{10\pi}{3}$

(d) $\dfrac{17\pi}{3}$

In Exercises 23–26, use a calculator to evaluate each trigonometric function. Round your answers to four decimal places.

23. (a) $\sin 10°$

(b) $\csc 10°$

24. (a) $\sec 225°$

(b) $\sec 135°$

25. (a) $\tan \dfrac{\pi}{9}$

(b) $\tan \dfrac{10\pi}{9}$

26. (a) $\cot(1.35)$

(b) $\tan(1.35)$

In Exercises 27–30, find two solutions of each equation. Give your answers in radians $(0 \le \theta < 2\pi)$. Do not use a calculator.

27. (a) $\cos \theta = \dfrac{\sqrt{2}}{2}$

(b) $\cos \theta = -\dfrac{\sqrt{2}}{2}$

28. (a) $\sec \theta = 2$

(b) $\sec \theta = -2$

29. (a) $\tan \theta = 1$

(b) $\cot \theta = -\sqrt{3}$

30. (a) $\sin \theta = \dfrac{\sqrt{3}}{2}$

(b) $\sin \theta = -\dfrac{\sqrt{3}}{2}$

In Exercises 31–38, solve the equation for θ $(0 \le \theta < 2\pi)$.

31. $2 \sin^2 \theta = 1$

32. $\tan^2 \theta = 3$

33. $\tan^2 \theta - \tan \theta = 0$

34. $2 \cos^2 \theta - \cos \theta = 1$

35. $\sec \theta \csc \theta = 2 \csc \theta$

36. $\sin \theta = \cos \theta$

37. $\cos^2 \theta + \sin \theta = 1$

38. $\cos \dfrac{\theta}{2} - \cos \theta = 1$

39. *Airplane Ascent*   An airplane leaves the runway climbing at an angle of $18°$ with a speed of 275 feet per second (see figure). Find the altitude a of the plane after 1 minute.

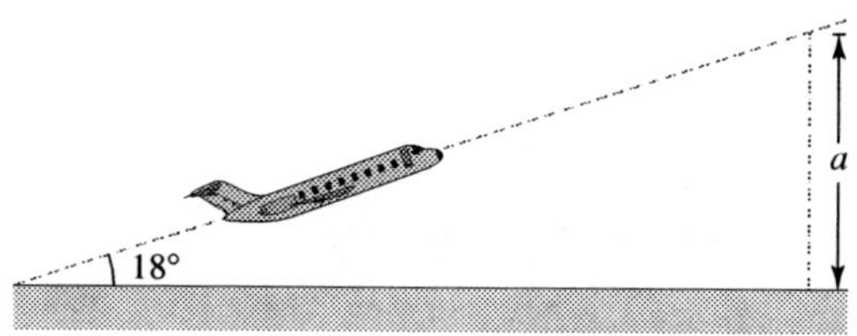

40. *Height of a Mountain*   While traveling across flat land, you notice a mountain directly in front of you. Its angle of elevation (to the peak) is $3.5°$. After you drive 13 miles closer to the mountain, the angle of elevation is $9°$. Approximate the height of the mountain.

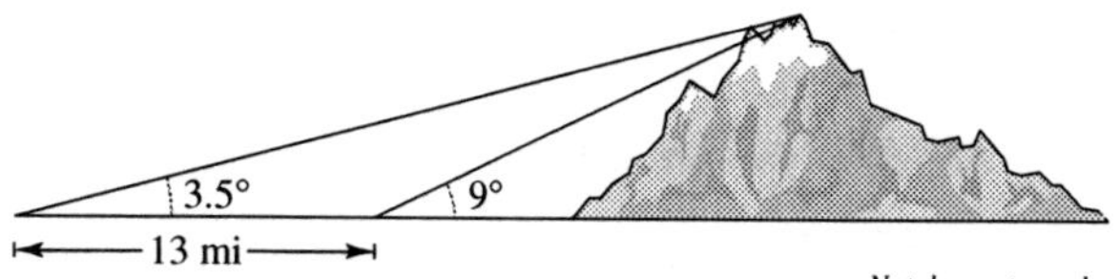

In Exercises 41–44, determine the period and amplitude of each function.

41. (a) $y = 2 \sin 2x$

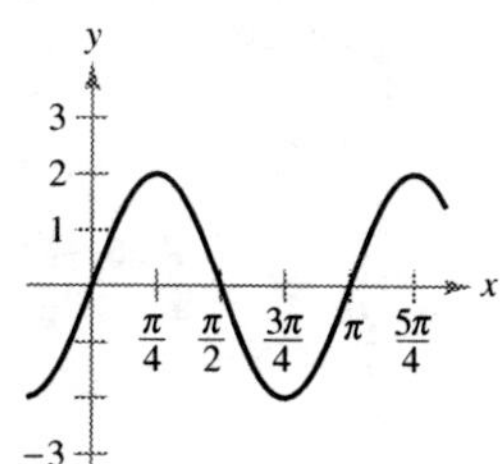

(b) $y = \dfrac{1}{2} \sin \pi x$

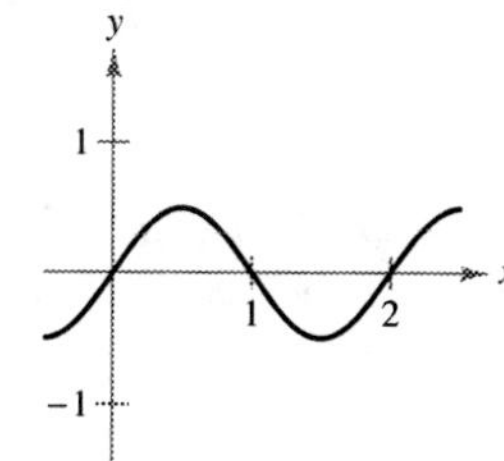

42. (a) $y = \dfrac{3}{2} \cos \dfrac{x}{2}$

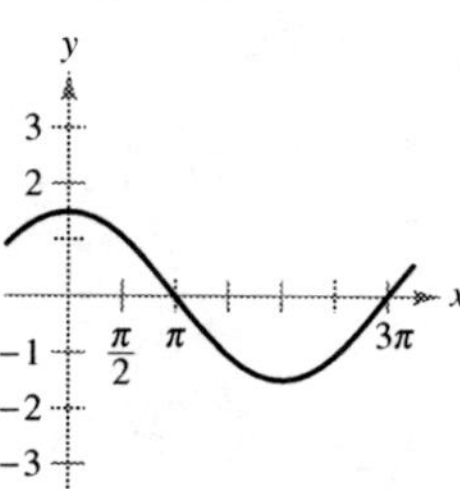

(b) $y = -2 \sin \dfrac{x}{3}$

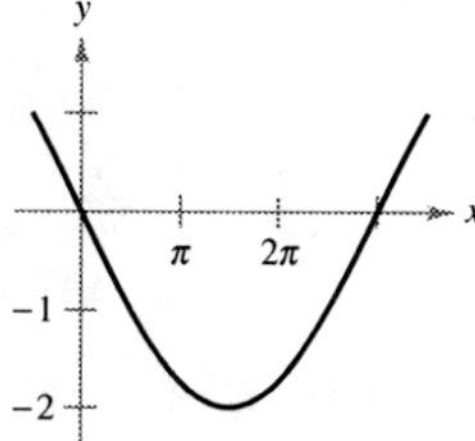

43. $y = 3 \sin 4\pi x$

44. $y = \dfrac{2}{3} \cos \dfrac{\pi x}{10}$

In Exercises 45–48, find the period of the function.

45. $y = 5 \tan 2x$

46. $y = 7 \tan 2\pi x$

47. $y = \sec 5x$

48. $y = \csc 4x$

Writing   In Exercises 49 and 50, use a graphing utility to graph each function f in the same viewing window for $c = -2$, $c = -1$, $c = 1$, and $c = 2$. Give a written description of the change in the graph caused by changing c.

49. (a) $f(x) = c \sin x$

    (b) $f(x) = \cos(cx)$

    (c) $f(x) = \cos(\pi x - c)$

50. (a) $f(x) = \sin x + c$

    (b) $f(x) = -\sin(2\pi x - c)$

    (c) $f(x) = c \cos x$

In Exercises 51–62, sketch the graph of the function.

51. $y = \sin \dfrac{x}{2}$

52. $y = 2 \cos 2x$

53. $y = -\sin \dfrac{2\pi x}{3}$

54. $y = 2 \tan x$

55. $y = \csc \dfrac{x}{2}$

56. $y = \tan 2x$

57. $y = 2 \sec 2x$

58. $y = \csc 2\pi x$

59. $y = \sin(x + \pi)$

60. $y = \cos\left(x - \dfrac{\pi}{3}\right)$

61. $y = 1 + \cos\left(x - \dfrac{\pi}{2}\right)$

62. $y = 1 + \sin\left(x + \dfrac{\pi}{2}\right)$

Graphical Reasoning   In Exercises 63 and 64, find a, b, and c such that the graph of the function matches the graph in the figure.

63. $y = a \cos(bx - c)$

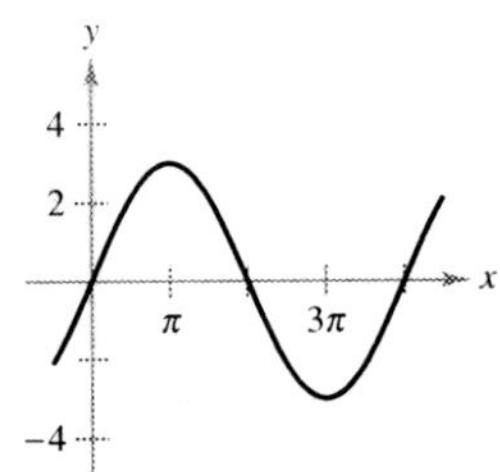

64. $y = a \sin(bx - c)$

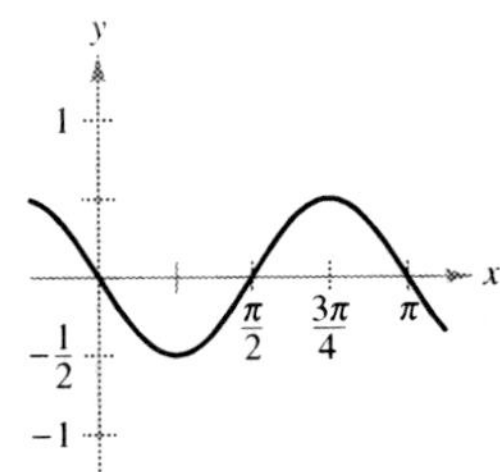

65. Think About It Sketch the graphs of $f(x) = \sin x$, $g(x) = |\sin x|$, and $h(x) = \sin(|x|)$. In general, how are the graphs of $|f(x)|$ and $f(|x|)$ related to the graph of f?

66. Think About It The model for the height h of a Ferris wheel car is

$$h = 51 + 50 \sin 8\pi t$$

where t is measured in minutes. (The Ferris wheel has a radius of 50 feet.) This model yields a height of 51 feet when $t = 0$. Alter the model so that the height of the car is 1 foot when $t = 0$.

67. Sales The monthly sales S (in thousands of units) of a seasonal product are modeled by

$$S = 58.3 + 32.5 \cos \frac{\pi t}{6}$$

where t is the time (in months), with $t = 1$ corresponding to January. Use a graphing utility to graph the model for S and determine the months when sales exceed 75,000 units.

68. Investigation Two trigonometric functions f and g have a period of 2, and their graphs intersect at $x = 5.35$.

    (a) Give one smaller and one larger positive value of x at which the functions have the same value.

    (b) Determine one negative value of x at which the graphs intersect.

    (c) Is it true that $f(13.35) = g(-4.65)$? Give a reason for your answer.

Pattern Recognition   In Exercises 69 and 70, use a graphing utility to compare the graph of f with the given graph. Try to improve the approximation by adding a term to $f(x)$. Use a graphing utility to verify that your new approximation is better than the original. Can you find other terms to add to make the approximation even better? What is the pattern? (In Exercise 69, sine terms can be used to improve the approximation, and in Exercise 70, cosine terms can be used.)

69. $f(x) = \dfrac{4}{\pi}\left(\sin \pi x + \dfrac{1}{3} \sin 3\pi x\right)$

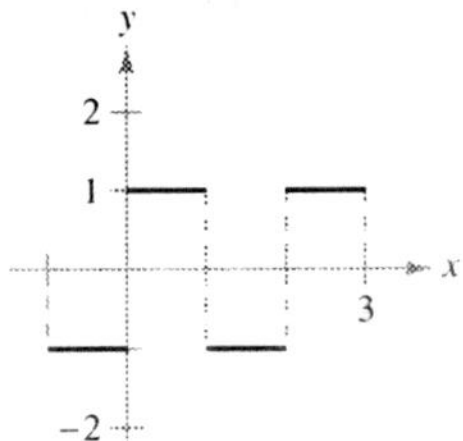

70. $f(x) = \dfrac{1}{2} - \dfrac{4}{\pi^2}\left(\cos \pi x + \dfrac{1}{9} \cos 3\pi x\right)$

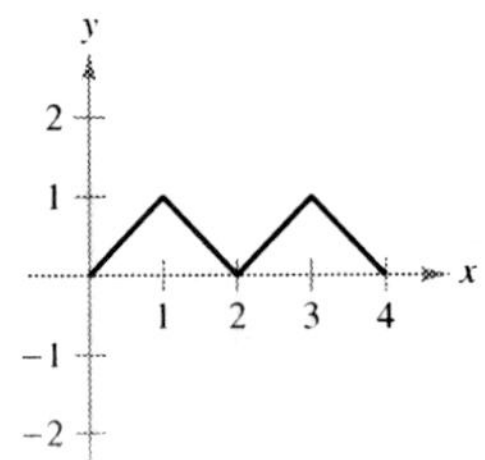

Answers to Odd-Numbered Exercises

Chapter 1

Section 1.1 (page 8)

1. b   **2.** d   **3.** a   **4.** c

5. 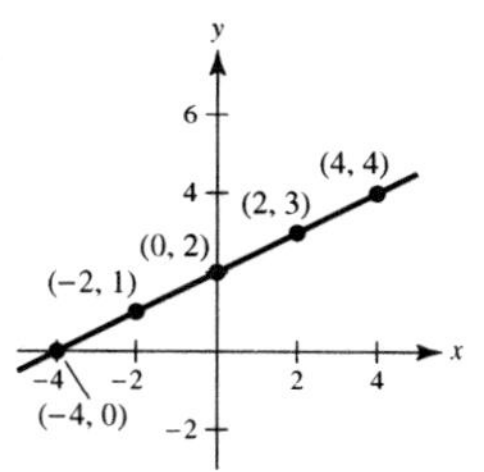

7. 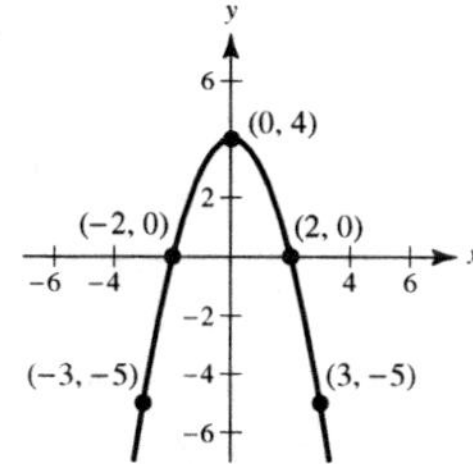

9. 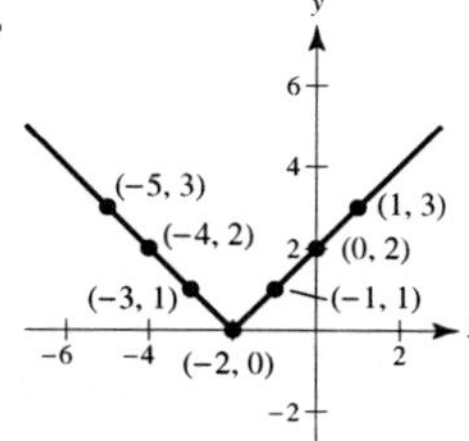

11. 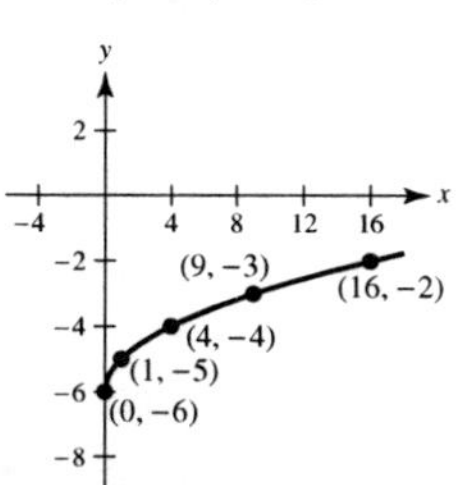

13. 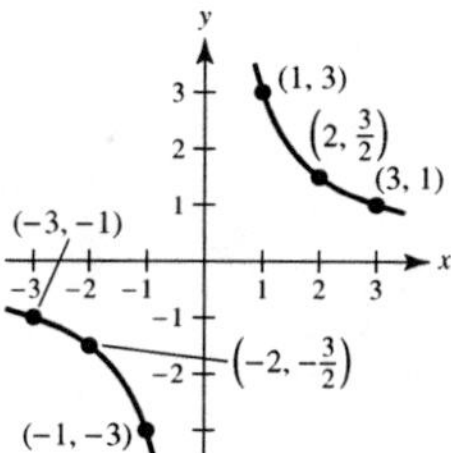

15. 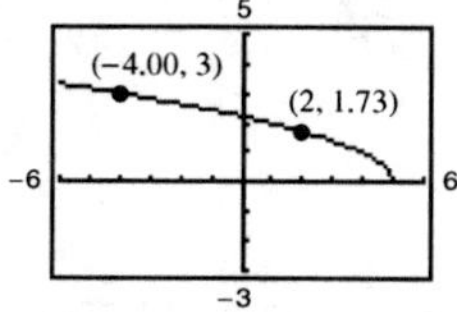

Xmin = -5
Xmax = 4
Xscl = 1
Ymin = -5
Ymax = 8
Yscl = 1

17. 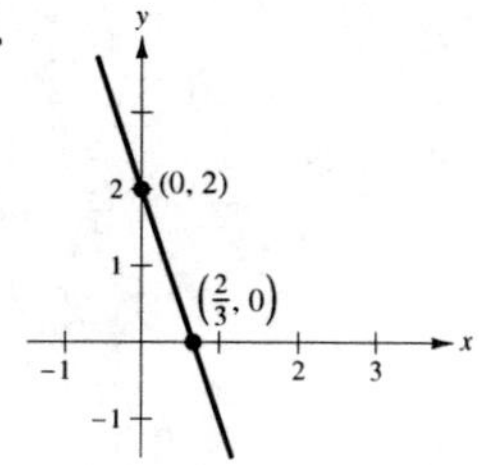

(a) $y \approx 1.73$

(b) $x = -4$

19. $(0, -5), \left(\frac{5}{2}, 0\right)$ **21.** $(0, -2), (-2, 0), (1, 0)$

23. $(0, 0), (4, 0), (-4, 0)$ **25.** $(4, 0)$ **27.** $(0, 0)$

29. Symmetric with respect to the y-axis

31. Symmetric with respect to the x-axis

33. Symmetric with respect to the origin   **35.** No symmetry

37. Symmetric with respect to the origin

39. Symmetric with respect to the y-axis

41.

Symmetry: none

43.

Symmetry: none

45. 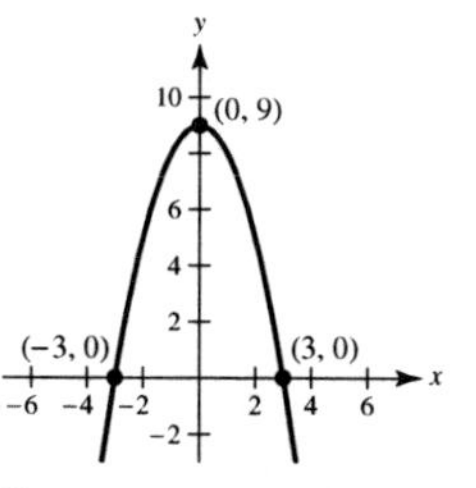

Symmetry: y-axis

47. 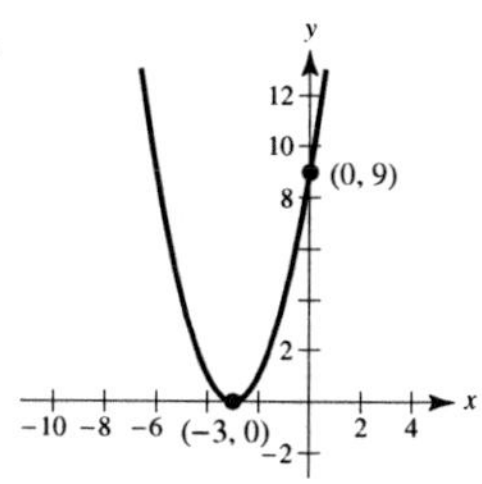

Symmetry: none

49. 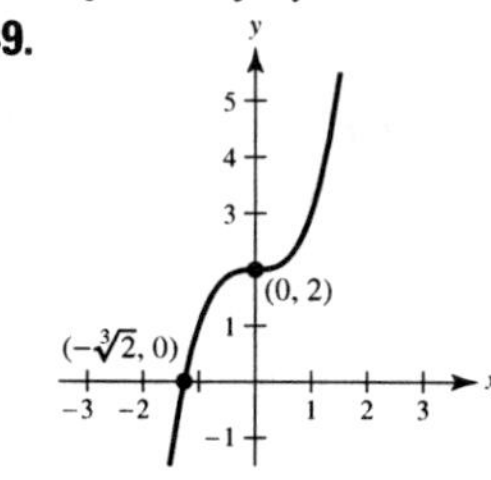

Symmetry: none

51. 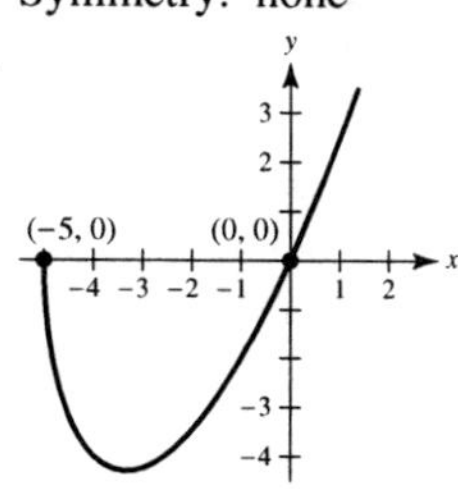

Symmetry: none

53. 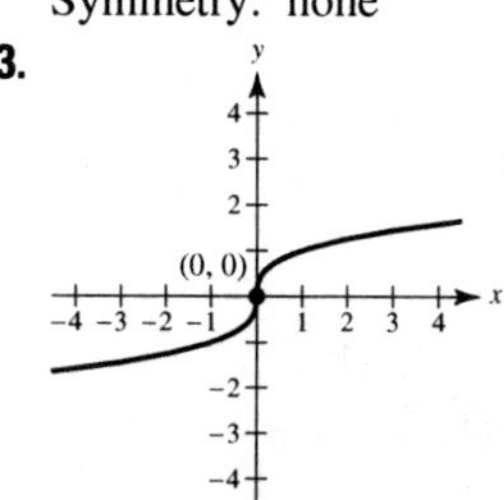

Symmetry: origin

55. 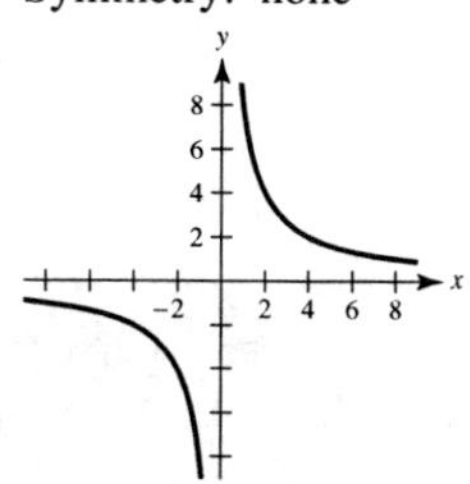

Symmetry: origin

57. 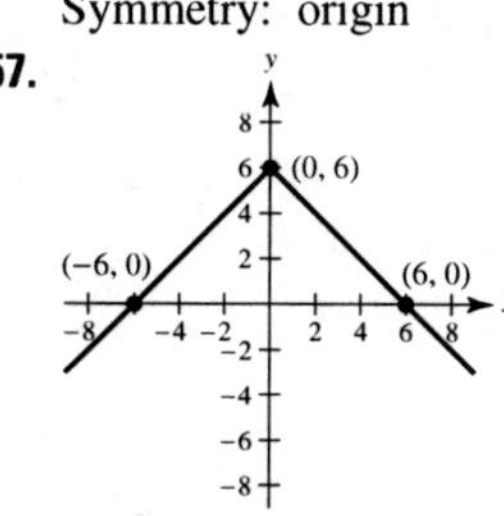

Symmetry: y-axis

59. 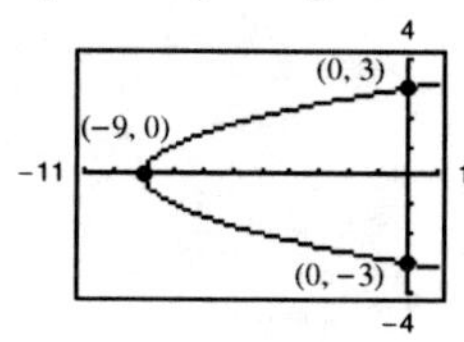

Symmetry: x-axis

61. 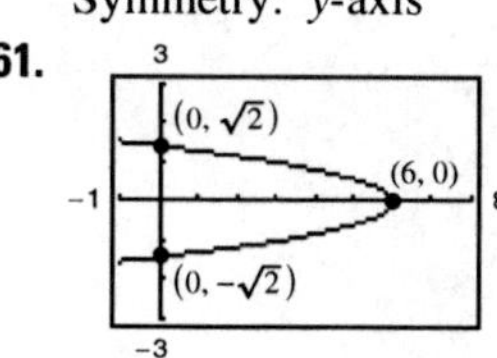

Symmetry: x-axis

63. $(3, 5)$ **65.** $(-1, 5), (2, 2)$ **67.** $(-1, -2), (2, 1)$

69. $(-1, -1), (0, 0), (1, 1)$ **71.** $(-1, -5), (0, -1), (2, 1)$

73. $(-2, 2), \left(-3, \sqrt{3}\right)$

75. (a) $y = -0.027t^2 + 5.73t + 26.9$

(b) 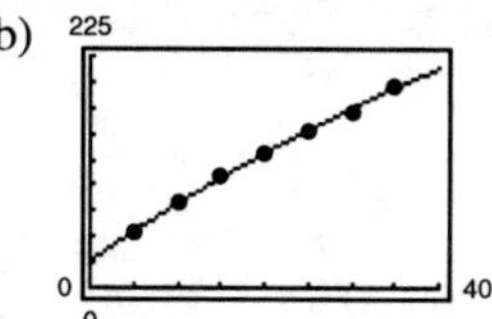   The model is a good fit for the data.

(c) 212.9

"""

77. $x \approx 3133$ units

79. Answers will vary. Sample answer: $y = (x + 4)(x - 3)(x - 8)$

81. (a) Proof   (b) Proof

83. False. $(4, -5)$ is not a point on the graph of $x = y^2 - 29$.

85. True   **87.** $x^2 + (y - 4)^2 = 4$

Section 1.2   (page 16)

1. $m = 1$ **3.** $m = 0$ **5.** $m = -12$

7. 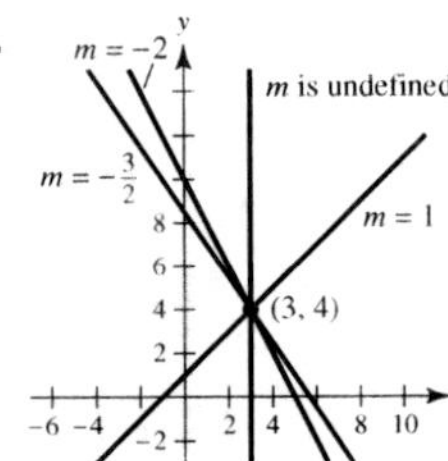

9. 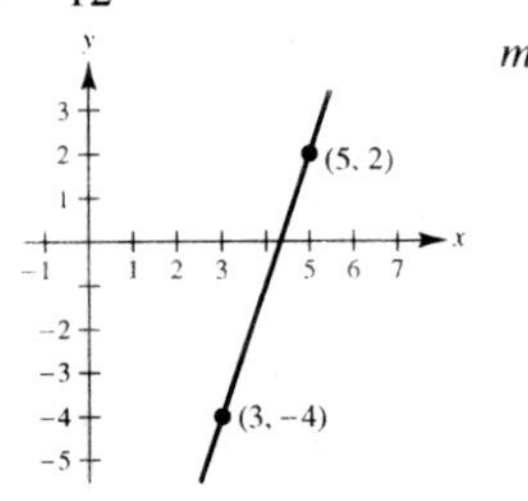 $m = 3$

11. 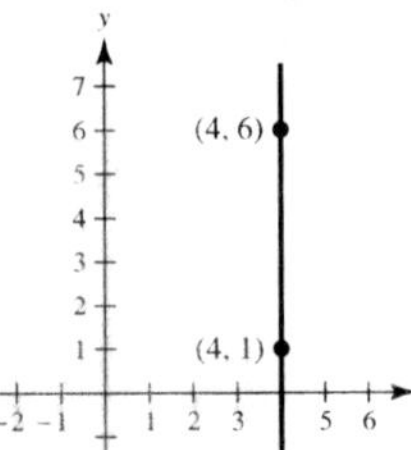

m is undefined.

13. 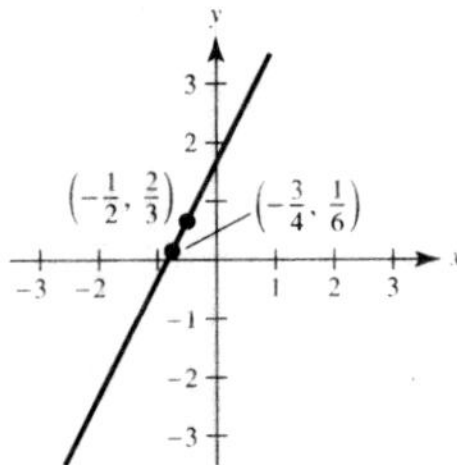

$m = 2$

15. Answers will vary. Sample answer: $(0, 2), (1, 2), (5, 2)$

17. Answers will vary. Sample answer: $(0, 10), (2, 4), (3, 1)$

19. (a) $m = \frac{1}{3}$ (b) $10\sqrt{10}$ ft

21. (a) 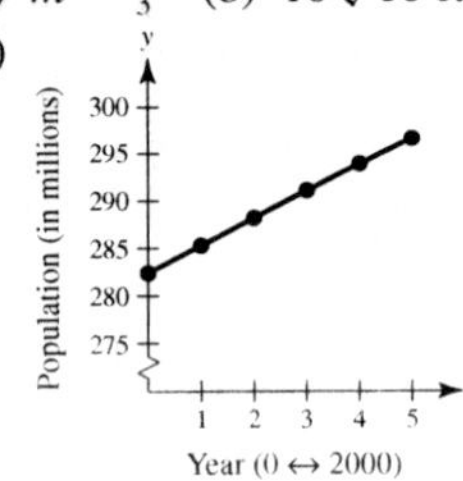   (b) Population increased least rapidly from 2004 to 2005.

23. $m = 4, (0, -3)$

25. $m = -\frac{1}{5}, (0, 4)$ **27.** m is undefined, no y-intercept

29. $3x - 4y + 12 = 0$ **31.** $2x - 3y = 0$

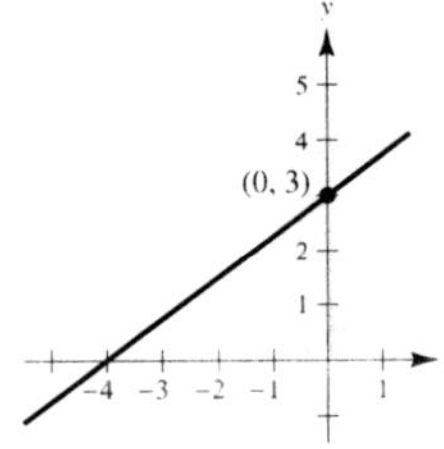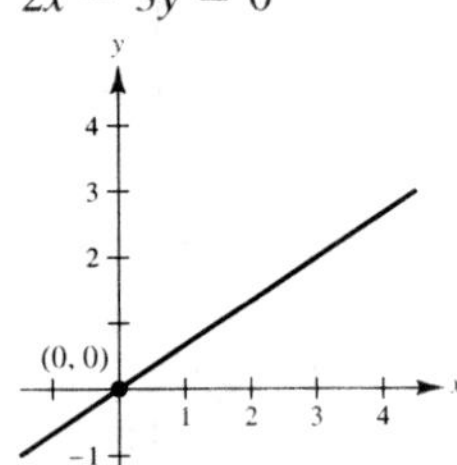

33. $3x - y - 11 = 0$ **35.** $2x - y = 0$

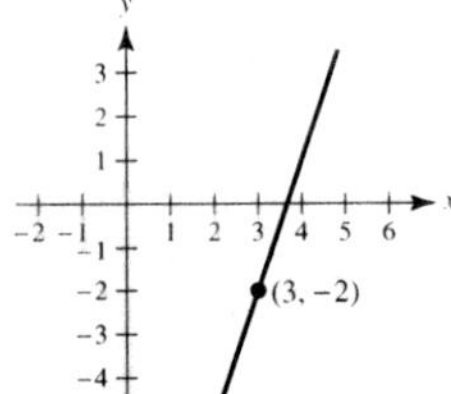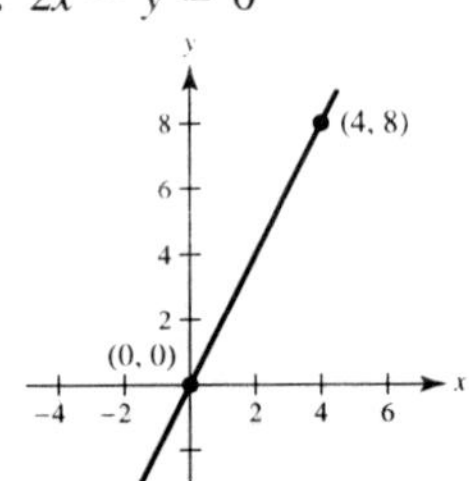

37. $2x - y - 3 = 0$

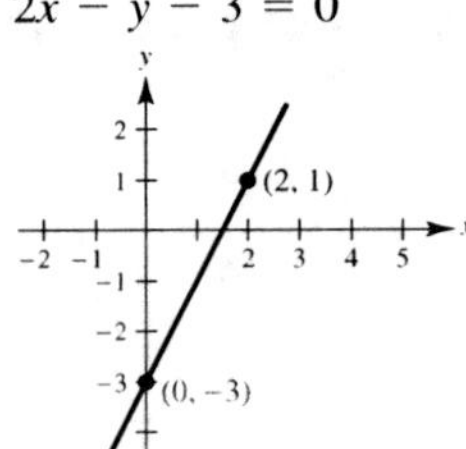

39. $8x + 3y - 40 = 0$

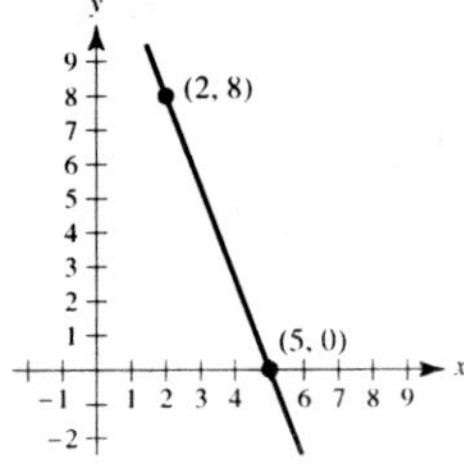

41. $x - 6 = 0$

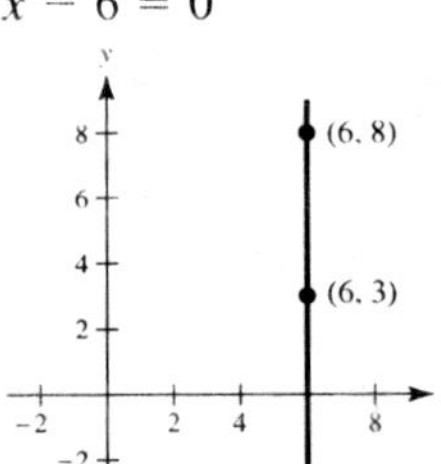

43. $22x - 4y + 3 = 0$

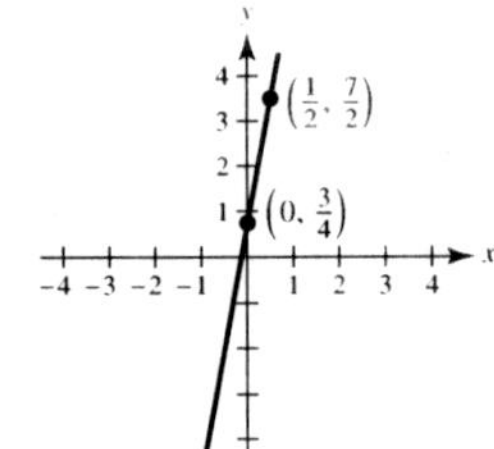

45. $x - 3 = 0$ **47.** $3x + 2y - 6 = 0$ **49.** $x + y - 3 = 0$

51. 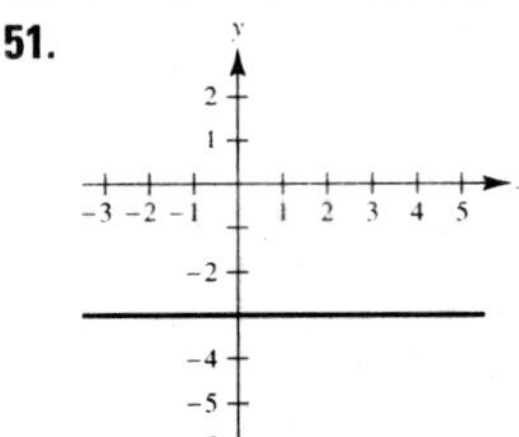

53. 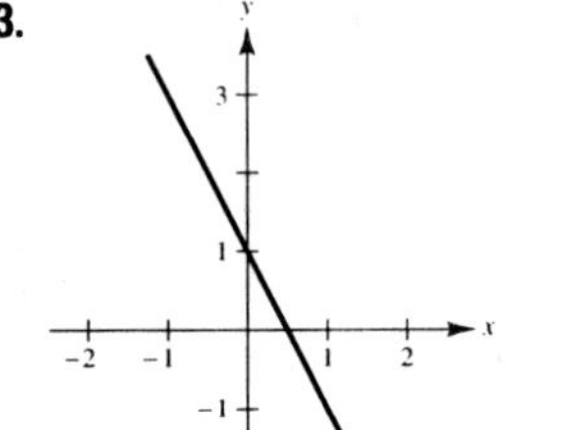

55. 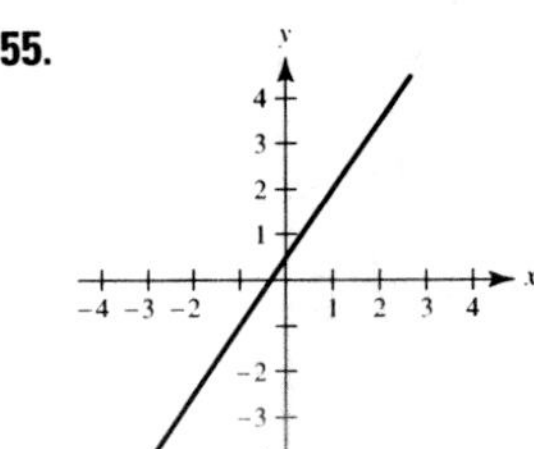

57. 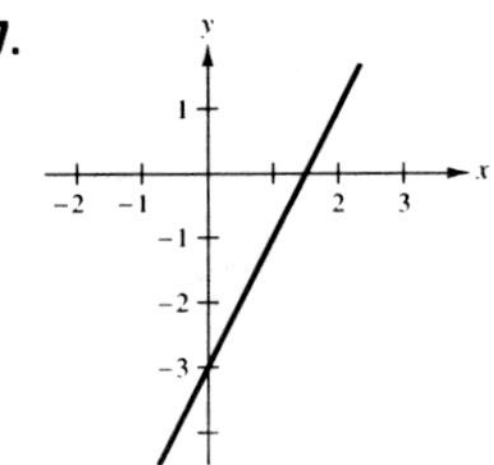

59. (a) 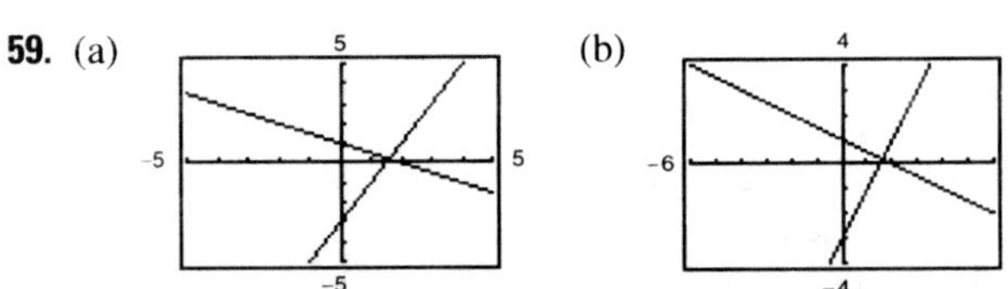 (b)

The lines in (a) do not appear perpendicular, but they do in (b) because a square setting is used. The lines are perpendicular.

61. (a) $x + 7 = 0$ (b) $y + 2 = 0$

63. (a) $2x - y - 3 = 0$ (b) $x + 2y - 4 = 0$

65. (a) $40x - 24y - 9 = 0$ (b) $24x + 40y - 53 = 0$

67. $V = 250t - 150$ **69.** $V = -1600t + 30,000$

71. $y = 2x$

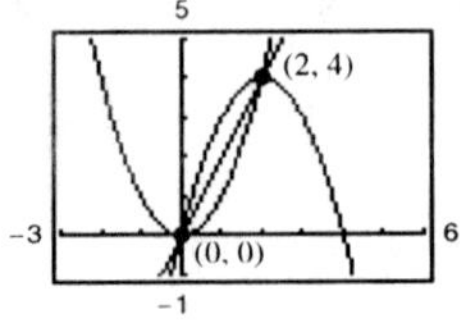

73. Not collinear, because $m_1 \neq m_2$

75. $\left(0, \dfrac{-a^2 + b^2 + c^2}{2c}\right)$ **77.** $\left(b, \dfrac{a^2 - b^2}{c}\right)$

79. $5F - 9C - 160 = 0$; $72°F \approx 22.2°C$

81. (a) $W_1 = 14.50 + 0.75x$, $W_2 = 11.20 + 1.30x$

(b) 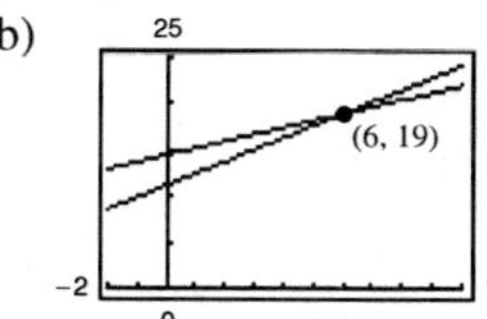

(c) When six units are produced, the wage for both options is $19.00 per hour. Choose option 1 if you think you will produce less than six units and choose option 2 if you think you will produce more than six units.

83. (a) $x = (1530 - p)/15$

(b) 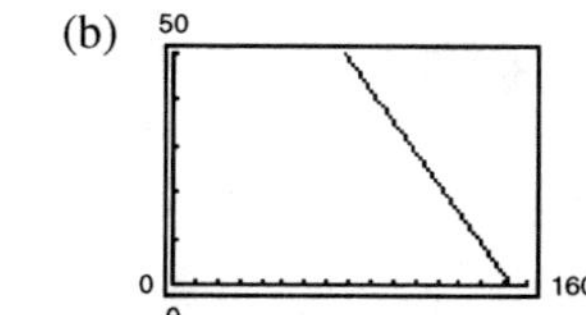   (c) 49 units

45 units

85. $12y + 5x - 169 = 0$ **87.** 2   **89.** $(5\sqrt{2})/2$ **91.** $2\sqrt{2}$

93. Proof   **95.** Proof   **97.** Proof   **99.** True

Section 1.3   (page 27)

1. (a) Domain of f: $[-4, 4]$; Range of f: $[-3, 5]$
Domain of g: $[-3, 3]$; Range of g: $[-4, 4]$

(b) $f(-2) = -1$; $g(3) = -4$

(c) $x = -1$ (d) $x = 1$ (e) $x = -1, x = 1,$ and $x = 2$

3. (a) -4 (b) -25 (c) $7b - 4$ (d) $7x - 11$

5. (a) 5   (b) 0   (c) 1   (d) $4 + 2t - t^2$

7. (a) 1   (b) 0   (c) $-\frac{1}{2}$

9. $3x^2 + 3x\,\Delta x + (\Delta x)^2$, $\Delta x \neq 0$

11. $\left(\sqrt{x - 1} - x + 1\right)/[(x - 2)(x - 1)]$

13. Domain: $(-\infty, \infty)$; Range: $[0, \infty)$

15. Domain: $[0, \infty)$; Range: $[0, \infty)$

17. Domain: All real numbers t such that $t \neq 4n + 2$, where n is an integer; Range: $(-\infty, -1] \cup [1, \infty)$

19. Domain: $(-\infty, 0) \cup (0, \infty)$; Range: $(-\infty, 0) \cup (0, \infty)$

21. Domain: $[0, 1]$

23. Domain: All real numbers x such that $x \neq 2n\pi$, where n is an integer

25. Domain: $(-\infty, -3) \cup (-3, \infty)$

27. (a) -1 (b) 2   (c) 6   (d) $2t^2 + 4$
Domain: $(-\infty, \infty)$; Range: $(-\infty, 1) \cup [2, \infty)$

29. (a) 4   (b) 0   (c) -2 (d) $-b^2$
Domain: $(-\infty, \infty)$; Range: $(-\infty, 0] \cup [1, \infty)$

31. 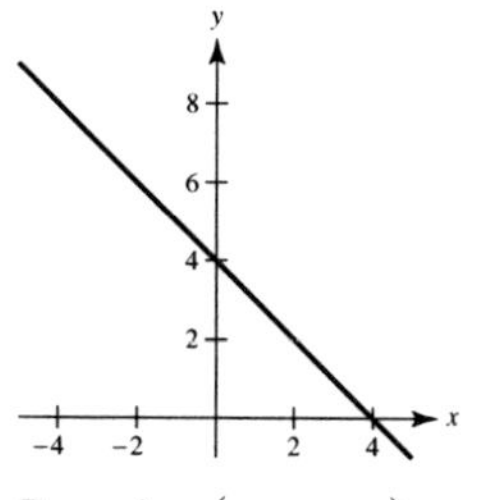
Domain: $(-\infty, \infty)$
Range: $(-\infty, \infty)$

33. 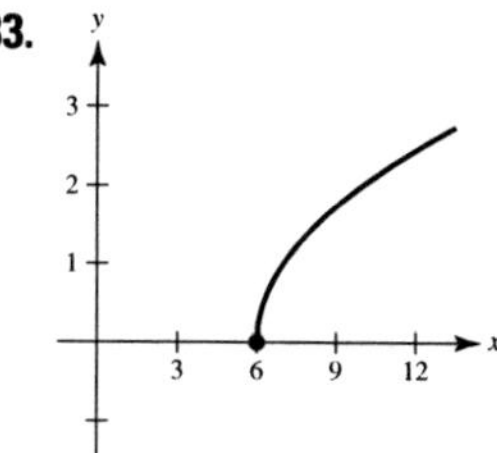
Domain: $[6, \infty)$
Range: $[0, \infty)$

35. 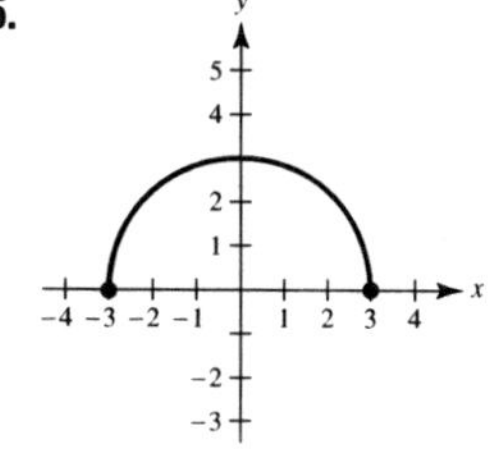
Domain: $[-3, 3]$
Range: $[0, 3]$

37. 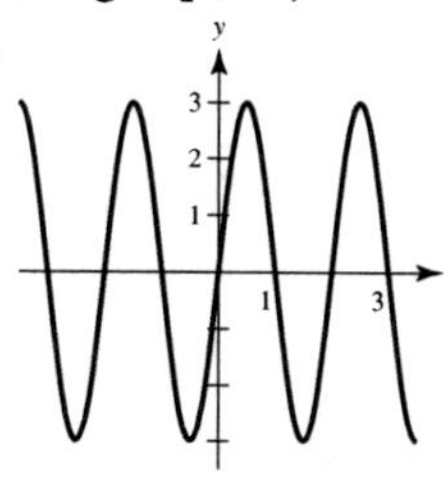
Domain: $(-\infty, \infty)$
Range: $[-3, 3]$

39. The student travels $\frac{1}{2}$ mile per minute during the first 4 minutes, is stationary for the next 2 minutes, and travels 1 mile per minute during the final 4 minutes.

41. y is not a function of x.   **43.** y is a function of x.

45. y is not a function of x.   **47.** y is not a function of x.

49. d   **50.** b   **51.** c   **52.** a   **53.** e   **54.** g

55. (a) 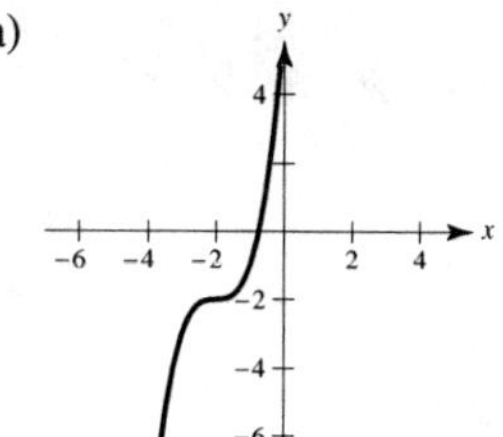   (b) 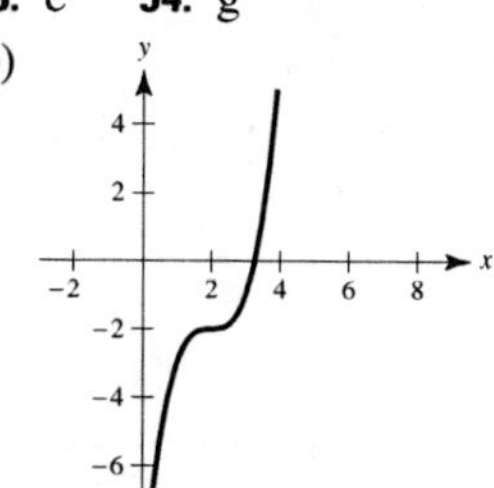

(c) 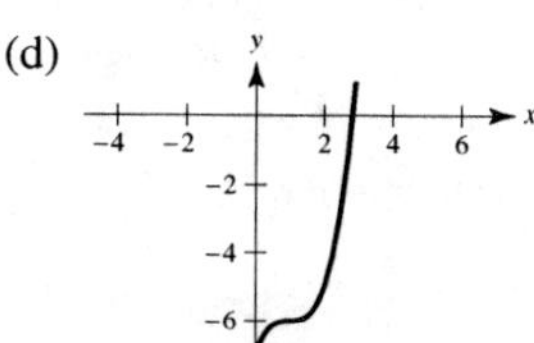   (d)

(e) 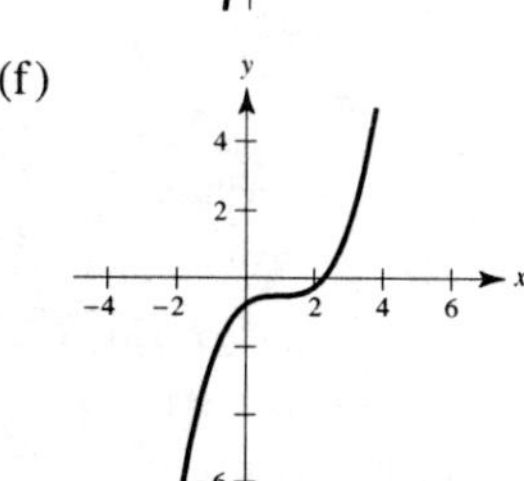   (f)

57. (a) 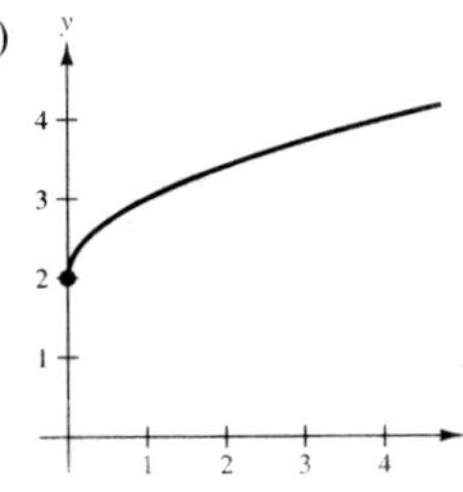   (b) 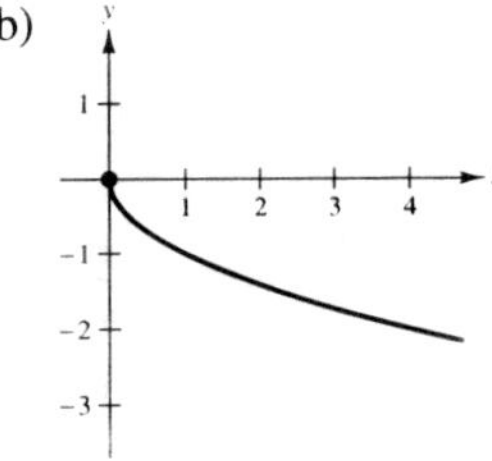

    Vertical translation   Reflection about the x-axis

(c) 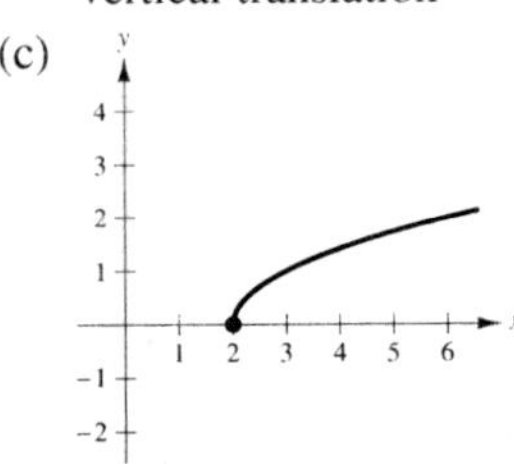

    Horizontal translation

59. (a) 0 (b) 0 (c) -1 (d) $\sqrt{15}$
    (e) $\sqrt{x^2 - 1}$ (f) $x - 1\ (x \geq 0)$

61. $(f \circ g)(x) = x$; Domain: $[0, \infty)$
    $(g \circ f)(x) = |x|$; Domain: $(-\infty, \infty)$
    No, their domains are different.

63. $(f \circ g)(x) = 3/(x^2 - 1)$; Domain: $(-\infty, -1) \cup (-1, 1) \cup (1, \infty)$
    $(g \circ f)(x) = (9/x^2) - 1$; Domain: $(-\infty, 0) \cup (0, \infty)$
    No

65. (a) 4 (b) -2
    (c) Undefined. The graph of g does not exist at $x = -5$.
    (d) 3 (e) 2
    (f) Undefined. The graph of f does not exist at $x = -4$.

67. Answers will vary.
    Sample answer: $f(x) = \sqrt{x}$; $g(x) = x - 2$; $h(x) = 2x$

69. Even   **71.** Odd   **73.** (a) $\left(\frac{3}{2}, 4\right)$ (b) $\left(\frac{3}{2}, -4\right)$

75. f is even. g is neither even nor odd. h is odd.

77. $f(x) = -5x - 6, -2 \leq x \leq 0$ **79.** $y = -\sqrt{-x}$

81. Answers will vary.   **83.** Answers will vary.
    Sample answer: Sample answer:

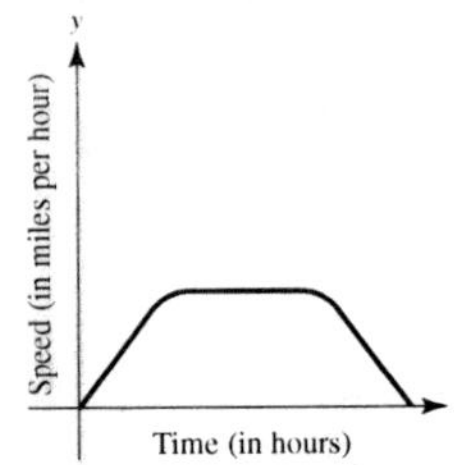   

85. $c = 25$

87. (a) $T(4) = 16°C$, $T(15) \approx 23°C$
    (b) The changes in temperature occur 1 hour later.
    (c) The temperatures are $1°$ lower.

89. (a) 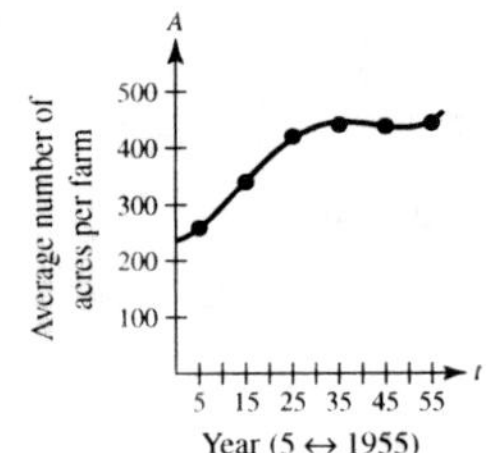   (b) $A(20) \approx 384$ acres/farm

91. $f(x) = |x| + |x - 2| = \begin{cases} 2x - 2, & \text{if } x \geq 2 \\ 2, & \text{if } 0 < x < 2 \\ -2x + 2, & \text{if } x \leq 0 \end{cases}$

93. Proof   **95.** Proof

97. (a) $V(x) = x(24 - 2x)^2, 0 < x < 12$
    (b) 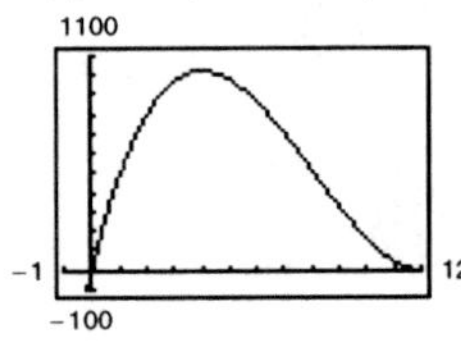

    $4 \times 16 \times 16$ cm

(c)

Height, x	Length and Width	Volume, V
1	$24 - 2(1)$	$1[24 - 2(1)]^2 = 484$
2	$24 - 2(2)$	$2[24 - 2(2)]^2 = 800$
3	$24 - 2(3)$	$3[24 - 2(3)]^2 = 972$
4	$24 - 2(4)$	$4[24 - 2(4)]^2 = 1024$
5	$24 - 2(5)$	$5[24 - 2(5)]^2 = 980$
6	$24 - 2(6)$	$6[24 - 2(6)]^2 = 864$

The dimensions of the box that yield a maximum volume are $4 \times 16 \times 16$ cm.

99. False. For example, if $f(x) = x^2$, then $f(-1) = f(1)$.

101. True   **103.** Putnam Problem A1, 1988

Section 1.4 (page 34)

1. Trigonometric   **3.** No relationship

5. (a) and (b)   **7.** (a) $d = 0.066F$

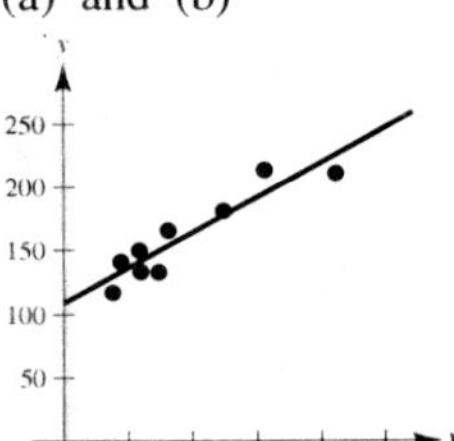   (b) 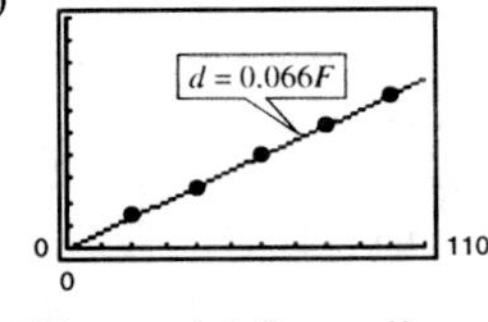

    Approximately linear   The model fits well.
    (c) 136   (c) 3.63 cm

9. (a) $y = 0.151x + 0.10$; $r \approx 0.880$
    (b) 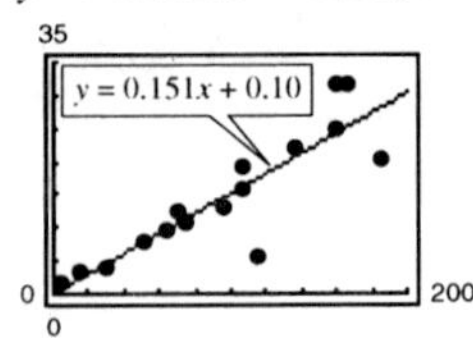

(c) Greater per capita energy consumption by a country tends to correspond to greater per capita gross national product of the country. The four countries that differ most from the linear model are Venezuela, South Korea, Hong Kong, and the United Kingdom.

(d) $y = 0.155x + 0.22$; $r \approx 0.984$

11. (a) $y_1 = 0.04040t^3 - 0.3695t^2 + 1.123t + 5.88$
$y_2 = 0.264t + 3.35$
$y_3 = 0.01439t^3 - 0.1886t^2 + 0.476t + 1.59$

(b) 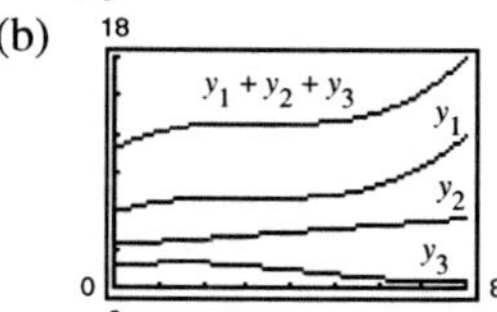   About 47.5 cents/mi

13. (a) $t = 0.002s^2 - 0.04s + 1.9$

(b) 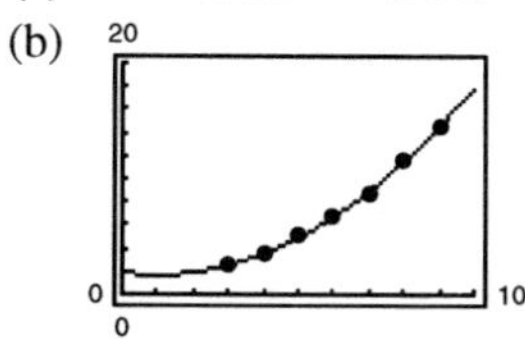

(c) According to the model, the times required to attain speeds of less than 20 miles per hour are all about the same.

(d) $t = 0.001s^2 + 0.02s + 0.1$

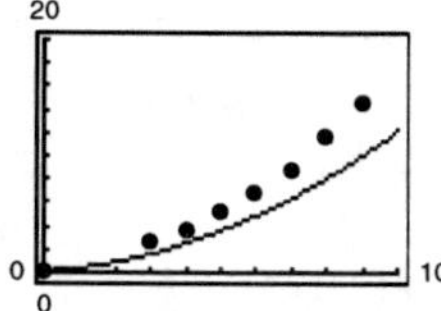

(e) No. From the graph in part (b), you can see that the model from part (a) follows the data more closely than the model from part (d).

15. (a) $y = -1.806x^3 + 14.58x^2 + 16.4x + 10$

(b) 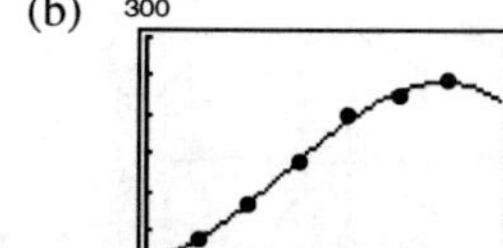   (c) 214 hp

17. (a) Yes. At time t there is one and only one displacement y.

(b) Amplitude: 0.35; Period: 0.5

(c) $y = 0.35 \sin(4\pi t) + 2$

(d) 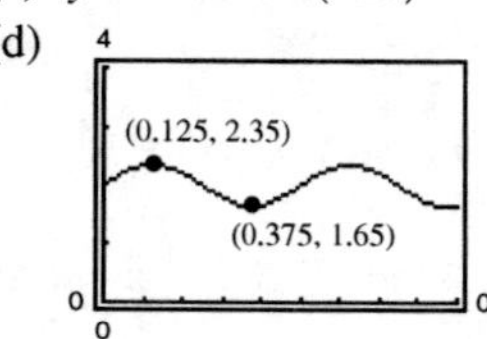

The model appears to fit the data well.

19. Answers will vary.   **21.** Putnam Problem A2, 2004

Section 1.5   (page 44)

1. (a) $f(g(x)) = 5[(x-1)/5] + 1 = x$
$g(f(x)) = [(5x+1) - 1]/5 = x$

(b) 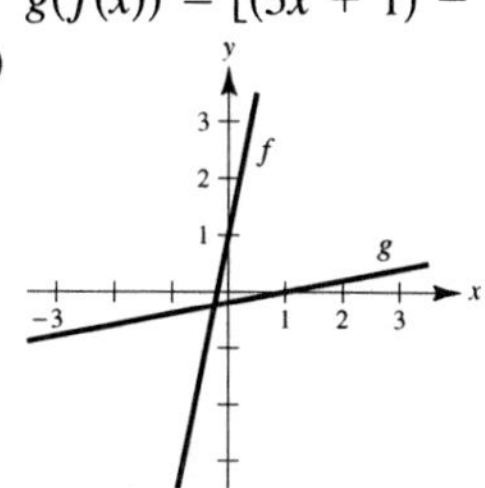

3. (a) $f(g(x)) = \left(\sqrt[3]{x}\right)^3 = x$; $g(f(x)) = \sqrt[3]{x^3} = x$

(b) 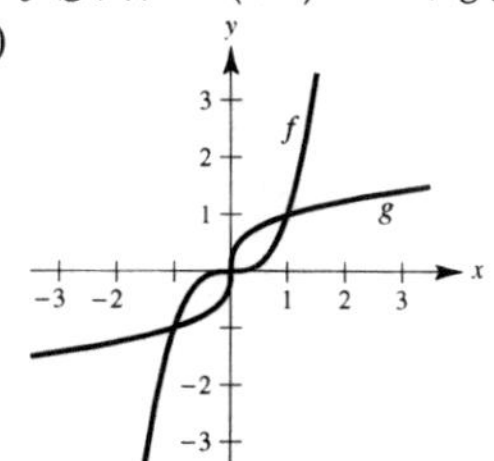

5. (a) $f(g(x)) = \sqrt{x^2 + 4} - 4 = x$
$g(f(x)) = \left(\sqrt{x - 4}\right)^2 + 4 = x$

(b) 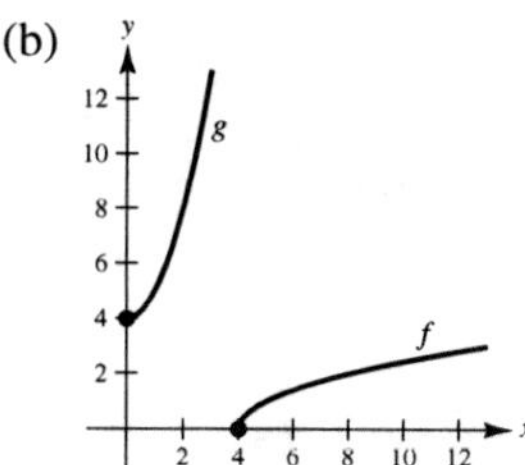

7. (a) $f(g(x)) = \dfrac{1}{1/x} = x$; $g(f(x)) = \dfrac{1}{1/x} = x$

(b) 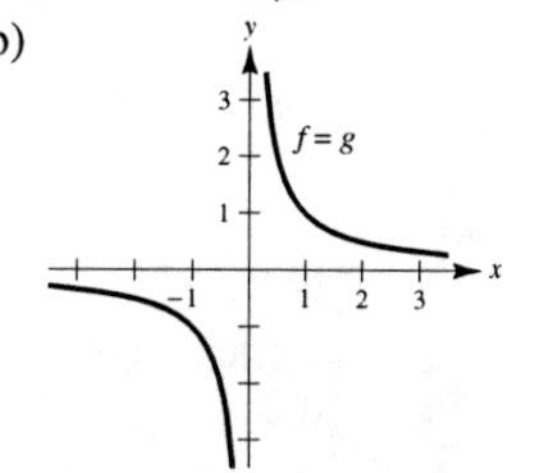

9. c   **10.** b   **11.** a   **12.** d   **13.** One-to-one, inverse exists.

15. Not one-to-one, inverse does not exist.

17. 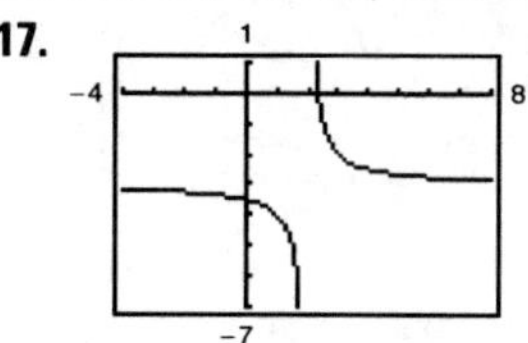

One-to-one, inverse exists.

19.

Not one-to-one, inverse does not exist.

21.

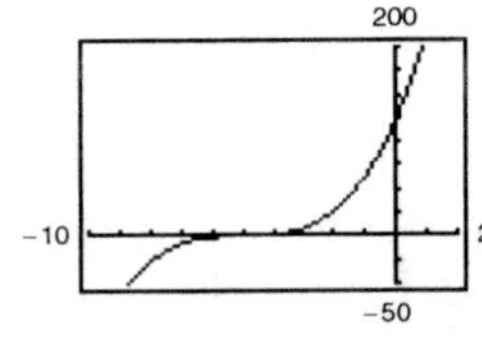

One-to-one, inverse exists.

23. One-to-one, inverse exists.

25. Not one-to-one, inverse does not exist.

27. One-to-one, inverse exists.

29. (a) $f^{-1}(x) = (x + 3)/2$

(b)

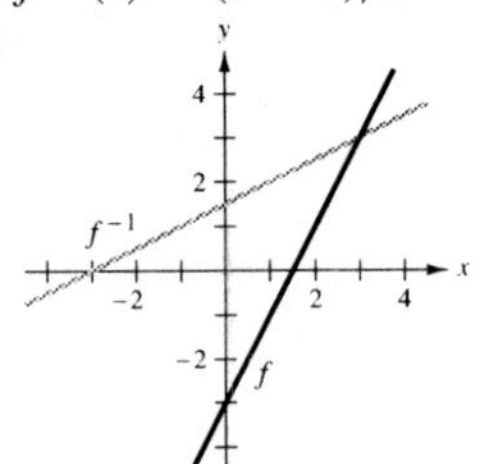

(c) f and f^{-1} are symmetric about $y = x$.

(d) Domain of f and f^{-1}: $(-\infty, \infty)$
Range of f and f^{-1}: $(-\infty, \infty)$

31. (a) $f^{-1}(x) = x^{1/5}$

(b)

(c) f and f^{-1} are symmetric about $y = x$.

(d) Domain of f and f^{-1}: $(-\infty, \infty)$
Range of f and f^{-1}: $(-\infty, \infty)$

33. (a) $f^{-1}(x) = x^2,\ x \geq 0$

(b)

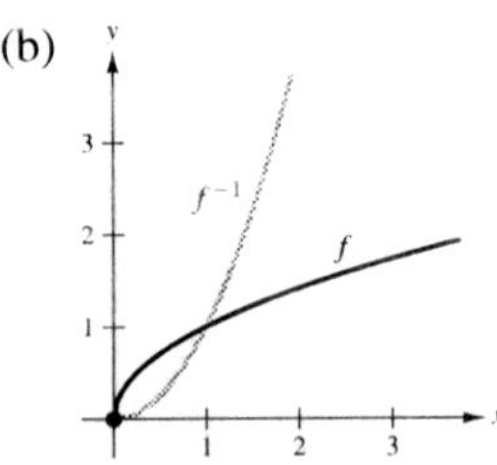

(c) f and f^{-1} are symmetric about $y = x$.

(d) Domain of f and f^{-1}: $[0, \infty)$
Range of f and f^{-1}: $[0, \infty)$

35. (a) $f^{-1}(x) = \sqrt{4 - x^2},\ 0 \leq x \leq 2$

(b)

(c) f and f^{-1} are symmetric about $y = x$.

(d) Domain of f and f^{-1}: $[0, 2]$
Range of f and f^{-1}: $[0, 2]$

37. (a) $f^{-1}(x) = x^3 + 1$

(b)

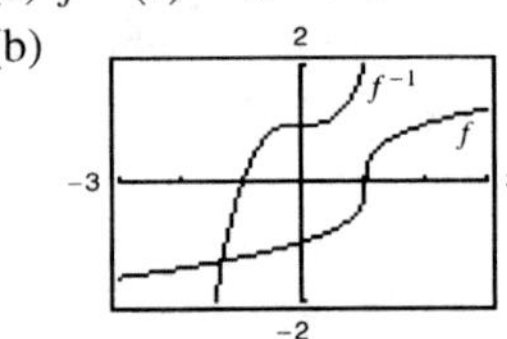

(c) f and f^{-1} are symmetric about $y = x$.

(d) Domain of f and f^{-1}: $(-\infty, \infty)$
Range of f and f^{-1}: $(-\infty, \infty)$

39. (a) $f^{-1}(x) = x^{3/2},\ x \geq 0$

(b)

(c) f and f^{-1} are symmetric about $y = x$.

(d) Domain of f and f^{-1}: $[0, \infty)$
Range of f and f^{-1}: $[0, \infty)$

41. (a) $f^{-1}(x) = \sqrt{7}x/\sqrt{1 - x^2},\ -1 < x < 1$

(b)

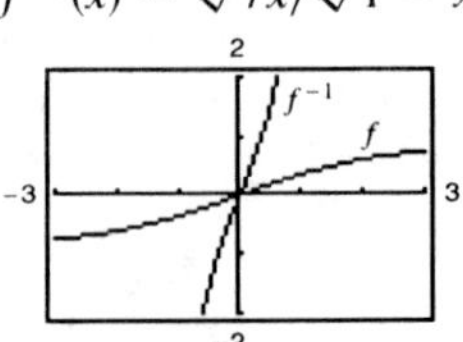

(c) f and f^{-1} are symmetric about $y = x$.

(d) Domain of f: $(-\infty, \infty)$
Range of f: $(-1, 1)$
Domain of f^{-1}: $(-1, 1)$
Range of f^{-1}: $(-\infty, \infty)$

43.

x	0	1	2	4
$f(x)$	1	2	3	4

x	1	2	3	4
$f^{-1}(x)$	0	1	2	4

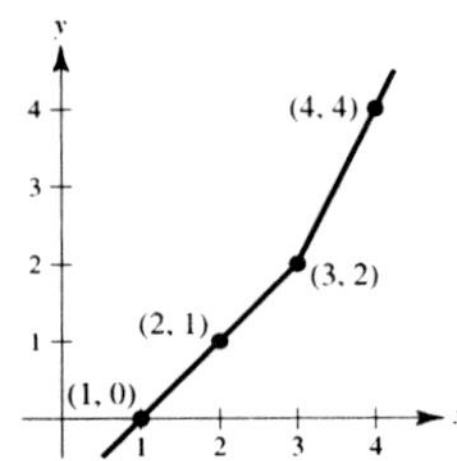

45. (a) Answers will vary.

(b) $y = \frac{20}{7}(80 - x)$
x: total cost
y: number of pounds of the less expensive commodity

(c) $[62.5, 80]$; The total cost will be between \$62.50 and \$80.00.

(d) 20 lb

47. $f^{-1}(x) = \begin{cases} (1 - \sqrt{1 + 16x^2})/(2x), & \text{if } x \neq 0 \\ 0, & \text{if } x = 0 \end{cases}$

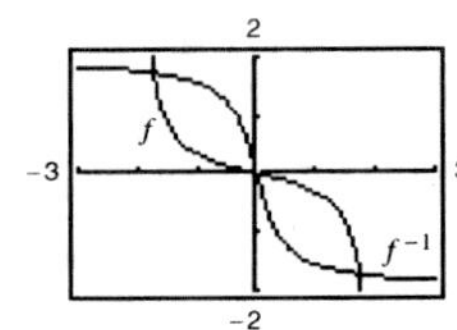

The graph of f^{-1} is a reflection of the graph of f in the line $y = x$.

49. (a) and (b)

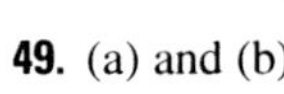
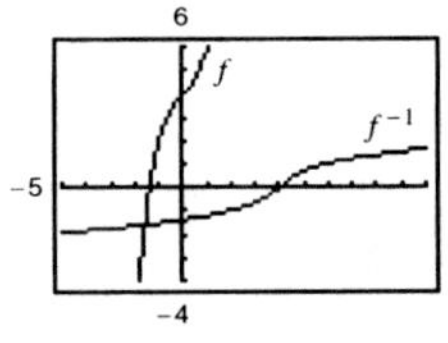

(c) Yes

51. (a) and (b)

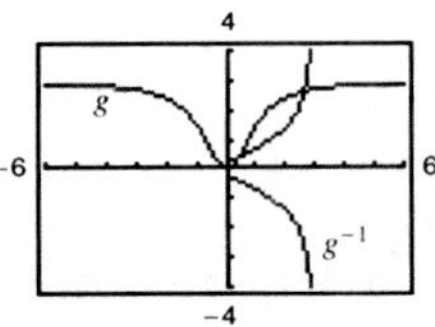

(c) No, it is not an inverse function. It does not pass the Vertical Line Test.

53. The function f passes the Horizontal Line Test on $[4, \infty)$, so it is one-to-one on $[4, \infty)$.

55. The function f passes the Horizontal Line Test on $(0, \infty)$, so it is one-to-one on $(0, \infty)$.

57. The function f passes the Horizontal Line Test on $[0, \pi]$, so it is one-to-one on $[0, \pi]$.

59. One-to-one
$f^{-1}(x) = x^2 + 2,\ x \geq 0$

61. One-to-one
$f^{-1}(x) = 2 - x,\ x \geq 0$

63. Answers will vary. Sample answer: $f^{-1}(x) = \sqrt{x} + 3,\ x \geq 0$

65. Answers will vary. Sample answer: $f^{-1}(x) = x - 3,\ x \geq 0$

67. (a)

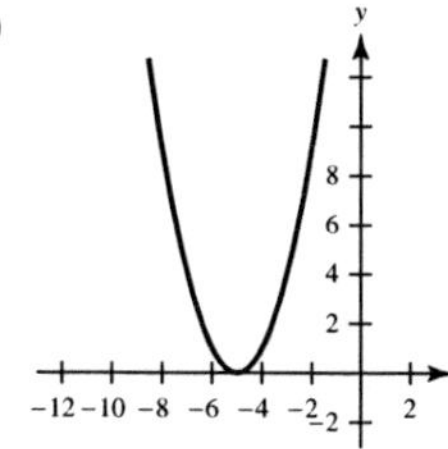

(b) Answers will vary.
Sample answer: $[-5, \infty)$
(c) $f^{-1}(x) = \sqrt{x} - 5$
(d) Domain of f^{-1}: $[0, \infty)$

69. (a)

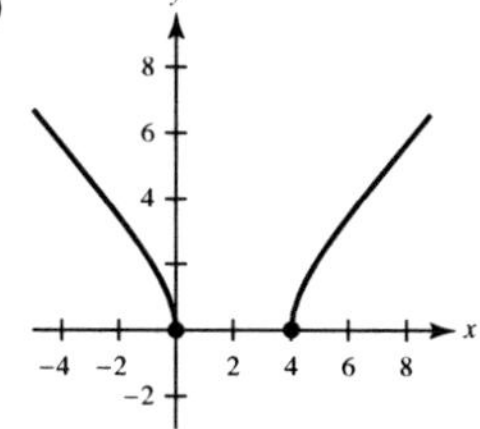

(b) Answers will vary.
Sample answer: $[4, \infty)$
(c) $f^{-1}(x) = 2 + \sqrt{x^2 + 4}$
(d) Domain of f^{-1}: $[0, \infty)$

71. (a)

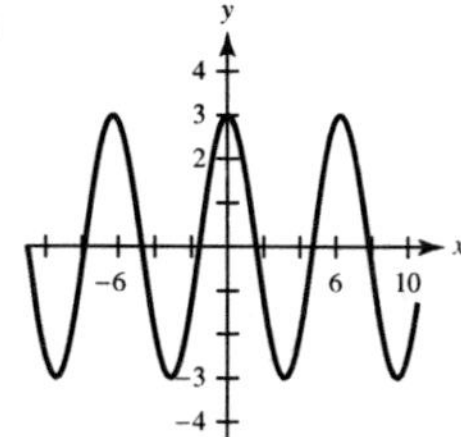

(b) Answers will vary.
Sample answer: $[0, \pi]$

(c) $f^{-1}(x) = \arccos\left(\dfrac{x}{3}\right)$

(d) Domain of f^{-1}: $[-3, 3]$

73. 1   **75.** $\dfrac{\pi}{6}$ **77.** 2   **79.** 32   **81.** 600

83. $(g^{-1} \circ f^{-1})(x) = \dfrac{x + 1}{2}$ **85.** $(f \circ g)^{-1}(x) = \dfrac{x + 1}{2}$

87. (a) f is one-to-one because it passes the Horizontal Line Test.
(b) $[-2, 2]$
(c) -4

89.

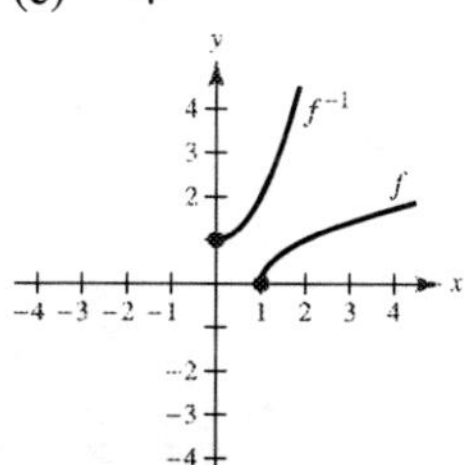

91. (a)

x	-1	-0.8	-0.6	-0.4	-0.2
y	-1.57	-0.93	-0.64	-0.41	-0.20

x	0	0.2	0.4	0.6	0.8	1
y	0	0.20	0.41	0.64	0.93	1.57

(b)

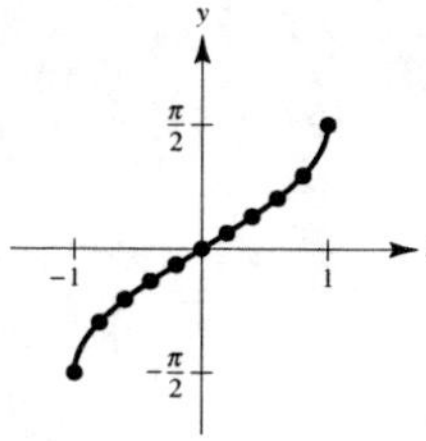

(c)

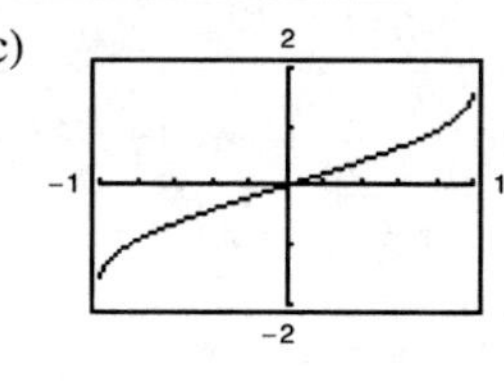

(d) Intercept: $(0, 0)$; Symmetry: origin

93. $\left(-\sqrt{2}/2, 3\pi/4\right), (1/2, \pi/3), \left(\sqrt{3}/2, \pi/6\right)$
95. $\pi/6$ **97.** $\pi/3$ **99.** $\pi/6$
101. $-\pi/4$ **103.** 2.50   **105.** 0.66
107. Let $y = f(x)$ be one-to-one. Solve for x as a function of y. Interchange x and y to get $y = f^{-1}(x)$. Let the domain of f^{-1} be the range of f. Verify that $f(f^{-1}(x)) = x$ and $f^{-1}(f(x)) = x$.
Sample answer: $f(x) = x^3$
$$y = x^3$$
$$x = \sqrt[3]{y}$$
$$y = \sqrt[3]{x}$$
$$f^{-1}(x) = \sqrt[3]{x}$$

109. Answers will vary. Sample answer: $y = x^4 - 2x^3$
111. If the domains were not restricted, then the trigonometric functions would not be one-to-one and hence would have no inverses.

113.

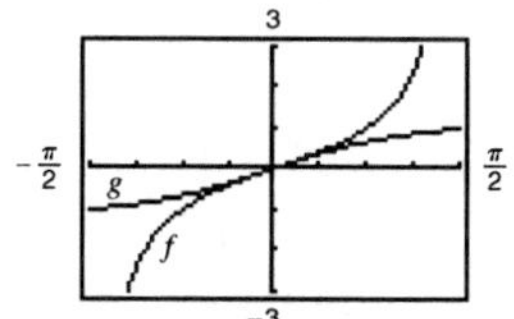

115. -0.1 **117.** (a) $\dfrac{1}{2}$ (b) $\dfrac{\sqrt{3}}{2}$ **119.** (a) $\dfrac{3}{5}$ (b) $\dfrac{5}{3}$

121. (a) $-\sqrt{3}$ (b) $-\dfrac{13}{5}$ **123.** x **125.** $\dfrac{\sqrt{1 - x^2}}{x}$

127. $\dfrac{1}{x}$ **129.** x **131.** $\sqrt{1 - 4x^2}$

133. $\dfrac{\sqrt{x^2 - 1}}{|x|}$ **135.** $\dfrac{\sqrt{x^2 - 9}}{3}$ **137.** $\dfrac{\sqrt{x^2 + 2}}{x}$

139. $x = \frac{1}{3}\left[\sin\left(\frac{1}{2}\right) + \pi\right] \approx 1.207$ **141.** $x = \frac{1}{3}$

143. $(0.7862, 0.6662)$ **145.** $\arcsin\left(\dfrac{9}{\sqrt{x^2 + 81}}\right)$

147. Answers will vary.

149.

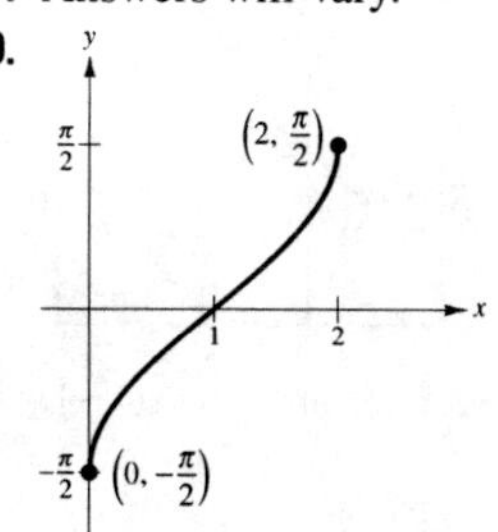

151.

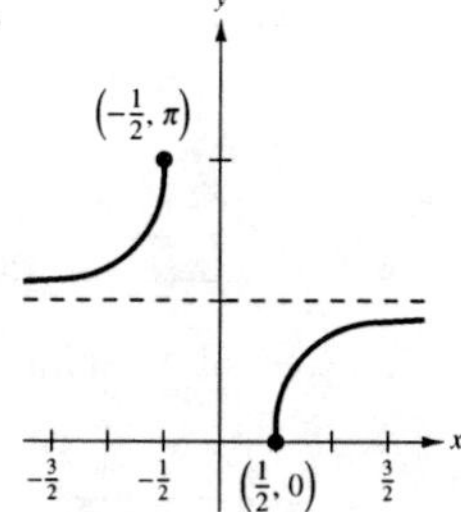

153. $f^{-1}(8) = -3$ **155.** Proof   **157.** Proof   **159.** Proof
161. False. Let $f(x) = x^2$.

163. False. $\arcsin^2 0 + \arccos^2 0 = \left(\dfrac{\pi}{2}\right)^2 \neq 1$

165. True   **167.** Answers will vary.   **169.** Proof

171. $f^{-1}(x) = \dfrac{-b - \sqrt{b^2 - 4ac + 4ax}}{2a}$

173. $ad - bc \neq 0$; $f^{-1}(x) = \dfrac{b - dx}{cx - a}$

Section 1.6   (page 54)

1. (a) 125   (b) 9   (c) $\frac{1}{9}$ (d) $\frac{1}{3}$

3. (a) 5^5 (b) $\frac{1}{5}$ (c) $\frac{1}{5}$ (d) 2^2

5. (a) e^6 (b) e^{12} (c) $\dfrac{1}{e^6}$ (d) e^2

7. $x = 4$ **9.** $x = 4$ **11.** $x = -5$ **13.** $x = -2$
15. $x = 2$ **17.** $x = 16$ **19.** $x = \ln 5 \approx 1.609$
21. $x = -\frac{5}{2}$ **23.** $2.7182805 < e$

25. 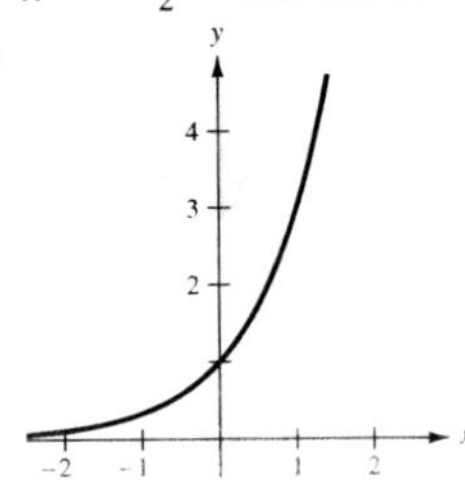   **27.** 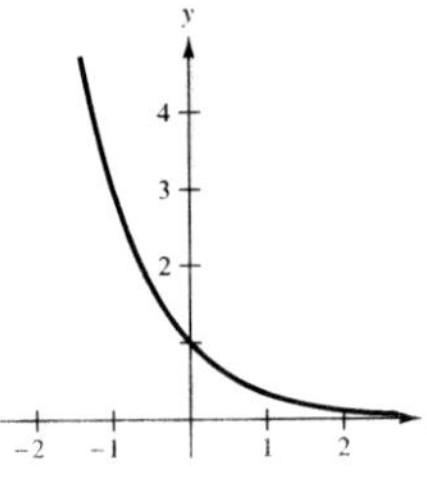

29. 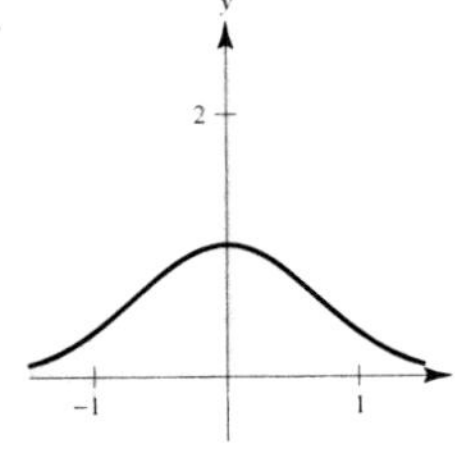   **31.** 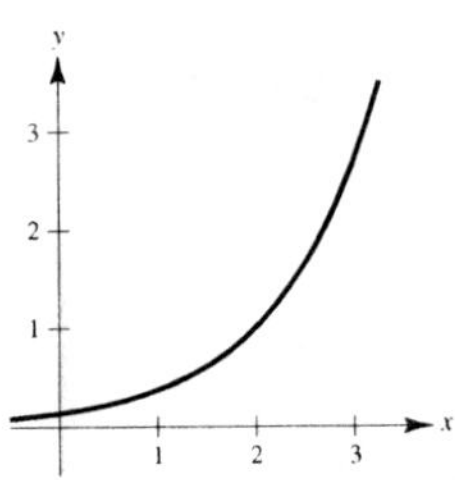

33. 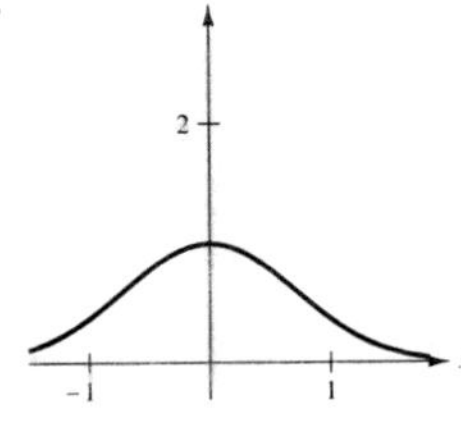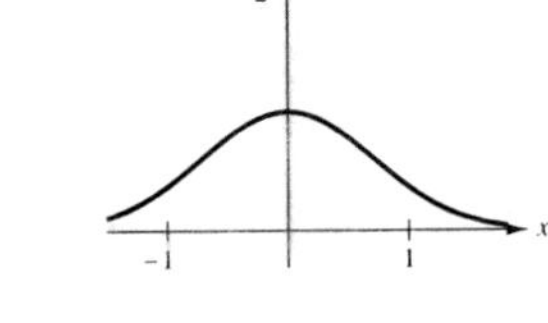   **35.** Domain: $(-\infty, \infty)$

37. Domain: $(-\infty, 0]$ **39.** Domain: $(-\infty, \infty)$

41. (a) 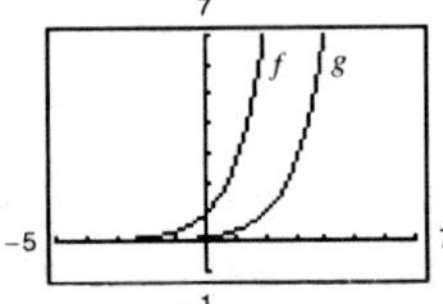   (b) 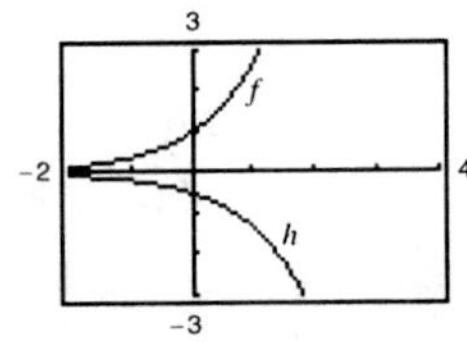

Translation two units to the right

Reflection in the x-axis and vertical shrink

(c) 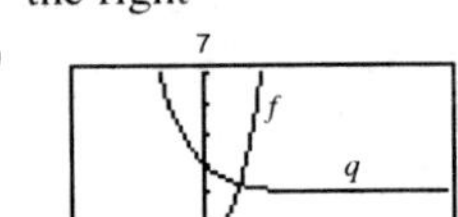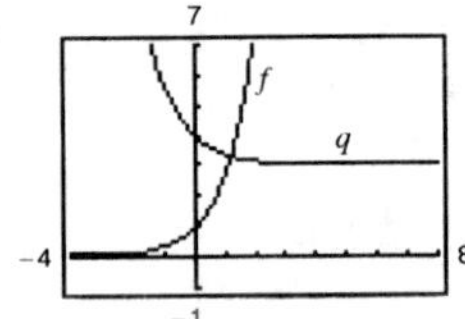

Reflection in the y-axis and translation three units upward

43. c   **44.** d   **45.** a   **46.** b   **47.** b   **48.** d
49. a   **50.** c   **51.** $y = 2(3^x)$
53. $\ln 1 = 0$ **55.** $e^{0.6931\ldots} = 2$

57. 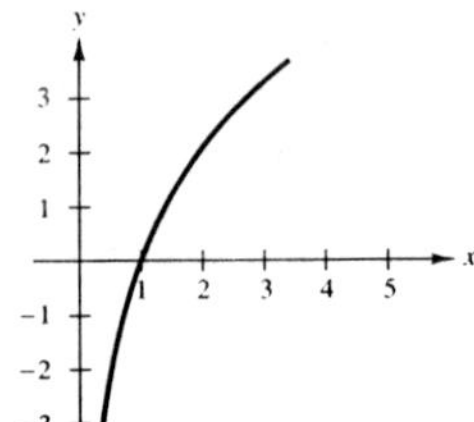   **59.** 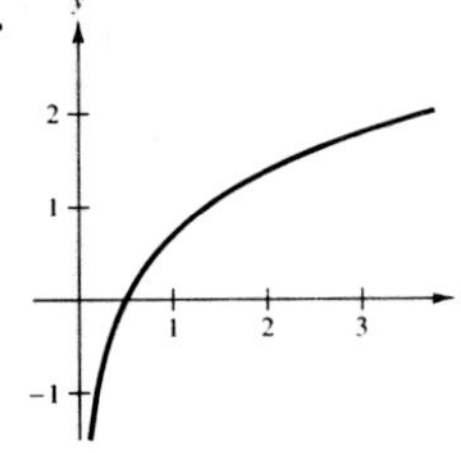

Domain: $x > 0$ Domain: $x > 0$

61. 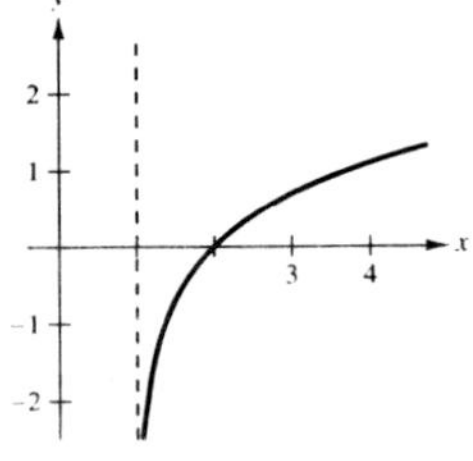

Domain: $x > 1$
63. $g(x) = -e^x - 8$ **65.** $g(x) = \ln(x - 5) - 1$
67. 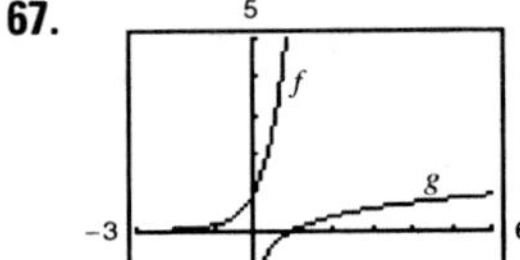   **69.** 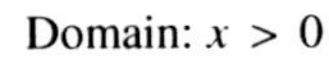

71. (a) $f^{-1}(x) = \dfrac{\ln x + 1}{4}$

(b) 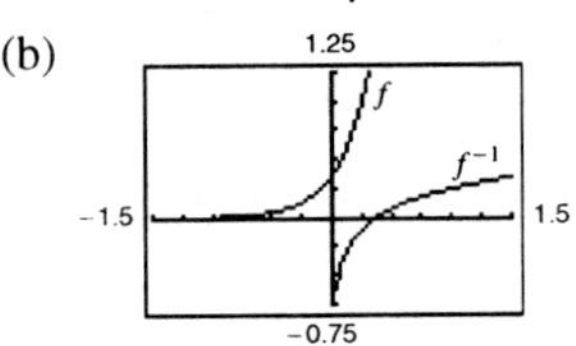

(c) Answers will vary.
73. (a) $f^{-1}(x) = e^{x/2} + 1$

(b) 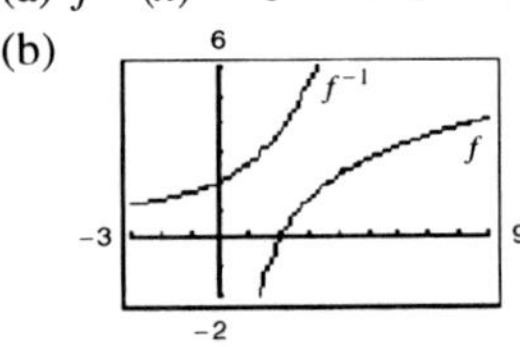

(c) Answers will vary.
75. x^2 **77.** $5x + 2$ **79.** $-1 + 2x$
81. (a) 1.7917   (b) -0.4055 (c) 4.3944   (d) 0.5493
83. Answers will vary.   **85.** Answers will vary.
87. $\ln x - \ln 4$ **89.** $\ln x + \ln y - \ln z$
91. $\ln x + \frac{1}{2}\ln(x^2 + 5)$ **93.** $\frac{1}{2}[\ln(x - 1) - \ln x]$
95. $2 + \ln 3$ **97.** $\ln(7x)$
99. $\ln \dfrac{x - 2}{x + 2}$ **101.** $\ln \sqrt[3]{\dfrac{x(x + 3)^2}{x^2 - 1}}$ **103.** $\ln \dfrac{9}{\sqrt{x^2 + 1}}$
105. (a) $x = 4$ (b) $x = \frac{3}{2}$
107. (a) $x = e^2 \approx 7.389$ (b) $x = \ln 4 \approx 1.386$
109. $x > \ln 5$ **111.** $e^{-2} < x < 1$

113. 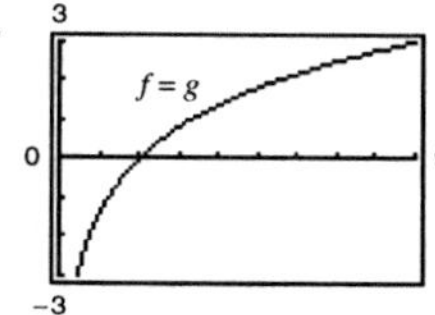

115. Proof

117. 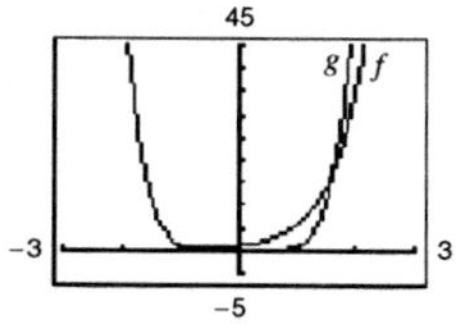

$(-0.7899, 0.2429)$,
$(1.6242, 18.3615)$,
and $(6, 46{,}656)$

As x increases, $f(x) = 6^x$ grows more rapidly.

119. (a) 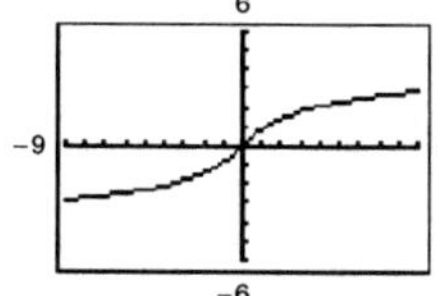   Domain: $(-\infty, \infty)$

(b) Proof

(c) $f^{-1}(x) = \dfrac{e^{2x} - 1}{2e^x}$

Review Exercises for Chapter 1   (page 57)

1. $(0, -8), \left(\tfrac{8}{5}, 0\right)$ **3.** $\left(0, \tfrac{3}{4}\right), (3, 0)$

5. Symmetric with respect to the y-axis

7. 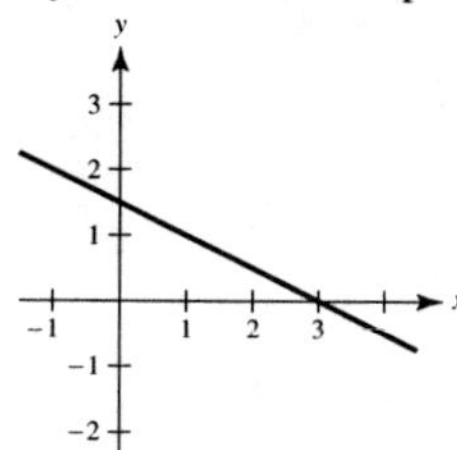

9. 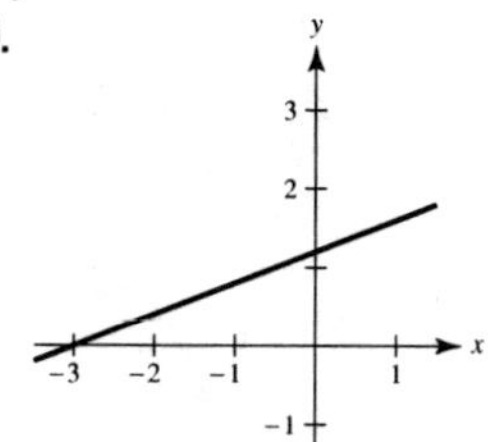

11. 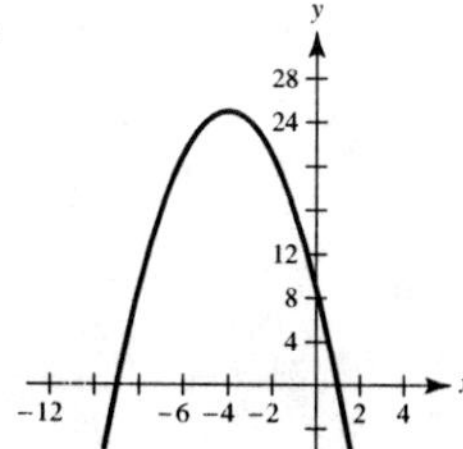

13. 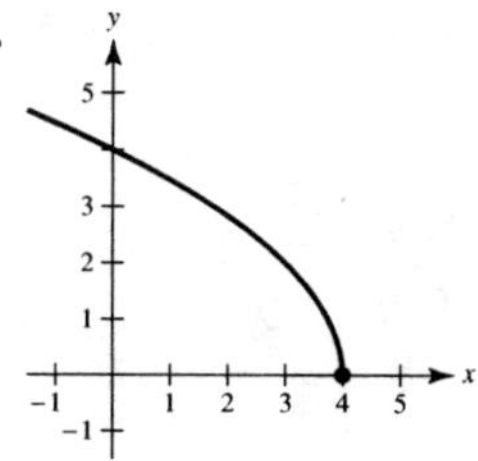

15. $(-2, 3)$

17. 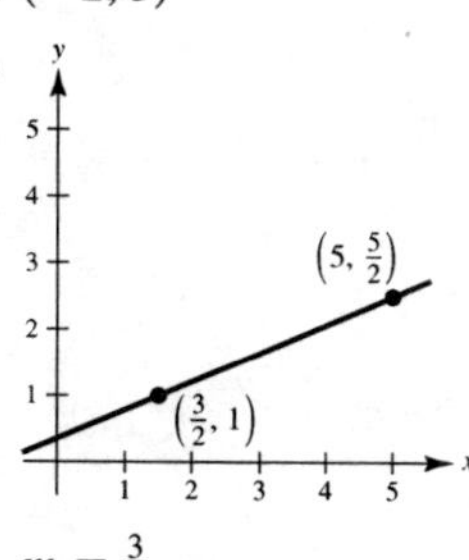

$m = \tfrac{3}{7}$

19. $t = \tfrac{1}{5}$

21. $7x - 4y - 41 = 0$

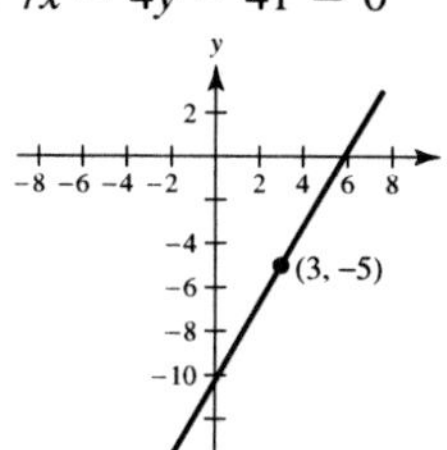

23. $2x + 3y + 6 = 0$

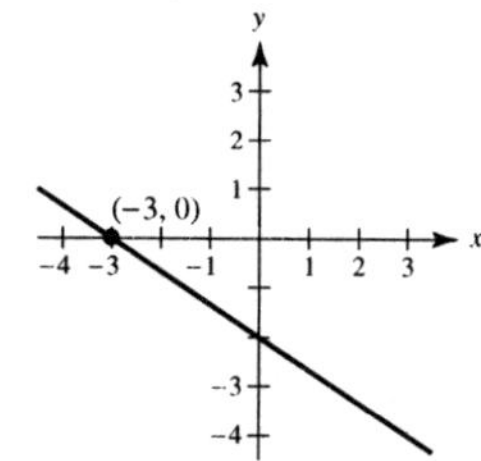

25. (a) $16y - 7x - 101 = 0$
(b) $3y - 5x - 30 = 0$
(c) $3y + 5x = 0$
(d) $x + 3 = 0$

27. $V = 12{,}500 - 850t$; $\$9950$

29. 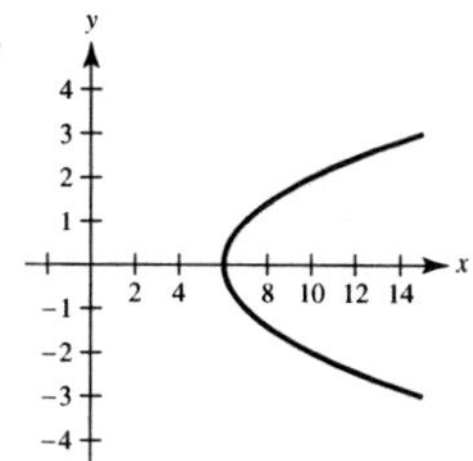

31. 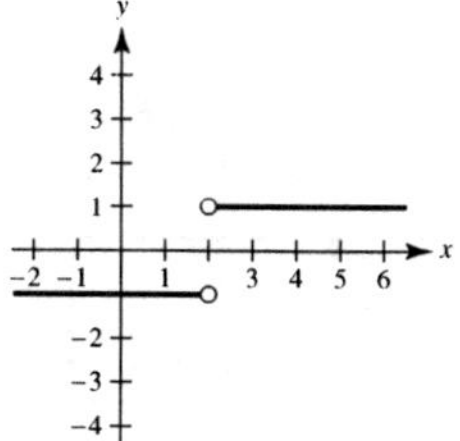

Not a function   Function

33. (a) Undefined   (b) $-1/(1 + \Delta x), \Delta x \neq 0, -1$

35. (a) Domain: $[-6, 6]$; Range: $[0, 6]$
(b) Domain: $(-\infty, 5) \cup (5, \infty)$; Range: $(-\infty, 0) \cup (0, \infty)$
(c) Domain: $(-\infty, \infty)$; Range: $(-\infty, \infty)$

37. (a) 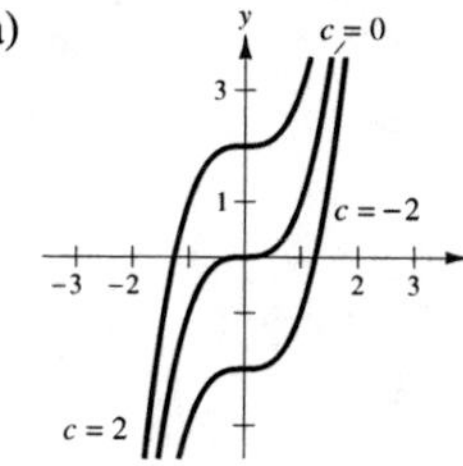   (b) 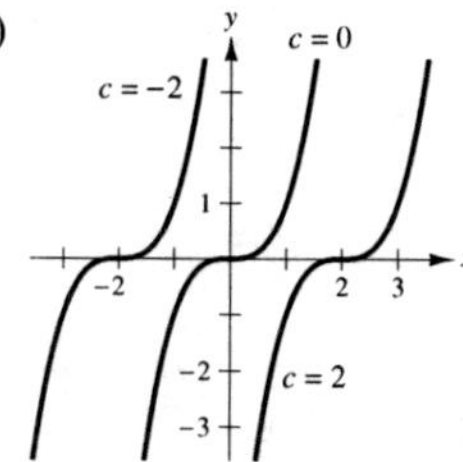

(c) 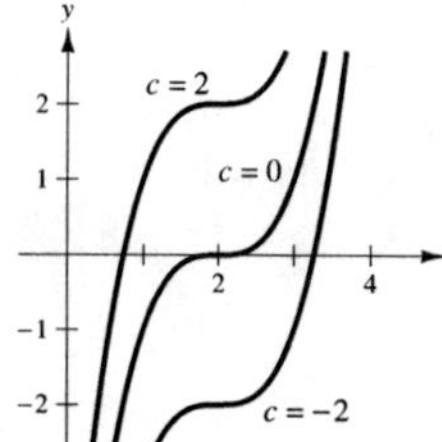   (d) 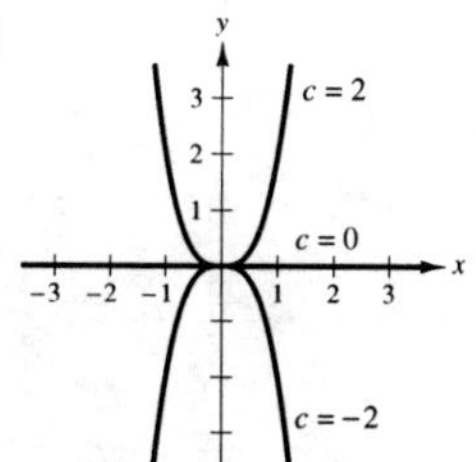

39. (a) Minimum degree: 3; Leading coefficient: negative
(b) Minimum degree: 4; Leading coefficient: positive
(c) Minimum degree: 2; Leading coefficient: negative
(d) Minimum degree: 5; Leading coefficient: positive

41. (a) Yes. For each time t there corresponds one and only one displacement y.
(b) Amplitude: 0.25; Period: 1.1
(c) $y = \tfrac{1}{4}\cos(5.7t)$
(d) 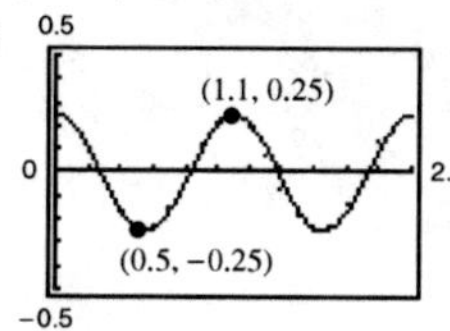   The model appears to fit the data well.

43. (a) $f^{-1}(x) = 2x + 6$

(b) 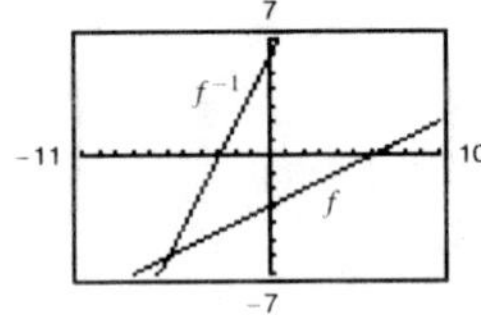

(c) Answers will vary.

45. (a) $f^{-1}(x) = x^2 - 1, \quad x \geq 0$

(b) 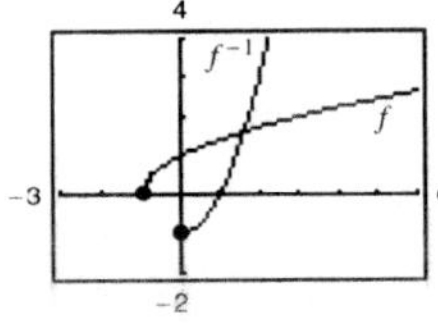

(c) Answers will vary.

47. (a) $f^{-1}(x) = x^3 - 1$

(b) 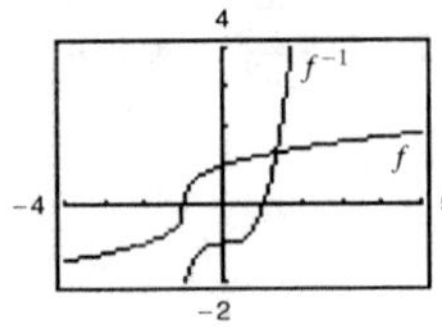

(c) Answers will vary.

49. 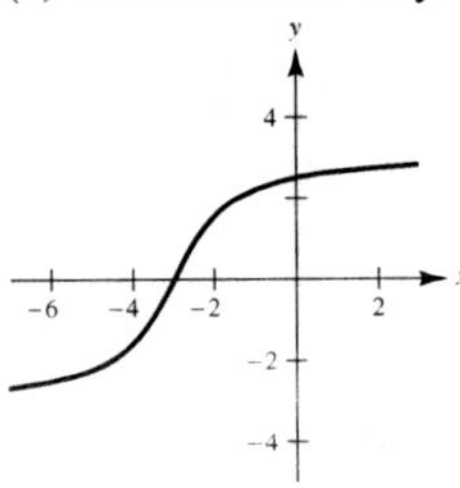   **51.** $\frac{1}{2}$ **53.** 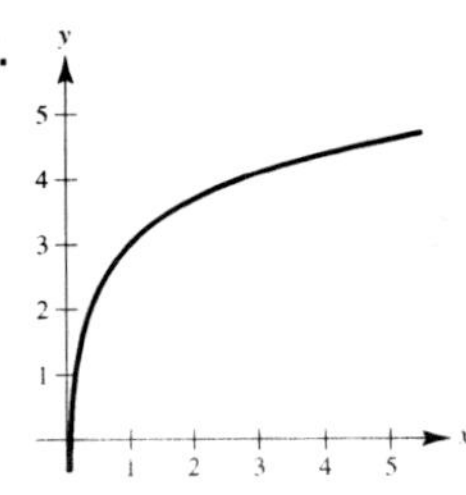

55. $\frac{1}{5}[\ln(2x + 1) + \ln(2x - 1) - \ln(4x^2 + 1)]$

57. $\ln\left(\dfrac{3\sqrt[3]{4 - x^2}}{x}\right)$ **59.** $x = e^4 - 1 \approx 53.598$

61. (a) $f^{-1}(x) = e^{2x}$

(b) 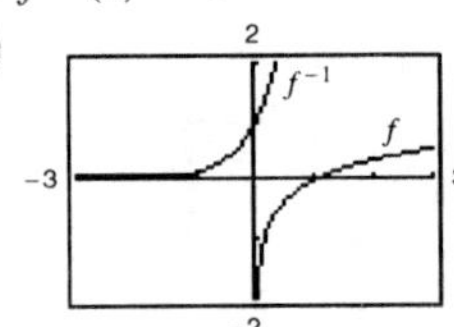   (c) Answers will vary.

63. 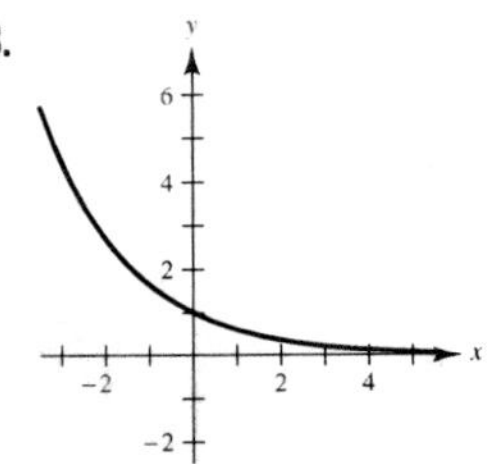

P.S. Problem Solving   (page 59)

1. (a) Center: $(3, 4)$; Radius: 5

(b) $y = -\frac{3}{4}x$

(c) $y = \frac{3}{4}x - \frac{9}{2}$

(d) $\left(3, -\frac{9}{4}\right)$

3. 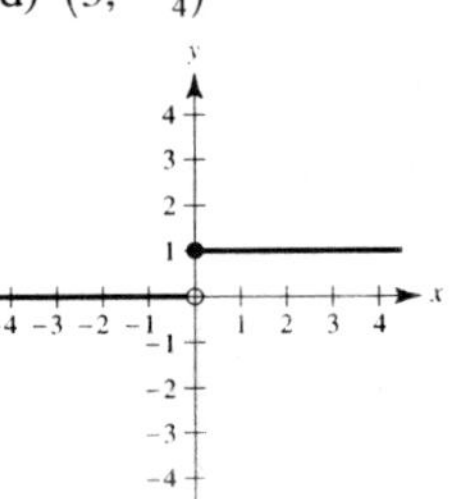

(a) 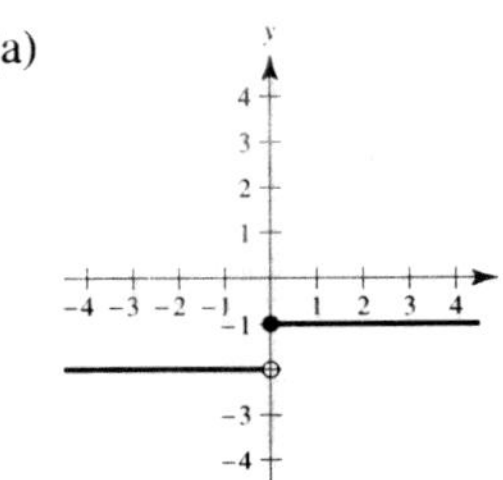   (b) 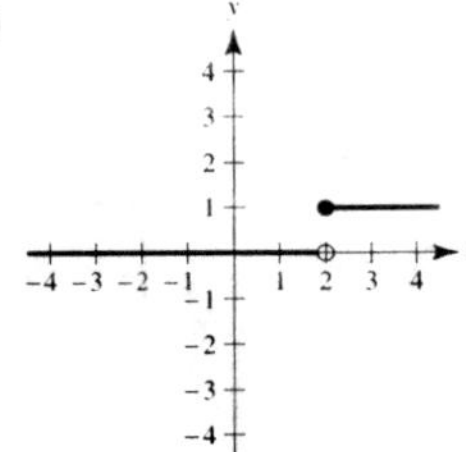

(c) 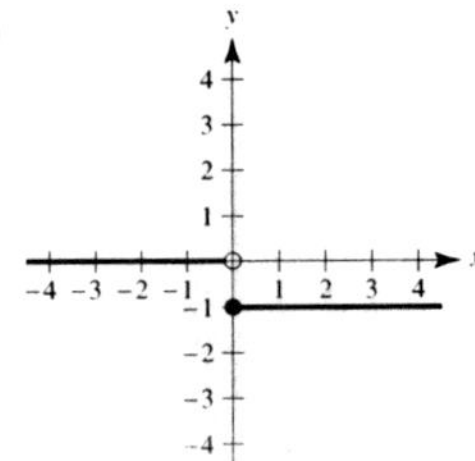   (d) 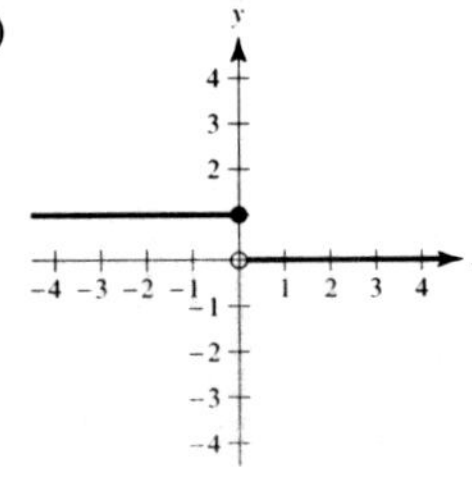

(e) 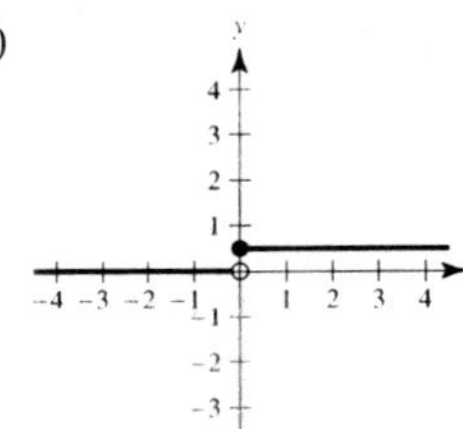   (f) 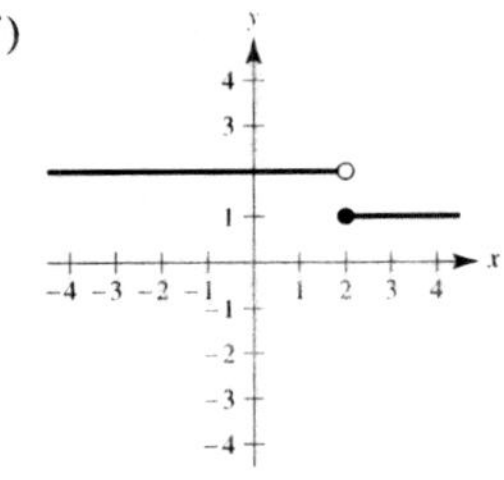

5. (a) $A(x) = x[(100 - x)/2]$; Domain: $(0, 100)$

(b) 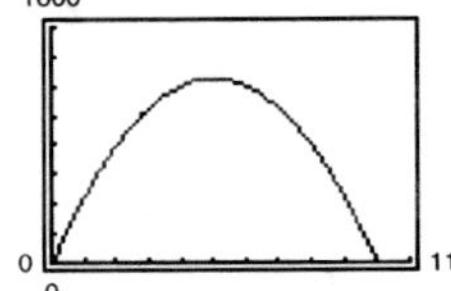

Dimensions 50 m $\times$ 25 m yield maximum area of 1250 m².

(c) 50 m $\times$ 25 m; Area = 1250 m²

7. $T(x) = \dfrac{2\sqrt{4 + x^2} + \sqrt{(3 - x)^2 + 1}}{4}$

9. (a) 5, less   (b) 3, greater   (c) 4.1, less

(d) $4 + h$ (e) 4; Answers will vary.

11. Using the definition of absolute value, you can rewrite the equation as

$$\begin{cases} 2y, & y > 0 \\ 0, & y \le 0 \end{cases} = \begin{cases} 2x, & x > 0 \\ 0, & x \le 0 \end{cases}.$$

For $x > 0$ and $y > 0$, you have $2y = 2x \to y = x$. For any $x \le 0$, y is any $y \le 0$. So, the graph of $y + |y| = x + |x|$ is as follows.

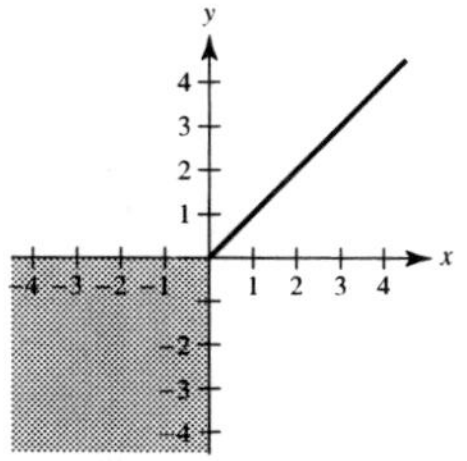

13. (a) $\left(x + \dfrac{4}{k-1} \right)^2 + y^2 = \dfrac{16k}{(k-1)^2}$

(b)

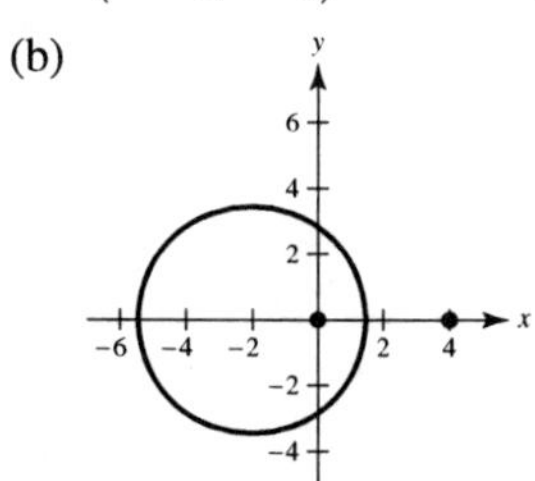

(c) As k becomes very large, $\dfrac{4}{k-1} \to 0$ and $\dfrac{16k}{(k-1)^2} \to 0$.

The center of the circle gets closer to $(0,0)$, and its radius approaches 0.

15. (a) Domain: $(-\infty, 1) \cup (1, \infty)$; Range: $(-\infty, 0) \cup (0, \infty)$

(b) $f(f(x)) = \dfrac{x-1}{x}$

Domain: $(-\infty, 0) \cup (0, 1) \cup (1, \infty)$

(c) $f(f(f(x))) = x$

Domain: $(-\infty, 0) \cup (0, 1) \cup (1, \infty)$

(d)

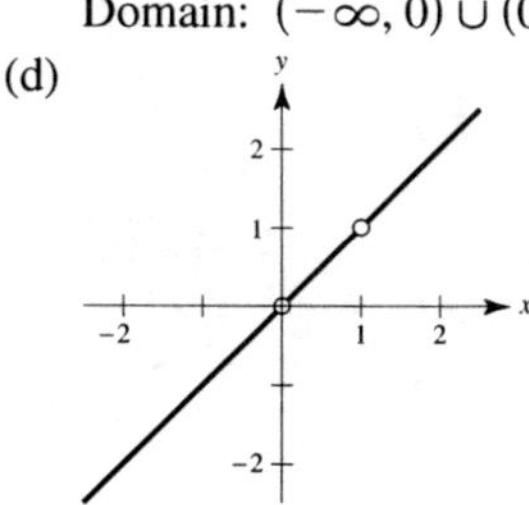

The graph is not a line because there are holes at $x = 0$ and $x = 1$.

Chapter 2

Section 2.1   (page 67)

1. Precalculus: 300 ft

3. Calculus: Slope of the tangent line at $x = 2$ is 0.16.

5. (a) Precalculus: 10 square units   (b) Calculus: 5 square units

7. (a)

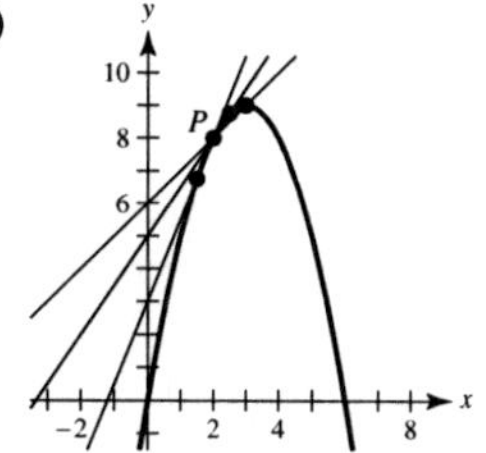

(b) $1; \frac{3}{2}; \frac{5}{2}$

(c) 2; Use points closer to P.

9. (a) Area ≈ 10.417; Area ≈ 9.145 (b) Use more rectangles.

11. (a) 5.66   (b) 6.11   (c) Increase the number of line segments.

Section 2.2   (page 74)

1.

x	3.9	3.99	3.999	4.001	4.01	4.1
$f(x)$	0.2041	0.2004	0.2000	0.2000	0.1996	0.1961

$$\lim_{x \to 4} \frac{x-4}{x^2 - 3x - 4} \approx 0.2000$$

3.

x	2.9	2.99	2.999
$f(x)$	-0.0641	-0.0627	-0.0625

x	3.001	3.01	3.1
$f(x)$	-0.0625	-0.0623	-0.0610

$$\lim_{x \to 3} \frac{[1/(x+1)] - (1/4)}{x-3} \approx -0.0625$$

5.

x	-0.1	-0.01	-0.001
$f(x)$	0.9983	0.99998	1.0000

x	0.001	0.01	0.1
$f(x)$	1.0000	0.99998	0.9983

$$\lim_{x \to 0} \frac{\sin x}{x} \approx 1.0000$$

7.

x	-0.1	-0.01	-0.001	0.001	0.01	0.1
$f(x)$	0.9516	0.9950	0.9995	1.0005	1.0050	1.0517

$$\lim_{x \to 0} \frac{e^x - 1}{x} \approx 1$$

9.

x	-0.1	-0.01	-0.001	0.001	0.01	0.1
$f(x)$	1.0536	1.0050	1.0005	0.9995	0.9950	0.9531

$$\lim_{x \to 0} \frac{\ln(x+1)}{x} \approx 1$$

11.

x	0.9	0.99	0.999	1.001	1.01	1.1
$f(x)$	0.2564	0.2506	0.2501	0.2499	0.2494	0.2439

$$\lim_{x\to 1}\frac{x-2}{x^2+x-6}\approx 0.2500$$

13.

x	0.9	0.99	0.999	1.001	1.01	1.1
$f(x)$	0.7340	0.6733	0.6673	0.6660	0.6600	0.6015

$$\lim_{x\to 1}\frac{x^4-1}{x^6-1}\approx 0.6666$$

15.

x	-0.1	-0.01	-0.001	0.001	0.01	0.1
$f(x)$	1.9867	1.9999	2.0000	2.0000	1.9999	1.9867

$$\lim_{x\to 0}\frac{\sin 2x}{x}\approx 2.0000$$

17. 1

19. Limit does not exist. The function approaches 1 from the right side of 2 but it approaches -1 from the left side of 2.

21. 0   **23.** 0

25. Limit does not exist. The function oscillates between 1 and -1 as x approaches 0.

27. (a) 2

(b) Limit does not exist. The function approaches 1 from the right side of 1 but it approaches 3.5 from the left side of 1.

(c) Value does not exist. The function is undefined at $x=4$.

(d) 2

29. $\lim_{x\to c} f(x)$ exists for all points on the graph except where $c=-3$.

31.

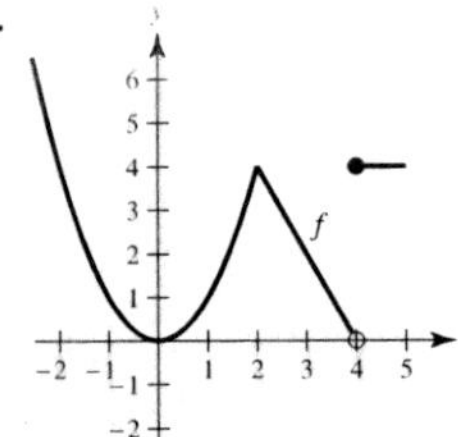

$\lim_{x\to c} f(x)$ exists for all points on the graph except where $c=4$.

33. Answers will vary.

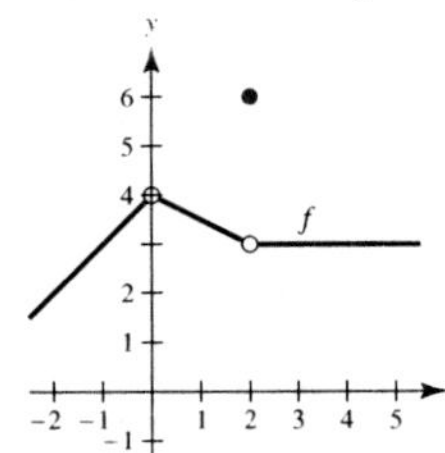

35. (a)

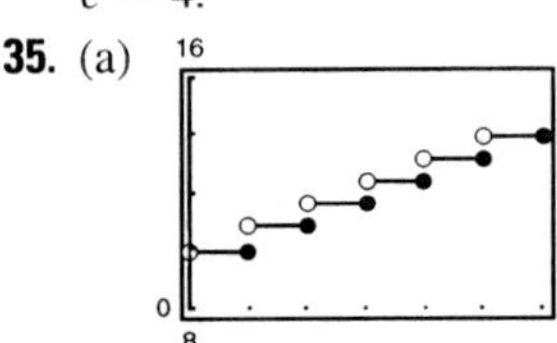

(b)

t	3	3.3	3.4	3.5	3.6	3.7	4
C	11.57	12.36	12.36	12.36	12.36	12.36	12.36

$$\lim_{t\to 3.5} C(t)=12.36$$

(c)

t	2	2.5	2.9	3	3.1	3.5	4
C	10.78	11.57	11.57	11.57	12.36	12.36	12.36

The limit does not exist because the limits from the right and left are not equal.

37. $\delta=0.4$ **39.** $\delta=\frac{1}{11}\approx 0.091$

41. $L=8$. Let $\delta=0.01/3\approx 0.0033$.

43. $L=1$. Let $\delta=0.01/5=0.002$.

45. 5; Proof   **47.** -3; Proof   **49.** 3; Proof   **51.** 0; Proof

53. 4; Proof   **55.** 2; Proof   **57.** 4

59.

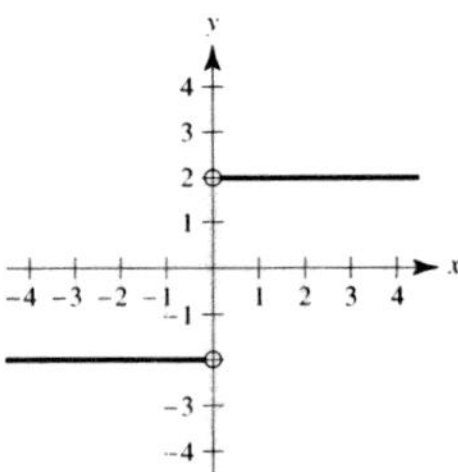

$$\lim_{x\to 4} f(x)=\tfrac{1}{6}$$

Domain: $[-5,4)\cup(4,\infty)$

The graph has a hole at $x=4$.

Answers will vary.

61.

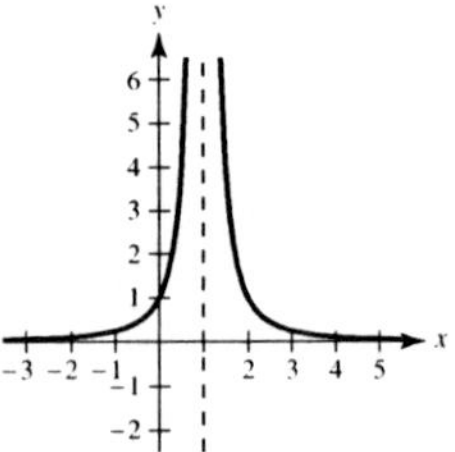

$$\lim_{x\to 9} f(x)=6$$

Domain: $[0,9)\cup(9,\infty)$

The graph has a hole at $x=9$.

Answers will vary.

63. Answers will vary. Sample answer: As x approaches 8 from either side, $f(x)$ becomes arbitrarily close to 25.

65. (i) The values of f approach different numbers as x approaches c from different sides of c.

(ii) The values of f increase or decrease without bound as x approaches c.

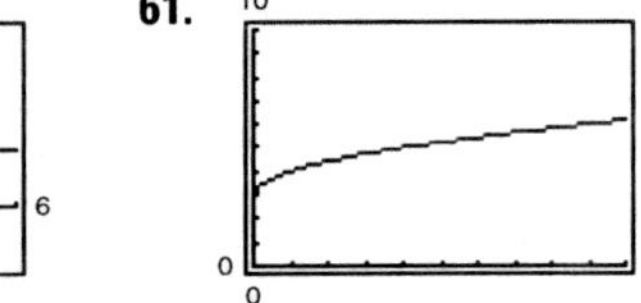

(iii) The values of f oscillate between two fixed numbers as x approaches c.

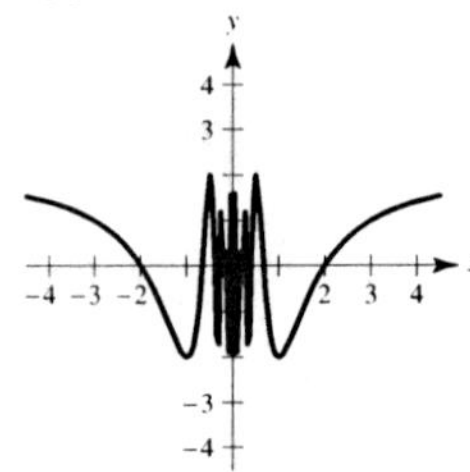

67. (a) $r = \dfrac{3}{\pi} \approx 0.9549$ cm

    (b) $\dfrac{5.5}{2\pi} \le r \le \dfrac{6.5}{2\pi}$, or approximately $0.8754 < r < 1.0345$

    (c) $\lim\limits_{r \to 3/\pi} 2\pi r = 6$; $\varepsilon = 0.5$; $\delta \approx 0.0796$

69. $\lim\limits_{x \to 0} f(x) \approx 2.7183$

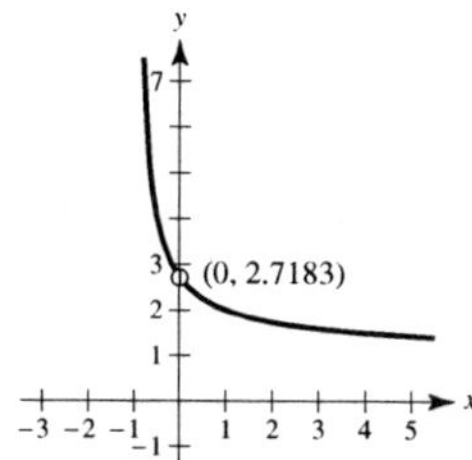

71.

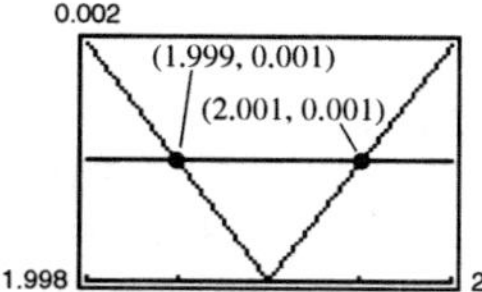

$\delta = 0.001$

$(1.999, 2.001)$

73. False. The existence or nonexistence of $f(x)$ at $x = c$ has no bearing on the existence of the limit of $f(x)$ as $x \to c$.

75. False. See Exercise 29.

77. Yes. As x approaches 0.25 from either side, $\sqrt{x}$ becomes arbitrarily close to 0.5.

79. $\lim\limits_{x \to 0} \dfrac{\sin nx}{x} = n$

81. Proof   **83.** Proof   **85.** Answers will vary.

87. Putnam Problem B1, 1986

Section 2.3   (page 87)

1. 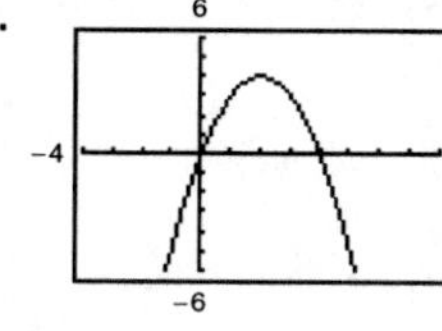   **3.** 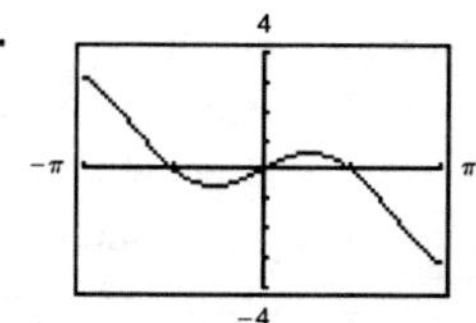

    (a) 0 (b) -5 (a) 0 (b) ≈ 0.52 or $\pi/6$

5. 8   **7.** -1 **9.** 7   **11.** 2   **13.** 1/2   **15.** 1/5

17. 7   **19.** 1   **21.** 1/2   **23.** 1   **25.** 1/2   **27.** -1

29. 1   **31.** $\ln 3 + e$ **33.** (a) 4 (b) 64 (c) 64

35. (a) 3 (b) 2 (c) 2   **37.** (a) 10 (b) 5 (c) 6 (d) 3/2

39. (a) 64 (b) 2 (c) 12 (d) 8

41. (a) -1 (b) -2

    $g(x) = \dfrac{x^2 - x}{x}$ and $f(x) = x - 1$ agree except at $x = 0$.

43. (a) 2 (b) 0

    $g(x) = \dfrac{x^3 - x}{x - 1}$ and $f(x) = x^2 + x$ agree except at $x = 1$.

45. -2

    $f(x) = \dfrac{x^2 - 1}{x + 1}$ and $g(x) = x - 1$ agree except at $x = -1$.

47. 12

    $f(x) = \dfrac{x^3 - 8}{x - 2}$ and $g(x) = x^2 + 2x + 4$ agree except at $x = 2$.

49. $-\dfrac{\ln 2}{8} \approx -0.0866$

    $f(x) = \dfrac{(x + 4)\ln(x + 6)}{x^2 - 16}$ and $g(x) = \dfrac{\ln(x + 6)}{x - 4}$ agree except at $x = -4$.

51. -1 **53.** 1/8   **55.** 5/6   **57.** 1/6   **59.** $\sqrt{5}/10$

61. $-1/9$ **63.** 2   **65.** $2x - 2$ **67.** 1/5   **69.** 0

71. 0   **73.** 0   **75.** 1   **77.** 1   **79.** 3/2

81.

≈ 0.354; $\dfrac{\sqrt{2}}{4}$

83.

≈ -0.250; $-\dfrac{1}{4}$

85.

≈ 3; 3

87.

≈ 0; 0

89.

≈ 1; 1

91. 3   **93.** $-1/(x + 3)^2$ **95.** 4

97.   **99.**

101.

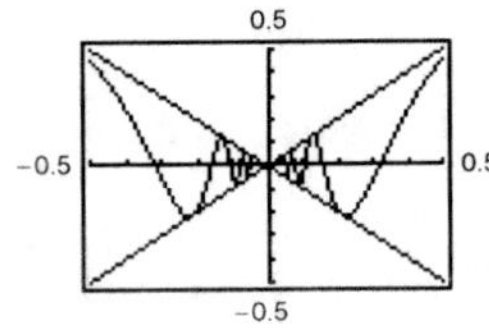

0

103. f and g agree at all but one point if c is a real number such that $f(x) = g(x)$ for all $x \neq c$.

105. An indeterminate form is obtained when the evaluation of a limit using direct substitution produces a meaningless fractional form, such as $\frac{0}{0}$.

107.

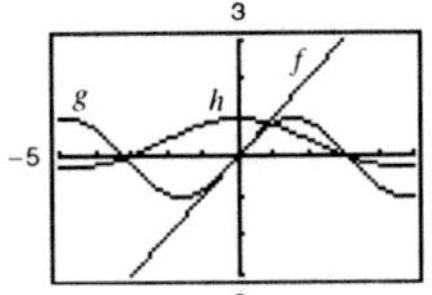

The magnitudes of $f(x)$ and $g(x)$ are approximately equal when x is "close to" 0. Therefore, their ratio is approximately 1.

109. -64 ft/sec (speed $= 64$ ft/sec)   **111.** -29.4 m/sec

113. Let $f(x) = 1/x$ and $g(x) = -1/x$.
$\lim\limits_{x \to 0} f(x)$ and $\lim\limits_{x \to 0} g(x)$ do not exist. However,
$$\lim_{x \to 0} \left[f(x) + g(x) \right] = \lim_{x \to 0} \left[\frac{1}{x} + \left(-\frac{1}{x} \right) \right] = \lim_{x \to 0} 0 = 0$$
and therefore does exist.

115. Proof   **117.** Proof   **119.** Proof

121. Answers will vary. Sample answer:
$$\text{Let } f(x) = \begin{cases} 4, & \text{if } x \geq 0 \\ -4, & \text{if } x < 0 \end{cases}$$

123. False. The limit does not exist because the function approaches 1 from the right side of 0 and approaches -1 from the left side of 0.

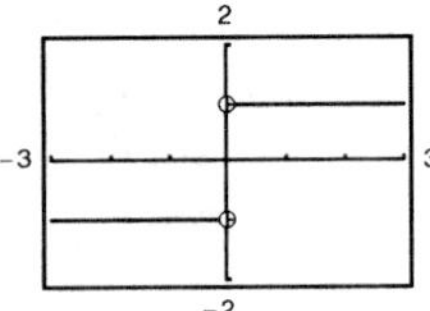

125. True. Theorem 2.7

127. False. The limit does not exist because $f(x)$ approaches 3 from the left side of 2 and approaches 0 from the right side of 2.

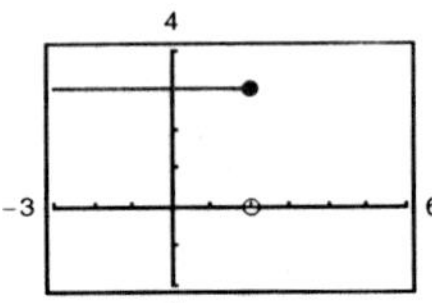

129. Proof

131. (a) All $x \neq 0, \dfrac{\pi}{2} + n\pi$

(b)

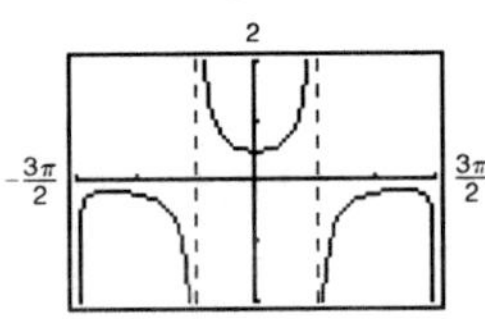

The domain is not obvious. The hole at $x = 0$ is not apparent from the graph.

(c) $\dfrac{1}{2}$ (d) $\dfrac{1}{2}$

133. The graphing utility was not set in *radian* mode.

Section 2.4   (page 98)

1. (a) 3   (b) 3   (c) 3; $f(x)$ is continuous on $(-\infty, \infty)$.

3. (a) 0   (b) 0   (c) 0; Discontinuity at $x = 3$

5. (a) -3 (b) 3   (c) Limit does not exist.
Discontinuity at $x = 2$

7. $\frac{1}{16}$ **9.** $\frac{1}{10}$

11. Limit does not exist. The function decreases without bound as x approaches -3 from the left.

13. -1 **15.** $-1/x^2$ **17.** 5/2   **19.** 2

21. Limit does not exist. The function decreases without bound as x approaches π from the left and increases without bound as x approaches π from the right.

23. 8

25. Limit does not exist. The function approaches 5 from the left side of 3 but approaches 6 from the right side of 3.

27. Limit does not exist. The function decreases without bound as x approaches 3 from the right.

29. ln 4   **31.** Discontinuous at $x = -2$ and $x = 2$

33. Discontinuous at every integer

35. Continuous on $[-7, 7]$ **37.** Continuous on $[-1, 4]$

39. Nonremovable discontinuity at $x = 0$

41. Continuous for all real x

43. Nonremovable discontinuities at $x = -2$ and $x = 2$

45. Continuous for all real x

47. Nonremovable discontinuity at $x = 1$
Removable discontinuity at $x = 0$

49. Continuous for all real x

51. Removable discontinuity at $x = -2$
Nonremovable discontinuity at $x = 5$

53. Nonremovable discontinuity at $x = -2$

55. Continuous for all real x

57. Nonremovable discontinuity at $x = 2$

59. Continuous for all real x

61. Nonremovable discontinuity at $x = 0$

63. Nonremovable discontinuities at integer multiples of $\pi/2$

65. Nonremovable discontinuity at each integer

67.

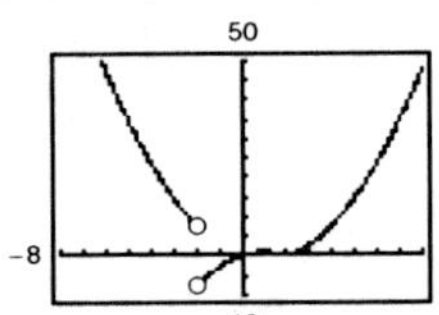

$\lim\limits_{x \to 0^+} f(x) = 0$
$\lim\limits_{x \to 0^-} f(x) = 0$
Discontinuity at $x = -2$

69. $a = 7$ **71.** $a = 2$ **73.** $a = -1, b = 1$ **75.** $a = -1$

77. Continuous for all real x

79. Nonremovable discontinuities at $x = 1$ and $x = -1$

81.

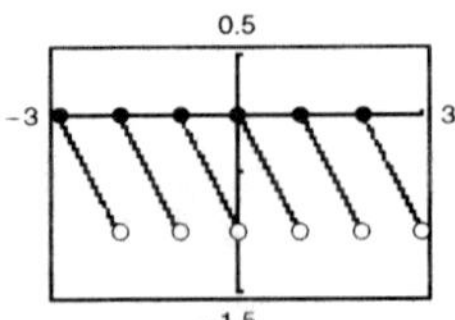

Nonremovable discontinuity at each integer

83.

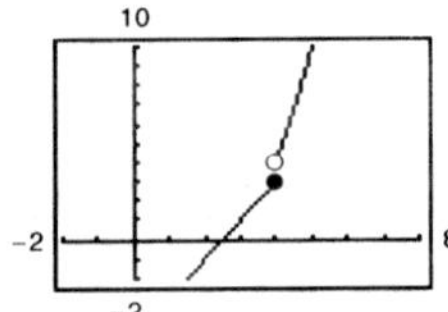

Nonremovable discontinuity at $x = 4$

85. Continuous on $(-\infty, \infty)$

87. Continuous on . . . , $(-6, -2), (-2, 2), (2, 6), \ldots$

89. 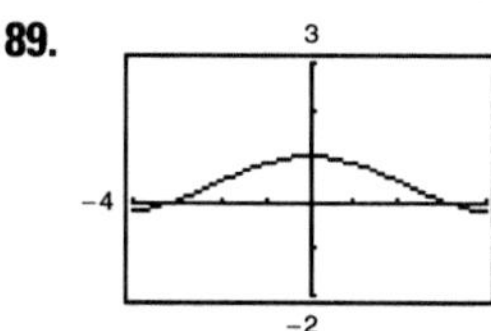
The graph has a hole at $x = 0$. The graph appears to be continuous, but the function is not continuous on $[-4, 4]$. It is not obvious from the graph that the function has a discontinuity at $x = 0$.

91. 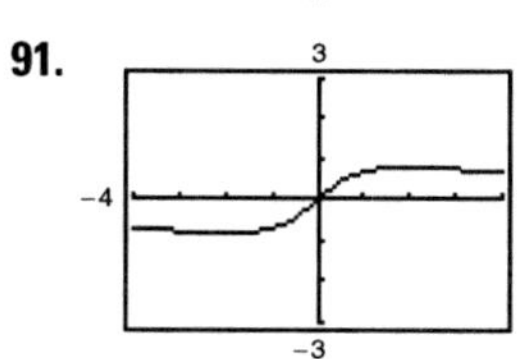
The graph has a hole at $x = 0$. The graph appears to be continuous, but the function is not continuous on $[-4, 4]$. It is not obvious from the graph that the function has a discontinuity at $x = 0$.

93. Because $f(x)$ is continuous on the interval $[1, 2]$, and $f(1) = 37/12$ and $f(2) = -8/3$, by the Intermediate Value Theorem there exists a real number c in $[1, 2]$ such that $f(c) = 0$.

95. Because $h(x)$ is continuous on the interval $[0, \pi/2]$, and $h(0) = -2$ and $h(\pi/2) \approx 0.9119$, by the Intermediate Value Theorem there exists a real number c in $[0, \pi/2]$ such that $f(c) = 0$.

97. 0.68, 0.6823   **99.** 0.56, 0.5636   **101.** 0.79, 0.7921

103. $f(3) = 11$ **105.** $f(2) = 4$

107. (a) The limit does not exist at $x = c$.
(b) The function is not defined at $x = c$.
(c) The limit exists, but it is not equal to the value of the function at $x = c$.
(d) The limit does not exist at $x = c$.

109. If f and g are continuous for all real x, then so is $f + g$ (Theorem 2.11, part 2). However, f/g might not be continuous if $g(x) = 0$. For example, let $f(x) = x$ and $g(x) = x^2 - 1$. Then f and g are continuous for all real x, but f/g is not continuous at $x = \pm 1$.

111. True

113. False. A rational function can be written as $P(x)/Q(x)$, where P and Q are polynomials of degree m and n, respectively. It can have at most n discontinuities.

115. $\lim\limits_{t \to 4^-} f(t) \approx 28$; $\lim\limits_{t \to 4^+} f(t) \approx 56$

At the end of day 3, the amount of chlorine in the pool is about 28 oz. At the beginning of day 4, the amount of chlorine in the pool is about 56 oz.

117. $C = \begin{cases} 0.40, & 0 < t \le 10 \\ 0.40 + 0.05[\![t - 9]\!], & t > 10, t \text{ is not an integer} \\ 0.40 + 0.05(t - 10), & t > 10, t \text{ is an integer} \end{cases}$

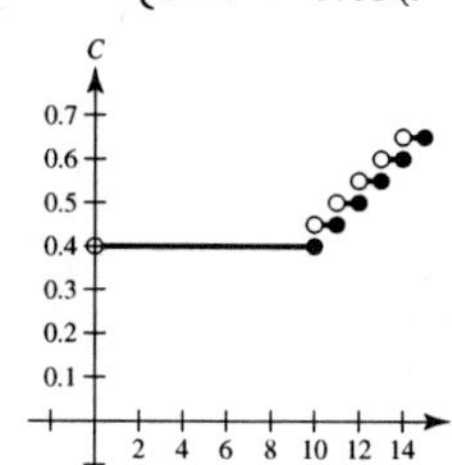
There is a nonremovable discontinuity at each integer greater than or equal to 10.

119. Proof   **121.** Proof   **123.** Proof

125. (a) 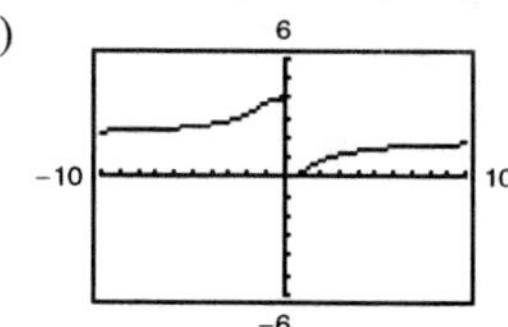

(b) Yes, there appears to be a limiting speed, and a possible cause is air resistance.

127. $c = \left(-1 \pm \sqrt{5}\right)/2$

129. Domain: $[-c^2, 0) \cup (0, \infty)$; Let $f(0) = 1/(2c)$.

131. $h(x)$ has a nonremovable discontinuity at every integer except 0.

133. (a) Domain: $(-\infty, 0) \cup (0, \infty)$

(b)

(c) $\lim\limits_{x \to 0^-} f(x) = 4$; $\lim\limits_{x \to 0^+} f(x) = 0$

(d) Answers will vary.

135. Putnam Problem A2, 1971

Section 2.5   (page 108)

1. $\lim\limits_{x \to 4^-} \dfrac{1}{x - 4} = -\infty$, $\lim\limits_{x \to 4^+} \dfrac{1}{x - 4} = \infty$

3. $\lim\limits_{x \to 4^-} \dfrac{1}{(x - 4)^2} = \infty$, $\lim\limits_{x \to 4^+} \dfrac{1}{(x - 4)^2} = \infty$

5. $\lim\limits_{x \to -2^-} 2\left|\dfrac{x}{x^2 - 4}\right| = \infty$, $\lim\limits_{x \to -2^+} 2\left|\dfrac{x}{x^2 - 4}\right| = \infty$

7. $\lim\limits_{x \to -2^-} \tan(\pi x/4) = \infty$, $\lim\limits_{x \to -2^+} \tan(\pi x/4) = -\infty$

9.

x	-3.5	-3.1	-3.01	-3.001
$f(x)$	0.31	1.64	16.6	167

x	-2.999	-2.99	-2.9	-2.5
$f(x)$	-167	-16.7	-1.69	-0.36

$\lim\limits_{x \to -3^-} f(x) = \infty$, $\lim\limits_{x \to -3^+} f(x) = -\infty$

11.

x	-3.5	-3.1	-3.01	-3.001
$f(x)$	3.8	16	151	1501

x	-2.999	-2.99	-2.9	-2.5
$f(x)$	-1499	-149	-14	-2.3

$\lim\limits_{x \to -3^-} f(x) = \infty$, $\lim\limits_{x \to -3^+} f(x) = -\infty$

13. $x = 0$ **15.** $x = \pm 2$ **17.** No vertical asymptote

19. $x = 2, x = -1$ **21.** $t = 0$ **23.** $x = -2, x = 1$

25. No vertical asymptote   **27.** $x = 1$ **29.** $t = -2$

31. $x = 0$

33. $x = \frac{1}{2} + n$, n is an integer.   **35.** $t = n\pi$, n is a nonzero integer.

37. Removable discontinuity at $x = -1$

39. Vertical asymptote at $x = -1$

41. Removable discontinuity at $x = -1$

43. ∞ **45.** ∞ **47.** ∞ **49.** $-\frac{1}{5}$ **51.** $\frac{1}{2}$ **53.** $-\infty$

55. ∞ **57.** $-\infty$ **59.** $-\infty$ **61.** Limit does not exist.

63. 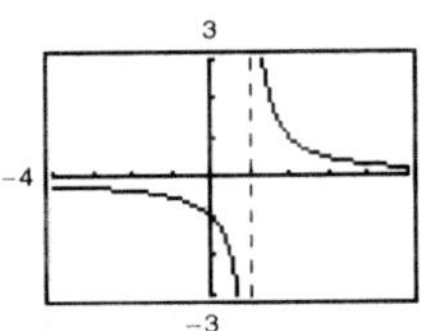   **65.** 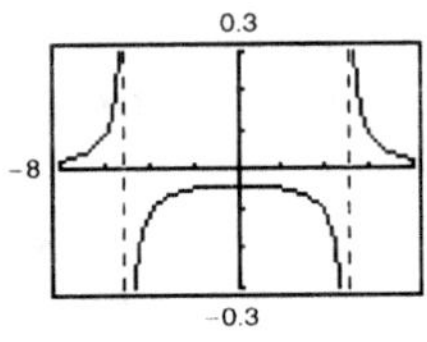

$\lim\limits_{x \to 1^+} f(x) = \infty$ $\lim\limits_{x \to 5} f(x) = -\infty$

67. Answers will vary.

69. Answers will vary. Sample answer: $f(x) = \dfrac{x - 3}{x^2 - 4x - 12}$

71. 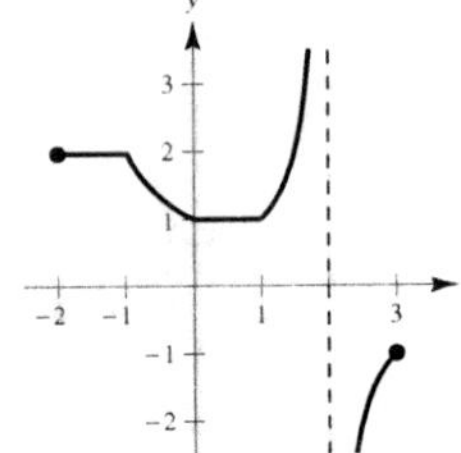   **73.** ∞

75. (a) $r = \frac{1}{3}(200\pi)$ ft/sec   (b) $r = 200\pi$ ft/sec

(c) $\lim\limits_{\theta \to (\pi/2)^-} [50\pi \sec^2 \theta] = \infty$

77. (a) Answers will vary. Domain: $x > 25$

(b)

x	30	40	50	60
y	150	66.667	50	42.857

Answers will vary.

(c) $\lim\limits_{x \to 25^+} \dfrac{25x}{x - 25} = \infty$

As x gets close to 25 mi/h, y becomes larger and larger.

79. (a) $A = 50 \tan \theta - 50\theta$; Domain: $(0, \pi/2)$

(b)

θ	0.3	0.6	0.9	1.2	1.5
$f(\theta)$	0.47	4.21	18.01	68.61	630.07

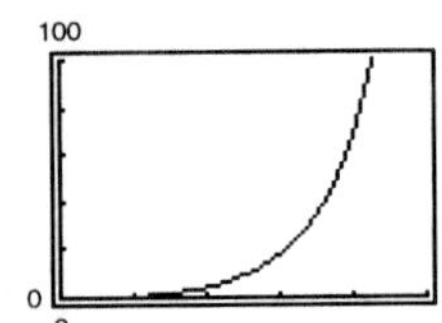   (c) $\lim\limits_{\theta \to \pi/2^-} A = \infty$

81. False. Let $f(x) = (x^2 - 1)/(x - 1)$.

83. False. Let $f(x) = \tan x$.

85. Answers will vary. Sample answer: Let $f(x) = \dfrac{1}{x^2}$, $g(x) = \dfrac{1}{x^4}$,

and $c = 0$.

87. Proof   **89.** Proof

Review Exercises for Chapter 2   (page 111)

1. Calculus   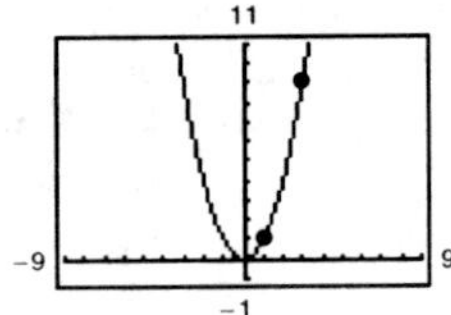   Estimate: 8.268

3.

x	-0.1	-0.01	-0.001
$f(x)$	-1.0526	-1.0050	-1.0005

x	0.001	0.01	0.1
$f(x)$	-0.9995	-0.9950	-0.9524

$\lim\limits_{x \to 0} f(x) = -1$

5.

x	-0.1	-0.01	-0.001
$f(x)$	0.8867	0.0988	0.0100

x	0.001	0.01	0.1
$f(x)$	-0.0100	-0.1013	-1.1394

$\lim\limits_{x \to 0} f(x) = 0$

7. (a) 4   (b) 5   **9.** (a) Limit does not exist.   (b) 0

11. 5; Proof   **13.** -3; Proof   **15.** 16   **17.** $\sqrt{6} \approx 2.45$

19. $-\frac{1}{4}$ **21.** $\frac{1}{2}$ **23.** -1 **25.** 75   **27.** 0   **29.** 1

31. $\sqrt{3}/2$ **33.** $-\frac{1}{2}$ **35.** $\frac{7}{12}$

37. (a)

x	1.1	1.01	1.001	1.0001
$f(x)$	0.5680	0.5764	0.5773	0.5773

$\lim\limits_{x \to 1^+} f(x) \approx 0.5773$

(b) 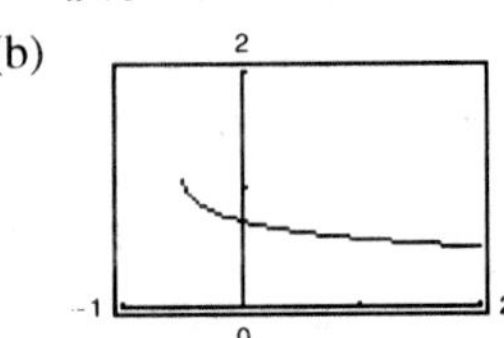 $\lim\limits_{x \to 1^+} f(x) \approx 0.5774$

(c) $\sqrt{3}/3$

39. -39.2 m/sec   **41.** -1 **43.** 0

45. Limit does not exist. The limit as t approaches 1 from the left is 2, whereas the limit as t approaches 1 from the right is 1.

47. Continuous for all real x

49. Continuous on $(k, k + 1)$ for all integers k

51. Continuous on $(-\infty, 1) \cup (1, \infty)$

53. Continuous on $(-\infty, 2) \cup (2, \infty)$

55. Continuous on $(-\infty, -1) \cup (-1, \infty)$

57. Continuous on $(2k, 2k + 2)$ for all integers k

59. Continuous on $(k, k + 1)$ for all integers k

61. $c = -\frac{1}{2}$ **63.** Proof

65. 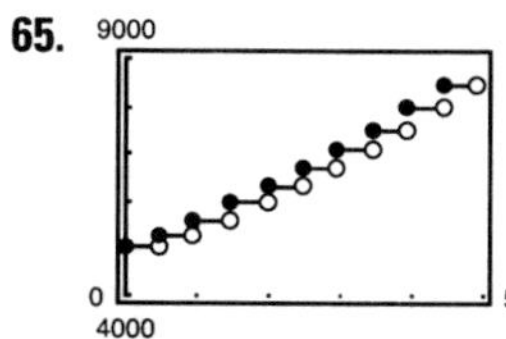

Nonremovable discontinuity every 6 months

67. $x = 0$ **69.** $x = 10$ **71.** $x = -5, x = 5$ **73.** $-\infty$

75. $\frac{1}{3}$ **77.** $-\infty$ **79.** $\frac{4}{5}$ **81.** ∞ **83.** $-\infty$

85. (a) 2

    (b) Yes, define as

$$f(x) = \begin{cases} \dfrac{\tan 2x}{x}, & x \neq 0 \\ 2, & x = 0 \end{cases}.$$

P.S. Problem Solving   (page 113)

1. (a) Perimeter $\triangle PAO = 1 + \sqrt{(x^2 - 1)^2 + x^2} + \sqrt{x^4 + x^2}$

       Perimeter $\triangle PBO = 1 + \sqrt{x^4 + (x - 1)^2} + \sqrt{x^4 + x^2}$

(b)

x	4	2	1
Perimeter $\triangle PAO$	33.0166	9.0777	3.4142
Perimeter $\triangle PBO$	33.7712	9.5952	3.4142
$r(x)$	0.9777	0.9461	1.0000

x	0.1	0.01
Perimeter $\triangle PAO$	2.0955	2.0100
Perimeter $\triangle PBO$	2.0006	2.0000
$r(x)$	1.0475	1.0050

(c) 1

3. (a) Area (hexagon) $= (3\sqrt{3})/2 \approx 2.5981$

       Area (circle) $= \pi \approx 3.1416$

       Area (circle) $-$ Area (hexagon) ≈ 0.5435

(b) $A_n = (n/2) \sin(2\pi/n)$

(c)

n	6	12	24	48	96
A_n	2.5981	3.0000	3.1058	3.1326	3.1394

(d) 3.1416 or π

5. (a) $m = -\frac{12}{5}$ (b) $y = \frac{5}{12}x - \frac{169}{12}$

    (c) $m_x = \dfrac{-\sqrt{169 - x^2} + 12}{x - 5}$

    (d) $\frac{5}{12}$; It is the same as the slope of the tangent line found in (b).

7. (a) Domain: $[-27, 1) \cup (1, \infty)$

    (b) 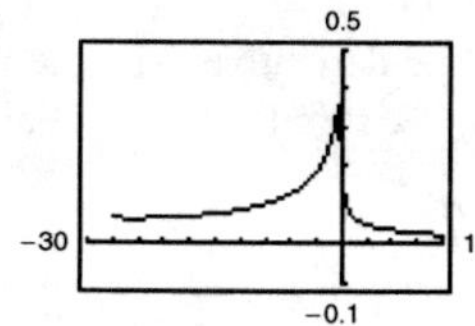   (c) $\frac{1}{14}$ (d) $\frac{1}{12}$

9. (a) g_1, g_4 (b) g_1 (c) g_1, g_3, g_4

11. 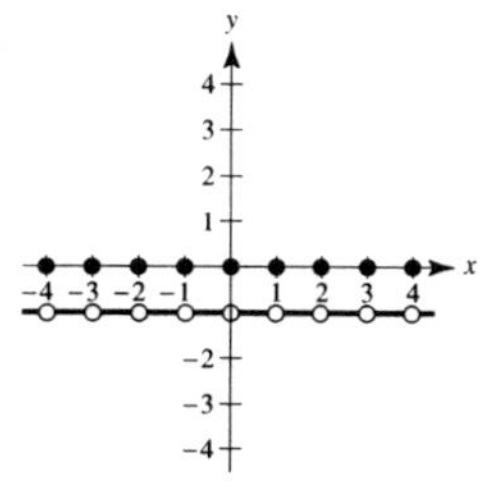

(a) $f(1) = 0$, $f(0) = 0$, $f\left(\frac{1}{2}\right) = -1$, $f(-2.7) = -1$

(b) $\lim\limits_{x \to 1^-} f(x) = -1$, $\lim\limits_{x \to 1^+} f(x) = -1$, $\lim\limits_{x \to 1/2} f(x) = -1$

(c) There is a discontinuity at each integer.

13. (a) 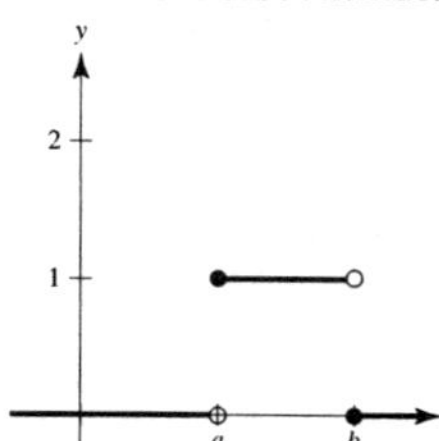

    (b) (i) $\lim\limits_{x \to a^+} P_{a,b}(x) = 1$

        (ii) $\lim\limits_{x \to a^-} P_{a,b}(x) = 0$

        (iii) $\lim\limits_{x \to b^+} P_{a,b}(x) = 0$

        (iv) $\lim\limits_{x \to b^-} P_{a,b}(x) = 1$

(c) Continuous for all positive real numbers except a and b

(d) The area under the graph of U, and above the x-axis, is 1.

Chapter 3

Section 3.1   (page 123)

1. (a) $m_1 = 0, m_2 = 5/2$ (b) $m_1 = -5/2, m_2 = 2$

3. 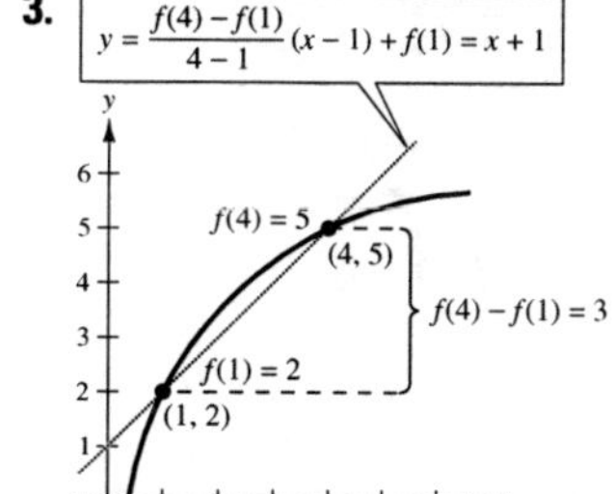

5. $m = -5$ **7.** $m = 4$

9. $m = 3$ **11.** $f'(x) = 0$ **13.** $f'(x) = -5$ **15.** $h'(s) = \frac{2}{3}$

17. $f'(x) = 2x + 1$ **19.** $f'(x) = 3x^2 - 12$

21. $f'(x) = \dfrac{-1}{(x - 1)^2}$ **23.** $f'(x) = \dfrac{1}{2\sqrt{x + 4}}$

25. (a) $y = 2x + 2$

    (b)

27. (a) $y = 12x - 16$

    (b)

29. (a) $y = \frac{1}{2}x + \frac{1}{2}$

    (b)

31. (a) $y = \frac{3}{4}x + 2$

    (b)

33. $y = 2x - 1$ **35.** $y = 3x - 2; y = 3x + 2$

37. $y = -\frac{1}{2}x + \frac{3}{2}$ **39.** b   **40.** d   **41.** a   **42.** c

43. $g(4) = 5; g'(4) = -\frac{5}{3}$

45. 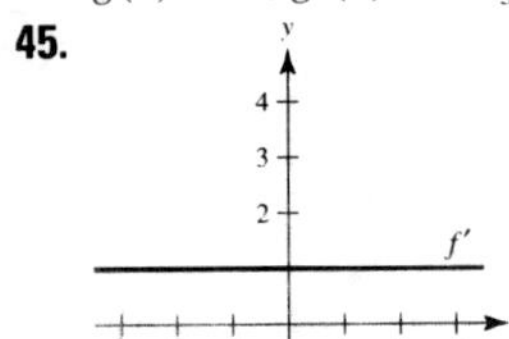   **47.** 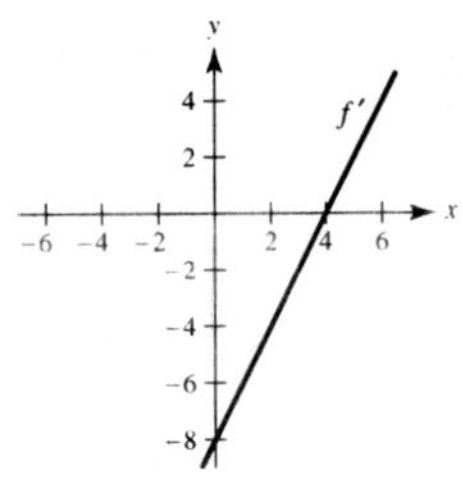

49. 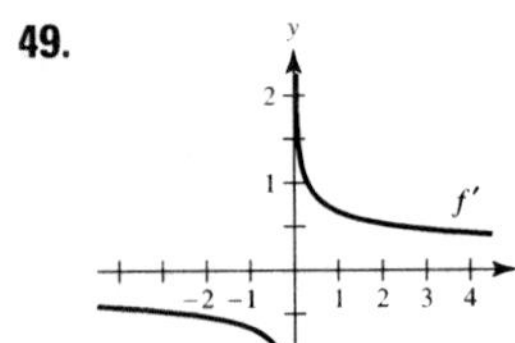   **51.** Answers will vary.
Sample answer: $y = -x$

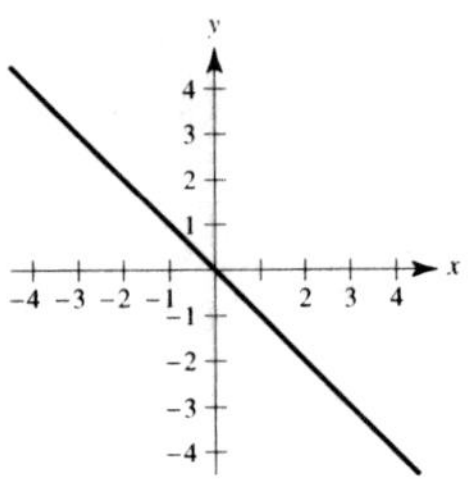

53. $f(x) = 5 - 3x$ **55.** $f(x) = -x^2$
    $c = 1$ $c = 6$

57. $f(x) = -3x + 2$ **59.** Answers will vary.
Sample answer: $f(x) = x^3$

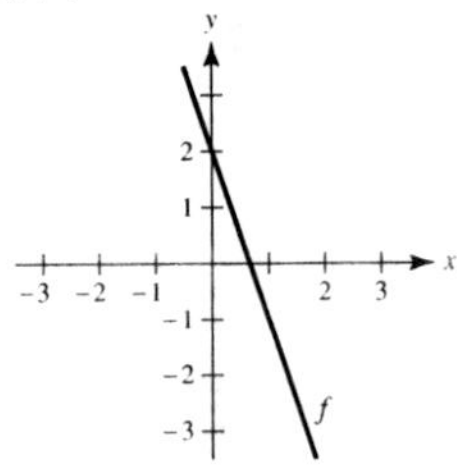   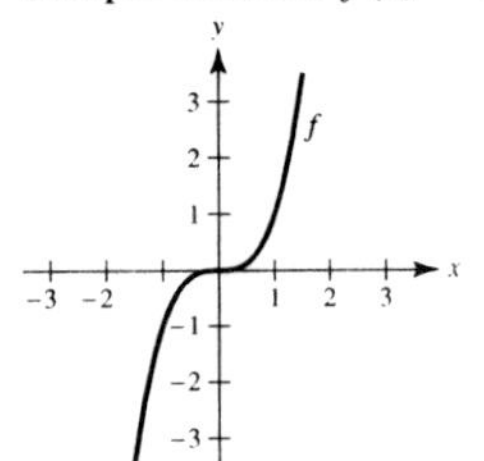

61. $y = 2x + 1; y = -2x + 9$

63. (a) 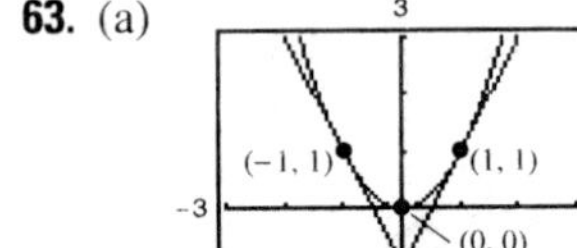

For this function, the slopes of the tangent lines are always distinct for different values of x.

(b) 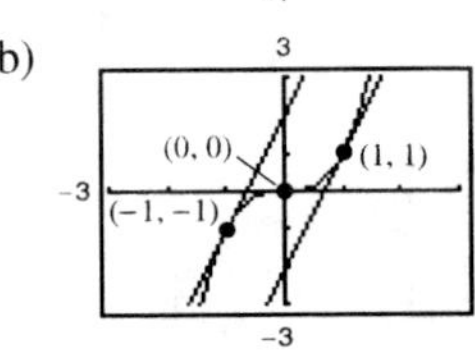

For this function, the slopes of the tangent lines are sometimes the same.

65. (a) 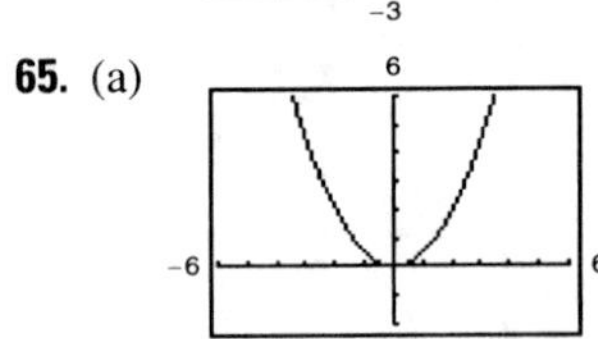

$f'(0) = 0, f'\left(\frac{1}{2}\right) = \frac{1}{2}, f'(1) = 1, f'(2) = 2$

(b) $f'\left(-\frac{1}{2}\right) = -\frac{1}{2}, f'(-1) = -1, f'(-2) = -2$

(c) 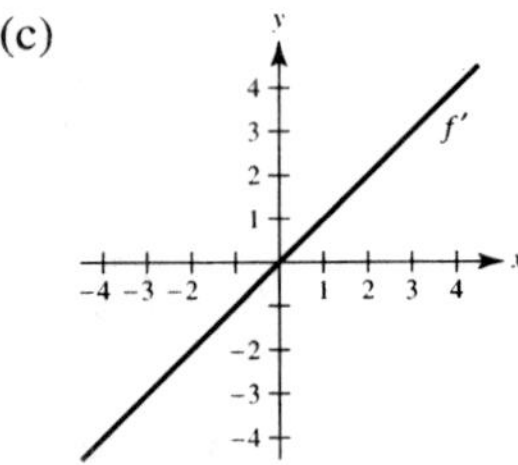

(d) $f'(x) = x$

67. 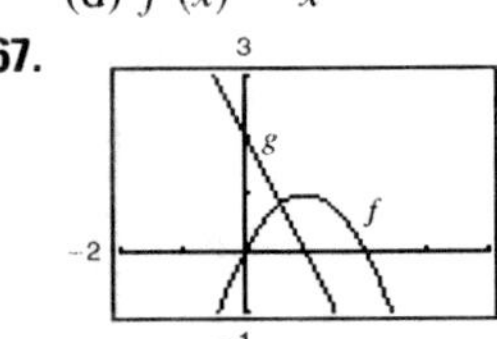 $g(x) \approx f'(x)$

69. $f(2) = 4; f(2.1) = 3.99; f'(2) \approx -0.1$

71. 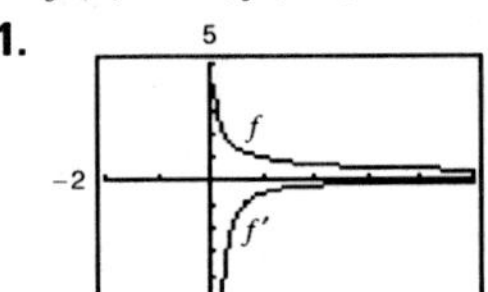   As x approaches infinity, the graph of f approaches a line of slope 0. Thus $f'(x)$ approaches 0.

73. 6   **75.** 4   **77.** $g(x)$ is not differentiable at $x = 0$.

79. $f(x)$ is not differentiable at $x = 6$.

81. $h(x)$ is not differentiable at $x = -7$.

83. $(-\infty, 3) \cup (3, \infty)$ **85.** $(-\infty, -4) \cup (-4, \infty)$

87. $(1, \infty)$

89. 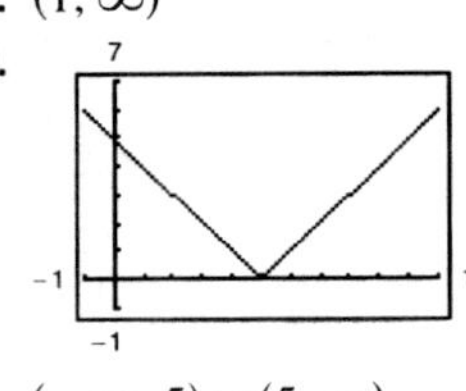   **91.** 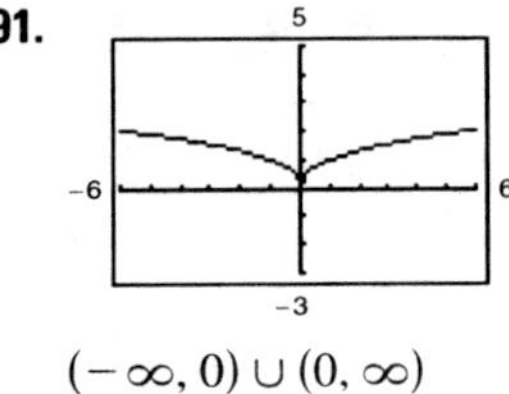

$(-\infty, 5) \cup (5, \infty)$ $(-\infty, 0) \cup (0, \infty)$

93. The derivative from the left is -1 and the derivative from the right is 1, so f is not differentiable at $x = 1$.

95. The derivatives from both the right and the left are 0, so $f'(1) = 0$.

97. f is differentiable at $x = 2$.

99. 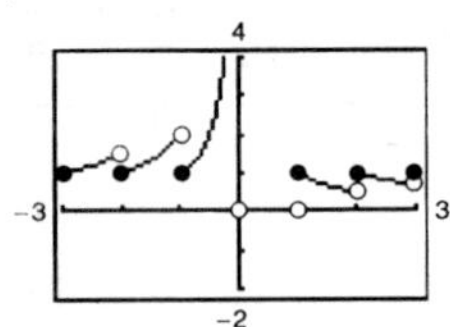

Yes, f is differentiable for all $x \neq n$, n is an integer.

101. False. The slope is $\lim\limits_{\Delta x \to 0} \dfrac{f(2 + \Delta x) - f(2)}{\Delta x}$.

103. False. For example: $f(x) = |x|$. The derivative from the left and the derivative from the right both exist but are not equal.

105. Proof

Section 3.2   (page 136)

1. (a) $\frac{1}{2}$ (b) 3   **3.** 0   **5.** $7x^6$ **7.** $-5/x^6$

9. $1/(5x^{4/5})$ **11.** 1 **13.** $-4t + 3$ **15.** $2x + 12x^2$

17. $3t^2 + 10t - 3$ **19.** $6 - 5e^x$

21. $\dfrac{\pi}{2}\cos\theta + \sin\theta$ **23.** $2x + \dfrac{1}{2}\sin x$ **25.** $\dfrac{1}{2}e^x - 3\cos x$

Function	Rewrite	Differentiate	Simplify
27. $y = \dfrac{5}{2x^2}$	$y = \dfrac{5}{2}x^{-2}$	$y' = -5x^{-3}$	$y' = -\dfrac{5}{x^3}$
29. $y = \dfrac{6}{(5x)^3}$	$y = \dfrac{6}{125}x^{-3}$	$y' = -\dfrac{18}{125}x^{-4}$	$y' = -\dfrac{18}{125x^4}$
31. $y = \dfrac{\sqrt{x}}{x}$	$y = x^{-1/2}$	$y' = -\dfrac{1}{2}x^{-3/2}$	$y' = -\dfrac{1}{2x^{3/2}}$

33. -2 **35.** 0 **37.** 3 **39.** $\frac{3}{4}$ **41.** $2t + 12/t^4$

43. $8x + 3$ **45.** $(x^3 - 8)/x^3$ **47.** $3x^2 + 1$

49. $\dfrac{1}{2\sqrt{x}} - \dfrac{2}{x^{2/3}}$ **51.** $\dfrac{4}{5s^{1/5}} - \dfrac{2}{3s^{1/3}}$

53. $\dfrac{3}{\sqrt{x}} - 5\sin x$ **55.** $\dfrac{-2}{x^3} - 2e^x$

57. (a) $5x + y + 3 = 0$ **59.** (a) $2x - y + 1 = 0$

(b) 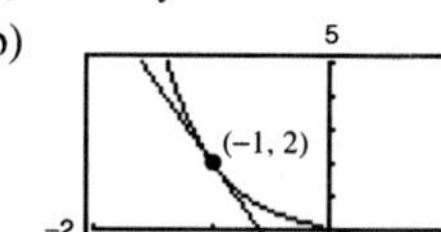   (b) 

61. $(-1, 2)$, $(0, 3)$, $(1, 2)$ **63.** No horizontal tangent lines

65. (π, π) **67.** $(\ln 4, 4 - 4\ln 4)$

69. $k = -1$, $k = -9$ **71.** $k = 3$ **73.** $k = 4/27$

75. 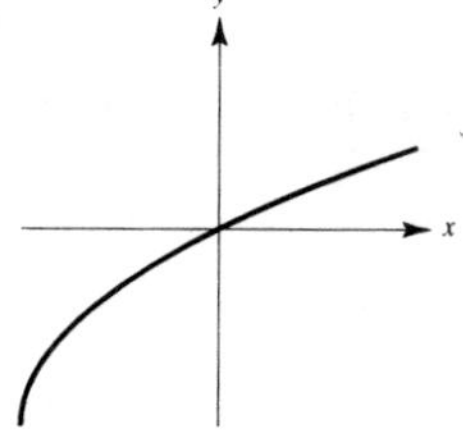

77. $g'(x) = f'(x)$

79. 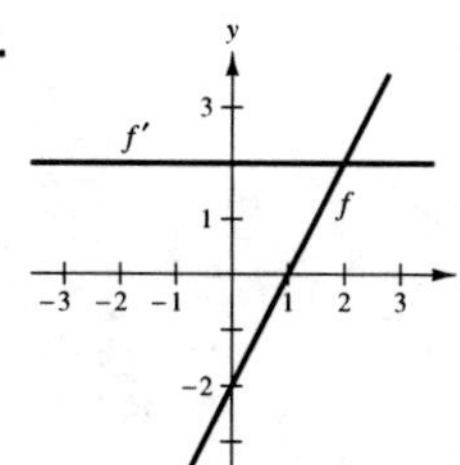

The rate of change of f is constant and therefore f' is a constant function.

81. 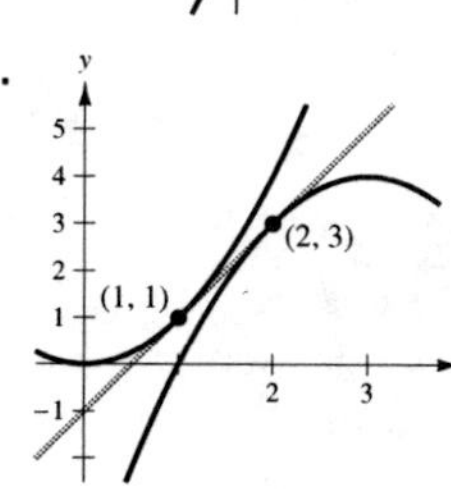

$y = 2x - 1$ $y = 4x - 4$

83. $f'(x) = 3 + \cos x \neq 0$ for all x.   **85.** $x - 4y + 4 = 0$

87. 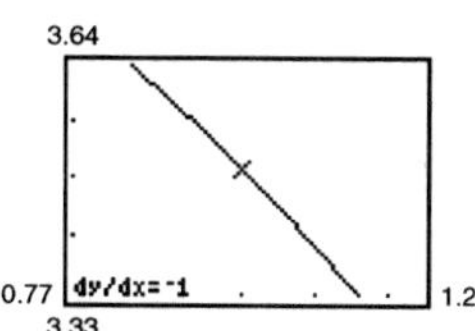

$f'(1)$ appears to be close to -1.

$f'(1) = -1$

89. (a) 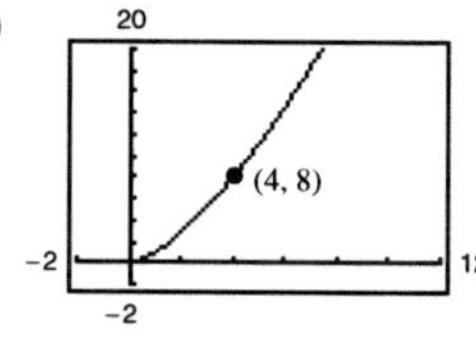

$(3.9, 7.7019)$; $S(x) = 2.981x - 3.924$

(b) $T(x) = 3(x - 4) + 8 = 3x - 4$

The slope (and equation) of the secant line approaches that of the tangent line at $(4, 8)$ as you choose points closer and closer to $(4, 8)$.

(c) 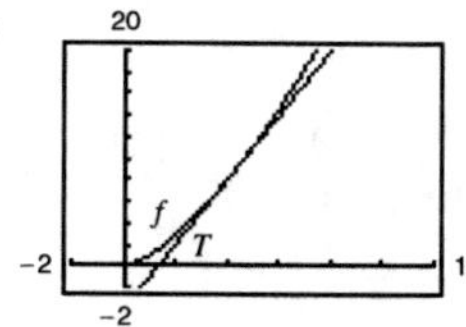

The approximation becomes less accurate.

(d)

Δx	-3	-2	-1	-0.5	-0.1	0
$f(4 + \Delta x)$	1	2.828	5.196	6.458	7.702	8
$T(4 + \Delta x)$	-1	2	5	6.5	7.7	8

Δx	0.1	0.5	1	2	3
$f(4 + \Delta x)$	8.302	9.546	11.180	14.697	18.520
$T(4 + \Delta x)$	8.3	9.5	11	14	17

91. False. Let $f(x) = x$ and $g(x) = x + 1$.

93. False. $dy/dx = 0$ **95.** True

97. Average rate: $\frac{1}{2}$

Instantaneous rates:

$f'(1) = 1$

$f'(2) = \frac{1}{4}$

99. Average rate: $e \approx 2.718$

Instantaneous rates:

$g'(0) = 1$

$g'(1) = 2 + e \approx 4.718$

101. (a) $s(t) = -16t^2 + 1362$; $v(t) = -32t$ (b) -48 ft/sec

(c) $s'(1) = -32$ ft/sec; $s'(2) = -64$ ft/sec

(d) $t = \dfrac{\sqrt{1362}}{4} \approx 9.226$ sec   (e) -295.242 ft/sec

103. $v(5) = 71$ m/sec; $v(10) = 22$ m/sec

105. 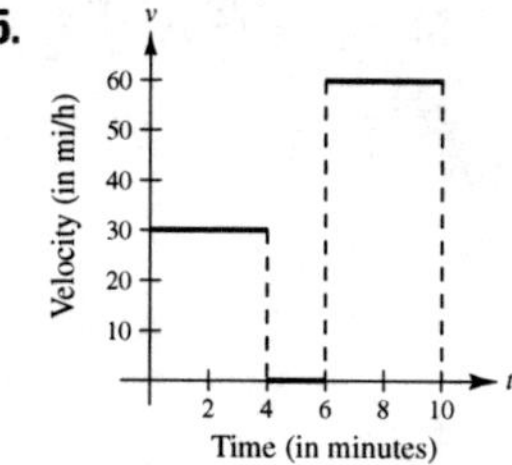

107. 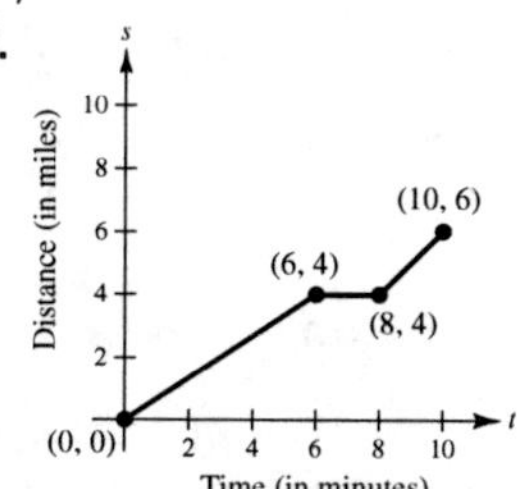

109. (a) $R(v) = 0.417v - 0.02$
   (b) $B(v) = 0.0056v^2 + 0.001v + 0.04$
   (c) $T(v) = 0.0056v^2 + 0.418v + 0.02$
   (d) 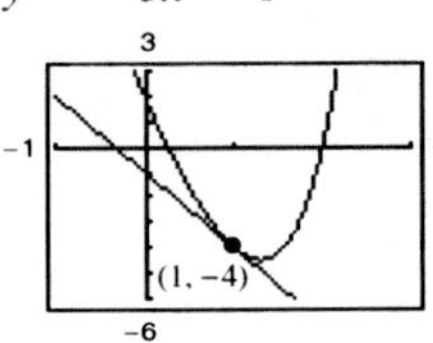
   (e) $T'(v) = 0.0112v + 0.418$
        $T'(40) = 0.866$
        $T'(80) = 1.314$
        $T'(100) = 1.538$
   (f) As speed increases, the total stopping distance increases at an increasing rate.

111. $V'(6) = 108 \text{ cm}^3/\text{cm}$ **113.** Answers will vary.

115. (a) The rate of change of the number of gallons of gasoline sold when the price is \$2.979
   (b) In general, the rate of change when $p = 2.979$ should be negative. As prices go up, sales go down.

117. $y = 2x^2 - 3x + 1$ **119.** $y = -9x,\ y = -\frac{9}{4}x - \frac{27}{4}$

121. $a = \frac{1}{3},\ b = -\frac{4}{3}$

123. $f_1(x) = |\sin x|$ is differentiable for all $x \neq n\pi$, n an integer.
   $f_2(x) = \sin|x|$ is differentiable for all $x \neq 0$.

Section 3.3 (page 147)

1. $2(2x^3 - 6x^2 + 3x - 6)$ **3.** $(1 - 5t^2)/(2\sqrt{t})$

5. $e^x(\cos x - \sin x)$ **7.** $(1 - x^2)/(x^2 + 1)^2$

9. $\dfrac{1 - 5x^3}{2\sqrt{x}(x^3 + 1)^2}$ **11.** $\dfrac{\cos x - \sin x}{e^x}$

13. $f'(x) = (x^3 + 4x)(6x + 2) + (3x^2 + 2x - 5)(3x^2 + 4)$
      $= 15x^4 + 8x^3 + 21x^2 + 16x - 20$
   $f'(0) = -20$

15. $f'(x) = \cos x - x \sin x$
   $f'\left(\dfrac{\pi}{4}\right) = \dfrac{\sqrt{2}}{8}(4 - \pi)$

17. $f'(x) = e^x(\cos x + \sin x)$
   $f'(0) = 1$

	Function	Rewrite	Differentiate	Simplify
19.	$y = \dfrac{x^2 + 3x}{7}$	$y = \dfrac{1}{7}x^2 + \dfrac{3}{7}x$	$y' = \dfrac{2}{7}x + \dfrac{3}{7}$	$y' = \dfrac{2x + 3}{7}$
21.	$y = \dfrac{6}{7x^2}$	$y = \dfrac{6}{7}x^{-2}$	$y' = -\dfrac{12}{7}x^{-3}$	$y' = -\dfrac{12}{7x^3}$
23.	$y = \dfrac{4x^{3/2}}{x}$	$y = 4x^{1/2},$ $x > 0$	$y' = 2x^{-1/2}$	$y' = \dfrac{2}{\sqrt{x}},$ $x > 0$

25. $\dfrac{3}{(x + 1)^2},\ x \neq 1$ **27.** $\dfrac{x^2 + 6x - 3}{(x + 3)^2}$ **29.** $\dfrac{3x + 1}{2x^{3/2}}$

31. $6s^2(s^3 - 2)$ **33.** $-(2x^2 - 2x + 3)/[x^2(x - 3)^2]$

35. $10x^4 - 8x^3 - 21x^2 - 10x - 30$ **37.** $-\dfrac{4xc^2}{(x^2 - c^2)^2}$

39. $t(t \cos t + 2 \sin t)$ **41.** $-(t \sin t + \cos t)/t^2$

43. $-e^x + \sec^2 x$ **45.** $\dfrac{1}{4t^{3/4}} - 6 \csc t \cot t$

47. $\frac{3}{2}\sec x(\tan x - \sec x)$ **49.** $\cos x \cot^2 x$

51. $x(x \sec^2 x + 2 \tan x)$ **53.** $2x \cos x + 2 \sin x + x^2 e^x + 2xe^x$

55. $\dfrac{e^x}{(8x^{3/2})(2x - 1)}$ **57.** $\dfrac{2x^2 + 8x - 1}{(x + 2)^2}$

59. $\dfrac{1 - \sin\theta + \theta \cos\theta}{(1 - \sin\theta)^2}$ **61.** $-4\sqrt{3}$ **63.** $\dfrac{1}{\pi^2}$

65. (a) $y = -3x - 1$
   (b) 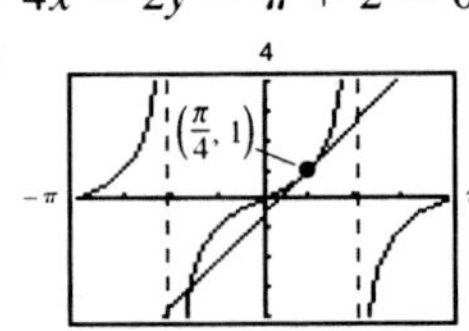

67. (a) $4x - 2y - \pi + 2 = 0$
   (b) 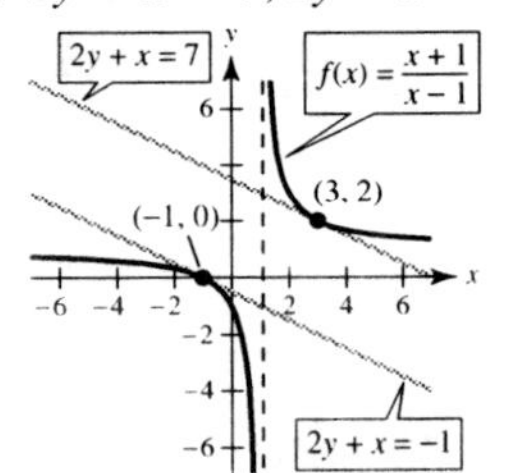

69. (a) $y = e(x - 1)$
   (b)

71. $2y + x - 4 = 0$ **73.** $25y - 12x + 16 = 0$

75. $(1, 1)$ **77.** $(3, 8e^{-3})$

79. $2y + x = 7;\ 2y + x = -1$

81. $f(x) + 2 = g(x)$ **83.** (a) $p'(1) = 1$ (b) $q'(4) = -1/3$

85. $(18t + 5)/(2\sqrt{t}) \text{ cm}^2/\text{sec}$

87. (a) About $-\$3.38$ per unit. The rate of change of the ordering and transportation cost C is decreasing at a rate of about \$3.38 per unit when the order size is 200 units.
   (b) \$0.00 per unit. The rate of change of the ordering and transportation cost C is not changing when the order size is 250 units.
   (c) About \$1.83 per unit. The rate of change of the ordering and transportation cost C is increasing at a rate of about \$1.83 per unit when the order size is 300 units.

89. 31.55 bacteria/hr   **91.** Proofs

93. (a) $q(t) = -0.0546t^3 + 2.529t^2 - 36.89t + 186.6$
      $v(t) = 0.0796t^3 - 2.162t^2 + 15.32t + 5.9$
   (b) 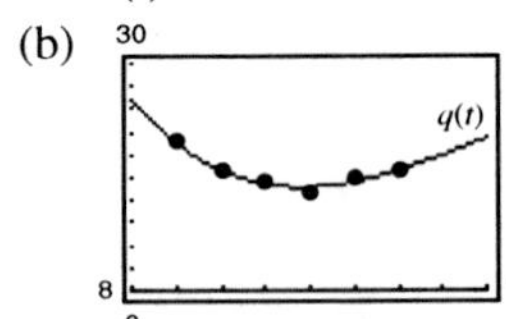 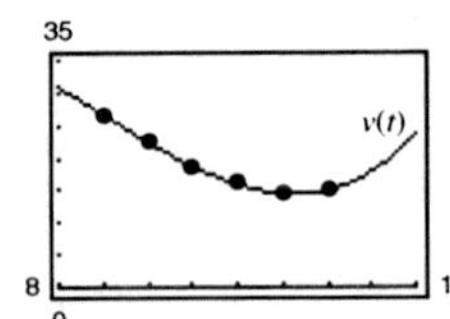
   (c) $A = \dfrac{0.0796t^3 - 2.162t^2 + 15.32t + 5.9}{-0.0546t^3 + 2.529t^2 - 36.89t + 186.6}$
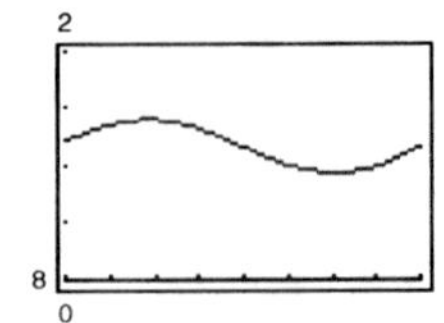
   A represents the average value (in billions of dollars) per one million personal computers.
   (d) $A'(t)$ represents the rate of change of the average value per one million personal computers for the given year.

95. $3/\sqrt{x}$ **97.** $2/(x-1)^3$ **99.** $2\cos x - x\sin x$
101. $e^x/x^3(x^2-2x+2)$ **103.** $2x$ **105.** $1/\sqrt{x}$
107. Answers will vary. Sample answer: $f(x) = (x-2)^2$

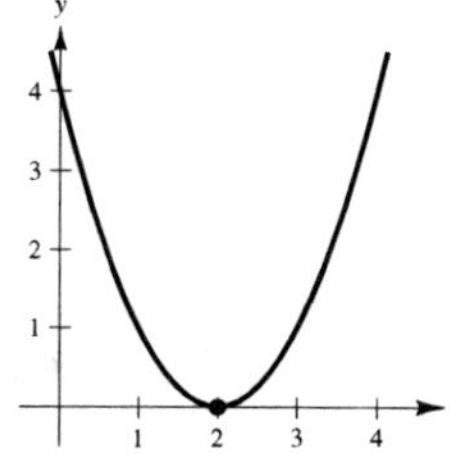

109. 0 **111.** -10
113.

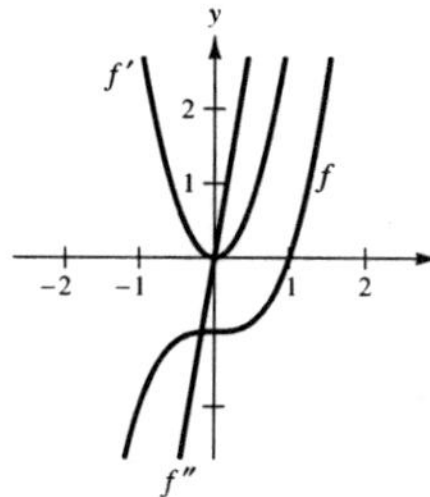

115.

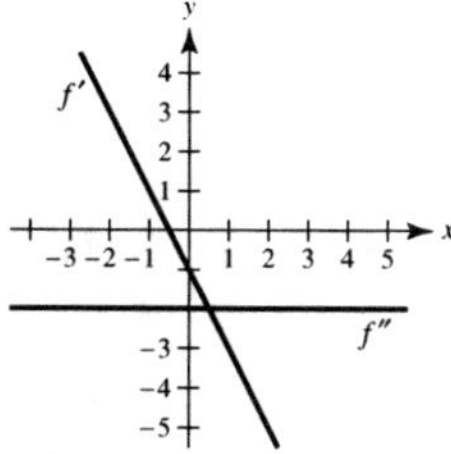

117.

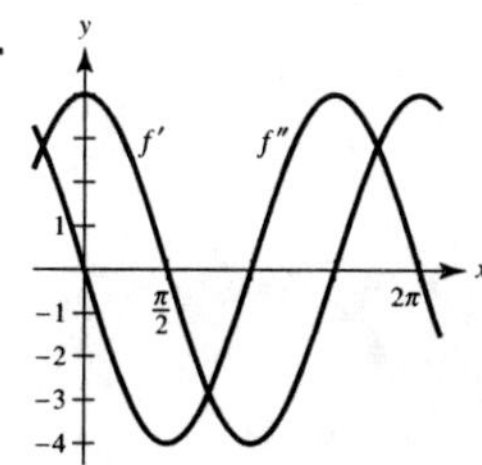

119. $v(3) = 27$ m/sec
$a(3) = -6$ m/sec^2
The speed of the object is decreasing.

121. $f^{(n)}(x) = n(n-1)(n-2)\cdots(2)(1) = n!$
123. (a) $f''(x) = g(x)h''(x) + 2g'(x)h'(x) + g''(x)h(x)$
$f'''(x) = g(x)h'''(x) + 3g'(x)h''(x) +$
$\qquad 3g''(x)h'(x) + g'''(x)h(x)$
$f^{(4)}(x) = g(x)h^{(4)}(x) + 4g'(x)h'''(x) + 6g''(x)h''(x) +$
$\qquad 4g'''(x)h'(x) + g^{(4)}(x)h(x)$

(b) $f^{(n)}(x) = g(x)h^{(n)}(x) + \dfrac{n!}{1!(n-1)!}g'(x)h^{(n-1)}(x) +$

$\qquad \dfrac{n!}{2!(n-2)!}g''(x)h^{(n-2)}(x) + \cdots +$

$\qquad \dfrac{n!}{(n-1)!1!}g^{(n-1)}(x)h'(x) + g^{(n)}(x)h(x)$

125. $n = 1$: $f'(x) = x\cos x + \sin x$
$n = 2$: $f'(x) = x^2\cos x + 2x\sin x$
$n = 3$: $f'(x) = x^3\cos x + 3x^2\sin x$
$n = 4$: $f'(x) = x^4\cos x + 4x^3\sin x$
General rule: $f'(x) = x^n\cos x + nx^{(n-1)}\sin x$
127. $y' = -1/x^2$, $y'' = 2/x^3$,
$x^3y'' + 2x^2y' = x^3(2/x^3) + 2x^2(-1/x^2)$
$\qquad = 2 - 2 = 0$
129. $y' = 2\cos x$, $y'' = -2\sin x$,
$y'' + y = -2\sin x + 2\sin x + 3 = 3$
131. False. $dy/dx = f(x)g'(x) + g(x)f'(x)$ **133.** True

135. True   **137.** $f(x) = 3x^2 - 2x - 1$
139. $f'(x) = 2|x|$; $f''(0)$ does not exist.   **141.** Proof

Section 3.4 (page 161)

	$y = f(g(x))$	$u = g(x)$	$y = f(u)$
1.	$y = (5x-8)^4$	$u = 5x-8$	$y = u^4$
3.	$y = \sqrt{x^3-7}$	$u = x^3-7$	$y = \sqrt{u}$
5.	$y = \csc^3 x$	$u = \csc x$	$y = u^3$
7.	$y = e^{-2x}$	$u = -2x$	$y = e^u$

9. $12(4x-1)^2$ **11.** $-108(4-9x)^3$ **13.** $-4x/[3(9-x^2)^{1/3}]$
15. $-1/(2\sqrt{5-t})$ **17.** $4x/\sqrt[3]{(6x^2+1)^2}$ **19.** $-x/\sqrt[4]{(9-x^2)^3}$
21. $-1/(x-2)^2$ **23.** $-2/(t-3)^3$ **25.** $-1/[2(x+2)^{3/2}]$

27. $2x(x-2)^3(3x-2)$ **29.** $\dfrac{1-2x^2}{\sqrt{1-x^2}}$ **31.** $\dfrac{1}{(x^2+1)^{3/2}}$

33. $\dfrac{-2(x+5)(x^2+10x-2)}{(x^2+2)^3}$

35. $20x(x^2+3)^9 + 2(x^2+3)^5 + 20x^2(x^2+3)^4 + 2x$

37. $\dfrac{1}{8\sqrt{x}\left(\sqrt{2+\sqrt{x}}\right)\left(\sqrt{2+\sqrt{2+\sqrt{x}}}\right)}$

39. $(1 - 3x^2 - 4x^{3/2})/[2\sqrt{x}(x^2+1)^2]$

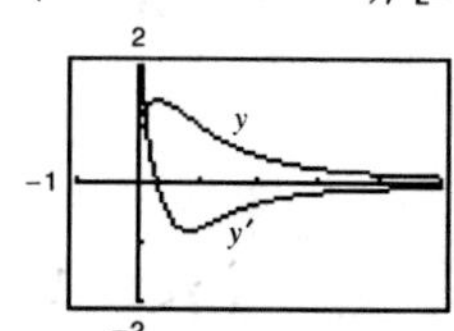

The zero of y' corresponds to the point on the graph of the function where the tangent line is horizontal.

41. $3t(t^2 + 3t - 2)/(t^2 + 2t - 1)^{3/2}$

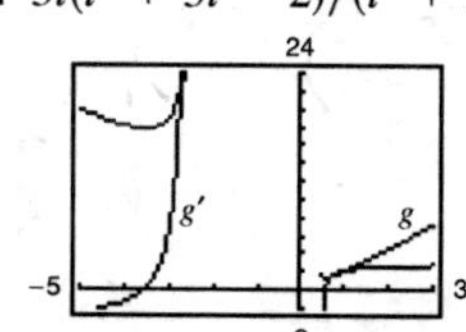

The zeros of $g'(t)$ correspond to the points on the graph of the function where the tangent line is horizontal.

43. $-\dfrac{\sqrt{\dfrac{x+1}{x}}}{2x(x+1)}$

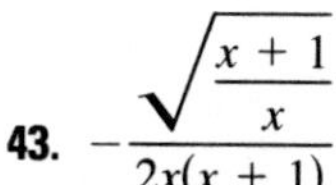

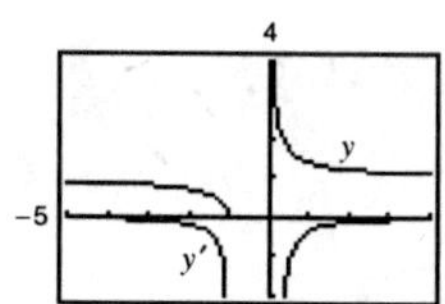

y' has no zeros.

45. $t/\sqrt{1+t}$

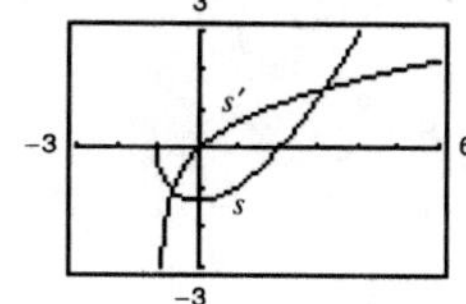

The zero of $s'(t)$ corresponds to the point on the graph of the function where the tangent line is horizontal.

47. $-[\pi x \sin(\pi x) + \cos(\pi x) + 1]/x^2$

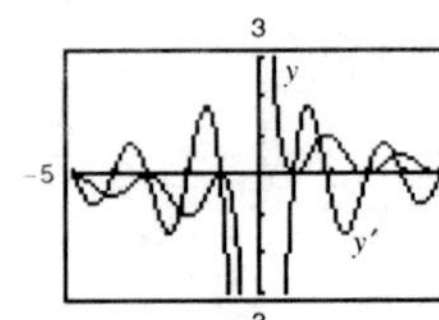

The zeros of y' correspond to the points on the graph of the function where the tangent lines are horizontal.

49. (a) 1 (b) 2; The slope of $\sin ax$ at the origin is a.

51. (a) 3 (b) -3 **53.** 3   **55.** 2   **57.** $-4\sin 4x$

59. $15\sec^2 3x$ **61.** $10\tan 5\theta \sec^2 5\theta$

63. $\sin 2\theta \cos 2\theta = \dfrac{1}{2}\sin 4\theta$ **65.** $\dfrac{1}{2\sqrt{x}} + 2x\cos(2x)^2$

67. $-\sin x \cos(\cos x)$ **69.** $2\sec^2 2x \cos(\tan 2x)$

71. $2e^{2x}$ **73.** $e^{\sqrt{x}}/(2\sqrt{x})$ **75.** $3(e^{-t} + e^t)^2(e^t - e^{-t})$

77. $2x$ **79.** $\dfrac{-2(e^x - e^{-x})}{(e^x + e^{-x})^2}$ **81.** $x^2 e^x$ **83.** $e^{-x}\left(\dfrac{1}{x} - \ln x\right)$

85. $2e^x \cos x$ **87.** $\dfrac{2}{x}$ **89.** $\dfrac{4(\ln x)^3}{x}$ **91.** $\dfrac{2x^2 - 1}{x(x^2 - 1)}$

93. $\dfrac{1 - x^2}{x(x^2 + 1)}$ **95.** $\dfrac{1 - 2\ln t}{t^3}$ **97.** $\dfrac{1}{1 - x^2}$

99. $\dfrac{\sqrt{x^2 + 1}}{x^2}$ **101.** $\cot x$ **103.** $-\tan x + \dfrac{\sin x}{\cos x - 1}$

105. $\dfrac{3\cos x}{(\sin x - 1)(\sin x + 2)}$ **107.** $2940(2 - 7x)^2$

109. $\dfrac{2}{(x - 6)^3}$ **111.** $2(\cos x^2 - 2x^2 \sin x^2)$ **113.** $3(6x + 5)e^{-3x}$

115. $\frac{6}{5}$ **117.** $-\frac{3}{5}$ **119.** -5 **121.** 0

123. (a) $8x - 5y - 7 = 0$ **125.** (a) $2x - y - 2\pi = 0$

(b) 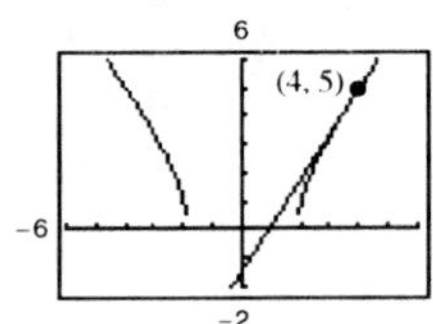   (b) 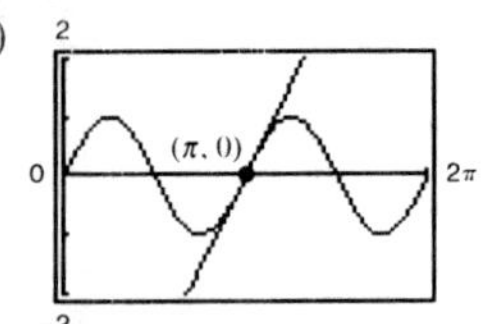

127. (a) $12x - y + 2 - 3\pi = 0$

(b) 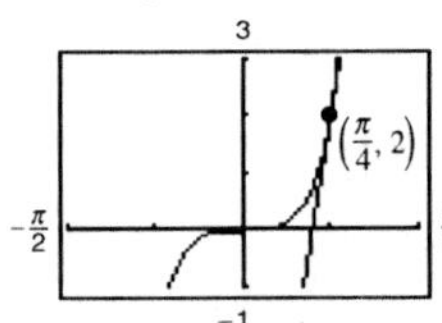

129. (a) $x + 2y - 8 = 0$

(b) 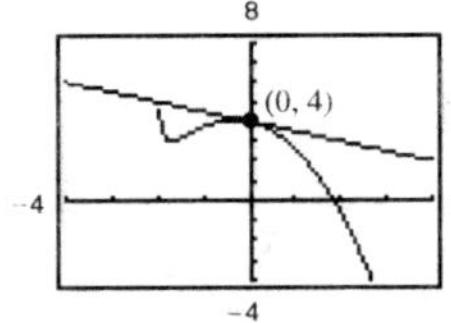

131. $(\ln 4)4^x$ **133.** $(\ln 5)5^{x-2}$ **135.** $t2^t(t\ln 2 + 2)$

137. $-2^{-\theta}[(\ln 2)\cos \pi\theta + \pi \sin \pi\theta]$ **139.** $1/[x(\ln 3)]$

141. $\dfrac{x - 2}{(\ln 2)x(x - 1)}$ **143.** $\dfrac{x}{(\ln 5)(x^2 - 1)}$

145. $\dfrac{5}{(\ln 2)t^2}(1 - \ln t)$

147. 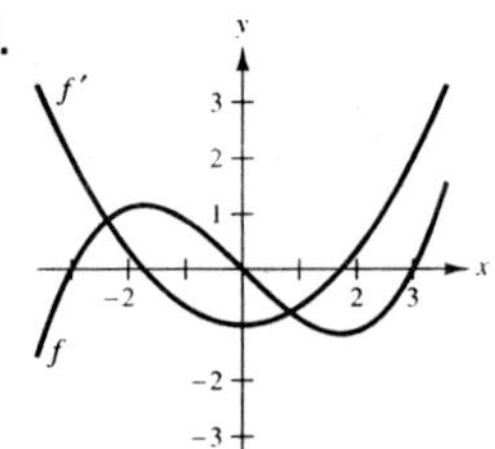

The zeros of f' correspond to the points where the graph of f has horizontal tangents.

149. 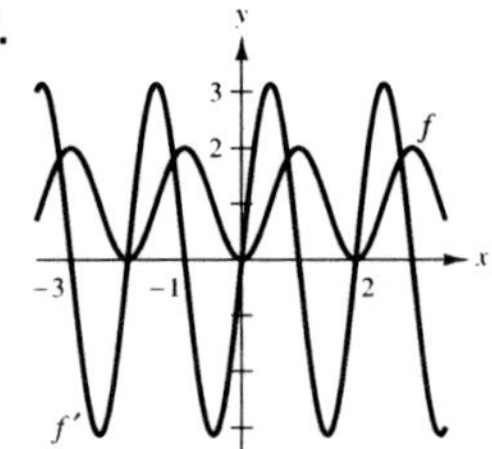

The zeros of f' correspond to the points where the graph of f has horizontal tangents.

151. $g'(x) = 3f'(3x)$ **153.** (a) 0 (b) Proof

155. (a) $g'(1/2) = -3$ **157.** (a) $s'(0) = 0$

(b) $3x + y - 3 = 0$ (b) $y = \frac{4}{3}$

(c) 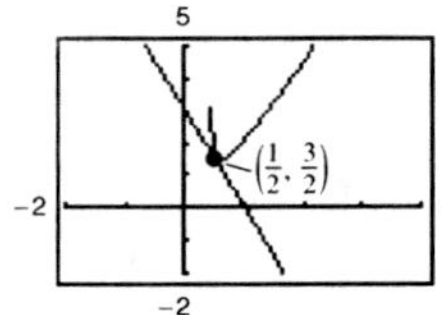   (c)

159. $3x + 4y - 25 = 0$

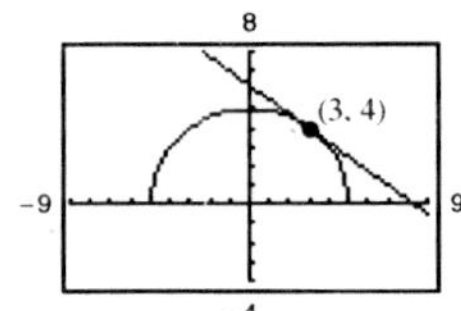

161. $\left(\dfrac{\pi}{6}, \dfrac{3\sqrt{3}}{2}\right), \left(\dfrac{5\pi}{6}, -\dfrac{3\sqrt{3}}{2}\right), \left(\dfrac{3\pi}{2}, 0\right)$ **163.** 24   **165.** 0

167. (a) 1.461 (b) -1.016

169. 0.2 rad, 1.45 rad/sec   **171.** 0.04224 cm/sec^2

173. 768π in.^{3}/sec

175. (a) 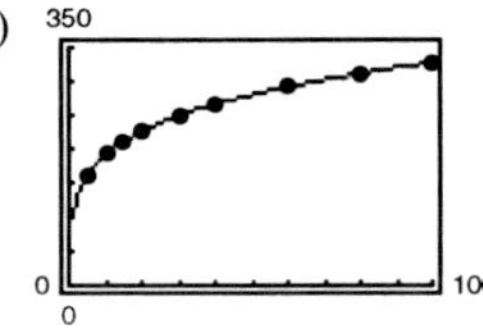

(b) $T'(10) \approx 4.75$ deg/lb/in.^{2}
    $T'(70) \approx 0.97$ deg/lb/in.^{2}

177. (a) \$48.79 (b) $C'(1) \approx 0.051P$, $C'(8) \approx 0.072P$

(c) $\ln 1.05$

179. (a) Yes; Answers will vary. (b) Yes; Answers will vary.

181. (a) Proof (b) Proof

183. $g'(x) = \dfrac{9x - 15}{|3x - 5|}$, $x \neq \dfrac{5}{3}$

185. $h'(x) = \dfrac{x}{|x|}\cos x - |x|\sin x$, $x \neq 0$

187. (a) $P_1(x) = 2\left(x - \dfrac{\pi}{4}\right) + 1$

$P_2(x) = 2\left(x - \dfrac{\pi}{4}\right)^2 + 2\left(x - \dfrac{\pi}{4}\right) + 1$

(b)

(c) P_2

(d) The accuracy worsens as you move away from $x = \pi/4$.

189. (a) $P_1(x) = x + 1$

$P_2(x) = \frac{1}{2}x^2 + x + 1$

(b) 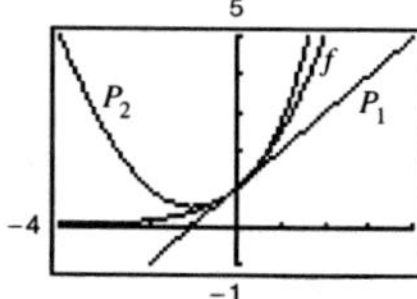

(c) P_2

(d) The accuracy worsens as you move away from $x = 0$.

191. False. $y' = \frac{1}{2}(1 - x)^{-1/2}(-1)$ **193.** True

195. Putnam Problem A1, 1967

Section 3.5   (page 171)

1. $-x/y$ **3.** $-\sqrt{y/x}$ **5.** $(y - 3x^2)/(2y - x)$

7. $\dfrac{10 - e^y}{xe^y + 3}$ **9.** $\dfrac{1 - 3x^2y^3}{3x^3y^2 - 1}$ **11.** $\dfrac{6xy - 3x^2 - 2y^2}{4xy - 3x^2}$

13. $\cos x/[4 \sin (2y)]$ **15.** $(\cos x - \tan y - 1)/(x \sec^2 y)$

17. $[y \cos(xy)]/[1 - x \cos(xy)]$ **19.** $2xy/(3 - 2y^2)$

21. $\dfrac{y(1 - 6x^2)}{1 + y}$

23. (a) $y_1 = \sqrt{64 - x^2};\ y_2 = -\sqrt{64 - x^2}$

(b) 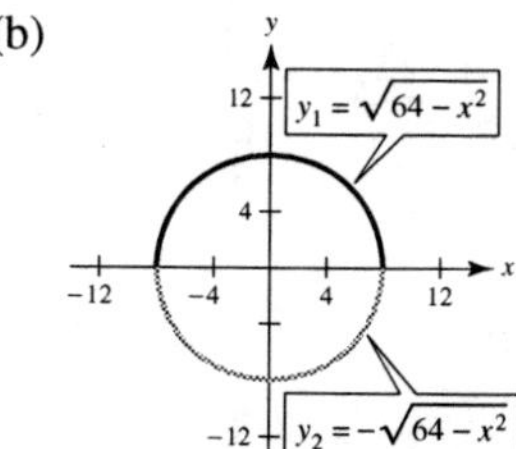

(c) $y' = \mp \dfrac{x}{\sqrt{64 - x^2}} = -\dfrac{x}{y}$

(d) $y' = -\dfrac{x}{y}$

25. (a) $y_1 = \frac{4}{5}\sqrt{25 - x^2};\ y_2 = -\frac{4}{5}\sqrt{25 - x^2}$

(b) 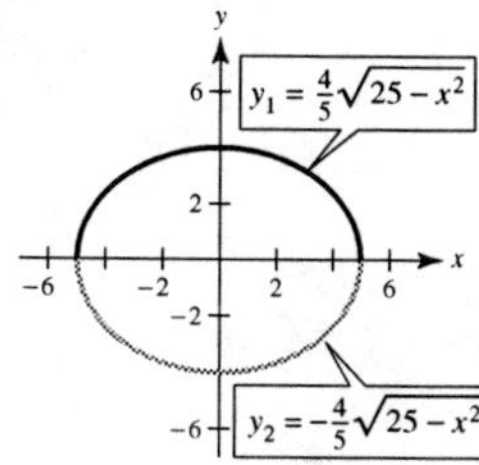

(c) $y' = \mp \dfrac{4x}{5\sqrt{25 - x^2}} = -\dfrac{16x}{25y}$

(d) $y' = -\dfrac{16x}{25y}$

27. $-\dfrac{y}{x}, -\dfrac{1}{6}$ **29.** $\dfrac{98x}{y(x^2 + 49)^2}$, Undefined   **31.** $-\sqrt[3]{\dfrac{y}{x}}, -\dfrac{1}{2}$

33. $-\sin^2(x + y)$ or $-\dfrac{x^2}{x^2 + 1}, 0$ **35.** $\dfrac{1 - 3ye^{xy}}{3xe^{xy}}, \dfrac{1}{9}$

37. $-\frac{1}{2}$ **39.** 0 **41.** $y = -\frac{9}{4}x + \frac{9}{2}$ **43.** $y = x - 1$

45. $y = -x + 7$ **47.** $y = -x + 2$

49. $y = \dfrac{\sqrt{3}x}{6} + \dfrac{8\sqrt{3}}{3}$ **51.** $y = -\dfrac{2}{11}x + \dfrac{30}{11}$

53. (a) $y = -2x + 4$ (b) Proof

55. $\cos^2 y, -\dfrac{\pi}{2} < y < \dfrac{\pi}{2}, \dfrac{1}{1 + x^2}$ **57.** $\dfrac{-4}{y^3}$

59. $-36/y^3$ **61.** $(3x)/(4y)$

63. 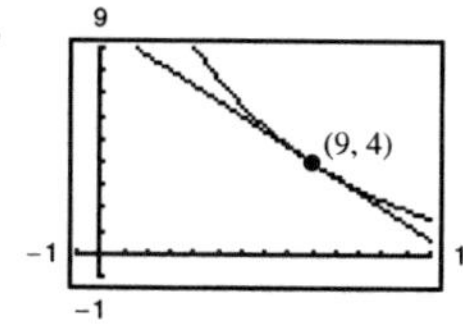

$2x + 3y - 30 = 0$

65. At $(4, 3)$:

Tangent line: $4x + 3y - 25 = 0$

Normal line: $3x - 4y = 0$

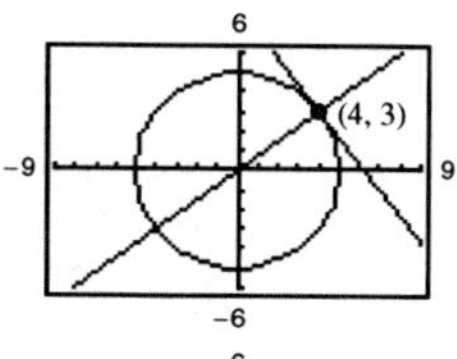

At $(-3, 4)$:

Tangent line: $3x - 4y + 25 = 0$

Normal line: $4x + 3y = 0$

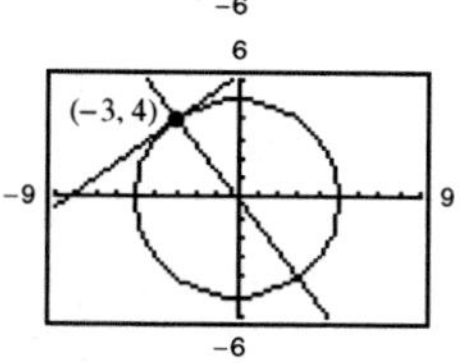

67. Proof

69. Horizontal tangents: $(-4, 0), (-4, 10)$

Vertical tangents: $(0, 5), (-8, 5)$

71. $\dfrac{2x^2 + 1}{\sqrt{x^2 + 1}}$ **73.** $\dfrac{3x^3 + 15x^2 - 8x}{2(x + 1)^3\sqrt{3x - 2}}$

75. $\dfrac{(2x^2 + 2x - 1)\sqrt{x - 1}}{(x + 1)^{3/2}}$ **77.** $2(1 - \ln x)x^{(2/x) - 2}$

79. $(x - 2)^{x+1}\left[\dfrac{x + 1}{x - 2} + \ln(x - 2)\right]$ **81.** $\dfrac{2x^{\ln x} \ln x}{x}$

83. 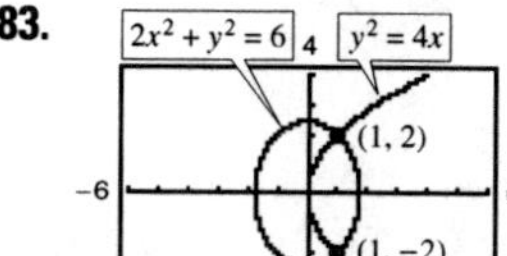   **85.** 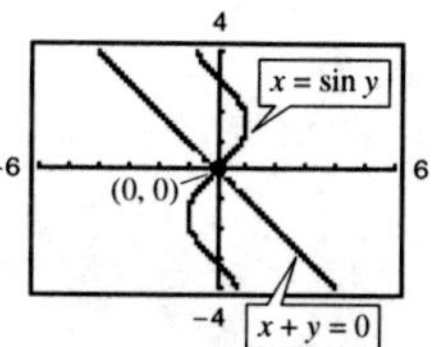

At $(1, 2)$:

Slope of ellipse: -1

Slope of parabola: 1

At $(1, -2)$:

Slope of ellipse: 1

Slope of parabola: -1

At $(0, 0)$:

Slope of line: -1

Slope of sine curve: 1

87. Derivatives: $\dfrac{dy}{dx} = -\dfrac{y}{x}, \dfrac{dy}{dx} = \dfrac{x}{y}$

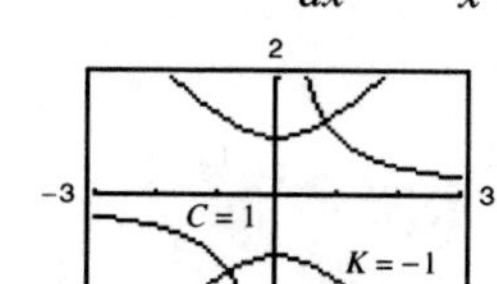   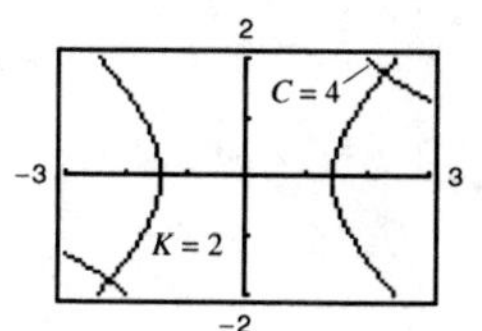

89. (a) $y\dfrac{dy}{dx} - 3x^3 = 0$ (b) $y\dfrac{dy}{dt} - 3x^3\dfrac{dx}{dt} = 0$

91. (a) $-\pi\sin(\pi y)(dy/dx) - 3\pi\cos(\pi x) = 0$

(b) $-\pi\sin(\pi y)\left(\dfrac{dy}{dt}\right) - 3\pi\cos(\pi x)\left(\dfrac{dx}{dt}\right) = 0$

93. Answers will vary. In the explicit form of a function, the variable is explicitly written as a function of x. In an implicit equation, the function is only implied by an equation. An example of an implicit function is $x^2 + xy = 5$. In explicit form, this equation would be $y = (5 - x^2)/x$.

95. 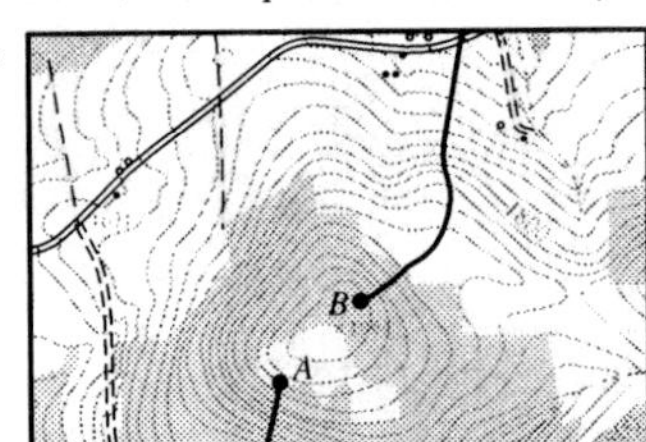 Use starting point B.

97. (a) 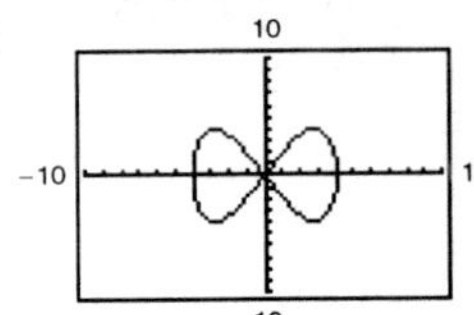

(b) 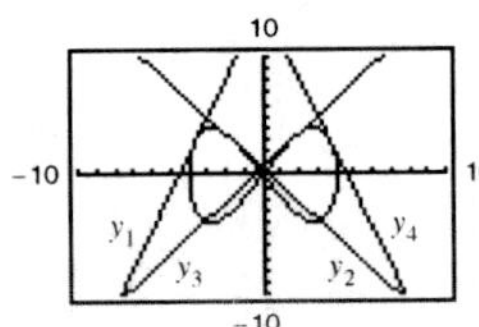

$y_1 = \tfrac{1}{3}\left[\left(\sqrt{7} + 7\right)x + \left(8\sqrt{7} + 23\right)\right]$

$y_2 = -\tfrac{1}{3}\left[\left(-\sqrt{7} + 7\right)x - \left(23 - 8\sqrt{7}\right)\right]$

$y_3 = -\tfrac{1}{3}\left[\left(\sqrt{7} - 7\right)x - \left(23 - 8\sqrt{7}\right)\right]$

$y_4 = -\tfrac{1}{3}\left[\left(\sqrt{7} + 7\right)x - \left(8\sqrt{7} + 23\right)\right]$

(c) $\left(8\sqrt{7}/7, 5\right)$

99. Proof   **101.** $(6, -8), (-6, 8)$

103. $y = -\dfrac{\sqrt{3}}{2}x + 2\sqrt{3},\ y = \dfrac{\sqrt{3}}{2}x - 2\sqrt{3}$

105. (a) $y = 2x - 6$

(b) 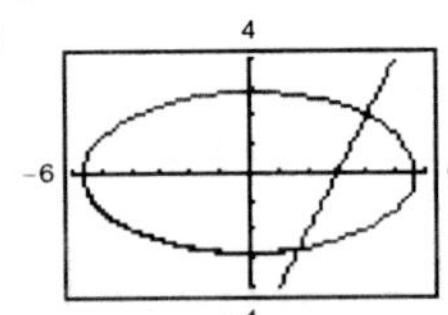   (c) $\left(\tfrac{28}{17}, -\tfrac{46}{17}\right)$

Section 3.6   (page 179)

1. $\dfrac{1}{27}$ **3.** $\dfrac{1}{5}$ **5.** $\dfrac{2\sqrt{3}}{3}$ **7.** -2

9. $f'\left(\tfrac{1}{2}\right) = \tfrac{3}{4},\ (f^{-1})'\left(\tfrac{1}{8}\right) = \tfrac{4}{3}$ **11.** $f'(5) = \tfrac{1}{2},\ (f^{-1})'(1) = 2$

13. (a) $y = \dfrac{\pi}{2}$

(b) 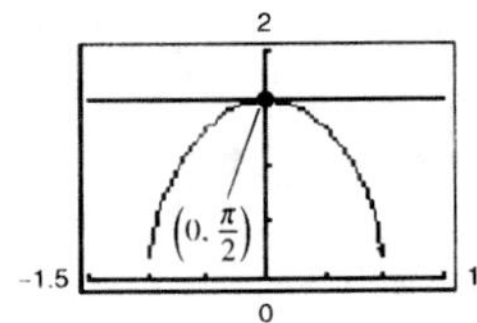

15. (a) $y = 3\sqrt{2}\,x + \dfrac{\pi}{4} - 1$

(b) 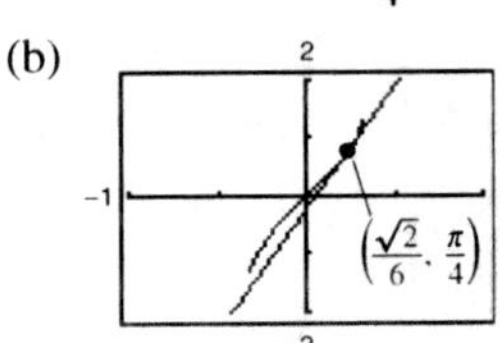

17. $-\dfrac{1}{11}$ **19.** $\dfrac{\pi + 2}{\pi}$ **21.** $\dfrac{1}{\sqrt{1 - (x + 1)^2}}$ **23.** $-\dfrac{3}{\sqrt{4 - x^2}}$

25. $\dfrac{e^x}{1 + e^{2x}}$ **27.** $\dfrac{3x - \sqrt{1 - 9x^2}\ \arcsin 3x}{x^2\sqrt{1 - 9x^2}}$

29. $-\dfrac{1}{(x + 1)\sqrt{1 - x^2}} - \dfrac{\arccos x}{(x + 1)^2}$ **31.** $-\dfrac{6}{1 + 36x^2}$

33. $-\dfrac{t}{\sqrt{1 - t^2}}$ **35.** $2\arccos x$ **37.** $\dfrac{1}{1 - x^4}$

39. $\dfrac{1}{(1 - t^2)^{3/2}}$ **41.** $\arcsin x$ **43.** $\dfrac{x^2}{\sqrt{16 - x^2}}$ **45.** $\dfrac{2}{(1 + x^2)^2}$

47. $y = \tfrac{1}{3}\left(4\sqrt{3}x - 2\sqrt{3} + \pi\right)$ **49.** $y = \tfrac{1}{4}x + (\pi - 2)/4$

51. $y = (2\pi - 4)x + 4$

53. $y = -2x + \left(\dfrac{\pi}{6} + \sqrt{3}\right)$

$y = -2x + \left(\dfrac{5\pi}{6} - \sqrt{3}\right)$

55. $P_1(x) = \dfrac{\pi}{6} + \dfrac{2\sqrt{3}}{3}\left(x - \dfrac{1}{2}\right)$

$P_2(x) = \dfrac{\pi}{6} + \dfrac{2\sqrt{3}}{3}\left(x - \dfrac{1}{2}\right) + \dfrac{2\sqrt{3}}{9}\left(x - \dfrac{1}{2}\right)^2$

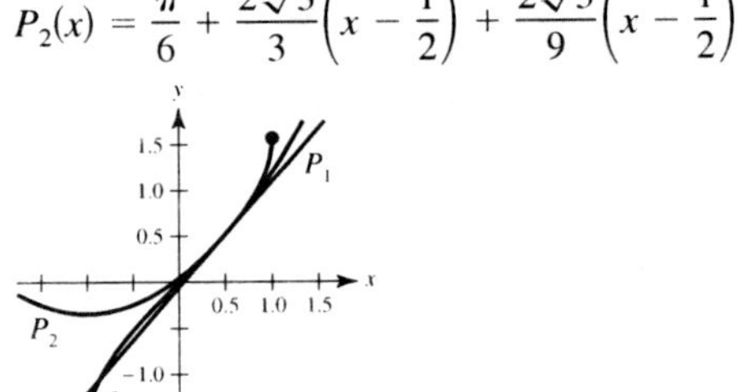

57. $P_1(x) = x;\ P_2(x) = x$

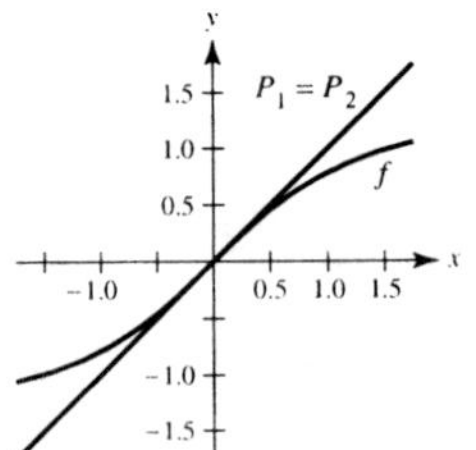

59. $y = [-2\pi x/(\pi + 8)] + 1 - [\pi^2/(2\pi + 16)]$

61. $y = -x + \sqrt{2}$

63. Many x-values yield the same y-value. For example, $f(\pi) = 0 = f(0)$. The graph is not continuous at $x = (2n - 1)\pi/2$, where n is an integer.

65. Theorem 3.17: Let f be differentiable on an interval I. If f has an inverse g, then g is differentiable at any x for which $f'(g(x)) \neq 0$. Moreover, $g'(x) = 1/f'(g(x)), f'(g(x)) \neq 0$.

67. (a) $\theta = \text{arccot}(x/5)$

(b) $x = 10$: 16 rad/hr; $x = 3$: 58.824 rad/hr

69. (a) $h(t) = -16t^2 + 256$; $t = 4$ sec
　(b) $t = 1$: -0.0520 rad/sec; $t = 2$: -0.1116 rad/sec
71. 0.015 rad/sec　**73.** Proofs　**75.** True　**77.** Proof
79. (a)

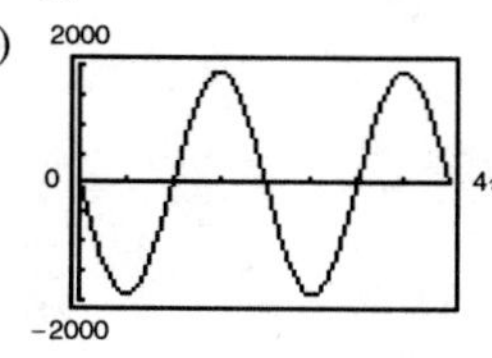

　　　　　　　　　　(b) Proof

81. Proof

Section 3.7　(page 187)

1. (a) $\frac{3}{4}$　(b) 20　**3.** (a) $-\frac{5}{8}$　(b) $\frac{3}{2}$
5. (a) -8 cm/sec　(b) 0 cm/sec　(c) 8 cm/sec
7. (a) 8 cm/sec　(b) 4 cm/sec　(c) 2 cm/sec
9. In a linear function, if x changes at a constant rate, so does y.
　However, unless $a = 1$, y does not change at the same rate as x.
11. $2(2x^3 + 3x)/\sqrt{x^4 + 3x^2 + 1}$
13. (a) 64π cm²/min　(b) 256π cm²/min
15. (a) Proof
　(b) When $\theta = \dfrac{\pi}{6}$, $\dfrac{dA}{dt} = \dfrac{\sqrt{3}}{8}s^2$. When $\theta = \dfrac{\pi}{3}$, $\dfrac{dA}{dt} = \dfrac{1}{8}s^2$.
　(c) If s and $d\theta/dt$ are constant, dA/dt is proportional to $\cos\theta$.
17. $\dfrac{3}{32\pi}$ m/min　**19.** (a) 144 cm²/sec　(b) 720 cm²/sec
21. $\dfrac{8}{405\pi}$ ft/min　**23.** (a) 12.5%　(b) $\dfrac{1}{144}$ m/min
25. (a) $-\frac{7}{12}$ ft/sec; $-\frac{3}{2}$ ft/sec; $-\frac{48}{7}$ ft/sec
　(b) $\frac{527}{24}$ ft²/sec　(c) $\frac{1}{12}$ rad/sec
27. Rate of vertical change: $\dfrac{1}{5}$ m/sec
　　Rate of horizontal change: $-\dfrac{\sqrt{3}}{15}$ m/sec
29. (a) -750 mi/h　(b) 30 min
31. $-50/\sqrt{85} \approx -5.42$ ft/sec　**33.** (a) $\frac{25}{3}$ ft/sec　(b) $\frac{10}{3}$ ft/sec
35. (a) 12 sec　(b) $\sqrt{3}/2$ m　(c) $\sqrt{5}\pi/120$ m/sec
37. Proof　**39.** $V^{0.3}\left(1.3p\dfrac{dV}{dt} + V\dfrac{dp}{dt}\right) = 0$　**41.** $\dfrac{1}{25}$ rad/sec
43. (a) $t = 65°$: $H \approx 99.8\%$　(b) -4.7%/hr
　　$t = 80°$: $H \approx 60.2\%$
45. (a) $\dfrac{dx}{dt} = -600\pi \sin\theta$
　(b)

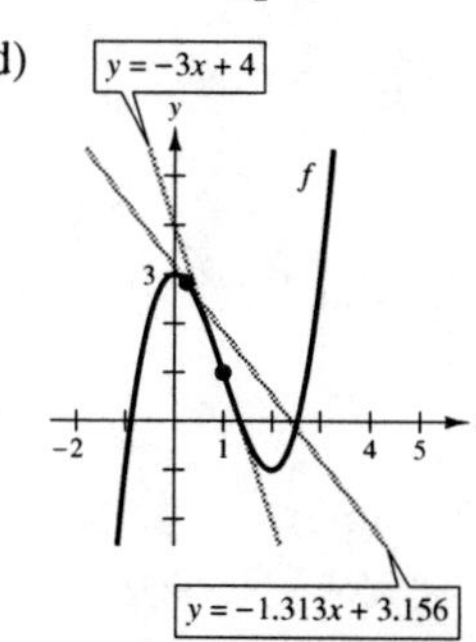

　(c) $\theta = \dfrac{\pi}{2} + n\pi$ (or $90° + n \cdot 180°$); $\theta = n\pi$ (or $n \cdot 180°$)
　(d) -300π cm/sec; $-300\sqrt{3}\pi$ cm/sec
47. $\dfrac{d\theta}{dt} = \dfrac{1}{25}\cos^2\theta$, $-\dfrac{\pi}{4} \le \theta \le \dfrac{\pi}{4}$

49. (a) $\frac{1}{2}$ rad/min　(b) $\frac{3}{2}$ rad/min　(c) 1.87 rad/min
51. -0.1808 ft/sec²
53. (a) $\dfrac{dy}{dt} = 3\dfrac{dx}{dt}$ means that y changes three times as fast as x changes.
　(b) y changes slowly when $x \approx 0$ or $x \approx L$. y changes more rapidly when x is near the middle of the interval.

Section 3.8　(page 195)

In Exercises 1 and 3, the following solutions may vary depending on the software or calculator used, and on rounding.

1.

n	x_n	$f(x_n)$	$f'(x_n)$	$\dfrac{f(x_n)}{f'(x_n)}$	$x_n - \dfrac{f(x_n)}{f'(x_n)}$
1	2.2000	-0.1600	4.4000	-0.0364	2.2364
2	2.2364	0.0015	4.4728	0.0003	2.2361

3.

n	x_n	$f(x_n)$	$f'(x_n)$	$\dfrac{f(x_n)}{f'(x_n)}$	$x_n - \dfrac{f(x_n)}{f'(x_n)}$
1	1.6	-0.0292	-0.9996	0.0292	1.5708
2	1.5708	0	-1	0	1.5708

5. -1.587　**7.** 0.682　**9.** 1.250, 5.000　**11.** 0.567
13. 0.900, 1.100, 1.900　**15.** 1.935　**17.** 0.569
19. 4.493　**21.** 0.567　**23.** 0.786
25. (a) Proof　(b) $\sqrt{5} \approx 2.236$; $\sqrt{7} \approx 2.646$
27. $f'(x_1) = 0$　**29.** $2 = x_1 = x_3 = \ldots$; $1 = x_2 = x_4 = \ldots$
31. 0.74　**33.** 1.12
35. (a)

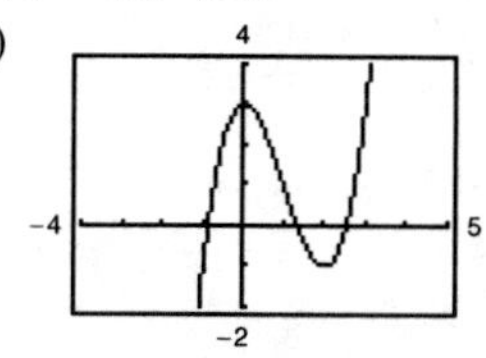

　　　　　　　　　(b) 1.347　(c) 2.532

　(d)

x-intercept of $y = -3x + 4$ is $\frac{4}{3}$.
x-intercept of $y = -1.313x + 3.156$ is approximately 2.404.

　(e) If the initial estimate $x = x_1$ is not sufficiently close to the desired zero of a function, the x-intercept of the corresponding tangent line to the function may approximate a second zero of the function.
37. Answers will vary.　**39.** Proof　**41.** 15.1, 26.8
43. False. Let $f(x) = (x^2 - 1)/(x - 1)$.
45. True　**47.** $x \approx 11.803$　**49.** 0.217

Review Exercises for Chapter 3   (page 197)

1. $2x - 4$ **3.** $\sqrt{x}/(2x)$ **5.** f is differentiable at all $x \neq 3$.

7.

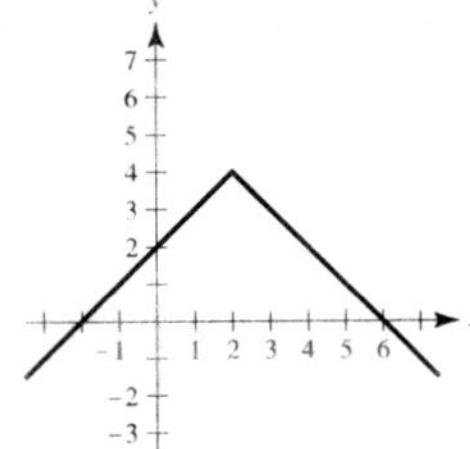

   (a) Yes

   (b) No, because the derivatives from the left and right are not equal.

9. $-\frac{3}{2}$

11. (a) $y = 3x + 1$

   (b)

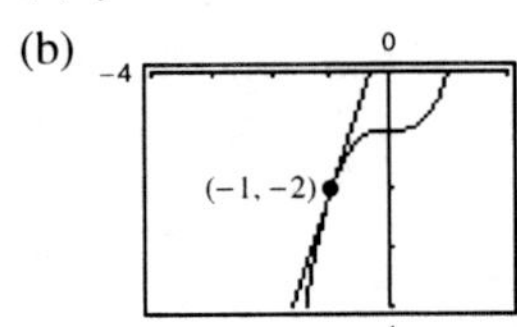

13. 8   **15.** 0   **17.** $8x^7$ **19.** $12t^3$ **21.** $3x(x - 2)$

23. $3/\sqrt{x} + 1/x^{2/3}$ **25.** $-4/(3t^3)$ **27.** $4 - 5\cos\theta$

29. $-3\sin t - 4e^t$

31.

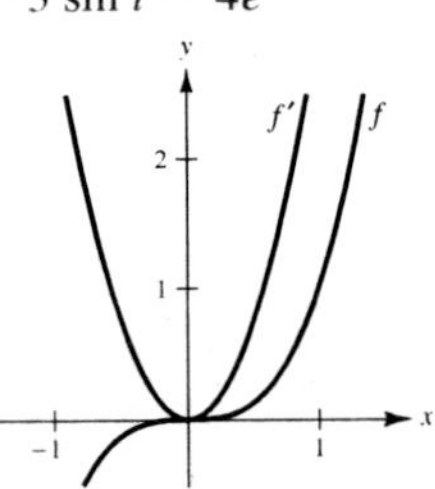

$f' > 0$ where the slopes of the tangent lines to the graph of f are positive.

33. (a) 50 vibrations/sec/lb

   (b) 33.33 vibrations/sec/lb

35. 1354.24 ft

37. (a)

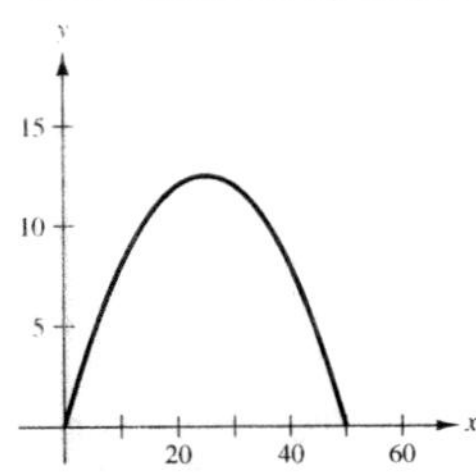

   (b) 50   (c) $x = 25$

   (d) $y' = 1 - 0.04x$

x	0	10	25	30	50
y'	1	0.6	0	-0.2	-1

   (e) $y'(25) = 0$

39. (a) $x'(t) = 2t - 3$ (b) $(-\infty, 1.5)$ (c) $x = -\frac{1}{4}$ (d) 1

41. $4(5x^3 - 15x^2 - 11x - 8)$ **43.** $\sqrt{x}\cos x + \sin x/(2\sqrt{x})$

45. $-(x^2 + 1)/(x^2 - 1)^2$ **47.** $(8x)/(9 - 4x^2)^2$

49. $\dfrac{4x^3\cos x + x^4\sin x}{\cos^2 x}$ **51.** $3x^2\sec x\tan x + 6x\sec x$

53. $-x\sin x$ **55.** $4e^x(x + 1)$ **57.** $y = 4x - 3$ **59.** $y = 0$

61. $v(4) = 20$ m/sec; $a(4) = -8$ m/sec^2

63. $-48t$ **65.** $\frac{225}{4}\sqrt{x}$ **67.** $6\sec^2\theta\tan\theta$ **69.** Proof

71. $x = \dfrac{3\pi}{4}, \dfrac{7\pi}{4}$ **73.** $\dfrac{2(x + 5)(-x^2 - 10x + 3)}{(x^2 + 3)^3}$

75. $s(s^2 - 1)^{3/2}(8s^3 - 3s + 25)$

77. $-45\sin(9x + 1)$ **79.** $\sin^2 x$ **81.** $\cos^3 x\sqrt{\sin x}$

83. $\dfrac{(x + 2)(\pi\cos\pi x) - \sin\pi x}{(x + 2)^2}$ **85.** $\dfrac{1}{4}te^{t/4}(t + 8)$

87. $\dfrac{e^{2x} - e^{-2x}}{\sqrt{e^{2x} + e^{-2x}}}$ **89.** $\dfrac{x(2 - x)}{e^x}$ **91.** $\dfrac{1}{2x}$

93. $\dfrac{1 + 2\ln x}{2\sqrt{\ln x}}$ **95.** $\dfrac{x}{(a + bx)^2}$ **97.** $\dfrac{1}{x(a + bx)}$

99. -2 **101.** 0

103. $t(t - 1)^4(7t - 2)$

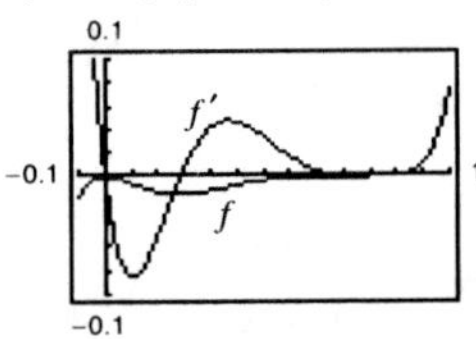

The zeros of f' correspond to the points on the graph of the function where the tangent line is horizontal.

105. $(x + 2)/(x + 1)^{3/2}$ **107.** $5/6(t + 1)^{1/6}$

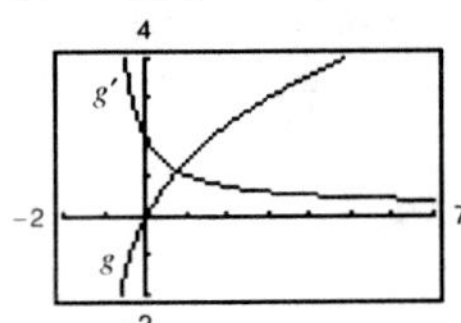 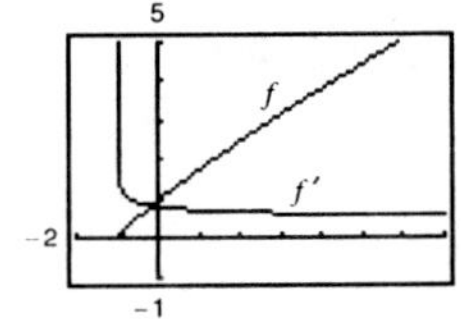

g' is not equal to zero for any x. f' has no zeros.

109. $-\sec^2\sqrt{1 - x}/(2\sqrt{1 - x})$

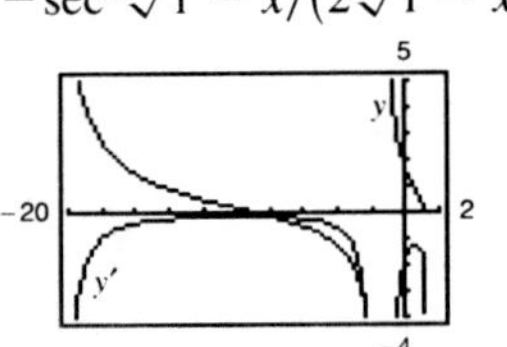

y' has no zeros.

111. $14 - 4\cos 2x$ **113.** $2\csc^2 x\cot x$

115. $[8(2t + 1)]/(1 - t)^4$

117. $18\sec^2 3\theta\tan 3\theta + \sin(\theta - 1)$

119. $x(6\ln x + 5)$

121. (a) $-18.667°$/hr   (b) $-7.284°$/hr

   (c) $-3.240°$/hr   (d) $-0.747°$/hr

123. (a) $h = 0$ is not in the domain of the function.

   (b) $h = 0.8627 - 6.4474\ln p$

   (c)

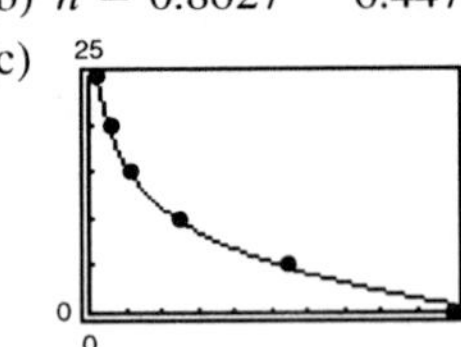

   (d) 2.72 km

   (e) 0.15 atm

   (f) $h = 5: \dfrac{dp}{dh} = -0.0816$ atm/km

      $h = 20: \dfrac{dp}{dh} = -0.008$ atm/km

As the altitude increases, the rate of change of pressure decreases.

125. $-\dfrac{2x + 3y}{3(x + y^2)}$ **127.** $-\dfrac{2x\sin x^2 + e^y}{xe^y}$

129. $\dfrac{2x - 9y}{9x - 32y}$ **131.** $\dfrac{y \sin x + \sin y}{\cos x - x \cos y}$

133. Tangent line: $3x + y - 10 = 0$
Normal line: $x - 3y = 0$

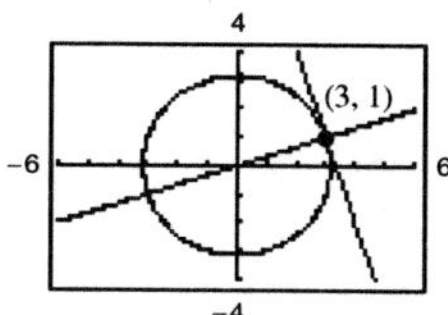

135. Tangent line: $xe^{-1} + y = 0$
Normal line: $xe - y - (e^2 + 1) = 0$

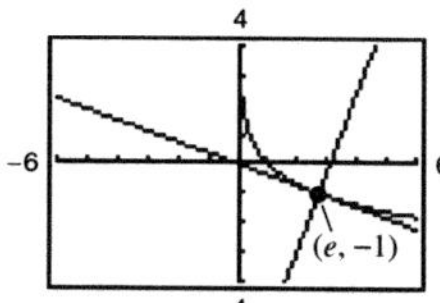

137. $\dfrac{x^3 + 8x^2 + 4}{(x + 4)^2 \sqrt{x^2 + 1}}$ **139.** $\dfrac{1}{3(\sqrt[3]{-3})^2} \approx 0.160$ **141.** $\dfrac{3}{4}$

143. $(1 - x^2)^{-3/2}$ **145.** $\dfrac{x}{|x|\sqrt{x^2 - 1}} + \operatorname{arcsec} x$

147. $(\arcsin x)^2$

149. (a) $2\sqrt{2}$ units/sec
(b) 4 units/sec
(c) 8 units/sec

151. $\frac{2}{25}$ m/min **153.** -38.34 m/sec

155. $-0.347, -1.532, 1.879$ **157.** 1.202

159. $-1.164, 1.453$

P.S. Problem Solving (page 201)

1. (a) $r = \frac{1}{2}$ (b) Center: $\left(0, \frac{5}{4}\right)$

3. (a) $P_1(x) = 1$ (b) $P_2(x) = 1 - \frac{1}{2}x^2$
(c)

x	-1.0	-0.1	-0.001	0	0.001
$\cos x$	0.5403	0.9950	1.000	1	1
$P_2(x)$	0.5000	0.9950	1.000	1	1

x	0.1	1.0
$\cos x$	0.9950	0.5403
$P_2(x)$	0.9950	0.5000

$P_2(x)$ is a good approximation of $f(x) = \cos x$ when x is very close to 0.
(d) $P_3(x) = x - \frac{1}{6}x^3$

5. $p(x) = 2x^3 + 4x^2 - 5$

7. (a) Graph $y_1 = \dfrac{1}{a}\sqrt{x^2(a^2 - x^2)}$ and $y_2 = -\dfrac{1}{a}\sqrt{x^2(a^2 - x^2)}$ as separate equations.
(b) Answers will vary.

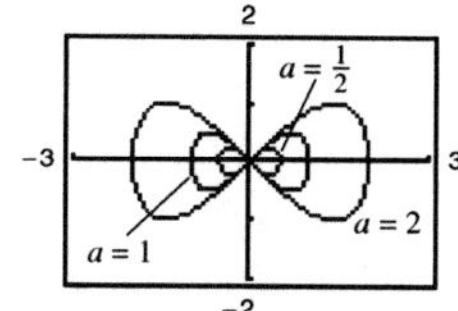

The intercepts will always be $(0, 0)$, $(a, 0)$, and $(-a, 0)$, and the maximum and minimum y-values appear to be $\pm\frac{1}{2}a$.
(c) $\left(\dfrac{a\sqrt{2}}{2}, \dfrac{a}{2}\right)$, $\left(\dfrac{a\sqrt{2}}{2}, -\dfrac{a}{2}\right)$, $\left(-\dfrac{a\sqrt{2}}{2}, \dfrac{a}{2}\right)$, $\left(-\dfrac{a\sqrt{2}}{2}, -\dfrac{a}{2}\right)$

9. $a = 1, b = \frac{1}{2}, c = -\frac{1}{2}$
$$f(x) = \frac{1 + \frac{1}{2}x}{1 - \frac{1}{2}x}$$

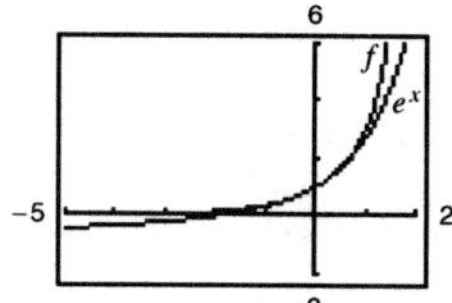

11. (a) 12 cm/sec (b) $49/\sqrt{17}$ cm/sec (c) $-4/17$ rad/sec

13. Proof

15. (a) $v(t) = -\frac{27}{5}t + 27$; $a(t) = -\frac{27}{5}$
(b) $t = 5$ sec; $s(5) = 73.5$ ft
(c) The acceleration due to gravity on Earth is greater in magnitude than that on the moon.

Chapter 4
Section 4.1 (page 209)

1. $f'(0) = 0$ **3.** $g'(2) = 0$ **5.** $f'(-2)$ is undefined.

7. 2, absolute maximum (and relative maximum)

9. 1, absolute maximum (and relative maximum);
2, absolute minimum (and relative minimum);
3, absolute maximum (and relative maximum)

11. $x = 0, 2$ **13.** $t = 8/3$ **15.** $x = \pi/3, \pi, 5\pi/3$

17. $t = \frac{1}{2}$ **19.** $x = 0$

21. Minimum: $(2, 1)$
Maximum: $(-1, 4)$

23. Minimum: $(1, -1)$
Maximum: $(4, 8)$

25. Minimum: $\left(-1, -\frac{5}{2}\right)$
Maximum: $(2, 2)$

27. Minimum: $(0, 0)$
Maximum: $(-1, 5)$

29. Minimum: $(1, -1)$
Maximum: $\left(0, -\frac{1}{2}\right)$

31. Minimum: $(0, 0)$
Maxima: $\left(-1, \frac{1}{4}\right)$ and $\left(1, \frac{1}{4}\right)$

33. Minimum: $\left(1/6, \sqrt{3}/2\right)$
Maximum: $(0, 1)$

35. Minimum: $(\pi, -3)$
Maxima: $(0, 3)$ and $(2\pi, 3)$

37. Minimum: $(0, 0)$
Maximum: $(-2, \arctan 4)$

39. Minimum: $\left(2, 5e^2 - e^4\right)$
Maximum: $\left(\ln \frac{5}{2}, \frac{25}{4}\right)$

41. Minima: $(0, 0)$ and $(\pi, 0)$
Maximum: $\left(3\pi/4, \left(\sqrt{2}/2\right)e^{3\pi/4}\right)$

43. (a) Minimum: $(0, -3)$
    Maximum: $(2, 1)$
(b) Minimum: $(0, -3)$
(c) Maximum: $(2, 1)$
(d) No extrema

45. 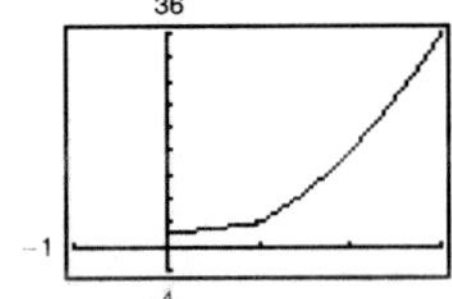
Minimum: $(0, 2)$
Maximum: $(3, 36)$

47. 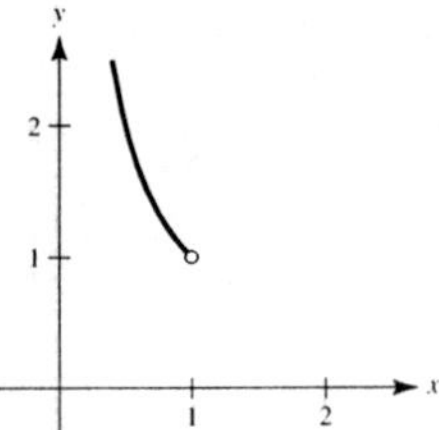
Minimum: $(4, 1)$

49. 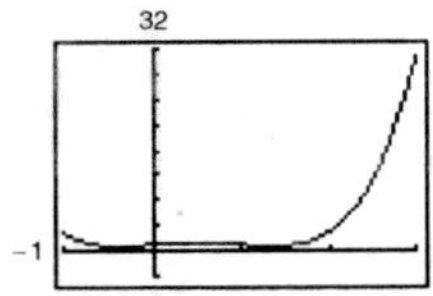
Minima: $\left((-\sqrt{3} + 1)/2, 3/4\right)$ and $\left((\sqrt{3} + 1)/2, 3/4\right)$
Maximum: $(3, 31)$

51. (a) 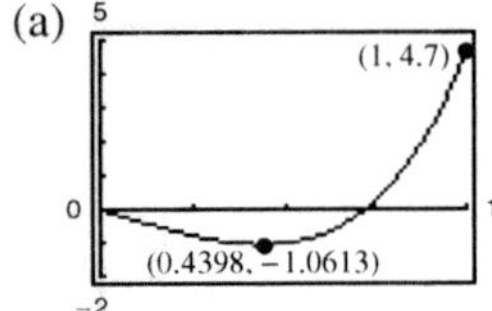
(b) Minimum: $(0.4398, -1.0613)$

53. (a) 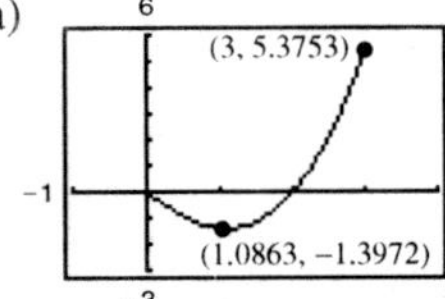
(b) Minimum: $(1.0863, -1.3972)$

55. (a) 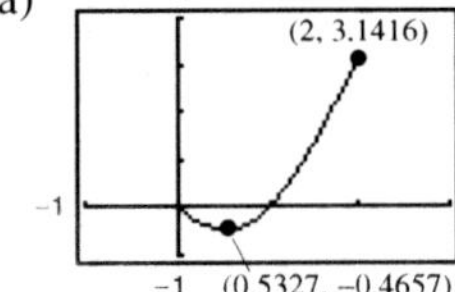
(b) Minimum: $(0.5327, -0.4657)$

57. Maximum: $\left|f''\left(\sqrt[3]{-10 + \sqrt{108}}\right)\right| = f''\left(\sqrt{3} - 1\right) \approx 1.47$

59. Maximum: $\left|f''(0)\right| = 1$ **61.** Maximum: $\left|f^{(4)}(0)\right| = \frac{56}{81}$

63. Answers will vary. Sample answer: Let $f(x) = 1/x$. f is continuous on $(0, 1)$ but does not have a maximum or minimum.

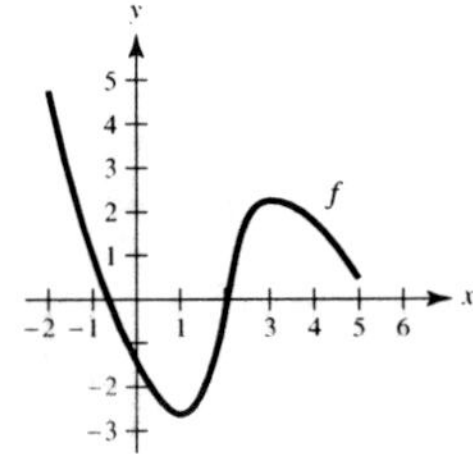

65. Answers will vary. Sample answer:

67. (a) Yes   (b) No   **69.** (a) No   (b) Yes

71. $dx/dt = (v^2 \cos 2\theta/16)\, d\theta/dt$
In the interval $[\pi/4, 3\pi/4]$, $\theta = \pi/4, 3\pi/4$ indicate minima for $|dx/dt|$ and $\theta = \pi/2$ indicates a maximum for $|dx/dt|$. This implies that the sprinkler waters longest when $\theta = \pi/4$ and $3\pi/4$. So, the lawn farthest from the sprinkler gets the most water.

73. True   **75.** True   **77.** Proof

79. (a) $A(-500, 45), B(500, 30)$
(b) $y = (3/40,000)x^2 - (3/200)x + 75/4$
(c)

x	-500	-400	-300	-200	-100
d	0	0.75	3	6.75	12

x	0	100	200	300	400	500
d	18.75	12	6.75	3	0.75	0

(d) Lowest point $\approx (100, 18)$; No

Section 4.2   (page 216)

1. $f(-1) = f(1) = 1$; f is not continuous on $(-1, 1)$.

3. $f(0) = f(2) = 0$; f is not differentiable on $(0, 2)$.

5. $(2, 0), (-1, 0); f'\left(\frac{1}{2}\right) = 0$ **7.** $(0, 0), (-4, 0); f'\left(-\frac{8}{3}\right) = 0$

9. $f'(-1) = 0$ **11.** $f'\left(\frac{3}{2}\right) = 0$

13. $f'\left((6 - \sqrt{3})/3\right) = 0; f'\left((6 + \sqrt{3})/3\right) = 0$

15. Not differentiable at $x = 0$ **17.** $f'\left(-2 + \sqrt{5}\right) = 0$

19. $f'\left(\sqrt{2}\right) = 0$ **21.** $f'(\pi/2) = 0; f'(3\pi/2) = 0$

23. $f'(0.249) \approx 0$ **25.** Not continuous on $[0, \pi]$

27. 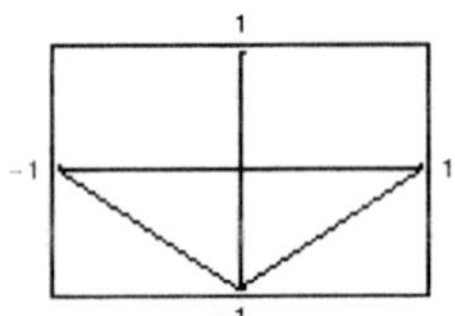
Rolle's Theorem does not apply.

29. 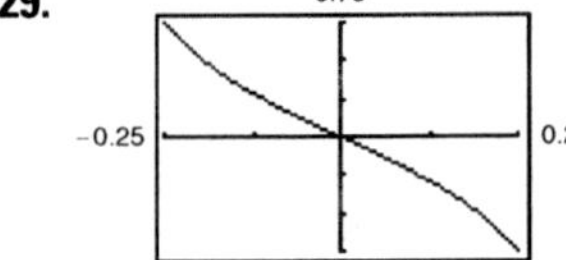
Rolle's Theorem does not apply.

31.

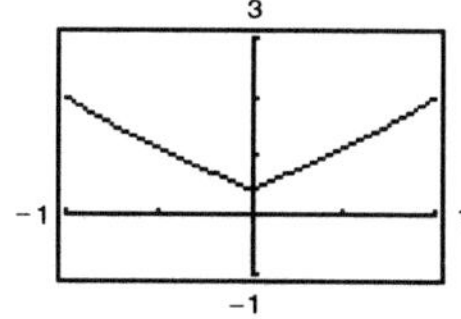

Rolle's Theorem does not apply.

33. (a) $f(1) = f(2) = 38$
(b) Velocity $= 0$ for some t in $(1, 2)$; $t = \frac{3}{2}$ sec

35.

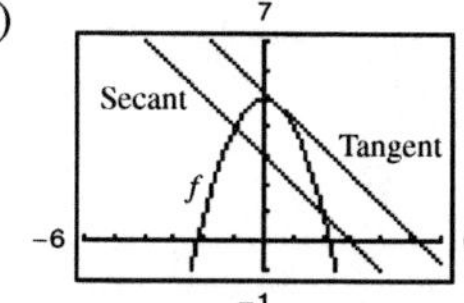

37. The function is not continuous on $[0, 6]$.
39. The function is not continuous on $[0, 6]$.
41. (a) Secant line: $x + y - 3 = 0$ (b) $c = \frac{1}{2}$
(c) Tangent line: $4x + 4y - 21 = 0$
(d)

43. $f'\left(-\frac{1}{2}\right) = -1$ **45.** $f'\left(1/\sqrt{3}\right) = 3, f'\left(-1/\sqrt{3}\right) = 3$
47. $f'\left(\frac{8}{27}\right) = 1$ **49.** f is not differentiable at $x = -\frac{1}{2}$.
51. $f'\left(\frac{\pi}{2}\right) = 0$ **53.** $f'\left(\ln \sqrt[3]{\dfrac{6}{1 - e^{-6}}}\right) = \dfrac{e^{-6} - 1}{2}$
55. $f'(4e^{-1}) = 2$
57. Secant line: $2x - 3y - 2 = 0$
Tangent line: $c = \left(-2 + \sqrt{6}\right)/2$, $2x - 3y + 5 - 2\sqrt{6} = 0$

59. Secant line: $x - 4y + 3 = 0$
Tangent line: $c = 4$, $x - 4y + 4 = 0$

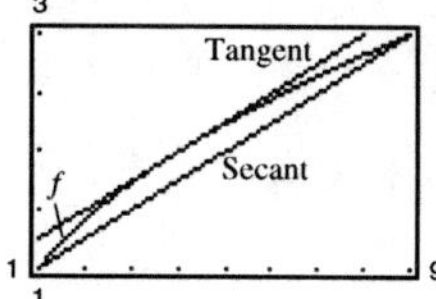

61. Secant line: $x + y - 2 = 0$
Tangent line: $c \approx 1.0161, x + y - 2.8161 = 0$

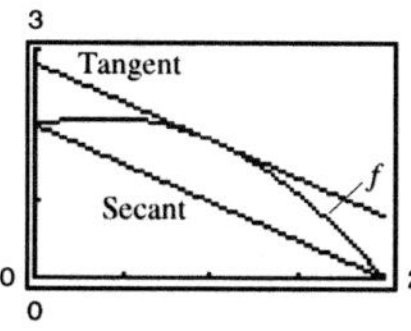

63. No. Let $f(x) = x^2$ on $[-1, 2]$.
65. No. $f(x)$ is not continuous on $[0, 1]$. So it does not satisfy the hypothesis of Rolle's Theorem.
67. By the Mean Value Theorem, there is a time when the speed of the plane must equal the average speed of 454.5 miles per hour. The speed was 400 miles per hour when the plane was accelerating to 454.5 miles per hour and decelerating from 454.5 miles per hour.
69. Proof
71. (a)

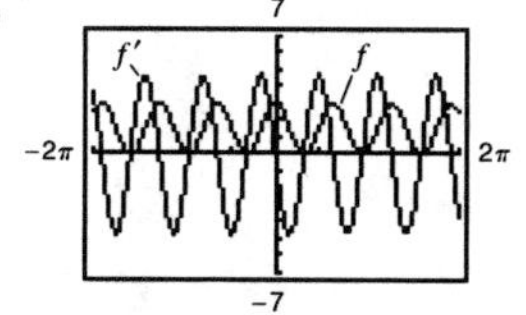

(b) Yes; yes
(c) Because $f(-1) = f(1) = 0$, Rolle's Theorem applies on $[-1, 1]$. Because $f(1) = 0$ and $f(2) = 3$, Rolle's Theorem does not apply on $[1, 2]$.
(d) $\lim\limits_{x \to 3^-} f'(x) = 0$; $\lim\limits_{x \to 3^+} f'(x) = 0$
73.

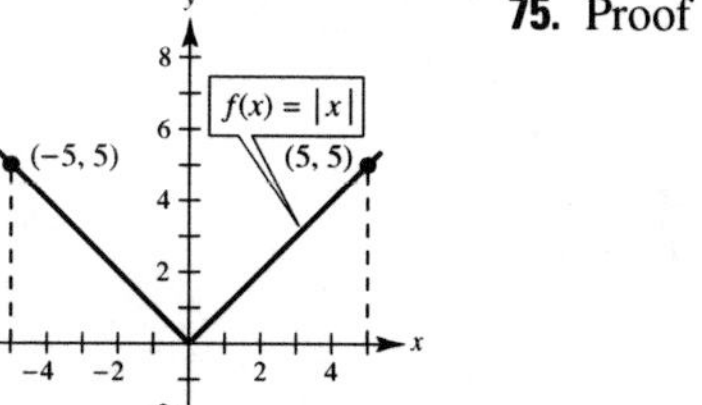

75. Proof   **77.** Proof
79. $a = 6, b = 1, c = 2$ **81.** $f(x) = 5$ **83.** $f(x) = x^2 - 1$
85. False. f is not continuous on $[-1, 1]$.   **87.** True
89–97. Proofs

Section 4.3   (page 226)

1. (a) $(0, 6)$ (b) $(6, 8)$
3. Increasing on $(3, \infty)$; Decreasing on $(-\infty, 3)$
5. Increasing on $(-\infty, -2)$ and $(2, \infty)$; Decreasing on $(-2, 2)$
7. Increasing on $(-\infty, -1)$; Decreasing on $(-1, \infty)$
9. Increasing on $(1, \infty)$; Decreasing on $(-\infty, 1)$
11. Increasing on $\left(-2\sqrt{2}, 2\sqrt{2}\right)$;
Decreasing on $\left(-4, -2\sqrt{2}\right), \left(2\sqrt{2}, 4\right)$
13. Increasing on $(0, \pi/2)$ and $(3\pi/2, 2\pi)$;
Decreasing on $(\pi/2, 3\pi/2)$
15. Increasing on $(0, 7\pi/6)$ and $(11\pi/6, 2\pi)$;
Decreasing on $(7\pi/6, 11\pi/6)$
17. Increasing on $\left(-\frac{1}{4}\ln 3, \infty\right)$;
Decreasing on $\left(-\infty, -\frac{1}{4}\ln 3\right)$

19. Increasing on $\left(\dfrac{2}{\sqrt{e}}, \infty\right)$;

  Decreasing on $\left(0, \dfrac{2}{\sqrt{e}}\right)$

21. Critical number: $x = 2$
  Increasing on $(2, \infty)$
  Decreasing on $(-\infty, 2)$
  Relative minimum: $(2, -4)$

23. Critical number: $x = 1$
  Increasing on $(-\infty, 1)$
  Decreasing on $(1, \infty)$
  Relative maximum: $(1, 5)$

25. Critical numbers: $x = -2, 1$
  Increasing on $(-\infty, -2)$ and $(1, \infty)$
  Decreasing on $(-2, 1)$
  Relative maximum: $(-2, 20)$
  Relative minimum: $(1, -7)$

27. Critical numbers: $x = -\frac{5}{3}, 1$
  Increasing on $\left(-\infty, -\frac{5}{3}\right)$ and $(1, \infty)$
  Decreasing on $\left(-\frac{5}{3}, 1\right)$
  Relative maximum: $\left(-\frac{5}{3}, \frac{256}{27}\right)$
  Relative minimum: $(1, 0)$

29. Critical numbers: $x = -1, 1$
  Increasing on $(-\infty, -1)$ and $(1, \infty)$
  Decreasing on $(-1, 1)$
  Relative maximum: $\left(-1, \frac{4}{5}\right)$
  Relative minimum: $\left(1, -\frac{4}{5}\right)$

31. Critical number: $x = 0$
  Increasing on $(-\infty, \infty)$
  No relative extrema

33. Critical number: $x = -2$
  Increasing on $(-2, \infty)$
  Decreasing on $(-\infty, -2)$
  Relative minimum: $(-2, 0)$

35. Critical number: $x = 5$
  Increasing on $(-\infty, 5)$
  Decreasing on $(5, \infty)$
  Relative maximum: $(5, 5)$

37. Critical numbers: $x = \pm\sqrt{2}/2$
  Discontinuity: $x = 0$
  Increasing on $(-\infty, -\sqrt{2}/2)$ and $(\sqrt{2}/2, \infty)$
  Decreasing on $\left(-\sqrt{2}/2, 0\right)$ and $\left(0, \sqrt{2}/2\right)$
  Relative maximum: $\left(-\sqrt{2}/2, -2\sqrt{2}\right)$
  Relative minimum: $\left(\sqrt{2}/2, 2\sqrt{2}\right)$

39. Critical number: $x = 0$
  Discontinuities: $x = -3, 3$
  Increasing on $(-\infty, -3)$ and $(-3, 0)$
  Decreasing on $(0, 3)$ and $(3, \infty)$
  Relative maximum: $(0, 0)$

41. Critical numbers: $x = -3, 1$
  Discontinuity: $x = -1$
  Increasing on $(-\infty, -3)$ and $(1, \infty)$
  Decreasing on $(-3, -1)$ and $(-1, 1)$
  Relative maximum: $(-3, -8)$
  Relative minimum: $(1, 0)$

43. Critical number: $x = 0$
  Increasing on $(-\infty, 0)$
  Decreasing on $(0, \infty)$
  No relative extrema

45. Critical number: $x = 1$
  Increasing on $(-\infty, 1)$
  Decreasing on $(1, \infty)$
  Relative maximum: $(1, 4)$

47. Critical number: $x = 2$
  Increasing on $(-\infty, 2)$
  Decreasing on $(2, \infty)$
  Relative maximum: $(2, e^{-1})$

49. Critical number: $x = 0$
  Decreasing on $[-1, 1]$
  No relative extrema

51. Critical number: $x = 1/\ln 3$
  Increasing on $(-\infty, 1/\ln 3)$
  Decreasing on $(1/\ln 3, \infty)$
  Relative maximum: $(1/\ln 3, (3^{-1/\ln 3})/\ln 3)$ or
      $(1/\ln 3, 1/(e \ln 3))$

53. Critical number: $x = 1/\ln 4$
  Increasing on $(1/\ln 4, \infty)$
  Decreasing on $(0, 1/\ln 4)$
  Relative minimum: $(1/\ln 4, (\ln(\ln 4) + 1)/\ln 4)$

55. No critical numbers
  Increasing on $(-\infty, \infty)$
  No relative extrema

57. No critical numbers
  Increasing on $(-\infty, 2)$ and $(2, \infty)$
  No relative extrema

59. (a) Increasing on $(0, \pi/6), (5\pi/6, 2\pi)$
       Decreasing on $(\pi/6, 5\pi/6)$
   (b) Relative maximum: $\left(\pi/6, (\pi + 6\sqrt{3})/12\right)$
       Relative minimum: $\left(5\pi/6, (5\pi - 6\sqrt{3})/12\right)$

61. (a) Increasing on $(0, \pi/4), (5\pi/4, 2\pi)$
       Decreasing on $(\pi/4, 5\pi/4)$
   (b) Relative maximum: $\left(\pi/4, \sqrt{2}\right)$
       Relative minimum: $\left(5\pi/4, -\sqrt{2}\right)$

63. (a) Increasing on $(\pi/4, \pi/2), (3\pi/4, \pi), (5\pi/4, 3\pi/2),$
       $(7\pi/4, 2\pi)$
       Decreasing on $(0, \pi/4), (\pi/2, 3\pi/4), (\pi, 5\pi/4),$
       $(3\pi/2, 7\pi/4)$
   (b) Relative maxima: $(\pi/2, 1), (\pi, 1), (3\pi/2, 1)$
       Relative minima: $(\pi/4, 0), (3\pi/4, 0), (5\pi/4, 0), (7\pi/4, 0)$

65. (a) Increasing on $(0, \pi/2)$, $(7\pi/6, 3\pi/2)$, $(11\pi/6, 2\pi)$
Decreasing on $(\pi/2, 7\pi/6)$, $(3\pi/2, 11\pi/6)$
(b) Relative maxima: $(\pi/2, 2)$, $(3\pi/2, 0)$
Relative minima: $(7\pi/6, -1/4)$, $(11\pi/6, -1/4)$

67. (a) $f'(x) = (2(9 - 2x^2))/\sqrt{9 - x^2}$
(b)

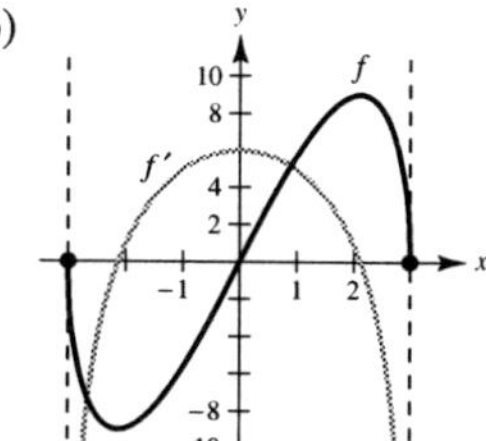

(c) Critical numbers: $x = \pm 3\sqrt{2}/2$
(d) $f' > 0$ on $\left(-3\sqrt{2}/2, 3\sqrt{2}/2\right)$
$f' < 0$ on $\left(-3, -3\sqrt{2}/2\right)$, $\left(3\sqrt{2}/2, 3\right)$
(e) f is increasing when f' is positive, and decreasing when f' is negative.

69. (a) $f'(t) = t(t \cos t + 2 \sin t)$
(b)

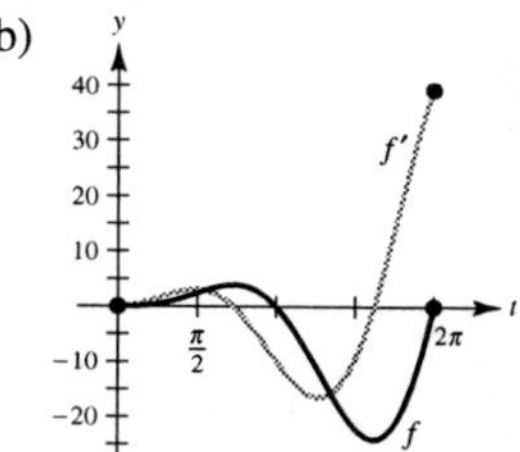

(c) Critical numbers: $t = 2.2889, 5.0870$
(d) $f' > 0$ on $(0, 2.2889)$, $(5.0870, 2\pi)$
$f' < 0$ on $(2.2889, 5.0870)$
(e) f is increasing when f' is positive, and decreasing when f' is negative.

71. (a) $f'(x) = (2x^2 - 1)/2x$
(b)

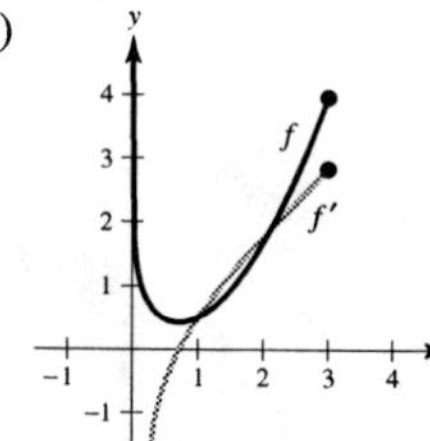

(c) Critical number: $x = \sqrt{2}/2$
(d) $f' > 0$ on $\left(\sqrt{2}/2, 3\right)$
$f' < 0$ on $\left(0, \sqrt{2}/2\right)$
(e) f is increasing when f' is positive, and decreasing when f' is negative.

73. $f(x)$ is symmetric with respect to the origin.
Zeros: $(0, 0)$, $\left(\pm\sqrt{3}, 0\right)$

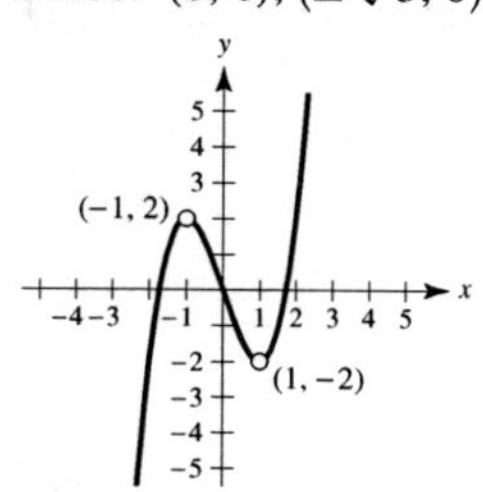

$g(x)$ is continuous on $(-\infty, \infty)$ and $f(x)$ has holes at $x = 1$ and $x = -1$.

75.

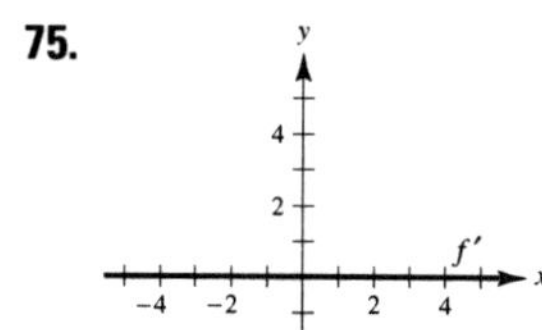

77.

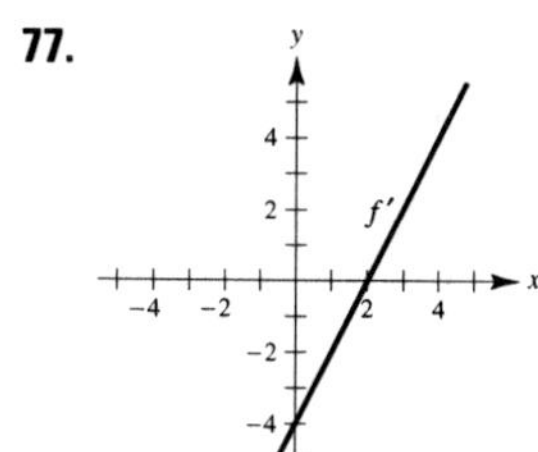

79.

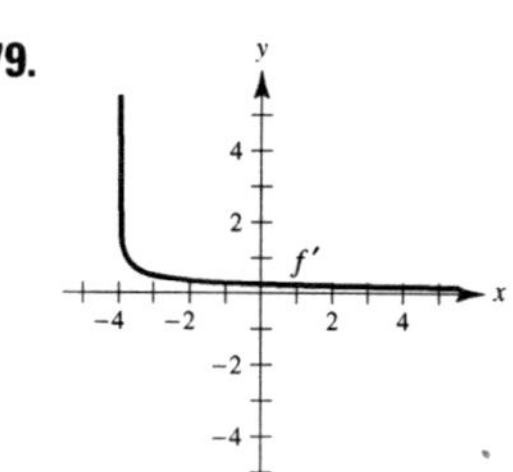

81. (a) Increasing on $(2, \infty)$; Decreasing on $(-\infty, 2)$
(b) Relative minimum: $x = 2$

83. (a) Increasing on $(-\infty, -1)$ and $(0, 1)$;
Decreasing on $(-1, 0)$ and $(1, \infty)$
(b) Relative maxima: $x = -1$ and $x = 1$
Relative minimum: $x = 0$

85. (a) Critical numbers: $x = -1, x = 1, x = 2$
(b) Relative maximum at $x = 1$, relative minimum at $x = 2$, and neither at $x = -1$

87. $g'(0) < 0$ **89.** $g'(-6) < 0$ **91.** $g'(0) > 0$

93. Answers will vary. Sample answer:

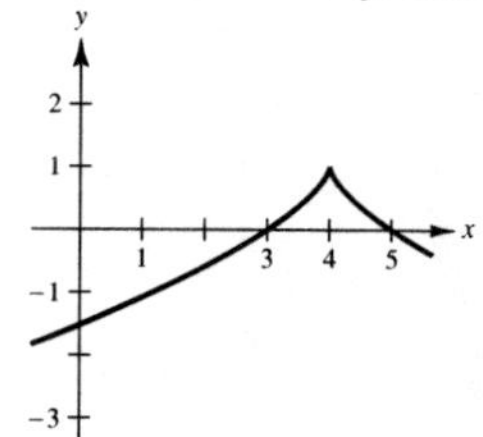

95. (a)

(b) Critical numbers: $x \approx -0.40$ and $x \approx 0.48$
(c) Relative maximum: $(0.48, 1.25)$
Relative minimum: $(-0.40, 0.75)$

97. (a) $s'(t) = 9.8(\sin \theta)t$; speed $= |9.8(\sin \theta)t|$
(b)

θ	0	$\pi/4$	$\pi/3$	$\pi/2$	$2\pi/3$	$3\pi/4$	π
$s'(t)$	0	$4.9\sqrt{2}t$	$4.9\sqrt{3}t$	$9.8t$	$4.9\sqrt{3}t$	$4.9\sqrt{2}t$	0

The speed is maximum at $\theta = \pi/2$.

99. (a)

x	0.5	1	1.5	2	2.5	3
$f(x)$	0.5	1	1.5	2	2.5	3
$g(x)$	0.48	0.84	1.00	0.91	0.60	0.14

$f(x) > g(x)$

(b) 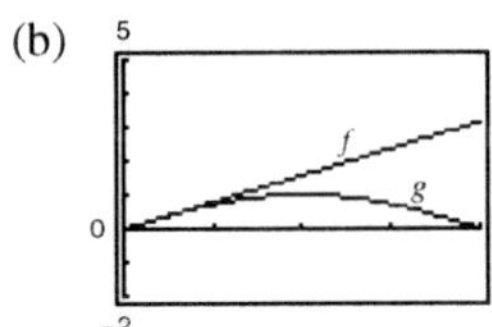

(c) Proof

$f(x) > g(x)$

101. $r = 2R/3$

103. (a) $v(t) = 6 - 2t$ (b) $[0, 3)$ (c) $(3, \infty)$ (d) $t = 3$

105. (a) $v(t) = 3t^2 - 10t + 4$

(b) $\left[0, (5 - \sqrt{13})/3\right)$ and $\left((5 + \sqrt{13})/3, \infty\right)$

(c) $\left((5 - \sqrt{13})/3, (5 + \sqrt{13})/3\right)$

(d) $t = (5 \pm \sqrt{13})/3$

107. Answers will vary.

109. (a) 3

(b) $a_3(0)^3 + a_2(0)^2 + a_1(0) + a_0 = 0$
$a_3(2)^3 + a_2(2)^2 + a_1(2) + a_0 = 2$
$3a_3(0)^2 + 2a_2(0) + a_1 = 0$
$3a_3(2)^2 + 2a_2(2) + a_1 = 0$

(c) $f(x) = -\frac{1}{2}x^3 + \frac{3}{2}x^2$

111. (a) 4

(b) $a_4(0)^4 + a_3(0)^3 + a_2(0)^2 + a_1(0) + a_0 = 0$
$a_4(2)^4 + a_3(2)^3 + a_2(2)^2 + a_1(2) + a_0 = 4$
$a_4(4)^4 + a_3(4)^3 + a_2(4)^2 + a_1(4) + a_0 = 0$
$4a_4(0)^3 + 3a_3(0)^2 + 2a_2(0) + a_1 = 0$
$4a_4(2)^3 + 3a_3(2)^2 + 2a_2(2) + a_1 = 0$

(c) $f(x) = \frac{1}{4}x^4 - 2x^3 + 4x^2$

113. True **115.** False. Let $f(x) = x^3$.

117. False. Let $f(x) = x^3$. There is a critical number at $x = 0$, but not a relative extremum.

119–123. Proofs **125.** Putnam Problem A3, 2003

Section 4.4 (page 235)

1. $f' > 0, f'' > 0$ **3.** $f' < 0, f'' < 0$

5. Concave upward: $(-\infty, \infty)$

7. Concave upward: $(-\infty, 1)$; Concave downward: $(1, \infty)$

9. Concave upward: $(-\infty, 2)$; Concave downward: $(2, \infty)$

11. Concave upward: $(-\infty, -2), (2, \infty)$
Concave downward: $(-2, 2)$

13. Concave upward: $(-\infty, -1), (1, \infty)$
Concave downward: $(-1, 1)$

15. Concave upward: $(-2, 2)$
Concave downward: $(-\infty, -2), (2, \infty)$

17. Concave upward: $(-\pi/2, 0)$; Concave downward: $(0, \pi/2)$

19. Points of inflection: $(-2, -8), (0, 0)$
Concave upward: $(-\infty, -2), (0, \infty)$
Concave downward: $(-2, 0)$

21. Point of inflection: $(2, 8)$
Concave downward: $(-\infty, 2)$
Concave upward: $(2, \infty)$

23. Points of inflection: $\left(\pm 2/\sqrt{3}, -20/9\right)$
Concave upward: $\left(-\infty, -2/\sqrt{3}\right), \left(2/\sqrt{3}, \infty\right)$
Concave downward: $\left(-2/\sqrt{3}, 2/\sqrt{3}\right)$

25. Points of inflection: $(2, -16), (4, 0)$
Concave upward: $(-\infty, 2), (4, \infty)$
Concave downward: $(2, 4)$

27. Concave upward: $(-3, \infty)$

29. Points of inflection: $\left(-\sqrt{3}/3, 3\right), \left(\sqrt{3}/3, 3\right)$
Concave upward: $\left(-\infty, -\sqrt{3}/3\right), \left(\sqrt{3}/3, \infty\right)$
Concave downward: $\left(-\sqrt{3}/3, \sqrt{3}/3\right)$

31. Point of inflection: $(2\pi, 0)$
Concave upward: $(2\pi, 4\pi)$
Concave downward: $(0, 2\pi)$

33. Concave upward: $(0, \pi), (2\pi, 3\pi)$
Concave downward: $(\pi, 2\pi), (3\pi, 4\pi)$

35. Points of inflection: $(\pi, 0), (1.823, 1.452), (4.46, -1.452)$
Concave upward: $(1.823, \pi), (4.46, 2\pi)$
Concave downward: $(0, 1.823), (\pi, 4.46)$

37. Point of inflection: $\left(\frac{3}{2}, e^{-2}\right)$
Concave upward: $(-\infty, 0), \left(0, \frac{3}{2}\right)$
Concave downward: $\left(\frac{3}{2}, \infty\right)$

39. Concave upward: $(0, \infty)$

41. Points of inflection:
$\left(-\left(\frac{1}{5}\right)^{5/8}, \arcsin\left(\sqrt{5}/5\right)\right), \left(\left(\frac{1}{5}\right)^{5/8}, \arcsin\left(\sqrt{5}/5\right)\right)$
Concave upward: $\left(-1, -\left(\frac{1}{5}\right)^{5/8}\right), \left(\left(\frac{1}{5}\right)^{5/8}, 1\right)$
Concave downward: $\left(-\left(\frac{1}{5}\right)^{5/8}, 0\right), \left(0, \left(\frac{1}{5}\right)^{5/8}\right)$

43. Relative minimum: $(5, 0)$ **45.** Relative maximum: $(3, 9)$

47. Relative maximum: $(0, 3)$
Relative minimum: $(2, -1)$

49. Relative minimum: $(3, -25)$

51. Relative maximum: $(2.4, 268.74)$
Relative minimum: $(0, 0)$

53. Relative minimum: $(0, -3)$

55. Relative maximum: $(-2, -4)$
Relative minimum: $(2, 4)$

57. No relative extrema, because f is nonincreasing.

59. Relative minimum: $\left(1, \frac{1}{2}\right)$

61. Relative minimum: (e, e) **63.** Relative minimum: $(0, 1)$

65. Relative minimum: $(0, 0)$
Relative maximum: $(2, 4e^{-2})$

67. Relative maximum: $(1/\ln 4, 4e^{-1}/\ln 2)$

69. Relative minimum: $(-1.272, 3.747)$
Relative maximum: $(1.272, -0.606)$

71. (a) $f'(x) = 0.2x(x - 3)^2(5x - 6)$
$f''(x) = 0.4(x - 3)(10x^2 - 24x + 9)$

(b) Relative maximum: $(0, 0)$
Relative minimum: $(1.2, -1.6796)$
Points of inflection: $(0.4652, -0.7048),$
$(1.9348, -0.9048), (3, 0)$

(c)

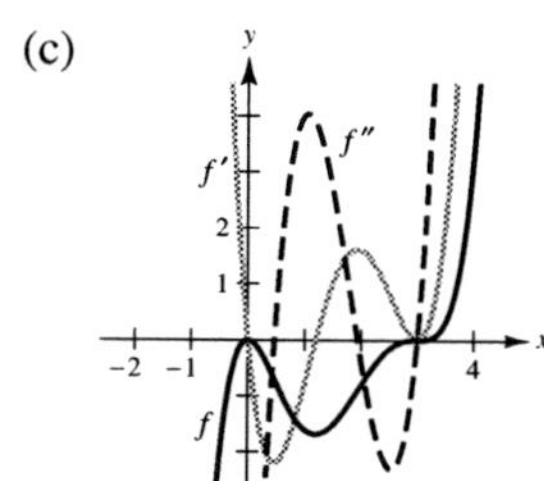

f is increasing when f' is positive, and decreasing when f' is negative. f is concave upward when f'' is positive, and concave downward when f'' is negative.

73. (a) $f'(x) = \cos x - \cos 3x + \cos 5x$

  $f''(x) = -\sin x + 3 \sin 3x - 5 \sin 5x$

(b) Relative maximum: $(\pi/2, 1.53333)$

  Points of inflection: $(0.5236, 0.2667)$, $(1.1731, 0.9637)$,

  $(1.9685, 0.9637)$, $(2.6180, 0.2667)$

(c)

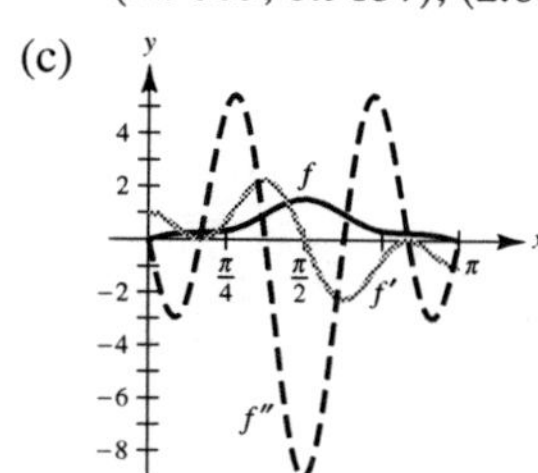

f is increasing when f' is positive, and decreasing when f' is negative. f is concave upward when f'' is positive, and concave downward when f'' is negative.

75. (a)

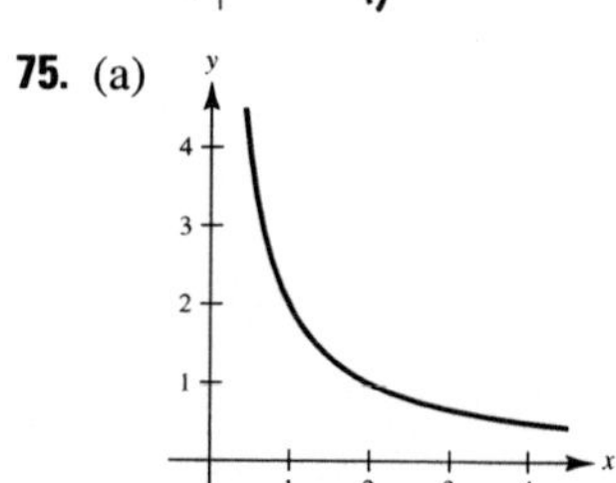

(b)

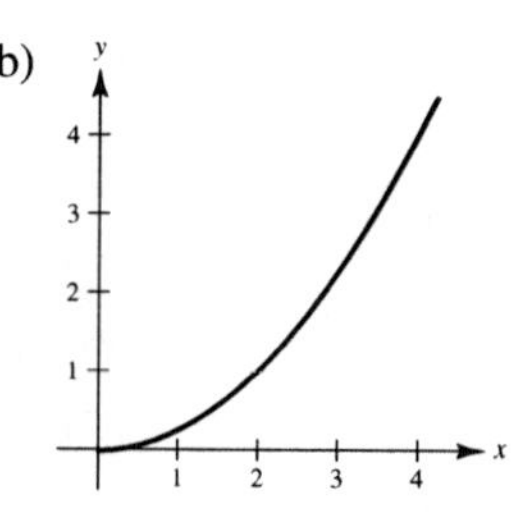

77. Answers will vary.

Sample answer: $f(x) = x^4$; $f''(0) = 0$, but $(0, 0)$ is not a point of inflection.

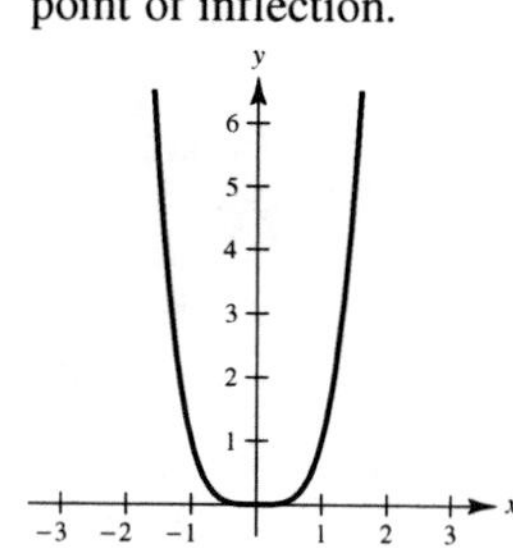

79.

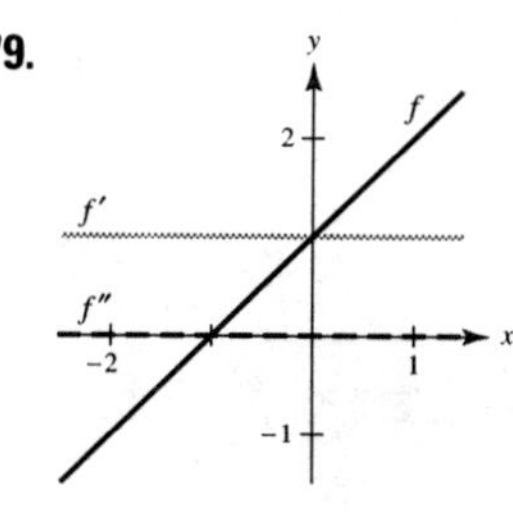

81.

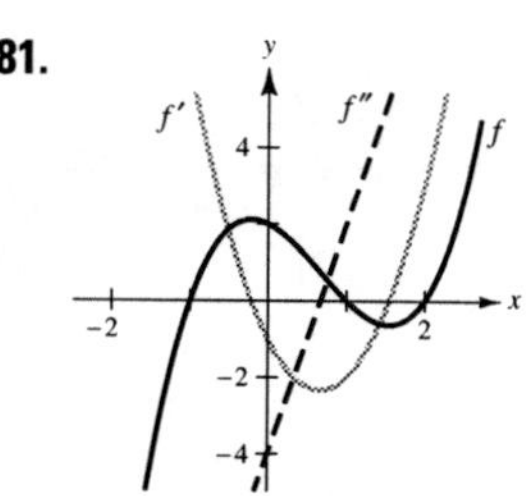

83.

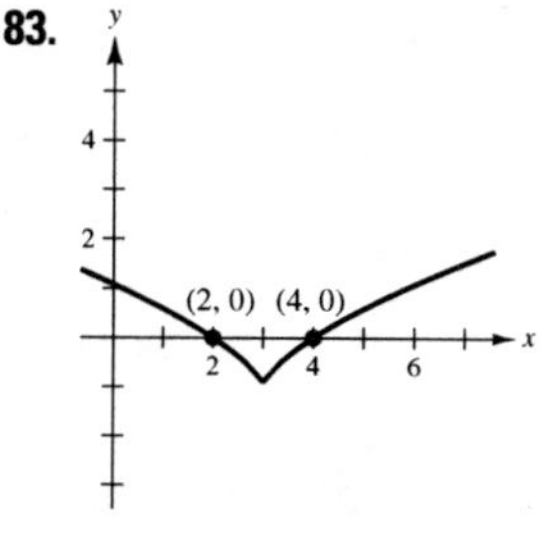

85.

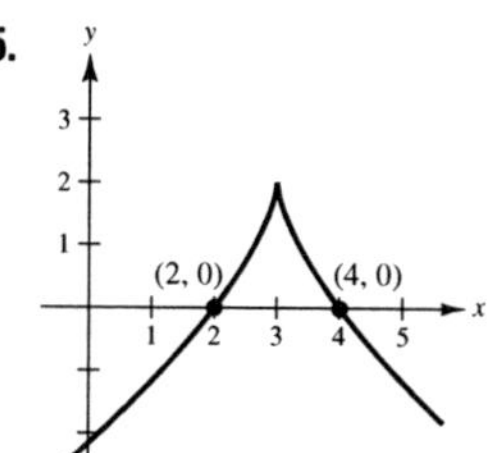

87. (a)

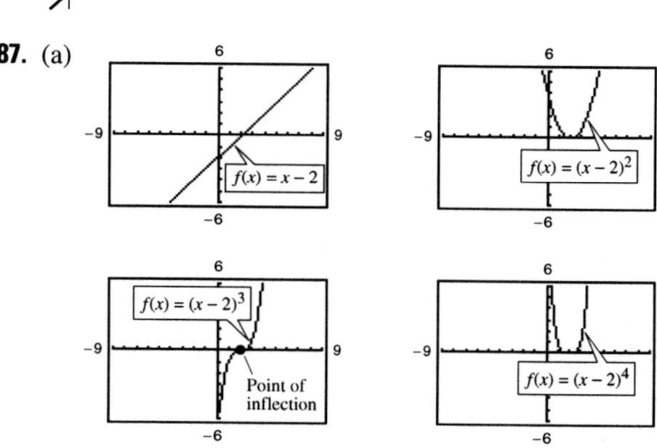

$f(x) = (x - 2)^n$ has a point of inflection at $(2, 0)$ if n is odd and $n \geq 3$.

(b) Answers will vary.

89. $f(x) = \frac{1}{2}x^3 - 6x^2 + \frac{45}{2}x - 24$

91. (a) $f(x) = \frac{1}{32}x^3 + \frac{3}{16}x^2$ (b) 2 miles from touchdown

93. $x = \left[(15 - \sqrt{33})/16\right]L \approx 0.578L$ **95.** $x = 100$ units

97. $P_1(x) = 2\sqrt{2}$

  $P_2(x) = 2\sqrt{2} - \sqrt{2}[x - (\pi/4)]^2$

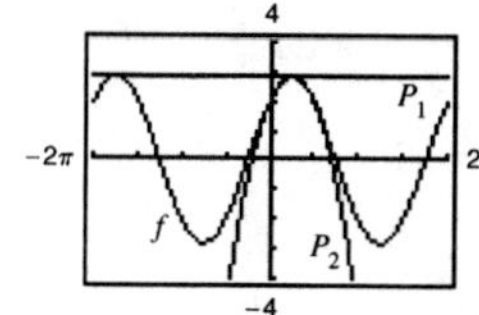

The values of f, P_1, and P_2 and their first derivatives are equal when $x = \pi/4$. The approximations worsen as you move away from $x = \pi/4$.

99. $P_1(x) = -\pi/4 + 1/2(x + 1)$

  $P_2(x) = -\pi/4 + 1/2(x + 1) + 1/4(x + 1)^2$

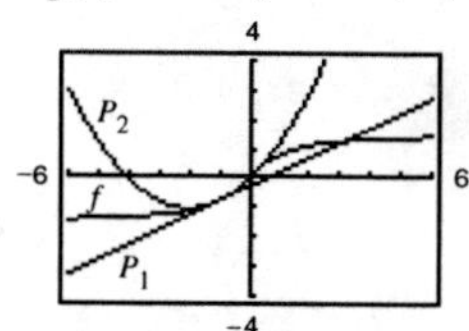

The values of f, P_1, and P_2 and their first derivatives are equal when $x = -1$. The approximations worsen as you move away from $x = -1$.

101. True   **103.** False. The maximum value is $\sqrt{13} \approx 3.60555$.

105. False. f is concave upward at $x = c$ if $f''(c) > 0$.   **107.** Proof

109.

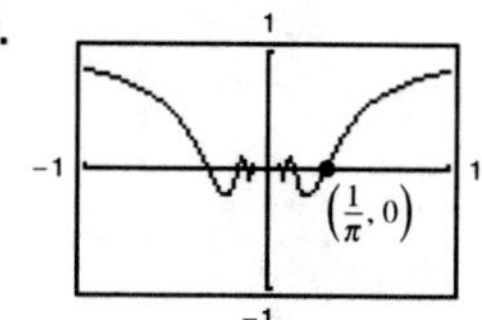

111. Proof

Section 4.5   (page 245)

1. f   **2.** c   **3.** d   **4.** a   **5.** b   **6.** e

7.

x	10^0	10^1	10^2	10^3
$f(x)$	7	2.2632	2.0251	2.0025

x	10^4	10^5	10^6
$f(x)$	2.0003	2.0000	2.0000

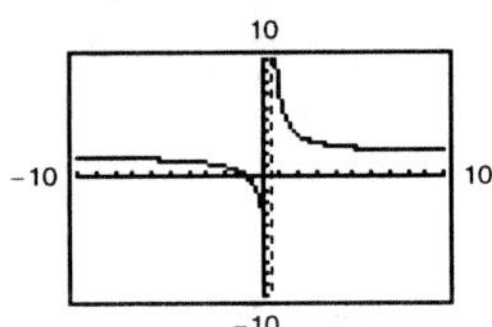

$$\lim_{x\to\infty}\frac{4x+3}{2x-1}=2$$

9.

x	10^0	10^1	10^2	10^3
$f(x)$	-2	-2.9814	-2.9998	-3.0000

x	10^4	10^5	10^6
$f(x)$	-3.0000	-3.0000	-3.0000

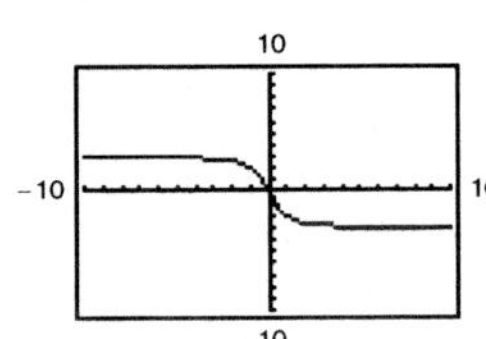

$$\lim_{x\to\infty}\frac{-6x}{\sqrt{4x^2+5}}=-3$$

11.

x	10^0	10^1	10^2	10^3
$f(x)$	4.5000	4.9901	4.9999	5.0000

x	10^4	10^5	10^6
$f(x)$	5.0000	5.0000	5.0000

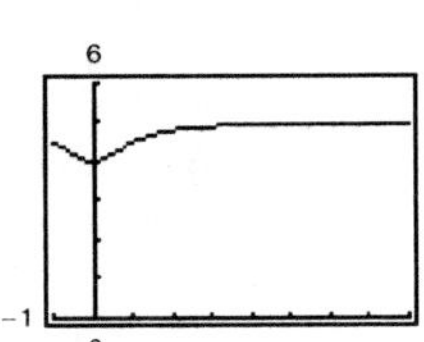

$$\lim_{x\to\infty}\left(5-\frac{1}{x^2+1}\right)=5$$

13. (a) ∞ (b) 5 (c) 0   **15.** (a) 0 (b) 1 (c) ∞
17. (a) 0 (b) $-\frac{2}{3}$ (c) $-\infty$ **19.** 4   **21.** $\frac{2}{3}$ **23.** 0
25. -1 **27.** -2 **29.** $\frac{1}{2}$ **31.** ∞ **33.** 0   **35.** 0
37. 2   **39.** 3   **41.** 0   **43.** $-\pi/2$
45. 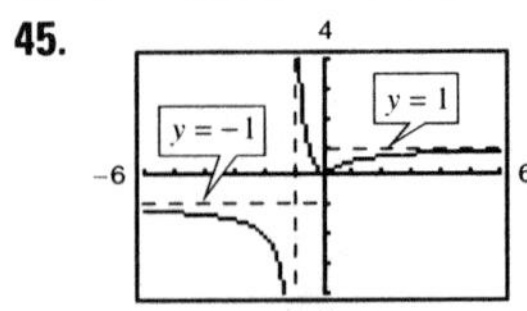   **47.** 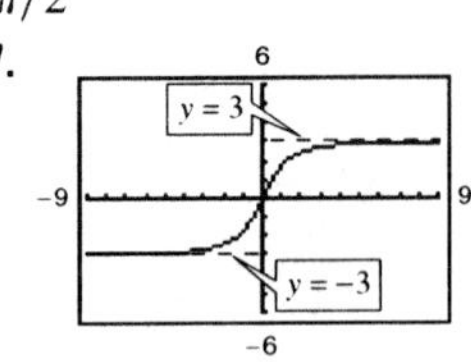

49. 1   **51.** 0   **53.** $\frac{1}{6}$

55.

x	10^0	10^1	10^2	10^3	10^4	10^5	10^6
$f(x)$	1.000	0.513	0.501	0.500	0.500	0.500	0.500

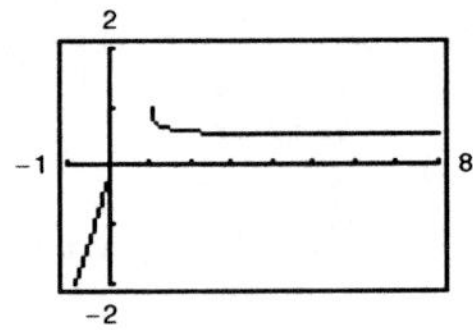

$$\lim_{x\to\infty}\left[x-\sqrt{x(x-1)}\right]=\tfrac{1}{2}$$

57.

x	10^0	10^1	10^2	10^3	10^4	10^5	10^6
$f(x)$	0.479	0.500	0.500	0.500	0.500	0.500	0.500

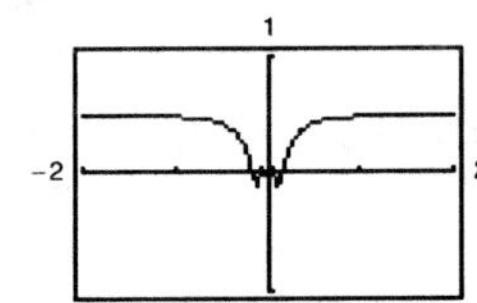

The graph has a hole at $x=0$.

$$\lim_{x\to\infty}x\sin\frac{1}{2x}=\frac{1}{2}$$

59. As x becomes large, $f(x)$ approaches 4.

61. Answers will vary. Sample answer:

$$\text{let } f(x)=\frac{-6}{0.1(x-2)^2+1}+6.$$

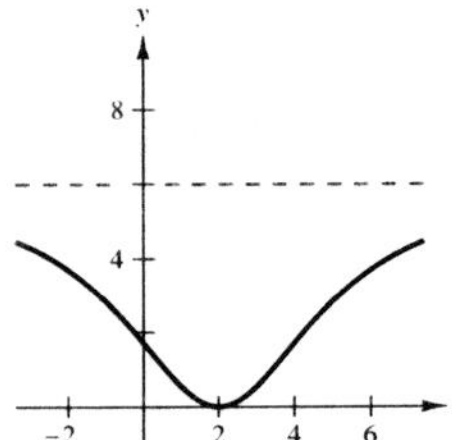

63. (a) 5 (b) -5

65. 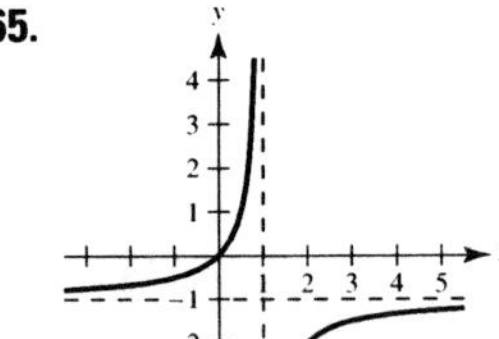   **67.** 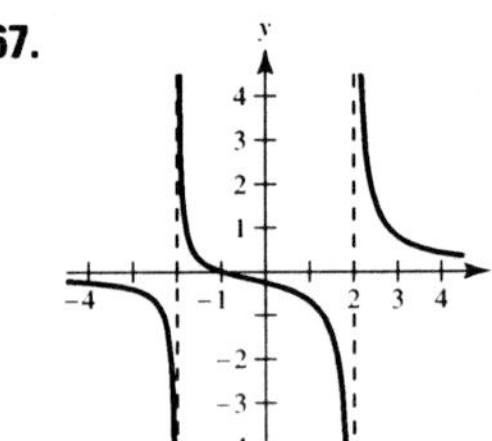

69. 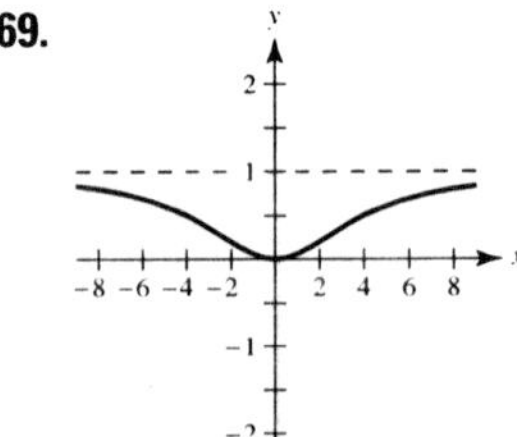   **71.** 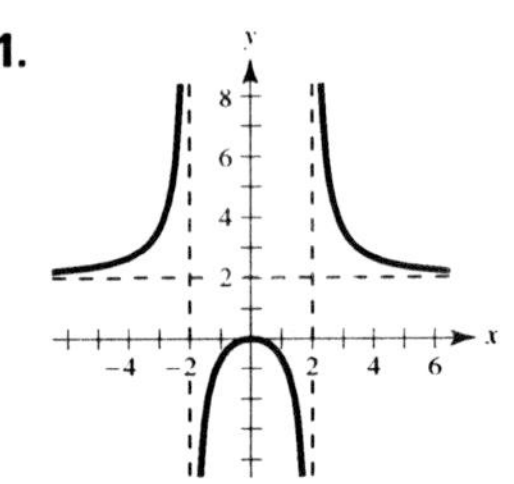

73. 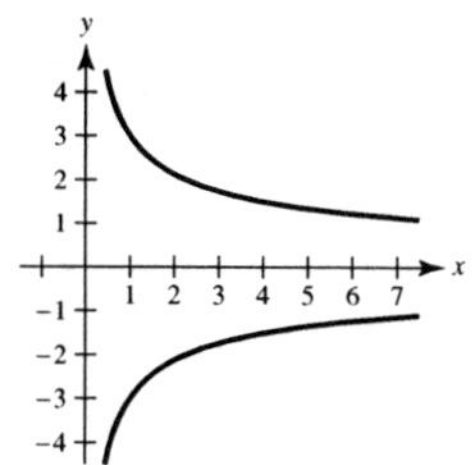

75. 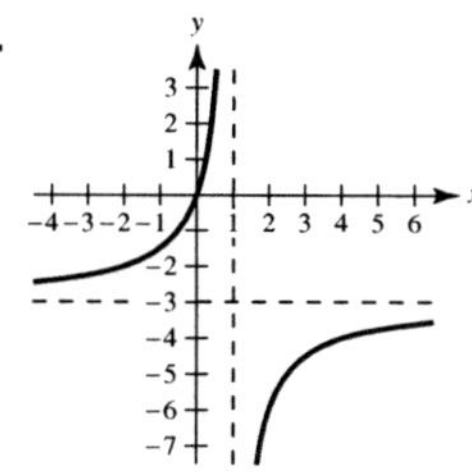

77. 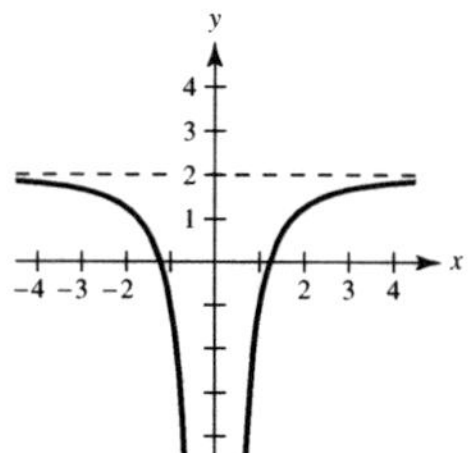

79. 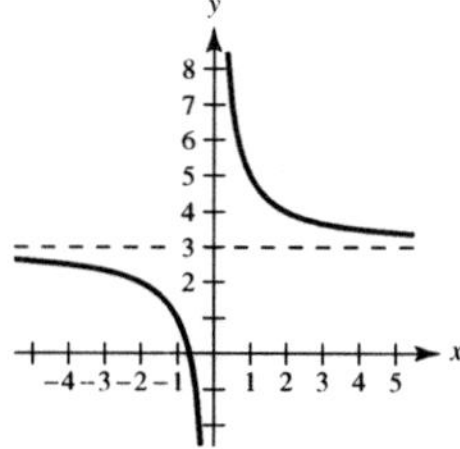

81. 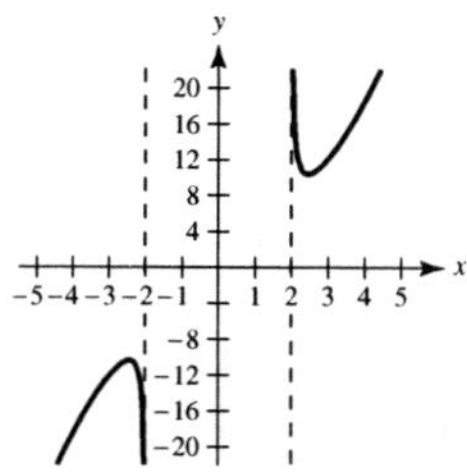

83. 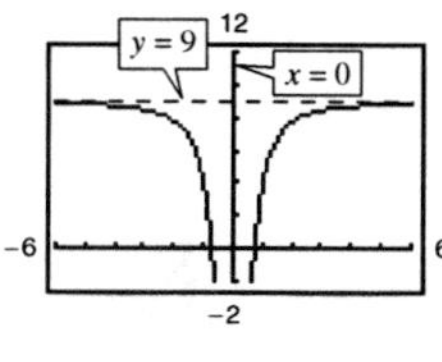

85. 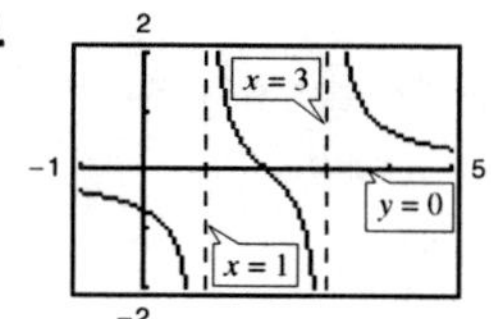

87. 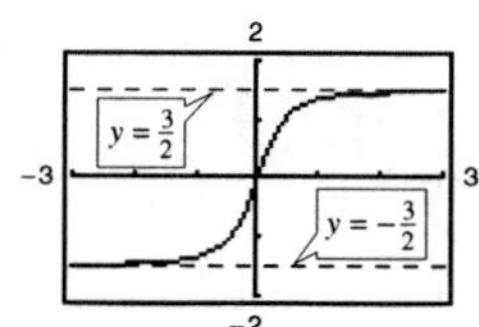

89. 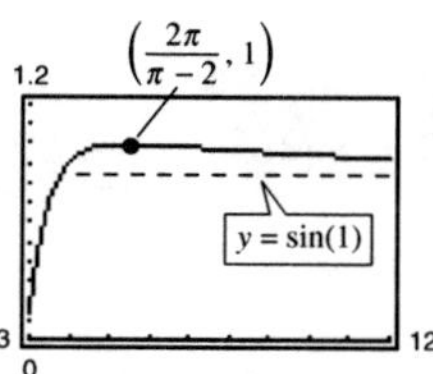

91. 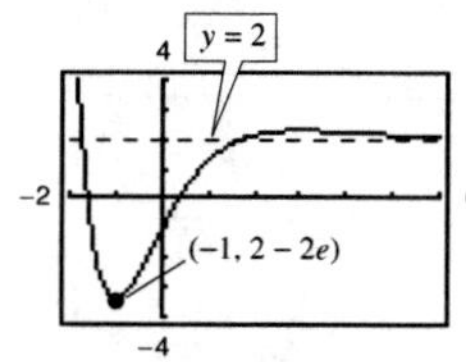

93. (a) 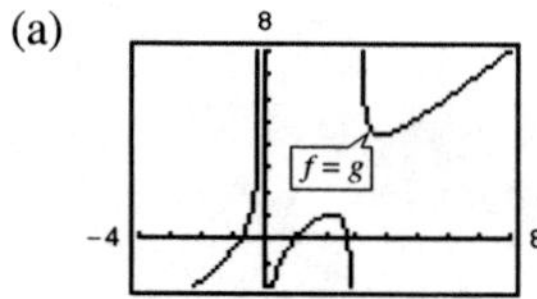

(c) 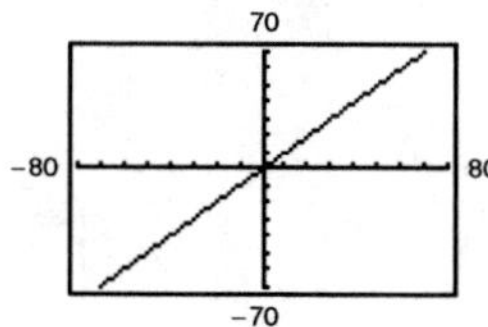

(b) Answers will vary.   The slant asymptote $y = x$

95. 100%   **97.** $\lim\limits_{t \to \infty} N(t) = +\infty$; $\lim\limits_{t \to \infty} E(t) = c$

99. (a) $T_1 = -0.003t^2 + 0.68t + 26.6$

(b) 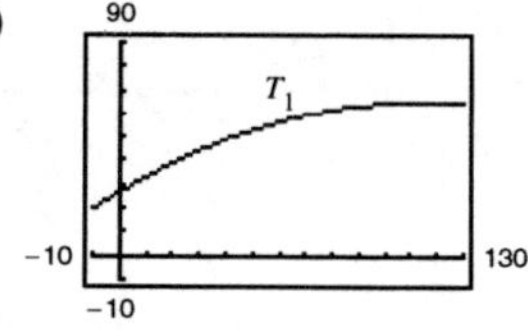

(c) 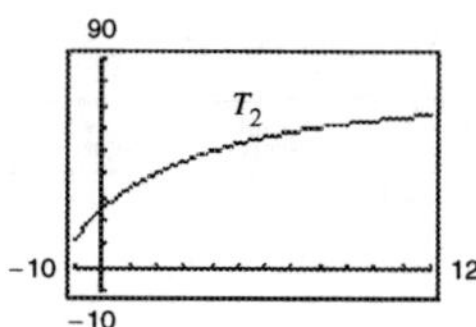

(d) $T_1(0) \approx 26.6°$, $T_2(0) \approx 25.0°$ (e) 86

(f) The limiting temperature is 86°.

No. T_1 has no horizontal asymptote.

101. (a) 7.1 million ft^3/acre

(b) $V'(20) \approx 0.077$

$V'(60) \approx 0.043$

103. (a) 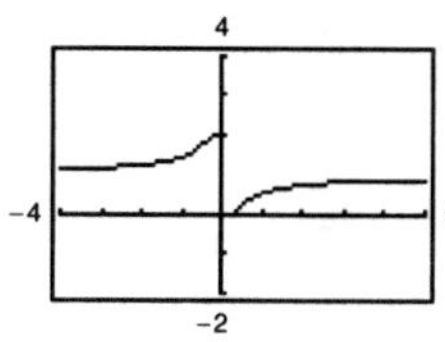

(b) Answers will vary.

105. (a) $d(m) = \dfrac{|3m + 3|}{\sqrt{m^2 + 1}}$

(b) 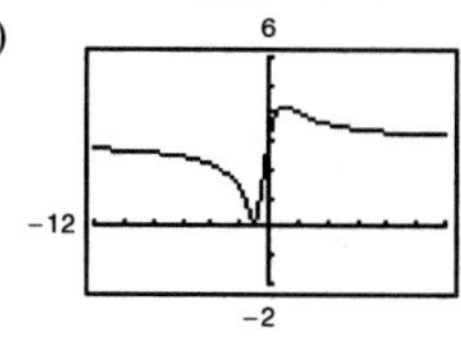

(c) $\lim\limits_{m \to \infty} d(m) = 3$

$\lim\limits_{m \to -\infty} d(m) = 3$

As m approaches $\pm\infty$, distance approaches 3.

107. (a) $\lim\limits_{x \to \infty} f(x) = 2$ (b) $x_1 = \sqrt{\dfrac{4 - 2\varepsilon}{\varepsilon}}$, $x_2 = -\sqrt{\dfrac{4 - 2\varepsilon}{\varepsilon}}$

(c) $\sqrt{\dfrac{4 - 2\varepsilon}{\varepsilon}}$ (d) $-\sqrt{\dfrac{4 - 2\varepsilon}{\varepsilon}}$

109. (a) Answers will vary. $M = \dfrac{5\sqrt{33}}{11}$

(b) Answers will vary. $M = \dfrac{29\sqrt{177}}{59}$

111–115. Proofs

117. False. Let $f(x) = \dfrac{2x}{\sqrt{x^2 + 2}}$.

Section 4.6   (page 255)

1. d   **2.** c   **3.** a   **4.** b

5. 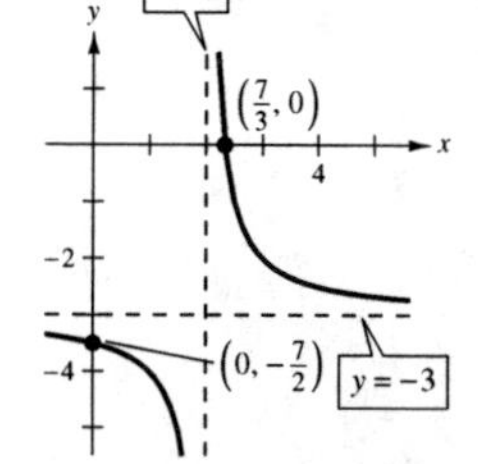

7. 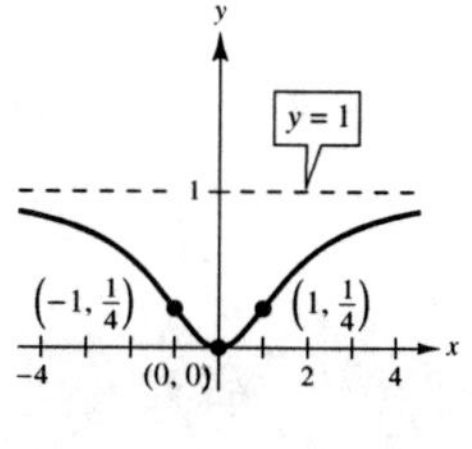

9. 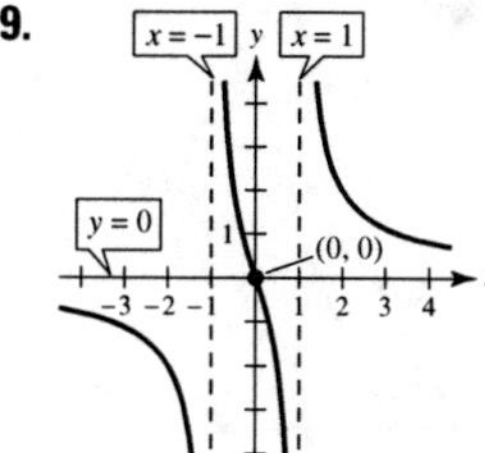

11. 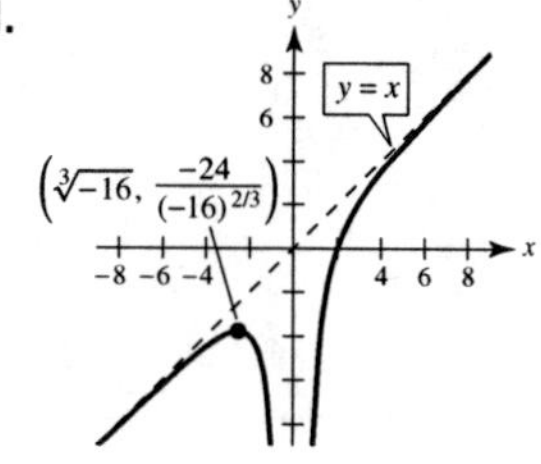

13. 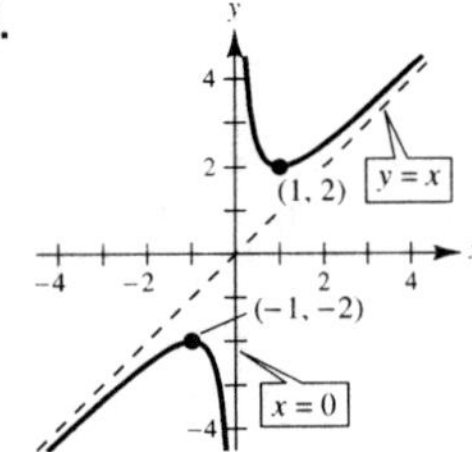

15. 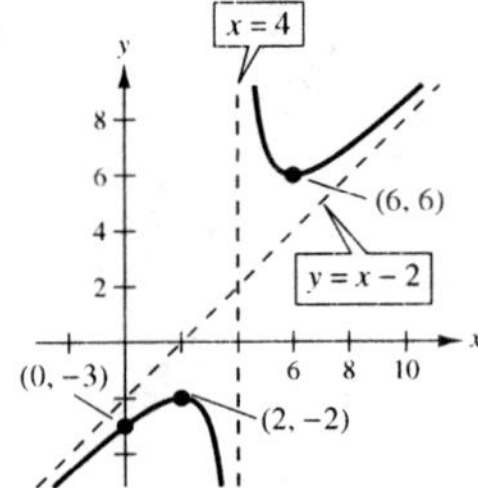

17. 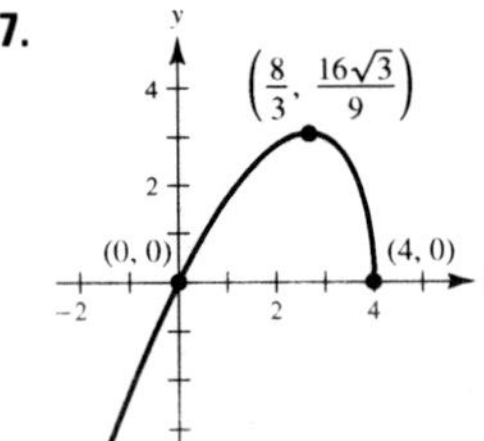

19. 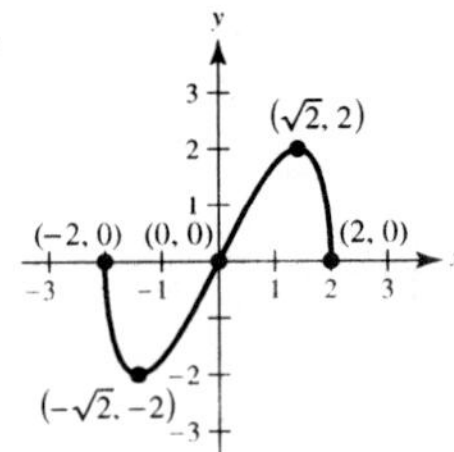

21. 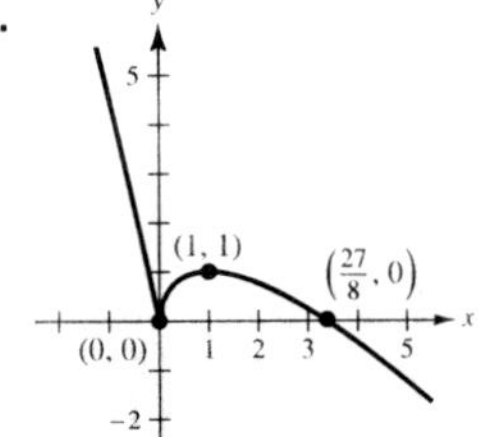

23. 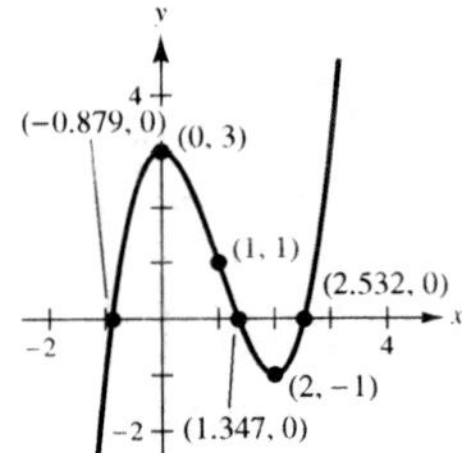

25. 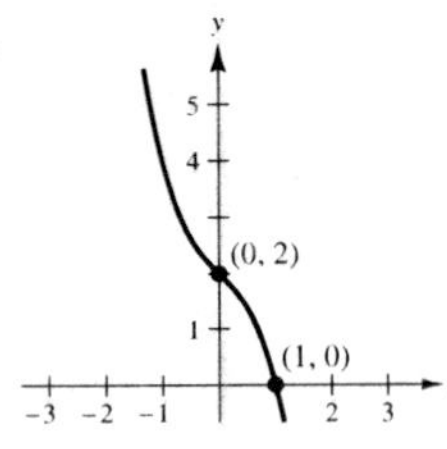

27. 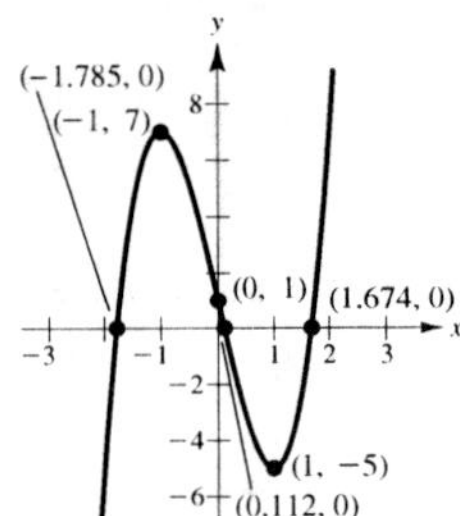

29. 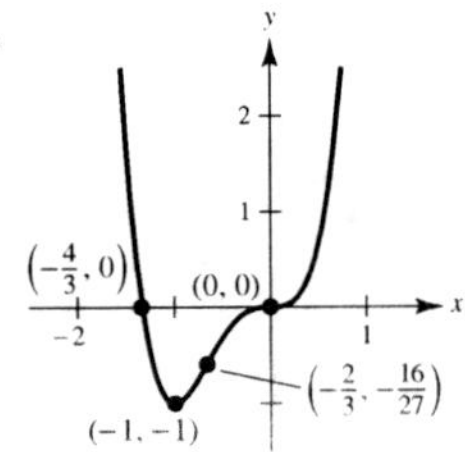

31. 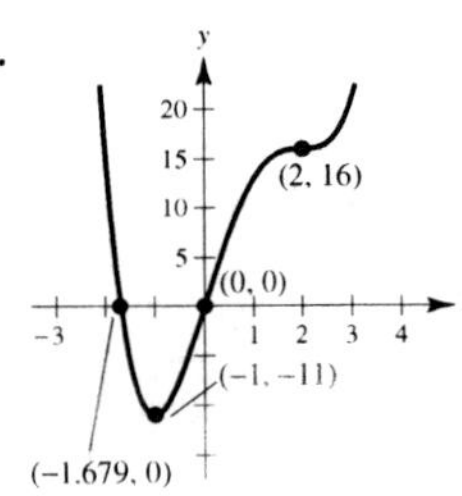

33. 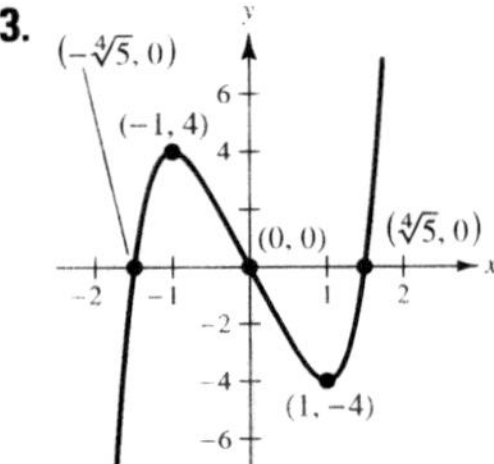

35. 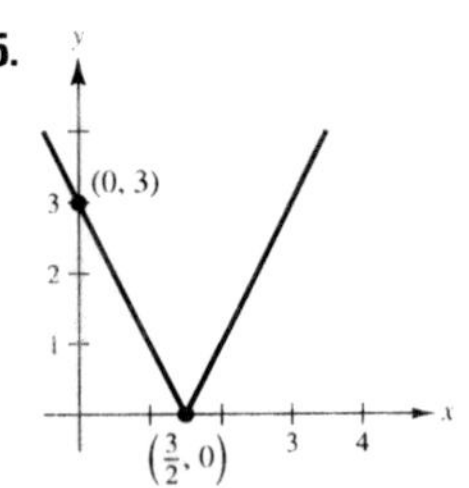

37. 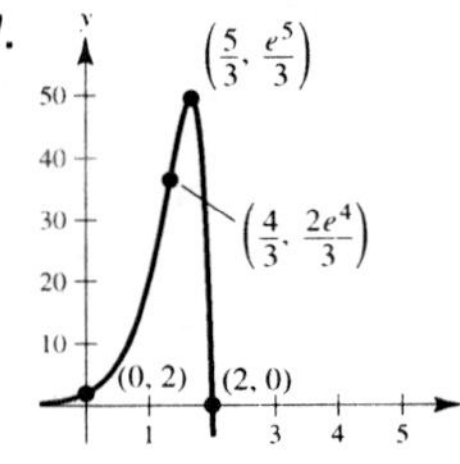

39. 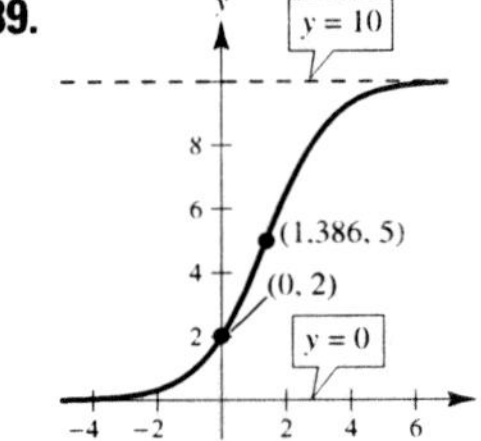

41. 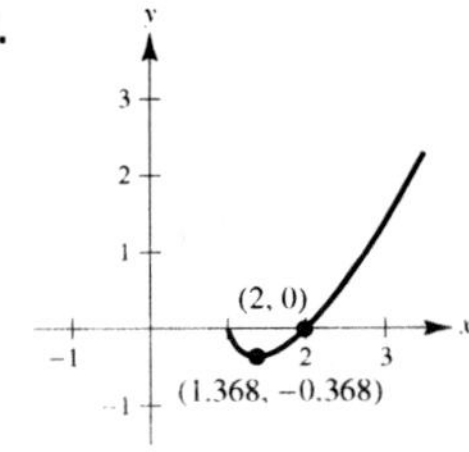

43. 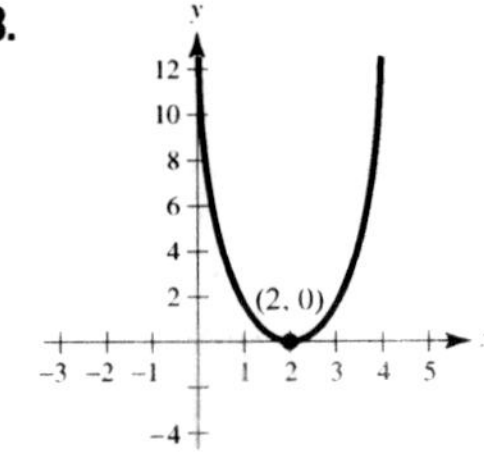

45. 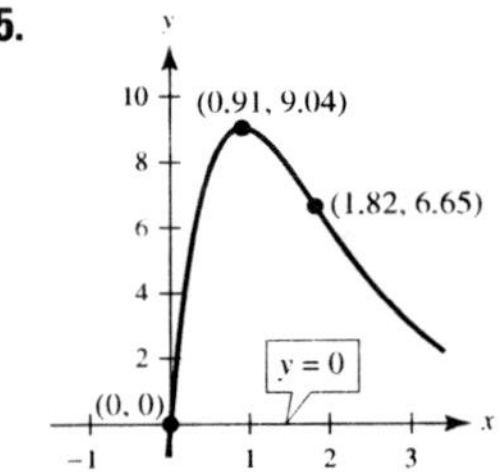

47. 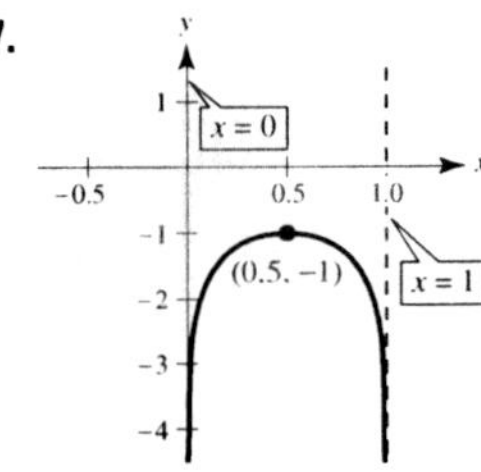

49. 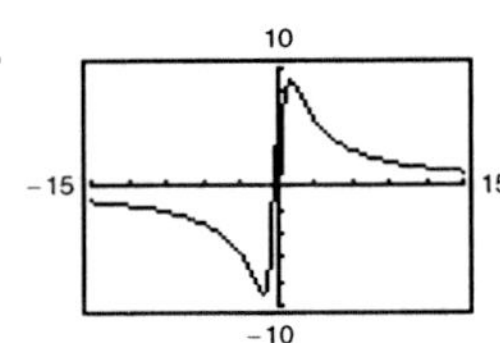

Minimum: $(-1.10, -9.05)$
Maximum: $(1.10, 9.05)$
Points of inflection:
$(-1.84, -7.86), (1.84, 7.86)$
Vertical asymptote: $x = 0$
Horizontal asymptote: $y = 0$

51. 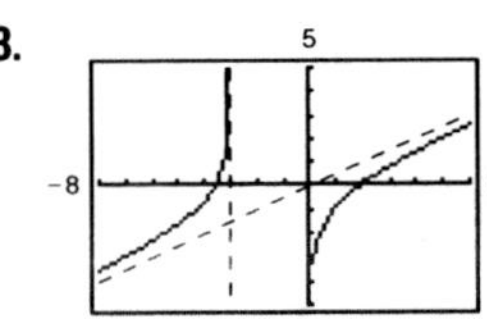

Point of inflection: $(0, 0)$
Horizontal asymptotes: $y = \pm 2$

53. 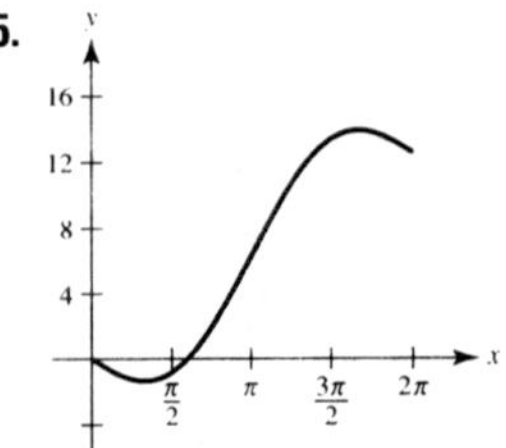

Vertical asymptotes: $x = -3,$
$x = 0$
Slant asymptote: $y = x/2$

55. 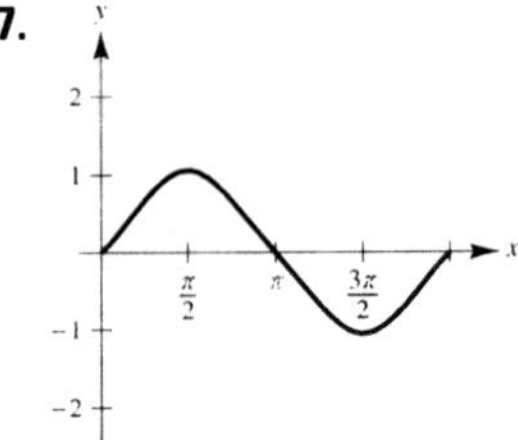

57.

59. 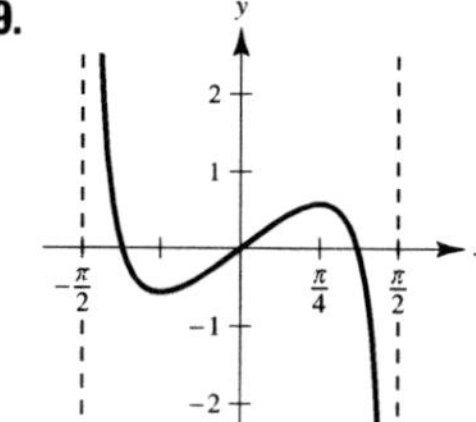

61. 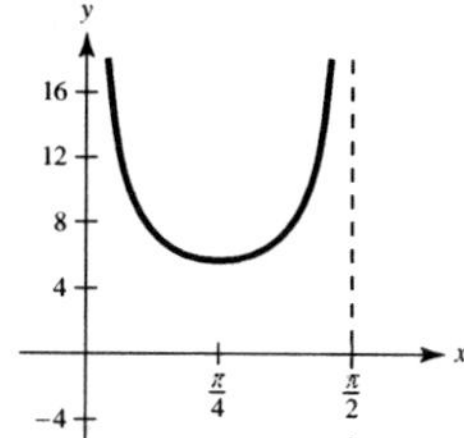

63. 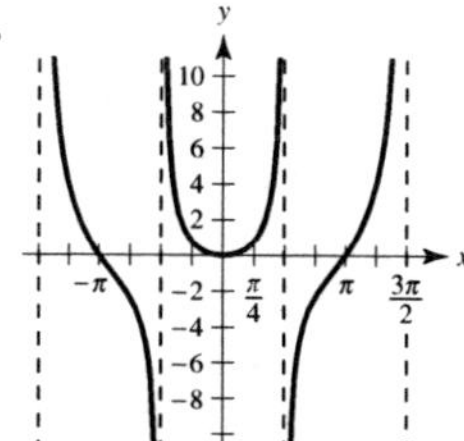

65. 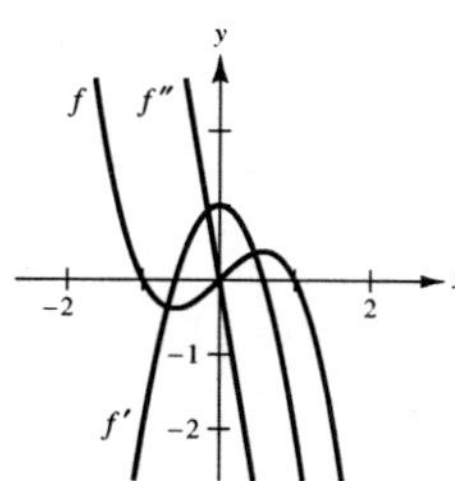

The zeros of f' correspond to the points where the graph of f has horizontal tangents. The zero of f'' corresponds to the point where the graph of f' has a horizontal tangent.

67. f is decreasing on $(2, 8)$ and therefore $f(3) > f(5)$.

69. 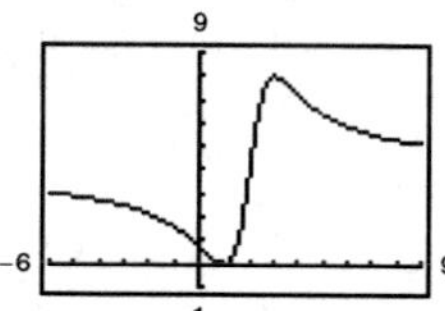

The graph crosses its horizontal asymptote $y = 4$. The graph of f does not cross its vertical asymptote $x = c$ because $f(c)$ does not exist.

71. 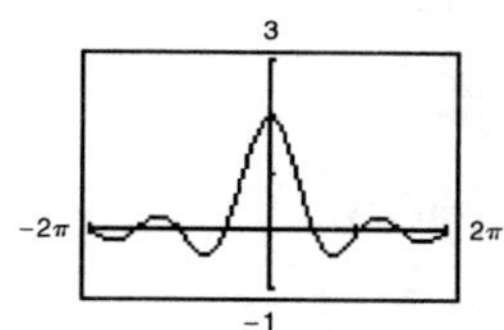

The graph has a hole at $x = 0$.
The graph crosses its horizontal asymptote $y = 0$.
The graph of a function f does not cross its vertical asymptote $x = c$ because $f(c)$ does not exist.

73. 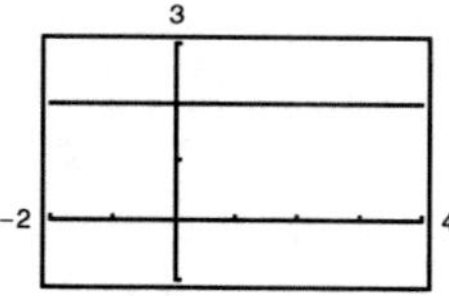

The graph has a hole at $x = 3$. The rational function is not reduced to lowest terms.

75. 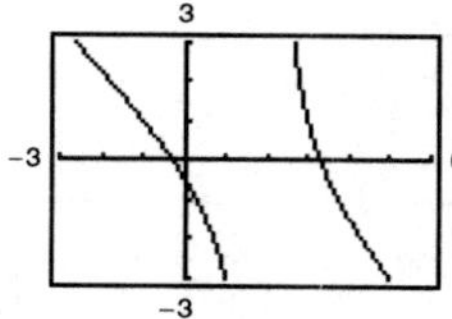

The graph appears to approach the line $y = -x + 1$, which is the slant asymptote.

77. 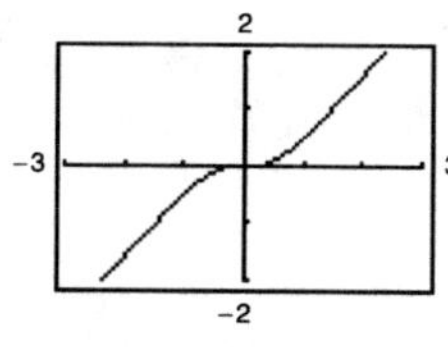

The graph appears to approach the line $y = x$, which is the slant asymptote.

79. 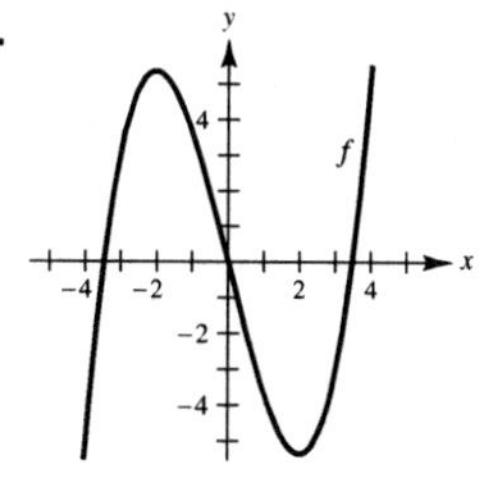

81. 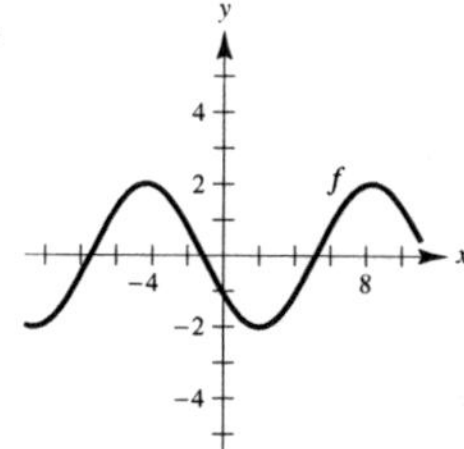

83. (a) 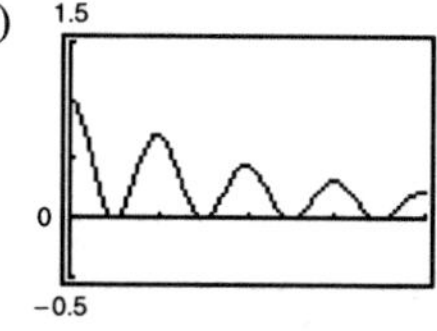

The graph has holes at $x = 0$ and at $x = 4$.
Visual approximation of critical numbers: $\frac{1}{2}, 1, \frac{3}{2}, 2, \frac{5}{2}, 3, \frac{7}{2}$

(b) $f'(x) = \dfrac{-x\cos^2(\pi x)}{(x^2 + 1)^{3/2}} - \dfrac{2\pi\sin(\pi x)\cos(\pi x)}{\sqrt{x^2 + 1}}$

Approximate critical numbers: $\frac{1}{2}, 0.97, \frac{3}{2}, 1.98, \frac{5}{2}, 2.98, \frac{7}{2}$
The critical numbers where maxima occur appear to be integers in part (a), but by approximating them using f', you see that they are not integers.

85. Answers will vary. Sample answer: $y = 1/(x - 3)$

87. Answers will vary. Sample answer:
$y = (3x^2 - 7x - 5)/(x - 3)$

89. (a) $f'(x) = 0$ for $x = \pm 2$; $f'(x) > 0$ for $(-\infty, -2), (2, \infty)$
$f'(x) < 0$ for $(-2, 2)$

(b) $f''(x) = 0$ for $x = 0$; $f''(x) > 0$ for $(0, \infty)$
$f''(x) < 0$ for $(-\infty, 0)$

(c) $(0, \infty)$

(d) f' is minimum for $x = 0$.
f is decreasing at the greatest rate at $x = 0$.

91. Answers will vary. Sample answer: The graph has a vertical asymptote at $x = b$. If a and b are both positive, or both negative, then the graph of f approaches ∞ as x approaches b, and the graph has a minimum at $x = -b$. If a and b have opposite signs, then the graph of f approaches $-\infty$ as x approaches b, and the graph has a maximum at $x = -b$.

93. (a) If n is even, f is symmetric with respect to the y-axis.
If n is odd, f is symmetric with respect to the origin.

(b) $n = 0, 1, 2, 3$ (c) $n = 4$

(d) When $n = 5$, the slant asymptote is $y = 2x$.

(e)

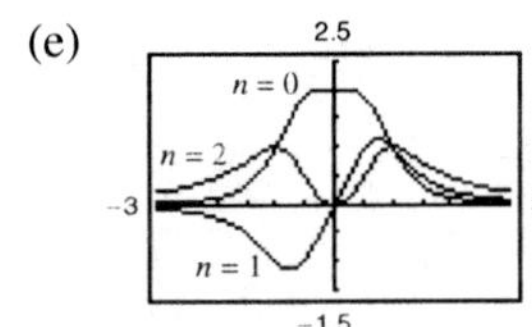 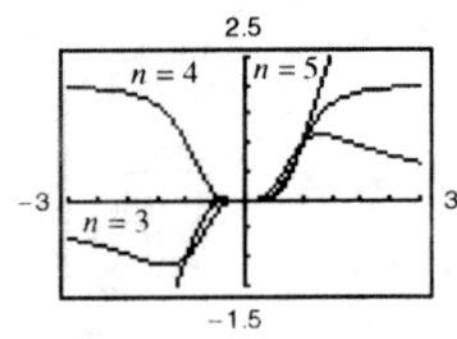

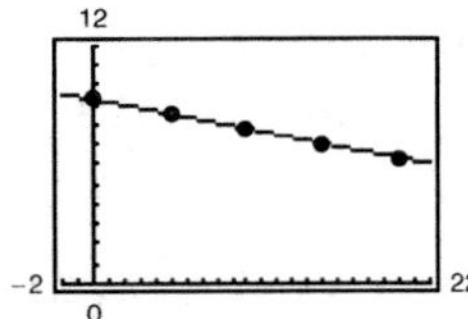

n	0	1	2	3	4	5
M	1	2	3	2	1	0
N	2	3	4	5	2	3

95. (a)

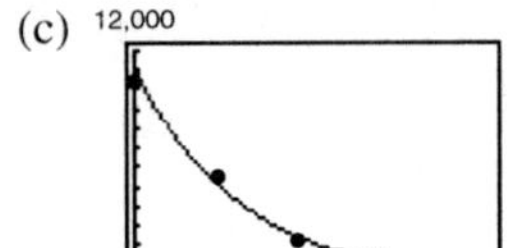

$$y = -0.1499h + 9.3018$$

(b) $P = 10{,}957.7e^{-0.1499h}$

(c)

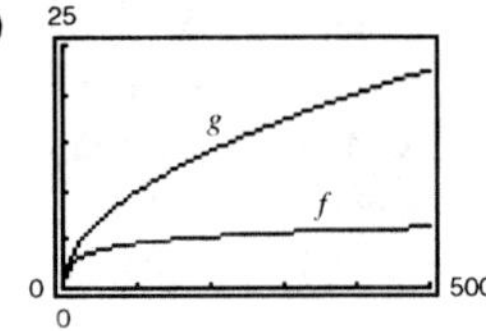

(d) $-776.3, -110.6$

97. (a)

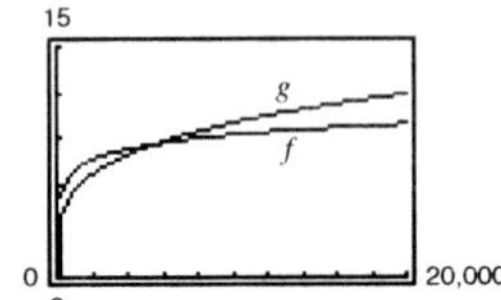

g; $f(x) = \ln x$ increases very slowly for "large" values of x.

(b)

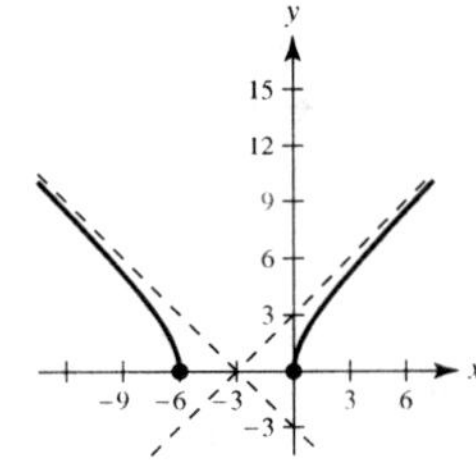

g; $f(x) = \ln x$ increases very slowly for "large" values of x.

99. $y = x + 3$, $y = -x - 3$

Section 4.7   (page 265)

1. (a) and (b)

First Number x	Second Number	Product P
10	$110 - 10$	$10(110 - 10) = 1000$
20	$110 - 20$	$20(110 - 20) = 1800$
30	$110 - 30$	$30(110 - 30) = 2400$
40	$110 - 40$	$40(110 - 40) = 2800$
50	$110 - 50$	$50(110 - 50) = 3000$
60	$110 - 60$	$60(110 - 60) = 3000$
70	$110 - 70$	$70(110 - 70) = 2800$
80	$110 - 80$	$80(110 - 80) = 2400$
90	$110 - 90$	$90(110 - 90) = 1800$
100	$110 - 100$	$100(110 - 100) = 1000$

(c) $P = x(110 - x)$

(d)

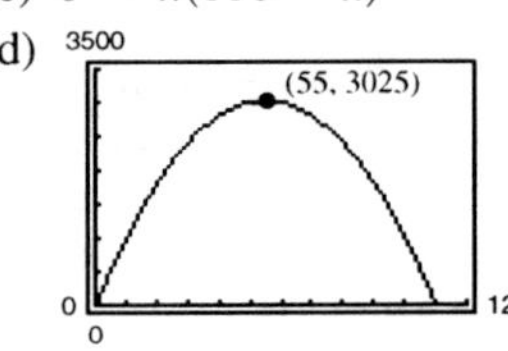

(e) 55 and 55

3. $S/2$ and $S/2$ **5.** 21 and 7   **7.** 54 and 27

9. $l = w = 20$ m   **11.** $l = w = 4\sqrt{2}$ ft

13. $(1, 1)$ **15.** $\left(\frac{7}{2}, \sqrt{\frac{7}{2}}\right)$

17. Dimensions of page: $\left(2 + \sqrt{30}\right)$ in. $\times \left(2 + \sqrt{30}\right)$ in.

19. $x = Q_0/2$ **21.** 700×350 m

23. (a) Answers will vary.

(b) $V_1 = 99$ in.^{3}, $V_2 = 125$ in.^{3}, $V_3 = 117$ in.^{3}

(c) $5 \times 5 \times 5$ in.

25. Rectangular portion: $16/(\pi + 4)$ ft $\times 32/(\pi + 4)$ ft

27. (a) $L = \sqrt{x^2 + 4 + \dfrac{8}{x - 1} + \dfrac{4}{(x - 1)^2}}$, $x > 1$

(b)

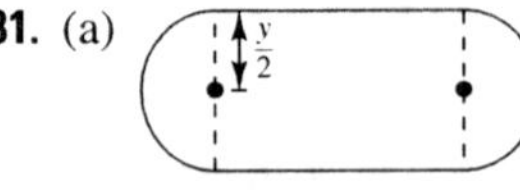

Minimum when $x \approx 2.587$

(c) $(0, 0), (2, 0), (0, 4)$

29. Length: $5\sqrt{2}$; Width: $5\sqrt{2}/2$

31. (a)

(b)

Length x	Width y	Area A
10	$2/\pi(100-10)$	$(10)(2/\pi)(100-10) \approx 573$
20	$2/\pi(100-20)$	$(20)(2/\pi)(100-20) \approx 1019$
30	$2/\pi(100-30)$	$(30)(2/\pi)(100-30) \approx 1337$
40	$2/\pi(100-40)$	$(40)(2/\pi)(100-40) \approx 1528$
50	$2/\pi(100-50)$	$(50)(2/\pi)(100-50) \approx 1592$
60	$2/\pi(100-60)$	$(60)(2/\pi)(100-60) \approx 1528$

The maximum area of the rectangle is approximately 1592 square meters.

(c) $A = \dfrac{2}{\pi}(100x - x^2), \quad 0 < x < 100$

(d) $\dfrac{dA}{dx} = \dfrac{2}{\pi}(100 - 2x)$
$= 0$ when $x = 50$
The maximum value is approximately 1592 when $x = 50$.

(e) 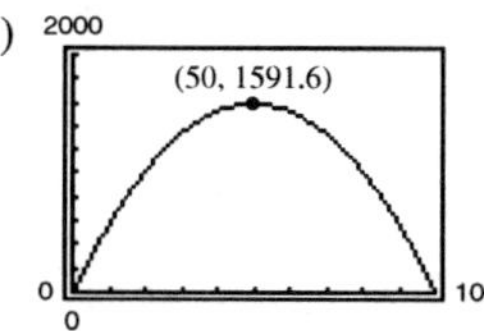

33. $18 \times 18 \times 36$ in.   **35.** $32\pi r^3/81$

37. No. The volume changes because the shape of the container changes when squeezed.

39. $r = \sqrt[3]{21/(2\pi)} \approx 1.50$ ($h = 0$, so the solid is a sphere.)

41. Side of triangle: $\dfrac{30}{9 + 4\sqrt{3}}$; Side of square: $\dfrac{10\sqrt{3}}{9 + 4\sqrt{3}}$

43. $w = \left(20\sqrt{3}\right)/3$ in., $h = \left(20\sqrt{6}\right)/3$ in.   **45.** $\theta = \pi/4$

47. $h = \sqrt{2}$ ft   **49.** One mile from the nearest point on the coast

51. Proof

53. 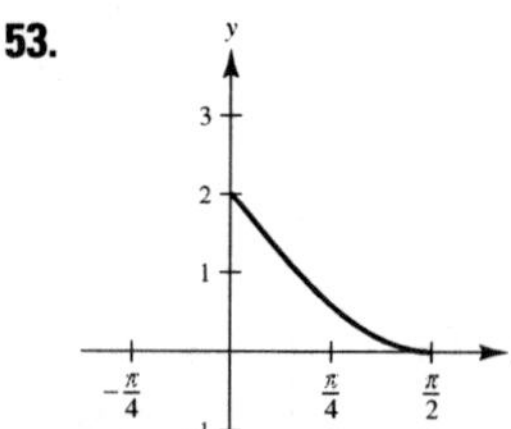

(a) Origin to y-intercept: 2
Origin to x-intercept: $\pi/2$

(b) $d = \sqrt{x^2 + (2 - 2\sin x)^2}$

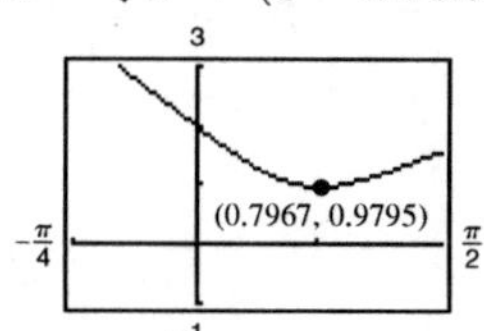

(c) Minimum distance is 0.9795 when $x \approx 0.7967$.

55. $\theta = \pi/3\left(6 - 2\sqrt{6}\right) \approx 1.15°$

57. 4045 units   **59.** $A = \sqrt{2}e^{-1/2}$ **61.** Answers will vary.

63. $y = \frac{64}{141}x$; $S_1 \approx 6.1$ mi   **65.** $y = \frac{3}{10}x$; $S_3 \approx 4.50$ mi

67. Putnam Problem A1, 1986

Section 4.8   (page 276)

1. $T(x) = 4x - 4$

x	1.9	1.99	2	2.01	2.1
$f(x)$	3.610	3.960	4	4.040	4.410
$T(x)$	3.600	3.960	4	4.040	4.400

3. $T(x) = 80x - 128$

x	1.9	1.99	2	2.01	2.1
$f(x)$	24.761	31.208	32	32.808	40.841
$T(x)$	24.000	31.200	32	32.800	40.000

5. $T(x) = (\cos 2)(x - 2) + \sin 2$

x	1.9	1.99	2	2.01	2.1
$f(x)$	0.946	0.913	0.909	0.905	0.863
$T(x)$	0.951	0.913	0.909	0.905	0.868

7. $\Delta y = 0.331$; $dy = 0.3$ **9.** $\Delta y = -0.039$; $dy = -0.040$

11. $6x \, dx$ **13.** $-\dfrac{3}{(2x-1)^2} \, dx$ **15.** $\dfrac{1 - 2x^2}{\sqrt{1 - x^2}} \, dx$

17. $x/(x^2 - 4) \, dx$ **19.** $(3 - \sin 2x) \, dx$

21. $-\pi \sin\left(\dfrac{6\pi x - 1}{2}\right) dx$ **23.** $\left(\arcsin x + \dfrac{x}{\sqrt{1 - x^2}}\right) dx$

25. (a) 0.9 (b) 1.04   **27.** (a) 1.05 (b) 0.98

29. (a) 8.035 (b) 7.95   **31.** $\pm\frac{5}{8}$ in.^{2}

33. $\pm 8\pi$ in.^{2} **35.** (a) $\frac{5}{6}$% (b) 1.25%

37. (a) $\pm 5.12\pi$ in.^{3} (b) $\pm 1.28\pi$ in.^{2} (c) 0.75%, 0.5%

39. 267.24, 3.1%   **41.** 80π cm^3

43. (a) 0.87% (b) 2.16%   **45.** 6407 ft

47. 9.97; Calculator: 9.97   **49.** 4.998; Calculator: 4.998

51. 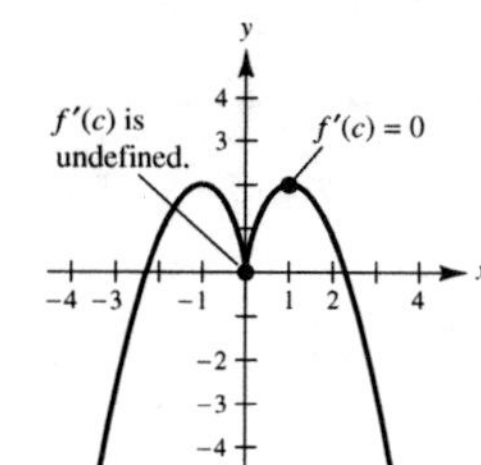

53. The value of dy becomes closer to the value of Δy as Δx decreases.

55. (a) $f(x) = \sqrt{x}$; $dy = \dfrac{1}{2\sqrt{x}} \, dx$

$f(4.02) \approx \sqrt{4} + \dfrac{1}{2\sqrt{4}}(0.02) = 2 + \dfrac{1}{4}(0.02)$

(b) $f(x) = \tan x$; $dy = \sec^2 x \, dx$
$f(0.05) \approx \tan 0 + \sec^2(0)(0.05) = 0 + 1(0.05)$

57. True   **59.** True

Review Exercises for Chapter 4   (page 278)

1. Let f be defined at c. If $f'(c) = 0$ or if f' is undefined at c, then c is a critical number of f.

3. Maximum: $(0, 0)$
   Minimum: $\left(-\frac{5}{2}, -\frac{25}{4}\right)$

5. Maximum: $(2\pi, 17.57)$
   Minimum: $(2.73, 0.88)$

7. $f(0) \neq f(4)$ **9.** Not continuous on $[-2, 2]$

11. (a)

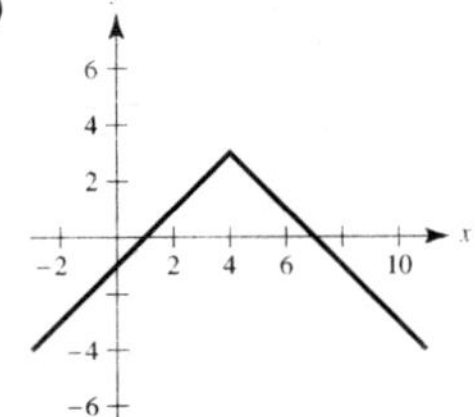

(b) f is not differentiable at $x = 4$.

13. $f'\left(\dfrac{2744}{729}\right) = \dfrac{3}{7}$ **15.** f is not differentiable at $x = 5$.

17. $f'(0) = 1$ **19.** $c = \dfrac{x_1 + x_2}{2}$

21. Critical number: $x = -\frac{3}{2}$
   Increasing on $\left(-\frac{3}{2}, \infty\right)$; Decreasing on $\left(-\infty, -\frac{3}{2}\right)$

23. Critical numbers: $x = 1, \frac{7}{3}$
   Increasing on $(-\infty, 1)$, $\left(\frac{7}{3}, \infty\right)$; Decreasing on $\left(1, \frac{7}{3}\right)$

25. Critical number: $x = 1$
   Increasing on $(1, \infty)$; Decreasing on $(0, 1)$

27. Critical number: $t = 2 - 1/\ln 2$
   Increasing on $(-\infty, 2 - 1/\ln 2)$
   Decreasing on $(2 - 1/\ln 2, \infty)$

29. Relative maximum: $\left(-\dfrac{\sqrt{15}}{6}, \dfrac{5\sqrt{15}}{9}\right)$
   Relative minimum: $\left(\dfrac{\sqrt{15}}{6}, -\dfrac{5\sqrt{15}}{9}\right)$

31. Minimum: $(2, -12)$

33. (a) $y = \frac{1}{4}$ in.; $v = 4$ in./sec   (b) Proof
   (c) Period: $\pi/6$; Frequency: $6/\pi$

35. $(3, -54)$; Concave upward: $(3, \infty)$;
   Concave downward: $(-\infty, 3)$

37. $(\pi/2, \pi/2)$, $(3\pi/2, 3\pi/2)$; Concave upward: $(\pi/2, 3\pi/2)$;
   Concave downward: $(0, \pi/2)$, $(3\pi/2, 2\pi)$

39. Relative minimum: $(-9, 0)$

41. Relative maxima: $\left(\sqrt{2}/2, 1/2\right)$, $\left(-\sqrt{2}/2, 1/2\right)$
   Relative minimum: $(0, 0)$

43.

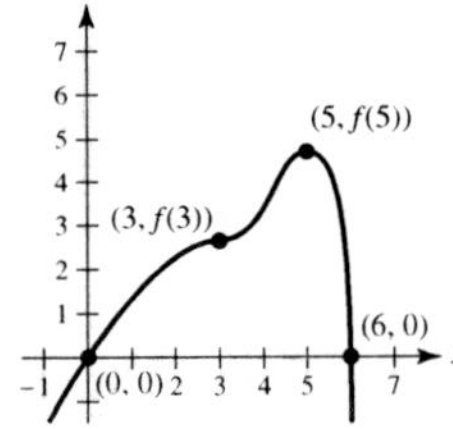

45. Increasing and concave down

47. (a) $D = 0.00430t^4 - 0.2856t^3 + 5.833t^2 - 26.85t + 87.1$
   (b)

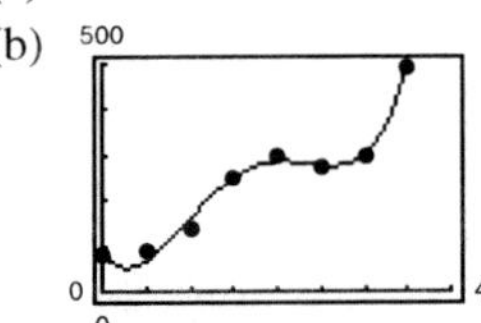

   (c) Maximum in 2005; Minimum in 1972   (d) 2005

49. 8   **51.** $\frac{2}{3}$ **53.** $-\infty$ **55.** 0   **57.** 6

59. Vertical asymptote: $x = 0$; Horizontal asymptote: $y = -2$

61. Vertical asymptote: $x = 4$; Horizontal asymptote: $y = 2$

63. Horizontal asymptotes: $y = 0$; $y = \frac{5}{3}$

65. Horizontal asymptote: $y = 0$

67.

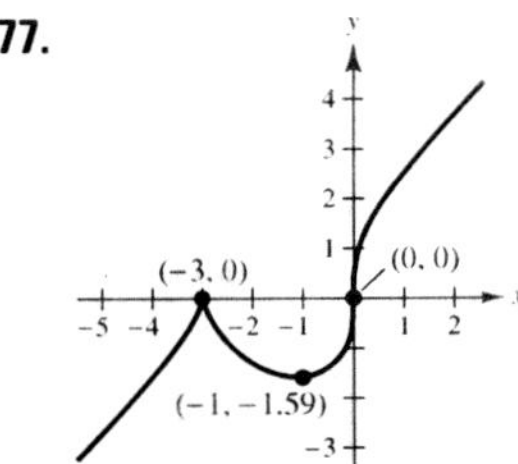

Vertical asymptote: $x = 0$
Relative minimum: $(3, 108)$
Relative maximum: $(-3, -108)$

69.

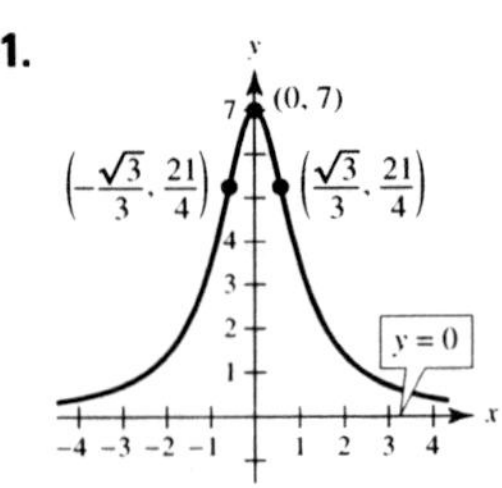

Horizontal asymptote: $y = 0$
Relative minimum: $(-0.155, -1.077)$
Relative maximum: $(2.155, 0.077)$

71.

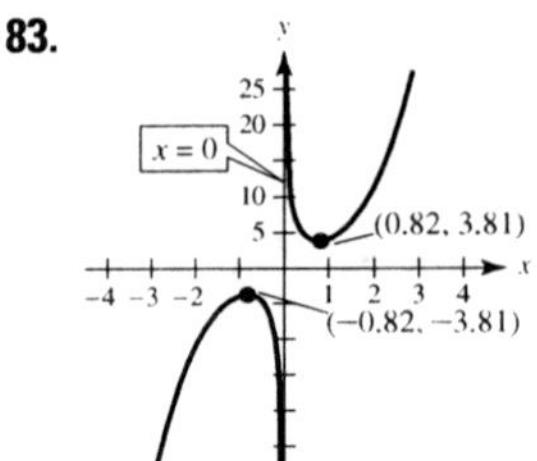

73.

75.

77.

79.

81.

83.

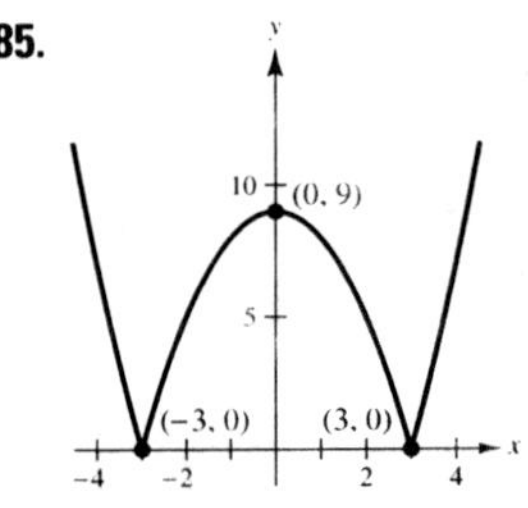

85.

87.

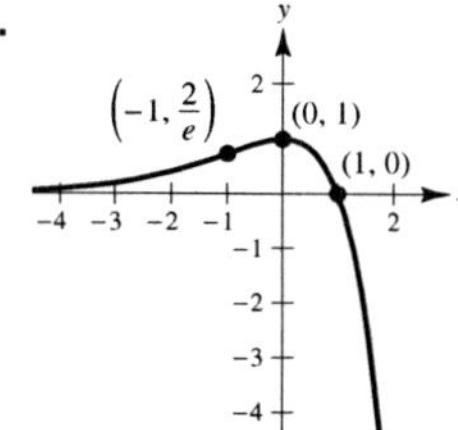

89.

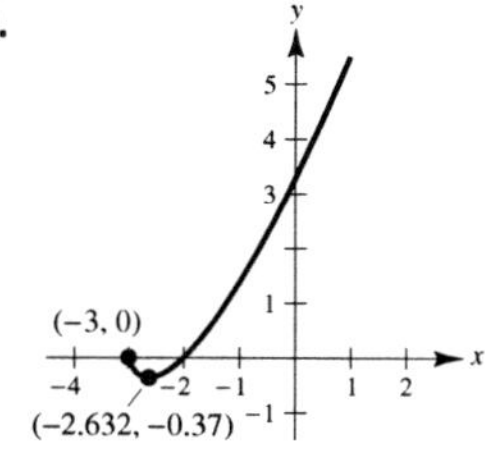

91.

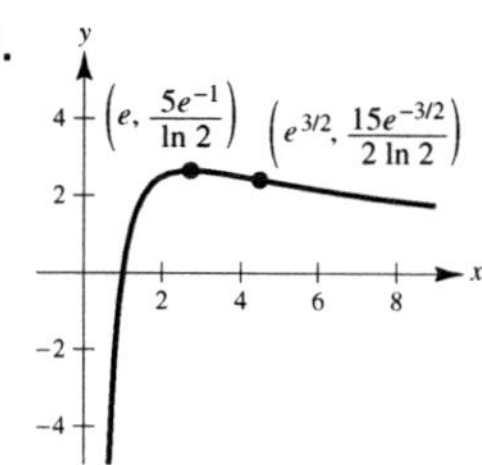

93.

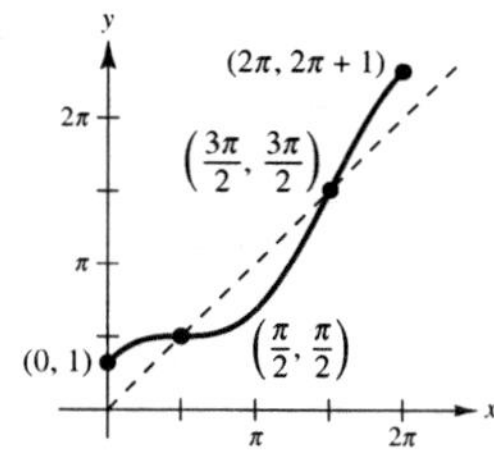

95. $\left(-\frac{\sqrt{2}}{2}, -2\sqrt{2} + 6\arctan\frac{\sqrt{2}}{2}\right)$

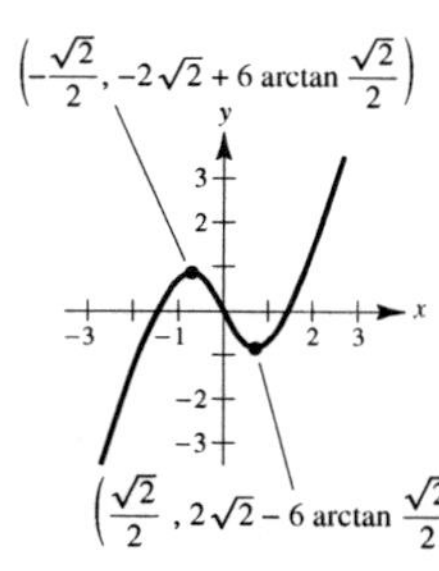

$\left(\frac{\sqrt{2}}{2}, 2\sqrt{2} - 6\arctan\frac{\sqrt{2}}{2}\right)$

97. Maximum: $(1, 3)$
Minimum: $(1, 1)$

99. $t \approx 4.92 \approx 4$:55 P.M.; $d \approx 64$ km
101. $(0, 0)$, $(5, 0)$, $(0, 10)$ **103.** Proof   **105.** 14.05 ft
107. $3(3^{2/3} + 2^{2/3})^{3/2} \approx 21.07$ ft   **109.** $v \approx 54.77$ mi/h
111. $dy = (1 - \cos x + x\sin x)\,dx$

113. $dS = \pm 1.8\pi \text{ cm}^2$, $\dfrac{dS}{S} \times 100 \approx \pm 0.56\%$

$dV = \pm 8.1\pi \text{ cm}^3$, $\dfrac{dV}{V} \times 100 \approx \pm 0.83\%$

P.S. Problem Solving   (page 281)

1. Choices of a may vary.

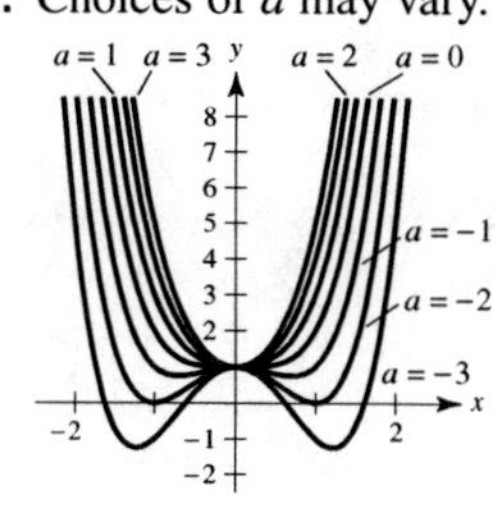

(a) One relative minimum at $(0, 1)$ for $a \geq 0$
(b) One relative maximum at $(0, 1)$ for $a < 0$
(c) Two relative minima for $a < 0$ when $x = \pm\sqrt{-a/2}$
(d) If $a < 0$, there are three critical points; if $a \geq 0$, there is only one critical point.

3. All c, where c is a real number   **5–7.** Proofs
9. About 9.19 ft
11. Minimum: $\left(\sqrt{2} - 1\right)d$; There is no maximum.
13. (a)

x	0	0.5	1	2
$\sqrt{1 + x}$	1	1.2247	1.4142	1.7321
$\frac{1}{2}x + 1$	1	1.25	1.5	2

(b) Proof

15. (a)–(c) Proofs
17. (a)

v	20	40	60	80	100
s	5.56	11.11	16.67	22.22	27.78
d	5.1	13.7	27.2	44.2	66.4

$d(s) = 0.071s^2 + 0.389s + 0.727$

(b) The distance between the back of the first vehicle and the front of the second vehicle is $d(s)$, the safe stopping distance. The first vehicle passes the given point in $5.5/s$ seconds, and the second vehicle takes $d(s)/s$ more seconds. So, $T = d(s)/s + 5.5/s$.

(c)

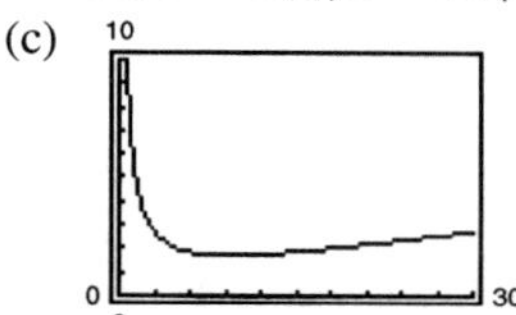

$s \approx 9.365$ m/sec

(d) $s \approx 9.365$ m/sec; 1.719 sec; 33.714 km/h   (e) 10.597 m

19. (a) $(0, \infty)$
(b) Answers will vary. Sample answer: $x = e^{\pi/2}$, $x = e^{(\pi/2) + 2\pi}$
(c) Answers will vary. Sample answer: $x = e^{-\pi/2}$, $x = e^{3\pi/2}$
(d) $[-1, 1]$ (e) $f'(x) = \dfrac{\cos(\ln x)}{x}$; Maximum $= e^{\pi/2}$

(f)

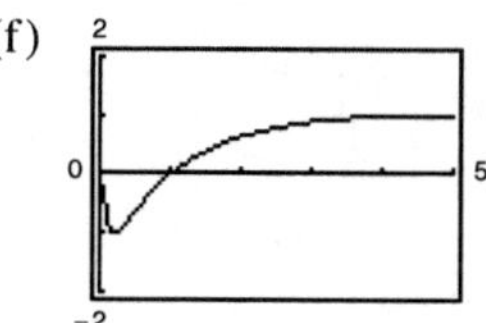

$\lim\limits_{x \to 0^+} f(x)$ seems to be $-\frac{1}{2}$. (This is incorrect.)

(g) The limit does not exist.

Chapter 5

Section 5.1   (page 291)

1. Proof   **3.** Proof   **5.** $y = 3t^3 + C$ **7.** $y = \frac{2}{5}x^{5/2} + C$

	Original Integral	Rewrite	Integrate	Simplify
9.	$\int \sqrt[3]{x}\,dx$	$\int x^{1/3}\,dx$	$\dfrac{x^{4/3}}{4/3} + C$	$\dfrac{3}{4}x^{4/3} + C$
11.	$\int \dfrac{1}{x\sqrt{x}}\,dx$	$\int x^{-3/2}\,dx$	$\dfrac{x^{-1/2}}{-1/2} + C$	$-\dfrac{2}{\sqrt{x}} + C$
13.	$\int \dfrac{1}{2x^3}\,dx$	$\dfrac{1}{2}\int x^{-3}\,dx$	$\dfrac{1}{2}\left(\dfrac{x^{-2}}{-2}\right) + C$	$-\dfrac{1}{4x^2} + C$

15. $\frac{1}{2}x^2 + 7x + C$ **17.** $\frac{1}{6}x^6 + x + C$
19. $\frac{2}{5}x^{5/2} + x^2 + x + C$ **21.** $-1/(4x^4) + C$
23. $\frac{2}{3}x^{3/2} + 12x^{1/2} + C = \frac{2}{3}x^{1/2}(x + 18) + C$
25. $x^3 + \frac{1}{2}x^2 - 2x + C$ **27.** $\frac{2}{7}y^{7/2} + C$ **29.** $x + C$
31. $5\sin x - 4\cos x + C$ **33.** $t + \csc t + C$
35. $-2\cos x - 5e^x + C$ **37.** $\tan\theta + \cos\theta + C$
39. $\tan y + C$ **41.** $x^2 - 4^x/\ln 4 + C$ **43.** $\frac{1}{2}x^2 - 5\ln|x| + C$

45. 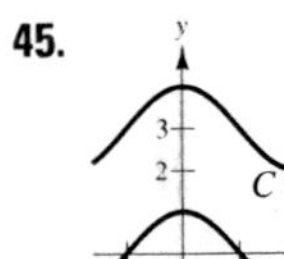

47. 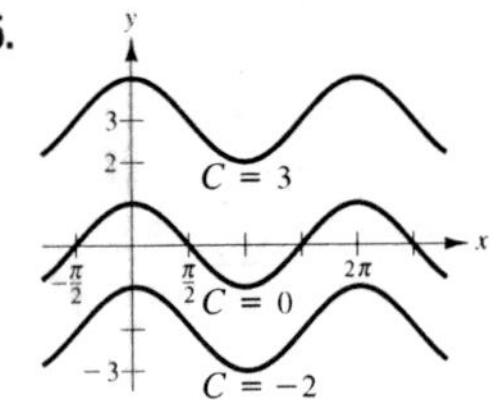

49. Answers will vary.
Sample answer:

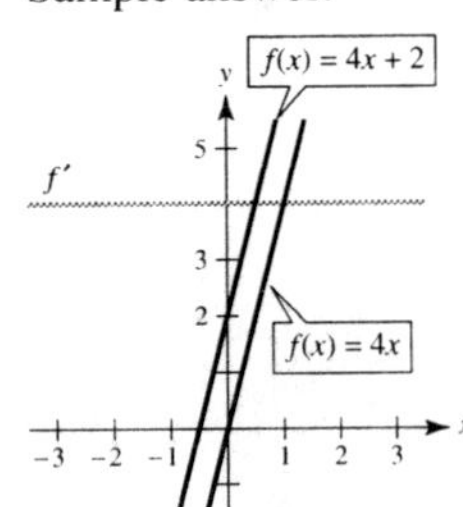

51. Answers will vary.
Sample answer:

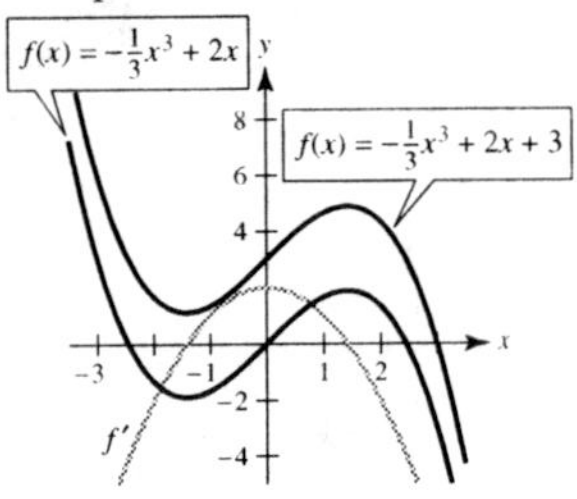

53. $y = x^2 - x + 1$ **55.** $y = \sin x + 4$

57. (a) Answers will vary.
Sample answer:

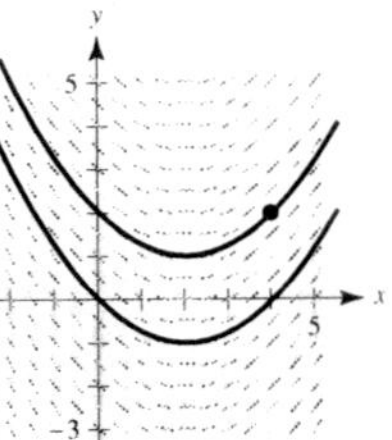

59. (a) Answers will vary.
Sample answer:

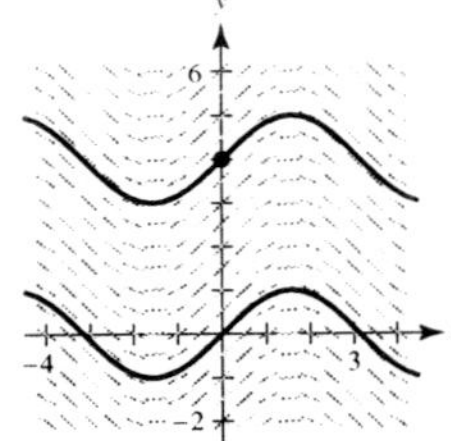

(b) $y = \frac{1}{4}x^2 - x + 2$

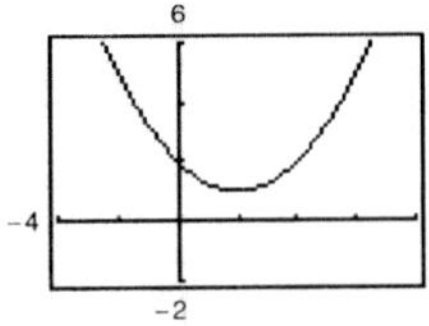

(b) $y = \sin x + 4$

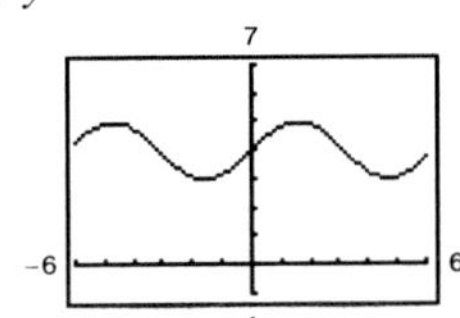

61. (a)

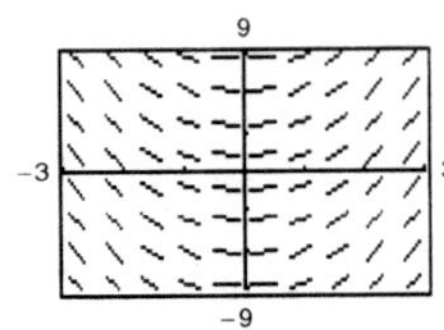

(b) $y = x^2 - 6$
(c)

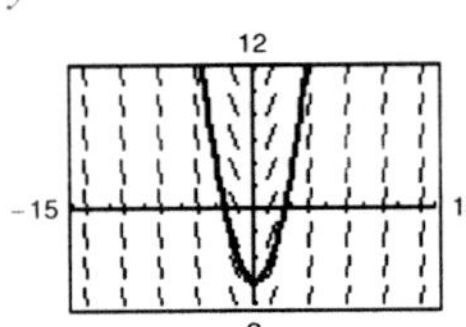

63. $f(x) = 3x^2 + 8$ **65.** $h(t) = 2t^4 + 5t - 11$
67. $f(x) = x^2 + x + 4$ **69.** $f(x) = -4\sqrt{x} + 3x$
71. $f(x) = e^x + x + 4$
73. When you evaluate the integral $\int f(x)\,dx$, you are finding a function $F(x)$ that is an antiderivative of $f(x)$. So, there is no difference.
75.

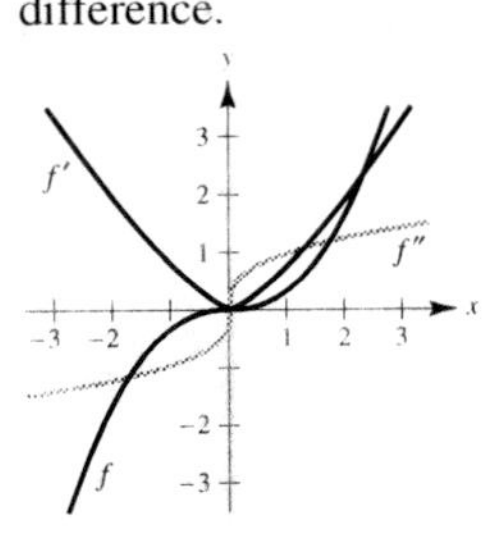

77. 62.25 ft   **79.** $v_0 \approx 187.617$ ft/sec
81. $v(t) = -9.8t + C_1 = -9.8t + v_0$
$\quad f(t) = -4.9t^2 + v_0 t + C_2 = -4.9t^2 + v_0 t + s_0$
83. 7.1 m   **85.** 320 m; -32 m/sec
87. (a) $v(t) = 3t^2 - 12t + 9$; $a(t) = 6t - 12$
$\quad$ (b) $(0, 1)$, $(3, 5)$ (c) -3
89. $a(t) = -1/(2t^{3/2})$; $x(t) = 2\sqrt{t} + 2$
91. (a) 1.18 m/sec^2 (b) 190 m
93. (a) 300 ft   (b) 60 ft/sec ≈ 41 mi/h
95. True   **97.** True
99. False. Let $f(x) = x$ and $g(x) = x + 1$.
101. $f(x) = \frac{1}{3}x^3 - 4x + \frac{16}{3}$
103. $f(x) = \begin{cases} x + 2, & 0 \le x < 2 \\ \frac{3}{2}x^2 - 2, & 2 \le x \le 5 \end{cases}$

$\quad f$ is not differentiable at $x = 2$, because the left- and right-hand derivatives at $x = 2$ do not agree.
105. Proof   **107.** Putnam Problem B2, 1991

Section 5.2   (page 303)

1. 75   **3.** $\dfrac{158}{85}$ **5.** $4c$ **7.** $\displaystyle\sum_{i=1}^{11} \frac{1}{5i}$ **9.** $\displaystyle\sum_{j=1}^{6} \left[7\left(\frac{j}{6}\right) + 5\right]$

11. $\dfrac{2}{n}\displaystyle\sum_{i=1}^{n}\left[\left(\frac{2i}{n}\right)^3 - \left(\frac{2i}{n}\right)\right]$ **13.** $\dfrac{3}{n}\displaystyle\sum_{i=1}^{n}\left[2\left(1 + \frac{3i}{n}\right)^2\right]$

15. 84   **17.** 1200   **19.** 2470   **21.** 12,040   **23.** 2930
25. (a)

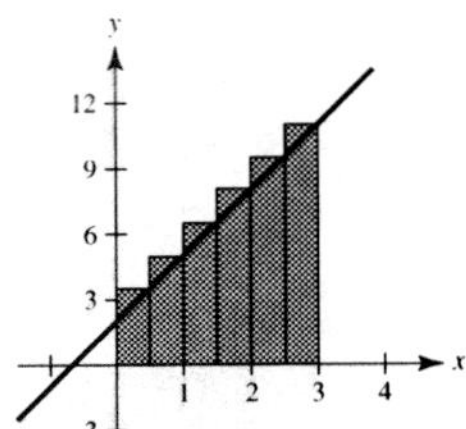

(b)

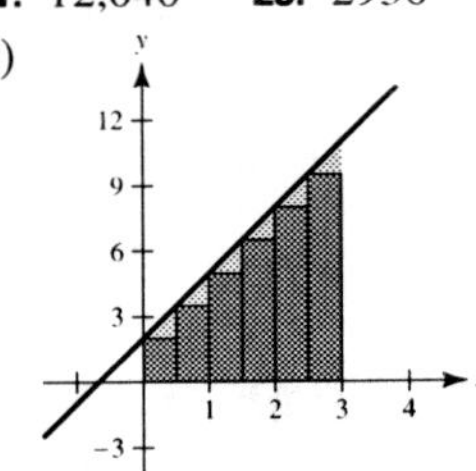

Area ≈ 21.75 $\qquad$ Area ≈ 17.25

27. $13 < $ (Area of region) < 15
29. $55 < $ (Area of region) < 74.5
31. $0.7908 < $ (Area of region) < 1.1835
33. The area of the shaded region falls between 12.5 square units and 16.5 square units.
35. The area of the shaded region falls between 7 square units and 11 square units.
37. $\frac{81}{4}$ **39.** 9   **41.** $A \approx S \approx 0.768$ **43.** $A \approx S \approx 0.746$
$\qquad\qquad\qquad\qquad A \approx s \approx 0.518 \qquad\qquad A \approx s \approx 0.646$
45. $(n + 2)/n$ $\qquad\qquad$ **47.** $[2(n + 1)(n - 1)]/n^2$
$\quad n = 10: S = 1.2 \qquad\qquad n = 10: S = 1.98$
$\quad n = 100: S = 1.02 \qquad\quad n = 100: S = 1.9998$
$\quad n = 1000: S = 1.002 \qquad n = 1000: S = 1.999998$
$\quad n = 10{,}000: S = 1.0002 \quad n = 10{,}000: S = 1.99999998$
49. $\displaystyle\lim_{n\to\infty}\left[\frac{12(n + 1)}{n}\right] = 12$ **51.** $\displaystyle\lim_{n\to\infty}\frac{1}{6}\left(\frac{2n^3 - 3n^2 + n}{n^3}\right) = \frac{1}{3}$
53. $\displaystyle\lim_{n\to\infty}\left[(3n + 1)/n\right] = 3$

55. (a) 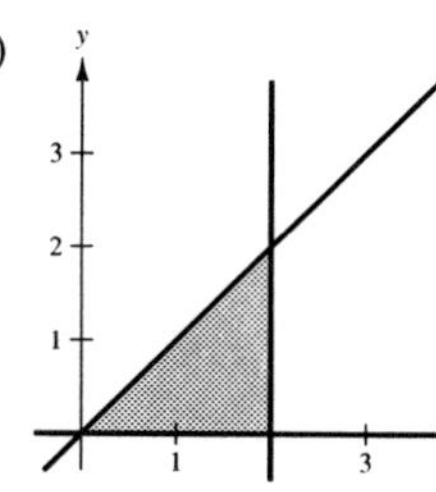

(b) $\Delta x = (2 - 0)/n = 2/n$

(c) $s(n) = \sum_{i=1}^{n} f(x_{i-1})\,\Delta x$
$$= \sum_{i=1}^{n}\left[(i-1)\left(\frac{2}{n}\right)\right]\left(\frac{2}{n}\right)$$

(d) $S(n) = \sum_{i=1}^{n} f(x_i)\,\Delta x$
$$= \sum_{i=1}^{n}\left[i\left(\frac{2}{n}\right)\right]\left(\frac{2}{n}\right)$$

(e)

n	5	10	50	100
$s(n)$	1.6	1.8	1.96	1.98
$S(n)$	2.4	2.2	2.04	2.02

(f) $\displaystyle\lim_{n\to\infty}\sum_{i=1}^{n}\left[(i-1)\left(\frac{2}{n}\right)\right]\left(\frac{2}{n}\right) = 2;\ \lim_{n\to\infty}\sum_{i=1}^{n}\left[i\left(\frac{2}{n}\right)\right]\left(\frac{2}{n}\right) = 2$

57. $A = 3$

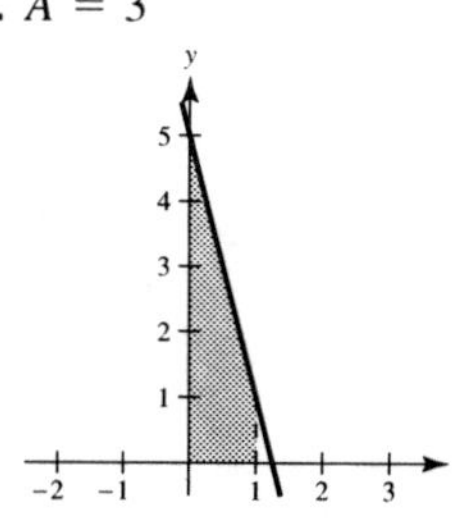

59. $A = \frac{7}{3}$

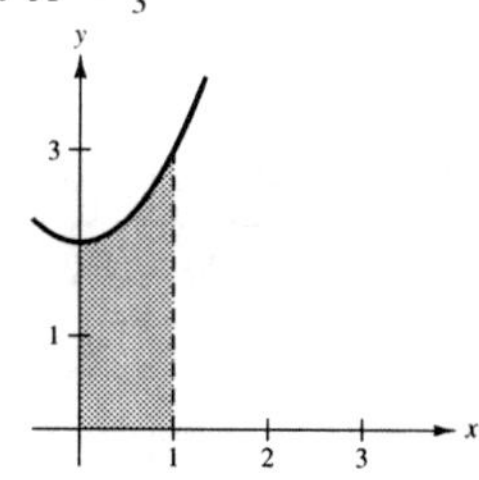

61. $A = 54$

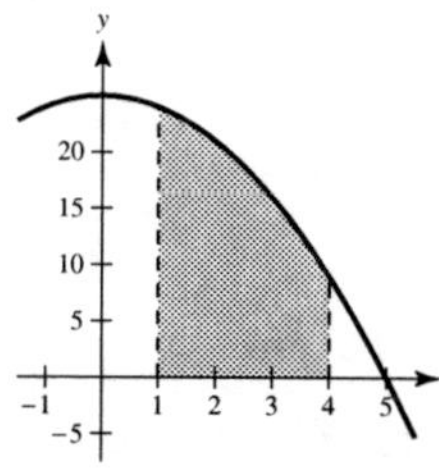

63. $A = 34$

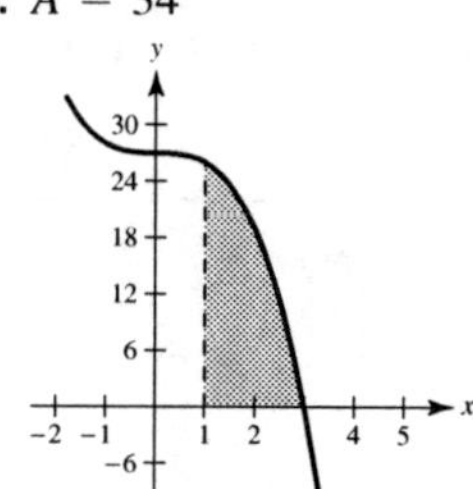

65. $A = \frac{2}{3}$

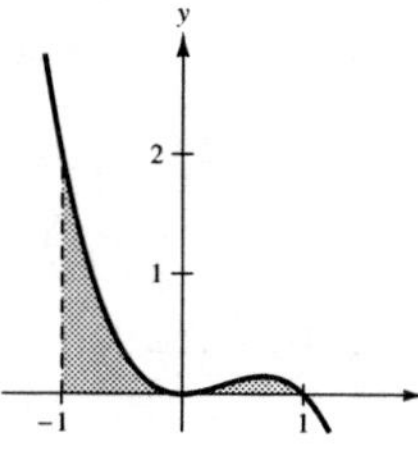

67. $A = 8$

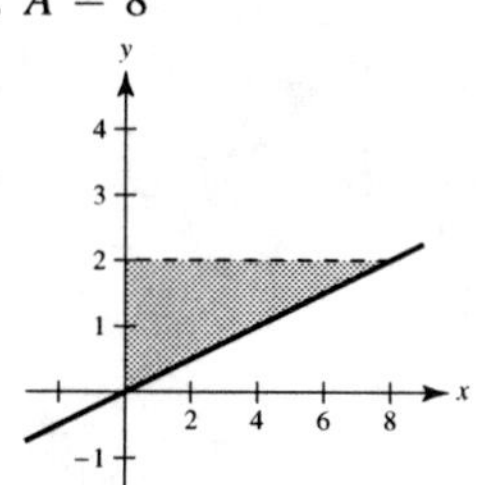

69. $A = \frac{125}{3}$

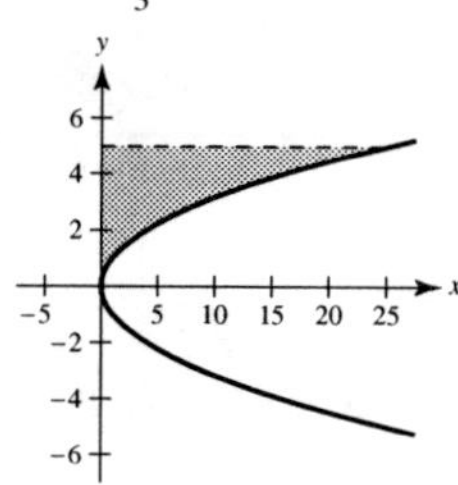

71. $A = \frac{44}{3}$

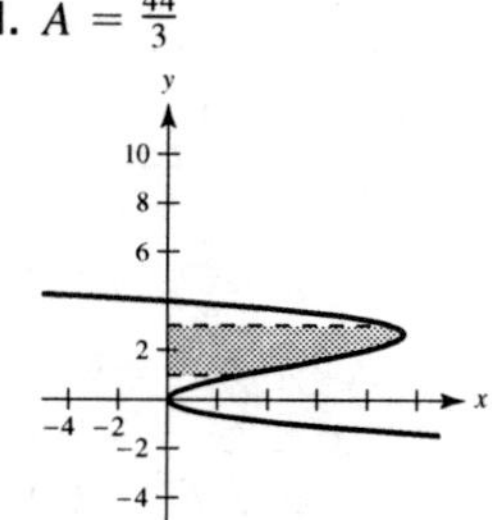

73. $\frac{69}{8}$ **75.** 0.345

77.

n	4	8	12	16	20
Approximate Area	5.3838	5.3523	5.3439	5.3403	5.3384

79.

n	4	8	12	16	20
Approximate Area	2.2223	2.2387	2.2418	2.2430	2.2435

81.

n	4	8	12	16	20
Approximate Area	4.0786	4.0554	4.0509	4.0493	4.0485

83. b

85. You can use the line $y = x$ bounded by $x = a$ and $x = b$. The sum of the areas of the inscribed rectangles in the figure below is the lower sum.

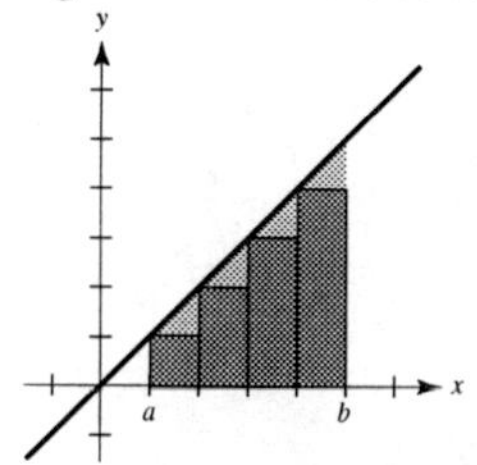

The sum of the areas of the circumscribed rectangles in the figure below is the upper sum.

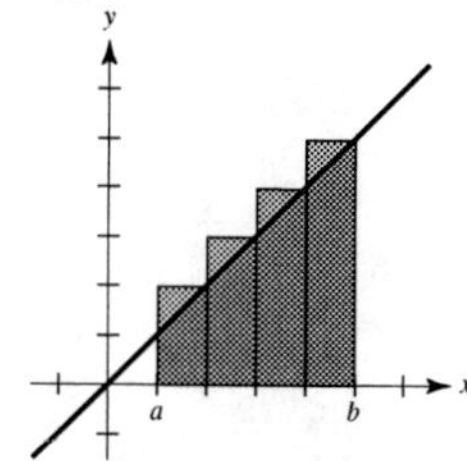

The rectangles in the first graph do not contain all of the area of the region, and the rectangles in the second graph cover more than the area of the region. The exact value of the area lies between these two sums.

87. (a) 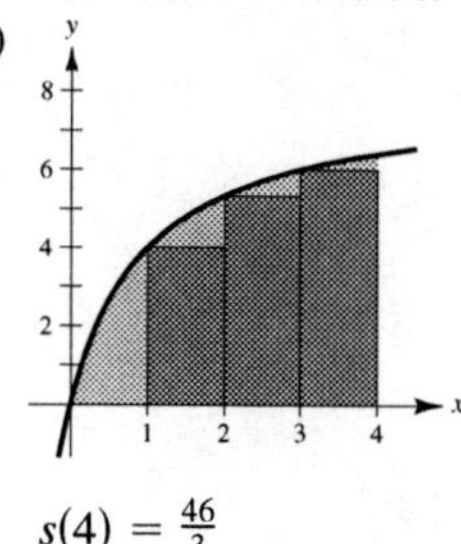

$s(4) = \frac{46}{3}$

(b) 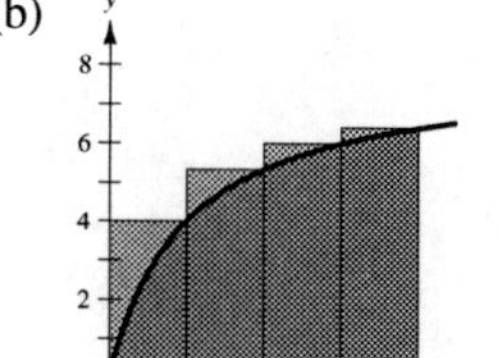

$S(4) = \frac{326}{15}$

(c) 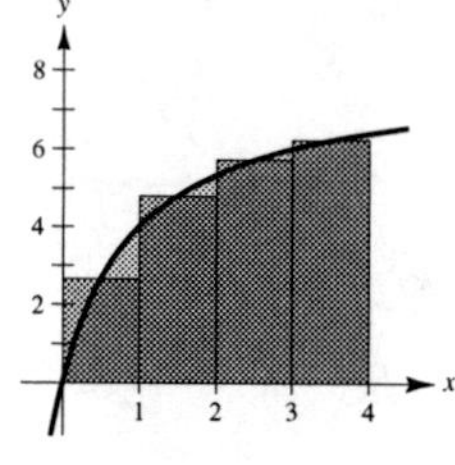

$M(4) = \frac{6112}{315}$

(d) Proof

(e)

n	4	8	20	100	200
$s(n)$	15.333	17.368	18.459	18.995	19.060
$S(n)$	21.733	20.568	19.739	19.251	19.188
$M(n)$	19.403	19.201	19.137	19.125	19.125

(f) Because f is an increasing function, $s(n)$ is always increasing and $S(n)$ is always decreasing.

89. True

91. Suppose there are n rows and $n + 1$ columns in the figure. The stars on the left total $1 + 2 + \cdots + n$, as do the stars on the right. There are $n(n + 1)$ stars in total. This means that $2[1 + 2 + \cdots + n] = n(n + 1)$. So $1 + 2 + \cdots + n = [n(n + 1)]/2$.

93. (a) $y = (-4.09 \times 10^{-5})x^3 + 0.016x^2 - 2.67x + 452.9$

(b) 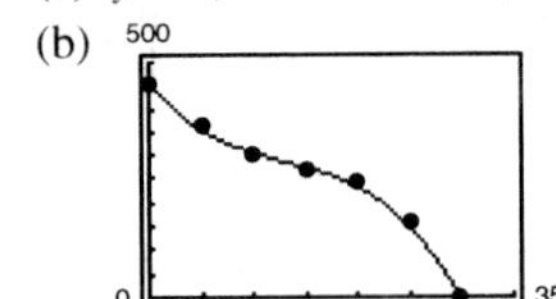

(c) $76{,}897.5 \text{ ft}^2$

95. Proof

Section 5.3   (page 314)

1. $2\sqrt{3} \approx 3.464$ **3.** 32   **5.** 0   **7.** $\frac{10}{3}$

9. $\displaystyle\int_{-1}^{5} (3x + 10)\, dx$ **11.** $\displaystyle\int_{0}^{3} \sqrt{x^2 + 4}\, dx$ **13.** $\displaystyle\int_{1}^{5} \left(1 + \frac{3}{x}\right) dx$

15. $\displaystyle\int_{0}^{4} 5\, dx$ **17.** $\displaystyle\int_{1}^{4} \frac{2}{x}\, dx$ **19.** $\displaystyle\int_{0}^{\pi/2} \cos x\, dx$ **21.** $\displaystyle\int_{0}^{2} y^3\, dy$

23. 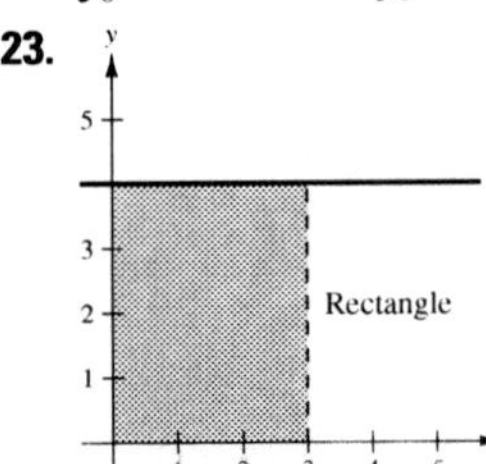

$A = 12$

25. 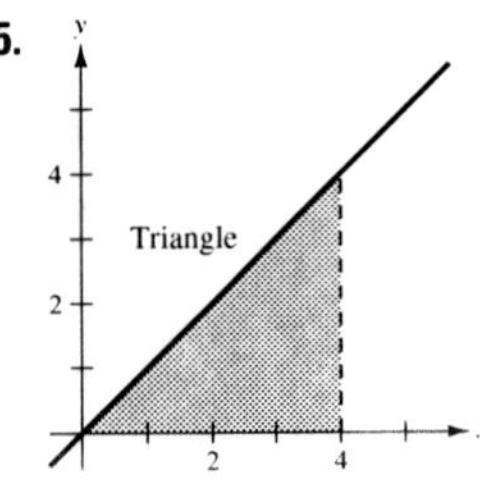

$A = 8$

27. 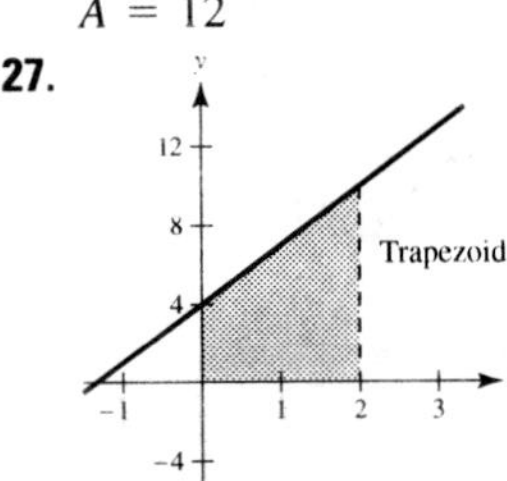

$A = 14$

29. 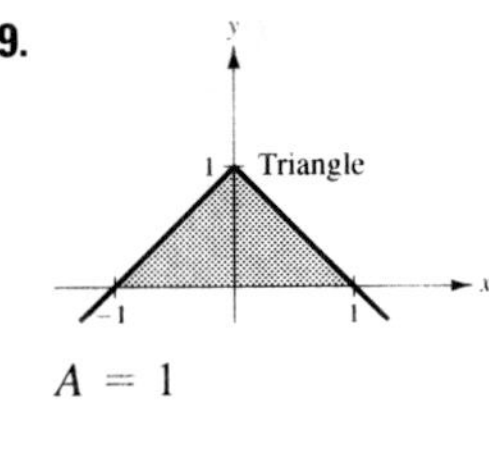

$A = 1$

31. 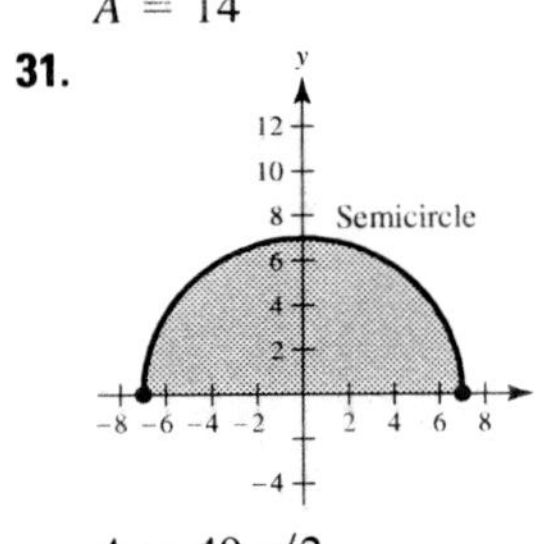

$A = 49\pi/2$

33. -6 **35.** 48   **37.** -12

39. 16   **41.** (a) 13   (b) -10 (c) 0   (d) 30

43. (a) 8   (b) -12 (c) -4 (d) 30   **45.** $-48, 88$

47. (a) $-\pi$ (b) 4   (c) $-(1 + 2\pi)$ (d) $3 - 2\pi$
(e) $5 + 2\pi$ (f) $23 - 2\pi$

49. (a) 14   (b) 4   (c) 8   (d) 0

51. 81   **53.** $\displaystyle\sum_{i=1}^{n} f(x_i)\, \Delta x > \int_{1}^{5} f(x)\, dx$

55. No. There is a discontinuity at $x = 4$.   **57.** a

59. d   **61.** c

63.

n	4	8	12	16	20
$L(n)$	3.6830	3.9956	4.0707	4.1016	4.1177
$M(n)$	4.3082	4.2076	4.1838	4.1740	4.1690
$R(n)$	3.6830	3.9956	4.0707	4.1016	4.1177

65.

n	4	8	12	16	20
$L(n)$	1.2833	1.1865	1.1562	1.1414	1.1327
$M(n)$	1.0898	1.0963	1.0976	1.0980	1.0982
$R(n)$	0.9500	1.0199	1.0451	1.0581	1.0660

67.

n	4	8	12	16	20
$L(n)$	0.5890	0.6872	0.7199	0.7363	0.7461
$M(n)$	0.7854	0.7854	0.7854	0.7854	0.7854
$R(n)$	0.9817	0.8836	0.8508	0.8345	0.8247

69. True   **71.** True

73. False: $\displaystyle\int_{0}^{2} (-x)\, dx = -2$ **75.** 272   **77.** Proof

79. No. No matter how small the subintervals, the number of both rational and irrational numbers within each subinterval is infinite and $f(c_i) = 0$ or $f(c_i) = 1$.

81. $a = -1$ and $b = 1$ maximize the integral.   **83.** $\frac{1}{3}$

Section 5.4   (page 329)

1. 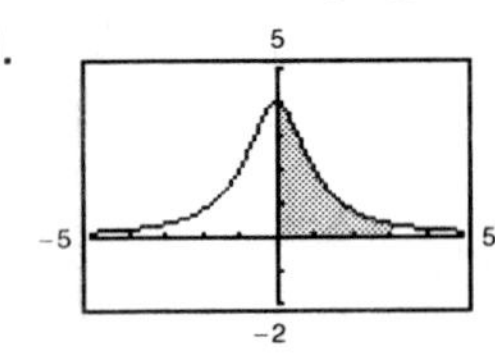

Positive

3. 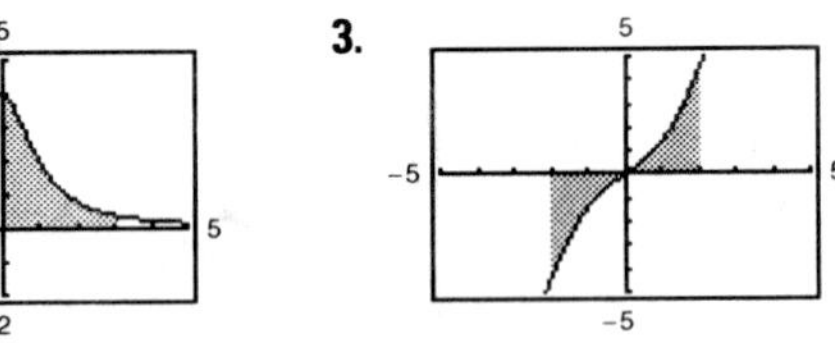

Zero

5. 12   **7.** -2 **9.** $-\frac{10}{3}$ **11.** $\frac{1}{3}$ **13.** $\frac{1}{2}$ **15.** $\frac{2}{3}$ **17.** -4

19. $-\frac{1}{18}$ **21.** $-\frac{27}{20}$ **23.** $\frac{25}{2}$ **25.** $\frac{64}{3}$ **27.** $\pi + 2$

29. $2\sqrt{3}/3$ **31.** $e^2 - 2$ **33.** 0   **35.** $(3/\ln 2) + 12$

37. $e - e^{-1}$ **39.** $\frac{1}{6}$ **41.** 1   **43.** $\frac{52}{3}$ **45.** 20   **47.** $\frac{32}{3}$

49. 4   **51.** $3\sqrt[3]{2}/2 \approx 1.8899$ **53.** $\frac{1444}{225} \approx 6.4178$

55. $\pm\arccos\sqrt{\pi}/2 \approx \pm 0.4817$ **57.** $3/\ln 4 \approx 2.1640$

59. Average value $= 6$ **61.** Average value $= \frac{1}{4}$
   $x = \pm\sqrt{3} \approx \pm 1.7321$ $x = \sqrt[3]{2}/2 \approx 0.6300$

63. Average value $= e - e^{-1} \approx 2.3504$
$x = \ln((e - e^{-1})/2) \approx 0.1614$

65. Average value $= 2/\pi$
$x \approx 0.690, \, x \approx 2.451$

67. About 540 ft

69. (a) 8 (b) $\frac{4}{3}$ (c) $\int_1^7 f(x)\,dx = 20$; Average value $= \frac{10}{3}$

71. (a) $F(x) = 500 \sec^2 x$ (b) $1500\sqrt{3}/\pi \approx 827$ N

73. About 0.5318 L

75. (a) $v = -0.00086t^3 + 0.0782t^2 - 0.208t + 0.10$
(b) 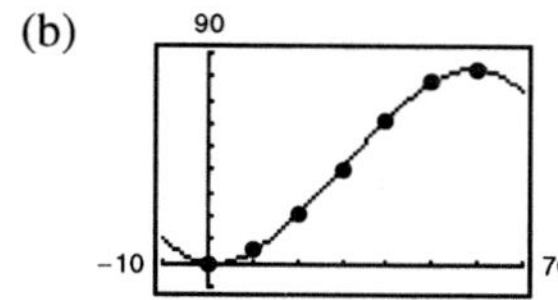 (c) 2475.6 m

77. $F(x) = 2x^2 - 7x$
$F(2) = -6$
$F(5) = 15$
$F(8) = 72$

79. $F(x) = -20/x + 20$
$F(2) = 10$
$F(5) = 16$
$F(8) = \frac{35}{2}$

81. $F(x) = \sin x - \sin 1$
$F(2) = \sin 2 - \sin 1 \approx 0.0678$
$F(5) = \sin 5 - \sin 1 \approx -1.8004$
$F(8) = \sin 8 - \sin 1 \approx 0.1479$

83. (a) $g(0) = 0, \, g(2) \approx 7, \, g(4) \approx 9, \, g(6) \approx 8, \, g(8) \approx 5$
(b) Increasing: $(0, 4)$; Decreasing: $(4, 8)$
(c) A maximum occurs at $x = 4$.
(d) 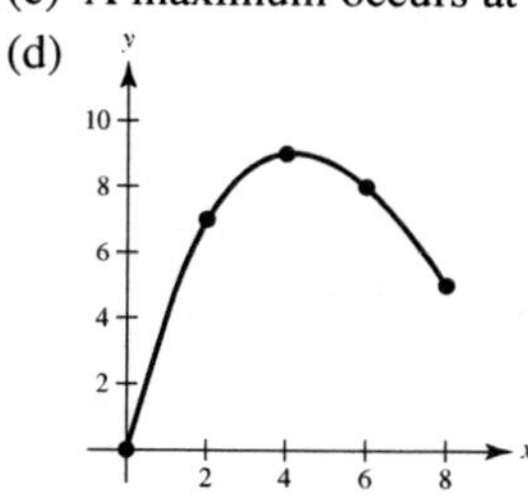

85. $\frac{1}{2}x^2 + 2x$ **87.** $\frac{3}{4}x^{4/3} - 12$ **89.** $\tan x - 1$

91. $e^x - e^{-1}$ **93.** $x^2 - 2x$ **95.** $\sqrt{x^4 + 1}$ **97.** $x \cos x$

99. 8 **101.** $\cos x \sqrt{\sin x}$ **103.** $3x^2 \sin x^6$

105. 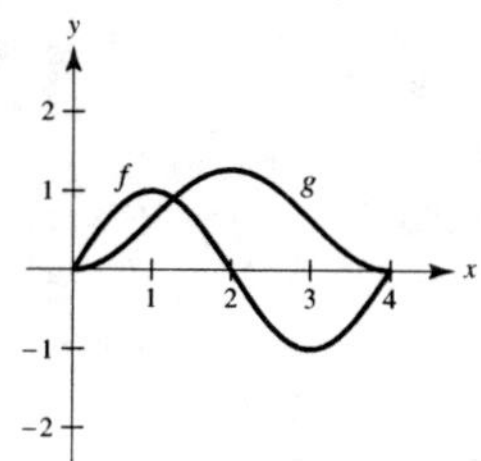

An extremum of g occurs at $x = 2$.

107. (a) $\frac{3}{2}$ ft to the right (b) $\frac{113}{10}$ ft **109.** (a) 0 ft (b) $\frac{63}{2}$ ft

111. (a) 2 ft to the right (b) 2 ft **113.** 28 units **115.** 8190 L

117. $f(x) = x^{-2}$ has a nonremovable discontinuity at $x = 0$.

119. $f(x) = \sec^2 x$ has a nonremovable discontinuity at $x = \pi/2$.

121. $2/\pi \approx 63.7\%$ **123.** True

125. $f'(x) = \dfrac{1}{(1/x)^2 + 1}\left(-\dfrac{1}{x^2}\right) + \dfrac{1}{x^2 + 1} = 0$
Because $f'(x) = 0$, $f(x)$ is constant.

127. (a) 0 (b) 0 (c) $xf(x) + \int_0^x f(t)\,dt$ (d) 0

Section 5.5 (page 342)

$\displaystyle\int f(g(x))g'(x)\,dx$	$u = g(x)$	$du = g'(x)\,dx$
1. $\displaystyle\int (8x^2 + 1)^2(16x)\,dx$	$8x^2 + 1$	$16x\,dx$
3. $\displaystyle\int \dfrac{x}{\sqrt{x^2 + 1}}\,dx$	$x^2 + 1$	$2x\,dx$
5. $\displaystyle\int \tan^2 x \sec^2 x\,dx$	$\tan x$	$\sec^2 x\,dx$

7. No **9.** Yes **11.** $\frac{1}{5}(1 + 6x)^5 + C$ **13.** $\frac{2}{3}(25 - x^2)^{3/2} + C$

15. $\frac{1}{12}(x^4 + 3)^3 + C$ **17.** $\frac{1}{15}(x^3 - 1)^5 + C$

19. $\frac{1}{3}(t^2 + 2)^{3/2} + C$ **21.** $-\frac{15}{8}(1 - x^2)^{4/3} + C$

23. $1/[4(1 - x^2)^2] + C$ **25.** $-1/[3(1 + x^3)] + C$

27. $-\sqrt{1 - x^2} + C$ **29.** $-\frac{1}{4}(1 + 1/t)^4 + C$ **31.** $\sqrt{2x} + C$

33. $\frac{2}{5}x^{5/2} + \frac{10}{3}x^{3/2} - 16x^{1/2} + C = \frac{1}{15}\sqrt{x}(6x^2 + 50x - 240) + C$

35. $\frac{1}{4}t^4 - 4t^2 + C$

37. $6y^{3/2} - \frac{2}{5}y^{5/2} + C = \frac{2}{5}y^{3/2}(15 - y) + C$

39. $2x^2 - 4\sqrt{16 - x^2} + C$ **41.** $-1/[2(x^2 + 2x - 3)] + C$

43. (a) Answers will vary.
Sample answer:
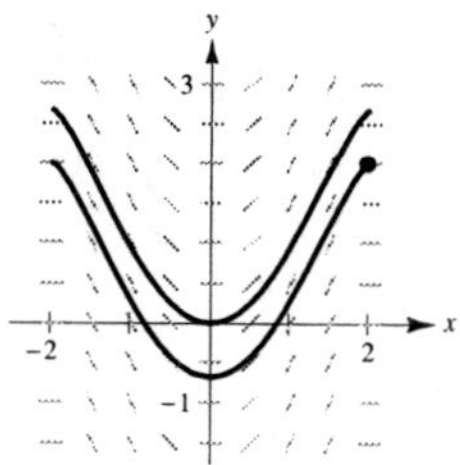
(b) $y = -\frac{1}{3}(4 - x^2)^{3/2} + 2$
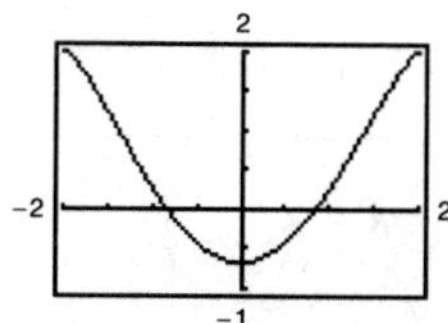

45. (a) Answers will vary.
Sample answer:
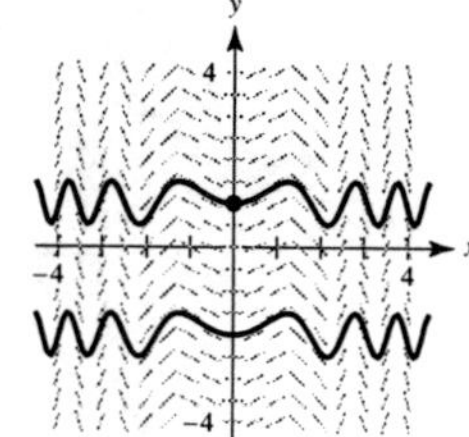
(b) $y = \frac{1}{2}\sin x^2 + 1$
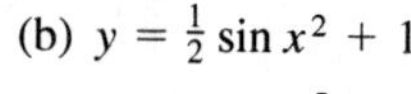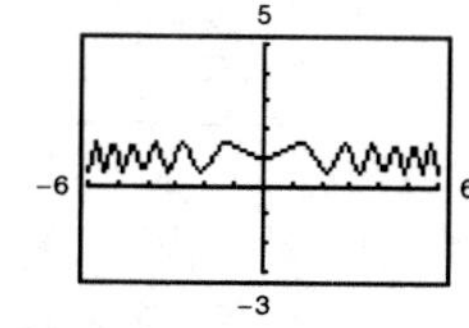

47. (a) Answers will vary.
Sample answer:
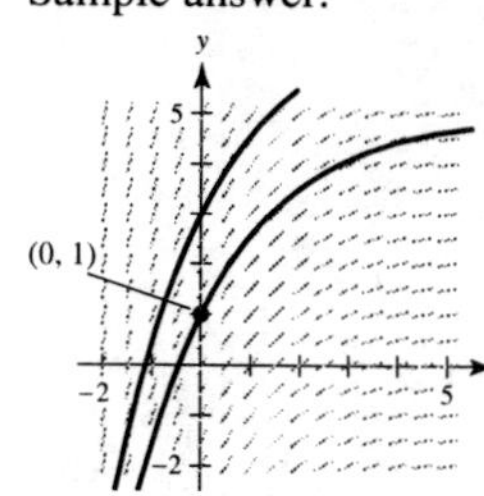
(b) $y = -4e^{-x/2} + 5$
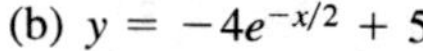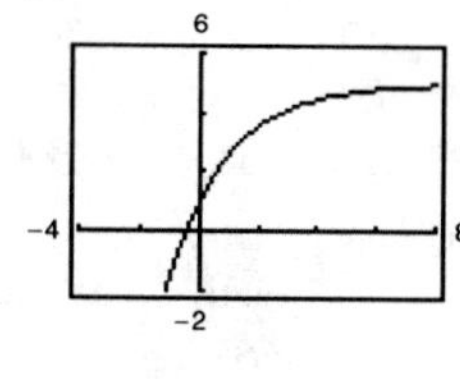

49. $-\cos(\pi x) + C$ **51.** $-\frac{1}{4}\cos 4x + C$ **53.** $-\sin(1/\theta) + C$

55. $e^{7x} + C$ **57.** $e^{x^2} + C$ **59.** $-\frac{1}{3}e^{-x^3} + C$

61. $\frac{1}{4}\sin^2 2x + C$ or $-\frac{1}{4}\cos^2 2x + C_1$ or $-\frac{1}{8}\cos 4x + C_2$

63. $\frac{1}{5}\tan^5 x + C$ **65.** $\frac{1}{2}\tan^2 x + C$ or $\frac{1}{2}\sec^2 x + C_1$

67. $-\cot x - x + C$ **69.** $\frac{1}{3}(e^x + 1)^3 + C$

71. $-\frac{2}{3}(1 - e^x)^{3/2} + C$ **73.** $-\frac{5}{2}e^{-2x} + e^{-x} + C$

75. $1/\pi\, e^{\sin \pi x} + C$ **77.** $-\tan(e^{-x}) + C$

79. $2/\ln 3\,(3^{x/2}) + C$ **81.** $-1/(2\ln 5)(5^{-x^2}) + C$

83. $f(x) = -\frac{1}{3}(4 - x^2)^{3/2} + 2$ **85.** $f(x) = 2\cos(x/2) + 4$

87. $f(x) = -8e^{-x/4} + 9$ **89.** $f(x) = \frac{1}{2}(e^x + e^{-x})$

91. $\frac{2}{5}(x + 6)^{3/2}(x - 4) + C$

93. $-\frac{2}{105}(1 - x)^{3/2}(15x^2 + 12x + 8) + C$

95. $\left(\sqrt{2x - 1}/15\right)(3x^2 + 2x - 13) + C$

97. $-x - 1 - 2\sqrt{x + 1} + C$ or $-\left(x + 2\sqrt{x + 1}\right) + C_1$

99. 0 **101.** $12 - \left(8\sqrt{2}/9\right)$ **103.** 2 **105.** $(e^2 - 1)/2e^2$

107. $e/3(e^2 - 1)$ **109.** $\frac{1}{2}$ **111.** $\frac{4}{15}$ **113.** $3\sqrt{3}/4$

115. $7/\ln 4$ **117.** $f(x) = (2x^3 + 1)^3 + 3$

119. $f(x) = \sqrt{2x^2 - 1} - 3$

121. $1209/28$ **123.** 4 **125.** $2\left(\sqrt{3} - 1\right)$

127. $e^5 - 1 \approx 147.413$ **129.** $2(1 - e^{-3/2}) \approx 1.554$

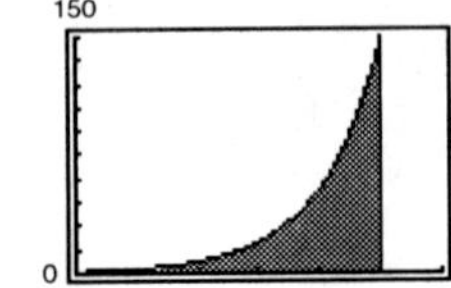 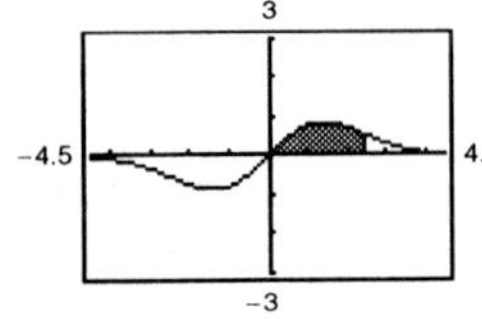

131. $\frac{14}{3}$ **133.** $\frac{144}{5}$

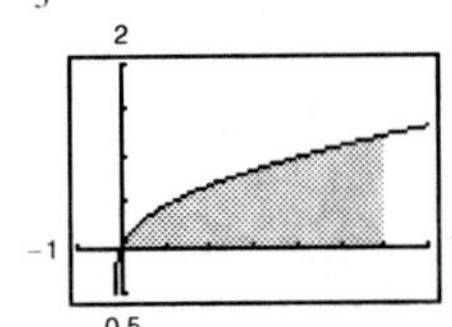 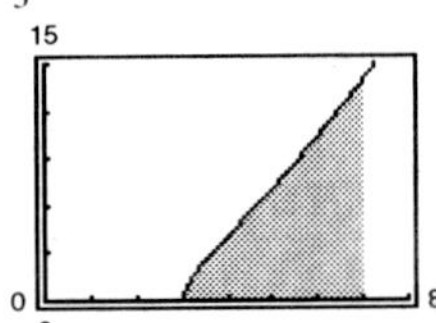

135. 9.21 **137.** $1 - e^{-1} \approx 0.632$

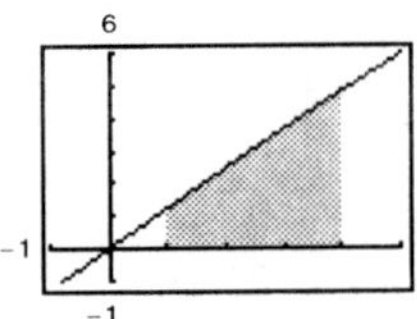 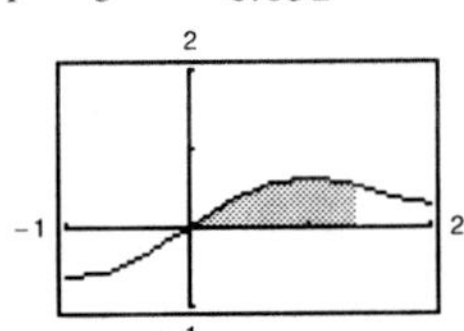

139. $\frac{272}{15}$ **141.** 0 **143.** (a) $\frac{64}{3}$ (b) $\frac{128}{3}$ (c) $-\frac{64}{3}$ (d) 64

145. $2\displaystyle\int_0^3 (4x^2 - 6)\, dx = 36$

147. If $u = 5 - x^2$, then $du = -2x\, dx$ and $\int x(5 - x^2)^3\, dx = -\frac{1}{2}\int(5 - x^2)^3(-2x)\, dx = -\frac{1}{2}\int u^3\, du.$

149. 16 **151.** $\$500{,}000$

153. (a) Relative minimum: $(6.7, 0.7)$ or July
        Relative maximum: $(1.3, 5.1)$ or February
     (b) 36.68 in.   (c) 3.99 in.

155. (a) 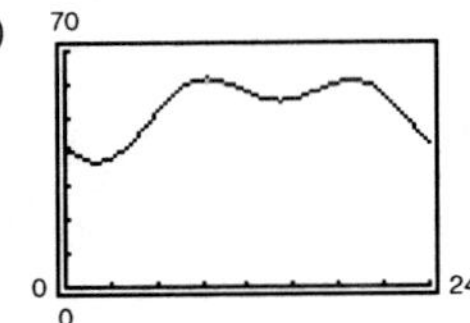   Maximum flow:
               $R \approx 61.713$ at $t = 9.36$

     (b) 1272 thousand gallons

157. (a) $P_{50,\,75} \approx 35.3\%$ (b) $b \approx 58.6\%$

159. 0.4772

161. (a) 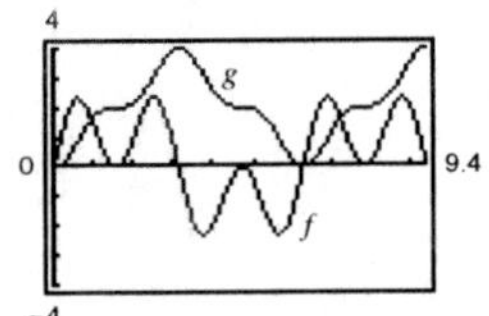   (b) g is nonnegative because the graph of f is positive at the beginning, and generally has more positive sections than negative ones.

     (c) The points on g that correspond to the extrema of f are points of inflection of g.

     (d) No, some zeros of f, such as $x = \pi/2$, do not correspond to extrema of g. The graph of g continues to increase after $x = \pi/2$ because f remains above the x-axis.

     (e) 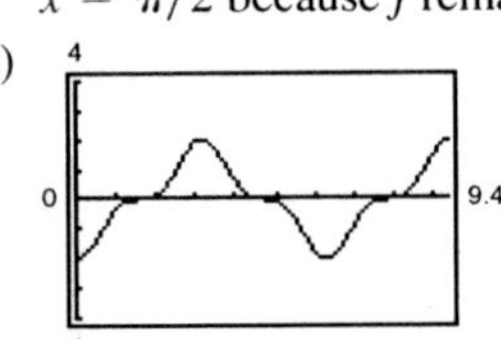

        The graph of h is that of g shifted 2 units downward.

163. (a) Proof   (b) Proof

165. False. $\displaystyle\int (2x + 1)^2\, dx = \frac{1}{6}(2x + 1)^3 + C$

167. True   **169.** True   **171.** Proof   **173.** Proof

175. Putnam Problem A1, 1958

Section 5.6   (page 352)

	Trapezoidal	*Simpson's*	*Exact*
1.	2.7500	2.6667	2.6667
3.	4.2500	4.0000	4.0000
5.	20.2222	20.0000	20.0000
7.	12.6640	12.6667	12.6667
9.	0.3352	0.3334	0.3333

	Trapezoidal	*Simpson's*	*Graphing Utility*
11.	3.2833	3.2396	3.2413
13.	0.3415	0.3720	0.3927
15.	0.5495	0.5483	0.5493
17.	-0.0975	-0.0977	-0.0977
19.	1.6845	1.6487	1.6479
21.	0.1940	0.1860	0.1858
23.	102.5553	93.3752	92.7437

25. Trapezoidal: First-degree (linear) polynomials
        Simpson's: Second-degree (quadratic) polynomials

27. (a) 1.500 (b) 0.000 **29.** (a) 0.1615 (b) 0.0066

31. (a) $n = 77$ (b) $n = 8$ **33.** (a) $n = 287$ (b) $n = 16$

35. (a) $n = 130$ (b) $n = 12$ **37.** (a) $n = 643$ (b) $n = 48$

39. (a) 24.5 (b) 25.67 **41.** Answers will vary.

43.

n	$L(n)$	$M(n)$	$R(n)$	$T(n)$	$S(n)$
4	0.8739	0.7960	0.6239	0.7489	0.7709
8	0.8350	0.7892	0.7100	0.7725	0.7803
10	0.8261	0.7881	0.7261	0.7761	0.7818
12	0.8200	0.7875	0.7367	0.7783	0.7826
16	0.8121	0.7867	0.7496	0.7808	0.7836
20	0.8071	0.7864	0.7571	0.7821	0.7841

45.

n	$L(n)$	$M(n)$	$R(n)$	$T(n)$	$S(n)$
4	0.7070	0.6597	0.6103	0.6586	0.6593
8	0.6833	0.6594	0.6350	0.6592	0.6593
10	0.6786	0.6594	0.6399	0.6592	0.6593
12	0.6754	0.6594	0.6431	0.6593	0.6593
16	0.6714	0.6594	0.6472	0.6593	0.6593
20	0.6690	0.6593	0.6496	0.6593	0.6593

47.

n	$L(n)$	$M(n)$	$R(n)$	$T(n)$	$S(n)$
4	2.5311	3.3953	4.3320	3.4316	3.4140
8	2.9632	3.4026	3.8637	3.4135	3.4074
10	3.0508	3.4037	3.7711	3.4109	3.4068
12	3.1094	3.4044	3.7100	3.4095	3.4065
16	3.1829	3.4050	3.6331	3.4080	3.4062
20	3.2272	3.4054	3.5874	3.4073	3.4061

49. 0.701   **51.** $\pi \approx 3.1416$ **53.** About 89,250 m²
55. Proof   **57.** Proof

Section 5.7 (page 360)

1. $5\ln|x| + C$ **3.** $\ln|x + 1| + C$ **5.** $\frac{1}{2}\ln|2x + 5| + C$
7. $\frac{1}{2}\ln|x^2 - 3| + C$ **9.** $\ln|x^4 + 3x| + C$
11. $x^2/2 - \ln(x^4) + C$ **13.** $\frac{1}{3}\ln|x^3 + 3x^2 + 9x| + C$
15. $\frac{1}{2}x^2 - 4x + 6\ln|x + 1| + C$ **17.** $\frac{1}{3}x^3 + 5\ln|x - 3| + C$
19. $\frac{1}{3}x^3 - 2x + \ln\sqrt{x^2 + 2} + C$ **21.** $\frac{1}{3}(\ln x)^3 + C$
23. $2\sqrt{x + 1} + C$ **25.** $2\ln|x - 1| - 2/(x - 1) + C$
27. $\sqrt{2x} - \ln\left|1 + \sqrt{2x}\right| + C$
29. $x + 6\sqrt{x} + 18\ln\left|\sqrt{x} - 3\right| + C$ **31.** $3\ln\left|\sin\frac{\theta}{3}\right| + C$
33. $-\frac{1}{2}\ln|\csc 2x + \cot 2x| + C$ **35.** $\frac{1}{3}\sin 3\theta - \theta + C$
37. $\ln|1 + \sin t| + C$ **39.** $\ln|\sec x - 1| + C$
41. $\ln|\cos(e^{-x})| + C$ **43.** $y = 4\ln|x| + C$

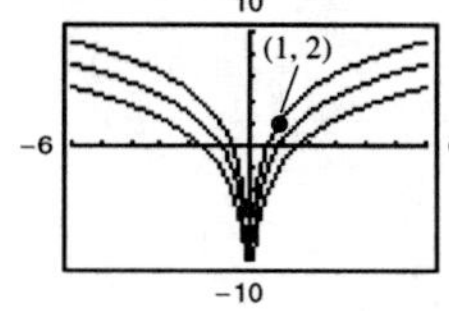

45. $y = -3\ln|2 - x| + C$ **47.** $s = -\frac{1}{2}\ln|\cos 2\theta| + C$

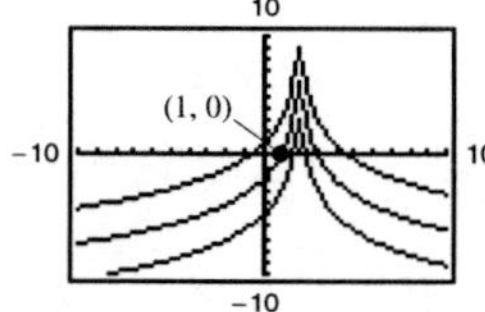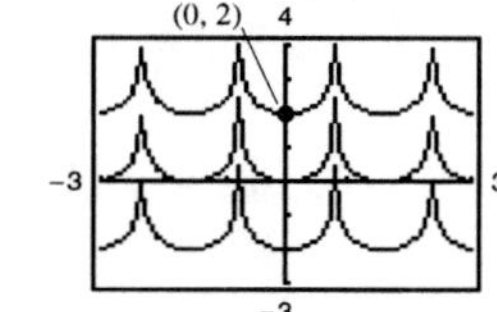

The graph has a hole at $x = 2$.
49. $f(x) = -2\ln x + 3x - 2$

51. (a)

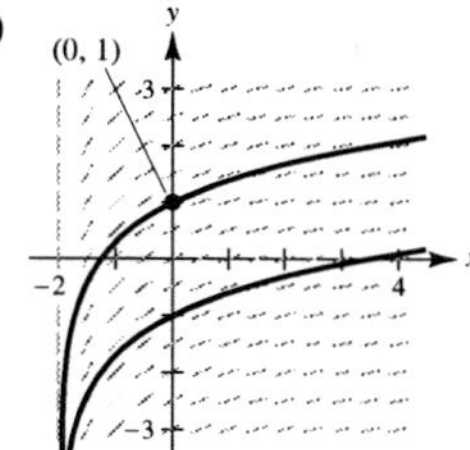

(b) $y = \ln[(x + 2)/2] + 1$

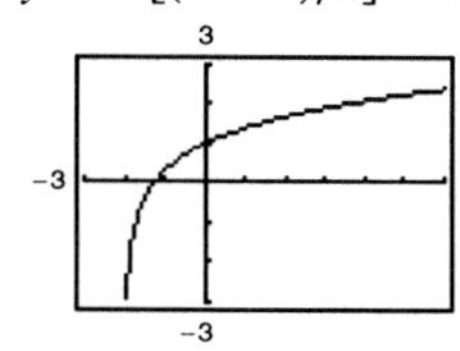

53. (a)

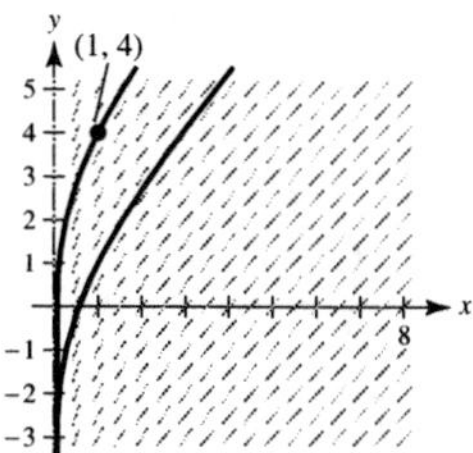

(b) $y = \ln x + x + 3$

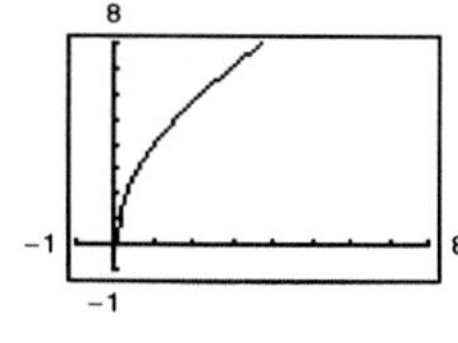

55. $\frac{5}{3}\ln 13 \approx 4.275$ **57.** $\frac{7}{3}$ **59.** $-\ln 3 \approx -1.099$
61. $\ln|(2 - \sin 2)/(1 - \sin 1)| \approx 1.929$
63. $2\left[\sqrt{x} - \ln\left(1 + \sqrt{x}\right)\right] + C$
65. $\ln\left[(\sqrt{x} - 1)/(\sqrt{x} + 1)\right] + 2\sqrt{x} + C$
67. $\ln(\sqrt{2} + 1) - \sqrt{2}/2 \approx 0.174$
69. $1/x$ **71.** $1/x$ **73.** d   **75.** $6\ln 3$ **77.** $\frac{1}{2}\ln 2$
79. $\frac{15}{2} + 8\ln 2 \approx 13.045$
81. $(12/\pi)\ln(2 + \sqrt{3}) \approx 5.03$
83. Trapezoidal Rule: 20.2   **85.** Trapezoidal Rule: 5.3368
    Simpson's Rule: 19.4667   Simpson's Rule: 5.3632
87. Power Rule   **89.** Log Rule   **91.** $x = 2$ **93.** Proof
95. $-\ln|\cos x| + C = \ln|1/\cos x| + C = \ln|\sec x| + C$
97. $\ln|\sec x + \tan x| + C = \ln\left|\dfrac{\sec^2 x - \tan^2 x}{\sec x - \tan x}\right| + C$
                        $= -\ln|\sec x - \tan x| + C$
99. 1   **101.** $1/(e - 1) \approx 0.582$
103. $P(t) = 1000(12\ln|1 + 0.25t| + 1); P(3) \approx 7715$
105. \$168.27   **107.** False. $\frac{1}{2}(\ln x) = \ln x^{1/2}$ **109.** True
111. (a)

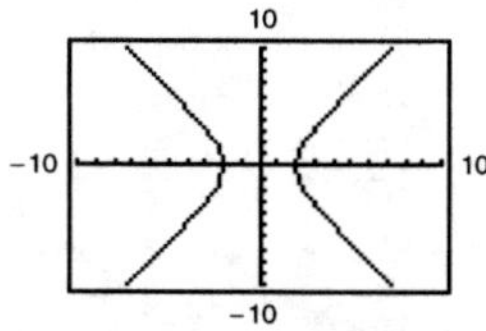

(b) Answers will vary. Sample answer: $y^2 = e^{-\ln x + \ln 4} = 4/x$

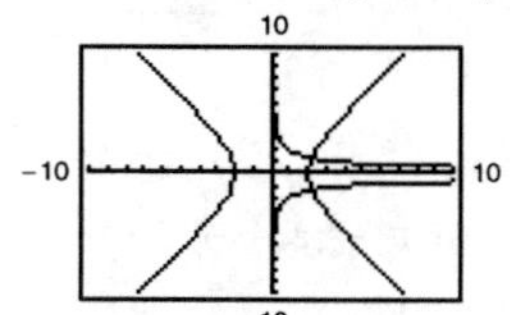

(c) Answers will vary.
113. Proof   **115.** Proof

Section 5.8   (page 368)

1. $\arcsin \dfrac{x}{3} + C$ **3.** $\dfrac{11}{6} \arctan \dfrac{x}{6} + C$ **5.** $\text{arcsec}|2x| + C$

7. $\arcsin(x + 1) + C$ **9.** $\frac{1}{2} \arcsin t^2 + C$

11. $\dfrac{1}{10} \arctan \dfrac{t^2}{5} + C$ **13.** $\dfrac{1}{4} \arctan\left(\dfrac{e^{2x}}{2}\right) + C$

15. $\arcsin\left(\dfrac{\tan x}{5}\right) + C$ **17.** $\dfrac{1}{2}x^2 - \dfrac{1}{2}\ln(x^2 + 1) + C$

19. $2 \arcsin \sqrt{x} + C$ **21.** $\frac{1}{2}\ln(x^2 + 1) - 3\arctan x + C$

23. $8 \arcsin[(x - 3)/3] - \sqrt{6x - x^2} + C$ **25.** $\pi/6$

27. $\pi/6$ **29.** $\frac{1}{2}(\sqrt{3} - 2) \approx -0.134$ **31.** $\frac{1}{5}\arctan\frac{3}{5} \approx 0.108$

33. $\arctan 5 - \dfrac{\pi}{4} \approx 0.588$ **35.** $\dfrac{\pi}{4}$ **37.** $\dfrac{1}{32}\pi^2 \approx 0.308$

39. $\pi/2$ **41.** $\ln|x^2 + 6x + 13| - 3\arctan[(x + 3)/2] + C$

43. $\arcsin[(x + 2)/2] + C$ **45.** $-\sqrt{-x^2 - 4x} + C$

47. $4 - 2\sqrt{3} + \frac{1}{6}\pi \approx 1.059$ **49.** $\frac{1}{2}\arctan(x^2 + 1) + C$

51. $2\sqrt{e^t - 3} - 2\sqrt{3}\arctan\left(\sqrt{e^t - 3}/\sqrt{3}\right) + C$

53. $\pi/6$ **55.** a and b   **57.** a, b, and c

59. No. This integral does not correspond to any of the basic integration rules.

61. $y = \arcsin(x/2) + \pi$

63. (a)   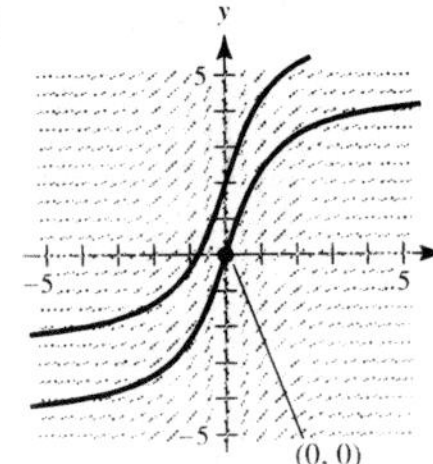   (b) $y = 3 \arctan x$ 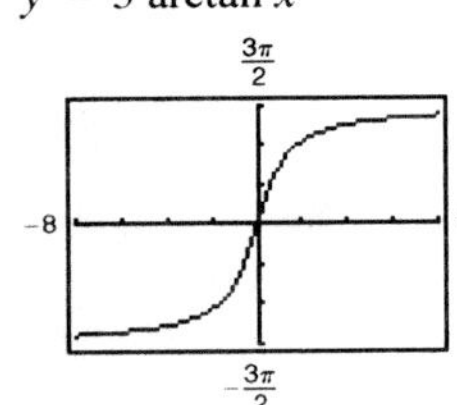

65. (a)   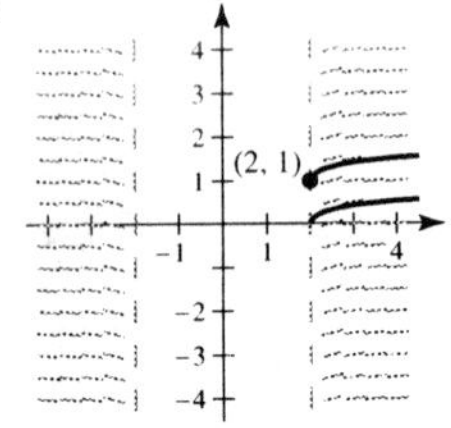   (b) $y = \frac{1}{2}\text{arcsec}(x/2) + 1$, $x \geq 2$ 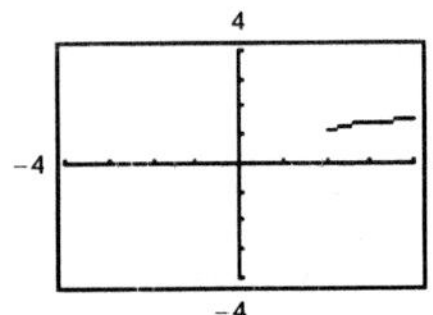

67.   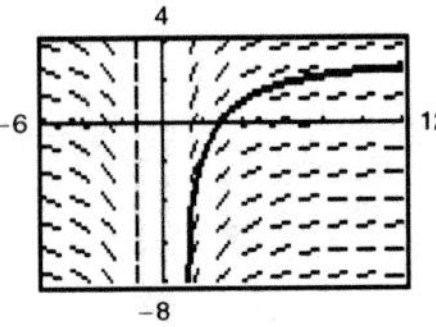   **69.**   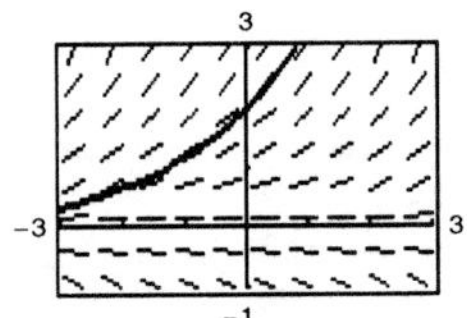

71. $\pi/3$ **73.** $\pi/8$ **75.** $3\pi/2$

77. (a) Proof   (b) $\ln(\sqrt{6}/2) + (9\pi - 4\pi\sqrt{3})/36$

79. (a)   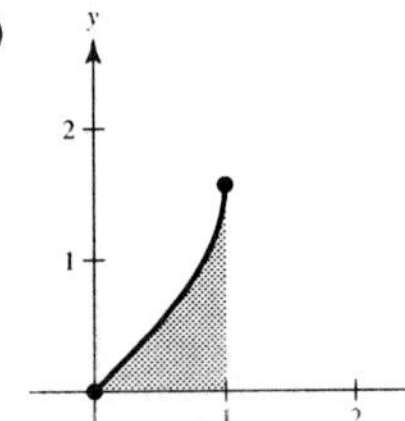   (b) 0.5708 (c) $(\pi - 2)/2$

81. (a) $F(x)$ represents the average value of $f(x)$ over the interval $[x, x + 2]$. Maximum at $x = -1$.
   (b) $x = -1$

83. False. $\displaystyle\int \dfrac{dx}{3x\sqrt{9x^2 - 16}} = \dfrac{1}{12}\text{arcsec}\dfrac{|3x|}{4} + C$

85. True   **87.** Proof   **89.** Proof

91. (a) $v(t) = -32t + 500$

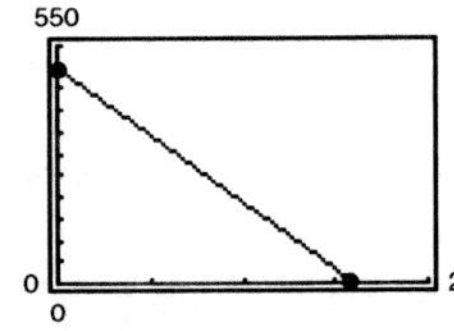

   (b) $s(t) = -16t^2 + 500t$; 3906.25 ft

   (c) $v(t) = \sqrt{\dfrac{32}{k}}\tan\left[\arctan\left(500\sqrt{\dfrac{k}{32}}\right) - \sqrt{32k}\,t\right]$

   (d) 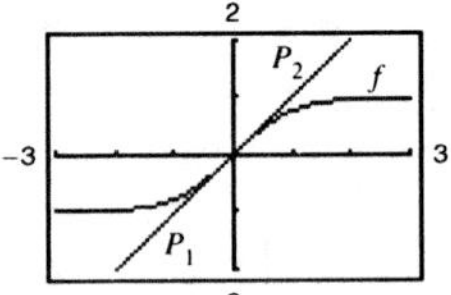   (e) 1088 ft
   (f) When air resistance is taken into account, the maximum height of the object is not as great.

   $t_0 = 6.86$ sec

Section 5.9   (page 379)

1. (a) 10.018 (b) -0.964 **3.** (a) $\frac{4}{3}$ (b) $\frac{13}{12}$

5. (a) 1.317 (b) 0.962 **7–15.** Proofs

17. $\cosh x = \sqrt{13}/2$; $\tanh x = 3\sqrt{13}/13$; $\text{csch } x = 2/3$; $\text{sech } x = 2\sqrt{13}/13$; $\coth x = \sqrt{13}/3$

19. $3 \cosh 3x$ **21.** $-10x[\text{sech}(5x^2)\tanh(5x^2)]$ **23.** $\coth x$

25. $\text{csch } x$ **27.** $\sinh^2 x$ **29.** $\text{sech } t$

31. $y = -2x + 2$ **33.** $y = 1 - 2x$

35. Relative maxima: $(\pm\pi, \cosh\pi)$; Relative minimum: $(0, -1)$

37. Relative maximum: $(1.20, 0.66)$
   Relative minimum: $(-1.20, -0.66)$

39. $y = a \sinh x$; $y' = a \cosh x$; $y'' = a \sinh x$; $y''' = a \cosh x$; Therefore, $y''' - y' = 0$.

41. $P_1(x) = x$; $P_2(x) = x$

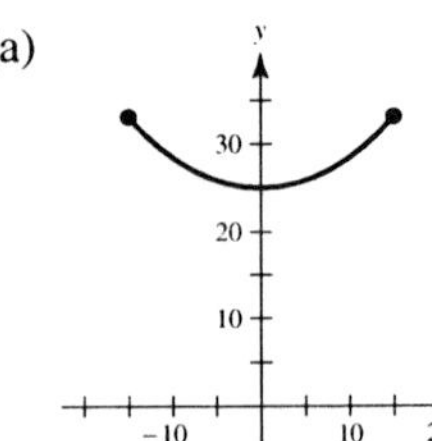

43. (a)   (b) 33.146 units; 25 units
   (c) $m = \sinh(1) \approx 1.175$

45. $\frac{1}{2}\sinh 2x + C$ **47.** $-\frac{1}{2}\cosh(1 - 2x) + C$

49. $\frac{1}{3}\cosh^3(x - 1) + C$ **51.** $\ln|\sinh x| + C$

53. $-\coth(x^2/2) + C$ **55.** $\text{csch}(1/x) + C$

57. $\frac{1}{2}\arctan x^2 + C$ **59.** $\ln(5/4)$ **61.** $\frac{1}{5}\ln 3$

63. $\frac{\pi}{4}$ **65.** $\dfrac{3}{\sqrt{9x^2-1}}$ **67.** $\dfrac{1}{2\sqrt{x}(1-x)}$ **69.** $|\sec x|$

71. $\dfrac{-2\,\operatorname{csch}^{-1}x}{|x|\sqrt{1+x^2}}$ **73.** $2\sinh^{-1}(2x)$ **75.** Answers will vary.

77. $\cosh x,\ \operatorname{sech} x$ **79.** ∞ **81.** 1   **83.** 0   **85.** 1

87. $\dfrac{\sqrt{3}}{18}\ln\left|\dfrac{1+\sqrt{3}x}{1-\sqrt{3}x}\right|+C$ **89.** $\ln\left(\sqrt{e^{2x}+1}-1\right)-x+C$

91. $2\sinh^{-1}\sqrt{x}+C=2\ln\left(\sqrt{x}+\sqrt{1+x}\right)+C$

93. $\dfrac{1}{4}\ln\left|\dfrac{x-4}{x}\right|+C$ **95.** $\dfrac{1}{2\sqrt{6}}\ln\left|\dfrac{\sqrt{2}(x+1)+\sqrt{3}}{\sqrt{2}(x+1)-\sqrt{3}}\right|+C$

97. $\ln\left(\dfrac{3+\sqrt{5}}{2}\right)$ **99.** $\dfrac{\ln 7}{12}$ **101.** $\dfrac{1}{4}\arcsin\left(\dfrac{4x-1}{9}\right)+C$

103. $-\dfrac{x^2}{2}-4x-\dfrac{10}{3}\ln\left|\dfrac{x-5}{x+1}\right|+C$

105. $8\arctan(e^2)-2\pi\approx 5.207$ **107.** $\frac{5}{2}\ln\left(\sqrt{17}+4\right)\approx 5.237$

109. (a) $\ln\left(\sqrt{3}+2\right)$ (b) $\sinh^{-1}\sqrt{3}$

111. $\frac{52}{31}$ kg   **113.** $-\sqrt{a^2-x^2}/x$ **115–123.** Proofs

125. Putnam Problem 8, 1939

Review Exercises for Chapter 5   (page 382)

1. 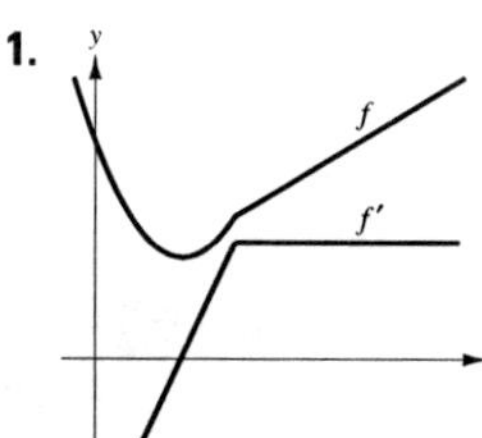   **3.** $\frac{4}{3}x^3+\frac{1}{2}x^2+3x+C$

5. $x^2/2-4/x^2+C$ **7.** $x^2+9\cos x+C$

9. $5x-e^x+C$ **11.** $5\ln|x|+C$ **13.** $y=1-3x^2$

15. 240 ft/sec   **17.** (a) 3 sec   (b) 144 ft   (c) $\frac{3}{2}$ sec   (d) 108 ft

19. (a) $\displaystyle\sum_{i=1}^{10}(2i-1)$ (b) $\displaystyle\sum_{i=1}^{n}i^3$ (c) $\displaystyle\sum_{i=1}^{10}(4i+2)$

21. $9.038 < (\text{Area of region}) < 13.038$

23. $A=15$ **25.** $A=12$

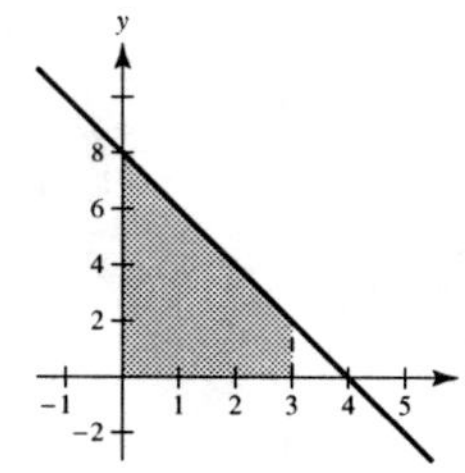   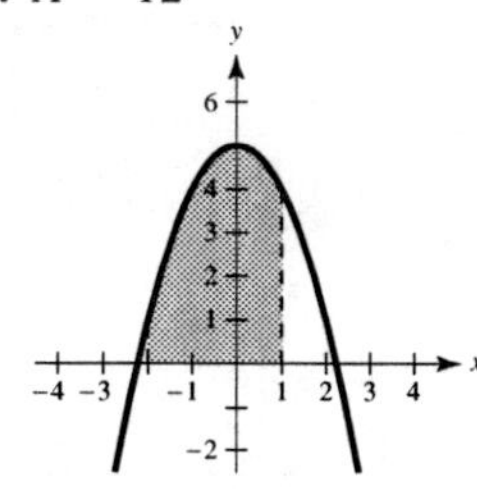

27. $\frac{27}{2}$ **29.** $\displaystyle\int_4^6 (2x-3)\,dx$

31. 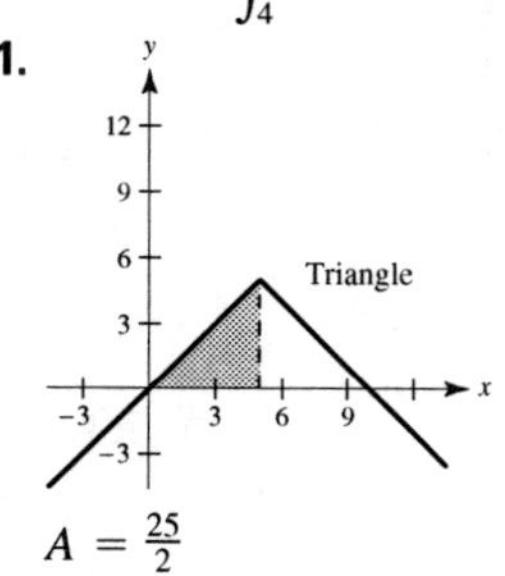

$A=\frac{25}{2}$

33. (a) 17   (b) 7   (c) 9   (d) 84   **35.** c   **37.** 56   **39.** 0

41. $\frac{422}{5}$ **43.** $\left(\sqrt{2}+2\right)/2$ **45.** e^2+1

47. 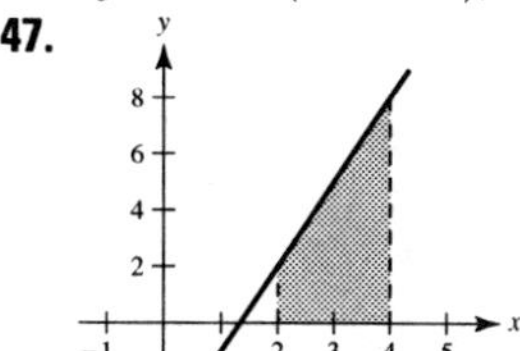   **49.** 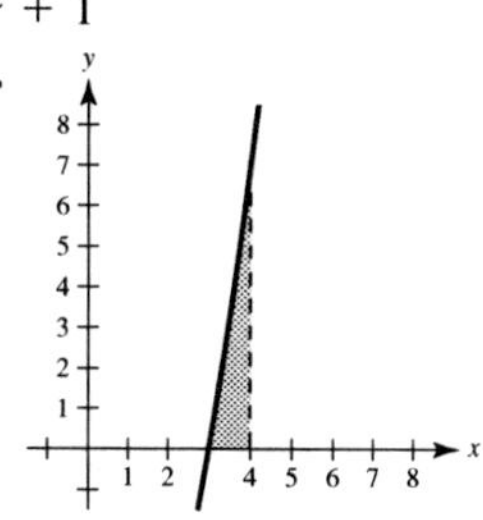

$A=10$ $A=\frac{10}{3}$

51. 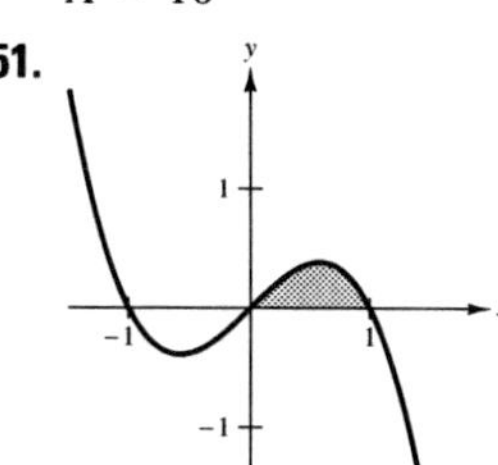   **53.** 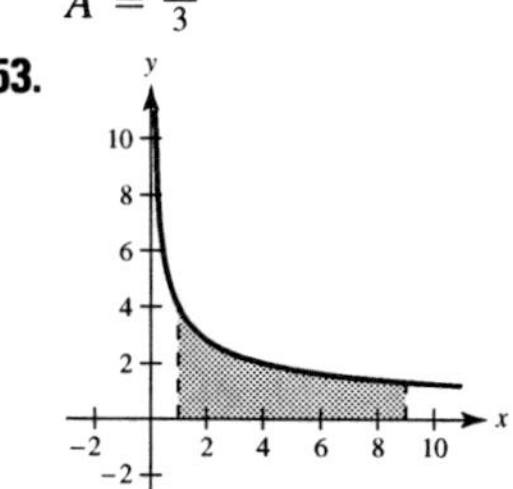

$A=\frac{1}{4}$ $A=16$

55. 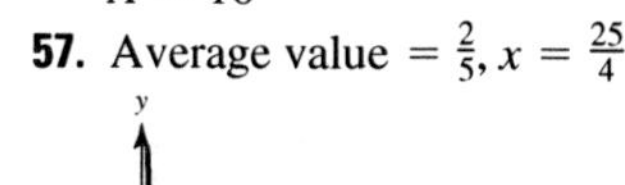   **57.** Average value $=\frac{2}{5}$, $x=\frac{25}{4}$

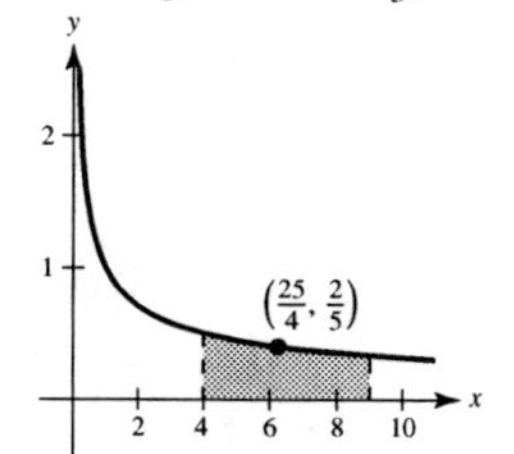

$A=2\ln 3\approx 2.1972$

59. $x^2\sqrt{1+x^3}$ **61.** x^2+3x+2

63. $-\frac{1}{7}x^7+\frac{9}{5}x^5-9x^3+27x+C$ **65.** $\frac{2}{3}\sqrt{x^3+3}+C$

67. $-\frac{1}{30}(1-3x^2)^5+C=(3x^2-1)^5/30+C$

69. $\frac{1}{4}\sin^4 x+C$ **71.** $-2\sqrt{1-\sin\theta}+C$

73. $\dfrac{\tan^{n+1}x}{n+1}+C,\ n\neq -1$ **75.** $\dfrac{1}{3\pi}(1+\sec\pi x)^3+C$

77. $-\frac{1}{6}e^{-3x^2}+C$ **79.** $\dfrac{1}{2\ln 5}(5^{(x+1)^2})+C$

81. $21/4$ **83.** 2   **85.** $28\pi/15$ **87.** 2

89. (a) $\int_0^{12}[2.880+2.125\sin(0.578t+0.745)]\,dt\approx 36.63$ in.
   (b) 2.22 in.

91. Trapezoidal Rule: 0.285   **93.** Trapezoidal Rule: 0.637
   Simpson's Rule: 0.284   Simpson's Rule: 0.685
   Graphing utility: 0.284   Graphing utility: 0.704

95. Trapezoidal Rule: 1.463
   Simpson's Rule: 1.494
   Graphing utility: 1.494

97. $\frac{1}{7}\ln|7x-2|+C$ **99.** $-\ln|1+\cos x|+C$

101. $3+\ln 2$ **103.** $\ln\left(2+\sqrt{3}\right)$ **105.** $\frac{1}{2}\ln(e^{2x}+e^{-2x})+C$

107. $\frac{1}{2}\arctan(e^{2x})+C$ **109.** $\frac{1}{2}\arcsin x^2+C$

111. $\ln\sqrt{16+x^2}+C$ **113.** $\frac{1}{4}[\arctan(x/2)]^2+C$

115. $y=A\sin\left(\sqrt{k/m}\,t\right)$ **117.** $2-\left[\left(\sinh\sqrt{x}\right)/(2\sqrt{x})\right]$

119. $\frac{1}{2}\ln\left(\sqrt{x^4-1}+x^2\right)+C$

P.S. Problem Solving   (page 385)

1. (a) $L(1) = 0$ (b) $L'(x) = 1/x$, $L'(1) = 1$
(c) $x \approx 2.718$ (d) Proof

3. (a)

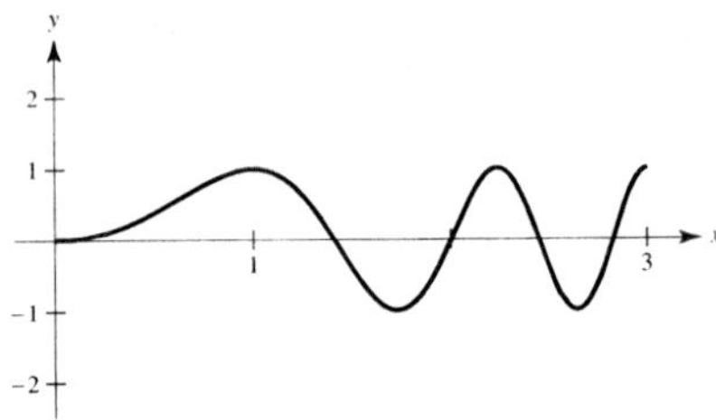

(b)

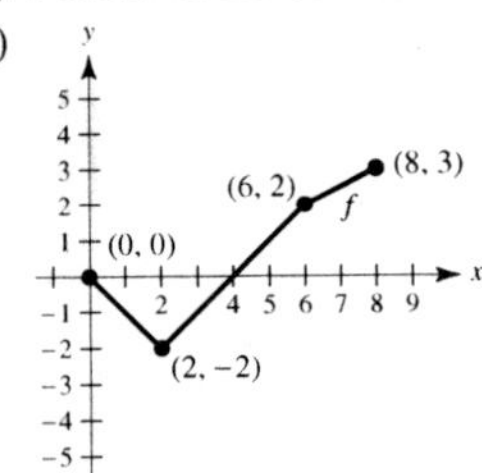

(c) Relative maxima at $x = \sqrt{2}, \sqrt{6}$
Relative minima at $x = 2, 2\sqrt{2}$
(d) Points of inflection at $x = 1, \sqrt{3}, \sqrt{5}, \sqrt{7}$

5. (a)

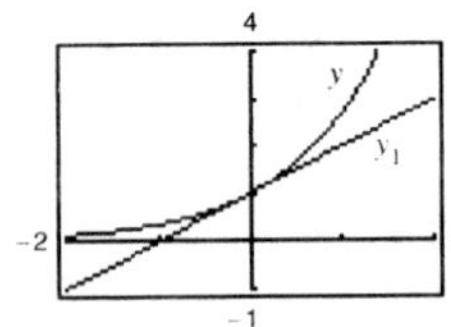

(b)

x	0	1	2	3	4	5	6	7	8
$F(x)$	0	$-\frac{1}{2}$	-2	$-\frac{7}{2}$	-4	$-\frac{7}{2}$	-2	$\frac{1}{4}$	3

(c) $x = 4, 8$ (d) $x = 2$

7. (a) 1.6758; Error of approximation ≈ 0.0071
(b) $\frac{3}{2}$ (c) Proof

9. Proof

11. $\displaystyle \lim_{n \to \infty} \sum_{i=1}^{n} \left(\frac{i}{n}\right)^5 \left(\frac{1}{n}\right) = \frac{1}{6}$ **13.** $1 \le \displaystyle\int_0^1 \sqrt{1 + x^4}\, dx \le \sqrt{2}$

15. Proof   **17.** $2 \ln\left(\frac{3}{2}\right) \approx 0.8109$

19. (a) (i)

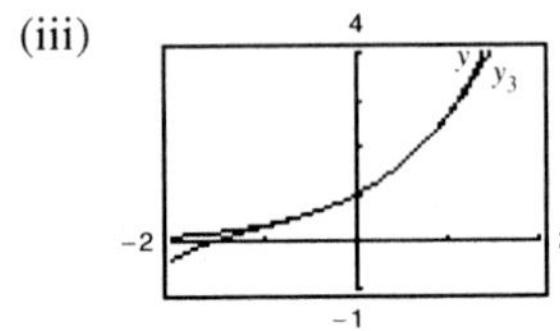

(ii)

(iii)

(b) $\displaystyle\sum_{n=0}^{\infty} \frac{x^n}{n!}$

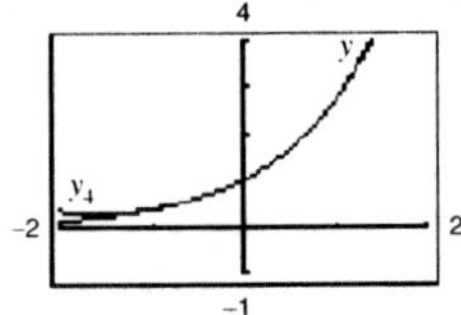

(c) $e^x = \displaystyle\sum_{n=0}^{\infty} \frac{x^n}{n!}$

Chapter 6

Section 6.1   (page 393)

1–11. Proofs   **13.** Not a solution   **15.** Solution
17. Solution   **19.** Solution   **21.** Not a solution
23. Solution   **25.** Not a solution   **27.** Not a solution
29. $y = 3e^{-x/2}$ **31.** $4y^2 = x^3$

33.

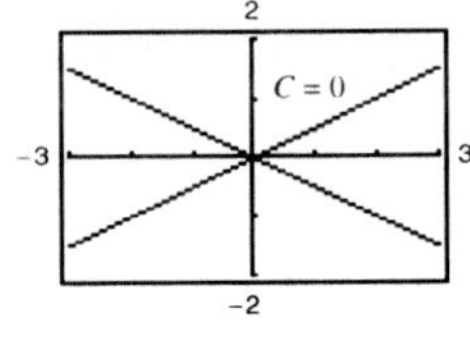

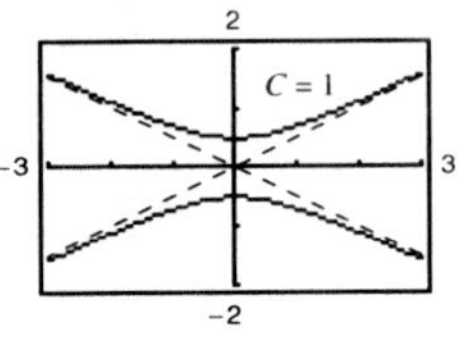

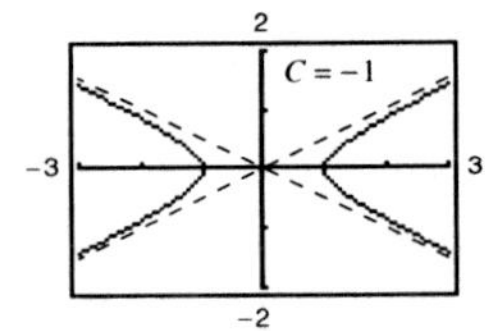

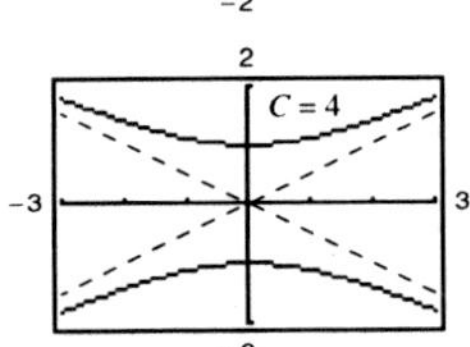

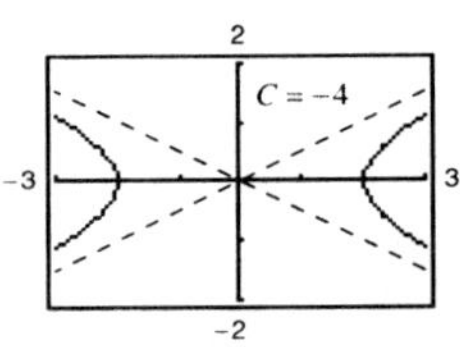

35. $y = 3e^{-2x}$ **37.** $y = 2 \sin 3x - \frac{1}{3}\cos 3x$
39. $y = -2x + \frac{1}{2}x^3$ **41.** $y = 2x^3 + C$
43. $y = \frac{1}{2}\ln(1 + x^2) + C$ **45.** $y = x - \ln x^2 + C$
47. $y = -\frac{1}{2}\cos 2x + C$
49. $y = \frac{2}{5}(x - 6)^{5/2} + 4(x - 6)^{3/2} + C$
51. $y = \frac{1}{2}e^{x^2} + C$

53.

x	-4	-2	0	2	4	8
y	2	0	4	4	6	8
dy/dx	-4	Undef.	0	1	$\frac{4}{3}$	2

55.

x	-4	-2	0	2	4	8
y	2	0	4	4	6	8
dy/dx	$-2\sqrt{2}$	-2	0	0	$-2\sqrt{2}$	-8

57. b   **58.** c   **59.** d   **60.** a

61. (a) and (b)

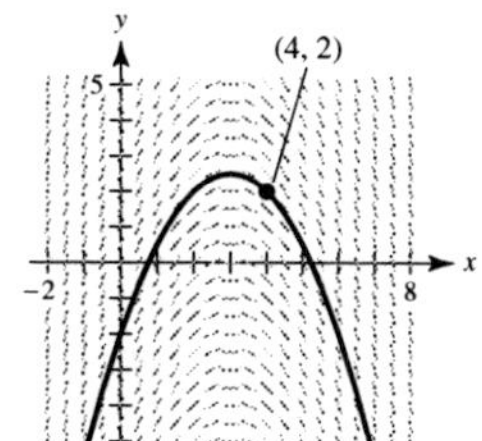

(c) As $x \to \infty$, $y \to -\infty$;
   as $x \to -\infty$, $y \to -\infty$

63. (a) and (b)

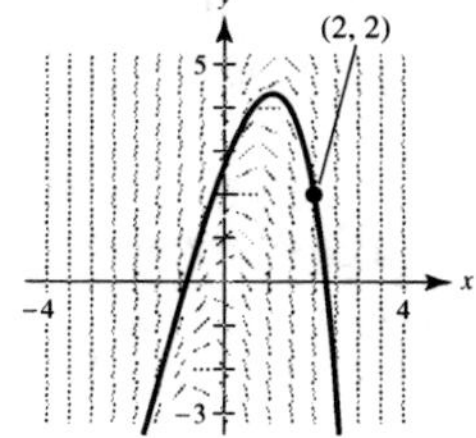

(c) As $x \to \infty$, $y \to -\infty$;
   as $x \to -\infty$, $y \to -\infty$

65. (a)

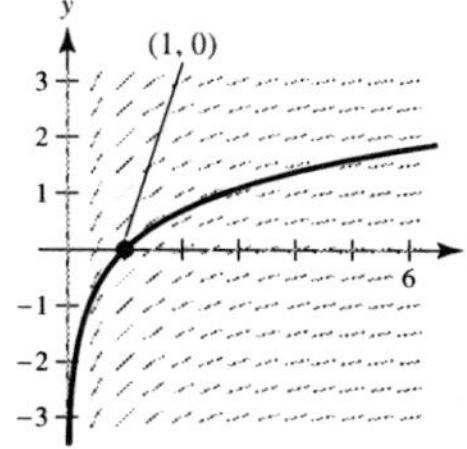

As $x \to \infty$, $y \to \infty$

(b)

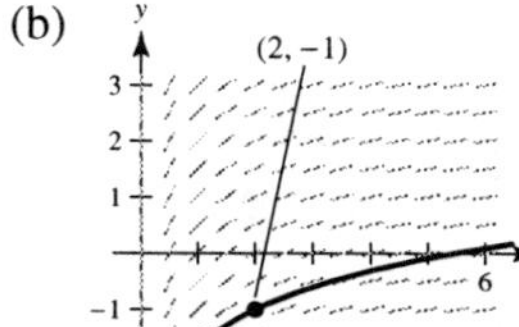

As $x \to \infty$, $y \to \infty$

67. (a) and (b)

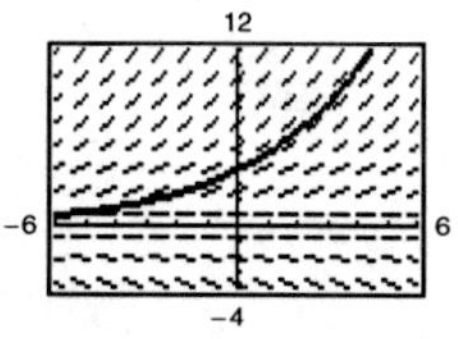

69. (a) and (b)

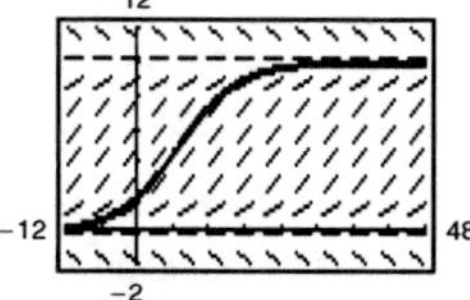

71. (a) and (b)

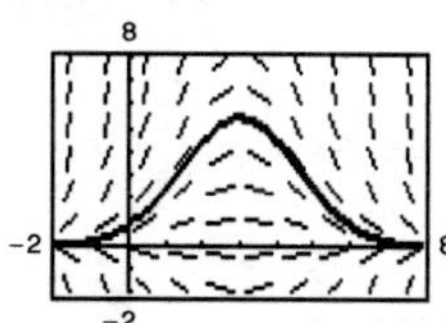

73.

n	0	1	2	3	4	5	6
x_n	0	0.1	0.2	0.3	0.4	0.5	0.6
y_n	2	2.2	2.43	2.693	2.992	3.332	3.715

n	7	8	9	10
x_n	0.7	0.8	0.9	1.0
y_n	4.146	4.631	5.174	5.781

75.

n	0	1	2	3	4	5	6
x_n	0	0.05	0.1	0.15	0.2	0.25	0.3
y_n	3	2.7	2.438	2.209	2.010	1.839	1.693

n	7	8	9	10
x_n	0.35	0.4	0.45	0.5
y_n	1.569	1.464	1.378	1.308

77.

n	0	1	2	3	4	5	6
x_n	0	0.1	0.2	0.3	0.4	0.5	0.6
y_n	1	1.1	1.212	1.339	1.488	1.670	1.900

n	7	8	9	10
x_n	0.7	0.8	0.9	1.0
y_n	2.213	2.684	3.540	5.958

79.

x	0	0.2	0.4	0.6	0.8	1
$y(x)$ (exact)	3.0000	3.6642	4.4755	5.4664	6.6766	8.1548
$y(x)$ ($h = 0.2$)	3.0000	3.6000	4.3200	5.1840	6.2208	7.4650
$y(x)$ ($h = 0.1$)	3.0000	3.6300	4.3923	5.3147	6.4308	7.7812

81.

x	0	0.2	0.4	0.6	0.8	1
$y(x)$ (exact)	0.0000	0.2200	0.4801	0.7807	1.1231	1.5097
$y(x)$ ($h = 0.2$)	0.0000	0.2000	0.4360	0.7074	1.0140	1.3561
$y(x)$ ($h = 0.1$)	0.0000	0.2095	0.4568	0.7418	1.0649	1.4273

83. (a) $y(1) = 112.7141°$; $y(2) = 96.3770°$; $y(3) = 86.5954°$
   (b) $y(1) = 113.2441°$; $y(2) = 97.0158°$; $y(3) = 87.1729°$
   (c) Euler's Method: $y(1) = 112.9828°$; $y(2) = 96.6998°$;
        $y(3) = 86.8863°$
      Exact solution: $y(1) = 113.2441°$; $y(2) = 97.0158°$;
        $y(3) = 87.1729°$
   The approximations are better using $h = 0.05$.

85. The general solution is a family of curves that satisfies the differential equation. A particular solution is one member of the family that satisfies given conditions.

87. Begin with a point (x_0, y_0) that satisfies the initial condition $y(x_0) = y_0$. Then, using a small step size h, calculate the point $(x_1, y_1) = (x_0 + h, y_0 + hF(x_0, y_0))$. Continue generating the sequence of points $(x_n + h, y_n + hF(x_n, y_n))$ or (x_{n+1}, y_{n+1}).

89. False. $y = x^3$ is a solution of $xy' - 3y = 0$, but $y = x^3 + 1$ is not a solution.

91. True

93. (a)

x	0	0.2	0.4	0.6	0.8	1
y	4	2.6813	1.7973	1.2048	0.8076	0.5413
y_1	4	2.56	1.6384	1.0486	0.6711	0.4295
y_2	4	2.4	1.44	0.864	0.5184	0.3110
e_1	0	0.1213	0.1589	0.1562	0.1365	0.1118
e_2	0	0.2813	0.3573	0.3408	0.2892	0.2303
r		0.4312	0.4447	0.4583	0.4720	0.4855

    (b) If h is halved, then the error is approximately halved because r is approximately 0.5.

    (c) The error will again be halved.

95. (a)   (b) $\lim\limits_{t \to \infty} I(t) = 2$

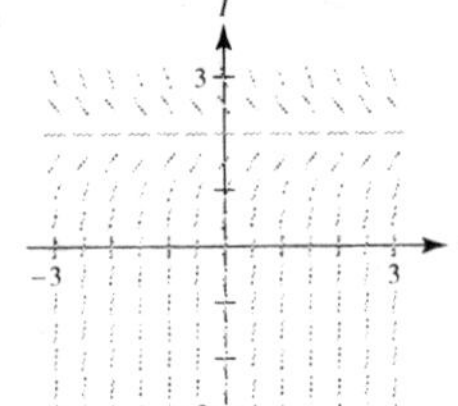

97. $\omega = \pm 4$ **99.** Putnam Problem 3, Morning Session, 1954

Section 6.2   (page 402)

1. $y = \frac{1}{2}x^2 + 3x + C$ **3.** $y = Ce^x - 3$ **5.** $y^2 - 5x^2 = C$

7. $y = Ce^{(2x^{3/2})/3}$ **9.** $y = C(1 + x^2)$

11. $dQ/dt = k/t^2$ **13.** $dN/ds = k(500 - s)$
    $Q = -k/t + C$ $N = -(k/2)(500 - s)^2 + C$

15. (a)   (b) $y = 6 - 6e^{-x^2/2}$

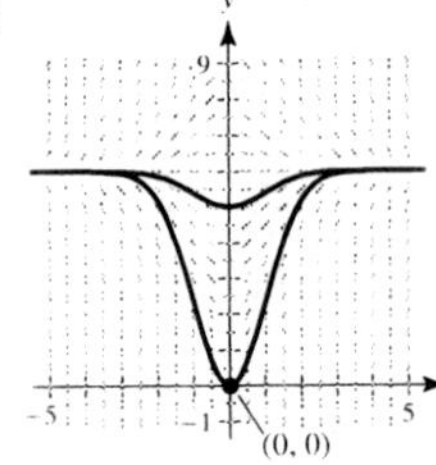
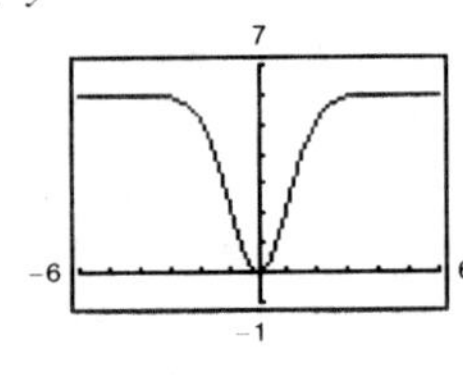

17. $y = \frac{1}{4}t^2 + 10$ **19.** $y = 10e^{-t/2}$

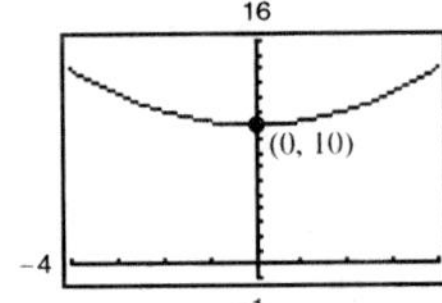
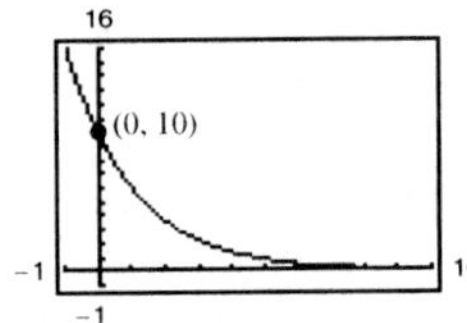

21. $dy/dx = ky$
    $y = 6e^{(1/4)\ln(5/2)x} \approx 6e^{0.2291x}$
    $y(8) \approx 37.5$

23. $dV/dt = kV$
    $V = 20{,}000e^{(1/4)\ln(5/8)t} \approx 20{,}000e^{-0.1175t}$
    $V(6) \approx 9882$

25. $y = (1/2)e^{[(\ln 10)/5]t} \approx (1/2)e^{0.4605t}$

27. $y = 5(5/2)^{1/4}e^{[\ln(2/5)/4]t} \approx 6.2872e^{-0.2291t}$

29. C is the initial value of y, and k is the proportionality constant.

31. Quadrants I and III; dy/dx is positive when both x and y are positive (Quadrant I) or when both x and y are negative (Quadrant III).

33. Amount after 1000 yr: 12.96 g;
    Amount after 10,000 yr: 0.26 g

35. Initial quantity: 7.63 g;
    Amount after 1000 yr: 4.95 g

37. Amount after 1000 yr: 4.43 g;
    Amount after 10,000 yr: 1.49 g

39. Initial quantity: 2.16 g;
    Amount after 10,000 yr: 1.62 g

41. 95.76%

43. Time to double: 11.55 yr; Amount after 10 yr: $7288.48

45. Annual rate: 8.94%; Amount after 10 yr: $1833.67

47. Annual rate: 9.50%; Time to double: 7.30 yr

49. $224,174.18   **51.** $61,377.75

53. (a) 10.24 yr (b) 9.93 yr (c) 9.90 yr (d) 9.90 yr

55. (a) 8.50 yr (b) 8.18 yr (c) 8.16 yr (d) 8.15 yr

57. (a) $P = 2.40e^{-0.006t}$ (b) 2.19 million
    (c) Because $k < 0$, the population is decreasing.

59. (a) $P = 5.66e^{0.024t}$ (b) 8.11 million
    (c) Because $k > 0$, the population is increasing.

61. (a) $P = 23.55e^{0.036t}$ (b) 40.41 million
    (c) Because $k > 0$, the population is increasing.

63. (a) $N = 100.1596(1.2455)^t$ (b) 6.3 h

65. (a) $N \approx 30(1 - e^{-0.0502t})$ (b) 36 days

67. (a) $P_1 \approx 181e^{0.01245t} \approx 181(1.01253)^t$
    (b) $P_2 = 182.3248(1.01091)^t$
    (c)   (d) 2011

    P_2 is a better approximation.

69. (a) 20 dB (b) 70 dB (c) 95 dB (d) 120 dB

71. 2024 ($t = 16$)   **73.** 379.2°F

75. False. The rate of growth dy/dx is proportional to y.

77. False. The prices are rising at a rate of 6.2% per year.

Section 6.3   (page 415)

1. $y^2 - x^2 = C$ **3.** $15y^2 + 2x^3 = C$ **5.** $r = Ce^{0.75s}$

7. $y = C(x + 2)^3$ **9.** $y^2 = C - 8\cos x$

11. $y = -\frac{1}{4}\sqrt{1 - 4x^2} + C$ **13.** $y = Ce^{(\ln x)^2/2}$

15. $y^2 = 4e^x + 5$ **17.** $y = e^{-(x^2+2x)/2}$

19. $y^2 = 4x^2 + 3$ **21.** $u = e^{(1 - \cos v^2)/2}$ **23.** $P = P_0e^{kt}$

25. $4y^2 - x^2 = 16$ **27.** $y = \frac{1}{3}\sqrt{x}$ **29.** $f(x) = Ce^{-x/2}$

31. Homogeneous of degree 3   **33.** Homogeneous of degree 3

35. Not homogeneous   **37.** Homogeneous of degree 0

39. $|x| = C(x - y)^2$ **41.** $|y^2 + 2xy - x^2| = C$

43. $y = Ce^{-x^2/2y^2}$ **45.** $e^{y/x} = 1 + \ln x^2$ **47.** $x = e^{\sin(y/x)}$

49. 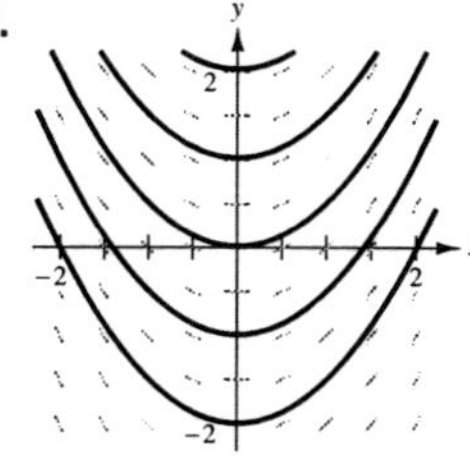

$y = \frac{1}{2}x^2 + C$

51. 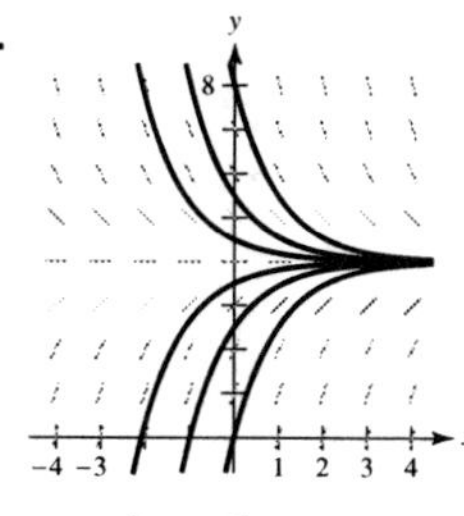

$y = 4 + Ce^{-x}$

53. (a) $y = 0.1602$ (b) $y = 5e^{-3x^2}$ (c) $y = 0.2489$
55. (a) $y = 3.0318$ (b) $y^3 - 4y = x^2 + 12x - 13$ (c) $y = 3$
57. 97.9% of the original amount
59. (a) $dy/dx = k(y - 4)$ (b) a   (c) Proof
60. (a) $dy/dx = k(x - 4)$ (b) b   (c) Proof
61. (a) $dy/dx = ky(y - 4)$ (b) c   (c) Proof
62. (a) $dy/dx = ky^2$ (b) d   (c) Proof
63. (a) $w = 1200 - 1140e^{-kt}$
(b) $w = 1200 - 1140e^{-0.8t}$ $w = 1200 - 1140e^{-0.9t}$

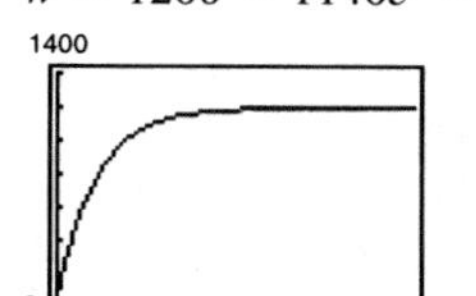

$w = 1200 - 1140e^{-t}$

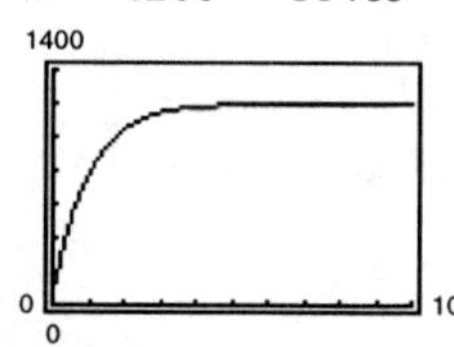

(c) 1.31 yr; 1.16 yr; 1.05 yr   (d) 1200 lb
65. Circles: $x^2 + y^2 = C$
Lines: $y = Kx$
Graphs will vary.

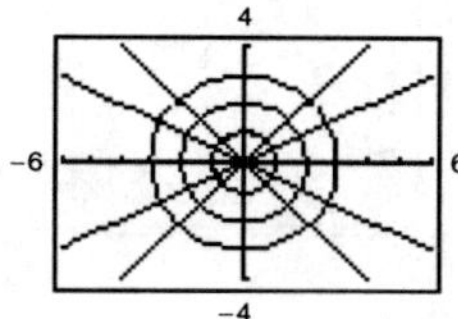

67. Parabolas: $x^2 = Cy$
Ellipses: $x^2 + 2y^2 = K$
Graphs will vary.

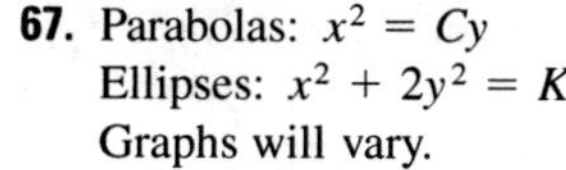
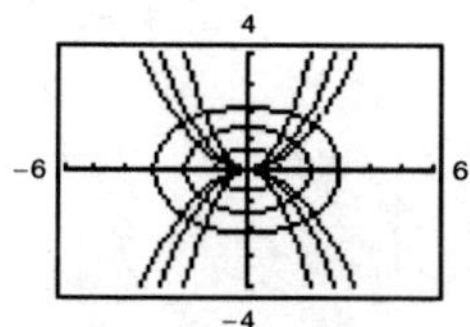

69. Curves: $y^2 = Cx^3$
Ellipses: $2x^2 + 3y^2 = K$
Graphs will vary.

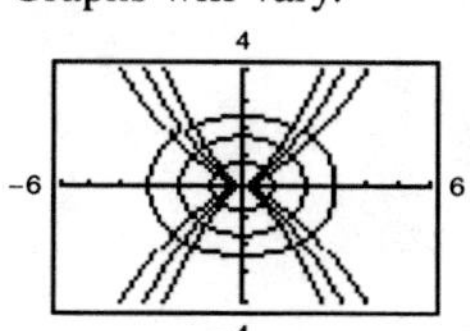

71. $N = 500/(1 + 4e^{-0.2452t})$

73. $y = 360/(8 + 41t)$

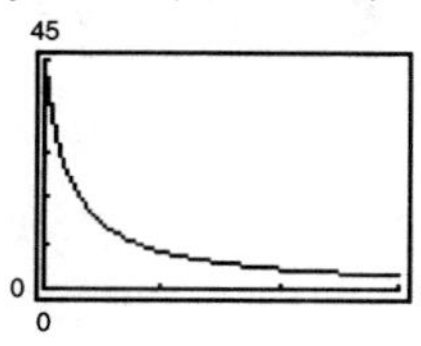

75. $y = 500e^{-1.6094e^{-0.1451t}}$

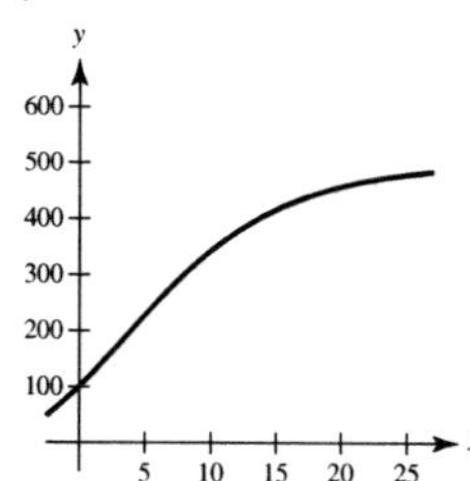

77. 34 beavers   **79.** 92%
81. (a) $Q = 25e^{-(1/20)t}$ (b) $t \approx 10.2$ min
83. (a) $y = Ce^{kt}$ (b) About 6.2 h   **85.** About 3.15 h
87. $P = Ce^{kt} - N/k$ **89.** $A = P/r(e^{rt} - 1)$
91. \$23,981,015.77
93. (a) 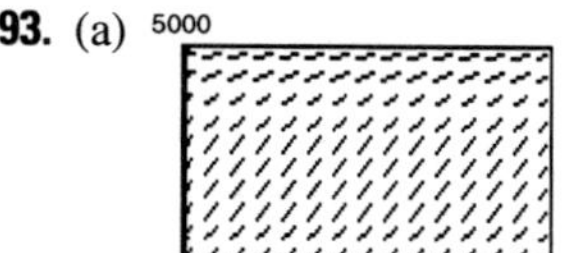

(b) As $t \to \infty$, $y \to L$.
(c) $y = 5000e^{-2.303e^{-0.02t}}$

(d) 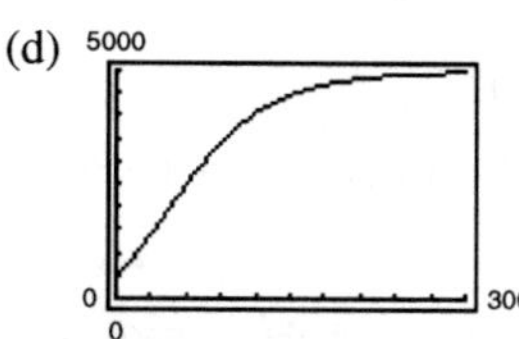

The graph is concave upward on $(0, 41.7)$ and concave downward on $(41.7, \infty)$.

95. Answers will vary.
97. Two families of curves are mutually orthogonal if each curve in the first family intersects each curve in the second family at right angles.
99. False. $y' = x/y$ is separable, but $y = 0$ is not a solution.
101. False. $f(tx, ty) \ne t^n f(x, y)$.
103. Putnam Problem A2, 1988

Section 6.4   (page 424)

1. d   **2.** a   **3.** b   **4.** c   **5.** $y(0) = 4$ **7.** $y(0) = \frac{12}{7}$
9. (a) 0.75   (b) 2100   (c) 70   (d) 4.49 yr
(e) $dP/dt = 0.75P(1 - P/2100)$
11. (a) 0.8   (b) 6000   (c) 1.2   (d) 10.65 yr
(e) $dP/dt = 0.8P(1 - (P/6000))$
13. (a) 3   (b) 100   **15.** (a) 0.1   (b) 250
(c) 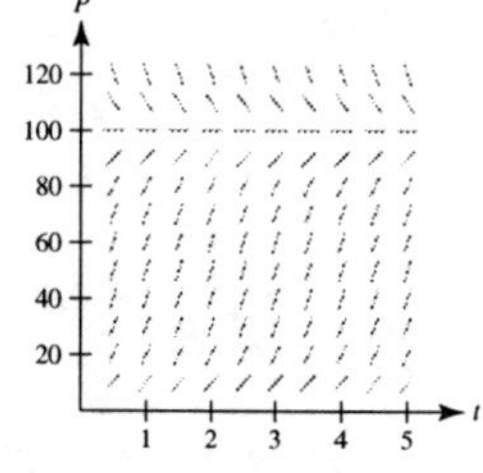

(d) 50   (d) 125
17. $y = 36/(1 + 8e^{-t})$; 34.16; 36
19. $y = 120/(1 + 14e^{-0.8t})$; 95.51; 120
21. c   **22.** d   **23.** b   **24.** a

25. (a) 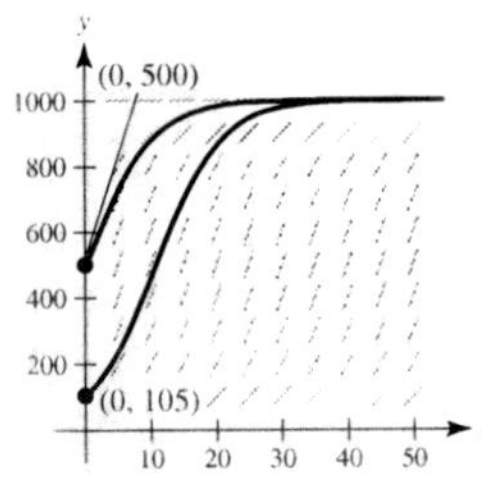

(b) $y = \dfrac{1000}{1 + (179/21)e^{-0.2t}}$

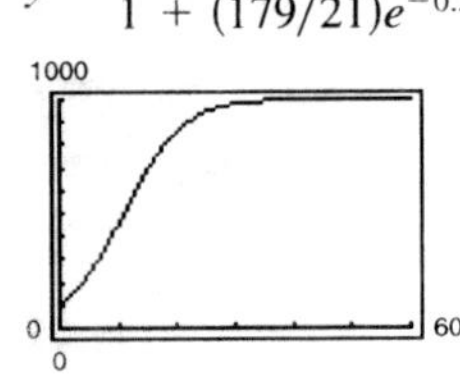

27. (a) 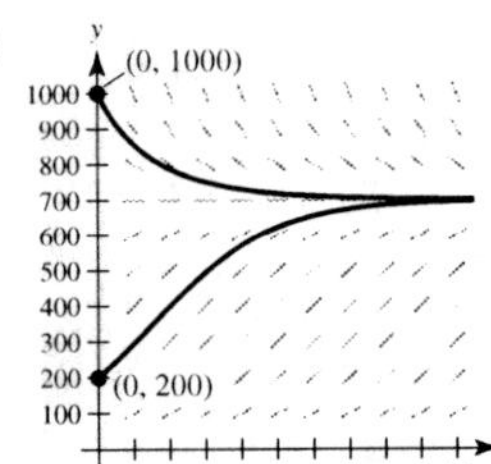

(b) $y = \dfrac{700}{1 - 0.3e^{-0.6t}}$

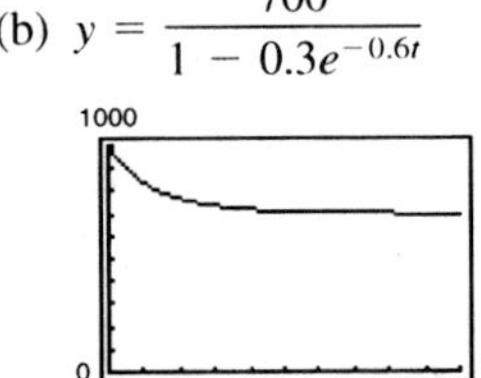

29. L represents the value that y approaches as t approaches infinity. L is the carrying capacity.

31. Yes. It can be written as $\dfrac{dy}{ky\left(1 - \dfrac{y}{L}\right)} = dt.$

33. (a) $P = \dfrac{200}{1 + 7e^{-0.2640t}}$ (b) 70 panthers   (c) 7.37 yr

(d) $dP/dt = 0.2640P(1 - P/200)$; 69.25 panthers   (e) 100 yr

35. False. $dy/dt < 0$ and the population decreases to approach L.

37. Proof

Section 6.5   (page 432)

1. Linear; can be written in the form $dy/dx + P(x)y = Q(x)$

3. Not linear; cannot be written in the form $dy/dx + P(x)y = Q(x)$

5. $y = 2x^2 + x + C/x$ **7.** $y = -16 + Ce^x$

9. $y = -1 + Ce^{\sin x}$ **11.** $y = (x^3 - 3x + C)/[3(x - 1)]$

13. $y = e^{x^3}(x + C)$

15. (a) Answers will vary.   (c) 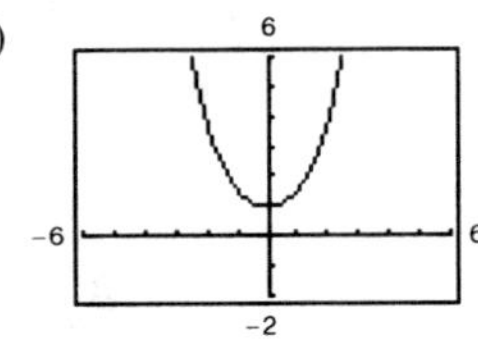

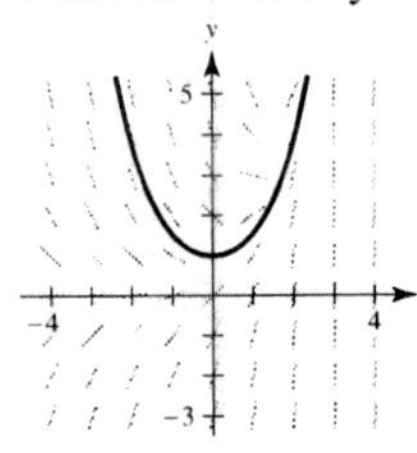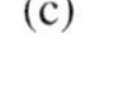

(b) $y = \frac{1}{2}(e^x + e^{-x})$

17. $y = 1 + 4/e^{\tan x}$ **19.** $y = \sin x + (x + 1)\cos x$

21. $xy = 4$ **23.** $y = -2 + x\ln|x| + 12x$

25. $1/y^2 = Ce^{2x^3} + \frac{1}{3}$ **27.** $y = 1/(Cx - x^2)$

29. $1/y^2 = 2x + Cx^2$ **31.** $y^{2/3} = 2e^x + Ce^{2x/3}$

33. (a) 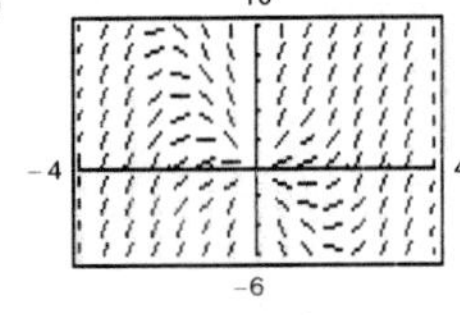   (c) 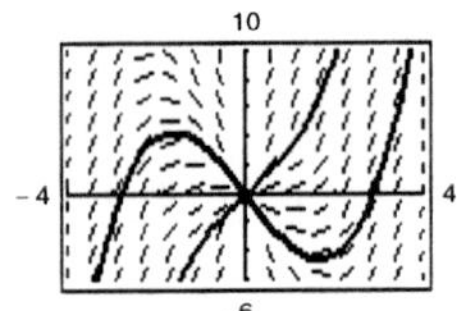

(b) $(-2, 4)$: $y = \frac{1}{2}x(x^2 - 8)$

$(2, 8)$: $y = \frac{1}{2}x(x^2 + 4)$

35. (a) 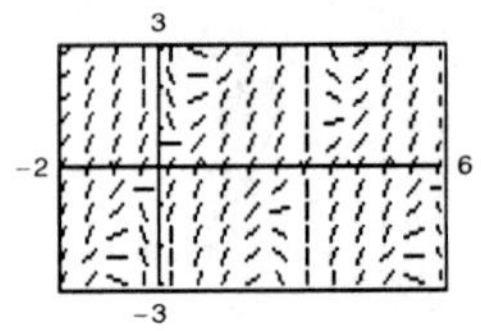

(b) $(1, 1)$: $y = (2\cos 1 + \sin 1)\csc x - 2\cot x$

$(3, -1)$: $y = (2\cos 3 - \sin 3)\csc x - 2\cot x$

(c) 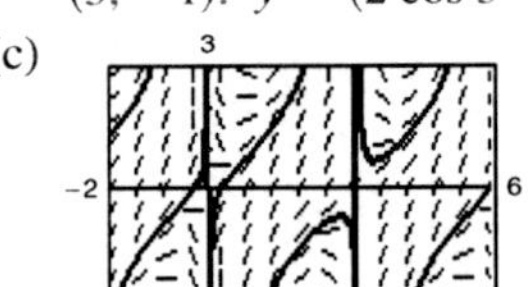

37. (a) $dQ/dt = q - kQ$ (b) $Q = q/k + (Q_0 - q/k)e^{-kt}$

(c) q/k

39. Proof

41. (a) $Q = 25e^{-t/20}$ (b) $-20\ln\left(\frac{3}{5}\right) \approx 10.2$ min   (c) 0

43. (a) $t = 50$ min   (b) $100 - 25/\sqrt{2} \approx 82.32$ lb

45. $v(t) = -159.47(1 - e^{-0.2007t})$; -159.47 ft/sec

47. $I = E_0/R + Ce^{-Rt/L}$

49. The term "first-order" refers to the first derivative.

51. $y' + P(x)y = Q(x)y^n$; Let $z = y^{1-n}$ $(n \neq 0, 1)$. Multiply by $(1 - n)y^{-n}$.

53. c   **54.** d   **55.** a   **56.** b   **57.** $2e^x + e^{-2y} = C$

59. $y = Ce^{-\sin x} + 1$ **61.** $x^3y^2 + x^4y = C$

63. $y = [e^x(x - 1) + C]/x^2$ **65.** $x^4y^4 - 2x^2 = C$

67. $y = \frac{12}{5}x^2 + C/x^3$ **69.** False. The equation contains $\sqrt{y}$.

Section 6.6   (page 440)

1. $dx/dt = 0.9x - 0.05xy$

$dy/dt = -0.6y + 0.008xy$; $(0, 0)$ and $(75, 18)$

3. $dx/dt = 0.5x - 0.01xy$

$dy/dt = -0.49y + 0.007xy$; $(0, 0)$ and $(70, 50)$

5. (a) 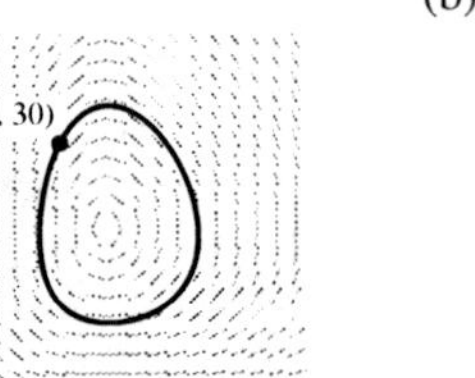   (b) 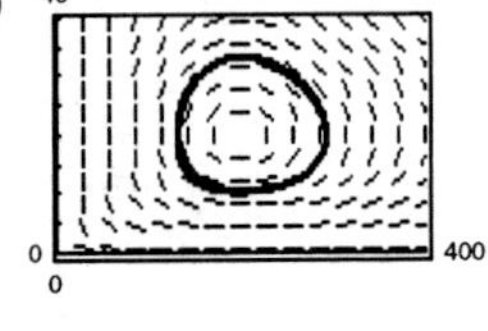

7. (a) $(40, 20)$ **9.** $(0, 0)$, $(50, 20)$

(b) 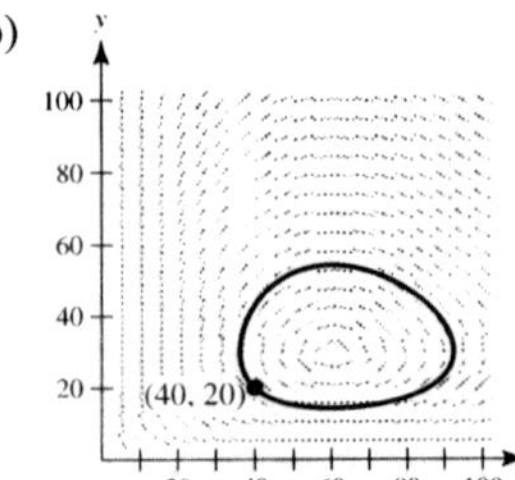

11. 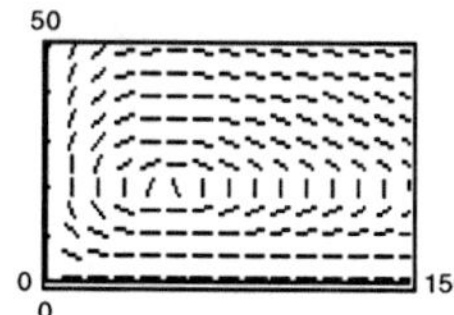

13. $(0, 0)$, $(10{,}000, 1250)$

15. 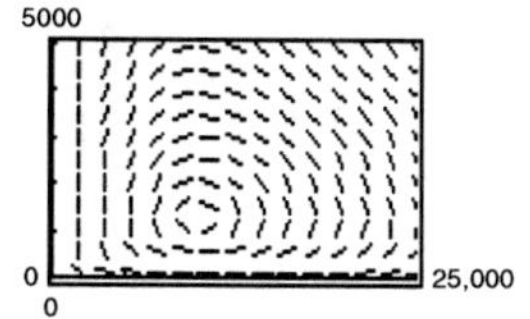

17. 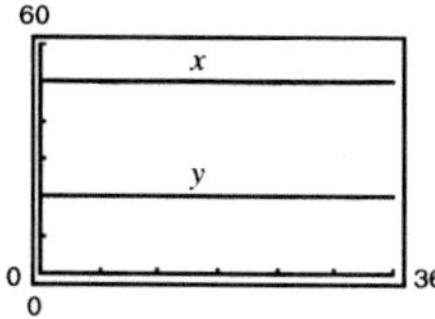   As t increases, both x and y are constant.

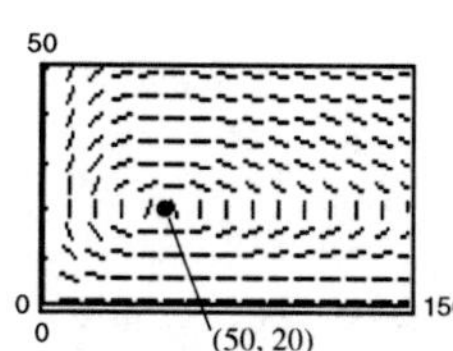

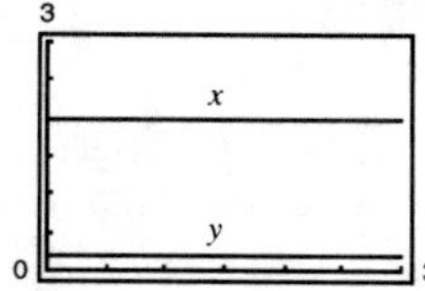   The solution curve reduces to a single point at $(50, 20)$.

19. $dx/dt = 2x - 3x^2 - 2xy$, $dy/dt = 2y - 3y^2 - 2xy$;
$(x, y) = (0, 0)$, $(2/5, 2/5)$, $(0, 2/3)$, and $(2/3, 0)$

21. $dx/dt = 0.15x - 0.6x^2 - 0.75xy$,
$dy/dt = 0.15y - 1.2y^2 - 0.45xy$;
$(x, y) = (0, 0)$, $(0, 1/8)$, $(1/4, 0)$, and $(3/17, 1/17)$

23. $(0, 0)$, $(0, 0.5)$, $(2, 0)$ and $(45/23, 4/23)$

25. $(0, 0)$, $(0, 0.5)$, $(2, 0)$ and $(-9/38, 17/19)$

27. 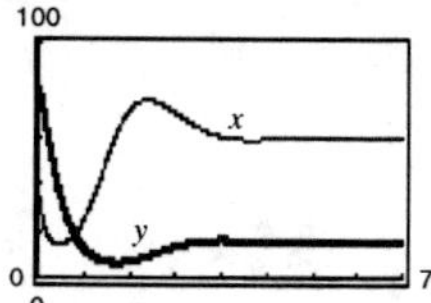   As t increases, both x and y are constant.

29. Yes. See bottom of page 437.

31. Use a critical point as the initial condition.

33. (a) $dx/dt = ax(1 - x/L)$. The equation is logistic.
(b) $dx/dt = 0.4x(1 - (x/100)) - 0.01xy$,
$dy/dt = -0.3y + 0.005xy$
Critical points: $(0, 0)$, $(60, 16)$, $(100, 0)$

(c) 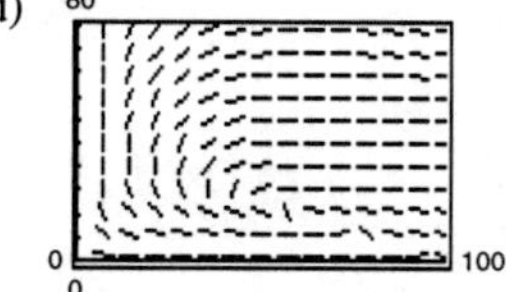   (d)

Answers will vary.

(e) 

Answers will vary.

Review Exercises for Chapter 6   (page 442)

1. Solution   **3.** $y = \frac{4}{3}x^3 + 7x + C$ **5.** $y = \frac{1}{2}\sin 2x + C$

7. $y = \frac{2}{5}(x - 5)^{5/2} + \frac{10}{3}(x - 5)^{3/2} + C$ **9.** $y = -e^{2-x} + C$

11.

x	-4	-2	0	2	4	8
y	2	0	4	4	6	8
dy/dx	-10	-4	-4	0	2	8

13. (a) and (b)   **15.** (a) and (b)

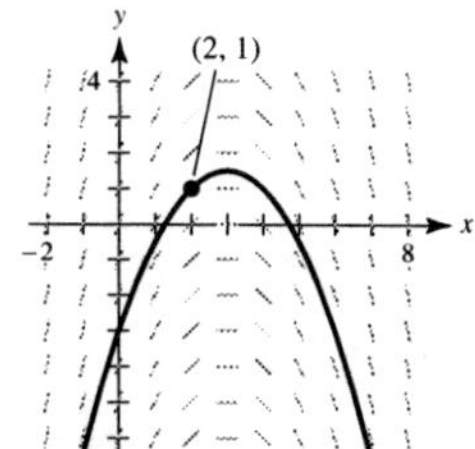   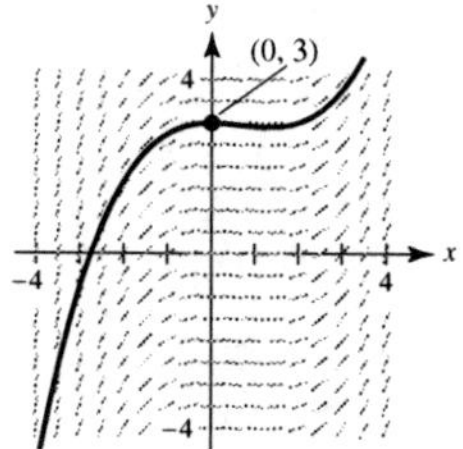

17. (a) and (b)   **19.** $y = 8x - \frac{1}{2}x^2 + C$

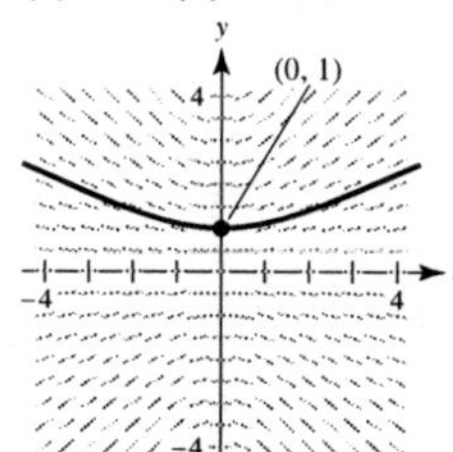

21. $y = -3 - 1/(x + C)$ **23.** $y = Ce^x/(2 + x)^2$

25. $y = \frac{3}{4}e^{0.379t}$ **27.** $y = 5e^{-0.680t}$ **29.** About 7.79 in.

31. (a) $S \approx 30e^{-1.7918/t}$ (b) 20,965 units

(c) 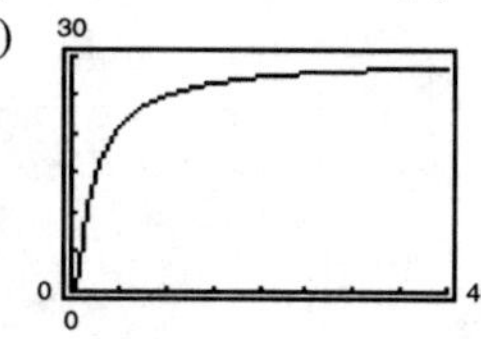

33. About 37.5 yr   **35.** $y = \frac{1}{5}x^5 + 7\ln|x| + C$

37. $y = Ce^{8x^2}$ **39.** $x/(x^2 - y^2) = C$

41. Proof; $y = -2x + \frac{1}{2}x^3$

43.

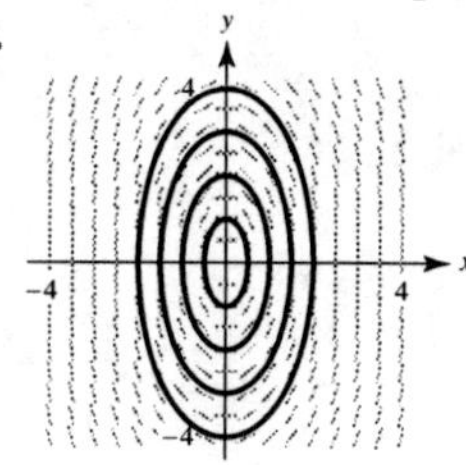

$4x^2 + y^2 = C$

45. (a) 0.55   (b) 5250   (c) 150   (d) 6.41 yr

(e) $\dfrac{dP}{dt} = 0.55P\left(1 - \dfrac{P}{5250}\right)$

47. $y = \dfrac{80}{1 + 9e^{-t}}$

49. (a) $P(t) = \dfrac{20,400}{1 + 16e^{-0.553t}}$ (b) 17,118 trout   (c) 4.94 yr

51. $dS/dt = k(L - S); S = L(1 - e^{-kt})$

53. $dP/dn = kP(L - P), P = CL/(e^{-Lkn} + C)$

55. (a) Answers will vary.   (b) $y = \frac{1}{3}(2e^{x/2} - 5e^{-x})$

(c)

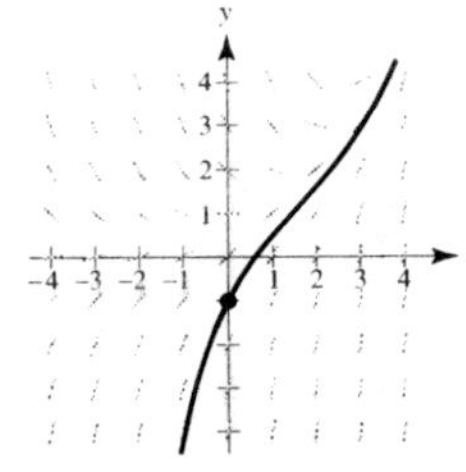

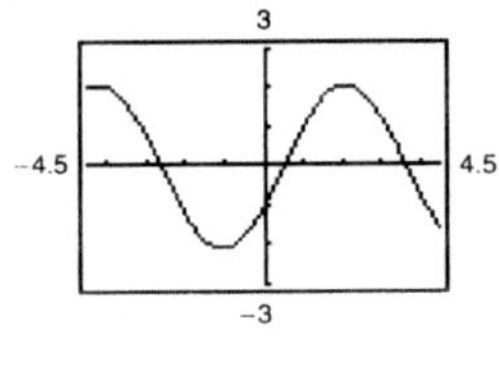

57. (a) Answers will vary.   (b) $y = -\cos x + 1.8305 \sin x$

(c)

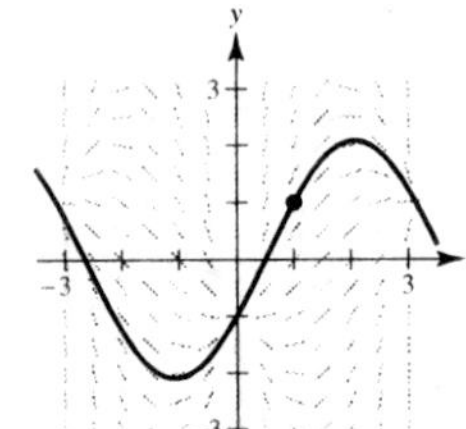

59. $y = -10 + Ce^x$ **61.** $y = e^{x/4}\left[\frac{1}{4}(x + C)\right]$

63. $y = (x + C)/(x - 2)$

65. $y = Ce^{3x} - \frac{1}{13}(2 \cos 2x + 3 \sin 2x)$

67. $y = e^{5x}/10 + Ce^{-5x}$ **69.** $y = 1/(1 + x + Ce^x)$

71. $y^{-2} = Cx^2 + 2/(3x)$

73. Answers will vary. Sample answer:

$(x^2 + 3y^2)\,dx - 2xy\,dy = 0; x^3 = C(x^2 + y^2)$

75. Answers will vary. Sample answer:

$x^3y' + 2x^2y = 1; x^2y = \ln|x| + C$

77. $A = P/r + (A_0 - (P/r))e^{rt}$

79. $t \approx 8.6$ yr

81. (a) Prey: $dx/dt = 0.4x - 0.04xy$

Predator: $dy/dt = -0.6y + 0.02xy$

(b) $(0, 0)$ and $(30, 10)$

(c)

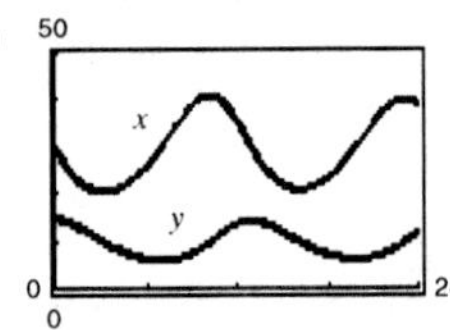

As t increases, both x and y oscillate.

83. (a) Species 1: $dx/dt = 15x - 2x^2 - 4xy$

Species 2: $dy/dt = 17y - 2y^2 - 4xy$

(b) $(0, 0)$, $(19/6, 13/6)$, $(0, 17/2)$, and $(15/2, 0)$

(c)

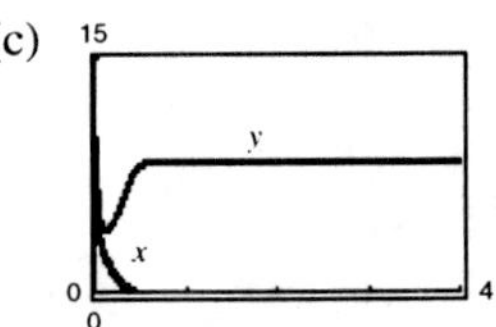

As t increases, x becomes extinct and y remains constant at approximately 8.5.

P.S. Problem Solving   (page 445)

1. (a) $y = 1/(1 - 0.01t)^{100}; T = 100$

(b) $y = 1 / \left[\left(\dfrac{1}{y_0}\right)^{\varepsilon} - k\varepsilon t\right]^{1/\varepsilon}$; Answers will vary.

3. (a) $dS/dt = kS(L - S); S = 100/(1 + 9e^{-0.8109t})$

(b) 2.7 months

(c)

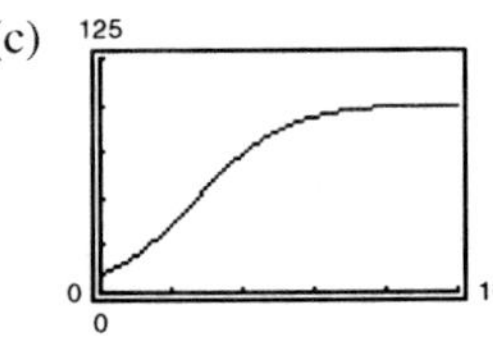

(d)

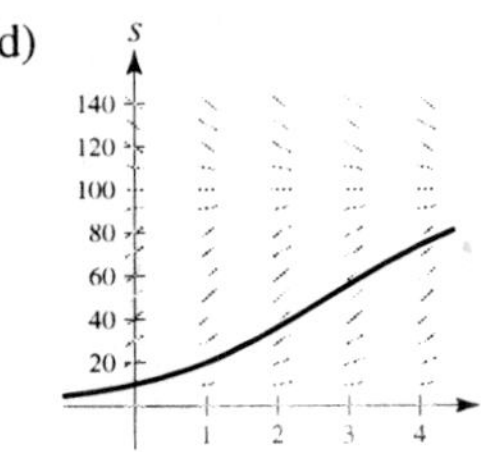

(e) Sales will decrease toward the line $S = L$.

5. Proof; The graph of the logistics function is just a shift of the graph of the hyperbolic tangent.

7. 1481.45 sec $\approx$ 24 min, 41 sec

9. 2575.95 sec $\approx$ 42 min, 56 sec

11. (a) $s = 184.21 - Ce^{-0.019t}$

(b)

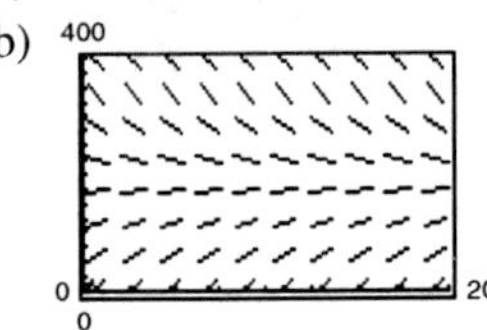

Answers will vary. Sample answer: The slope field seems to have a horizontal asymptote.

(c) As $t \to \infty$, $Ce^{-0.019t} \to 0$, and $s \to 184.21$.

13. (a) $C = 0.6e^{-0.25t}$ (b) $C = 0.6e^{-0.75t}$

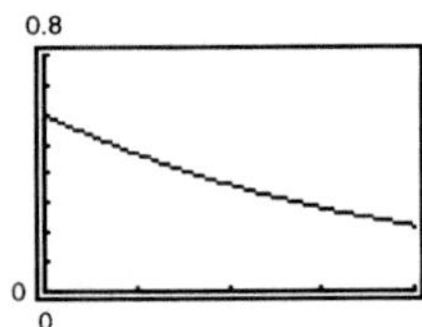

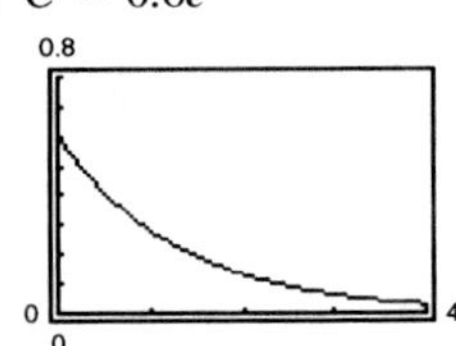

Chapter 7
Section 7.1   (page 454)

1. $-\displaystyle\int_0^6 (x^2 - 6x)\,dx$ **3.** $\displaystyle\int_0^3 (-2x^2 + 6x)\,dx$

5. $-6\displaystyle\int_0^1 (x^3 - x)\,dx$

7. 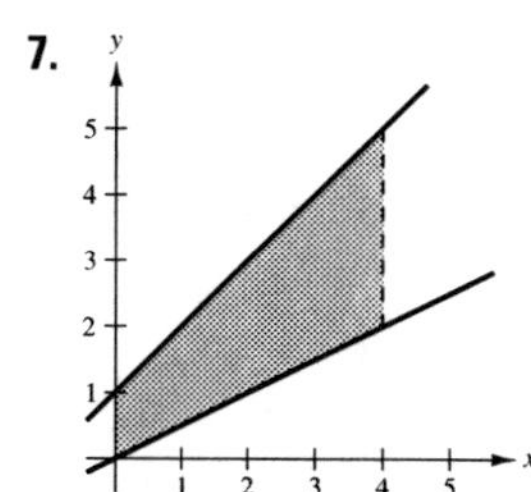

9. 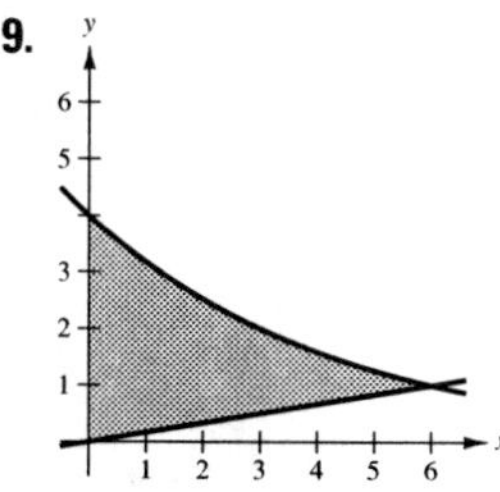

11. 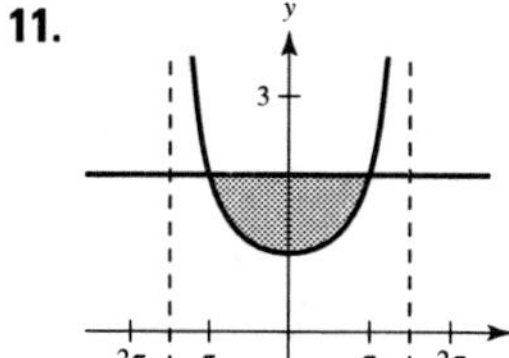

13. 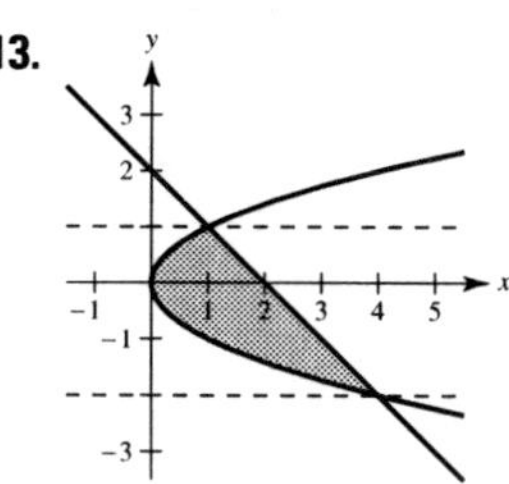

15. d

17. (a) $\frac{125}{6}$ (b) $\frac{125}{6}$ (c) Integrating with respect to y; Answers will vary.

19. 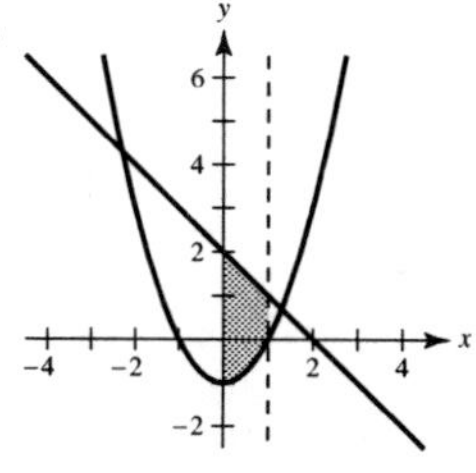

$\frac{13}{6}$

21. 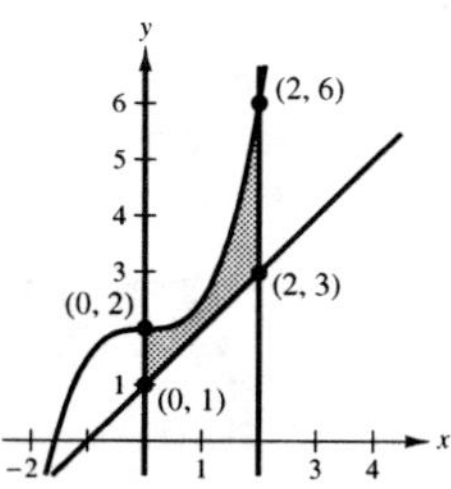

2

23. 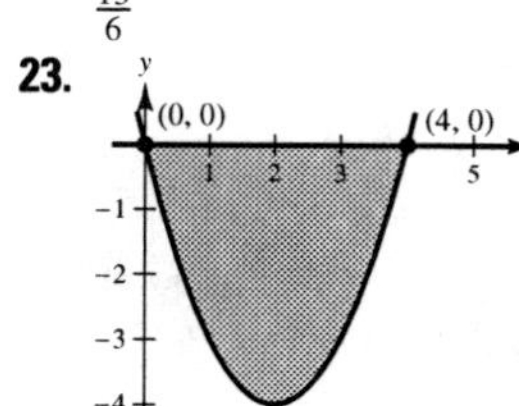

$\frac{32}{3}$

25. 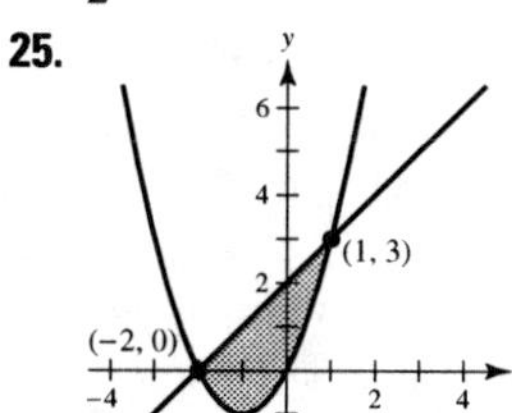

$\frac{9}{2}$

27. 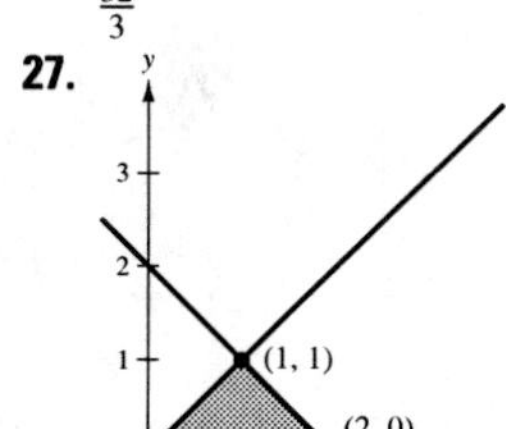

1

29. 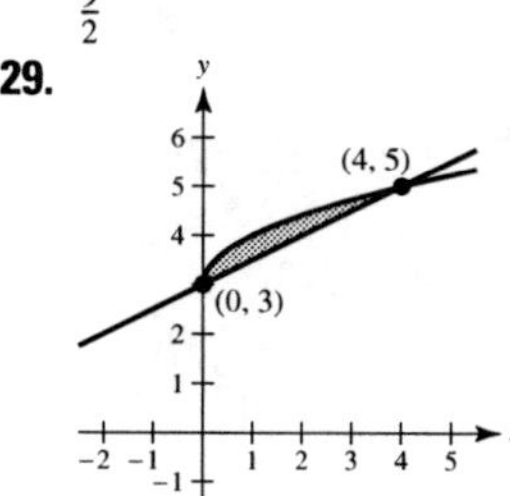

$\frac{4}{3}$

31. 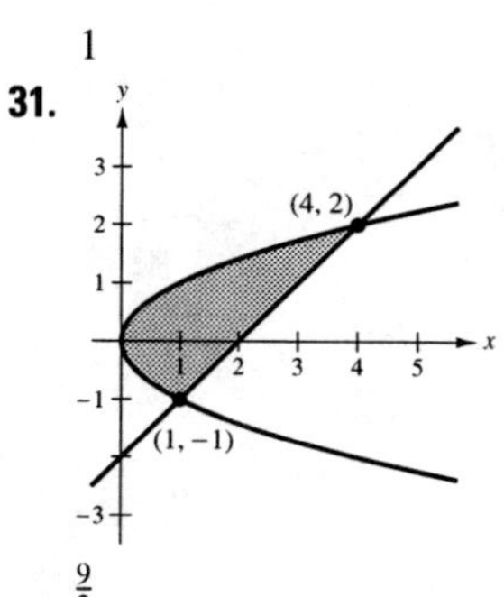

$\frac{9}{2}$

33. 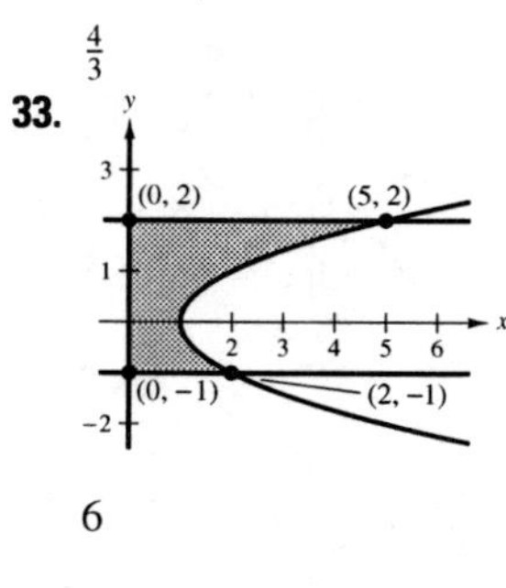

6

35. 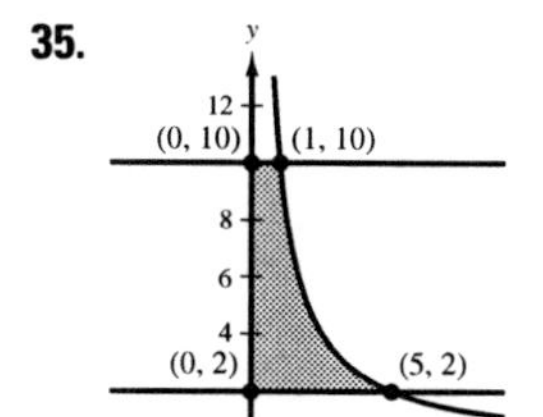

$10 \ln 5 \approx 16.094$

37. (a) 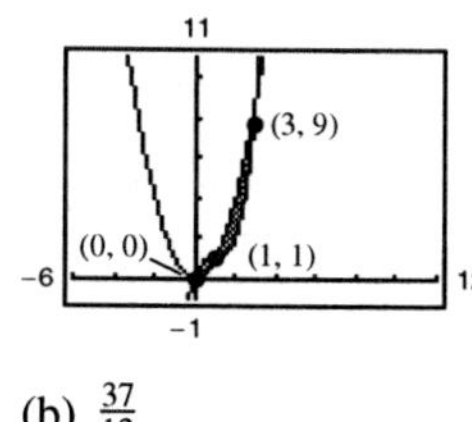

(b) $\frac{37}{12}$

39. (a) 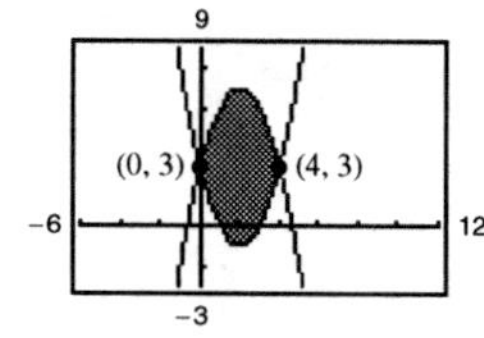

(b) $\frac{64}{3}$

41. (a) 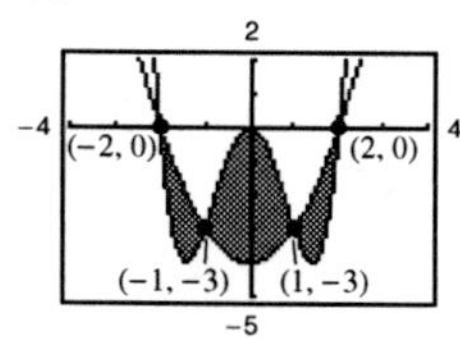

(b) 8

43. (a) 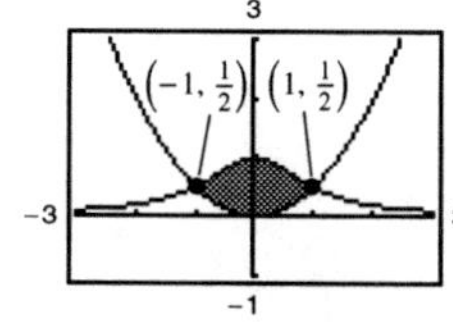

(b) $\pi/2 - 1/3 \approx 1.237$

45. (a) 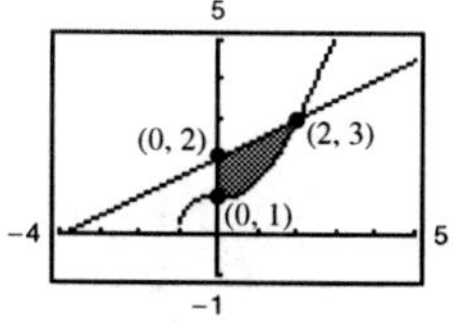

(b) ≈ 1.759

47. 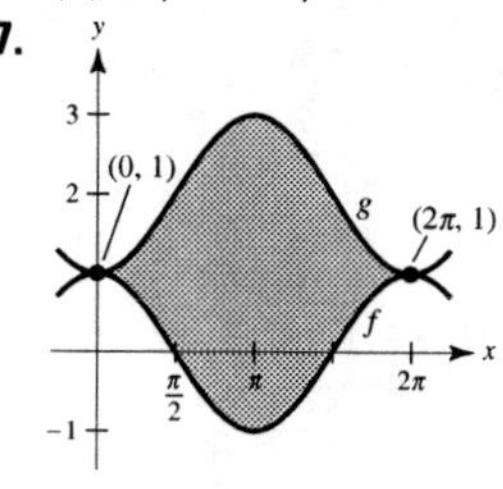

$4\pi \approx 12.566$

49. 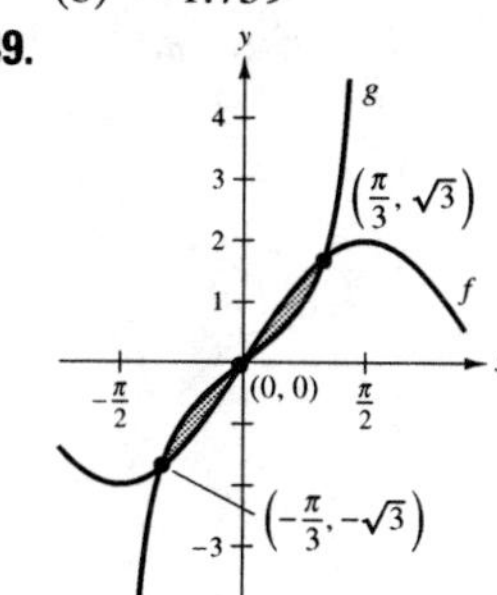

$2(1 - \ln 2) \approx 0.614$

51. 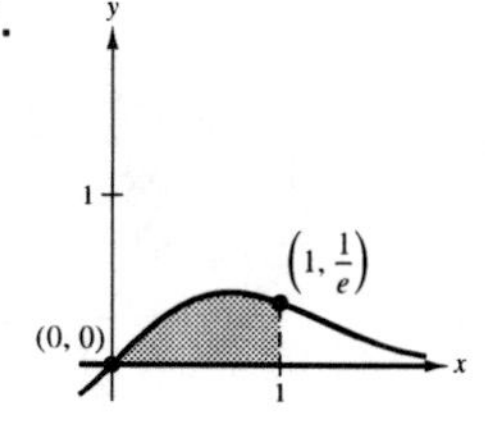

$(1/2)(1 - 1/e) \approx 0.316$

53. (a) 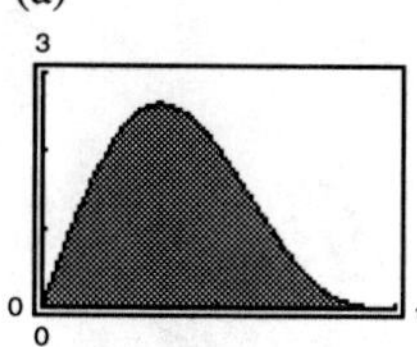

(b) 4

55. (a) 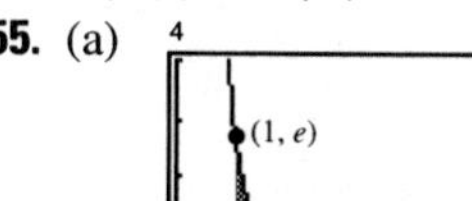

(b) About 1.323

57. (a) 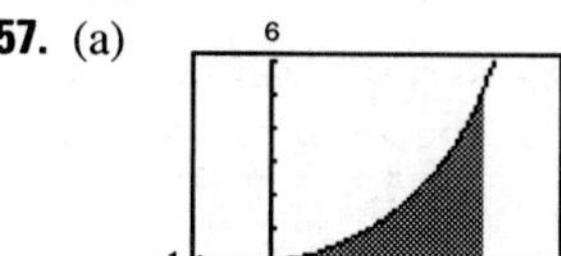

(b) The function is difficult to integrate.

(c) About 4.7721

59. (a)

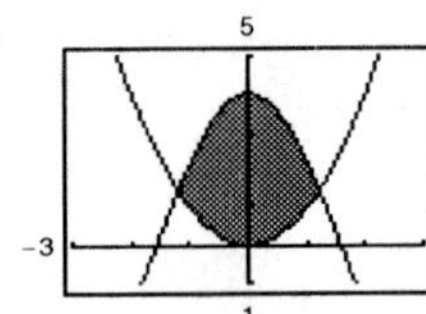

(b) The intersections are difficult to find.

(c) About 6.3043

61. $F(x) = \frac{1}{4}x^2 + x$

(a) $F(0) = 0$

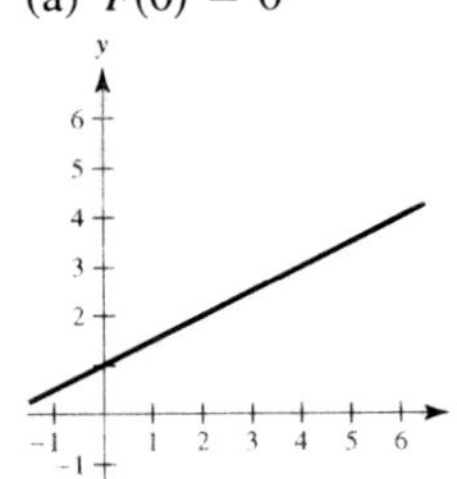

(b) $F(2) = 3$

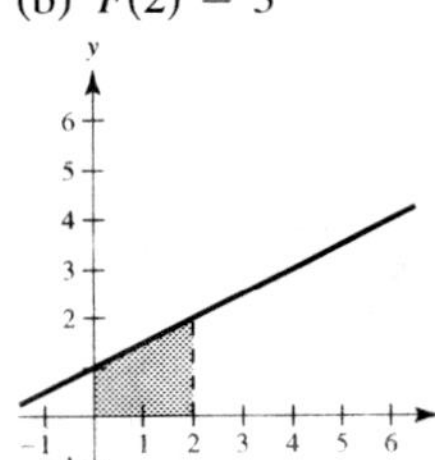

(c) $F(6) = 15$

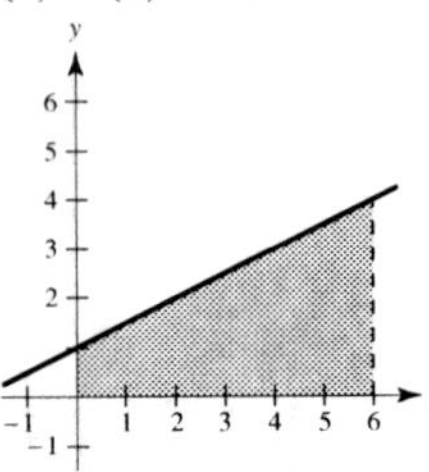

63. $F(\alpha) = (2/\pi)[\sin(\pi\alpha/2) + 1]$

(a) $F(-1) = 0$

(b) $F(0) = 2/\pi \approx 0.6366$

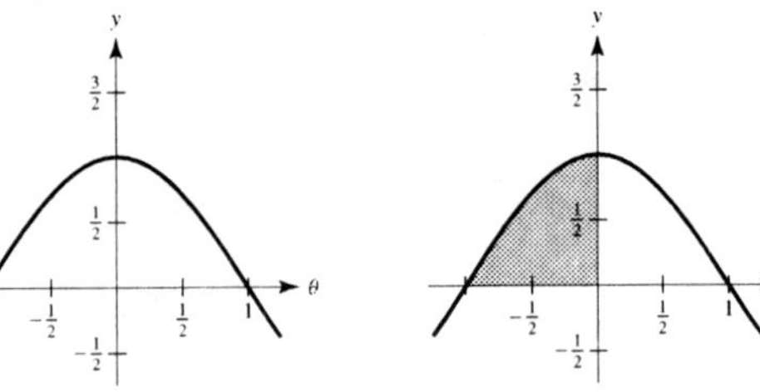

(c) $F(1/2) = (\sqrt{2} + 2)/\pi \approx 1.0868$

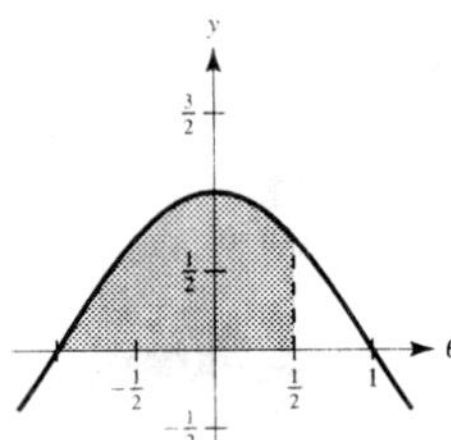

65. 14   **67.** 16

69. Answers will vary. Sample answers:

(a) About 966 ft^2 (b) About 1004 ft^2

71. $A = \dfrac{3\sqrt{3}}{4} - \dfrac{1}{2} \approx 0.7990$ **73.** $\displaystyle\int_{-2}^{1} [x^3 - (3x - 2)]\,dx = \frac{27}{4}$

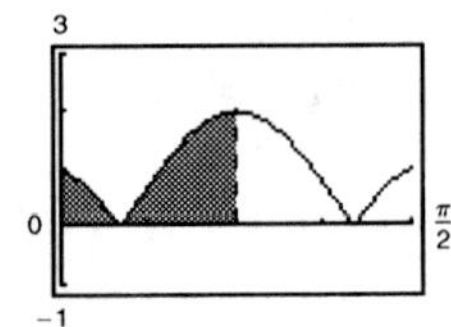

75. $\displaystyle\int_{0}^{1} \left[\frac{1}{x^2 + 1} - \left(-\frac{1}{2}x + 1 \right) \right] dx \approx 0.0354$

77. Answers will vary. Example: $x^4 - 2x^2 + 1 \le 1 - x^2$ on $[-1, 1]$

$$\int_{-1}^{1} [(1 - x^2) - (x^4 - 2x^2 + 1)]\,dx = \frac{4}{15}$$

79. Offer 2 is better because the cumulative salary (area under the curve) is greater.

81. (a) The integral $\int_{0}^{5} [v_1(t) - v_2(t)]\,dt = 10$ means that the first car traveled 10 more meters than the second car between 0 and 5 seconds.

The integral $\int_{0}^{10} [v_1(t) - v_2(t)]\,dt = 30$ means that the first car traveled 30 more meters than the second car between 0 and 10 seconds.

The integral $\int_{20}^{30} [v_1(t) - v_2(t)]\,dt = -5$ means that the second car traveled 5 more meters than the first car between 20 and 30 seconds.

(b) No. You do not know when both cars started or the initial distance between the cars.

(c) The car with velocity v_1 is ahead by 30 meters.

(d) Car 1 is ahead by 8 meters.

83. $b = 9(1 - 1/\sqrt[3]{4}) \approx 3.330$ **85.** $a = 4 - 2\sqrt{2} \approx 1.172$

87. Answers will vary. Sample answer: $\frac{1}{6}$

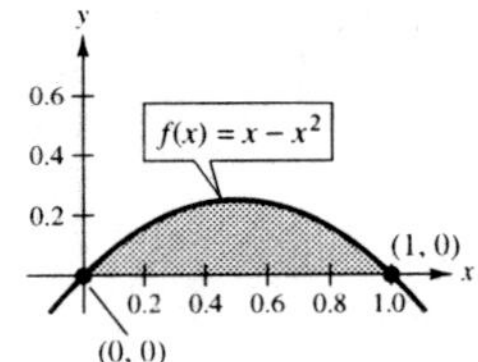

89. (a) $(-2, -11)$, $(0, 7)$ (b) $y = 9x + 7$

(c) 3.2, 6.4, 3.2; The area between the two inflection points is the sum of the areas between the other two regions.

91. \$6.825 billion

93. (a) $y = 0.0124x^2 - 0.385x + 7.85$

(b)

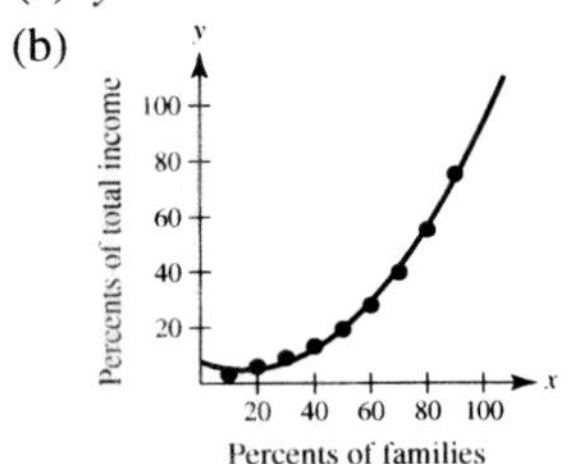

(c)

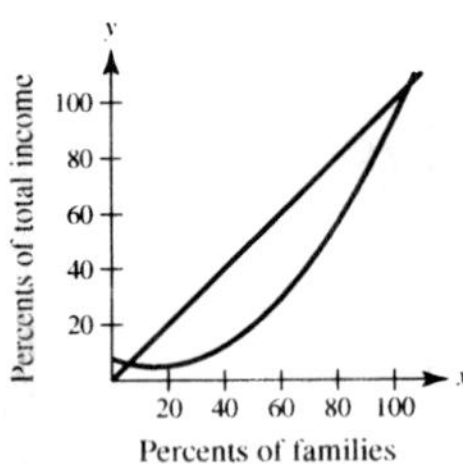

(d) About 2006.7

95. $\frac{16}{3}(4\sqrt{2} - 5) \approx 3.503$

97. (a) About 6.031 m^2 (b) About 12.062 m^3 (c) 60,310 lb

99. True

101. False. Let $f(x) = x$ and $g(x) = 2x - x^2$. f and g intersect at $(1, 1)$, the midpoint of $[0, 2]$, but

$$\int_{a}^{b} [f(x) - g(x)]\,dx = \int_{0}^{2} [x - (2x - x^2)]\,dx = \frac{2}{3} \ne 0.$$

103. $\sqrt{3}/2 + 7\pi/24 + 1 \approx 2.7823$

105. Putnam Problem A1, 1993

Section 7.2 (page 465)

1. $\pi \displaystyle\int_0^1 (-x + 1)^2\,dx = \dfrac{\pi}{3}$ **3.** $\pi \displaystyle\int_1^4 \left(\sqrt{x}\right)^2 dx = \dfrac{15\pi}{2}$

5. $\pi \displaystyle\int_0^1 [(x^2)^2 - (x^5)^2]\,dx = \dfrac{6\pi}{55}$ **7.** $\pi \displaystyle\int_0^4 \left(\sqrt{y}\right)^2 dy = 8\pi$

9. $\pi \displaystyle\int_0^1 (y^{3/2})^2\,dy = \dfrac{\pi}{4}$

11. (a) $9\pi/2$ (b) $\left(36\pi\sqrt{3}\right)/5$ (c) $\left(24\pi\sqrt{3}\right)/5$
  (d) $\left(84\pi\sqrt{3}\right)/5$

13. (a) $32\pi/3$ (b) $64\pi/3$ **15.** 18π

17. $\pi\left(48\ln 2 - \frac{27}{4}\right) \approx 83.318$ **19.** $124\pi/3$ **21.** $832\pi/15$

23. $\pi \ln 5$ **25.** $2\pi/3$ **27.** $(\pi/2)(1 - 1/e^2) \approx 1.358$

29. $277\pi/3$ **31.** 8π **33.** $\pi^2/2 \approx 4.935$

35. $(\pi/2)(e^2 - 1) \approx 10.036$ **37.** 1.969 **39.** 15.4115

41. $\pi/3$ **43.** $2\pi/15$ **45.** $\pi/2$ **47.** $\pi/6$

49. (a) The area appears to be close to 1 and therefore the volume (area squared $\times\,\pi$) is near 3.

51. A sine curve on $[0, \pi/2]$ revolved about the x-axis

53. The parabola $y = 4x - x^2$ is a horizontal translation of the parabola $y = 4 - x^2$. Therefore, their volumes are equal.

55. (a) This statement is true. Explanations will vary.
  (b) This statement is false. Explanations will vary.

57. 18π **59.** Proof   **61.** $\pi r^2 h[1 - (h/H) + h^2/(3H^2)]$

63.

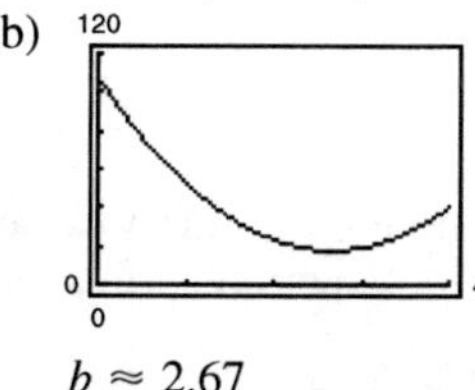

$\pi/30$

65. (a) 60π (b) 50π

67. (a) $V = \pi\left(4b^2 - \frac{64}{3}b + \frac{512}{15}\right)$
  (b)
  (c) $b = \frac{8}{3} \approx 2.67$

$b \approx 2.67$

69. (a) ii; right circular cylinder of radius r and height h
  (b) iv; ellipsoid whose underlying ellipse has the equation $(x/b)^2 + (y/a)^2 = 1$
  (c) iii; sphere of radius r
  (d) i; right circular cone of radius r and height h
  (e) v; torus of cross-sectional radius r and other radius R

71. (a) $\frac{81}{10}$ (b) $\frac{9}{2}$ **73.** $\frac{16}{3}r^3$ **75.** $V = \frac{4}{3}\pi(R^2 - r^2)^{3/2}$

77. 19.7443 **79.** (a) $\frac{2}{3}r^3$ (b) $\frac{2}{3}r^3 \tan\theta$; As $\theta \to 90°$, $V \to \infty$.

Section 7.3 (page 474)

1. $2\pi \displaystyle\int_0^2 x^2\,dx = \dfrac{16\pi}{3}$ **3.** $2\pi \displaystyle\int_0^4 x\sqrt{x}\,dx = \dfrac{128\pi}{5}$

5. $2\pi \displaystyle\int_0^3 x^3\,dx = \dfrac{81}{2}\pi$ **7.** $2\pi \displaystyle\int_0^2 x(4x - 2x^2)\,dx = \dfrac{16\pi}{3}$

9. $2\pi \displaystyle\int_0^2 x(x^2 - 4x + 4)\,dx = \dfrac{8\pi}{3}$

11. $2\pi \displaystyle\int_2^4 x\sqrt{x - 2}\,dx = \dfrac{128\pi}{15}\sqrt{2}$

13. $2\pi \displaystyle\int_0^1 x\left(\dfrac{1}{\sqrt{2\pi}}e^{-x^2/2}\right)dx = \sqrt{2\pi}\left(1 - \dfrac{1}{\sqrt{e}}\right) \approx 0.986$

15. $2\pi \displaystyle\int_0^2 y(2 - y)\,dy = \dfrac{8\pi}{3}$

17. $2\pi\left[\displaystyle\int_0^{1/2} y\,dy + \int_{1/2}^1 y\left(\dfrac{1}{y} - 1\right)dy\right] = \dfrac{\pi}{2}$

19. $2\pi\left[\displaystyle\int_0^8 y^{4/3}\,dy\right] = \dfrac{768\pi}{7}$

21. $2\pi \displaystyle\int_0^2 y(4 - 2y)\,dy = 16\pi/3$ **23.** 64π **25.** 16π

27. Shell method; it is much easier to put x in terms of y rather than vice versa.

29. (a) $128\pi/7$ (b) $64\pi/5$ (c) $96\pi/5$

31. (a) $\pi a^3/15$ (b) $\pi a^3/15$ (c) $4\pi a^3/15$

33. (a)

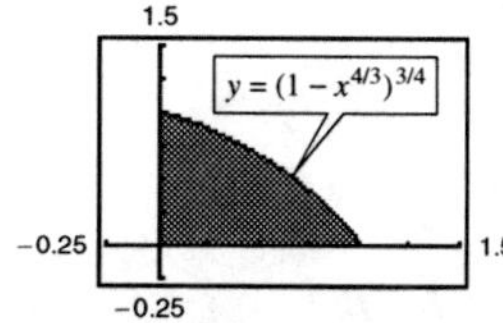

35. (a)

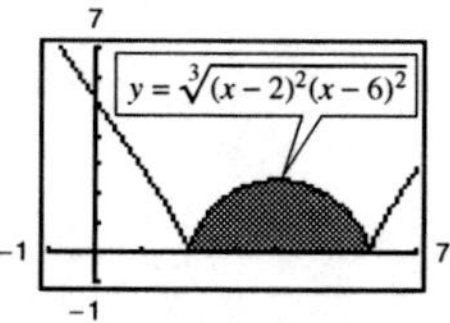

(b) 1.506 (b) 187.25

37. d   **39.** a, c, b

41. Both integrals yield the volume of the solid generated by revolving the region bounded by the graphs of $y = \sqrt{x - 1}$, $y = 0$, and $x = 5$ about the x-axis.

43. (a) The rectangles would be vertical.
  (b) The rectangles would be horizontal.

45. Diameter $= 2\sqrt{4 - 2\sqrt{3}} \approx 1.464$ **47.** $4\pi^2$

49. (a) Region bounded by $y = x^2$, $y = 0$, $x = 0$, $x = 2$
  (b) Revolved about the y-axis

51. (a) Region bounded by $x = \sqrt{6 - y}$, $y = 0$, $x = 0$
  (b) Revolved about $y = -2$

53. (a) Proof (b) (i) $V = 2\pi$ (ii) $V = 6\pi^2$

55. Proof

57. (a) $R_1(n) = n/(n + 1)$ (b) $\displaystyle\lim_{n\to\infty} R_1(n) = 1$
  (c) $V = \pi ab^{n+2}[n/(n + 2)]$; $R_2(n) = n/(n + 2)$
  (d) $\displaystyle\lim_{n\to\infty} R_2(n) = 1$
  (e) As $n \to \infty$, the graph approaches the line $x = b$.

59. (a) and (b) About $121{,}475$ ft^3 **61.** $c = 2$

63. (a) $64\pi/3$ (b) $2048\pi/35$ (c) $8192\pi/105$

Section 7.4 (page 485)

1. (a) and (b) 17 **3.** $\frac{5}{3}$ **5.** $\frac{2}{3}\left(2\sqrt{2} - 1\right) \approx 1.219$

7. $5\sqrt{5} - 2\sqrt{2} \approx 8.352$ **9.** 309.3195

11. $\ln\left[\left(\sqrt{2} + 1\right)/\left(\sqrt{2} - 1\right)\right] \approx 1.763$

13. $\frac{1}{2}(e^2 - 1/e^2) \approx 3.627$ **15.** $\frac{76}{3}$

17. (a)

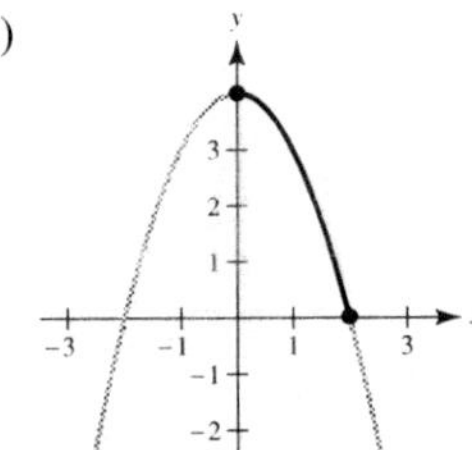

(b) $\displaystyle\int_0^2 \sqrt{1 + 4x^2}\, dx$

(c) About 4.647

19. (a)

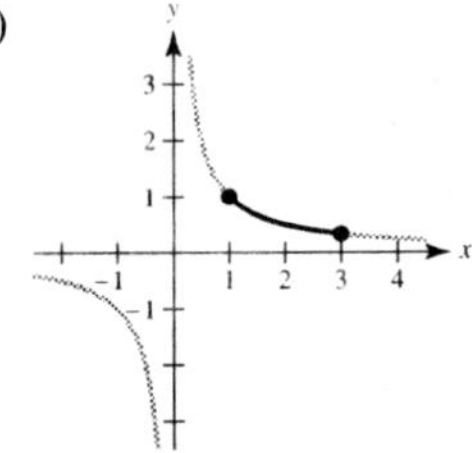

(b) $\displaystyle\int_1^3 \sqrt{1 + \frac{1}{x^4}}\, dx$

(c) About 2.147

21. (a)

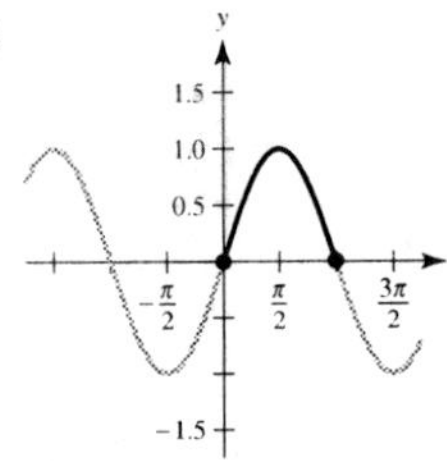

(b) $\displaystyle\int_0^\pi \sqrt{1 + \cos^2 x}\, dx$

(c) About 3.820

23. (a)

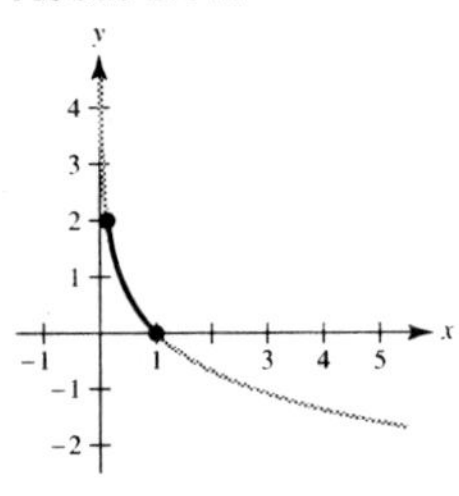

(b) $\displaystyle\int_0^2 \sqrt{1 + e^{-2y}}\, dy$

$\displaystyle = \int_{e^{-2}}^1 \sqrt{1 + \frac{1}{x^2}}\, dx$

(c) About 2.221

25. (a)

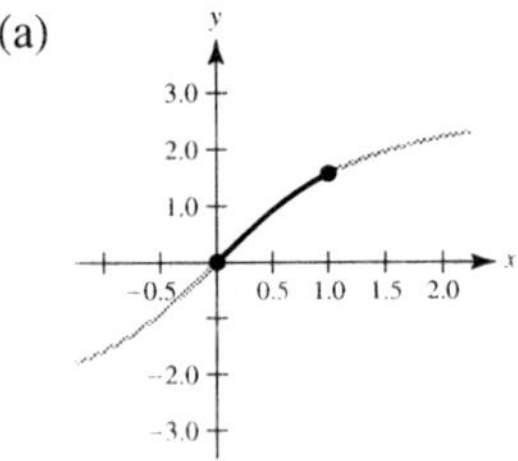

(b) $\displaystyle\int_0^1 \sqrt{1 + \left(\frac{2}{1 + x^2}\right)^2}\, dx$

(c) About 1.871

27. b   **29.** (a) 64.125   (b) 64.525   (c) 64.666   (d) 64.672

31. $20[\sinh 1 - \sinh(-1)] \approx 47.0$ m   **33.** About 1480

35. $3 \arcsin \frac{2}{3} \approx 2.1892$

37. $\displaystyle 2\pi \int_0^3 \frac{1}{3}x^3 \sqrt{1 + x^4}\, dx = \frac{\pi}{9}\left(82\sqrt{82} - 1\right) \approx 258.85$

39. $\displaystyle 2\pi \int_1^2 \left(\frac{x^3}{6} + \frac{1}{2x}\right)\left(\frac{x^2}{2} + \frac{1}{2x^2}\right) dx = \frac{47\pi}{16} \approx 9.23$

41. $\displaystyle 2\pi \int_{-1}^1 2\, dx = 8\pi \approx 25.13$

43. $\displaystyle 2\pi \int_1^8 x \sqrt{1 + \frac{1}{9x^{4/3}}}\, dx = \frac{\pi}{27}\left(145\sqrt{145} - 10\sqrt{10}\right) \approx 199.48$

45. $\displaystyle 2\pi \int_0^2 x \sqrt{1 + \frac{x^2}{4}}\, dx = \frac{\pi}{3}\left(16\sqrt{2} - 8\right) \approx 15.318$

47. 14.424

49. A rectifiable curve is a curve with a finite arc length.

51. The integral formula for the area of a surface of revolution is derived from the formula for the lateral surface area of the frustum of a right circular cone. The formula is $S = 2\pi r L$, where $r = \frac{1}{2}(r_1 + r_2)$, which is the average radius of the frustum, and L is the length of a line segment on the frustum. The representative element is $2\pi f(d_i)\sqrt{1 + (\Delta y_i/\Delta x_i)^2}\,\Delta x_i$.

53. (a)

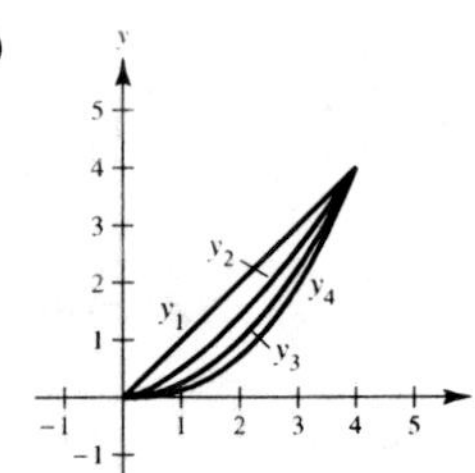

(b) y_1, y_2, y_3, y_4

(c) $s_1 \approx 5.657;\ s_2 \approx 5.759;$
$\quad s_3 \approx 5.916;\ s_4 \approx 6.063$

55. 20π **57.** $6\pi\left(3 - \sqrt{5}\right) \approx 14.40$

59. (a) Answers will vary. Sample answer: 5207.62 in.^{3}

(b) Answers will vary. Sample answer: 1168.64 in.^{2}

(c) $r = 0.0040y^3 - 0.142y^2 + 1.23y + 7.9$

(d) 5279.64 in.^{3}; 1179.5 in.^{2}

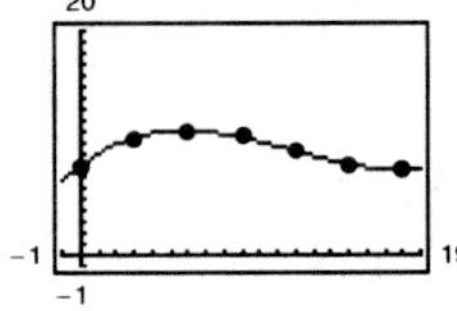

61. (a) $\pi(1 - 1/b)$ (b) $\displaystyle 2\pi \int_1^b \sqrt{x^4 + 1}/x^3\, dx$

(c) $\displaystyle \lim_{b\to\infty} V = \lim_{b\to\infty} \pi(1 - 1/b) = \pi$

(d) Because $\displaystyle \frac{\sqrt{x^4 + 1}}{x^3} > \frac{\sqrt{x^4}}{x^3} = \frac{1}{x} > 0$ on $[1, b]$,

you have $\displaystyle \int_1^b \frac{\sqrt{x^4 + 1}}{x^3}\, dx > \int_1^b \frac{1}{x}\, dx = \Big[\ln x\Big]_1^b = \ln b$

and $\displaystyle \lim_{b\to\infty} \ln b \to \infty$. So, $\displaystyle \lim_{b\to\infty} 2\pi \int_1^b \frac{\sqrt{x^4 + 1}}{x^3}\, dx = \infty$.

63. (a)

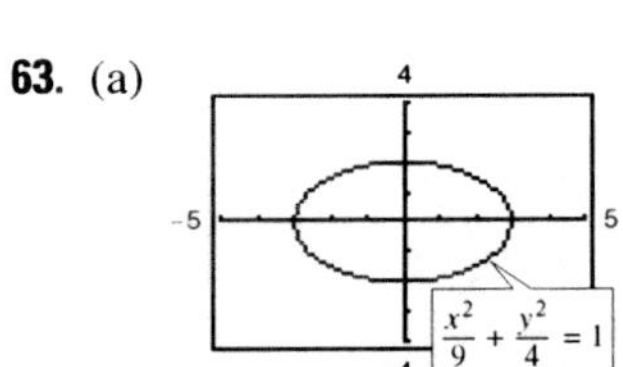

(b) $\displaystyle \int_0^3 \sqrt{1 + \frac{4x^2}{81 - 9x^2}}\, dx$

(c) You cannot evaluate this definite integral because the integrand is not defined at $x = 3$. Simpson's Rule will not work for the same reason.

65. Fleeing object: $\frac{2}{3}$ unit

Pursuer: $\displaystyle \frac{1}{2}\int_0^1 \frac{x + 1}{\sqrt{x}}\, dx = \frac{4}{3} = 2\left(\frac{2}{3}\right)$

67. $384\pi/5$ **69.** Proof   **71.** Proof

Section 7.5   (page 495)

1. 2000 ft-lb   **3.** 896 N-m

5. 40.833 in.-lb $\approx$ 3.403 ft-lb   **7.** 8750 N-cm = 87.5 N-m

9. 160 in.-lb $\approx$ 13.3 ft-lb   **11.** 37.125 ft-lb

13. (a) 487.805 mile-tons $\approx 5.151 \times 10^9$ ft-lb

(b) 1395.349 mile-tons $\approx 1.473 \times 10^{10}$ ft-lb

15. (a) 2.93×10^4 mile-tons $\approx 3.10 \times 10^{11}$ ft-lb

(b) 3.38×10^4 mile-tons $\approx 3.57 \times 10^{11}$ ft-lb

17. (a) 2496 ft-lb   (b) 9984 ft-lb   **19.** $470,400\pi$ N-m

21. 2995.2π ft-lb   **23.** $20,217.6\pi$ ft-lb   **25.** 2457π ft-lb

27. 600 ft-lb   **29.** 450 ft-lb   **31.** 168.75 ft-lb

33. If an object is moved a distance D in the direction of an applied constant force F, then the work W done by the force is defined as $W = FD$.

35. The situation in part (a) requires more work. There is no work required for part (b) because the distance is 0.

37. (a) 54 ft-lb   (b) 160 ft-lb   (c) 9 ft-lb   (d) 18 ft-lb

39. $2000 \ln(3/2) \approx 810.93$ ft-lb   **41.** $3k/4$

43. 3249.4 ft-lb   **45.** 10,330.3 ft-lb

Section 7.6   (page 506)

1. $\bar{x} = -\frac{4}{3}$ **3.** $\bar{x} = 12$ **5.** (a) $\bar{x} = 16$ (b) $\bar{x} = -2$

7. $x = 6$ ft   **9.** $(\bar{x}, \bar{y}) = \left(\frac{10}{9}, -\frac{1}{9}\right)$ **11.** $(\bar{x}, \bar{y}) = \left(2, \frac{48}{25}\right)$

13. $M_x = \rho/3,\, M_y = 4\rho/3,\quad (\bar{x}, \bar{y}) = (4/3, 1/3)$

15. $M_x = 4\rho,\, M_y = 64\rho/5,\quad (\bar{x}, \bar{y}) = (12/5, 3/4)$

17. $M_x = \rho/35,\, M_y = \rho/20,\quad (\bar{x}, \bar{y}) = (3/5, 12/35)$

19. $M_x = 99\rho/5,\, M_y = 27\rho/4,\quad (\bar{x}, \bar{y}) = (3/2, 22/5)$

21. $M_x = 192\rho/7,\, M_y = 96\rho,\quad (\bar{x}, \bar{y}) = (5, 10/7)$

23. $M_x = 0,\, M_y = 256\rho/15,\quad (\bar{x}, \bar{y}) = (8/5, 0)$

25. $M_x = 27\rho/4,\, M_y = -27\rho/10,\quad (\bar{x}, \bar{y}) = (-3/5, 3/2)$

27. $A = \displaystyle\int_0^2 (2x - x^2)\, dx = \frac{4}{3}$

$M_x = \displaystyle\int_0^2 \left(\frac{2x + x^2}{2}\right)(2x - x^2)\, dx = \frac{32}{15}$

$M_y = \displaystyle\int_0^2 x(2x - x^2)\, dx = \frac{4}{3}$

29. $A = \displaystyle\int_0^3 (2x + 4)\, dx = 21$

$M_x = \displaystyle\int_0^3 \left(\frac{2x + 4}{2}\right)(2x + 4)\, dx = 78$

$M_y = \displaystyle\int_0^3 x(2x + 4)\, dx = 36$

31. 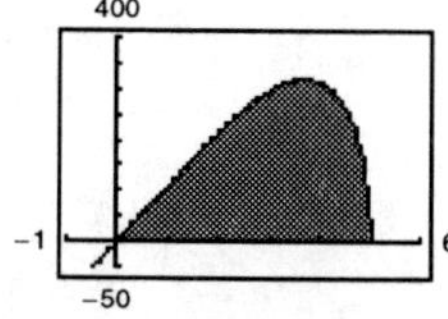

$(\bar{x}, \bar{y}) = (3.0, 126.0)$

33. 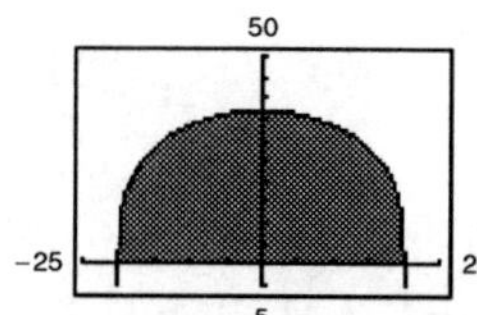

$(\bar{x}, \bar{y}) = (0, 16.2)$

35. $(\bar{x}, \bar{y}) = \left(\dfrac{b}{3}, \dfrac{c}{3}\right)$ **37.** $(\bar{x}, \bar{y}) = \left(\dfrac{(a + 2b)c}{3(a + b)}, \dfrac{a^2 + ab + b^2}{3(a + b)}\right)$

39. $(\bar{x}, \bar{y}) = (0, 4b/(3\pi))$

41. (a) 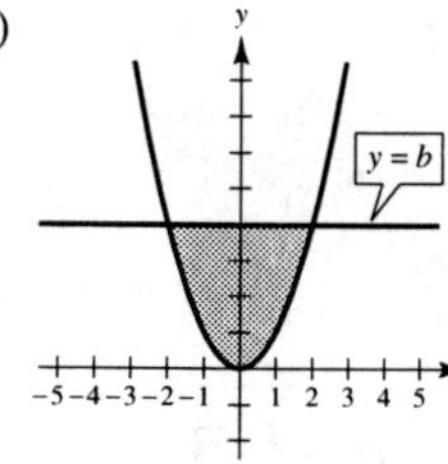

(b) $\bar{x} = 0$ by symmetry

(c) $M_y = \displaystyle\int_{-\sqrt{b}}^{\sqrt{b}} x(b - x^2)\, dx = 0$ because $x(b - x^2)$ is an odd function.

(d) $\bar{y} > b/2$ because the area is greater for $y > b/2$.

(e) $\bar{y} = (3/5)b$

43. (a) $(\bar{x}, \bar{y}) = (0, 12.98)$

(b) $y = (-1.02 \times 10^{-5})x^4 - 0.0019x^2 + 29.28$

(c) $(\bar{x}, \bar{y}) = (0, 12.85)$

45. 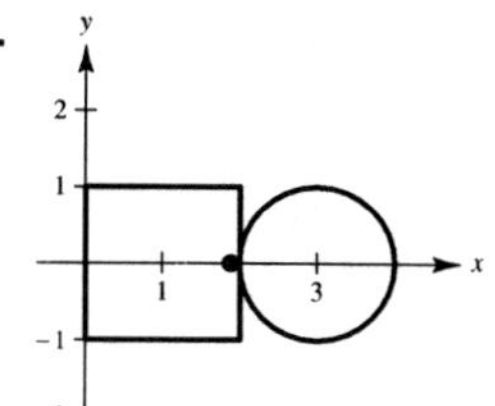

$(\bar{x}, \bar{y}) = \left(\dfrac{4 + 3\pi}{4 + \pi}, 0\right)$

47. $(\bar{x}, \bar{y}) = \left(0, \dfrac{135}{34}\right)$

49. $(\bar{x}, \bar{y}) = \left(\dfrac{2 + 3\pi}{2 + \pi}, 0\right)$ **51.** $160\pi^2 \approx 1579.14$

53. $128\pi/3 \approx 134.04$

55. The center of mass $(\bar{x}, \bar{y})$ is $\bar{x} = M_y/m$ and $\bar{y} = M_x/m$, where:
1. $m = m_1 + m_2 + \cdots + m_n$ is the total mass of the system.
2. $M_y = m_1 x_1 + m_2 x_2 + \cdots + m_n x_n$ is the moment about the y-axis.
3. $M_x = m_1 y_1 + m_2 y_2 + \cdots + m_n y_n$ is the moment about the x-axis.

57. See Theorem 7.1 on page 505.   **59.** $(\bar{x}, \bar{y}) = (0, 2r/\pi)$

61. $(\bar{x}, \bar{y}) = \left(\dfrac{n + 1}{n + 2}, \dfrac{n + 1}{4n + 2}\right)$; As $n \to \infty$, the region shrinks toward the line segments $y = 0$ for $0 \le x \le 1$ and $x = 1$ for $0 \le y \le 1$; $(\bar{x}, \bar{y}) \to \left(1, \dfrac{1}{4}\right)$.

Section 7.7   (page 513)

1. 1497.6 lb   **3.** 4992 lb   **5.** 748.8 lb   **7.** 1123.2 lb

9. 748.8 lb   **11.** 1064.96 lb   **13.** 117,600 N

15. 2,381,400 N   **17.** 2814 lb   **19.** 6753.6 lb   **21.** 94.5 lb

23. Proof   **25.** Proof   **27.** 960 lb

29. Answers will vary. Sample answer (using Simpson's Rule): 3010.8 lb

31. 8213.0 lb

33. $3\sqrt{2}/2 \approx 2.12$ ft; The pressure increases with increasing depth.

35. Because you are measuring total force against a region between two depths.

Review Exercises for Chapter 7   (page 515)

1. 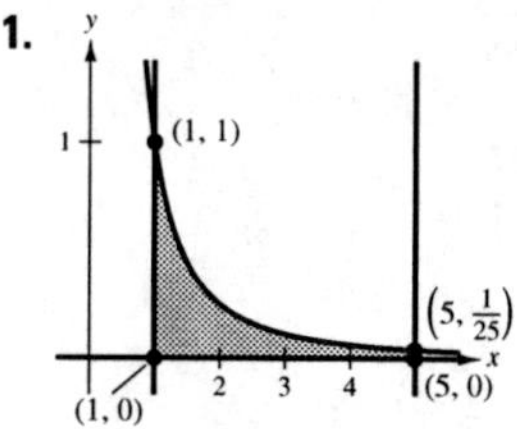

4/5

3. 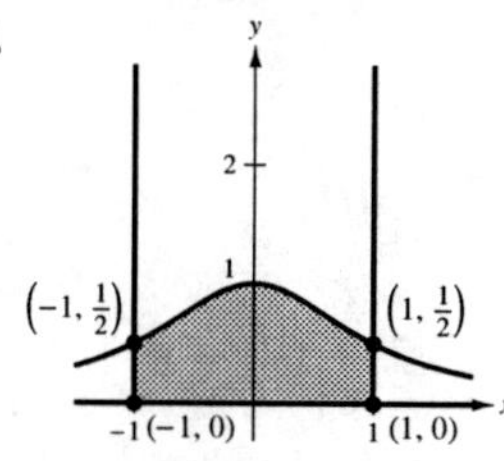

$\pi/2$

5. 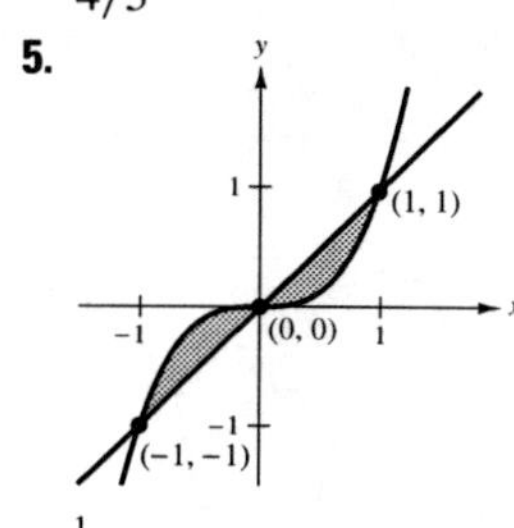

$\dfrac{1}{2}$

7. 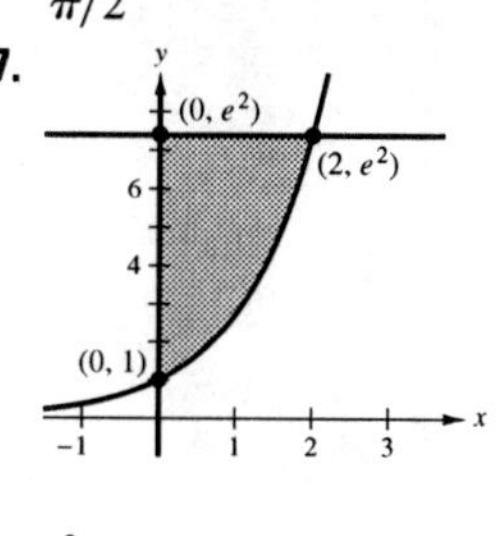

$e^2 + 1$

9. 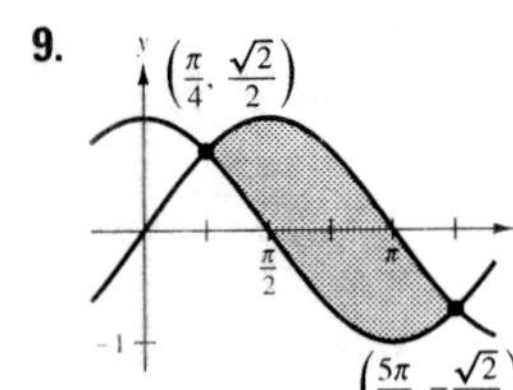

$2\sqrt{2}$

11. 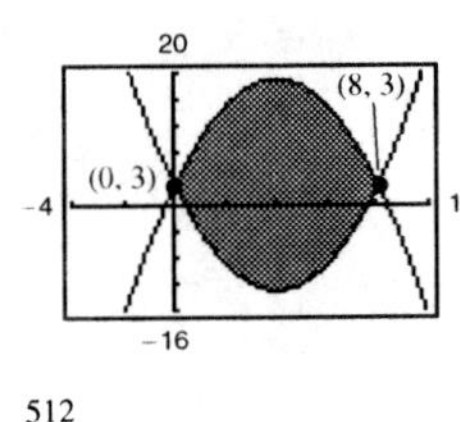

$\dfrac{512}{3}$

13. 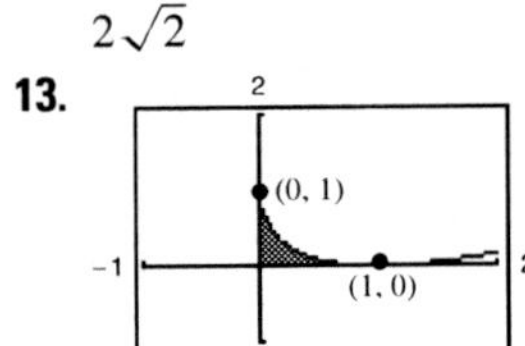

$\dfrac{1}{6}$

15. $\displaystyle\int_0^2 [0 - (y^2 - 2y)]\,dy = \int_{-1}^0 2\sqrt{x+1}\,dx = \dfrac{4}{3}$

17. $\displaystyle\int_0^2 \left[1 - \left(1 - \dfrac{x}{2}\right)\right]dx + \int_2^3 [1 - (x-2)]\,dx$

$\qquad = \displaystyle\int_0^1 [(y+2) - (2-2y)]\,dy = \dfrac{3}{2}$

19. (a) 9920 ft^2 (b) $10{,}413\frac{1}{3}$ ft^2

21. (a) 9π (b) 18π (c) 9π (d) 36π

23. (a) 64π (b) 48π **25.** $\pi^2/4$

27. $(4\pi/3)(20 - 9\ln 3) \approx 42.359$

29. (a) $\dfrac{4}{15}$ (b) $\dfrac{\pi}{12}$ (c) $\dfrac{32\pi}{105}$ **31.** 1.958 ft

33. $\frac{8}{15}\left(1 + 6\sqrt{3}\right) \approx 6.076$ **35.** 4018.2 ft

37. 15π **39.** 62.5 in.-lb ≈ 5.208 ft-lb

41. $122{,}980\pi$ ft-lb ≈ 193.2 foot-tons   **43.** 200 ft-lb

45. $a = 15/4$ **47.** $(\bar{x}, \bar{y}) = (a/5, a/5)$ **49.** $(\bar{x}, \bar{y}) = (0, 2a^2/5)$

51. $(\bar{x}, \bar{y}) = \left(\dfrac{2(9\pi + 49)}{3(\pi + 9)}, 0\right)$

53. Let D = surface of liquid; ρ = weight per cubic volume.

$$F = \rho\int_c^d (D - y)[f(y) - g(y)]\,dy$$

$$= \rho\left[\int_c^d D[f(y) - g(y)]\,dy - \int_c^d y[f(y) - g(y)]\,dy\right]$$

$$= \rho\left[\int_c^d [f(y) - g(y)]\,dy\right]\left[D - \dfrac{\int_c^d y[f(y) - g(y)]\,dy}{\int_c^d [f(y) - g(y)]\,dy}\right]$$

$$= \rho(\text{area})(D - \bar{y})$$

$$= \rho(\text{area})(\text{depth of centroid})$$

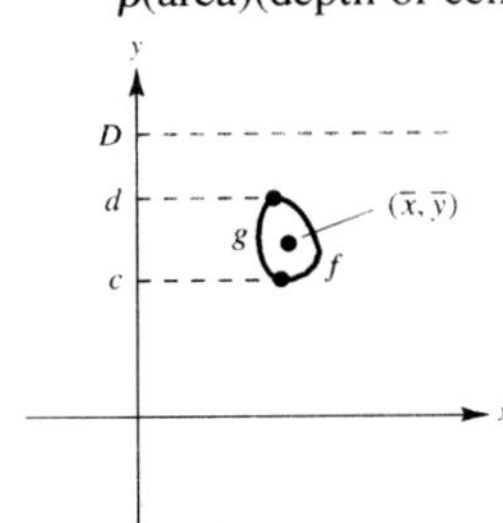

P.S. Problem Solving   (page 517)

1. 3   **3.** $y = 0.2063x$

5. 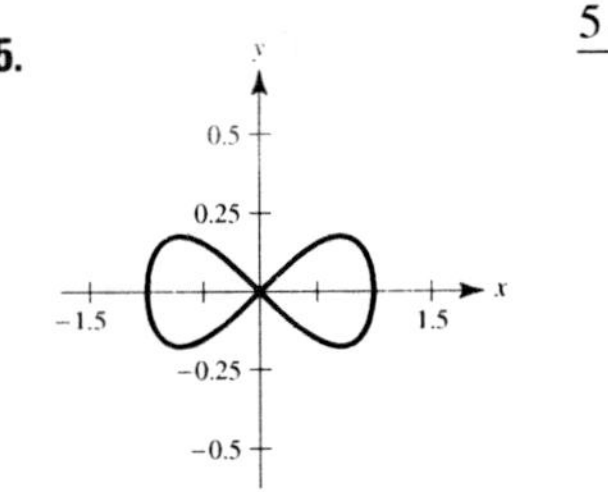

$\qquad\dfrac{5\sqrt{2}\pi}{3}$

7. $V = 2\pi\left[d + \frac{1}{2}\sqrt{w^2 + l^2}\right]lw$ **9.** $f(x) = 2e^{x/2} - 2$

11. 89.3%

13. 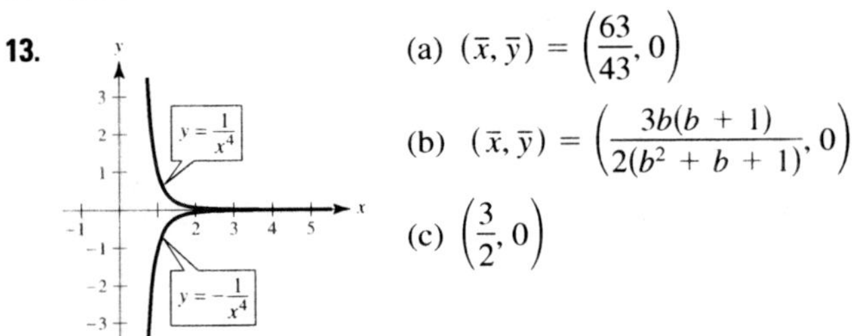

(a) $(\bar{x}, \bar{y}) = \left(\dfrac{63}{43}, 0\right)$

(b) $(\bar{x}, \bar{y}) = \left(\dfrac{3b(b+1)}{2(b^2 + b + 1)}, 0\right)$

(c) $\left(\dfrac{3}{2}, 0\right)$

15. Consumer surplus: 1600; Producer surplus: 400

17. Wall at shallow end: 9984 lb

Wall at deep end: 39,936 lb

Side wall: $19{,}968 + 26{,}624 = 46{,}592$ lb

Chapter 8

Section 8.1   (page 524)

1. b   **3.** c

5. $\displaystyle\int u^n\,du$ **7.** $\displaystyle\int \dfrac{du}{u}$ **9.** $\displaystyle\int \dfrac{du}{\sqrt{a^2 - u^2}}$

$\quad u = 5x - 3, n = 4$ $\quad u = 1 - 2\sqrt{x}$ $\quad u = t, a = 1$

11. $\displaystyle\int \sin u\,du$ **13.** $\displaystyle\int e^u\,du$ **15.** $2(x - 5)^7 + C$

$\quad u = t^2$ $\quad u = \sin x$

17. $-7/[6(z - 10)^6] + C$ **19.** $\frac{1}{2}v^2 - 1/[6(3v - 1)^2] + C$

21. $-\frac{1}{3}\ln|-t^3 + 9t + 1| + C$ **23.** $\frac{1}{2}x^2 + x + \ln|x - 1| + C$

25. $\ln(1 + e^x) + C$ **27.** $\dfrac{x}{15}(48x^4 + 200x^2 + 375) + C$

29. $\sin(2\pi x^2)/(4\pi) + C$ **31.** $-(1/\pi)\csc \pi x + C$

33. $\frac{1}{11}e^{11x} + C$ **35.** $2\ln(1 + e^x) + C$ **37.** $(\ln x)^2 + C$

39. $-\ln(1 - \sin x) + C = \ln|\sec x(\sec x + \tan x)| + C$

41. $\csc \theta + \cot \theta + C = (1 + \cos \theta)/\sin \theta + C$

43. $-\frac{1}{4}\arcsin(4t + 1) + C$ **45.** $\frac{1}{2}\ln|\cos(2/t)| + C$

47. $6\arcsin[(x - 5)/5] + C$ **49.** $\frac{1}{4}\arctan[(2x + 1)/8] + C$

51. $\arcsin\left[(x + 2)/\sqrt{5}\right] + C$

53. (a) 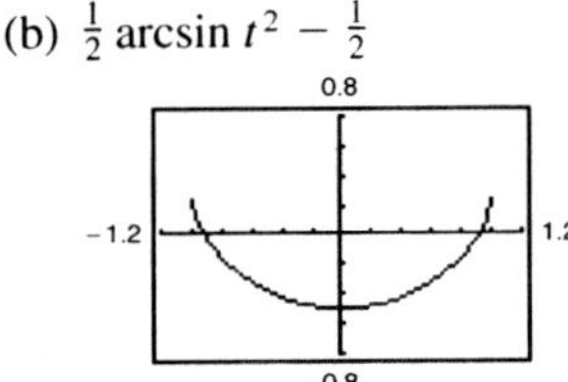   (b) $\frac{1}{2}\arcsin t^2 - \frac{1}{2}$

55.

(a)

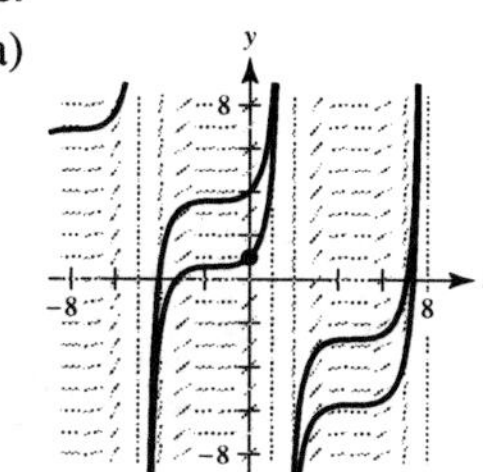

(b) $2 \tan x + 2 \sec x - x - 1 + C$

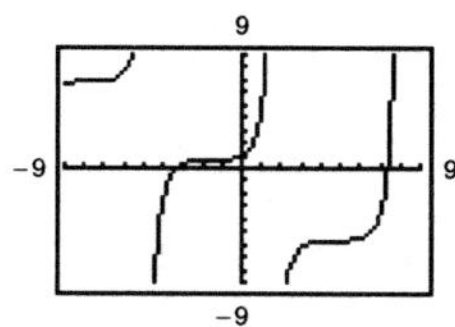

57. $y = 4e^{0.8x}$ **59.** $y = \frac{1}{2}e^{2x} + 10e^x + 25x + C$

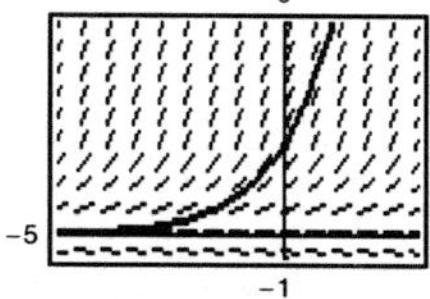

61. $r = 10 \arcsin e^t + C$ **63.** $y = \frac{1}{2}\arctan(\tan x/2) + C$

65. $\frac{1}{2}$ **67.** $\frac{1}{2}(1 - e^{-1}) \approx 0.316$ **69.** 8   **71.** $\pi/18$

73. $18\sqrt{6}/5 \approx 8.82$ **75.** $\frac{3}{2}\ln\left(\frac{34}{9}\right) + \frac{2}{3}\arctan\left(\frac{5}{3}\right) \approx 2.68$

77. $\frac{4}{3} \approx 1.333$

79. $\frac{1}{3}\arctan\left[\frac{1}{3}(x + 2)\right] + C$ **81.** $\tan \theta - \sec \theta + C$

Graphs will vary.   Graphs will vary.

Example: Example:

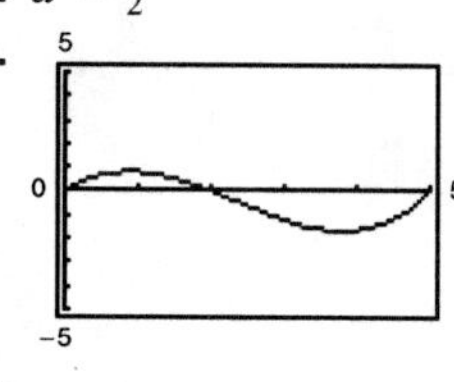   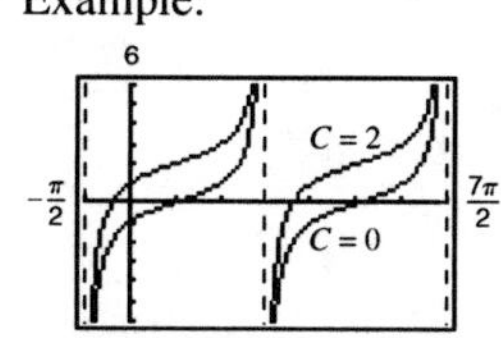

One graph is a vertical   One graph is a vertical

translation of the other.   translation of the other.

83. Power Rule: $\int u^n \, du = \frac{u^{n+1}}{n+1} + C$; $u = x^2 + 1$, $n = 3$

85. Log Rule: $\int \frac{du}{u} = \ln|u| + C$; $u = x^2 + 1$

87. $a = \sqrt{2}, b = \frac{\pi}{4}$; $-\frac{1}{\sqrt{2}}\ln\left|\csc\left(x + \frac{\pi}{4}\right) + \cot\left(x + \frac{\pi}{4}\right)\right| + C$

89. $a = \frac{1}{2}$

91.

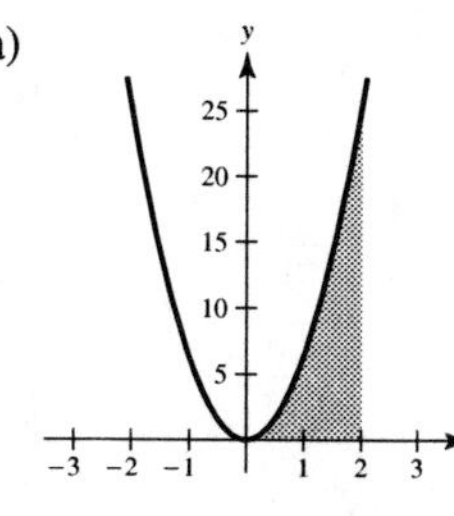

Negative; more area below the x-axis than above

93. a

95. (a)

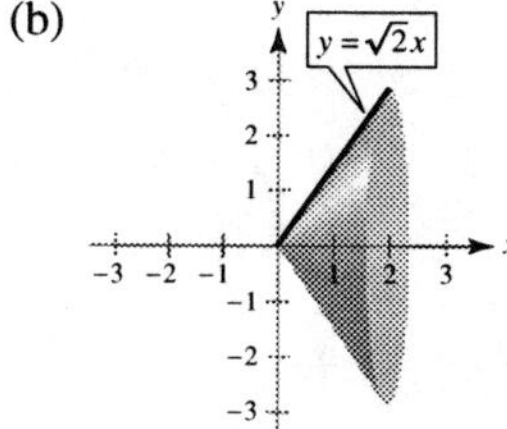

(b)

(c)

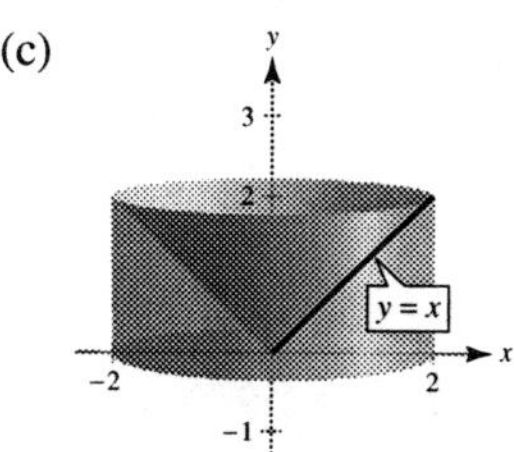

97. (a) $\pi(1 - e^{-1}) \approx 1.986$ (b) $b = \sqrt{\ln\left(\frac{3\pi}{3\pi - 4}\right)} \approx 0.743$

99. $\ln(\sqrt{2} + 1) \approx 0.8814$

101. $(8\pi/3)(10\sqrt{10} - 1) \approx 256.545$

103. $\frac{1}{3}\arctan 3 \approx 0.416$ **105.** About 1.0320

107. (a) $\frac{1}{3}\sin x(\cos^2 x + 2)$

(b) $\frac{1}{15}\sin x(3\cos^4 x + 4\cos^2 x + 8)$

(c) $\frac{1}{35}\sin x(5\cos^6 x + 6\cos^4 x + 8\cos^2 x + 16)$

(d) $\int \cos^{15} x \, dx = \int (1 - \sin^2 x)^7 \cos x \, dx$

You would expand $(1 - \sin^2 x)^7$.

109. Proof

Section 8.2   (page 533)

1. $u = x, dv = e^{2x} dx$ **3.** $u = (\ln x)^2, dv = dx$

5. $u = x, dv = \sec^2 x \, dx$ **7.** $\frac{1}{16}x^4(4 \ln x - 1) + C$

9. $\frac{1}{9}\sin 3x - \frac{1}{3}x \cos 3x + C$ **11.** $-\frac{1}{16e^{4x}}(4x + 1) + C$

13. $e^x(x^3 - 3x^2 + 6x - 6) + C$

15. $\frac{1}{3}e^{x^3} + C$ **17.** $\frac{1}{4}[2(t^2 - 1)\ln|t + 1| - t^2 + 2t] + C$

19. $\frac{1}{3}(\ln x)^3 + C$ **21.** $e^{2x}/[4(2x + 1)] + C$

23. $(x - 1)^2 e^x + C$ **25.** $\frac{2}{15}(x - 5)^{3/2}(3x + 10) + C$

27. $x \sin x + \cos x + C$

29. $(6x - x^3)\cos x + (3x^2 - 6)\sin x + C$

31. $-t \csc t - \ln|\csc t + \cot t| + C$

33. $x \arctan x - \frac{1}{2}\ln(1 + x^2) + C$

35. $\frac{1}{5}e^{2x}(2 \sin x - \cos x) + C$

37. $\frac{1}{5}e^{-x}(2 \sin 2x - \cos 2x) + C$ **39.** $y = \frac{1}{2}e^{x^2} + C$

41. $y = \frac{2}{5}t^2\sqrt{3 + 5t} - \frac{8t}{75}(3 + 5t)^{3/2} + \frac{16}{1875}(3 + 5t)^{5/2} + C$

$\qquad = \frac{2}{625}\sqrt{3 + 5t}(25t^2 - 20t + 24) + C$

43. $\sin y = x^2 + C$

45. (a)

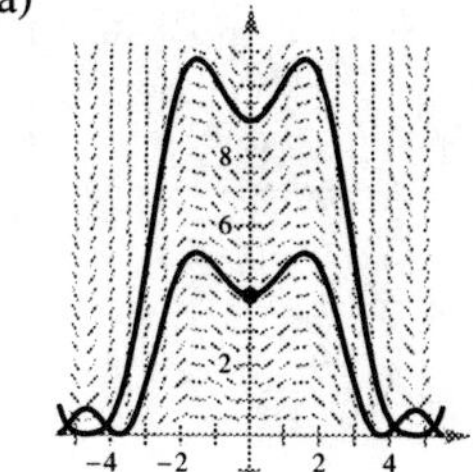

(b) $2\sqrt{y} - \cos x - x \sin x = 3$

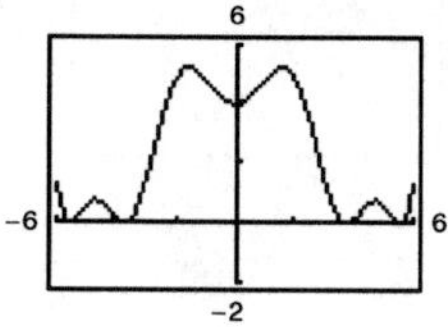

47.

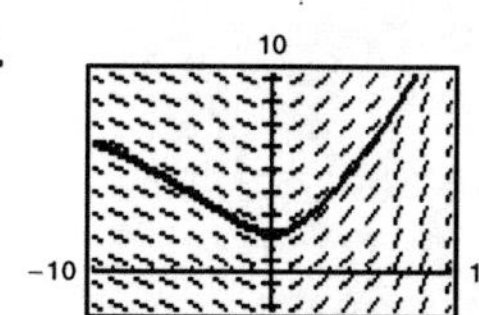

49. $2e^{3/2} + 4 \approx 12.963$

51. $\frac{\pi}{8} - \frac{1}{4} \approx 0.143$

53. $(\pi - 3\sqrt{3} + 6)/6 \approx 0.658$

55. $\frac{1}{2}[e(\sin 1 - \cos 1) + 1] \approx 0.909$

57. $\frac{4}{3}\sqrt{2}\ln 2 - \frac{8}{9}\sqrt{2} + \frac{4}{9} \approx 0.494$

59. $8 \operatorname{arcsec} 4 + \sqrt{3}/2 - \sqrt{15}/2 - 2\pi/3 \approx 7.380$

61. $(e^{2x}/4)(2x^2 - 2x + 1) + C$

63. $(3x^2 - 6)\sin x - (x^3 - 6x)\cos x + C$

65. $x \tan x + \ln|\cos x| + C$ **67.** $2(\sin\sqrt{x} - \sqrt{x}\cos\sqrt{x}) + C$

69. $\frac{128}{15}$ **71.** $\frac{1}{2}(x^4 e^{x^2} - 2x^2 e^{x^2} + 2e^{x^2}) + C$

73. $\frac{1}{2}x[\cos(\ln x) + \sin(\ln x)] + C$ **75.** Product Rule

77. In order for the integration by parts technique to be efficient, you want dv to be the most complicated portion of the integrand and you want u to be the portion of the integrand whose derivative is a function simpler than u. If you let $u = \sin x$, then du is not a simpler function.

79. (a) $-(e^{-4t}/128)(32t^3 + 24t^2 + 12t + 3) + C$

(b) Graphs will vary. Example: (c) One graph is a vertical translation of the other.

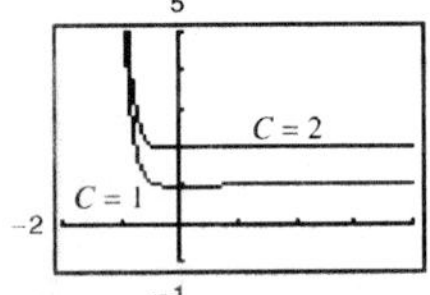

81. (a) $\frac{1}{13}(2e^{-\pi} + 3) \approx 0.2374$

(b) Graphs will vary. Example: (c) One graph is a vertical translation of the other.

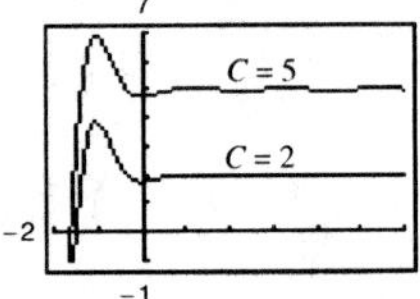

83. $\frac{2}{5}(2x - 3)^{3/2}(x + 1) + C$ **85.** $\frac{1}{3}\sqrt{4 + x^2}(x^2 - 8) + C$

87. $n = 0$: $x(\ln x - 1) + C$

$n = 1$: $\frac{1}{4}x^2(2\ln x - 1) + C$

$n = 2$: $\frac{1}{9}x^3(3\ln x - 1) + C$

$n = 3$: $\frac{1}{16}x^4(4\ln x - 1) + C$

$n = 4$: $\frac{1}{25}x^5(5\ln x - 1) + C$

$$\int x^n \ln x \, dx = \frac{x^{n+1}}{(n+1)^2}[(n+1)\ln x - 1] + C$$

89–93. Proofs

95. $\frac{1}{36}x^6(6\ln x - 1) + C$ **97.** $\frac{1}{13}e^{2x}(2\cos 3x + 3\sin 3x) + C$

99.

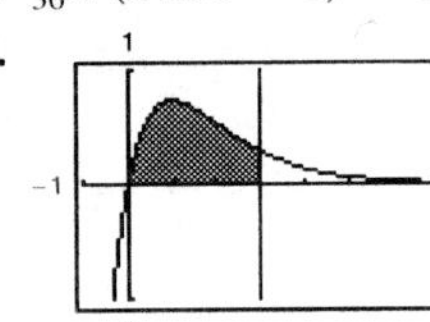

$2 - \dfrac{8}{e^3} \approx 1.602$

101.

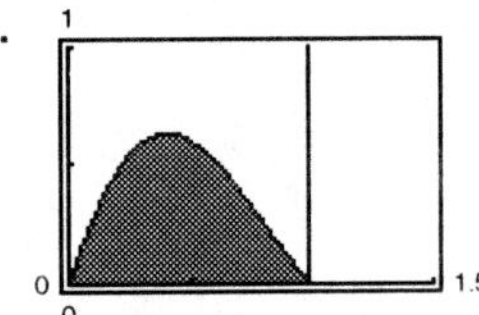

$\dfrac{\pi}{1 + \pi^2}\left(\dfrac{1}{e} + 1\right) \approx 0.395$

103. (a) 1   (b) $\pi(e - 2) \approx 2.257$ (c) $\frac{1}{2}\pi(e^2 + 1) \approx 13.177$

(d) $\left(\dfrac{e^2 + 1}{4}, \dfrac{e - 2}{2}\right) \approx (2.097, 0.359)$

105. In Example 6, we showed that the centroid of an equivalent region was $(1, \pi/8)$. By symmetry, the centroid of this region is $(\pi/8, 1)$.

107. $[7/(10\pi)](1 - e^{-4\pi}) \approx 0.223$ **109.** \$931,265

111. Proof   **113.** $b_n = [8h/(n\pi)^2]\sin(n\pi/2)$

115. Shell: $V = \pi\left[b^2 f(b) - a^2 f(a) - \displaystyle\int_a^b x^2 f'(x)\, dx\right]$

Disk: $V = \pi\left[b^2 f(b) - a^2 f(a) - \displaystyle\int_{f(a)}^{f(b)} [f^{-1}(y)]^2\, dy\right]$

Both methods yield the same volume because $x = f^{-1}(y)$, $f'(x)\,dx = dy$, if $y = f(a)$ then $x = a$, and if $y = f(b)$ then $x = b$.

117. (a) $y = \frac{1}{4}(3\sin 2x - 6x\cos 2x)$

(b)

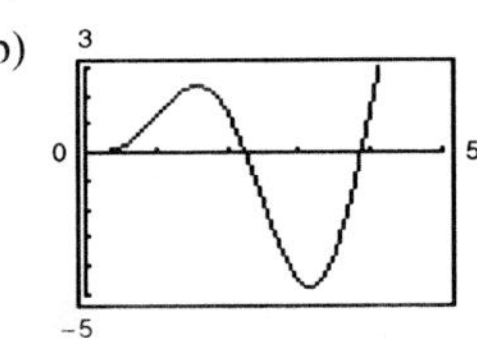

(c) You obtain the following points.

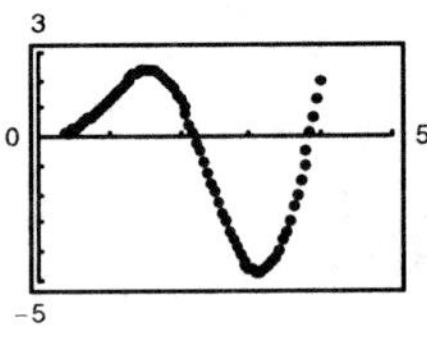

n	x_n	y_n
0	0	0
1	0.05	0
2	0.10	7.4875×10^{-4}
3	0.15	0.0037
4	0.20	0.0104
$\vdots$	$\vdots$	$\vdots$
80	4.00	1.3181

(d) You obtain the following points.

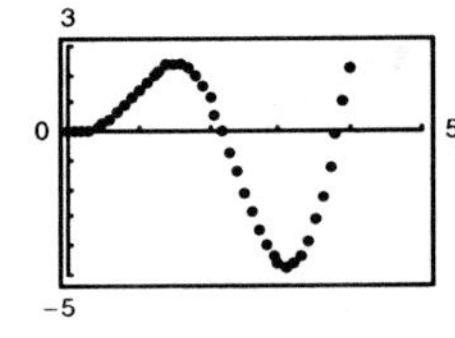

n	x_n	y_n
0	0	0
1	0.1	0
2	0.2	0.0060
3	0.3	0.0293
4	0.4	0.0801
$\vdots$	$\vdots$	$\vdots$
40	4.0	1.0210

119. The graph of $y = x\sin x$ is below the graph of $y = x$ on $[0, \pi/2]$.

Section 8.3   (page 542)

1. c   **2.** a   **3.** d   **4.** b   **5.** $-\frac{1}{6}\cos^6 x + C$

7. $\frac{1}{16}\sin^8 2x + C$ **9.** $-\frac{1}{3}\cos^3 x + \frac{1}{5}\cos^5 x + C$

11. $-\frac{1}{3}(\cos 2\theta)^{3/2} + \frac{1}{7}(\cos 2\theta)^{7/2} + C$

13. $\frac{1}{12}(6x + \sin 6x) + C$

15. $\frac{3}{8}\alpha + \frac{1}{12}\sin 6\alpha + \frac{1}{96}\sin 12\alpha + C$

17. $\frac{1}{8}(2x^2 - 2x\sin 2x - \cos 2x) + C$ **19.** $\frac{16}{35}$

21. $63\pi/512$ **23.** $5\pi/32$ **25.** $\frac{1}{7}\ln|\sec 7x + \tan 7x| + C$

27. $\frac{1}{15}\tan 5x(3 + \tan^2 5x) + C$

29. $\left(\sec \pi x \tan \pi x + \ln|\sec \pi x + \tan \pi x|\right)/(2\pi) + C$

31. $\frac{1}{2}\tan^4(x/2) - \tan^2(x/2) - 2\ln|\cos(x/2)| + C$

33. $\frac{1}{2}\tan^2 x + C$ **35.** $\frac{1}{3}\tan^3 x + \frac{1}{5}\tan^5 x + C$

37. $\frac{1}{24}\sec^6 4x + C$ **39.** $\frac{1}{7}\sec^7 x - \frac{1}{5}\sec^5 x + C$

41. $\ln|\sec x + \tan x| - \sin x + C$

43. $(12\pi\theta - 8\sin 2\pi\theta + \sin 4\pi\theta)/(32\pi) + C$

45. $y = \frac{1}{9}\sec^3 3x - \frac{1}{3}\sec 3x + C$

47. (a)   (b) $y = \frac{1}{2}x - \frac{1}{4}\sin 2x$

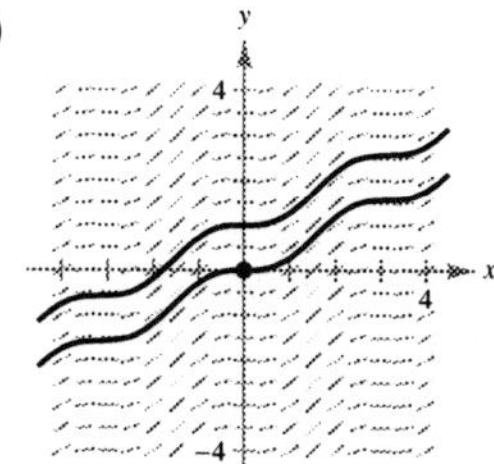   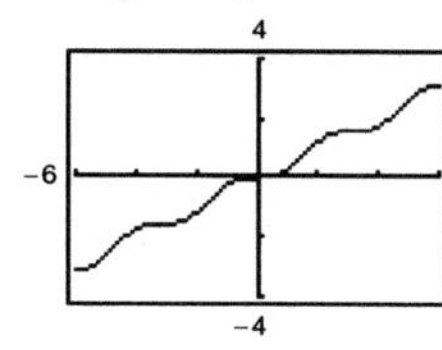

49.   **51.** $\frac{1}{16}(2\sin 4x + \sin 8x) + C$

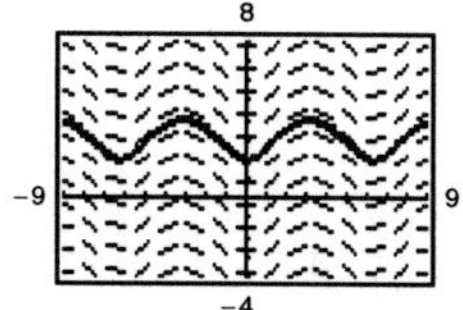

53. $\frac{1}{12}(3\cos 2x - \cos 6x) + C$ **55.** $\frac{1}{8}(2\sin 2\theta - \sin 4\theta) + C$

57. $\frac{1}{4}(\ln|\csc^2 2x| - \cot^2 2x) + C$ **59.** $-\frac{1}{2}\cot 2x - \frac{1}{6}\cot^3 2x + C$

61. $\ln|\csc t - \cot t| + \cos t + C$

63. $\ln|\csc x - \cot x| + \cos x + C$ **65.** $t - 2\tan t + C$

67. π **69.** $3(1 - \ln 2)$ **71.** $\ln 2$ **73.** 4

75. $\frac{1}{16}(6x + 8\sin x + \sin 2x) + C$

Graphs will vary. Example:

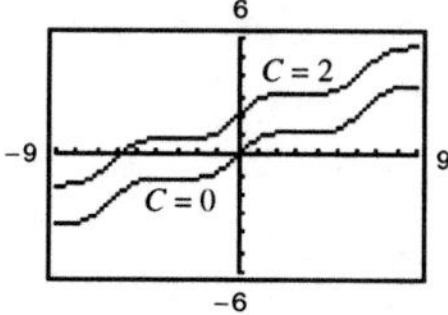

77. $\Big[\sec^3 \pi x \tan \pi x +$

$\frac{3}{2}\big(\sec \pi x \tan \pi x + \ln|\sec \pi x + \tan \pi x|\big)\Big]/(4\pi) + C$

Graphs will vary. Example:

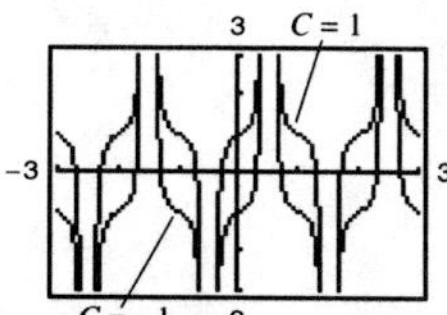

79. $(\sec^5 \pi x)/(5\pi) + C$ **81.** $2\sqrt{2}/7$ **83.** $3\pi/16$

Graphs will vary. Example:

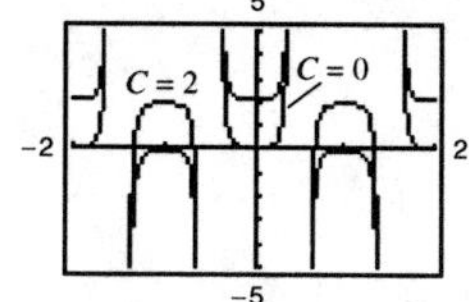

85. (a) Save one sine factor and convert the remaining factors to cosines. Then expand and integrate.

(b) Save one cosine factor and convert the remaining factors to sines. Then expand and integrate.

(c) Make repeated use of the power reducing formulas to convert the integrand to odd powers of the cosine. Then proceed as in part (b).

87. (a) $\frac{1}{2}\sin^2 x + C$ (b) $-\frac{1}{2}\cos^2 x + C$

(c) $\frac{1}{2}\sin^2 x + C$ (d) $-\frac{1}{4}\cos 2x + C$

The answers are all the same, they are just written in different forms. Using trigonometric identities, you can rewrite each answer in the same form.

89. (a) $\frac{1}{18}\tan^6 3x + \frac{1}{12}\tan^4 3x + C_1$, $\frac{1}{18}\sec^6 3x - \frac{1}{12}\sec^4 3x + C_2$

(b)   (c) Proof

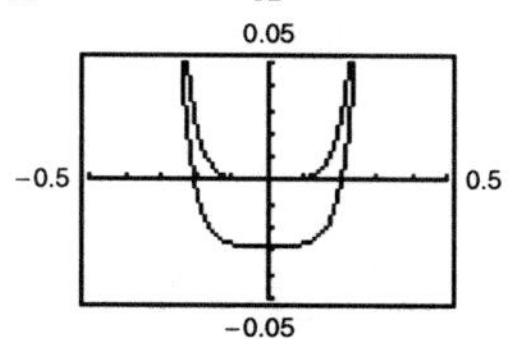

91. $\frac{1}{3}$ **93.** 1 **95.** $2\pi(1 - \pi/4) \approx 1.348$

97. (a) $\pi^2/2$ (b) $(\bar{x}, \bar{y}) = (\pi/2, \pi/8)$ **99–101.** Proofs

103. $-\frac{1}{15}\cos x(3\sin^4 x + 4\sin^2 x + 8) + C$

105. $\dfrac{5}{6\pi}\tan\dfrac{2\pi x}{5}\left(\sec^2\dfrac{2\pi x}{5} + 2\right) + C$

107. (a) $H(t) \approx 57.72 - 23.36\cos(\pi t/6) - 2.75\sin(\pi t/6)$

(b) $L(t) \approx 42.04 - 20.91\cos(\pi t/6) - 4.33\sin(\pi t/6)$

(c)   The maximum difference is at $t \approx 4.9$, or late spring.

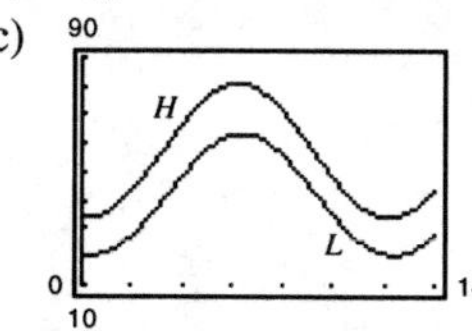

109. Proof

Section 8.4   (page 551)

1. $x = 3\tan\theta$ **3.** $x = 4\sin\theta$ **5.** $x/\left(16\sqrt{16 - x^2}\right) + C$

7. $4\ln\left|(4 - \sqrt{16 - x^2})/x\right| + \sqrt{16 - x^2} + C$

9. $\ln\left|x + \sqrt{x^2 - 25}\right| + C$

11. $\frac{1}{15}(x^2 - 25)^{3/2}(3x^2 + 50) + C$

13. $\frac{1}{3}(1 + x^2)^{3/2} + C$ **15.** $\frac{1}{2}[\arctan x + x/(1 + x^2)] + C$

17. $\frac{1}{2}x\sqrt{9 + 16x^2} + \frac{9}{8}\ln\left|4x + \sqrt{9 + 16x^2}\right| + C$

19. $\frac{25}{4}\arcsin(2x/5) + \frac{1}{2}x\sqrt{25 - 4x^2} + C$

21. $\sqrt{x^2 + 36} + C$ **23.** $\arcsin(x/4) + C$

25. $4\arcsin(x/2) + x\sqrt{4 - x^2} + C$ **27.** $\ln\left|x + \sqrt{x^2 - 4}\right| + C$

29. $-\dfrac{(1 - x^2)^{3/2}}{3x^3} + C$ **31.** $-\dfrac{1}{3}\ln\left|\dfrac{\sqrt{4x^2 + 9} + 3}{2x}\right| + C$

33. $3/\sqrt{x^2 + 3} + C$ **35.** $\frac{1}{3}(1 + e^{2x})^{3/2} + C$

37. $\frac{1}{2}\left(\arcsin e^x + e^x\sqrt{1 - e^{2x}}\right) + C$

39. $\frac{1}{4}\left[x/(x^2 + 2) + (1/\sqrt{2})\arctan(x/\sqrt{2})\right] + C$

41. $x\,\text{arcsec}\,2x - \frac{1}{2}\ln\left|2x + \sqrt{4x^2 - 1}\right| + C$

43. $\arcsin[(x - 2)/2] + C$

45. $\sqrt{x^2 + 6x + 12} - 3\ln\left|\sqrt{x^2 + 6x + 12} + (x + 3)\right| + C$

47. (a) and (b) $\sqrt{3} - \pi/3 \approx 0.685$

49. (a) and (b) $9(2 - \sqrt{2}) \approx 5.272$

51. (a) and (b) $-(9/2)\ln(2\sqrt{7}/3 - 4\sqrt{3}/3 - \sqrt{21}/3 + 8/3)$
$+ 9\sqrt{3} - 2\sqrt{7} \approx 12.644$

53. $\sqrt{x^2 - 9} - 3\arctan(\sqrt{x^2 - 9}/3) + 1$

55. $\frac{1}{2}(x - 15)\sqrt{x^2 + 10x + 9}$
$+ 33\ln|\sqrt{x^2 + 10x + 9} + (x + 5)| + C$

57. $\frac{1}{2}(x\sqrt{x^2 - 1} + \ln|x + \sqrt{x^2 - 1}|) + C$

59. (a) Let $u = a\sin\theta$, $\sqrt{a^2 - u^2} = a\cos\theta$, where
$-\pi/2 \le \theta \le \pi/2$.

(b) Let $u = a\tan\theta$, $\sqrt{a^2 + u^2} = a\sec\theta$, where
$-\pi/2 < \theta < \pi/2$.

(c) Let $u = a\sec\theta$, $\sqrt{u^2 - a^2} = \tan\theta$ if $u > a$ and
$\sqrt{u^2 - a^2} = -\tan\theta$ if $u < -a$, where $0 \le \theta < \pi/2$
or $\pi/2 < \theta \le \pi$.

61. Trigonometric substitution: $x = \sec\theta$ **63.** True

65. False: $\displaystyle\int_0^{\sqrt{3}} \frac{dx}{(1 + x^2)^{3/2}} = \int_0^{\pi/3} \cos\theta\, d\theta$

67. πab **69.** (a) $5\sqrt{2}$ (b) $25(1 - \pi/4)$ (c) $r^2(1 - \pi/4)$

71. $6\pi^2$ **73.** $\ln\left[\dfrac{5(\sqrt{2} + 1)}{\sqrt{26} + 1}\right] + \sqrt{26} - \sqrt{2} \approx 4.367$

75. Length of one arch of sine curve: $y = \sin x$, $y' = \cos x$

$$L_1 = \int_0^{\pi} \sqrt{1 + \cos^2 x}\, dx$$

Length of one arch of cosine curve: $y = \cos x$, $y' = -\sin x$

$$L_2 = \int_{-\pi/2}^{\pi/2} \sqrt{1 + \sin^2 x}\, dx$$

$$= \int_{-\pi/2}^{\pi/2} \sqrt{1 + \cos^2(x - \pi/2)}\, dx, \quad u = x - \pi/2,\, du = dx$$

$$= \int_{-\pi}^{0} \sqrt{1 + \cos^2 u}\, du = \int_0^{\pi} \sqrt{1 + \cos^2 u}\, du = L_1$$

77. (a) 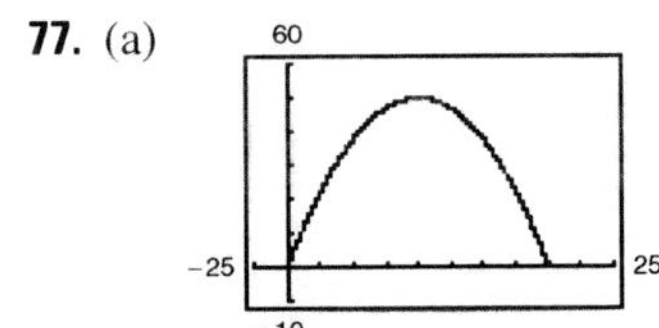   (b) 200

(c) $100\sqrt{2} + 50\ln[(\sqrt{2} + 1)/(\sqrt{2} - 1)] \approx 229.559$

79. $(0, 0.422)$ **81.** $(\pi/32)[102\sqrt{2} - \ln(3 + 2\sqrt{2})] \approx 13.989$

83. (a) 187.2π lb (b) $62.4\pi d$ lb

85. Proof   **87.** $12 + 9\pi/2 - 25\arcsin(3/5) \approx 10.050$

89. Putnam Problem A5, 2005

Section 8.5   (page 561)

1. $\dfrac{A}{x} + \dfrac{B}{x - 8}$ **3.** $\dfrac{A}{x} + \dfrac{Bx + C}{x^2 + 10}$ **5.** $\dfrac{A}{x} + \dfrac{B}{x - 6}$

7. $\frac{1}{6}\ln|(x - 3)/(x + 3)| + C$ **9.** $\ln|(x - 1)/(x + 4)| + C$

11. $\frac{3}{2}\ln|2x - 1| - 2\ln|x + 1| + C$

13. $5\ln|x - 2| - \ln|x + 2| - 3\ln|x| + C$

15. $x^2 + \frac{3}{2}\ln|x - 4| - \frac{1}{2}\ln|x + 2| + C$

17. $1/x + \ln|x^4 + x^3| + C$

19. $2\ln|x - 2| - \ln|x| - 3/(x - 2) + C$

21. $\ln|(x^2 + 1)/x| + C$

23. $\frac{1}{6}\left[\ln|(x - 2)/(x + 2)| + \sqrt{2}\arctan(x/\sqrt{2})\right] + C$

25. $\frac{1}{16}\ln|(4x^2 - 1)/(4x^2 + 1)| + C$

27. $\ln|x + 1| + \sqrt{2}\arctan[(x - 1)/\sqrt{2}] + C$

29. $\ln 3$ **31.** $\frac{1}{2}\ln(8/5) - \pi/4 + \arctan 2 \approx 0.557$

33. $y = 5\ln|x - 5| - 5x/(x - 5) + 30$

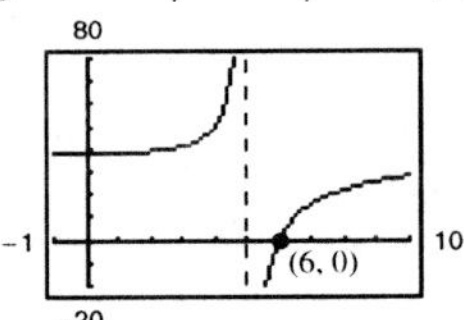

35. $y = (\sqrt{2}/2)\arctan(x/\sqrt{2}) - 1/[2(x^2 + 2)] + 5/4$

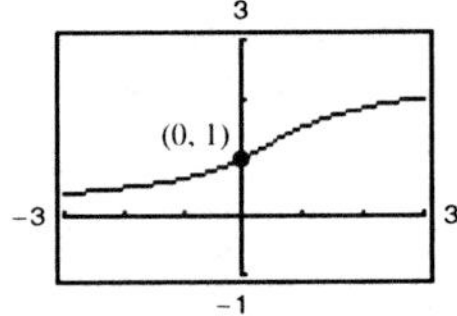

37. $y = \ln|x - 2| + \frac{1}{2}\ln|x^2 + x + 1|$
$- \sqrt{3}\arctan[(2x + 1)/\sqrt{3}] - \frac{1}{2}\ln 13$
$+ \sqrt{3}\arctan(7/\sqrt{3}) + 10$

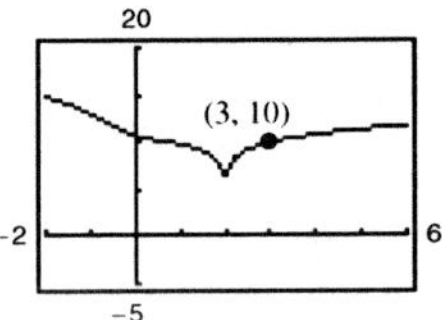

39. $y = \frac{1}{10}\ln|(x - 5)/(x + 5)| + \frac{1}{10}\ln 6 + 2$

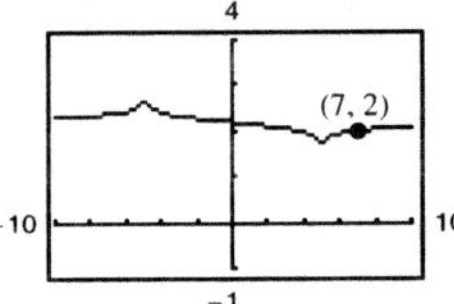

41. $\ln\left|\dfrac{\cos x}{\cos x - 1}\right| + C$ **43.** $\ln\left|\dfrac{\sin x}{1 + \sin x}\right| + C$

45. $\ln\left|\dfrac{\tan x + 2}{\tan x + 3}\right| + C$ **47.** $\frac{1}{5}\ln\left|\dfrac{e^x - 1}{e^x + 4}\right| + C$

49. $2\sqrt{x} + 2\ln\left|\dfrac{\sqrt{x} - 2}{\sqrt{x} + 2}\right| + C$ **51–53.** Proofs

55. $y = \dfrac{3}{2}\ln\left|\dfrac{2 + x}{2 - x}\right| + 3$ **57.** First divide x^3 by $(x - 5)$.

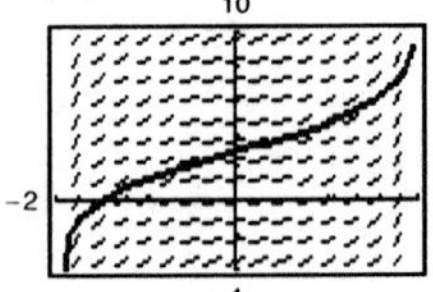

59. $12\ln(\frac{9}{8}) \approx 1.4134$ **61.** $6 - \frac{7}{4}\ln 7 \approx 2.5947$

63. 4.90 or \$490,000

65. $V = 2\pi(\arctan 3 - \frac{3}{10}) \approx 5.963$; $(\bar{x}, \bar{y}) \approx (1.521, 0.412)$

67. $x = n[e^{(n+1)kt} - 1]/[n + e^{(n+1)kt}]$ **69.** $\pi/8$

Section 8.6 (page 567)

1. $-\frac{1}{2}x(10 - x) + 25 \ln|5 + x| + C$

3. $\frac{1}{2}\left[e^x\sqrt{e^{2x} + 1} + \ln\left(e^x + \sqrt{e^{2x} + 1}\right)\right] + C$

5. $-\sqrt{1 - x^2}/x + C$

7. $\frac{1}{24}(3x + \sin 3x \cos 3x + 2 \cos^3 3x \sin 3x) + C$

9. $-2\left(\cot\sqrt{x} + \csc\sqrt{x}\right) + C$ **11.** $x - \frac{1}{2}\ln(1 + e^{2x}) + C$

13. $\frac{1}{64}x^8(8 \ln x - 1) + C$

15. (a) and (b) $\frac{1}{27}e^{3x}(9x^2 - 6x + 2) + C$

17. (a) and (b) $\ln|(x + 1)/x| - 1/x + C$

19. $\frac{1}{2}\left[(x^2 + 1) \operatorname{arccsc}(x^2 + 1) + \ln\left(x^2 + 1 + \sqrt{x^4 + 2x^2}\right)\right] + C$

21. $\sqrt{x^2 - 4}/(4x) + C$ **23.** $\frac{4}{25}[\ln|2 - 5x| + 2/(2 - 5x)] + C$

25. $e^x \arccos(e^x) - \sqrt{1 - e^{2x}} + C$

27. $\frac{1}{2}(x^2 + \cot x^2 + \csc x^2) + C$

29. $\left(\sqrt{2}/2\right)\arctan\left[(1 + \sin\theta)/\sqrt{2}\right] + C$

31. $-\sqrt{2 + 9x^2}/(2x) + C$

33. $\frac{1}{4}\left(2\ln|x| - 3\ln|3 + 2\ln|x||\right) + C$

35. $(3x - 10)/[2(x^2 - 6x + 10)] + \frac{3}{2}\arctan(x - 3) + C$

37. $\frac{1}{2}\ln\left|x^2 - 3 + \sqrt{x^4 - 6x^2 + 5}\right| + C$

39. $-\frac{1}{3}\sqrt{4 - x^2}(x^2 + 8) + C$

41. $2/(1 + e^x) - 1/[2(1 + e^x)^2] + \ln(1 + e^x) + C$

43. $\frac{1}{2}(e - 1) \approx 0.8591$ **45.** $\frac{32}{5}\ln 2 - \frac{31}{25} \approx 3.1961$

47. $\pi/2$ **49.** $\pi^3/8 - 3\pi + 6 \approx 0.4510$ **51–55.** Proofs

57. $y = -2\sqrt{1 - x}/\sqrt{x} + 7$

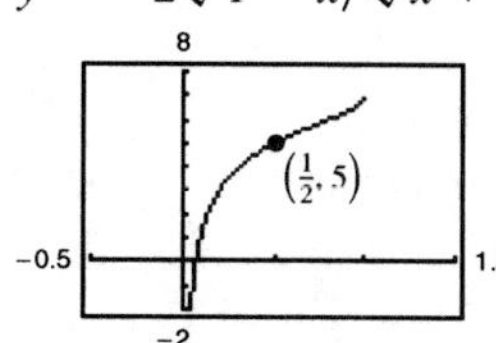

59. $y = \frac{1}{2}[(x - 3)/(x^2 - 6x + 10) + \arctan(x - 3)]$

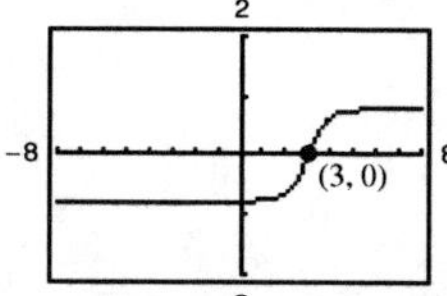

61. $y = -\csc\theta + \sqrt{2} + 2$

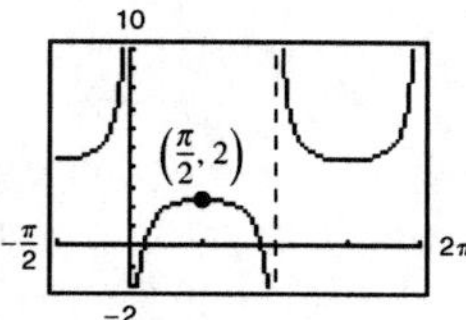

63. $\dfrac{1}{\sqrt{5}}\ln\left|\dfrac{2\tan(\theta/2) - 3 - \sqrt{5}}{2\tan(\theta/2) - 3 + \sqrt{5}}\right| + C$ **65.** $\ln 2$

67. $\frac{1}{2}\ln(3 - 2\cos\theta) + C$ **69.** $-2\cos\sqrt{\theta} + C$ **71.** $4\sqrt{3}$

73. (a) $\displaystyle\int x \ln x \, dx = \frac{1}{2}x^2 \ln x - \frac{1}{4}x^2 + C$

$\displaystyle\int x^2 \ln x \, dx = \frac{1}{3}x^3 \ln x - \frac{1}{9}x^3 + C$

$\displaystyle\int x^3 \ln x \, dx = \frac{1}{4}x^4 \ln x - \frac{1}{16}x^4 + C$

(b) $\displaystyle\int x^n \ln x \, dx = x^{n+1} \ln x/(n + 1) - x^{n+1}/(n + 1)^2 + C$

75. False. Substitutions may first have to be made to rewrite the integral in a form that appears in the table.

77. $32\pi^2$ **79.** 1919.145 ft-lb

81. (a) $V = 80 \ln\left(\sqrt{10} + 3\right) \approx 145.5$ ft^3

$W = 11,840 \ln\left(\sqrt{10} + 3\right) \approx 21{,}530.4$ lb

(b) $(0, 1.19)$

83. (a) $k = 30/\ln 7 \approx 15.42$ **85.** Putnam Problem A3, 1980

(b)

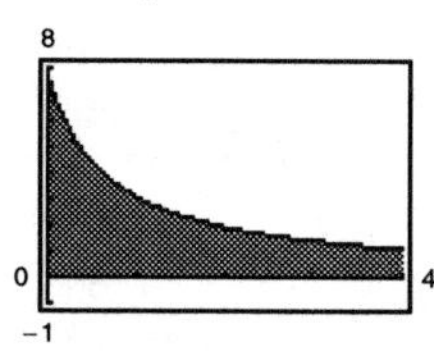

Section 8.7 (page 576)

1.

x	-0.1	-0.01	-0.001	0.001	0.01	0.1
$f(x)$	1.3177	1.3332	1.3333	1.3333	1.3332	1.3177

$\frac{4}{3}$

3.

x	1	10	10^2	10^3	10^4	10^5
$f(x)$	0.9900	90,483.7	3.7×10^9	4.5×10^{10}	0	0

0

5. $\frac{3}{8}$ **7.** $\frac{1}{8}$ **9.** $\frac{5}{3}$ **11.** 4 **13.** 0 **15.** 2

17. ∞ **19.** $\frac{11}{4}$ **21.** $\frac{3}{5}$ **23.** 1 **25.** $\frac{5}{4}$ **27.** ∞

29. 0 **31.** 1 **33.** 0 **35.** 0 **37.** ∞

39. $\frac{5}{9}$ **41.** 1 **43.** ∞

45. (a) Not indeterminate **47.** (a) $0 \cdot \infty$

(b) ∞ (b) 1

(c) 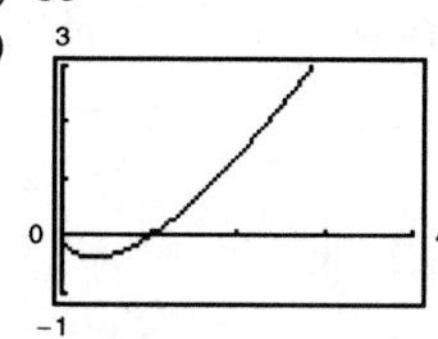 (c) 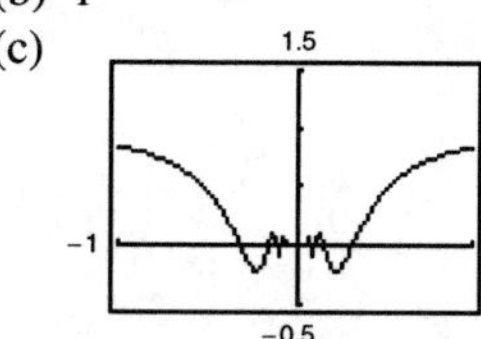

49. (a) Not indeterminate **51.** (a) ∞^0

(b) 0 (b) 1

(c) 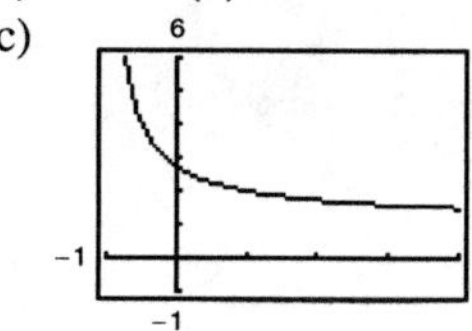 (c) 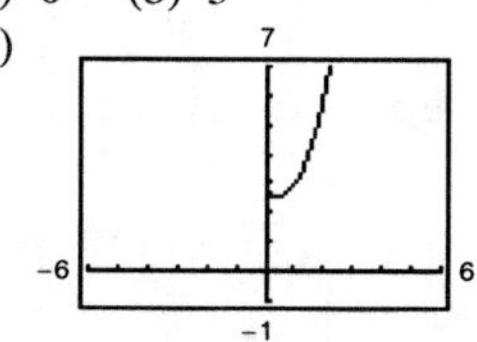

53. (a) 1^∞ (b) e **55.** (a) 0^0 (b) 3

(c) (c)

57. (a) 0^0 (b) 1
(c) 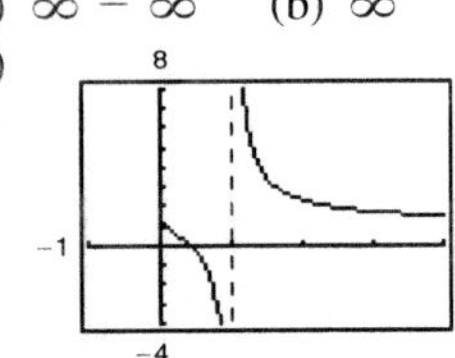

59. (a) $\infty - \infty$ (b) $-\frac{3}{2}$
(c) 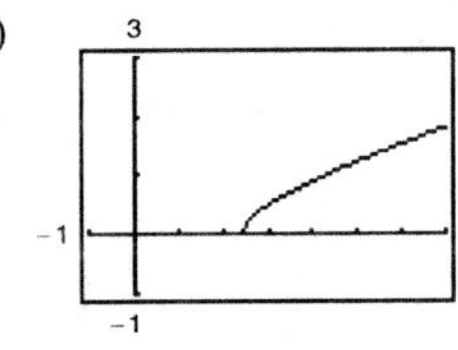

61. (a) $\infty - \infty$ (b) ∞
(c) 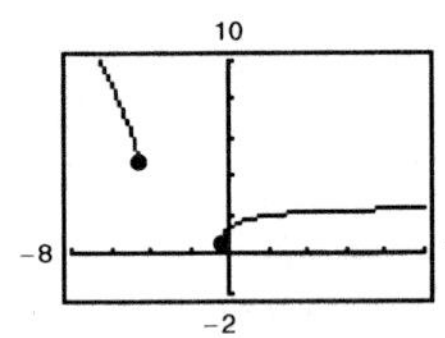

63. (a) 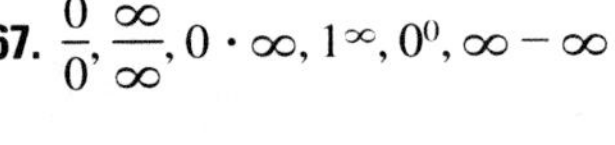
(b) $\frac{1}{2}$

65. (a) 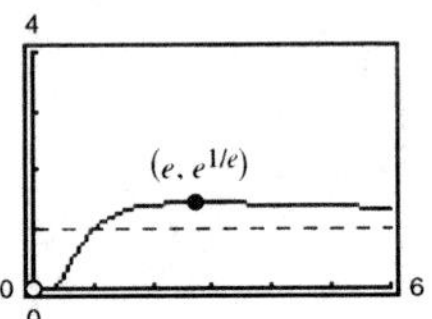
(b) $\frac{5}{2}$

67. $\dfrac{0}{0}, \dfrac{\infty}{\infty}, 0 \cdot \infty, 1^\infty, 0^0, \infty - \infty$

69. Answers will vary. Examples:
(a) $f(x) = x^2 - 25,\ g(x) = x - 5$
(b) $f(x) = (x - 5)^2,\ g(x) = x^2 - 25$
(c) $f(x) = x^2 - 25,\ g(x) = (x - 5)^3$

71.

x	10	10^2	10^4	10^6	10^8	10^{10}
$\dfrac{(\ln x)^4}{x}$	2.811	4.498	0.720	0.036	0.001	0.000

73. 0   **75.** 0   **77.** 0

79. Horizontal asymptote:
$y = 1$
Relative maximum: $(e, e^{1/e})$

81. Horizontal asymptote:
$y = 0$
Relative maximum: $(1, 2/e)$

83. Limit is not of the form $0/0$ or ∞/∞.
85. Limit is not of the form $0/0$ or ∞/∞.
87. Limit is not of the form $0/0$ or ∞/∞.

89. (a) $\displaystyle \lim_{x \to \infty} \frac{x}{\sqrt{x^2 + 1}} = \lim_{x \to \infty} \frac{\sqrt{x^2 + 1}}{x} = \lim_{x \to \infty} \frac{x}{\sqrt{x^2 + 1}}$

Applying L'Hôpital's Rule twice results in the original limit, so L'Hôpital's Rule fails.

(b) 1
(c) 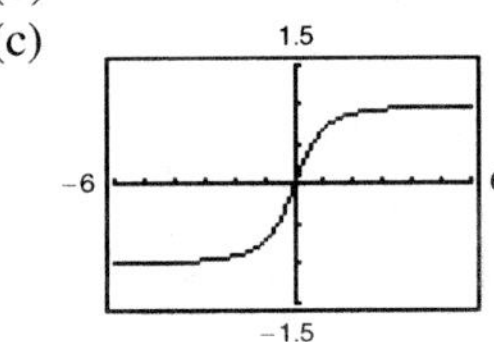

91. 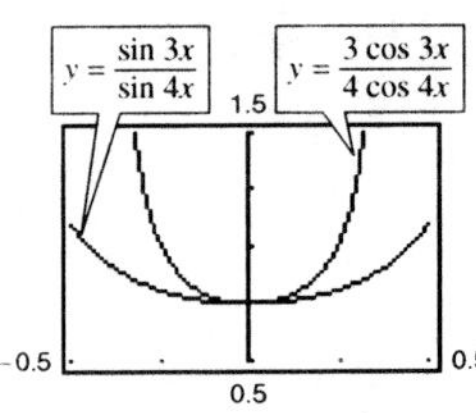

As $x \to 0$, the graphs get closer together (they both approach 0.75).

By L'Hôpital's Rule,
$$\lim_{x \to 0} \frac{\sin 3x}{\sin 4x} = \lim_{x \to 0} \frac{3 \cos 3x}{4 \cos 4x} = \frac{3}{4}.$$

93. $v = 32t + v_0$ **95.** Proof   **97.** $c = \frac{2}{3}$ **99.** $c = \pi/4$

101. False: L'Hôpital's Rule does not apply, because $\displaystyle \lim_{x \to 0}(x^2 + x + 1) \neq 0$.   **103.** True

105. $\frac{3}{4}$ **107.** $\frac{4}{3}$ **109.** $a = 1,\ b = \pm 2$ **111.** Proof

113. 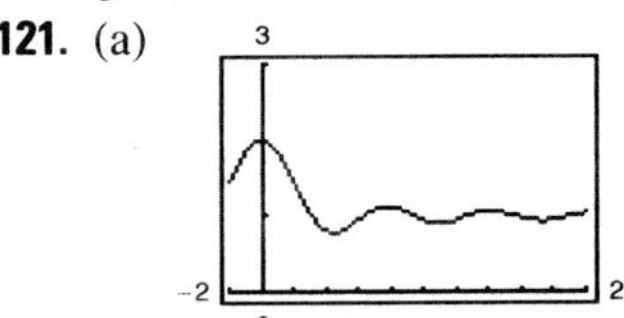
$g'(0) = 0$

115. (a) $0 \cdot \infty$ (b) 0
117. Proof
119. (a)–(c) 2

121. (a) 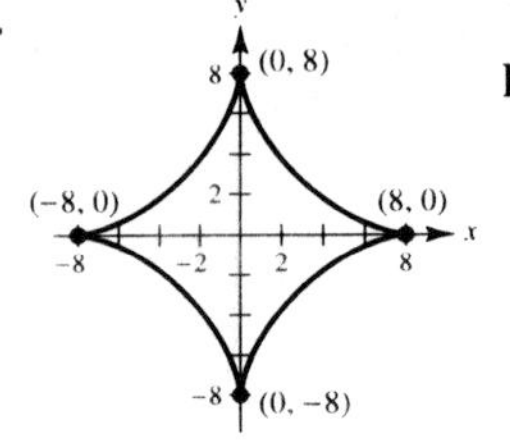   (b) $\displaystyle \lim_{x \to \infty} h(x) = 1$
(c) No

123. Putnam Problem A1, 1956

Section 8.8   (page 587)

1. Improper; $0 \le \frac{3}{5} \le 1$ **3.** Not improper; continuous on $[0, 1]$
5. Not improper; continuous on $[0, 2]$
7. Improper; infinite limits of integration
9. Infinite discontinuity at $x = 0$; 4
11. Infinite discontinuity at $x = 1$; diverges
13. Infinite limit of integration; $\frac{1}{4}$
15. Infinite discontinuity at $x = 0$; diverges
17. Infinite limit of integration; converges to 1   **19.** $\frac{1}{2}$
21. Diverges   **23.** Diverges   **25.** 2   **27.** $\frac{1}{2}$
29. $1/[2(\ln 4)^2]$ **31.** π **33.** $\pi/4$ **35.** Diverges
37. Diverges   **39.** 6   **41.** $-\frac{1}{4}$ **43.** Diverges   **45.** $\pi/3$
47. $\ln(2 + \sqrt{3})$ **49.** 0   **51.** $\pi/6$ **53.** $2\pi\sqrt{6}/3$
55. $p > 1$ **57.** Proof   **59.** Diverges   **61.** Converges
63. Converges   **65.** Diverges   **67.** Diverges   **69.** Converges
71. An integral with infinite integration limits, an integral with an infinite discontinuity at or between the integration limits
73. The improper integral diverges.   **75.** e **77.** π
79. (a) 1   (b) $\pi/2$ (c) 2π
81. 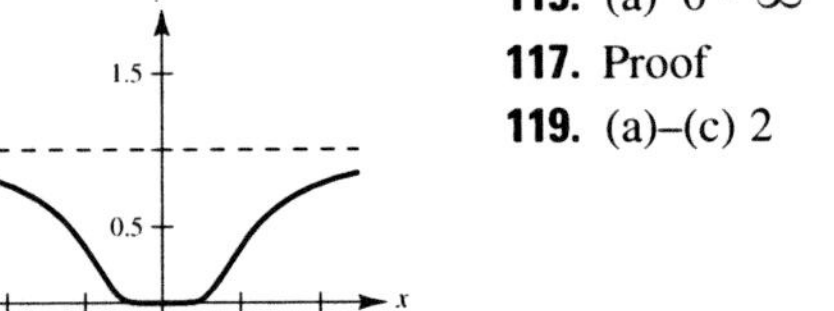

Perimeter $= 48$

83. $8\pi^2$ **85.** (a) $W = 20{,}000$ mile-tons   (b) 4000 mi
87. (a) Proof   (b) $P = 43.53\%$ (c) $E(x) = 7$
89. (a) \$757,992.41   (b) \$837,995.15   (c) \$1,066,666.67
91. $P = \left[2\pi NI\left(\sqrt{r^2 + c^2} - c\right)\right]/\left(kr\sqrt{r^2 + c^2}\right)$
93. False. Let $f(x) = 1/(x + 1)$.   **95.** True
97. (a) and (b) Proofs

(c) The definition of the improper integral $\displaystyle\int_{-\infty}^{\infty}$ is not $\displaystyle\lim_{a\to\infty}\int_{-a}^{a}$ but rather if you rewrite the integral that diverges, you can find that the integral converges.

99. (a) $\displaystyle\int_{1}^{\infty}\frac{1}{x^n}\,dx$ will converge if $n > 1$ and diverge if $n \le 1$.

(b) 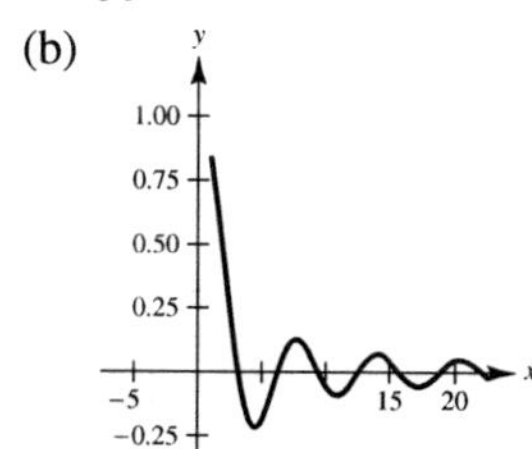   (c) Converges

101. (a) $\Gamma(1) = 1, \Gamma(2) = 1, \Gamma(3) = 2$ (b) Proof
(c) $\Gamma(n) = (n - 1)!$
103. $1/s, \; s > 0$ **105.** $2/s^3, \; s > 0$ **107.** $s/(s^2 + a^2), \; s > 0$
109. $s/(s^2 - a^2), \; s > |a|$
111. (a) 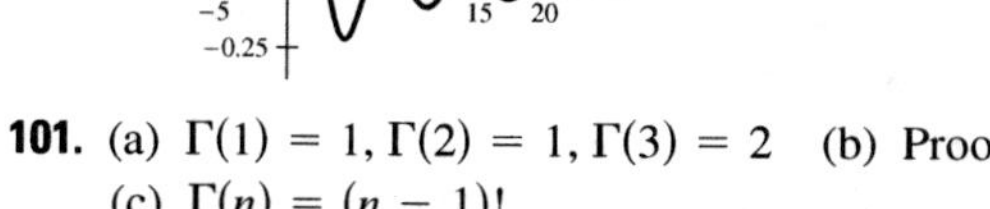   (b) About 0.2525
(c) 0.2525; same by symmetry

113. $c = 1; \ln(2)$
115. $8\pi[(\ln 2)^2/3 - (\ln 4)/9 + 2/27] \approx 2.01545$
117. $\displaystyle\int_{0}^{1} 2\sin(u^2)\,du; \; 0.6278$
119. (a) 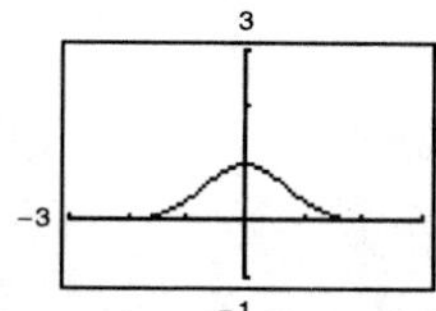   (b) Proof

Review Exercises for Chapter 8   (page 591)

1. $\frac{1}{3}(x^2 - 36)^{3/2} + C$ **3.** $\frac{1}{2}\ln|x^2 - 49| + C$
5. $\ln(2) + \frac{1}{2} \approx 1.1931$ **7.** $100\arcsin(x/10) + C$
9. $\frac{1}{9}e^{3x}(3x - 1) + C$
11. $\frac{1}{13}e^{2x}(2\sin 3x - 3\cos 3x) + C$
13. $\frac{2}{15}(x - 1)^{3/2}(3x + 2) + C$
15. $-\frac{1}{2}x^2\cos 2x + \frac{1}{2}x\sin 2x + \frac{1}{4}\cos 2x + C$
17. $\frac{1}{16}\left[(8x^2 - 1)\arcsin 2x + 2x\sqrt{1 - 4x^2}\right] + C$
19. $\sin(\pi x - 1)[\cos^2(\pi x - 1) + 2]/(3\pi) + C$
21. $\frac{2}{3}[\tan^3(x/2) + 3\tan(x/2)] + C$ **23.** $\tan\theta + \sec\theta + C$
25. $3\pi/16 + \frac{1}{2} \approx 1.0890$ **27.** $3\sqrt{4 - x^2}/x + C$
29. $\frac{1}{3}(x^2 + 4)^{1/2}(x^2 - 8) + C$ **31.** π

33. (a), (b), and (c) $\frac{1}{3}\sqrt{4 + x^2}(x^2 - 8) + C$
35. $6\ln|x + 3| - 5\ln|x - 4| + C$
37. $\frac{1}{4}[6\ln|x - 1| - \ln(x^2 + 1) + 6\arctan x] + C$
39. $x - \frac{64}{11}\ln|x + 8| + \frac{9}{11}\ln|x - 3| + C$
41. $\frac{1}{25}[4/(4 + 5x) + \ln|4 + 5x|] + C$ **43.** $1 - \sqrt{2}/2$
45. $\frac{1}{2}\ln|x^2 + 4x + 8| - \arctan[(x + 2)/2] + C$
47. $\ln|\tan \pi x|/\pi + C$ **49.** Proof
51. $\frac{1}{8}(\sin 2\theta - 2\theta\cos 2\theta) + C$
53. $\frac{4}{3}[x^{3/4} - 3x^{1/4} + 3\arctan(x^{1/4})] + C$
55. $2\sqrt{1 - \cos x} + C$ **57.** $\sin x\ln(\sin x) - \sin x + C$
59. $\frac{5}{2}\ln|(x - 5)/(x + 5)| + C$
61. $y = x\ln|x^2 + x| - 2x + \ln|x + 1| + C$ **63.** $\frac{1}{5}$
65. $\frac{1}{2}(\ln 4)^2 \approx 0.961$ **67.** π **69.** $\frac{128}{15}$
71. $(\bar{x}, \bar{y}) = (0, 4/(3\pi))$ **73.** 3.82   **75.** 0   **77.** ∞ **79.** 1
81. $1000e^{0.09} \approx 1094.17$ **83.** Converges; $\frac{32}{3}$ **85.** Diverges
87. Converges; 1   **89.** Converges; $\pi/4$
91. (a) \$6,321,205.59   (b) \$10,000,000
93. (a) 0.4581   (b) 0.0135

P.S. Problem Solving   (page 593)

1. (a) $\frac{4}{3}, \frac{16}{15}$ (b) Proof   **3.** $\ln 3$ **5.** Proof
7. (a) 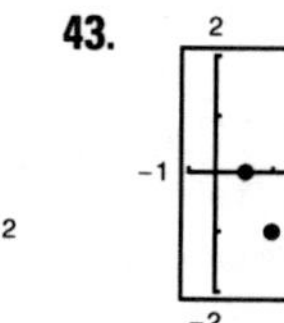   (b) $\ln 3 - \frac{4}{5}$
(c) $\ln 3 - \frac{4}{5}$

Area ≈ 0.2986

9. $\ln 3 - \frac{1}{2} \approx 0.5986$ **11.** Proof   **13.** About 0.8670
15. (a) ∞ (b) 0   (c) $-\frac{2}{3}$
The form $0 \cdot \infty$ is indeterminant.
17. $\dfrac{1/12}{x} + \dfrac{1/42}{x - 3} + \dfrac{1/10}{x - 1} + \dfrac{111/140}{x + 4}$
19. Proof   **21.** About 0.0158

Chapter 9
Section 9.1   (page 604)

1. $3, 9, 27, 81, 243$ **3.** $-\frac{1}{4}, \frac{1}{16}, -\frac{1}{64}, \frac{1}{256}, -\frac{1}{1024}$
5. $1, 0, -1, 0, 1$ **7.** $-1, -\frac{1}{4}, \frac{1}{9}, \frac{1}{16}, -\frac{1}{25}$ **9.** $5, \frac{19}{4}, \frac{43}{9}, \frac{77}{16}, \frac{121}{25}$
11. $3, 4, 6, 10, 18$ **13.** $32, 16, 8, 4, 2$
15. c   **16.** a   **17.** d   **18.** b   **19.** b   **20.** c
21. a   **22.** d   **23.** 14, 17; add 3 to preceding term
25. 80, 160; multiply preceding term by 2.
27. $\frac{3}{16}, -\frac{3}{32}$; multiply preceding term by $-\frac{1}{2}$
29. $11 \cdot 10 \cdot 9 = 990$ **31.** $n + 1$ **33.** $1/[(2n + 1)(2n)]$
35. 5   **37.** 2   **39.** 0
41.

43.

Converges to 1   Diverges

45. Converges to -1 **47.** Converges to 0
49. Diverges   **51.** Converges to $\frac{3}{2}$ **53.** Converges to 0
55. Converges to 0   **57.** Converges to 0   **59.** Converges to 0
61. Diverges   **63.** Converges to 0   **65.** Converges to 0
67. Converges to 1   **69.** Converges to e^k **71.** Converges to 0
73. Answers will vary. Sample answer: $3n - 2$
75. Answers will vary. Sample answer: $n^2 - 2$
77. Answers will vary. Sample answer: $(n + 1)/(n + 2)$
79. Answers will vary. Sample answer: $(n + 1)/n$
81. Answers will vary. Sample answer: $n/[(n + 1)(n + 2)]$
83. Answers will vary. Sample answer:
$$\frac{(-1)^{n-1}}{1 \cdot 3 \cdot 5 \cdots (2n - 1)} = \frac{(-1)^{n-1}2^n n!}{(2n)!}$$
85. Answers will vary. Sample answer: $(2n)!$
87. Monotonic, bounded   **89.** Monotonic, bounded
91. Not monotonic, bounded   **93.** Monotonic, bounded
95. Not monotonic, bounded   **97.** Not monotonic, bounded
99. (a) $\left|5 + \dfrac{1}{n}\right| \le 6 \Rightarrow$ bounded   (b)

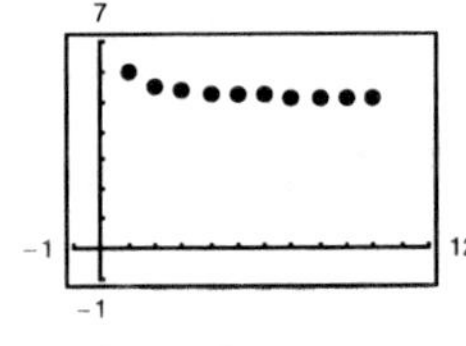

$a_n > a_{n+1} \Rightarrow$ monotonic

So, $\{a_n\}$ converges.

Limit $= 5$

101. (a) $\left|\dfrac{1}{3}\left(1 - \dfrac{1}{3^n}\right)\right| < \dfrac{1}{3} \Rightarrow$ bounded

$a_n < a_{n+1} \Rightarrow$ monotonic

So, $\{a_n\}$ converges.

(b)

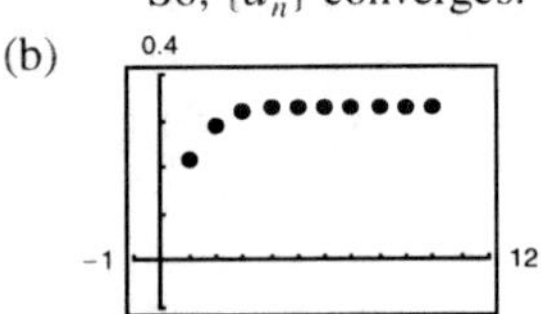

Limit $= \frac{1}{3}$

103. $\{a_n\}$ has a limit because it is bounded and monotonic; since $2 \le a_n \le 4, 2 \le L \le 4$.
105. (a) No; $\lim\limits_{n \to \infty} A_n$ does not exist.

(b)

n	1	2	3	4
A_n	\$10,045.83	\$10,091.88	\$10,138.13	\$10,184.60

n	5	6	7
A_n	\$10,231.28	\$10,278.17	\$10,325.28

n	8	9	10
A_n	\$10,372.60	\$10,420.14	\$10,467.90

107. No. A sequence is said to converge when its terms approach a real number.
109. The graph on the left represents a sequence with alternating signs because the terms alternate from being above the x-axis to being below the x-axis.

111. (a) $\$4,500,000,000(0.8)^n$

(b)

Year	1	2
Budget	\$3,600,000,000	\$2,880,000,000

Year	3	4
Budget	\$2,304,000,000	\$1,843,200,000

(c) Converges to 0
113. (a) $a_n = -5.364n^2 + 608.04n + 4998.3$

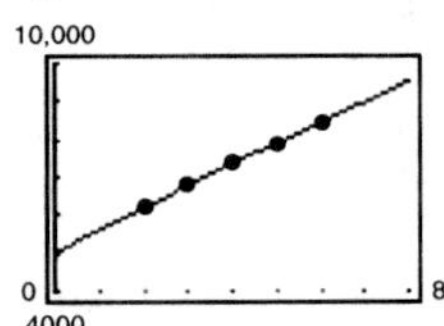

(b) \$11,522.4 billion
115. (a) $a_9 = a_{10} = 1,562,500/567$ (b) Decreasing
(c) Factorials increase more rapidly than exponentials.
117. 1, 1.4142, 1.4422, 1.4142, 1.3797, 1.3480; Converges to 1
119. True   **121.** True   **123.** True
125. (a) 1, 1, 2, 3, 5, 8, 13, 21, 34, 55, 89, 144
(b) 1, 2, 1.5, 1.6667, 1.6, 1.6250, 1.6154, 1.6190, 1.6176, 1.6182
(c) Proof   (d) $\rho = \left(1 + \sqrt{5}\right)/2 \approx 1.6180$
127. (a) 1.4142, 1.8478, 1.9616, 1.9904, 1.9976
(b) $a_n = \sqrt{2 + a_{n-1}}$ (c) $\lim\limits_{n \to \infty} a_n = 2$
129. (a) Proof   (b) Proof   (c) $\lim\limits_{n \to \infty} a_n = \left(1 + \sqrt{1 + 4k}\right)/2$
131. (a) Proof   (b) Proof
133. (a) Proof
(b)

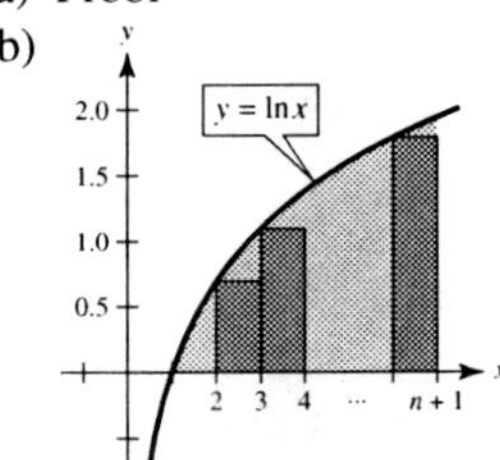

(c) Proof   (d) Proof
(e) $\dfrac{\sqrt[20]{20!}}{20} \approx 0.4152;$
$\dfrac{\sqrt[50]{50!}}{50} \approx 0.3897;$
$\dfrac{\sqrt[100]{100!}}{100} \approx 0.3799$

135. Proof
137. Answers will vary. Sample answer: $a_n = (-1)^n$
139. Proof   **141.** Putnam Problem A1, 1990

Section 9.2   (page 614)

1. 1, 1.25, 1.361, 1.424, 1.464
3. 3, -1.5, 5.25, -4.875, 10.3125
5. 3, 4.5, 5.25, 5.625, 5.8125
7. $\{a_n\}$ converges, $\Sigma\, a_n$ diverges
9. Geometric series: $r = \frac{7}{6} > 1$
11. Geometric series: $r = 1.055 > 1$ **13.** $\lim\limits_{n \to \infty} a_n = 1 \ne 0$
15. $\lim\limits_{n \to \infty} a_n = 1 \ne 0$ **17.** $\lim\limits_{n \to \infty} a_n = \frac{1}{2} \ne 0$ **19.** c; 3
20. b; 3   **21.** a; 3   **22.** d; 3   **23.** f; $\frac{34}{9}$ **24.** e; $\frac{5}{3}$
25. Geometric series: $r = \frac{5}{6} < 1$
27. Geometric series: $r = 0.9 < 1$

29. Telescoping series: $a_n = 1/n - 1/(n+1)$; Converges to 1.

31. (a) $\frac{11}{3}$

(b)

n	5	10	20	50	100
S_n	2.7976	3.1643	3.3936	3.5513	3.6078

(c) 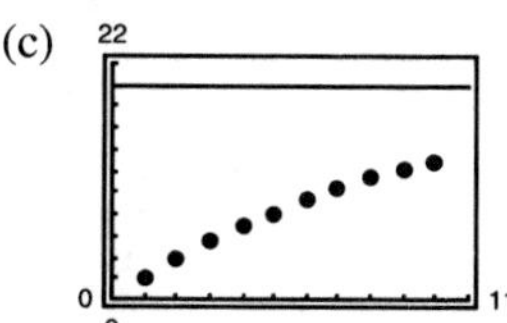

(d) The terms of the series decrease in magnitude relatively slowly, and the sequence of partial sums approaches the sum of the series relatively slowly.

33. (a) 20

(b)

n	5	10	20	50	100
S_n	8.1902	13.0264	17.5685	19.8969	19.9995

(c) 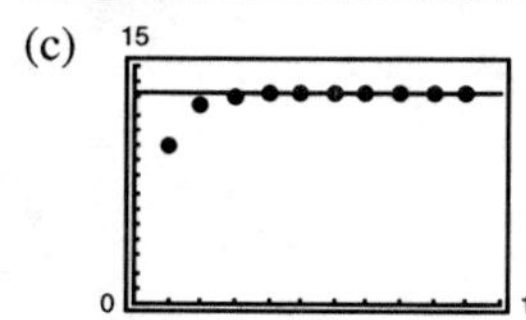

(d) The terms of the series decrease in magnitude relatively slowly, and the sequence of partial sums approaches the sum of the series relatively slowly.

35. (a) $\frac{40}{3}$

(b)

n	5	10	20	50	100
S_n	13.3203	13.3333	13.3333	13.3333	13.3333

(c)

(d) The terms of the series decrease in magnitude relatively rapidly, and the sequence of partial sums approaches the sum of the series relatively rapidly.

37. 2   **39.** $\frac{3}{4}$ **41.** $\frac{3}{4}$ **43.** 4   **45.** $\frac{10}{9}$ **47.** $\frac{9}{4}$ **49.** $\frac{1}{2}$

51. $\dfrac{\sin(1)}{1 - \sin(1)}$ **53.** (a) $\displaystyle\sum_{n=0}^{\infty} \frac{4}{10}(0.1)^n$ (b) $\dfrac{4}{9}$

55. (a) $\displaystyle\sum_{n=0}^{\infty} \frac{81}{100}(0.01)^n$ (b) $\dfrac{9}{11}$

57. (a) $\displaystyle\sum_{n=0}^{\infty} \frac{3}{40}(0.01)^n$ (b) $\dfrac{5}{66}$ **59.** Diverges   **61.** Diverges

63. Converges   **65.** Converges   **67.** Diverges

69. Converges   **71.** Diverges   **73.** Diverges   **75.** Diverges

77. See definitions on page 608.

79. The series given by

$$\sum_{n=0}^{\infty} ar^n = a + ar + ar^2 + \cdots + ar^n + \cdots, a \neq 0$$

is a geometric series with ratio r. When $0 < |r| < 1$, the series converges to the sum $\displaystyle\sum_{n=0}^{\infty} ar^n = \frac{a}{1 - r}$.

81. The series in (a) and (b) are the same. The series in (c) is different unless $a_1 = a_2 = \cdots = a$ is constant.

83. $-2 < x < 2$; $x/(2 - x)$ **85.** $0 < x < 2$; $(x - 1)/(2 - x)$

87. $-1 < x < 1$; $1/(1 + x)$

89. x: $(-\infty, -1) \cup (1, \infty)$; $x/(x - 1)$ **91.** $c = (\sqrt{3} - 1)/2$

93. Neither statement is true. The formula is valid for $-1 < x < 1$.

95. (a) x (b) $f(x) = 1/(1 - x)$, $|x| < 1$

(c) 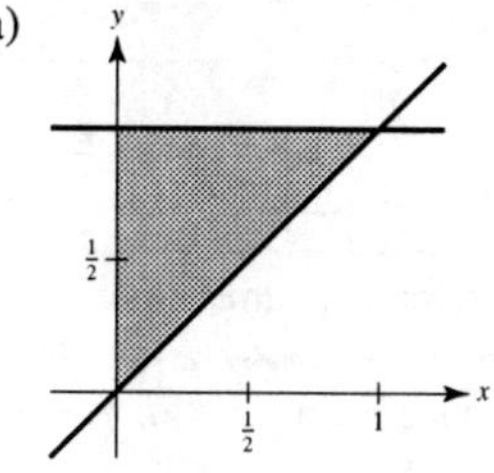

Answers will vary.

97. 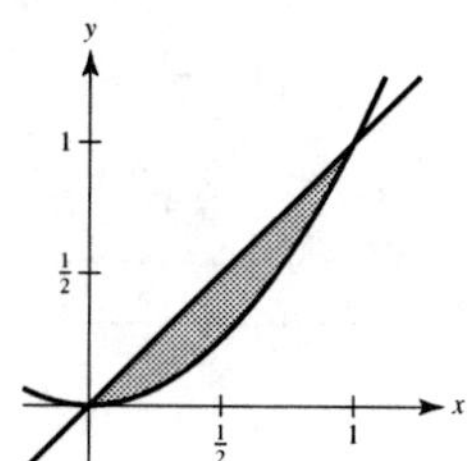

Horizontal asymptote: $y = 6$
The horizontal asymptote is the sum of the series.

99. The required terms for the two series are $n = 100$ and $n = 5$, respectively. The second series converges at a higher rate.

101. $160{,}000(1 - 0.95^n)$ units

103. $\displaystyle\sum_{i=0}^{\infty} 200(0.75)^i$; Sum = \$800 million

105. 152.42 feet   **107.** $\dfrac{1}{8}$; $\displaystyle\sum_{n=0}^{\infty} \frac{1}{2}\left(\frac{1}{2}\right)^n = \frac{1/2}{1 - 1/2} = 1$

109. (a) $-1 + \displaystyle\sum_{n=0}^{\infty}\left(\frac{1}{2}\right)^n = -1 + \frac{a}{1 - r} = -1 + \frac{1}{1 - 1/2} = 1$

(b) No   (c) 2

111. (a) 126 in.^{2} (b) 128 in.^{2}

113. The \$2,000,000 sweepstakes has a present value of \$1,146,992.12. After accruing interest over the 20-year period, it attains its full value.

115. (a) \$5,368,709.11   (b) \$10,737,418.23   (c) \$21,474,836.47

117. (a) \$14,773.59   (b) \$14,779.65

119. (a) \$91,373.09   (b) \$91,503.32   **121.** \$4,751,275.79

123. False. $\displaystyle\lim_{n\to\infty} \frac{1}{n} = 0$, but $\displaystyle\sum_{n=1}^{\infty} \frac{1}{n}$ diverges.

125. False. $\displaystyle\sum_{n=1}^{\infty} ar^n = \left(\frac{a}{1 - r}\right) - a$ The formula requires that the geometric series begins with $n = 0$.

127. True   **129.** Proof

131. Answers will vary. Example: $\displaystyle\sum_{n=0}^{\infty} 1$, $\displaystyle\sum_{n=0}^{\infty}(-1)$

133–137. Proofs

139. (a) 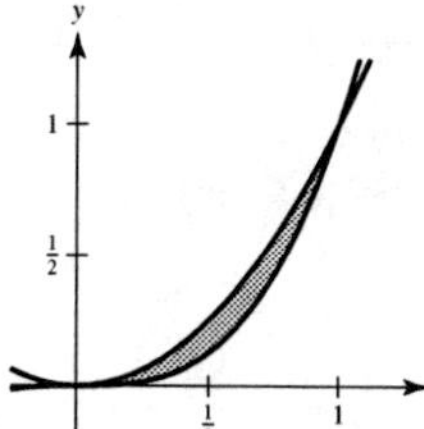

(b) $\displaystyle\int_0^1 (1 - x)\, dx = \frac{1}{2}$

$\displaystyle\int_0^1 (x - x^2)\, dx = \frac{1}{6}$

$\displaystyle\int_0^1 (x^2 - x^3)\, dx = \frac{1}{12}$

(c) $a_n = \dfrac{1}{n} - \dfrac{1}{n + 1}$ and $\displaystyle\sum_{n=1}^{\infty} a_n = 1$; The sum of all the shaded regions is the area of the square, 1.

141. H = half-life of the drug

n = number of equal doses

P = number of units of the drug

t = equal time intervals

The total amount of the drug in the patient's system at the time the last dose is given is

$T_n = P + Pe^{kt} + Pe^{2kt} + \cdots + Pe^{(n-1)kt}$

where $k = -(\ln 2)/H$. One time interval after the last dose is administered is given by

$T_{n+1} = Pe^{kt} + Pe^{2kt} + Pe^{3kt} + \cdots + Pe^{nkt}$

and so on. Because $k < 0$, $T_{n+s} \to 0$ as $s \to \infty$.

143. Putnam Problem A1, 1966

Section 9.3   (page 622)

1. Diverges   **3.** Converges   **5.** Converges   **7.** Converges

9. Diverges   **11.** Diverges   **13.** Diverges   **15.** Converges

17. Converges   **19.** Converges   **21.** Diverges

23. Diverges   **25.** Diverges   **27.** $f(x)$ is not positive for $x \geq 1$.

29. $f(x)$ is not always decreasing.   **31.** Converges   **33.** Diverges

35. Diverges   **37.** Diverges   **39.** Converges   **41.** Converges

43. c; diverges   **44.** f; diverges   **45.** b; converges

46. a; diverges   **47.** d; converges   **48.** e; converges

49. (a)

n	5	10	20	50	100
S_n	3.7488	3.75	3.75	3.75	3.75

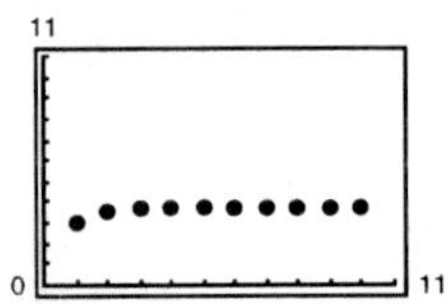

The partial sums approach the sum 3.75 very quickly.

(b)

n	5	10	20	50	100
S_n	1.4636	1.5498	1.5962	1.6251	1.635

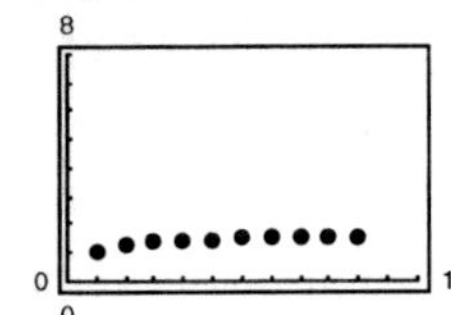

The partial sums approach the sum $\pi^2/6 \approx 1.6449$ more slowly than the series in part (a).

51. See Theorem 9.10 on page 619. Answers will vary. For example, convergence or divergence can be determined for the series $\displaystyle\sum_{n=1}^{\infty} \frac{1}{n^2 + 1}$.

53. No. Because $\displaystyle\sum_{n=1}^{\infty} \frac{1}{n}$ diverges, $\displaystyle\sum_{n=10,000}^{\infty} \frac{1}{n}$ also diverges. The convergence or divergence of a series is not determined by the first finite number of terms of the series.

55.

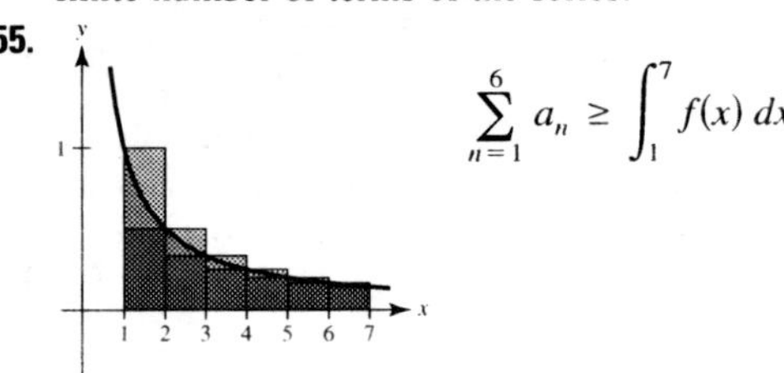

$$\sum_{n=1}^{6} a_n \geq \int_1^7 f(x)\, dx \geq \sum_{n=2}^{7} a_n$$

57. $p > 1$ **59.** $p > 1$ **61.** $p > 1$ **63.** Diverges

65. Converges   **67.** Proof

69. $S_6 \approx 1.0811$ **71.** $S_{10} \approx 0.9818$ **73.** $S_4 \approx 0.4049$

$R_6 \approx 0.0015$; $R_{10} \approx 0.0997$; $R_4 \approx 5.6 \times 10^{-8}$

75. $N \geq 7$ **77.** $N \geq 2$ **79.** $N \geq 1000$

81. (a) $\displaystyle\sum_{n=2}^{\infty} \frac{1}{n^{1.1}}$ converges by the p-Series Test because $1.1 > 1$.

$\displaystyle\sum_{n=2}^{\infty} \frac{1}{n \ln n}$ diverges by the Integral Test because $\displaystyle\int_2^{\infty} \frac{1}{x \ln x}\, dx$ diverges.

(b) $\displaystyle\sum_{n=2}^{\infty} \frac{1}{n^{1.1}} = 0.4665 + 0.2987 + 0.2176 + 0.1703$

$+\, 0.1393 + \cdots$

$\displaystyle\sum_{n=2}^{\infty} \frac{1}{n \ln n} = 0.7213 + 0.3034 + 0.1803 + 0.1243$

$+\, 0.0930 + \cdots$

(c) $n \geq 3.431 \times 10^{15}$

83. (a) Let $f(x) = 1/x$. f is positive, continuous, and decreasing on $[1, \infty)$.

$S_n - 1 \leq \displaystyle\int_1^n \frac{1}{x}\, dx = \ln n$

$S_n \geq \displaystyle\int_1^{n+1} \frac{1}{x}\, dx = \ln(n + 1)$

So, $\ln(n + 1) \leq S_n \leq 1 + \ln n$.

(b) $\ln(n + 1) - \ln n \leq S_n - \ln n \leq 1$.

Also, $\ln(n + 1) - \ln n > 0$ for $n \geq 1$. So, $0 \leq S_n - \ln n \leq 1$, and the sequence $\{a_n\}$ is bounded.

(c) $a_n - a_{n+1} = [S_n - \ln n] - [S_{n+1} - \ln(n + 1)]$

$= \displaystyle\int_n^{n+1} \frac{1}{x}\, dx - \frac{1}{n + 1} \geq 0$

So, $a_n \geq a_{n+1}$.

(d) Because the sequence is bounded and monotonic, it converges to a limit, γ.

(e) 0.5822

85. (a) Diverges   (b) Diverges

(c) $\displaystyle\sum_{n=2}^{\infty} x^{\ln n}$ converges for $x < 1/e$.

87. Diverges   **89.** Converges   **91.** Converges   **93.** Diverges

95. Diverges   **97.** Converges

Section 9.4   (page 630)

1. (a)

(b) $\displaystyle\sum_{n=1}^{\infty} \frac{6}{n^{3/2}}$; Converges

(c) The magnitudes of the terms are less than the magnitudes of the terms of the p-series. Therefore, the series converges.

(d) The smaller the magnitudes of the terms, the smaller the magnitudes of the terms of the sequence of partial sums.

3. Converges   **5.** Diverges   **7.** Converges   **9.** Diverges
11. Converges   **13.** Converges   **15.** Diverges   **17.** Diverges
19. Converges   **21.** Converges   **23.** Converges
25. Diverges   **27.** Diverges   **29.** Diverges; p-Series Test

31. Converges; Direct Comparison Test with $\displaystyle\sum_{n=1}^{\infty} \left(\frac{1}{5}\right)^{n}$

33. Diverges; nth-Term Test · **35.** Converges; Integral Test

37. $\displaystyle\lim_{n\to\infty} \frac{a_n}{1/n} = \lim_{n\to\infty} na_n$

$\displaystyle\lim_{n\to\infty} na_n \neq 0$, but is finite.

The series diverges by the Limit Comparison Test.

39. Diverges   **41.** Converges

43. $\displaystyle\lim_{n\to\infty} n\left(\frac{n^3}{5n^4 + 3}\right) = \frac{1}{5} \neq 0$

So, $\displaystyle\sum_{n=1}^{\infty} \frac{n^3}{5n^4 + 3}$ diverges.

45. Diverges   **47.** Converges

49. Convergence or divergence is dependent on the form of the general term for the series and not necessarily on the magnitudes of the terms.

51. See Theorem 9.13 on page 628. Answers will vary. For example,

$\displaystyle\sum_{n=2}^{\infty} \frac{1}{\sqrt{n-1}}$ diverges because $\displaystyle\lim_{n\to\infty} \frac{1/\sqrt{n-1}}{1/\sqrt{n}} = 1$ and

$\displaystyle\sum_{n=2}^{\infty} \frac{1}{\sqrt{n}}$ diverges (p-series).

53. (a) Proof

(b)

n	5	10	20	50	100
S_n	1.1839	1.2087	1.2212	1.2287	1.2312

(c) 0.1226   (d) 0.0277

55. False. Let $a_n = 1/n^3$ and $b_n = 1/n^2$.

57. True   **59.** True   **61.** Proof   **63.** $\displaystyle\sum_{n=1}^{\infty} \frac{1}{n^2}, \sum_{n=1}^{\infty} \frac{1}{n^3}$

65–71. Proofs   **73.** Putnam Problem B4, 1988

Section 9.5   (page 638)

1. d   **2.** f   **3.** a   **4.** b   **5.** e   **6.** c

7. (a)

n	1	2	3	4	5
S_n	1.0000	0.6667	0.8667	0.7238	0.8349

n	6	7	8	9	10
S_n	0.7440	0.8209	0.7543	0.8131	0.7605

(b)

(c) The points alternate sides of the horizontal line $y = \pi/4$ that represents the sum of the series. The distances between the successive points and the line decrease.

(d) The distance in part (c) is always less than the magnitude of the next term of the series.

9. (a)

n	1	2	3	4	5
S_n	1.0000	0.7500	0.8611	0.7986	0.8386

n	6	7	8	9	10
S_n	0.8108	0.8312	0.8156	0.8280	0.8180

(b)

(c) The points alternate sides of the horizontal line $y = \pi^2/12$ that represents the sum of the series. The distances between the successive points and the line decrease.

(d) The distance in part (c) is always less than the magnitude of the next term of the series.

11. Converges   **13.** Converges   **15.** Diverges
17. Converges   **19.** Diverges   **21.** Converges   **23.** Diverges
25. Diverges   **27.** Diverges   **29.** Converges
31. Converges   **33.** Converges   **35.** Converges
37. $0.7305 \leq S \leq 0.7361$ **39.** $2.3713 \leq S \leq 2.4937$
41. (a) 7 terms (Note that the sum begins with $n = 0$.)   (b) 0.368
43. (a) 3 terms (Note that the sum begins with $n = 0$.)   (b) 0.842
45. (a) 1000 terms   (b) 0.693   **47.** 10   **49.** 7
51. Converges absolutely   **53.** Converges absolutely
55. Converges absolutely   **57.** Converges conditionally
59. Diverges   **61.** Converges conditionally
63. Converges absolutely   **65.** Converges absolutely
67. Converges conditionally   **69.** Converges absolutely
71. An alternating series is a series whose terms alternate in sign.
73. $|S - S_N| = |R_N| \leq a_{N+1}$
75. Graph (b). The partial sums alternate above and below the horizontal line representing the sum.
77. True   **79.** $p > 0$
81. Proof; The converse is false. For example: Let $a_n = 1/n$.
83. $\displaystyle\sum_{n=1}^{\infty} \frac{1}{n^2}$ converges, hence so does $\displaystyle\sum_{n=1}^{\infty} \frac{1}{n^4}$.
85. (a) No; $a_{n+1} \leq a_n$ is not satisfied for all n. For example, $\frac{1}{9} < \frac{1}{8}$.
     (b) Yes; 0.5
87. Converges; p-Series Test   **89.** Diverges; nth-Term Test
91. Converges; Geometric Series Test

93. Converges; Integral Test
95. Converges; Alternating Series Test
97. The first term of the series is 0, not 1. You cannot regroup series terms arbitrarily.
99. Putnam Problem 2, afternoon session, 1954

Section 9.6   (page 647)

1–3. Proofs   **5.** d   **6.** c   **7.** f   **8.** b   **9.** a   **10.** e
11. (a) Proof
(b)

n	5	10	15	20	25
S_n	9.2104	16.7598	18.8016	19.1878	19.2491

(c) 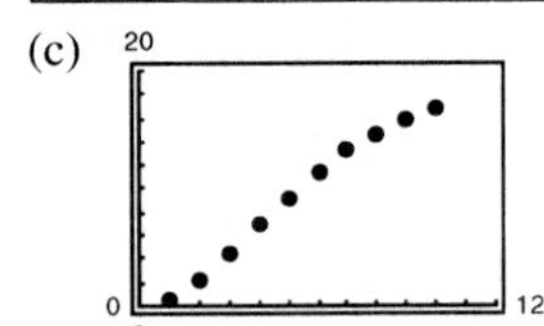   (d) 19.26

(e) The more rapidly the terms of the series approach 0, the more rapidly the sequence of partial sums approaches the sum of the series.

13. Converges   **15.** Diverges   **17.** Diverges
19. Converges   **21.** Diverges   **23.** Converges
25. Diverges   **27.** Converges   **29.** Converges
31. Diverges   **33.** Converges   **35.** Converges
37. Converges   **39.** Diverges   **41.** Converges
43. Diverges   **45.** Converges   **47.** Converges
49. Converges   **51.** Converges; Alternating Series Test
53. Converges; p-Series Test   **55.** Diverges; nth-Term Test
57. Diverges; Geometric Series Test
59. Converges; Limit Comparison Test with $b_n = 1/2^n$
61. Converges; Direct Comparison Test with $b_n = 1/3^n$
63. Converges; Ratio Test   **65.** Converges; Ratio Test
67. Converges; Ratio Test   **69.** a and c   **71.** a and b
73. $\sum_{n=0}^{\infty} \dfrac{n+1}{7^{n+1}}$ **75.** (a) 9   (b) -0.7769

77. Diverges; $\lim\limits_{n\to\infty} \left| \dfrac{a_{n+1}}{a_n} \right| > 1$

79. Converges; $\lim\limits_{n\to\infty} \left| \dfrac{a_{n+1}}{a_n} \right| < 1$

81. Diverges; $\lim a_n \neq 0$ **83.** Converges   **85.** Converges
87. $(-3, 3)$ **89.** $(-2, 0]$ **91.** $x = 0$
93. See Theorem 9.17 on page 641.
95. No; the series $\sum\limits_{n=1}^{\infty} \dfrac{1}{n + 10,000}$ diverges.

97. Absolutely; by Theorem 9.17   **99–105.** Proofs
107. (a) Diverges   (b) Converges   (c) Converges
(d) Converges for all integers $x \geq 2$
109. Answers will vary.
111. Putnam Problem 7, morning session, 1951

Section 9.7   (page 658)

1. d   **2.** c   **3.** a   **4.** b

5. $P_1 = -\frac{1}{2}x + 6$

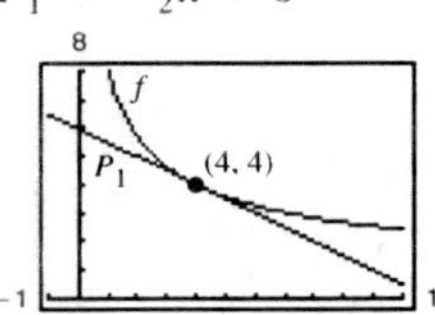

P_1 is the first-degree Taylor polynomial for f at 4.

7. $P_1 = \sqrt{2}\,x + \sqrt{2}(4 - \pi)/4$

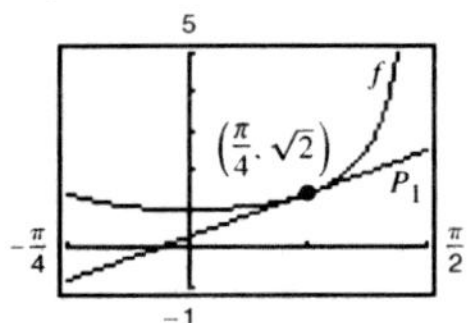

P_1 is the first-degree Taylor polynomial for f at $\pi/4$.

9. 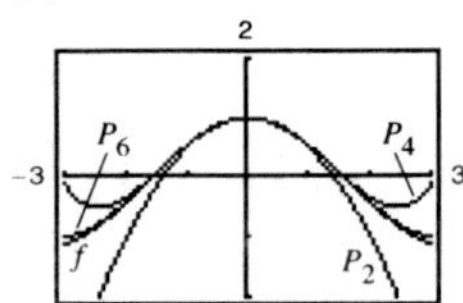

x	0	0.8	0.9	1	1.1
$f(x)$	Error	4.4721	4.2164	4.0000	3.8139
$P_2(x)$	7.5000	4.4600	4.2150	4.0000	3.8150

x	1.2	2
$f(x)$	3.6515	2.8284
$P_2(x)$	3.6600	3.5000

11. (a) 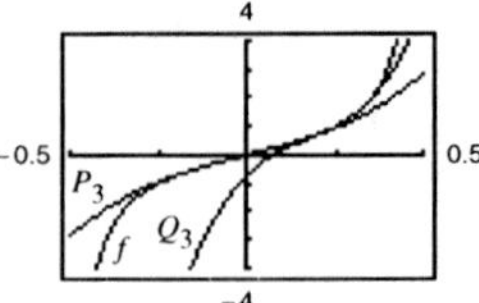

(b) $f^{(2)}(0) = -1$ $P_2^{(2)}(0) = -1$
$f^{(4)}(0) = 1$ $P_4^{(4)}(0) = 1$
$f^{(6)}(0) = -1$ $P_6^{(6)}(0) = -1$

(c) $f^{(n)}(0) = P_n^{(n)}(0)$

13. $1 + 3x + \frac{9}{2}x^2 + \frac{9}{2}x^3 + \frac{27}{8}x^4$
15. $1 - \frac{1}{2}x + \frac{1}{8}x^2 - \frac{1}{48}x^3 + \frac{1}{384}x^4$
17. $x - \frac{1}{6}x^3 + \frac{1}{120}x^5$ **19.** $x + x^2 + \frac{1}{2}x^3 + \frac{1}{6}x^4$
21. $1 - x + x^2 - x^3 + x^4 - x^5$ **23.** $1 + \frac{1}{2}x^2$
25. $2 - 2(x - 1) + 2(x - 1)^2 - 2(x - 1)^3$
27. $2 + \frac{1}{4}(x - 4) - \frac{1}{64}(x - 4)^2 + \frac{1}{512}(x - 4)^3$
29. $\ln 2 + \frac{1}{2}(x - 2) - \frac{1}{8}(x - 2)^2 + \frac{1}{24}(x - 2)^3 - \frac{1}{64}(x - 2)^4$

31. (a) $P_3(x) = \pi x + \dfrac{\pi^3}{3}x^3$

(b)
$$Q_3(x) = 1 + 2\pi\left(x - \frac{1}{4}\right) + 2\pi^2\left(x - \frac{1}{4}\right)^2 + \frac{8\pi^3}{3}\left(x - \frac{1}{4}\right)^3$$

33. (a)

x	0	0.25	0.50	0.75	1.00
$\sin x$	0	0.2474	0.4794	0.6816	0.8415
$P_1(x)$	0	0.25	0.50	0.75	1.00
$P_3(x)$	0	0.2474	0.4792	0.6797	0.8333
$P_5(x)$	0	0.2474	0.4794	0.6817	0.8417

(b) 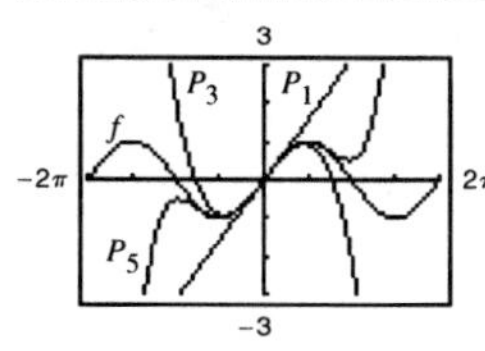

(c) As the distance increases, the polynomial approximation becomes less accurate.

35. (a) $P_3(x) = x + \frac{1}{6}x^3$

(b)

x	-0.75	-0.50	-0.25	0	0.25
$f(x)$	-0.848	-0.524	-0.253	0	0.253
$P_3(x)$	-0.820	-0.521	-0.253	0	0.253

x	0.50	0.75
$f(x)$	0.524	0.848
$P_3(x)$	0.521	0.820

(c) 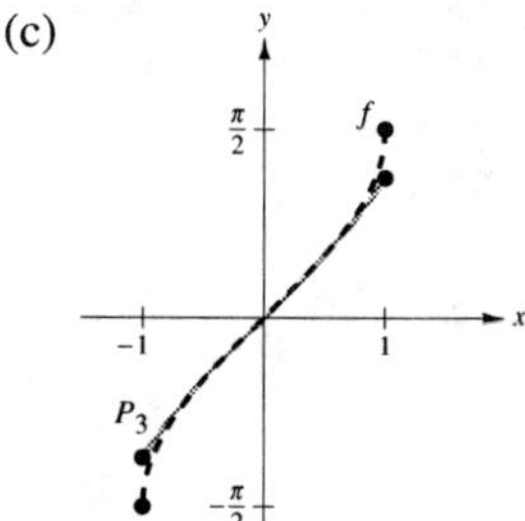

37. 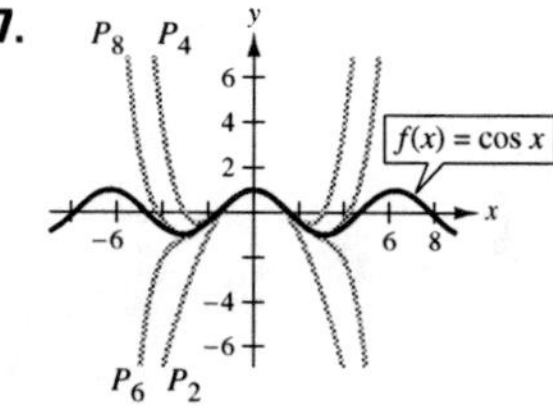

39. 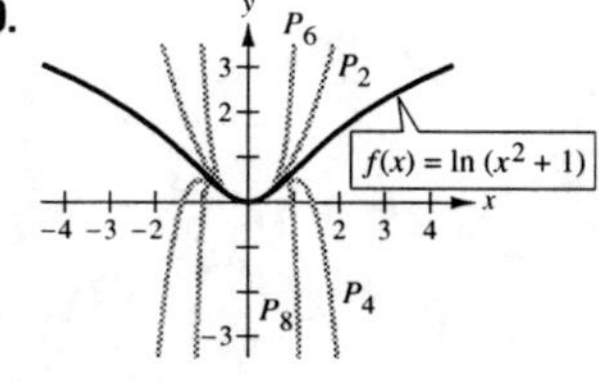

41. 4.3984   **43.** 0.7419   **45.** $R_4 \le 2.03 \times 10^{-5}$; 0.000001

47. $R_3 \le 7.82 \times 10^{-3}$; 0.00085   **49.** 3   **51.** 5

53. $n = 9$; $\ln(1.5) \approx 0.4055$ **55.** $n = 16$; $e^{-\pi(1.3)} \approx 0.01684$

57. $-0.3936 < x < 0$ **59.** $-0.9467 < x < 0.9467$

61. The graph of the approximating polynomial P and the elementary function f both pass through the point $(c, f(c))$, and the slope of the graph of P is the same as the slope of the graph of f at the point $(c, f(c))$. If P is of degree n, then the first n derivatives of f and P agree at c. This allows for the graph of P to resemble the graph of f near the point $(c, f(c))$.

63. See "Definitions of nth Taylor Polynomial and nth Maclaurin Polynomial" on page 652.

65. As the degree of the polynomial increases, the graph of the Taylor polynomial becomes a better and better approximation of the function within the interval of convergence. Therefore, the accuracy is increased.

67. (a) $f(x) \approx P_4(x) = 1 + x + (1/2)x^2 + (1/6)x^3 + (1/24)x^4$
$g(x) \approx Q_5(x) = x + x^2 + (1/2)x^3 + (1/6)x^4 + (1/24)x^5$
$Q_5(x) = xP_4(x)$

(b) $g(x) \approx P_6(x) = x^2 - x^4/3! + x^6/5!$

(c) $g(x) \approx P_4(x) = 1 - x^2/3! + x^4/5!$

69. (a) $Q_2(x) = -1 + (\pi^2/32)(x + 2)^2$

(b) $R_2(x) = -1 + (\pi^2/32)(x - 6)^2$

(c) No. Horizontal translations of the result in part (a) are possible only at $x = -2 + 8n$ (where n is an integer) because the period of f is 8.

71. Proof

73. As you move away from $x = c$, the Taylor polynomial becomes less and less accurate.

Section 9.8   (page 668)

1. 0   **3.** 2   **5.** $R = 1$ **7.** $R = \frac{1}{4}$ **9.** $R = \infty$

11. $(-4, 4)$ **13.** $(-1, 1]$ **15.** $(-\infty, \infty)$ **17.** $x = 0$

19. $(-4, 4)$ **21.** $(-5, 13]$ **23.** $(0, 2]$ **25.** $(0, 6)$

27. $\left(-\frac{1}{2}, \frac{1}{2}\right)$ **29.** $(-\infty, \infty)$ **31.** $(-1, 1)$ **33.** $x = 3$

35. $R = c$ **37.** $(-k, k)$ **39.** $(-1, 1)$

41. $\displaystyle\sum_{n=1}^{\infty} \frac{x^{n-1}}{(n-1)!}$ **43.** $\displaystyle\sum_{n=1}^{\infty} \frac{x^{2n-1}}{(2n-1)!}$

45. (a) $(-3, 3)$ (b) $(-3, 3)$ (c) $(-3, 3)$ (d) $[-3, 3)$

47. (a) $(0, 2]$ (b) $(0, 2)$ (c) $(0, 2)$ (d) $[0, 2]$

49. c; $S_1 = 1$, $S_2 = 1.33$ **50.** a; $S_1 = 1$, $S_2 = 1.67$

51. b; diverges   **52.** d; alternating

53. b   **54.** c   **55.** d   **56.** a

57. A series of the form

$$\sum_{n=0}^{\infty} a_n(x - c)^n = a_0 + a_1(x - c) + a_2(x - c)^2 + \cdots$$
$$+ a_n(x - c)^n + \cdots$$

is called a power series centered at c, where c is a constant.

59. 1. A single point   2. An interval centered at c
3. The entire real line

61. Answers will vary.

63. (a) For $f(x)$: $(-\infty, \infty)$; For $g(x)$: $(-\infty, \infty)$
(b) Proof (c) Proof (d) $f(x) = \sin x$; $g(x) = \cos x$

65–69. Proofs

71. (a) Proof (b) Proof
(c)   (d) 0.92

73.

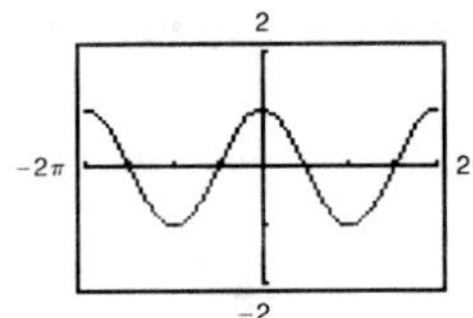

$f(x) = \cos x$

75.

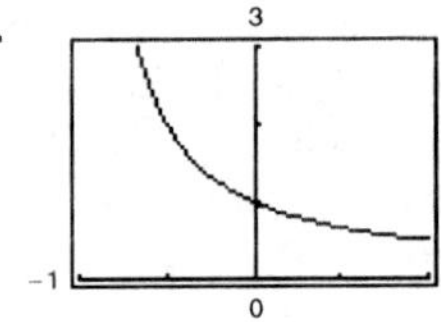

$f(x) = 1/(1 + x)$

77. (a) $\frac{8}{3}$ (b) $\frac{8}{13}$

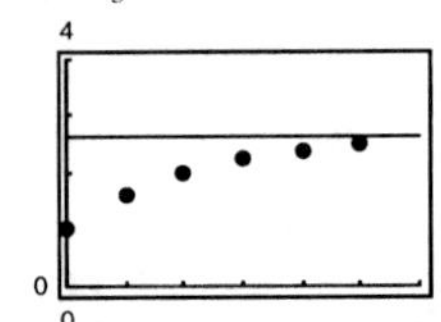

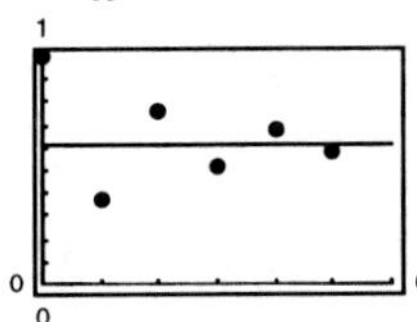

(c) The alternating series converges more rapidly. The partial sums of the series of positive terms approach the sum from below. The partial sums of the alternating series alternate sides of the horizontal line representing the sum.

(d)

M	10	100	1000	10,000
N	5	14	24	35

79. False. Let $a_n = (-1)^n/(n2^n)$ **81.** True **83.** Proof
85. (a) $(-1, 1)$ (b) $f(x) = (c_0 + c_1x + c_2x^2)/(1 - x^3)$
87. Proof

Section 9.9 (page 676)

1. $\sum_{n=0}^{\infty} \frac{x^n}{4^{n+1}}$ **3.** $\sum_{n=0}^{\infty} \frac{3(-1)^n x^n}{4^{n+1}}$ **5.** $\sum_{n=0}^{\infty} \frac{(x-1)^n}{2^{n+1}}$ **7.** $\sum_{n=0}^{\infty} (3x)^n$

$\qquad\qquad\qquad\qquad\qquad\qquad\qquad\quad (-1, 3) \qquad\qquad \left(-\frac{1}{3}, \frac{1}{3}\right)$

9. $-\frac{5}{9} \sum_{n=0}^{\infty} \left[\frac{2}{9}(x+3)\right]^n$ **11.** $\sum_{n=0}^{\infty} \frac{(-1)^n 2^{n+1} x^n}{3^{n+1}}$

$\left(-\frac{15}{2}, \frac{3}{2}\right) \qquad\qquad\qquad \left(-\frac{3}{2}, \frac{3}{2}\right)$

13. $\sum_{n=0}^{\infty} \left[\frac{1}{(-3)^n} - 1\right]x^n$ **15.** $\sum_{n=0}^{\infty} x^n[1 + (-1)^n] = 2\sum_{n=0}^{\infty} x^{2n}$

$(-1, 1) \qquad\qquad\qquad\qquad (-1, 1)$

17. $2\sum_{n=0}^{\infty} x^{2n}$ **19.** $\sum_{n=1}^{\infty} n(-1)^n x^{n-1}$ **21.** $\sum_{n=0}^{\infty} \frac{(-1)^n x^{n+1}}{n+1}$

$(-1, 1) \qquad\quad (-1, 1) \qquad\qquad\quad (-1, 1]$

23. $\sum_{n=0}^{\infty} (-1)^n x^{2n}$ **25.** $\sum_{n=0}^{\infty} (-1)^n (2x)^{2n}$

$(-1, 1) \qquad\qquad\quad \left(-\frac{1}{2}, \frac{1}{2}\right)$

27.

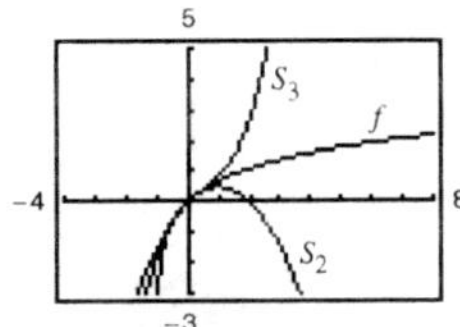

x	0.0	0.2	0.4	0.6	0.8	1.0
S_2	0.000	0.180	0.320	0.420	0.480	0.500
$\ln(x+1)$	0.000	0.182	0.336	0.470	0.588	0.693
S_3	0.000	0.183	0.341	0.492	0.651	0.833

29. (a)

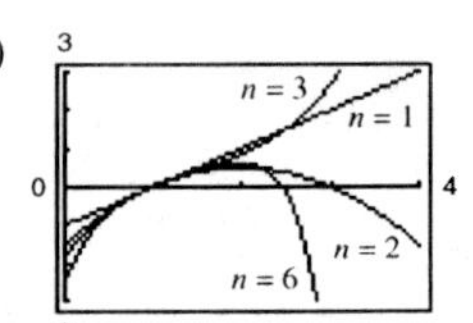

(b) $\ln x, 0 < x \le 2, R = 1$
(c) -0.6931
(d) $\ln(0.5)$; The error is approximately 0.

31. c **32.** d **33.** a **34.** b **35.** 0.245 **37.** 0.125

39. $\sum_{n=1}^{\infty} nx^{n-1}, \ -1 < x < 1$ **41.** $\sum_{n=0}^{\infty} (2n+1)x^n, \ -1 < x < 1$

43. $E(n) = 2$. Because the probability of obtaining a head on a single toss is $\frac{1}{2}$, it is expected that, on average, a head will be obtained in two tosses.

45. Because $\frac{1}{1+x} = \frac{1}{1-(-x)}$, substitute $(-x)$ into the geometric series.

47. Because $\frac{5}{1+x} = 5\left(\frac{1}{1-(-x)}\right)$, substitute $(-x)$ into the geometric series and then multiply the series by 5.

49. Proof **51.** (a) Proof (b) 3.14
53. $\ln \frac{3}{2} \approx 0.4055$; See Exercise 21.
55. $\ln \frac{7}{5} \approx 0.3365$; See Exercise 53.
57. $\arctan \frac{1}{2} \approx 0.4636$; See Exercise 56.
59. $f(x) = \arctan x$ is an odd function (symmetric to the origin).
61. The series in Exercise 56 converges to its sum at a lower rate because its terms approach 0 at a much lower rate.
63. The series converges on the interval $(-5, 3)$ and perhaps also at one or both endpoints.
65. $\sqrt{3}\pi/6$ **67.** $S_1 = 0.3183098862$, $1/\pi \approx 0.3183098862$

Section 9.10 (page 687)

1. $\sum_{n=0}^{\infty} \frac{(2x)^n}{n!}$ **3.** $\frac{\sqrt{2}}{2} \sum_{n=0}^{\infty} \frac{(-1)^{n(n+1)/2}}{n!}\left(x - \frac{\pi}{4}\right)^n$

5. $\sum_{n=0}^{\infty} (-1)^n (x-1)^n$ **7.** $\sum_{n=0}^{\infty} \frac{(-1)^n (x-1)^{n+1}}{n+1}$

9. $\sum_{n=0}^{\infty} \frac{(-1)^n (3x)^{2n+1}}{(2n+1)!}$ **11.** $1 + x^2/2! + 5x^4/4! + \cdots$

13–15. Proofs **17.** $\sum_{n=0}^{\infty} (-1)^n (n+1)x^n$

19. $1 + \sum_{n=1}^{\infty} \frac{1 \cdot 3 \cdot 5 \cdots (2n-1)x^n}{2^n n!}$

21. $\frac{1}{2}\left[1 + \sum_{n=1}^{\infty} \frac{(-1)^n 1 \cdot 3 \cdot 5 \cdots (2n-1)x^{2n}}{2^{3n} n!}\right]$

23. $1 + \frac{x}{2} + \sum_{n=2}^{\infty} \frac{(-1)^{n+1} 1 \cdot 3 \cdot 5 \cdots (2n-3)x^n}{2^n n!}$

25. $1 + \frac{x^2}{2} + \sum_{n=2}^{\infty} \frac{(-1)^{n+1} 1 \cdot 3 \cdot 5 \cdots (2n-3)x^{2n}}{2^n n!}$

27. $\sum_{n=0}^{\infty} \frac{x^{2n}}{2^n n!}$ **29.** $\sum_{n=1}^{\infty} \frac{(-1)^{n-1} x^n}{n}$ **31.** $\sum_{n=0}^{\infty} \frac{(-1)^n (3x)^{2n+1}}{(2n+1)!}$

33. $\sum_{n=0}^{\infty} \frac{(-1)^n 4^{2n} x^{2n}}{(2n)!}$ **35.** $\sum_{n=0}^{\infty} \frac{(-1)^n x^{3n}}{(2n)!}$ **37.** $\sum_{n=0}^{\infty} \frac{x^{2n+1}}{(2n+1)!}$

39. $\frac{1}{2}\left[1 + \sum_{n=0}^{\infty} \frac{(-1)^n (2x)^{2n}}{(2n)!}\right]$ **41.** $\sum_{n=0}^{\infty} \frac{(-1)^n x^{2n+2}}{(2n+1)!}$

43. $\begin{cases} \displaystyle\sum_{n=0}^{\infty} \dfrac{(-1)^n x^{2n}}{(2n+1)!}, & x \neq 0 \\ \qquad\qquad 1, & x = 0 \end{cases}$ **45.** Proof

47. $P_5(x) = x + x^2 + \frac{1}{3}x^3 - \frac{1}{30}x^5$

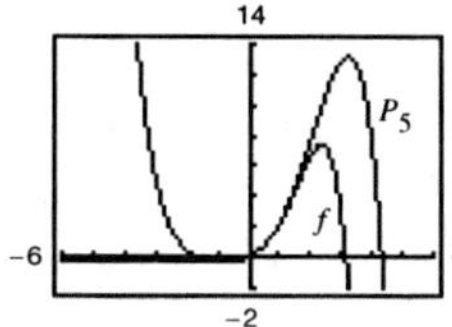

49. $P_5(x) = x - \frac{1}{2}x^2 - \frac{1}{6}x^3 + \frac{3}{40}x^5$

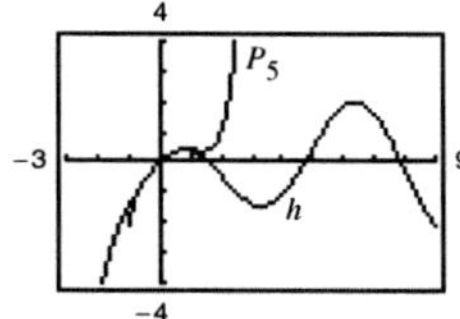

51. $P_4(x) = x - x^2 + \frac{5}{6}x^3 - \frac{5}{6}x^4$

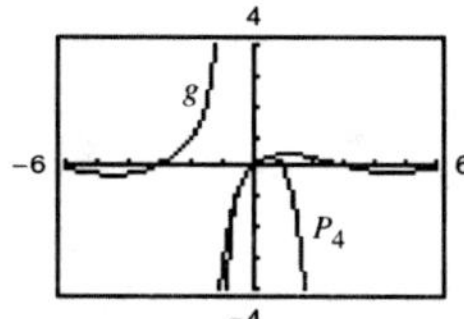

53. c; $f(x) = x \sin x$ **54.** d; $f(x) = x \cos x$ **55.** a; $f(x) = xe^x$

56. b; $f(x) = x^2 \left(\dfrac{1}{1+x}\right)$ **57.** $\displaystyle\sum_{n=0}^{\infty} \dfrac{(-1)^{(n+1)} x^{2n+3}}{(2n+3)(n+1)!}$

59. 0.6931 **61.** 7.3891 **63.** 0 **65.** 1 **67.** 0.8075
69. 0.9461 **71.** 0.4872 **73.** 0.2010 **75.** 0.7040
77. 0.3412

79. $P_5(x) = x - 2x^3 + \frac{2}{3}x^5$

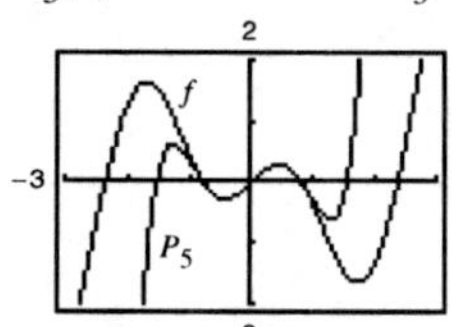
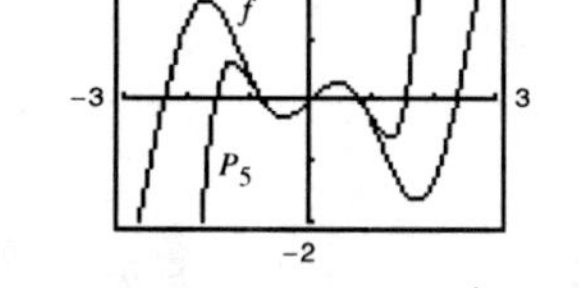

$\left[-\frac{3}{4}, \frac{3}{4}\right]$

81. $P_5(x) = (x-1) - \frac{1}{24}(x-1)^3 + \frac{1}{24}(x-1)^4 - \frac{71}{1920}(x-1)^5$

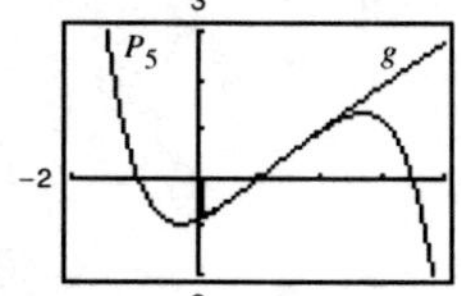

$\left[\frac{1}{4}, 2\right]$

83. See "Guidelines for Finding a Taylor Series" on page 682.

85. The binomial series is given by

$$(1+x)^k = 1 + kx + \frac{k(k-1)}{2!}x^2 + \frac{k(k-1)(k-2)}{3!}x^3 + \cdots.$$

The radius of convergence is $R = 1$.

87. Proof

89. (a)

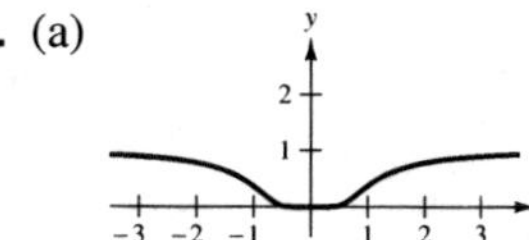

(b) Proof

(c) $\displaystyle\sum_{n=0}^{\infty} 0x^n = 0 \neq f(x)$

91. Proof **93.** 10 **95.** -0.0390625

97. $\displaystyle\sum_{n=0}^{\infty} \binom{k}{n} x^n$ **99.** Proof

Review Exercises for Chapter 9 (page 690)

1. $a_n = 1/(n! + 1)$ **3.** a **4.** c **5.** d **6.** b

7.

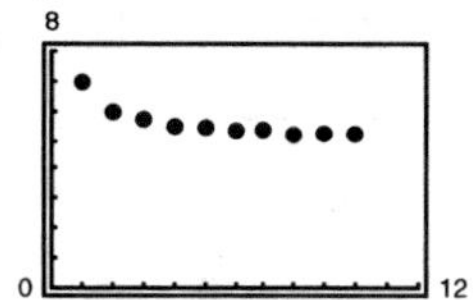

Converges to 5

9. Converges to 3 **11.** Diverges **13.** Converges to 0
15. Converges to 0 **17.** Converges to 0

19. (a)

n	1	2	3	4
A_n	\$8100.00	\$8201.25	\$8303.77	\$8407.56

n	5	6	7	8
A_n	\$8512.66	\$8619.07	\$8726.80	\$8835.89

(b) \$13,148.96

21. (a)

n	5	10	15	20	25
S_n	13.2	113.3	873.8	6648.5	50,500.3

(b)

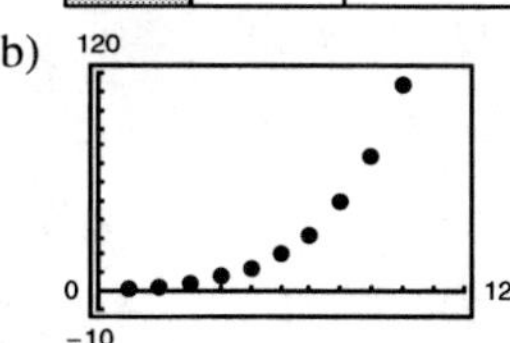

23. (a)

n	5	10	15	20	25
S_n	0.4597	0.4597	0.4597	0.4597	0.4597

(b)

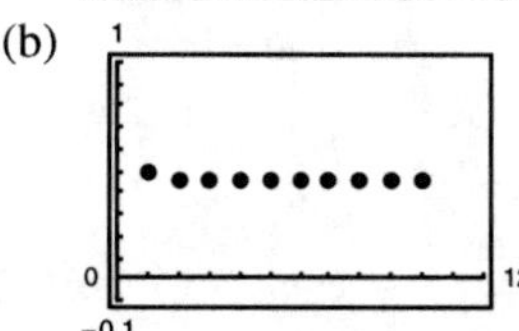

25. 3 **27.** 5.5 **29.** (a) $\displaystyle\sum_{n=0}^{\infty} (0.09)(0.01)^n$ (b) $\frac{1}{11}$

31. Diverges **33.** Diverges **35.** $45\frac{1}{3}$ m **37.** \$7630.70
39. Converges **41.** Diverges **43.** Diverges
45. Converges **47.** Diverges **49.** Converges
51. Converges **53.** Diverges **55.** Diverges
57. Converges **59.** Diverges
61. (a) Proof

(b)

n	5	10	15	20	25
S_n	2.8752	3.6366	3.7377	3.7488	3.7499

(c)

(d) 3.75

63. (a)

N	5	10	20	30	40
$\sum_{n=1}^{N} \dfrac{1}{n^2}$	1.4636	1.5498	1.5962	1.6122	1.6202
$\displaystyle\int_{N}^{\infty} \dfrac{1}{x^2}\,dx$	0.2000	0.1000	0.0500	0.0333	0.0250

(b)

N	5	10	20	30	40
$\sum_{n=1}^{N} \dfrac{1}{n^5}$	1.0367	1.0369	1.0369	1.0369	1.0369
$\displaystyle\int_{N}^{\infty} \dfrac{1}{x^5}\,dx$	0.0004	0.0000	0.0000	0.0000	0.0000

The series in part (b) converges more rapidly. This is evident from the integrals that give the remainders of the partial sums.

65. $P_3(x) = 1 - 3x + \frac{9}{2}x^2 - \frac{9}{2}x^3$ **67.** 0.996 **69.** 0.559

71. (a) 4 (b) 6 (c) 5 (d) 10   **73.** $(-10, 10)$

75. $[1, 3]$ **77.** Converges only at $x = 2$ **79.** Proof

81. $\displaystyle\sum_{n=0}^{\infty} \frac{2}{3}\left(\frac{x}{3}\right)^n$ **83.** $\displaystyle\sum_{n=0}^{\infty} \frac{2}{9}(n+1)\left(\frac{x}{3}\right)^n$

85. $f(x) = \dfrac{3}{3 - 2x},\ \left(-\dfrac{3}{2}, \dfrac{3}{2}\right)$ **87.** $\dfrac{\sqrt{2}}{2}\displaystyle\sum_{n=0}^{\infty} \frac{(-1)^{n(n+1)/2}}{n!}\left(x - \frac{3\pi}{4}\right)^n$

89. $\displaystyle\sum_{n=0}^{\infty} \frac{(x \ln 3)^n}{n!}$ **91.** $-\displaystyle\sum_{n=0}^{\infty} (x + 1)^n$

93. $1 + x/5 - 2x^2/25 + 6x^3/125 - 21x^4/625 + \cdots$

95. $\ln \frac{5}{4} \approx 0.2231$ **97.** $e^{1/2} \approx 1.6487$ **99.** $\cos \frac{2}{3} \approx 0.7859$

101. The series in Exercise 45 converges to its sum at a lower rate because its terms approach 0 at a lower rate.

103. (a)–(c) $1 + 2x + 2x^2 + \dfrac{4}{3}x^3$

105. $\displaystyle\sum_{n=0}^{\infty} \frac{(-1)^n x^{2n+1}}{(2n + 1)(2n + 1)!}$ **107.** $\displaystyle\sum_{n=0}^{\infty} \frac{(-1)^n x^{n+1}}{(n + 1)^2}$ **109.** 0

P.S. Problem Solving (page 693)

1. (a) 1 (b) Answers will vary. Example: $0, \frac{1}{3}, \frac{2}{3}$ (c) 0

3. Proof   **5.** (a) Proof (b) Yes (c) Any distance

7. For $a = b$, the series converges conditionally. For no values of a and b does the series converge absolutely.

9. 665,280   **11.** (a) Proof (b) Diverges

13. Proof   **15.** (a) Proof (b) Proof

17. (a) The height is infinite. (b) The surface area is infinite.
  (c) Proof

Chapter 10
Section 10.1 (page 706)

1. h   **2.** a   **3.** e   **4.** b   **5.** f   **6.** g   **7.** c   **8.** d

9. Vertex: $(0, 0)$
  Focus: $(-2, 0)$
  Directrix: $x = 2$

11. Vertex: $(-5, 3)$
  Focus: $\left(-\frac{21}{4}, 3\right)$
  Directrix: $x = -\frac{19}{4}$

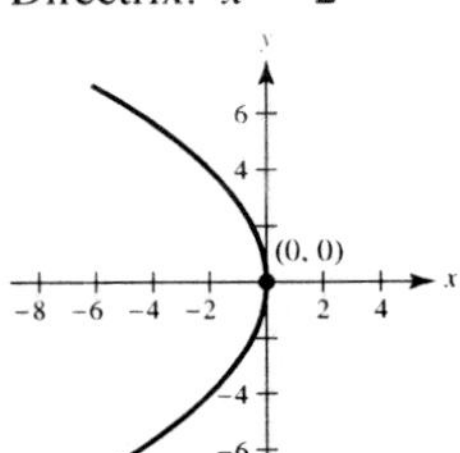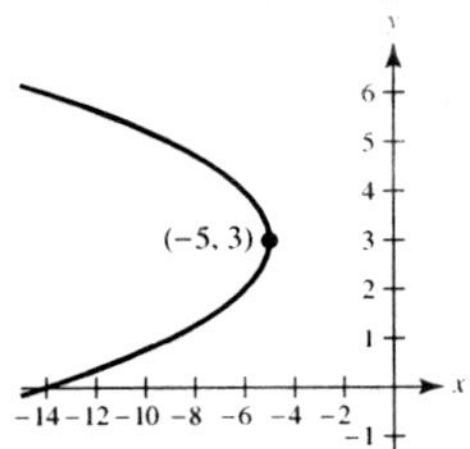

13. Vertex: $(-1, 2)$
  Focus: $(0, 2)$
  Directrix: $x = -2$

15. Vertex: $(-2, 2)$
  Focus: $(-2, 1)$
  Directrix: $y = 3$

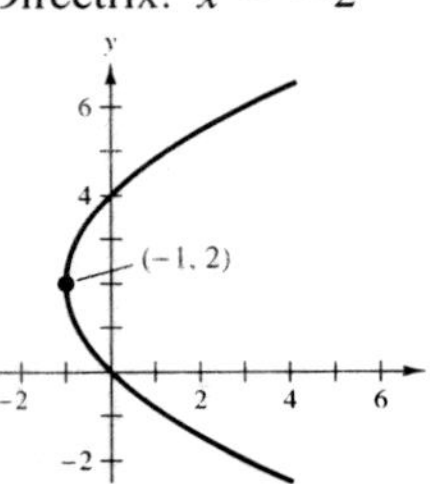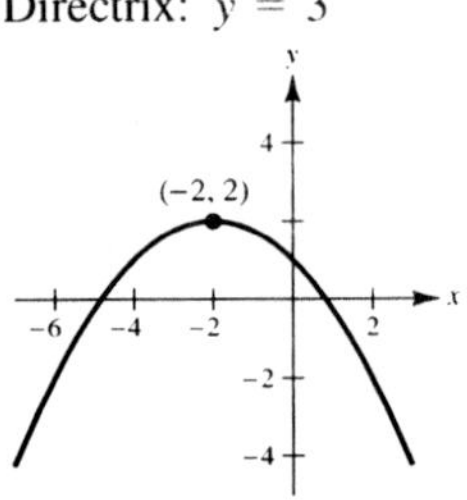

17. Vertex: $\left(\frac{1}{4}, -\frac{1}{2}\right)$
  Focus: $\left(0, -\frac{1}{2}\right)$
  Directrix: $x = \frac{1}{2}$

19. Vertex: $(-1, 0)$
  Focus: $(0, 0)$
  Directrix: $x = -2$

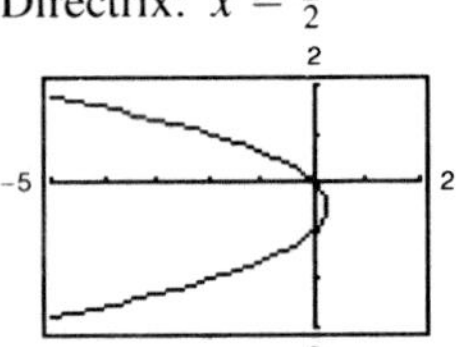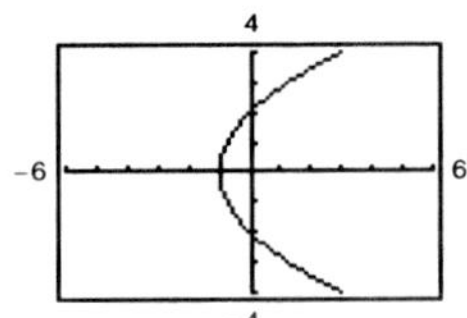

21. $y^2 - 8y + 8x - 24 = 0$ **23.** $x^2 - 32y + 160 = 0$

25. $x^2 + y - 4 = 0$ **27.** $5x^2 - 14x - 3y + 9 = 0$

29. Center: $(0, 0)$
  Foci: $\left(0, \pm\sqrt{15}\right)$
  Vertices: $(0, \pm 4)$
  $e = \sqrt{15}/4$

31. Center: $(3, 1)$
  Foci: $(3, 4), (3, -2)$
  Vertices: $(3, 6), (3, -4)$
  $e = \frac{3}{5}$

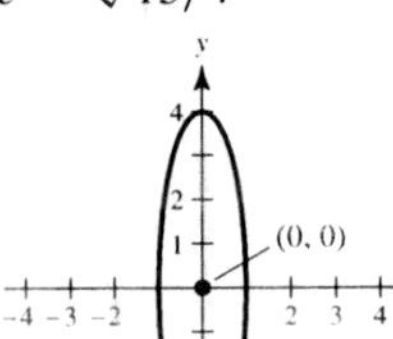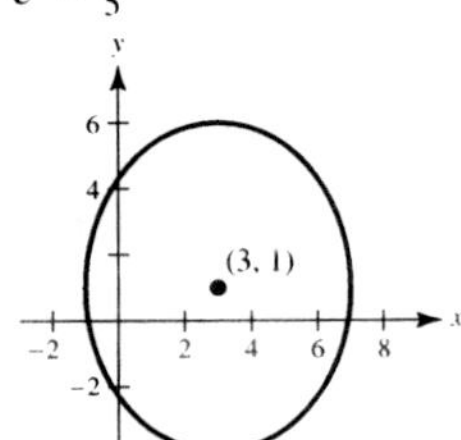

33. Center: $(-2, 3)$
Foci: $\left(-2, 3 \pm \sqrt{5}\right)$
Vertices: $(-2, 6), (-2, 0)$
$e = \sqrt{5}/3$

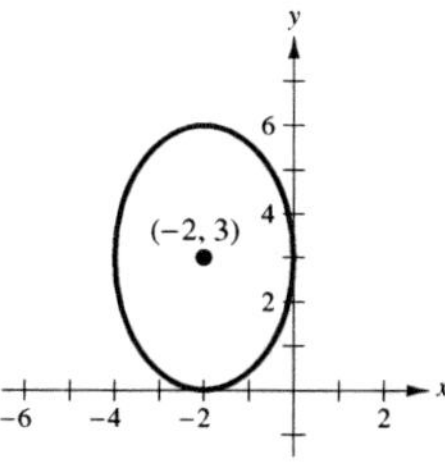

35. Center: $\left(\frac{1}{2}, -1\right)$
Foci: $\left(\frac{1}{2} \pm \sqrt{2}, -1\right)$
Vertices: $\left(\frac{1}{2} \pm \sqrt{5}, -1\right)$
To obtain the graph, solve
for y and get

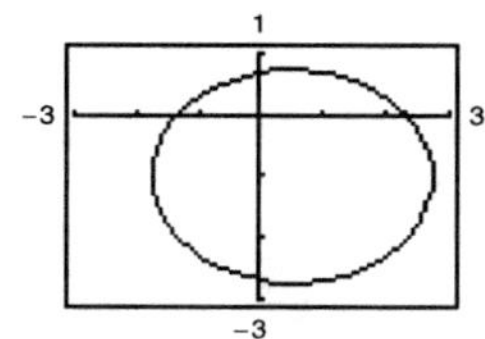

$y_1 = -1 + \sqrt{(57 + 12x - 12x^2)/20}$ and
$y_2 = -1 - \sqrt{(57 + 12x - 12x^2)/20}$.
Graph these equations in the same viewing window.

37. Center: $\left(\frac{3}{2}, -1\right)$
Foci: $\left(\frac{3}{2} \pm \sqrt{2}, -1\right)$
Vertices: $\left(-\frac{1}{2}, -1\right), \left(\frac{7}{2}, -1\right)$
To obtain the graph, solve for y
and get

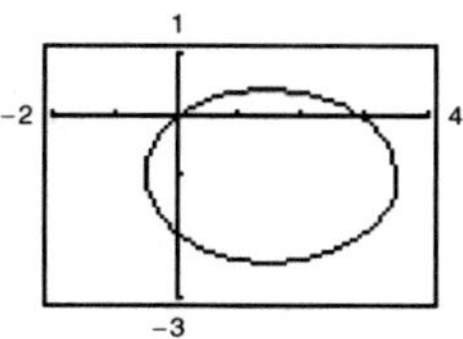

$y_1 = -1 + \sqrt{(7 + 12x - 4x^2)/8}$ and
$y_2 = -1 - \sqrt{(7 + 12x - 4x^2)/8}$.
Graph these equations in the same viewing window.

39. $x^2/36 + y^2/11 = 1$ **41.** $(x - 3)^2/9 + (y - 5)^2/16 = 1$
43. $x^2/16 + 7y^2/16 = 1$

45. Center: $(0, 0)$
Foci: $\left(0, \pm \sqrt{10}\right)$
Vertices: $(0, \pm 1)$

47. Center: $(1, -2)$
Foci: $\left(1 \pm \sqrt{5}, -2\right)$
Vertices: $(-1, -2), (3, -2)$

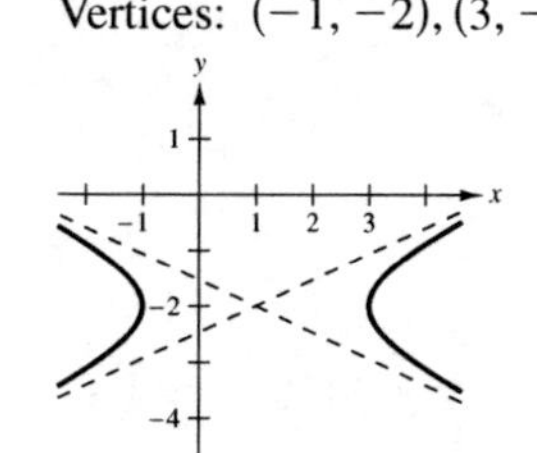

49. Center: $(2, -3)$
Foci: $\left(2 \pm \sqrt{10}, -3\right)$
Vertices: $(1, -3), (3, -3)$

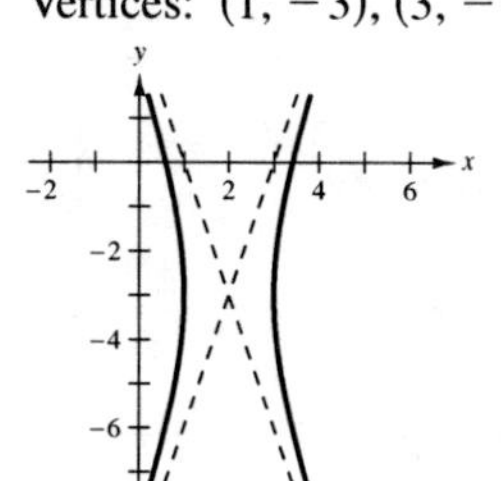

51. Degenerate hyperbola
Graph is two lines
$y = -3 \pm \frac{1}{3}(x + 1)$
intersecting at $(-1, -3)$.

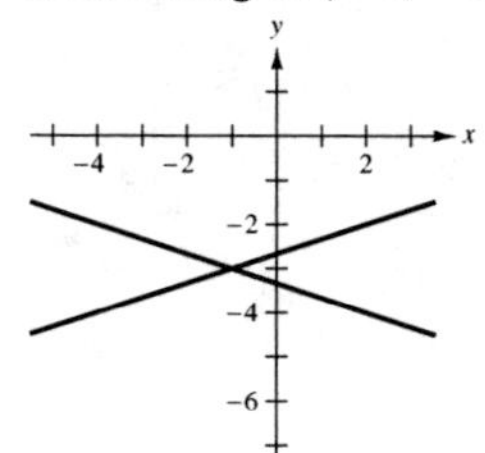

53. Center: $(1, -3)$
Foci: $\left(1, -3 \pm 2\sqrt{5}\right)$
Vertices: $\left(1, -3 \pm \sqrt{2}\right)$
Asymptotes:
$y = \frac{1}{3}x - \frac{1}{3} - 3$;
$y = -\frac{1}{3}x + \frac{1}{3} - 3$

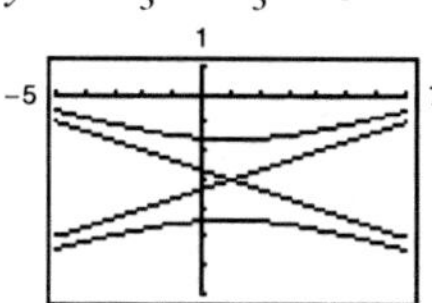

55. Center: $(1, -3)$
Foci: $\left(1 \pm \sqrt{10}, -3\right)$
Vertices: $(-1, -3), (3, -3)$
Asymptotes:
$y = \sqrt{6}\,x/2 - \sqrt{6}/2 - 3$;
$y = -\sqrt{6}\,x/2 + \sqrt{6}/2 - 3$

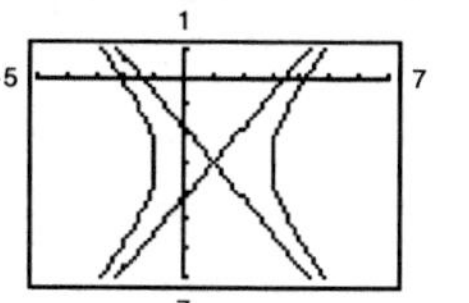

57. $x^2/1 - y^2/25 = 1$ **59.** $y^2/9 - (x - 2)^2/(9/4) = 1$
61. $y^2/4 - x^2/12 = 1$ **63.** $(x - 3)^2/9 - (y - 2)^2/4 = 1$
65. (a) $\left(6, \sqrt{3}\right): 2x - 3\sqrt{3}\,y - 3 = 0$
$\left(6, -\sqrt{3}\right): 2x + 3\sqrt{3}\,y - 3 = 0$
(b) $\left(6, \sqrt{3}\right): 9x + 2\sqrt{3}\,y - 60 = 0$
$\left(6, -\sqrt{3}\right): 9x - 2\sqrt{3}\,y - 60 = 0$
67. Ellipse   **69.** Parabola   **71.** Circle
73. Circle   **75.** Hyperbola
77. (a) A parabola is the set of all points (x, y) that are equidistant
from a fixed line and a fixed point not on the line.
(b) For directrix $y = k - p$: $(x - h)^2 = 4p(y - k)$
For directrix $x = h - p$: $(y - k)^2 = 4p(x - h)$
(c) If P is a point on a parabola, then the tangent line to the
parabola at P makes equal angles with the line passing
through P and the focus, and with the line passing through P
parallel to the axis of the parabola.
79. (a) A hyperbola is the set of all points (x, y) for which the
absolute value of the difference between the distances from
two distinct fixed points is constant.
(b) Transverse axis is horizontal: $\dfrac{(x - h)^2}{a^2} - \dfrac{(y - k)^2}{b^2} = 1$

Transverse axis is vertical: $\dfrac{(y - k)^2}{a^2} - \dfrac{(x - h)^2}{b^2} = 1$

(c) Transverse axis is horizontal:
$y = k + (b/a)(x - h)$ and $y = k - (b/a)(x - h)$
Transverse axis is vertical:
$y = k + (a/b)(x - h)$ and $y = k - (a/b)(x - h)$
81. $\frac{9}{4}$ m   **83.** $y = 2ax_0 x - ax_0^2$ **85.** (a) Proof   (b) Proof
87. $x_0 = 2\sqrt{3}/3$; Distance from hill: $2\sqrt{3}/3 - 1$
89. $[16(4 + 3\sqrt{3} - 2\pi)]/3 \approx 15.536 \text{ ft}^2$
91. (a) $y = (1/180)x^2$

(b) $10\left[2\sqrt{13} + 9 \ln\left(\dfrac{2 + \sqrt{13}}{3}\right)\right] \approx 128.4 \text{ m}$

93.

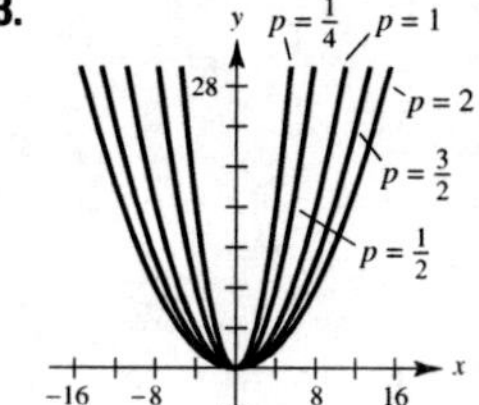

As p increases, the graph of
$x^2 = 4py$ gets wider.

95. (a) $L = 2a$

(b) The thumbtacks are located at the foci, and the length of string is the constant sum of distances from the foci.

97. 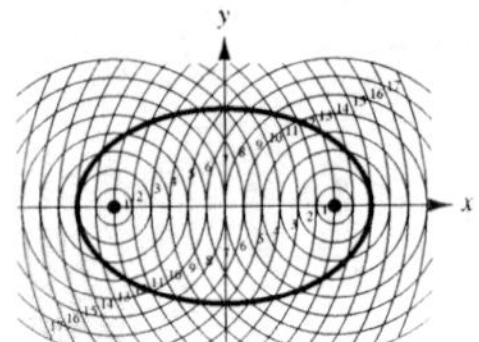

99. Proof   **101.** $e \approx 0.1776$

103. $e \approx 0.9671$ **105.** $\left(0, \frac{25}{3}\right)$

107. Minor-axis endpoints: $(-6, -2), (0, -2)$
Major-axis endpoints: $(-3, -6), (-3, 2)$

109. (a) Area $= 2\pi$

(b) Volume $= 8\pi/3$

Surface area $= [2\pi(9 + 4\sqrt{3}\pi)]/9 \approx 21.48$

(c) Volume $= 16\pi/3$

Surface area $= \dfrac{4\pi\left[6 + \sqrt{3}\ln\left(2 + \sqrt{3}\right)\right]}{3} \approx 34.69$

111. 37.96   **113.** 40   **115.** $(x-6)^2/9 - (y-2)^2/7 = 1$

117. 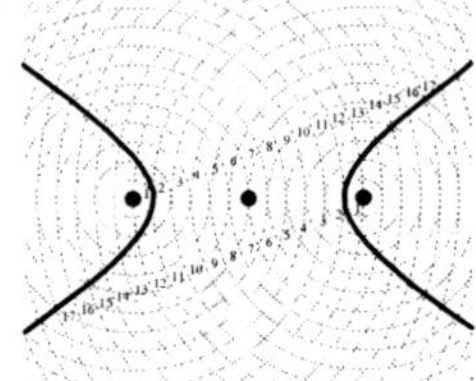

119. Proof

121. $x = \left(-90 + 96\sqrt{2}\right)/7 \approx 6.538$
$y = \left(160 - 96\sqrt{2}\right)/7 \approx 3.462$

123. There are four points of intersection.

At $\left(\dfrac{\sqrt{2}\,ac}{\sqrt{2a^2 - b^2}}, \dfrac{b^2}{\sqrt{2}\sqrt{2a^2 - b^2}}\right)$, the slopes of the tangent lines are $y'_e = -c/a$ and $y'_h = a/c$.

Because the slopes are negative reciprocals, the tangent lines are perpendicular. Similarly, the curves are perpendicular at the other three points of intersection.

125. False. See the definition of a parabola.   **127.** True

129. True   **131.** Putnam Problem B4, 1976

Section 10.2   (page 718)

1. (a)

t	0	1	2	3	4
x	0	1	$\sqrt{2}$	$\sqrt{3}$	2
y	3	2	1	0	-1

(b) and (c)   (d) $y = 3 - x^2, x \geq 0$

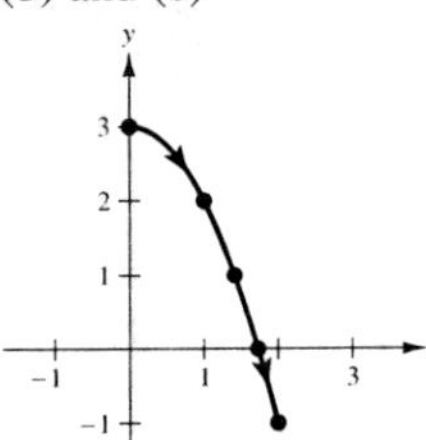

3. 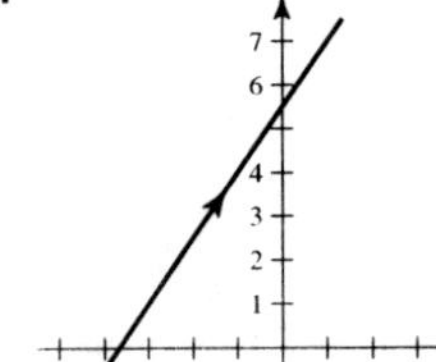

$3x - 2y + 11 = 0$

5. 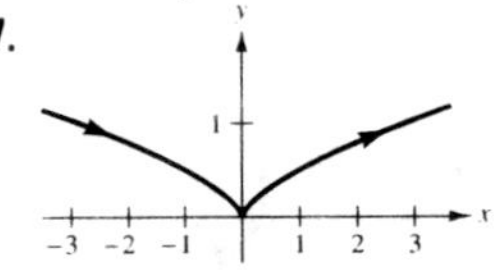

$y = (x - 1)^2$

7. 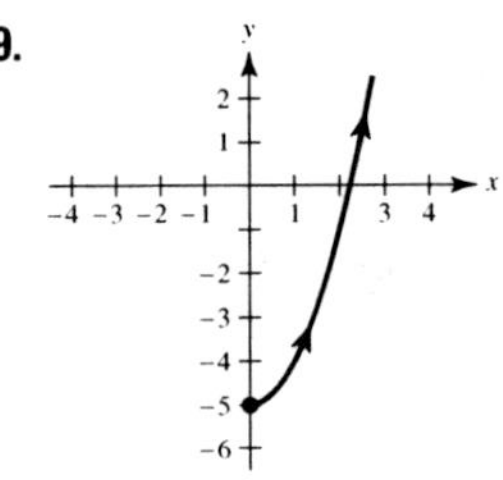

$y = \frac{1}{2}x^{2/3}$

9. 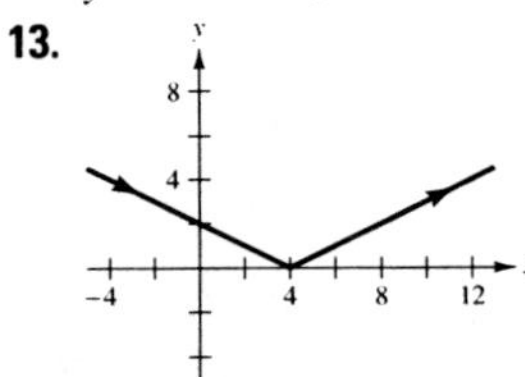

$y = x^2 - 5, \quad x \geq 0$

11. 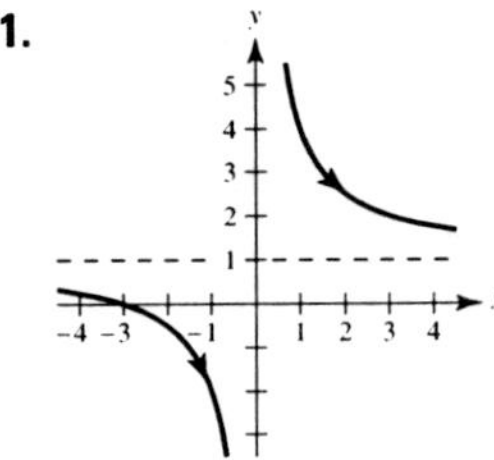

$y = (x + 3)/x$

13. 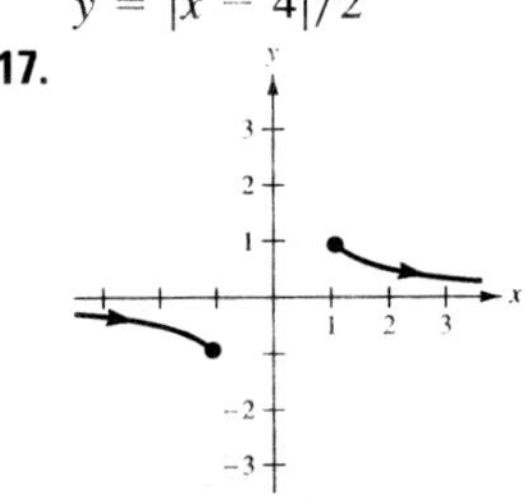

$y = |x - 4|/2$

15. 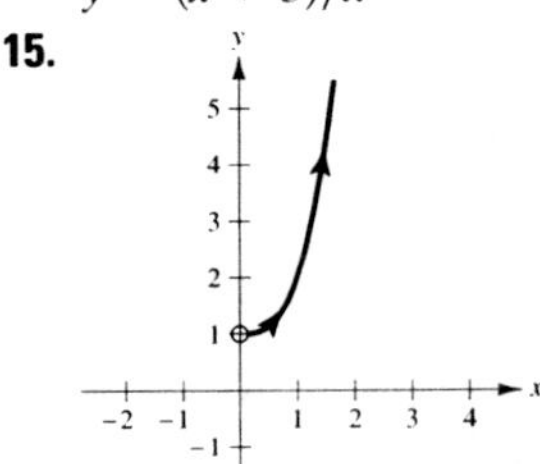

$y = x^3 + 1, \quad x > 0$

17. 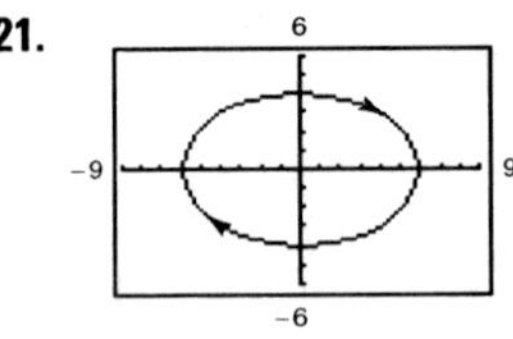

$y = 1/x, \quad |x| \geq 1$

19. 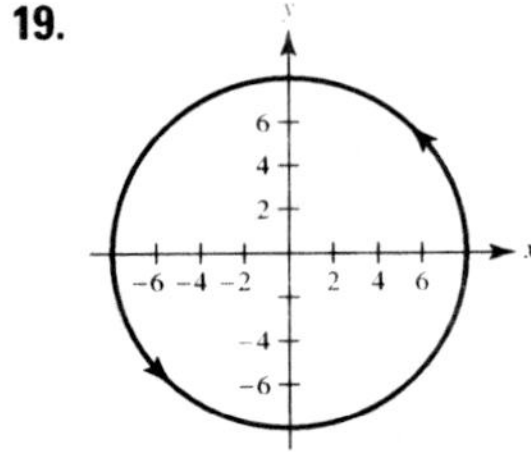

$x^2 + y^2 = 64$

21. 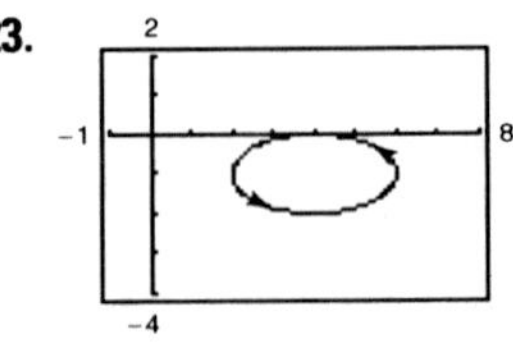

$\dfrac{x^2}{36} + \dfrac{y^2}{16} = 1$

23.

$\dfrac{(x-4)^2}{4} + \dfrac{(y+1)^2}{1} = 1$

25. 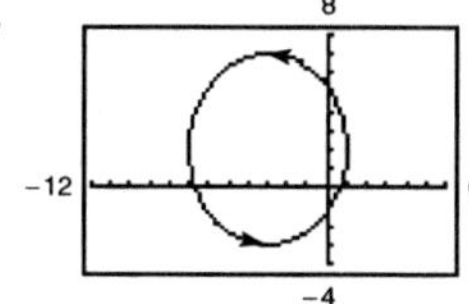

27. 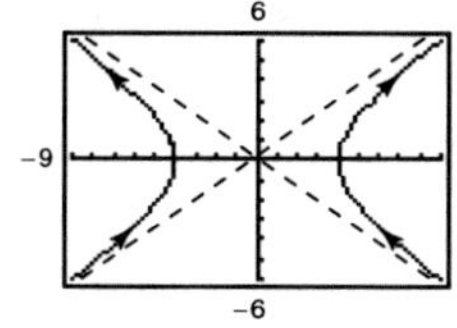

$$\frac{(x+3)^2}{16}+\frac{(y-2)^2}{25}=1$$

$$\frac{x^2}{16}-\frac{y^2}{9}=1$$

29. 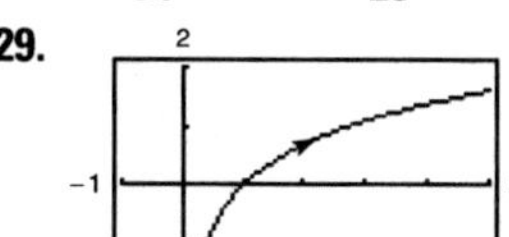

31. 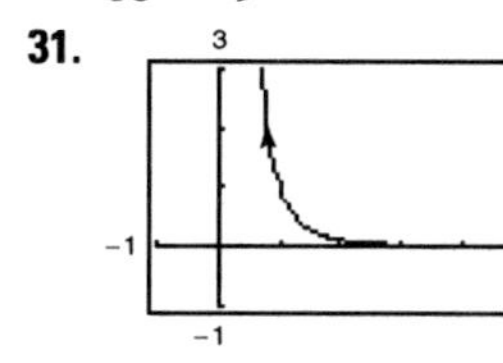

$$y=\ln x$$

$$y=1/x^3,\quad x>0$$

33. Each curve represents a portion of the line $y=2x+1$.

	Domain	Orientation	Smooth
(a)	$-\infty < x < \infty$	Up	Yes
(b)	$-1 \le x \le 1$	Oscillates	No, $\dfrac{dx}{d\theta}=\dfrac{dy}{d\theta}=0$ when $\theta=0,\pm\pi,\pm2\pi,\ldots$
(c)	$0 < x < \infty$	Down	Yes
(d)	$0 < x < \infty$	Up	Yes

35. (a) and (b) represent the parabola $y=2(1-x^2)$ for $-1 \le x \le 1$. The curve is smooth. The orientation is from right to left in part (a) and in part (b).

37. (a) 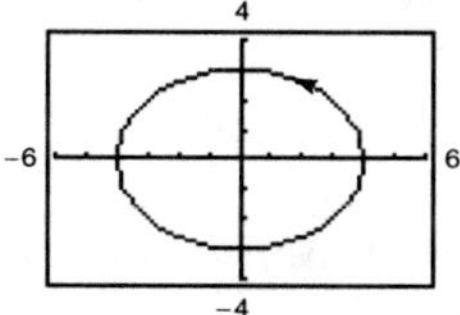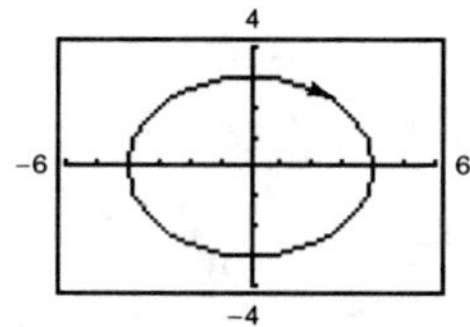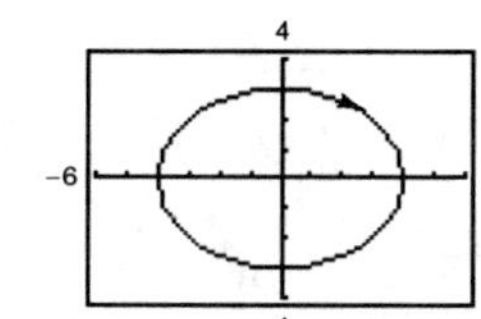

(b) The orientation is reversed.   (c) The orientation is reversed.
(d) Answers will vary. For example,

$$x=2\sec t \qquad x=2\sec(-t)$$
$$y=5\sin t \qquad y=5\sin(-t)$$

have the same graphs, but their orientations are reversed.

39. $y-y_1=\dfrac{y_2-y_1}{x_2-x_1}(x-x_1)$ **41.** $\dfrac{(x-h)^2}{a^2}+\dfrac{(y-k)^2}{b^2}=1$

43. $x=4t$
$y=-7t$
(Solution is not unique.)

45. $x=3+2\cos\theta$
$y=1+2\sin\theta$
(Solution is not unique.)

47. $x=10\cos\theta$
$y=6\sin\theta$
(Solution is not unique.)

49. $x=4\sec\theta$
$y=3\tan\theta$
(Solution is not unique.)

51. $x=t$
$y=6t-5;$
$x=t+1$
$y=6t+1$
(Solution is not unique.)

53. $x=t$
$y=t^3;$
$x=\tan t$
$y=\tan^3 t$
(Solution is not unique.)

55. $x=t+3,\ y=2t+1$ **57.** $x=t,\ y=t^2$

59. 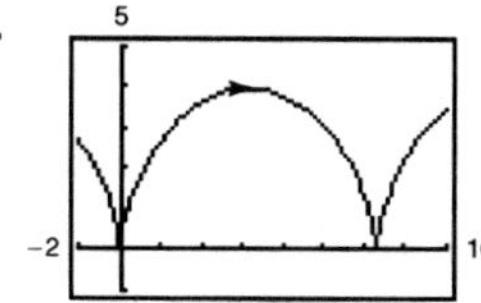

61. 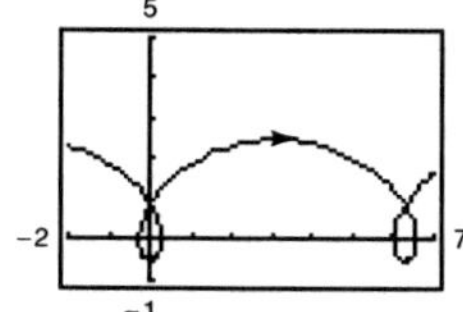

Not smooth at $\theta=2n\pi$ Smooth everywhere

63. 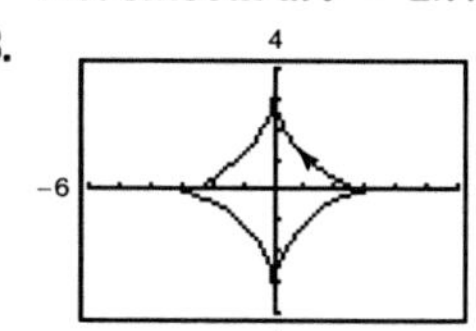

65. 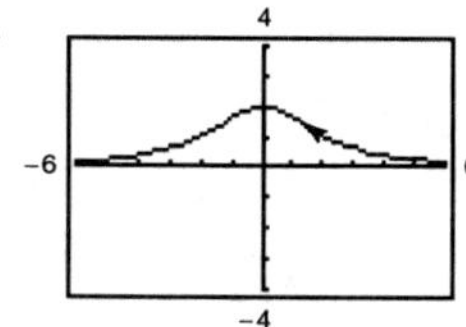

Not smooth at $\theta=\frac{1}{2}n\pi$ Smooth everywhere

67. Each point (x,y) in the plane is determined by the plane curve $x=f(t),y=g(t)$. For each t, plot (x,y). As t increases, the curve is traced out in a specific direction called the orientation of the curve.

69. $x=a\theta-b\sin\theta;\ y=a-b\cos\theta$

71. False. The graph of the parametric equations is the portion of the line $y=x$ when $x\ge 0$.

73. True

75. (a) $x=\left(\frac{440}{3}\cos\theta\right)t;\ y=3+\left(\frac{440}{3}\sin\theta\right)t-16t^2$

(b) 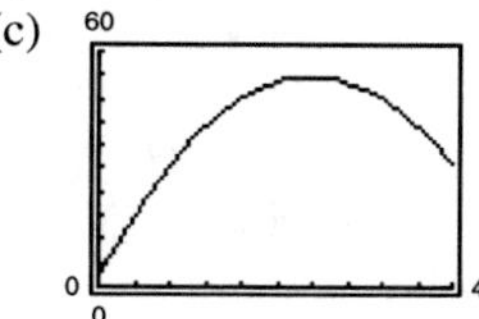   (c)

Not a home run   Home run

(d) $19.4°$

Section 10.3   (page 727)

1. $-3/t$ **3.** -1

5. $\dfrac{dy}{dx}=\dfrac{3}{4},\dfrac{d^2y}{dx^2}=0$; Neither concave upward nor concave downward

7. $dy/dx=2t+3,\ d^2y/dx^2=2$
At $t=-1,dy/dx=1,\ d^2y/dx^2=2$; Concave upward

9. $dy/dx=-\cot\theta,\ d^2y/dx^2=-(\csc\theta)^3/4$
At $\theta=\pi/4,dy/dx=-1,\ d^2y/dx^2=-\sqrt{2}/2$;
Concave downward

11. $dy/dx=2\csc\theta,\ d^2y/dx^2=-2\cot^3\theta$
At $\theta=\pi/6,dy/dx=4,\ d^2y/dx^2=-6\sqrt{3}$;
Concave downward

13. $dy/dx=-\tan\theta,\ d^2y/dx^2=\sec^4\theta\csc\theta/3$
At $\theta=\pi/4,dy/dx=-1,\ d^2y/dx^2=4\sqrt{2}/3$;
Concave upward

15. $\left(-2/\sqrt{3},3/2\right):\ 3\sqrt{3}x-8y+18=0$
$(0,2):\ y-2=0$
$\left(2\sqrt{3},1/2\right):\ \sqrt{3}x+8y-10=0$

17. $(0,0):\ 2y-x=0$
$(-3,-1):\ y+1=0$
$(-3,3):\ 2x-y+9=0$

19. (a) and (d)

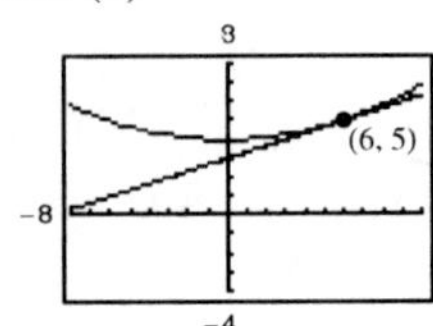

(b) At $t = 1$, $dx/dt = 6$, $dy/dt = 2$, and $dy/dx = 1/3$.
(c) $y = \frac{1}{3}x + 3$

21. (a) and (d)

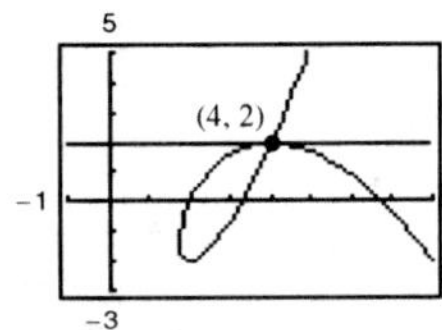

(b) At $t = -1$, $dx/dt = -3$, $dy/dt = 0$, and $dy/dx = 0$.
(c) $y = 2$

23. $y = \pm\frac{3}{4}x$ **25.** $y = 3x - 5$ and $y = 1$

27. Horizontal: $(1, 0), (-1, \pi), (1, -2\pi)$
Vertical: $(\pi/2, 1), (-3\pi/2, -1), (5\pi/2, 1)$

29. Horizontal: $(4, 0)$
Vertical: None

31. Horizontal: $(5, -2), (3, 2)$
Vertical: None

33. Horizontal: $(0, 3), (0, -3)$
Vertical: $(3, 0), (-3, 0)$

35. Horizontal: $(5, -1), (5, -3)$
Vertical: $(8, -2), (2, -2)$

37. Horizontal: None
Vertical: $(1, 0), (-1, 0)$

39. Concave downward: $-\infty < t < 0$
Concave upward: $0 < t < \infty$

41. Concave upward: $t > 0$

43. Concave downward: $0 < t < \pi/2$
Concave upward: $\pi/2 < t < \pi$

45. $\displaystyle\int_1^3 \sqrt{4t^2 - 3t + 9}\, dt$ **47.** $\displaystyle\int_{-2}^2 \sqrt{e^{2t} + 4}\, dt$

49. $4\sqrt{13} \approx 14.422$ **51.** $70\sqrt{5} \approx 156.525$

53. $\sqrt{2}\left(1 - e^{-\pi/2}\right) \approx 1.12$

55. $\frac{1}{12}\left[\ln\left(\sqrt{37} + 6\right) + 6\sqrt{37}\right] \approx 3.249$ **57.** $6a$ **59.** $8a$

61. (a)

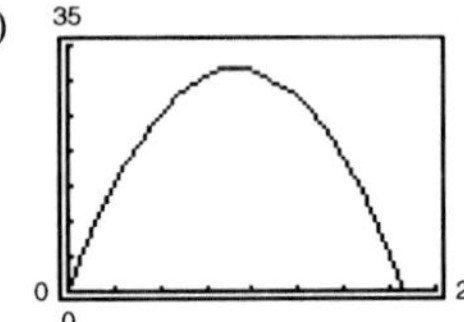

(b) 219.2 ft
(c) 230.8 ft

63. (a)

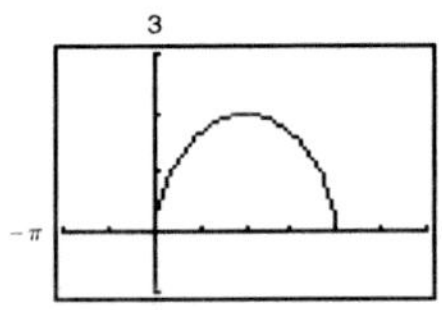

(b) $(0, 0), \left(4\sqrt[3]{2}/3, 4\sqrt[3]{4}/3\right)$
(c) About 6.557

65. (a)

(b) The average speed of the particle on the second path is twice the average speed of the particle on the first path.
(c) 4π

67. $S = 2\pi\displaystyle\int_0^4 \sqrt{10}(t + 2)\, dt = 32\pi\sqrt{10} \approx 317.907$

69. $S = 2\pi\displaystyle\int_0^{\pi/2} \left(\sin\theta\cos\theta\sqrt{4\cos^2\theta + 1}\right) d\theta = \dfrac{\left(5\sqrt{5} - 1\right)\pi}{6}$
≈ 5.330

71. (a) $27\pi\sqrt{13}$ (b) $18\pi\sqrt{13}$ **73.** 50π **75.** $12\pi a^2/5$

77. See Theorem 10.7, Parametric Form of the Derivative, on page 721.

79. 6

81. (a) $S = 2\pi\displaystyle\int_a^b g(t)\sqrt{\left(\dfrac{dx}{dt}\right)^2 + \left(\dfrac{dy}{dt}\right)^2}\, dt$

(b) $S = 2\pi\displaystyle\int_a^b f(t)\sqrt{\left(\dfrac{dx}{dt}\right)^2 + \left(\dfrac{dy}{dt}\right)^2}\, dt$

83. Proof   **85.** $3\pi/2$ **87.** d   **88.** b   **89.** f   **90.** c

91. a   **92.** e   **93.** $\left(\frac{3}{4}, \frac{8}{5}\right)$ **95.** 288π

97. (a) $dy/dx = \sin\theta/(1 - \cos\theta)$; $d^2y/dx^2 = -1/[a(\cos\theta - 1)^2]$
(b) $y = \left(2 + \sqrt{3}\right)\left[x - a\left(\pi/6 - \frac{1}{2}\right)\right] + a\left(1 - \sqrt{3}/2\right)$
(c) $\left(a(2n + 1)\pi, 2a\right)$
(d) Concave downward on $(0, 2\pi), (2\pi, 4\pi)$, etc.
(e) $s = 8a$

99. Proof

101. (a)

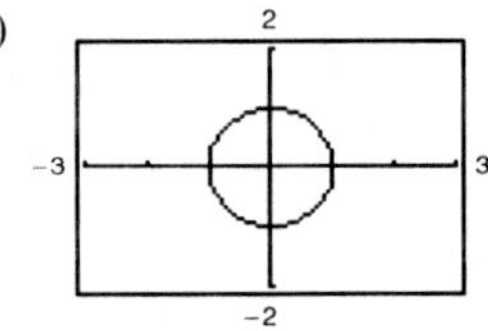

(b) Circle of radius 1 and center at $(0, 0)$ except the point $(-1, 0)$
(c) As t increases from -20 to 0, the speed increases, and as t increases from 0 to 20, the speed decreases.

103. False: $\dfrac{d^2y}{dx^2} = \dfrac{\dfrac{d}{dt}\left[\dfrac{g'(t)}{f'(t)}\right]}{f'(t)} = \dfrac{f'(t)g''(t) - g'(t)f''(t)}{[f'(t)]^3}$.

105. About 982 ft

Section 10.4   (page 738)

1.

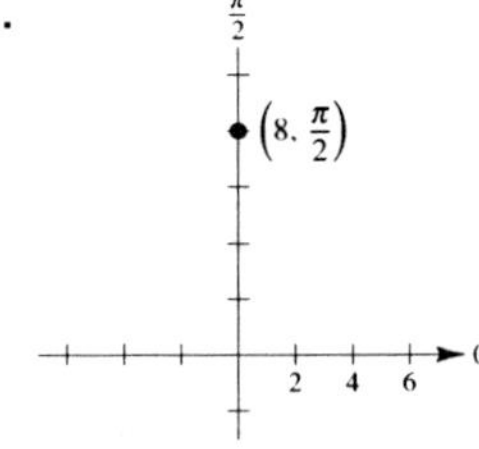

$\left(8, \frac{\pi}{2}\right)$

$(0, 8)$

3.

$\left(-4, -\frac{3\pi}{4}\right)$

$\left(2\sqrt{2}, 2\sqrt{2}\right) \approx (2.828, 2.828)$

5.

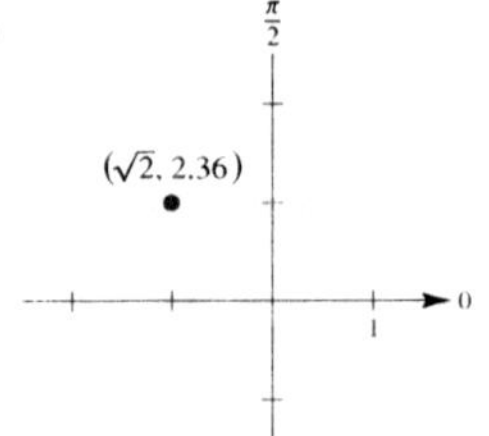

$\left(\sqrt{2}, 2.36\right)$

$(-1.004, 0.996)$

7.

$(-4.95, -4.95)$

9.
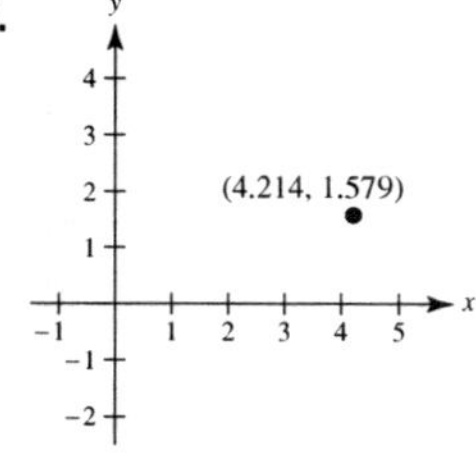
(4.214, 1.579)

11.
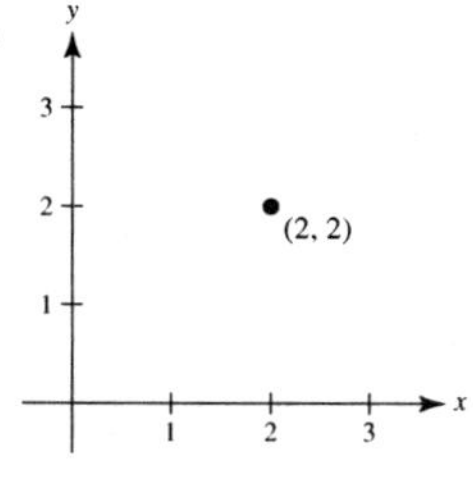
(2, 2)

$$\left(2\sqrt{2},\ \pi/4\right),\ \left(-2\sqrt{2},\ 5\pi/4\right)$$

13.
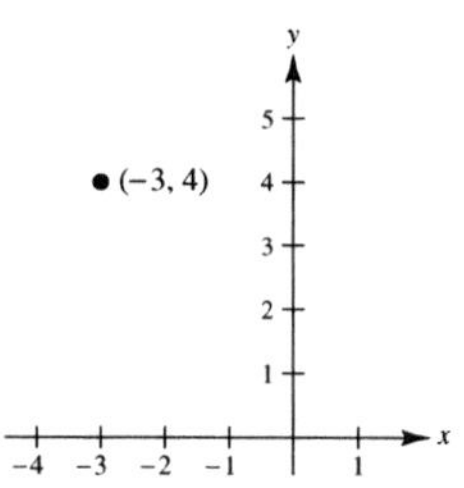
(−3, 4)

$(5, 2.214),\ (-5, 5.356)$

15.
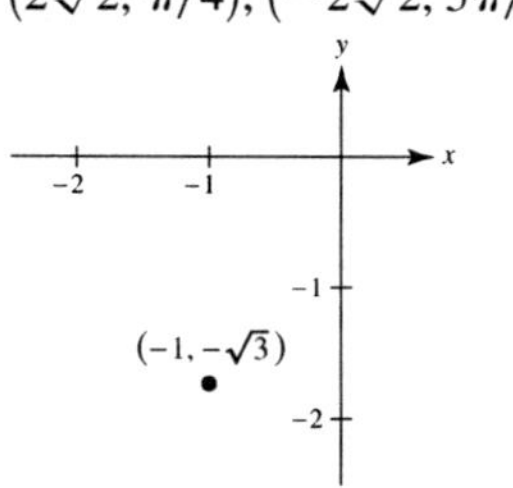
$(-1, -\sqrt{3})$

$(2, 4\pi/3),\ (-2, \pi/3)$

17. $(3.606, -0.588)$ **19.** $(3.052, 0.960)$

21. (a)
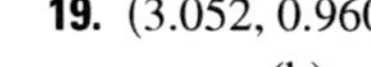
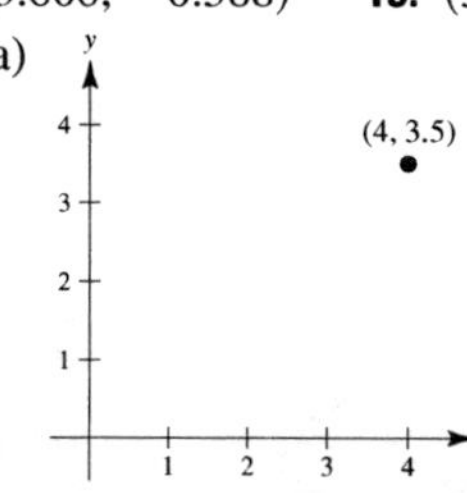
(4, 3.5)

(b)
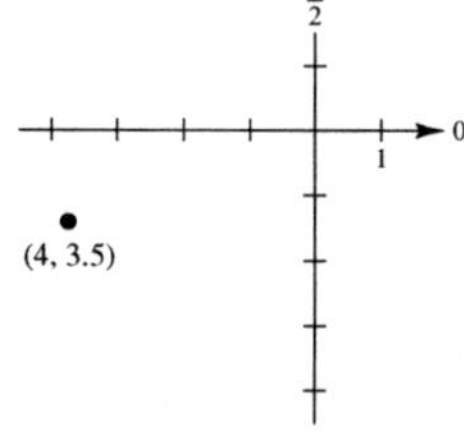
(4, 3.5)

23. c   **24.** b   **25.** a   **26.** d

27. $r = 3$
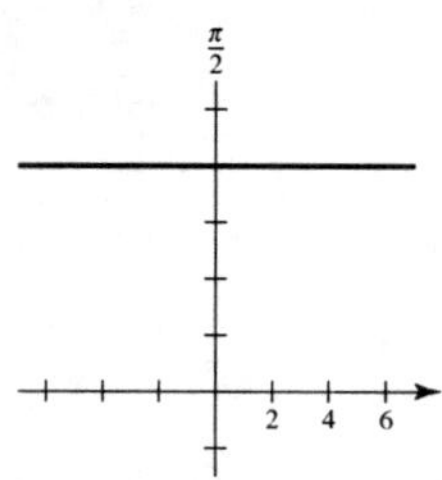

29. $r = a$
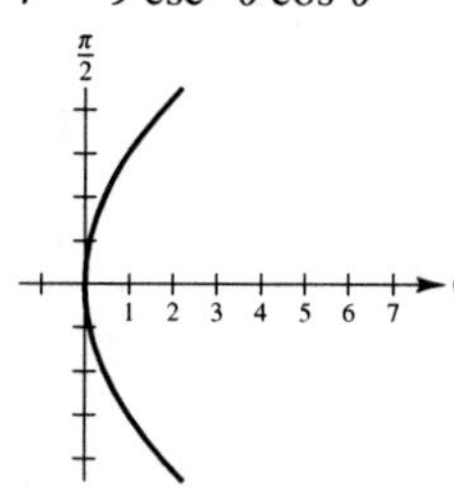

31. $r = 8 \csc \theta$

33. $r = \dfrac{-2}{3 \cos \theta - \sin \theta}$

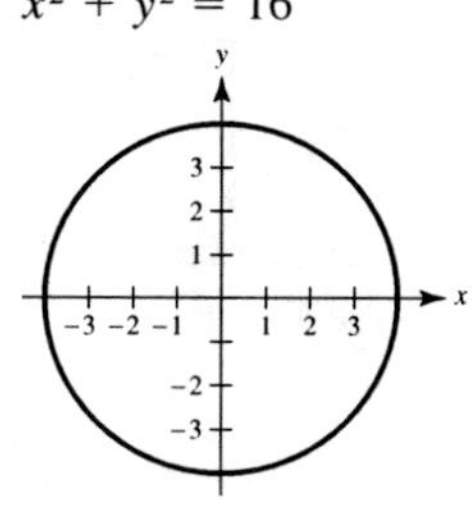

35. $r = 9 \csc^2 \theta \cos \theta$

37. $x^2 + y^2 = 16$

39. $x^2 + y^2 - 3y = 0$
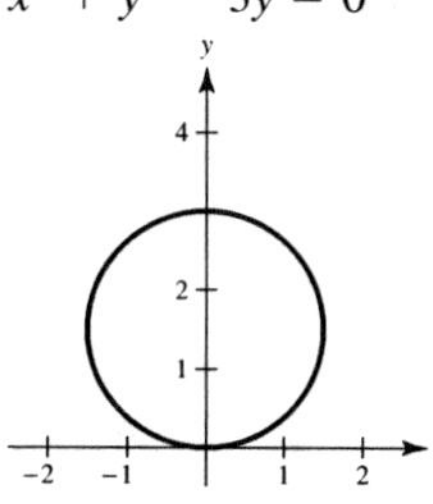

41. $\sqrt{x^2 + y^2} = \arctan\left(y/x\right)$
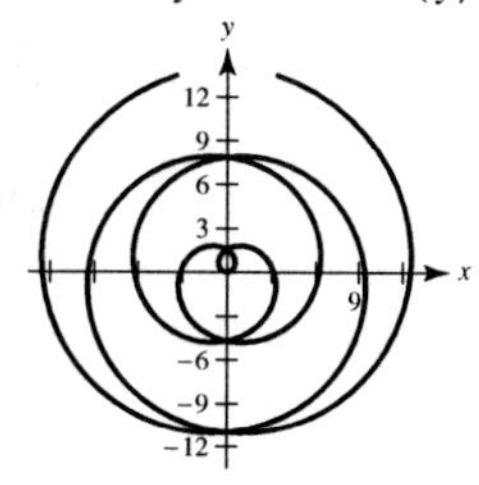

43. $x - 3 = 0$
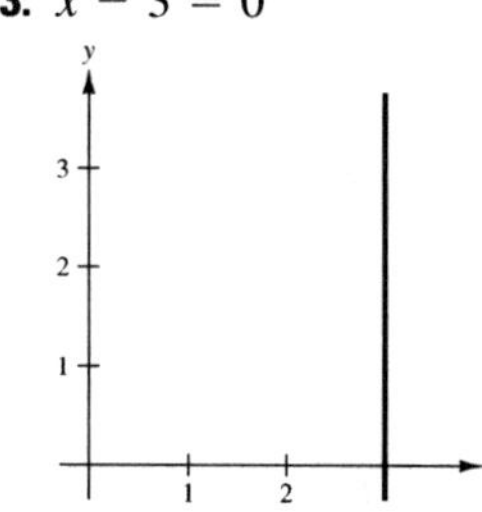

45. $x^2 - y = 0$
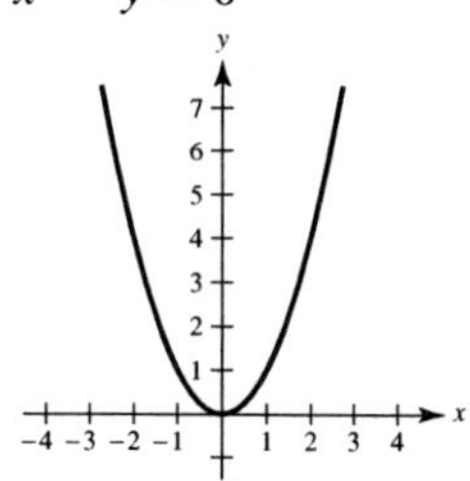

47.
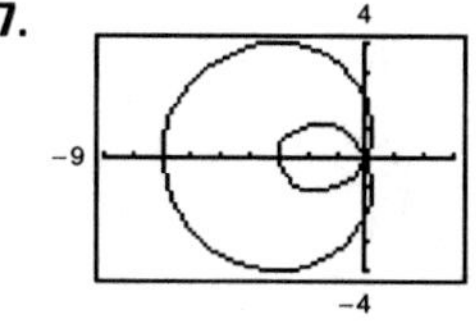
$0 \le \theta < 2\pi$

49.
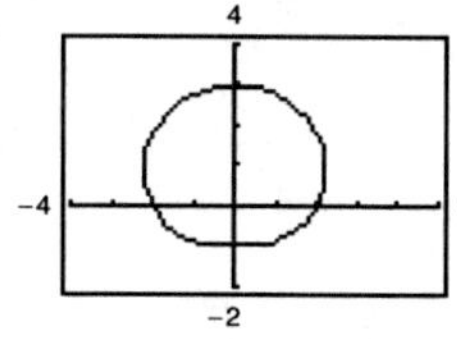
$0 \le \theta < 2\pi$

51.
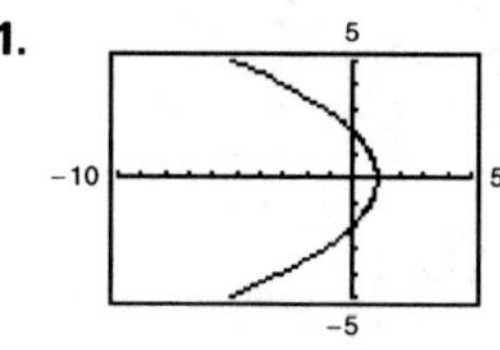
$-\pi < \theta < \pi$

53.
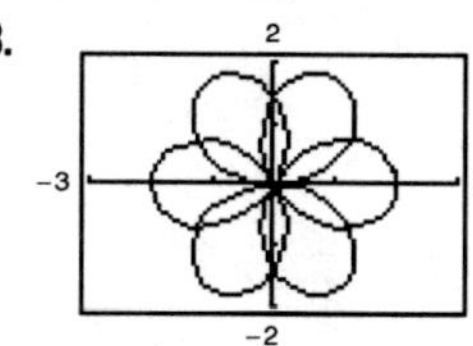
$0 \le \theta < 4\pi$

55.
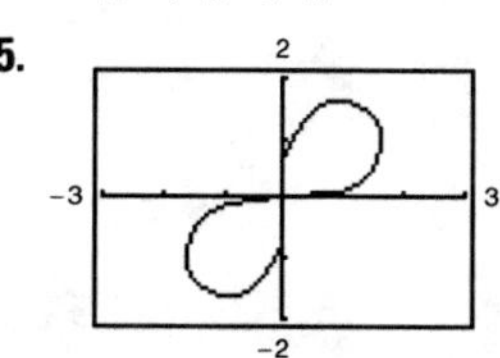
$0 \le \theta < \pi/2$

57. $(x - h)^2 + (y - k)^2 = h^2 + k^2$
Radius: $\sqrt{h^2 + k^2}$
Center: (h, k)

59. $\sqrt{17}$ **61.** About 5.6

63. $\dfrac{dy}{dx} = \dfrac{2 \cos \theta (3 \sin \theta + 1)}{6 \cos^2 \theta - 2 \sin \theta - 3}$

$(5, \pi/2):\ dy/dx = 0$
$(2, \pi):\ dy/dx = -2/3$
$(-1, 3\pi/2):\ dy/dx = 0$

65. (a) and (b)
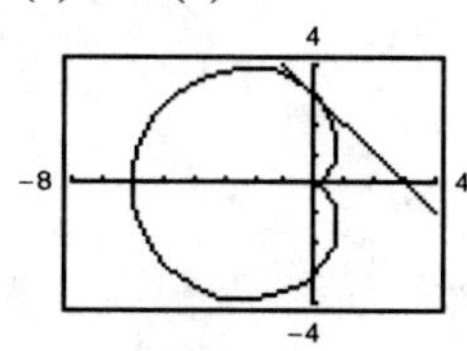
(c) $dy/dx = -1$

67. (a) and (b)
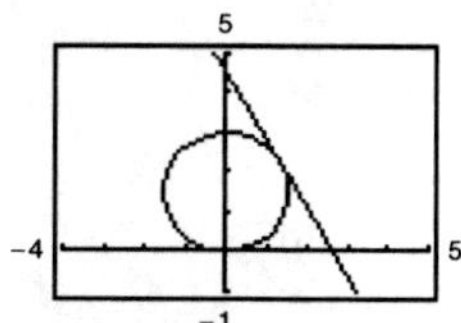
(c) $dy/dx = -\sqrt{3}$

69. Horizontal: $(2, 3\pi/2), (\frac{1}{2}, \pi/6), (\frac{1}{2}, 5\pi/6)$
Vertical: $(\frac{3}{2}, 7\pi/6), (\frac{3}{2}, 11\pi/6)$

71. $(5, \pi/2), (1, 3\pi/2)$

73. 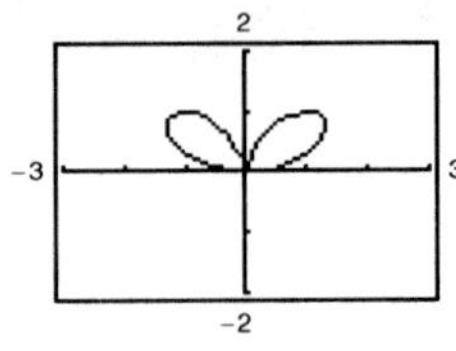   **75.** 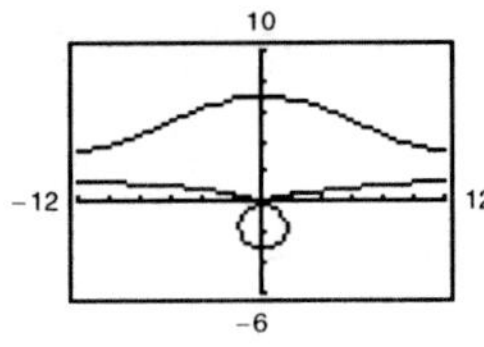

$(0, 0), (1.4142, 0.7854),$ $(7, 1.5708), (3, 4.7124)$
$(1.4142, 2.3562)$

77. 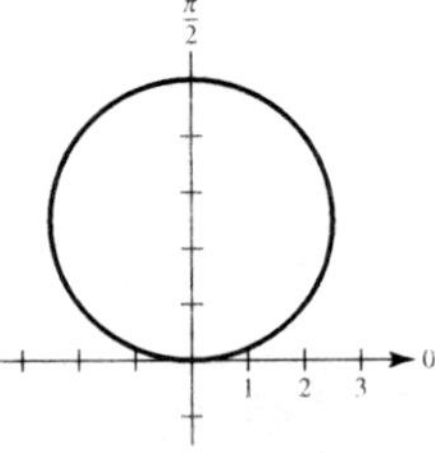   **79.** 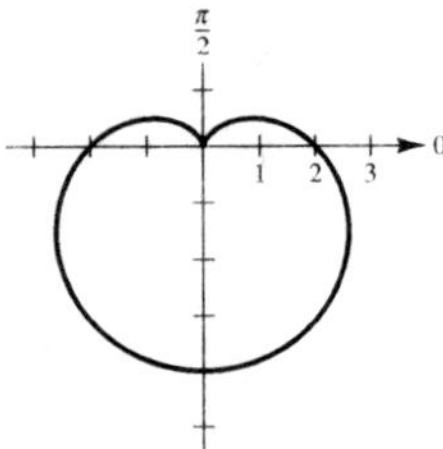

$\theta = 0$ $\theta = \pi/2$

81. 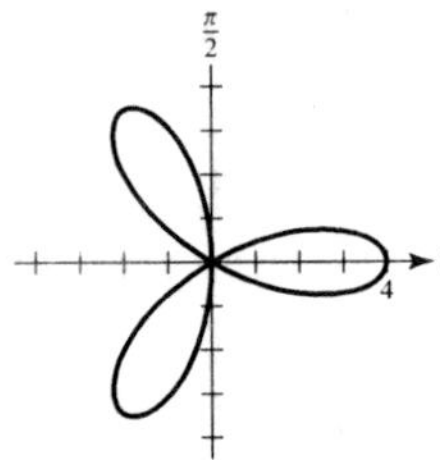   **83.** 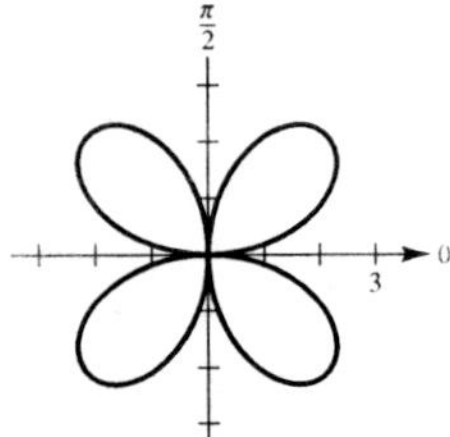

$\theta = \pi/6, \pi/2, 5\pi/6$ $\theta = 0, \pi/2$

85. 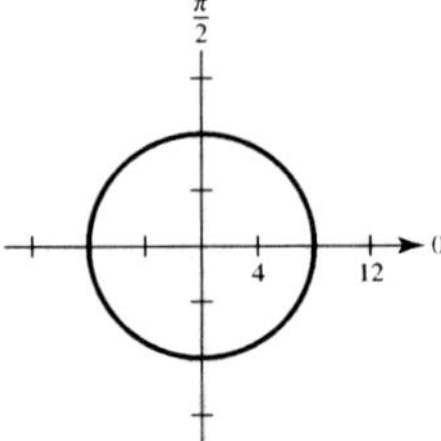   **87.** 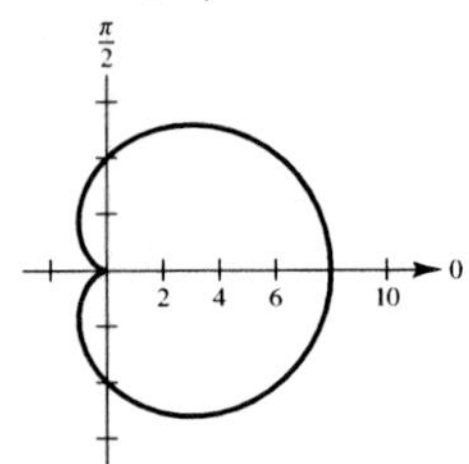

89. 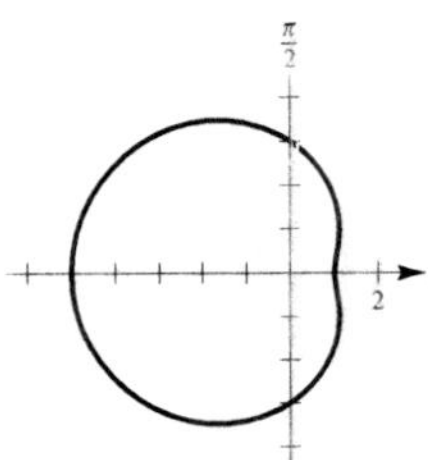   **91.** 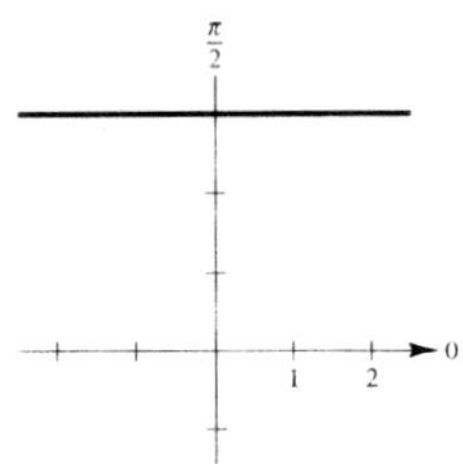

93. 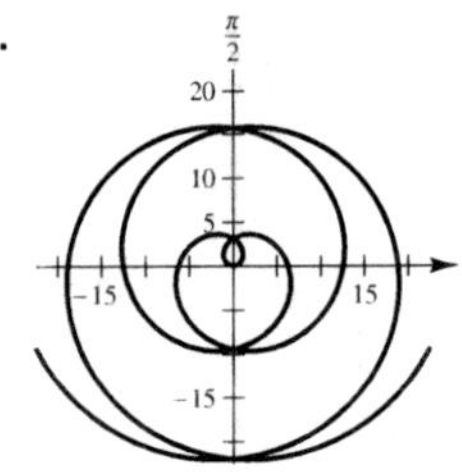   **95.** 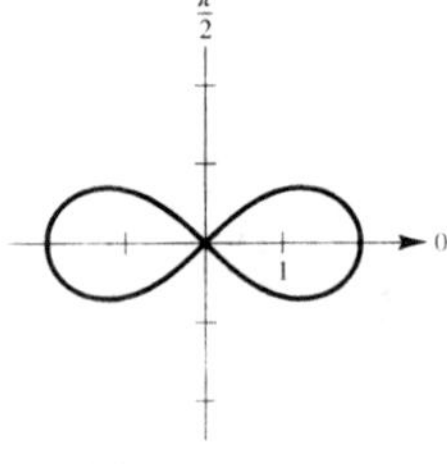

97. 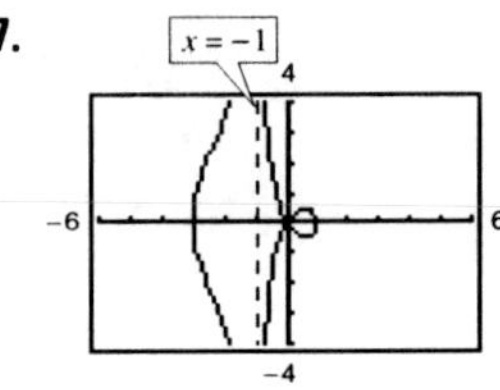

99. 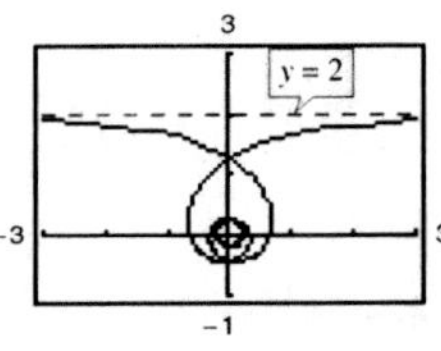

101. The rectangular coordinate system is a collection of points of the form (x, y), where x is the directed distance from the y-axis to the point and y is the directed distance from the x-axis to the point. Every point has a unique representation.

The polar coordinate system is a collection of points of the form (r, θ), where r is the directed distance from the origin O to a point P and θ is the directed angle, measured counterclockwise, from the polar axis to the segment $\overline{OP}$. Polar coordinates do not have unique representations.

103. Slope of tangent line to graph of $r = f(\theta)$ at (r, θ) is
$$\frac{dy}{dx} = \frac{f(\theta)\cos\theta + f'(\theta)\sin\theta}{-f(\theta)\sin\theta + f'(\theta)\cos\theta}.$$
If $f(\alpha) = 0$ and $f'(\alpha) \neq 0$, then $\theta = \alpha$ is tangent at the pole.

105. (a) 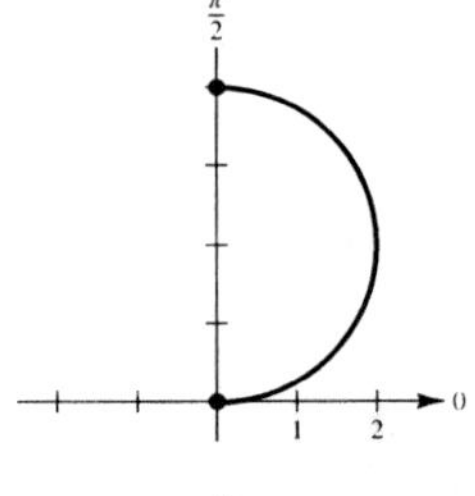   (b) 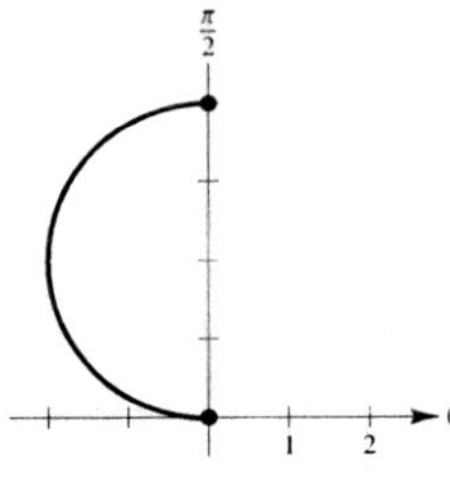

(c)

107. Proof

109. (a) $r = 2 - \sin(\theta - \pi/4)$ (b) $r = 2 + \cos\theta$
$$= 2 - \frac{\sqrt{2}(\sin\theta - \cos\theta)}{2}$$
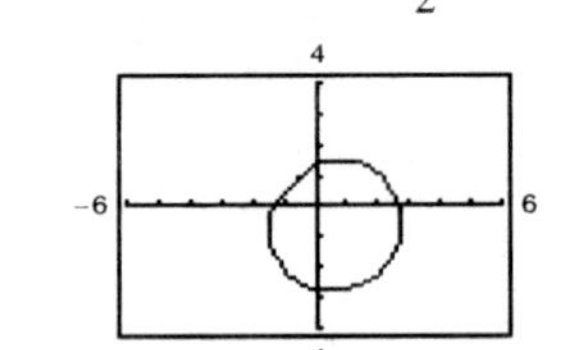   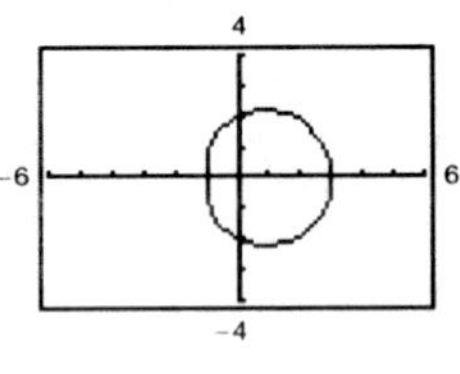

(c) $r = 2 + \sin\theta$ (d) $r = 2 - \cos\theta$
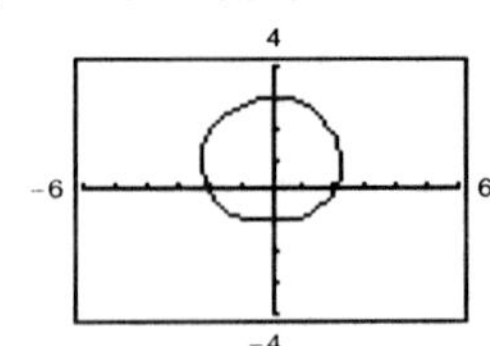   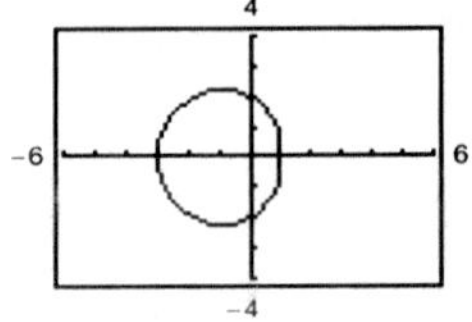

111. (a) 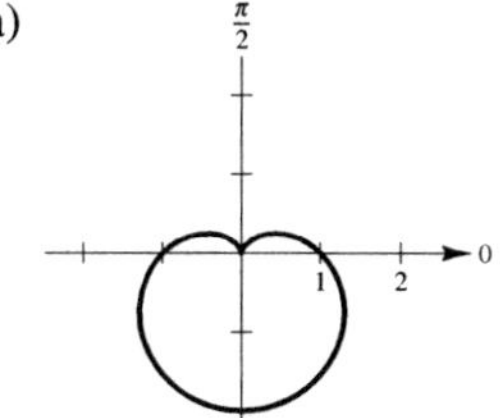 (b) 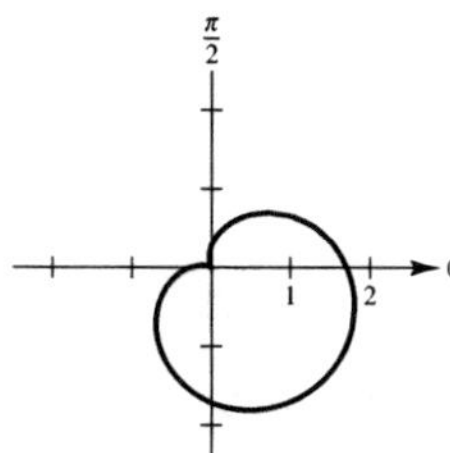

113. 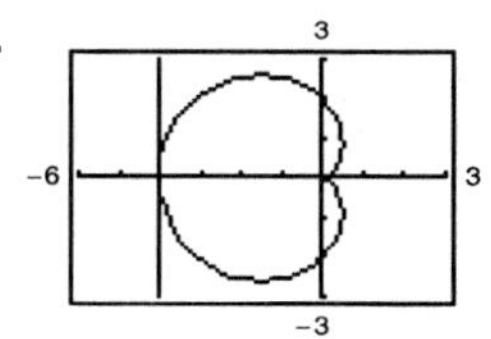

$\psi = \pi/2$

115. 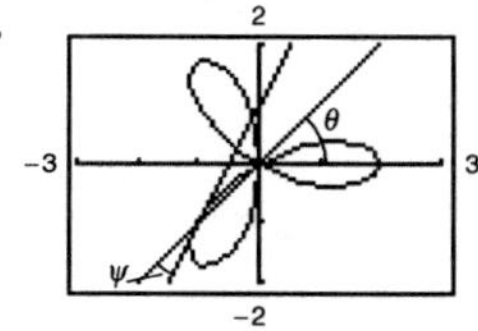

$\psi = \arctan \frac{1}{3} \approx 18.4°$

117. 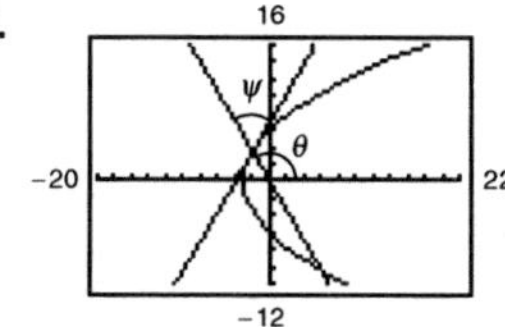

$\psi = \pi/3,\ 60°$

119. True **121.** True

Section 10.5 (page 747)

1. $8 \displaystyle\int_0^{\pi/2} \sin^2\theta\, d\theta$ **3.** $\dfrac{1}{2} \displaystyle\int_{\pi/2}^{3\pi/2} (3 - 2\sin\theta)^2\, d\theta$ **5.** 9π

7. $\pi/3$ **9.** $\pi/8$ **11.** $3\pi/2$ **13.** 27π **15.** 4

17. 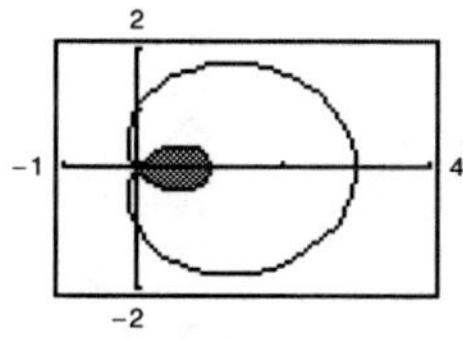

$(2\pi - 3\sqrt{3})/2$

19. 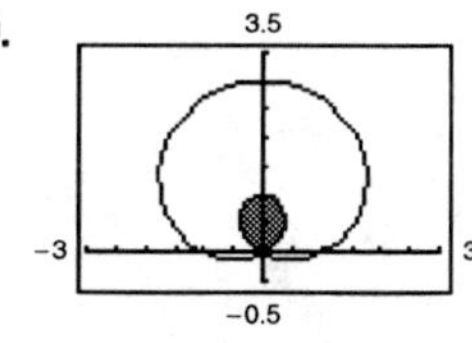

$(2\pi - 3\sqrt{3})/2$

21. 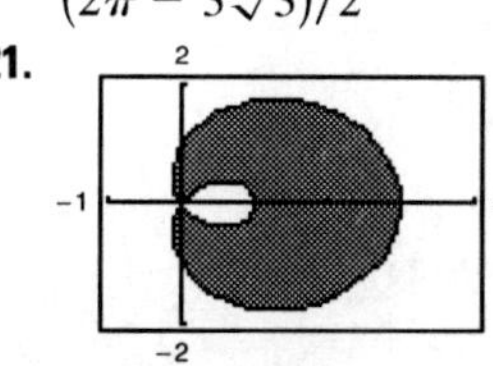

$\pi + 3\sqrt{3}$

23. 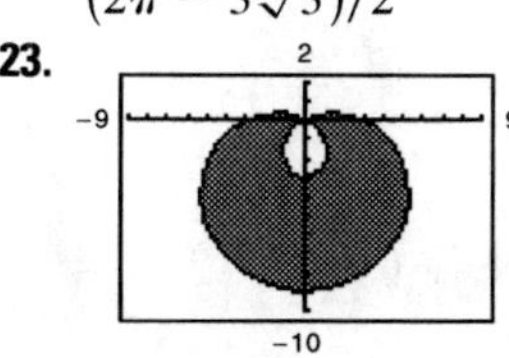

$9\pi + 27\sqrt{3}$

25. $(1, \pi/2),\ (1, 3\pi/2),\ (0, 0)$

27. $\left(\dfrac{2 - \sqrt{2}}{2}, \dfrac{3\pi}{4}\right), \left(\dfrac{2 + \sqrt{2}}{2}, \dfrac{7\pi}{4}\right),\ (0, 0)$

29. $\left(\dfrac{3}{2}, \dfrac{\pi}{6}\right), \left(\dfrac{3}{2}, \dfrac{5\pi}{6}\right),\ (0, 0)$ **31.** $(2, 4),\ (-2, -4)$

33. $(1, \pi/12),\ (1, 5\pi/12),\ (1, 7\pi/12),\ (1, 11\pi/12),$
$(1, 13\pi/12),\ (1, 17\pi/12),\ (1, 19\pi/12),\ (1, 23\pi/12)$

35. 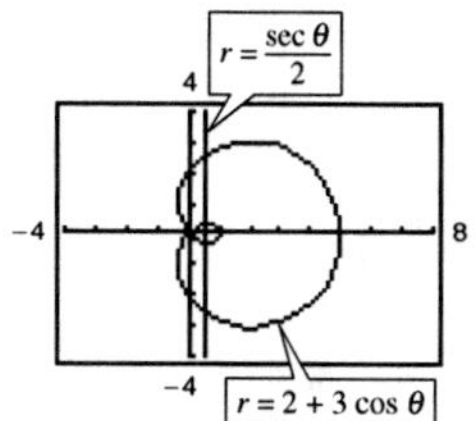

$(-0.581, \pm 2.607),$
$(2.581, \pm 1.376)$

37. 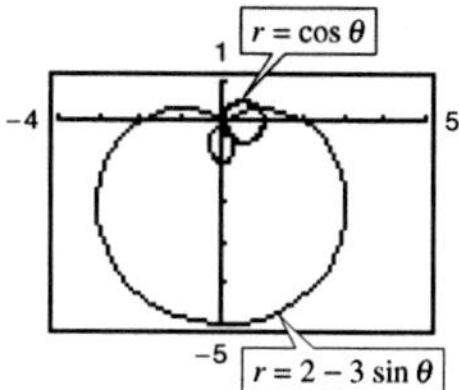

$(0, 0),\ (0.935, 0.363),$
$(0.535, -1.006)$

The graphs reach the pole at different times (θ-values).

39. 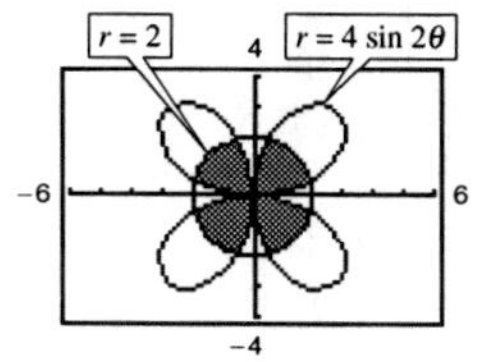

$\dfrac{4}{3}(4\pi - 3\sqrt{3})$

41. 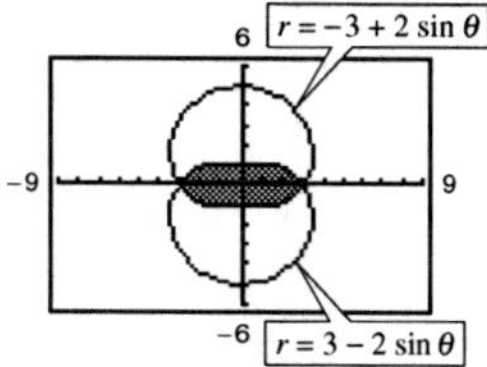

$11\pi - 24$

43. 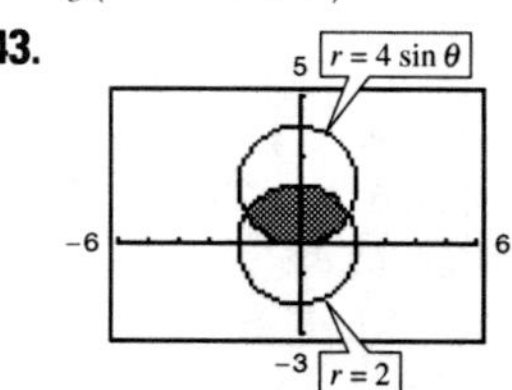

$\dfrac{2}{3}(4\pi - 3\sqrt{3})$

45. 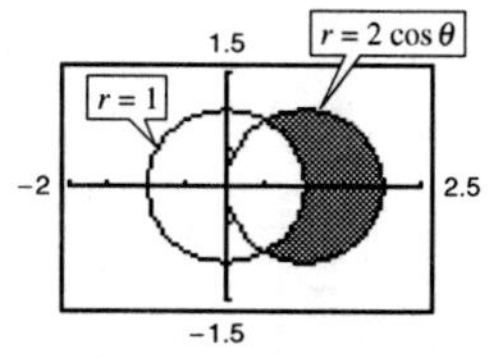

$\pi/3 + \sqrt{3}/2$

47. $5\pi a^2/4$ **49.** $(a^2/2)(\pi - 2)$

51. (a) $(x^2 + y^2)^{3/2} = ax^2$

(b) 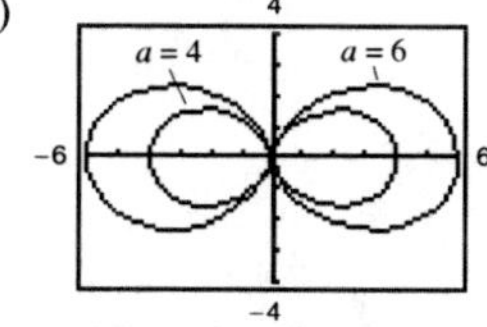

(c) $15\pi/2$

53. The area enclosed by the function is $\pi a^2/4$ if n is odd and is $\pi a^2/2$ if n is even.

55. 16π **57.** 4π **59.** 8

61. 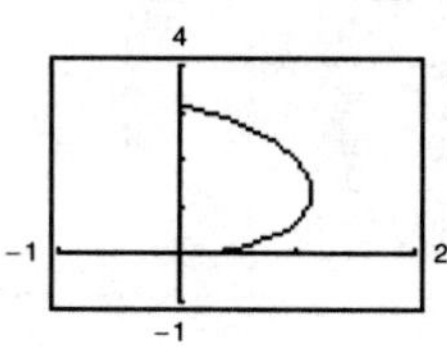

About 4.16

63. 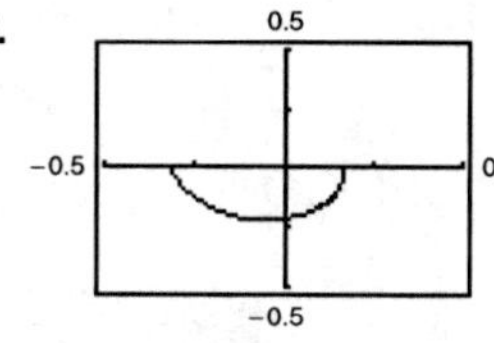

About 0.71

65. 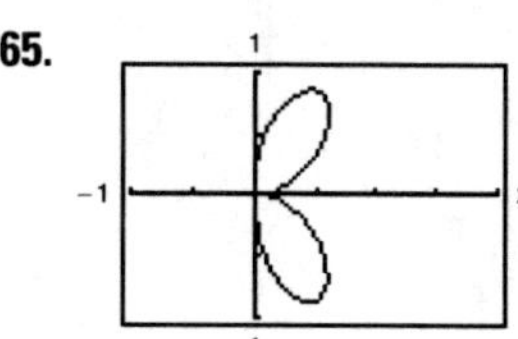

About 4.39

67. 36π

69. $\dfrac{2\pi\sqrt{1 + a^2}}{1 + 4a^2}(e^{\pi a} - 2a)$ **71.** 21.87

73. You will only find simultaneous points of intersection. There may be intersection points that do not occur with the same coordinates in the two graphs.

75. (a) $S = 2\pi \displaystyle\int_\alpha^\beta f(\theta)\sin\theta\sqrt{f(\theta)^2 + f'(\theta)^2}\,d\theta$

(b) $S = 2\pi \displaystyle\int_\alpha^\beta f(\theta)\cos\theta\sqrt{f(\theta)^2 + f'(\theta)^2}\,d\theta$

77. $40\pi^2$

79. (a) 16π

(b)

θ	0.2	0.4	0.6	0.8	1.0	1.2	1.4
A	6.32	12.14	17.06	20.80	23.27	24.60	25.08

(c) and (d) For $\frac{1}{4}$ of area ($4\pi \approx 12.57$): 0.42

For $\frac{1}{2}$ of area ($8\pi \approx 25.13$): 1.57($\pi/2$)

For $\frac{3}{4}$ of area ($12\pi \approx 37.70$): 2.73

(e) No. The results do not depend on the radius. Answers will vary.

81. Circle

83. (a) 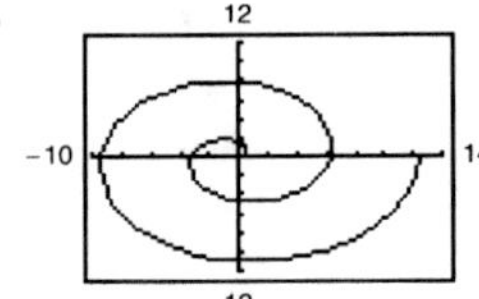   The graph becomes larger and more spread out. The graph is reflected over the y-axis.

(b) $(an\pi, n\pi)$ where $n = 1, 2, 3, \ldots$.

(c) About 21.26   (d) $4/3\pi^3$

85. $r = \sqrt{2}\cos\theta$

87. False. The graphs of $f(\theta) = 1$ and $g(\theta) = -1$ coincide.

89. Proof

Section 10.6   (page 755)

1. 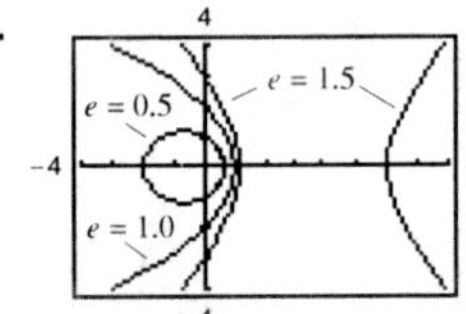

(a) Parabola   (b) Ellipse

(c) Hyperbola

3. 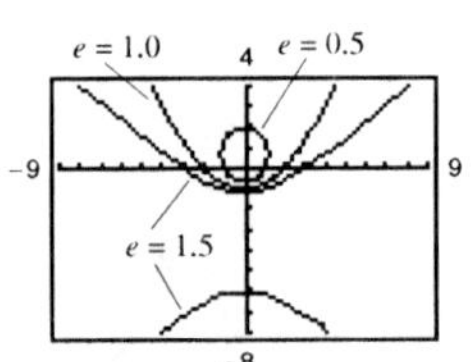

(a) Parabola   (b) Ellipse

(c) Hyperbola

5. (a) 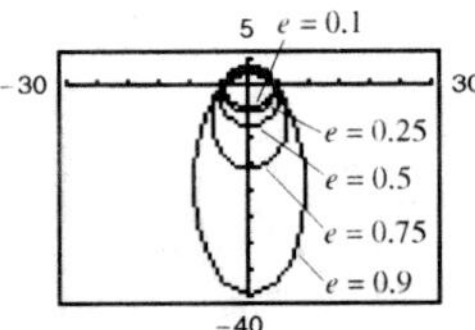   (b) 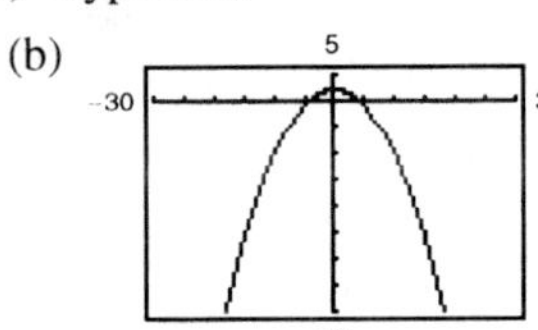

Ellipse   Parabola

As $e \to 1^-$, the ellipse becomes more elliptical, and as $e \to 0^+$, it becomes more circular.

(c) 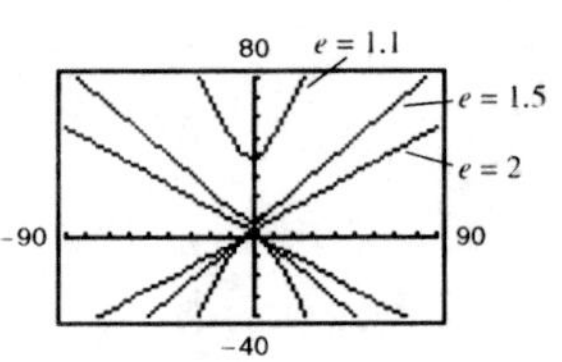   Hyperbola

As $e \to 1^+$, the hyperbola opens more slowly, and as $e \to \infty$, it opens more rapidly.

7. c   **8.** f   **9.** a   **10.** e   **11.** b   **12.** d

13. $e = 1$

Distance $= 1$

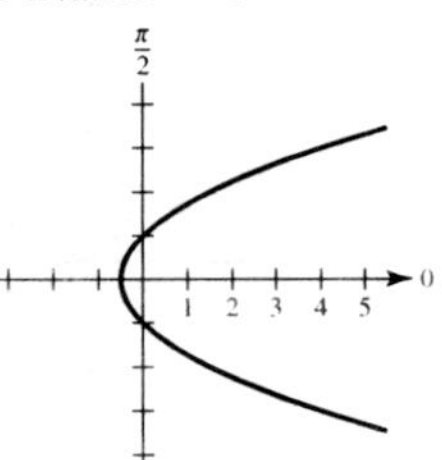

Parabola

15. $e = 1$

Distance $= 4$

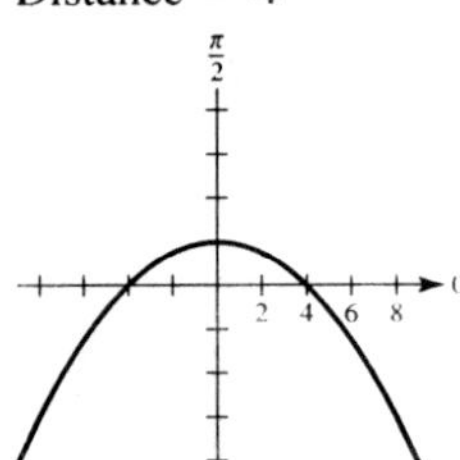

Parabola

17. $e = \frac{1}{2}$

Distance $= 6$

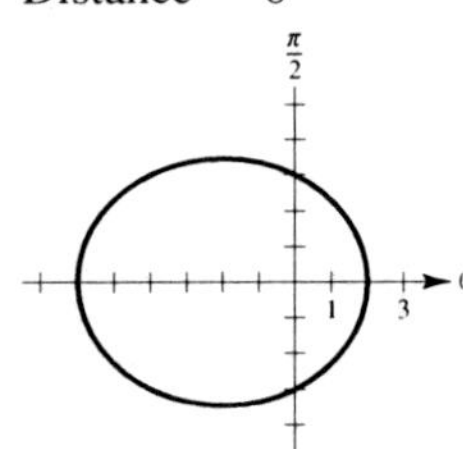

Ellipse

19. $e = \frac{1}{2}$

Distance $= 4$

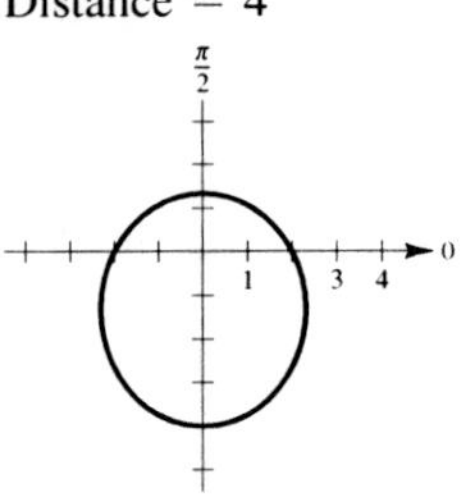

Ellipse

21. $e = 2$

Distance $= \frac{5}{2}$

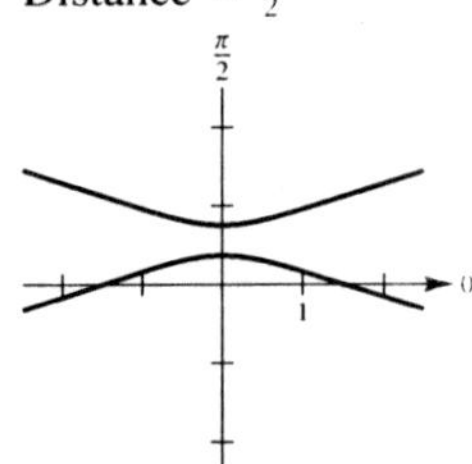

Hyperbola

23. $e = 3$

Distance $= \frac{1}{2}$

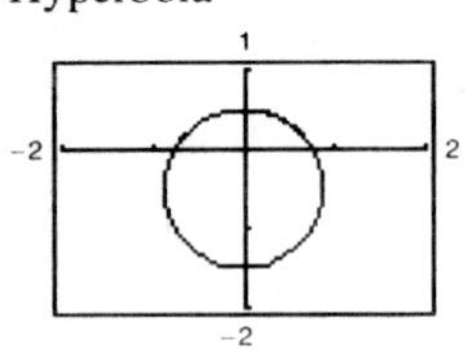

Hyperbola

25. $e = \frac{1}{2}$

Distance $= 50$

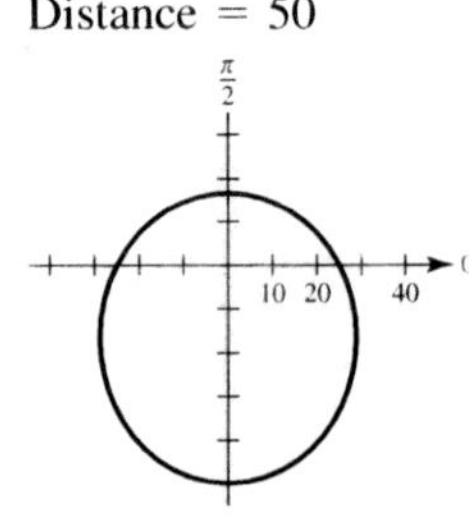

Ellipse

27. 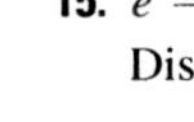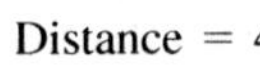

Ellipse

$e = \frac{1}{2}$

29. 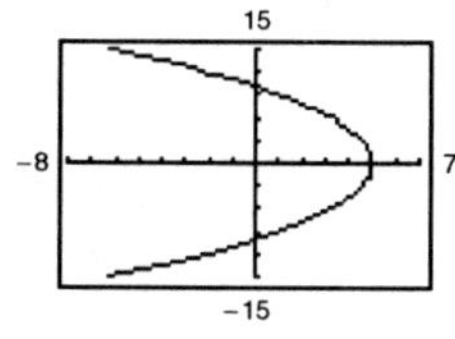

Parabola
$e = 1$

31. 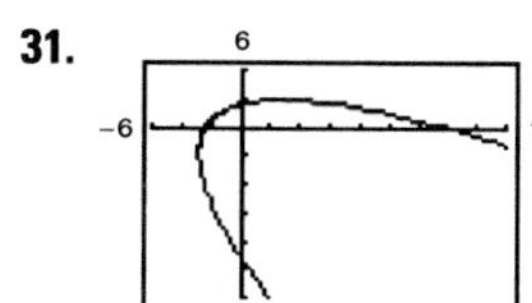

Rotated $\pi/4$ radian
counterclockwise.

33. 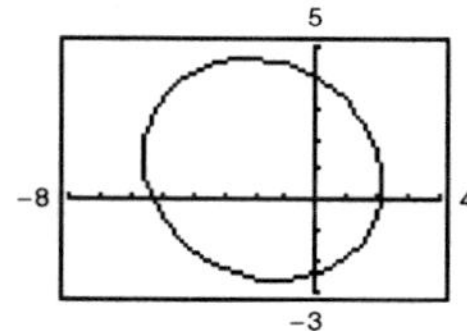

Rotated $\pi/6$ radian clockwise.

35. $r = \dfrac{8}{8 + 5 \cos\left(\theta + \dfrac{\pi}{6}\right)}$

37. $r = 3/(1 - \cos\theta)$ **39.** $r = 1/(2 + \sin\theta)$
41. $r = 2/(1 + 2\cos\theta)$ **43.** $r = 2/(1 - \sin\theta)$
45. $r = 16/(5 + 3\cos\theta)$ **47.** $r = 9/(4 - 5\sin\theta)$
49. $r = 4/(2 + \cos\theta)$
51. If $0 < e < 1$, the conic is an ellipse.
 If $e = 1$, the conic is a parabola.
 If $e > 1$, the conic is a hyperbola.
53. If the foci are fixed and $e \to 0$, then $d \to \infty$. To see this, compare
 the ellipses

$$r = \frac{1/2}{1 + (1/2)\cos\theta}, \ e = \frac{1}{2}, \ d = 1 \text{ and}$$

$$r = \frac{5/16}{1 + (1/4)\cos\theta}, \ e = \frac{1}{4}, \ d = \frac{5}{4}.$$

55. Proof

57. $r^2 = \dfrac{9}{1 - (16/25)\cos^2\theta}$ **59.** $r^2 = \dfrac{-16}{1 - (25/9)\cos^2\theta}$

61. About 10.88   **63.** 3.37   **65.** $\dfrac{7979.21}{1 - 0.9372\cos\theta}$; 11,015 mi

67. $r = \dfrac{149{,}558{,}278.0560}{1 - 0.0167\cos\theta}$ **69.** $r = \dfrac{4{,}497{,}667{,}328}{1 - 0.0086\cos\theta}$
 Perihelion: 147,101,680 km   Perihelion: 4,459,317,200 km
 Aphelion: 152,098,320 km   Aphelion: 4,536,682,800 km
71. Answers will vary. Sample answers:
 (a) 3.591×10^{18} km^2; 9.322 yr
 (b) $\alpha \approx 0.361 + \pi$; Larger angle with the smaller ray to
  generate an equal area
 (c) Part (a): 1.583×10^9 km; 1.698×10^8 km/yr
  Part (b): 1.610×10^9 km; 1.727×10^8 km/yr
73. Proof
75. Let $r_1 = ed/(1 + \sin\theta)$ and $r_2 = ed/(1 - \sin\theta)$.

 The points of intersection of r_1 and r_2 are $(ed, 0)$ and (ed, π).
 The slopes of the tangent lines to r_1 are -1 at $(ed, 0)$ and 1 at
 (ed, π). The slopes of the tangent lines to r_2 are 1 at $(ed, 0)$ and
 -1 at (ed, π). Therefore, at $(ed, 0)$, $m_1 m_2 = -1$, and at (ed, π),
 $m_1 m_2 = -1$, and the curves intersect at right angles.

Review Exercises for Chapter 10   (page 758)

1. e   **2.** c   **3.** b   **4.** d   **5.** a   **6.** f
7. Circle
 Center: $\left(\frac{1}{2}, -\frac{3}{4}\right)$
 Radius: 1

9. Hyperbola
 Center: $(-4, 3)$
 Vertices: $\left(-4 \pm \sqrt{2}, 3\right)$

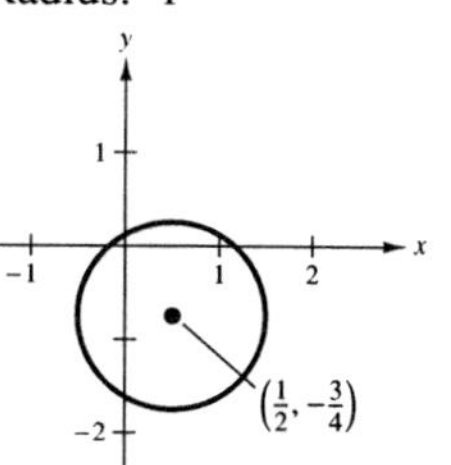
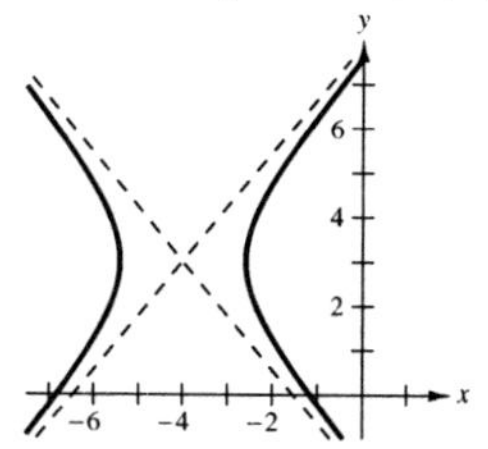

11. Ellipse
 Center: $(2, -3)$
 Vertices: $\left(2, -3 \pm \sqrt{2}/2\right)$

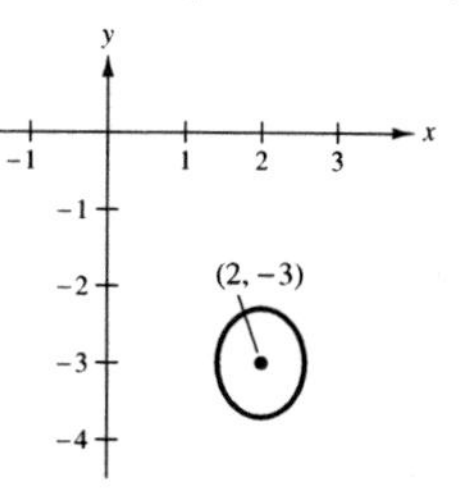

13. $y^2 - 4y - 12x + 4 = 0$ **15.** $(x - 1)^2/36 + y^2/20 = 1$
17. $x^2/49 - y^2/32 = 1$ **19.** About 15.87
21. $4x + 4y - 7 = 0$ **23.** (a) $(0, 50)$ (b) About 38,294.49

25. 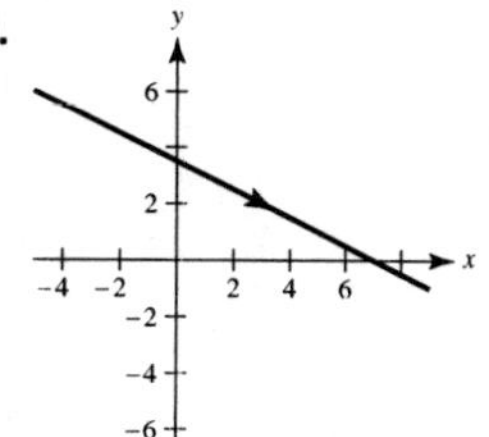

$x + 2y - 7 = 0$

27. 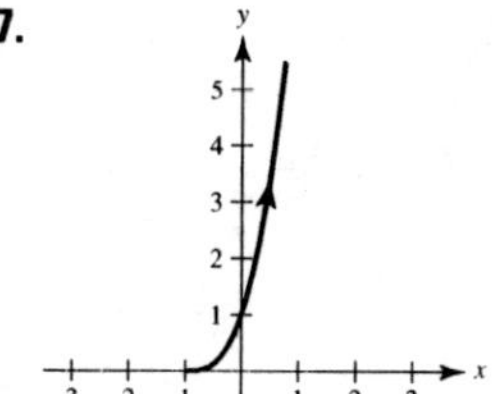
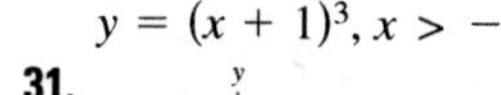

$y = (x + 1)^3, x > -1$

29. 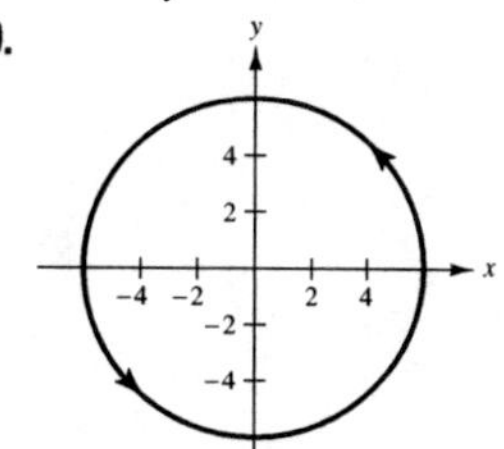

$x^2 + y^2 = 36$

31. 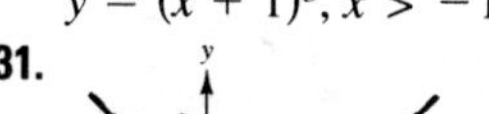
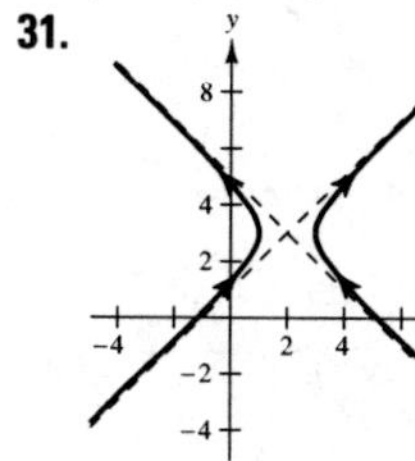

$(x - 2)^2 - (y - 3)^2 = 1$

33. Answers will vary. Sample answer:
 $x = 5t - 2$
 $y = 6 - 4t$
35. $x = 4\cos\theta - 3$
 $y = 4 + 3\sin\theta$

37. 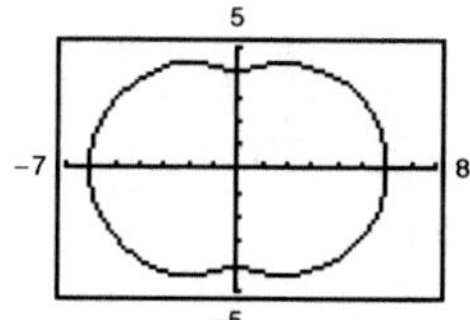

39. (a) $dy/dx = -\frac{4}{5}$;
Horizontal tangents: None
(b) $y = (-4x + 13)/5$
(c) 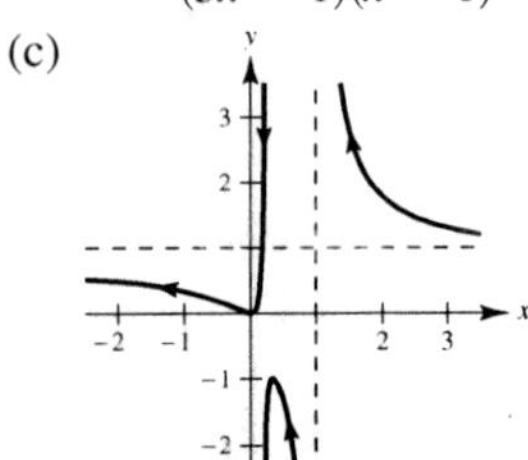

41. (a) $dy/dx = -2t^2$;
Horizontal tangents: None
(b) $y = 3 + 2/x$
(c) 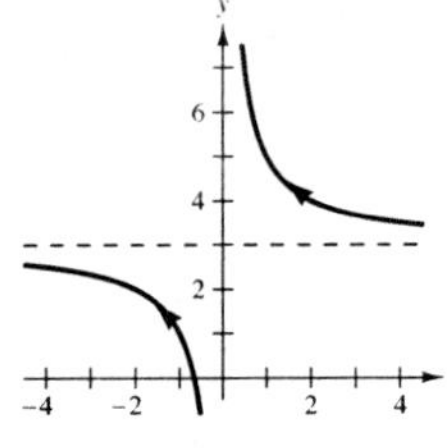

43. (a) $\dfrac{dy}{dx} = \dfrac{(t-1)(2t+1)^2}{t^2(t-2)^2}$;
Horizontal tangent: $\left(\dfrac{1}{3}, -1\right)$
(b) $y = \dfrac{4x^2}{(5x-1)(x-1)}$
(c) 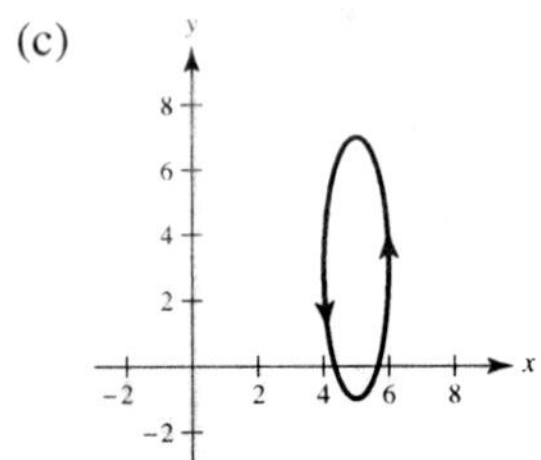

45. (a) $\dfrac{dy}{dx} = -4\cot\theta$;
Horizontal tangents: $(5, 7), (5, -1)$
(b) $(x-5)^2 + \dfrac{(y-3)^2}{16} = 1$
(c)

47. (a) $\dfrac{dy}{dx} = -4\tan\theta$;
Horizontal tangents: None
(b) $x^{2/3} + (y/4)^{2/3} = 1$
(c) 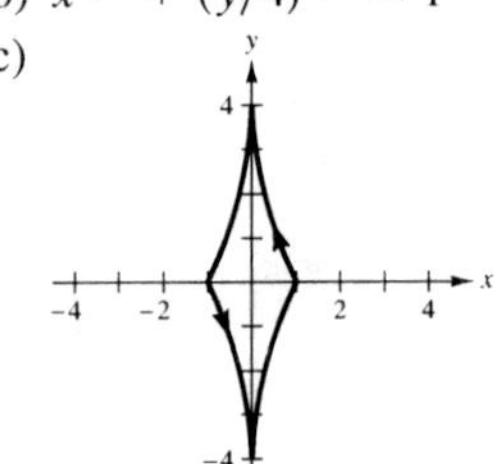

49. Horizontal: $(5, 0)$
Vertical: None

51. Horizontal: $(2, 2), (2, 0)$
Vertical: $(4, 1), (0, 1)$

53. (a) and (c)
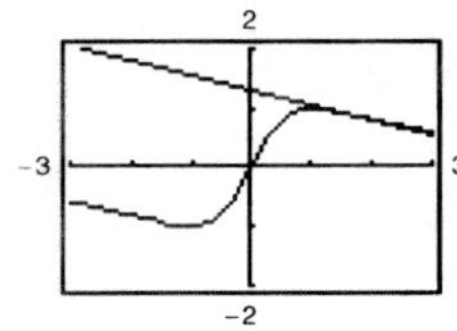
(b) $dx/d\theta = -4, dy/d\theta = 1, dy/dx = -\frac{1}{4}$

55. $\frac{1}{2}\pi^2 r$

57. (a) $s = 12\pi\sqrt{10} \approx 119.215$ **59.** $A = 3\pi$
(b) $s = 4\pi\sqrt{10} \approx 39.738$

61. 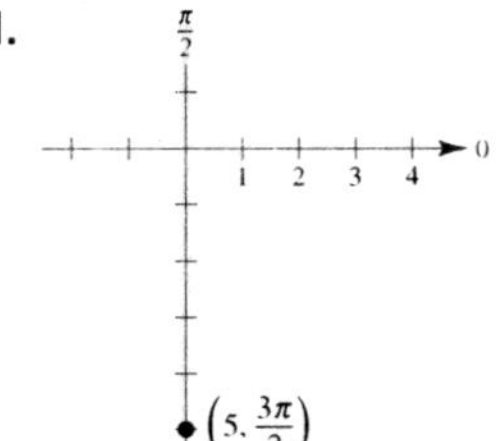
Rectangular: $(0, -5)$

63. 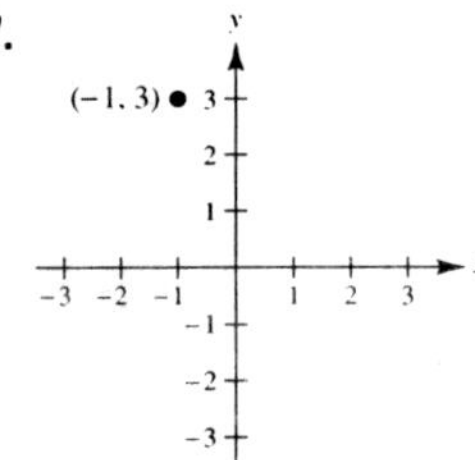
Rectangular: $(0.0187, 1.7320)$

65. 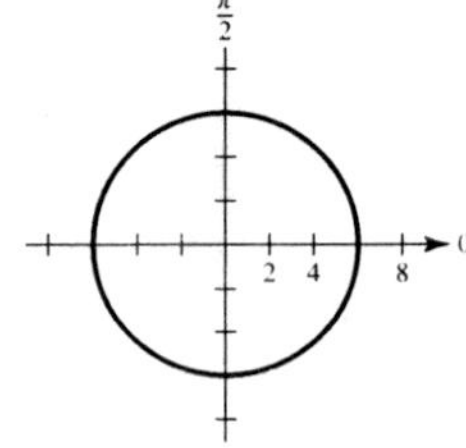

67. 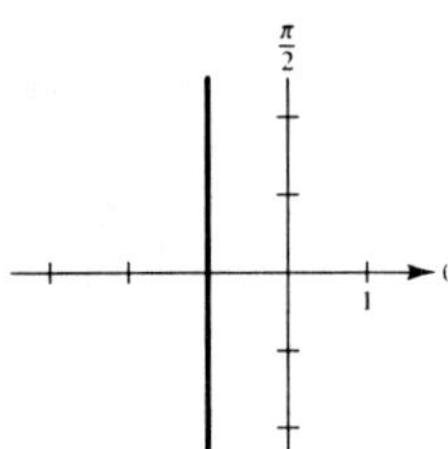

$\left(4\sqrt{2}, \dfrac{7\pi}{4}\right), \left(-4\sqrt{2}, \dfrac{3\pi}{4}\right)$ $\left(\sqrt{10}, 1.89\right), \left(-\sqrt{10}, 5.03\right)$

69. $x^2 + y^2 - 3x = 0$ **71.** $(x^2 + y^2 + 2x)^2 = 4(x^2 + y^2)$
73. $(x^2 + y^2)^2 = x^2 - y^2$ **75.** $y^2 = x^2[(4-x)/(4+x)]$
77. $r = a\cos^2\theta\sin\theta$ **79.** $r^2 = a^2\theta^2$
81. Circle   **83.** Line
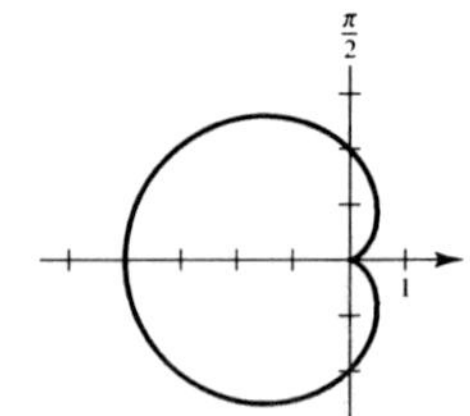   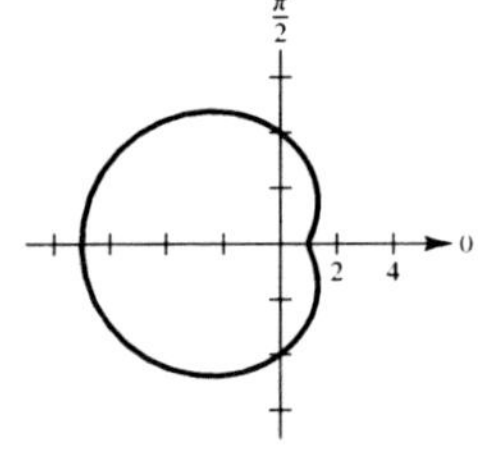

85. Cardioid   **87.** Limaçon

89. Rose curve

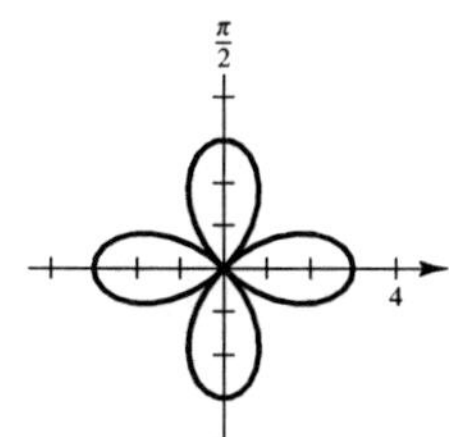

91. Rose curve

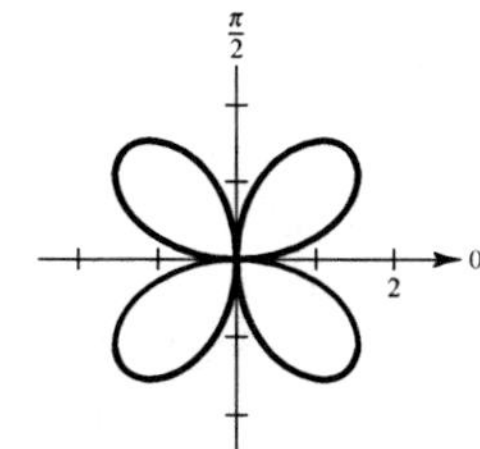

117. Parabola

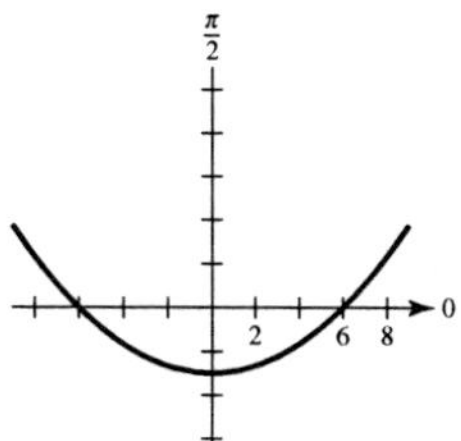

119. Ellipse

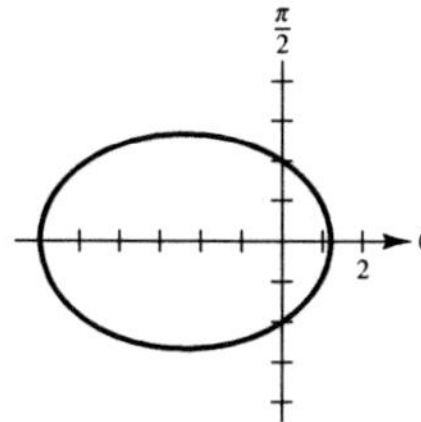

93.

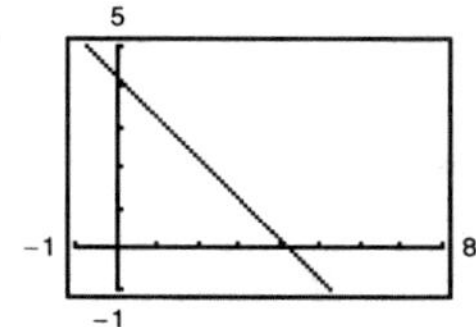

95.

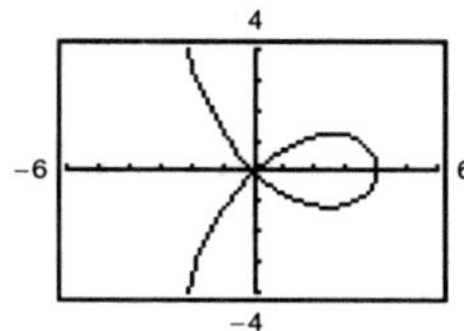

121. Hyperbola

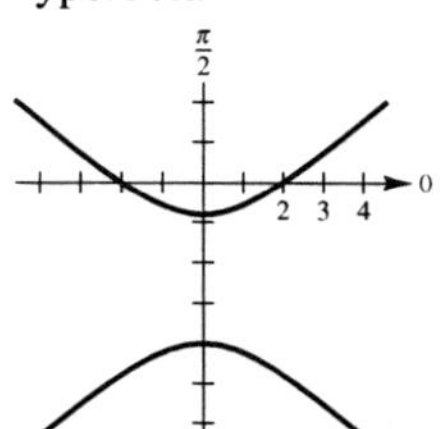

123. $r = 10 \sin \theta$

97. (a) $\theta = \pm \pi/3$

(b) Vertical: $(-1, 0)$, $(3, \pi)$, $\left(\frac{1}{2}, \pm 1.318\right)$

Horizontal: $(-0.686, \pm 0.568)$, $(2.186, \pm 2.206)$

(c)

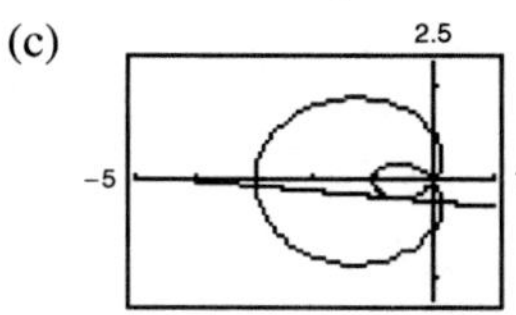

125. $r = 4/(1 - \cos \theta)$ **127.** $r = 5/(3 - 2 \cos \theta)$

P.S. Problem Solving   (page 761)

1. (a)

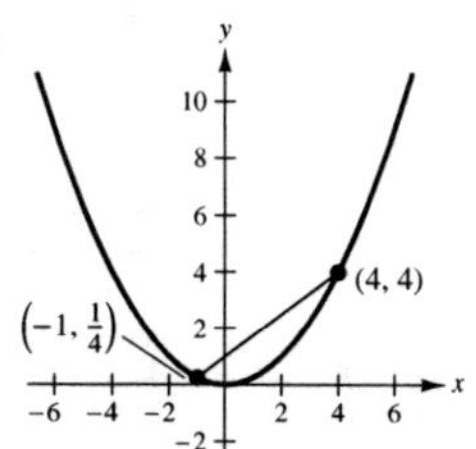

3. Proof

(b) and (c) Proofs

5. (a) $r = 2a \tan \theta \sin \theta$

(b) $x = 2at^2/(1 + t^2)$

$y = 2at^3/(1 + t^2)$

(c) $y^2 = x^3/(2a - x)$

7. (a) $y^2 = x^2[(1 - x)/(1 + x)]$

(b) $r = \cos 2\theta \cdot \sec \theta$

(c)

99. Proof   **101.** $\dfrac{9\pi}{20}$ **103.** $\dfrac{9\pi}{2}$ **105.** 4

107. $\left(1 + \dfrac{\sqrt{2}}{2}, \dfrac{3\pi}{4}\right)$, $\left(1 - \dfrac{\sqrt{2}}{2}, \dfrac{7\pi}{4}\right)$, $(0, 0)$

109.

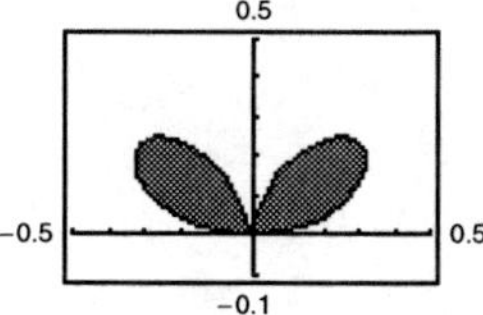

$$A = 2\left(\frac{1}{2}\right)\int_0^{\pi/2} \sin^2 \theta \cos^4 \theta \, d\theta \approx 0.10$$

111.

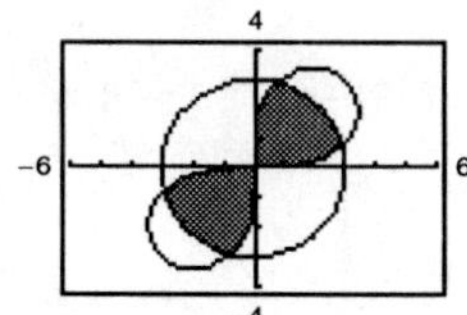

$$A = 2\left[\frac{1}{2}\int_0^{\pi/12} 18 \sin 2\theta + \frac{1}{2}\int_{\pi/12}^{5\pi/12} 9 \, d\theta + \frac{1}{2}\int_{5\pi/12}^{\pi/2} 18 \sin 2\theta \, d\theta\right]$$
$$\approx 1.2058 + 9.4248 + 1.2058 = 11.8364$$

113. $4a$

115. $S = 2\pi\displaystyle\int_0^{\pi/2} (1 + 4 \cos \theta) \sin \theta \sqrt{17 + 8 \cos \theta} \, d\theta$

$= 34\pi\sqrt{17}/5 \approx 88.08$

(d) $y = x$, $y = -x$

(e) $\left(\dfrac{\sqrt{5} - 1}{2}, \pm \dfrac{\sqrt{5} - 1}{2}\sqrt{-2 + \sqrt{5}}\right)$

9. (a)

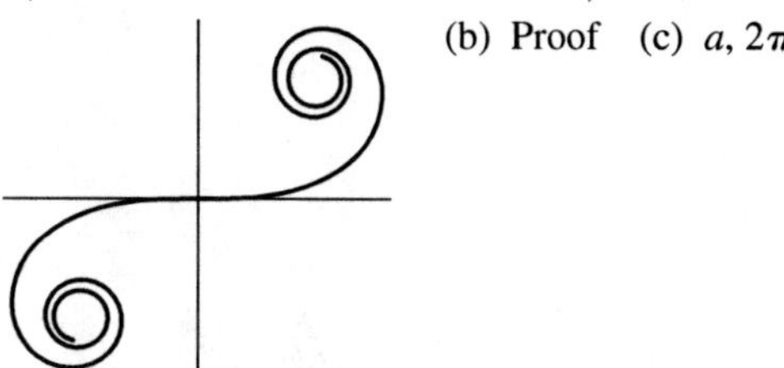

(b) Proof   (c) $a, 2\pi$

Generated by Mathematica

11. $A = \frac{1}{2}ab$ **13.** $r^2 = 2\cos 2\theta$

15. (a) First plane: $x_1 = \cos 70°(150 - 375t)$
$$y_1 = \sin 70°(150 - 375t)$$
Second plane: $x_2 = \cos 45°(450t - 190)$
$$y_2 = \sin 45°(190 - 450t)$$
(b) $\{[\cos 45°(450t - 190) - \cos 70°(150 - 375t)]^2$
$$+ [\sin 45°(190 - 450t) - \sin 70°(150 - 375t)]^2\}^{1/2}$$
(c) 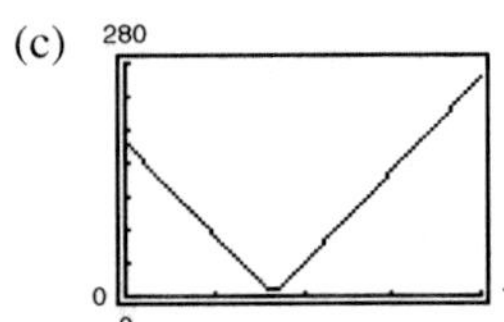

0.4145 h; Yes

17. 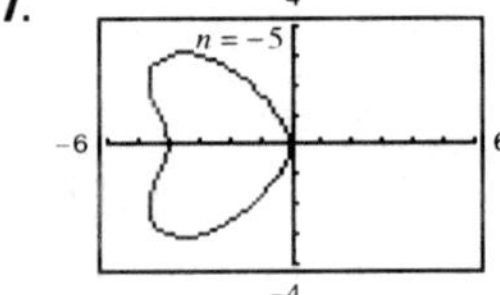 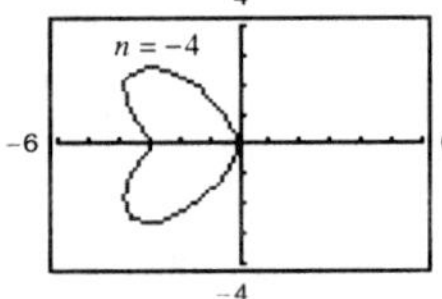

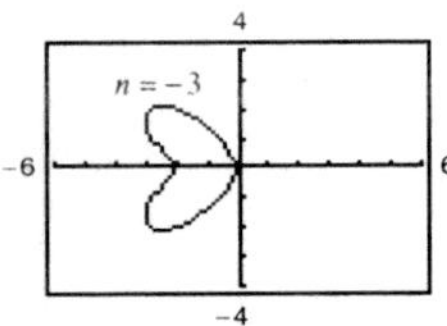 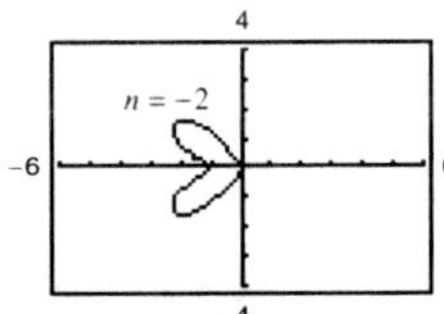

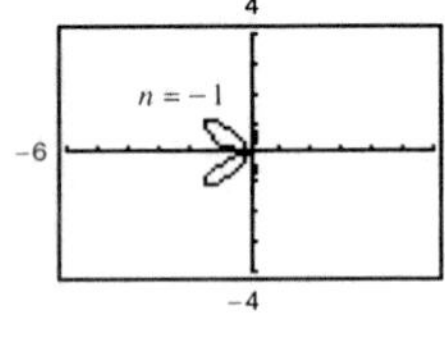 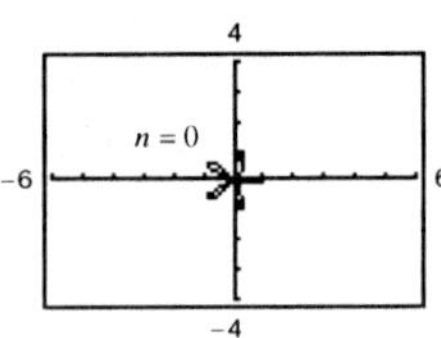

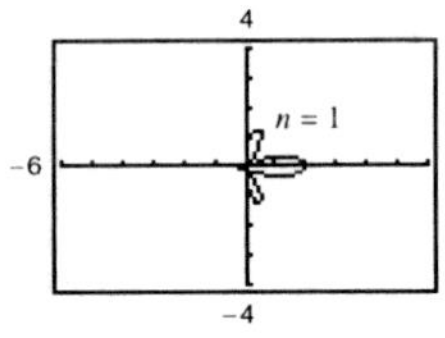 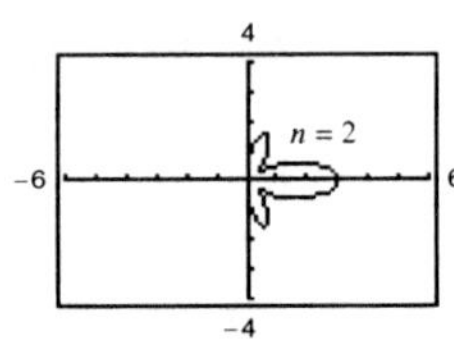

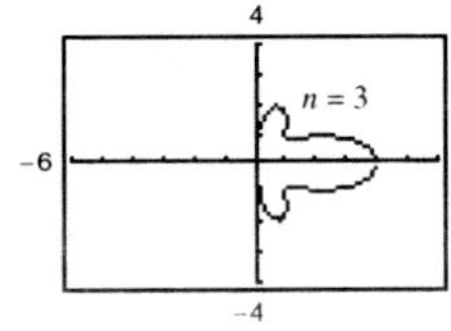 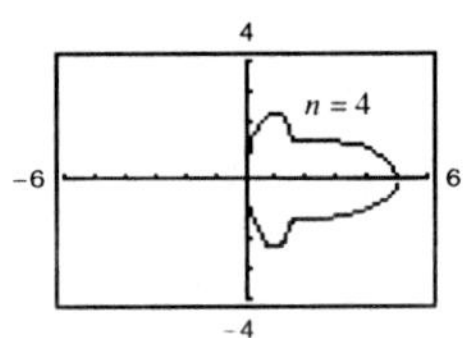

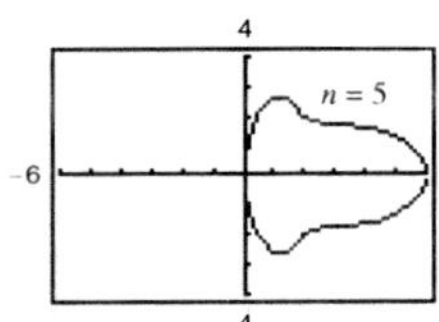

$n = 1, 2, 3, 4, 5$ produce "bells"; $n = -1, -2, -3, -4, -5$ produce "hearts."

Appendix C

Appendix C.1   (page A31)

1. Rational   **3.** Irrational   **5.** Rational   **7.** Rational
9. Rational   **11.** $\frac{4}{11}$ **13.** $\frac{11}{37}$
15. (a) True   (b) False   (c) True   (d) False
(e) False   (f) False
17. x is greater than -3 and less than 3.
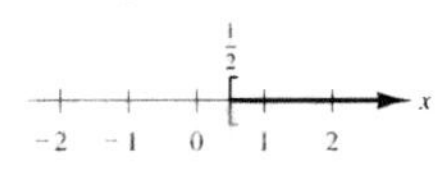   The interval is bounded.
19. x is no more than 5.
    The interval is unbounded.
21. $y \geq 4, [4, \infty)$ **23.** $0.03 < r \leq 0.07, (0.03, 0.07]$
25. $x \geq \frac{1}{2}$ **27.** $-\frac{1}{2} < x < \frac{7}{2}$

29. $x > 6$ **31.** $-1 < x < 1$

33. $x \geq 13, x \leq -7$ **35.** $a - b < x < a + b$

37. $-3 < x < 2$ **39.** $0 < x < 3$

41. $-3 \leq x \leq 1$ **43.** $-3 \leq x \leq 2$

45. $4, -4, 4$ **47.** (a) $-51, 51, 51$ (b) $51, -51, 51$
49. $|x| \leq 2$ **51.** $|x - 2| > 2$
53. (a) $|x - 12| \leq 10$ (b) $|x - 12| \geq 10$
55. 1   **57.** (a) 14   (b) 10
59. $x \geq 36$ units   **61.** $x \leq 41$ or $x \geq 59$
63. (a) $\frac{355}{113} > \pi$ (b) $\frac{22}{7} > \pi$ **65.** b
67. False. The reciprocal of 2 is $\frac{1}{2}$, which is not an integer.
69. True   **71.** True   **73–79.** Proofs
81. $|-3 - 1| > |-3| - |1|; |3 - 1| = |3| - |1|$; Proof

Appendix C.2   (page A38)

1. (a)   **3.** (a)

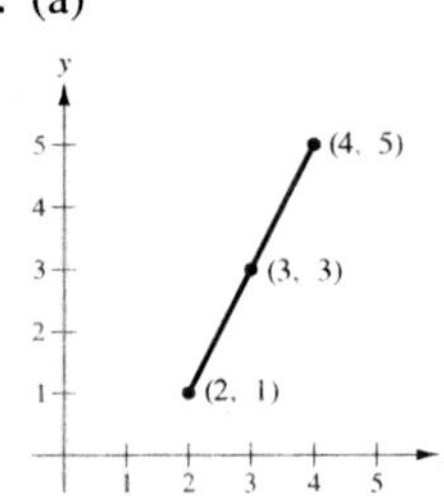

(b) $2\sqrt{5}$ (b) $2\sqrt{10}$
(c) $(3, 3)$ (c) $\left(-\frac{1}{2}, -2\right)$

5. (a)

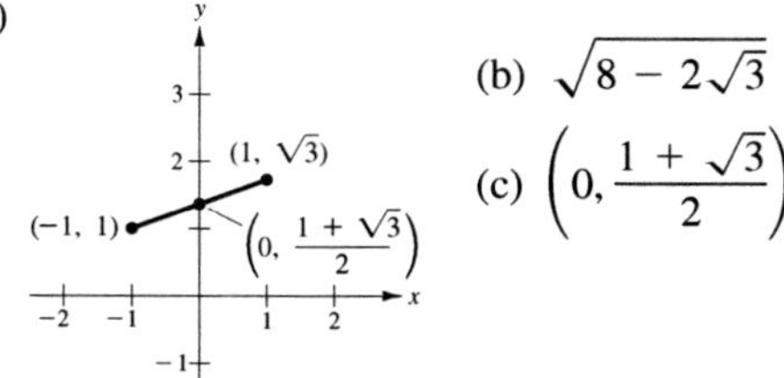

(b) $\sqrt{8 - 2\sqrt{3}}$

(c) $\left(0, \dfrac{1 + \sqrt{3}}{2}\right)$

7. Quadrant II   **9.** Quadrants I and III

11. Right triangle:
$d_1 = \sqrt{45},\ d_2 = \sqrt{5}$
$d_3 = \sqrt{50}$
$(d_1)^2 + (d_2)^2 = (d_3)^2$

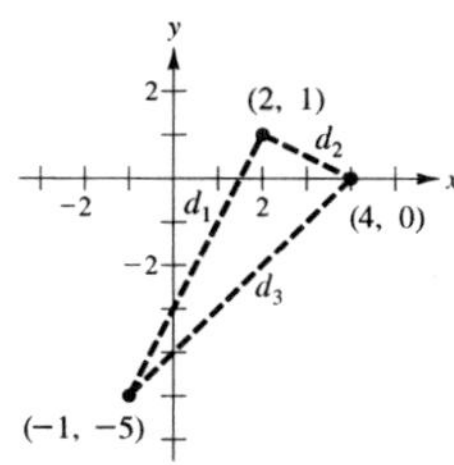

13. Rhombus: the length of each side is $\sqrt{5}$.

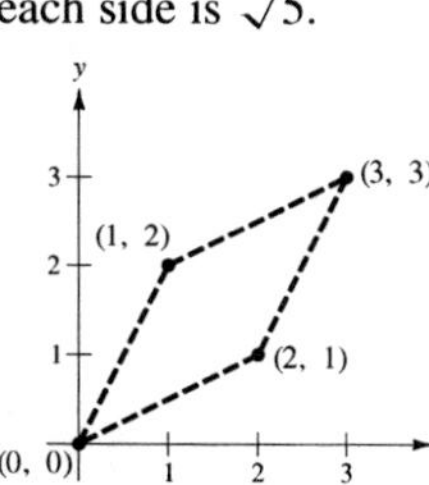

15.

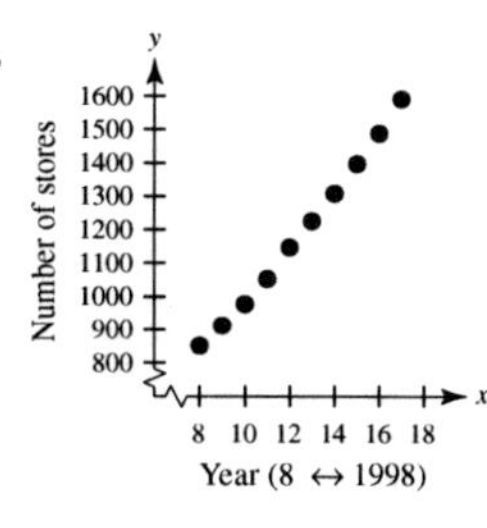

17. $d_1 = 2\sqrt{5},\ d_2 = \sqrt{5},\ d_3 = 3\sqrt{5}$
Collinear, because $d_1 + d_2 = d_3$.

19. $d_1 = \sqrt{2},\ d_2 = \sqrt{13},\ d_3 = 5$
Not collinear, because $d_1 + d_2 > d_3$.

21. $x = \pm 3$ **23.** $y = \pm\sqrt{55}$

25. $\left(\dfrac{3x_1 + x_2}{4}, \dfrac{3y_1 + y_2}{4}\right),\ \left(\dfrac{x_1 + x_2}{2}, \dfrac{y_1 + y_2}{2}\right),$

$\left(\dfrac{x_1 + 3x_2}{4}, \dfrac{y_1 + 3y_2}{4}\right)$

27. c   **28.** b   **29.** a   **30.** d   **31.** $x^2 + y^2 - 9 = 0$

33. $x^2 + y^2 - 4x + 2y - 11 = 0$

35. $x^2 + y^2 + 2x - 4y = 0$

37. $x^2 + y^2 - 6x - 4y + 3 = 0$ **39.** $x^2 + y^2 = 26{,}000^2$

41. $(x - 1)^2 + (y + 3)^2 = 4$ **43.** $(x - 1)^2 + (y + 3)^2 = 0$

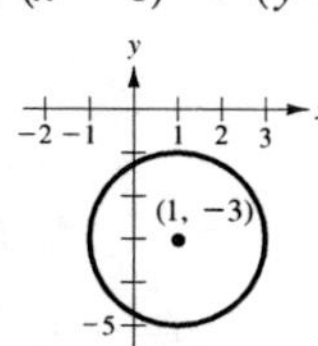

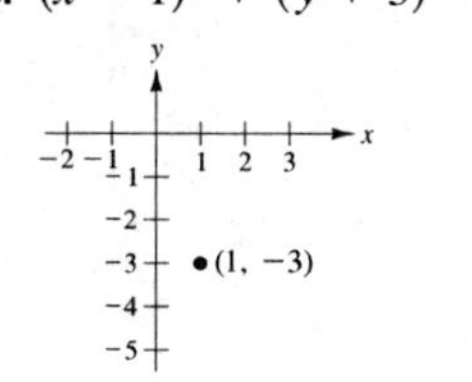

45. $\left(x - \tfrac{1}{2}\right)^2 + \left(y - \tfrac{1}{2}\right)^2 = 2$ **47.** $\left(x + \tfrac{1}{2}\right)^2 + \left(y + \tfrac{5}{4}\right)^2 = \tfrac{9}{4}$

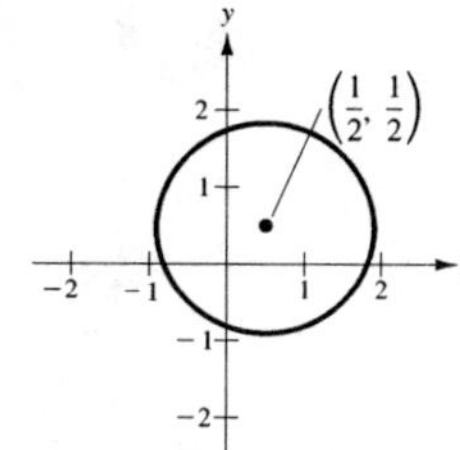

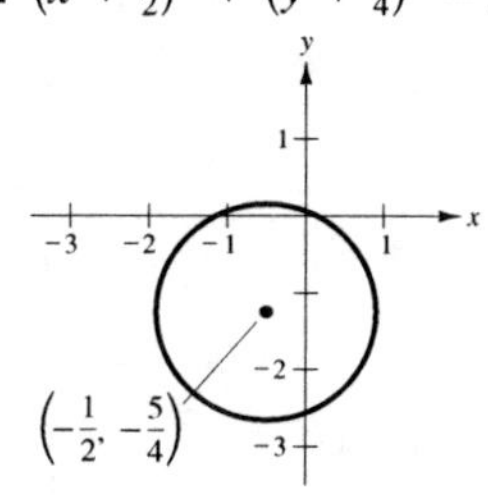

49.

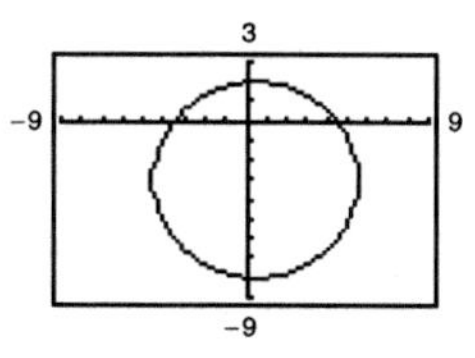

51.

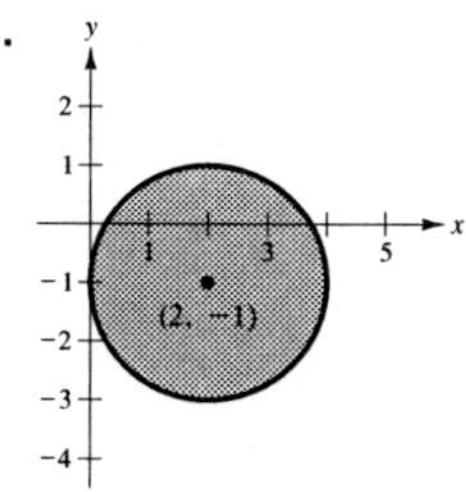

53. Proof   **55.** True   **57.** True   **59.** Proof   **61.** Proof

Appendix C.3 (page A48)

1. (a) $396°,\ -324°$ (b) $240°,\ -480°$

3. (a) $\dfrac{19\pi}{9},\ -\dfrac{17\pi}{9}$ (b) $\dfrac{10\pi}{3},\ -\dfrac{2\pi}{3}$

5. (a) $\dfrac{\pi}{6},\ 0.524$ (b) $\dfrac{5\pi}{6},\ 2.618$

(c) $\dfrac{7\pi}{4},\ 5.498$ (d) $\dfrac{2\pi}{3},\ 2.094$

7. (a) $270°$ (b) $210°$ (c) $-105°$ (d) $-151.1°$

9.

r	8 ft	15 in.	85 cm	24 in.	$\dfrac{12{,}963}{\pi}$ mi
s	12 ft	24 in.	63.75π cm	96 in.	8642 mi
θ	1.5	1.6	$\dfrac{3\pi}{4}$	4	$\dfrac{2\pi}{3}$

11. (a) $\sin\theta = \tfrac{4}{5}$ $\csc\theta = \tfrac{5}{4}$ (b) $\sin\theta = -\tfrac{5}{13}$ $\csc\theta = -\tfrac{13}{5}$
     $\cos\theta = \tfrac{3}{5}$ $\sec\theta = \tfrac{5}{3}$ $\cos\theta = -\tfrac{12}{13}$ $\sec\theta = -\tfrac{13}{12}$
     $\tan\theta = \tfrac{4}{3}$ $\cot\theta = \tfrac{3}{4}$ $\tan\theta = \tfrac{5}{12}$ $\cot\theta = \tfrac{12}{5}$

13. (a) Quadrant III   (b) Quadrant IV

15. $\dfrac{\sqrt{3}}{2}$ **17.** $\dfrac{4}{3}$

19. (a) $\sin 60° = \dfrac{\sqrt{3}}{2}$ (b) $\sin 120° = \dfrac{\sqrt{3}}{2}$

     $\cos 60° = \dfrac{1}{2}$ $\cos 120° = -\dfrac{1}{2}$

     $\tan 60° = \sqrt{3}$ $\tan 120° = -\sqrt{3}$

(c) $\sin\dfrac{\pi}{4} = \dfrac{\sqrt{2}}{2}$ (d) $\sin\dfrac{5\pi}{4} = -\dfrac{\sqrt{2}}{2}$

     $\cos\dfrac{\pi}{4} = \dfrac{\sqrt{2}}{2}$ $\cos\dfrac{5\pi}{4} = -\dfrac{\sqrt{2}}{2}$

     $\tan\dfrac{\pi}{4} = 1$ $\tan\dfrac{5\pi}{4} = 1$

21. (a) $\sin 225° = -\dfrac{\sqrt{2}}{2}$ (b) $\sin(-225°) = \dfrac{\sqrt{2}}{2}$

     $\cos 225° = -\dfrac{\sqrt{2}}{2}$ $\cos(-225°) = -\dfrac{\sqrt{2}}{2}$

     $\tan 225° = 1$ $\tan(-225°) = -1$

(c) $\sin \dfrac{5\pi}{3} = -\dfrac{\sqrt{3}}{2}$ (d) $\sin \dfrac{11\pi}{6} = -\dfrac{1}{2}$

$\cos \dfrac{5\pi}{3} = \dfrac{1}{2}$ $\cos \dfrac{11\pi}{6} = \dfrac{\sqrt{3}}{2}$

$\tan \dfrac{5\pi}{3} = -\sqrt{3}$ $\tan \dfrac{11\pi}{6} = -\dfrac{\sqrt{3}}{3}$

23. (a) 0.1736   (b) 5.7588   **25.** (a) 0.3640   (b) 0.3640

27. (a) $\theta = \dfrac{\pi}{4}, \dfrac{7\pi}{4}$ (b) $\theta = \dfrac{3\pi}{4}, \dfrac{5\pi}{4}$

29. (a) $\theta = \dfrac{\pi}{4}, \dfrac{5\pi}{4}$ (b) $\theta = \dfrac{5\pi}{6}, \dfrac{11\pi}{6}$

31. $\theta = \dfrac{\pi}{4}, \dfrac{3\pi}{4}, \dfrac{5\pi}{4}, \dfrac{7\pi}{4}$ **33.** $\theta = 0, \dfrac{\pi}{4}, \pi, \dfrac{5\pi}{4}$

35. $\theta = \dfrac{\pi}{3}, \dfrac{5\pi}{3}$ **37.** $\theta = 0, \dfrac{\pi}{2}, \pi$ **39.** 5099 ft

41. (a) Period: π (b) Period: 2   **43.** Period: $\dfrac{1}{2}$

Amplitude: 2   Amplitude: $\dfrac{1}{2}$ Amplitude: 3

45. Period: $\dfrac{\pi}{2}$ **47.** Period: $\dfrac{2\pi}{5}$

49. (a) 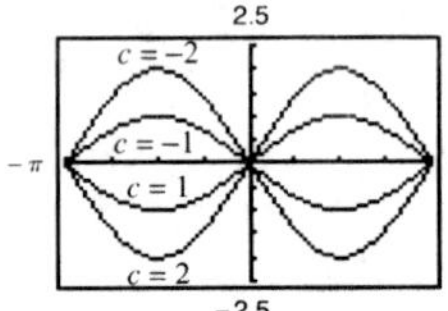   (b) 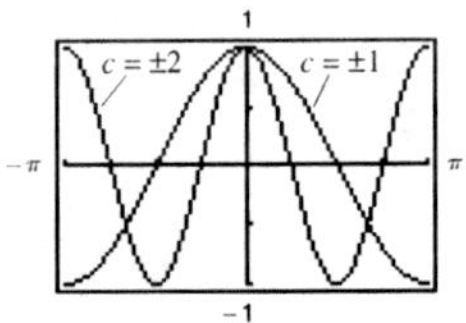

Change in amplitude   Change in period

(c) 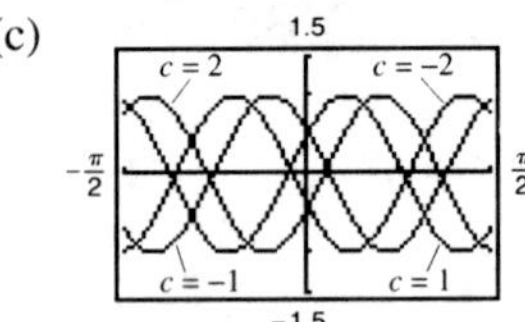

Horizontal translation

51. 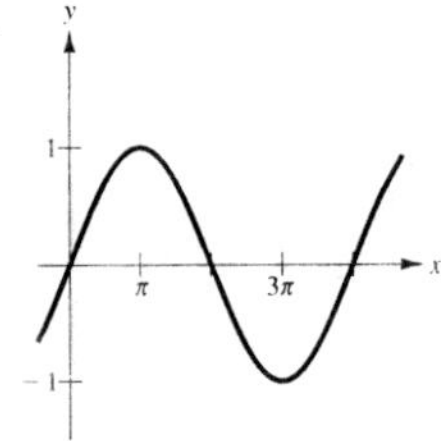   **53.** 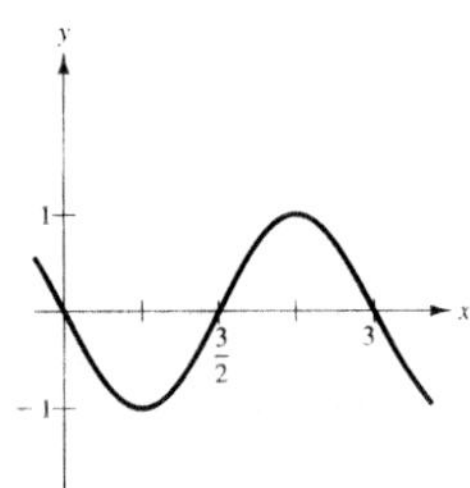

55. 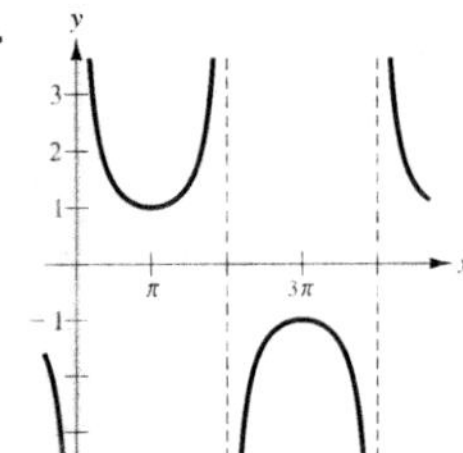   **57.** 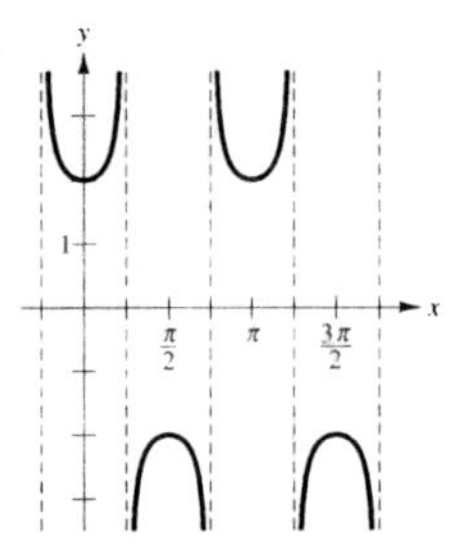

59. 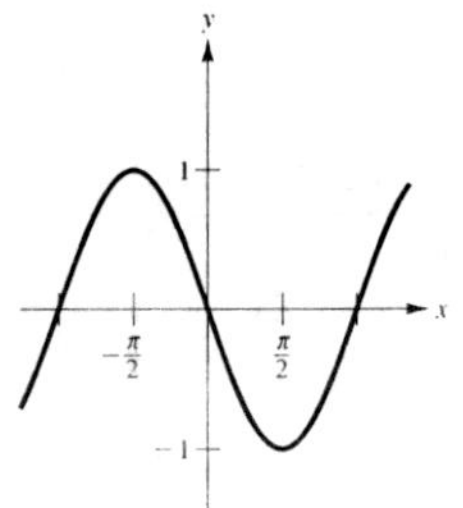   **61.** 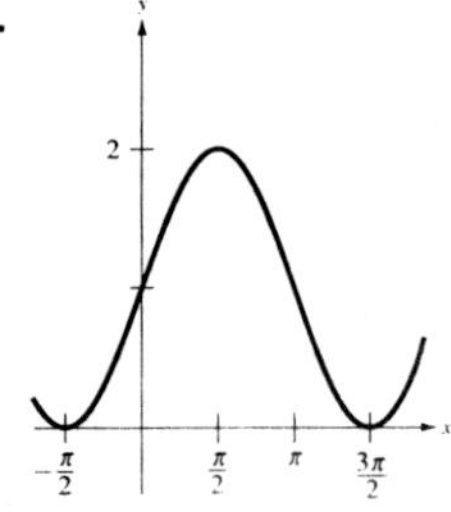

63. $a = 3, b = \dfrac{1}{2}, c = \dfrac{\pi}{2}$

65. 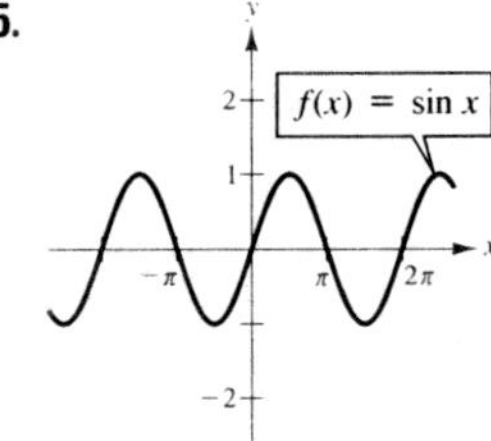

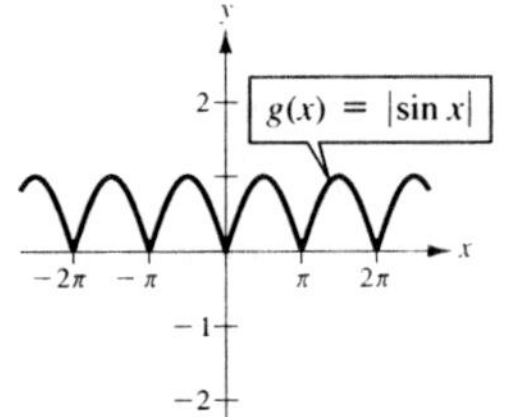

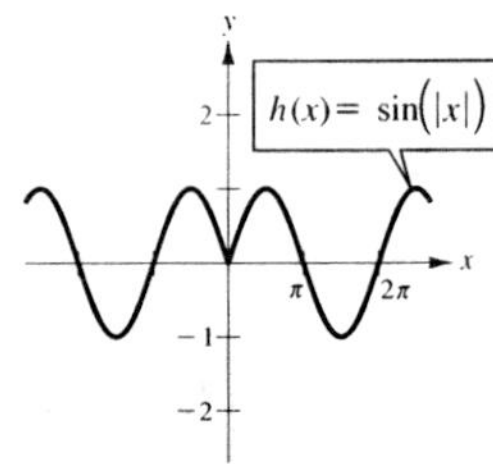

The graph of $|f(x)|$ will reflect any parts of the graph of $f(x)$ below the x-axis about the x-axis. The graph of $f(|x|)$ will reflect the part of the graph of $f(x)$ right of the y-axis about the y-axis.

67. 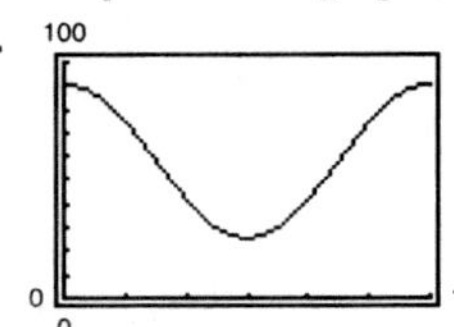

January, November, December

69. $f(x) = \dfrac{4}{\pi}\left(\sin \pi x + \dfrac{1}{3}\sin 3\pi x + \dfrac{1}{5}\sin 5\pi x + \cdots\right)$

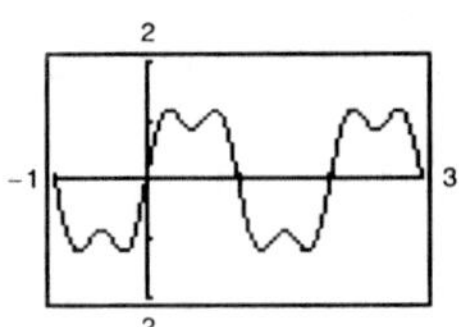

Index

APPENDIX F

Business and Economic Applications

Business and Economics Applications

Previously, you learned that one of the most common ways to measure change is with respect to time. In this section, you will study some important rates of change in economics that are not measured with respect to time. For example, economists refer to **marginal profit, marginal revenue,** and **marginal cost** as the rates of change of the profit, revenue, and cost with respect to the number of units produced or sold.

Summary of Business Terms and Formulas

Basic Terms | *Basic Formulas*

x is the number of units produced (or sold).

p is the price per unit.

R is the total revenue from selling x units. $\qquad R = xp$

C is the total cost of producing x units.

$\overline{C}$ is the average cost per unit. $\qquad \overline{C} = \dfrac{C}{x}$

P is the total profit from selling x units. $\qquad P = R - C$

The **break-even point** is the number of units for which $R = C$.

Marginals

$$\frac{dR}{dx} = \text{Marginal revenue} \approx \textit{extra} \text{ revenue from selling one additional unit}$$

$$\frac{dC}{dx} = \text{Marginal cost} \approx \textit{extra} \text{ cost of producing one additional unit}$$

$$\frac{dP}{dx} = \text{Marginal profit} \approx \textit{extra} \text{ profit from selling one additional unit}$$

In this summary, note that marginals can be used to approximate the *extra* revenue, cost, or profit associated with selling or producing one additional unit. This is illustrated graphically for marginal revenue in Figure F.1.

Marginal revenue

1 unit

Extra revenue for one unit

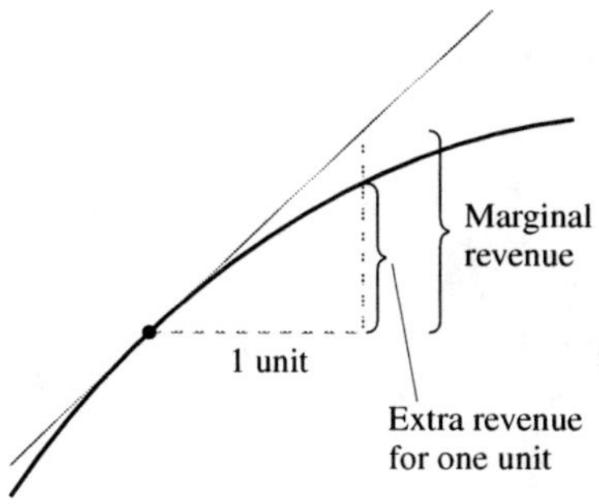

A revenue function

Figure F.1

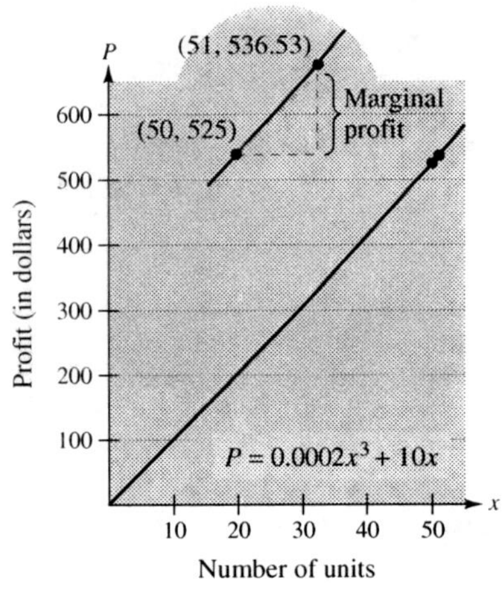

Marginal profit is the extra profit from selling one additional unit.
Figure F.2

EXAMPLE 1   Using Marginals as Approximations

A manufacturer determines that the profit P (in dollars) derived from selling x units of an item is given by

$$P = 0.0002x^3 + 10x.$$

a. Find the marginal profit for a production level of 50 units.

b. Compare this with the actual gain in profit obtained by increasing production from 50 to 51 units. (See Figure F.2.)

Solution

a. Because the profit is $P = 0.0002x^3 + 10x$, the marginal profit is given by the derivative

$$\frac{dP}{dx} = 0.0006x^2 + 10.$$

When $x = 50$, the marginal profit is

$$\frac{dP}{dx} = (0.0006)(50)^2 + 10 \qquad \text{Marginal profit for } x = 50$$

$$= \$11.50.$$

b. For $x = 50$ and 51, the actual profits are

$$P = (0.0002)(50)^3 + 10(50)$$

$$= 25 + 50$$

$$= \$525.00$$

$$P = (0.0002)(51)^3 + 10(51)$$

$$= 26.53 + 510$$

$$= \$536.53.$$

So, the additional profit obtained by increasing the production level from 50 to 51 units is

$$536.53 - 525.00 = \$11.53. \qquad \text{Extra profit for one unit}$$

The profit function in Example 1 is unusual in that the profit continues to increase as long as the number of units sold increases. In practice, it is more common to encounter situations in which sales can be increased only by lowering the price per item. Such reductions in price ultimately cause the profit to decline.

The number of units x that consumers are willing to purchase at a given price p per unit is defined as the **demand function**

$$p = f(x). \qquad \text{Demand function}$$

EXAMPLE 2   Finding a Demand Function

A business sells 2000 items per month at a price of \$10 each. It is estimated that monthly sales will increase by 250 items for each \$0.25 reduction in price. Find the demand function corresponding to this estimate.

Solution   From the given estimate, x increases 250 units each time p drops \$0.25 from the original cost of \$10. This is described by the equation

$$x = 2000 + 250\left(\frac{10 - p}{0.25}\right)$$

$$= 12{,}000 - 1000p$$

or

$$p = 12 - \frac{x}{1000}, \quad x \geq 2000. \qquad \text{Demand function}$$

The graph of the demand function is shown in Figure F.3.

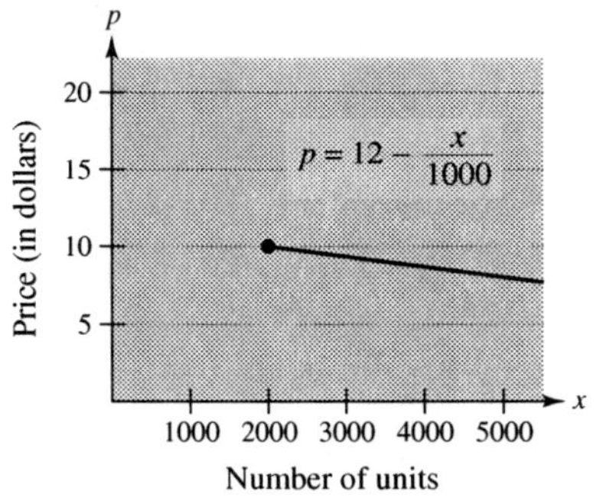

A demand function p
Figure F.3

EXAMPLE 3   Finding the Marginal Revenue

A fast-food restaurant has determined that the monthly demand for its hamburgers is

$$p = \frac{60{,}000 - x}{20{,}000}.$$

Find the increase in revenue per hamburger (marginal revenue) for monthly sales of 20,000 hamburgers. (See Figure F.4.)

Solution   Because the total revenue is given by $R = xp$, you have

$$R = xp = x\left(\frac{60{,}000 - x}{20{,}000}\right) = \frac{1}{20{,}000}(60{,}000x - x^2).$$

By differentiating, you can find the marginal revenue to be

$$\frac{dR}{dx} = \frac{1}{20{,}000}(60{,}000 - 2x).$$

When $x = 20{,}000$, the marginal revenue is

$$\frac{dR}{dx} = \frac{1}{20{,}000}[60{,}000 - 2(20{,}000)]$$

$$= \frac{20{,}000}{20{,}000}$$

$$= \$1 \text{ per unit.}$$

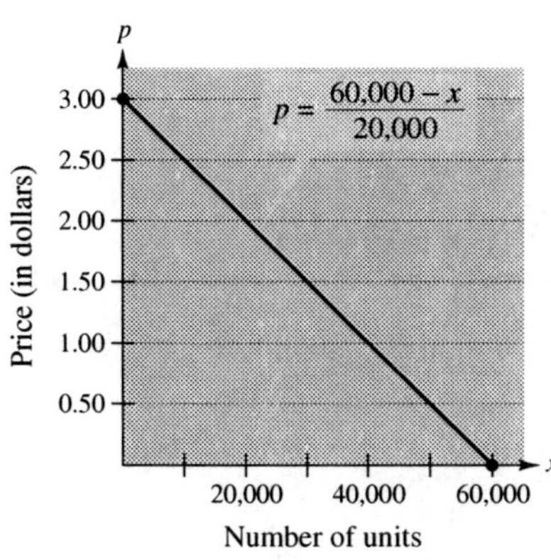

As the price decreases, more hamburgers are sold.
Figure F.4

NOTE   The demand function in Example 3 is typical in that a high demand corresponds to a low price, as shown in Figure F.4.

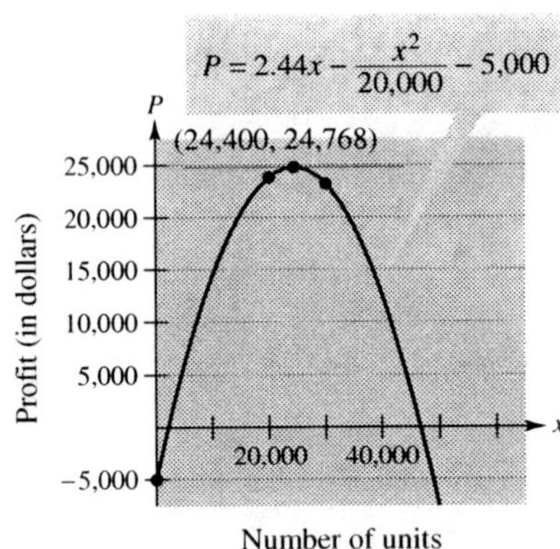

$$P = 2.44x - \frac{x^2}{20,000} - 5,000$$

The maximum profit corresponds to the point where the marginal profit is 0. When more than 24,400 hamburgers are sold, the marginal profit is negative—increasing production beyond this point will *reduce* rather than increase profit.

Figure F.5

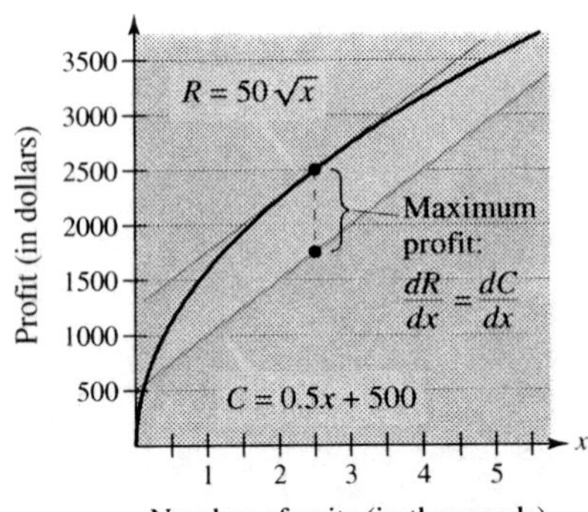

Maximum profit occurs when $\dfrac{dR}{dx} = \dfrac{dC}{dx}$.

Figure F.6

EXAMPLE 4 Finding the Marginal Profit

Suppose that in Example 3 the cost C (in dollars) of producing x hamburgers is

$$C = 5000 + 0.56x, \quad 0 \le x \le 50,000.$$

Find the total profit and the marginal profit for 20,000, 24,400, and 30,000 units.

Solution Because $P = R - C$, you can use the revenue function in Example 3 to obtain

$$P = \frac{1}{20,000}(60,000x - x^2) - 5000 - 0.56x$$

$$= 2.44x - \frac{x^2}{20,000} - 5000.$$

So, the marginal profit is

$$\frac{dP}{dx} = 2.44 - \frac{x}{10,000}.$$

The table shows the total profit and the marginal profit for each of the three indicated demands. Figure F.5 shows the graph of the profit function.

Demand	20,000	24,400	30,000
Profit	\$23,800	\$24,768	\$23,200
Marginal profit	\$0.44	\$0.00	−\$0.56

EXAMPLE 5 Finding the Maximum Profit

In marketing an item, a business has discovered that the demand for the item is represented by

$$p = \frac{50}{\sqrt{x}}. \qquad \text{Demand function}$$

The cost C (in dollars) of producing x items is given by $C = 0.5x + 500$. Find the price per unit that yields a maximum profit (see Figure F.6).

Solution From the given cost function, you obtain

$$P = R - C = xp - (0.5x + 500). \qquad \text{Primary equation}$$

Substituting for p (from the demand function) produces

$$P = x\left(\frac{50}{\sqrt{x}}\right) - (0.5x + 500) = 50\sqrt{x} - 0.5x - 500.$$

Setting the marginal profit equal to 0

$$\frac{dP}{dx} = \frac{25}{\sqrt{x}} - 0.5 = 0$$

yields $x = 2500$. From this, you can conclude that the maximum profit occurs when the price is

$$p = \frac{50}{\sqrt{2500}} = \frac{50}{50} = \$1.00.$$

NOTE   To find the maximum profit in Example 5, the profit function, $P = R - C$, was differentiated and set equal to 0. From the equation

$$\frac{dP}{dx} = \frac{dR}{dx} - \frac{dC}{dx} = 0$$

it follows that the maximum profit occurs when the marginal revenue is equal to the marginal cost, as shown in Figure F.6.

EXAMPLE 6   Minimizing the Average Cost

A company estimates that the cost C (in dollars) of producing x units of a product is given by $C = 800 + 0.04x + 0.0002x^2$. Find the production level that minimizes the average cost per unit.

Solution   Substituting from the given equation for C produces

$$\overline{C} = \frac{C}{x} = \frac{800 + 0.04x + 0.0002x^2}{x} = \frac{800}{x} + 0.04 + 0.0002x.$$

Setting the derivative $d\overline{C}/dx$ equal to 0 yields

$$\frac{d\overline{C}}{dx} = -\frac{800}{x^2} + 0.0002 = 0$$

$$x^2 = \frac{800}{0.0002} = 4,000,000 \implies x = 2000 \text{ units.}$$

See Figure F.7.

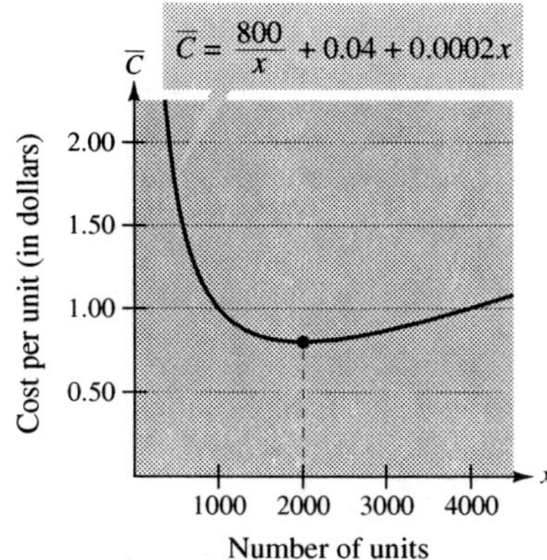

Minimum average cost occurs when $\dfrac{d\overline{C}}{dx} = 0$.

Figure F.7

EXERCISES FOR APPENDIX F

1. Think About It   The figure shows the cost C of producing x units of a product.

  (a) What is $C(0)$ called?

  (b) Sketch a graph of the marginal cost function.

  (c) Does the marginal cost function have an extremum? If so, describe what it means in economic terms.

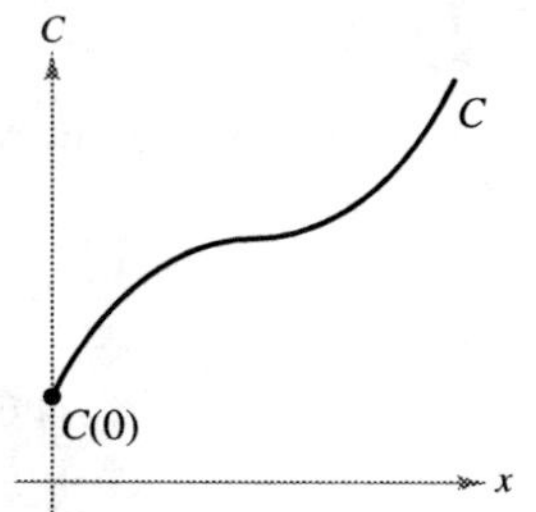

Figure for 1

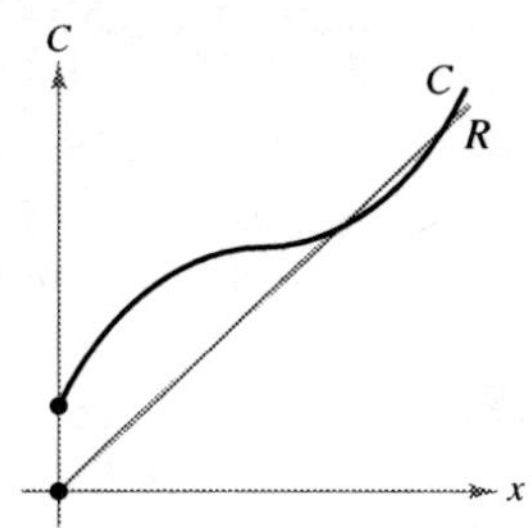

Figure for 2

2. Think About It   The figure shows the cost C and revenue R for producing and selling x units of a product.

  (a) Sketch a graph of the marginal revenue function.

  (b) Sketch a graph of the profit function. Approximate the position of the value of x for which profit is maximum.

In Exercises 3–6, find the number of units x that produces a maximum revenue R.

  3. $R = 900x - 0.1x^2$

  4. $R = 600x^2 - 0.02x^3$

  5. $R = \dfrac{1{,}000{,}000x}{0.02x^2 + 1800}$

  6. $R = 30x^{2/3} - 2x$

In Exercises 7–10, find the number of units x that produces the minimum average cost per unit $\overline{C}$.

  7. $C = 0.125x^2 + 20x + 5000$

  8. $C = 0.001x^3 - 5x + 250$

  9. $C = 3000x - x^2\sqrt{300 - x}$

  10. $C = \dfrac{2x^3 - x^2 + 5000x}{x^2 + 2500}$

In Exercises 11–14, find the price per unit p (in dollars) that produces the maximum profit P.

Cost Function	Demand Function
11. $C = 100 + 30x$	$p = 90 - x$
12. $C = 2400x + 5200$	$p = 6000 - 0.4x^2$
13. $C = 4000 - 40x + 0.02x^2$	$p = 50 - \dfrac{x}{100}$
14. $C = 35x + 2\sqrt{x - 1}$	$p = 40 - \sqrt{x - 1}$

Average Cost   **In Exercises 15 and 16, use the cost function to find the value of x at which the average cost is a minimum. For that value of x, show that the marginal cost and average cost are equal.**

15. $C = 2x^2 + 5x + 18$ **16.** $C = x^3 - 6x^2 + 13x$

17. Prove that the average cost is a minimum at the value of x where the average cost equals the marginal cost.

18. *Maximum Profit*   The profit P for a company is

$$P = 230 + 20s - \tfrac{1}{2}s^2$$

where s is the amount (in hundreds of dollars) spent on advertising. What amount of advertising produces a maximum profit?

19. *Numerical, Graphical, and Analytic Analysis*   The cost per unit for the production of a radio is $60. The manufacturer charges $90 per unit for orders of 100 or less. To encourage large orders, the manufacturer reduces the charge by $0.15 per radio for each unit ordered in excess of 100 (for example, there would be a charge of $87 per radio for an order size of 120).

(a) Analytically complete six rows of a table such as the one below. (The first two rows are shown.)

x	Price	Profit
102	$90 - 2(0.15)$	$102[90 - 2(0.15)] - 102(60) = 3029.40$
104	$90 - 4(0.15)$	$104[90 - 4(0.15)] - 104(60) = 3057.60$

(b) Use a graphing utility to generate additional rows of the table. Use the table to estimate the maximum profit. (*Hint:* Use the *table* feature of the graphing utility.)

(c) Write the profit P as a function of x.

(d) Use calculus to find the critical number of the function in part (c) and find the required order size.

(e) Use a graphing utility to graph the function in part (c) and verify the maximum profit from the graph.

20. *Maximum Profit*   A real estate office handles 50 apartment units. When the rent is $720 per month, all units are occupied. However, on the average, for each $40 increase in rent, one unit becomes vacant. Each occupied unit requires an average of $48 per month for service and repairs. What rent should be charged to obtain a maximum profit?

21. *Minimum Cost*   A power station is on one side of a river that is $\frac{1}{2}$-mile wide, and a factory is 6 miles downstream on the other side. It costs $12 per foot to run power lines over land and $16 per foot to run them underwater. Find the most economical path for the transmission line from the power station to the factory.

22. *Maximum Revenue*   When a wholesaler sold a product at $25 per unit, sales were 800 units per week. After a price increase of $5, the average number of units sold dropped to 775 per week. Assume that the demand function is linear, and find the price that will maximize the total revenue.

23. *Minimum Cost*   The ordering and transportation cost C (in thousands of dollars) of the components used in manufacturing a product is

$$C = 100\left(\frac{200}{x^2} + \frac{x}{x + 30}\right), \quad 1 \le x$$

where x is the order size (in hundreds). Find the order size that minimizes the cost. (*Hint:* Use Newton's Method or the *zero* feature of a graphing utility.)

24. *Average Cost*   A company estimates that the cost C (in dollars) of producing x units of a product is

$$C = 800 + 0.4x + 0.02x^2 + 0.0001x^3.$$

Find the production level that minimizes the average cost per unit. (*Hint:* Use Newton's Method or the *zero* feature of a graphing utility.)

25. *Revenue*   The revenue R for a company selling x units is

$$R = 900x - 0.1x^2.$$

Use differentials to approximate the change in revenue if sales increase from $x = 3000$ to $x = 3100$ units.

26. *Analytic and Graphical Analysis*   A manufacturer of fertilizer finds that the national sales of fertilizer roughly follow the seasonal pattern

$$F = 100{,}000\left\{1 + \sin\left[\frac{2\pi(t - 60)}{365}\right]\right\}$$

where F is measured in pounds. Time t is measured in days, with $t = 1$ corresponding to January 1.

(a) Use calculus to determine the day of the year when the maximum amount of fertilizer is sold.

(b) Use a graphing utility to graph the function and approximate the day of the year when sales are minimum.

27. *Modeling Data*   The table shows the monthly sales G (in thousands of gallons) of gasoline at a gas station in 2004. The time in months is represented by t, with $t = 1$ corresponding to January.

t	1	2	3	4	5	6
G	8.91	9.18	9.79	9.83	10.37	10.16

t	7	8	9	10	11	12
G	10.37	10.81	10.03	9.97	9.85	9.51

A model for these data is

$$G = 9.90 - 0.64 \cos\left(\frac{\pi t}{6} - 0.62\right).$$

 (a) Use a graphing utility to plot the data and graph the model.

(b) Use the model to approximate the month when gasoline sales were greatest.

(c) What factor in the model causes the seasonal variation in sales of gasoline? What part of the model gives the average monthly sales of gasoline?

(d) Suppose the gas station added the term $0.02t$ to the model. What does the inclusion of this term mean? Use this model to estimate the maximum monthly sales in the year 2008.

28. *Airline Revenues*   The annual revenue R (in millions of dollars) for an airline for the years 1995–2004 can be modeled by

$$R = 4.7t^4 - 193.5t^3 + 2941.7t^2 - 19,294.7t + 52,012$$

where $t = 5$ corresponds to 1995.

(a) During which year (between 1995 and 2004) was the airline's revenue the least?

(b) During which year was the revenue the greatest?

(c) Find the revenues for the years in which the revenue was the least and greatest.

(d) Use a graphing utility to confirm the results in parts (a) and (b).

29. *Modeling Data*   The manager of a department store recorded the quarterly sales S (in thousands of dollars) of a new seasonal product over a period of 2 years, as shown in the table, where t is the time in quarters, with $t = 1$ corresponding to the winter quarter of 2002.

t	1	2	3	4	5	6	7	8
S	7.5	6.2	5.3	7.0	9.1	7.8	6.9	8.6

(a) Use a graphing utility to plot the data.

(b) Find a model of the form $S = a + bt + c \sin \beta t$ for the data. (*Hint:* Start by finding β. Next, use a graphing utility to find $a + bt$. Finally, approximate c.)

(c) Use a graphing utility to graph the model with the data and make any adjustments necessary to obtain a better fit.

(d) Use the model to predict the maximum quarterly sales in the year 2006.

30. *Think About It*   Match each graph with the function it best represents—a demand function, a revenue function, a cost function, or a profit function. Explain your reasoning. [The graphs are labeled (a), (b), (c), and (d).]

(a)
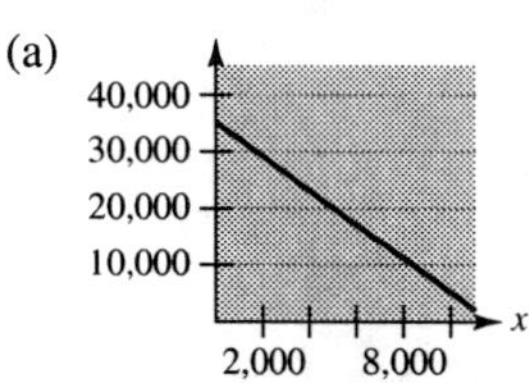

(b)
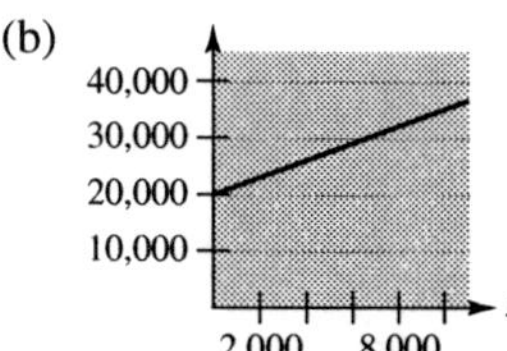

(c)
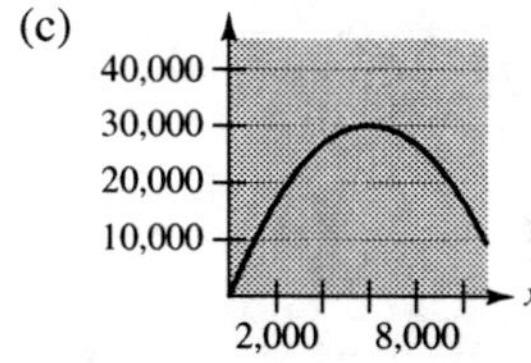

(d)
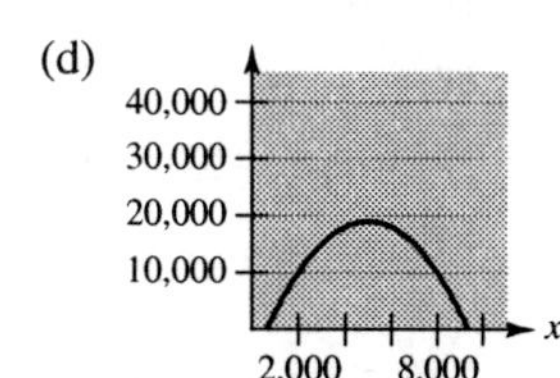

Elasticity   The relative responsiveness of consumers to a change in the price of an item is called the *price elasticity of demand*. If $p = f(x)$ is a differentiable demand function, the price elasticity of demand is

$$\eta = \frac{p/x}{dp/dx}.$$

For a given price, if $|\eta| < 1$, the demand is *inelastic*, and if $|\eta| > 1$, the demand is *elastic*. In Exercises 31–34, find η for the demand function at the indicated x-value. Is the demand elastic, inelastic, or neither at the indicated x-value?

31. $p = 400 - 3x$
  $x = 20$

32. $p = 5 - 0.03x$
  $x = 100$

33. $p = 400 - 0.5x^2$
  $x = 20$

34. $p = \dfrac{500}{x + 2}$
  $x = 23$

ALGEBRA

Factors and Zeros of Polynomials

Let $p(x) = a_n x^n + a_{n-1} x^{n-1} + \cdots + a_1 x + a_0$ be a polynomial. If $p(a) = 0$, then a is a *zero* of the polynomial and a solution of the equation $p(x) = 0$. Furthermore, $(x - a)$ is a *factor* of the polynomial.

Fundamental Theorem of Algebra

An *n*th degree polynomial has n (not necessarily distinct) zeros. Although all of these zeros may be imaginary, a real polynomial of odd degree must have at least one real zero.

Quadratic Formula

If $p(x) = ax^2 + bx + c$, and $0 \leq b^2 - 4ac$, then the real zeros of p are $x = \left(-b \pm \sqrt{b^2 - 4ac}\right)/2a$.

Special Factors

$$x^2 - a^2 = (x - a)(x + a) \qquad\qquad x^3 - a^3 = (x - a)(x^2 + ax + a^2)$$

$$x^3 + a^3 = (x + a)(x^2 - ax + a^2) \qquad\qquad x^4 - a^4 = (x^2 - a^2)(x^2 + a^2)$$

Binomial Theorem

$$(x + y)^2 = x^2 + 2xy + y^2 \qquad\qquad (x - y)^2 = x^2 - 2xy + y^2$$

$$(x + y)^3 = x^3 + 3x^2 y + 3xy^2 + y^3 \qquad\qquad (x - y)^3 = x^3 - 3x^2 y + 3xy^2 - y^3$$

$$(x + y)^4 = x^4 + 4x^3 y + 6x^2 y^2 + 4xy^3 + y^4 \qquad\qquad (x - y)^4 = x^4 - 4x^3 y + 6x^2 y^2 - 4xy^3 + y^4$$

$$(x + y)^n = x^n + nx^{n-1} y + \frac{n(n - 1)}{2!} x^{n-2} y^2 + \cdots + nxy^{n-1} + y^n$$

$$(x - y)^n = x^n - nx^{n-1} y + \frac{n(n - 1)}{2!} x^{n-2} y^2 - \cdots \pm nxy^{n-1} \mp y^n$$

Rational Zero Theorem

If $p(x) = a_n x^n + a_{n-1} x^{n-1} + \cdots + a_1 x + a_0$ has integer coefficients, then every *rational zero* of p is of the form $x = r/s$, where r is a factor of a_0 and s is a factor of a_n.

Factoring by Grouping

$$acx^3 + adx^2 + bcx + bd = ax^2(cx + d) + b(cx + d) = (ax^2 + b)(cx + d)$$

Arithmetic Operations

$$ab + ac = a(b + c) \qquad \frac{a}{b} + \frac{c}{d} = \frac{ad + bc}{bd} \qquad \frac{a + b}{c} = \frac{a}{c} + \frac{b}{c}$$

$$\frac{\left(\dfrac{a}{b}\right)}{\left(\dfrac{c}{d}\right)} = \left(\frac{a}{b}\right)\left(\frac{d}{c}\right) = \frac{ad}{bc} \qquad \frac{\left(\dfrac{a}{b}\right)}{c} = \frac{a}{bc} \qquad \frac{a}{\left(\dfrac{b}{c}\right)} = \frac{ac}{b}$$

$$a\left(\frac{b}{c}\right) = \frac{ab}{c} \qquad \frac{a - b}{c - d} = \frac{b - a}{d - c} \qquad \frac{ab + ac}{a} = b + c$$

Exponents and Radicals

$$a^0 = 1, \quad a \neq 0 \qquad (ab)^x = a^x b^x \qquad a^x a^y = a^{x+y} \qquad \sqrt{a} = a^{1/2} \qquad \frac{a^x}{a^y} = a^{x-y} \qquad \sqrt[n]{a} = a^{1/n}$$

$$\left(\frac{a}{b}\right)^x = \frac{a^x}{b^x} \qquad \sqrt[n]{a^m} = a^{m/n} \qquad a^{-x} = \frac{1}{a^x} \qquad \sqrt[n]{ab} = \sqrt[n]{a}\,\sqrt[n]{b} \qquad (a^x)^y = a^{xy} \qquad \sqrt[n]{\frac{a}{b}} = \frac{\sqrt[n]{a}}{\sqrt[n]{b}}$$

FORMULAS FROM GEOMETRY

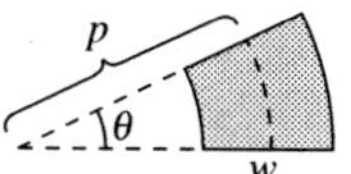

Triangle

$h = a \sin \theta$

$\text{Area} = \dfrac{1}{2}bh$

(Law of Cosines)

$c^2 = a^2 + b^2 - 2ab \cos \theta$

Sector of Circular Ring

(p = average radius,

w = width of ring,

θ in radians)

$\text{Area} = \theta p w$

Right Triangle

(Pythagorean Theorem)

$c^2 = a^2 + b^2$

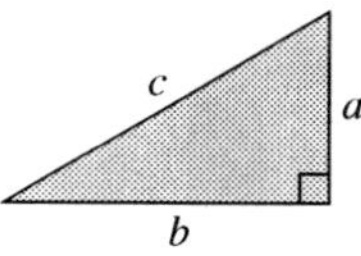

Ellipse

$\text{Area} = \pi a b$

$\text{Circumference} \approx 2\pi \sqrt{\dfrac{a^2 + b^2}{2}}$

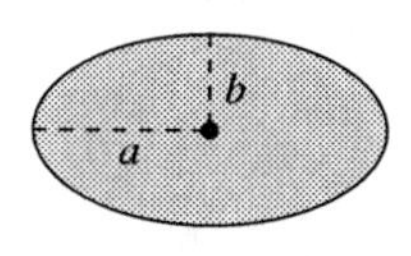

Equilateral Triangle

$h = \dfrac{\sqrt{3}\,s}{2}$

$\text{Area} = \dfrac{\sqrt{3}\,s^2}{4}$

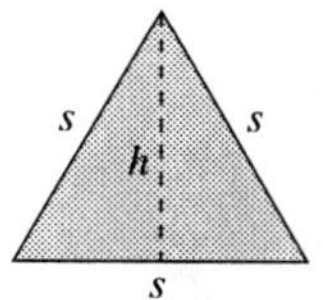

Cone

(A = area of base)

$\text{Volume} = \dfrac{Ah}{3}$

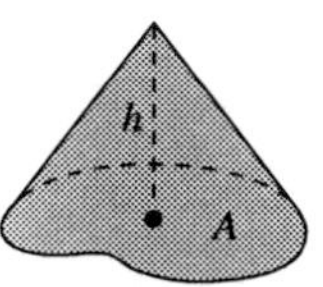

Parallelogram

$\text{Area} = bh$

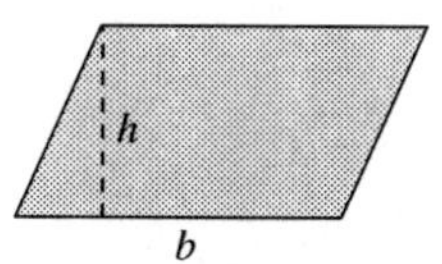

Right Circular Cone

$\text{Volume} = \dfrac{\pi r^2 h}{3}$

$\text{Lateral Surface Area} = \pi r \sqrt{r^2 + h^2}$

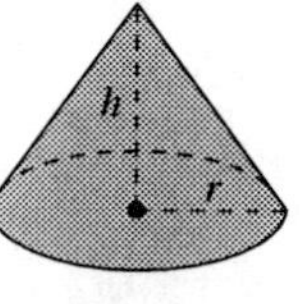

Trapezoid

$\text{Area} = \dfrac{h}{2}(a + b)$

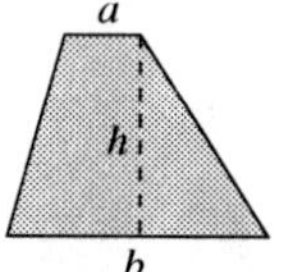 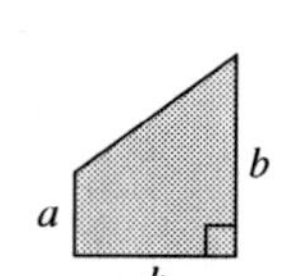

Frustum of Right Circular Cone

$\text{Volume} = \dfrac{\pi(r^2 + rR + R^2)h}{3}$

$\text{Lateral Surface Area} = \pi s(R + r)$

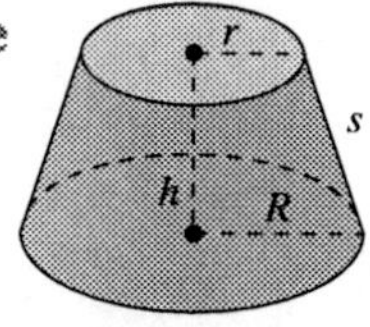

Circle

$\text{Area} = \pi r^2$

$\text{Circumference} = 2\pi r$

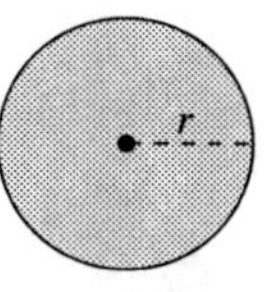

Right Circular Cylinder

$\text{Volume} = \pi r^2 h$

$\text{Lateral Surface Area} = 2\pi r h$

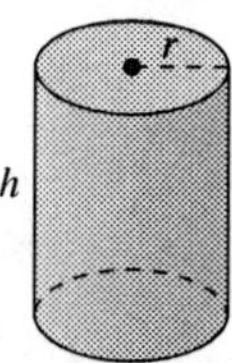

Sector of Circle

(θ in radians)

$\text{Area} = \dfrac{\theta r^2}{2}$

$s = r\theta$

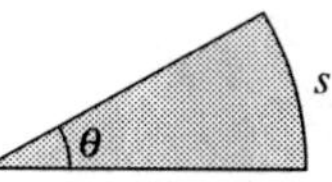

Sphere

$\text{Volume} = \dfrac{4}{3}\pi r^3$

$\text{Surface Area} = 4\pi r^2$

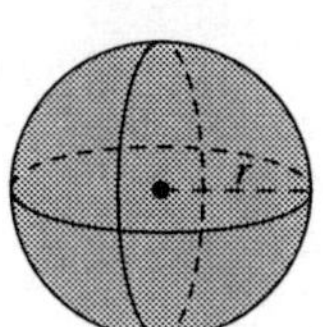

Circular Ring

(p = average radius,

w = width of ring)

$\text{Area} = \pi(R^2 - r^2)$

$\qquad = 2\pi p w$

 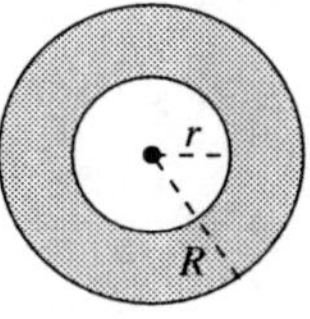 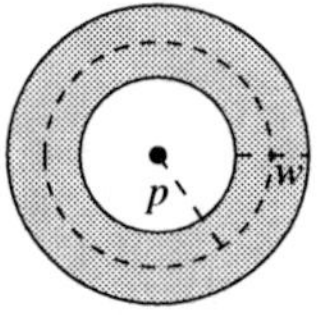

Wedge

(A = area of upper face,

B = area of base)

$A = B \sec \theta$

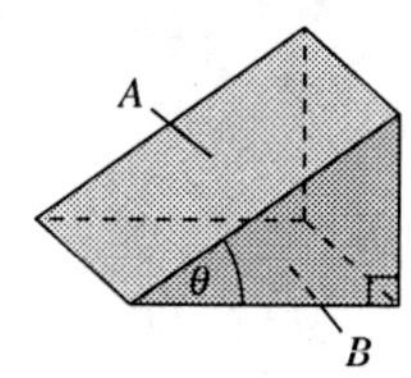